Nanotechnology for Cancer Therapy

Nanotechnology for Cancer Therapy

Edited by
Mansoor M. Amiji

CRC Press
Taylor & Francis Group
Boca Raton London New York

CRC Press is an imprint of the
Taylor & Francis Group, an informa business

CRC Press
Taylor & Francis Group
6000 Broken Sound Parkway NW, Suite 300
Boca Raton, FL 33487-2742

© 2007 by Taylor & Francis Group, LLC
CRC Press is an imprint of Taylor & Francis Group, an Informa business

First issued in paperback 2019

No claim to original U.S. Government works

ISBN 13: 978-0-367-45327-5 (pbk)
ISBN 13: 978-0-8493-7194-3 (hbk)

Library of Congress Cataloging-in-Publication Data

Nanotechnology in cancer therapy / edited by Mansoor M. Amiji.
    p. ; cm.
"A CRC title."
Includes bibliographical references and index.
ISBN-13: 978-0-8493-7194-3 (hardcover : alk. paper)
ISBN-10: 0-8493-7194-5 (hardcover : alk. paper)
  1. Cancer. 2. Nanotechnology. 3. Drug targeting. 4. Antineoplastic agents. I. Amiji, Mansoor M.
  [DNLM: 1. Neoplasms--therapy. 2. Drug Delivery Systems. 3. Nanotechnology. 4.
Neoplasms--diagnosis. QZ 266 N186 2007]

RC262.N22 2007
616.99'4061--dc22   2006024885

Visit the Taylor & Francis Web site at
http://www.taylorandfrancis.com

and the CRC Press Web site at
http://www.crcpress.com

Dedication

———

To the loving memory of my dad,
Mustafa Amiji, and to my mom, Shirin Amiji

Foreword

We are at the threshold of a new era in cancer treatment and diagnosis, brought about by the convergence of two disciplines—materials engineering and life sciences—that 30 years ago might have been difficult to envision. The product of this curious marriage, nanobiotechnology, is yielding many surprises and fostering many hopes in the drug-development space. Nanoparticles, engineered to exquisite precision using polymers, metals, lipids, and carbon, have been combined with molecular targeting, molecular imaging, and therapeutic techniques to create a powerful set of tools in the fight against cancer. The unique properties of nanomaterials enable selective drug delivery to tumors, novel treatment methods, intraoperative imaging guides to surgery, highly sensitive imaging agents for early tumor detection, and real-time monitoring of response to treatment.

Using nanotechnology, it may be possible to leap over many of the hurdles of cancer drug delivery that have confounded conventional drugs. These hurdles include hindered access to the central nervous system through the blood–brain barrier, sequestration by the reticulo-endothelial system, inability to penetrate the interior of solid tumors, and overcoming multi-drug resistance mechanisms. These obstacles can be mitigated by manipulating the size, surface charge, hydrophilicity, and attached targeting ligands of a therapeutic nanoparticle.

The novelty of using nanotechnology for medical applications presents its own challenges, similar in many ways to the challenges faced by the introduction of the first synthetic protein-based drugs. In the early 1980s, protein-based drugs offered an entirely new approach to targeting and treating disease. They could be "engineered" to be highly specific and even offered the capability of separating the targeting and therapeutic functions of a drug, such as when monoclonal antibodies are conjugated to cytotoxic agents. But monoclonal antibodies and recombinant protein-based drugs necessitated a different approach to lead optimization, metabolism, and toxicity screening from that of small-molecule drugs. Factors such as stability, immunogenicity, and species specificity had to be considered and tested in ways that were not familiar to the developers of traditional drugs.

Nanotechnology-based drugs will catalyze a reevaluation of optimization, metabolism, and toxicity-screening protocols. Interactions between novel materials and biological pathways are largely unknown but are widely suspected to depend heavily on physical and chemical characteristics such as particle size, particle size distribution, surface area, surface chemistry (including charge and hydrophobicity), shape, and aggregation state—features not usually scrutinized for traditional drugs. Standard criteria for physicochemical characterization will have to be established before safety testing can yield interpretable and reproducible results.

Would nanotechnology-based drugs be successful? All drugs have to run an obstacle course through physiological and biochemical barriers. For monoclonal antibody drugs, only a minute portion of an intravenously administered drug reaches its target. However, drugs based on nanomaterials, including nanoparticles, have unique properties that enable them to specifically bind to and penetrate solid tumors. Size and surface chemistries can be manipulated to facilitate extravasation through tumor vasculature, or the therapeutic agent can be encapsulated in polymer micelles or liposomes to prevent degradation and increase circulation half-life to improve the odds of it reaching the target.

This level of functional engineering has not been available to either small-molecule or protein-based drugs. For these drugs, modification of one characteristic, such as solubility or charge, can have dramatic effects on other essential characteristics, such as potency or target specificity. Nanoparticles, however, introduce a much higher degree of modularity and offer at least four advantages over the antibody conjugates: (1) the delivery of a larger therapeutic payload per target recognition event, enhancing *potency*; (2) the ability to carry multiple targeting agents, enhancing *selectivity*; (3) the ability to carry multiple therapeutic agents, enabling *targeted combination therapies*; and (4) the ability to *bypass physiological and biological barriers*.

We would all like to hasten the day when chemotherapy—the administration of non-specific, toxic anti-cancer agents—is relegated to medical history. Improvements in diagnostic screening and the development of drugs that target specific biological pathways have begun to turn the tide and contribute to a slow, yet consistent decline in death rates. While confirming that early diagnosis and targeted therapies are pointing us in the right direction, the progress is still incremental. Nanotechnology may be our best hope for overcoming many of the barriers faced by today's drugs in the battle against cancer. The combined creative forces of engineering, chemistry, physics, and biology will beget new and hopefully transformative options to old, intractable problems.

Piotr Grodzinski, Ph.D.
National Cancer Institute

Preface

With parallel breakthroughs occurring in molecular biology and nanoscience/technology, the newly recognized research thrust on "nanomedicine" is expected to have a revolutionary impact on the future of healthcare. To advance nanotechnology research for cancer prevention, diagnosis, and treatment, the United States National Cancer Institute (NCI) established the Alliance for Nanotechnology in Cancer in September 2004 and has pledged $144.3 million in the next five years (for details, visit http://nano.cancer.gov). Among the approaches for exploiting developments in nanotechnology for cancer molecular medicine, nanoparticles offer some unique advantages as sensing, delivery, and image enhancement agents. Several varieties of nanoparticles are available, including polymeric conjugates and nanoparticles, micelles, dendrimers, liposomes, and nanoassemblies.

This book focuses specifically on nanoscientific and nanotechnological strategies that are effective and promising for imaging and treatment of cancer. Among the various approaches considered, nanotechnology offers the best promise for targeted delivery of drugs and genes to the tumor site and alleviation of the side effects of chemotherapeutic agents. Multifunctional nanosystems offer tremendous opportunity for combining more than one drug or using drug and imaging agents. The expertise of world-renowned academic and industrial researchers is brought together here to provide a comprehensive treatise on this subject.

The book is composed of thirty-eight chapters divided into seven sections that address the specific nanoplatforms used for imaging and delivery of therapeutic molecules. Section 1 focuses on the rationale and fundamental understanding of targeting strategies, including pharmacokinetic considerations for delivery to tumors in vivo, multifunctional nanotherapeutics, boron neutron capture therapy, and the discussion on nanotechnology characterization for cancer therapy, as well as guidance from the U.S. Food and Drug Administration on approval of nanotechnology products. Section 2 focuses on polymeric conjugates used for tumor-targeted imaging and delivery, including special consideration on the use of imaging to evaluate therapeutic efficacy. In Section 3, polymeric nanoparticle systems are discussed with emphasis on biodegradable, long-circulating nanoparticles for passive and active targeting. Section 4 focuses on polymeric micellar assemblies, where sophisticated chemistry is applied for the development of novel nanosystems that can provide efficient delivery to tumors. Many of the micellar delivery systems are undergoing clinical trials in Japan and other countries across the globe. Dendritic nanostructures used for cancer imaging and therapy are discussed in Section 5. Section 6 focuses on the oldest nanotechnology for cancer therapy—liposome-based delivery systems—with emphasis on surface modification to enhance target efficiency and temperature-responsive liposomes. Lastly, Section 7 focuses on other lipid nanosystems used for targeted delivery of cancer therapy, including nanoemulsions that can cross biological barriers, solid-lipid nanoparticles, lipoprotein nanoparticles, and DQAsomes for mitochondria-specific delivery.

Words cannot adequately express my admiration and gratitude to all of the contributing authors. Each chapter is written by a world-renowned authority on the subject, and I am deeply grateful for their willingness to participate in this project. I am also extremely grateful to Dr. Piotr Grodzinski for providing the Foreword. Drs. Fredika Robertson and Mauro Ferrari have done a superb job in laying the foundation by providing a chapter entitled "Introduction and Rationale for Nanotechnology in Cancer." I am grateful to Professor Kinam Park at Purdue University, Professor Robert Langer at MIT, and Professor Vladimir Torchilin at Northeastern University, who have been my mentors and collaborators, as well as many other researchers from academia and industry. Special thanks are due to the postdoctoral associates and graduate students in my laboratory at Northeastern University who have been the "soldiers in the trenches" in our quest to use nanotechnology for the targeted delivery of drugs and genes to solid tumors. Lastly, I am deeply grateful to the wonderful people at Taylor & Francis-CRC Press, including Stephen Zollo, Patricia Roberson, and many others, who have made the concept of this book into reality.

Any comments and constructive criticisms of the book can be sent to the editor at m.amiji@neu.edu.

Editor

Dr. Mansoor M. Amiji received his undergraduate degree in pharmacy from Northeastern University in 1988 and his PhD in pharmaceutics from Purdue University in 1992. His areas of specialization include polymeric biomaterials, advanced drug delivery systems, and nanomedical technologies.

Dr. Amiji's research interests include the synthesis of novel polymeric materials for medical and pharmaceutical applications; surface modification of cationic polymers by the complexation-interpenetration method to develop biocompatible materials; the preparation and characterization of polymeric membranes and microcapsules with controlled permeability properties for medical and pharmaceutical applications; target-specific drug and vaccine delivery systems for gastrointestinal tract infections; localized delivery of cytotoxic and anti-angiogenic drugs for solid tumors in novel biodegradable polymeric nanoparticles; intracellular delivery systems for drugs and genes using target-specific, long-circulating, biodegradable polymeric nanoparticles; and gold and iron-gold core-shell nanoparticles for biosensing, imaging, and delivery applications. His research has received sustained funding from the National Institutes of Health (NIH), the National Science Foundation (NSF), foundations, and local industries.

Dr. Amiji is Professor and Associate Chair of the Pharmaceutical Sciences Department and Co-Director of the Northeastern University Nanomedicine Education and Research Consortium (NERC). The NERC oversees a doctoral training grant in nanomedicine science and technology that was co-funded by the NIH and NSF. He has two published books, *Applied Physical Pharmacy* and *Polymeric Gene Delivery: Principles and Applications*, along with numerous manuscript publications. He has also received a number of awards, including the 2003 Eurand Award for Innovative Oral Drug Delivery Research, Third Prize.

Dr. Amiji has supervised the research efforts of over 50 postdoctoral associates, doctoral and master's level graduate students, and undergraduate honors students over the last 13 years. His teaching responsibilities include the Doctor of Pharmacy (PharmD) program and graduate programs (MS and PhD) in pharmaceutical sciences, biotechnology, and nanomedicine.

Contributors

Hamidreza Montazeri Aliabadi
Department of Pharmacy and Pharmaceutical
Sciences
University of Alberta
Edmonton, Alberta, Canada

Christine Allen
Department of Pharmaceutical Sciences
and
Department of Chemical Engineering and
Applied Chemistry
University of Toronto
Toronto, Ontario, Canada

Ashootosh V. Ambade
Department of Chemistry
University of Massachusetts at Amherst
Amherst, Massachusetts

Mansoor M. Amiji
Department of Pharmaceutical Sciences
Northeastern University
Boston, Massachusetts

Joseph M. Backer
SibTech, Inc.
Newington, Connecticut

Marina V. Backer
SibTech, Inc.
Newington, Connecticut

You Han Bae
Department of Pharmaceutics and
Pharmaceutical Chemistry
University of Utah
Salt Lake City, Utah

Lajos P. Balogh
Department of Radiation Medicine
and
Department of Cellular Stress and
Cancer Biology
Roswell Park Cancer Institute
Buffalo, New York

Rolf F. Barth
Department of Pathology
The Ohio State University
Columbus, Ohio

Reina Bendayan
Department of Pharmaceutical Sciences
University of Toronto
Toronto, Ontario, Canada

Tania Betancourt
The University of Texas at Austin
Austin, Texas

D. Bhadra
Department of Pharmaceutical Sciences
Dr. H.S. Gour University
Sagar, India

S. Bhadra
Department of Pharmaceutical Sciences
Dr. H.S. Gour University
Sagar, India

Sangeeta N. Bhatia
Harvard–MIT Division of Health Sciences
and Technology
Massachusetts Institute of Technology
Cambridge, Massachusetts

Sarathi V. Boddapati
Department of Pharmaceutical Sciences
Northeastern University
Boston, Massachusetts

Lisa Brannon-Peppas
The University of Texas at Austin
Austin, Texas

Mark Butler
Department of Chemical Engineering
and Applied Chemistry
University of Toronto
Toronto, Ontario, Canada

John C. Byrd
Division of Hematology–Oncology
NCI Comprehensive Cancer Center
The Ohio State University
Columbus, Ohio

Robert B. Campbell
Department of Pharmaceutical Sciences
Northeastern University
Boston, Massachusetts

Shing-Ming Cheng
Department of Pharmaceutical Sciences
Northeastern University
Boston, Massachusetts

Gerard G. M. D'Souza
Department of Pharmaceutical Sciences
Northeastern University
Boston, Massachusetts

Marina A. Dobrovolskaia
Nanotechnology Characterization Laboratory
NCI Frederick
Frederick, Maryland

Amber Doiron
The University of Texas at Austin
Austin, Texas

P. K. Dubey
Department of Pharmaceutical Sciences
Dr. H. S. Gold University
Sagar, India

Omid C. Farokhzad
Department of Anesthesiology, Perioperative
 and Pain Medicine
Brigham and Women's Hospital and Harvard
 Medical School
Boston, Massachusetts

Mauro Ferrari
The Brown Foundation Institute of
 Molecular Medicine
M. D. Anderson Cancer Center
and
Alliance for Nanohealth
The University of Texas
Houston, Texas

Alberto A. Gabizon
Oncology Institute
Shaare Zedek Medical Center and
 Hebrew University
Jerusalem, Israel

Jinming Gao
Simmons Comprehensive Cancer Center
University of Texas Southwestern
 Medical Center
Dallas, Texas

Hamidreza Ghandehari
Department of Pharmaceutical Sciences
University of Maryland
Baltimore, Maryland

Todd J. Harris
Harvard–MIT Division of Health Sciences
 and Technology
Massachusetts Institute of Technology
Cambridge, Massachusetts

Mitsuru Hashida
Department of Drug Delivery Research
Graduate School of Pharmaceutical Sciences
Kyoto University
Sakyo-ku, Kyoto, Japan

Yuriko Higuchi
Department of Drug Delivery Research
Graduate School of Pharmaceutical Sciences
Kyoto University
Sakyo-ku, Kyoto, Japan

Kang Moo Huh
Department of Polymer Science
 and Engineering
Chungnam National University
Yuseong-Gu, Daejeon, South Korea

Edward F. Jackson
Division of Diagnostic Imaging
The University of Texas M. D. Anderson
 Cancer Center
Houston, Texas

N. K. Jain
Department of Pharmaceutical Sciences
Dr. H.S. Gour University
Sagar, India

Eun-Kee Jeong
Department of Radiology
University of Utah
Salt Lake City, Utah

Ji Hoon Jeong
Department of Pharmaceutics and
    Pharmaceutical Chemistry
University of Utah
Salt Lake City, Utah

Sangyong Jon
Department of Life Science
Gwangju Institute of Science and Technology
Gwangju, South Korea

Shigeru Kawakami
Department of Drug Delivery Research
Graduate School of Pharmaceutical Sciences
Kyoto University
Sakyo-ku, Kyoto, Japan

Mohamed K. Khan
Department of Radiation Medicine
and
Department of Cellular Stress and
    Cancer Biology
Roswell Park Cancer Institute
Buffalo, New York

Chalermchai Khemtong
Simmons Comprehensive Cancer Center
University of Texas Southwestern
    Medical Center
Dallas, Texas

Sun Hwa Kim
Department of Biological Sciences
Korea Advanced Institute of Science
    and Technology
Daejeon, South Korea

Tae-il Kim
School of Chemistry and Molecular
    Engineering
Seoul National University
Seoul, South Korea

Sushma Kommareddy
Department of Pharmaceutical Sciences
Northeastern University
Boston, Massachusetts

Vinod Labhasetwar
Departments of Pharmaceutical Sciences and
    Biochemistry and Molecular Biology
University of Nebraska Medical Center
Omaha, Nebraska

Andras G. Lacko
Departments of Molecular Biology and
    Immunology
University of North Texas Health Science
    Center
Fort Worth, Texas

Robert Langer
Department of Chemical Engineering
Massachusetts Institute of Technology
Cambridge, Massachusetts

Afsaneh Lavasanifar
Department of Pharmacy and Pharmaceutical
    Sciences
University of Alberta
Edmonton, Alberta, Canada

Eun Seong Lee
Amorepacific Corporation/R&D Center
Giheung-gu Yongin-si Gyeonggi-do
South Korea

Helen Lee
Department of Pharmaceutical Sciences
University of Toronto
Toronto, Ontario, Canada

Robert J. Lee
NCI Comprehensive Cancer Center
and
NSF Nanoscale Science and
    Engineering Center
The Ohio State University
Columbus, Ohio

Sang Cheon Lee
Nanomaterials Application Division
Korea Institute of Ceramic Engineering and
    Technology
Guemcheon-gu, Seoul, South Korea

Chun Li
M. D. Anderson
Cancer Center
The University of Texas
Houston, Texas

Yongqiang Li
Department of Pharmaceutical Sciences
University of Toronto
Toronto, Ontario, Canada

Bruce R. Line
Department of Radiology, and Greenebaum
    Cancer Center
University of Maryland
Baltimore, Maryland

Jubo Liu
Department of Pharmaceutical Sciences
University of Toronto
Toronto, Ontario, Canada

Zheng-Rong Lu
Departments of Pharmaceutics and
    Pharmaceutical Chemistry
University of Utah
Salt Lake City, Utah

S. Mahor
Department of Pharmaceutical Sciences
Dr. H. S. Gour University
Sagar, India

Walter J. McConathy
Department of Internal Medicine
University of North Texas Health Science
    Center
Fort Worth, Texas

Scott E. McNeil
Nanotechnology Characterization Laboratory
NCI Frederick
Frederick, Maryland

T. J. Miller
Center for Drug Evaluation and Research
Food and Drug Administration
Silver Spring, Maryland

Amitava Mitra
Department of Pharmaceutical Sciences and
    Center for Nanomedicine and
    Cellular Delivery
University of Maryland
Baltimore, Maryland

S. M. Moghimi
Molecular Targeting and Polymer
    Toxicology Group
School of Pharmacy
University of Brighton
Brighton, United Kingdom

Randall J. Mrsny
Center for Drug Delivery/Biology
Welsh School of Pharmacy
Cardiff University
Cardiff, United Kingdom

R. S. R. Murthy
Pharmacy Department
The M.S. University of Baroda
Vadodara, India

Natarajan Muthusamy
NCI Comprehensive Cancer Center
The Ohio State University
Columbus, Ohio

Anjan Nan
Department of Pharmaceutical Sciences and
    Center for Nanomedicine and
    Cellular Delivery
University of Maryland
Baltimore, Maryland

Maya Nair
Departments of Molecular Biology and
    Immunology
University of North Texas Health
    Science Center
Fort Worth, Texas

Norased Nasongkla
Simmons Comprehensive Cancer Center
University of Texas Southwestern
    Medical Center
Dallas, Texas

David Needham
Department of Mechanical Engineering
    and Materials Science
Duke University
Durham, North Carolina

Tooru Ooya
Department of Materials Science
Japan Advanced Institute of Science
    and Technology
Tatsunokuchi, Ishikawa, Japan

Jong-Sang Park
Department of Chemistry and Molecular
    Engineering
Seoul National University
Seoul, South Korea

Kinam Park
Departments of Pharmaceutics and
    Biomedical Engineering
Purdue University
West Lafayette, Indiana

Tae Gwan Park
Department of Biological Sciences
Korea Advanced Institute of Science and
    Technology
Daejeon, South Korea

Anil K. Patri
Nanotechnology Characterization Laboratory
NCI Frederick
Frederick, Maryland

Ana Ponce
Department of Biomedical Engineering
Duke University
Durham, North Carolina

Natalya Rapoport
Department of Bioengineering
University of Utah
Salt Lake City, Utah

Mike Andrew Rauth
Department of Pharmaceutical Sciences
University of Toronto
Toronto, Ontario, Canada

L. Harivardan Reddy
Pharmacy Department
The M.S. University of Baroda
Vadodara, India

Fredika M. Robertson
College of Medicine and Comprehensive
    Cancer Center
The Ohio State University
Columbus, Ohio

N. Sadrieh
Center for Drug Evaluation and Research
Food and Drug Administration
Silver Spring, Maryland

Sanjeeb K. Sahoo
Department of Pharmaceutical Sciences
Nebraska Medical Center
Omaha, Nebraska
and
Institute of Life Sciences
Bhubaneswar, India

Elamprakash N. Savariar
Department of Chemistry
University of Massachusetts at Amherst
Amherst, Massachusetts

Dinesh B. Shenoy
Department of Pharmaceutical Sciences
Northeastern University
Boston, Massachusetts and Novavax, Inc.
Malvern, Pennsylvania

Patrick Lim Soo
Department of Pharmaceutical Sciences
University of Toronto
Toronto, Ontario, Canada

Stephan T. Stern
Nanotechnology Characterization
    Laboratory
NCI Frederick
Frederick, Maryland

S. (Thai) Thayumanavan
Department of Chemistry
University of Massachusetts at Amherst
Amherst, Massachusetts

Sandip B. Tiwari
Department of Pharmaceutical Sciences
Northeastern University
Boston, Massachusetts

Werner Tjarks
College of Pharmacy
The Ohio State University
Columbus, Ohio

Vladimir P. Torchilin
Department of Pharmaceutical Sciences
Northeastern University
Boston, Massachusetts

Anagha Vaidya
Departments of Pharmaceutics and
  Pharmaceutical Chemistry
University of Utah
Salt Lake City, Utah

Geoffrey von Maltzahn
Harvard–MIT Division of Health Sciences
  and Technology
Massachusetts Institute of Technology
Cambridge, Massachusetts

S. P. Vyas
Department of Pharmaceutical Sciences
Dr. H. S. Gour University
Sagar, India

Sidney Wallace
Division of Diagnostic Imaging
The University of Texas M. D. Anderson
Cancer Center
Houston, Texas

Yanli Wang
Departments of Pharmaceutics and
  Pharmaceutical Chemistry
University of Utah
Salt Lake City, Utah

Volkmar Weissig
Department of Pharmaceutical Sciences
Northeastern University
Boston, Massachusetts

Ho Lun Wong
Department of Pharmaceutical Sciences
University of Toronto
Toronto, Ontario, Canada

Gong Wu
Department of Pathology
The Ohio State University
Columbus, Ohio

Xiao Yu Wu
Department of Pharmaceutical Sciences
University of Toronto
Toronto, Ontario, Canada

Xiao-Bing Xiong
Faculty of Pharmacy and Pharmaceutical
  Sciences
University of Alberta
Edmonton, Alberta, Canada

Weilian Yang
Department of Pathology
The Ohio State University
Columbus, Ohio

Furong Ye
Departments of Pharmaceutics and
  Pharmaceutical Chemistry
University of Utah
Salt Lake City, Utah

Guodong Zhang
M. D. Anderson
Cancer Center
The University of Texas
Houston, Texas

Xiaobin B. Zhao
NCI Comprehensive Cancer Center
The Ohio State University
Columbus, Ohio

Table of Contents

Section 1
Nanotechnology and Cancer. 1

Chapter 1   Introduction and Rationale for Nanotechnology
in Cancer Therapy 3

Fredika M. Robertson and Mauro Ferrari

Chapter 2   Passive Targeting of Solid Tumors: Pathophysiological
Principles and Physicochemical Aspects of
Delivery Systems. 11

S. M. Moghimi

Chapter 3   Active Targeting Strategies in Cancer with a Focus on
Potential Nanotechnology Applications 19

Randall J. Mrsny

Chapter 4   Pharmacokinetics of Nanocarrier-Mediated
Drug and Gene Delivery 43

Yuriko Higuchi, Shigeru Kawakami, and Mitsuru Hashida

Chapter 5   Multifunctional Nanoparticles for Cancer Therapy. 59

Todd J. Harris, Geoffrey von Maltzahn, and Sangeeta N. Bhatia

Chapter 6   Neutron Capture Therapy of Cancer: Nanoparticles
and High Molecular Weight Boron Delivery Agents. 77

*Gong Wu, Rolf F. Barth, Weilian Yang, Robert J. Lee, Werner Tjarks, Marina V. Backer, and
Joseph M. Backer*

Chapter 7   Preclinical Characterization of Engineered Nanoparticles
Intended for Cancer Therapeutics 105

Anil K. Patri, Marina A. Dobrovolskaia, Stephan T. Stern, and Scott E. McNeil

Chapter 8   Nanotechnology: Regulatory Perspective for Drug
Development in Cancer Therapeutics 139

N. Sadrieh and T. J. Miller

Section 2
Polymer Conjugates. 157

Chapter 9   Polymeric Conjugates for Angiogenesis Targeted Tumor
Imaging and Therapy. 159

Amitava Mitra, Anjan Nan, Bruce R. Line, and Hamidreza Ghandehari

Chapter 10 Poly(L-Glutamic Acid): Efficient Carrier of Cancer Therapeutics and
Diagnostics 185

Guodong Zhang, Edward F. Jackson, Sidney Wallace, and Chun Li

Chapter 11 Noninvasive Visualization of In Vivo Drug Delivery
of Paramagnetic Polymer Conjugates with MRI. 201

Zheng-Rong Lu, Yanli Wang, Furong Ye, Anagha Vaidya, and Eun-Kee Jeong

Section 3
Polymeric Nanoparticles 213

Chapter 12 Polymeric Nanoparticles for Tumor-Targeted Drug
Delivery. 215

Tania Betancourt, Amber Doiron, and Lisa Brannon-Peppas

Chapter 13 Long-Circulating Polymeric Nanoparticles for Drug and
Gene Delivery to Tumors 231

Sushma Kommareddy, Dinesh B. Shenoy, and Mansoor M. Amiji

Chapter 14 Biodegradable PLGA/PLA Nanoparticles for Anti-cancer
Therapy. 243

Sanjeeb K. Sahoo and Vinod Labhasetwar

Chapter 15 Poly(Alkyl Cyanoacrylate) Nanoparticles for Delivery
of Anti-Cancer Drugs 251

R. S. R. Murthy and L. Harivardhan Reddy

Chapter 16 Aptamers and Cancer Nanotechnology 289

Omid C. Farokhzad, Sangyong Jon, and Robert Langer

Section 4
Polymeric Micelles 315

Chapter 17 Polymeric Micelles for Formulation of Anti-Cancer Drugs. 317

Helen Lee, Patrick Lim Soo, Jubo Liu, Mark Butler, and Christine Allen

Chapter 18 PEO-Modified Poly(L-Amino Acid) Micelles for
Drug Delivery 357

Xiao-Bing Xiong, Hamidreza Montazeri Aliabadi, and Afsaneh Lavasanifar

Chapter 19 Hydrotropic Polymer Micelles for Cancer Therapeutics 385

Sang Cheon Lee, Kang Moo Huh, Tooru Ooya, and Kinam Park

Chapter 20 Tumor-Targeted Delivery of Sparingly-Soluble Anti-Cancer
Drugs with Polymeric Lipid-Core Immunomicelles 409

Vladimir P. Torchilin

Chapter 21   Combined Cancer Therapy by Micellar-Encapsulated
Drugs and Ultrasound 421

Natalya Rapoport

Chapter 22   Polymeric Micelles Targeting Tumor pH 443

Eun Seong Lee and You Han Bae

Chapter 23   cRGD-Encoded, MRI-Visible Polymeric Micelles
for Tumor-Targeted Drug Delivery 465

Jinming Gao, Norased Nasongkla, and Chalermchai Khemtong

Chapter 24   Targeted Antisense Oligonucleotide Micellar
Delivery Systems 477

Ji Hoon Jeong, Sun Hwa Kim, and Tae Gwan Park

Section 5
Dendritic Nanocarriers 487

Chapter 25   Dendrimers as Drug and Gene Delivery Systems 489

Tae-il Kim and Jong-Sang Park

Chapter 26   Dendritic Nanostructures for Cancer Therapy 509

Ashootosh V. Ambade, Elamprakash N. Savariar, and S. (Thai) Thayumanavan

Chapter 27   PEGylated Dendritic Nanoparticulate Carriers of
Anti-Cancer Drugs 523

D. Bhadra, S. Bhadra, and N. K. Jain

Chapter 28   Dendrimer Nanocomposites for Cancer Therapy 551

Lajos P. Balogh and Mohamed K. Khan

Section 6
Liposomes 593

Chapter 29   Applications of Liposomal Drug Delivery Systems
to Cancer Therapy 595

Alberto A. Gabizon

Chapter 30   Positively-Charged Liposomes for Targeting
Tumor Vasculature 613

Robert B. Campbell

Chapter 31   Cell Penetrating Peptide (CPP)–Modified Liposomal
Nanocarriers for Intracellular Drug and Gene Delivery 629

Vladimir P. Torchilin

Chapter 32 RGD-Modified Liposomes for Tumor Targeting 643

P. K. Dubey, S. Mahor, and S. P. Vyas

Chapter 33 Folate Receptor-Targeted Liposomes for
Cancer Therapy 663

Xiaobin B. Zhao, Natarajan Muthusamy, John C. Byrd, and Robert J. Lee

Chapter 34 Nanoscale Drug Delivery Vehicles for Solid Tumors:
A New Paradigm for Localized Drug Delivery
Using Temperature Sensitive Liposomes 677

David Needham and Ana Ponce

Section 7
Other Lipid Nanostructures 721

Chapter 35 Nanoemulsion Formulations for Tumor-Targeted Delivery 723

Sandip B. Tiwari and Mansoor M. Amiji

Chapter 36 Solid Lipid Nanoparticles for Antitumor Drug Delivery 741

Ho Lun Wong, Yongqiang Li, Reina Bendayan, Mike Andrew Rauth, and Xiao Yu Wu

Chapter 37 Lipoprotein Nanoparticles as Delivery Vehicles
for Anti-Cancer Agents 777

Andras G. Lacko, Maya Nair, and Walter J. McConathy

Chapter 38 DQAsomes as Mitochondria-Targeted Nanocarriers
for Anti-Cancer Drugs 787

Shing-Ming Cheng, Sarathi V. Boddapati, Gerard G. M. D'Souza, and Volkmar Weissig

Index 803

Section 1

Nanotechnology and Cancer

1 Introduction and Rationale for Nanotechnology in Cancer Therapy

Fredika M. Robertson and Mauro Ferrari

CONTENT

References 8

Although there have been significant advances in defining the fundamentals of cancer biology over the past 25 years, this has not translated into similar clinical advances in cancer therapeutics. One area that holds great promise for making such advances is the area defined as *cancer nanotechnology*, which involves the intersection of a variety of disciplines, including engineering, materials science, chemistry, and physics with cancer biology. This multidisciplinary convergence has resulted in the creation of devices and/or materials that are themselves or have essential components in the 1–1000-nm range for at least one dimension and holds the possibility of rapidly advancing the state of cancer therapeutics and tumor imaging.

This newly developing area of "nanohealth" may ultimately allow detection of human tumors at the very earliest stages, regardless of the location of the primary tumor and/or metastases, and may provide approaches to more effectively destroy tumors as well as their associated vascular supplies with fewer adverse side effects. The chapters within this book describe the most recent cutting-edge approaches in nanotechnology that are focused on overcoming the multiple barriers that have, in the past, blocked the successful treatment and ultimate eradication of human cancers.

The majority of nanotechnology-based devices useful for cancer therapeutics have been defined as nanovectors, which are injectable nanoscale delivery systems.[1,2] Nanovectors offer the promise of providing breakthrough solutions to the problems of optimizing efficacy of therapeutic agents while simultaneously diminishing the deleterious side-effects that commonly accompany the use of both single chemotherapeutic agents as well as multimodality therapeutic regimens.

In general, nanovectors are comprised of at least three constituents, which include a core material, a therapeutic and/or imaging "payload," and a biological surface modification, which aids in both appropriate biodistribution and selective localization of the nanovector and its cytotoxic and/or imaging agent. Although the first type of molecules used to enhance the selective localization and delivery of nanovectors were antibodies, more sophisticated recognition systems have been devised as a result of our expanding knowledge base in cancer biology.

For example, while the biological and molecular characteristics of human tumors of different origins continue to be defined and exploited for cancer therapeutics and tumor imaging, the significance of blood vessels that develop around actively growing tumors as potential therapeutic targets has only been widely recognized and found clinical utility within the past decade.[3,4] The process of tumor-associated angiogenesis is now known to be an essential component of tumor expansion and metastasis. This realization and the accompanied potential for development of successful cancer therapeutics, as well as for tumor imaging strategies based on tumor-associated vasculature, has opened new doors for the application of nanotechnology in cancer therapeutics. Additionally, the molecules that drive the process of tumor angiogenesis may provide a means to gauge the timing and extent of individual patient responses to cancer treatment, which can also be monitored using nanovector approaches. The importance of tumor neovasculature in cancer nano-technology is highlighted throughout the chapters in *Nanotechnology for Cancer Therapy*.

One of the most important characteristics of nanovectors is their ability to be functionalized to overcome barriers that block access of agents used for treatment of tumors and for imaging of tumors and their associated vasculature. These biological barriers are numerous and complex. One such barrier is the blood–brain barrier, which prevents access to brain malignancies, compounding the difficulties in their successful treatment. One example of the potential utility of nanovectors in overcoming this biobarrier, which is critical to treatment of malignant brain tumors, is the use of nanoparticles in combination with boron neutron capture therapy (BCNT). The current approaches used for BCNT combining nanoparticles and the potential for this nanotechnology-based approach to treatment of brain tumors is described in detail in a chapter in *Nanotechnology for Cancer Therapy*. Additional biobarriers that must be overcome include epithelial–endothelial cell barriers, the barriers presented by the markedly tortuous structures that are characteristic of angiogenic vasculature associated with tumors, as well as the barrier set up by the rapid uptake of nanovectors by resident macrophages within the reticuloendothelial system (RES) which may prevent nano-vectors from reaching their targeted location.

To achieve breakthrough advances in cancer therapeutics, there are two related and essential components which must be addressed. The first issue in successful use of nanovectors is recognition of the tumor and the second is the ability of the nanovector to reach the site of the tumor and associated blood vessels. The goal is to preferentially achieve high concentrations of a specific chemotherapeutic agent, a tumor imaging agent, and/or gene therapies at the site(s) of tumors and associated vasculature. In addition, nanovectors must be able to deliver an active agent to achieve effective anti-tumor treatment, or tumor imaging, which is essential for tumor diagnosis and for monitoring the extent and timing of an individual patient's response to anti-tumor therapy.

The first nanotechnology-based approach to be used as a means of delivering cancer chemotherapy was liposomes, which is a type of nanovector made of lipids surrounding a water core. Liposomes are the simplest form of nanovector and their utility is based on the significant difference in endothelial structures—defined as *fenestrations*—between normal vasculature and tumor-associated vessels. The increase in fenestrations in tumor neovasculature allows the prefer-ential concentration of liposome-encapsulated anti-tumor agent in close proximity to the local tumor site, a phenomenon defined as enhanced penetration and retention (EPR), which is considered to be a characteristic of passive targeting of tumors. Both the passive targeting of solid tumors as well as approaches for active targeting of nanovectors such as liposomes are described in the following chapters.

The first studies to demonstrate the proof of principle that liposomes could be used as a nanovector for effective delivery of an active cancer therapeutic was based on the encapsulation of the anthracycline antibiotic doxorubicin, which is an effective and commonly used first line chemotherapeutic agent for both leukemia and solid tumors. Although use of doxorubicin is associated with dose limiting cardiotoxicity, when anionic liposomes were used to encapsulate doxorubicin, preclinical studies found that the anti-tumor action of doxorubicin was significantly enhanced and the cardiotoxicity was significantly diminished.[5,6] Liposome encapsulated doxoru-bicin has now been clearly demonstrated safe and efficacious clinically in such malignancies as breast and ovarian cancer,[7–9] allowing for greater amounts of this agent to be administered than the recommended lifetime cumulative dose of free doxorubicin, which was 450–550 mg/m^2.

Importantly, liposomes have been shown to be an effective means of delivering a diverse group of anti-tumor agents such as doxorubicin, as well as the poorly soluble drug paclitaxel.[10]

As evidenced by the chapters in this volume, interest in improving the use of liposomes for enhancing the selective localization and delivery of cancer therapeutics to tumors and tumor-associated blood vessels remains an active area of investigation. Because liposomes can be modified with respect to composition, charge, and external coating, liposomes remain a versatile nanotechnology platform, historically serving as the prototypic nanovector.

An approach that was first used with liposome nanovectors that has found utility with other types of nanovectors is "PEGylation," a modification of liposome surface characteristics using poly(ethylene glycol) (PEG), resulting in what has been defined as "sterically stabilized liposomes." This PEG modification of liposomes provides protection against uptake by resident macrophages within the RES biobarrier, thus increasing the circulation time of liposome-encapsulated anti-tumor agent, resulting in significantly increased therapeutic efficacy.[11]

This approach to overcoming rapid uptake and destruction by resident macrophages within the RES by PEGylation has been used alone or in tandem with other modifications of liposomes to aid in avoiding or overcoming biobarriers and to more selectively localize nanovectors. For example, based on the recognition that a wide variety of solid tumors as well as hematologic malignancies express amplified levels of the folate receptor on their cell surface, liposome nanovectors have been developed to target folate receptors.[12,13] Using the strategy of biomarker-based targeting, folate receptor targeted liposomes have been shown to not only overcome multidrug resistance associated with P-glycoprotein that often occurs in human tumors, but also have proven to be antiangiogenic.[13] Another means of more selectively localizing liposome-encapsulated anti-tumor agents to tumor-associated vasculature is to add a cationic charge,[10,14] which has been shown to double the accumulation of anti-tumor drug encapsulated into liposomes selectively within vessels surrounding tumors. Other approaches to biomolecular targeting have included the use of specific proteins such as vasoactive intestinal peptide (VIP)[15] to target gastric tumors as well as signature RGD amino acid sequences expressed on the surface of endothelial cells that are associated with expression of $\alpha_v\beta_3$ integrin on angiogenic endothelium.[16]

In addition to serving as cancer drug delivery systems, liposomes have been shown to be an effective means for delivery of other agents such as genes and antisense oligonucleotides, and would allow access of such entities as small interfering RNA.[13,17] As an example, peptides such as cell penetrating peptides (CPPs) have been used to modify liposomes, allowing them to be used as a means for intracellular drug delivery and delivery of proteins such as antibodies and genes, as well as providing a window for cellular imaging.[17]

In addition to liposomes as an example of the prototype of nanovectors, there are many others now available in the nanotechnology toolbox that are being investigated for use in cancer therapeutics and tumor imaging. Polymer-based nanovectors are a specific area of interest in cancer therapeutics and a number of polymer-based nanovector systems are described in the following chapters. For the purposes of this volume, polymer-based nanovectors have been subdivided into several categories, including polymer conjugates, polymeric nanoparticles, and polymeric micelles.

Polymeric conjugates that have been investigated for use as nanovector systems for delivery of anti-tumor agents include the water soluble biocompatible polymer N-(2-hydroxypropyl)methacrylamide (HPMA). This polymeric conjugate has been targeted based on overexpression of hyaluronan (HA) receptors that are present on cancer cells. HPMA–HA polymeric conjugate drug delivery nanovector systems have been developed to carry a doxorubicin "payload" and have been shown to have selectivity for endocytosis of the targeted polymeric nanovector to breast, ovarian, and colon tumor cells, compared to an HPMA polymer with doxorubicin but lacking the HA targeting conjugate.[18,19] Copolymer–peptide conjugates have also been developed as nanovectors using HPMA in combination with RGD targeting peptides,[20] which molecularly target radiotherapeutic agents to tumor-associated vasculature, resulting in both antiangiogenic as well as anti-tumor activity. In addition to serving as nanovectors for delivery of chemotherapy and radiotherapy, copolymer nanovectors show promise as agents to deliver biodegradable blood-pool contrast agents for use with tumor imaging methodologies. Other copolymers that have been used with gadolinium chelates for improvements in MRI based contrast agents include poly(L-glutamic acid)[21] as well as (Gd-DTPA)-cystine copolymers (GDCP) and (Gd-DTPA)-cystine diethyl ester copolymers (GDCEP).[22,23] As with liposome nanovectors, polymer conjugates can be targeted to a specific site, can carry a multitude of agents useful for cancer chemotherapeutics, radiotherapy, or tumor imaging, and their surfaces can be modified using different biological surface modifiers to enhance selectivity of localization of the nanovector.

Studies focusing on polymeric nanoparticles as nanovectors have included the use of poly(lactic-co-glycolic acid) (PLGA) and polylactide (PLA) that can be selectively targeted against such surface receptors as the transferrin receptor on breast tumor cells, as well as can be optimized to avoid biobarriers such as the RES. In addition, polymeric nanoparticles can be altered to enhance the intracellular drug concentration once delivered. Polymeric nanoparticles are also mutable for enhancement of retention at the local tumor site and to combine anti-tumor agents with antiangiogenesis strategies.[24–26] In addition to delivery of anti-tumor agents that are antiangiogenic, nanovectors such as PLGA and PLA polymers as well as PEGylated gelatin nanoparticles have been developed. In the case of PEGylated gelatin nanovectors, these have been improved by thiolation to optimize their time in circulation and to enhance delivery of payloads such as non-viral DNA for gene therapy.[25,27–29]

Relatively new polymeric nanovectors recently shown to have potential utility in cancer nanotechnology are bioconjugates composed of PLGA polymers and nucleic acid ligands defined as aptamers.[30] Because these nanovectors can be PEGylated and further modified with selectively targeted approaches directed against such proteins as prostate specific antigen (PSA) which are present in abundance on the surface of prostatic tumor cells, these nanoparticle–aptamer bioconjugates represent a new class of polymeric nanoparticles with promise as multifunctional nanovectors. Polymeric aptamer bioconjugates are currently being evaluated for their optimization for use in a large number of systems that may lead to in vivo studies in the near future.[31]

Another type of nanovector discussed in Nanotechnology for Cancer Therapy includes polymeric micelles, which can be defined as self-assembling nanosized colloidal particles with a hydrophobic core and hydrophilic shell.[32] Polymeric micelles have been commonly used in the field of drug delivery for over two decades. The advantages of these nanovectors include the ability to encapsulate agents with poor aqueous solubility profiles such as has been observed with the anti-tumor agent paclitaxel[33] as well as with the photodynamic therapy agent, meso-tetratphenylporphine (TPP).[34] Other advantages associated with the use of polymeric micelles include the ability to control both drug targeting using immunotargeting[34] and rates of drug release.[32,35] The efficacy of agents encapsulated within polymeric micelle nanovectors occurs via the passive targeting method defined as enhanced penetration and retention (EPR), which results in diminished adverse side effects usually observed in normal tissues by administration of cytotoxic cancer chemotherapeutic agents.

In addition to the utility of polymeric micelles for delivery of anti-cancer agents,[32–35] these nanovectors are also described as an effective means for the combined delivery of anti-tumor agents as well as simultaneous tumor imaging using focused ultrasound imaging. This makes it possible to non-invasively penetrate tumors that are typically difficult to access, such as with ovarian cancer.[36] Polymeric micelles have also found usefulness as multifunctional nanovectors that are both visible to MRI imaging modalities using RGD-based targeting of the tumor-associated vasculature combined with drug delivery. Other approaches using polymeric micelles have focused on targeting the acidic extracellular pH of the tumor microenvironment.[37] Polymeric micelles have also been evaluated for delivery of antisense oligonucleotides using folate receptor targeting strategies.[38] Although polymeric micelle nanovectors have been used as strategies for drug delivery, it is clear that they continue to have promise in cancer nanotechnology.

Another class of nanovectors described in the following chapters is dendrimers, which are self assembling synthetic polymers possessing tunable nanoscale dimensions that assume conformations that are dependent upon the microenvironment. The development of polymeric micelles preceded the development of dendrimers. However, due to the characteristics of dendrimers as "perfectly branched monodispersed macromolecules,"[39] this class of nanovectors may be suitable as exquisitely sensitive carriers for defined release of anti-tumor agents, as well as very useful for targeted delivery of such molecules as contrast agents for tumor imaging.

Dendritic nanocarriers have been evaluated for their capability to serve as nanovectors for delivery of drugs and for plasmid DNA, suggesting that dendrimer nanovectors may be useful for gene therapy.[40] Like liposomes, dendrimers can also be modified by PEGylation[41] to increase their bioavailability and half-life in the circulation. Dendrimers have been shown to be suitable for enhancing the solubility of drugs, and to have suitable profiles for controlled delivery as well as selective delivery of drugs such as flurbiprofen that are active within a local site of inflammation.[42]

Although dendrimers that have been evaluated for potential utility in cancer therapeutics have primarily been composed of polyamidoamine (PAMAM), there are also gold/dendrimer nanocomposites[43] and silver/dendrimer nanocomposite materials[44] that have been developed and are currently being evaluated for their comparative chemical and biological characteristics with potential utility as biomarkers, as well as for use in cancer therapeutics, tumor imaging, and radiotherapy.

The varied nature of nanovectors covered in *Nanotechnology for Cancer Therapy* is illustrated by such novel classes of nanocarriers as DQAsomes, which are mitochondrial-targeted nanovectors capable of delivery of anti-cancer drugs as well as carriers of gene therapy directed against this organelle.[44,45] The development of this nanovector delivery approach is based on the relatively recent realization that a number of genetic disorders associated with defects in oxidative metabolism are associated with genetic defects in mitochondrial DNA, with a significant lack of available therapies. While the majority of nanovectors target tumors and/or tumor-associated vasculature, the DQAsomes are composed of derivatives of the self-assembling mitochondriotropic bola-amphiphile dequalinium chloride, which forms cationic vesicles which can bind and transport DNA to mitochondria in living mammalian cells.[44]

Another approach using selective targeting of more classically used characteristics of cancer cells includes the development of lipoprotein nanoparticles as delivery vehicles for anti-cancer therapy.[46] These high density lipoprotein (rHDL) nanoparticles are composed of phosphatidylcholine, apolipoprotein A-1, cholesterol, and cholesteryl esters with encapsulated paclitaxel. They were shown to be useful as a nanovector for anti-tumor drug delivery based on the ability of cancer cells to selectively acquire HDL, while the use of these nanovectors decreased the side effects normally observed with this type of anti-tumor therapy.

Given the wide variety of nanovectors and nanoplatforms, *Nanotechnology for Cancer Therapy* cannot possibly cover each nanoscale application that has potential for use in anti-tumor therapy and tumor imaging. One nanoplatform that should be mentioned is the so-called "quantum dots" (Q-dots) that have recently received wide attention. Although cadmium and selenium crystalline nanovectors were first used over two decades ago, Q-dots have recently received attention for their utility as tunable multicolor nanoparticles that can be used for in vivo tumor imaging in animal models of cancer as well as for multiplex detection of mRNAs and proteins. Although the current composition of the Q-dots nanovectors prohibits their clinical use for cancer therapy, there is significant promise for the potential for development of biocompatible nanovectors which are tunable to wavelengths useful for tumor imaging. They may provide the ability to selectively target tumors as well as serve as multifunctional nanovector devices for cancer therapy, imaging, diagnostics, and monitoring of response to anti-tumor therapy.

Although the development of nanoplatforms for cancer therapeutics is a critical component to the successful use of nanovectors for anti-tumor therapy and tumor imaging, there are other necessary components for nanoplatforms to gain clinical acceptance. One such parameter is the necessary infrastructure to characterize nanoplatforms and to ensure that the rapid translation of discoveries in this area proceeds from preclinical to clinical use. The recognition by the National Cancer Institute (NCI) that nanodevices and nanomaterials will be critical in making significant advances in imaging, diagnosis, and treatment of human tumors has lead to the development of The Nanotechnology Characterization Laboratory (NCL). This laboratory provides critical support to the NCI's Alliance in Nanotechnology for Cancer. The role and function of the services provided by NCL is supported by the NCI, and works in concert with the National Institute of Standards and Technology (NIST) and the U.S. Food and Drug Administration (FDA). The NCL provides a valuable service to those investigators working in the area of cancer nanotechnology and performs preclinical efficacy and toxicity testing of nanoscale devices and materials. The goal of the NCL is to support the rapid acceleration of preclinical development of nanovectors to support IND application and subsequent clinical trials.

The continued development and ultimate success of nanotechnology-based strategies for cancer therapeutics, imaging, and tracking patient therapeutic response will be dependent upon a number of parameters. A multidisciplinary approach to achieving success in cancer nanotechnology is absolutely necessary. While remaining grounded in our basic knowledge of the known targets that are critical for localization of multifunctional nanovectors to the tumor and surrounding tissue, researchers must be aware of new targets that have been identified as being critically involved in the response of tumors and tumor vasclulature to anti-tumor agents, and to recognize and take full advantage of known targets such as surface receptors that are amplified on tumor cells as well as recognize and incorporate newer targets such as mitochondrial DNA and HDL uptake. For example, while researchers continue to identify novel means to functionalize nanovectors to eliminate biobarriers that historically have limited access to tumors and their associated angiogenic vasculature and stroma, we will still need to produce and test multifunctional nanovectors capable of selective localization and delivery of agents that function both as cytotoxic agents as well as imaging agents.

Because the size of nanodevices and nanomaterials is similar to that of naturally occurring components of cells, including surface receptors, organelles such as mitochondria, lipoproteins, and other, as yet unidentified molecules such as siRNA, microRNA, genes, and unknown proteins, nanodevices and nanomaterials can easily interface with biological molecules. The mutability of nanotechnology platforms has allowed their rapid introduction into the field of basic cancer biology and the specialty of oncology. The approaches described in the chapters within *Nanotechnology for Cancer Therapy* range from liposomes, which are among the first and the simplest nanovectors to be used for delivery of cytotoxic drugs, to advanced nanoscale devices for use as imaging contrast agents in conjunction with ultrasound and magnetic resonance imaging (MRI). Each of the nanotechnology approaches described in this book will be absolutely necessary to achieve significant breakthroughs in cancer therapeutics. It is truly an exciting time to be an integral part of this field as these discoveries in cancer nanotechnology are made and put into practice with the goal to use nanovectors and nanotechnology platforms to significantly impact cancer morbidity and mortality in the next decade.

REFERENCES

1. Ferrari, M., Nanovector therapeutics, *Curr. Opin. Chem. Biol.*, 9, 343–346, 2005.
2. Ferrari, M., Cancer nanotechnology: Opportunities and challenges, *Nat. Rev. Cancer*, 5(3), 161–171, 2005.
3. Folkman, J., Fundamental concepts of the angiogenic process, *Curr. Mol. Med.*, 3(7), 643–651, 2003.
4. Ferrara, N. and Kerbel, R. S., Angiogenesis as a therapeutic target, *Nature*, 438, 967–974, 2005.
5. Treat, J., Greenspan, A., Forst, D., Sanchez, J. A., Ferrans, V. J., Potkul, L. A., Woolley, P. V., and Rahman, A., Antitumor activity of liposome-encapsulated doxorubicin in advanced breast cancer: Phase II study, *J. Natl. Cancer Inst.*, 82(21), 1706–1710, 1990.

6. Forssen, E. A. and Tokes, Z. A., Improved therapeutic benefits of doxorubicin by entrapment in anionic liposomes, *Cancer Res.*, 43(2), 546–550, 1983.

7. Straubinger, R. M., Lopez, N. G., Debs, R. J., Hong, K., and Papahadjopoulos, D., Liposome-based therapy of human ovarian cancer: Parameters determining potency of negatively charged and antibody-targeted liposomes, *Cancer Res.*, 48(18), 5237–5245, 1988.

8. Robert, N. J., Vogel, C. L., Henderson, I. C., Sparano, J. A., Moore, M. R., Silverman, P., Overmoyer, B. A. et al., The role of the liposomal anthracyclines and other systemic therapies in the management of advanced breast cancer, *Semin. Oncol.*, 31(6 Suppl 13), 106–146, 2004.

9. Ewer, M. S., Martin, F. J., and Henderson, Use of anionic liposomes for the reduction of chronic doxorubicin-induced cardiotoxicity, *Proc. Natl. Acad. Sci. USA*, 78(3), 1873–1877, 1981.

10. Campbell, R. B., Balasubramanian, S. V., and Straubinger, R. M., Influence of cationic lipids on the stability and membrane properties of paclitaxel-containing liposomes, *J. Pharm. Sci.*, 90(8), 1091–1105, 2001.

11. Gabizon, A., Shmeeda, H., Horowitz, A. T., and Zalipsky, S., Tumor cell targeting of liposome-entrapped drugs with phospholipid-anchored folic acid-PEG conjugates, *Adv. Drug Deliv. Rev.*, 56(8), 1177–1192, 2004.

12. Gabizon, A., Horowitz, A. T., Goren, D., Tzemach, D., Shmeeda, H., and Zalipsky, S., In vivo fate of folate-targeted polyethylene-glycol liposomes in tumor-bearing mice, *Clin. Cancer Res.*, 9(17), 6551–6559, 2003.

13. Pan, X. and Lee, R. J., Tumour-selective drug delivery via folate receptor-targeted liposomes, *Expert Opin. Drug Deliv.*, 1(1), 7–17, 2004.

14. Campbell, R. B., Fukumura, D., Brown, E. B., Mazzola, L. M., Izumi, Y., Jain, R. K., Torchilin, V. P., and Munn, L. L., Cationic charge determines the distribution of liposomes between the vascular and extravascular compartments of tumors, *Cancer Res.*, 62(23), 6831–6836, 2002.

15. Sethi, V., Onyuksel, H., and Rubinstein, I., Liposomal vasoactive intestinal peptide, *Methods Enzymol.*, 391, 377–395, 2005.

16. Dubey, P. K., Mishra, V., Jain, S., Mahor, S., and Vyas, S. P., Liposomes modified with cyclic RGD peptide for tumor targeting, *J. Drug Target.*, 12(5), 257–264, 2004.

17. Gupta, B., Levchenko, T. S., and Torchilin, V. P., Intracellular delivery of large molecules and small particles by cell-penetrating proteins and peptides, *Adv. Drug Deliv. Rev.*, 57(4), 637–651, 2005.

18. Luo, Y. and Prestwich, G. D., Cancer-targeted polymeric drugs, *Curr. Cancer Drug Targets*, 2(3), 209–226, 2002.

19. Luo, Y., Bernshaw, N. J., Lu, Z. R., Kopecek, J., and Prestwich, G. D., Targeted delivery of doxorubicin by HPMA copolymer-hyaluronan bioconjugates, *Pharm. Res.*, 19(4), 396–402, 2002.

20. Mitra, A., Nan, A., Papadimitriou, J. C., Ghandehari, H., and Line, B. R., Polymer–peptide conjugates for angiogenesis targeted tumor radiotherapy, *Nucl. Med. Biol.*, 33(1), 43–52, 2006.

21. Wen, X., Jackson, E. F., Price, R. E., Kim, E. E., Wu, Q., Wallace, S., Charnsangavej, C., Gelovani, J. G., and Li, C., Synthesis and characterization of poly(L-glutamic acid) gadolinium chelate: a new biodegradable MRI contrast agent, *Bioconjug. Chem.*, 15(6), 1408–1415, 2004.

22. Wang, X., Feng, Y., Ke, T., Schabel, M., and Lu, Z. R., Pharmacokinetics and tissue retention of (Gd-DTPA)-cystamine copolymers, a biodegradable macromolecular magnetic resonance imaging contrast agent, *Pharm. Res.*, 22(4), 596–602, 2005.

23. Zong, Y., Wang, X., Goodrich, K. C., Mohs, A. M., Parker, D. L., and Lu, Z. R., Contrast-enhanced MRI with new biodegradable macromolecular Gd(III) complexes in tumor-bearing mice, *Magn. Reson. Med.*, 53(4), 835–842, 2005.

24. Brannon-Peppas, L. and Blanchette, J. O., Nanoparticle and targeted systems for cancertherapy, *Adv. Drug Deliv. Rev.*, 56(11), 1649–1659, 2004.

25. Labhasetwar, V., Nanotechnology for drug and gene therapy: the importance of understanding molecular mechanisms of delivery, *Curr. Opin. Biotechnol.*, 16(6), 674–680, 2005.

26. Sahoo, S. K. and Labhasetwar, V., Enhanced antiproliferative activity of transferrin-conjugated paclitaxel-loaded nanoparticles is mediated via sustained intracellular drug retention, *Mol. Pharm.*, 2(5), 373–383, 2005.

27. Kommareddy, S., Tiwari, S. B., and Amiji, M. M., Long-circulating polymeric nanovectors for tumor-selective gene delivery, *Technol. Cancer Res. Treat.*, 4(6), 615–625, 2005.

28. Kommareddy, S. and Amiji, M., Preparation and evaluation of thiol-modified gelatin nanoparticles for intracellular DNA delivery in response to glutathione, *Bioconjug. Chem.*, 16(6), 1423–1432, 2005.
29. Kaul, G. and Amiji, M., Tumor-targeted gene delivery using poly(ethylene glycol)-modified gelatin nanoparticles: in vitro and in vivo studies, *Pharm. Res.*, 22(6), 951–961, 2005.
30. Farokhzad, O. C., Jon, S., Khademhosseini, A., Tran, T. N., Lavan, D. A., and Langer, R., Nanoparticle–aptamer bioconjugates: a new approach for targeting prostate cancer cells, *Cancer Res.*, 64(21), 7668–7672, 2004.
31. Farokhzad, O. C., Khademhosseini, A., Jon, S., Hermmann, A., Cheng, J., Chin, C., Kiselyuk, A., Teply, B., Eng, G., and Langer, R., Microfluidic system for studying the interaction of nanoparticles and microparticles with cells, *Anal. Chem.*, 77(17), 5453–5459, 2005.
32. Torchilin, V. P., Lipid-core micelles for targeted drug delivery, *Curr. Drug Deliv.*, 2(4), 319–327, 2005.
33. Zeng, F., Liu, J., and Allen, C., Synthesis and characterization of biodegradable poly(ethylene glycol)-block-poly(5-benzyloxy-trimethylene carbonate) copolymers for drug delivery, *Biomacromolecules*, 5(5), 1810–1817, 2004.
34. Roby, A., Erdogan, S., and Torchilin, V. P., Solubilization of poorly soluble PDT agent, mesotetraphenylporphin, in plain or immunotargeted PEG-PE micelles results in dramatically improved cancer cell killing in vitro, *Eur. J. Pharm. Biopharm.*, 62(3), 235–240, 2006.
35. Aliabadi, H. M. and Lavasanifar, A., Polymeric micelles for drug delivery, *Expert Opin. Drug Deliv.*, 3(1), 139–162, 2006.
36. Rapoport, N., Combined cancer therapy by micellar-encapsulated drug and ultrasound, *Int. J. Pharm.*, 277(1,2), 155–162, 2004.
37. Gao, Z. G., Lee, D. H., Kim, D. I., and Bae, Y. H., Doxorubicin loaded pH-sensitive micelletargeting acidic extracellular pH of human ovarian A2780 tumor in mice, *J. Drug Target.*, 13(7), 391–397, 2005.
38. Jeong, J. H., Kim, S. H., Kim, S. W., and Park, T. G., In vivo tumor targeting of ODN-PEG-folic acid/PEI polyelectrolyte complex micelles, *J. Biomater. Sci. Polym. Ed.*, 16(11), 1409–1419, 2005.
39. Ambade, A. V., Savariar, E. N., and Thayumanavan, S., Dendrimeric micelles for controlled drug release and targeted delivery, *Mol. Pharm.*, 2(4), 264–272, 2005.
40. Lee, J. H., Lim, Y. B., Choi, J. S., Lee, Y., Kim, T. I., Kim, H. J., Yoon, J. K., Kim, K., and Park, J. S., Polyplexes assembled with internally quaternized PAMAM-OH dendrimer and plasmid DNA have a neutral surface and gene delivery potency, *Bioconjug. Chem.*, 14(6), 1214–1221, 2003.
41. Bhadra, D., Bhadra, S., Jain, P., and Jain, N. K., Pegnology: a review of PEG-ylated systems, *Pharmazie*, 57(1), 5–29, 2002.
42. Asthana, A., Chauhan, A. S., Diwan, P. V., and Jain, N. K., Poly(amidoamine) (PAMAM) dendritic nanostructures for controlled site-specific delivery of acidic anti-inflammatory active ingredient, *AAPS Pharm. Sci. Tech.*, 6(3), E536–E542, 2005.
43. Khan, M. K., Nigavekar, S.S, Mine, L. D., Kariapper, M. S., Nair, B. M., Lesniak, W. G., and Balogh, L. P., In vivo biodistribution of dendrimers and dendrimer nanocomposites: Implications for cancer imaging and therapy, *Technol. Cancer Res. Treat.*, 4(6), 603–613, 2005.
44. Lesniak, W., Bielinska, A. U., Sun, K., Janczak, K. W., Shi, X., Baker, J. R., and Balogh, L. P., Silver/dendrimer nanocomposites as biomarkers: Fabrication, characterization, in vitro toxicity, and intracellular detection, *Nano Lett.*, 5(11), 2123–2130, 2005.
45. D'Souza, G. G., Boddapati, S. V., and Weissig, V., Mitochondrial leader sequence: Plasmid DNA conjugates delivered into mammalian cells by DQAsomes co-localized with mitochondria, *Mitochondrion*, 5(5), 352–358, 2005.
46. D'Souza, G. G., Rammohan, R., Cheng, S. M., Torchilin, V. P., and Weissig, V., DQAsomemediated delivery of plasmid DNA toward mitochondria in living cells, *J. Controlled Release*, 92(1,2), 189–197, 2003.

2 Passive Targeting of Solid Tumors: Pathophysiological Principles and Physicochemical Aspects of Delivery Systems

S. M. Moghimi

CONTENTS

2.1 Introduction 11
2.2 Barriers to Extravasation..12
2.3 Selected Delivery Systems...12
    2.3.1 Liposomes...12
    2.3.2 Polymeric Nanoparticles 14
    2.3.3 Nanotechnology-Derived Nanoparticles 15
    2.3.4 Macromolecular and Related Delivery 16
2.4 Conclusions 16
References 17

2.1 INTRODUCTION

A solid tumor comprises two major cellular components: the tumor parenchyma and the stroma; the latter incorporating the vasculature and other supporting cells. As the tumor grows, in order to meet the metabolic requirements of an expanding population of tumor cells, the pre-existing blood vessels become subject to intense angiogenic pressure. Several factors produced by tumor cells and infiltrating immune-competent effector cells in the tumor parenchyma are believed to signal the development of new capillaries from the pre-existing vessels by capillary sprouting and/or dysregulated intussusceptive microvascular growth.[1] Further, in many solid tumors, endothelial cells destined to create new vessels are recruited not only from nearby vessels, but also to a significant extent from precursor cells within the bone marrow (so-called endothelial progenitor cells), a process referred to as "vasculogenesis."[2]

Scanning electron microscopy of microvascular corrosion casts has allowed visualization of the geometry of blood vessel architecture in solid tumors. From these studies, it has become apparent that tumor blood vessels are highly irregular and show gross architectural changes that differ from those in normal organs and from newly formed blood vessels, such as those found in wound healing and in other angiogenic sites.[1,3] For instance, the thickness of a tumor blood vessel wall is poorly correlated to its diameter. Therefore, despite the large size of some tumor vessels, the tumor blood flows is chaotic, with high flow rates in some segments and stagnation in others.[1] Also, the blood flow may temporarily change direction within individual tumor vessels.

Further, the structure and organization of the endothelial cells, pericytes, and vascular basement membrane of tumor vessels are all abnormal.[1,3–6] One consistent abnormality of tumor blood vessels is their high permeability to macromolecules, arising from irregularly shaped and loosely inter-connected endothelial cells (where the size of fenestrae often ranges from 200 to 2000 nm) and their less frequent and intimate association with pericytes and the vascular basement membrane.[1,3–5] Marked variability has been noted in endothelial permeability among different tumors, different vessels within the same tumor, and during tumor growth, regression, and relapse. The extent of tumor blood vessel permeability is also controlled by the host microenvironment, and increases with the histological grade and malignant potential of tumors.[7,8]

Given the potency and toxicity of modern pharmacological agents, tissue selectivity is a major issue. In the delivery of chemotherapeutic agents to solid tumors this is particularly critical, since the therapeutic window for these agents is often small and the dose–response curve steep. Therefore, the idea of exploiting the well-documented vascular abnormalities of tumors, restricting penetration into normal tissue interstitium while allowing freer access to that of the tumor, becomes particularly attractive.

2.2 BARRIERS TO EXTRAVASATION

As a consequence of temporal and spatial heterogeneity in tumor blood flow, solid tumors usually contain well-perfused, rapidly growing regions, and poorly perfused, often necrotic areas.[1,3] As in normal tissues, diffusive and convective forces govern the movement of molecules into the inter-stitium of tumors. However, diffusion is believed to play a minor role in the movement of solutes across the endothelial barrier in comparison with bulk fluid flow. Examination of pressure gradients in experimental tumors has suggested that the movement of macromolecules and particulate materials out of the tumor blood vessels and into the extra-vascular compartment is remarkably limited. This has been attributed to a higher-than-expected interstitial pressure, in part due to a lack of functional lymphatic drainage, coupled with lower intra-vascular pressure.[3] In addition, inter-stitial pressure tends to be higher at the center of solid tumors, diminishing towards the periphery, creating a mass flow movement of fluid away from the central region of the tumor.[3] For example, the measured interstitial fluid pressure in invasive breast ductal carcinoma was 29 ± 3 mm Hg, compared with 3.0 ± 0.8 mm Hg in patients with benign tumors and -3.0 ± 0.1 mm Hg in patients with normal breast parenchyma.[9] Nevertheless, the lower interstitial pressure in the periphery still permits adequate extravasation of fluid and macromolecules.

These pathophysiological characteristics have serious implications for the systemic delivery of not only low-molecular-weight and macromolecular agents, but also particulate delivery vehicles. Simply enhancing the plasma half-life of these agents (e.g., long-circulating carriers) will not necessarily lead to an increase in therapeutic effect.[10] Furthermore, distribution, organization, and relative levels of collagen, decorin, and hyaluronan also impede the diffusion of extravasated macromolecules and particulate systems in tumors.[11] Thus, diffusion of macromolecules and particles will vary with tumor types, anatomical locations, and possibly by factors that influence extracellular matrix composition and/or structure.

2.3 SELECTED DELIVERY SYSTEMS

2.3.1 LIPOSOMES

Liposomes are perhaps the best studied vehicles in cancer drug delivery, capable of either increasing the drug concentration in solid tumors and/or limiting drug exposure to critical target sites such as bone marrow and myocardium.[10,12] For example, Myocet™ is a liposomal formulation of doxorubicin (an inhibitor of topoisomerase II) approximately 190 nm in size that was approved by the European Agency for the Evaluation of Medicinal Products (EMEA) in 2000 for the treatment of metastatic breast cancer. This formulation provides a limited degree of prolonged circulation when compared with doxorubicin in the free form. Myocet™ releases more than half of its associated doxorubicin within a few hours of administration and 90% within 24 h. Similar to intravenously injected nanoparticulate systems, liposomes are rapidly intercepted by macrophages of the reticuloendothelial system.[10] Hepatic deposition of Myocet™ could lead to gradual release of the cytotoxic agent back to the systemic circulation (a macrophage depot system), as well as induction of Kupffer cell apoptosis.[10] Following apoptosis, restoration of Kupffer cells may take up to two weeks.[13] A potentially harmful effect is the occurrence of bacteriemia during the period of Kupffer cell deficiency. Although Myocet™ administration decreases the frequency of cardiotoxicity and neutropenia compared with free drug,[14] there is still controversy as to whether liposomal encapsulation exhibits equivalent efficacy to doxorubicin.[15]

Macrophage deposition of intravenously administered liposomes can be markedly minimized either by bilayer or surface modification.[10,16] Regulatory approved examples include DaunoXome®, a daunorubicin-encapsulated liposome 45 nm in size with a rigid bilayer for HIV-related Kaposi's sarcoma, and Doxil®/Caelyx®, a poly(ethylene glycol)-grafted rigid vesicle of 100-nm diameter with encapsulated doxorubicin for HIV-related Kaposi's sarcoma and refractory ovarian carcinoma. As a result of their small size, rigid bilayer, and hydrophilic surface display (as in the case of Doxil®), these formulations exhibit poor surface opsonization, a process that limits vesicle recognition by macrophages in contact with the blood and consequently prolongs their residency time within the vasculature.[10,16] For instance, Doxil® has a biphasic circulation half-life of 84 min and 46 h in humans. In addition, Doxil® also has a high drug loading capacity; here doxorubicin is loaded actively by an ammonium sulfate gradient (as doxorubicin sulfate) yielding liposomes with a high content of doxorubicin aggregates, which remain highly stable within the vasculature with minimum drug loss.[17] Therefore, it is not surprising to see that such liposomal formulations exhibit favorable pharmacokinetics when compared with the free drug. For example, the area under the curve after a dose of 50 mg/m^2 doxorubicin encapsulated in long-circulating liposomes is approximately 300-fold greater than that of free doxorubicin.[17] In addition, clearance and volume of distribution are reduced by at least 250- and 60-fold, respectively.[17] However, as a result of their prolonged circulation times, alternative toxic reactions have been reported with such vehicles. The most notable dose-limiting toxicity associated with continuous infusion of Doxil® is palmar–plantar erythrodysesthesis.[17]

Following extravasation into solid tumors, long-circulating liposomes often distribute heterogeneously in perivascular clusters that do not move significantly and poorly interact with cancer cells.[18] Therefore, the efflux of drug must follow the process of liposome extravasation at a rate that maintains free drug levels in the therapeutic range. The rate of drug release from liposomes not only depends on the composition of the interstitial fluid surrounding tumors but also on the drug type and encapsulation procedures. The importance of the latter is highlighted by the observation that extravasated long-circulating cisplatin-containing liposomes (where cisplatin is loaded passively) lack anti-tumor activity, whereas cisplatin in free form is capable of inserting cytotoxicity.[19] This is in contrast to the effective anti-tumor property of the same liposomal lipid composition containing entrapped doxorubicin. It is believed that nonspecific chemical disruption or collapse of the liposomal pH gradient, that is used to load liposomes actively with doxorubicin, may trigger doxorubicin release.[17]

Long-circulating liposomes have the capability to deliver between 3 and 10 times more drug to solid tumors compared with the administered drug in its free form. If the entrapped drugs are released from extravasated liposomes, it is very likely that these vesicles inherently overcome a certain degree of multidrug resistance by the tumor cells. Thus, tumor regression is to be expected with tumors exhibiting a low resistance factor. With tumors exhibiting higher resistance levels, due to over-expression of energy-dependent efflux pumps such as P-glycoprotein and multidrug-resistance-related protein, alternative approaches are necessary. One effective strategy is to use long-circulating temperature-sensitive liposomes in conjugation with hyperthermia, but this approach has limited applicability for visceral and widespread malignancies.[20] Others have elaborated on biochemical triggers such as cleaveable poly(ethylene glycol)-phospholipid conjugates to generate fusion competent vesicles[21] and enzyme-mediated liposome destabilization and pore formation.[22–24] Examples of the latter include long-circulating liposomes with attached protease-sensitive haemolysin[22] and pro-drug ether liposomes (e.g., vesicles containing phospholipids with a nonhydrolyzable ether bond in the 1-position), which are susceptible to degradation by secretory phospholipases.[23,24] For instance, the level of secretory phospholipases (such as the secretory phospholipase A2) is dramatically elevated within the interstitium of various tumors. Secretory phospholiapse A2 not only acts as a trigger resulting in the release of encapsulated cytotoxic drugs from pro-drug ether liposomes, but also generates highly cytotoxic lysolipids that destabilizes plasma membrane of tumor cells, thereby enhancing their permeability to cytotoxic drugs.[24]

There are several approaches that exploit active targeting of long-circulating liposomes to tumor cells, where receptor-mediated internalization is strongly believed to bypass tumor cell multidrug-efflux pumps.[10,16,17,25] These strategies utilize tumor-specific monoclonal antibodies or their internalizing epitopes, or ligands, such as folic acid, which are attached to the distal end of the poly(ethylene glycol) chains expressed on the surface of long-circulating liposomes. Nevertheless, with such approaches the delivery part is still passive and relies on liposome extravasation.

2.3.2 POLYMERIC NANOPARTICLES

Abraxane™ is the only example of a regulatory approved (FDA, USA) nanoparticle formulation for intravenous drug delivery in cancer patients. It is paclitaxel bound to albumin nanoparticles, with a mean diameter of 130 nm, for use in individuals with metastatic breast cancer who have failed combination chemotherapy or relapse within 6 months of adjuvant chemotherapy. This formulation overcomes poor solubility of paclitaxel in the blood and allows patients to receive 50% more paclitaxel per dose over a 30-min period.[26,27] Unlike Cremophor® EL/ethanol or Tween® 80-solubilized taxanes, acute hypersensitivity reactions, which are secondary to complement activation, have yet to be reported following Abraxane™ infusion. Albumin nanoparticles seem to interact with gp60 receptors present on tumor blood vessels that transport the nanoparticles into tumor interstitial spaces by transcytosis, a process that may partly contribute to the effectiveness of Abraxane™. However, hepatic deposition (Kupffer cell capture) and processing of a significant fraction of albumin nanoparticles are most likely to occur. Indeed, after a 30 min infusion of 260 mg/m^2 doses of Abraxane™, faecal excretion accounted for approximately 20% of the administered dose (ABRAXIS Oncology, A division of American Pharmaceutical Partners, Inc., Schaumburg, IL 60173, USA; 2005), thus supporting a role for hepatic handling and biliary excretion of albumin nanoparticles (or its components).

Nanoparticles assembled from synthetic polymers have also received much attention in cancer drug delivery.[28] One interesting example is doxorubicin-loaded poly(alkyl cyanoacrylate) (PACA) nanoparticles. In vitro studies have indicated that PACA nanoparticles can overcome drug resistance in tumor cells expressing multidrug-resistance-1-type efflux pumps.[29] The mechanism of action is related to adherence of PACA nanoparticles to tumor cell plasma membrane, which initiates particle degradation and provides a concentration gradient for doxorubicin, and diffusion of doxorubicin across the plasma membrane following formation of an ion pair between the positively charged doxorubicin and the negatively charged cyanoacrylic acid (a nanoparticle degradation product).[29] These observations clearly indicate that drug release and nanoparticle degradation must occur simultaneously, yielding an appropriate size complex with correct physicochemical properties for diffusion across the plasma membrane. Further developments with PACA nanoparticles include preparations that contain doxorubicin within the particle core and cyclosphorin, an inhibitor of the P-glycoprotein, at the surface.[30] Similar to liposomes, long-circulating versions of PACA nanoparticles have also been engineered for passive as well as active targeting to solid tumors.[31]

2.3.3 NANOTECHNOLOGY-DERIVED NANOPARTICLES

Nanotechnology is a cross-disciplinary field, which involves the ability to design and exploit the unique properties that emerge from man-made materials ranging in size from 1 to greater than 100 nm.[16] Indeed, the physical and chemical properties of materials—such as porosity, electrical conductivity, light emission, and magnetism—can significantly improve or radically change as their size is scaled down to small clusters of atoms. These advances are beginning to have a paradigm-shifting impact not least in experimental (e.g., thermal tumor killing) and diagnostic oncology.[16,32] Examples include superparamagnetic iron oxide nanocrystals, quantum dots (QDs), inorganic nanoparticles, and composite nanoshells. The surfaces of these entities are amenable to modification with synthetic polymers (to afford long-circulating properties) and/or to targeting ligands. However, a key problem with these technologies is toxicity and is discussed elsewhere.[16]

Iron oxide nanocrystals are formed from an inner core of hexagonally shaped iron oxide particles of approximately 5 nm, which express correlated electron behavior; at a high enough temperature, they are superparamagnetic.[33,34] In addition, dextran or synthetic polymers such as poly(ethyleneglycol) surround the crystal core. Indeed, it is the combination of the small size and surface characteristics that allow iron oxide nanocrystals, once injected into the blood stream, to bypass rapid detection by the body's defence cells and accumulate in tumor sites by extravasation. Therefore, they are useful for patient selection, detection of tumor progression, and tracking of the effectiveness of anti-tumor treatment regimens by magnetic resonance imaging (MRI). These approaches can be extended for site-specific imaging of tumor vasculature with targeting ligands. In addition, iron oxide nanocrystals can slowly extravasate from the vasculature into the interstitial spaces, from which they are transported to lymph nodes by way of lymphatic vessels.[34] Within lymph nodes they are captured by local macrophages, and their intracellular accumulation shortens the spin relaxation process of nearby protons detectable by MRI. On magnetic resonance images, those node regions accumulating iron oxide appear dark relative to surrounding tissues. Indeed, iron oxide nanocrystals can distinguish between normal and tumor-bearing nodes and reactive and metastatic nodes.[34]

QDs are made of semiconductors like silicon and gallium arsenide.[35,36] In these particles there are discrete electronic energy levels (valance band and conduction band), but the spacing of the electronic energy levels (band gap) can be precisely controlled through variation in size. When a photon, with higher energy than the energy of the band gap, hits a QD, an electron is promoted from valance band into the conduction band, leaving a hole behind. Electrons emit their excess energy as light when they recombine with holes. Since optical response is due to the excitation of single electron-hole pairs, the size and shape of QDs can be tailored to fluoresce specific colors. The ability of QDs to tune broad wavelength together with their photostability is of paramount importance in biological labeling.[35,36] Indeed, QDs stay lit much longer than conventional dyes used for imaging and tagging purposes and therefore have the potential to improve the resolution of tumor cells to the single cell level by optical imaging as well as determining heterogeneity among cancer cells in a solid tumor.[37–39] Unlike QDs, where optical response is due to the excitation of single electron hole pairs, in metallic nanoparticles (e.g., gold) incident light can couple to the plasmon excitation of the metal. This involves the light-induced motion of all valence electrons.[36] Therefore, the type of plasmon that exists on a surface of a metallic nanoparticle is directly related to its shape and curvature; so it is possible to make a wide range of light scatterers that can be detected at different wavelengths. Composite nanoshells consist of a spherical dielectric core (e.g., silica) surrounded by a thin metal shell (e.g., gold). Again, by controlling the relative thickness of the core and shell layers of the composite nanoparticle, the plasmon resonance and the resultant optical absorption properties can be tuned from near-UV to the mid-infrared. Of particular interest is the ability of near-infrared light (700–1000 nm) to penetrate through tissue at depths of a few cm with minimal heat generation and tissue damage. Thus, a recent study demonstrated rapid irreversible photothermal ablation of tumor tissue in vivo following administration of near-infrared-absorbing silica–gold nanoshells in combination with an extracorporeal low-power diode laser.[40]

2.3.4 MACROMOLECULAR AND RELATED DELIVERY

Polymer-based drug delivery systems also favorably alter the pharmacokinetics and biodistribution of conjugated drugs and accumulate in tumor interstitium following extravasation.[41] Examples include SMANCS (a conjugate of the polymer styrene-*co*-maleic acid/anhydride and neocarzinostatin for treatment of hepatocellular carcinoma), conjugates of various cytotoxic agents (e.g., paclitaxel, doxorubicin, platinate, and campthothecin) with polyglutamate and nonbiodegradable hydroxypropyl methacrylamide.

Other related polymer-based systems in cancer drug delivery include micelles[42,43] and dendrimers.[44] For example, Pluronics® (copolymers of ethylene oxide and propylene oxide) are capable of forming micelles, and some members of Pluronic copolymers can overcome multidrug resistance.[42] However, it is becoming clear that Pluronic copolymers can induce complement activation, even at concentrations below their critical micelle concentration, which may increase the risk of pseudoallergy in sensitive patients.[45]

Dendrimers are highly branched macromolecules with controlled near monodisperse three-dimensional architecture emanating from a central core.[44] Polymer growth starts from a central core molecule and growth occurs in an outward direction by a series of polymerization reactions. Hence, precise control over size can be achieved by the extent polymerization, starting from a few nanometers. Cavities in the core structure and folding of the branches create cages and channels. The surface groups of dendrimers are amenable to modification and can be tailored for specific applications. Therapeutic and diagnostic agents are usually attached to surface groups on dendrimers by chemical modification. For example, a recent study has used tagged-dendrimers for in vivo evaluation of tumor-associated matrix metalloproteinase-7 (matrilysin) activity.[46]

Other macromolecular systems for cancer targeting and treatment include various forms of monoclonal and bispecific monoclonal antibodies against tumor-associated antigens.[47] These can further be coupled to drugs, toxins, enzymes (as in antibody-directed enzyme pro-drug therapy), cytokines, radionuclides, etc.

2.4 CONCLUSIONS

The chaotic blood flow in tumor vasculature and the heterogeneous vascular permeability of tumor blood vessels are among the key barriers controlling passive delivery of macromolecular and particulate delivery systems into the interstitium of solid tumors. Already compromised by abnormal hydrostatic pressure gradients, compressive mechanical forces generated by tumor cell proliferation cause intratumoral vessels to compress and collapse, thus creating further barriers for passive targeting. Interestingly, tumor-specific cytotoxic therapy, reducing tumor cell number, may result in more efficient delivery by decompressing these same vessels,[48] but this enhanced perfusion could provide a route for metastasis. Pathophysiologoical barriers, however, are not fully developed in micrometastases, and also pose a lesser problem in the diagnostic oncology as well as in drug delivery to well-perfused and low-pressure regions in larger tumors. Some of the problems may possibly be overcome by design of long-circulating multifunctional carriers (carriers that contain appropriate combinations of cytotoxic agents, diagnostic, and barrier-avoiding components) with biochemical triggering mechanisms.[16] The vascular barrier of the solid tumor is also its Achilles' heel; the nutritionally demanding tumor cells are entirely dependent upon a functional vasculature.

For this reason, interest has also been focused on the concept of tumor vasculature as a target rather than a barrier and is reviewed in this compendium and elsewhere.[49–51]

REFERENCES

1. Munn, L. L., Aberrant vascular architecture in tumors and its importance in drug-based therapies, *Drug Discov. Today*, 8, 396, 2003.
2. Lyden, D. et al., Impaired recruitment of bone-marrow-derived endothelial and haematopoietic precursor cells blocks, tumor angiogenesis, and growth, *Nat. Med.*, 7, 1194, 2001.
3. Jain, R. K., Delivery of molecular medicine to solid tumors: Lessons from in vivo imaging of gene expression and function, *J. Control. Release*, 74, 7, 2001.
4. Dvorak, H. F. et al., Identification and characterization of the blood vessels of solid tumors that are leaky to circulating macromolecules, *Am. J. Pathol.*, 133, 95, 1988.
5. Hashizume, H. et al., Openings between defective endothelial cells explain tumor vessel leakiness, *Am. J. Pathol.*, 156, 1363, 2000.
6. Morikawa, S. et al., Abnormalities in pericytes on blood vessels and endothelial sprouts in tumors, *Am. J. Pathol.*, 160, 985, 2002.
7. Fukumura, D. et al., Effect of host microenvironment on the microcirculation of human colon adeno-carcinoma, *Am. J. Pathol.*, 151, 679, 1997.
8. Daldrup, H. et al., Correlation of dynamic contrast-enhanced MR imaging with histologic tumor grade: Comparison of macromolecular and small-molecular contrast media, *Am. J. Roentgenol.*, 171, 941, 1998.
9. Nathanson, S. D. and Nelson, L., Interstitial fluid pressure in breast cancer, benign breast conditions, and breast parenchyma, *Ann. Surg. Oncol.*, 1, 333, 1994.
10. Moghimi, S. M., Hunter, A. C., and Murray, J. C., Long-circulating and target-specific nanoparticles: Theory to practice, *Pharmacol. Rev.*, 53, 283, 2001.
11. Pluen, A. et al., Role of tumor–host interactions in interstitial diffusion of macromolecules: Cranial vs. subcutaneous tumors, *Proc. Natl Acad. Sci. USA*, 98, 4628, 2001.
12. Torchilin, V. P., Recent advances with liposomes as pharmaceutical carriers, *Nat. Rev. Drug Discov.*, 4, 145, 2005.
13. Daemen, T. et al., Liposomal doxorubicin-induced toxicity: Depletion and impairment of phagocytic activity of liver macrophages, *Int. J. Cancer*, 61, 716, 1995.
14. Batist, G. et al., Reduced cardiotoxicity and preserved antitumor efficacy of liposome-encapsulated doxorubicin and cyclophosphamide compared with conventional doxorubicin and cyclophosphamide in a randomized, multicenter trial of metastatic breast cancer, *J. Clin. Oncol.*, 19, 1444, 2001.
15. Williams, G., Cortazar, P., and Pazdur, R., Developing drugs to decrease the toxicity of chemotherapy, *J. Clin. Oncol.*, 19, 3439, 2001.
16. Moghimi, S. M., Hunter, A. C., and Murray, J. C., Nanomedicine: Current status and future prospects, *FASEB J.*, 19, 311, 2005.
17. Gabizon, A., Shmeeda, H., and Barenholz, Y., Pharmacokinetics of pegylated liposomal doxorubicin: Review of animal and human studies, *Clin. Pharmacokinet.*, 42, 419, 2003.
18. Yuan, F. et al., Microvascular permeability and interstitial penetration of sterically stabilized (stealth) liposomes in a human tumor xenograft, *Cancer Res.*, 54, 3352, 1994.
19. Bandak, S. et al., Pharmacological studies of cisplatin encapsulated in long-circulating liposomes in mouse tumor models, *Anti-Cancer Drugs*, 10, 911, 1999.
20. Huang, S. K. et al., Liposomes and hyperthermia in mice: Increased tumor uptake and therapeutic efficacy of doxorubicin in sterically stabilized liposomes, *Cancer Res.*, 54, 2186, 1994.
21. Zalipsky, S. et al., New detachable poly(ethylene glycol) conjugates: Cysteine–cleavable lipopolymers regenerating natural phospholipid, diacylphosphatidylethanolamine, *Bioconjug. Chem.*, 10, 703, 1999.
22. Provoda, C. J. and Lee, K.-D., Bacterial pore-forming hemolysins and their use in the cytosolic delivery of macromolecules, *Adv. Drug Deliv. Rev.*, 41, 209, 2000.
23. Andresen, T. L., Jensen, S. S., and Jørgensen, K., Advanced strategies in liposomal cancer therapy: Problems and prospects of active and tumor specific drug release, *Prog. Lipid Res.*, 44, 68, 2005.

24. Andresen, T. L. et al., Triggered activation and release of liposomal prodrugs and drugs in cancer tissue by secretory phospholipase A2, *Curr. Drug Deliv.*, 2, 353, 2005.
25. Allen, T. M., Ligand-targeted therapeutics in anti-cancer therapy, *Nat. Rev. Cancer*, 2, 750, 2002.
26. Bartels, C. L. and Wilson, A. F., How does a novel formulation of paclitaxel affect drug delivery in metastatic breast cancer?, *US Pharm.*, 29, HS18, 2004.
27. Garber, K., Improved paclitaxel formulation hints at new chemotherapy approach, *J. Natl Cancer Inst.*, 96, 90, 2004.
28. Brigger, I., Dubernet, C., and Couvreur, P., Nanoparticles in cancer therapy and diagnosis, *Adv. Drug Deliv. Rev.*, 54, 631, 2002.
29. Colin de Verdiere, A. et al., Reversion of multidrug resistance with polyalkylcyanoacrylate nanoparticles: Towards a mechanism of action, *Br. J. Cancer*, 76, 198, 1997.
30. Soma, C. E. et al., Reversion of multidrug resistance by co-encapsulation of doxorubicin and cyclosporin A in polyalkylcyanoacrylate nanoparticles, *Biomaterials*, 21, 1, 2000.
31. Stella, B. et al., Design of folic acid-conjugated nanoparticles for drug targeting, *J. Pharm. Sci.*, 89, 1452, 2000.
32. Ferrari, M., Cancer nanotechnology: Opportunities and challenges, *Nat. Rev. Cancer*, 5, 161, 2005.
33. Perez, J. M., Josephson, L., and Weissleder, R., Use of magnetic nanoparticles as nanosensors to probe for molecular interactions, *Chem. Bio. Chem.*, 5, 261, 2004.
34. Moghimi, S. M. and Bonnemain, B., Subcutaneous and intravenous delivery of diagnostic agents to the lymphatic system: Applications in lymphoscintigraphy and indirect lymphography, *Adv. Drug Deliv. Rev.*, 37, 295, 1999.
35. Medintz, I. L. et al., Quantum dot bioconjugates for imaging, labelling, and sensing, *Nat. Mater.*, 4, 435, 2005.
36. Alivisatos, P., The use of nanocrystals in biological detection, *Nat. Biotechnol.*, 22, 47, 2003.
37. Stroh, M. et al., Quantum dots spectrally distinguish multiple species within the tumor milieu in vivo, *Nat. Med.*, 11, 678, 2005.
38. Wu, X. et al., Immunofluorescent labeling of cancer marker Her2 and other cellular targets with semiconductor quantum dots, *Nat. Biotechnol.*, 21, 41, 2003.
39. Gao, X. et al., In vivo cancer targeting and imaging with semiconductor quantum dots, *Nat. Biotechnol.*, 22, 969, 2004.
40. Hirsch, L. R. et al., Nanoshell-mediated near-infrared thermal therapy of tumors under magnetic resonance guidance, *Proc. Natl Acad. Sci. USA*, 100, 13549, 2003.
41. Duncan, R., The dawning era of polymer therapeutics, *Nat. Rev. Drug Discov.*, 2, 347, 2003.
42. Oh, K. T., Bronich, T. K., and Kabanov, A. V., Micellar formulations for drug delivery based on mixtures of hydrophobic and hydrophilic Pluronic® block copolymers, *J. Control. Release*, 94, 411, 2004.
43. Adams, M. L., Lavasanifar, A., and Kwon, G. S., Amphiphilic block copolymers for drug delivery, *J. Pharm. Sci.*, 92, 1343, 2003.
44. Haag, R., Supramolecular drug-delivery systems based on polymeric core-shell architectures, *Angew. Chem. Int. Ed.*, 43, 278, 2004.
45. Moghimi, S. M. et al., Causative factors behind poloxamer 188 (Pluronic F68 Flocor)-induced complement activation in human sera: a protective role against poloxamer-mediated complement activation by elevated serum lipoprotein levels, *Biochim. Biophys. Acta-Mol. Basis Dis.*, 103, 1689, 2004.
46. McIntyre, J. O. et al., Development of a novel fluorogenic proteolytic beacon for in vivo detection and imaging of tumor-associated matrix metalloproteinase-7 activity, *Biochem. J.*, 377, 617, 2004.
47. Kosterink, J. G. W. et al., Strategies for specific drug targeting to tumor cells. *Vol. 12 of Drug Targeting, Organ-Specific Strategies*, Molema, G., and Meijer, D. K. F., Eds., Wiley-VCH, Weinheim pp.199–232, 2001, chapter 8.
48. Padera, T. P. et al., Cancer cells compress intratumor vessels, *Nature*, 427, 695, 2004.
49. Pasqualini, R., Arap, W., and McDonald, D. M., Probing the structural and molecular diversity of tumor vasculature, *Trend Mol. Med.*, 8, 563, 2002.
50. Murray, J. C. and Moghimi, S. M., Endothelial cells as therapeutic targets in cancer: New biology and novel delivery systems, *Crit. Rev. Ther. Drug Carrier Syst.*, 20, 139, 2003.
51. Reynolds, A. R., Moghimi, S. M., and Hodivala-Dilke, K., Nanoparticle-mediated gene delivery to tumor neovasculature, *Trend Mol. Med.*, 9, 2, 2003.

3 Active Targeting Strategies in Cancer with a Focus on Potential Nanotechnology Applications

Randall J. Mrsny

CONTENTS

3.1 Introduction 19
3.2 Nanoparticle Characteristics 20
    3.2.1 Composition and Biocompatibility 22
    3.2.2 Derivitization 23
    3.2.3 Detection 23
3.3 Nanoparticle Targeting 24
    3.3.1 Inherent Targeting 24
    3.3.2 Complicating Aspects 25
    3.3.3 Safety Issues 26
3.4 Targeting to Cancer Cells 27
    3.4.1 Cell Surface Properties 28
    3.4.2 Metabolic Properties 29
3.5 Targeting to Tumors 30
    3.5.1 Aberrant Vasculature 30
    3.5.2 Metabolic Environment 32
3.6 Fate of Nanoparticles 32
    3.6.1 Site of Initial Application 33
    3.6.2 Distribution to Specific Organs 34
    3.6.3 Intracellular Uptake and Fate 34
3.7 Conclusions 35
References 37

3.1 INTRODUCTION

Nanotechnology has recently become a buzzword in several scientific fields, including the area of drug delivery,[1] and a variety of nanomedicine opportunities have recently been reviewed.[2] The concept of nanoparticles as drug delivery vehicles is not new as reviews on the subject were published nearly 20 years ago.[3] That nanoparticles could be selectively targeted by coating with monoclonal antibodies provided an early, ground-breaking proof-of-concept that these materials might one day be effective tools for the diagnosis and treatment of cancers. At their inception, all new technologies

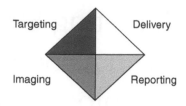

FIGURE 3.1 Integrated potential applications of nanoparticles as visualized by the National Cancer Institute of the NIH. Nanoparticles, because of their general capacity for multiple modifications as well as their inherent properties, can have multiple functions relevant to targeting cancers for therapeutic and/or diagnostic purposes.

appear to have a plethora of possibilities because limitations have not yet become apparent. Subsequent studies that examine the limits of various applications then provide a menu of feasible applications. This menu can change as new developments of the technology are realized that allow one to address additional applications. This has certainly been the case for nanotechnology applications that involve active targeting to cancers for diagnostic and therapeutic purposes.[4]

Although nanoparticles have shown tremendous promise in facilitating the targeted delivery of therapeutics and diagnostics to cancers, the composition and size of the particles have inherent physical and chemical properties that can compromise their capability to localize to and/or treat cancers because of non-selective cell and tissue uptake. Therefore, active targeting to cancers using nanoparticles requires consideration of not only unique properties of the cancer that allow for specific targeting but also attention to issues that minimize non-selective delivery to uninvolved regions of the body. Methods to overcome functional barriers that limit uptake of materials or delivery to cancer cells must be also considered. Even with proper consideration of these issues, successful disposition and targeting to cancers is quite a challenge. Certainly, lack of consideration of these issues can dramatically increase the risk of serious negative outcomes, particularly when targeted nanoparticles contain potent cytotoxic agents.

Advantages and disadvantages of specific nanoparticles as well as methods to potentially correct shortcomings of nanoparticle targeting to cancers will be discussed in this chapter. Several of the topics raised in this chapter will also be discussed in much greater depth in other chapters in this text. Specifically, two major strategies of cancer targeting related to nanotechnology opportunities will be addressed: targeting to cancer cells and targeting to tumors. Attention will be paid to similarities as well as differences for these two strategies. Targeting cancer cells and tumors for diagnostic purposes as well as therapy will also be examined. Several applications for nanoparticles have been outlined by the National Cancer Institute (http://nano.cancer.gov) regarding unmet medical needs in the areas of targeting, imaging, delivery, and reporting agents for cancer diagnosis and treatment (Figure 3.1). General cellular responses to and fates of nanoparticles that can affect these potential applications will be discussed as critical aspects of ultimate clinical success.

3.2 NANOPARTICLE CHARACTERISTICS

Clinical success of nanoparticle-based diagnostics and therapeutics requires proper matching of particle characteristics. The characteristics of nanoparticles are critically dependent upon the materials used to prepare the nanoparticle. Nanoparticles can now be readily prepared from a wide range of inorganic and organic materials in a range of sizes from two to several hundred nanometers (nm) in diameter. Put in perspective, human cells are typically 10,000–20,000 nm in diameter. The plasma membrane of these cells is 6 nm in thickness. In most cases, nanoparticles can be generated to have narrow and defined size ranges. Other chapters in this text will focus on the physical and chemical characteristics of nanoparticles made from various materials as well as methods for their production. Although nanoparticles can be prepared from a wide variety of materials (inorganic salts, lipids, synthetic organic polymers, polymeric forms of amino acids, nucleic acids, etc.), this chapter will primarily focus on those prepared from materials that would be considered sufficiently safe for repeated systemic administrations and/or would be perceived to have an acceptable safety profile that would warrant use in man. In general, it is desirable for nanoparticles to be either readily metabolized or sufficiently broken down to produce only non-toxic metabolites that can be safely excreted. Indeed, tremendous advances have been made in controlling the chemical nature, degradable characteristics, and dimensions of nanoparticles.

Many of the initial studies examining nanoparticles as delivery tools used particles prepared from materials such as polyalkylcyanoacrylates (PAA).[5] The extreme stability of PAA is both a positive and a negative. PAA nanoparticles will not be degraded prior to reaching a tissue or cell target site; however, once they reach that site, it is unlikely that they will be efficiently metabolized. Therefore, PAA nanoparticles have been extremely useful for initial studies of nanomaterials for cancer targeting, but an inability to clear PAA nanoparticles presents uncertainties as to their ultimate toxicological fate. Concerns over repeated PAA nanoparticle administrations in man and the need for more acceptable materials were highlighted early on.[3] One of the biggest concerns regarding poorly metabolized nanoparticles is that of accumulation and the potential sequelae associated with such an outcome. In some cases where a limited number of exposures would occur, one could consider the use of materials that are not readily metabolized by the body. In the case of certain cancer applications, it might be possible to use materials that otherwise would be considered to have an unacceptable safety signal following repeat dosing or that have the potential to accumulate. Therefore, rationales exist for the potential application of nanoparticles prepared from a wide range of materials, even those that, at first glance, would be considered unacceptable.

Methods of production and composition define nanoparticle characteristics; these characteristics define potential issues (and opportunities) related to biocompatibility, derivitization, and detection. Nanoparticles can be prepared from a singular subunit that is chemically coupled and organized in a defined (e.g., dendrimers) or in a more random (e.g., polylactic acid) manner. Although these materials would not have a defined core, they can be impregnated with compatible materials and/or chemically modified at their surface. Materials such as glyconanoparticles would provide one approach where a distinct core with radiating ligands could be positioned using linkers. In such a case, the solid core, used to anchor each linker used for the attachment of targeting ligands, could be used to deliver a therapeutic or diagnostic payload. Liposomes are an example of nanoshell structures that can be loaded internally as well as impregnated within the shell. Many types of nanomaterials fall into one of these three general structural architectures (Figure 3.2).

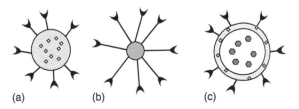

(a)   (b)   (c)

FIGURE 3.2 General schema for three types of nanoparticle structures. (a) Nanoparticles can be formed from one type of material that can be impregnated with therapeutic or lipophilic imaging reagents (open diamonds) and modified with targeting ligands (crescents) positioned by chemical coupling through linker moieties. (b) Metal (or similar) cores (circles can be modified through a linker-targeting ligand system to generate another type of nanoparticle structure. In this case, it might be possible to use elaborated linkers as an environment compatible for incorporation of therapeutic or imaging reagents. (c) Shell-type nanoparticles such as liposomes where an aqueous compartment is enclosed by a bilayer of phospholipids can also be used for the targeted delivery of hydrophilic therapeutic or imaging reagents (filled hexagons).

3.2.1 COMPOSITION AND BIOCOMPATIBILITY

Biological organic polymers can be formed from amino acids to form peptides and proteins. Some proteins have, by themselves, been used as nanoparticle delivery systems. Indeed, the protein ferritin functions as a coated nanoparticle. This molecule is ~ 12 nm in diameter and can carry hydrous ferrous oxide (~ 5–7 nm in diameter) during its role as an iron storage system for the body.[6] Alternately, proteins can be used to generate nanoparticles for carrier applications; gelatin has been used to prepare nanoparticles.[7] Albumin nanoparticles have been described.[8] A 13-MDa ribonucleoprotein, termed a vault, has also been identified as a nanomaterial that could be used to deliver therapeutic and diagnostic agents.[9] Materials prepared from nucleic acids also have the potential to act as nanoparticle carriers. In general, nanoparticles prepared from biological materials would be biocompatible as a result of obvious elimination mechanisms.

Nanoparticles can also be prepared from biological materials that are found in the body but are not typically organized as nanoparticle-size polymers; complex mixtures of polysaccharides, poly-lysine, and poly(D,L-lactic and glycolic acids; PLGA) have been prepared in a variety of sizes and in formats that allow ligand coupling with targeting moieties as well as diagnostic or therapeutic agents.[10] Synthetic organic polymers such as PLGA have been used to produce resorbable sutures, providing nanoparticles that will produce a sustained release of its contents. PLGA nanoparticles have been used to deliver wild-type p53 protein to cancer cells.[11] Such polymers have been studied for the development of nanoparticle delivery vehicles.[12–14] Although such nanoparticles would be considered relatively safe because of their biocompatibility, particles prepared from these types of materials can initiate inflammatory responses at sites of accumulation or deposit.[15] A number of new synthetic organic polymer materials are being examined for their capacity to generate nanoparticles useful for drug encapsulation and targeting.[16] In general, organic polymer-based nanoparticles have provided promising results in the field of cancer targeting for diagnostics and therapy with the added capability of sustained release in some cases.[17]

Dendrimer structures are prepared from a series of repetitive chemical steps that perpetually increase their size as additional shells are added. Varying the seed molecule can produce nanoparticle structures of spherical or more elongated shapes. Dendrimer-based nanoparticle structures are inherently different from other organic-based nanoparticles prepared as linear sequences of subunits (e.g., amino acids used to form proteins) or from subunits that can randomly branch (e.g., polysaccharides). With the flexibility of chemistries that have now been described to prepare dendrimers, these materials show tremendous promise to prepare highly defined nanoparticles that can be targeted to cancers for therapeutic and diagnostic applications. To date, polyamidoamine (PAMAM) dendrimers have been the most extensively studied family of dendrimers that can be used to deliver anti-cancer drugs.[18]

A variety of inorganic materials can be used to generate nanoparticles for drug delivery that might be applied to cancer and tumor targeting. For example, iron oxide (Fe_2O_3) nanoparticles can be used to deliver anti-cancer agents.[19] Fe_2O_3 nanoparticles targeted to a cancer can become hot enough in an applied oscillating magnetic field to kill cells. Calcium phosphate precipitates can also be also made into nanoparticles. Although calcium phosphate precipitates can be metabolized over time and would be considered biocompatible, these materials can act as a potent adjuvant, potentially enhancing their application to target the delivery of cancer antigens.[20] In this way, calcium phosphate nanoparticles are similar to another inorganic salt precipitate, aluminum hydroxide (alum), that is currently approved as an adjuvant for human vaccines.[21] Semiconductor nanocrystals (quantum dots) are another example of inorganic nanoparticles. Quantum dots have exceptional characteristics for in vivo imaging and diagnostic applications.[22,23] However, some materials used to generate quantum dots such as CdSe can release toxic Cd^{2+} ions that alter ion channel function and lead to cell death when sufficient levels are reached.[24] Therefore, some of the inorganic materials used to generate nanoparticles may have significant biocompatibility issues.

Lipids such as phospholipids and cholesterol can be used to generate single- or multi-lamellar spheres (liposomes) in the nanometer-size range. Liposome-based nanocapsules that can be loaded with diagnostic or therapeutic agents for the targeted delivery to cancers[25] and enhancement strategies for lipid-based nanoparticle-mediated tumor targeting have been reviewed.[26] Liposomes were some of the first nanoparticle structures extensively evaluated for cancer targeting. Early studies using liposomes highlighted issues associated with recognition and clearance by cells of the reticuloendothelial system (RES) that remove particulates from the systemic circulation.[27] Methods of masking liposomes from the RES such as modification with poly(ethylene glycol) (PEG) have been successfully used to limit RES clearance and increase circulating half-lives in serum.[28]

Solid lipid nanoparticles, nanostructure lipid carriers, and lipid–drug conjugate nanoparticles have also been described for the drug delivery strategies that could be applied to cancer diagnosis and/or therapy.[29] Lipid-based nanospheres can be sterically stabilized by the incorporation of artificial lipid derivatives that can be cross-linked. Subsequently, stabilized lipid-based nanospheres can be targeted to cancers using a conjugated antibody.[30] Such covalent modifications can improve the stability of lipid-based nanoparticles but can also reduce the biocompatible natures of these materials by modifying their clearance from the body. Apolipoprotein E-containing liposomes have also been prepared as a carrier for a lipophilic prodrug of daunorubicin as a means of targeting cancer cells that overexpress the receptor for low-density lipoproteins (LDL).[31]

3.2.2 DERIVITIZATION

Because of their chemical and physical characteristics, nanoparticles exhibit inherent cellular targeting and uptake characteristics. Size and surface charge seem to be the two prominent characteristics that affect inherent nanoparticle targeting and cellular uptake. Because inherent targeting mechanisms may not provide the targeting or delivery characteristics desired, methods to modify nanoparticles with targeting agents can be critical. Although some nanoparticle materials are composed of materials with functional groups useful for chemical coupling, others are not. Such nanoparticle systems must be either modified to allow chemical coupling or doped with reagents that can be used for this modification. A number of coupling strategies have also been worked out that allow for efficient functionalization of nanomaterials through both reversible and irreversible chemistries.[14,32] These modifications allow for the coupling of antibodies, receptor ligands, and other potential targeting agents. Similar to the concerns associated with composition of the nanoparticle itself, any modification through chemical derivitization must also be considered with regard to generating materials with unacceptable toxicity or neutralization of the function of the nanoparticle or its targeting element.

Nanoparticles have the advantage that they can be modified with multiple ligands to enhance their targeting selectivity and/or allow for simultaneous delivery of diagnostic and therapeutic agents.[33] It is important to appreciate the relative size of components used to construct and derivitize nanoparticles. For example, a quantum dot may be only 10 nm in diameter. Targeting that sized particle with an antibody might require the attachment of an IgG antibody that is roughly equal in size. By comparison, a fluorescent material that might be useful for localization of a targeted nanoparticle such as green fluorescent protein (GFP) is about 5 nm. Derivitization strategies for the construction of targeted nanoparticles must incorporate a consideration of potential steric conflicts for incorporation of targeting, detection, and therapeutic components. Other chapters in this text will extensively examine derivitization technologies for nanoparticles.

3.2.3 DETECTION

Most, if not all, nanoparticle structures currently investigated for delivery of cancer therapeutics also have the capacity to be detected or modified to contain a detectable agent that could be simultaneously used for cancer diagnosis. For example, PAMAM folate-dendrimers that contain contrast media for detection by magnetic resonance imaging (MRI) are effectively targeted to cancer cells that overexpress the high affinity folate receptor.[34] Gadolinium ion–dendrimer nanoparticles are readily detected by CT imaging and appear to provide several advantages over previous methodologies.[35] PAMAM dendrimer nanoparticles covalently coupled with a fluorescent label can be visualized as sites of increased retention.[36] Any strategy to produce nanoparticles that combines diagnostic and therapeutic elements, however, must consider potential conflicting aspects. Introduction of some heterocyclic anti-cancer molecules may act to quench fluorescent properties and the capacity to detect fluorescent labels.

Some nanoparticles are particularly promising for cancer diagnosis because of their exceptional properties of detection using current radiographic and magnetic methods. Some new materials being prepared as nanoparticles will provide the potential for visualization using novel imaging methods that may lead to greater selectivity of signal and reduce false positives.[37] For example, quantum dots associated with metastatic cancer cells can be visualized using fluorescence emission-scanning microscopy.[38] Similarly, lipid-encapsulated liquid perfluorocarbon contrast media at the site of a tumor can be detected by ultrasonic acoustic transmission.[39] It is also possible to functionalize materials such as quantum dots to incorporate materials that can be photo-activated to enhance their activity or detection.[23]

3.3 NANOPARTICLE TARGETING

Nanoparticles can be designed in a variety of ways to achieve targeted delivery. Some targeting strategies rely upon inherent properties of the particle, in particular, its composition, size, and surface properties. Furthermore, the particle itself can either be the agent being delivered, or it can be prepared to carry a cargo for delivery. Cargo release from the nanoparticles can occur while the nanoparticle is still relatively intact or through its decomposition. A number of methods have been described to integrate and retain cargo components within nanoparticles and these, in general, match to chemical or physical characteristics of the cargo with those of the material used to generate the nanostructure. For example, positively charged cargo can be held within the nanoparticle through interactions with an internal network such as a polyanionic polymer that resembles the organization of secretory granules synthesized by cells.[40] Alternately, organized complexes akin to coacervates proposed to participate in cell structure evolution can be formed between cargo and particle matrix.[40] Therefore, for some cancer-targeting strategies, one should consider not only compatibility of the nanoparticle carrier with its cargo but also degradation events that might affect temporal aspects of particle stability and cargo release.

Some nanoparticles can be designed or delivered in such a way as to produce a default targeting event; other nanoparticles must be decorated on their surface to produce a targeted structure. Topical application of a nanoparticle system at the target site may be all that is required for a successful outcome. Such a simple approach is not typically sufficient for effective targeting of many cancers. Successful targeting may require reduction of inherent targeting tendencies for the material(s) used to prepare the nanoparticle. Depending upon the physical and chemical nature of the nanoparticle and the mode of administration, there can also be complicating factors that affect the effectiveness of the targeting method. Inherent targeting and complicating factors associated with some nanoparticles used for a targeted delivery can impart safety issues that must also be considered. With such characteristics, it is easy to see why active targeting of nanoparticles to cancers can be both complicated by competing biological events as well as facilitated by these same properties.[41]

3.3.1 INHERENT TARGETING

The RES is composed of a series of sentinel cells located in several highly perfused organs, including the liver and spleen.[27] Nanoparticles can be rapidly cleared from the blood if they are recognized by RES cells in a non-selective fashion, typically before achievement of effective targeting.[13,42] In some instances, this inherent targeting can provide a means to selectively delivery materials.[43] In most cases, it is possible to modify the physical and chemical characteristics of nanoparticles to reduce their default uptake by the RES.[44] Methods to avoid the RES will be addressed in depth in other chapters in this text. In general, these measures follow principles initially outlined in the development of stealth liposomes that provided a means of extending the circulating half-life of a nanoparticle. Although PEG molecules of various lengths coupled using various chemistries[45] are frequently used in this approach, heparan sulfate glycosaminoglycans (HSGs) have also been shown to provide a protective coating that reduces immune detection.[46] Interestingly, HSGs might be shed at tumors by tumor-associated heparanase activity.

Another inherent targeting aspect of nanoparticles relates to the nature of tumor-associated vasculature. In general, nanoparticles smaller than 20 nm have the ability to transit out of blood vessels. Solid tumors grow rapidly; tumor-associated endothelial cells are continually bathed by a plethora of cancer cell-secreted growth factors. In turn, endothelial cells sprout new vessels to provide needed nutrients for the continued growth of the tumor. This cancer cell-endothelial cell relationship, however, leads to the establishment of a poorly organized vasculature that, under the constant drive of growth factor stimulation, fails to organize into a mature vascular bed. Therefore, tumor-associated vascular beds are poorly organized and more leaky that normal vasculature. Nanoparticles will inherently target to tumors as exudates through leaky vasculature. This phenomenon, referred to as the enhanced permeability and retention (EPR) effect,[47] will be covered extensively in other chapters in this text.

Finally, peculiar surface properties of certain nanomaterials might affect their inherent interactions that could act to detract from a targeted delivery strategy. For example, some polyanionic dendrimers can be taken up by cells and act within those cells to interfere with replication of human immunodeficiency virus (HIV) that is considered to be the causative agent of AIDS.[48] Although, from such studies, it is unclear if these dendrimers interact with the host cell or the pathogen to block their interaction; such a finding points to the potential for nanoparticles to interact with structures that might affect their cellular properties or cell function. In some cases, a nanoparticle with inherent capacity to interact with a cell or tissue might provide an added advantage of using that material for a specific indication. In other cases, such an inherent capacity to bind to or recognize cell or tissue components might highlight potential distractive aspects of that material for certain indications.

3.3.2 COMPLICATING ASPECTS

By their eponymous descriptor, nanoparticles have physical dimensions in the nanometer size scale similar to viruses and other materials that are either recognized by the body as pathogens or are elements associated with an infective event. Toll-like receptors (TLR) present on monocytes, leukocytes, and dendritic cells play a critical role in innate immunity with the capacity to recognize organized patterns present on viruses and bacteria.[49] TLR proteins are present on the surface of cells in the lung, spleen, prostate, liver, and kidney. Because the patterned surfaces of nanoparticles can look like pathogen components recognized by TLR proteins such as DNA, RNA, and repeating proteins like flagellin, it is possible that a number of cell types might non-selectively interact with some nanoparticles. If such an interaction occurs, there are several potential outcomes that might produce complicating aspects for nanoparticle targeting to cancers. Nanoparticle materials might be immediately recognized and cleared by cells of the innate immune system, limiting the usefulness of even their initial application. Alternately, only a fraction of the applied nanoparticles might engage TLR that would not significantly affect the effectiveness of the administration. Recognition of even a small fraction of the administered nanoparticles, however, might lead to immune events that diminish the effectiveness of subsequent administrations. Nanoparticles that the naïve body initially tolerates may become a focus of immune responses upon repeated exposure.

Oppositely, one could envisage how potential recognition by TLR proteins might be beneficial in certain applications. In particular, cancer cells or tumors sites that express a particular TLR might actually provide a unique targeting aspect for nanoparticles prepared from the right material. In fact, some studies have been described using empty RNA virus capsules from cowpea mosaic virus as biological nanoparticles for delivery.[50] Because tumors can also contain a number of immune cells with some that might express TLR, the potential exists to have inherent targeting of nanoparticles to cancers through these tumor-associated cells. Such a circumstance could reduce the complexity of construction for a targeted nanoparticle complex. Such a suggestion has solid grounding in the extensive use of nanoparticles as adjuvants used for vaccination.[51] Additionally, nanoparticles such as liposomes can be selectively directed specifically to Langerhans cells in the skin as a means of enhancing the delivery of antigens to these professional antigen presenting cells.[52]

Recognition of nanoparticles by TLR proteins may or may not act to stimulate potential target cells. If stimulation occurs, the result can include release of cytokines and a variety of potent regulatory molecules. Induction of pro-inflammatory cytokines may provide an undesirable outcome because cancerous states appear to be motivated by inflammatory events.[53] The potential for a beneficial or negative impact of such outcomes would require case-by-case scrutiny. By themselves, nanoparticles have been shown to stimulate several cell-signaling pathways, including those that drive the release of pro-inflammatory cytokines such as interleukin (IL)-6 and IL-8.[54,55] The incorporation of some therapeutic or diagnostic agents into nanoparticles can further enhance this potential for an inflammatory outcome.[56] Additionally, a variety of nanoparticles, including fullerenes and quantum dots, has been shown to stimulate the creation of reactive oxygen species, and this effect can be enhanced by exposure to UV light that might be used for visualization and/or activation (reviewed in Oberdorster, Oberdorster, and Oberdorster[57]). As discussed at the beginning of this section, the potential for such events may be desirable for induction of selected events, e.g., immunization against cancer cell antigens. Nanoparticles can also be prepared to release cytokines that might affect an anti-tumor immune events event.[58]

3.3.3 Safety Issues

Although generalizations can be made for a particular targeted nanoparticle delivery system, specific issues will arise for each type of payload it contains and for each indication. It will be important to balance potential safety concerns for using nanoparticles with their potential benefits for reducing a safety concern that occurs without their use for comparable (or even improved) efficacy. Nanoparticles can tremendously reduce toxicity by sequestering cytotoxic materials from non-specific tissues and organs of the body until reaching the cancer site.[59] Polymeric nanoparticles have been loaded with tamoxifen for targeted delivery to breast cancer cells to improve the efficacy to safety quotient relative to direct administration of this cytotoxic agent.[60] Coupling of doxorubicin-loaded liposomes to antibodies or antibody fragments that can enhance targeting to cancer cells appears quite promising as a means of further improving the efficacy to toxicity profile for this chemotherapeutic.[61] PEGylated liposomal doxorubicin has improved tolerability with similar efficacy compared to free drug.[62] Liposomal formulations of anthracyclines appear to improve the cardiotoxicity profile of this anti-neoplastic.[63] PAMAM dendrimers conjugated to cisplatin not only improved the water solubility characteristics of this chemotherapeutic but also improve its toxicity profile.[64] PAMAM dendrimers have also been used to increase the effectiveness of a radioimmunotherapy approach by increasing the specific accumulation of radioactive atoms at a tumor site as well as improving selectivity of biodistribution.[65]

As previously mentioned, potential inherent targeting of some nanoparticles may initiate cellular responses, following recognition by immune cell surface receptor systems. Outcomes from such events could produce safety concerns, particularly for repeated exposures. Data suggest additional safety concerns related to the material(s) used in nanoparticle preparation as well as nanoparticle size for some routes of administration. Recent studies have called into question the safety of repeated aerosolized exposure of carbon nanomaterials,[66] and nanoparticles have been shown to have a higher inflammatory potential per given mass than do larger particles.[57] Such particles may provide an efficient means to actively target lung cancers that is facilitated by the natural properties of these materials. In general, safety issues related to using nanoparticles to target cancers for treatment and diagnosis will be specific for the material used to prepare the particle: its size, the type of targeting mechanism it utilizes, its fate at the targeted site, and its non-targeted distribution and elimination pattern from the body.

3.4 TARGETING TO CANCER CELLS

In a very simplified sense, cancers occur through dysregulation of normal cell function. The body is designed to correct tissue defects following damage, to expand selected immune cells in response to a pathogen, and to compensate for perceived deficiencies in one cell type by altering the phenotype of another such as in stem cell recruitment. Each of these processes is mediated by (normally) regulated events that allow cells to lose their differentiated phenotype with restrained growth characteristics and acquire a replication-driven phenotype. A lack of re-differentiation into a growth-restrained, differentiated phenotype is the paradigm of cancer. Repair of an epithelial wound is a good example of this phenomenon (Figure 3.3). It is the lack of re-differentiation and continued responsiveness of these cells to growth factors and growth activators that supports and maintains the cancer phenotype. Extensive genomic differences between differentiated and non-differentiated forms of the same cell account for the differences observed between these two cell phenotypes.[67]

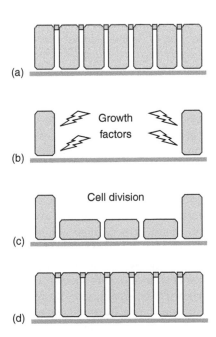

FIGURE 3.3 Loss and recovery of barrier function associated with normal epithelia. (a) Epithelia express tight junctions (▫) and associate at their base with a complex of proteins known as the basement membrane (▬▬▬▬). (b) Damage to epithelia increases responses to growth factors and results in differentiated epithelial cells that can freely divide. (c) Cell division continues until damaged area is covered. (d) Cell–cell contacts such as adherens junctions, tight junctions, and gap junctions have been shown to suppress growth and stimulate re-differentiation.

Many of the differences between replicating and non-replicating cells are associated with surface and metabolic properties that can be used to discriminate between differentiated (normal) and de-differentiated (cancer) cells. One of the biggest concerns using this information to target cancer cells is that non-cancerous cells undergoing normal and necessary repair processes may transiently express these same targets. For example, herceptin is an antibody that binds to her2/neu receptors that are over expressed on the surfaces of cancer cells. Unfortunately, this antibody can also target normal cardiac cells undergoing repair induced by the actions of a common anti-cancer agent, doxorubicin; patients on doxorubicin treatment are placed on an extended washout period prior to exposure to herceptin. Therefore, nanoparticles having a cytotoxic capacity and targeted using the herceptin antibody could result in cardiomyopathy. Fortunately, the high, transient, systemic levels of doxorubicin associated with direct administration can be muted by administration in nanomaterials such as liposomes, reducing the risk of cardiomyopathy.[68] That many surface and metabolic properties are the same for both cancer cells of a particular cell type and the undifferentiated form of that cell type during normal cell function must be appreciated as one examines potential cancer cell-targeting strategies for nanoparticles.

The general issues raised above concerning safety aspects of targeting cancer cells highlight concerns of selecting a strategy that properly accounts for unique cell surface properties and metabolic activities of cancer cells relative to normal cells. All too frequently, normal cells can undergo processes (e.g., wound repair) that will transiently transform them into a cell with surface properties or metabolic characteristics indistinguishable from a cancer cell. In some aspects, these altered surface properties and metabolic activities are intertwined. Alterations in metabolic properties can induce increased expression of nutrient uptake pathways and catabolic proteins that assist in nutrient absorption to sustain the accelerated growth rate of cancer cells. Because some of these components are present at the cell surface, a cancer cell's composition and profile are modified by their presence. From an opposing perspective, increased surface expression of components such as growth factor receptors will shift the metabolic activity of a cancer cell following activation of that receptor (either constitutive activation or ligand-induced activation).

3.4.1 CELL SURFACE PROPERTIES

In general, one thinks of targeting cancer cells by use of a highly specific surface material that absolutely identifies only that cancer cell within the entire body. Some studies, however, have demonstrated that cancer cells growing in different sites of the body can have altered surface properties that could facilitate non-specific nanoparticle binding and uptake; colloidal iron hydroxide (CIH) nanoparticle association and uptake appears to be enhanced for transformed cells,[69] and CIH nanoparticles show differences in cell surface interactions following transformation of chick embryo fibroblasts.[70] Such observations suggest that generic cell-surface charge differences might provide a targeting strategy for nanoparticles that could be, even if not highly discriminating, used to enrich nanoparticle delivery to cancer cells through non-specific associations. In combination with other targeting strategies, surface charge differences might provide a useful adjunct. For example, some cancer cells express unique sets of surface enzymes that might be useful to activate a prodrug once a nanoparticle has been localized to the surface of a cancer cell. In this regard, a number of proteases have been shown to be significantly up-regulated by oncogenic conversion.[71] A proof of concept for this type of specific application has been described using a polymer-based fluorogenic substrate PB-M7VIS that serves as a selective proteobeacon.[36]

A number of overexpressed growth factor receptors have been used to selectively target cancer cell surfaces, and the description of many of these targets has been reviewed.[72] With regard to targeting nanoparticles, it is important to remember that once engaged by the targeting ligand, some of these surface components are internalized whereas others will remain at the cell surface. Matching the type of nanoparticle material and its potential cargo with the likely fate of the targeted structure can be critical to optimizing the desired outcome. It is also possible that once bound, ligand–nanoparticle complexes could lead to receptor internalization that might not occur by the presence of the targeting ligand alone. The basis for this difference might come from the potential for nanoparticles to contain a coordinated ligand matrix to sequester of cell-surface receptors in a manner that facilitates internalization.

Antibody-based targeting of nanoparticles to solid tumors has been a highly promising strategy that is augmented by the enhanced vascular permeability (EPR effect) of solid tumors. For example, an antibody-directed (anti-p185HER2) liposome loaded with an anti-neoplastic can be an effective cancer therapeutic approach.[73] Blood cell-based cancers (e.g., leukemias and lymphomas) can also be targeted by nanoparticles as a way to reduce unwanted systemic side effects. Nanoparticles could be targeted to T-cell leukemia cells using an antibody to a surface cluster of differentiation (CD) antigen, CD3, on the surface of lymphocytes.[74] B-cell lymphomas can be targeted by anti-idiotypic antibodies specific for the unique monoclonal antibody expressed by each individual cancer.[75] In both of these cases, the potential for these B- and T-cell-derived cancer cells to actively take up particles on their surfaces as part of their normal function in antigen surveillance and presentation might facilitate and even augment the desired outcome using a targeted nanoparticle.

Efficient, targeted delivery of gene therapy elements and/or antigens to antigen presentation cells (APC) has long been a goal for the induction of anti-cancer cell immune responses. Nanoparticles provide an exciting possibility to achieve this goal. Coating nanoparticles with mannan facilitates their uptake by APCs such as macrophages and dendritic cells that acts to target these materials to local-draining lymph nodes following their administration.[76] Such an approach is likely to provide additional synergy in APC activation because polymer nanoparticles are efficiently phagocytosed by dendritic cells.[77] Many APCs express LDL-type receptors, and molecules that interact with this class of receptors could be a means of targeting as well.[78] Interestingly, LDL receptors can be an attractive targeting strategy for cancers because many tumors of different origins express elevated levels of this receptor.[79] Therefore, LDL-based nanoparticles could be useful in targeting cancer.[31]

It might also be possible to intentionally alter the surfaces of cancer cells to improve nanoparticle targeting. Cells could be transfected with a protein that expresses the appropriate acceptor peptide recognized by a surface-applied bacterial biotin ligase.[80] Although there would be multiple issues to overcome prior to clinical application, this approach outlines one strategy where cancer cells might be altered to express a unique surface structure such as biotin that could be very selectively targeted. Reversed-response targeting might also be performed using discriminating cell-surface properties. Hepatocytes can be targeted using nanocapsules decorated with the surface antigen of hepatitis B virus (SAgHBV).[81] Because liver cells may lose their capacity to bind SAgHBV following oncogenic conversion, this targeting strategy could be used to deliver cytoprotective materials to normal hepatocytes and enhance the efficacy of chemotherapeutics aimed at liver cancers. Similarly, hepatocytes exclusively express high affinity cell-surface receptors for asialoglycoproteins, and this ligand–receptor system has been used to target albumin nanoparticles to non-cancer cells of the liver.[8] Nanoparticles coated with galactose might also be used to target the liver.[82]

3.4.2 Metabolic Properties

One of the most detrimental aspects of cancer cells, their high rate of proliferation, can also be considered their Achilles' heel. As proliferation rate increases, metabolic requirements follow accordingly. This places cancer cells in a precarious position where a blockade of critical metabolic steps can lead to cytotoxic outcomes; this is the basis from a number of currently approved anti-cancer agents that function as metabolic poisons.[83] There are two obvious approaches that could be used for targeting using this characteristic of cancer cells: ligands that emulate the nutrient, vitamins or co-factors, and antibodies that recognize these surface transport elements. Nanoparticles coated with a ligand for one of these receptors such as folate can be used to target to cancer cells.[84]

Folate has been used to target dendrimers[85,86] and iron oxide nanoparticles.[87] Folate receptor-targeted lipid nanoparticles for the delivery of a lipophilic paclitaxel prodrug have shown promising pre-clinical outcomes.[88] Cancer cells can also overexpress transferrin receptors (REF), and transferrin-conjugated gold nanoparticle uptake by cells has been demonstrated.[89] A transferrin-modified cyclodextrin polymeric nanoparticle has been described that could be used to deliver genetic material to cancer cells.[90]

Increased nutrient uptake associated with increased requirements for amino acids and nucleic acids may also provide a targeting strategy for nanoparticles. That many classical anti-cancer agents function by interfering with amino acid or nucleic acid incorporation into polymer structures should provide an important template for strategies to use nanoparticle technologies to effectively target cancer cells. For example, pancreatic cancer cell lines appear to overexpress the peptide transporter system PepT1,[91] possibly providing a growth advantage by its ability to provide additional amino acid uptake. Nanoparticles decorated with (or composed of) ligands recognized by this uptake pathway may provide an important targeting opportunity. Such an approach can be used for similar target-specific delivery of other molecules. It is important to remember that materials such as amino acids and nucleic acids, unlike co-factors discussed above that are used more as part of catalytic cellular events, are required in stoiciometric amounts for cell growth. Targeting strategies using uptake processes against vitamins and co-factors will have the added benefit of potentially depriving cancer cells of these critical materials whereas targeting strategies using amino acids and nucleic acids will probably not affect the overall influx of these materials and their incorporation into nascent polymers required for continued cell growth.

A growing bank of experimental and clinical data has provided strong evidence that chronic inflammation can drive epithelial cell populations into an oncogenic phenotype.[92] It is this pre-neoplastic character that may act to alter the metabolic character of cells that might be useful for targeting nanoparticles. Such a targeting strategy would make use of transitions in cellular function in response to pro-inflammatory signals (Figure 3.4). In this regard, one could envisage ligand-directed targeting to inflammatory sites as well as activation of nanoparticles (or their components) for localized delivery at these sites by the presence of unique enzymatic activities. Additionally, one could contemplate nanoparticles that deliver cancer prevention agents that work through suppression of inflammatory events. A ligand peptide that binds endothelial vascular adhesion molecule-1 (VCAM-1) on the surfaces of inflamed vessels has been used to target nanoparticles.[93] One major concern with using such metabolic processes for targeting nanoparticles is that inflammatory events are a common and essential function of the body, and they are not necessarily associated with pre-cancerous or cancerous states. Some nanoparticles can induce an inflammatory response. Therefore, the method selected for such a nanoparticle-targeting strategy most keep this in mind. It might be possible to utilize additional cancer-targeting mechanisms (e.g., the EPR effect) to augment inflammation-based strategies.

3.5 TARGETING TO TUMORS

Nanoparticle targeting strategies involving peculiar and unique properties of tumors have been described. Two of these properties will be discussed: aberrant vasculature and the unique metabolic environment produced by inefficient blood supply resulting from the aberrant vasculature. These targeting strategies have strong similarities to those described for targeting cancer cells, taking advantage of unique endothelial cell-surface properties in tumors or the cells' peculiar environment as they respond to the metabolically overactive environment of a tumor.

3.5.1 ABERRANT VASCULATURE

Tumor neovasculature is a promising site for targeting nanoparticles for both diagnosis and therapy. Once a tumor reaches a size of greater than ~ 1 cm^3, vascular assistance to deliver oxygen and

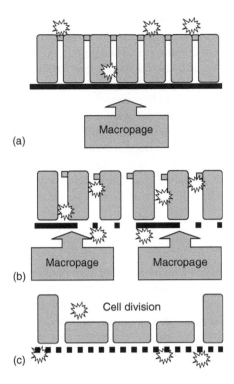

FIGURE 3.4 Effect of chronic pro-inflammatory stimulus on epithelial barrier patency. (a) A variety of stimuli (✹) can incite the release of pro-inflammatory cytokine (e.g., TNFα and IFNγ) from macrophages and other cells. Genetic predisposition can enhance these responses. (b) Released pro-inflammatory cytokines act to open tight junction (TJ) structures (▫), allowing entry of additional activating stimuli that, in turn, attract more cells associated with inflammation. (c) Chronic inflammation leads to breakdown of basement membrane, loss of TJ function, and disorganization of the epithelia characteristic with a pre-neoplastic state.

remove by-products is required for cell survival. Cancer cells secrete a variety of growth factors that stimulate the formation of nascent blood and lymph vessels. Tumor-associated vascular beds have unique surface properties that are associated with their rapid growth characteristics, resulting in aberrant vascular beds that have been used to design a combined vascular imaging and therapy approach using nanoparticles.[94] Non-specific targeting of nanoparticles (10–500 nm in diameter) to solid tumors through this EPR capacity of solid tumors[47] can provide a means to enrich the localization of an anti-neoplastic agent to a tumor when it is coupled to a nanoparticle compared to its free form.[95] The EPR effect has also been used to increase localization through inherent targeting[96] of long-circulating liposomes.[97]

PAMAM dendrimers useful for boron neutron capture therapy (BNCT) of cancers have been targeted to tumor vasculature by attachment of vascular endothelial growth factor (VEGF) that acts to target VEGF receptors that are frequently overexpressed on tumor neovasculature.[98] Targeting VEGF receptors Flk or Flt on tumor-associated endothelial cells could also be effective; this has been done with a complex material composed of anti-Flk-1 antibody-coated ^{90}Y-labeled nanoparticles[99]. PECAM (or CD31) is highly expressed on the surface of endothelial cells present in immature vasculature. Platelet endothelial cell adhesion molecule (PECAM) up-regulation occurs following VEGF stimulation of endothelial cells. The presence of such endothelial surface markers also provides the opportunity to target nanoparticles that contain DNA for gene therapy applications,[100] release anti-angiogenesis agents as well as chemotherapeutics,[101] and antigens/agents to stimulate an anti-cancer cell immune response.[101,102] Vascular cell adhesion molecule-1 (VCAM-1) is a marker for inflammation of the endothelial and has been used to target nanoparticles to these sites for magento-optical imaging.[103]

Differences in tissue-specific endothelial surface properties might be used for targeting nano-particles.[104] In one such example, a peptide that binds to membrane dipeptidase on lung endothelial cells can be used to target nanocrystals to the lung.[105] Doxorubicin-loaded nanoparticles have been targeted to tumor vasculature by surface decoration with a cyclic arginine–glycine–aspartic acid (RGD) peptide that binds to the cell adhesion molecule integrin $\alpha_v\beta_3$ on the surface of endothelial cells.[106] A peptide that interacts with the lymphatic vessel marker podoplanin can be used to target nanocrystals to lymph vessels and some tumor cells.[105] Cationic nanoparticles, containing genes that can block endothelial cell signaling that were selectively targeted to tumor vasculature by coupling to an integrin $\alpha_v\beta_3$ ligand, were shown to produce endothelial apoptosis and tumor regression.[107]

3.5.2 Metabolic Environment

Solid tumors typically grow at such a rapid pace that vascular function fails to keep pace with local metabolic requirements. This situation, leading to reduced oxygen tension and a depressed pH as a result of a lack of waste acid clearance, is associated with the onset of necrotic cores of large tumors. A number of studies have shown that cancer cells become adapted to successful growth in these difficult conditions, providing unique characteristics that might be exploited for tumor targeting. For example, nanoparticles will accumulate at inflammation sites along the bowel when administered orally.[108,109] Because these inflammatory sites are typically associated with a slightly acidic environment, a similar targeting strategy may be possible for tumors where the pH has also been depressed. Nanoparticles targeted to cancer cells might also be modified at low oxygen tension to either release or activate a payload. Frequently, cancer cells rely more on glycolysis than oxidative metabolism for energy production. This finding is consistent with the reduced oxygen tension of poorly perfused tumors. These conditions would result in the release of acidic glycolytic end-products that are not effectively cleared by the sluggish blood flow through tumor vascular beds. Therefore, targeting that takes advantage of a slightly depressed local pH that might be found in some tumors could be an attractive design strategy.

3.6 FATE OF NANOPARTICLES

Successful targeting of nanoparticles to cancers or tumors may involve overcoming multiple biological, physiological, and physical barriers. Site of initial application can be critical. For example, nanoparticles administered into the gut or lung would initially confront epithelial barriers. Metabolic events or cellular responses at an injection site represent another initial barrier to targeted nanoparticle delivery. Once nanoparticles have entered the body, their size, shape, or surface characteristics can initiate events that present a second barrier to targeted delivery—misdirection of the material away from its targeted site through undesired interactions. Access and/or enriched distribution to specific organs or regions of the body may be critical for successful nanoparticle targeting. Finally, once nanoparticles have reached a targeted site, metabolic or physical aspects of the cancer cell or tumor might limit their effectiveness. Here, the intracellular uptake and fate may dictate the potential success of each nanoparticle approach. Events at each of these barriers act in a cumulative fashion to limit the success of any nanoparticle-based targeting strategy. Obviously, the overall fate of nanoparticles might be improved by using materials that are not affected by these barriers or by modification of nanoparticles in ways that can neutralize these barrier issues. Matching the size and composition profile of a nanoparticle delivery system with the targeting strategy allows for an optimized approach that takes advantage of default targeting events as much as possible.

As previously discussed, it is critical that inherent targeting mechanisms associated with a particular nanoparticle does not overwhelm or work in concert with any applied targeting strategy. Once at the target site, the fate of the nanoparticle can significantly affect its potential to provide the desired outcome. The delivery of hybridization-competent antisense oligonucleotides (ODNs) targeted to a cancer cell or tumor would not provide the desired outcome unless this material is efficiently internalized. ODNs covalently conjugated to anionic dendrimers have been shown to effectively deliver through an endocytosis process and down-regulate epidermal growth factor receptor expression in cancer cells.[110] Other targeting strategies may not provide the desired outcome if the nanoparticle is internalized by cells. For example, enzyme-coupled nanoparticles targeted to a tumor that would activate a prodrug could fail to provide a desired outcome. Therefore, the potential for nanoparticles to have a successful outcome of targeting cancer cells or tumors requires favorable events at natural barriers of the body and their distribution within the body, but also their fate at the targeted site.

3.6.1 Site of Initial Application

Several extracellular barriers exist for the administration and targeted delivery of nanoparticles. Initial entry into the body represent on obvious barrier. This is not an issue for situations where the cancer cell or tumor targeted is readily accessible by a topical application. Such a situation, however, is rather rare. Entry into the body across mucosal surfaces such as those in the gut or lung is typically very inefficient. Even viral particles are not very successful at this approach with most relying upon infecting cells of the barrier from the apical exposure by only a few viral particles that can replicate inside the cells to allow the basolateral (systemic) release of large numbers of progeny. Viral particle entry at apical surfaces of epithelial cells is decreased by physical barriers such as secreted mucus as well as proteases and other enzymatic barriers. Extracellular (acellular) matrix environments that viruses might encounter after systemic infection could similarly act to diminish cellular targeting and entry. Man-made nanoparticle delivery systems are likely to be impeded by these same physical and biological barriers at epithelial surfaces and within the body. Reduced surface exposure of highly charged or protruding structures is commonly used by viruses to minimize the impact of these extracellular barriers on viral infectivity. Similar considerations may facilitate optimization of nanoparticle delivery strategies.

There are several common methods for administering materials to the body: injection or application to an epithelial surface (skin, intestine, lung, etc.) of the body. Nanoparticles can be absorbed into the skin after topical application.[111] Although nanoparticles can be taken up through appendages of the skin (sweat gland ducts, hair follicles) following topical application,[112] microneedles can be used to dramatically increase the efficiency of their uptake into and across skin.[113] The intravitreous injection of nanoparticles results in transretinal movement with a preferential localization in retinal pigment epithelial cells,[114] allowing for a sustained delivery strategy to the inner eye.

Nanoparticles can be absorbed from the lumen of the gut, but this absorption is inefficient.[115] A number of factors have been examined related to regulation of nanoparticle uptake from the gut lumen.[116] Nanometer-sized liposomes enter into the intestinal mucosa better than larger, multi-lamellar liposomes, and this uptake can be improved by coating with a mucoadhesive polymer such as chitosan.[117] It is interesting that lipid-based materials absorbed from the gut partition into the lymphatic system and studies have suggested that these particles have remarkable access to the hepatocytes.[118] One way to potentially improve nanoparticle uptake from the gut is to PEGylate these materials in a manner that selectively increases binding to the intestinal mucosa rather than the stomach wall.[119] Additionally, anionic PAMAM dendrimers have been shown to rapidly cross the intestinal mucosa in vitro and may provide a method to improving oral delivery of nanoparticles.[120] Cationic dendrimers also show a transcytosis capability in vitro; in general, cationic dendrimers are more cytotoxic than anionic dendrimers, but this characteristic can be reduced by additional surface modifications using lipids.[121] Formulation studies have been performed to identify optimal methods for aerosol delivery of nanoparticles to the lung.[122] In general, the uptake of nanoparticles at the lung or gut surface occurs, but the efficiency of this uptake is dramatically improved by incorporation of a specific uptake mechanism. Even without a specific uptake mechanism, an appreciable amount of nanoparticle absorption can occur at these sites if they are sufficiently stable and are not removed by clearance mechanisms.

A large number of studies have been performed to assess nanoparticle absorption following inhalation exposure, and concerns over the safety of such an approach for drug delivery have been raised.[123] Nanoparticles deposited in the airways appear to be taken up through transcytosis pathways that allow the passage of these materials across epithelial and endothelial cells to reach the blood and lymphatics.[57] Surface properties of nanoparticles greatly affect the capacity of nanoparticles to be taken into cells through the process of endocytosis and uptake following pulmonary deposition. As part of the respiratory tree, intranasal administration of nanoparticles can potentially provide a route into the brain.

3.6.2 DISTRIBUTION TO SPECIFIC ORGANS

One of the most difficult challenges of administering cytotoxic chemotherapeutics involves unwanted exposure to non-cancer cells and to tissue and organ compartments not involved with the disease. Nanoparticles carrying a chemotherapeutic can reduce the undesirable distribution of such compounds as they are restricted from some compartments of the body such as the brain.[59] Oppositely, nanoparticles can be modified, e.g., conjugation to chelators, to acquire the capacity to transport across the blood–brain barrier or BBB.[124] Alternately, nanoparticles coupled to certain protein ligands such as apolipoprotein E can be used to target and transport across the BBB.[125] In both cases, these nanoparticle delivery approaches lead to the unique distributions of materials that must be cleared and/or metabolized. Therefore, one consequence of targeting cancers cells is that the fate of these materials, by accessing and localizing to sites where cancer cells reside, may be affected that could affect their overall safety as well as efficacy.

Targeting to some cancers may require overcoming additional hurdles beyond interaction with specific cancer cell or tumor components. Some tumors are located in difficult-to-reach sites such as the brain and testes. Accessing these sites from the systemic vasculature requires that nanoparticle materials must first avoid systemic clearance by the RES and have the capacity to move across either the blood–brain or blood–testes barrier. Whereas some types of nanoparticles can keep materials out of compartments of the body such as the brain,[59] other types of nanoparticles may provide access to this difficult-to-reach compartment.[126] In general, surface characteristics that can be altered through chemical modifications can be used to regulate the targeted delivery of nano-particles to specific sites within the body such as the brain (reviewed in Olivier[14]). It has even been reported that coating nanoparticles with polysorbate 80 can facilitate brain targeting by enhancing their interaction with brain microvasculature.[127] Alternately, intranasal delivery of macromolecules has been suggested to move in a retrograde fashion into the brain following uptake at the olfactory epithelium.[128] Polylactic acid–PEG nanoparticles have been shown to transport across the nasal mucosa[129] and could, theoretically, also provide some access to the brain because both materials appear to transport via a transcytosis mechanism.

3.6.3 INTRACELLULAR UPTAKE AND FATE

Particularly in the case of cancer therapeutics, some targeted delivery may require access to intra-cellular sites such as the nucleus or mitochondria. In general, nanoparticles with diameters less than 50 nm can easily enter most cells. Successful cellular uptake of nanoparticle systems targeted to cancer cells and/or tumors, however, frequently depends upon the balance of mechanisms that act to clear nanoparticles from the circulation and mechanisms that allow for their retention in this compartment. Several clearance mechanisms exist, and loss of nanoparticles from the circulation appears to be dominated by macrophage uptake, following complement activation or surface opsinization.[130]

Some agents such as commonly used chemotherapeutics are capable of moving efficiently across the plasma membrane of cells and into the cytoplasm that allows access to intracellular organelles if the material can be released locally to the target site. A nanoparticle-based delivery may require the release of a cargo either at the cell surface or after internalization of the nanoparticle by the target cell. Uptake of nanoparticles into cells can occur through a clathrin-coated endocytosis event,[131] through caveolae structures,[132] or through uptake mechanisms that do not appear to involve clathrin or caveolae.[133] Differences in surface properties and nanoparticle size will likely dictate the predominant route of entry into a cell. Such uptake mechanisms into cells following nanoparticle targeting to a cancer cell or tumor can involve vesicular trafficking to acidified, protease, and nuclease-enriched lysosomal compartments within the cell.

Unless the nanoparticle carrying a chemotherapeutic agent can release it prior to the trafficking of these vesicles to destructive (lysosomal) pathways or it can avoid such a pathway once inside the cell, the effectiveness of the absorbed material may be dramatically reduced. Also, some new classes of anti-cancer agents have poor membrane permeability properties and would not readily leave the endosome after uptake. Furthermore, exposure of these materials to lysosomal environments would destroy their biological activity; nucleic acid- or peptide/protein-based therapeutics capable of marking a cancer cell for clearance by the immune system would be examples of this type of approach. In these cases, proper selection of the nanoparticle composition and characteristics allows these materials to escape the fate of this default uptake event. There are endogenous properties of some materials as well as the capacity to include specific intracellular targeting agents that can be matched with the intracellular delivery desired for the material being targeted.

Following specific (or non-specific) targeting of nanoparticles to a cell, a number of events act to traffic these material within the cell. In general, nanoparticles that have associated with a specific cell-surface target are internalized through an endocytosis process that produces the formation of intracellular vesicles, containing the nanoparticle. Based upon their physical and chemical characteristics, most nanoparticles reaching these endosomal vesicles are likely delivered to lysosomes within cells where they would be metabolized or retained. Therefore, nanoparticles would not readily access the cytoplasm of target cells. An ultimate fate of lysosomal structures within targeted cells is not necessarily a problem. Many diagnostics have already performed their function and/or can still be detected within this compartment. In the case of therapeutics, many of these may have already been released from the nanoparticle prior to its arrival at the lysosome, and/or the materials are stable in this hostile environment and continue to act upon the target cell from this location. However, there are a number of potential therapeutic compounds that will be inactivated by this outcome and that require additional delivery events to achieve their optimal function on the target cell. Oligonucleotide delivery to tumors is one such example where the therapeutic must access the target cell cytoplasm for its desired action.[134]

Reduction in the rate of endocytosis of nanoparticles can be achieved by coating them with proteins such as lactoferrin or ceruloplasmin that act to retain the material at the cell surface.[135] One approach to facilitate nanoparticle delivery to the cytoplasm of a target cell is to covalently attach the membrane penetrating TAt peptide derived from HIV-1 to the surface of these nanoparticles.[136] Because nanoparticles of several compositions have been shown to target to the mitochondria,[137,138] these materials may access the cell's cytoplasm to reach this organelle. Such an outcome may be driven by the physico-chemical characteristics of the nanoparticle with relation to the unique proton and ion gradients found in mitochondria. Modification of the nanoparticle to affect this inherent targeting may be important.

3.7 CONCLUSIONS

Nanoparticles provide a range of new opportunities to increase the targeting of currently approved diagnostic and therapeutic agents to cancers. Improvements in targeting can lead not only to increased efficiency of these agents but also to increased signal-to-noise ratios for diagnostics and better efficacy to toxicity ratios for therapeutics. Currently, a whole new spectrum of biopharmaceuticals and biotechnological agents for cancer diagnosis and therapy are also being developed. Some of these materials require special formulation technologies to overcome drug-associated problems. Although nanoparticles offer improved profiles for some currently approved diagnostic and therapeutic agents, many biotechnology-based materials absolutely require some method of delivery that compensates for their poor stability or non-selective activity in a systemic setting. Nanoparticles offer a set of new opportunities for the development of these agents.[29]

Nanoparticles can be prepared in such a way as to have diagnostic or therapeutic agents integrated into them in ways that either freely releases the agents or that requires decomposition of the nanoparticle for the release to occur. Because of the inherent nature of small (nanometer-sized) structures, the body can identify and respond to these as foreign. Such a response can by suppressed by incorporation of agents that might suppress undesirable responses, or the application can be matched to the nanoparticle to make use of these natural responses. It is even possible to modify these natural responses to better match the desired clinical outcome. Specific components used to prepare nanoparticles can affect not only their stability in the body but also their capacity to be absorbed across natural barriers of the body (e.g., BBB) as well as the inherent systemic distribution of the nanoparticle that might compete with or complement efforts to selective targeting strategies.

Without the current fanfare related to nanotechnologies, nanoparticles have been used to selectively target a number of organs of the body for a number of years. Nanoparticle colloids were shown to have contrast media properties that related to the unique surface properties of cells in specific organs of the body.[139] Deviations from normal function such as oncogenic transformation can lead to changes in a cell's surface properties and its capacity to interact with nanomaterials. For example, gadolinium-based nanoparticles are taken up by hepatocytes, and by the decreased function and density of cancer cells in the liver tumors, a reduced level of uptake of these particles can be used to identify tumors using T1-weighted images obtained from MRI.[42]

Throughout this chapter, there has been frequent referral to viral infection events as a paradigm for cellular and intracellular targeting strategies for nanoparticles. Indeed, these materials are very successful models for nanoparticle targeting because they have developed mechanisms to discriminate between the various cells of the body (e.g., tropism for only cells of the intestinal tract) and can deliver labile (polynucleic acids) payloads that dramatically affect cell function. In response to these nanoparticle invaders, host cells have established intricate and complicated mechanisms that block viral infectivity and cellular actions. Such evolutionary pressures have led to the incorporation of intricate and novel methods by viruses to effectively combat these protective systems established by host cells. It is into this environment where the virus nanoparticles and host cells have battled back and forth for millennia that efforts to use nanoparticles to deliver agents to cancers for diagnosis and/or therapy must be framed.

Finally, it is important to sound a precautionary note for the potential to over-engineer nanoparticles. Nanoparticles provide a platform that can potentially be used to simultaneously function in targeting therapeutic molecules as well a reporter and/or imaging agents. Elegant studies have been performed with nanoparticles modified three, four, or even five times with materials that promoted active targeting and/or reduced non-specific targeting as well as corrected undesirable properties of residence, biodistribution, and stability. Such tour-de-force efforts would be considered unrealistic by pharmaceutical companies for scaling to a process that would gain approval from regulatory agencies. Clinical success can be demonstrated for a nanoparticle system only if it can get to the clinic; a viable production process is critical to this development path. Therefore, successful applications of nanoparticles in target cancers for therapy and diagnosis will require designing systems where the inherent activities and distribution of nanoparticle size and composition allow for minimal modifications that will translate into production process steps.

REFERENCES

1. Kayser, O., Lemke, A., and Hernandez-Trejo, N., The impact of nanobiotechnology on the development of new drug delivery systems, *Curr. Pharm. Biotechnol.*, 6(1), 3–5, 2005.
2. Duncan, R., The dawning era of polymer therapeutics, *Nat. Rev. Drug Discov.*, 2(5), 347–360, 2003.
3. Douglas, S. J., Davis, S. S., and Illum, L., Nanoparticles in drug delivery, *Crit. Rev. Ther. Drug Carrier Syst.*, 3(3), 233–261, 1987.
4. Moghimi, S. M., Hunter, A. C., and Murray, J. C., Nanomedicine: Current status and future prospects, *FASEB J.*, 19(3), 311–330, 2005.
5. Zimmer, A., Antisense oligonucleotide delivery with polyhexylcyanoacrylate nanoparticles as carriers, *Methods*, 18(3), 286–295, 1999, see also p. 322.
6. Li, H. and Qian, Z. M., Transferrin/transferrin receptor-mediated drug delivery, *Med. Res. Rev.*, 22(3), 225–250, 2002.
7. Balthasar, S. et al., Preparation and characterisation of antibody modified gelatin nanoparticles as drug carrier system for uptake in lymphocytes, *Biomaterials*, 26(15), 2723–2732, 2005.
8. Mao, S. J. et al., Uptake of albumin nanoparticle surface modified with glycyrrhizin by primary cultured rat hepatocytes, *World J. Gastroenterol.*, 11(20), 3075–3079, 2005.
9. Kickhoefer, V. A. et al., Engineering of vault nanocapsules with enzymatic and fluorescent properties, *Proc. Natl. Acad. Sci. USA*, 102(12), 4348–4352, 2005.
10. Benns, J. M. and Kim, S. W., Tailoring new gene delivery designs for specific targets, *J. Drug Target.*, 8(1), 1–12, 2000.
11. Prabha, S. and Labhasetwar, V., Nanoparticle-mediated wild-type p53 gene delivery results in sustained antiproliferative activity in breast cancer cells, *Mol. Pharmacol.*, 1(3), 211–219, 2004.
12. Bala, I., Hariharan, S., and Kumar, M. N., PLGA nanoparticles in drug delivery: The state of the art, *Crit. Rev. Ther. Drug Carrier Syst.*, 21(5), 387–422, 2004.
13. Bejjani, R. A. et al., Nanoparticles for gene delivery to retinal pigment epithelial cells, *Mol. Vis.*, 11, 124–132, 2005.
14. Olivier, J. C., Drug transport to brain with targeted nanoparticles, *NeuroRx*, 2(1), 108–119, 2005.
15. Daugherty, A. L. et al., Pharmacological modulation of the tissue response to implanted polylactic-co-glycolic acid microspheres, *Eur. J. Pharm. Biopharm.*, 44(1637), 89–102, 1997.
16. Lemarchand, C. et al., Novel polyester–polysaccharide nanoparticles, *Pharm. Res.*, 20(8), 1284–1292, 2003.
17. Mainardes, R. M. *et al.*, Colloidal carriers for ophthalmic drug delivery, *Curr. Drug Targets*, 6(3), 363–371, 2005.
18. Kojima, C. et al., Synthesis of polyamidoamine dendrimers having poly(ethylene glycol) grafts and their ability to encapsulate anti-cancer drugs, *Bioconjugate Chem.*, 11(6), 910–917, 2000.
19. Jain, T. K. et al., Iron oxide nanoparticles for sustained delivery of anti-cancer agents, *Mol. Pharmacol.*, 2(3), 194–205, 2005.
20. He, Q. et al., Calcium phosphate nanoparticle adjuvant, *Clin. Diagn. Lab. Immunol.*, 7(6), 899–903, 2000.
21. Petrovsky, N. and Aguilar, J. C., Vaccine adjuvants: Current state and future trends, *Immunol. Cell Biol.*, 82(5), 488–496, 2004.
22. Gao, X. et al., In vivo molecular and cellular imaging with quantum dots, *Curr. Opin. Biotechnol.*, 16(1), 63–72, 2005.
23. Michalet, X. et al., Quantum dots for live cells, in vivo imaging, and diagnostics, *Science*, 307(5709), 538–544, 2005.
24. Kirchner, C. et al., Cytotoxicity of colloidal CdSe and CdSe/ZnS nanoparticles, *Nano Lett.*, 5(2), 331–338, 2005.
25. Ruysschaert, T. et al., Liposome-based nanocapsules, *IEEE Trans. Nanobioscience*, 3(1), 49–55, 2004.
26. Shenoy, V. S., Vijay, I. K., and Murthy, R. S., Tumour targeting: Biological factors and formulation advances in injectable lipid nanoparticles, *J. Pharm. Pharmacol.*, 57(4), 411–422, 2005.
27. Moghimi, S. M. and Hunter, A. C., Capture of stealth nanoparticles by the body's defences, *Crit. Rev. Ther. Drug Carrier Syst.*, 18(6), 527–550, 2001.

28. Allen, C. *et al.*, Controlling the physical behavior and biological performance of liposome formulations through use of surface grafted poly(ethylene glycol), *Biosci. Rep.*, 22(2), 225–250, 2002.

29. Muller, R. H. and Keck, C. M., Challenges and solutions for the delivery of biotech drugs—A review of drug nanocrystal technology and lipid nanoparticles, *J. Biotechnol.*, 113(1–3), 151–170, 2004.

30. Emanuel, N. et al., Preparation and characterization of doxorubicin-loaded sterically stabilized immunoliposomes, *Pharm. Res.*, 13(3), 352–359, 1996.

31. Versluis, A. J. et al., Stable incorporation of a lipophilic daunorubicin prodrug into apolipoprotein E-exposing liposomes induces uptake of prodrug via low-density lipoprotein receptor in vivo, *J. Pharmacol. Exp. Ther.*, 289(1), 1–7, 1999.

32. Nobs, L. et al., Poly(lactic acid) nanoparticles labeled with biologically active Neutravidin for active targeting, *Eur. J. Pharm. Biopharm.*, 58(3), 483–490, 2004.

33. Gref, R. et al., Surface-engineered nanoparticles for multiple ligand coupling, *Biomaterials*, 24(24), 4529–4537, 2003.

34. Konda, S. D. et al., Specific targeting of folate-dendrimer MRI contrast agents to the high affinity folate receptor expressed in ovarian tumor xenografts, *Magma*, 12(2–3), 104–113, 2001.

35. Kobayashi, H. and Brechbiel, M. W., Dendrimer-based macromolecular MRI contrast agents: Characteristics and application, *Mol. Imaging*, 2(1), 1–10, 2003.

36. McIntyre, J. O. et al., Development of a novel fluorogenic proteolytic beacon for in vivo detection and imaging of tumour-associated matrix metalloproteinase-7 activity, *Biochem. J.*, 377(Pt 3), 617–628, 2004.

37. Morawski, A. M., Lanza, G. A., and Wickline, S. A., Targeted contrast agents for magnetic resonance imaging and ultrasound, *Curr. Opin. Biotechnol.*, 16(1), 89–92, 2005.

38. Voura, E. B. et al., Tracking metastatic tumor cell extravasation with quantum dot nanocrystals and fluorescence emission-scanning microscopy, *Nat. Med.*, 10(9), 993–998, 2004.

39. Marsh, J. N. et al., Improvements in the ultrasonic contrast of targeted perfluorocarbon nanoparticles using an acoustic transmission line model, *IEEE Trans. Ultrason. Ferroelectr. Freq. Control*, 49(1), 29–38, 2002.

40. Kiser, P. F., Wilson, G., and Needham, D., A synthetic mimic of the secretory granule for drug delivery, *Nature*, 394(6692), 459–462, 1998.

41. Brannon-Peppas, L. and Blanchette, J. O., Nanoparticle and targeted systems for cancer therapy, *Adv. Drug Deliv. Rev.*, 56(11), 1649–1659, 2004.

42. Mintorovitch, J. and Shamsi, K., Eovist injection and Resovist injection: Two new liver-specific contrast agents for MRI, *Oncology (Williston Park)*, 14(6 Suppl. 3), 37–40, 2000.

43. Perez, R. V. et al., Selective targeting of Kupffer cells with liposomal butyrate augments portal venous transfusion-induced immunosuppression, *Transplantation*, 65(10), 1294–1298, 1998.

44. Kim, B. K. et al., Hydrophilized poly(lactide-*co*-glycolide) nanospheres with poly(ethylene oxide)–Poly(propylene oxide)–poly(ethylene oxide) triblock copolymer, *J. Microencapsul.*, 21(7), 697–707, 2004.

45. Veronese, F. M. and Pasut, G., PEGylation, successful approach to drug delivery, *Drug Discov. Today*, 10(21), 1451–1458, 2005.

46. Cole, C. et al., Tumor-targeted, systemic delivery of therapeutic viral vectors using hitchhiking on antigen-specific T cells, *Nat. Med.*, 11(10), 1073–1081, 2005.

47. Maeda, H. et al., Tumor vascular permeability and the EPR effect in macromolecular therapeutics: A review, *J. Control. Release*, 65(1–2), 271–284, 2000.

48. Witvrouw, M. et al., Polyanionic (i.e., polysulfonate) dendrimers can inhibit the replication of human immunodeficiency virus by interfering with both virus adsorption and later steps (reverse transcriptase/integrase) in the virus replicative cycle, *Mol. Pharmacol.*, 58(5), 1100–1108, 2000.

49. Moynagh, P. N., TLR signalling and activation of IRFs: Revisiting old friends from the NF-kappaB pathway, *Trends Immunol.*, 26(9), 469–476, 2005.

50. Rae, C. S. et al., Systemic trafficking of plant virus nanoparticles in mice via the oral route, *Virology*, 343(2), 224–235, 2005.

51. Aucouturier, J., Dupuis, L., and Ganne, V., Adjuvants designed for veterinary and human vaccines, *Vaccine*, 19(17–19), 2666–2672, 2001.

52. McGuire, M. J. et al., A library-selected, Langerhans cell-targeting peptide enhances an immune response, *DNA Cell Biol.*, 23(11), 742–752, 2004.

53. Mizukami, Y. et al., Induction of interleukin-8 preserves the angiogenic response in HIF-1alpha-deficient colon cancer cells, *Nat. Med.*, 11(9), 992–997, 2005.

54. Donaldson, K. and Tran, C. L., Inflammation caused by particles and fibers, *Inhal. Toxicol.*, 14(1), 5–27, 2002.

55. Wottrich, R., Diabate, S., and Krug, H. F., Biological effects of ultrafine model particles in human macrophages and epithelial cells in mono- and co-culture, *Int. J. Hyg. Environ. Health*, 207(4), 353–361, 2004.

56. Gopalan, B. et al., Nanoparticle based systemic gene therapy for lung cancer: Molecular mechanisms and strategies to suppress nanoparticle-mediated inflammatory response, *Technol. Cancer Res. Treat.*, 3(6), 647–657, 2004.

57. Oberdorster, G., Oberdorster, E., and Oberdorster, J., Nanotoxicology: An emerging discipline evolving from studies of ultrafine particles, *Environ. Health Perspect.*, 113(7), 823–839, 2005.

58. Segura, S. et al., Potential of albumin nanoparticles as carriers for interferon gamma, *Drug Dev. Ind. Pharm.*, 31(3), 271–280, 2005.

59. Hamaguchi, T. et al., NK105, a paclitaxel-incorporating micellar nanoparticle formulation, can extend in vivo antitumour activity and reduce the neurotoxicity of paclitaxel, *Br. J. Cancer*, 92(7), 1240–1246, 2005.

60. Shenoy, D. B. and Amiji, M. M., Poly(ethylene oxide)-modified poly(epsilon-caprolactone) nanoparticles for targeted delivery of tamoxifen in breast cancer, *Int. J. Pharm.*, 293(1–2), 261–270, 2005.

61. Park, J. W., Benz, C. C., and Martin, F. J., Future directions of liposome- and immunoliposome-based cancer therapeutics, *Semin. Oncol.*, 31(6 Suppl. 13), 196–205, 2004.

62. Rose, P. G., Pegylated liposomal doxorubicin: Optimizing the dosing schedule in ovarian cancer, *Oncologist*, 10(3), 205–214, 2005.

63. Robert, N. J. *et al.*, The role of the liposomal anthracyclines and other systemic therapies in the management of advanced breast cancer, *Semin. Oncol.*, 31(6 Suppl. 13), 106–146, 2004.

64. Malik, N., Evagorou, E. G., and Duncan, R., Dendrimer-platinate: A novel approach to cancer chemotherapy, *Anti-cancer Drugs*, 10(8), 767–776, 1999.

65. Kobayashi, H. et al., Monoclonal antibody–dendrimer conjugates enable radiolabeling of antibody with markedly high specific activity with minimal loss of immunoreactivity, *Eur. J. Nucl. Med.*, 27(9), 1334–1339, 2000.

66. Lam, C. W. et al., Pulmonary toxicity of single-wall carbon nanotubes in mice 7 and 90 days after intratracheal instillation, *Toxicol. Sci.*, 77(1), 126–134, 2004.

67. Liu, J. J. et al., Multiclass cancer classification and biomarker discovery using GA-based algorithms, *Bioinformatics*, 21(11), 2691–2697, 2005.

68. Ewer, M. S. et al., Cardiac safety of liposomal anthracyclines, *Semin. Oncol.*, 31(6 Suppl. 13), 161–181, 2004.

69. Harlos, J. P. and Weiss, L., Differences in the peripheries of Lewis lung tumor cells growing in different sites in the mouse, *Int. J. Cancer*, 32(6), 745–750, 1983.

70. Subjeck, J. R., Weiss, L., and Warren, L., Colloidal iron hydroxide-binding to the surfaces of chick embryo fibroblasts transformed by wild-type and a temperature-sensitive mutant of Rous sarcoma virus, *J. Cell. Physiol.*, 91(3), 329–334, 1977.

71. Lynch, C. C. et al., MMP-7 promotes prostate cancer-induced osteolysis via the solubilization of RANKL, *Cancer Cell*, 7(5), 485–496, 2005.

72. Carter, P., Smith, L., and Ryan, M., Identification and validation of cell surface antigens for antibody targeting in oncology, *Endocr. Relat. Cancer*, 11(4), 659–687, 2004.

73. Park, J. W. *et al.*, Development of anti-p185HER2 immunoliposomes for cancer therapy, *Proc. Natl Acad. Sci. USA*, 92(5), 1327–1331, 1995.

74. Dinauer, N. *et al.*, Selective targeting of antibody-conjugated nanoparticles to leukemic cells and primary T-lymphocytes, *Biomaterials*, 26(29), 5898–5906, 2005.

75. Brown, S. L., Miller, R. A., and Levy, R., Antiidiotype antibody therapy of B-cell lymphoma, *Semin. Oncol.*, 16(3), 199–210, 1989.

76. Cui, Z., Han, S. J., and Huang, L., Coating of mannan on LPD particles containing HPV E7 peptide significantly enhances immunity against HPV-positive tumor, *Pharm. Res.*, 21(6), 1018–1025, 2004.

77. Elamanchili, P. et al., Characterization of poly(D,L- lactic-*co*-glycolic acid) based nanoparticulate system for enhanced delivery of antigens to dendritic cells, *Vaccine*, 22(19), 2406–2412, 2004.

78. Mrsny, R. J. et al., Bacterial toxins as tools for mucosal vaccination, *Drug Discov. Today*, 7(4), 247–258, 2002.

79. Vitols, S., Uptake of low-density lipoprotein by malignant cells—Possible therapeutic applications, *Cancer Cells*, 3(12), 488–495, 1991.

80. Howarth, M. et al., Targeting quantum dots to surface proteins in living cells with biotin ligase, *Proc. Natl Acad. Sci. USA*, 102(21), 7583–7588, 2005.

81. Yu, D. et al., The specific delivery of proteins to human liver cells by engineered bio-nanocapsules, *FEBS J.*, 272(14), 3651–3660, 2005.

82. Popielarski, S. R., Pun, S. H., and Davis, M. E., A nanoparticle-based model delivery system to guide the rational design of gene delivery to the liver. 1. Synthesis and characterization, *Bioconjugate Chem.*, 16(5), 1063–1070, 2005.

83. Ma, M. K. and McLeod, H. L., Lessons learned from the irinotecan metabolic pathway, *Curr. Med. Chem.*, 10(1), 41–49, 2003.

84. Rossin, R. et al., ^{64}Cu-labeled folate-conjugated shell cross-linked nanoparticles for tumor imaging and radiotherapy: Synthesis, radiolabeling, and biologic evaluation, *J. Nucl. Med.*, 46(7), 1210–1218, 2005.

85. Kukowska-Latallo, J. F. et al., Nanoparticle targeting of anti-cancer drug improves therapeutic response in animal model of human epithelial cancer, *Cancer Res.*, 65(12), 5317–5324, 2005.

86. Thomas, T. P. et al., Targeting and inhibition of cell growth by an engineered dendritic nanodevice, *J. Med. Chem.*, 48(11), 3729–3735, 2005.

87. Sonvico, F. et al., Folate-conjugated iron oxide nanoparticles for solid tumor targeting as potential specific magnetic hyperthermia mediators: Synthesis, physicochemical characterization, and in vitro experiments, *Bioconjugate Chem.*, 16(5), 1181–1188, 2005.

88. Stevens, P. J., Sekido, M., and Lee, R. J., A folate receptor-targeted lipid nanoparticle formulation for a lipophilic paclitaxel prodrug, *Pharm. Res.*, 21(12), 2153–2157, 2004.

89. Yang, P. H. et al., Transferrin-mediated gold nanoparticle cellular uptake, *Bioconjugate Chem.*, 16(3), 494–496, 2005.

90. Bellocq, N. C. et al., Transferrin-containing, cyclodextrin polymer-based particles for tumor-targeted gene delivery, *Bioconjugate Chem.*, 14(6), 1122–1132, 2003.

91. Gonzalez, D. E. et al., An oligopeptide transporter is expressed at high levels in the pancreatic carcinoma cell lines AsPc-1 and Capan-2, *Cancer Res.*, 58(3), 519–525, 1998.

92. Luo, J. L., Kamata, H., and Karin, M., IKK/NF-kappaB signaling: Balancing life and death—A new approach to cancer therapy, *J. Clin. Invest.*, 115(10), 2625–2632, 2005.

93. Kelly, K. A. et al., Detection of vascular adhesion molecule-1 expression using a novel multimodal nanoparticle, *Circ. Res.*, 96(3), 327–336, 2005.

94. Li, K. C., Guccione, S., and Bednarski, M. D., Combined vascular targeted imaging and therapy: A paradigm for personalized treatment, *J. Cell. Biochem. Suppl.*, 39, 65–71, 2002.

95. Torchilin, V. P., Drug targeting, *Eur. J. Pharm. Sci.*, 11(Suppl. 2), S81–S91, 2000.

96. Gabizon, A. A., Selective tumor localization and improved therapeutic index of anthracyclines encapsulated in long-circulating liposomes, *Cancer Res.*, 52(4), 891–896, 1992.

97. Allen, T. M., Long-circulating (sterically stabilized) liposomes for targeted drug delivery, *Trends Pharmacol. Sci.*, 15(7), 215–220, 1992.

98. Backer, M. V. et al., Vascular endothelial growth factor selectively targets boronated dendrimers to tumor vasculature, *Mol. Cancer Ther.*, 4(9), 1423–1429, 2005.

99. Li, L. et al., A novel antiangiogenesis therapy using an integrin antagonist or anti-Flk-1 antibody coated ^{90}Y-labeled nanoparticles, *Int. J. Radiat. Oncol. Biol. Phys.*, 58(4), 1215–1227, 2004.

100. Konstan, M. W. et al., Compacted DNA nanoparticles administered to the nasal mucosa of cystic fibrosis subjects are safe and demonstrate partial to complete cystic fibrosis transmembrane regulator reconstitution, *Hum. Gene Ther.*, 15(12), 1255–1269, 2004.

101. Sengupta, S. et al., Temporal targeting of tumour cells and neovasculature with a nanoscale delivery system, *Nature*, 436(7050), 568–572, 2005.

102. Tanaka, K. et al., Intratumoral injection of immature dendritic cells enhances antitumor effect of hyperthermia using magnetic nanoparticles, *Int. J. Cancer*, 116(4), 624–633, 2005.

103. Tsourkas, A. et al., In vivo imaging of activated endothelium using an anti-VCAM-1 magnetooptical probe, *Bioconjugate Chem.*, 16(3), 576–581, 2005.
104. Pasqualini, R. and Ruoslahti, E., Organ targeting in vivo using phage display peptide libraries, *Nature*, 380(6572), 364–366, 1996.
105. Akerman, M. E. et al., Nanocrystal targeting in vivo, *Proc. Natl Acad. Sci. USA*, 99(20), 12617–12621, 2002.
106. Bibby, D. C. et al., Pharmacokinetics and biodistribution of RGD-targeted doxorubicin-loaded nanoparticles in tumor-bearing mice, *Int. J. Pharm.*, 293(1–2), 281–290, 2005.
107. Hood, J. D. et al., Tumor regression by targeted gene delivery to the neovasculature, *Science*, 296(5577), 2404–2407, 2002.
108. Lamprecht, A. et al., Biodegradable nanoparticles for targeted drug delivery in treatment of inflammatory bowel disease, *J. Pharmacol. Exp. Ther.*, 299(2), 775–781, 2001.
109. Lamprecht, A. et al., A pH-sensitive microsphere system for the colon delivery of tacrolimus containing nanoparticles, *J. Control. Release*, 104(2), 337–346, 2005.
110. Hussain, M. et al., A novel anionic dendrimer for improved cellular delivery of antisense oligonucleotides, *J. Control. Release*, 99(1), 139–155, 2004.
111. Santos Maia, C. et al., Drug targeting by solid lipid nanoparticles for dermal use, *J. Drug Target.*, 10(6), 489–495, 2002.
112. Alvarez-Roman, R. et al., Skin penetration and distribution of polymeric nanoparticles, *J. Control. Release*, 99(1), 53–62, 2004.
113. Prausnitz, M. R. et al., Microneedles for transdermal drug delivery, *Adv. Drug Deliv. Rev.*, 56(5), 581–587, 2004.
114. Bourges, J. L. et al., Ocular drug delivery targeting the retina and retinal pigment epithelium using polylactide nanoparticles, *Invest. Ophthalmol. Vis. Sci.*, 44(8), 3562–3569, 2003.
115. Florence, A. T., Issues in oral nanoparticle drug carrier uptake and targeting, *J. Drug Target.*, 12(2), 65–70, 2004.
116. Florence, A. T. et al., Factors affecting the oral uptake and translocation of polystyrene nanoparticles: Histological and analytical evidence, *J. Drug Target.*, 3(1), 65–70, 1995.
117. Takeuchi, H. et al., Effectiveness of submicron-sized, chitosan-coated liposomes in oral administration of peptide drugs, *Int. J. Pharm.*, 303(1–2), 160–170, 2005.
118. Hu, J. et al., A remarkable permeability of canalicular tight junctions might facilitate retrograde, non-viral gene delivery to the liver via the bile duct, *Gut*, 54(10), 1473–1479, 2005.
119. Yoncheva, K., Lizarraga, E., and Irache, J. M., Pegylated nanoparticles based on poly(methyl vinyl ether-*co*-maleic anhydride): Preparation and evaluation of their bioadhesive properties, *Eur. J. Pharm. Sci.*, 24(5), 411–419, 2005.
120. Wiwattanapatapee, R. et al., Anionic PAMAM dendrimers rapidly cross adult rat intestine in vitro: A potential oral delivery system?, *Pharm. Res.*, 17(8), 991–998, 2000.
121. Jevprasesphant, R. et al., Engineering of dendrimer surfaces to enhance transepithelial transport and reduce cytotoxicity, *Pharm. Res.*, 20(10), 1543–1550, 2003.
122. Sham, J. O. et al., Formulation and characterization of spray-dried powders containing nanoparticles for aerosol delivery to the lung, *Int. J. Pharm.*, 269(2), 457–467, 2004.
123. Borm, P. J. and Kreyling, W., Toxicological hazards of inhaled nanoparticles—Potential implications for drug delivery, *J. Nanosci. Nanotechnol.*, 4(5), 521–531, 2004.
124. Liu, G. et al., Nanoparticle and other metal chelation therapeutics in Alzheimer disease, *Biochim. Biophys. Acta*, 1741(3), 246–252, 2005.
125. Muller, R. H. and Keck, C. M., Drug delivery to the brain—Realization by novel drug carriers, *J. Nanosci. Nanotechnol.*, 4(5), 471–483, 2004.
126. Garcia-Garcia, E. et al., A relevant in vitro rat model for the evaluation of blood–brain barrier translocation of nanoparticles, *Cell. Mol. Life Sci.*, 62(12), 1400–1408, 2005.
127. Sun, W. et al., Specific role of polysorbate 80 coating on the targeting of nanoparticles to the brain, *Biomaterials*, 25(15), 3065–3071, 2004.
128. Thorne, R. G. et al., Quantitative analysis of the olfactory pathway for drug delivery to the brain, *Brain Res.*, 692(1–2), 278–282, 1995.
129. Vila, A. et al., Transport of PLA–PEG particles across the nasal mucosa: Effect of particle size and PEG coating density, *J. Control. Release*, 98(2), 231–244, 2004.

130. Moghimi, S. M. and Szebeni, J., Stealth liposomes and long circulating nanoparticles: Critical issues in pharmacokinetics, opsonization and protein-binding properties, *Prog. Lipid Res.*, 42(6), 463–478, 2003.

131. Ma, Z. and Lim, L. Y., Uptake of chitosan and associated insulin in Caco-2 cell monolayers: A comparison between chitosan molecules and chitosan nanoparticles, *Pharm. Res.*, 20(11), 1812–1819, 2003.

132. Gumbleton, M. et al., Targeting caveolae for vesicular drug transport, *J. Control. Release*, 87(1–3), 139–151, 2003.

133. Qaddoumi, M. G. et al., Clathrin and caveolin-1 expression in primary pigmented rabbit conjunctival epithelial cells: Role in PLGA nanoparticle endocytosis, *Mol. Vis.*, 9, 559–568, 2003.

134. Dass, C. R., Oligonucleotide delivery to tumours using macromolecular carriers, *Biotechnol. Appl. Biochem.*, 40(Pt 2), 113–122, 2004.

135. Gupta, A. K. and Curtis, A. S., Lactoferrin and ceruloplasmin derivatized superparamagnetic iron oxide nanoparticles for targeting cell surface receptors, *Biomaterials*, 25(15), 3029–3040, 2004.

136. de la Fuente, J. M. and Berry, C. C., Tat Peptide as an efficient molecule to translocate gold nanoparticles into the cell nucleus, *Bioconjugate Chem.*, 16(5), 1176–1180, 2005.

137. Li, N. et al., Ultrafine particulate pollutants induceoxidative stress and mitochondrial damage, *Environ. Health Perspect.*, 111(4), 455–460, 2003.

138. Savic, R. et al., Micellar nanocontainers distribute to defined cytoplasmic organelles, *Science*, 300(5619), 615–618, 2003.

139. Fischer, H. W., Improvement in radiographic contrast media through the development of colloidal or particulate media: An analysis, *J. Theor. Biol.*, 67(4), 653–670, 1977.

4 Pharmacokinetics of Nanocarrier-Mediated Drug and Gene Delivery

Yuriko Higuchi, Shigeru Kawakami, and Mitsuru Hashida

CONTENTS

4.1 Introduction 44
4.2 Characteristics of the Structure and Pharmacokinetic Properties
of the Nanocarrier 44
   4.2.1 Linear Polymers 44
      4.2.1.1 Their History and Characteristics 44
      4.2.1.2 Pharmacokinetic Properties and Applications of Drug Carriers 45
   4.2.2 Polymeric Micelles..46
      4.2.2.1 Their History and Characteristics 46
      4.2.2.2 Pharmacokinetic Properties and Applications as Drug Carriers 46
   4.2.3 Dendrimers 47
      4.2.3.1 Their History and Characteristics 47
      4.2.3.2 Pharmacokinetic Properties and Applications as Drug Carriers 47
   4.2.4 Liposomes..48
      4.2.4.1 Their History and Characteristics 48
      4.2.4.2 Pharmacokinetic Properties and Applications of Drug Carriers 49
4.3 Pharmacokinetic Analysis and Tissue Distribution Characteristics
of Macromolecules 50
   4.3.1 Effect of Size on the Pharmacokinetic Properties 51
   4.3.2 Effect of Charge on the Pharmacokinetic Properties 51
   4.3.3 Effect of Modification for Cell-Selective Delivery 51
4.4 Nanocarriers for Gene Delivery..51
   4.4.1 Polymers 52
   4.4.2 Liposomes..52
      4.4.2.1 Cationic Liposomes for Lung Targeting 52
      4.4.2.2 Mannose Liposomes for Liver NPC Targeting 53
      4.4.2.3 Galactose Liposomes for Liver PC Targeting....................53
      4.4.2.4 Folate Receptor-Mediated Targeting 54
References 54

4.1 INTRODUCTION

The recent development of nanoscale technologies is beginning to change the foundations of disease prevention, diagnosis, and treatment. Nanotechnology has allowed improvement of a novel drug carrier, nanocarrier, involving the nanoscale size and capable of targeting different cells in the body. These include polymeric micelles, dendrimers, and liposomes. The focus is on anti-tumor drugs and, although various anti-tumor drugs have been developed for cancer chemotherapy, such drugs often cause severe side effects related to the high cytotoxicity to tumor cells, and they are often toxic to normal cells. Therefore, a carefully designed nanocarrier is required to deliver anti-tumor drugs to their target sites for effective chemotherapy.

Tissue accumulation and cellular uptake of externally administered agents are generally determined by their physicochemical and biological properties and the anatomical and physiological properties of the body or tissues. As far as cancer therapy is concerned, site-specific delivery of drugs by carrier is broadly categorized as passive and active targeting. Passive targeting is the method determined by the physicochemical properties of the carrier relative to the anatomical and physiological characteristics of the tissue. From the 1970s forward, it has been demonstrated that macromolecule–drug conjugates accumulate in high concentrations in solid tumors and produce a therapeutic effect on the cancer.[1-4] This phenomenon is dependent on the anatomical and physiological characteristics of the solid tumor tissue, including a large vascular permeability, high interstitial diffusivity, and a lack of lymphatic drainage.[5,6] As far as this effect is concerned, Matsumura and Maeda proposed the concept of an enhanced permeability and retention effect (EPR effect) of macromolecules in solid tumors.[7,8] On the other hand, active targeting refers to alterations in the drug carrier, using specific interactions such as ligand–receptor and antigen–antibody phenomena. This approach involves designing carrier materials such as lipids and polymers and the combination of materials that enable site-specific drug delivery at a cellular or sub-cellular level.

Because the therapeutic effect of anti-tumor drugs is closely related to their pharmacokinetic properties, it is important for the rational design of a nanocarrier for targeted delivery to understand the pharmacokinetics of the nanocarrier in relation to the physicochemical and biological properties of the drugs involved.[9,10] In this chapter, site-specific delivery using nanocarriers is discussed based on pharmacokinetic considerations. As far as tissue distribution is concerned, because a nanocarrier possesses several features common to the macromolecule, the pharmacokinetics properties of the macromolecule are also summarized.

4.2 CHARACTERISTICS OF THE STRUCTURE AND PHARMACOKINETIC PROPERTIES OF THE NANOCARRIER

Recently, progress in nanotechnology has allowed the development of several types of nanocarriers capable of delivering drugs to target tissue and cells. Figure 4.1 summarizes typical types of nanocarriers and their characteristic structures. Because each type of nanocarrier has a unique structure and physicochemical characteristics, it exhibits unique pharmacokinetic properties in the body. In order to develop a strategy for establishing a nanocarrier system, it is necessary to understand these pharmacokinetic properties.

4.2.1 Linear Polymers

4.2.1.1 Their History and Characteristics

In recent decades, the use of polymers as carriers of both covalently bound and physically entrapped drugs has been widely explored. The larger hydrodynamic volume of polymers contributes to the

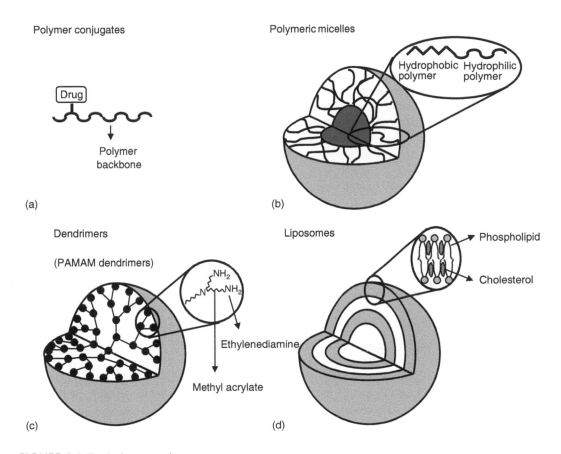

FIGURE 4.1 Typical nanocarriers.

increase in the plasma half-life of drug–polymer conjugates, increasing the probability of accumulation of the therapeutic agent in the tumor tissue because of the EPR effect. Drug–polymer conjugates (Figure 4.1) exhibit improved water solubility and reduced toxicity as a result of accumulation in the target tissue, and they protect the drug from enzymatic degradation or hydrolysis. However, drugs covalently bound to a polymer lose their pharmacological activity because a drug cannot interact with the target site because of steric constraints. In this respect, in 1975, Ringesdorf[11] was the first to propose a model for the rational design of a polymer conjugate with a drug that consisted of five components: the polymeric backbone, the drug, the spacer, the targeting group, and the solubilizing agent. The polymeric carrier can be either an inert or a biodegradable polymer. The drug can be fixed directly or via a spacer group onto the polymer backbone that is about 5–40 kDa. The proper selection of this spacer offers the possibility of controlling the site and the rate of release of the active drug from the conjugate by hydrolytic or enzymatic cleavage.

4.2.1.2 Pharmacokinetic Properties and Applications of Drug Carriers

To date, poly[N-(2-hydroxypropyl)methacrylamide] (PHPMA) is the most advanced example of a synthetic polymer used as a drug carrier in both basic research and clinical applications reported by Duncan[12] and Kocheck[13]. The biocompatibility of this polymer has been well characterized,[14] and it was found to have an inert character when it was injected in the blood stream.[15,16] A PHPMA copolymer conjugated with doxorubicin as an anti-tumor agent linked via a peptidyl linker

(Gly–Phe–Leu–Gly) was developed. The PHPMA copolymer conjugated with doxorubicin is stable in plasma[17] and has been shown to be concentrated within solid tumor models.[18,19] It is then cleaved intracellulary by lysosomal cysteine proteinases,[20] thereby allowing intratumoral drug release. Preclinical investigations have shown that a PHPMA copolymer conjugated with doxorubicin has radically different pharmacokinetics compared to free doxorubicin with a distribution plasma half-life that is increased from 5 min to 1 h.[21] The stable peptidyl linker also ensures that little or no free doxorubicin is released into the circulation following intravenous injection, increasing the therapeutic index of the conjugate. Besides the prolonged circulation in blood and the high solid tumor accumulation, it is also possible to attach targeting moieties such as antibody[22] and saccharides[23] to PHPMA. Another polymer conjugate, poly(ethylene glycol) (PEG), is the most basic and widely used polymer. Stylene maleic anhydride conjugated with the anti-tumor protein, neocarzinostatin (SMANCS), has been studied and is already used for the treatment of hepatocellular carcinoma.[24] Furthermore, by attaching homing devices as a cell-specific uptake enhancer, the biodistribution could be altered and cellular targeting of drugs conjugated with polymer could be achieved.

4.2.2 POLYMERIC MICELLES

4.2.2.1 Their History and Characteristics

Polymeric micelles that are prepared from amphiphilic block or graft copolymers with a spherical core and a shell with a carrier size of 10–100 nm have been undergoing investigation since 1984 and have been studied more actively since 1990.[25–27] The term *block* refers to the linear architecture of the copolymer in which the end of one segment is covalently joined to the head of the other segment to give a diblock AB type (Figure 4.1) or multiple block (AB_n) type copolymers. On the other hand, graft copolymers have a comb-like structure with hydrophilic segments attached on the side of the cationic segments. Although the influence of the copolymer architecture on biological activity has not yet been clarified, both block and graft copolymers can form polymeric micelles because of their amphiphilic character. Interactions between the polymer chains that serve as the driving force for micelle formation include hydrophobic, electrostatic, and π–π interactions and hydrogen bonding. Because hydrophobic drugs can be stably trapped in the hydrophobic core of the polymeric micelles and exhibit water-solubility, polymeric micelles are attractive carriers for hydrophobic drugs. Moreover, electrostatic (ionic) interactions involving the hydrophilic surface of polymeric micelles may also be applicable to macromolecules possessing many electrostatic (ionic) charges, e.g., DNA.[28]

4.2.2.2 Pharmacokinetic Properties and Applications as Drug Carriers

Early reports showed that the anti-cancer drug, doxorubicin, could be incorporated into a PEG-polyaspartate block copolymer.[29,30] PEG constituted the outer shell of the micelle that conferred a stealth property on the drug. After intravenous injection, doxorubicin incorporated in polymeric micelles was less avidly taken up by the reticuloendothelial system (RES) and could be retained in the circulation for a longer time compared with an intravenous injection of doxorubicin alone.[29] The in vivo anti-tumor activity against Colon-26 solid tumors showed that there was critical suppression of tumor growth and a prolonged life span for doxorubicin incorporated in polymeric micelles when administered to mice.[30]

Another type of polymeric micelle composed poly(ethylene oxide–aspartate)block copolymer and doxorubicin conjugate also exhibited a sustained circulation in blood,[31] reduced uptake by the RES, and more than 100 times higher accumulation in tumors in Colon-26 tumor-bearing mice.[32] Another report described the tumor-targeting delivery of paclitaxel by block copolymers producing both anti-tumor activity and a reduction in a major side effect of paclitaxel, neurotoxicity.[33]

In a previous study, all-*trans* retinoic acid (ATRA) incorporated polymeric micelles produced a higher AUC and lower hepatic clearance than free ATRA, demonstrating escape from the RES.[34]

Polymeric micelles have also been used in an active targeting approach, using the fact that certain pathological processes are associated with local temperature increase and/or acidosis. The use of polymeric micelles prepared from thermo-[35,36] or pH-sensitive[37] block co-polymers allows the disintegration of micelles and release of the incorporated drugs specifically at the site of interest. Additionally, the micelles can be targeted by attaching to their surface a vector molecule such as an antibody, peptide, lectin, saccharide, hormone, or some low-molecular weight compounds if this vector molecule binds to ligands characteristic of the site of interest.

4.2.3 Dendrimers

4.2.3.1 Their History and Characteristics

Dendrimers are a new class of polymeric materials discovered in the 1980s that are highly branched, monodisperse macromolecules. These hyperbranched molecules were described by Tomalia et al.[38] and named *dendrimers*. The word *dendrimer* comes from the Greek *dendron*, meaning a tree or a branch, and *meros* meaning part. At the same time, Vogtle and his co-workers[39] published the first report of the synthesis of polymers with a dendritic structure that they called cascade molecules. Newkome et al.[40] independently reported the synthesis of similar macromolecules that they called *arboros* from the Latin word *arbor*, meaning a tree.

Although linear polymers are highly polydisperse with a wide range of molecular weights because of their automatic assembly, dendrimers are mono dispersed or low-polydispersed that is necessary for homogeneous and reproducible pharmacokinetic properties. Therefore, dendrimers that have well-defined structures and flexible functions are good candidates for drug carriers.

Poly(amidoamine) (PAMAM) dendrimers (Figure 4.1) are the first complete dendrimer family to be synthesized, characterized, and commercialized. Dendrimers have a molecular architecture characterized by regular, dendric branching with radial symmetry. The molecular weights are similar to those of proteins, and their size ranges from 1 to 10 nm with each generation, or layer, of polymer adding 1 nm to the diameter of the molecule.[41] Lower generation dendrimers have a highly asymmetric shape and possess more open structures compared with those of a higher generation. As the chains growing from the core molecule become longer and more branched, dendrimers adopt a globular structure. Dendrimers become densely packed as they extend out to the periphery that results in a closed membrane-like structure. When a critical branched state is reached, dendrimers cannot continue to grow because of a lack of space. This is called the starburst effect. In the case of PAMAM dendrimer synthesis, this was observed after the 10th generation that contains 6141 monomer units and has a diameter of about 12.4 nm.[41]

4.2.3.2 Pharmacokinetic Properties and Applications as Drug Carriers

As far as dendrimers' pharmacokinetics are concerned, the biodistribution of [^{14}C] labeled PAMAM dendrimers after intravenous injection was determined for generation 3, 5, and 7 PAMAM dendrimers.[42] At each generation, the biodistribution characteristics were different. Generation 3 dendrimers showed the highest accumulation in kidney (15% of dose/g tissue over 48 h); however, generation 5 and 7 dendrimers preferentially accumulated in the pancreas (peak level 32% dose/g tissue at 24 h and 20% of dose/g tissue at 2 h, respectively). In addition, generation 7 dendrimers showed extremely high urinary excretion with values of 46 and 74% of dose/g tissue at 2 and 4 h, respectively. These results suggested that the biodistribution of a dendrimer depended on the generation because physicochemical properties were determined by the generation. Other reports described the biodistribution of [^{3}H]-labeled neutral and positively charged generation 5 PAMAM dendrimers in a tumor-bearing mouse model.[43] Both neutral and positively charged dendrimers showed a similar biodistribution trend 1 h after intravenous injection; the highest level was found in the lungs, kidney, and liver (positive, 28–6.1% of dose/g tissue; neutral, 5.3–2.9% of dose/g tissue) followed by the tumor, spleen, and pancreas (positive, 3.3–2.6% of dose/g, tissue; neutral 3.4–0.92% of dose/g tissue). Differences between neutral and positively charged dendrimers were observed for accumulation in the lung. A high lung accumulation of cationic dendrimer was found compared with conventional cationic liposomes, and this may be explained by the stacking to the capillary vessels of the lung. Moreover, [14C]-labeling of the dendrimer results in partial neutralization of the dendrimer. Therefore, different biodistribution properties were observed in the lung and liver accumulation between [14C]-and [3H]-labeled dendrimers. These results suggest that the biodistribution properties of radio-labeled dendrimers have to be examined carefully, taking into consideration the surface charge caused by the radio-labeling. As far as tumor accumulation is concerned, this may be explained by the EPR effect.

Drug delivery applications of dendrimers have been studied by Szoka and co-workers.[44] The biodistribution of doxorubicin covalently bound via a hydrazone linkage with a polyester dendritic scaffold based on 2,2-bis(hydroxymethyl)propanoic acid was examined. After intravenous injection, doxorubicin was extracted from the tumor and organs. Compared with free doxorubicin, the serum half-life was significantly increased. Therefore, this dendrimer would be a promising drug carrier for cancer therapy.

4.2.4 LIPOSOMES

4.2.4.1 Their History and Characteristics

Liposomes that are vesicles of lipid bilayers were first prepared in the 1960s. Liposome encapsulation has been proven useful for reducing the toxicity of drugs and increasing the solubility of hydrophobic drugs. However, liposomes tend to be trapped by the RES after intravenous injection. Therefore, allowing escape from the RES is one of the key strategies for developing cell-specific carriers because liposomes escaping from the RES accumulate in tumor tissue by a passive mechanism. As far as methods of escaping from the RES are concerned, Allen et al. in 1987[45] were the first to show that ganglioside and sphingomyelin (SM) synergistically acted to dramatically reduce the rate and extent of uptake of liposomes by the RES; therefore, this type of liposome is called a stealth liposome. In 1990, Klibanov et al.[46] demonstrated that PEG-liposomes prepared as poly (ethylene glycol)–phosphatidylethanolamine conjugates (PEG–PE) could also reduce uptake by the RES. Because PEG is easy to prepare and relatively cheap, and it has controllable molecular weight and is able to bind to lipids or proteins, it can be used for active targeting such as glycosylated liposomes[47] and immuno-liposomes.[48]

On the other hand, as far as active targeting is concerned, the receptor-mediated endocytosis systems present in various cell types would be useful, and a number of gene delivery systems were developed in the 1990s. Ligand–receptor recognition is an attractive tool for the development of cell-specific targeting systems; i.e., galactose to asialoglycoprotein receptors expressed on hepatocytes parenchymal cell (PC),[49] mannose to mannose receptors expressed on macrophages, sinusoidal endothelial cells, Kupffer cells, and dendritic cells,[50] folate receptors expressed on cancer cells,[51] and so on. As far as the size effect is concerned, it is possible to prepare liposomes of the size required by filtration. Because capillary vessels in a human tumor inoculated into SCID mice are permeable even to liposomes up to 400 nm in diameter,[52] liposomes with a diameter less than 50–200 nm are typically used. On the other hand, with regard to the effect of the surface charge of liposomes, it is known that strongly cationic liposomes highly accumulate in the lung after intravenous injection. Table 4.1 summarizes liposomes for drug delivery.

TABLE 4.1
In Vivo Targeted Drug Delivery

System	Method	Target Tissue (Cell)	Reference
EPR (prolongation of circulating)	Sphingomyelin modification	Tumor	45
			53
			55
			56
	PEG modification	Tumor	46
			57
			58
			59
Asialoglycoprotein receptor mediated	Galactose modification	Liver (hepatocyte)	62
			63
			64
			61
			65
Mannose receptor mediated	Mannose modification	Liver (non-parenchymal cell)	66
			67
			71

4.2.4.2 Pharmacokinetic Properties and Applications of Drug Carriers

SM-liposomes described by Allen et al. dramatically reduced the uptake by the RES and exhibited a prolonged circulation in blood and, consequently, accumulated in tumor tissue.[45] Gabizon and Papahadjopoulos[53] have examined a number of other natural and synthetic lipids to prolong the circulation time and lead to preferential uptake by tumor cells in vivo. In these respects, SM-containing liposomes would be an attractive option as a tumor targeting carrier for highly lipophilic anti-tumor drugs. The preparation of SM containing emulsion formulations has been also described to prolong the blood retention of encapsulated [^{3}H] labeled ATRA.[54] Furthermore, after the intravenous injection of vincristine encapsulated in SM-liposomes, the plasma vincristine level was 7-fold higher than that following administration in the form of bare liposomes, and the half-life of the elimination from the circulation was prolonged.[55,56] The improved circulation lifetime of vincristine in SM-liposomes correlated with the increased vincristine accumulation in subcutaneous solid A431 human xenograft tumors. Moreover, treatment with vincristine in SM-liposomes delayed the increase in tumor mass.[55]

PEG-liposomes are the most widely studied stealth liposomes. This occurs because the presence of PEG protects liposomes from the interaction with opsonins in the blood plasma and prevents their rapid uptake by the RES.[46,57,58] Therefore, doxorubicin encapsulated into PEG liposomes exhibited a dramatically increase in circulating time in blood and much better therapeutic efficacy as compared with the free drug. Therefore, they are currently being used in the clinic as a component of Doxil®.[59]

On the other hand, as far as active targeting is concerned, galactose modification could be an attractive tool for PC targeting of lipophilic drugs. Recently, a novel galactosylated cholesterol derivative, Gal-C4-Chol, that could be stably incorporated into liposomes by means of its lipophilic anchor and allowed the introduction of galactose residues on the surface of the liposomes was synthesized.[60,61] After intravenous injection, galactosylated liposomes prepared using Gal-C4-Chol were preferentially taken up by the liver.[62] In addition, prostaglandin E[63] and probucol[64] were investigated as model lipophilic drugs incorporated in galactosylated liposomes prepared using Gal-C4-Chol, and they were found to be effectively taken up by PC. Recently, other types of galactosylated liposome were developed.[65]

Using a mannosylated cholesterol derivative (Man-C4-Chol) that was synthesized by the same method as Gal-C4-Chol, mannosylated liposomes (Man-liposomes) for non-parenchymal cell (NPC) and macrophages targeting were prepared.[66–68] As far as cancer immunotherapy was concerned, macrophages, Kupffer cells, and dendric cells are attractive targets. These cells expressed a large number of mannose receptors on the cell surface and, therefore, Man-liposomes would be ideal carriers for cancer therapy. Previous studies have shown that Kupffer cells (contained in NPC) can be activated to a tumor reducible state by the administration of immuno-modulators such as muramyl dipeptide.[69] However, muramyl dipeptide parentally administered in free form is rapidly cleared from the body and excreted in the urine.[70] It was demonstrated that active targeting of muramyl dipeptide to liver non-parenchymal cells by Man-liposomes resulted in more effective inhibition of liver metastasis than delivery of muramyl dipeptide by liposomes without mannose.[71] Moreover, treatment with muramyl dipeptide incorporated in Man-liposomes increased the survival of tumor-bearing mice.[71]

4.3 PHARMACOKINETIC ANALYSIS AND TISSUE DISTRIBUTION CHARACTERISTICS OF MACROMOLECULES

The pharmacokinetic characteristics are generally determined by the physicochemical properties of the molecule such as the size, molecular weight, and surface charge.[3] Moreover, specific interactions such as ligand–receptor or antibody–antigen have a major effect on the characteristics of the molecule after intravenous injection.[4] The former factors are closely related to passive targeting, and the latter factors are closely related to active targeting. Figure 4.2 summarizes the

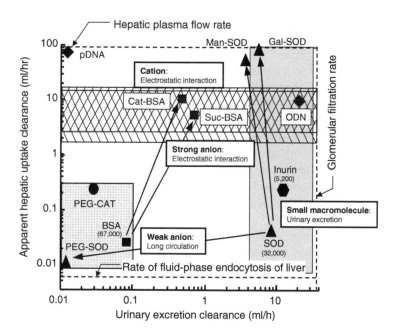

FIGURE 4.2 Effect of physicochemical characteristics of macromolecules on their hepatic uptake and urinary excretion in mice. The number in parenthesis is the molecular weight. Abbreviations in this figure are as follows: SOD, superoxide dismutase; Man-SOD, mannosylated superoxide dismutase; Gal-SOD, galactosy-lated superoxide dismutase; PEG-SOD, pegylated superoxide dismutase; Cat-BSA, cationized bovine serum albumin; PEG-CAT, pegylated catalase; ODN, oligonucleotide. (From Kawabata, K., Takakura, Y., and Hashida, M., *Pharm. Res.*, 12, 1995; Hashida, M., et al., *J. Control. Release*, 41,1996; Fujita, T., et al., *J. Pharmacol. Exp. Theor.*, 263, 1992; Yamasaki, Y., et al., *J. Pharmacol. Exp. Theor.*, 301, 2002; Yabe, Y., et al., *J. Pharmacol. Exp. Theor.*, 289, 1999.)

effect of size, charge, and ligand modification of macromolecules on hepatic uptake clearance and urinary excretion clearance that essentially decide the pharmacokinetic properties of the macromolecule. Understanding the effect of the physicochemical properties of macromolecules on their tissue distribution allows theoretical design delivery systems, involving nanocarriers.

4.3.1 Effect of Size on the Pharmacokinetic Properties

After intravenous injection, the kidney plays an important role in the disposition of macromolecules circulating in the blood. Macromolecules with a molecular weight of less than 50,000 (approximately 6 nm in diameter) are susceptible to glomerular filtration and are excreted in the urine.[72–75] Therefore, low molecular weight drugs and oligonucleotides with a lower molecular weight undergo rapid glomerular filtration by the kidney.[76,77] On the other hand, pDNA is too large to be filtered without degradation.[78]

4.3.2 Effect of Charge on the Pharmacokinetic Properties

Electric charge is also an important factor in determining the pharmacokinetic properties of macromolecules.[79] Because the glomerular capillary walls also function as a charge-selective barrier having negative charges, positively charged macromolecules exhibit higher glomerular permeability than anionic macromolecules of similar molecular weights. Larger molecular weight cationic macromolecules escaping glomerular filtration mainly accumulate in the liver, kidney, and lung.[80–82] This phenomenon can be explained by following factors:

1. Because the cell surface is negatively charged in general, cationic molecules tend to electrostatically bind to it.
2. Because the liver and kidney have fenestrated capillaries, macromolecules can cross the endothelial barrier.
3. Direct interaction with the lung endothelial cells.
4. Embolization of aggregates of the complex with negatively charged blood components.

On the other hand, strongly negatively charged macromolecules are taken up by liver non-parenchymal cells through liver fenestrae by scavenger receptor-mediated endocytosis.[78,83] Neutral or weakly anionic macromolecules with a molecular size that does not allow glomerular filtration such as polyethylene glycol and serum albumin posses a very weak affinity for the cell surface, and they are cleared very slowly.[7,8,84,85] Consequently, such neutral or weakly anionic macromolecules have a longer elimination half-life compared with the approximately same sized macromolecule with a strong negative or positive charge.

4.3.3 Effect of Modification for Cell-Selective Delivery

Ligand modification for active targeting using receptor-mediated uptake markedly affects the pharmacokinetic properties of macromolecules.[4] For example, pharmacokinetic studies of receptor-mediated targeting to liver parenchymal cells via galactose recognition have been carried out based on the clearance concept. Galactose modification increased the hepatic uptake of macromolecules of different molecular weights.[3] The clearance was almost the same as the hepatic flow rate in mice, showing that it is almost identical to the theoretical maximum value.

4.4 NANOCARRIERS FOR GENE DELIVERY

In addition to traditional drugs, pDNA has recently also been used as a drug for cancer therapy; for example, antigen-encoding pDNA has a potential application in DNA vaccine therapy for cancer. Moreover, new techniques to inhibit target gene expression through transcriptional regulation without any changes in the functions of other genes using synthetic oligonucleotides such as antisense DNA, decoy oligonucleotide, and siRNA have been also considered as a novel anti-tumor chemotherapy. However, because pDNA and synthetic oligonucleotides are easily degraded or metabolized, there is no cell selectivity, and enough transgene expression or inhibition of target gene expression for effective therapy could not be achieved by the administration of naked pDNA or oligonucleotides. Therefore, for effective therapy, it is essential to develop a novel carrier that makes it possible to deliver such drugs to the target tissue or cells.

4.4.1 POLYMERS

Polymeric micelles prepared from cationic copolymer could be a novel gene carrier. After preparing a pDNA complex (standard size 5,000–7,000 bp), the mean particle of the complex is about 50–200 nm. pDNA entrapped in polymeric micelles is highly resistant to nuclease degradation.[28]

Dendrimers are also an attractive carrier for gene delivery because they can interact on an electrostatic charge basis with biologically relevant polyanions such as nucleic acids[86] and pDNA[87] because their surfaces are covered with primary amino groups. Moreover, chemical or biological modification of the surface of dendrimers would make them more effective drug or gene carriers.

4.4.2 LIPOSOMES

Liposomes with cationic charge on their surface are novel carriers for genes. Several kinds of liposomes for gene delivery have been developed depending on passive or active targeting mechanisms. Table 4.2 summarizes liposomes for gene delivery.

4.4.2.1 Cationic Liposomes for Lung Targeting

After intravenous administration of pDNA complex, the immediate effect on the complex of erythrocytes is to induce aggregation.[88] Large aggregates (>400 nm) may be readily entrapped in the lung capillary. Mahato et al. demonstrated that cationic liposome/[^{32}P] pDNA complexes (lipoplexes) were rapidly cleared from the blood circulation and accumulated extensively in the lung and liver.[89] In addition, [^{32}P] lipoplexes were predominantly taken up by the liver NPC and

TABLE 4.2
In Vivo Targeted Gene Delivery

System	Liposome	Target Tissue (Cell)	Reference
Capillary trapping	Cationic-liposome	Lung	90
			89
			91
			93
Asialoglycoprotein receptor madiated	Asialofetuin-liposome	Liver (hepatocyte)	96
	Galactose-liposome	Liver (hepatocyte) (hepatocyte)	97
			60
			98
Mannose receptor mediated	Mannose-liposome	Liver (non-parenchymal cell)	66
			94
			95
Folate receptor mediated	Folate-liposome	Tumor cell	99
			100

this uptake was inhibited by the pre-administration of dextran sulfate, suggesting the involvement of a phagocytic process. However, gene expression after intravenous administration of lipoplexes was extremely high in the lung.[90–92] It is well known that IFNβ exhibits anti-tumor activity. Taking this into consideration, intravenously injected pDNA encoding IFNβ complexed with cationic liposome was found to significantly inhibit established metastatic lung tumors in CT-26 cells inoculated into mice.[93]

4.4.2.2 Mannose Liposomes for Liver NPC Targeting

Because Man-C4-Chol has a cationic charge, a cationic charge could also be induced on the surface of a Man-liposome. Therefore, Man-liposomes are also an attractive potential gene carrier because of this cationic charge that would allow electrical interaction with the gene. As far as in vivo gene transfection is concerned, the highest gene expression was observed in the liver after intravenous injection of pDNA/Man-liposome complexes (Man-lipoplex) in mice.[66–68] In addition, gene transfer by a Man-liposome was mannose receptor-mediated because the gene expression with Man-lipoplex in the liver was significantly reduced by pre-dosing with mannosylated bovine serum albumin. Because macrophages in the liver and spleen exist at the endothelial cells intervals and could make contact with the complex without passing through the sinusoid (100–200 nm), cell-selective gene transfection could be achieved by intravenous administration of Man-lipoplex. Therefore, a mannosylated gene carrier would be effective for the NPC-selective gene transfection system even after intravenous administration.[94] As far as the application to cancer therapy using targeted gene delivery by mannosylated liposomes is concerned, DNA vaccine therapy is suitable because antigen encoded pDNA can be efficiently transfected into dendritic cells that expressed a large number of mannose receptors. Recently, Hattori et al.[95] demonstrated that the targeted delivery of a DNA vaccine by Man-liposomes is a potent method for DNA vaccine therapy.

4.4.2.3 Galactose Liposomes for Liver PC Targeting

PC exclusively expresses large numbers of high affinity cell-surface receptors that can bind asialoglycoprotein and internalize to the cell interior. In order to achieve PC-specific gene transfection, galactose moieties are introduced into the cationic liposomes. The galactosylation of liposomes can be achieved by coating with either glycoproteins or galactose conjugated synthetic lipids. Hara et al.[96] reported that asialofetuin-labeled liposomes encapsulating pDNA were taken up by asialoglycoprotein receptor-mediated endocytosis using cultured PC and showed the highest hepatic gene expression after intraportal injection with a preloading of EDTA. Gal-C4-Chol that possess a similar structure to Man-C4-Chol was synthesized.[58] In vivo gene transfer was examined by optimizing the pharmacokinetics and physicochemical properties.[97] The radioactivity in the liver from the Gal-liposome/[^{32}P] pDNA complex (Gal-lipoplexes) was about 75% of the dose even 1 min after intraportal administration. The hepatic gene expression of Gal-lipoplexes was more than a 10-fold greater than that of lipoplexes with bare cationic liposomes. When gene expression was examined at the intrahepatic cellular level, the gene expression of PC of Gal-lipoplexes was significantly higher than that of liver NPC. On the other hand, the gene expression of PC and NPC of lipoplexes was almost identical. In addition, asialoglycoprotein receptor-mediated endocytosis was also confirmed by the inhibitory effect of pre-dosing an excess amount of galactosylated bovine serum albumin. It was previously reported that the lipoplexes interact with erythrocytes after intravenous administration.[88] Recently, the presence of an essential amount of sodium chloride (NaCl) during the formation of Gal-lipoplexes stabilizing the complexes in accordance with the surface charge regulation (SCR) theory was demonstrated.[98] The transfection activity in hepatocytes of SCR Gal-lipoplexes was significantly higher than that of conventional complexes.

4.4.2.4 Folate Receptor-Mediated Targeting

The folate receptor is known to be over expressed in a large fraction of human tumors, but it is only minimally distributed in normal tissues. Therefore, the folate receptor has also been used as a tumor-targeting ligand for several drug delivery systems. Recently, Hofland et al. synthesized folate-PEG-lipid derivatives to prepare folate-modified cationic liposomes.[99] After an intravenous injection of folate–conjugated liposome complex, gene expression in the tumors was not changed, whereas gene expression in the lung was reduced as compared with the conventional complex. However, after intraperitoneal injection in a murine disseminated peritoneal tumor model, folate–conjugated liposome complex formulations produced approximately a 10-fold increase in tumor-associated gene expression compared with conventional complex.[100]

REFERENCES

1. Sezaki, H., Takakura, Y., and Hashida, M., Soluble macromolecular carriers for the delivery of antitumour drugs, *Adv. Drug Deliv. Rev.*, 3, 247, 1989.
2. Sezaki, H. and Hashida, M., Macromolecule–drug conjugates in targeted cancer chemotherapy, *Crit. Rev. Theor. Drug Carrier Syst.*, 1, 1, 1984.
3. Takakura, Y. and Hashida, M., Macromolecular carrier systems for targeted drug delivery: Pharmacokinetic considerations on biodistribution, *Pharm. Res.*, 13, 820, 1996.
4. Hashida, M., Kawakami, S., and Yamashita, F., Lipid carrier systems for targeted drug and gene delivery, *Chem. Pharm. Bull.*, 53, 871, 2005.
5. Takakura, Y. and Hashida, M., Macromolecular drug carrier systems in cancer chemotherapy: Macromolecular prodrugs, *Crit. Rev. Oncol. Hematol.*, 18, 207, 1995.
6. Jain, R. K., Transport of molecules across tumor vasculature, *Cancer Metastasis Rev.*, 6, 559, 1987.
7. Matsumura, Y. and Maeda, H., A new concept for macromolecular therapeutics in cancer chemotherapy: Mechanism of tumoritropic accumulation of proteins and the anti-tumor agent smancs, *Cancer Res.*, 46, 6387, 1986.
8. Maeda, H. and Matsumura, Y., Tumoritropic and lymphotropic principles of macromolecular drugs, *Crit. Rev. Theor. Drug Carrier Syst.*, 6, 193, 1989.
9. Takakura, Y. et al., Influence of physicochemical properties on pharmacokinetics of non-viral vectors for gene delivery, *J. Drug Target.*, 10, 99, 2002.
10. Nishikawa, M. and Hashida, M., Pharmacokinetics of anti-cancer drugs, plasmid DNA, and their delivery systems in tissue-isolated perfused tumors, *Adv. Drug Deliv. Rev.*, 40, 19, 1999.
11. Ringsdorf, H., Structure and properties of pharmacologically active polymers, *J. Polym. Sci. Polym. Symp.*, 51, 135, 1975.
12. Duncan, R. et al., Polymers containing enzymatically degradable bonds 7. Design of oligopeptide sidechains in poly(*N*-(2-hydroxypropyl) metacylamide) copolymers to promote efficient degradation by lysosomal enzymes, *Makromol. Chem.*, 184, 1997–2008, 1983.
13. Kopecek, J. and Bazilova, H., Poly[*N*-(hydroxypropyl)-methacrylamide]—1. Radical polymerisation and copolymerization, *Eur. Polym. J.*, 9, 7, 1973.
14. Rihova, B. et al., Biocompatibility of *N*-(2-hydroxypropyl) methacrylamide copolymers containing adriamycin, *Biomaterials*, 10, 335, 1989.
15. Cartlidge, S. A. et al., Soluble, crosslinked *N*-(2-hydroxypropyl) methacrylamide copolymers as potential drug carriers 2: Effect of molecular weight on blood clearance and body distribution in the rat after intravenous administration. Distribution of unfractionated copolymer after intraperitoneal, subcutaneous or oral administration, *J. Control Release*, 4, 253, 1987.
16. Seymour, L. W. et al., Effect of molecular weight of *N*-(2-hydroxypropyl)methacrylamide copolymers on body distribution and rate of excretion after subcutaneous, intraperitoneal, and intravenous administration to rats, *J. Biomed. Mater. Res.*, 21, 1341, 1987.
17. Rejmanova, P. et al., Stability in rat plasma and serum of lysosomally degradable oligopeptide sequences in *N*-(2-hydroxypropyl)methacrylamide) copolymers, *Biomaterials*, 6, 45, 1985.
18. Seymour, L. W. et al., Tumour tropism and anti-cancer efficacy of polymer-based doxorubicin prodrugs in the treatment of subcutaneous murine B16F10 melanoma, *Br. J. Cancer*, 70, 636, 1994.

19. Cassidy, J. et al., Clinical trials of nimodipine as a potential neuroprotector in ovarian cancer patients treated with cisplatin, *Cancer Chemother. Pharmacol.*, 44, 161, 1998.

20. Duncan, R. et al., Degradation of side-chains of *N*-(2-hydroxypropyl)methacrylamide) copolymers by lysosomal thiolproteinases, *Biosci. Rep.*, 2, 1041, 1982.

21. Seymour, L. W. et al., Pharmacokinetics of polymer-bound Adriamycin, *Biochem. Pharmacol.*, 39, 1125, 1990.

22. Flanagan, P. A. et al., Evaluation of protein-*N*-(2-hydroxypropyl) methacrylamide copolymer conjugates as targetable drug carriers 2. Body distribution of conjugates containing transferrin, anti-transferrin receptor antibody or anti-Thy 1.2 antibody and effectiveness of transferrin-containing daunomycin conjugates against mouse L 1210 leukaemia in vivo, *J. Control Release*, 18, 25, 1992.

23. Wedge, S. R., Duncan, R., and Kopeckova, P., Comparison of the liver subcelluar distribution of free daunomycin and that bound to galactosamine targeted *N*-(2-hydroxypropyl)methacrylamide copolymers, following intravenous administration in the rat, *Br. J. Cancer*, 63, 546, 1991.

24. Maeda, H., Sawa, T., and Konno, T., Mechanism of tumor-targeted delivery of macromolecular drugs, including the EPR effect in solid tumor and clinical overview of the prototype polymeric drug SMANCS, *J. Control Release*, 74, 47, 2001.

25. Yokoyama, M., Block copolymers as drug carriers, *Crit. Rev. Theor. Drug Carrier Syst.*, 9, 213, 1992.

26. Kataoka, K. et al., Block copolymer micelles as vehicles for drug delivery, *J. Control Release*, 24, 119, 1993.

27. Torchilin, V. P., Structure and design of polymeric surfactant-based drug delivery systems, *J. Control Release*, 73, 137, 2001.

28. Kakizawa, Y. and Kataoka, K., Block copolymer micelles for delivery of gene and related compounds, *Adv. Drug Deliv. Rev.*, 54, 203, 2002.

29. Yokoyama, M. et al., Selective delivery of adriamycin to a solid tumor using a polymeric micelle carrier system, *J. Drug Target.*, 7, 171, 1999.

30. Yokoyama, M. et al., Toxicity and anti-tumor activity against solid tumors of micelle-forming polymeric anti-cancer drug and its extremely long circulation in blood, *Cancer Res.*, 51, 3229, 1991.

31. Kwon, G. et al., Biodistribution of micelle-forming polymer–drug conjugates, *Pharm. Res.*, 10, 970, 1993.

32. Kwon, G. et al., Enhanced tumor accumulation and prolonged circulation times of micelle-forming poly (ethylene oxide–aspartate) block copolymer adriamycin conjugates, *J. Control Release*, 29, 17, 1994.

33. Hamaguchi, T. et al., NK105, a paclitaxel-incorporating micellar nanoparticle formulation, can extend in vivo antitumour activity and reduce the neurotoxicity of paclitaxel, *Br. J. Cancer*, 92, 1240, 2005.

34. Kawakami, S. et al., Biodistribution characteristics of all-trans retinoic acid incorporated in liposomes and polymeric micelles following intravenous administration, *J. Pharm. Sci.*, 94, 2606, 2005.

35. Kohori, F. et al., Preparation and characterization of thermally responsive block copolymer micelles comprising poly(*N*-isopropylacrylamide-*b*-DL-lactide), *J. Control Release*, 55, 87, 1998.

36. Chung, J. E. et al., Effect of molecular architecture of hydrophobically modified poly(*N*-isopropylacrylamide) on the formation of thermoresponsive core-shell micellar drug carriers, *J. Control Release*, 53, 119, 1998.

37. Le Garrec, D. et al., Optimizing pH-responsive polymeric micelles for drug delivery in a cancer photodynamic therapy model, *J. Drug Target.*, 10, 429, 2002.

38. Tomalia, D. A. et al., A new class of polymers: Starburst-dendritic macromolecules, *Polym. J.*, 17, 117, 1985.

39. Buhleier, E., Wehner, W., and Vogtle, F., *Synthesis*, 405, 155, 1978.

40. Newkome, G. R. et al., Cascade molecules: A new approach to micelles, A[27]-arborol, *J. Org. Chem.*, 50, 2003–2004, 1985.

41. Svenson, S. and Tomalia, D. A., Dendrimers in biomedical applications-reflections on the field, *Adv. Drug Deliv. Rev.*, 57, 2106, 2005.

42. Roberts, J. C., Bhalgat, M. K., and Zera, R. T., Preliminary biological evaluation of polyamidoamine (PAMAM) Starburst dendrimers, *J. Biomed. Mater. Res.*, 30, 53, 1996.

43. Nigavekar, S. S. et al., ^{3}H dendrimer nanoparticle organ/tumor distribution, *Pharm. Res.*, 21, 476, 2004.

44. Padilla De Jesus, O. L. et al., Polyester dendritic systems for drug delivery applications: In vitro and in vivo evaluation, *Bioconjug. Chem.*, 13, 453, 2002.

45. Allen, T. M. and Chonn, A., Large unilamellar liposomes with low uptake into the reticuloendothelial system, *FEBS Lett.*, 223, 42, 1987.

46. Klibanov, A. L. et al., Amphipathic polyethyleneglycols effectively prolong the circulation time of liposomes, *FEBS Lett.*, 268, 235, 1990.

47. Managit, C. et al., Targeted and sustained drug delivery using PEGylated galactosylated liposomes, *Int. J. Pharm.*, 266, 77, 2003.

48. Maruyama, K., Kennel, S. J., and Huang, L., Lipid composition is important for highly efficient target binding and retention of immunoliposomes, *Proc. Natl Acad. Sci. U.S.A*, 87, 5744, 1990.

49. Ashwell, G. and Harford, J., Carbohydrate-specific receptors of the liver, *Annu. Rev. Biochem.*, 51, 531, 1982.

50. Kuiper, J., *The Liver: Biology and Pthobiology*, 3rd ed., Raven Press, New York, 1994.

51. Weitman, S. D. et al., Distribution of the folate receptor GP38 in normal and malignant cell lines and tissues, *Cancer Res.*, 52, 3396, 1992.

52. Yuan, F. et al., Vascular permeability in a human tumor xenograft: Molecular size dependence and cutoff size, *Cancer Res.*, 55, 3752, 1995.

53. Gabizon, A. and Papahadjopoulos, D., Liposome formulations with prolonged circulation time in blood and enhanced uptake by tumors, *Proc. Natl Acad. Sci. U.S.A*, 85, 6949, 1988.

54. Takino, T. et al., Long circulating emulsion carrier systems for highly lipophilic drugs, *Biol. Pharm. Bull.*, 17, 121, 1994.

55. Webb, M. S. et al., Sphingomyelin-cholesterol liposomes significantly enhance the pharmacokinetic and therapeutic properties of vincristine in murine and human tumour models, *Br. J. Cancer*, 72, 896, 1995.

56. Krishna, R. et al., Liposomal and nonliposomal drug pharmacokinetics after administration of liposome-encapsulated vincristine and their contribution to drug tissue distribution properties, *J. Pharmacol. Exp. Theor.*, 298, 1206, 2001.

57. Blume, G. and Cevc, G., Liposomes for the sustained drug release in vivo, *Biochim. Biophys. Acta*, 1029, 91, 1990.

58. Gabizon, A. A., Barenholz, Y., and Bialer, M., Prolongation of the circulation time of doxorubicin encapsulated in liposomes containing a polyethylene glycol-derivatized phospholipid: Pharmacokinetic studies in rodents and dogs, *Pharm. Res.*, 10, 703, 1993.

59. Gabizon, A. A., Pegylated liposomal doxorubicin: Metamorphosis of an old drug into a new form of chemotherapy, *Cancer Invest.*, 19, 424, 2001.

60. Kawakami, S. et al., Asialoglycoprotein receptor-mediated gene transfer using novel galactosylated cationic liposomes, *Biochem. Biophys. Res. Commun.*, 252, 78, 1998.

61. Kawakami, S. et al., Novel galactosylated liposomes for hepatocyte-selective targeting of lipophilic drugs, *J. Pharm. Sci.*, 90, 105, 2001.

62. Kawakami, S. et al., Biodistribution characteristics of mannosylated, fucosylated, and galactosylated liposomes in mice, *Biochim. Biophys. Acta*, 1524, 258, 2000.

63. Kawakami, S. et al., Targeted delivery of prostaglandin E1 to hepatocytes using galactosylated liposomes, *J. Drug Target.*, 8, 137, 2000.

64. Hattori, Y. et al., Controlled biodistribution of galactosylated liposomes and incorporated probucol in hepatocyte-selective drug targeting, *J. Control Release*, 69, 369, 2000.

65. Wang, S. N. et al., Synthesis of a novel galactosylated lipid and its application to the hepatocyte-selective targeting of liposomal doxorubicin, *Eur. J. Pharm. Biopharm.*, 62, 32, 2006.

66. Kawakami, S. et al., Mannose receptor-mediated gene transfer into macrophages using novel mannosylated cationic liposomes, *Gene Theor.*, 7, 292, 2000.

67. Kawakami, S. et al., The effect of lipid composition on receptor-mediated in vivo gene transfection using mannosylated cationic liposomes in mice, *S.T.P. Pharm. Sci.*, 11, 117, 2001.

68. Kawakami, S. et al., Effect of cationic charge on receptor-mediated transfection using mannosylated cationic liposome/plasmid DNA complexes following the intravenous administration in mice, *Pharmazie*, 59, 405, 2004.

69. Taniyama, T. and Holden, H. T., Direct augmentation of cytolytic activity of tumor-derived macrophages and macrophage cell lines by muramyl dipeptide, *Cell. Immunol.*, 48, 369, 1979.
70. Fox, A. and Fox, K., Rapid elimination of a synthetic adjuvant peptide from the circulation after systemic administration and absence of detectable natural muramyl peptides in normal serum at current analytical limits, *Infect. Immunol.*, 59, 1202, 1991.
71. Opanasopit, P. et al., Inhibition of liver metastasis by targeting of immunomodulators using mannosylated liposome carriers, *J. Control Release*, 80, 283, 2002.
72. Brenner, B. M., Hostetter, T. H., and Humes, H. D., Glomerular permselectivity: Barrier function based on discrimination of molecular size and charge, *Am. J. Physiol.*, 234, F455, 1978.
73. Maack, T. et al., Renal filtration, transport, and metabolism of low-molecular-weight proteins: A review, *Kidney Int.*, 16, 251, 1979.
74. Mihara, K. et al., Disposition characteristics of protein drugs in the perfused rat kidney, *Pharm. Res.*, 10, 823, 1993.
75. Mihara, K. et al., Disposition characteristics of model macromolecules in the perfused rat kidney, *Biol. Pharm. Bull.*, 16, 158, 1993.
76. Sawai, K. et al., Renal disposition characteristics of oligonucleotides modified at terminal linkages in the perfused rat kidney, *Antisense Res. Dev.*, 5, 279, 1995.
77. Sawai, K. et al., Disposition of oligonucleotides in isolated perfused rat kidney: Involvement of scavenger receptors in their renal uptake, *J. Pharmacol. Exp. Theor.*, 279, 284, 1996.
78. Kawabata, K., Takakura, Y., and Hashida, M., The fate of plasmid DNA after intravenous injection in mice: Involvement of scavenger receptors in its hepatic uptake, *Pharm. Res.*, 12, 825, 1995.
79. Hashida, M. et al., Pharmacokinetics and targeted delivery of proteins and genes, *J. Control Release*, 41, 91, 1996.
80. Fujita, T. et al., Targeted delivery of human recombinant superoxide dismutase by chemical modification with mono- and polysaccharide derivatives, *J. Pharmacol. Exp. Theor.*, 263, 971, 1992.
81. Mahato, R. I. et al., In vivo disposition characteristics of plasmid DNA complexed with cationic liposomes, *J. Drug Target.*, 3, 149, 1995.
82. Ma, S. F. et al., Cationic charge-dependent hepatic delivery of amidated serum albumin, *J. Control Release*, 102, 583, 2005.
83. Yamasaki, Y. et al., Pharmacokinetic analysis of in vivo disposition of succinylated proteins targeted to liver nonparenchymal cells via scavenger receptors: Importance of molecular size and negative charge density for in vivo recognition by receptors, *J. Pharmacol. Exp. Theor.*, 301, 467, 2002.
84. Fujita, T. et al., Control of in vivo fate of albumin derivatives utilizing combined chemical modification, *J. Drug Target.*, 2, 157, 1994.
85. Yabe, Y. et al., Targeted delivery and improved therapeutic potential of catalase by chemical modification: Combination with superoxide dismutase derivatives, *J. Pharmacol. Exp. Theor.*, 289, 1176, 1999.
86. Bielinska, A. et al., Regulation of in vitro gene expression using antisense oligonucleotides or antisense expression plasmids transfected using starburst PAMAM dendrimers, *Nucleic Acids Res.*, 24, 2176, 1996.
87. Tang, M. X., Redemann, C. T., and Szoka, F. C., In vitro gene delivery by degraded polyamidoamine dendrimers, *Bioconjug. Chem.*, 7, 703, 1996.
88. Sakurai, F. et al., Interaction between DNA–cationic liposome complexes and erythrocytes is an important factor in systemic gene transfer via the intravenous route in mice: The role of the neutral helper lipid, *Gene Theor.*, 8, 677, 2001.
89. Mahato, R. I. et al., Physicochemical and pharmacokinetic characteristics of plasmid DNA/cationic liposome complexes, *J. Pharm. Sci.*, 84, 1267, 1995.
90. Zhu, N. et al., Systemic gene expression after intravenous DNA delivery into adult mice, *Science*, 261, 209, 1993.
91. Song, Y. K. et al., Characterization of cationic liposome-mediated gene transfer in vivo by intravenous administration, *Hum. Gene Theor.*, 8, 1585, 1997.
92. Uyechi, L. S. et al., Mechanism of lipoplex gene delivery in mouse lung: Binding and internalization of fluorescent lipid and DNA components, *Gene Theor.*, 8, 828, 2001.
93. Sakurai, F. et al., Therapeutic effect of intravenous delivery of lipoplexes containing the interferon-beta gene and poly I: poly C in a murine lung metastasis model, *Cancer Gene Theor.*, 10, 661, 2003.

94. Yamada, M. et al., Tissue and intrahepatic distribution and subcellular localization of a mannosy-
    lated lipoplex after intravenous administration in mice, *J. Control Release*, 98, 157, 2004.
95. Hattori, Y. et al., Enhancement of immune responses by DNA vaccination through targeted gene
    delivery using mannosylated cationic liposome formulations following intravenous administration in
    mice, *Biochem. Biophys. Res. Commun.*, 317, 992, 2004.
96. Hara, T. et al., Receptor-mediated transfer of pSV2CAT DNA to a human hepatoblastoma cell line
    HepG2 using asialofetuin-labeled cationic liposomes, *Gene*, 159, 167, 1995.
97. Kawakami, S. et al., In vivo gene delivery to the liver using novel galactosylated cationic liposomes,
    Pharm. Res., 17, 306, 2000.
98. Fumoto, S. et al., Enhanced hepatocyte-selective in vivo gene expression by stabilized galactosy-
    lated liposome/plasmid DNA complex using sodium chloride for complex formation, *Mol. Theor.*,
    10, 719, 2004.
99. Hofland, H. E. et al., Folate-targeted gene transfer in vivo, *Mol. Theor.*, 5, 739, 2002.
100. Reddy, J. A. et al., Folate-targeted, cationic liposome-mediated gene transfer into disseminated
    peritoneal tumors, *Gene Theor.*, 9, 1542, 2002.

5 Multifunctional Nanoparticles for Cancer Therapy

Todd J. Harris, Geoffrey von Maltzahn, and Sangeeta N. Bhatia

CONTENTS

5.1 Introduction 59
5.2 Modular Functionalities at the Biosynthetic Interface 60
    5.2.1 Targeting.. 60
    5.2.2 Imaging Agents 61
    5.2.3 Sensing 64
    5.2.4 Therapeutic Payloads 66
    5.2.5 Remote Actuation.. 69
5.3 Challenges in Integrating Multiple Functionalities and Future Directions 70
References 70

5.1 INTRODUCTION

The use of nanoparticles in cancer therapy is attractive for several reasons: they exhibit unique pharmacokinetics, including minimal renal filtration; they have high surface-to-volume ratios enabling modification with various surface functional groups that home, internalize, or stabilize; and they may be constructed from a wide range of materials used to encapsulate or solubilize therapeutic agents for drug delivery or to provide unique optical, magnetic, and electrical properties for imaging and remote actuation. The topology of a nanoparticle—core, coating, and surface functional groups—makes it particularly amenable to modular design, whereby features and functional moieties may be interchanged or combined. Although many functionalities of nanoparticles have been demonstrated, including some clinically approved drug formulations and imaging agents,[3,8] the consolidation of these into multifunctional nanoparticles capable of targeting, imaging, and delivering therapeutics is an exciting area of research that holds great promise for cancer therapy in the future.

Figure 5.1[1] schematically depicts a hypothetical multifunctional particle that has been engineered to include many features such as the ability to target tumors, evade uptake by the reticuloendothelial system (RES), protect therapeutics that can be released on demand, act as sensors of tumor responsiveness, and provide image contrast to visualize sites of disease and monitor disease progression. Some of these features, such as targeting, leverage biological machinery. Others are derived synthetically and enable external probing or manipulation that is otherwise not feasible in biological systems. In this chapter, we review both bio-inspired and synthetic nanoparticle functionalities that have been used in cancer therapy and address both current efforts and future opportunities to combine these into multifunctional devices.

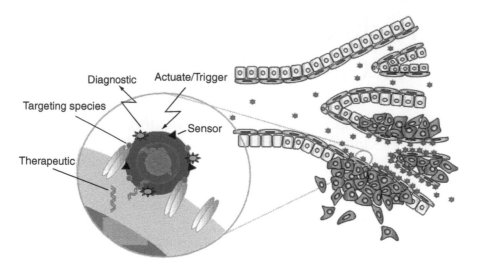

FIGURE 5.1 Schematic depiction of a multifunctional nanoparticle. A hypothetical nanoparticle targets the tumor, senses and reports molecular signatures, and delivers a therapeutic in response to an external or biological trigger. (From Ruoslahti, E., *Cancer Cell*, 2, 97–98, 2002.)

5.2 MODULAR FUNCTIONALITIES AT THE BIOSYNTHETIC INTERFACE

5.2.1 TARGETING

The ability to physically target diseased cells to receive therapeutics while avoiding residual uptake in other tissues has long been a goal in cancer therapy.[9–12] The homing of stem cells to a tissue niche, or the susceptibility of one cell over another to viral infection, demonstrates that biomolecular recognition can be used to direct species to specific extracellular and intracellular sites. The microenvironment of the tumor including cell-surface markers, extracellular matrix, soluble factors, and proteases, as well as the tumor's unique architecture and transport properties, may be exploited for targeting.[13–17]

Both passive and active targeting have been utilized for nanoparticle delivery. Passive targeting relies upon the unique pharmacokinetics of nanoparticles including minimal renal clearance and enhanced permeability and retention (EPR) through the porous angiogenic vessels in the tumor.[18,19] Surface attachment of polymers such as poly(ethylene glycol) (PEG) and poly(ethylene oxide) (PEO) enables nanoparticles to avoid uptake by mononuclear phagocytes in the liver, spleen, and lymph nodes, thereby improving accumulation in the tumor.[20–22] Active targeting relies on ligand-directed binding of nanoparticles to receptors expressed in the tumor. Binding of ligands to the vasculature can occur immediately, as it is directly accessible to nanoparticles circulating in the blood. Over longer time periods, particles extravasate into the tissues where receptors expressed on cancer cells and in the interstitium may be used for localization.[23–25]

Many candidate tumor markers have been described, some of which bind known ligands such as arginine–glycine–aspartic acid (RGD)-binding $_V\beta_3$ and $_V\beta_5$ integrins expressed on the surface of angiogenic blood vessels, and folic acid-binding receptors on the surface of cancer cells. These and others have been attached to the surface of various nanoparticle cores to deliver them to tumors.[17,26–28] Monoclonal antibodies have also been used extensively for targeting. These can be isolated with high affinity for tumor markers and are useful for targeting receptors of unknown or low affinity ligands.[29,30] Novel screens for discovering tumor homing ligands have been developed using phage and bacterial display as well as libraries of aptamers, peptides, polymers, and small molecules.[31,32] These techniques may be used to isolate targeting ligands, even when their target receptor is unknown. For example, the 34 amino acid, cationic peptide F3, which has been used to deliver quantum dots to tumor endothelium, was uncovered initially by a blind-page display screen in a breast cancer xenograft model and later found to bind cell surface nucleolin expressed on tumor endothelium and cancer cells.[6,33,34]

Although extracellular targeting to the tumor is sufficient for many modes of imaging and drug delivery, intracellular delivery of nanoparticles into the cytosol is essential for some applications. For example, nanoparticles carrying membrane-impermeable cargo that perform their function in the cytosol, such as siRNA, antisense DNA, peptides, and other drugs, are minimally effective if delivered extracellularly or sequestered in the endosome.[35] Protein and peptide motifs capable of translocating nanoparticles into the cytoplasm have been borrowed from mechanisms of viral transfection. Two important classes of translocating domains include polycationic sequences and membrane fusion domains. Attaching the short polycationic sequence of HIV's TAT protein, amino acid residues 48–57, to a nanoparticle facilitates its adsorption on a cell surface and subsequent internalization into the cell.[36,37] This peptide has been used to internalize dextran coated iron-oxide nanoparticles into T-cells in vitro, which were subsequently used to monitor T-cell trafficking in tumors with MRI.[38] Use of this peptide for intracellular delivery in vivo is limited by the adverse effect that polycationic sequences have on nanoparticle circulation time and RES uptake.[39] The amphiphilic domain derived from the N-terminus of the influenza protein hemagluttinin (HA2) is a membrane fusion peptide that destabilizes the endosome at low pH and facilitates viral escape into the cytosol.[40] Variations of this peptide with improved infectivity have also been synthesized.[41] Influenza-derived peptides have been used to enhance the delivery of liposomes as well as 100 nm poly-L-lysine particles. Although the peptide modification of these particles improves endosomal escape over unmodified particles, the transfection efficiency still remains well below that of intact viruses.[42,43]

Another level of targeting can occur after translocation of nanoparticles into the cytosol to direct nanoparticles to specific sub-cellular structures. Using peptide localization sequences, fluorescent quantum dots have been targeted to the nucleus and the mitochondria (Figure 5.2).[2] Several other localization sequences exist and could be used to traffic nanoparticles to the endoplasmic reticulum, golgi apparatus, or peroxisomes. Although work in this area has been focused on organelle labeling, the potential for delivering therapeutic nanoparticles to sub-cellular structures is possible. Such nanoparticles could sense sub-cellular aspects of disease or specifically intervene for more potent treatment or eradication of cancer cells (i.e., free-radical-mediated mitochondrial damage to induce apoptosis).

5.2.2 Imaging Agents

Imaging cancer is crucial for guiding decisions about treatment and for monitoring the efficacy of administered therapies. The use of nanoparticles for image contrast and enhancement has enabled improvements in cancer imaging by conventional modalities, such as magnetic resonance imaging (MRI) and ultrasound, and has also established new techniques such as optical-based imaging for cancer detection.[39,44,45] Targeted imaging agents that can identify specific biomarkers have the potential to improve detection, classification, and treatment of cancer with minimal invasiveness and reduced costs.

The use of nanoparticles in cancer imaging has already demonstrated clinical efficacy in detecting liver cancer and staging lymph node metastasis noninvasively.[3,46] Superparamagnetic iron-oxide nanoparticles disrupt local magnetic field gradients in tissues, causing a detectable signal void in MRI. Dextran coated iron-oxide nanoparticles administered intravenously get phagocytosed by normal macrophages of the liver and lymph and the failure of these tissues to darken after iron-oxide administration identifies invading cancer cells. Directly targeting these magnetic nanoparticles to cancer cells has also been demonstrated. For example, herceptin mAb and folic acid on the surface of iron-oxide nanoparticles enable MRI-based molecular imaging of their respective targets

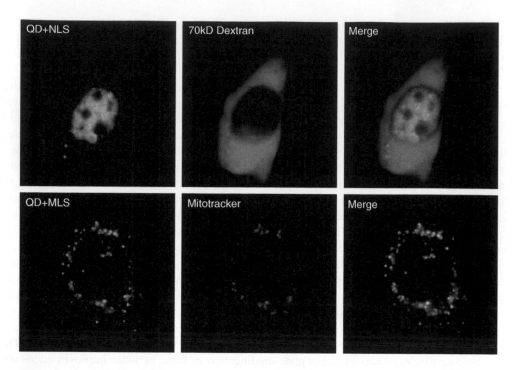

FIGURE 5.2 (See color insert following page 522.) Labeling of intracellular targets with peptide labeled quantum dots. Quantum dots (QD) modified with PEG and a nuclear localization sequence (NLS) or a mitochondrial localization sequence (MLS) were shown to target the nucleus or mitochondria of cells respectively. Seventy kilo Dalton PEG distributed in the cytoplasm contrasts nuclear localization while MitoTracker colocalizes with mitochondrial localization. (From Derfus, A. M., Chan, W. C. W., and Bhatia, S. N., *Advanced Materials*, 16, 961, 2004.)

in tumors.[47,48] Other nanoparticle cores including dendrimers, micelles, and liposomes modified with paramagnetic gadolinium have also been used for tumor targeted MRI contrast.[48–50]

Gold nanoshells offer a promising alternative to MRI probes by providing contrast for optical imaging.[45] These nanoparticles are constructed from a dielectric core (silicon) and a metallic conducting shell (gold). By varying the dimension of the core and shell, the plasmon resonance of these particles can be engineered to either absorb or scatter wavelengths of light, from UV to infrared. Particles that are tailored to scatter light in the near-infrared, where tissues have minimal absorbance, have been used to enhance imaging modalities such as reflectance confocal microscopy and optical coherence tomography (OCT).[51] Although the penetration of optical techniques does not approach that of CT or MRI, imaging features is possible at depths of a few centimeters. Gold colloids have also been used for optical contrast, but these lack the inherent tunability of nanoshells. The conjugation of optical contrast agents to antibodies has been used for the molecular imaging of the EGFR receptor on early cervical precancers and for Her2+ breast carcinoma cells in mice.[52,53]

Fluorescent nanoparticles offer another useful tool to enhance optical detection. These probes are identified easily in microscopy and are useful for tracking the biodistribution of nanoparticles in experimental models. Fluorescent semiconductor nanocrystals, quantum dots, have been used to show ligand-mediated nanoparticle targeting to distinct features in the tumor.[6] Three different phage display-derived peptides were used to specifically target these nanocrystals to tumor blood vessels, tumor lymphatics, or lung endothelium (Figure 5.3). Functionalization of these nanoparticles with PEG eliminated detectable accumulation in RES organs, including the liver (Figure 5.4). Quantum dots have a distinct advantage over conventional fluorophores because

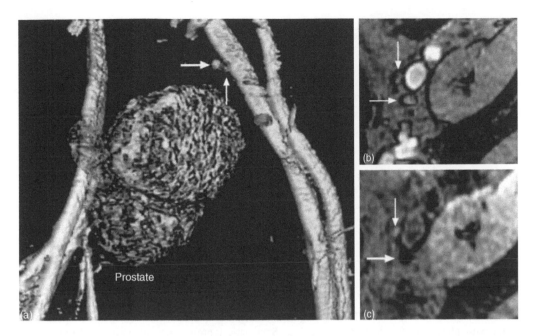

FIGURE 5.3 (See color insert following page 522.) Noninvasive detection of lymph node metastasis with iron-oxide nanoparticles and MRI. (a) A three-dimensional reconstruction using nanoparticle-enhanced MRI of the prostate. Metastatic lymph nodes are in red and normal nodes are in green. (b) Conventional MRI image shows similar signal intensity from two adjacent nodes. (c) Nanoparticle-enhanced MRI shows a decreased signal in the normal node from macrophage uptake (thick arrow), but not in the metastatic node (thin arrow). (From Harisinghani, M. G. and Weissleder, R., *Plos Medicine*, 1, 202–209, 2004.)

of their size-tunable excitation and emission profiles, narrow bandwidths, and high photo-stability.[54,55] Using nanocrystals that fluoresce in the near-infrared could extend their utility to clinical settings,[56] though a key limitation has been their potential toxicity because they are formulated from heavy metals.[57] Efforts to make these of nontoxic materials are ongoing (Figure 5.4). Alternative fluorescent nanoparticle probes have been developed including fluorescently tagged dendrimers and fluorophore-embedded silica nanoparticles.[58–60]

Nanoparticle formulations that provide contrast for other imaging modalities including ultrasound and CT have been described. Perfluorocarbon emulsion nanoparticles composed of lipid-encapsulated perfluorocarbon liquid, about 250 nm diameter, are effective in giving echo contrast.[61] Air-entrapping liposomes formulated from freeze-drying techniques have also been developed to give ultrasound contrast.[62,63] These agents passively distribute in RES organs and areas of angiogenenis enabling enhanced imaging of these features. Bismuth sulfide nanoparticles can be used as contrast agents for CT imaging, giving blood pool contrast similar to that of iodine but at lower concentrations. These can also be used to image lymph nodes after phagocytic uptake.[64] As with other imaging modalities, the attachment of appropriate ligands to nanoparticle-based CT and ultrasound contrast agents could be used for molecular imaging.

An interesting extension of the imaging agents described above is their combination into multimodal imaging nanoparticles. The combination of fluorescence and magnetic properties within a single particle has been used for dual optical and MRI imaging. Fluorescence is used to determine the localization of targeted imaging agents down to specific micro-structures inside and outside cells and magnetic domains provide three-dimensional whole-body imaging capabilities with MRI.[59,65] These magneto-fluorescent nanoparticles could be effective for image guided surgeries where MRI is used to locate cancer in the body and fluorescence is used to more precisely

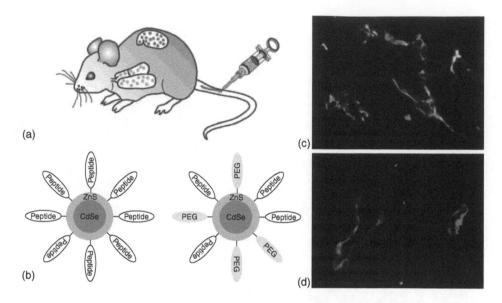

FIGURE 5.4 (See color insert following page 522.) Targeting quantum dots (QDs) to site-specific endothelium with phage display-derived peptides. (a) Schematic representation of co-injected red and green quantum dots that home to tumor and lung vasculature respectively after intravenous injection. (b) Schematic representation of peptide-coated QDs and peptide-coated PEG-QDs. (c) QDs labeled with the tumor endothelium homing peptide F3 co-localize with a blood vessel marker. (d) QDs labeled with the tumor lymphatic homing peptide Lyp-1 highlight the endothelium but do not colocalize with a blood vessel marker. (From Akerman, M. E., Chan, W. C. W., Laakkonen, P., Bhatia, S. N., and Ruoslahti, E., *Proceedings of the National Academy of Sciences of the United States of America*, 99, 12617–12621, 2002.)

delineate tumor borders during resection. Other dual-imaging probes have been described including: perfluorocarbon emulsions tagged with gadolinium for combined ultrasound and MRI.[66]

5.2.3 SENSING

Functionalities that undergo chemical alterations in response to enzymatic activity or other properties such as pH or oxygen could be used as sensors to report information about the status of the tumor or efficacy of treatment. Many nanoparticle-based sensors that respond to biological triggers including proteases, DNAses, proteins, peroxidase, pH, and others have been demonstrated in vitro.[67–72] These generally rely on assembly or disassembly of inorganic nanocrystals including: gold nanoparticles, nanoshells, or nanorods, which undergo a shift in their plasmon resonance when aggregated; iron-oxide nanoparticles have enhanced T2 relaxivity when clustered; and fluorescent quencher-based nanoparticle systems that dequench after triggered release.

A system using cleavable polymeric shielding of self-assembling nanoparticles has been proposed as a mechanism for translating nanoparticle-based enzyme sensors to in vivo use (Figure 5.5).[73] Self-assembling, complementary iron-oxide nanoparticles are rendered latent with PEG polymers linked to the nanoparticle surface by protease-cleavable substrates that serve to both inhibit assembly and stabilize the particles in serum. Upon proteolytic removal of PEG polymers by MMP-2 expressing cancer cells, nanoparticles assemble and acquire amplified magnetic properties that can be detected with MRI. In the future, similar to thrombin-driven self-assembly of fibrin and platelets at sites of endothelial injury, this system may allow the hyper-active proteolytic processes of cancers to drive the self-assembly of nanoparticles in regions of cancer angiogenesis, invasion, and metastasis in vivo. Due to its modular design, this system can easily be modified for a number of detection schemes by substituting the complementary binding pairs, cleavable substrates

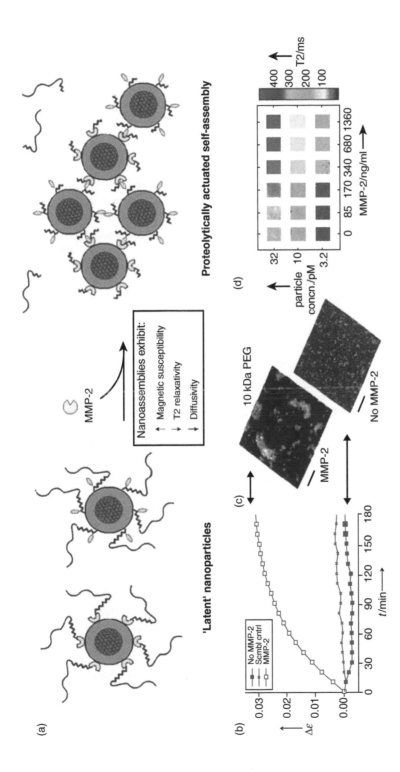

FIGURE 5.5 (See color insert following page 522.) Protease-triggered self-assembling nanoparticles with polymer-shielded coatings. (a) Schematic representation of nanoparticles that self-assemble after protease-mediated cleavage of PEG chains reveals complementary moieties (neutravidin and biotin). (b) Iron-oxide nanoparticles with cleavable linkers assemble in the presence of MMP-2, as measured by changes their light extinction, while particles with noncleavable scrambled peptides do not. (c) Atomic force micrographs of particles incubated with MMP-2 shows detectable aggregation (scale bars are 500 nm). (d) T2 maps of particles in solution using a 4.7 T MRI demonstrate enhanced T2 relaxivity for increasing concentrations of MMP-2.

(e.g., glycans, lipids, oligonucleotides), or multivalent nanoparticle cores (e.g., gold, quantum dot, dendrimer).

5.2.4 THERAPEUTIC PAYLOADS

The use of nanoparticulate drug carriers can address many critical challenges in drug delivery including: improving drug solubility and stability; extending drug half-lives in the blood; reducing adverse effects in nontarget organs; and concentrating drugs at the disease site.[74] Drugs may be dispersed in a matrix, encapsulated in a vesicle, dissolved in a hydrophobic core, or attached to the surface of a nanoparticle. Several nanoparticle-based drug delivery systems including liposomes, polymeric nanoparticles, dendrimers, ceramic based carriers, micelles, and others have been used to carry small molecule, peptide, and oligonucleotide therapeutic agents.[24,75] Many promising anti-cancer drugs fail to make it to the clinic because of poor solubility or high collateral toxicity at therapeutic levels, thus motivating the need for these carriers in cancer therapy.

Liposomes have been the most extensively utilized nanoparticle-based carriers for delivering anti-cancer drugs. First described decades ago, these submicron-sized carriers consist of amphiphilic lipids assembled to form vesicles that can encapsulate drugs.[76] Liposome-encapsulated doxorubicin is a clinically approved nanoparticle formulation used for chemotherapy.[77] The surface of this nanocarrier is PEGylated to reduce rapid uptake by phagocytic cells and extend the drug circulation time for better therapeutic efficacy. Several other liposome-encapsulated chemotherapeutic drugs have been described, with many in clinical trials.[78] Active targeting of these liposomes through the attachment of antibodies and various ligands has also been demonstrated.[30,79] Drug loaded liposomes with encapsulated or surface-functionalized gadolinium or fluorophores have been used to simultaneously image tumors during nanoparticle-targeted drug delivery.[80–82]

Biodegradable polymer nanocarriers have also been investigated as a means of encapsulating drugs and releasing them over time. Both poly dl-lactide co-glycolide (PLGA) and polylactide (PLA) nanoparticles have been formed that immobilize drugs dispersed in their matrix and release them upon degradation.[24,83] Other polymers, including polyethyleneimine (PEI), polylysine, and cyclodextrin-containing polymers, are used to condense DNA or RNA into nanoparticle carriers that can be targeted to cancer cells for gene or siRNA delivery.[35] Polymeric micelles consist of amphiphilic block copolymers that self-assemble into a water-soluble nanoparticle with a hydrophobic core. These can be used to encapsulate water-insoluble drugs such as doxorubicin and adriamycin and targeted to tumors.[84–86] Polymersomes are another variation of polymer-based nanoparticulate vesicles that self-assemble from amphiphilic block copolymers.[87] These have been used to encapsulate doxorubicin with well-controlled release over several days.[88,89]

Another class of nanoparticle-based drug carriers are dendrimers. These consist of a network of branching chemical bonds around an inner core. One of the more popular dendrimers, polyamidoamine dendrimers (PAMAMs), are nonimmunogenic, water-soluble, and possess terminal amine functional groups for conjugation of a variety of surface moieties.[90] Their inner core can been used to encapsulate anti-cancer drugs such as doxorubicin and methotrexate.[91] Drugs may also be conjugated to the dendrimer surface along with ligands for targeting.[92,93] A dendrimer functionalized with FITC, folic acid, and methoxetrate has been synthesized to have imaging, targeting, and drug delivery capabilities (Figure 5.6).[4,7,94] The synthesis of these conjugates in a scalable and reproducible manner has been described for potential clinical applications.[4]

Other nanoparticulate carriers, including nanoemulsions, drug nanocrystals, and polyelectrolyte carriers, have been developed. Nanoemulsions are formed by dissolving a drug in a lipid, cooling the solution under high pressure, and using homogenization to form solid nanoparticle lipid carriers at body temperature. Homogenization techniques can also be used to form crystalline nanosuspensions of drugs.[74] These formulations increase drug solubility and control release kinetics of the drug in the blood and at the tumor site. Polyelectrolyte carriers formed by the

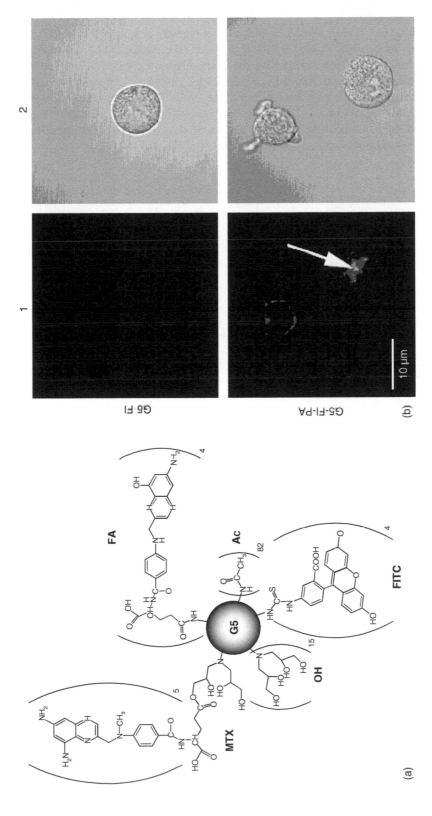

FIGURE 5.6 Multifunctional dendrimers for targeting, imaging, and drug delivery. (a) Schematic of a multifunctional dendrimer labeled with fluorescein (FITC) for imaging, folic acid (FA) for targeting, methotrexate (MTX) for therapeutic delivery, and alcohol (OH) and acetylated (Ac) moieties for particle stabilization. (b) Confocal images of FITC-labeled dendrimers incubated over cells with and without a targeting antibody, anti-PSMA (PA). (From Majoros, I. J., Thomas, T. P., Mehta, C. B., and Baker, J. R., *Journal of Medicinal Chemistry*, 48, 5892–5899, 2005; Thomas, T. P. et al., *Biomacromolecules*, 5, 2269–2274, 2004.)

layer-by-layer absorption of polycationic and polyanionic moieties can be used to encapsulate therapeutic cargo, particularly larger agents such as peptides and oligonucleotides.[95]

A clever combination of drug-release modalities was recently demonstrated by the creation of a dual drug-release nanoparticle having a PLGA polymer core encapsulating doxorubicin and a PEG-lipid block copolymer shell loaded with the combretastatin (Figure 5.7).[5] The lipophilic anti-angiogenesis drug, combretastatin, intercalates in the nanoparticle membrane and releases rapidly upon association with tumor endothelial cells, while the slower-releasing doxorubicin increases cytotoxic killing of tumor cells for a prolonged time after the vasculature shuts down. This novel system demonstrates the feasibility of integrating multiple functionalities of drug delivery on a single nanoparticle to enhance therapeutic efficacy.

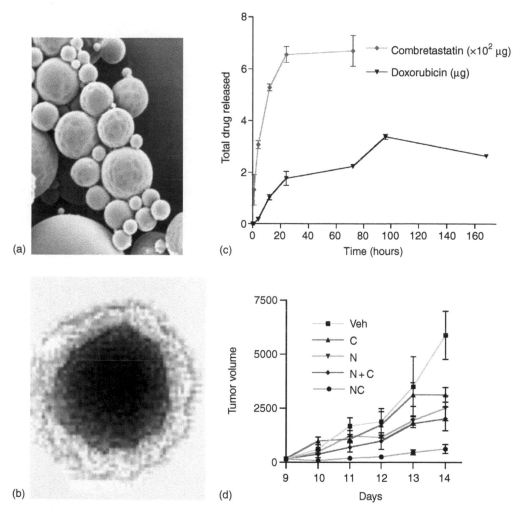

FIGURE 5.7 Dual drug-release nanoparticle for combined anti-angiogenisis and anti-cancer treatment. (a) Scanning electron micrograph showing nanocores prepared from doxorubicin-coupled PLGA. The nanocores are encapsulated inside a lipid coat, which is also loaded with an anti-angiogenesis agent. (b) Cross section of a nanocell with the dark nanocore. The lipid coat is surface-modified through pegylation, which confers stealth characteristics to the nanocell from the RES. (c) The composition of the nanocell enables a spatiotemporal release of the two agents in an acidic pH mimicking the tumor environment, as shown in the graph. (d) In vivo studies using F10 melanoma clearly show that the spatiotemporal release from the nanocells (NC) achieves better outcome than the doxorubicin-loaded nanocore or the lipid-entrapped combretastatin (C) alone, or combinations of both (N+C).[5]

5.2.5 REMOTE ACTUATION

Temporal and spatial control of therapeutic administration is important for eliminating off-target toxicity and achieving optimal delivery. Temporally controlled release profiles may be designed into nanoparticle carriers mentioned previously and spatial control can be improved with targeting. However, off-target effects, including eventual accumulation of nanoparticles in RES organs, limit many aspects of these methods of control. The ability to trigger the therapeutic activity of administered nanoparticles remotely could be a valuable tool for localizing treatments to a diseased site. Many inorganic nanocrystals and nanoemulsions used for imaging contrast absorb electromagnetic or ultrasonic energy that can also be used to remotely heat or trigger drug delivery.

Thermal ablation of tumors by nanoparticles that absorb external energy has been demonstrated both with iron-oxide nanoparticles and gold nanoshells. Superparamagnetic iron-oxide nanoparticles under the influence of an alternating electromagnetic (EM) field heat by *Brownian relaxation*, where heat is generated by the rotation of particles in the field, and *Neel relaxation*, where the magnetic domains are moved away from their easy axis with the resultant energy being deposited as heat in the solution.[96,97] Nanoparticle concentrations of 0.1–1% are required to achieve critical temperatures for tumor ablation.[98,99] Ongoing work to increase the absorption of magnetic nanoparticles using clinically safe RF frequencies and to increase the concentration of particles that can be targeted to the tumor may extend the utility of this technique. Alternatively, near-infrared-absorbing gold nanoshells targeted to the tumor can be used to thermally ablate the cancer cells upon illumination with a high intensity laser.[100,101] This technique can be applied to solid tumors in close proximity to the skin, but cannot be applied to deeper lesions because of tissue absorbance.[53,100,101] By synthesizing nanoshells with a plasmon resonance that has both absorption *and* scattering profiles, these nanoparticles may be capable of both heating and imaging tumors.[53]

Remotely-triggered release of a therapy by heating is a promising extension of the use of nanoparticles that can absorb external energy. An example of this has been demonstrated with a model drug linked to an iron-oxide nanoparticle via a heat-labile tether that is released and diffuses into the peripheral tissue after irradiation with RF energy.[98] By modifying the susceptibility of the linker, it is possible to tune the release profile over a range of temperatures and to enable repeated administrations. The iron core of these drug-releasing nanoparticles can be used simultaneously for imaging with MRI. Additionally, the magnetic properties of these nanoparticles can be manipulated by magnetic field gradients to target sites near externally- or internally-placed magnets.[102]

Drug activation using EM energy has been explored extensively with photodynamic therapy (PDT). PDT agents, when irradiated by light, produce reactive oxygen species that are toxic to cells. Agents such as porphyrins have been conjugated to various nanoparticle cores including dendrimers, liposomes, and polymers.[103,104] When excited by light, these nanoparticles can produce enough reactive oxygen species to kill tumor cells.[60] The inherent fluorescent properties of many PDT agents enable simultaneous imaging with therapeutic delivery. A multifunctional nanoparticle platform combining MRI contrast and photodynamic therapy has been used to target, image, and treat brain cancer in a rat model.[105] In the future, integrating these nanoparticles with peptides capable of targeting tumors *and* subcellularly localizing them to the nuclei or mitochondria of tumor cells may enhance the therapeutic efficacy of these treatments.

Other forms of externally applied energy such as ultrasound and x-ray radiation provide alternative mechanisms to achieve remote actuation. Acoustic energy has been shown to enhance the delivery of lipid drugs from a perfluorocarbon emulsion targeted to cell membranes and from doxorubicin-loaded polymeric micelles.[106,107] Atomically dense nanoparticles have been shown to increase the absorption of x-ray radiation, enhancing their destructive effect in surrounding tissue.[108] There is potential for simultaneous imaging and therapeutic delivery with these particles also.

5.3 CHALLENGES IN INTEGRATING MULTIPLE FUNCTIONALITIES AND FUTURE DIRECTIONS

Although remotely actuated nanoparticle cores such as iron-oxide and metal nanoshells naturally lend themselves to dual-imaging and therapeutic applications, the combination of imaging and other functionalities using other nanoparticle cores can be challenging. There are inherent trade-offs when combining many functional groups into one nanoparticle. In many cases, a limited number of attachment sites are available on the particle surface, making it difficult to couple several functional groups in sufficient concentration for each to function. Moreover, some groups may interact to sterically shield or alter the activity of one another when combined in close proximity. Multiple functional moieties on a nanoparticle may also reduce colloidal stability or adversely affect its in vivo pharmacokinetics. With significant characterization and fine tuning, dendrimers that combine targeting, imaging, and therapeutic moieties on their surface have been synthesized successfully.[4] Similar efforts will be necessary to achieve other multifunctional nanoparticles with decorated-surface moieties.

An alternative strategy to consolidate multiple functionalities onto a single particle is to use core-shell architecture. In this case, an outer shell with one functionality, such as targeting, may be unveiled to reveal an inner core that performs a secondary function such as endosomal escape or drug release. This has been demonstrated with the conjugation of targeting moieties or protective PEG groups on the surface of dendrimers or polymers via acid-labile chemistries that degrade in the lower pH of the endosome and unveil endosomal escape mechanisms on the particle core.[109,110] This has also been demonstrated with protease-cleavable linkers that release protective polymers on the surface of complementary nanoparticles to initiate their self-assembly.[73]

The synthesis of nanoparticles with polar domains is another strategy that could be used to incorporate multiple functionalities on a single particle. Janus nanoparticles—named for Janus, the Roman God of doorways typically depicted with faces on the front and back of his head—have been engineered with two chemically distinct hemispheres or surfaces. These nanoparticles may be spherical (with opposing faces of unique composition), dumbbell-shaped (with two equal-sized spheres linked together), snowman-shaped, and may have other morphologies as well.[95,111] The creation of nanoparticles with spatially separated chemical domains is a step towards replicating the controlled polarity exhibited in nature across many length scales. Separate hemispheres may be used to isolate and organize functional domains on nanoparticles such that they may simultaneously carry targeting molecules, endosomal escape domains, sensing moieties, hydrophilic and hydrophobic therapeutics, or contrast agents that otherwise might be mutually inhibitory if randomly incorporated. Moreover, there may be specific applications for which the polarity and anisotropy of Janus nanoparticles have benefit, such as real-time detection of oriented binding events, targeted bridging of multiple components at a tumor cell, directed drug delivery, or guided self-assembly.

Although there have been many exciting advances in the application of nanoparticles for cancer imaging and treatment, the true power of these materials will be in their ability to interact with disease processes intelligently. The modular design of functionalities that target, sense, signal, and treat and the ongoing efforts to consolidate these into single nanoparticle platforms is one way in which such 'smart' materials are being developed. The further elucidation of complex biological processes in tumorogenesis, the discovery of nanomaterials with other novel properties, and the consolidation of biological and synthetic machinery in these materials in new and elegant ways are key factors that will determine their future success in cancer therapy.

REFERENCES

1. Ruoslahti, E., Antiangiogenics meet nanotechnology, *Cancer Cell*, 2, 97–98, 2002.
2. Derfus, A. M., Chan, W. C. W., and Bhatia, S. N., Intracellular delivery of quantum dots for live cell labeling and organelle tracking, *Advanced Materials*, 16, 961–966, 2004.

3. Harisinghani, M. G. and Weissleder, R., Sensitive, noninvasive detection of lymph node metastases, *Plos Medicine*, 1, 202–209, 2004.

4. Majoros, I. J., Thomas, T. P., Mehta, C. B., and Baker, J. R., Poly(amidoamine) dendrimer-based multifunctional engineered nanodevice for cancer therapy, *Journal of Medicinal Chemistry*, 48, 5892–5899, 2005.

5. Sengupta, S. et al., Temporal targeting of tumour cells and neovasculature with a nanoscale delivery system, *Nature*, 436, 568–572, 2005.

6. Akerman, M. E., Chan, W. C. W., Laakkonen, P., Bhatia, S. N., and Ruoslahti, E., Nanocrystal targeting in vivo, *Proceedings of the National Academy of Sciences of the United States of America*, 99, 12617–12621, 2002.

7. Thomas, T. P. et al., In vitro targeting of synthesized antibody-conjugated dendrimer nanoparticles, *Biomacromolecules*, 5, 2269–2274, 2004.

8. Gordon, A. N. et al., Recurrent epithelial ovarian carcinoma: A randomized phase III study of pegylated liposomal doxorubicin versus topotecan, *Journal of Clinical Oncology*, 19, 3312–3322, 2001.

9. Ruoslahti, E., Drug targeting to specific vascular sites, *Drug Discovery Today*, 7, 1138–1143, 2002.

10. Allen, T. M., Charrois, G. J. R., and Sapra, P., Recent advances in passively and actively targeted liposomal drug delivery systems for the treatment of cancer, *Abstracts of Papers of the American Chemical Society*, 226, U458, 2003.

11. Allen, T. M. and Cullis, P. R., Drug delivery systems: Entering the mainstream, *Science*, 303, 1818–1822, 2004.

12. Wickham, T. J., Targeting adenovirus, *Gene Therapy*, 7, 110–114, 2000.

13. Ruoslahti, E. and Rajotte, D., An address system in the vasculature of normal tissues and tumors, *Annual Review of Immunology*, 18, 813–827, 2000.

14. Jain, R. K., Delivery of molecular and cellular medicine to solid tumors, *Advanced Drug Delivery Reviews*, 46, 149–168, 2001.

15. Jain, R. K., Delivery of molecular and cellular medicine to solid tumors, *Journal of Controlled Release*, 53, 49–67, 1998.

16. Jain, R. K., Delivery of molecular and cellular medicine to solid tumors, *Microcirculation — London*, 4, 3–23, 1997.

17. Satchi-Fainaro, R. et al., Targeting angiogenesis with a conjugate of HPMA copolymer and TNP-470, *Nature Medicine*, 10, 255–261, 2004.

18. Matsumura, Y. and Maeda, H., A new concept for macromolecular therapeutics in cancer-chemotherapy—mechanism of tumoritropic accumulation of proteins and the antitumor agent Smancs, *Cancer Research*, 46, 6387–6392, 1986.

19. Tabata, T., Murakami, Y., and Ikada, Y., Tumor accumulation of poly(vinyl alcohol) of different sizes after intravenous injection, *Journal of Controlled Release*, 50, 123–133, 1998.

20. Moghimi, S. M., Hunter, A. C., and Murray, J. C., Long-circulating and target-specific nanoparticles: Theory to practice, *Pharmacological Reviews*, 53, 283–318, 2001.

21. Moghimi, S. M. and Hunter, A. C., Recognition by macrophages and liver cells of opsonized phospholipid vesicles and phospholipid headgroups, *Pharmaceutical Research*, 18, 1–8, 2001.

22. Nicolazzi, C. et al., Anionic polyethyleneglycol lipids added to cationic lipoplexes increase their plasmatic circulation time, *Journal of Controlled Release*, 88, 429–443, 2003.

23. Ruoslahti, E., Specialization of tumour vasculature, *Nature Reviews Cancer*, 2, 83–90, 2002.

24. Sahoo, S. K. and Labhasetwar, V., Nanotech approaches to delivery and imaging drug, *Drug Discovery Today*, 8, 1112–1120, 2003.

25. Wickline, S. A. and Lanza, G. M., Nanotechnology for molecular imaging and targeted therapy, *Circulation*, 107, 1092–1095, 2003.

26. Stella, B. et al., Design of folic acid-conjugated nanoparticles for drug targeting, *Journal of Pharmaceutical Sciences*, 89, 1452–1464, 2000.

27. Lockman, P. R. et al., Brain uptake of thiamine-coated nanoparticles, *Journal of Controlled Release*, 93, 271–282, 2003.

28. Dubey, P. K., Mishra, V., Jain, S., Mahor, S., and Vyas, S. P., Liposomes modified with cyclic RGD peptide for tumor targeting, *Journal of Drug Targeting*, 12, 257–264, 2004.

29. Kirpotin, D. et al., Sterically stabilized Anti-HER2 immunoliposomes: Design and targeting to human breast cancer cells in vitro, *Biochemistry*, 36, 66–75, 1997.

30. Li, L. Y. et al., A novel antiangiogenesis therapy using an integrin antagonist or anti-FLK-1 antibody coated Y-90-labeled nanoparticles, *International Journal of Radiation Oncology Biology Physics*, 58, 1215–1227, 2004.

31. Farokhzad, O. C. et al., Nanopartide-aptamer bioconjugates: A new approach for targeting prostate cancer cells, *Cancer Research*, 64, 7668–7672, 2004.

32. Allen, T. M., Sapra, P., Moase, E., Moreira, J., and Iden, D., Adventures in targeting, *Journal of Liposome Research*, 12, 5–12, 2002.

33. Christian, S. et al., Nucleolin expressed at the cell surface is a marker of endothelial cells in angiogenic blood vessels, *Journal of Cell Biology*, 163, 871–878, 2003.

34. Porkka, K., Laakkonen, P., Hoffman, J. A., Bernasconi, M., and Ruoslahti, E., A fragment of the HMGN2 protein homes to the nuclei of tumor cells and tumor endothelial cells in vivo, *Proceedings of the National Academy of Sciences of the United States of America*, 99, 7444–7449, 2002.

35. Pack, D. W., Hoffman, A. S., Pun, S., and Stayton, P. S., Design and development of polymers for gene delivery, *Nature Reviews Drug Discovery*, 4, 581–593, 2005.

36. Frankel, A. D. and Pabo, C. O., Cellular uptake of the tat protein from human immunodeficiency virus, *Cell*, 55, 1189–1193, 1988.

37. Wadia, J. S., Stan, R. V., and Dowdy, S. F., Transducible TAT-HA fusogenic peptide enhances escape of TAT-fusion proteins after lipid raft macropinocytosis, *Nature Medicine*, 10, 310–315, 2004.

38. Kircher, M. F. et al., In vivo high resolution three-dimensional imaging of antigen-specific cytotoxic T-lymphocyte trafficking to tumors, *Cancer Research*, 63, 6838–6846, 2003.

39. Weissleder, R., Bogdanov, A., Neuwelt, E. A., and Papisov, M., Long-circulating iron-oxides for Mr-imaging, *Advanced Drug Delivery Reviews*, 16, 321–334, 1995.

40. Wagner, E., Application of membrane-active peptides for nonviral gene delivery, *Advanced Drug Delivery Reviews*, 38, 279–289, 1999.

41. Plank, C., Oberhauser, B., Mechtler, K., Koch, C., and Wagner, E., The influence of endosome-disruptive peptides on gene-transfer using synthetic virus-like gene-transfer systems, *Journal of Biological Chemistry*, 269, 12918–12924, 1994.

42. Mastrobattista, E., Crommelin, D. J. A., Wilschut, J., and Storm, G., Targeted liposomes for delivery of protein-based drugs into the cytoplasm of tumor cells, *Journal of Liposome Research*, 12, 57–65, 2002.

43. Kakudo, T. et al., Transferrin-modified liposomes equipped with a pH-sensitive fusogenic peptide: An artificial viral-like delivery system, *Biochemistry*, 43, 5618–5628, 2004.

44. Lanza, G. M. et al., Novel paramagnetic contrast agents for molecular imaging and targeted drug delivery, *Current Pharmaceutical Biotechnology*, 5, 495–507, 2004.

45. West, J. L. and Halas, N. J., Engineered nanomaterials for biophotonics applications: Improving sensing, imaging, and therapeutics, *Annual Review of Biomedical Engineering*, 5, 285–292, 2003.

46. Stark, D. D. et al., Superparamagnetic iron-oxide—clinical-application as a contrast agent for MR imaging of the liver, *Radiology*, 168, 297–301, 1988.

47. Choi, H., Choi, S. R., Zhou, R., Kung, H. F., and Chen, I. W., Iron oxide nanoparticles as magnetic resonance contrast agent for tumor imaging via folate receptor-targeted delivery, *Academic Radiology*, 11, 996–1004, 2004.

48. Huh, Y. M. et al., In vivo magnetic resonance detection of cancer by using multifunctional magnetic nanocrystals, *Journal of the American Chemical Society*, 127, 12387–12391, 2005.

49. Wang, S. J., Brechbiel, M., and Wiener, E. C., Characteristics of a new MRI contrast agent prepared from polypropyleneimine dendrimers, generation 2, *Investigative Radiology*, 38, 662–668, 2003.

50. Lanza, G. M. et al., Molecular imaging and targeted drug delivery with a novel, ligand-directed paramagnetic nanoparticle technology, *Academic Radiology*, 9, S330–S331, 2002.

51. Loo, C. et al., Nanoshell-enabled photonics-based imaging and therapy of cancer, *Technology in Cancer Research & Treatment*, 3, 33–40, 2004.

52. Sokolov, K. et al., Real-time vital optical imaging of precancer using anti-epidermal growth factor receptor antibodies conjugated to gold nanoparticles, *Cancer Research*, 63, 1999–2004, 2003.

53. Loo, C., Lowery, A., Halas, N., West, J., and Drezek, R., Immunotargeted nanoshells for integrated cancer imaging and therapy, *Nano Letters*, 5, 709–711, 2005.

54. Parak, W. J. et al., Biological applications of colloidal nanocrystals, *Nanotechnology*, 14, R15–R27, 2003.

55. Gao, X. H., Cui, Y. Y., Levenson, R. M., Chung, L. W. K., and Nie, S. M., In vivo cancer targeting and imaging with semiconductor quantum dots, *Nature Biotechnology*, 22, 969–976, 2004.

56. Kim, S. et al., Near-infrared fluorescent type II quantum dots for sentinel lymph node mapping, *Nature Biotechnology*, 22, 93–97, 2004.

57. Derfus, A. M., Chan, W. C. W., and Bhatia, S. N., Probing the cytotoxicity of semiconductor quantum dots, *Nano Letters*, 4, 11–18, 2004.

58. Choi, Y. and Baker, J. R., Targeting cancer cells with DNA-assembled dendrimers—a mix and match strategy for cancer, *Cell Cycle*, 4, 669–671, 2005.

59. Lu, Y., Yin, Y. D., Mayers, B. T., and Xia, Y. N., Modifying the surface properties of superparamagnetic iron oxide nanoparticles through a sol–gel approach, *Nano Letters*, 2, 183–186, 2002.

60. Roy, I. et al., Ceramic-based nanoparticles entrapping water-insoluble photosensitizing anti-cancer drugs: A novel drug–carrier system for photodynamic therapy, *Journal of the American Chemical Society*, 125, 7860–7865, 2003.

61. Lanza, G. M. et al., A novel site-targeted ultrasonic contrast agent with broad biomedical application, *Circulation*, 94, 3334–3340, 1996.

62. Alkan-Onyuksel, H. et al., Development of inherently echogenic liposomes as an ultrasonic contrast agent, *Journal of Pharmaceutical Sciences*, 85, 486–490, 1996.

63. Huang, S. L. et al., Improving ultrasound reflectivity and stability of echogenic liposomal dispersions for use as targeted ultrasound contrast agents, *Journal of Pharmaceutical Sciences*, 90, 1917–1926, 2001.

64. Rabin, O., Perez, J. M., Grimm, J., Wojtkiewicz, G., and Weissleder, R., An x-ray computed tomography imaging agent based on long-circulating bismuth sulphide nanoparticles, *Nature Materials*, 5, 118–122, 2006.

65. Kelly, K. A. et al., Detection of vascular adhesion molecule-1 expression using a novel multimodal nanoparticle, *Circulation Research*, 96, 327–336, 2005.

66. Morawski, A. M., Lanza, G. A., and Wickline, S. A., Targeted contrast agents for magnetic resonance imaging and ultrasound, *Current Opinion in Biotechnology*, 16, 89–92, 2005.

67. Perez, J. M., Josephson, L., O'Loughlin, T., Hogemann, D., and Weissleder, R., Magnetic relaxation switches capable of sensing molecular interactions, *Nature Biotechnology*, 20, 816–820, 2002.

68. Perez, J. M., Simeone, F. J., Saeki, Y., Josephson, L., and Weissleder, R., Viral-induced self-assembly of magnetic nanoparticles allows the detection of viral particles in biological media, *Journal of the American Chemical Society*, 125, 10192–10193, 2003.

69. Perez, J. M., Simeone, F. J., Tsourkas, A., Josephson, L., and Weissleder, R., Peroxidase substrate nanosensors for MR imaging, *Nano Letters*, 4, 119–122, 2004.

70. Georganopoulou, D. G. et al., Nanoparticle-based detection in cerebral spinal fluid of a soluble pathogenic biomarker for Alzheimer's disease, *Proceedings of the National Academy of Sciences of the United States of America*, 102, 2273–2276, 2005.

71. Mirkin, C. A., Letsinger, R. L., Mucic, R. C., and Storhoff, J. J., A DNA-based method for rationally assembling nanoparticles into macroscopic materials, *Nature*, 382, 607–609, 1996.

72. Stevens, M. M., Flynn, N. T., Wang, C., Tirrell, D. A., and Langer, R., Coiled-coil peptide-based assembly of gold nanoparticles, *Advanced Materials*, 16, 915–918, 2004.

73. Harris, T. J., von Maltzahn, G., Derfus, A. M., Ruoslahti, E., and Bhatia, S. N., Proteolytic actuation of nanoparticle self-assembly, Angewandte Chemie-International Edition, In press.

74. Muller, R. H. and Keck, C. M., Challenges and solutions for the delivery of biotech drugs—a review of drug nanocrystal technology and lipid nanoparticles, *Journal of Biotechnology*, 113, 151–170, 2004.

75. Duncan, R., The dawning era of polymer therapeutics, *Nature Reviews Drug Discovery*, 2, 347–360, 2003.

76. Sessa, G. and Weissman, G., Phospholipid spherules (liposomes) as a model for biological membranes, *Journal of Lipid Research*, 9(3): 310–318, 1968.

77. Gabizon, A. and Martin, F., Polyethylene glycol coated (pegylated) liposomal doxorubicin—rationale for use in solid tumours, *Drugs*, 54, 15–21, 1997.

78. Campbell, R. B., Balasubramanian, S. V., and Straubinger, R. M., Influence of cationic lipids on the stability and membrane properties of paclitaxel-containing liposomes, *Journal of Pharmaceutical Sciences*, 90, 1091–1105, 2001.

79. Sapra, P. and Allen, T. M., Improved outcome when B-cell lymphoma is treated with combinations of immunoliposomal anti-cancer drugs targeted to both the CD19 and CD20 epitopes, *Clinical Cancer Research*, 10, 4893, 2004.

80. Torchilin, V. P. et al., p-nitrophenylcarbonyl-PEG-PE-liposomes: Fast and simple attachment of specific ligands, including monoclonal antibodies, to distal ends of PEG chains via p-nitrophenyl-carbonyl groups, *Biochimica et Biophysica Acta-Biomembranes*, 1511, 397–411, 2001.

81. Park, J. W. et al., Tumor targeting using anti-her2 immunoliposomes, *Journal of Controlled Release*, 74, 95–113, 2001.

82. Miyamoto, M. et al., Preparation of gadolinium-containing emulsions stabilized with phosphatidyl-choline-surfactant mixtures for neutron-capture therapy, *Chemical & Pharmaceutical Bulletin*, 47, 203–208, 1999.

83. Panyam, J. and Labhasetwar, V., Biodegradable nanoparticles for drug and gene delivery to cells and tissue, *Advanced Drug Delivery Reviews*, 55, 329–347, 2003.

84. Nakanishi, T. et al., Development of the polymer micelle carrier system for doxorubicin, *Journal of Controlled Release*, 74, 295–302, 2001.

85. Yokoyama, M. et al., Selective delivery of adiramycin to a solid tumor using a polymeric micelle carrier system, *Journal of Drug Targeting*, 7, 171–186, 1999.

86. Torchilin, V. P., Structure and design of polymeric surfactant-based drug delivery systems, *Journal of Controlled Release*, 73, 137–172, 2001.

87. Discher, B. M. et al., Polymersomes: Tough vesicles made from diblock copolymers, *Science*, 284, 1143–1146, 1999.

88. Ahmed, F. and Discher, D. E., Self-porating polymersomes of PEG-PLA and PEG-PCL: Hydrolysis-triggered controlled release vesicles, *Journal of Controlled Release*, 96, 37–53, 2004.

89. Xu, J. P., Ji, J., Chen, W. D., and Shen, J. C., Novel biomimetic polymersomes as polymer thera-peutics for drug delivery, *Journal of Controlled Release*, 107, 502–512, 2005.

90. Patri, A. K., Majoros, I. J., and Baker, J. R., Dendritic polymer macromolecular carriers for drug delivery, *Current Opinion in Chemical Biology*, 6, 466–471, 2002.

91. Kojima, C., Kono, K., Maruyama, K., and Takagishi, T., Synthesis of polyamidoamine dendrimers having poly(ethylene glycol) grafts and their ability to encapsulate anti-cancer drugs, *Bioconjugate Chemistry*, 11, 910–917, 2000.

92. Patri, A. K. et al., Synthesis and in vitro testing of J591 antibody-dendrimer conjugates for targeted prostate cancer therapy, *Bioconjugate Chemistry*, 15, 1174–1181, 2004.

93. Tripathi, P. K. et al., Dendrimer grafts for delivery of 5-flurouracil, *Pharmazie*, 57, 261–264, 2002.

94. Quintana, A. et al., Design and function of a dendrimer-based therapeutic nanodevice targeted to tumor cells through the folate receptor, *Pharmaceutical Research*, 19, 1310–1316, 2002.

95. Vinogradov, S. V., Batrakova, E. V., and Kabanov, A. V., Nanogels for oligonucleotide delivery to the brain, *Bioconjugate Chemistry*, 15, 50–60, 2004.

96. Jordan, A., Scholz, R., Wust, P., Fahling, H., and Felix, R., Magnetic fluid hyperthermia (MFH): Cancer treatment with AC magnetic field induced excitation of biocompatible superparamagnetic nanoparticles, *Journal of Magnetism and Magnetic Materials*, 201, 413–419, 1999.

97. Pankhurst, Q. A., Connolly, J., Jones, S. K., and Dobson, J., Applications of magnetic nanoparticles in biomedicine, *Journal of Physics D—Applied Physics*, 36, R167–R181, 2003.

98. Derfus, A. M. and Bhatia, S. N., Unpublished data.

99. Rabin, Y., Is intracellular hyperthermia superior to extracellular hyperthermia in the thermal sense? *International Journal of Hyperthermia*, 18, 194–202, 2002.

100. Hirsch, L. R. et al., Nanoshell-mediated near-infrared thermal therapy of tumors under magnetic resonance guidance, *Proceedings of the National Academy of Sciences of the United States of America*, 100, 13549–13554, 2003.

101. O'Neal, D. P., Hirsch, L. R., Halas, N. J., Payne, J. D., and West, J. L., Photo-thermal tumor ablation in mice using near infrared-absorbing nanoparticles, *Cancer Letters*, 209, 171–176, 2004.

102. Plank, C. et al., The magnetofection method: Using magnetic force to enhance gene delivery, *Biological Chemistry*, 384, 737–747, 2003.

103. Nishiyama, N. et al., Light-harvesting ionic dendrimer porphyrins as new photosensitizers for photodynamic therapy, *Bioconjugate Chemistry*, 14, 58–66, 2003.

104. Konan, Y. N., Gurny, R., and Allemann, E., State of the art in the delivery of photosensitizers for photodynamic therapy, *Journal of Photochemistry and Photobiology B—Biology*, 66, 89–106, 2002.

105. Kopelman, R. et al., Multifunctional nanoparticle platforms for in vivo MRI enhancement and photodynamic therapy of a rat brain cancer, *Journal of Magnetism and Magnetic Materials*, 293, 404–410, 2005.

106. Crowder, K. C. et al., Sonic activation of molecularly-targeted nanoparticles accelerates trans-membrane lipid delivery to cancer cells through contact-mediated mechanisms: Implications for enhanced local drug delivery, *Ultrasound in Medicine and Biology*, 31, 1693–1700, 2005.

107. Gao, Z. G., Fain, H. D., and Rapoport, N., Controlled and targeted tumor chemotherapy by micellar-encapsulated drug and ultrasound, *Journal of Controlled Release*, 102, 203–222, 2005.

108. Hainfeld, J. F., Slatkin, D. N., and Smilowitz, H. M., The use of gold nanoparticles to enhance radiotherapy in mice, *Physics in Medicine and Biology*, 49, N309–N315, 2004.

109. Haag, R., Supramolecular drug-delivery systems based on polymeric core-shell architectures, *Angewandte Chemie-International Edition*, 43, 278–282, 2004.

110. Murthy, N., Campbell, J., Fausto, N., Hoffman, A. S., and Stayton, P. S., Design and synthesis of pH-responsive polymeric carriers that target uptake and enhance the intracellular delivery of oligonucleotides, *Journal of Controlled Release*, 89, 365–374, 2003.

111. Forster, S., Abetz, V., and Muller, A. H. E., Polyelectrolyte block copolymer micelles, *Polyelectrolytes with Defined Molecular Architecture Ii*, 166, 173–210, 2004.

6 Neutron Capture Therapy of Cancer: Nanoparticles and High Molecular Weight Boron Delivery Agents

*Gong Wu, Rolf F. Barth, Weilian Yang, Robert J. Lee,
Werner Tjarks, Marina V. Backer, and Joseph M. Backer*

CONTENTS

6.1 Introduction 78
6.2 General Requirements for Boron Delivery Agents 78
6.3 Low-Molecular-Weight Delivery Agents 79
6.4 High-Molecular-Weight Boron Delivery Agents 79
6.5 Dendrimer-Related Delivery Agents 80
    6.5.1 Properties of Dendrimers 80
    6.5.2 Boronated Dendrimers Linked to Monoclonal Antibodies 80
        6.5.2.1 Boron Clusters Directly Linked to mAb 80
        6.5.2.2 Attachment of Boronated Dendrimers to mAb 81
    6.5.3 Boronated Dendrimers Delivered by Receptor Ligands 81
        6.5.3.1 Epidermal Growth Factors (EGF) 81
        6.5.3.2 Folate Receptor-Targeting Agents 83
        6.5.3.3 Vascular Endothelial Growth Factor (VEGF) 83
    6.5.4 Other Boronated Dendrimers 85
6.6 Liposomes as Boron Delivery Agents 86
    6.6.1 Overview of Liposomes 86
    6.6.2 Boron Delivery by Nontargeted Liposomes 86
        6.6.2.1 Liposomal Encapsulation of Sodium Borocaptate and
                Boronophenylalanine 86
        6.6.2.2 Liposomal Encapsulation of Other Boranes and Carboranes 87
    6.6.3 Boron Delivery by Targeted Liposomes 89
        6.6.3.1 Immunoliposomes 89
        6.6.3.2 Folate Receptor-Targeted Liposomes 90
        6.6.3.3 EGFR-Targeted Liposomes 91
6.7 Boron Delivery by Dextrans 92

Experimental studies described in this report were supported by N.I.H. grants 1 R01 CA098945 (R.F. Barth), 1 R01 CA79758 (W. Tjarks), and Department of Energy grants DE-FG02-98ER62595 (R.F. Barth) and DE-FG-2-02FR83520 (J.M. Backer).

6.8   Other Macromolecules Used for Delivering Boron Compounds 93
6.9   Delivery of Boron-Containing Macromolecules to Brain Tumors 93
       6.9.1   General Considerations 93
       6.9.2   Drug-Transport Vectors 93
       6.9.3   Direct Intracerebral Delivery 94
       6.9.4   Convection-Enhanced Delivery 94
6.10   Clinical Considerations and Conclusions 95
6.11   Summary 96
Acknowledgments 96
References 96

6.1 INTRODUCTION

After decades of intensive research, high-grade gliomas, and specifically glioblastoma multiforme (GBM), are still extremely resistant to all current forms of therapy including surgery, chemotherapy, radiotherapy, immunotherapy and gene therapy.[1–5] The five-year survival rate of patients diagnosed with GBM in the United States is less than a few percent[6,7] despite aggressive treatment using combinations of therapeutic modalities. This is due to the infiltration of malignant cells beyond the margins of resection and their spread into both gray and white matter by the time of surgical resection.[8,9] High-grade gliomas are histologically complex and heterogeneous in their cellular composition. Recent molecular genetic studies of gliomas have shown how complex the development of these tumors is.[10] Glioma cells and their neoplastic precursors have biologic properties that allow them to evade a tumor associated host immune response,[11] and biochemical properties that allow them to invade the unique extracellular environment of the brain.[12,13] Consequently, high-grade supratentorial gliomas must be regarded as a whole-brain disease.[14] The inability of chemo- and radiotherapy to cure patients with high-grade gliomas is due to their failure to eradicate micro-invasive tumor cells within the brain. To successfully treat these tumors, strategies must be developed that can selectively target malignant cells with little or no effect on normal cells and tissues adjacent to the tumor. Boron neutron capture therapy (BNCT) is based on the nuclear capture and fission reactions that occur when non radioactive boron-10 is irradiated with low-energy thermal neutrons to yield high-linear-energy-transfer (LET) alpha particles (^{4}He) and recoiling lithium-7 (^{7}Li) nuclei. For BNCT to be successful, a sufficient number of ^{10}B atoms (approximately 10^9 atoms/cell) must be delivered selectively to the tumor and enough thermal neutrons must be absorbed by them to sustain a lethal ^{10}B(n,α) ^{7}Li-capture reaction. The destructive effects of these high-energy particles are limited to boron-containing cells. BNCT primarily has been used to treat high-grade gliomas,[15,16] and either cutaneous primaries[17] or cerebral metastases of melanoma.[18] More recently, it also has been used to treat patients with head-and-neck[19,20] and metastatic liver cancer.[21,22] BNCT is a biologically rather than physically targeted type of radiation treatment. If sufficient amounts of ^{10}B and thermal neutrons can be delivered to the target volume, the potential exists to destroy tumor cells dispersed in the normal tissue parenchyma. Readers interested in more in depth coverage of other topics related to BNCT are referred to several recent reviews[23–25] and monographs.[15,24] This review will focus on boron-containing macromolecules and nanovehicles as boron delivery agents.

6.2 GENERAL REQUIREMENTS FOR BORON DELIVERY AGENTS

A successful boron delivery agent should have (1) no or minimal systemic toxicity with rapid clearance from blood and normal tissues, (2) high tumor (approximately 20 µg ^{10}B/g) and low normal tissue uptake, (3) high tumor to brain (T:Br) and tumor to blood (T:Bl) concentration ratios (greater than 3–4:1), and (4) persistence in the tumor for a sufficient period of time to carry out BNCT. At this time, no single boron delivery agent fulfills all of these criteria. However, as a result of new synthetic techniques and increased knowledge of the biological and biochemical requirements for an effective agent, multiple new boron agents have emerged and these are described in a special issue of Anti-Cancer Agents in Medical Chemistry (6, 2, 2006). The major challenge in their development has been the requirement for specific tumor targeting to achieve boron concentrations sufficient to deliver therapeutic doses of radiation to the tumor with minimal normal-tissue toxicity. The selective destruction of GBM cells in the presence of normal cells represents an even greater challenge compared to malignancies at other anatomic sites.

6.3 LOW-MOLECULAR-WEIGHT DELIVERY AGENTS

In the 1950s and early 1960s clinical trials of BNCT were carried out using boric acid and some of its derivatives as delivery agents. These simple chemical compounds had poor tumor retention, attained low T:Br ratios and were nonselective.[26,27] Among the hundreds of low-molecular-weight (LMW) boron-containing compounds that were synthesized, two appeared to be promising. One, based on arylboronic acids,[28] was (L)-4-dihydroxy-borylphenylalanine, referred to as boronophenylalanine or BPA (Figure 6.1, 1). The second, a polyhedral borane anion, was sodium mercaptoundecahydro-closo-dodecaborate,[29] more commonly known as sodium borocaptate or BSH (2). These two compounds persisted longer in animal tumors compared with related molecules, attained T:Br and T:Bl boron ratios greater than 1 and had low toxicity. [10]B-enriched BPA, complexed with fructose to improve its water solubility, and BSH, have been used clinically for BNCT of brain, as well as extracranial tumors. Although their selective accumulation in tumors is not ideal, the safety of these two drugs following intravenous (i.v.) administration has been well established.[30,31]

6.4 HIGH-MOLECULAR-WEIGHT BORON DELIVERY AGENTS

High-molecular-weight (HMW) delivery agents usually contain a stable boron group or cluster linked via a hydrolytically stable bond to a tumor-targeting moiety, such as monoclonal antibodies (mAbs) or LMW receptor-targeting ligands. Examples of these include epidermal growth factor

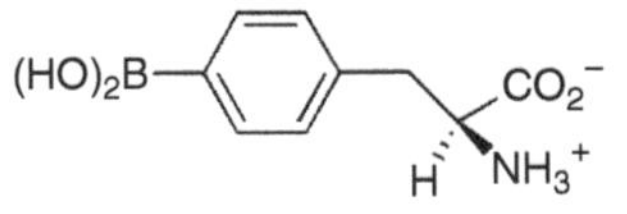

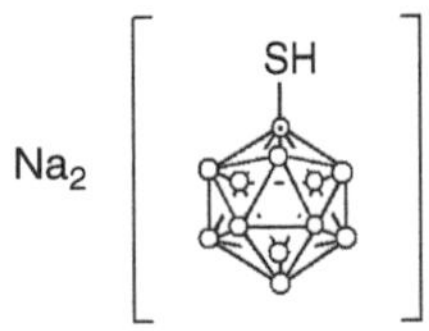

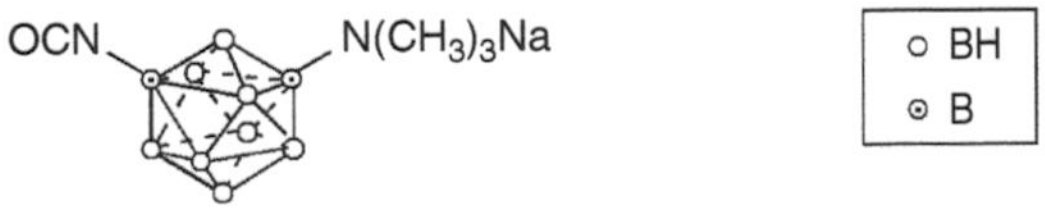

FIGURE 6.1 Structures of two compounds used clinically for boron neutron capture therapy (BNCT), 4-dihydroxyborylphenylalanine or boronophenylalanine (BPA, **1**), and disodium undecahydro-*closo*-dodecaborate or sodium borocaptate (BSH, **2**) and the isocyanato polyhedral borane (**3**), which has been used to heavily boronate dendrimers.

(EGF) and the mAb cetuximab (IMC-C225) to target the EGF receptor (EGFR) or its mutant isoform EGFRvIII, which are overexpressed in a variety of malignant tumors including gliomas, and squamous-cell carcinomas of the head and neck.[32] Agents that are to be administered systemically should be water soluble (WS), but lipophilicity is important to cross the blood–brain barrier (BBB) and diffuse within the brain and the tumor. There should be a favorable differential in boron concentrations between tumor and normal brain, thereby enhancing their tumor specificity. Their amphiphilic character is not as crucial for LMW agents that target-specific biological transport systems and/or are incorporated into nanovehicles such as liposomes. Molecular weight also is an important factor because it determines the rate of diffusion both within the brain and the tumor. Detailed reviews of the state-of-the-art of compound development for BNCT have been published.[33,34] In this review, we will focus on boron containing macromolecules and liposomes as delivery agents for BNCT, and how they can be most effectively administered.

6.5 DENDRIMER-RELATED DELIVERY AGENTS

6.5.1 PROPERTIES OF DENDRIMERS

Dendrimers are synthetic polymers with a well-defined globular structure. As shown in Figure 6.2, they are composed of a core molecule, repeat units that have three or more functionalities, and reactive surface groups.[35,36] Two techniques have been used to synthesize these macromolecules: divergent growth outwards from the core,[37] or convergent growth from the terminal groups inwards towards the core.[36,38] Regular and repeated branching at each monomer group gives rise to a symmetric structure and pattern to the entire globular dendrimers. Dendrimers are an attractive platform for macromolecular imaging and gene delivery because of their low cytotoxicity and their multiple types of reactive terminal groups.[36,39–44]

6.5.2 BORONATED DENDRIMERS LINKED TO MONOCLONAL ANTIBODIES

6.5.2.1 Boron Clusters Directly Linked to mAb

Monoclonal antibodies (mAb) have been attractive targeting agents for delivering radionuclides,[45] drugs,[46–50] toxins,[51] and boron to tumors.[52–55] Prior to the introduction of dendrimers as boron carriers, boron compounds were directly attached to mAbs.[53,54] It has been calculated that

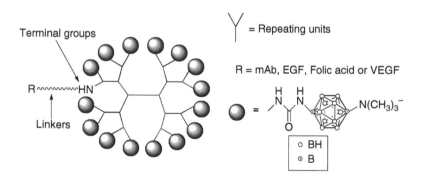

FIGURE 6.2 Structure of a boronated PAMAM dendrimer that has been linked to targeting moieties. PAMAM dendrimers consist of a core, repeating polyamido amino units, and reactive terminal groups. Each successively higher generation of PAMAM dendrimer has a geometrically incremental number of terminal groups. Boronated dendrimers have been prepared by reacting them with water-soluble isocyanato polyhedral boranes and subsequently attaching them to targeting moieties by means of heterobifunctional linkers.

approximately 10^9 ^{10}B atoms per cell (approximately 20 µg/g tumor) must be delivered to kill tumor cells.[55,56] Based on the assumption of 10^6 antigenic receptor sites per cell, approximately 50–100 boron cage structures of carboranes, or polyhedral borane anions and their derivatives must be linked to each mAb molecule to deliver the required amount of boron for NCT. The attachment of such a large number of boron cages to a mAb may result in precipitation of the bioconjugate or a loss of its immunological activity. Solubility can be improved by inserting a water-soluble gluco-namide group into the protein-binding boron cage compounds, thereby enhancing their water solubility.[57] This modification makes it possible to incorporate up to 1100 boron atoms into a human gamma globulin (HGG) molecule without any precipitation. Other approaches to enhance solubility include the use of negatively charged carboranes[58] or polyhedral borane anions,[59] as well as the insertion of carbohydrate groups.[60,61] A major limitation of using an agent containing a single boron cage is that a large number of sites must be modified to deliver 10^3 boron atoms per molecule of antibody and this can reduce its immunoreactivity activity. Alam et al. showed that attachment of an average of 1300 boron atoms to mAb 17-1A, which is directed against human colorectal carcinoma cells, resulted in a 90% loss of its immunoreactivity.[62]

6.5.2.2 Attachment of Boronated Dendrimers to mAb

Dendrimers are one of the most attractive polymers that have been used as boron carriers due to their well-defined structure and large number of reactive terminal groups. Depending on the antigen site density, approximately 1000 boron atoms need to be attached per molecule of dendrimer and subsequently linked to the mAb. In our first study, second- and fourth-generation polyamido amino (PAMAM or "starburst") dendrimers, which have 12 and 48 reactive terminal amino groups, respectively, were reacted with the water-soluble isocyanato polyhedral borane [Na(CH$_3$)$_3$NB$_{10}$H$_8$NCO] (3).[63,64] The boronated dendrimer then was linked to the mAb IB16-6, which is directed against the murine B16 melanoma, by means of two heterobifunctional linkers, m-maleimido-benzoyl-N-hydroxysulfosuccinimide ester (sulfo-MBS) and N-succinimidyl 3-(2-pyridyldithio) propionate (SPDP).[63,65] However, following i.v. administration, large amounts of the bioconjugate accumulated in the liver, and spleen and it was concluded that random conjugation of boronated dendrimers to a mAb could alter its binding affinity and biodistribution. To minimize the loss of mAb reactivity, a fifth-generation PAMAM dendrimer was boronated with the same polyhedral borane anion, and more recently it was site-specifically linked to the anti-EGFR mAb cutuximab (or IMC-C225) or the EGFRvIII-specific mAb L8A4 (Figure 6.3). Cetuximab was linked via glycosidic moieties in the Fc region by means of two heterobifunctional reagents, SPDP and N-(k-maleimidoundecanoic acid) hydrazide (KMUH).[66,67] The resulting bioconjugate, designated C225-G5-B$_{1100}$, contained approximately 1100 boron atoms per molecule of cetuximab and retained its aqueous solubility in 10% DMSO and its in vitro and in vivo immunoreactivity. As determined by a competitive binding assay, there was a less than 1 log unit decrease in affinity for EGFR (+) glioma cell line F98$_{EGFR}$, compared to that of unmodified cetuximab.[66] In vivo bio-distribution studies, carried out 24 h after intratumoral (i.t) administration of the bioconjugate, demonstrated that 92.3 µg/g of boron was retained in rats bearing F98$_{EGFR}$ gliomas, compared to 36.5 µg/g in EGFR (−) F98 parental tumors and 6.7 µg/g in normal brain.[67]

6.5.3 BORONATED DENDRIMERS DELIVERED BY RECEPTOR LIGANDS

6.5.3.1 Epidermal Growth Factors (EGF)

Due to its increased expression in a variety of tumors, including high-grade gliomas, and its low or undetectable expression in normal brain, EGFR is an attractive target for cancer therapy.[68–70] As described above, targeting of EGFR has been carried out using either mAbs or alternatively, as described in this section, EGF, which is a single-chain, 53-mer heat and acid stable polypeptide.

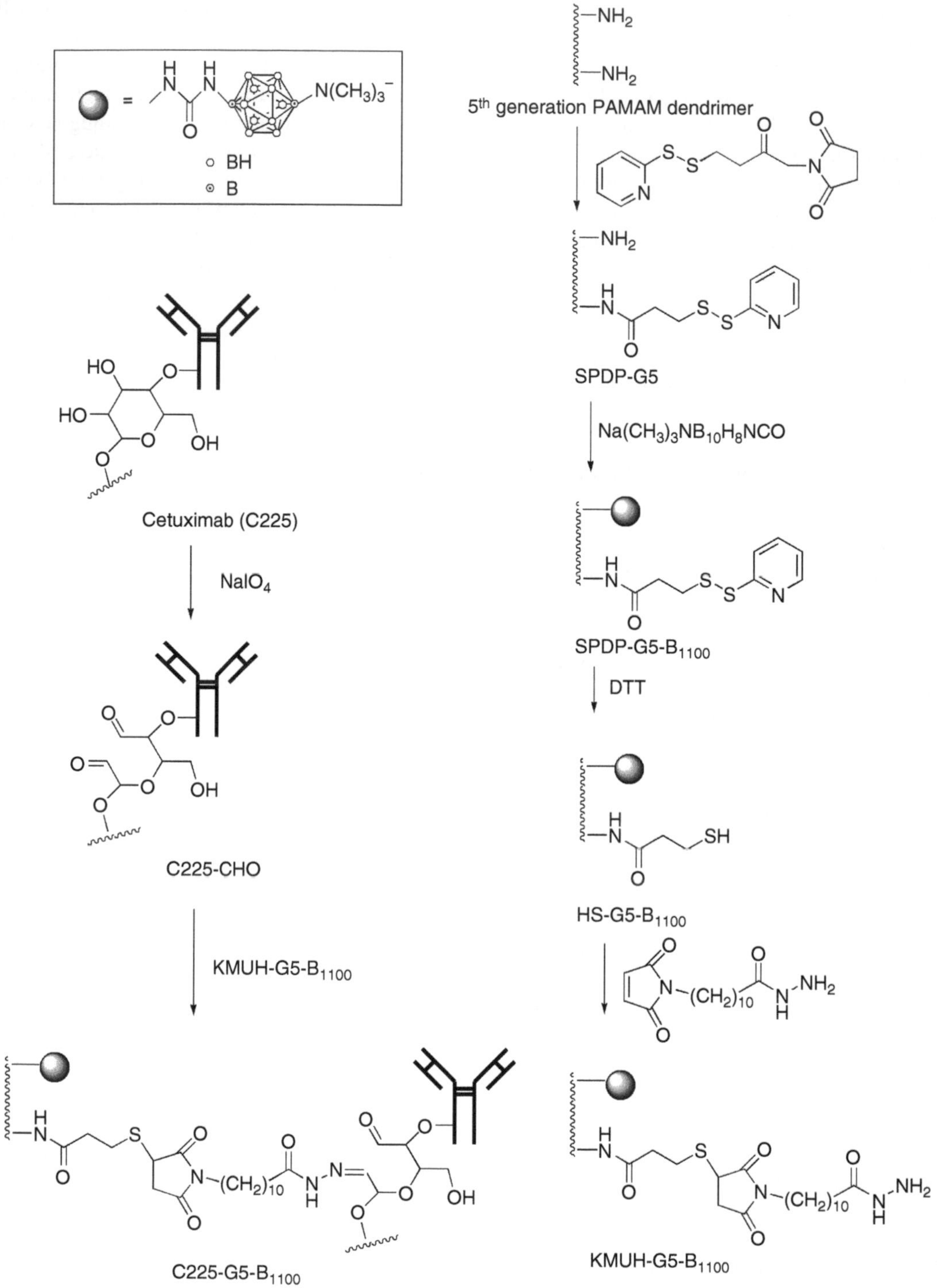

FIGURE 6.3 Conjugation scheme for linkage of a boron-containing dendrimer to the monoclonal antibody, C225 (cetuximab), which is directed against EGFR. (From Wu, G. et al., *Bioconjug. Chem.*, 15, 185, 2004; Barth, R. F. et al., *Appl. Radiat. Isot.* 61, 899, 2004. With permission.)

It binds to a transmembrane glycoprotein with tyrosine kinase activity, which triggers dimerization and internalization.[71,72] Because the EGF boron bioconjugates have a much smaller MW than mAb conjugates, they should be capable of more rapid and effective tumor targeting than has been observed with mAbs.[67,73]

The procedure used to conjugate EGF to a boronated dendrimer was slightly different from that used to boronate mAbs. A fourth-generation PAMAM dendrimer was reacted with the isocyanato polyhedral borane $Na(CH_3)_3NB_{10}H_8NCO$. Next, reactive thiol groups were introduced into the boronated dendrimer using SPDP, and EGF was derivatized with sulfo-MBS. The reaction of thiol groups of the derivatized, boronated starburst dendrimer (BSD) with maleimide groups produced a stable BSD–EGF bioconjugate, which contained approximately 960 atoms of boron per molecule of EGF.[74] The BSD–EGF initially was bound to the cell surface membrane and then was endocytosed, which resulted in accumulation of boron in lysosomes.[74] Subsequently, in vitro and in vivo studies were carried out to evaluate the potential efficacy of the bioconjugate as a boron delivery agent for BNCT.[73] As will be described in more detail later on in this review, therapy studies demonstrated that $F98_{EGFR}$-glioma-bearing rats that received either boronated EGF or mAb by either direct i.t. injection or convection enhanced delivery into the brain had a longer mean survival time (MST) than animals bearing F98 parental tumors following BNCT.[75–77]

6.5.3.2 Folate Receptor-Targeting Agents

Folate receptor (FR) is overexpressed on a variety of human cancers, including those originating in ovary, lung, breast, endometrium and kidney.[78–80] Folic acid (FA) is a vitamin that is transported into cells via FR mediated endocytosis. It has been well documented that the attachment of FA via its γ-carboxylic function to other molecules does not alter its endocytosis by FR-expressing cells.[81] FR targeting has been used successfully to deliver protein toxins, chemotherapeutic, radioimaging, therapeutic and MRI contrast agents,[82] liposomes,[83] gene transfer vectors,[84] antisense oligonucleotides,[85] ribozymes, and immunotherapeutic agents to FR-positive cancers.[86] To deliver boron compounds, FA was conjugated to heavily boronated third generation PAMAM dendrimers containing polyethylene glycol (PEG).[87] PEG was introduced into the bioconjugate to reduce its uptake by the reticuloendothelial system (RES), and more specifically, the liver and spleen. It was observed that folate linked to third generation PAMAM dendrimers containing 12–15 decaborate clusters and 1–1.5 PEG_{2000} units had the lowest hepatic uptake in C57Bl/6 mice (7.2–7.7% injected dose [I.D.]/g liver). In vitro studies using FR (+) KB cells demonstrated receptor-dependent uptake of the bioconjugate. Biodistribution studies with this conjugate, carried out in C57Bl/6 mice bearing subcutaneous (s.c.) implants of the FR (+) murine sarcoma 24JK-FBP, demonstrated selective tumor uptake (6.0% I.D./g tumor), but there was high hepatic (38.8% I.D./g) and renal (62.8% I.D./g) uptake.[87]

6.5.3.3 Vascular Endothelial Growth Factor (VEGF)

There is preclinical and clinical evidence indicating that angiogenesis plays a major role in the growth and dissemination of malignant tumors.[88,89] Inhibition of angiogenesis has yielded promising results in a number of experimental animal tumor models.[90,91] The most important molecular targets have been vascular endothelial growth factor (VEGF) and its receptor (VEGFR).[92,93] We recently have constructed a human $VEGF_{121}$ isoform fused to a novel 15-aa cysteine-containing tag designed for site-specific conjugation of therapeutic and diagnostic agents[94] (Figure 6.4). A boronated fifth-generation PAMAM dendrimer (BD), tagged with a near-infrared (IR) fluorescent dye Cy5, was conjugated using the heterobifunctional reagent sulfo-LC–SPDP to BD/Cy5 via reactive SH-groups generated in the VEGF fusion protein by mild dithiothreitol (DTT) treatment. The bioconjugate, designated VEGF–BD/Cy5, contained 1050–1100 boron atoms per dimeric VEGF molecule. VEGF–BD/Cy5 retained the in vitro

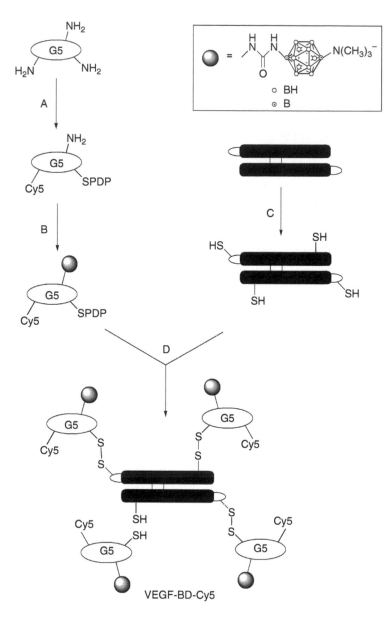

FIGURE 6.4 Preparation of a vascular endothelial growth factor (VEGF) receptor-targeting nanovehicle (VEGF–BD/Cy5) (A). A fifth-generation (G5) PAMAM dendrimer initially is reacted with SPDP and the near-IR dye, Cy5-NHS, (A) dissolved in dimethylformamide/methanol mixture for 1 h. Following this, it is reacted with the polyisocynato polyhedral borane, $Na(CH_3)_3–NB_{10}H_8NCO$, in a carbonate buffer (pH 9.0)/9% acetone mixture for 24 h (B). Then a 1.5-fold molar excess of DTT is added to the VEGF monomer in a solution containing 20 mmol/L NaOAc (pH 6.5), 0.5 mol/L urea and 0.5 mol/L NDSB-221 and incubated at 4°C for 16 h (C). Finally, the boronated dendrimer is incubated with the targeting vehicle for 15 min to produce the VEGF–BD/Cy5 (D). (From Backer, M. V. et al., *Mol. Cancer Ther.*, 4, 1423, 2005.)

functional activity of $VEGF_{121}$, which was similar to that of native VEGF. Tagging the bioconjugate with Cy5 dye permitted near-IR fluorescence imaging of its in vitro and in vivo uptake. In vitro uptake was determined by incubating VEGF–BD/Cy5 with VEGFR-2 overexpressing PAE/KDR cells and in vivo uptake was evaluated in mice bearing s.c. tumor implants. In vitro, the bioconjugate localized in the perinuclear region and in vivo it primarily localized in regions of active angiogenesis. Furthermore, depletion of VEGFR-2 overexpressing cells from the tumor vasculature with VEGF-toxin fusion protein significantly decreased bioconjugate uptake, indicating that these cells were the primary targets of VEGF–BD/Cy5. These studies demonstrated that VEGF–BD/Cy5 potentially could be used as a diagnostic agent.[94] Further studies are planned to evaluate its therapeutic efficacy using the F98 rat glioma model, which we have used extensively to evaluate both LMW and HMW boron delivery agents.[95]

6.5.4 Other Boronated Dendrimers

An alternative method to deliver boron compounds by means of dendrimers is to incorporate carborane cages within the dendrimer (Figure 6.5). Matthew et al. have reported that 4, 8, or 16 carboranes could be inserted into an aliphatic polyester dendrimer by means of a highly effective

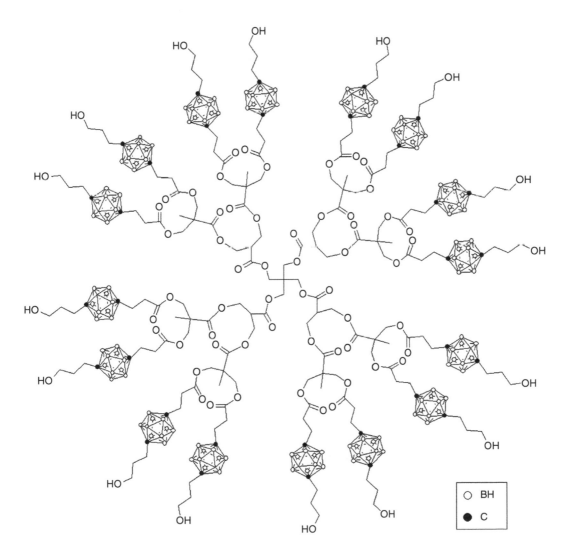

FIGURE 6.5 Structure of a barborane-containing aliphatic polyester dendrimer. Carborane cages were incorporated into the interior of the dendrimer structure and the peripheral hydrophilic groups improved water solubility and were available for modification.

synthon, a bifunctional carborane derivative with an acid group and a benzyl ether protected alcohol.[96] The procedure employed a divergent synthesis with high yield. The resulting polyhydroxylated dendrimer was WS with a minimum ratio of eight hydroxyl groups per carborane cage. Carborane containing dendrimers potentially could be used as boron delivery agents for BNCT because it is possible to control the number of carborane moieties and overall solubility.

6.6 LIPOSOMES AS BORON DELIVERY AGENTS

6.6.1 OVERVIEW OF LIPOSOMES

Liposomes are biodegradable, nontoxic vesicles that have been used to deliver both hydrophilic and hydrophobic agents (Figure 6.6).[97] Both classical and PEGylated ("stealth") liposomes can increase the amounts of anti-cancer drugs that can be delivered to solid tumors by passive targeting. Rapidly growing solid tumors have increased permeability to nanoparticles due to increased capillary pore size. These can range from 100 to 800 nm. In comparison, endothelial pore size of normal tissues, which are impermeable to liposomes, can range in size from 60 to 80 nm. In addition, tumors lack efficient lymphatic drainage, and consequently, clearance of extravasated liposomes is slow.[98] Modification of the liposomal surface by PEGylation or attachment of antibodies or receptor ligands, will improve their selective targeting and increase their circulation time.[98]

6.6.2 BORON DELIVERY BY NONTARGETED LIPOSOMES

6.6.2.1 Liposomal Encapsulation of Sodium Borocaptate and Boronophenylalanine

Liposomes have been extensively evaluated as nanovehicles for the delivery of boron compounds for NCT.[99,100] In vitro and in vivo studies have demonstrated that they can effectively and selectively deliver large quantities of boron to tumors and that the compounds delivered by liposomes have a longer tumor retention time. BPA is an amino acid analogue that is preferentially taken up by cells with increased metabolic activity, such as tumor cells of varying histopathologic types

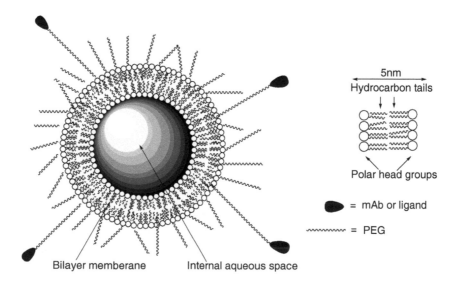

FIGURE 6.6 Schematic diagram of the structure of a liposome that has an aqueous core and a lipid bilayer membrane. The latter is composed of polar head groups with hydrocarbon tails. The liposomal surface can be modified by PEGylation to prolong its circulation time and linked to either a mAb or a ligand for targeting.

including melanomas,[31,101] gliomas[102], and squamous cell carcinomas.[103,104] Because of its low aqueous solubility, BPA has been used as a fructose complex, which has permitted it to be administered i.v. rather than orally.[105,106] Following i.v. administration of BPA, which had been incorporated into conventional liposomes, there was rapid elimination by the RES with very low blood boron concentrations at 3 h. In contrast, if BPA was incorporated into liposomes composed of distearoyl phosphoethanolamine (DSPE)–PEG, therapeutically effective tumor boron concentrations (greater than 20 µg/g) were seen at 3 h and at 6 h, indicating that PEG-liposomes had evaded the RES.[107] In addition, BPA has been incorporated into the lipid bilayer of liposomes, composed of positively charged lipid 1,2-dioleoyl-3-trimethylammonium-propane (DOTAP) and the zwitterionic lipid, 1,2-dioleoyl-sn-glycerol-3-phosphoethanolamine (DOPE).[108] Cationic liposomes have been widely used as carriers of biomolecules that specifically target the cell nucleus,[109] which would be advantageous for BNCT. Another clinically used drug, BSH, has been incorporated into liposomes composed of dipalmitoylphosphatidylcholine (DPPC)/Chol in a 1:1 molar ratio with and without PEG stabilization.[110] The average diameter of liposomes containing BSH was in the range of 100–110 nm. Both types of liposomes resulted in a significant improvement in their circulation time compared to that of free BSH. At 24 h following i.v. injection of PEG-liposomes, 19% of the I.D. of boron was in the blood compared to 7% following formulation of BSH in conventional liposomes. The mean percent uptake by the liver and spleen was not significantly different for the two types of liposomes. However, the blood to RES ratios were higher for PEG-liposomes at all time points indicating that a higher fraction of the injected dose (I.D.) of BSH was still in the blood. Ji et al. have reported that there were no significant differences in the in vitro uptake by 9L gliosarcoma cells of free BSH versus a liposomal formulation after 16 h incubation. However, cellular persistence was increased at 12 and 24 h for BSH-loaded liposomes.[111] BSH also has been incorporated into transferrin (TF) conjugated PEG liposomes (TF–PEG liposomes),[112] which then were taken up by cells via TF receptor-mediated endocytosis. Intravenous administration of this formulation increased boron retention at the tumor site compared with PEG liposomes, bare liposomes or free BSH and suppressed tumor growth following BNCT. These results suggest that TF targeted liposomes might be useful as intracellular targeting vehicles.

6.6.2.2 Liposomal Encapsulation of Other Boranes and Carboranes

Polyhedral boranes[34] and carboranes[113,114] are another class of boron compounds that have been used for NCT. They contain multiple boron atoms per molecule, are resistant to metabolic degradation, and are lipophilic, thereby permitting easier penetration of the tumor cell membrane.[113] In addition to BSH, carboranylpropylamine (CPA, Figure 6.7, **4**)[115] has been incorporated into conventional and PEGylated liposomes by active loading, using a transmembrane pH gradient.[115,116] Although as many as 13,000 molecules of CPA were loaded into liposomes having a mean average diameter of 100 nm, there was no in vitro toxicity to both the glioblastoma cell line SK-MG-1 and normal human peripheral blood lymphocytes. Borane anions, such as $B_{10}H_{10}^{2-}$, $B_{12}H_{11}SH^{2-}$, $B_{20}H_{17}OH^{4-}$, and $B_{20}H_{19}^{3-}$, and the normal form and photoisomer of $B_{20}H_{18}^{2-}$, also have been encapsulated into small unilamellar vesicles with mean diameters of 70 nm or less.[117–120] These were composed of the pure synthetic phospholipids, distearoyl phosphatidylcholine, and cholesterol.[117] Although encapsulation efficiencies were only 2–3%, following i.v. injection, these liposomes were selectively delivered to the tumors in mice bearing the EMT6 mammary carcinoma and had attained boron concentrations of greater than 15 µg boron/g, with a T:Bl ratio of greater than 3. Two isomers of $B_{20}H_{18}^{2-}$ attained boron concentrations of 13.6 and 13.9 µg/g with T:Bl ratios of 3.3 and 12, respectively, at 48 h following administration. High boron retention and prolongation of their circulation time was observed due to the interaction with intracellular components after it had been released from liposomes within tumor cells.[117]

To examine the effect of charge and substitution on the retention of boranes, two isomers $[B_{20}H_{17}NH_3]^{3-}$ and $[1-(2'-B_{10}H_9)-2-NH_3B_{10}H_8]^{3-}$ (**5**) were prepared from the polyhedral

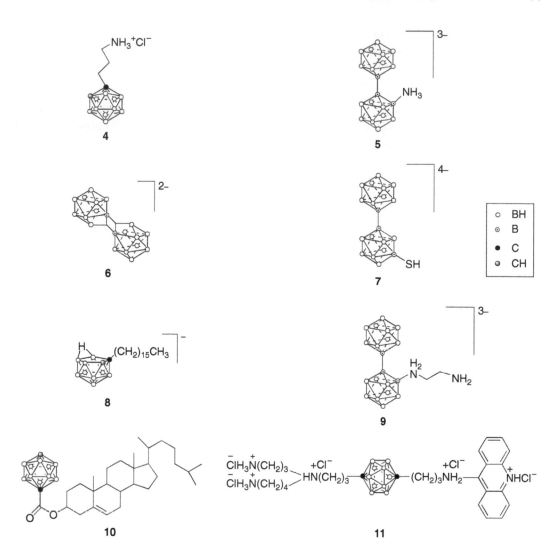

FIGURE 6.7 Structures of hydrophilic and lipophilic boron-containing compounds that have been incorporated into liposomes. Carboranylpropylamine (CPA, **4**), $[1\text{-}(2'\text{-}B_{10}H_9)\text{-}2\text{-}NH_3B_{10}H_8]^{3-}$ (**5**), $[n\text{-}B_{20}H_{18}]^{2-}$ (**6**), $[B_{20}H_{17}SH]^{4-}$ (**7**), $Na_3[a2\text{-}B_{20}H_{17}NH_2CH_2CH_2NH_2]$ (**9**) and boronated water soluble acridine (WSA, **11**) were encapsulated into the aqueous core. $K[nido\text{-}7\text{-}CH_3(CH_2)_{15}\text{-}7,8\text{-}C_2B_9H_{11}]$ (**8**) and cholesteryl 1,12-dicarba-*closo*-dodecaboranel-carboxylate (**10**) were incorporated into the lipid bilayer of liposomes.

borane anion $[n\text{-}B_{20}H_{18}]^{2-}$ (**6**).[118,119] The sodium salts of these two isomers had been encapsulated within small unilamellar liposomes, composed of distearoyl phosphatidylcholine/cholesterol at a 1:1 ratio. Both isomers of $[B_{20}H_{17}NH_3]^{3-}$ had excellent tumor uptake and selectivity in EMT 6 tumor bearing mice, even at very low I.D.s, and this resulted in peak tumor boron concentrations of 30–40 μg B/g and a T:Bl ratio of approximately 5. Due to low boron retention of liposomal $Na_3[B_{20}H_{19}]$ and $Na_4[e^2\text{-}B_{20}H_{17}OH]$ and rapid clearance of liposomal $[2\text{-}NH_3B_{10}H_9]^-$, the enhanced retention of liposomal $Na_3[ae\text{-}B_{20}H_{17}NH_3]$ was not due to the anionic charge or substitution in the borane cage. Rather, it could be attributed to their facile intracellular oxidation to an extremely reactive NH_3-substituted $[n\text{-}B_{20}H_{18}]^{2-}$ electrophilic anion, $[B_{20}H_{17}NH_3]^-$. Another anion $[ae\text{-}B_{20}H_{17}NH_3]^{3-}$ also was encapsulated into liposomes prepared with 5% PEG-2000-distearoyl phosphatidylethanolamine as a constituent of the membrane. These liposomes had longer in vivo circulation times, which resulted in continued accumulation of boron in the tumor over the entire 48 h time period, and reached a maximum concentration of 47 µg B/g tumor.

$[B_{20}H_{17}SH]^{4-}$ (**7**), a thiol derivative of $[B_{20}H_{18}]^{4-}$, possesses a reactive thiol substituent and this can be oxidized into the more reactive $[B_{20}H_{17}SH]^{2-}$ anion. Both of these were considered to be essential for high tumor boron retention[120] and they have been encapsulated into small, unilamellar liposomes. Biodistribution was determined after i.v. injection into BALB/c mice bearing EMT6 tumors. At low I.D.s, tumor boron concentrations increased throughout the duration of the experiment, resulting in a maximum concentration of 47 µg B/g tumor at 48 h, which corresponded to 22.2% I.D./g and a T:Bl ratio of 7.7. This was the most promising of the polyhedral borane anions that had been investigated for liposomal delivery. Although they were able to deliver adequate amounts of boron to tumor cells, their application to BNCT has been limited due to their low incorporation efficiency (approximately 3%).

Lipophilic boron compounds incorporated into the lipid bilayer would be an alternative approach. Small unilamellar vesicles composed of 3:3:1 ratio of distearoylphosphatidylcholine, cholesterol and $K[nido\text{-}7\text{-}CH_3(CH_2)_{15}\text{-}7,8\text{-}C_2B_9H_{11}]$ (**8**) in the lipid bilayer and $Na_3[a2\text{-}B_{20}H_{17}$ $NH_2CH_2CH_2NH_2]$ (**9**) in the aqueous core were produced as a delivery agents for NCT mediated synovectomy.[121] Biodistribution studies were carried out in Louvain rats that had a collagen-induced arthritis. The maximum synovial boron concentration was 29 µg/g tissue at 30 h and this had only decreased to 22 µg/g at 96 h following i.v. administration. The prolonged retention by synovium provided sufficient time for extensive clearance of boron from other tissues so that at 96 h the synovium to blood (Syn:Bl) ratio was 3.0. To accelerate blood clearance, serum stability of the liposomes was lowered by increasing the proportion of $K[nido\text{-}7\text{-}CH_3(CH_2)_{15}\text{-}7,8\text{-}C_2B_9H_{11}]$ embedded in the lipid bilayer. Liposomes were formulated with a 3:3:2 ratio of DSPC to Ch to $K[nido\text{-}7\text{-}CH_3(CH_2)_{15}\text{-}7,8\text{-}C_2B_9H_{11}]$ in the lipid bilayer and $Na_3[a2\text{-}B_{20}H_{17}NH_2CH_2CH_2NH_2]$ was encapsulated in the aqueous core. The boron concentration in the synovium reached a maximum of 26 µg/g at 48 h with a Syn:Bl ratio of 2, following which it slowly decreased to 14 µg/g at 96 h at which time the Syn:Bl ratio was 7.5.[121]

Another method to deliver hydrophilic boron containing compounds would be to incorporate them into cholesterol to target tumor cells expressing amplified low density lipoprotein (LDL) receptors.[122–124] Glioma cells, which absorb more cholesterol, have been reported to take up more LDL than the corresponding normal tissue cells.[125–127] The cellular uptake of liposomal cholesteryl 1,12-dicarba-closo-dodecaboranel-carboxylate (**10**) by two fast growing human glioma cell lines, SF-763 and SF-767, was mediated via the LDL receptor and was much higher than that of human neurons. The cellular boron concentration was approximately 10–11 times greater than that required for BNCT.[128]

6.6.3 Boron Delivery by Targeted Liposomes

To improve the specificity of liposomally encapsulated drugs and to increase the amount of boron delivered, targeting moieties have been attached to the surface of liposomes. These could be any molecules that selectively recognized and bound to target antigens or receptors that were over expressed on neoplastic cells or tumor-associated neovasculature. These have included either intact mAb molecules or fragments, LMW, naturally occurring or synthetic ligands such as peptides or receptor-binding-ligands such as EGF. To date, liposomes linked to mAbs or their fragments,[129] EGF,[130] folate,[131] and transferin,[112] have been the most extensively studied as targeting moieties (Figure 6.8).

6.6.3.1 Immunoliposomes

The murine anticarcinoembryonic antigen (CEA) mAb 2C-8 has been conjugated to large multi-lamellar liposomes containing ^{10}B compounds.[132,133] The maximum number of ^{10}B atoms attached

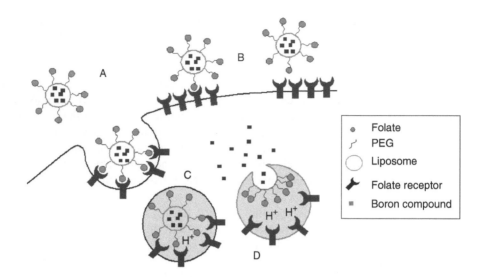

FIGURE 6.8 Proposed mechanism of intracellular boron delivery based on receptor-mediated tumor cell targeting. Liposomes loaded with a boron compound (A) are conjugated to a targeting ligand (e.g., folate, transferrin (TF), or anti-EGFR antibody). These bind to receptors on the cell surface (B), following which they are internalized by receptor mediated endocytosis (C) into the acidified endosomal/lysosomal compartment. The boron compound then is released (D) into the cytosol by liposomal degradation, endosome/lysosome disruption or liposome–endosome/lysome membrane fusion.

per molecule of mAb was approximately 1.2×10^4. These immunoliposomes bound selectively to the human pancreatic carcinoma cell line, AsPC-1 that overexpressed CEA. Incubating the immunoliposomes with either MRKnu/nu-1 or AsPC-1 tumor cells, suppressed in vitro tumor cell growth following thermal neutron irradiation.[134] This was dependent upon the liposomal concentration of the ^{10}B-compound and on the number of molecules of mAb conjugated to the liposomes. Immunoliposomes containing either $(Et_4N)_2B_{10}H_{10}$ and linked to the mAb MGb 2, directed against human gastric cancer[135,136] or water-soluble boronated acridine (WSA, **11**) linked to trastuzumab, directed against HER-2, have been prepared and evaluated in vitro.[129] There was specific binding and high uptake of these immunoliposomes, which delivered a sufficient amount of ^{10}B to produce a tumoricidal effect following thermal neutron irradiation.

6.6.3.2 Folate Receptor-Targeted Liposomes

A highly ionized boron compound, $Na_3B_{20}H_{17}NH_3$, was incorporated into liposomes by passive loading.[131,137,138] This showed high in vitro uptake by the FR expressing human cell line KB (American Type Culture Collection CCL 17), which originally was thought to be derived from a squamous cell carcinoma of the mouth, and subsequently was shown to be identical to HeLa cells, as determined by isoenyzyme markers, DNA fingerprinting, and karyotypic analysis. KB tumor-bearing mice that received either FR-targeted or nontargeted control liposomes had equivalent tumor boron values (approximately 85 µg/g), which attained a maximum at 24 h, while the T:Bl ratio reached a maximum at 72 h. Additional studies were carried out with the lipophilic boron compound, $K[nido\text{-}7\text{-}CH_3(CH_2)_{15}\text{-}7,8\text{-}C_2B_9H_{11}]$. This was incorporated into large unilamellar vesicles, approximately 200 nm in diameter, which were composed of egg PC/Chol/K[$nido\text{-}7\text{-}CH_3(CH_2)_{15}\text{-}7,8\text{-}C_2B_9H_{11}$] at a 2:2:1, mol/mol ratio, and an additional 0.5 mol% of folate–PEG–DSPE or PEG–DSPE for the FR-targeted or nontargeted liposomal formulations.[139] The boron uptake by FR-overexpressing KB cells, treated with these targeted liposomes, was approximately 10 times greater compared with those treated with control liposomes. In addition, BSH and

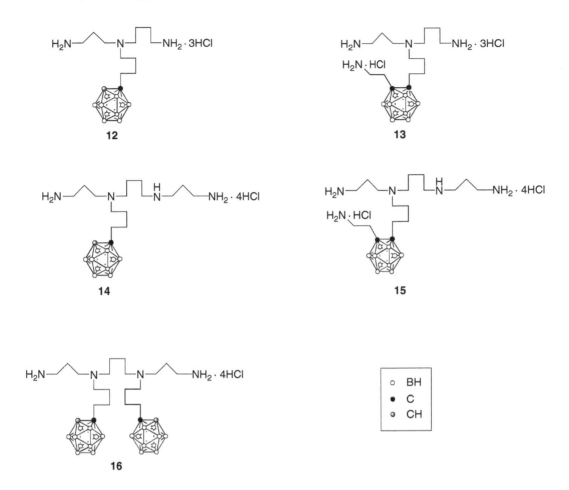

FIGURE 6.9 Structures of five weakly basic boronated polyamines encapsulated in FR-targeting liposomes. Two spermidine (**12**, **13**) and three spermine derivatives (**14–16**) that contain hydrophilic amine groups and lipophilic carboranyl cages had DNA-binding properties.

five weakly basic boronated polyamines were evaluated (Figure 6.9). Two of these were the spermidine derivatives N^5-(4-carboranylbutyl)spermidine·3HCl (**12**) and N^5-[4-(2-aminoethyl-o-carboranyl)butyl] spermidine·4HCl (**13**). Three were the spermine derivatives N^5-(4-o-carboranyl-butyl) spermine·4HCl (**14**), N^5-[4-(2-aminoethyl-o-carboranyl) butyl] spermine·5HCl (**15**), and N^5,N^{10}-bis(4-o-carboranylbutyl) spermine·4HCl (**16**). These were incorporated into liposomes by a pH-gradient-driven remote loading method with varying loading efficiencies that were influenced by the specific trapping agent and the structure of the boron compound. Greater loading efficiencies were obtained with lower molecular-weight boron derivatives, using ammonium sulfate as the trapping agent, compared to those obtained with sodium citrate.

6.6.3.3 EGFR-Targeted Liposomes

Acridine is a WS, DNA-intercalator. Its boronated derivative WSA was incorporated into liposomes composed of EGF-conjugated lipids. Their surface contained approximately 5 mol% PEG and 10–15 molecules of EGF, and 10^4–10^5 of WSA molecules were encapsulated. These liposomes had EGFR-specific cellular binding to cultured human glioma cells[130,140] and were internalized following specific binding to the receptor. Following internalization, WSA primarily was localized in the cytoplasm, and had high cellular retention with 80% of the boron remaining cell-associated after 48 h.[141]

6.7 BORON DELIVERY BY DEXTRANS

Dextrans are glucose polymers that consist mainly of a linear α-1,6-glucosidic linkage with some degree of branching via a 1,3-linkage.[142,143] Dextrans have been used extensively as drug and protein carriers to increase drug circulation time.[144,145] In addition, native or chemically-modified dextrans have been used for passive targeting to tumors, the RES or active receptor-specific cellular targeting. To link boron compounds to dextrans,[146] β-decachloro-o-carborane derivatives, in which one of the carbon atoms was substituted by –CH₂CHOHCH₂–O–CH₂CH=CH₂, were epoxidized and then subsequently bound to dextran with a resulting boron content of 4.3% (w/w).[147] The modified dextran then could be attached to tumor-specific antibodies.[147–150] BSH was covalently coupled to dextran derivatives by two methods.[151] In the first method, dextran was activated with 1-cyano-4-(dimethylamino)pyridine (CDAP) and subsequently coupled with 2-aminoethyl pyridyl disulfide. Then, thiolated dextran was linked to BSH in a disulfide exchange reaction. A total of 10–20 boron cages were attached to each dextran chain. In the second method, dextran was derivatized to a multiallyl derivative (Figure 6.10, **17**), which was reacted with BSH in a free-radical-initiated addition reaction. Using this method, 100–125 boron cages could be attached per dextran chain, suggesting that this derivative might be a promising template for the development of other HMW delivery agents. In the second method, designed to target EGFR overexpressing cells, EGF and BSH were covalently linked to a 70 kDa dextran (**18**).[152–154] Bioconjugates, having a small number of BSH molecules, attained maximum in vitro binding at 4 h with the human glioma cell line U-343

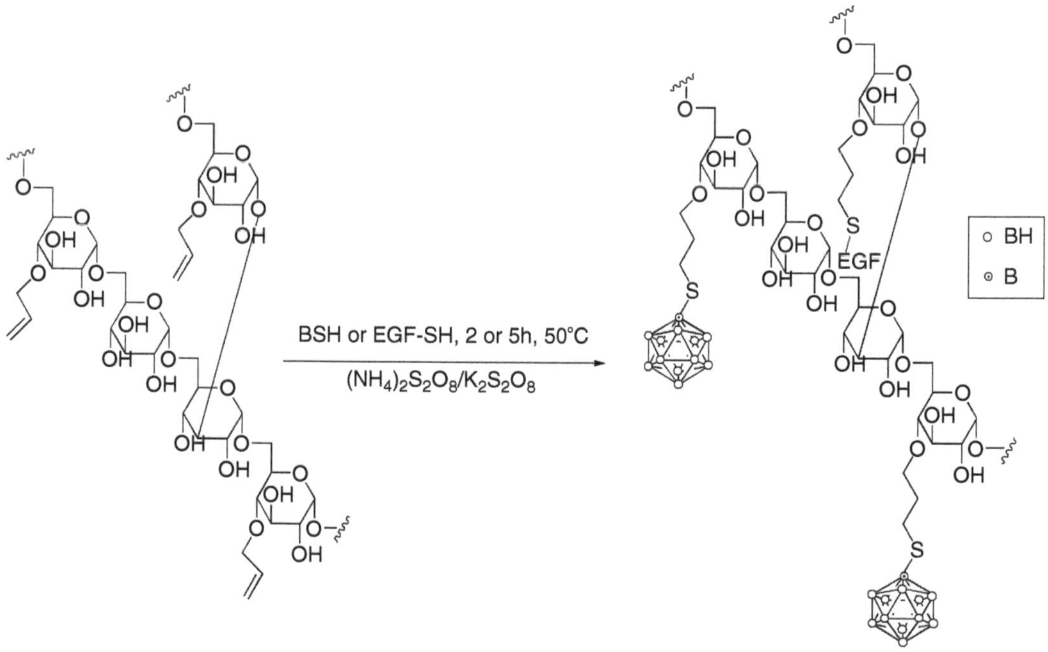

17   **18**

FIGURE 6.10 Preparation of EGF-targeted, boronated dextrans. The bioconjugate was prepared by a free-radical-initiated addition reaction between multiallyl dextran derivatives and BSH or thiolated EGF at 50°C using K₂S₂O₈ as an initiator.

MGaC12:6. In contrast, there was a slow increase of binding over 24 h for those having a large number of BSH molecules. Although most of the bioconjugates were internalized, in vitro retention was low, as was in vivo uptake following i.v. injection into nude mice bearing s.c. implants of Chinese hamster ovary (CHO) cells transfected with the human gene encoding EGFR (designated CHO–EGFR). However, following i.t. injection, boron uptake was higher with CHO–EGFR(+) tumors compared to wildtype EGFR(−) CHO tumors.[155]

6.8 OTHER MACROMOLECULES USED FOR DELIVERING BORON COMPOUNDS

Polylysine is another polymer having multiple reactive amino groups that has been used as a platform for the delivery of boron compounds.[53,156] The protein-binding polyhedral boron derivatives, isocyanatoundecahydro-closo-dodecaborate ($B_{12}H_{11}NCO^{2-}$), was linked to polylysine and subsequently to the anti-B16 melanoma mAb IB16-6 using two heterobifunctional linkers, SPDP and sulfo-MBS. The bioconjugate had an average of 2700 boron atoms per molecule and retained 58% of the immunoreactivity of the native antibody, as determined by a semiquantitative immuno-fluorescent assay or by ELISA. Other bioconjugates prepared by this method had greater than 1000 boron atoms per molecule of antibody and retained 40–90% of the immunoreactivity of the native antibody.[53] Using another approach, site-specific linkage of boronated polylysine to the carbohydrate moieties of anti-TSH antibody resulted in a bioconjugate that had approximately 6×10^3 boron atoms with retention of its immunoreactivity.[156]

A streptavidin/biotin system also has been developed to specifically deliver boron to tumors. Biotin was linked to a mAb and streptavidin was attached to the boron containing moiety. The indirect linking of boron to the mAb minimized loss of its immunoreactivity. BSH was attached to poly-(D-glutamate D-lysine) (poly-GL) via a heterobifunctional agent.[157] This boronated poly-GL then was activated by a carbodiimide reagent and in turn reacted with streptavidin. Another approach employed a streptavidin mutant that had 20 cysteine residues per molecule. BSH was conjugated via sulfhydryl-specific bifunctional reagents to incorporate approximately 230 boron atoms/molecule.[158] A closomer species with an icosahedral dodecaborate core and twelve pendant anionic nido-7,8-carborane groups was developed as a new class of unimolecular nanovehicles for evaluation as a delivery agent for BNCT.[159]

6.9 DELIVERY OF BORON-CONTAINING MACROMOLECULES TO BRAIN TUMORS

6.9.1 General Considerations

Drug delivery to brain tumors is dependent upon (1) the plasma concentration profile of the drug, which depends upon the amount and route of administration; (2) the ability of the agent to traverse the BBB; (3) blood flow within the tumor; and (4) the lipophilicity of the drug. In general, a high steady-state blood concentration will maximize brain uptake, while rapid clearance will reduce it, except in the case of intra-arterial (i.a.) drug administration. Although the i.v. route currently is being used clinically to administer both BSH and BPA, this may not be ideal for boron containing macromolecules and other strategies must be employed to improve their delivery.

6.9.2 Drug-Transport Vectors

One approach to improve brain tumor uptake of boron compounds has been to conjugate them to a drug-transport vector by means of receptor-specific transport systems.[160,161] Proteins such as insulin, insulin-like growth factor (IGF), TF,[162] and leptin can traverse the BBB. BSH encapsulated in TF–PEG liposomes had a prolonged residence time in the circulation and low RES uptake in tumor-bearing mice, resulting in enhanced extravasation of the liposomes into the tumor and concomitant internalization by receptor-mediated endocytosis.[163,164] Mice that received BSH containing TF-liposomes following by BNCT had a significant prolongation in survival time compared to those that received PEG-liposomes, bare liposomes and free BSH, thereby establishing proof-of-principle for transcytosis of a boron containing nanovehicle.[112]

6.9.3 DIRECT INTRACEREBRAL DELIVERY

Studies carried out by us have clearly demonstrated that the systemic route of administration is not suitable for delivery of boronated EGF or mAbs to glioma-bearing rats.[75,165] Intravenous injection of technetium-99m labeled EGF to rats bearing intracerebral implants of the C6 rat glioma, which had been genetically engineered to express the human EGFR gene, resulted in 0.14% I.D. localizing in the tumor. Intracarotid (i.c.) injection with or without BBB disruption increased the tumor uptake from 0.34 to 0.45% I.D./g, but based even on the most optimistic assumptions the amount of boron that could be delivered to the tumor by i.v. injection, this would have been inadequate for BNCT.[165] Direct i.t. injection of boronated EGF (BSD–EGF), on the other hand, resulted in tumor boron concentrations of 22 µg/g compared to 0.01 µg/g following i.v. injection and almost identical boron uptake values were obtained using the F98$_{EGFR}$ glioma model.[77] This was produced by transfecting F98 glioma cells with the gene encoding human EGFR. Based on our biodistribution results, therapy studies were initiated with the F98$_{EGFR}$ glioma in syngeneic Fischer rats. F98$_{EGFR}$ glioma bearing rats that received BSD–EGF i.t. had a MST of 45 ± d compared to 33 ± 2 d in animals that had EGFR(−) wildtype F98 gliomas. Because it is unlikely that any single boron delivery agent will be able to target all tumor cells, the combination of i.t. administration of BSD–EGF with i.v. injection of BPA was evaluated. This resulted in further increase in MST to 57 ± 8 d compared to 39 ± 2 d for i.v. BPA alone.[73] These data provide proof-of-principle for the idea of using a combination of LMW and HMW boron delivery agents.

6.9.4 CONVECTION-ENHANCED DELIVERY

Convection-Enhanced Delivery (CED), by which therapeutic agents are directly infused into the brain, is an innovative method to increase their uptake and distribution.[166–168] Under normal physiological conditions, interstitial fluids move through the brain by both convection and diffusion. Diffusion of a drug in tissue depends upon its molecular weight, ionic charge and its concentration gradient within normal tissue and the tumor. The higher the molecular weight of the drug, the more positively charged the ionic species and the lower its concentration, the slower its diffusion. For example, diffusion of antibody into a tumor requires 3 d to diffuse 1 mm from the point of origin. Unlike diffusion, however, convection or "bulk" flow results from a pressure gradient that is independent of the molecular weight of the substance. CED potentially can improve the targeting of both LMW and HMW molecules, as well as liposomes, to the central nervous system by applying a pressure gradient to establish bulk flow during interstitial infusion. The volume of distribution (V_d) is a linear function of the volume of the infusate (V_i). CED has been used to efficiently deliver drugs and HMW agents such as mAbs and toxin fusion proteins to brain tumors.[168–170] CED can provide more homogenous dispersion of the agent and at higher concentrations than otherwise would be attainable by i.v. injection.[165] For example, in our own studies, CED of ^{125}I-labeled EGF to F98$_{EGFR}$ glioma-bearing rats resulted in 47% I.D./g of the bioconjugate localizing in the tumor compared to 10% I.D./g in normal brain at 24 h following administration. The corresponding boron values were 22 and 2.9–4.9 µg/g, respectively.[76] Based on these results, therapy studies were initiated. F98$_{EGFR}$ glioma-bearing rats that received BD–EGF by CED had a MST of 53 ± 13 d compared to 40 ± 5 d for animals that received BPA i.v.[73] Similar studies have been carried out using either boronated cetuximab (IMC-C225) or the mAb L8A4,[171,172] which is specifically directed against the tumor-specific mutant isoform, EGFRvIII, and comparable results were obtained.[173] Direct intracerebral administration of these and other HMW agents by CED has opened up the possibility that they actually could be used clinically, since CED is being used to administer radiolabeled antibodies, toxin fusion proteins, and gene vectors to patients with GBM. It is only a matter of time before this approach also will be used to deliver both LMW and HMW boron-containing agents for NCT.

6.10 CLINICAL CONSIDERATIONS AND CONCLUSIONS

In this review we have focused on HMW boron delivery agents and nanovehicles that potentially could be used clinically for targeting intra- and extracranial tumors. Studies carried out in glioma bearing rats have demonstrated that boronated EGF and the mAb, cetuximab, both of which bind to EGFR, selectively targeted receptor ($+$) tumors following direct i.c. delivery. Furthermore, following BNCT, a significant increase in MST was observed, and this was further enhanced if BPA was administered in combination with the HMW agents. These studies provide proof-of-principle first for the potential utility of HMW agents, and second, the therapeutic gain associated with the combination of HMW and LMW boron delivery agents.

There is a question as to whether or not any of these agents will ever be used clinically. Critical *issues* must be addressed if BNCT is to ever become a useful modality for the treatment of cancer. *Large* clinical trials, preferably randomized, must be carried out to convincingly demonstrate the efficacy of BPA and BSH, the two drugs that currently are being used. Once efficacy has been established, studies with HMW EGFR targeting agents could move forward.

Both direct i.t. injection[174,175] and CED[170,176–179] have been used clinically to deliver mAbs and toxin fusion proteins to patients who have had surgical resection of their brain tumors. These studies provide a strong clinical rationale for the direct intracerebral delivery of HMW agents. Initially, the primary focus should be on determining the safety of administering them to patients prior to surgical resection of their brain tumors. After this has been established, then biodistribution studies could be carried out in patients scheduled to undergo surgical resection of their brain tumors. The patients' tumor and normal tissues would then be analyzed for their boron content, and if there was evidence of preferential tumor localization with boron concentrations in the range 10–20 µg/g and normal brain concentrations of less than 5 µg/g, then therapy studies could be undertaken.

Because there is considerable variability in EGFR expression in gliomas, it is highly unlikely that any single agent will be able to deliver the requisite amount of boron to all tumor cells, and that HMW agents would need to be used in combination with BPA/BSH. This general plan would also be applicable to the other HMW delivery agents and nanovehicles that have been discussed in this review. The joining together of chemistry and nanotechnology[180,181] represents a major step forward for the development of effective boron delivery agents for NCT. Nanovehicles offer the possibility of tumor targeting with enhanced boron payloads. Potentially, this could solve the central problem of how to selectively deliver large number of boron atoms to individual cancer cells.

As can be seen from the preceding discussion, the development of HMW boron delivery agents must proceed in step with strategies to optimize their delivery and an appreciation as to how they would be used clinically. Intracerebral delivery has been used in clinically advanced settings, but nuclear reactors, which currently are the only source of neutrons for BNCT, would not be conducive to this. Therefore, the development of accelerator-neutron sources,[182] which could be easily sited in hospitals, is especially important. This also would facilitate the initiation of large scale clinical trials at selected centers that treat large numbers of patients with brain tumors and would permit evaluation of new boron delivery agents.

In conclusion, there is a plethora of HMW boron delivery agents that have been designed and synthesized. The challenge is to move from experimental animal studies to clinical biodistribution studies, a step which has yet to be taken.

6.11 SUMMARY

BNCT is based on the nuclear capture and fission reactions that occur when nonradioactive boron-10 is irradiated with low energy thermal neutrons to yield LET alpha particles (^{4}He) and recoiling lithium-7 (^{7}Li) nuclei. For BNCT to be successful, a sufficient number of ^{10}B atoms (approximately 10^9 atoms/cell) must be selectively delivered to the tumor and enough thermal neutrons must be absorbed by them to sustain a lethal ^{10}B(n,α) ^{7}Li capture reaction. BNCT primarily has been used to treat patients with brain tumors, and more recently those with head-and-neck cancer. Two LMW boron delivery agents currently are being used clinically, BSH and BPA. However, a variety of HMW agents consisting of macromolecules and nanovehicles have been developed. This review focuses on the latter, which includes mAbs, dendrimers, liposomes, dextrans, polylysine, avidin, FA, and both epidermal and vascular endothelial growth factors (EGF and VEGF). Procedures for introducing boron atoms into these HMW agents and their chemical properties are discussed. In vivo studies on their biodistribution are described, and the efficacy of a subset of them, those which have been used for BNCT of tumors in experimental animals, will be discussed.

Because brain tumors currently are the primary candidates for treatment by BNCT, delivery of these HMW agents across the BBB presents a special challenge. Various routes of administration are discussed, including receptor-facilitated transcytosis following i.v. administration, direct i.t injection and convection-enhanced delivery in which a pump is used to apply a pressure gradient to establish bulk flow of the HMW agent during interstitial infusion. Finally, we have concluded with a discussion relating to issues that must be addressed if these HMW agents are to be used clinically.

ACKNOWLEDGMENTS

We thank Dr. Achintya Bandyopadhyaya for suggestions on figures and Mrs. Beth Kahl for secretarial assistance. We thank Bentham Science Publishers, Ltd. for granting copyright release to republish the text and figures in this chapter.

REFERENCES

1. Berger, M. S., Malignant astrocytomas: Surgical aspects, *Semin. Oncol.*, 21, 172, 1994.
2. Gutin, P. H. and Posner, J. B., Neuro-oncology: Diagnosis and management of cerebral gliomas—past, present, and future, *Neurosurgery*, 47, 1, 2000.
3. Parney, I. F. and Chang, S. M., Current chemotherapy for glioblastoma, *Cancer J.*, 9, 149, 2003.
4. Paul, D. B. and Kruse, C. A., Immunologic approaches to therapy for brain tumors, *Curr. Neurol. Neurosci. Rep.*, 1, 238, 2001.
5. Rainov, N. G. and Ren, H., Gene therapy for human malignant brain tumors, *Cancer J.*, 9, 180, 2003.
6. Curran, W.J., Jr., et al., Recursive partitioning analysis of prognostic factors in three Radiation Therapy Oncology Group malignant glioma trials, *J. Natl Cancer Inst.*, 85, 704, 1993.
7. Lacroix, M. et al., A multivariate analysis of 416 patients with glioblastoma multiforme: Prognosis, extent of resection, and survival, *J. Neurosurg.*, 95, 190, 2001.
8. Hentschel, S. J. and Lang, F. F., Current surgical management of glioblastoma, *Cancer J.*, 9, 113, 2003.
9. Laws, E.R., Jr. and Shaffrey, M. E., The inherent invasiveness of cerebral gliomas: Implications for clinical management, *Int. J. Dev. NeuroSci.*, 17, 413, 1999.
10. Ware, M. L., Berger, M. S., and Binder, D. K., Molecular biology of glioma tumorigenesis, *Histol. Histopathol.*, 18, 207, 2003.

11. Parney, I. F., Hao, C., and Petruk, K. C., Glioma immunology and immunotherapy, *Neurosurgery*, 46, 778, 2000.

12. Kaczarek, E. et al., Dissecting glioma invasion: Interrelation of adhesion, migration and intercellular contacts determine the invasive phenotype, *Int. J. Dev. NeuroSci.*, 17, 625, 1999.

13. Nutt, C. L., Matthews, R. T., and Hockfield, S., Glial tumor invasion: A role for the upregulation and cleavage of BEHAB/brevican, *Neuroscientist*, 7, 113, 2001.

14. Halperin, E. C., Burger, P. C., and Bullard, D. E., The fallacy of the localized supratentorial malignant glioma, *Int. J. Radiat. Oncol. Biol. Phys.*, 15, 505, 1988.

15. Barth, R. F., A critical assessment of boron neutron capture therapy: An overview, *J. Neurooncol.*, 62, 1, 2003.

16. Nakagawa, Y. et al., Clinical review of the Japanese experience with boron neutron capture therapy and a proposed strategy using epithermal neutron beams, *J. Neurooncol.*, 62, 87, 2003.

17. Wadabayashi, N. et al., Selective boron accumulation in human ocular melanoma vs. surrounding eye components after ^{10}B1-*p*-boronophenylalanine administration. Prerequisite for clinical trial of neutron-capture therapy, *Melanoma Res.*, 4, 185, 1994.

18. Busse, P. M. et al., A critical examination of the results from the Harvard-MIT NCT program phase I clinical trial of neutron capture therapy for intracranial disease, *J. Neurooncol.*, 62, 111, 2003.

19. Kato, I. et al., Effectiveness of BNCT for recurrent head and neck malignancies, *Appl. Radiat. Isot.*, 61, 1069, 2004.

20. Rao, M. et al., BNCT of 3 cases of spontaneous head and neck cancer in feline patients, *Appl. Radiat. Isot.*, 61, 947, 2004.

21. Koivunoro, H. et al., BNCT dose distribution in liver with epithermal D–D and D–T fusion-based neutron beams, *Appl. Radiat. Isot.*, 61, 853, 2004.

22. Pinelli, T. et al., TAOrMINA: from the first idea to the application to the human liver, In *Research and Development in Neutron Capture Therapy*, Sauerwein, M. W., Moss, R., Wittig, A. et al., Eds., Modduzzi Editore, International Proceedings Division, Bologna, pp. 1065–1072, 2002.

23. Coderre, J. A. et al., Boron neutron capture therapy: Cellular targeting of high linear energy transfer radiation, *Technol. Cancer Res. Treat.*, 2, 355, 2003.

24. Barth, R. F. et al., Boron neutron capture therapy of cancer: Current status and future prospects, *Clin. Cancer Res.*, 11, 3987, 2005.

25. Zamenhof, R. G. et al., Eleventh World Congress on Neutron Capture Therapy, *Appl. Radiat. Isot.*, 61, 731, 2004.

26. Farr, L. E. et al., Neutron capture therapy with boron in the treatment of glioblastoma multiforme, *Am. J. Roentgenol. Radium Ther. Nucl. Med.*, 71, 279, 1954.

27. Goodwin, J. T. et al., Pathological study of eight patients with glioblastoma multiforme treated by neutron-capture therapy using boron 10, *Cancer*, 8, 601, 1955.

28. Snyder, H. R., Reedy, A. J., and Lennarz, W. J., Synthesis of aromatic boronic acids. Aldehydo boronic acids and a boronic acid analog of tyrosine, *J. Am. Chem. Soc.*, 80, 835, 1958.

29. Soloway, A. H., Hatanaka, H., and Davis, M. A., Penetration of brain and brain tumor. VII. Tumor-binding sulfhydryl boron compounds, *J. Med. Chem.*, 10, 714, 1967.

30. Hatanaka, H. and Nakagawa, Y., Clinical results of long-surviving brain tumor patients who underwent boron neutron capture therapy, *Int. J. Radiat. Oncol. Biol. Phys.*, 28, 1061, 1994.

31. Mishima, Y., Selective thermal neutron capture therapy of cancer cells using their specific metabolic activities—melanoma as prototype, In *Cancer Neutron Capture Therapy*, Mishima, Y., Ed., Plenum Press, New York, pp. 1–26, 1996.

32. Ang, K. K. et al., Impact of epidermal growth factor receptor expression on survival and pattern of relapse in patients with advanced head and neck carcinoma, *Cancer Res.*, 62, 7350, 2002.

33. Hawthorne, M. F. and Lee, M. W., A critical assessment of boron target compounds for boron neutron capture therapy, *J. Neurooncol.*, 62, 33, 2003.

34. Soloway, A. H. et al., The chemistry of neutron capture therapy, *Chem. Rev.*, 98, 1515, 1998.

35. Gillies, E. R. and Frechet, J. M. J., Dendrimers and dendritic polymers in drug delivery, *Drug Discov. Today*, 10, 35, 2005.

36. Esfand, R. and Tomalia, D. A., Poly(amidoamine) (PAMAM) dendrimers: From biomimicry to drug delivery and biomedical applications, *Drug Discov. Today*, 6, 427, 2001.

37. Tomalia, D.A., Birth of a new macromolecular architecture: Dendrimers as quantized building blocks for nanoscale synthetic organic chemistry, *Aldrichimica ACTA*, 37, 39, 2004

38. Verheyde, B., Maes, W., and Dehaen, W., The use of 1,3,5-triazines in dendrimer synthesis, *Mater. Sci. Eng., C*, 18, 243, 2001.

39. McCarthy, T. D. et al., Dendrimers as drugs: Discovery and preclinical and clinical development of dendrimer-based microbicides for HIV and STI prevention, *Mol. Pharmacol.*, 2, 312, 2005.

40. Venditto, V. J., Regino, C. A., and Brechbiel, M. W., PAMAM dendrimer based macromolecules as improved contrast agents, *Mol. Pharmacol.*, 2, 302, 2005.

41. Ambade, A. V., Savariar, E. N., and Thayumanavan, S., Dendrimeric micelles for controlled drug release and targeted delivery, *Mol. Pharmacol.*, 2, 264, 2005.

42. Majoros, I. J. et al., Poly(amidoamine) dendrimer-based multifunctional engineered nanodevice for cancer therapy, *J. Med. Chem.*, 48, 5892, 2005.

43. Boas, U. and Heegaard, P. M., Dendrimers in drug research, *Chem. Soc. Rev.*, 33, 43, 2004.

44. Klajnert, B. and Bryszewska, M., Dendrimers: Properties and applications, *Acta Biochim. Pol.*, 48, 199, 2001.

45. Sharkey, R. M. and Goldenberg, D. M., Perspectives on cancer therapy with radiolabeled monoclonal antibodies, *J. Nucl. Med.*, 46(Suppl 1), 115S, 2005.

46. Jaracz, S. et al., Recent advances in tumor-targeting anticancer drug conjugates, *Bioorg. Med. Chem.*, 13, 5043, 2005.

47. Garnett, M. C., Targeted drug conjugates: Principles and progress, *Adv. Drug Deliv. Rev.*, 53, 171, 2001.

48. Chari, R. V., Targeted delivery of chemotherapeutics: Tumor-activated prodrug therapy, *Adv. Drug Deliv. Rev.*, 31, 89, 1998.

49. Hamblett, K. J. et al., Effects of drug loading on the antitumor activity of a monoclonal antibody drug conjugate, *Clin. Cancer Res.*, 10, 7063, 2004.

50. Trail, P. A., King, H. D., and Dubowchik, G. M., Monoclonal antibody drug immunoconjugates for targeted treatment of cancer, *Cancer Immunol. Immunother.*, 52, 328, 2003.

51. Fracasso, G. et al., Immunotoxins and other conjugates: Preparation and general characteristics, *Mini-Rev. Med. Chem.*, 4, 545, 2004.

52. Liu, L. et al., Bispecific antibodies as targeting agents for boron neutron capture therapy of brain tumors, *J. Hematother.*, 4, 477, 1995.

53. Alam, F. et al., Boron neutron capture therapy: Linkage of a boronated macromolecule to monoclonal antibodies directed against tumor-associated antigens, *J. Med. Chem.*, 32, 2326, 1989.

54. Barth, R. F. et al., Neutron capture using boronated monoclonal antibody directed against tumor-associated antigens, *Cancer Detect. Prev.*, 5, 315, 1982.

55. Alam, F., Barth, R. F., and Soloway, A. H., Boron containing immunoconjugates for neutron capture therapy of cancer and for immunocyto chemistry, *Antibody Immunoconjugates Radiopharm.*, 2, 145, 1989.

56. Tolpin, E. I. et al., Boron neutron capture therapy of cerebral gliomas. II. Utilization of the blood–brain barrier and tumor-specific antigens for the selective concentration of boron in gliomas, *Oncology*, 32, 223, 1975.

57. Sneath, R. L. et al., Protein-binding polyhedral boranes, *J. Med. Chem.*, 19, 1290, 1976.

58. Varadarajan, A. et al., Conjugation of phenyl isothiocyanate derivatives of carborane to antitumor antibody and in vivo localization of conjugates in nude mice, *Bioconjug. Chem.*, 2, 102, 1991.

59. Takahashi, T. et al., Preliminary study for application of anti-alpha-fetoprotein monoclonal antibody to boron-neutron capture therapy, *Jpn. J. Exp. Med.*, 57, 83, 1987.

60. Compostella, F. et al., Synthesis of glycosyl carboranes with different linkers between the sugar and the boron cage moieties, In *Research and Development in Neutron Capture Therapy. Proceedings of the 10th International Congress on Neutron Capture Therapy*, Saverwein, W., Moss, R., and Wittig, A., Eds., Monduzzi Editore, Bologna, p.8, 2002.

61. Giovenzana, G. B. et al., Synthesis of carboranyl derivatives of alkynyl glycosides as potential BNCT agents, *Tetrahedron*, 55, 14123, 1999.

62. Alam, F. et al., Dicesium *N*-succinimidyl 3-(undecahydro-closo-dodecaboranyldithio)propionate, a novel heterobifunctional boronating agent, *J. Med. Chem.*, 28, 522, 1985.

63. Barth, R. F. et al., Boronated starburst dendrimer-monoclonal antibody immunoconjugates: evaluation as a potential delivery system for neutron capture therapy, *Bioconjug. Chem.*, 5, 58, 1994.
64. Liu, L. et al., Critical evaluation of bispecific antibodies as targeting agents for boron neutron capture therapy of brain tumors, *Anticancer Res.*, 16, 2581, 1996.
65. Barth, R. F. et al., In vivo distribution of boronated monoclonal antibodies and starburst dendrimers, In *Adv Neutron Capture Ther*, Soloway, A. H., Barth, R. F., and Carpenter, D.E., Eds., Plenum Press, New York, p. 35, 1993.
66. Wu, G. et al., Site-specific conjugation of boron-containing dendrimers to anti-EGF receptor monoclonal antibody cetuximab (IMC-C225) and its evaluation as a potential delivery agent for neutron capture therapy, *Bioconjug. Chem.*, 15, 185, 2004.
67. Barth, R. F. et al., Neutron capture therapy of epidermal growth factor (+) gliomas using boronated cetuximab (IMC-C225) as a delivery agent, *Appl. Radiat. Isot.*, 61, 899, 2004.
68. Arteaga, C. L., Overview of epidermal growth factor receptor biology and its role as a therapeutic target in human neoplasia, *Semin. Oncol.*, 29, 3, 2002.
69. Mendelson, J. and Beselga, J., Epidermal growth factor receptor targeting in cancer, *Seminars Oncol.*, 33, 369, 2006.
70. Pal, S. K. and Pegram, M., Epidermal growth factor receptor and signal transduction: potential targets for anti-cancer therapy, *Anticancer Drugs*, 16, 483, 2005.
71. Normanno, N. et al., The ErbB receptors and their ligands in cancer: an overview, *Curr. Drug Targets*, 6, 243, 2005.
72. Jorissen, R. N. et al., Epidermal growth factor receptor: mechanisms of activation and signaling, *Exp. Cell Res.*, 284, 31, 2003.
73. Yang, W. et al., Boronated epidermal growth factor as a delivery agent for neutron capture therapy of EGF receptor positive gliomas, *Appl. Radiat. Isot.*, 61, 981, 2004.
74. Capala, J. et al., Boronated epidermal growth factor as a potential targeting agent for boron neutron capture therapy of brain tumors, *Bioconjug. Chem.*, 7, 7, 1996.
75. Yang, W. et al., Intratumoral delivery of boronated epidermal growth factor for neutron capture therapy of brain tumors, *Cancer Res.*, 57, 4333, 1997.
76. Yang, W. et al., Convection-enhanced delivery of boronated epidermal growth factor for molecular targeting of EGF receptor-positive gliomas, *Cancer Res.*, 62, 6552, 2002.
77. Barth, R. F. et al., Molecular targeting of the epidermal growth factor receptor for neutron capture therapy of gliomas, *Cancer Res.*, 62, 3159, 2002.
78. Reddy, J. A., Allagadda, V. M., and Leamon, C. P., Targeting therapeutic and imaging agents to folate receptor positive tumors, *Curr. Pharm. Biotechnol.*, 6, 131, 2005.
79. Leamon, C. P. and Reddy, J. A., Folate-targeted chemotherapy, *Adv. Drug Deliv. Rev.*, 56, 1127, 2004.
80. Sudimack, J. and Lee, R. J., Targeted drug delivery via the folate receptor, *Adv. Drug Deliv. Rev.*, 41, 147, 2000.
81. Paulos, C. M. et al., Ligand binding and kinetics of folate receptor recycling in vivo: impact on receptor-mediated drug delivery, *Mol. Pharmacol.*, 66, 1406, 2004.
82. Reddy, J. A. and Low, P. S., Folate-mediated targeting of therapeutic and imaging agents to cancers, *Crit. Rev. Ther. Drug Carrier Syst.*, 15, 587, 1998.
83. Stephenson Stacy, M. et al., Folate receptor-targeted liposomes as possible delivery vehicles for boron neutron capture therapy, *Anticancer Res.*, 23, 3341, 2003.
84. Ward, C. M., Folate-targeted non-viral DNA vectors for cancer gene therapy, *Curr. Opin. Mol. Ther.*, 2, 182, 2000.
85. Gottschalk, S. et al., Folate receptor mediated DNA delivery into tumor cells: Protosomal disruption results in enhanced gene expression, *Gene Ther.*, 1, 185, 1994.
86. Hilgenbrink, A. R. and Low, P. S., Folate receptor-mediated drug targeting: from therapeutics to diagnostics, *J. Pharm. Sci.*, 94, 2135, 2005.
87. Shukla, S. et al., Synthesis and biological evaluation of folate receptor-targeted boronated PAMAM dendrimers as potential agents for neutron capture therapy, *Bioconjug. Chem.*, 14, 158, 2003.
88. Benouchan, M. and Colombo, B. M., Anti-angiogenic strategies for cancer therapy (Review), *Int. J. Oncol.*, 27, 563, 2005.

89. Tortora, G., Melisi, D., and Ciardiello, F., Angiogenesis: a target for cancer therapy, *Curr. Pharm. Des.*, 10, 11, 2004.

90. Brekken, R. A., Li, C., and Kumar, S., Strategies for vascular targeting in tumors, *Int. J. Cancer*, 100, 123, 2002.

91. Gaya, A. M. and Rustin, G. J., Vascular disrupting agents: a new class of drug in cancer therapy, *Clin. Oncol. (R. Coll. Radiol.)*, 17, 277, 2005.

92. Bergsland, E. K., Vascular endothelial growth factor as a therapeutic target in cancer, *Am. J. Health Syst. Pharm.*, 61, S4, 2004.

93. Hicklin, D. J. and Ellis, L. M., Role of the vascular endothelial growth factor pathway in tumor growth and angiogenesis, *J. Clin. Oncol.*, 23, 1011, 2005.

94. Backer, M. V. et al., Vascular endothelial growth factor selectively targets boronated dendrimers to tumor vasculature, *Mol. Cancer Ther.*, 4, 1423, 2005.

95. Barth, R. F., Yang, W., and Coderre, J. A., Rat brain tumor models to assess the efficacy of boron neutron capture therapy: a critical evaluation, *J. Neurooncol.*, 62, 61, 2003.

96. Parrott, M. C. et al., Synthesis and properties of carborane-functionalized aliphatic polyester dendrimers, *J. Am. Chem. Soc.*, 127, 12081, 2005.

97. Park, J. W., Benz, C. C., and Martin, F. J., Future directions of liposome- and immunoliposome-based cancer therapeutics, *Semin. Oncol.*, 31, 196, 2004.

98. Sapra, P. and Allen, T. M., Ligand-targeted liposomal anticancer drugs, *Prog. Lipid Res.*, 42, 439, 2003.

99. Carlsson, J. et al., Ligand liposomes and boron neutron capture therapy, *J. Neurooncol.*, 62, 47, 2003.

100. Hawthorne, M. F. and Shelly, K., Liposomes as drug delivery vehicles for boron agents, *J. Neurooncol.*, 33, 53, 1997.

101. Mishima, Y. et al., Treatment of malignant melanoma by single thermal neutron capture therapy with melanoma-seeking ^{10}B-compound, *Lancet*, 2, 388, 1989.

102. Coderre, J. A. et al., Selective delivery of boron by the melanin precursor analogue *p*-boronophenylalanine to tumors other than melanoma, *Cancer Res.*, 50, 138, 1990.

103. Ono, K. et al., The combined effect of boronophenylalanine and borocaptate in boron neutron capture therapy for SCCVII tumors in mice, *Int. J. Radiat. Oncol. Biol. Phys.*, 43, 431, 1999.

104. Obayashi, S. et al., Delivery of (10)boron to oral squamous cell carcinoma using boronophenylalanine and borocaptate sodium for boron neutron capture therapy, *Oral Oncol.*, 40, 474, 2004.

105. Yoshino, K. et al., Improvement of solubility of *p*-boronophenylalanine by complex formation with monosaccharides, *Strahlenther. Onkol.*, 165, 127, 1989.

106. Ryynanen, P. M. et al., Models for estimation of the ^{10}B concentration after BPA-fructose complex infusion in patients during epithermal neutron irradiation in BNCT, *Int. J. Radiat. Oncol. Biol. Phys.*, 48, 1145, 2000.

107. Pavanetto, F. et al., Boron-loaded liposomes in the treatment of hepatic metastases: preliminary investigation by autoradiography analysis, *Drug Deliv.*, 7, 97, 2000.

108. Martini, S. et al., Boronphenylalanine insertion in cationic liposomes for boron neutron capture therapy, *Biophys. Chem.*, 111, 27, 2004.

109. Smyth Templeton, N. et al., Cationic liposomes as in vivo delivery vehicles, *Curr. Med. Chem.*, 10, 1279, 2003.

110. Mehta, S. C., Lai, J. C., and Lu, D. R., Liposomal formulations containing sodium mercaptoundecahydrododecaborate (BSH) for boron neutron capture therapy, *J. Microencapsul.*, 13, 269, 1996.

111. Ji, B. et al., Cell culture and animal studies for intracerebral delivery of borocaptate in liposomal formulation, *Drug Deliv.*, 8, 13, 2001.

112. Maruyama, K. et al., Intracellular targeting of sodium mercaptoundecahydrododecaborate (BSH) to solid tumors by transferrin-PEG liposomes, for boron neutron-capture therapy (BNCT), *J. Control. Release*, 98, 195, 2004.

113. Valliant, J. F. et al., The medicinal chemistry of carboranes, *Coord. Chem. Rev.*, 232, 173, 2002.

114. Hawthorne, M. F., The role of chemistry in the development of cancer therapy by the boron-neutron capture reaction, *Angew. Chem. Int. Ed. Engl.*, 105, 997, 1993.

115. Moraes, A. M., Santanaand, M. H. A., and Carbonell, R. G., Preparation and characterization of liposomal systems entrapping the boronated compound *o*-carboranylpropylamine, *J. Microencapsul.*, 16, 647, 1999.

116. Moraes, A. M., Santana, M. H. A., and Carbonell, R. G., Characterization of liposomal systems entrapping boron-containing compounds in response to pH gradients, In *Biofunctional Membranes, Proceedings of the International Conference on Biofunctional Membranes*, Butterfield, A., Ed., Plenum Press, New York, p.259, 1996.

117. Shelly, K. et al., Model studies directed toward the boron neutron-capture therapy of cancer: boron delivery to murine tumors with liposomes, *Proc. Natl. Acad. Sci. USA*, 89, 9039, 1992.

118. Feakes, D. A. et al., [$Na_3B_{20}H_{17}NH_3$]: Synthesis and liposomal delivery to murine tumors, *Proc. Natl. Acad. Sci. USA*, 91, 3029, 1994.

119. Hawthorne, M. F., Shelly, K., and Li, F., The versatile chemistry of the $B_{20}H_{18}]^{2-}$ ions: novel reactions and structural motifs, *Chem. Commun. (Camb.)*, 547, 2002.

120. Feakes, D. A. et al., Synthesis and in vivo murine evaluation of [$Na_41-(1'-B_{10}H_9)-6-SHB_{10}H_8$] as a potential agent for boron neutron capture therapy, *Proc. Natl Acad. Sci. USA*, 96, 6406, 1999.

121. Watson-Clark, R. A. et al., Model studies directed toward the application of boron neutron capture therapy to rheumatoid arthritis: boron delivery by liposomes in rat collagen-induced arthritis, *Proc. Natl. Acad. Sci. USA*, 95, 2531, 1998.

122. Feakes, D. A., Spinler, J. K., and Harris, F. R., Synthesis of boron-containing cholesterol derivatives for incorporation into unilamellar liposomes and evaluation as potential agents for BNCT, *Tetrahedron*, 55, 11177, 1999.

123. Thirumamagal, B. T. S., Zhao, X. B., Bandyopadhyaya, A. K., Narayanasamy, S., Johnsamuel, J., Tiwari, R., Golightly, D. W., Patel, V., Jehning, B. T., Backer, M. V., Barth, R. F., Lee, R. J., Backer, J. M., and Tjarks, W., Receptor-targeted liposomal delivery of boron-containing cholesterol mimics for boron neutron capture therapy (BNCT), *Bioconjug. Chem.*, 2006, in press.

124. Pan, G., Oie, S., and Lu, D. R., Uptake of the carborane derivative of cholesteryl ester by glioma cancer cells is mediated through LDL receptors, *Pharm. Res.*, 21, 1257, 2004.

125. Maletinska, L. et al., Human glioblastoma cell lines: levels of low-density lipoprotein receptor and low-density lipoprotein receptor-related protein, *Cancer Res.*, 60, 2300, 2000.

126. Nygren, C. et al., Increased levels of cholesterol esters in glioma tissue and surrounding areas of human brain, *Br. J. Neurosurg.*, 11, 216, 1997.

127. Leppala, J. et al., Accumulation of ^{99m}Tc-low-density lipoprotein in human malignant glioma, *Br. J. Cancer*, 71, 383, 1995.

128. Peacock, G. et al., In vitro uptake of a new cholesteryl carborane ester compound by human glioma cell lines, *J. Pharm. Sci.*, 93, 13, 2004.

129. Wei, Q., Kullberg, E. B., and Gedda, L., Trastuzumab-conjugated boron-containing liposomes for tumor-cell targeting; development and cellular studies, *Int. J. Oncol.*, 23, 1159, 2003.

130. Bohl Kullberg, E. et al., Development of EGF-conjugated liposomes for targeted delivery of boronated DNA-binding agents, *Bioconjug. Chem.*, 13, 737, 2002.

131. Stephenson, S. M. et al., Folate receptor-targeted liposomes as possible delivery vehicles for boron neutron capture therapy, *Anticancer Res.*, 23, 3341, 2003.

132. Yanagie, H. et al., Boron neutron capture therapy using [10]B entrapped anti-CEA immunoliposome, *Hum. Cell*, 2, 290, 1989.

133. Yanagie, H. et al., Application of boronated anti-CEA immunoliposome to tumour cell growth inhibition in in vitro boron neutron capture therapy model, *Br. J. Cancer*, 63, 522, 1991.

134. Yanagie, H. et al., Inhibition of growth of human breast cancer cells in culture by neutron capture using liposomes containing [10]B, *Biomed. Pharmacother.*, 56, 93, 2002.

135. Xu, L., Boron neutron capture therapy of human gastric cancer by boron-containing immunoliposomes under thermal neutron irradiation (in Chinese), *Zhonghua Yi Xue Za Zhi.*, 71, 568, 1991.

136. Xu, L., Zhang, X. Y., and Zhang, S. Y., In vitro and in vivo targeting therapy of immunoliposomes against human gastric cancer (in Chinese), *Zhonghua Yi Xue Za Zhi.*, 74, 83, 1994.

137. Pan, X. Q., Wang, H., and Lee, R. J., Boron delivery to a murine lung carcinoma using folate receptor-targeted liposomes, *Anticancer Res.*, 22, 1629, 2002.

138. Pan, X. Q. et al., Boron-containing folate receptor-targeted liposomes as potential delivery agents for neutron capture therapy, *Bioconjug. Chem.*, 13, 435, 2002.

139. Sudimack, J. J. et al., Folate receptor-mediated liposomal delivery of a lipophilic boron agent to tumor cells in vitro for neutron capture therapy, *Pharm. Res.*, 19, 1502, 2002.

140. Bohl Kullberg, E. et al., Introductory experiments on ligand liposomes as delivery agents for boron neutron capture therapy, *Int. J. Oncol.*, 23, 461, 2003.

141. Kullberg, E. B., Nestor, M., and Gedda, L., Tumor-cell targeted epiderimal growth factor liposomes loaded with boronated acridine: uptake and processing, *Pharm. Res.*, 20, 229, 2003.

142. Mehvar, R., Dextrans for targeted and sustained delivery of therapeutic and imaging agents, *J. Control. Release*, 69, 1, 2000.

143. Mehvar, R., Recent trends in the use of polysaccharides for improved delivery of therapeutic agents: pharmacokinetic and pharmacodynamic perspectives, *Curr. Pharm. Biotechnol.*, 4, 283, 2003.

144. Chau, Y., Tan, F. E., and Langer, R., Synthesis and characterization of dextran-peptide–methotrexate conjugates for tumor targeting via mediation by matrix metalloproteinase II and matrix metalloproteinase IX, *Bioconjug. Chem.*, 15, 931, 2004.

145. Zhang, X. and Mehvar, R., Dextran–methylprednisolone succinate as a prodrug of methylprednisolone: Plasma and tissue disposition, *J. Pharm. Sci.*, 90, 2078, 2001.

146. Larsson, B., Gabel, D., and Borner, H. G., Boron-loaded macromolecules in experimental physiology: tracing by neutron capture radiography, *Phys. Med. Biol.*, 29, 361, 1984.

147. Gabel, D. and Walczyna, R., B-Decachloro-*o*-carborane derivatives suitable for the preparation of boron-labeled biological macromolecules, *Z. Naturforsch., C*, 37, 1038, 1982.

148. Pettersson, M. L. et al., In vitro immunological activity of a dextran-boronated monoclonal antibody, *Strahlenther. Onkol.*, 165, 151, 1989.

149. Ujeno, Y. et al., The enhancement of thermal-neutron induced cell death by 10-boron dextran, *Strahlenther. Onkol.*, 165, 201, 1989.

150. Pettersson, M. L. et al., Immunoreactivity of boronated antibodies, *J. Immunol. Methods*, 126, 95, 1990.

151. Holmberg, A. and Meurling, L., Preparation of sulfhydrylborane-dextran conjugates for boron neutron capture therapy, *Bioconjug. Chem.*, 4, 570, 1993.

152. Carlsson, J. et al., Strategy for boron neutron capture therapy against tumor cells with over-expression of the epidermal growth factor-receptor, *Int. J. Radiat. Oncol. Biol. Phys.*, 30, 105, 1994.

153. Gedda, L. et al., Development and in vitro studies of epidermal growth factor-dextran conjugates for boron neutron capture therapy, *Bioconjug. Chem.*, 7, 584, 1996.

154. Mehta, S. C. and Lu, D. R., Targeted drug delivery for boron neutron capture therapy, *Pharm. Res.*, 13, 344, 1996.

155. Olsson, P. et al., Uptake of a boronated epidermal growth factor-dextran conjugate in CHO xenografts with and without human EGF-receptor expression, *Anti-Cancer Drug Des.*, 13, 279, 1998.

156. Novick, S. et al., Linkage of boronated polylysine to glycoside moieties of polyclonal antibody; boronated antibodies as potential delivery agents for neutron capture therapy, *Nucl. Med. Biol.*, 29, 159, 2002.

157. Ferro, V. A., Morris, J. H., and Stimson, W. H., A novel method for boronating antibodies without loss of immunoreactivity, for use in neutron capture therapy, *Drug Des. Discov.*, 13, 13, 1995.

158. Sano, T., Boron-enriched streptavidin potentially useful as a component of boron carriers for neutron capture therapy of cancer, *Bioconjug. Chem.*, 10, 905, 1999.

159. Thomas, J. and Hawthorne, M. F., Dodeca(carboranyl)-substituted closomers: Toward unimolecular nanoparticles as delivery vehicles for BNCT, *Chem. Commun. (Camb.)*, 1884, 2001.

160. Pardridge, W. M., Vector-mediated drug delivery to the brain, *Adv. Drug Deliv. Rev.*, 36, 299, 1999.

161. Pardridge, W. M., Blood–brain barrier biology and methodology, *J. Neurovirol.*, 5, 556, 1999.

162. Hatakeyama, H. et al., Factors governing the in vivo tissue uptake of transferrin-coupled polyethylene glycol liposomes in vivo, *Int. J. Pharm.*, 281, 25, 2004.

163. Yanagie, H. et al., Accumulation of boron compounds to tumor with polyethylene–glycol binding liposome by using neutron capture autoradiography, *Appl. Radiat. Isot.*, 61, 639, 2004.

164. Maruyama, K. et al., Targetability of novel immunoliposomes modified with amphipathic poly (ethylene glycol)s conjugated at their distal terminals to monoclonal antibodies, *Biochim. Biophys. Acta*, 1234, 74, 1995.

165. Yang, W. et al., Evaluation of systemically administered radiolabeled epidermal growth factor as a brain tumor targeting agent, *J. Neurooncol.*, 55, 19, 2001.

166. Bobo, R. H. et al., Convection-enhanced delivery of macromolecules in the brain, *Proc. Natl Acad. Sci. USA*, 91, 2076, 1994.

167. Groothuis, D. R., The blood–brain and blood–tumor barriers: A review of strategies for increasing drug delivery, *Neurooncology*, 2, 45, 2000.

168. Vogelbaum, M. A., Convection enhanced delivery for the treatment of malignant gliomas: symposium review, *J. Neurooncol.*, 73, 57, 2005.

169. Husain, S. R. and Puri, R. K., Interleukin-13 receptor-directed cytotoxin for malignant glioma therapy: from bench to bedside, *J. Neurooncol.*, 65, 37, 2003.

170. Kunwar, S., Convection enhanced delivery of IL13-PE38QQR for treatment of recurrent malignant glioma: presentation of interim findings from ongoing phase 1 studies, *Acta Neurochir. Suppl.*, 88, 105, 2003.

171. Wikstrand, C. J. et al., Monoclonal antibodies against EGFRvIII are tumor specific and react with breast and lung carcinomas and malignant gliomas, *Cancer Res.*, 55, 3140, 1995.

172. Wikstrand, C. J. et al., Cell surface localization and density of the tumor-associated variant of the epidermal growth factor receptor, EGFRvIII, *Cancer Res.*, 57, 4130, 1997.

173. Barth, R. F., Wu, G., Kawabata, S., Sferra, T. J., Bandyopadhyaya, A. K., Tjarks, W., Ferketich, A. K., Binns, P. J., Riley, K. J., Coderre, J. A., Ciesielski, M. J., Fenstermaker, R. A., Wikstrand, C. J. et al., Molecular targeting and treatment of EGFRvIII positive gliomas using boronated monoclonal antibody L8A4, *Clin. Cancer Res.*, 12, 3792, 2006.

174. Cokgor, I. et al., Phase I trial results of iodine-131-labeled antitenascin monoclonal antibody 81C6 treatment of patients with newly diagnosed malignant gliomas, *J. Clin. Oncol.*, 18, 3862, 2000.

175. Akabani, G. et al., Dosimetry and radiographic analysis of 131I-labeled anti-tenascin 81C6 murine monoclonal antibody in newly diagnosed patients with malignant gliomas: a phase II study, *J. Nucl. Med.*, 46, 1042, 2005.

176. Laske, D. W., Youle, R. J., and Oldfield, E. H., Tumor regression with regional distribution of the targeted toxin TF-CRM107 in patients with malignant brain tumors, *Nat. Med.*, 3, 1362, 1997.

177. Sampson, J. H. et al., Progress report of a Phase I study of the intracerebral microinfusion of a recombinant chimeric protein composed of transforming growth factor (TGF)-alpha and a mutated form of the Pseudomonas exotoxin termed PE-38 (TP-38) for the treatment of malignant brain tumors, *J. Neurooncol.*, 65, 27, 2003.

178. Weber, F. et al., Safety, tolerability, and tumor response of IL4-Pseudomonas exotoxin (NBI-3001) in patients with recurrent malignant glioma, *J. Neurooncol.*, 64, 125, 2003.

179. Weber, F. W. et al., Local convection enhanced delivery of IL4-Pseudomonas exotoxin (NBI-3001) for treatment of patients with recurrent malignant glioma, *Acta Neurochir. Suppl.*, 88, 93, 2003.

180. Ferrari, M., Nanovector therapeutics, *Curr. Opin. Chem. Biol.*, 9, 343, 2005.

181. Ferrari, M., Cancer nanotechnology: opportunities and challenges, *Nat. Rev. Cancer*, 5, 161, 2005.

182. Blue, T. E. and Yanch, J. C., Accelerator-based epithermal neutron sources for boron neutrom capture therapy of brain tumors, *J. Neurooncol.*, 62, 19, 2003.

7 Preclinical Characterization of Engineered Nanoparticles Intended for Cancer Therapeutics

Anil K. Patri, Marina A. Dobrovolskaia, Stephan T. Stern, and Scott E. McNeil

CONTENTS

7.1 Introduction 106
    7.1.1 Physicochemical Characterization 106
    7.1.2 In Vitro Characterization 107
    7.1.3 In Vivo Characterization 107
7.2 Physicochemical Characterization 107
    7.2.1 Physicochemical Characterization Strategies 108
    7.2.2 Instrumentation 108
        7.2.2.1 Spectroscopy 108
        7.2.2.2 Chromatography 109
        7.2.2.3 Microscopy 109
        7.2.2.4 Size and Size Distribution 110
        7.2.2.5 Surface Characteristics 112
        7.2.2.6 Functionality 112
        7.2.2.7 Composition and Purity 113
        7.2.2.8 Stability 114
7.3 In Vitro Pharmacological and Toxicological Assessments 115
    7.3.1 Special Considerations 115
    7.3.2 In Vitro Target-Organ Toxicity 116
    7.3.3 Cytotoxicity 117
    7.3.4 Oxidative Stress... 118
    7.3.5 Apoptosis and Mitochondrial Dysfunction 119
    7.3.6 Proteomics and Toxicogenomics 121
7.4 In Vivo Pharmacokinetic and Toxicological Assessments 121
7.5 Immunotoxicity 124

This project has been funded in whole or in part with federal funds from the National Cancer Institute, National Institutes of Health, under contract N01-CO-12400. The content of this publication does not necessarily reflect the views or policies of the Department of Health and Human Services, nor does mention of trade names, commercial products, or organizations imply endorsement by the U.S. Government.

7.5.1   Applicability of Standard Immunological Methods for Nanoparticle Evaluation
        and Challenges Specific to Nanoparticle Characterization 125
        7.5.1.1   Blood Contact Properties 125
7.5.2   Immunogenicity 128
7.6   Conclusions 128
References 129

7.1 INTRODUCTION

As discussed in other chapters of this book, nanotechnology offers the research community the potential to significantly transform cancer diagnostics and therapeutics. Our ability to manipulate the biological and physicochemical properties of nanomaterial allows for more efficient drug targeting and delivery, resulting in greater potency and specificity, and decreased adverse side effects. The combinatorial possibilities of these multifunctional platforms have been the focus of considerable research and funding, but realization of their use in clinical trials is highly dependent on rigorous preclinical characterization to meet regulatory provisions and elucidated structure–activity relationships (SARs). A rational characterization strategy for biomedical nanoparticles contains three essential components (see Figure 7.1): physicochemical characterization, in vitro assays, and in vivo studies.

7.1.1 PHYSICOCHEMICAL CHARACTERIZATION

One of the major criticisms of early biomedical nanotechnology research was the general lack of physicochemical characterization that did not allow for the meaningful interpretation of resulting data or inter-laboratory comparisons. Traditional small molecule drugs are characterized by data that contribute to the chemistry, manufacturing and controls (CMC) section of the investigational new drug (IND) application with Food and Drug Administration (FDA), which include their molecular weight, chemical composition, identity, purity, solubility, and stability. The instrumentation to ascertain these properties has been well established, and the techniques are standardized. Engineered nanomaterials have dimensions between small molecules and bulk materials and often exhibit different physical and chemical properties than their counterparts. These physical and chemical

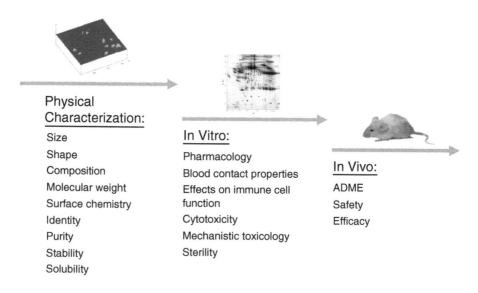

FIGURE 7.1 An assay cascade for preclinical characterization of nanomaterials.

properties influence the biological activity of nanoparticles and may depend on parameters such as particle size, size distribution, surface area, surface charge, surface functionality, shape, and aggregation state. Additionally, because most nanoparticle concepts are multifunctional, the distribution of targeting, imaging, and therapeutic components can also have dramatic effects on nanoparticle biological activity. There is a need to establish and standardize techniques to define these nanoparticle attributes.

There is now ample evidence that size and surface characteristics can dramatically affect nanoparticle behavior in biological systems.[1–4] For instance, a decrease in particle size leads to an exponential increase in surface area per unit mass, and an attendant increase in the availability of reactive groups on the surface. Nanoparticles with cationic surface character have a notably increased ability to cross the blood–brain barrier compared to nanoparticles with anionic surfaces.[5] In general, surface area, rather than mass, provides a better fit of dose–response relationships in toxicity studies for particles of various sizes.[6,7] Physicochemical characterization of properties, such as size, surface area, surface chemistry, and aggregation/agglomeration state, can provide the basis for better understanding of SARs.

7.1.2 IN VITRO CHARACTERIZATION

In vitro assays enable the isolation and analysis of specific biological and mechanistic pathways under controlled conditions. While many in vitro assays for nanomaterials will be similar to those used for traditional drugs, others will address mechanisms more specific to nanoparticles, such as oxidative stress. Noncellular assays measuring processes, such as protein adsorption, will also be an important accompaniment to cell-based assays. For example, monitoring the profile of serum proteins that absorb to nanoparticles in an in vitro environment may further our understanding of how nanoparticles interact with components of the reticuloendothelial system (RES) in vivo.[8,9] Additionally, proteomics and toxicogenomics can be employed to potentially identify biomarkers of toxicity related to nanomaterial exposure.[10]

7.1.3 IN VIVO CHARACTERIZATION

In vivo studies must be conducted to better understand the safety and behavior of nanoparticles in a living organism. As with any new chemical entity (NCE), the nanoparticle formulations' pharmacological and toxicological properties (i.e., ADME/Tox) need to be thoroughly characterized. In vivo studies should include examination of nanoparticles' effects on various organs and systems, such as the liver, heart, kidney, and immune system.

In this chapter, we outline the scientific rationale underlying the development of an assay cascade, with special attention paid to the selection and adaptations of assays and analytical protocols needed to extract meaningful efficacy and safety data from nanomaterials. These are presented in the following four sections: (1) physicochemical characterization, (2) in vitro pharmacology and toxicology assessment, (3) in vivo pharmacology and toxicology assessment, and (4) immunotoxicity. Standardized characterization of nanomaterials will facilitate better inter-laboratory comparison of results and will enhance the quality of scientific data submitted by investigators in support of their IND applications.

7.2 PHYSICOCHEMICAL CHARACTERIZATION

The physicochemical characteristics of nanomaterials can affect their cellular uptake, binding to blood proteins, access to target sites, and ability to cause damage to cells and tissue.[11] Standard methods for physicochemical characterization of nanomaterials will provide the basis for rational product development as well as consistent and interpretable results for tests of efficacy and safety. Few examples of standard characterization criteria exist in the literature, and there is as yet no consensus as to what measurement criteria are appropriate for any given nanomaterial product. However, it is clear that the diversity and complexity of nanomaterials used in biomedical applications dictates a more comprehensive and strategic approach to characterization than has been applied to date.

There are many varieties of nanomaterials currently being investigated for biomedical applications, especially for cancer diagnosis, imaging, and targeted drug delivery. These nanomaterials may be classified under several broad categories:

- Organic nanoparticles (e.g., dendrimers, polymers, functionalized fullerenes)
- Inorganic nanoparticles and organic–inorganic hybrids (e.g., iron oxide core particles, quantum dots)
- Liposomes and other biological nanomaterials

Each category of nanomaterial has a distinctly different composition that gives rise to different physical properties, such as solubility, stability, surface characteristics, and functional capabilities. Even within a single category, there can be a tremendous variety of product compositions, each with unique physical and chemical characteristics, and each requiring a different strategy for measuring those characteristics. This section examines the various categories of nanoparticles, and the tools and instrumentation available to address physicochemical characterization.

7.2.1 PHYSICOCHEMICAL CHARACTERIZATION STRATEGIES

Successful characterization strategies will enable one to begin associating the physicochemical properties of a nanomaterial with its in vivo behavior (i.e., SARs). This is an important step in the development of any material used for medical applications. For small molecules, the basis of most traditional drugs, the characterization techniques have been well established and standardized to determine their attributes, such as melting point, boiling point, molecular weight and structure, identity, composition, solubility, purity, and stability. These characteristics are measured and adequately defined using elemental analysis, mass spectrometry (MS), nuclear magnetic resonance (NMR), ultraviolet-visible (UV–vis) spectrophotometry, infrared (IR) spectroscopy, high-performance liquid chromatography (HPLC), gas chromatography (GC), capillary electrophoresis (CE), polarimetry, and other common analytical methods. Each of these individual techniques provides unique information about the sample, while together they provide the foundation for product quality control, manufacturing, and regulatory approval.

Many of the techniques used to characterize small molecules apply to nanomaterials. However, due to the composite nature of nanomaterials, the definition and measurement of these attributes can be quite different. To fully understand the attributes of a nanomaterial, additional characterizations are needed, such as size, surface chemistry, surface area, polydispersity, and zeta potential (see Figure 7.2). A comprehensive analysis of these properties is necessary to better understand in vivo effects and to allow for greater consistency and reproducibility in their preparation. The requirements set by regulatory bodies for quality control and consistency of biomedical nanomaterials are likely to be as stringent as those for small molecule preparations, but the path to verifying quality will require a more sophisticated approach. At the core of this analysis is an array of tools and instrumentation that are particularly well suited to measuring the properties of nanomaterials.

7.2.2 INSTRUMENTATION

7.2.2.1 Spectroscopy

Many traditional analytical methods can be applied to the characterization of nanomaterials. For example, NMR is extensively used to characterize dendrimers, polymers, and fullerenes derivatives, and provides unique information on the structure, purity, and functionality.[12–14]

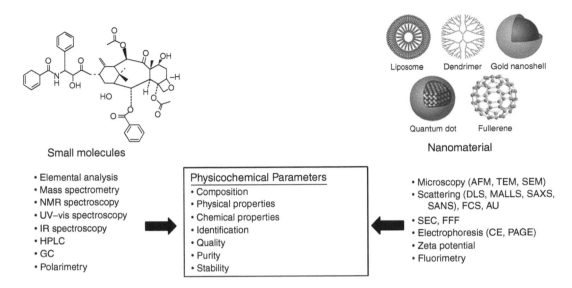

FIGURE 7.2 Physicochemical characterization methods and instrumentation for small molecules and nanotechnology.

In addition, the average number of terminal capping groups, number of small molecule ligands, and drugs in a multifunctional nanomaterial can be ascertained by comparing the integration values with chemical shifts unique to the ligands. UV–vis absorption spectrophotometry is also extensively used to identify and quantify the chromophore present in the preparation by using its extinction coefficient. Spectrofluorimetry is used in cases where the material has inherent fluorescence (such as quantum dots) or labeled with a fluorescence probe. Matrix-assisted laser desorption ionization time-of-flight (MALDI-TOF) MS is used extensively for macromolecules, dendrimers, and polymers to determine the molecular weight and utilize novel matrices to minimize the fragmentation of the macromolecule before reaching the detector. In the case of lower generation dendrimers, the presence of impurities, incomplete reaction, and reaction byproducts can be easily determined using MS.

7.2.2.2 Chromatography

Liquid chromatography methods such as analytical HPLC and size-exclusion chromatography (SEC; also called gel-permeation chromatography or GPC) utilize a column to separate components of a mixture in a liquid mobile phase based on their interaction with a solid stationary phase. The eluents are passed through UV–vis and fluorescence detectors with a flow cell where the absorbance and fluorescence is recorded to determine the purity of the sample. Although these techniques are suitable for stable polymers, dendrimers,[15] functionalized fullerenes, and protein- and peptide-based nanomaterial, they are not suitable for particles that degrade under experimental conditions or have excessive nonspecific binding to the solid matrix.

7.2.2.3 Microscopy

Scanning probe microscopy (SPM) techniques can be employed to measure the size, topography, composition, and structural properties of nanoparticles. Related techniques such as scanning tunneling microscopy (STM), electric field gradient microscopy (EFM), scanning thermal microscopy, and magnetic field microscopy (MFM) combined with atomic force microscopy (AFM), can be used to investigate the structural, electronic, thermal, and magnetic properties of a nanomaterial. AFM uses a nanoscale probe to detect the inter-atomic forces and interactions between the probe and the material being analyzed and is capable of determining size and shape within a spatial resolution of a few angstroms.[16] Apart from the ability to measure the particle size in a dry state as well as in aqueous and physiological conditions, AFM is a useful tool to probe the interaction of nanoparticles with supported lipid bilayers. This technique has been successfully used to compare nanoparticle interactions in in vitro cell assays.[17,18] The ability to image under physiological conditions makes AFM a powerful tool for the characterization of nanoparticles in a dynamic, biological context. A variant of this method, molecular recognition force microscopy (MRFM), can be employed to study the specific ligand–receptor interactions between nanoparticles and their biological targets.

Optical microscopy techniques are useful at the micron scale and are extensively used for imaging structural features. Fluorescence and confocal microscopy may be used to determine cellular binding and internalization of fluorescent-labeled nanoparticles[19] or those that are inherently fluorescent, such as quantum dots. But a more precise analysis of nanomaterial size and other direct measurements of physical properties will require a more sophisticated and specialized set of microscopic and spectroscopic techniques.

Scanning electron microscopy (SEM) provides information on the size, size distribution, shape, and density of nanomaterials. Transmission electron microscopy (TEM) and high-resolution TEM are more powerful than SEM in providing details at the atomic scale and can yield information regarding the crystal structure, quality, and grain size. TEM can be coupled with other characterization tools, such as electron energy loss spectrometry (EELS) or energy dispersive x-ray spectrometry (EDS), to provide additional information on the electronic structure and elemental composition of nanomaterials. Samples for TEM are evaluated dry or in a frozen state, under high-vacuum conditions. Nanoparticles analyzed by this instrument must therefore be stable under these extreme conditions. Additionally, while considered a gold standard of microscopic characterization methods, TEM requires a great deal of skill and time to obtain good data. In principle, when establishing characterization protocols, TEM can be used to validate characterization methods that are easier to use on a routine basis. Further description of analytical technologies as they apply to the measurement of specific nanomaterial properties is provided in the following sections.

7.2.2.4 Size and Size Distribution

Size is one of the critical parameters that dictate the absorption, biodistribution, and route of elimination for biomedical nanomaterials.[20] Generally, nanoparticles with dimensions of less than 5–10 nm are rapidly cleared after systemic administration, while particles from 10 to 70 nm in diameter may penetrate capillary walls throughout the body.[21,22] Larger particles 70–200 nm often remain in circulation for extended times.[22,23] This general correlation of biodistribution and elimination with respect to size may vary greatly depending on nanoparticle surface characteristics.

Specifically in cancer applications, size is an important factor in the accumulation of therapeutic nanomaterials in tumors, usually as a result of enhanced permeation and retention (EPR), caused by local defects in the vasculature and poor lymphatic drainage.[24] Particle size can be precisely tuned to take advantage of this phenomenon and passively target and deliver a therapeutic payload to tumors.[25–27]

Depending on the category of the nanomaterial, synthesis and scale-up can be problematic. Most biomedical nanomaterials for therapeutic and diagnostic applications are complex and involve some combination of molecular self-assembly, encapsulation, *and/*or the use of *nano*-sized metal or polymer cores, surfactants and/or proteins to impart solubility and functionality. Due to inherent variability in the manufacturing process, one rarely achieves a monodisperse, homogeneous product. It is therefore important to ascertain the precise size, size distribution, and polydispersity index (PDI) of the material. There are several techniques available to assess these parameters, including electron microscopy, AFM, and light scattering. Light scattering techniques can measure overall size and polydispersity of the particles. TEM is powerful in ascertaining the homogeneity of nanoparticles with encapsulated metals and in determining core size. With knowledge of nanoparticle geometry and size, surface area can also be estimated.

For biological applications, it is important to measure the physical characteristics of the nano-material in isotonic solution at physiological pH and temperature. The hydrodynamic size can be measured under these conditions using dynamic light scattering (DLS) (also known as photon correlation spectroscopy [PCS] and quasi elastic light scattering [QELS]) and analytical ultracen-trifugation (AU). In a DLS experiment, the effects of Brownian motion (particle movement caused by random collisions in solution) provide information on particle size and size distribution. The sample is illuminated with a laser, and the intensity fluctuations in the scattered light are analyzed and related to the size of the suspended particles. This technique is useful in determining whether the nanomaterial is monodisperse in size distribution. These data are influenced by the viscosity and the temperature of the medium, since Brownian motion depends on these factors. The pH of the medium and salt concentration may also affect the degree of agglomeration in some samples. With DLS, sample preparation is easy, the measurement is quick, and data are reproducible on larger sample volumes compared to microscopy techniques; however, better standardization of procedures, conditions, and data analysis tools will be required. Static light scattering provides information on molar mass and root-mean-squared (rms) radius for fractionated or monodisperse samples. One limitation of light scattering instruments is the inability to measure the size when the nanoparticles absorbs in the wavelength of the laser being used. Small-angle x-ray scattering (SAXS)[28] and small-angle neutron scattering (SANS)[29] can be used to measure the size, shape and orientation of components. Due to their cost and infrastructure requirements, there is limited availability of these instruments. For fluorescent nanomaterials such as quantum dots, size can be measured using fluorescence correlation spectroscopy (FCS).[30]

The hydrodynamic size of nanoparticles can also be measured with AU, which is traditionally used to measure the size of proteins.[31] The instrument spins the protein sample solution under high vacuum at a controlled speed and temperature while recording concentration distribution at set times. Even though this technique is designed to measure the size of proteins in solution, it has potential applications in the measurement of the hydrodynamic size of nanoparticles samples that are stable under the experimental conditions. Fractionation using SEC separates stable polymers into individual components and helps in the determination of the PDI. In the case of unfractionated samples, batch mode measurement provides averaged quantities such as weight-averaged molar mass and z-average rms radius. This technique is especially useful when combined with a refractive index detector to obtain absolute molecular weight for very high molecular weight polymers where traditional MS methods fail.

In cases where the separation and fractionation of nanomaterial is not possible using a column with a stationary phase, such as when the nanomaterial may interact with the column packing material and render it unstable, asymmetric-flow field flow fractionation (AFFF) is useful.[32] In AFFF, separation occurs when the sample passes through a narrow channel with a cross-flow through a porous semi-permeable membrane. The faster moving smaller particles rise to the top of the flow and come out first followed by larger particles that stay closer to the membrane and migrate more slowly. One advantage in this method is that there is no stationary phase in the separation: the sample injected comes out intact with little loss of material due to nonspecific binding. This feature is particularly useful for less stable nanoparticles such as liposomes, or for polymer- or protein-coated metal nanoparticles that would otherwise interfere with the performance of a traditional GPC column. The efficiency of separation for AFFF is not as good as with GPC, but there have been recent improvements in instrumentation that are closing the gap in performance. For both GPC and AFFF, the quantity and hydrodynamic size of the nanoparticles are detected in eluted peaks by measuring absorbance, refractive index, and light scattering.

In addition to size, the shape of a nanoparticle may affect its distribution and absorption in the body. Spherical, tubular, plate-like, or nano-porous materials of the same composition can vary significantly in their surface energy, biological activity, and access to different physiological structures, such as cell walls, capillary vessels, etc. Methods such as AFM, SEM, TEM, and STM can be used to determine the distribution of shape in a nanoparticle preparation.

7.2.2.5 Surface Characteristics

Surface characteristics contribute to the nanoparticle's solubility, aggregation tendency, ability to traverse biological barriers (such as a cell wall), biocompatibility, and targeting ability. The nanoparticle surface is also responsible for interaction and binding with plasma proteins in vivo, which in turn may alter the nanoparticle's distribution and pharmacokinetics. For multifunctional nanoparticles, modifying agents are often attached to the surface to bind to receptors in target tissues and organs. The presence of charged functionalities on the nanoparticle surface may increase nonspecific uptake, making the preparation less effective in targeting. It has been shown that dendrimer nanoparticles displaying positively charged amine groups on their surface can be significantly more hemolytic and cytotoxic than nanoparticles displaying negatively charged carboxylates.[20] The negatively charged nanoparticles were also cleared more slowly from the blood compared to positively charged species, following intravenous administration to rats.[20] Another potential effect of surface charge is to alter a nanoparticle's ability to penetrate the blood–brain barrier. Studies have shown that for emulsifying wax nanoparticles, anionic surfaces were superior to neutral or cationic surfaces for penetration of the blood–brain barrier.[33]

Surface characteristics can be tuned to improve receptor binding, reduce toxicity, or alter biodistribution. For example, when the above-mentioned dendrimers were acetylated to neutralize exposed surface charges, the toxic effects of the nanoparticles were also neutralized.[20,34] Surface properties can also lead to toxicity through interaction with molecular oxygen, leading to oxidative stress and inflammation. Electron capture at the surface of the nanoparticle results in the formation of the superoxide radical, which can set off a cascade of reactions (e.g., through Fenton reaction or disumation) to generate reactive oxygen species (ROS). ROS generation has been studied extensively for inhaled nanoparticles,[11,35] and has been observed in engineered nanoparticles such as fullerenes, single walled nanotubes (SWNTs), and quantum dots.[6,36–44] Studies have shown a direct correlation between nanoparticle surface area and ROS-generating capacity and inflammatory effects.[11]

The nature and integrity of nanomaterial surfaces must be established through analytical measurements to ensure product quality and account for surface-dependent effects on biodistribution and toxicity. Potentiometric titrations provide crucial information on the net charge of a nanoparticle, and include zeta potential analysis, which provides information on the net charge and distribution under physiological conditions. Polyacrylamide gel electrophoresis (PAGE) analysis of dendrimers and other nanopolymers yields information on the molecular weight and the polydispersity of nanoparticles (such as trailing generations in dendrimer populations) based on their migration through the gel under an electric field. PAGE is also a powerful tool in the qualitative analysis of bioconjugates of nanomaterials with DNA, oligonucleotides, antibodies, and other ligands. Further analysis of the surface charge distribution and polydispersity of nanomaterials can be conducted using CE. MS is also effective in ascertaining the number and distribution of charges, especially for smaller and purer nanoparticles with known molecular weight.

7.2.2.6 Functionality

Analysis of the functional components of nanomaterials, such as targeting, imaging, and therapeutic agents, is critical to understand the in vivo efficacy of the preparation. Characteristic features of functional components include their quantity, distribution, orientation, and activity. For targeting agents, a key advantage of their use in nanoparticles is their ability to provide increased avidity to the target due to polyvalency. The level of polyvalency and activity of targeting agents can be monitored using surface plasmon resonance (SPR) to measure the rate constants for nanoparticle association and dissociation. During preclinical development, the affinity of nanomaterial preparations for their target molecule/receptor can be analyzed using SPR and compared to data obtained for binding to cellular receptors in culture.[45]

The average number of targeting agents per nanoparticle has to be optimized for both solubility and binding affinity. Affinity chromatography or SEC can be employed with some nanoparticles to separate nanoparticles with targeting agents from those without targeting agents. In nanoparticles containing antibodies[19,46] or proteins, quantification can be achieved using an enzyme-linked immunosorbent assay (ELISA) or bicinchoninic acid assay (BCA) if the inherent property of the nanoparticle itself does not interfere with the assay. In the case of dendrimers, NMR has been successfully applied to analyze the average number of targeting agents by comparing the integration values of the signals associated with the targeting agents to those belonging to the dendrimer. This is still an averaged technique that cannot distinguish the distribution of targeting agent density on a population of nanoparticles.

For targeted drug delivery applications, it is obviously important for both the targeting and therapeutic agents to be on the same particle. If the therapeutic has UV–vis absorption, it can be quantified using UV–vis spectroscopy with the extinction coefficient of the drug. HPLC analysis is possible in some cases to evaluate the amount of the drug present in a known amount of material, after isolating the drug from the sample.

7.2.2.7 Composition and Purity

Biomedical nanomaterials can be comprised of a wide variety of substances, including polymers, metals and metal oxides, lipids and other organic compounds, and large biomolecules such as protein or DNA. In most cases, the nanomaterials combine two or more of these substances, such as in a core or shell of a particle, and in encapsulated or conjugated material. Analysis of chemical composition will be critical for confirming the purity and homogeneity of nanomaterial product preparations.

Elemental analysis, such as CHN analysis, is most often used to ascertain the purity of small molecules. For nanomaterials, elemental analysis can be used to determine the composition and ratios of different elements present in the sample. For example, this technique can be used to determine the amount of linker present, if a unique element (such as sulfur) has been employed in the synthesis. In the case of core–shell metal nanoparticles, the ratio of core to shell material ratios can be determined.

Atomic absorption (AA) and atomic emission (AE) spectroscopies can also be utilized to determine the composition of nanomaterials. For imaging applications using iron oxide nanoparticles or gadolinium (Gd)-based chelates, composition analysis is very important to quantify metals present in the preparation which influence imaging efficacy. Inductively coupled plasmon optical emission spectroscopy (ICP-OES) is very sensitive to determine the amount of Gd in such contrast agent conjugates.[47] Specific T1/T2 relaxivities of magnetic resonance contrast agents can, of course, be assessed under in vitro conditions in the actual MRI instrument.

The purity of synthetic small molecules can be determined with a high degree of certainty since the analyte usually consists of a single component. With nanomaterials, purity must be determined in the context of multiple layered, conjugated, and encapsulated components. Purity analysis must account for the presence of solvents, free metals and chelates, unconjugated therapeutic or other agents, precursors, dimers, etc., that result in artifacts and side products of the preparation.[48] Characterization of the inhomogeneity in ligand distribution is very important for efficacy as well as testing batch-to-batch reproducibility.[49] Proper methods and techniques to detect the presence of all these entities are required to ensure the purity and quality of nanomaterial preparations and to further expand our understanding of SARs.

7.2.2.8 Stability

The ability of multifunctional nanoparticles to combine targeting, therapeutic, and imaging modalities is a key aspect of their versatility and anticipated clinical impact.[50–52] With such complex compositions, the stability of all the components in nanoparticles is essential to their biological function. Premature release of any of the components from the composite preparation may render it ineffective. For example, in a nanodelivery system containing a targeting agent and a drug, the nanoparticles with the drug cannot bind to the desired targeting site if the targeting agent is prematurely cleaved or released. If the drug is prematurely released, even if the nanoparticle reaches its target, there will no longer be a therapeutic benefit.[53] For this reason, it is important to determine the in vitro functional component stability under physiological conditions.

For a nanomaterial providing targeted or timed-release drug delivery with an encapsulated drug, the release profile should be determined at different ionic strength, pH, and temperature conditions. Examples of such conditions include the stability at pH 7.4, in buffers such as phosphate buffered saline (PBS), and serum at 37°C. There are many nanoparticle designs being pursued which incorporate the selective release of components triggered by an external stimulus after targeted delivery. If a therapeutic attached to a nanoparticle uses a cleavable linkage, the efficiency of release should be determined under the expected cleavage conditions.[54]

In cases where a metal complex is used (for example, a Gd chelate for enhanced MRI contrast), the stability constants for the encapsulation or complexation should be determined, since any release of free heavy metal will increase the in vivo toxicity of the preparation.[55] The potential in vivo application of quantum dots has raised some concerns that the CdSe core might be exposed by the breakdown of its protective polymer or inorganic shell, releasing the highly toxic heavy metal Cd^{2+} ions into the bloodstream.[56] The quantum dot shells have been designed to be protective, but their long-term stability (e.g., susceptibility to Cd leaching) has not been established. Studies conducted on primary hepatocytes in vitro suggest that CdSe core quantum dots may be acutely toxic under certain conditions.[57] Other studies suggest that under physiological conditions, appropriately coated quantum dots do not expose the host organism to toxic levels of the core material.[58–60] Apparently conflicting evidence as to the safety of quantum dots highlights the necessity of clearly and objectively establishing the stability of these nanoparticles under physiological conditions using standardized methodologies.

It is also important to determine the stability of the nanoparticle under nonphysiological conditions to account for the effects of short-term and long-term storage, lyophilization, ultrafiltration, thermal exposure, pH variation, freeze–thawing, and exposure to light.

In summary, adequate physicochemical characterization of nanomaterials should be included as an essential requirement for preclinical characterization. Just as molecular characterization forms the basis of dosing and toxicity studies for small molecule therapeutics and diagnostic compounds, physicochemical characterization provides the foundation for dosing and toxicity studies for nanomaterials intended for clinical applications. Standardized protocols are being established by Standards Developing Organizations, such as the International Standards Organization (ISO) and American Society for Testing and Materials (ASTM), for characterizing the many types of biomedical nanomaterials being developed today for human use. Additionally, standardized reference material (SRM) will enable analytical technologies to be calibrated and protocols to be tested for consistency and to facilitate inter-laboratory comparisons.

To better control for the results of in vivo studies of nanomaterial absorption, distribution, metabolism, elimination, and toxicity, it will be necessary to examine the material in the same physicochemical state as would be found under physiological conditions. Particle-specific attributes that should be evaluated include surface characteristics, chemical composition, shape, size, and ligand dispersity. Additional properties that are influenced by experimental conditions include solubility, stability, protein binding, and aggregation state. Knowing the exact physiological conditions in different tissues and organs and developing a means to either replicate those conditions or measure physicochemical properties in situ is a significant challenge. But continued studies in this area will provide further data to elucidate the linkages between physicochemical characteristics of nanomaterials and their biological effects (i.e., SARs).

7.3 IN VITRO PHARMACOLOGICAL AND TOXICOLOGICAL ASSESSMENTS

Prior to filing an IND or investigational device exemption (IDE) application with the FDA and subsequent clinical testing in humans, a new product must be adequately studied for efficacy and safety using animal models. The cost- and labor-intensiveness of these in vivo studies impel drug and device researchers to make use of predictive in vitro methodologies wherever technology permits. In vitro models can serve as an initial assessment of a nanomaterial's efficacy and absorption, distribution, metabolism, elimination, and toxicity (ADME/Tox), allowing a more strategic approach to animal studies. Used iteratively with in vivo studies, the two approaches can inform each other and help narrow investigations of the physiological and biochemical pathways that contribute to ADME/Tox behavior.

A variety of cell-based in vitro systems are available, including perfused organs, tissue slices, cell cultures based on a single cell line or combination of cell lines, and primary cell preparations freshly derived from organ and tissue sources. In vitro models allow examination of biochemical mechanisms under controlled conditions, including specific toxicological pathways that may occur in target organs and tissues. Examples of mechanistic toxicological endpoints assessed in vitro include inhibition of protein synthesis and microtubule injury. These mechanistic endpoints can provide information not only as to the potential mechanisms of cell death, but also can identify compounds that may cause chronic toxicities that often results from sublethal mechanisms that may not cause overt toxicity in cytotoxicity assays. Common mechanistic paradigms associated with nanoparticle toxicity include oxidative stress, apoptosis, and mitochondrial dysfunction. Due to the nanoparticle- and approach-specific nature of pharmacology studies, it is beyond the scope of this chapter to discuss pharmacological assay specifics. Where appropriate models exist, chemotherapeutic efficacy can be examined in vitro. In certain cases, targeting of chemotherapeutic agents may be demonstrated as well, using optimized treatment/wash-out schemes in cell lines expressing the targeted receptor. Though nanoparticle metabolism or enzyme induction has yet to be demonstrated, certain nanomaterials with attractive chemistries may be subjected to phase-I/II metabolism and induction studies using cell-based, microsomal, and/or recombinant enzyme systems.

7.3.1 SPECIAL CONSIDERATIONS

Many of the standard methods used to evaluate biocompatibility of new molecular and chemical entities are fully applicable to nanoparticles. However, existing test protocols may require further development and laboratory validation before they become available for routine testing. Careful attention must be paid to potential sources of interference with analytical endpoints that may lead to false-positive or false-negative results. Nanoparticle interference could result from: interference with assay spectral measurements; inhibition/enhancement of enzymatic reactions[61,62]; and absorption of reagents to nanoparticle surfaces. In the event of nanoparticle interference, additional sample preparation steps or alternative methods may be required.

When evaluating the results of in vitro assays, it is important to recognize that dose–response relationships will not always follow a classical linear pattern. These atypical dose–response relationships have previously been attributed to shifts between the different mechanisms underlying the measured response.[63] In the case of nanomaterials, it is also important to bear in mind that concentration-dependent changes in the physical state (e.g., aggregation state, degree of protein binding) may also result in apparent nonlinearity.

Another key consideration when evaluating the results of nanoparticle research is the impact of dose metric (e.g., mass, particle number, surface area), sample preparation (e.g., sonication), and experimental conditions (e.g., exposure to light) on the interpretation of results. For example, surface area or particle number may be a more appropriate metric than mass when comparing data generated for different sized particles. This has been shown to be the case for 20- and 250-nm titanium dioxide nanoparticles, in which lung inflammation in rats, as assessed by percentage of neutrophils in lung lavage fluid, correlated with total surface area rather than mass.[64] The importance of experimental conditions in study design is highlighted by an investigation of functionalized fullerenes, demonstrating that the cytotoxicity of dendritic and malonic acid functionalized fullerenes to human T-lymphocytes in vitro is enhanced by photoexcitation.[41] The standardization of these experimental variables should limit inter-laboratory variability and make data generated more comparable.

7.3.2 In Vitro Target-Organ Toxicity

A recently published report from the International Life Sciences Institute Research Foundation/Risk Science Institute Nanomaterial Toxicity Screening Working Group[73] recommends the inclusion of several specific in vitro assays in a standard protocol of safety assessment. Much of the report focused on toxicity screening for environmental exposure to nanoparticles and thus emphasized environmentally relevant exposure routes. However, in addition to the in vitro examination of so-called portal-of-entry tissues, the report expressed the need for inclusion of potential target organs. The liver and kidney were selected as ideal candidates for these initial in vitro target organ toxicity studies, since preliminary investigations (discussed below) have identified these as the primary organs involved in the accumulation, processing, and eventual clearance of nanoparticles.

The liver has been identified in many studies as the primary organ responsible for reticuloendothelial capture of nanoparticles, often due to phagocytosis by Kupffer cells.[65–67] Fluorescein isothiocyanate-labeled polystyrene nanoparticles and radiolabeled dendrimers, for example, are rapidly cleared from the systemic circulation by hepatic uptake following intravenous injection.[4,68] Hepatic uptake has also been shown to be a primary mechanism of hepatic clearance for parenterally administered fullerenes, dendrimers, and quantum dots.[20,69,70] In addition to hepatic accumulation, nanoparticles have also been shown to have a detrimental effect on liver function ex vivo and alter hepatic morphology. Hepatocytes isolated from rats intravenously administered polyalkylcyanoacrylate nanoparticles had diminished secretion of albumin and decreased glucose production.[71] This alteration in albumin synthesis was also observed in freshly isolated rat hepatocytes exposed to polyalkylcyanoacrylate nanoparticles. Dendrimers have also been shown to cause liver injury. Repeated dosing of mice with polyamidoamine (PAMAM) dendrimers resulted in vacuolization of the hepatic parenchyma, suggesting lysosomal dysfunction.[72]

Sprague–Dawley rat hepatic primary cells and human hepatoma Hep-G2, selected for in vitro hepatic target organ toxicity assays, have a long history of use in toxicological evaluation.[73–75] Hep-G2 cells were chosen since they are a readily available hepatocyte cell line with high metabolic activity.[76] Rat hepatic primary cells were also chosen for toxicological studies, since hepatic primary cells in culture are more reflective of in vivo hepatocytes with regard to enzyme expression and specialized functions. One survey found rat hepatic primary cells up to ten times more sensitive to model hepatotoxic agents than established hepatic cell lines.[77] Rat hepatic primaries represent a suitable alternative to human hepatic primary cells, which are scarce, costly, and suffer from interindividual variability.

Preliminary pharmacokinetic studies of parenterally administered, radiolabeled carbon nanotubes, dendritic fullerenes, and low generation dendrimers in rodents have identified urinary excretion as the principal mechanism of clearance.[78–80] A variety of engineered nanoparticles, including actinomycin D-loaded isobutylcyanoacrylate, doxorubicin-loaded cyanoacrylate, and dendrimer nanoparticles, have also been shown to distribute to renal tissue following parenteral administration in rodents.[68,81,82] In the case of the doxorubicin-loaded cyanoacrylate nanoparticles, doxorubicin renal distribution was increased due to capture of the nanoparticles by glomerular mesangial cells. This resulted in a shift in the primary target organ from the heart to the kidney. Doxorubicin-induced renal injury presented as a severe proteinuria. Kidney injury has been demonstrated for other nanomaterials as well. Nano-zinc particles, for example, caused severe histological alterations in murine kidneys and Q-dots were shown to be cytotoxic to African green monkey kidney cells.[83,84]

The porcine renal proximal tubule cell line, LLC-PK1, was selected as a representative kidney cell line, since it is readily available through ATCC and has been used extensively in nephrotoxicity screening and mechanistic studies.[85] The SD rat hepatic primary, LLC-PK1 and Hep-G2 cells are adherent, which can simplify sample preparation, and can be propagated in a 96-well plate format suitable for high-throughput screening. The 96-well format allows for detailed concentration–response curves and multiple controls to be run on the same microplate. These cell lines were subjected to a variety of in vitro assays, described below, to evaluate cytotoxicity, mechanistic toxicology, and pharmacology. These assays have been selected primarily for their superior performance, convenience, and adaptability in evaluating this new class of biomedical agent.

7.3.3 Cytotoxicity

Cell viability of adherent cell lines can be assessed by a variety of methods.[86] These methods fall broadly into four categories, assays that measure: (1) loss of membrane integrity; (2) loss of metabolic activity; (3) loss of monolayer adherence; and (4) cell cycle analysis. Data generated using these various viability assays can be used to identify cell lines susceptible to nanoparticle toxicity and potentially give clues as to the type (i.e., cytostatic/cytotoxic) and location of cellular injury. Many of the cytotoxicity assays discussed below are available as commercial kits. These kits should be used whenever feasible since they provide an extra level of quality control.

1. Membrane integrity assays are particularly important as a measure of cellular damage, since there is evidence that some cationic nanoparticles, such as amine terminated dendrimers, exhibit toxic effects by disrupting the cell membrane.[87] Examples of assays that measure membrane integrity include the trypan blue exclusion assay and lactate dehydrogenase (LDH) leakage assay, which measures the presence of LDH released into the media through cell lysis.[88,89] The LDH leakage assay was selected because of its sensitivity and suitability for the high-throughput, 96-well plate format.
2. Examples of assays which measure metabolic activity include tetrazolium dye reduction, ATP, and 3H-thymidine incorporation assays. The 3-(4,5-dimethyl-2-thiazolyl)-2,5-diphenyl-2H-tetrazolium bromide (MTT) reduction assay was chosen for measurement of metabolic activity in the assay cascade, since it does not use radioactivity, and historically has been proven sensitive and reliable. MTT is a yellow water-soluble tetrazolium dye that is metabolized by live cells to water insoluble, purple formazan crystals. The formazan can be dissolved in DMSO and quantified by measuring the absorbance of the solution at 550 nm. Comparisons between the spectra of samples from nanoparticle treated and untreated cells can provide a relative estimate of cytotoxicity.[90]

   The MTT assay requires a solubilization step that is not required for the newer generation of tetrazolium dyes that form water-soluble formazans (e.g., XTT). However, these analogs require an intermediate electron acceptor that is often unstable, adding to assay variability. Furthermore, the net negative charge of these newer analogs limits cellular uptake, resulting in extracellular reduction.[91] MTT, with a net positive charge, readily crosses cell membranes and is reduced intracellularly, primarily in the mitochondria. Because nanoparticles have been shown to interact with cell membranes and could potentially interfere with the reduction of the newer generation analog via trans-plasma membrane electron transport, the traditional MTT assay would appear to be a better choice to assess cellular viability in nanoparticle cytotoxicity experiments. Analytes that are antioxidants, or are substrate/inhibitors of drug efflux pumps, have been shown to interfere with the MTT assay.[92,93] Functionalized fullerenes, which have not identified as efflux pump inhibitors or substrates, but do possess potent antioxidant activity, have been observed in our laboratory to cause MTT assay interference, resulting in enhanced MTT reduction and overestimation of cell viability (unpublished data).

3. Loss of monolayer adherence to plating surfaces is often used as a marker of cytotoxicity. Monolayer adherence is commonly measured by staining for total protein, following fixation of adherent cells. This simple assay is often a very sensitive indicator of loss of cell viability.[55] The sulforhodamine B total protein staining assay was selected for determination of monolayer adherence. Advantages of this assay include the ability to store the fixed, stained microplates for extended periods prior to measurement, making the assay especially suitable for high throughput.[94]

4. Cell cycle analysis is conducted using propidium iodide staining of DNA and flow cytometry.[95] Flow cytometric can be used as a screening test for toxicity of chemicals. This method can determine the effect of nanoparticle treatment on cell cycle progression, as well as cell death. Cell cycle effects have been shown for a variety of nanoparticles. For instance, carbon nanotubes have been shown to cause G1 cell cycle arrest in human embryonic kidney cells, with a corresponding decrease in expression of G1-associated cdks and cyclins.[96]

7.3.4 OXIDATIVE STRESS

The generation of free radicals by nanomaterials is well documented.[97,98] In most cases, the studied material was of ambient or industrial origin (quartz, carbon black, metal fumes, and diesel exhaust particles). However, engineered nanomaterials, such as fullerenes and polystyrene nanoparticles, have been shown to generate oxidative stress as well.[40,99,100] Lovric et al., for example, determined ROS to play an important role in cytotoxicity of quantum dots that have lost their protective coating.[101] The unique surface chemistries, large surface area, and redox active or catalytic contaminants (e.g., metals, quinones) of nanoparticles can facilitate ROS generation.[102] For example, fullerenes can perform electron transfer (phase-I pathway) or energy transfer (phase-II pathway) reactions with molecular oxygen following photoexcitation,[44] resulting in the formation of the superoxide anion radical or singlet oxygen, respectively. The superoxide anion radical can then undergo further reactions, such as dismutation and Fenton chemistry, to generate additional ROS species (e.g., ˙OH), resulting in cellular injury (see Scheme 7.1).[103] Evidence of fullerene-induced oxidative stress includes lipid peroxidation in the brains of exposed fish and treated rat liver microsomes.[40,104] Additional biomarkers of oxidative stress include a decrease in the reduced glutathione/oxidized glutathione ratio (GSH/GSSG), DNA fragmentation, and protein carbonyls.[105]

Biomarkers of nanoparticle-induced oxidative stress measured in our laboratory include ROS, lipid peroxidation products, and GSH/GSSG ratio. The fluorescent dichlorodihydroflourescein (DCFH) assay is used for measurement of ROS, such as hydrogen peroxide.[106] DCFH-DA is a ROS probe that undergoes intracellular deacetylation, followed by ROS-mediated oxidation to a fluorescent species, with excitation 485 nm and emission 530 nm. DCFH-DA can be used to measure ROS generation in the cytoplasm and cellular organelles, such as the mitochondria. The

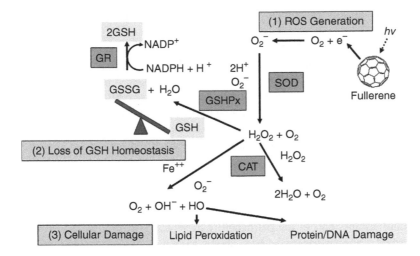

SCHEME 7.1 (1) Photoexcited fullerenes can perform electron-transfer reactions with molecular dioxygen to form the superoxide anion radical (O_2^-). Superoxide can then undergo superoxide dismutase (SOD)-catalyzed dismutation to hydrogen peroxide (H_2O_2). H_2O_2 is a substrate for catalase (CAT)-and glutathione peroxidase (GSHPx)-catalyzed detoxification reactions. (2) The oxidation of glutathione (GSH) to form oxidized glutathione (GSSG) during detoxification of H_2O_2 can result in a loss of glutathione homeostasis. GSH can be regenerated by glutathione reductase (GR). (3) Alternatively, hydrogen peroxide can undergo transition metal (Fe^{++})-catalyzed Fenton chemistry to form the highly reactive hydroxyl radical ($HO^\cdot$) that is capable of initiating lipid peroxidation and DNA/protein oxidation.

thiobarbituric acid reactive substances (TBARS) assay is used for measurement of lipid peroxidation products, such as lipid hydroperoxides and aldehydes. A molondialdehyde (MDA) standard curve is used for quantitation. MDA, a lipid peroxidation product, combines with thiobarbituric acid in a 1:2 ratio to form a fluorescent adduct, that is measured at 521 nm (excitation) and 552 nm (emission). TBARS are expressed as MDA equivalents.[107] The dithionitrobenzene (DTNB) assay is used for evaluation of glutathione homeostasis. In the DTNB assay, reduced GSH interacts with 5,5′-dithiobis(2-nitrobenzoic acid) (DTNB) to form the colored product 2-nitro-5-thiobenzoic acid, which is measured at 415 nm, and GSSG. GSSG is then reduced by glutathione reductase to form reduced GSH, which is again measured by the preceding method. Pretreatment with thiol-masking reagent, 1-methyl-4-vinyl-pyridinium trifluoromethane sulfonate, prevents GSH measurement, resulting in measurement of GSSG alone.[108]

7.3.5 Apoptosis and Mitochondrial Dysfunction

Nanoparticle-induced cell death can occur by either necrosis or apoptosis, processes that can be distinguished both morphologically and biochemically. Morphologically, apoptosis is characterized by perinuclear partitioning of condensed chromatin and budding of the cell membrane to form apoptotic bodies, whereas necrosis is characterized by cellular swelling (oncosis) and blebbing of the cell membrane.[109] In vitro studies have demonstrated the ability of nanoparticles, such as dendrimers and carbon nanotubes, to induce apoptosis.[110–112] In vitro exposure of macrophage-like mouse RAW 264.7 cells to cationic dendrimers led to apoptosis confirmed by morphological observation and the evidence of DNA cleavage.[112] Pretreatment of cells with a general caspase inhibitor (zVAD-fmk) reduced the apoptotic effect of the cationic dendrimer.[112] Apoptosis has also been observed in cultured human embryonic kidney cells (HEK293) and T lymphocytes treated with single walled carbon nanotubes, and in MCF-7 breast cancer cells treated with quantum dots.[101,110,113]

Apoptosis in mammalian cells can be initiated by four potential pathways: (1) mitochondrial pathway, (2) Death receptor-mediated pathway, (3) ER-mediated pathway, and (4) Granzyme B-mediated pathway.[114] Our laboratory has focused on caspase-3 activation in liver and kidney cells as a biomarker of apoptosis, since this a downstream event in all the classical apoptotic signaling pathways and can be measured using a fluorometric protease assay. This assay quantifies caspase-3 activation in vitro by measuring the cleavage of DEVD-7-amino-4-trifluoromethyl coumarin (AFC) to free AFC that emits a yellow-green fluorescence ($\lambda_{max} = 505$ nm).[115] This initial apoptosis screen can then be followed by additional analysis, as cellular morphology studies using nuclear staining techniques to detect perinuclear chromatin, or agarose gel electrophoresis to detect DNA laddering.[116]

Evidence supports a role for ROS in generation of the mitochondrial permeability transition via oxidation of thiol components of the permeability transition pore complex.[117] As discussed in the preceding sections, nanoparticles have been shown to induce oxidative stress, and thus this ROS-mediated pathway for induction of the mitochondrial permeability transition is a plausible apoptotic mechanism for nanomaterials. For instance, ambient ultrafine particulates have been shown to translocate to the mitochondria of RAW 264.7 murine macrophage cells, cause structural damage, and altered mitochondrial permeability.[98] A subsequent study demonstrated that mitochondrial dysfunction and apoptosis in the RAW 264.7 cells could be induced by polar compounds fractionated from ultrafine particles, suggesting that the mitochondrial dysfunction caused by ultrafines was the result of redox cycling of quinone contaminants on the surface of the particle.[118] This link between oxidative stress, mitochondrial dysfunction, and apoptosis has also been observed for man-made nanoparticles. For example, metal and quantum dot engineered nanoparticles have both been shown to induce oxidative stress, mitochondrial dysfunction, and apoptosis in various in vitro models.[101,119] Water-soluble, derivatized fullerenes, which have been shown to accumulate in the mitochondria of HS 68 human fibroblast cells, have also been shown to induce apoptosis in U251 human glioma cells.[120,121] While this derivatized fullerene-induced apoptosis in the glioma cell line did not involve oxidative stress, mitochondrial dysfunction was not measured and cannot be ruled out. Mitochondrial dysfunction and apoptosis have also been observed in a human gastric carcinoma cell line exposed to chitosan nanoparticles.[122] Taken together, these observations support a role for mitochondrial dysfunction and oxidative stress in nanoparticle-induced apoptosis. Apart from apoptosis, mitochondrial dysfunction has long been associated with necrotic cell death, and represents a potential necrotic mechanism of nanoparticle-induced injury as well.[123]

Mitochondrial dysfunction can result from several mechanisms in addition to opening of the permeability transition pore complex, including uncoupling of oxidative phosphorylation, damage to mitochondrial DNA, disruption of the electron transport chain, and inhibition of fatty acid β-oxidation.[124] Methods used to detect mitochondrial dysfunction include measurement of ATPase activity (via luciferin–luciferase reaction), oxygen consumption (via polarographic technique), morphology (via electron microscopy), and membrane potential (via fluorescent probe analysis).[125] Our laboratory measured loss of mitochondrial membrane potential in rat hepatic primaries, and Hep-G2 and LLC-PK1 cell lines, using the 5,5′,6,6′-tetrachloro-1,1′,3,3′-tetraethyl-benzimidazolcarbocyanine iodide (JC-1) assay, which is a convenient assay that does not require mitochondrial isolation or use specialized equipment.[126] This fluorescent dye partitions into the mitochondrial matrix as a result of the membrane potential. Concentration of JC-1 in the matrix results in aggregation that fluoresces at 590 nm (red). Upon loss of membrane potential, the dye dissipates from the matrix and can be measured, in its monomer state at emission 527 nm (green). The proportion of green to red fluorescence reflects the degree of mitochondrial membrane depolarization.

7.3.6 PROTEOMICS AND TOXICOGENOMICS

Proteomics and toxicogenomics are useful tools for identifying the mechanisms underlying toxicity.[127] Using gel electrophoresis in combination with MS identification, or gene microarray technology, the expression of specific pathway-responsive genes, such as Phase-II enzymes for oxidative stress, or cytokines for inflammation, can be identified. The delineation of these toxic pathways could help further refine the in vitro and in vivo study of nanomaterials, potentially leading to the development of novel biomarkers that could then be used in clinical and occupational toxicology studies. Proteomic and genomic research on biomedically relevant nanomaterials is presently underway, using a series of human hepatocyte, kidney, and immunological primary cells.

7.4 IN VIVO PHARMACOKINETIC AND TOXICOLOGICAL ASSESSMENTS

As is the case with any NCE, a thorough understanding of the properties that govern biocompatability is necessary to allow transition of nanomaterials to human clinical trials. Although in vitro toxicology studies can be informative, the obvious caveat is that phenomenon observed in vitro may not materialize in vivo due to differences in biological response or nanoparticle concentrations. Therefore, nanoparticle safety and therapeutic efficacy can only be definitively assessed by rigorous in vivo testing. This phase is guided in part by insights obtained from the physicochemical and in vitro characterization programs.

The primary goal of in vivo studies is to evaluate nanomaterials' pharmacokinetics, safety, and efficacy (see Table 7.1) in the most appropriate animal models. Preclinical toxicological and pharmacokinetic studies are conducted in accordance with the FDA regulatory guidance for IND and IDE submission. It is not within the scope of this chapter to review this regulatory guidance; instead, the reader is directed to the regulatory chapter of this text. While it is generally agreed that the current in vivo pharmacological and toxicological endpoints used for devices and small molecule drugs should be appropriate in assessing the safety and efficacy of biomedical nanomaterials, the qualities of nanomaterials that lend themselves to biomedical application, such as

---

TABLE 7.1
In Vivo Pharmacological and Toxicological Assessment of Nanomatersials

Category	Assessment
Initial disposition study	Tissue distribution
	Clearance mechanisms
	Half-life
	Systemic exposure (plasma AUC)
Immunotoxicity	28-day screen
	Immunogenicity (repeat-dose toxicity study)
	Hypersensitivity
	Immunostimulation
	Immunosuppression
Dose-range finding toxicity	NOAEL
	STD10
Good laboratory practices studies	PK/ADME
	Expanded single dose acute toxicity
	Repeated dose toxicity
Efficacy	Targeting
	Therapeutic
	Imaging

macromolecular structure and polydispersity, could also be problematic with respect to preclinical characterization, as will be discussed below. Early efforts are focused on identifying and standardizing the analytical and toxicological methodologies that are unique to nanoparticle preclinical characterization.

Several factors can influence the clinical viability of a new cancer diagnostic or therapeutic agent, aside from economic feasibility. These include: (1) demonstrated advantage over the current market standard; (2) appropriate administration route/schedule/elimination half-life; and (3) favorable safety profile.

1. The in vivo characterization phase includes an assessment of nanoparticle imaging and/or therapeutic efficacy in animal models that most closely approximate the human disease state. For instance, targeting efficacy is addressed by comparing a nanoparticle distribution profile with that of a nontargeted nanoparticle from the same class; however, this approach may be prone to ambiguities due to passive targeting via the EPR effects. For those particles with imaging components, the signal enhancement and tissue distribution profile is monitored using the appropriate magnetic resonance, ultrasound, optical, or positron emission tomography imaging instrumentation. In all cases, the approach will be compared against the present market standard to provide comparable data regarding efficacy, pharmacokinetics, and safety.

2. Animal studies of nanoparticles have rarely utilized the oral administration route, but those that have used that route have demonstrated poor intestinal absorption. For example, 98% of orally administered, PEG-functionalized fullerenes are eliminated in the feces within 48 h. Oral bioavailability of polystyrene nanoparticles were similarly poor, with less than 7% of 3000-, 1000-, 100-, 50-, and 3-nm sized particles absorbed.[128] Due to this extremely low nanoparticle oral bioavailability, the majority of biomedical nanoparticle formulations encountered undoubtedly will be intended for parenteral administration. Since diagnostic and chemotherapeutic regimens are typically of short duration, the inconvenience of intravenous administration does not appear to be a significant hurdle for eventual clinical transition.

   To be a successful diagnostic or therapeutic agent, nanoparticles should be eliminated from the body in a reasonable timeframe. However, studies have shown that optimum passive targeting of tumors, by the EPR effect, also requires that nanoparticle agents remain in the systemic circulation for prolonged periods.[129] Therefore, there must be a balance between systemic residence and clearance. At present, avoidance of the RES system and eventual urinary clearance appears to be a formidable obstacle for many of the current approaches, such as iron oxide MRI contrast agents and lipidic nanoparticle drug delivery agents, which have been shown to undergo capture by organs of the RES and remain for extended periods.[130,131] The primary mechanism of nanoparticle clearance, as discussed previously, appears to be glomerular filtration, which is governed by charge, molecular weight, and degree of protein binding.[132,133] A good case study of the importance of molecular weight in mediating glomerular filtration, by Gillies and colleagues, demonstrated that dendrimer–polyethylene oxide complexes greater than 40 kDa were cleared less readily than lower molecular weight species. Because timely clearance is an important drug attribute, accurate and thorough disposition studies are required.

   The single greatest obstacle for nanoparticle disposition studies is analytical methodology to quantitate nanoparticle concentrations in biological matrices, such as plasma and tissues. Due to their macromolecular and polydispersed nature, nanomaterials do not lend themselves to quantitation by traditional methods, such as HPLC and LC/MS, and may require alternative methods such as radiolabeling and scintillation counting. Since many biomedical nanoparticles will be multifunctional, and have imaging components, the imaging functionality of the particle could potentially be used for in vivo quantitation. In cases where no imaging component is present, the nanoparticles could be tagged with an appropriate probe to allow for imaging quantitation. The labeling method utilized may alter the surface properties of a nanoparticle, and thus affect the tissue distribution profile; comparison of alternative labeling methods would help identify consensus behavior. In any event, the use of imaging would first require validation using traditional methods such as LC/MS or scintillation to ensure image intensity could correlate to nanoparticle concentration in a linear fashion. Since many nanomaterials are electrondense, electron microscopy might also be used for tracking tissue distribution.

3. The objectives of the preclinical toxicological studies are to identify target organs of toxicity and to aid in the selection of starting doses for phase-I human clinical trials. Toward this end, toxicity studies seek to determine dose ranges causing (1) no adverse effects (NOAEL) and (2) life-threatening toxicity (i.e., severe toxic dose 10% [STD_{10}]). Studies are performed in two mammalian species, rodent and nonrodent, with rats the preferred rodent species, since they exhibit the greatest concordance with human toxicities.[134] Nanoparticle formulations are administered according to the intended clinical treatment cycle, with regard to schedule, duration, route, and formulation. Necropsy, performed on animals showing signs of morbidity during the study and at study termination, includes comprehensive hematology, histopathology, and clinical chemistry (see Figure 7.3). Several preclinical studies suggest a key role for reticuloendothelial organs, such as liver, kidney, and bone marrow, in the uptake of nanoparticles from the systemic circulation.[67,69] Therefore, these organs should receive special attention with regard to functional and histopathological evaluation. A review of the limited in vivo safety data available for nanoparticles supports the scrutinizing of these tissues, as there are several examples of RES organ injury, including hepatotoxicity and nephrotoxicity. For example, intravenous administration of cationic PAMAM dendrimers at low doses has

Histopathology			Hematology
Brain	Pancreas	Salivary gland	Erthrocyte count (RBC)
Lymph node	Esophagus	Parathyroid	Hemoglobin (HGB)
Thyroid	Trachea	Adrenal	Hematocrit (HCT)
Pituitary	Heart	Kidney	Mean corpuscular volume (MCV)
Thymus	Gall bladder	Liver	Mean corpuscular hemoglobin (MCH)
Spleen	Lung	Duodenum	Mean corpuscular hemoglobin conc. (MCHC)
Ileum	Rectum	Stomach	Platelet count (Plate)
Cecum	Colon	Jejunum	Reticulocyte count (RETIC)
Lymph node	Epididymis	Ovary	Total leukocyte count (WBC)
Prostate	Seminal vesicle	Testis	Differential leukocyte count
Urinary bladder	Uterus	Eye	Nucleated red blood cell count
Hardian gland	Nasal sections	Femur	
Femur	Vertebra	Spinal cord	
Mammary gland	Skin/subcuits	Tongue	

Clinical Chemistry		
BUN	AST	ALT
GGT	GLUC	Creatinine
total protein	Albumin	Globulin
A/G	Sodium	Potassium
Chloride		

FIGURE 7.3 Panel of tissues and chemistries to be assessed for comprehensive toxicology.

been shown to cause liver injury, as determined by histopathology and elevations in serum alanin aminotransferase when administered intravenously to mice; larger doses of the same cationic dendrimer were lethal to 100% of the mice.[135] Nephrotoxicity and hepatotoxicity, as determined by histopathology and serum enzyme markers, have both been observed in mice treated orally with nano-zinc.[72,84]

One of the potential advantages of nanoparticle drug formulations is an improved safety profile as a result of targeted therapy or elimination of toxic solubilization agents. For example, Baker and colleagues have shown that methotrexate-conjugated PAMAM dendrimers containing a folate receptor targeting ligand are more efficacious, and less toxic, than unformulated methotrexate against a murine human epithelial cancer model.[34] Abraxane is an example of a nanoparticle formulation of paclitaxel, presently on the market, that takes advantage of the solubilizing effect of albumin, eliminating the need for the toxic vehicle Cremophor EL.[136] Phase-III studies have shown enhanced therapeutic response of abraxane compared to Cremophor-formulated paclitaxel, while side effects, such as myelosuppression and peripheral neuropathy, were significantly reduced. As discussed above, nanoparticle drug formulations are compared against the unformulated drug to determine if the improvement in safety profile is realized.

7.5 IMMUNOTOXICITY

A growing body of evidence suggests that immunotoxicity provides a considerable contribution to onset and development of various disorders, including cancer and autoimmune diseases.[137–139] Nevertheless, it was not until recently that this relatively new field of toxicology emerged as an important interface between the fields of novel drug design and pharmacology. Recognition of immunosuppressive properties of new pharmaceuticals during early drug development phase is very important to eliminate potentially dangerous substances from the drug pipelines. For example, treatment of patients diagnosed with Crohn's disease and rheumatoid arthritis with inflix-imab and etanercept (both drugs represent neutralizing anti-TNF antibodies) resulted in increased incidence of tuberculosis and histoplasmosis.[140–143] Although these data did not result in withdrawal of any of the products from the pharmaceutical market, they helped initiate the strategy of preparing patients for anti-TNF therapy by screening for, and treatment of, latent tuberculosis prior to admin-istration of anti-TNF medications. Immunosuppression caused by pharmaceuticals can also lead to the development of lymphomas and acute leukemia.[144–146] Undesirable immunostimulation caused by pharmacological intervention include immunogenicity, hypersensitivity, and increased risk of autoimmune response. The standard toxicology endpoints employed for safety assessment of new pharmaceuticals primarily rely on clinical chemistry and histopathological evaluation of immune organs and were developed several decades ago. Currently, there is an increasing demand for the development of new methods for immunotoxicity assessment because of drug candidates' more complex structure as well as the application of new technologies in their manufacturing. The introduction of new molecular and immune cell biology methods into the immunotoxicology assessment framework is not a trivial and straightforward process. It requires not only scrupulous validation and standardization of the new techniques but also demonstration of the physiological relevance for the proposed battery of assays. These processes are expensive, time-consuming, and necessitate cooperation across the various pharmaceutical industry players. Unlike traditional drugs, multifunctional nanomaterials combine both chemistry-based and biotechnology-derived com-ponents, and therefore their characterization using standard methodologies requires adjustments and/or modification of classical experimental protocols. Below we will attempt to summarize data on critical aspects of immunotoxicological evaluation of nanomaterials and examine challenges in the application of standard methodologies for the assessments of nanoparticle safety to the immune system.

7.5.1 Applicability of Standard Immunological Methods for Nanoparticle Evaluation and Challenges Specific to Nanoparticle Characterization

Immunotoxicological evaluation of new drug candidates includes studies on both immunosuppression and immunostimulation, and is applicable to nanomaterials intended for use as drug candidates and/or drug delivery platforms. Short-term in vitro assays are being developed to allow for quick evaluation of nanoparticles' biocompatibility. The in vitro immunotoxicity assay cascade includes the following methods: analysis of plasma protein binding by two-dimensianol polyacrylamide gel electrophoresis (PAGE), hemolysis, platelet aggregation, coagulation, complement activation, colony forming unit-granulocyte macrophage (CFU-GM), leukocyte proliferation, phagocytosis, cytokine secretion by macrophages, chemotaxis, oxidative burst, and evaluation of cytotoxic activity of NK cells.[147] In addition to these methods, our in vitro tests include sterility assessment based on pyrogen contamination test (L-amebocyte lysate (LAL) assay) and evaluation of microbiological contamination. The assay cascade is based on several regulatory documents recommended buy the FDA for immunotoxicological evaluation of new investigational drugs, medical devices, and biotechnology, derived pharmaceuticals,[148–152] as well as on ASTM and ISO standards developed for characterization of blood contact properties of medical devices.[153–155] The aim of the in vitro immunoassay cascade is to provide quick evaluation of nanomaterials of interest prior to initiation of more thorough in vivo studies. Challenges specific for immunotoxicity assessment of nanoparticulate materials are summarized below.

7.5.1.1 Blood Contact Properties

One important aspect of nanoparticle used for medical applications is the assurance that they will not cause toxicity to blood elements when injected into a patient.

Hemolysis (i.e., damage to red blood cells) can lead to life-threatening conditions such as anemia, hypertension, arrhythmia, and renal failure. In our laboratory we have developed a protocol to evaluate hemolytic properties of nanoparticles based on the existing ASTM International standard used to characterize other materials.[147,156] We have identified several problems when applying existing protocol for nanoparticles characterization. For example, colloidal gold nanoparticles with size 5–50 nm have absorbance at 535 nm, which overlaps with the assay wavelength of 540 nm. Removal of these particles by centrifugation was required prior to sample evaluation for the presence of plasma-free hemoglobin to avoid false-positive results. Though it worked well for gold particles with size 10–50 nm, centrifugation may be problematic for other nanoparticles. For example, small colloidal gold particles with a size of 5 nm require higher centrifugation force to be removed from the supernatant. Hemoglobin has a size of 5 nm; therefore, one cannot exclude the possibility that ultracentrifugation of supernatant may pellet hemoglobin along with the gold particles and thus result in a false-negative result. Ultracentrifugation is not feasible for fullerenes or dendrimer particles. Analysis of polystyrene particles of 20, 50, and 80 nm revealed another complication. We found that particle preparation damages red blood cells; the damage is caused by the surfactant used during particle manufacturing, and is detected for 20- and 50-nm particles. For 80-nm particles, the hemolysis assay showed false-negative results due to the adsorption of hemoglobin by the particles. When we applied dialysis to remove surfactant, 50-nm particles revealed same phenomenon as 80-nm particles, i.e., they caused hemolysis, but adsorbed hemoglobin resulting in a false-negative result. Another potential problem with the application of standard hemolysis protocol for nanoparticles characterization is that metal-containing particles may oxidize hemoglobin and result in a change in assay OD responses. Therefore, assay procedures may require slight modifications depending on the particle type. In general, inclusion of particle test samples without blood allowed for quick assessment of the potential particle interference with the assay. In some instances, deduction of the result generated for blood-free particle control from that obtained for blood-plus particle sample was possible and allowed estimation of particle potential to cause damage to erythrocytes.

Blood coagulation. Blood coagulation may be affected by nanomaterials. For example, modification of surface chemistry has been shown to improve immunological compatibility at the particle–blood interface: application of polyvinyl chloride resin particles resulted in $19 \pm 4\%$ decrease in platelet count, indicating platelet adhesion/aggregation and increased blood coagulation time; the same particle preparation coated with PEG affected neither the platelet count nor elements of coagulation cascade.[157] Similarly, folate-coated and PEG-coated Gd nanoparticles did not aggregate platelets or activate neutrophils.[158] The evaluation of nanoparticle effects on blood coagulation includes studies on platelet function and coagulation factors. The in vitro cascade includes platelet aggregation assay and four coagulation assays measuring prothrombin time, activated partial thromboplastin time, thrombin time, and reptilase time.

Interaction with plasma proteins. High-resolution, 2D PAGE may be the method of choice to investigate plasma protein adsorption by nanoparticles. 2D PAGE has been used in several labs to isolate and identify plasma proteins adsorbed on the surface of stealth polycyanoacrylate particles[8], liposomes[159,160], solid lipid[161], and iron oxide nanoparticles.[162] Proteins commonly identified on several types of nanomaterials include antithrombin, C3 component of complement, alpha-2-macroglobulin, haptoglobin, plasminogen, immunoglobulins, albumin, fibrinogen, apolipoprotein, and transthyretin; albumin, immunoglobulins, and fibrinogen are the most abundant. Studies using this approach revealed that surface chemistry is important for protein adsorption. For example, Peracchia et al. demonstrated that coating with PEG results in approximately a fourfold reduction in protein binding by polycyanoacrylate particles.[163] Gessner et al. prepared polystyrene latex model nanoparticles with different surface charges. This study demonstrated that increasing surface charge density results in a quantitative increase in plasma protein adsorption, but did not show significant differences in the qualitative composition of the absorbed protein mixture.[164]

One of the most important step in this procedure is the separation of particles from plasma after incubation is complete. Ultracentrifugation was shown to be successful for isolation of iron oxide, solid lipid particles, and some polymer-based particles.[8,160,162] Gel filtration was applicable to liposomes,[159] solid lipid, and iron oxide nanoparticles. Thode and colleagues compared four methods for isolation of iron oxide particles, i.e., ultracentrifugation, static filtration, magnetic separation, and gel filtration. Depending on the method used for particle separation from bulk plasma, different quantities of the same proteins and different species of proteins were identified on the particles of the same size and surface chemistry.[162] For example, albumin was the predominant protein if static filtration and gel filtration were employed, while small quantities of this protein were found after ultracentrifugation; it was almost undetectable when magnetic separation was used. There was no difference in isolation of fibrinogen among the four methods. Comparable quantities of IgG gamma-chain were isolated using ultracentrifugation, static filtration, and magnetic separation, while gel filtration appeared to be inefficient in isolation of this protein.[162] Attention has to be paid to the sample preparation to avoid artificial protein adsorption due to desorption during the separation, for example. Other critical steps are the number of washes to remove an excess of bulk plasma, the type of wash buffer, and a buffer to dislodge protein from the particle surface. In our lab we also found that using polypropylene low-retention tubes and pipette tips is crucial for isolation of particle-specific proteins (unpublished data).

Complement activation and phagocytosis. Following intravenous administration, nanoscale drug carriers may suffer a drawback in that they may be taken up by cells of the mononuclear phagocytic system. Consequently, such uptake facilitates clearance of nanoparticulate carriers and associated drugs, thus leading to a decrease in drug efficacy.[165] The initial adsorption of plasma proteins such as components of complement and immunoglobulins promotes nanoparticles clearance.[166,167] Therefore, the investigation of a nanoparticle's ability to interact with and activate a complement and uptake by mononuclear cells seems to be one of the key assays in the preclinical characterization cascade. Classical immunoassays used to evaluate complement activation, such as the total hemolytic complement assay (CH50) and the alternative pathway (rabbit CH50 or APCH50), are based on the hemolysis of rabbit erythrocytes. These hemolytic assays can be used to measure functional activity of specific components of either pathway. The main challenge in applying these assays for nanomaterial characterization is the ability of nanoparticles per se to lyse RBCs, thus generating false-positive results. To overcome this limitation, the approach for evaluation of complement activation includes two techniques. One is a qualitative yes or no rapid screen for the presence of C3 cleavage products using western blot. The second assay is a quantitative evaluation of samples found positive at an initial screen for the presence of C4a, C3a, and C5a components of complement using a flow cytometry-based multiplex array.

The difficulties with application of standard phagocytosis assay are: (1) light microscopy used in traditional phagocytosis assay[168] is not applicable to nanoparticles due to their smaller size; (2) when light microscopy is substituted with TEM, visualization of particle may be complicated since their size is similar to that of cell organelles resulting in ambiguous interpretation of TEM data, thus TEM is limited to electron dense metal containing particles (see Figure 7.4); (3) labeling of nanoparticles with fluorescent tags[9,169] is a superior approach for visualizing internalized particles, but should be avoided in preclinical tests as chemical attachment of fluorophore may create a new molecular entity with properties widely divergent from those of the original particle; (4) application of luminol to detect phagocytosed particles[8] provides an exceptional technique which can overcome all limitations listed above; however, it is not free of applicability reservations as well (e.g., nanoparticles may interfere with activation of luminol once it is internalized, etc.).

CFU-GM assays. CFU-GM assays allow for the evaluation of potential nanoparticles interference with growth and differentiation of bone marrow stem cells into granulocyte and macrophages. This assay may provide valuable information for development of anti-cancer nanotechnology platforms. Myelosuppression is a very common dose-limiting toxicity associated with the use of oncology cytotoxic drugs. Incorporation of such drugs into nanoparticle carriers targeted to specific cancer cells may help to reduce toxicity to normal tissues, including bone marrow, and should be considered during initial characterization of nanocarriers.

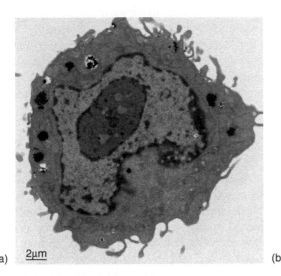
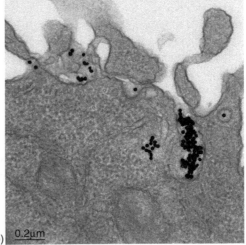

(a) 2μm   (b) 0.2μm

FIGURE 7.4 Study of internalization of 30-nm colloidal gold nanoparticles by murine macrophage cell line RAW 264.7 by TEM. (a) Analysis of single cell. (b) Zoom-in analysis of the selected area in the same cell.

7.5.2 IMMUNOGENICITY

This issue of immune stimulation by pharmaceuticals came into the forefront when biotechnology-derived products, especially recombinant proteins, moved toward clinical trials. Today it is evident that the immune system can effectively recognize biological therapeutics as foreign substances and build up a multi-level immune response against them. A number of factors result in the immune system responding to the administration of a pharmacological product, such as structure, formulation, folding architecture, but also degradation byproducts.[170] In addition, the route of administration and the dosage were shown to influence the staging and the amplitude of the immune system's response. In general, immune responses to biological products could be classified as benign in the sense that they affect only the pharmacological efficacy of the administered compound. The greatest concern is the robust immune response to certain biotechnology-derived products that are fraught with serious clinical consequences that may even result in a fatal outcome due to specific recognition and elimination of the patient's endogenous growth factors critical for survival.[171–173] For example, in the case of thrombopoietin, the immune response may result in the production of neutralizing antibodies, causing inhibition of the endogenous thrombopoietin with subsequent development of thrombocytopenia.[173] The patient's immune response to recombinant erythropoietin product Eprex® has been reported to induce pure red cell aplasia.[171,172,174] In the latter example, cross-reactivity tests indicated that antibodies generated against Eprex® could also neutralize other forms of erythropoietin products such as Epogen®, NeoRecormon®, and Aranesp® and suggested that antibodies are directed against some specific conformation of the erythropoietin active site.[170] Although incidence of the acute pure red cell aplasia remains relatively rare, the long-term implications are of great concern, as over half of patients who developed the auto-antibody remained transfusion dependent. The potential of using multifunctional nanoparticles for medical applications raises a key question of whether nanoparticle materials by themselves can induce an anti-nanoparticle immune response, stimulate allergic reactions, or trigger synthesis of nanopreparation-specific IgE. One can expect that the generation of antibodies to nanoparticles will ultimately affect only efficacy of the particle-based product. Of greatest concern will be the immune response to particles functionalized with growth factors, receptors or other biological molecules, which would result in the formation of antibodies, neutralizing the effect of these biological molecules and leading to potential exclusion of both particle-linked and endogenous proteins, akin to similar effects observed with biotechnology-derived pharmaceuticals. There are a limited number of studies on immunogenicity of nanomaterials. A few of them have shown that nanoparticles may both induce nanoparticles specific immune response and act as adjuvants.[175–178] Although preclinical animal studies may not be predictive to human immune response, available data described above do suggest that the immune system can recognize and build an immune response against some nanoparticles. Therefore, evaluation of nanoparticle antigenicity is seen as an important step during preclinical development. Other immunogenicity characterization should include evaluation of a nanoparticles' ability to act as an adjuvant and to induce allergic reactions.

7.6 CONCLUSIONS

The urgency to eliminate the pain and suffering associated with cancer is fueling research into novel therapies at a blistering pace. Through exquisitely targeted and multifunctional approaches, nanotechnology in particular holds great promise for enhancing cancer therapy. This promise, however, will never be realized if the safety of the nanomaterial is not demonstrated to allay public concern and if the regulatory structure is not in place to allow proper evaluation of the science. Without such a framework, the return on investment will be uncertain and progress will undoubtedly be stalled. The effort to develop a standardized set of protocols to characterize nanomaterials and their biological effects will provide a foundation for future regulation and will lead to a body of knowledge that will guide the design of safer nanotechnology products.

The biologic activity and toxicity of nanoscale particles are dependent on many parameters not typically examined for conventional small molecule therapeutics: size, shape, surface chemistry, stability of outer coating, agglomeration state, etc., and many conventional properties, such as stability or biodistribution, must be analyzed using a very different set of protocols and/or instrumentation. Emerging data from studies on nanoparticles engineered for medical use will build toward a consensus of instrumentation and experimental methodology needed to reproducibly determine the pharmacology and safety of these novel products.

The greatest challenge may not be in the development of new screening technologies, but in the ability to promulgate an accepted set of characterization protocols throughout the Nano Bio industry. The Nanotechnology Characterization Laboratory at the National Cancer Institute is working together with the FDA, NIST, and other regulatory and standards-setting organizations to establish the standard assays and technologies needed for timely delivery of safe and effective nanotechnology products.

REFERENCES

1. Furumoto, K., Nagayama, S., Ogawara, K., Takakura, Y., Hashida, M., Higaki, K., and Kimura, T., Hepatic uptake of negatively charged particles in rats: Possible involvement of serum proteins in recognition by scavenger receptor, *Journal of Controlled Release*, 97, 133–141, 2004.
2. Oberdorster, G., Toxicology of ultrafine particles: In vivo studies, *Philosophical Transactions of the Royal Society of London Series A—Mathematical Physical and Engineering Sciences*, 358, 2719–2739, 2000.
3. Ogawara, K., Yoshida, M., Higaki, K., Kimura, T., Shiraishi, K., Nishikawa, M., Takakura, Y., and Hashida, M., Hepatic uptake of polystyrene microspheres in rats: Effect of particle size on intrahepatic distribution, *Journal of Controlled Release*, 59, 15–22, 1999.
4. Ogawara, K., Yoshida, M., Kubo, J., Nishikawa, M., Takakura, Y., Hashida, M., Higaki, K., and Kimura, T., Mechanisms of hepatic disposition of polystyrene microspheres in rats: Effects of serum depend on the sizes of microspheres, *Journal of Controlled Release*, 61, 241–250, 1999.
5. Fenart, L., Casanova, A., Dehouck, B., Duhem, C., Slupek, S., Cecchelli, R., and Betbeder, D., Evaluation of effect of charge and lipid coating on ability of 60-nm nanoparticles to cross an in vitro model of the blood–brain barrier, *Journal of Pharmacological and Experimental Therapeutics*, 291, 1017–1022, 1999.
6. Brown, D. M., Wilson, M. R., MacNee, W., Stone, V., and Donaldson, K., Size-dependent proinflammatory effects of ultrafine polystyrene particles: a role for surface area and oxidative stress in the enhanced activity of ultrafines, *Toxicology and Applied Pharmacology*, 175, 191–199, 2001.
7. Donaldson, K. and Tran, C. L., Inflammation caused by particles and fibers, *Inhalation Toxicology*, 14, 5–27, 2002.
8. Gref, R., Luck, M., Quellec, P., Marchand, M., Dellacherie, E., Harnisch, S., Blunk, T., and Muller, R. H., 'Stealth' corona-core nanoparticles surface modified by polyethylene glycol (PEG): Influences of the corona (PEG chain length and surface density) and of the core composition on phagocytic uptake and plasma protein adsorption, *Colloids and Surfaces B*, 18, 301–313, 2000.
9. Leroux, J. C., Gravel, P., Balant, L., Volet, B., Anner, B. M., Allemann, E., Doelker, E., and Gurny, R., Internalization of poly(D,L-Lactic Acid) nanoparticles by isolated human-leukocytes and analysis of plasma-proteins adsorbed onto the particles, *Journal of Biomedical Materials Research*, 28, 471–481, 1994.
10. Reynolds, L. J. and Richards, R. J., Can toxicogenomics provide information on the bioreactivity of diesel exhaust particles? *Toxicology*, 165, 145–152, 2001.
11. Oberdorster, G., Oberdorster, E., and Oberdorster, J., Nanotoxicology: An emerging discipline evolving from studies of ultrafine particles, *Environmental Health Perspectives*, 113, 823–839, 2005.
12. Van de Coevering, R., Kreiter, R., Cardinali, F., van Koten, G., Nierengarten, J. F., and Gebbink, R., An octa-cationic core–shell dendrimer as a molecular template for the assembly of anionic fullerene derivatives, *Tetrahedron Letters*, 46, 3353–3356, 2005.

13. Tomalia, D. A., Naylor, A. M., and Goddard, W. A., Starburst dendrimers—molecular-level control of size shape, surface-chemistry, topology, and flexibility from atoms to macroscopic matter, *Angewandte Chemie-International Edition in English*, 29, 138–175, 1990.

14. Tomalia, D. A., Baker, H., Dewald, J., Hall, M., Kallos, G., Martin, S., Roeck, J., Ryder, J., and Smith, P., A new class of polymers—starburst-dendritic macromolecules, *Polymer Journal*, 17, 117–132, 1985.

15. Islam, M. T., Majoros, I. J., and Baker, J. R., HPLC analysis of PAMAM dendrimer based multifunctional devices, *Journal of Chromatography B*, 822, 21–26, 2005.

16. Binnig, G., Quate, C. F., and Gerber, C., Atomic force microscope, *Physical Review Letters*, 56, 930–933, 1986.

17. Mecke, A., Lee, D. K., Ramamoorthy, A., Orr, B. G., and Holl, M. M. B., Synthetic and natural polycationic polymer nanoparticles interact selectively with fluid-phase domains of DMPC lipid bilayers, *Langmuir*, 21, 8588–8590, 2005.

18. Mecke, A., Majoros, I. J., Patri, A. K., Baker, J. R., Holl, M. M. B., and Orr, B. G., Lipid bilayer disruption by polycationic polymers: The roles of size and chemical functional group, *Langmuir*, 21, 10348–10354, 2005.

19. Patri, A. K., Myc, A., Beals, J., Thomas, T. P., Bander, N. H., and Baker, J. R., Synthesis and in vitro testing of J591 antibody–dendrimer conjugates for targeted prostate cancer therapy, *Bioconjugate Chemistry*, 15, 1174–1181, 2004.

20. Malik, N., Wiwattanapatapee, R., Klopsch, R., Lorenz, K., Frey, H., Weener, J. W., Meijer, E. W., Paulus, W., and Duncan, R., Dendrimers: Relationship between structure and biocompatibility in vitro, and preliminary studies on the biodistribution of I-125-labelled polyamidoamine dendrimers in vivo, *Journal of Controlled Release*, 65, 133–148, 2000.

21. Hawley, A. E., Davis, S. S., and Illum, L., Targeting of colloids to lymph nodes: Influence of lymphatic physiology and colloidal characteristics, *Advanced Drug Delivery Reviews*, 17, 129–148, 1995.

22. Stolnik, S., Illum, L., and Davis, S. S., Long circulating microparticulate drug carriers, *Advanced Drug Delivery Reviews*, 16, 195–214, 1995.

23. Ishida, O., Maruyama, K., Sasaki, K., and Iwatsuru, M., Size-dependent extravasation and interstitial localization of polyethyleneglycol liposomes in solid tumor-bearing mice, *International Journal of Pharmaceutics*, 190, 49–56, 1999.

24. Maeda, H., Wu, J., Sawa, T., Matsumura, Y., and Hori, K., Tumor vascular permeability and the EPR effect in macromolecular therapeutics: A review, *Journal of Controlled Release*, 65, 271–284, 2000.

25. Brigger, I., Dubernet, C., and Couvreur, P., Nanoparticles in cancer therapy and diagnosis, *Advanced Drug Delivery Reviews*, 54, 631–651, 2002.0

26. Kawasaki, E. S. and Player, A., Nanotechnology, nanomedicine, and the development of new, effective therapies for cancer, *Nanomedicine: Nanotechnology, Biology and Medicine*, 1, 101–109, 2005.

27. Torchilin, V. P., Drug targeting, *European Journal of Pharmaceutical Sciences*, 11, S81–S91, 2000.

28. Prosa, T. J., Bauer, B. J., and Amis, E. J., From stars to spheres: A SAXS analysis of dilute dendrimer solutions, *Macromolecules*, 34, 4897–4906, 2001.

29. Nisato, G., Ivkov, R., and Amis, E. J., Structure of charged dendrimer solutions as seen by small-angle neutron scattering, *Macromolecules*, 32, 5895–5900, 1999.

30. Tsay, J. M., Doose, S., and Weiss, S., Rotational and translational diffusion of peptide-coated CdSe/CdS/ZnS nanorods studied by fluorescence correlation spectroscopy, *Journal of the American Chemical Society*, 128, 1639–1647, 2006.

31. Calabretta, M., Jamison, J. A., Falkner, J. C., Liu, Y., Yuhas, B. D., Matthews, K. S., and Colvin, V. L., Analytical ultracentrifugation for characterizing nanocrystals and their bioconjugates, *Nano Letters*, 5, 963–967, 2005.

32. Giddings, J. C., Field-flow fractionation: Analysis of macromolecular, colloidal, and particulate materials, *Science*, 260, 1456–1465, 1993.

33. Lockman, P. R., Koziara, J. M., Mumper, R. J., and Allen, D. D., Nanoparticle surface charges alter blood–brain barrier integrity and permeability, *Journal of Drug Targeting*, 12, 635–641, 2004.

34. Kukowska-Latallo, J. F., Candido, K. A., Cao, Z., Nigavekar, S. S., Majoros, I. J., Thomas, T. P., Balogh, L. P., Khan, M. K., and Baker, J. R., Nanoparticle targeting of anticancer drug improves therapeutic response in animal model of human epithelial cancer, *Cancer Research*, 65, 5317–5324, 2005.

35. Nel, A., Atmosphere: Enhanced: Air pollution-related illness: Effects of particles, *Science*, 308, 804–806, 2005.

36. Brown, D. M., Stone, V., Findlay, P., MacNee, W., and Donaldson, K., Increased inflammation and intracellular calcium caused by ultrafine carbon black is independent of transition metals or other soluble components, *Occupational and Environmental Medicine*, 57, 685–691, 2000.

37. Derfus, A. M., Chan, W. C. W., and Bhatia, S. N., Probing the cytotoxicity of semiconductor quantum dots, *Nano Letters*, 4, 11–18, 2004.

38. Joo, S. H., Feitz, A. J., and Waite, T. D., Oxidative degradation of the carbothioate herbicide, molinate, using nanoscale zero-valent iron, *Environmental Science & Technology*, 38, 2242–2247, 2004.

39. Nagaveni, K., Sivalingam, G., Hegde, M. S., and Madras, G., Photocatalytic degradation of organic compounds over combustion-synthesized nano-TiO_2, *Environmental Science & Technology*, 38, 1600–1604, 2004.

40. Oberdorster, E., Manufactured nanomaterials (fullerenes, C60) induce oxidative stress in the brain of juvenile largemouth bass, *Environmental Health Perspectives*, 112, 1058–1062, 2004.

41. Rancan, F., Rosan, S., Boehm, F., Cantrell, A., Brellreich, M., Schoenberger, H., Hirsch, A., and Moussa, F., Cytotoxicity and photocytotoxicity of a dendritic C(60) mono-adduct and a malonic acid C(60) tris-adduct on Jurkat cells, *Journal of Photochemistry and Photobiology B*, 67, 157–162, 2002.

42. Sayes, C., Fortner, J., Guo, W., Lyon, D., Boyd, A., Ausman, K., Tao, Y., Sitharaman, B., Wilson, L., Hughes, J., West, J., and Colvin, V. L., The differential cytotoxicity of water-soluble fullerenes, *Nano Letters*, 4, 1881–1887, 2004.

43. Wilson, M. R., Lightbody, J. H., Donaldson, K., Sales, J., and Stone, V., Interactions between ultrafine particles and transition metals in vivo and in vitro, *Toxicology and Applied Pharmacology*, 184, 172–179, 2002.

44. Yamakoshi, Y., Umezawa, N., Ryu, A., Arakane, K., Miyata, N., Goda, Y., Masumizu, T., and Nagano, T., Active oxygen species generated from photoexcited fullerene (C60) as potential medicines: O2-* versus 1O2, *Journal of the American Chemical Society*, 125, 12803–12809, 2003.

45. Sonvico, F., Mornet, S., Vasseur, S., Dubernet, C., Jaillard, D., Degrouard, J., Hocbeke, J., Duguet, E., Colombo, P., and Couvreur, P., Folate-conjugated iron oxide nanoparticles for solid tumor targeting as potential specific magnetic hyperthermia mediators: synthesis, physicochemical characterization, and in vitro experiments, *Bioconjugate Chemistry*, 16, 1181–1188, 2005.

46. Thomas, T. P., Patri, A. K., Myc, A., Myaing, M. T., Ye, J. Y., Norris, T. B., and Baker, J. R., In vitro targeting of synthesized antibody-conjugated dendrimer nanoparticles, *Biomacromolecules*, 5, 2269–2274, 2004.

47. Kobayashi, H. and Brechbiel, M. W., Nano-sized MRI contrast agents with dendrimer cores, *Advanced Drug Delivery Reviews*, 57, 2271–2286, 2005.

48. Shi, X., Lesniak, W., Islam, M. T., MuNiz, M. C., Balogh, L. P., and Baker, J. J. R., Comprehensive characterization of surface-functionalized poly(amidoamine) dendrimers with acetamide, hydroxyl, and carboxyl groups, *Colloids and Surfaces A*, 272, 139–150, 2006.

49. Shi, X., Majoros, I. J., Patri, A. K., Bi, X., Islam, M. T., Desai, A., Ganser, T. R., and Baker, J. R., Molecular heterogeneity analysis of poly(amidoamine) dendrimer-based mono- and multifunctional nanodevices by capillary electrophoresis, *Analyst*, 131, 374–381, 2006.

50. Ferrari, M., Cancer nanotechnology: Opportunities and challenges, *Nature Reviews Cancer*, 5, 161–171, 2005.

51. Patri, A. K., Majoros, I. J., and Baker, J. R., Dendritic polymer macromolecular carriers for drug delivery, *Current Opinion in Chemical Biology*, 6, 466–471, 2002.

52. Portney, N. G. and Ozkan, M., Nano-oncology: drug delivery, imaging, and sensing, *Analytical and Bioanalytical Chemistry*, 384, 620–630, 2006.

53. Patri, A. K., Kukowska-Latallo, J. F., and Baker, J. R., Targeted drug delivery with dendrimers: Comparison of the release kinetics of covalently conjugated drug and non-covalent drug inclusion complex, *Advanced Drug Delivery Reviews*, 57, 2203–2214, 2005.

54. Gillies, E. R. and Frechet, J. M., pH-responsive copolymer assemblies for controlled release of doxorubicin, *Bioconjugate Chemistry*, 16, 361–368, 2005.
55. Brucher, E., Kinetic stabilities of gadolinium(III) chelates used as MRI contrast agents, *Topics in Current Chemistry*, 221, 103–122, 2002.
56. Hardman, R., A toxicologic review of quantum dots: Toxicity depends on physicochemical and environmental factors, *Environmental Health Perspectives*, 114, 165–172, 2006.
57. Derfus, A. M., Chan, W. C. W., and Bhatia, S. N., Probing the cytotoxicity of semiconductor quantum dots, *Nano Letters*, 4, 11–18, 2004.
58. Dubertret, B., Skourides, P., Norris, D. J., Noireaux, V., Brivanlou, A. H., and Libchaber, A., In vivo imaging of quantum dots encapsulated in phospholipid micelles, *Science*, 298, 1759–1762, 2002.
59. Hoshino, A., Hanaki, K., Suzuki, K., and Yamamoto, K., Applications of T-lymphoma labeled with fluorescent quantum dots to cell tracing markers in mouse body, *Biochemical and Biophysical Research Communications*, 314, 46–53, 2004.
60. Voura, E. B., Jaiswal, J. K., Mattoussi, H., and Simon, S. M., Tracking metastatic tumor cell extravasation with quantum dot nanocrystals and fluorescence emission-scanning microscopy, *Nature Medicine*, 10, 993–998, 2004.
61. Ueng, T. H., Kang, J. J., Wang, H. W., Cheng, Y. W., and Chiang, L. Y., Suppression of microsomal cytochrome P450-dependent monooxygenases and mitochondrial oxidative phosphorylation by fullerenol, a polyhydroxylated fullerene C60, *Toxicology Letters*, 93, 29–37, 1997.
62. Shcharbin, D., Jokiel, M., Klajnert, B., and Bryszewska, M., Effect of dendrimers on pure acetylcholinesterase activity and structure, *Bioelectrochemistry*, 68, 56–59, 2006.
63. Slikker, W., Andersen, M. E., Bogdanffy, M. S., Bus, J. S., Cohen, S. D., Conolly, R. B., David, R. M., Doerrer, N. G., Dorman, D. C., Gaylor, D. W., Hattis, D., Rogers, J. M., Setzer, R. W., Swenberg, J. A., and Wallace, K., Dose-dependent transitions in mechanisms of toxicity: Case studies, *Toxicology and Applied Pharmacology*, 201, 226–294, 2004.
64. Oberdorster, G., Finkelstein, J. N., Johnston, C., Gelein, R., Cox, C., Baggs, R., and Elder, A. C., Acute pulmonary effects of ultrafine particles in rats and mice, *Research Report (Health Effects Institute)*, 5–74, 2000. (disc 75–86)
65. Bazile, D. V., Ropert, C., Huve, P., Verrecchia, T., Marlard, M., Frydman, A., Veillard, M., and Spenlehauer, G., Body distribution of fully biodegradable [14C]-poly(lactic acid) nanoparticles coated with albumin after parenteral administration to rats, *Biomaterials*, 13, 1093–1102, 1992.
66. Cagle, D. W., Kennel, S. J., Mirzadeh, S., Alford, J. M., and Wilson, L. J., In vivo studies of fullerene-based materials using endohedral metallofullerene radiotracers, *Proceedings of the National Academy of Sciences of the United States of America*, 96, 5182–5187, 1999.
67. Ogawara, K., Furumoto, K., Takakura, Y., Hashida, M., Higaki, K., and Kimura, T., Surface hydrophobicity of particles is not necessarily the most important determinant in their in vivo disposition after intravenous administration in rats, *Journal of Controlled Release*, 77, 191–198, 2001.
68. Nigavekar, S. S., Sung, L. Y., Llanes, M., El-Jawahri, A., Lawrence, T. S., Becker, C. W., Balogh, L., and Khan, M. K., H-3 dendrimer nanoparticle organ/tumor distribution, *Pharmaceutical Research*, 21, 476–483, 2004.
69. Ballou, B., Lagerholm, B. C., Ernst, L. A., Bruchez, M. P., and Waggoner, A. S., Noninvasive imaging of quantum dots in mice, *Bioconjugate Chemistry*, 15, 79–86, 2004.
70. Gharbi, N., Pressac, M., Hadchouel, M., Szwarc, H., Wilson, S. R., and Moussa, F., [60]Fullerene is a powerful antioxidant in vivo with no acute or subacute toxicity, *Nano Letters*, 5, 2578–2585, 2005.
71. Fernandezurrusuno, R., Fattal, E., Porquet, D., Feger, J., and Couvreur, P., Evaluation of liver toxicological effects induced by polyalkylcyanoacrylate nanoparticles, *Toxicology and Applied Pharmacology*, 130, 272–279, 1995.
72. Roberts, J. C., Bhalgat, M. K., and Zera, R. T., Preliminary biological evaluation of polyamidoamine (PAMAM) starburst dendrimers, *Journal of Biomedical Materials Research*, 30, 53–65, 1996.
73. Oberdorster, G., Maynard, A., Donaldson, K., Castranova, V., Fitzpatrick, J., Ausman, K., Carter, J., Karn, B., Kreyling, W., Lai, D., Olin, S., Monteiro-Riviere, N., Warheit, D., and Yang, H., Principles for characterizing the potential human health effects from exposure to nanomaterials: elements of a screening strategy, *Particle and Fibre Toxicology*, 2, 8, 2005.

74. Dierickx, P., Prediction of human acute toxicity by the hep G2/24-hour/total protein assay, with protein measurement by the CBQCA method, *Alternatives to Laboratory Animals*, 33, 207–213, 2005.

75. Knasmuller, S., Parzefall, W., Sanyal, R., Ecker, S., Schwab, C., Uhl, M., Mersch-Sundermann, V., Williamson, G., Hietsch, G., Langer, T., Darroudi, F., and Natarajan, A. T., Use of metabolically competent human hepatoma cells for the detection of mutagens and antimutagens, *Mutation Research*, 402, 185–202, 1998.

76. Urani, C., Doldi, M., Crippa, S., and Camatini, M., Human-derived cell lines to study xenobiotic metabolism, *Chemosphere*, 37, 2785–2795, 1998.

77. Wang, K., Shindoh, H., Inoue, T., and Horii, I., Advantages of in vitro cytotoxicity testing by using primary rat hepatocytes in comparison with established cell lines, *Journal of Toxicological Sciences*, 27, 229–237, 2002.

78. Gharbi, N., Pressac, M., Tomberli, V., Da Ros, T., Brettreich, M., Hadchouel, M., Arbeille, B., Trivin, F., Ceolin, R., Hirsch, A., Prato, M., Szwarc, H., Bensasson, R., and Moussa, F., In *Fullerenes 2000: Functionalized Fullerenes*, Maggini, M., Martin, N., and Guldi, D. M., Eds., vol. 9, The Electrochemical Society, Pennington, NJ, 2000.

79. Lee, C. C., MacKay, J. A., Frechet, J. M., and Szoka, F. C., Designing dendrimers for biological applications, *Nature Biotechnology*, 23, 1517–1526, 2005.

80. Wang, H., Wang, J., Deng, X., Sun, H., Shi, Z., Gu, Z., Liu, Y., and Zhao, Y., Biodistribution of carbon single-wall carbon nanotubes in mice, *Journal of Nanoscience and Nanotechnology*, 4, 1019–1024, 2004.

81. Manil, L., Couvreur, P., and Mahieu, P., Acute renal toxicity of doxorubicin (adriamycin)-loaded cyanoacrylate nanoparticles, *Pharmaceutical Research*, 12, 85–87, 1995.

82. Manil, L., Davin, J. C., Duchenne, C., Kubiak, C., Foidart, J., Couvreur, P., and Mahieu, P., Uptake of nanoparticles by rat glomerular mesangial cells in-vivo and in-vitro, *Pharmaceutical Research*, 11, 1160–1165, 1994.

83. Shiohara, A., Hoshino, A., Hanaki, K., Suzuki, K., and Yamamoto, K., On the cyto-toxicity caused by quantum dots, *Microbiology and Immunology*, 48, 669–675, 2004.

84. Wang, B., Feng, W. Y., Wang, T. C., Jia, G., Wang, M., Shi, J. W., Zhang, F., Zhao, Y. L., and Chai, Z. F., Acute toxicity of nano- and micro-scale zinc powder in healthy adult mice, *Toxicology Letters*, 161, 115–123, 2006.

85. Toutain, H. and Morin, J. P., Renal proximal tubule cell cultures for studying drug-induced nephro-toxicity and modulation of phenotype expression by medium components, *Renal Failure*, 14, 371–383, 1992.

86. Mickuviene, I., Kirveliene, V., and Juodka, B., Experimental survey of non-clonogenic viability assays for adherent cells in vitro, *Toxicology In Vitro*, 18, 639–648, 2004.

87. Hong, S., Bielinska, A. U., Mecke, A., Keszler, B., Beals, J. L., Shi, X., Balogh, L., Orr, B. G., Baker, J. R., and Banaszak Holl, M. M., Interaction of poly(amidoamine) dendrimers with supported lipid bilayers and cells: hole formation and the relation to transport, *Bioconjugate Chemistry*, 15, 774–782, 2004.

88. Decker, T. and Lohmann-Matthes, M. L., A quick and simple method for the quantitation of lactate dehydrogenase release in measurements of cellular cytotoxicity and tumor necrosis factor (TNF) activity, *Journal of Immunological Methods*, 115, 61–69, 1988.

89. Korzeniewski, C. and Callewaert, D. M., An enzyme-release assay for natural cytotoxicity, *Journal of Immunological Methods*, 64, 313–320, 1983.

90. Alley, M. C., Scudiero, D. A., Monks, A., Hursey, M. L., Czerwinski, M. J., Fine, D. L., Abbott, B. J., Mayo, J. G., Shoemaker, R. H., and Boyd, M. R., Feasibility of drug screening with panels of human tumor cell lines using a microculture tetrazolium assay, *Cancer Research*, 48, 589–601, 1988.

91. Berridge, M. V., Herst, P. M., and Tan, A. S., Tetrazolium dyes as tools in cell biology: New insights into their cellular reduction, *Biotechnology Annual Review*, 11, 127–152, 2005.

92. Natarajan, M., Mohan, S., Martinez, B. R., Meltz, M. L., and Herman, T. S., Antioxidant compounds interfere with the 3, *Cancer Detection and Prevention*, 24, 405–414, 2000.

93. Vellonen, K. S., Honkakoski, P., and Urtti, A., Substrates and inhibitors of efflux proteins interfere with the MTT assay in cells and may lead to underestimation of drug toxicity, *European Journal of Pharmaceutical Sciences*, 23, 181–188, 2004.

94. Voigt, W., Sulforhodamine B assay and chemosensitivity, *Methods in Molecular Medicine*, 110, 39–48, 2005.

95. Tuschl, H. and Schwab, C. E., Flow cytometric methods used as screening tests for basal toxicity of chemicals, *Toxicology In Vitro*, 18, 483–491, 2004.

96. Cui, D., Tian, F., Ozkan, C. S., Wang, M., and Gao, H., Effect of single wall carbon nanotubes on human HEK293 cells, *Toxicology Letters*, 155, 73–85, 2005.

97. Tao, F., Gonzalez-Flecha, B., and Kobzik, L., Reactive oxygen species in pulmonary inflammation by ambient particulates, *Free Radical Biology and Medicine*, 35, 327–340, 2003.

98. Li, N., Sioutas, C., Cho, A., Schmitz, D., Misra, C., Sempf, J., Wang, M., Oberley, T., Froines, J., and Nel, A., Ultrafine particulate pollutants induce oxidative stress and mitochondrial damage, *Environmental Health Perspectives*, 111, 455–460, 2003.

99. Fernandez-Urrusuno, R., Fattal, E., Rodrigues, J. M., Feger, J., Bedossa, P., and Couvreur, P., Effect of polymeric nanoparticle administration on the clearance activity of the mononuclear phagocyte system in mice, *Journal of Biomedical Materials Research*, 31, 401–408, 1996.

100. Fernandez-Urrusuno, R., Fattal, E., Feger, J., Couvreur, P., and Therond, P., Evaluation of hepatic antioxidant systems after intravenous administration of polymeric nanoparticles, *Biomaterials*, 18, 511–517, 1997.

101. Lovric, J., Cho, S. J., Winnik, F. M., and Maysinger, D., Unmodified cadmium telluride quantum dots induce reactive oxygen species formation leading to multiple organelle damage and cell death, *Chemical Biology*, 12, 1227–1234, 2005.

102. Risom, L., Moller, P., and Loft, S., Oxidative stress-induced DNA damage by particulate air pollution, *Mutation Research*, 592, 119–137, 2005.

103. Grisham, M. B. and McCord, J. M., Chemistry and cytotoxicity of reactive oxygen metabolites, In *Physiology of Oxygen Radicals*, Taylor, A. E., Matalon, S., and Ward, P., Eds., American Physiology Society, Bethesda, MD, pp. 1–18, 1986.

104. Kamat, J. P., Devasagayam, T. P. A., Priyadarsini, K. I., and Mohan, H., Reactive oxygen species mediated membrane damage induced by fullerene derivatives and its possible biological implications, *Toxicology*, 155, 55–61, 2000.

105. Dotan, Y., Lichtenberg, D., and Pinchuk, I., Lipid peroxidation cannot be used as a universal criterion of oxidative stress, *Progress in Lipid Research*, 43, 200–227, 2004.

106. Black, M. J. and Brandt, R. B., Spectrofluorometric analysis of hydrogen peroxide, *Analytical Biochemistry*, 58, 246–254, 1974.

107. Dubuisson, M. L., de Wergifosse, B., Trouet, A., Baguet, F., Marchand-Brynaert, J., and Rees, J. F., Antioxidative properties of natural coelenterazine and synthetic methyl coelenterazine in rat hepatocytes subjected to *tert*-butyl hydroperoxide-induced oxidative stress, *Biochemical Pharmacology*, 60, 471–478, 2000.

108. Shaik, I. H. and Mehvar, R., Rapid determination of reduced and oxidized glutathione levels using a new thiol-masking reagent and the enzymatic recycling method: application to the rat liver and bile samples, *Analytical and Bioanalytical Chemistry*, 385 (1), 105–113, 2006.

109. Van Cruchten, S. and Van den Broeck, W., Morphological and biochemical aspects of apoptosis, oncosis and necrosis, *Anatomia, Histologia, Embryologia: Journal of Veterinary Medicine Series C*, 31, 214–223, 2002.

110. Bottini, M., Bruckner, S., Nika, K., Bottini, N., Bellucci, S., Magrini, A., Bergamaschi, A., and Mustelin, T., Multi-walled carbon nanotubes induce T lymphocyte apoptosis, *Toxicology Letters*, 160, 121–126, 2006.

111. Cui, D., Tian, F., Ozkan, C. S., Wang, M., and Gao, H., Effect of single wall carbon nanotubes on human HEK293 cells, *Toxicology Letters*, 155, 73–85, 2005.

112. Kuo, J. H., Jan, M. S., and Chiu, H. W., Mechanism of cell death induced by cationic dendrimers in RAW 264.7 murine macrophage-like cells, *Journal of Pharmacy and Pharmacology*, 57, 489–495, 2005.

113. Cui, D., Tian, F., Ozkan, C. S., Wang, M., and Gao, H., Effect of single wall carbon nanotubes on human HEK293 cells, *Toxicology Letters*, 155, 73–85, 2005.

114. Wang, Z. B., Liu, Y. Q., and Cui, Y. F., Pathways to caspase activation, *Cell Biology International*, 29, 489–496, 2005.

115. Gurtu, V., Kain, S. R., and Zhang, G., Fluorometric and colorimetric detection of caspase activity associated with apoptosis, *Analytical Biochemistry*, 251, 98–102, 1997.

116. Loo, D. T. and Rillema, J. R., Measurement of cell death, *Methods in Cell Biology*, 57, 251–264, 1998.
117. Le Bras, M., Clement, M. V., Pervaiz, S., and Brenner, C., Reactive oxygen species and the mitochondrial signaling pathway of cell death, *Histology and Histopathology*, 20, 205–219, 2005.
118. Xia, T., Korge, P., Weiss, J. N., Li, N., Venkatesen, M. I., Sioutas, C., and Nel, A., Quinones and aromatic chemical compounds in particulate matter induce mitochondrial dysfunction: implications for ultrafine particle toxicity, *Environmental Health Perspectives*, 112, 1347–1358, 2004.
119. Hussain, S. M., Hess, K. L., Gearhart, J. M., Geiss, K. T., and Schlager, J. J., In vitro toxicity of nanoparticles in BRL 3A rat liver cells, *Toxicology In Vitro*, 19, 975–983, 2005.
120. Foley, S., Crowley, C., Smaihi, M., Bonfils, C., Erlanger, B. F., Seta, P., and Larroque, C., Cellular localisation of a water-soluble fullerene derivative, *Biochemical and Biophysical Research Communications*, 294, 116–119, 2002.
121. Isakovic, A., Markovic, Z., Todorovic-Markovic, B., Nikolic, N., Vranjes-Djuric, S., Mirkovic, M., Dramicanin, M., Harhaji, L., Raicevic, N., Nikolic, Z., and Trajkovic, V., Distinct cytotoxic mechanisms of pristine versus hydroxylated fullerene, *Toxicological Sciences*, 91 (1), 173–183, 2006.
122. Qi, L. F., Xu, Z. R., Li, Y., Jiang, X., and Han, X. Y., In vitro effects of chitosan nanoparticles on proliferation of human gastric carcinoma cell line MGC803 cells, *World Journal of Gastroenterology*, 11, 5136–5141, 2005.
123. Hirsch, T., Susin, S. A., Marzo, I., Marchetti, P., Zamzami, N., and Kroemer, G., Mitochondrial permeability transition in apoptosis and necrosis, *Cell Biology and Toxicology*, 14, 141–145, 1998.
124. Amacher, D. E., Drug-associated mitochondrial toxicity and its detection, *Current Medicinal Chemistry*, 12, 1829–1839, 2005.
125. Gogvadze, V., Orrenius, S., and Zhivotovsky, B., Analysis of mitochondrial dysfunction during cell death. In *Current Protocols in Toxicology*, Wiley, New York, pp. 10.1–2.10.27, 2004.
126. Guo, W.-X., Pye, Q. N., Williamson, K. S., Stewart, C. A., Hensley, K. L., Kotake, Y., Floyd, R. A., and Broyles, R. H., Mitochondrial dysfunction in choline deficiency-induced apoptosis in cultured rat hepatocytes, *Free Radical Biology and Medicine*, 39, 641–650, 2005.
127. Merrick, B. A. and Madenspacher, J. H., Complementary gene and protein expression studies and integrative approaches in toxicogenomics, *Toxicology and Applied Pharmacology*, 207, 189–194, 2005.
128. Jani, P., Halbert, G. W., Langridge, J., and Florence, A. T., Nanoparticle uptake by the rat gastrointestinal mucosa: Quantitation and particle size dependency, *Journal of Pharmacy and Pharmacology*, 42, 821–826, 1990.
129. Duncan, R., Polymer conjugates for tumour targeting and intracytoplasmic delivery. The EPR effect as a common gateway? *Pharmaceutical Science & Technology Today*, 2, 441–449, 1999.
130. Briley-Saebo, K., Hustvedt, S. O., Haldorsen, A., and Bjornerud, A., Long-term imaging effects in rat liver after a single injection of an iron oxide nanoparticle based MR contrast agent, *Journal of Magnetic Resonance Imaging*, 20, 622–631, 2004.
131. Cahouet, A., Denizot, B., Hindre, F., Passirani, C., Heurtault, B., Moreau, M., Le Jeune, J., and Benoit, J., Biodistribution of dual radiolabeled lipidic nanocapsules in the rat using scintigraphy and gamma counting, *International Journal of Pharmaceutics*, 242, 367–371, 2002.
132. Deen, W. M., Lazzara, M. J., and Myers, B. D., Structural determinants of glomerular permeability, *American Journal of Physiology—Renal Physiology*, 281, F579–F596, 2001.
133. Walton, K., Dorne, J. L., and Renwick, A. G., Species-specific uncertainty factors for compounds eliminated principally by renal excretion in humans, *Food and Chemical Toxicology*, 42, 261–274, 2004.
134. Olson, H., Betton, G., Robinson, D., Thomas, K., Monro, A., Kolaja, G., Lilly, P., Sanders, J., Sipes, G., Bracken, W., Dorato, M., Van Deun, K., Smith, P., Berger, B., and Heller, A., Concordance of the toxicity of pharmaceuticals in humans and in animals, *Regulatory Toxicology and Pharmacology*, 32, 56–67, 2000.
135. Neerman, M. F., Zhang, W., Parrish, A. R., and Simanek, E. E., In vitro and in vivo evaluation of a melamine dendrimer as a vehicle for drug delivery, *International Journal of Pharmaceutics*, 281, 129–132, 2004.
136. Harries, M., Ellis, P., and Harper, P., Nanoparticle albumin-bound paclitaxel for metastatic breast cancer, *Journal of Clinical Oncology*, 23, 7768–7771, 2005.
137. Descotes, J., Importance of immunotoxicity in safety assessment: a medical toxicologist's perspective, *Toxicology Letters*, 149, 103–108, 2004.

138. Dobrovolskaia, M. A. and Kozlov, S. V., Inflammation and cancer: When NF-kappa B amalgamates the perilous partnership, *Current Cancer Drug Targets*, 5, 325–344, 2005.

139. Merk, H. F., Sachs, B., and Baron, J., The skin: Target organ in immunotoxicology of small-molecular-weight compounds, *Skin Pharmacology and Applied Skin Physiology*, 14, 419–430, 2001.

140. Myers, A., Clark, J., and Foster, H., Tuberculosis and treatment with infliximab, *New England Journal of Medicine*, 346, 625–626, 2002.

141. Keane, J., Gershon, S., Wise, R. P., Mirabile-Levens, E., Kasznica, J., Schwieterman, W. D., Siegel, J. N., and Braun, M. M., Tuberculosis associated with infliximab, a tumor necrosis factor (alpha)-neutralizing agent, *New England Journal of Medicine*, 345, 1098–1104, 2001.

142. Zhang, Z., Correa, H., and Begue, R. E., Tuberculosis and treatment with infliximab, *New England Journal of Medicine*, 346, 623–626, 2002.

143. Lim, W. S., Powell, R. J., and Johnston, I. D., Tuberculosis and treatment with infliximab, *New England Journal Medicine*, 346, 623–626, 2002.

144. Nart, D., Nalbantgil, S., Yagdi, T., Yilmaz, F., Hekimgil, M., Yuce, G., and Hamulu, A., Primary cardiac lymphoma in a heart transplant recipient, *Transplant Proceedings*, 37, 1362–1364, 2005.

145. Caillard, S., Pencreach, E., Braun, L., Marcellin, L., Jaegle, M. L., Wolf, P., Parissiadis, A., Hannedouche, T., Gaub, M. P., and Moulin, B., Simultaneous development of lymphoma in recipients of renal transplants from a single donor: Donor origin confirmed by human leukocyte antigen staining and microsatellite analysis, *Transplantation*, 79, 79–84, 2005.

146. Karakus, S., Ozyilkan, O., Akcali, Z., Demirhan, B., and Haberal, M., Acute myeloid leukemia 4 years after Kaposi's sarcoma in a renal transplant recipient, *Onkologie*, 27, 163–165, 2004.

147. http://ncl.cancer.gov/working_assay-cascade.asp

148. FDA/CDER, Guidance for industry. Immunotoxicology evaluation of investigational new drugs, 2002, http://www.fda.gov/Cder/guidance/.

149. FDA/CDER, ICH S8: Immunotoxicity studies for human pharmaceuticals (draft), 2004, http://www.fda.gov/Cder/guidance/.

150. FDA/CBER/CDER, Guidance for industry. Developing medical imaging drug and biological products. Part 1: Conducting safety assessments, 2004, http://www.fda.gov/Cder/guidance/.

151. FDA/CDER/CBER, Guidance for industry. ICH S6: Preclinical safety evaluation of biotechnology-derived pharmaceuticals, 1997, http://www.fda.gov/Cder/guidance/.

152. FDA/CDRH, Guidance for industry and FDA reviewers. Immunotoxicity testing guidance, May 1999, http://www.fda.gov/cdrh/guidance.html.

153. ANSI/AAMI/ISO, 10993-4: Biological evaluation of medical devices. Part 4: Selection of tests for interaction with blood, 2002, http://www.iso.org/iso/en/CatalogueListPage.CatalogueList.

154. ASTM, Standard practice. F1906-98: Evaluation of immune responses in biocompatibility testing using ELISA tests, lymphocyte proliferation, and cell migration, 2003, http://www.astm.org/cgibin/SoftCart.exe/NEWSITE_JAVASCRIPT/index.shtml?L+mystore+zdrn2970+1157995562.

155. ASTM, Standard practice. F748-98: Selecting generic biological test methods for materials and devices, http://www.astm.org/cgibin/SoftCart.exe/NEWSITE_JAVASCRIPT/index.shtml?L+mystore+zdrn2970+1157995562.

156. ASTM, F756-00: Standard practice for assessment of hemolytic properties of materials, 2000, http://www.astm.org/cgibin/SoftCart.exe/NEWSITE_JAVASCRIPT/index.shtml?L+mystore+zdrn2970+1157995562.

157. Balakrishnan, B., Kumar, D. S., Yoshida, Y., and Jayakrishnan, A., Chemical modification of poly(vinyl chloride) resin using poly(ethylene glycol) to improve blood compatibility, *Biomaterials*, 26, 3495–3502, 2005.

158. Oyewumi, M. O., Yokel, R. A., Jay, M., Coakley, T., and Mumper, R. J., Comparison of cell uptake, biodistribution and tumor retention of folate-coated and PEG-coated gadolinium nanoparticles in tumor-bearing mice, *Journal of Controlled Release*, 95, 613–626, 2004.

159. Diederichs, J. E., Plasma protein adsorption patterns on liposomes: establishment of analytical procedure, *Electrophoresis*, 17, 607–611, 1996.

160. Harnisch, S. and Muller, R. H., Plasma protein adsorption patterns on emulsions for parenteral administration: establishment of a protocol for two-dimensional polyacrylamide electrophoresis, *Electrophoresis*, 19, 349–354, 1998.

161. Goppert, T. M. and Muller, R. H., Alternative sample preparation prior to two-dimensional electrophoresis protein analysis on solid lipid nanoparticles, *Electrophoresis*, 25, 134–140, 2004.

162. Thode, K., Luck, M., Semmler, W., Muller, R. H., and Kresse, M., Determination of plasma protein adsorption on magnetic iron oxides: sample preparation, *Pharmaceutial Research*, 14, 905–910, 1997.

163. Peracchia, M. T., Harnisch, S., Pinto-Alphandary, H., Gulik, A., Dedieu, J. C., Desmaele, D., d'Angelo, J., Muller, R. H., and Couvreur, P., Visualization of in vitro protein-rejecting properties of PEGylated stealth (R) polycyanoacrylate nanoparticles, *Biomaterials*, 20, 1269–1275, 1999.

164. Gessner, A., Lieske, A., Paulke, B. R., and Muller, R. H., Influence of surface charge density on protein adsorption on polymeric nanoparticles: Analysis by two-dimensional electrophoresis, *European Journal of Pharmaceutics and Biopharmaceutics*, 54, 165–170, 2002.

165. Moghimi, S. M., Hunter, A. C., and Murray, J. C., Long-circulating and target-specific nanoparticles: Theory to practice, *Pharmacological Reviews*, 53, 283–318, 2001.

166. Blunk, T., Hochstrasser, D. F., Sanchez, J. C., Muller, B. W., and Muller, R. H., Colloidal carriers for intravenous drug targeting: Plasma protein adsorption patterns on surface-modified latex particles evaluated by two-dimensional polyacrylamide gel electrophoresis, *Electrophoresis*, 14, 1382–1387, 1993.

167. Diluzio, N. R. and Zilversmit, D. B., Influence of exogenous proteins on blood clearance and tissue distribution of colloidal gold, *American Journal of Physiology*, 180, 563–565, 1955.

168. Campbell, P. A., Canono, B. P., and Drevets, D. A., Measurement of bacterial ingestion and killing by macrophages, *Current Protocols in Immunology*, Suppl. 12, 14.6.1–14.6.13, 1994.

169. Prabha, S., Zhou, W. Z., Panyam, J., and Labhasetwar, V., Size-dependency of nanoparticle-mediated gene transfection: studies with fractionated nanoparticles, *International Journal of Pharmaceutics*, 244, 105–115, 2002.

170. Chamberlain, P. and Mire-Sluis, A. R., An overview of scientific and regulatory issues for the immunogenicity of biological products, *Developments in Biological Standardization (Basel)*, 112, 3–11, 2003.

171. Casadevall, N., Antibodies against rHuEPO: native and recombinant, *Nephrology Dialysis Transplantation*, 11(Suppl. 5), 42–47, 2002.

172. Casadevall, N., Nataf, J., Viron, B., Kolta, A., Kiladjian, J. J., Martin-Dupont, P., Michaud, P., Papo, T., Ugo, V., Teyssandier, I., Varet, B., and Mayeux, P., Pure red-cell aplasia and antierythropoietin antibodies in patients treated with recombinant erythropoietin, *New England Journal of Medicine*, 346, 469–475, 2002.

173. Koren, E., Zuckerman, L. A., and Mire-Sluis, A. R., Immune responses to therapeutic proteins in humans—clinical significance, assessment and prediction, *Current Pharmaceutical Biotechnology*, 3, 349–360, 2002.

174. Swanson, S. J., Ferbas, J., Mayeux, P., and Casadevall, N., Evaluation of methods to detect and characterize antibodies against recombinant human erythropoietin, *Nephron Clinical Practice*, 96, c88–c95, 2004.

175. Andreev, S. M., Babakhin, A. A., Petrukhina, A. O., Romanova, V. S., Parnes, Z. N., and Petrov, R. V., Immunogenetic and allergenic activities conjugates of fullere with amino acids and protein, *Doklady Akademii Nauk*, 370, 261–264, 2000.

176. Braden, B. C., Goldbaum, F. A., Chen, B. X., Kirschner, A. N., Wilson, S. R., and Erlanger, B. F., X-ray crystal structure of an anti-Buckminsterfullerene antibody Fab fragment: Biomolecular recognition of C-60, *Proceedings of the National Academy of Sciences of the United States of America*, 97, 12193–12197, 2000.

177. Chen, B. X., Wilson, S. R., Das, M., Coughlin, D. J., and Erlanger, B. F., Antigenicity of fullerenes: antibodies specific for fullerenes and their characteristics, *Proceedings of the National Academy of Sciences of the United States of America*, 95, 10809–10813, 1998.

178. Masalova, O. V., Shepelev, A. V., Atanadze, S. N., Parnes, Z. N., Romanova, V. S., Vol'pina, O. M., Semiletov Iu, A., and Kushch, A. A., Immunostimulating effect of water-soluble fullerene derivatives—perspective adjuvants for a new generation of vaccine, *Doklady Akademii Nauk*, 369, 411–413, 1999.

8 Nanotechnology: Regulatory Perspective for Drug Development in Cancer Therapeutics

N. Sadrieh and T. J. Miller

CONTENTS

8.1 Introduction 139
8.2 FDA Experience with Nanotechnology Product Applications......................... 140
8.3 FDA Definition of Nanotechnology 143
8.4 FDA Initiatives in Nanotechnology 143
8.5 Research at the FDA on Nanotechnology 145
8.6 Regulatory Considerations for Nanotechnology Products 145
    8.6.1 Jurisdiction of Nanotechnology Products 145
    8.6.2 Existing FDA Regulations That Apply to Nanomaterial-Containing Products 146
8.7 Scientific Considerations for the Development of Nanotechnology Products 147
    8.7.1 Characterization of Nanoparticles... 147
    8.7.2 Safety Considerations.. 149
8.8 Conclusions 149
8.9 Trademark Listing 151
References 151

8.1 INTRODUCTION

In the last decade, remarkable scientific progress in the development of novel, nanoscale technologies has generated rising expectations for this enabling technology to advance our efforts in the prevention, diagnosis and treatment of human disease. These nanotechnologies, commonly referred to as nanomedicines when applied to the treatment, diagnosis, monitoring, and control of biological systems, incorporate a multitude of diverse and often unique nanosized products including new and/or improved therapeutic drugs, diagnostic/surgical devices, and targeted drug delivery systems.[1] The list of nanoparticle medicines approved and/or currently being developed is quite large; these products range from early forms of drug-encapsulating liposomes (e.g., liposomes containing doxorubicin for treatment of breast and ovarian cancer) and simple polymeric structures (e.g., actinomycin D adsorbed on poly(methylcryanoacrylate) (PMCA) and poly(ethylcryanoacrylate) (PECA) nanosized particles)[2] to more intricate multifunctional "nanoplatforms" and diagnostic devices, including polymeric dendrimers, nanocrystal quantum dots, carbon-based full-erenes, and nanoshells, to name a few (see Table 8.1). The vast majority of these products are being developed to improve early detection of cancer and/or to provide alternative clinical therapeutic approaches to fighting cancer, some of these products have been approved by the Food and Drug Administration for clinical use.[75]

Although there have been many articles published on the in vitro toxicity of certain nanomaterials, very little in vivo safety data evaluating the general toxicity of these novel products is available in the published literature. Many in vitro assays (cell-based, organelle-based, and cell/organelle-free systems) have been used to evaluate the cytotoxicity of nanomaterials to help predict the relative safety and general biocompatibility of the nanoproducts in vivo. Several such studies include viability assays in vascular endothelial cells exposed to C_{60};[76] fullerene induction/inhibition of oxidative stress and lipid peroxidation in isolated microsomes, dermal fibroblasts, and cultured cortical neurons;[77–79] cytotoxicity of metal oxide nanoparticles to BRL 3A liver cells;[80] and macrophage apoptosis after exposure to carbon nanomaterials and ultrafine particles.[81–83] Although a vast amount of useful information has been obtained from these various in vitro models, the relevance and application of these findings to human drug safety remain uncertain until validated with in vivo studies with identical nanomaterials.

Targeted in vivo efficacy data is readily available for many of the novel, potential nanotherapeutics; however, necessary in vivo data indicating mechanism of action, general pharmacology/toxicity profiles, and optimal product characterization efforts rarely appear in the published literature. Much of our current understanding of the biocompatibility of nanomaterials comes from focused inhalation and dermal studies using nanosized products and ultrafine contaminants, as a model of environmental and occupational risk.[84–87] However, there are questions regarding the level of applicability of these studies to human drug safety assessment due to the variability in size and purity of the chemical or structural composition reported or not reported in these studies, the route of primary exposure, and the complex differences in the structure and function of the affected organs.[88] In addition, other in vivo studies, which identify intrinsic anti-oxidant and oxidative stress-inducing properties of unsubstituted fullerenes in a rodent and fish species, respectively, have important but limited value in drug discovery in the absence of systemic, whole animal toxicology data[67,89].

While the enthusiasm and expectations for success in utilizing nanomaterials and nanometer-sized structures in biomedical applications grow, so do the concerns regarding the potential impact of such products on the environment and human health.[88] The absence of available safety information for a vast majority of these novel nanosized materials and nanoparticlized drugs in biological systems only further supports the need for additional research into any potential toxic mechanisms in humans. The questions that appear most often in deciding the safety of any nanoscale material include:

- Are nanomaterials and nanotechnology new to the United States Food and Drug Administration (FDA)?
- How do existing regulations ensure the development of safe and effective nanotechnology-based drugs and how is the FDA preparing to deal with nanotechnology?
- What are the scientific considerations that raise unique concerns for nanotechnology-based therapeutics?

8.2 FDA EXPERIENCE WITH NANOTECHNOLOGY PRODUCT APPLICATIONS

Historically, the FDA has encountered and approved many products with particulate materials characteristic of a nanomedicine or nanoproduct (e.g., small size, mechanism of delivery, increased

TABLE 8.1
Nanomedicines: Drugs/Devices FDA Approved and in Development

Name	Structure	Company	Size (nm)	Phase	Indication	Reference[a,b]
Magnevist[®c]	PIO	Shering AG	<1	APP	MRI tumor imaging	3–5
Feridex[®c]	SPIO	Adv.Magnetics	120–180	APP	MRI tumor imaging	5–9
Rapamune[®c]	Nanocrystal drug	Wyeth	100–1,000	APP	Immunosuppressant agent	10–12
Emend[®c]	Nanocrystal drug	Merck	100–1,000	APP	Nausea	13,14
Doxil[®c]	Liposome-drug encapsulation	Ortho Biotech	≈100	APP	Metastatic ovarian cancer	15–17
Abraxane[®c]	Nanoparticulate albumin	American Pharmaceutical Partners	≈130	APP	Non-small-cell lung cancer, breast cancer, others	18–20
SilvaGard[®]	Silver nanoparticles	AcryMed	≈10	APP	Antimicrobial surface treatment (medical devices)	21–23
AmBisome[®c]	Liposome-drug encapsulation	Gilead Science	45–80	APP	Fungal infections	24–26
Leunesse[®]	Solid lipid nanoparticles	Aano-therapeutics	N/A	On market	Cosmetics	N/A
NB-001[d]	Nanoemulsion	NanoBio Corp	≈150	Phase II	Topical microbicide for herpes labialis	27–30
VivaGel[®]	Dendrimer	Starpharma Holdings Ltd.	N/A	Phase I	Vaginal microbicide (STI and HIV prevention)	31–34
INGN-401[c]	Drug–liposome complex	Introgen Therapeutics	N/A	Phase I	Metastatic lung cancer	35–38
Combidex[®c]	USPIO	Advanced Magnetics	20–50	NDA filed	MRI tumor imaging	39–42
IT-101[c]	Drug–polymer complex	Insert Therapeutics	N/A	IND filed	Metastatic solid tumors	43,44
Dendrimers[c]	Polymeric spheres, multifunctional	Dendritech Nanotech., Starpharma	>1 to <500	IDV	Diagnostic, contrast agents, microbicide, drug delivery, B neutron capture	45–54
Nanoshells[c]	Metal shell, SiO_2 core	Nanospectra	100–200	IDV	Photo-thermal ablation therapy /imaging tumors	55–58
Fullerenes[c]	Carbon sphere, C_{60}, multifunctional	C Sixty	1	IDV	MRI contrast agent, antiviral, antioxidant	59–67
Quantum dots[c]	CdSe[a] nanocrystal	Evident Technologies	2–10	IDV	Optical imaging, tumor imaging, photosensitizer	68–74

APP, approved; IDV, in development; PIO, paramagnetic iron oxide; SPIO, superparamagnetic iron oxide; USPIO, ultrasmall superparamagnetic iron oxide; STI, sexually transmitted infection; HIV, human immunodeficiency virus; N/A, not available.

[a] Most typical semiconductor metal used, however can also be composed of nearly any other semiconductor metal (e.g., CdS, CdTe, ZnS, PbS, etc.).

[b] References do not necessarily refer to the product described, and may reflect general properties about the product.

[c] Drugs with proposed or proven diagnostic or therapeutic benefit to cancer patients.

[d] Unable to locate peer-reviewed publications. References listed are press releases found during internet search.

surface area, specialized function related to size and increased surface area, etc.). Although the field of nanomedicine may be new, the submission of applications for products containing novel dosage forms or drug delivery systems, including nanomaterials, nanoparticlized drugs and medical devices, is not particularly new to this agency. Products are reviewed on a product-by-product basis, and as such, various concepts of risk management (i.e., risk identification, risk analysis, risk control, etc.) are often implemented to support the drug review process.

The FDA is aware of several FDA regulated products that employ nanotechnology. However, to date, few manufacturers of regulated products have claimed the use of nanotechnology in the manufacture of their products or made any nanotechnology claims for the finished product. The FDA is aware that a few cosmetic products and sunscreens claim to contain nanoparticles to increase the product stability, modify release of ingredients, or change the opacity of the product by using nanoparticles of titanium dioxide and zinc oxide. Similarly, several drugs such as Gd-DTPA (Magnevist®, tumor contrast agent),[3-5] ferumoxides (Feridex®, tumor contrast agent),[5-9] liposomal doxorubicin (Doxil®, chemotherapeutic),[15-17] and albumin-bound paclitaxel (Abraxane®, chemotherapeutic),[18-20] to name a few, have all been approved for human clinical use in the diagnosis and treatment of a variety of patient tumors and metastatic cancers. For public information regarding FDA approval letters, label information, and review packages for each of these drugs and others listed in Table 8.1, please visit the following website: http://www.accessdata.fda.gov/scripts/cder/drugsatfda/.

Of the many drugs listed in Table 8.1, tumor contrast agents comprise some of the earliest nanosized products to receive FDA approval for human clinical use. Critical to therapeutic outcome, diagnostic imaging of nanosized contrast agents using non-invasive magnetic resonance imaging (MRI) and positron emission tomography (PET), have shown advanced utility for early detection of small tumors, metastatic lymph nodes, and altered tumor microenvironments.[90] The leading MR contrast agent, Magnevist®, a paramagnetic, gadolinium-based contrast medium, and Feridex®, a superparamagnetic iron oxide contrast agent, received primary FDA approval for clinical use in 1988 and 1996, respectively. Each contrast agent has a different mean particle size (<1 nm and 120–180 nm, respectively), is approved for different clinical indications, and demonstrates product-specific adverse sensitivity and pharmacology/toxicology profiles in patients.[3-9]

Another early nanomedicine, liposomal encapsulated doxorubicin (Doxil®), is regulated by the FDA and has been available in the clinic for treatment of various cancers since 1995. Drug incorporation inside the hydrophilic core or within the hydrophobic phospholipid bilayer coat of liposomes, has been shown to improve drug solubility, enhance drug transfer into cells and tissues, facilitate organ avoidance, and modify drug release profiles, minimizing toxicity.[91] The liposomal formulation of the popular anthracyline, doxorubicin, which is commonly used to treat metastatic breast and ovarian cancer, is reported to have diminished cardiotoxicity and enhanced therapeutic efficacy compared to the free form of the drug.[15-17] This increased efficacy is most likely due to the passive targeting of solid tumors through the enhanced permeability and retention effect inherent to tumor vasculature and aberrant tumor morphology.[17] The approximate diameter of the doxorubicin–liposomal product is reported to be 100 nm, near the size limits described in the FDA definition of a nanomedicine.

Nanoparticlized, albumin-bound paclitaxel, Abraxane®, is one of the first FDA-approved (January, 2005) cancer drugs that was submitted to the FDA with specific claims regarding nano-related characteristics that improved this particular formulation over previous submissions.[18-20] Initially identified as (ABI-007), this chemotherapeutic agent that is approved for the treatment of metastatic breast cancer eliminates the excipient Cremophor® by overcoming intrinsic insolubility with nanoparticlization of the active paclitaxel ingredient. Although widely used in many drug formulations, Cremophor has been associated with an adverse hypersensitivity reaction that limited the amount of paclitaxel that could be safely administered.[20] Unlike the previous three agents, the approximate diameter of Abraxane is 130 nm, slightly larger than the requirements described in the National Nanotechnology Initiative (NNI) definition of nanotechnology (<100 nm), and thus originally was determined not to fit the size requirement of a nanomedicine. However, the fact that the exact particle size exceeds 100 nm may not necessarily preclude Abraxane and other similarly-sized or larger products from being considered nanotechnology.

8.3 FDA DEFINITION OF NANOTECHNOLOGY

As described on the NNI website (http://www.nano.gov/index.html), "Nanotechnology is the understanding and control of matter at dimensions of roughly 1–100 nm, where unique phenomena enable novel applications. Encompassing nanoscale science, engineering and technology, nanotechnology involves imaging, measuring, modeling, and manipulating matter at this length scale." Similarly, in a recent publication by the European Science Foundation on Nanomedicine,[92] nanotechnology is defined as the following: "In case of these devices, nanoscale objects were defined as molecules or devices within the size range of one to hundreds of nanometers that are the active component or object, even within the frame-work of a larger micro-size device or at a macro-interface." A discussion regarding the definition of nanotechnology is currently an active topic among various organizations. However, while a definition is always useful, the FDA has traditionally dealt with products on a case-by-case basis. This approach is likely to continue for nanotechnology products, regardless of the definition. However, we are actively involved in the various organizations and committees dealing with issues regarding terminology, such as the American Society for Testing and Materials (ASTM) International.

The definition of nanotechnology at the FDA has been consistent with the definition of nanotechnology as reported by the NNI. As such, the FDA currently defines nanotechnology with the following definition:

1. The existence of materials or products at the atomic, molecular, or macromolecular levels, where at least one dimension that affects the functional behavior of the drug/device product is in the length scale range of approximately 1–100 nm.
2. The creation and use of structures, devices, and systems that have novel properties and functions because of their small size.
3. The ability to control or manipulate the product on the atomic scale.

8.4 FDA INITIATIVES IN NANOTECHNOLOGY

The FDA is responsible for protecting the health of the public by assuring the safety, efficacy and security of human and veterinary drugs, biological products, medical devices, our nation's food supply, cosmetics, and products that emit radiation. The FDA is also responsible for advancing the public's health by helping to speed innovations that make medicines and foods more effective, safer and more affordable; and by helping the public get the accurate, science-based information they need to use medicines and foods to improve their health. Towards this end, the FDA needs to proactively pursue those innovations that would result in a significant enhancement of public health. Therefore, it falls directly within the mandate of its mission for the FDA to understand and help to speed innovations that promise to improve many products that will be regulated by the FDA, such as those resulting from nanotechnology.

In response to a slowdown in innovating medical therapies submitted to the FDA for approval, the FDA released the Critical Path Initiative (http://www.fda.gov/oc/initiatives/criticalpath/whitepaper.html) in March 2004. The report describes the urgent need to modernize the medical product development process—the Critical Path—to make product development more predictable and less costly. The Critical Path Initiative at the FDA is a serious attempt to bring attention and focus to the need for targeted scientific efforts to modernize the techniques and methods used to evaluate the safety, efficacy, and quality of medical products as they move from product selection and design to mass manufacture. It is intended to integrate new science into the regulatory process. As such, a list of opportunities for the Critical Path Initiative was published by the FDA, and nanotechnology was identified as one of the elements under the Critical Path Initiative. Figure 8.1 illustrates the three dimensions of the Critical Path (safety, medical utility and industrialization).

However, the Food and Drug Administration is only one of several government agencies that currently regulates and evaluates a wide range of products that utilize nanotechnology and/or contain nanomaterials, including foods, cosmetics, drugs, devices, and veterinary products. As a member of both the National Science and Technology Council (NSTC) Committee on Technology, and the Nanoscale Science, Engineering and Technology (NSET) Working Group on Nanomaterials Environmental and Health Implications (NEHI), the FDA coordinates with other government agencies, including the National Institute for Environmental Health Sciences (NIEHS), the National Toxicology Program (NTP), and the National Institute of Occupational Safety and Health (NIOSH) to develop novel strategies for the characterization of and safety evaluation standards for nanomaterials. The FDA also works closely with the National Cancer Institute (NCI) and the National Institute of Standards and Technology (NIST) in the Interagency Oncology Task Force, Nanotechnology subcommittee, to pool knowledge and resources to facilitate the development and approval of new cancer drugs and to address the use of nanotechnology in cancer therapies and diagnostics. Within the FDA, the Office of Science and Health Coordination manages the regular interaction and discussion of nanotechnology between the various agency centers and offices, and the nanotechnology working groups within each center, and address the concerns about the regulation of nanosized drugs and materials submitted to the individual centers.

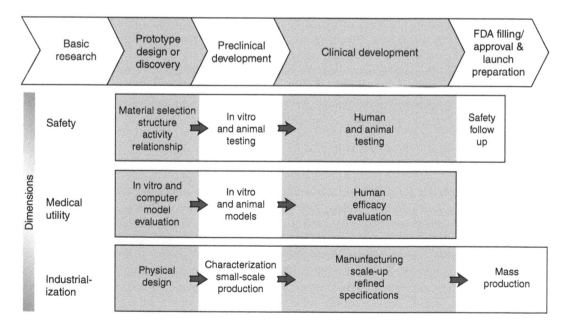

FIGURE 8.1 A highly generalized description of activities that must be successfully completed at different points and in different dimensions along the Critical Path. Many of these activities are highly complex—whole industries are devoted to supporting them. Not all the described activities are performed for every product, and many activities have been omitted for the sake of simplicity. (From http://www.fda.gov/oc/initiatives/criticalpath/whitepaper. html.)

8.5 RESEARCH AT THE FDA ON NANOTECHNOLOGY

Although the FDA is a regulatory agency, it is also engaged in a number of research activities. However, the FDA recognizes that science supported by research provides the foundation for sound regulation. As such, the FDA is interested in research projects that can address specific regulatory needs. As an example, FDA has a grants program in support of orphan products research and development. The FDA does not have a grants program to support other research in non-FDA laboratories. Nevertheless, in several of its Centers the FDA conducts research to understand the characteristics of nanomaterials and nanotechnology processes. Research interests include any areas related to the use of nanoproducts that the FDA needs to consider in the regulation of these products.

Although many FDA Centers are engaged in some nanotechnology research activities, only some of the specific projects that are currently ongoing are listed below. The research projects, which happen to be conducted by the Center for Drug Evaluation and Research (CDER), include:

1. Particle size determination in marketed sunscreens with TiO2 and ZnO nanoparticles. Sunscreens are considered drugs in the U.S., and although some undergo a premarket FDA review, others are marketed under over-the-counter (OTC) monographs that outline the active ingredients that the FDA has found to be safe and effective, as well as labeling that is truthful and not misleading. TiO_2 and ZnO used in sunscreens are therefore manufactured according to OTC monographs, and these monographs do not specify particle size. As a result, a manufacturer can formulate a sunscreen using particles of TiO_2 and ZnO that may or may not be in the nano size range. Therefore, to assess the particle size of TiO_2 and ZnO in currently marketed sunscreens, CDER has undertaken a research project, in collaboration with NIST.

2. Another project focuses on manufacturing various nanoformulations in our laboratories and characterizing the physical and chemical properties of the nanoparticles in these formulations, including evaluating the effects of excipients on the measured parameters, the effects of preparation methodology on the measured parameters, and the impact (if any) that process and formulation variables may have on nanotechnology product characteristics and stability.

3. Another project is studying the in vivo effects of selected nanoparticles, using validated animal models. The toxicity of the same nanoparticles will also be evaluated, using various in vitro systems.

8.6 REGULATORY CONSIDERATIONS FOR NANOTECHNOLOGY PRODUCTS

8.6.1 JURISDICTION OF NANOTECHNOLOGY PRODUCTS

There has been much interest in the question of whether nanotechnology products will be considered drugs, devices, biologics or a combination of the three for regulatory purposes and assignment of work. This has been extensively discussed internally within the FDA and the current thinking is that many of these products will be considered combination products.

Combination products (i.e., drug–device, drug–biologic, and device–biologic products) are increasingly incorporating cutting edge, novel technologies that hold great promise for advancing patient care. For example, innovative drug delivery devices have the potential to make treatments safer, more effective, or more convenient to patients. For example, drug-eluting cardiovascular stents have the potential to reduce the need for surgery by preventing the restenosis that sometimes occurs following stent implantation. Drugs and biologics can be used in combination to potentially enhance the safety and/or effectiveness of either product used alone. Biologics are being incorporated into novel orthopedic implants to help facilitate the regeneration of bone required to permanently stabilize the implants.

Because combination products involve components that would normally be regulated under different types of regulatory authorities, and frequently by different FDA centers, they also raise challenging regulatory, policy, and review management issues. A number of criticisms have been raised regarding the FDA's regulation of combination products. These include concerns about the consistency, predictability, and transparency of the assignment process; issues related to the management of the review process when two (or more) FDA centers have review responsibilities for a combination product; lack of clarity about the postmarket regulatory controls applicable to combination products; and lack of clarity regarding certain Agency policies, such as when applications to more than one agency center are needed.

To address these concerns, FDA's Office of Combination Products (OCP) was established on December 24, 2002, as required by the Medical Device User Fee and Modernization Act of 2002 (MDUFMA). The law gives the office broad responsibilities covering the regulatory life cycle of drug–device, drug–biologic, and device–biologic combination products. However, the primary regulatory responsibilities for, and oversight of, specific combination products will remain in one of three product centers—the Center for Drug Evaluation and Research, the Center for Biologics Evaluation and Research, or the Center for Devices and Radiological Health—to which they are assigned.

A combination product is assigned to an agency center or alternative organizational component that will have primary jurisdiction for its premarket review and regulation. Under section 503(g)(1) of the act, assignment to a center with primary jurisdiction, or a lead center, is based on a determination of the "primary mode of action" (PMOA) of the combination product. For example, if the PMOA of a device–biological combination product is attributable to the biological product, the agency component responsible for premarket review of that biological product would have primary jurisdiction for the combination product.

A final rule defining the PMOA of a combination product was published in the August 25, 2005 edition of the *Federal Register*, and is available at http://www.fda.gov/oc/combination/. The final rule defines PMOA as "the single mode of action of a combination product that provides the most important therapeutic action of the combination product."

8.6.2 Existing FDA Regulations That Apply to Nanomaterial-Containing Products

The FDA has specific guidance/requirements in place for most products. These guidance documents can be accessed on the FDA's website (http://www.fda.gov/oc/industry/default.htm). These requirements apply to all products, whether they contain nanomaterials or not. Existing requirements are therefore considered to be adequate at this time for most nanotechnology pharmaceutical products.

In the following section, there will be a brief description of the current preclinical requirements for the submission of most drugs, to highlight why we believe that our current regulations are adequate for the evaluation of the types of nanotechnology products that have been reported as being "close" to submission to the agency as an investigational new drug (IND).

Within CDER, the preclinical requirements for approval to market pharmaceutical products include both short-term and long-term toxicity testing in rodent and non-rodent species. Specifically, the types of studies conducted by pharmaceutical manufacturers prior to New Drug Application (NDA) submission include pharmacology (mechanism of action), safety pharmacology (functional studies in various organ systems, most notably the cardiovascular system), absorption, distribution, metabolism, and excretion (ADME), genotoxicity (potential to cause mutations in both in vivo and in vitro assays), developmental toxicity (to assess effects on reproduction, fertility and lactation), irritation studies (to assess local irritation effects), immunotoxicology (to assess effects on the components of the immune system), carcinogenicity (to assess the capacity of drugs to induce tumors in animal models) and other possible studies (specific studies for a product being developed). The current battery of tests, as listed above, is therefore considered to be adequate because of the following reasons: the studies use high dose multiples of the drug (usually three doses, one that results in low, one that results in medium and one that results in high toxicity), at least two animal species are used (usually one rodent and one non-rodent), extensive histopathology is conducted on most organs (where most organs are removed, fixed, stained and viewed under the microscope to assess if there are structural changes that may have resulted from damage to the organs), functional tests are conducted to assess if there are effects on specific organ systems (such as the heart, brain, respiratory system, reproductive system, immune system, etc.) and drug treatments in animals can be for extended periods of time (up to two years for carcinogenicity studies).

Additional studies might be requested based on drug-specific considerations and on a case-by-case basis. Therefore, as for other drugs, nanotechnology products will be handled on a case-by-case basis, depending on the characteristics of the particular product being developed. As research provides additional understanding with regards to the toxicity of nanomaterials, the FDA may require additional toxicological tests to ensure the safety of the products it regulates.

8.7 SCIENTIFIC CONSIDERATIONS FOR THE DEVELOPMENT OF NANOTECHNOLOGY PRODUCTS

Ensuring the safety of nanomaterials in drug applications is one of the most important concerns for regulators and drug developers. However, as important as safety is, it is also important to adequately characterize the nanomaterials used in drug products. Proper characterization and manufacturing procedures will ensure that a consistent product is being used, both during preclinical and clinical development, and after approval. Some of the questions that have been raised internally regarding the assurance of safety of nanomaterial-containing products, as well as questions regarding characterization, are listed below. It is not within the scope of this chapter to address questions regarding manufacturing and scale-up; however, it should be noted that this is an issue that needs serious consideration, because unique characteristics of nanotechnology products are likely to require unique manufacturing procedures.

8.7.1 CHARACTERIZATION OF NANOPARTICLES

It has become clear to most of those involved in research or development of potential nanotechnology applications that it is only possible to assess the safety of nanomaterials when the material under study is adequately characterized. If two or more laboratories are working on what they think is the same material, then in order for the data from one laboratory to be compared with data from another laboratory, there needs to be some confirmation that in fact the nanomaterials being studied are actually identical. It has been mentioned by many that for nanomaterials, small differences in product characteristics (such as size, surface charge, hydrophobicity, or a number of other attributes) may impact product behavior and thus result in vastly different safety profiles. Published data show that slight modifications of a nanoparticle have resulted in different toxicity profiles. For example, several in vitro studies with Starburst® polyamidoamine (PAMAM) dendrimers and Caco-2 cells have shown decreased cytotoxicity of cationic dendrimers upon addition of lauroyl or polyethylene glycol (PEG) to the surface of the macromolecules.[93-94] Similar findings of diminished toxicity were observed in J774 macrophages upon conjugation of polysaccharides to polymer nanoparticles and in fibroblasts and CHO cells with the addition of PEG–silica to the surface of CdSe and CdSe/ZnS nanocrystals.[95-96]

Although much less toxicity data is available in vivo, it is evident that surface chemistry is of similar importance to the toxico- and pharmacokinetics, toxico- and pharmacodynamics, and overall stability of the nanomaterials in vivo.[97-99] When injected into rodents, several generations of dendrimers with different surface configurations displayed contrasting organ distribution patterns and plasma clearance profiles in vivo, typically depositing within the liver, kidney, and pancreas.[98–100] PEG-conjugation to the dendrimer particle surface significantly increased the blood half-life and diminished the liver accumulation of dendrimer particles.[99] Similarly, quantum dots coated with a complex polymer with carbon alkyl side chains demonstrated greater in vivo stability and less genotoxicity than those coated with a simpler polymer or lipid coating.[101] Based on the findings of these studies and the collection of pulmonary inhalation data with environmental ultrafines and other in vitro cytotoxicity results obtained with different nanomaterials (e.g., quantum dots, fullerenes, and nanoshells, etc.), it appears that a thorough characterization of the nanomaterial is essential to the interpretation and understanding of toxicity data collected from in vitro and in vivo studies.

To summarize some of the ongoing discussions within CDER's nanotechnology working group, a list of questions has been put together regarding the characterization of nanomaterial-containing products. At this time, it appears that there are no adequate answers to these questions. However, it may be that in the future, to satisfy regulatory requirements for nanomaterial-containing products, the questions listed below will need to be addressed by product sponsors.

1. What are the forms in which particles are presented to the organism, the cells and the organelles? Are these particles soluble or insoluble, are the nanomaterials organic or inorganic molecules, are the nanomaterials described as nanoemulsions, nanocrystal colloid dispersions, liposomes, nanoparticles that are combination products (drug–device, drug–biologic, drug–device–biologic), or something else? Does the product fall within the currently established definition of nanotechnology?

2. What are the tools that can be used to characterize the properties of the nanoparticles? Can standard tools that are used of other types of products (for example, spectroscopic tools) be used to characterize nanomaterial-containing products, or do these products require specialized tools that are not widely available for product characterization (atomic force microscopy, tunneling electron microscopy, or other specialized modalities)?

3. Are there validated assays to detect and quantify nanoparticles in the drug product and in tissues?

4. How can long- and short-term stability of nanomaterials be assessed? Specifically, how can the stability of nanomaterials be assessed in various environments (buffer, blood, plasma, and tissues)?

5. What are the critical physical and chemical properties of nanomaterials in a drug product and how do residual solvents, processing variables, impurities and excipients impact these properties?

6. How do physical characteristics impact product quality and performance? Within what range can these physical properties be modified without significantly affecting product quality and performance?

7. What are the critical steps in the scale-up and manufacturing process for nanotechnology products? How could manufacturing procedures for nanomaterials differ from those for other types of drugs?

8. How are issues concerning the characterization and manufacturing procedures for nanotechnology products assessed, when considering the development of "personalized therapies"? In this context, personalized therapies refer to those multifunctional products that will be "custom made" for specific patients, depending on issues such as the expression of a certain receptor on a tumor or the response of the tumor to a particular chemotherapeutic agent. For these types of therapies, what should be the level of characterization of the nanomaterials? Does the preclinical testing need to be repeated for each modification of the multifunctional molecule, or can there be limited testing to

cover certain aspects of the product behavior, such as the ADME, the toxicology (a bridging study for the acute and/or repeat dose toxicology studies)? What aspects of the chemistry, manufacturing and controls (CMC) studies need to be repeated and what should be the extent of physical characterization?

8.7.2 SAFETY CONSIDERATIONS

Other than characterization, the other major consideration has to do with safety. However, it is only within the context of adequately characterized products that safety studies provide value. Therefore, although safety is instinctively the primary concern, adequate characterization of nanomaterials has to precede any studies on safety.

As particle size decreases, there may be size-specific effects. These effects may impact safety if the materials are more reactive. It has been mentioned in numerous discussions that, as particle size decreases, reactivity is expected to rise. The increased reactivity combined with the decreased size has led us to pose many questions. Some of these questions are listed below:

1. Will nanoparticles gain access to tissues and cells that normally would be bypassed by larger particles?
2. If nanoparticles enter cells, what effects do they have on cellular and tissue functions? Are the effects transient or permanent? Might there be different effects in different cell types, depending on specific tissue targets or receptors?
3. Once nanoparticles enter tissues, how long do they remain there and where do they concentrate?
4. What are the mechanisms by which nanoparticles are cleared from tissues and blood and how can their clearance be measured?
5. What are some route-specific issues for nanomaterials, and how are these issues different than for other types of materials that are not in the nanoscale?
6. With regards to ADME, specific questions include:
   a. What are the differences in the ADME profile for nanoparticles versus larger particles of the same drug?
   b. Are current methods used for measuring drug levels in blood and tissues adequate for assessing levels of nanoparticles (appropriateness of method, limits of detection)?
   c. How accurate are mass balance studies, especially if the levels of drug administered are very low; i.e., can 100% of the amount of the administered drug be accounted for?
   d. How is clearance of targeted nanoparticles accurately assessed?
   e. If nanoparticles concentrate in a particular tissue, how can clearance be accurately determined?
   f. Can nanoparticles be successfully labeled for ADME studies?

8.8 CONCLUSIONS

In this chapter, we have tried to highlight some of the steps being taken by the FDA to meet the challenges of nanotechnology products. The FDA has engaged in an open discussion with numerous groups and organizations, has participated in and sponsored workshops, has established internal working groups, has outlined a path for the review of combination nanotechnology products and has initiated research projects in nanotechnology. All these efforts have been geared towards a better understanding of the science and towards assessing the adequacy of existing regulations for the types of products that are expected to be regulated by the FDA.

At this point, and based on internal discussions, the FDA feels that its current regulations are adequate to address the types of nanotechnology products that are being developed as drugs or drug delivery systems. The reasons for this have been outlined in this chapter. Although there have been requests by investigators and developers of nanotechnology products for the FDA to issue nanotechnology-specific guidances, no new guidance documents applicable specifically to nanotechnology products will be issued at this time. This is because it is felt that all existing Guidance documents apply to these products. All products being developed, whether nanotechnology-based or not, are unique and therefore likely to have unique issues requiring consideration. As such, and as for all other products reviewed by the FDA, nanotechnology products will be reviewed on a case-by-case basis.

As shown in Figure 8.2, the drug development process is multiphasic and along the path, there are numerous opportunities for meetings between the sponsor and the FDA. The first meeting is the pre-IND meeting, and it is highly recommended that sponsors of nanotechnology products take the opportunity to meet with the FDA at this early stage. An early meeting with the FDA can avoid unnecessary or unfocussed studies that could result in a waste of sponsor resources. During early meetings, sponsors can present, with minimal data, their preclinical plans to the FDA and obtain guidance on how to focus the studies submitted in the IND package. This type of interaction between the FDA and the sponsor will be valuable to both parties and will pave the way for the most efficient drug development plan.

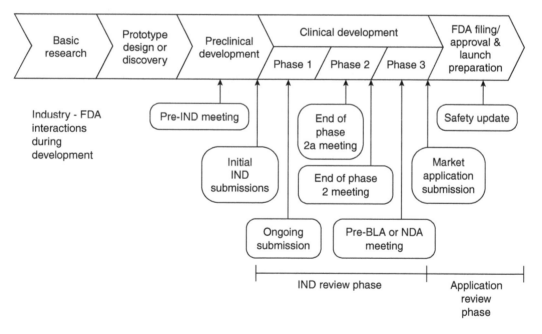

FIGURE 8.2 This figure depicts the extensive industry–FDA interactions that occur during product development, using the drug development process as a specific example.* Developers often meet with the agency before submitting an investigational new drug (IND) application to discuss early development plans. An IND must be filed and cleared by the FDA before human testing can commence in the United States. During the clinical phase, there are ongoing submissions of new protocols and results of testing. Developers often request additional meetings to get FDA agreement on the methods proposed for evaluation of safety or efficacy, and also on manufacturing issues. *Note: Clinical drug development is conventionally divided into three phases. This is not the case for medical device development. (From www.fda.gov/oc/initiatives/criticalpath/whitepaper.html.)

8.9 TRADEMARK LISTING

1. Magnevist® (Schering Aktiengesellschaft, Berlin, Germany)
2. Feridex® (Advanced Magnetics, Inc., Cambridge, Massachusetts, U.S.A.)
3. Rapamune® (Wyeth, Madison, New Jersey, U.S.A.)
4. Emend® (Merck & CO., Inc., Whitehouse Station, New Jersey, U.S.A.)
5. Doxil® (Ortho Biotech Products, L.P., Bridgewater, New Jersey, U.S.A.)
6. Abraxane® (American Pharmaceutical Partners, Inc., Schaumburg, Illinois, U.S.A.)
7. SilvaGard® (AcryMed, Beaverton, Oregon, U.S.A.)
8. AmBisome® (Gilead Sciences, Inc., Foster City, California, U.S.A.)
9. Leunesse® (Nanotherapeutics, Inc., Alachua, Florida, U.S.A.)
10. VivaGel® (Starpharma Holdings, Ltd., Melbourne, Australia)
11. Combidex® (Advanced Magnetics, Inc., Cambridge, Massachusetts, U.S.A.)
12. Cremophor® (BASF Aktiengesellschaft, CO., Ludwigshafen, Germany)
13. Starburst® (Dendritic Nanotechnologies, Inc., Mount Pleasant, Michigan, U.S.A.)

REFERENCES

1. Moghimi, S. M., Hunter, A. C., and Murray, J. C., Nanomedicine: Current status and future prospects, *FASEB Journal*, 19 (311), 330, 2005.
2. Brigger, I., Dubernet, C., and Couvreur, P., Nanoparticles in cancer therapy and diagnosis, *Advanced Drug Delivery Review*, 54 (5), 631–651, 2002.
3. Kieffer, S. A., Gadopentetate dimeglumine: Observations on the clinical research process, *Radiology*, 174 (1), 7–8, 1990.
4. Zhang, H. L., Ersoy, H., and Prince, M. R., Effects of gadopentetate dimeglumine and gadodiamide on serum calcium, magnesium, and creatinine measurements, *Journal of Magnetic Resonance Imaging*, 23, 383–387, 2006.
5. de Lussanet, Q. G., Backes, W. H., Griffioen, A. W., van Engleshoven, J. M., and Beets-Tan, R. G., Gadopentenate dimeglumine versus ultra-small superparamagnetic iron oxide for dynamic contrast-enhanced MR imaging of tumor angiogenesis in human colon carcinoma in mice, *Radiology*, 229, 429–438, 2003.
6. Frank, J. A., Zywicke, H., Jordan, E. K., Mitchell, J., Lewis, B. K., Miller, B., Bryant, L. H. Jr., and Bulte, J. W., Magnetic intracellular labeling of mammalian cells by combining (FDA approved) superparamagnetic iron oxide MR contrast agents and commonly used transfection agents, *Academy of Radiology*, 9 (Suppl. 2), S484–S487, 2002.
7. Shapiro, E. M., Skrtic, S., Sharer, K., Hill, J. M., Dunbar, C. E., and Koretsky, A. P., MRI detection of single particles for cellular imaging, *Proceedings of the National Academy of Sciences of the United States of America*, 101 (30), 10901–10906, 2004.
8. Daldrup-Link, H. E., Rudelius, M., Oostendorp, R. A., Settles, M., Piontek, G., Metz, S., Rosenbrock, H., Keller, U., Heinzmann, U., Rummeny, E. J., Schlegel, J., and Link, T. M., Targeting of hematopoietic progenitor cells with MR contrast agents, *Radiology*, 228 (3), 760–767, 2003.
9. Kostura, L., Kraitchman, D. L., Macka, A. M., Pittenger, M. F., and Bulte, J. W., Feridex labeling of mesenchymal stem cells inhibits chondrogenesis but not adipogenesis or osteogenesis, *NMR in Biomedicine*, 17 (7), 513–517, 2004.
10. LaVan, D. A., Lynn, D. M., and Langer, R., Moving smaller in drug discovery and delivery, *National Review of Drug Discovery*, 1, 77–84, 2002.
11. Miller, J. L., Sirolimus approved with renal transplant indication, *American Journal of Health-System Pharmacy*, 56 (21), 2177–2178, 1999.
12. Aspeslet, L. J. and Yatscoff, R. W., Requirements for therapeutic drug monitoring of sirolimus, an immunosuppressive agents used in renal transplant, *Clinical Therapeutics*, 22 (Suppl. B), B86–B92, 2000.
13. Patel, L. and Lindley, C., Aprepitant-a novel NK1-receptor antagonist, *Expert Opinion on Pharmacotherapy*, 4 (12), 2279–2296, 2003.

14. Wu, Y., Loper, A., Landis, E., Hettrick, L., Novak, L., Lynn, K., Chen, C., Thompson, K., Higgins, R., Batra, U., Shelukar, S., Kwei, G., and Storey, D., The role of biopharmaceutics in the development of a clinical nanoparticle formulation of MK-0869: A beagle dog model predicts improved bioavailability and diminished food effect on absorption in human, *International Journal of Pharmaceutics*, 285 (1–2), 135–146, 2004.

15. Markman, M., Kennedy, A., Webster, K., Peterson, G., Kulp, B., and Belinson, J., Phase 2 trial of liposomal doxorubicin (40 mg/m(2)) in platinum/paclitaxel refractory ovarian fallopian tube cancers and primary carcinoma of the peritoneum, *Gynecologic Oncology*, 78, 369–372, 2000.

16. Frykman, G., Williams, G., and Pazdur, R., Conflicting phase II efficacy date for Doxil, *Journal of Clinical Oncology*, 19 (2), 596–597, 2001.

17. Laginha, K. M., Verwoert, S., Charrois, G. J. R., and Allen, T. M., Determination of doxorubicin levels in whole tumor and tumor nuclei in murine breast cancer tumors, *Clinical Cancer Research*, 11 (19), 6944–6949, 2005.

18. Harries, M., Ellis, P., and Harper, P., Nanoparticle albumin-bound paclitaxel for metastatic breast cancer, *Journal of Clinical Oncology*, 23 (31), 7768–7771, 2005.

19. Nyman, D. W., Campbell, K. J., Hersh, E., Long, K., Richardson, K., Trieu, V., Desai, N., Hawkins, M. J., and Von Hoff, D. D., Phase I and pharmacokinetics trial of ABI-007, a novel nanoparticle formulation of paclitaxel in patients with advanced nonhematologic malignancies, *Journal of Clinical Oncology*, 23 (31), 7785–7793, 2005.

20. Sparreboom, A., Scripture, C. D., Trieu, V., Williams, P. J., De, T., Yang, A., Beals, B., Figg, W. D., Hawkins, M., and Desai, N., Comparative preclinical and clinical pharmacokinetics of a cremaphor-free nanoparticle albumin-bound paclitaxel (ABI-007) and Paclitaxel formulated in cremaphor (Taxol), *Clinical Cancer Research*, 11 (11), 4136–4143, 2005.

21. Chaw, K. C., Manimaran, M., and Tay, F. E., Role of silver ions in destabilization of intermolecular adhesion forces measured by atomic force microscopy in staphylococcus epidermis biofilms, *Antimicrobial Agents and Chemotherapy*, 49 (12), 4853–4859, 2005.

22. Warriner, R. and Burrell, R., Infection and the chronic wound: A focus on silver, *Advanced Skin Wound Care*, 18 (Suppl. 1), 2–12, 2005.

23. Baker, C., Pradhan, A., Pakstis, L., Pochan, D. J., and Shah, S. I., Synthesis and antibaceterial properties of silver nanoparticles, *Journal of Nanoscience and Nanotechnology*, 5 (2), 244–249, 2005.

24. Takemoto, K., Yamamoto, Y., Ueda, Y., Sumita, Y., Yoshida, K., and Niki, Y., Comparative study on the efficacy of Ambisome and Fungizone in a mouse model of pulmonary aspergillosis, *Journal of Antimicrobial Chemotherapy*, 57 (4), 724–731, 2006.

25. Cetin, H., Yalaz, M., Akisu, M., Hilmioglu, S., Metin, D., and Kultursay, N., The efficacy of two different lipid-based amphotericin B in neonatal Candida septicemia, *Pediatrics International*, 47 (6), 676–680, 2005.

26. Barratt, G. and Legrand, P., Comparison of the efficacy and pharmacology of formulations of amphotericin B used in treatment of leishmaniasis, *Current Opinion in Infectious Diseases*, 18 (6), 527–530, 2005.

27. The Healthcare Sales and Marketing Network. NanoBio® corporation announces FDA approval for phase II clinical trials for herpes labialis treatment. http://salesandmarketingnetwork.com/news_-release.php?ID = 14361 (accessed February 24, 2004).

28. NewsRX. FDA approves phase II clinical trials for herpes labialis treatment. http://www.newsrx.com/article.php?articleID = 87034 (accessed February 24, 2004).

29. BioSpace. NanoBio announces FDA approval for phase II clinical trials for herpes labialis treatment. http://www.biospace.com/news_story.aspx?StoryID = 15257520&full = 1 (accessed February 24, 2004).

30. Nanotsunami. NanoBio® completes successful phase 2 herpes trial, prepares for phase 3 trial. http://www.voyle.net/Nano%20Medicine%202005/Medicine%202005-0087.htm (accessed July 22, 2005).

31. Abner, S. R., Guenthner, P. C., Guarner, J., Hancock, K. A., Cummins, J. E. Jr., Fink, A., Gilmore, G. T., Staley, C., Ward, A., Ali, O., Binderow, S., Cohen, S., Grohskopf, L. A., Paxton, L., Hart, C. E., and Dezzutti, C. S., A human colorectal explant culture to evaluate topical microbicides for the prevention of HIV infection, *Journal of Infectious Diseases*, 192 (9), 1545–1556, 2005.

32. Gong, E., Matthews, B., McCarthy, T., Chu, J., Holan, G., Raff, J., and Sacks, S., Evaluation of dendrimer SPL7013, a lead microbicide candidate against herpes simplex virus, *Antiviral Research*, 68 (3), 139–146, 2005.

33. McCarthy, T. D., Karellas, P., Henderson, S. A., Giannis, M., O'Keefe, D. G., Heery, G., Paull, J. R. A., Matthews, B. R., and Holan, G., Dendrimers as drugs, discovery and preclinical and clinical development of dendrimer-based microbicides for HIV and STI prevention, *Molecular Pharmacology*, 2 (4), 312–318, 2005.

34. Jiang, Y. H., Emau, P., Cairns, J. S., Flanary, L., Morton, W. R., McCarthy, T. D., and Tsai, C. C., SPL7013 gel as topical microbicide for prevention of vaginal transmission of SHIV89.6P in macaques, *AIDS Research and Human Retroviruses*, 21 (3), 207–213, 2005.

35. Ito, I., Ji, L., Tanaka, F., Saito, Y., Gopalan, B., Branch, C. D., Xu, K., Atkinson, E. N., Bekele, B. N., Stephens, L. C., Minna, J. D., Roth, J. A., and Ramesh, R., Liposomal vector mediated delivery of the 3p FUS1 gene demonstrates potent antitumor activity against human lung cancer in vivo, *Cancer Gene Therapy*, 11 (11), 733–739, 2004.

36. Smalltimes. Study: Nano therapy kills tumors, extends life. http://www.smalltimes.com/document_-display.cfm?section_id = 53&document_id = 8429 (accessed November 11, 2004).

37. Nanotechnology News. Introgen's nanoparticle tumor treatment increases survival rate by 70%. http://www.azonano.com/news.asp?newsID = 406 (accessed November 12, 2004).

38. Nanotsunami. Introgen's nanoparticle cancer therapy INGN 401 demonstrates promise in the treatment of lung cancer. http://www.voyle.net/Nano%20Medicine%202005/Medicine%202005-0052.htm (accessed May 17, 2005).

39. Bourrinet, P., Bengele, H. H., Bonnemain, B., Dencausse, A., Idee, J. M., Jacobs, P. M., and Lewis, J. M., Preclinical safety and pharmacokinetic profile of ferumoxtran-10, an ultrasmall superparamagnetic iron oxide magnetic resonance contrast agent, *Investigative Radiology*, 41 (3), 313–324, 2006.

40. Saksena, M. A., Saokar, A., and Harisinghani, M. G., Lymphotropic nanoparticle enhanced MR imaging (LNMRI) technique for lymph node imaging, *European Journal of Radiology*, 58 (3), 367–374, 2006.

41. Harisinghani, M. G., Saksena, M. A., Hahn, P. F., King, B., Kim, J., Torabi, M. T., and Weissleder, R., Ferumoxtran-10-enhanced MR lymphangiography: Does contrast-enhanced imaging alone suffice for accurate lymph node characterization?, *American Journal of Roentgenology*, 186 (1), 144–148, 2006.

42. Metz, S., Lohr, S., Settles, M., Beer, A., Woertler, K., Rummeny, E. J., and Daldrup-Link, H. E., Ferumoxtran-10-enhanced MR imaging of the bone marrow before and after conditioning therapy in patients with non-Hodgkin lymphomas, *European Radiology*, 16 (3), 598–607, 2006.

43. Schluep, T., Hwang, J., Cheng, J., Heidel, J. D., Bartlett, D. W., Hollister, B., and Davis, M. E., Preclinical efficacy of the camptothecin-polymer conjugate IT-101 in multiple cancer models, *Clinical Cancer Research*, 12 (5), 1606–1614, 2006.

44. Schluep, T., Cheng, J., Khin, K. T., and Davis, M. E., Pharmacokinetics and biodistribution of the camptothecin–polymer conjugate IT-101 in rats and tumor-bearing mice, *Cancer Chemotherapy and Pharmacology*, 57, 654–662, 2006.

45. Newkome, G. R., Yao, Z., Baker, G. R., and Gupta, V. K., Cascade molecules: A new approach to micelles. A (27)-Arborol, *Journal of Organic Chemistry*, 50, 2003–2004, 1985.

46. Tomalia, D. A., Baker, H., Dewald, J. R., Hall, M., Kallos, G., Maris, S., Roeck, J., Ryder, J., and Smith, P. A., New class of polymers: Starburst–dendritic macromolecules, *Polymer Journal (Tokyo)*, 17, 117–123, 1985.

47. Wiener, E. C., Brechbiel, M. W., Brothers, H., Magin, R. L., Gansow, O. A., Tomalia, D. A., and Lauterbur, P. C., Dendrimer-based metal chelates: A new class of magnetic resonance imaging contrast agents, *Magnetic Resonance in Medicine*, 31 (1), 1–8, 1994.

48. Grinstaff, M. W., Biodendrimers: New polymeric biomaterials for tissue engineering, *Chemistry*, 8, 2839–2846, 2002.

49. Bourne, N., Stanberry, L. R., Kern, E. R., Holan, G., Matthews, B., and Berstein, D. I., Dendrimers, a new class of candidate topic microbicides with activity against herpes simplex virus infestation, *Antimicrobial Agents in Chemotherapy*, 44, 2471–2474, 2000.

50. Zanini, D. and Roy, R., Practical synthesis of PAMAM αthiosialodendrimers for probing multivalent carbohydrate–lectin binding properties, *Journal of Organic Chemistry*, 63, 3486–3491, 1998.
51. Zhuo, R. X., Du, B., and Lu, Z. R., In vitro release of 5-fluoroacil with cyclic core dendritic polymer, *Journal of Controlled Release*, 57, 249–257, 1999.
52. Liu, M., Kono, K., and Frechet, J. M. J., Water-soluble dendritic unimolecular micelles: Their potential as drug delivery agents, *Journal of Controlled Release*, 65, 121–131, 2000.
53. Kobayashi, H. and Brechbiel, M. W., Dendrimer-based nanosized MRI contrast agents, *Current Pharmaeutical Technology*, 5, 539–549, 2004.
54. Konda, S. D., Wang, S., Brechbiel, M., and Wiener, E. C., Biodistribution of a 153 Gd-folate dendrimer generation = 4, in mice with folate-receptor positive and negative ovarian tumor xenografts, *Investigative Radiology*, 37, 199–204, 2002.
55. Hirsch, L. R., Stafford, R. J., Bankson, J. A., Sershen, S. R., Rivers, B., Price, R. E., Hazle, J. D., Halas, N. J., and West, J. L., Nanoshell-mediated nar-infared thermal therapy of tumors under magnetic resonance guidance, *Proceedings of the National Academy of Sciences of the United States of America*, 100, 13549–13554, 2003.
56. Loo, C., Lin, A., Hirsch, L., Lee, M. H., Barton, J., Halas, N., West, J., and Drezek, R., Nanoshell-enabled photonics-based imaging and therapy of cancer, *Technology Cancer Research and Treatment*, 3, 33–40, 2004.
57. Loo, C., Lowery, A., Halas, N., West, J., and Drezek, R., Immunotargeted nanoshells for integrated cancer imaging and therapy, *Nano Letters*, 5, 709–711, 2005.
58. O'Neal, D. P., Hirsch, L. R., Halas, N., Payne, J. D., and West, J. L., Photo-thermal tumor ablation in mice using near-infrared-absorbing nanoparticles, *Cancer Letters*, 209 (2), 171–176, 2004.
59. de Jong, W. H., Roszek, B., and Geertsman, R. E., Nanotechnology in medical applications: Possible risks for human health. RIVM Report 265001002, Bilthoven, 2005.
60. Haddon, R. C., Hebard, A. F., Rosseinsky, M. J., Murphy, D. W., Duclos, S. J., Lyons, K. B., Miller, B., Rosamilia, J. M., Fleming, R. M., Kortan, A. R., Glarum, S. H., Makjiha, A. V., Muller, A. J., Eick, R. H., Zahurak, S. M., Tycko, R., Dabbagh, G., and Thiel, F. A., Conducting films of C60 and C70 by alkali-metal doping, *Nature*, 350, 320–322, 1991.
61. Bosi, S., Da Ros, T., Spalluto, G., and Prato, M., Fullerene derivatives: An attractive tool for biological applications, *European Journal of Medical Chemistry*, 38, 913–923, 2003.
62. Friedman, S. H., Decamp, D. L., Sijbesma, R. P., Srdanov, G., Wudl, F., and Kenyon, G. L., Inhibition of the HIV-1 protease by fullerene derivatives: Model building studies and experimental verification, *Journal of the American Chemical Society*, 115, 6506–6509, 1993.
63. Schiniazi, R. F., Sijbesma, R., Srdanov, G., Hill, C. L., and Wudl, F., Synthesis and virucidal activity of a water soluble, configurationally stable, derivatized C60 fullerene, *Antimicrobial Agents and Chemotherapy*, 37, 1707–1710, 1993.
64. Chueh, S. C., Lai, K. M., Lee, M. S., Chiang, L. Y., Ho, T., and Chen, S. C., Decrease of free radical level in organ perfusate by a novel water soluble carbon-sixty hexa(sulfobutyl)fullerenes, *Transplant Proceedings*, 31, 1976–1977, 1999.
65. Lin, A. M., Chyi, B. Y., Wang, S. D., Yu, H., Kanakamma, P. P., Luh, T. Y., Chou, C. K., and Ho, L. T., Carboxyfullerene prevents iron-induced oxidative stress in rat brain, *Journal of Neurochemistry*, 72, 1634–1640, 1999.
66. Cagle, D. W., Kennel, S. J., Mirzadeh, S., Alford, J. M., and Wilson, L. J., In vivo studies of fullerene-based materials using endohedral metallofullerene radiotracers, *Proceedings of the National Academy of Sciences of the United States of America*, 96, 5182–5187, 1999.
67. Gharbi, N., Pressac, M., Hadchouel, M., Szwarc, H., Wilson, S. R., and Moussa, F., (60)Fullerene is a powerful antioxidant in vivo with no acute or subacute toxicity, *Nano Letters*, 5 (12), 2578–2585, 2005.
68. Bruchez, M. Jr., Moronne, M., Gin, P., Weiss, S., and Alivisatos, A. P., Semiconductor nanocrystals as fluorescent biological labels, *Science*, 281, 2013–2016, 1998.
69. Sukhanova, A., Venteo, L., Devy, J., Artemyev, M., Oleinikov, V., Pluot, M., and Nabiev, I., Highly stable fluorescent nanocrystals as a novel class of labels for immunohistochemical analysis of paraffin embedded tissue sections, *Laboratory Investigation*, 82, 1259–1261, 2002.

70. Sukhanova, A., Devy, J., Venteo, L., Kaplan, H., Artemyev, M., Oleinikov, V., Klinov, D., Pluot, M., Cohen, J. K., and Nabiev, I., Biocompatible fluorescent nanocrystals for immunolabeling of membrane proteins and cells, *Analytical Biochemistry*, 324, 60–67, 2004.

71. Lidke, D. S., Nagy, P., Heintzmann, R., Arndt-Jovin, D. J., Post, J. N., Freco, H. E., Jares-Erijman, E. A., and Jovin, T. M., Quantum dot ligands provide new insights into erbB/HER receptor mediated signal transduction, *Nature Biotechnology*, 22, 198–203, 2004.

72. Akerman, M. E., Chan, W. C., Laakkonen, P., Bhatia, S. N., and Ruoslahti, E., Nanocrystal targeting in vivo, *Proceedings of the National Academy of Sciences of the United States of America*, 99, 12617–12621, 2002.

73. Kim, S., Lim, Y. T., Soltesz, E. G., De Grand, A. M., Lee, J., Nakayama, A., Parker, J. A., Mihalijevic, T., Laurence, R. G., Dor, D. M., Cohn, L. H., Bawendi, M. G., and Frangioni, J. V., Near-infrared fluorescent type II quantum dots for sentinel lymph node mapping, *Nature Biotechnology*, 22, 93–97, 2003.

74. Ballou, B., Lagerholm, B. C., Ernst, L. A., Bruchez, M. P., and Waggoner, A. S., Noninvasive imaging of quantum dots in mice, *Bioconjugate Chemistry*, 15, 79–86, 2004.

75. Duncan, R., Nanomedicine gets clinical, *Materials Today*, 8 (8 Suppl. 1), 16–17, 2005.

76. Yamawaki, H. and Iwai, N., Cytotoxicity of water-soluble fullerene in vascular endothelial cells, *American Journal of Physiology. Cell Physiology*, 290 (6), C1495–C1502, 2006.

77. Kamat, J. P., Devasagayam, T. P. A., Priyadarsini, K. I., and Mohan, H., Reactive oxygen species mediated membrane damage induced by fullerene derivatives and its possible biological implications, *Toxicology*, 155, 55–61, 2000.

78. Sayes, C. M., Gobin, A. M., Ausman, K. D., Mendez, J., West, J. L., and Colvin, V. L., Nano-C60 cytotoxicity is due to lipid peroxidation, *Biomaterials*, 26, 7587–7595, 2005.

79. Dugan, L. L., Gabrielsen, J. K., Yu, S. P., Lin, T. S., and Choi, D. W., Buckminsterfullerenol free radical scavengers reduce excitotoxic and apoptotic death of cultured cortical neurons, *Neurobiology of Disease*, 3, 129–135, 1996.

80. Hussain, S. M., Hess, K. L., Gearhart, J. M., Geiss, K. T., and Shlager, J. J., In vitro toxicity of nanoparticles in BRL 3A rat liver cells, *Toxicology In Vitro*, 19, 975–983, 2005.

81. Jia, F., Wang, H., Yan, L., Wang, X., Pei, R., Yan, T., Zhao, Y., and Guo, X., Cytotoxicity of carbon nanomaterials: Single-wall nanotube, multi-wall nanotube, and fullerene, *Environmental Science & Technology*, 39, 1378–1383, 2005.

82. Renwick, L. C., Donaldson, K., and Clouter, A., Impairment of alveolar macrophage phagocytosis by ultrafine particles, *Toxicology and Applied Pharmacology*, 172, 119–127, 2001.

83. Moller, W., Hofer, T., Ziesenis, A., Karg, E., and Heyder, J., Ultrafine particles cause cytoskeletal dysfunctions in macrophages, *Toxicology and Applied Pharmacology*, 182, 197–207, 2002.

84. Oberdorster, G., Toxicology of ultrafine particles: In vivo studies, *Philosophical Transactions of the Royal Society of London, Series A*, 358, 2719–2740, 2000.

85. Brown, D. M., Wilson, M. R., Macnee, W., Stone, V., and Donaldson, K., Size-dependent proinflammatory effects of ultrafine polystyrene particles: A role for surface area and oxidative stress in the enhanced activity of ultrafines, *Toxicology and Applied Pharmacology*, 175, 191–199, 2001.

86. Johnston, C. H., Finkelstein, J. N., Mercer, P., Corson, N., Gelein, R., and Oberdorster, G., Pulmonary effects induced by ultrafine PTFE particles, *Toxicology and Applied Pharmacology*, 168, 208–215, 2000.

87. Oberdorster, G., Oberdorster, E., and Oberdorster, J., Nanotoxicology: An emerging discipline evolving from studies of ultrafine particles, *Environmental Health Perspectives*, 113 (7), 823–839, 2005.

88. Holsapple, M. P. and Lehman-McKeeman, L. D., Forum series: Research strategies for safety evaluation of nanomaterials, *Toxicological Sciences*, 87 (2), 315, 2005.

89. Oberdorster, E., Manufactured nanomaterials (fullerenes, C60) induce oxidative stress in the brain of juvenile largemouth bass, *Environmental Health Perspectives*, 112 (10), 1058–1062, 2004.

90. Sullivan, D. C. and Ferrari, M., Nanotechnology and tumor imaging: Seizing an opportunity, *Molecular Imaging*, 3 (4), 364–369, 2004.

91. Portney, N. G. and Ozkan, M., Nano-oncology: Drug delivery, imaging, sensing, *Analytical and Bioanalytical Chemistry*, 384, 620–630, 2006.

92. Nanomedicine, an ESF-European Medical Research Councils, (EMRC) forward look report. http://www.esf.org/, 2005.

93. El-Sayed, M., Ginski, M., Rhodes, C., and Ghandehari, H., Transepithelial transport or poly(amidoamine) dendrimers across Caco-2 cell monolayers, *Journal of Controlled Release*, 81, 355–365, 2002.

94. Jevprasesphant, R., Penny, J., Jalal, R., Attwood, D., McKeown, N. B., and D'Emanuele, A. D., The influence of surface modification on the cytotoxicity of PAMAM dendrimers, *International Journal of Pharmaceutics*, 252, 263–266, 2003.

95. Lemarchand, C., Gref, R., Pasirani, C., Garcion, E., Petri, B., Muller, R., Costantini, D., and Couvreur, P., Influence of polysaccharide coating on the interactions of nanoparticles with biological systems, *Biomaterials*, 27, 108–118, 2006.

96. Kirchner, C., Liedl, T., Kudera, S., Pellegrino, T., Javier, A. M., Gaub, H. E., Stolzle, S., Fertig, N., and Parak, W. J., Cytotoxicity of colloidal CdSe and CdSe/ZnS nanoparticles, *Nano Letters*, 5 (2), 331–338, 2005.

97. Neerman, M. F., Zhang, W., Parrish, A. R., and Simanek, E. E., In vitro and in vivo evaluation of a melamine dendrimer as a vehicle for drug delivery, *International Journal of Pharmaceutics*, 281, 129–132, 2004.

98. Nigavekar, S., Sung, L. Y., Llanes, M., El-Jawahri, A., Lawrence, R. S., Becker, C. W., Balogh, L., and Khan, M. K., 3H Dendrimer nanoparticle organ/tumor distribution, *Pharmaceutical Research*, 21 (3), 476–483, 2004.

99. Malik, N., Wiwattanapatapee, R., Klopsch, R., Lorenz, K., Frey, H., Weener, J. W., Meijer, E. W., Paulus, W., and Duncan, R., Relationship between structure and biocompatibility in vitro, and preliminary studies on the biodistribution of 125I-labeled polyamidoamine dendrimers in vivo, *Journal of Controlled Release*, 65, 133–148, 2000.

100. Roberts, J. C., Bhalgat, M. K., and Zera, R. T., Preliminary biological evaluation of polyamidoamine (PAMAM) starburst dendrimers, *Journal of Biomedical Materials Research*, 30 (1), 53–65, 1996.

101. Hardman, R., A toxicological review of quantum dots: Toxicity depends on physiochemical and environmental factors, *Environmental Health Perspectives*, 114 (2), 165–172, 2006.

Section 2

Polymer Conjugates

9 Polymeric Conjugates for Angiogenesis-Targeted Tumor Imaging and Therapy

Amitava Mitra, Anjan Nan, Bruce R. Line, and Hamidreza Ghandehari

CONTENTS

9.1 Introduction 159
9.2 Tumor Angiogenesis 161
9.3 Angiogenesis Markers 162
    9.3.1 VEGF and VEGF Receptors 162
    9.3.2 Aminopeptidases 162
    9.3.3 Matrix Metalloproteinases 164
    9.3.4 Integrins 166
        9.3.4.1 RGD Multimers 167
9.4 Polymeric Conjugates for Angiogenesis Targeting 168
    9.4.1 Polymer-Based Antiangiogenic Gene Therapy 169
    9.4.2 Polymer-Based Nuclear Imaging and Radiotherapy 170
9.5 Summary and Future Directions 175
9.6 List of Abbreviations 175
Acknowledgment 177
References 177

9.1 INTRODUCTION

Despite significant progress in the surgical treatment, radiotherapy, and chemotherapy of solid tumors (e.g., breast, colon, lung, and prostate), too few patients survive long term. This critical unmet need has been the long term focus of a wide range of therapies directed at the malignant cancer cell phenotype, its dysfunctional apoptotic signaling pathways, and survival adaptations. To enhance drug effectiveness and decrease non-specific toxicities, many of these agents have been designed to specifically target tumor cell antigens, receptors, and tumor microenvironmental markers.

In contrast to direct targeting of specific tumor cell markers, targeting endothelial cells supporting tumor angiogenesis provides an alternative, broadly applicable method for cancer diagnosis and therapy.[1–5] All growing tumors require an augmented blood supply, so angiogenesis is a critical common denominator for therapeutic attack. Two distinct types of vessel targeted therapies have evolved.[6] The first general group, the anti-angiogenic agents, inhibit tumor-induced neovascularization by preventing proliferation, migration, and differentiation of endothelial cells.[7–9] Anti-angiogenic drugs have been carried through substantial development to clinical trials.[7,10–12] Bevacuzimab (Avastin), an anti-vascular endothelial growth factor (VEGF) monoclonal antibody, has been approved for the treatment of colorectal cancer.[13] The second type, the antivascular agents, cause rapid and selective occlusion of existing tumor blood vessels, leading to tumor ischemia and extensive hemorrhagic necrosis.[5,14,15] Although still in preclinical development, antivascular-targeted therapeutics produce a characteristic pattern of necrosis after administration to mice and rats with solid tumors.[16,17] They cause a widespread central necrosis that can extend to as much as 95% of the tumor.[16–18]

Indeed, over the past 5–10 years, there has been a rapid development of small molecular therapies directed against neovascular angiogenesis. These molecules are typically smaller than 2 kDa and show very low to absent immunogenicity, high target affinity, rapid targeting, and fast blood clearance. Small molecular agents have been directed at a number of angiogenesis-related targets and have been coupled to moieties to support scintigraphic, optical, and magnetic resonance imaging (MRI).[19,20] Other modifications have supported delivery of drugs, tumor enzyme-cleaved prodrugs, and therapeutic radionuclides.[21] This rich foundation provides the infrastructure upon which polymer-based conjugates have allowed further strides. Well-designed polymer conjugates demonstrate improved pharmacokinetics with prolonged blood residence, lower non-specific tissue penetration, and increased passive tumor localization because of enhanced permeability, and retention; i.e., the EPR effect.[22] Polymer conjugates bearing targeting ligands often demonstrate high tumor affinity via multivalent interactions.[23] Relative to smaller molecular platforms, they can also carry higher diagnostic and therapeutic agent payloads (Figure 9.1).

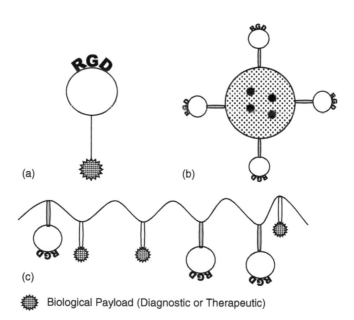

FIGURE 9.1 Schematic of angiogenesis-targeted conjugates. (a) Conjugate of cyclic RGD peptide with a therapeutic or a diagnostic agent. Chemotherapeutic or antiangiogenic drugs can be linked via biodegradable linkers, whereas imaging agents or radionuclides are generally attached via non-biodegradable linkers. (b) Particulate systems such as liposomes and nanoparticles are conjugated to cyclic RGD peptides via a non-biodegradable linker, and the therapeutic or diagnostic agents are encapsulated. (c) Polymeric conjugates have cyclic RGD peptides linked onto the side chains via non-biodegradable linker, whereas biodegradable linkers can be used for drugs and non-biodegradable linkers for imaging agents or radionuclides.

Recent advances provide a clear indication that polyvalent, moderate molecular weight ligand-polymer conjugates will provide the platform to target angiogenesis and deliver a broad range of effective therapies. This chapter presents an overview of tumor angiogenesis, strategies for targeting angiogenic vasculature, and current developments in angiogenesis-targeted polymeric conjugates for diagnosis and therapy.

9.2 TUMOR ANGIOGENESIS

For tumors to grow and metastasize, they must secure an expanded blood supply by taking over existing blood vessels and promoting angiogenesis, i.e., the sprouting of new blood vessels from existing vessels.[2,3] Except in wound healing or in the female reproductive tissues, only 0.01% of normal endothelial cells are involved in angiogenesis.[24,25] In contrast, angiogenesis is both essential for tumor growth beyond 1–2 mm size and is highly specific for neoplasia.[26,27] Indeed, a highly significant statistical correlation has been shown between microvessel density and clinical stage, histopathology stage, and disease-specific survival.[28,29]

The induction of angiogenesis—the angiogenic switch—is an important early event in tumor progression and propagation.[30,31] The switch to an angiogenic phenotype is tightly regulated by a balance between endogenous anti-angiogenic[32–34] and pro-angiogenic[35,36] molecules. Studies in transgenic mice suggest that the switch occurs early in tumor development as a response to local stresses such as hypoxia and low pH.[37,38] Hypoxia increases levels of hypoxia-inducible factors (HIF) in tumor cells and in surrounding stromal cells. HIF then promotes the transcription of genes for VEGF and several other pro-angiogenic factors.[38,39]

During normal angiogenesis such as during reproduction, development, and wound healing, new vessels rapidly mature and stabilize.[2,4] By contrast, tumor blood vessels often fail to mature and are characterized by structures that are leaky, tortuous, and irregular in shape.[40] The high levels of growth factors within a tumor support a continual state of vascular growth, regression, and regrowth. Many lack functional pericytes and are exceptionally permeable because of the presence of fenestrae, transcellular holes, and a lack of a complete basement membrane.[40–44] These ultrastructural changes lead to substantial leaks of tissue fluids into the tumor microenvironment and increased interstitial fluid pressure (IFP). Tumor IFP begins to increase as soon as the host vessels become leaky in response to angiogenic molecules such as VEGF. The importance of this transvascular flux is compounded by the impaired lymphatic system characteristic of cancer tissues. Hence, hyperpermeable angiogenic tumor vessels allow preferential extravasation of circulating macromolecules, and once in the interstitium, they are retained there by a lack of intratumoral lymphatic drainage. This enhanced permeability and retention (EPR) effect[22] results in intratumoral penetration and retention of macromolecules. It is one important factor explaining the delivery of macromolecular anti-cancer agents to the tumor microenvironment.

In addition to the angiogenic factors that enhance the passive delivery of macromolecules via the EPR effect, angiogenic tumor vessels also display a variety of markers (Table 9.1) that can be the focus of active molecular targeting. These markers provide several targeting advantages relative to direct targeting of tumor cell.[45,46] First, endothelial cells are immediately accessible to intravenously delivered ligands. In contrast, the increased interstitial pressure associated with the tumor microenvironment can limit the access of macromolecular agents to tumor-associated targets. Second, endothelial cells are derived from normal, genetically stable host vessels. Hence, they present a target with characteristics that do not fluctuate unlike the genomic instability of most cancer tissues. Third, because large numbers of tumor cells rely on each blood vessel, anti-tumor effects will be amplified by vascular bed injury. Endothelial cell damage leads to vascular coagulation and downstream hypoxic death of approximately 100-fold more cancer cells. Fourth, endothelial targets are common to all solid tumors, whereas most other tumor cell associated markers are cancer phenotype specific.

TABLE 9.1
Some Ligands and Their Target Receptors with Potential for Targeted Delivery to Tumor Angiogenesis

Target	Targeting Ligand[a]	Reference
$\alpha_V\beta_3$ integrin	RGD	53
Aminopeptidase N (CD13)	NGR	54
Vascular endothelial growth factor	Anti-VEGF antibody (Bevacizumab)	13
$\alpha_V\beta_3$ integrin	Anti-$\alpha_V\beta_3$ integrin antibody (Vitaxin)	193
MMP-2 and MMP-9	HWGF	73
Aminopeptidase A	CPRECESIC	67
Cell surface nucleolin	F3 (N-terminal fragment of human high mobility group protein 2)	194, 195
Heparan sulfates	CGKRK, CDTRL	196
Unknown	SMSIARL	197
Kallikrein-9	CSRPRRSEC	196
Platelet-derived growth factor receptor-β	CRGRRST	198

[a] Single amino acid codes for peptides (see list of abbreviations at end of chapter for details).

9.3 ANGIOGENESIS MARKERS

A relatively large number of endothelial cell surface proteins, ligands, and receptors can distinguish tumor angiogenic vessels from normal vasculature (Table 9.1). These markers on tumor endothelial cells have been identified by in vitro techniques (e.g., protein electrophoresis, mRNA-based serial analysis of gene expression, microarray analysis)[47] and in vivo techniques (e.g., biopanning of phage display libraries).[48,49] Many of these targets are the subject of reviews by Ruoshlathi and others.[43,44,47,50]

The identification of angiogenesis markers has spurred the development of technologies to enhance early cancer diagnosis, therapy planning, and therapeutic intervention. The development of diagnostic agents is mainly focused in the areas of nuclear scintigraphy, MRI, and near-infrared (NIR) imaging using matrix metalloproteinase (MMP) inhibitors and $\alpha_V\beta_3$ integrin antagonists (Table 9.2).[19,20,51,52] Tumor vascular targeting peptides have been used to deliver chemotherapy (Table 9.3) given that they selectively target tumor angiogenesis[53,54] and are internalized once bound to cell surface receptors.[55,56] The most important classes of angiogenic targets and pertinent imaging and therapy studies are briefly discussed below.

9.3.1 VEGF and VEGF Receptors

VEGF has become one of the most important targets for antivascular therapy. It is an endothelial cell-specific mitogen that is a potent inducer of angiogenesis. It stimulates endothelial cell growth and acts through a family of closely related receptor tyrosine kinases, the most important of which is VEGFR-2.[36,57,58] VEGF receptor inhibitors have been shown to have anti-tumor effects.[59,60] The most clinically developed is a humanized anti-VEGF antibody (Bevacizumab) that has been approved by the U.S. Food and Drug Administration as a first-line therapy for metastatic colorectal cancer.[13]

9.3.2 Aminopeptidases

Aminopeptidase N (APN) (also known as CD13) is a membrane, spanning 140 kD cell surface protein[61] that plays a role in tumor invasion.[62,63] APN has been successfully targeted using peptides

TABLE 9.2
Conjugates for Angiogenesis Targeted Imaging

Compound	Label	Target	Tumor Model	Reference
Nuclear imaging				
C(RGDfY)	I-125	αVβ3 integrin	Murine osteosarcoma	117
Gluco-RGD	I-125	αVβ3 integrin	M21 human melanoma	120
Galacto-RGD	F-18	αVβ3 integrin	M21 human melanoma	121
c(RGDf(N-Me)V)	F-18	αVβ3 integrin	DLD-1 human colon adenocarcinoma	199
DOTA-c(RGDyK)	Cu-64	αVβ3 integrin	MDA-MB-435 human mammary carcinoma	118
HYNIC-RGD4C	Tc-99m	αVβ3 integrin	Human renal adenocarcinoma	200
FB-c(RGDyK)$_2$	F-18	αVβ3 integrin	U87MG human glioblastoma	136
DOTA-c(RGDyK)$_2$	Cu-64	αVβ3 integrin	MDA-MB-435 human mammary carcinoma	135
HYNIC-E-c(RGDfK)$_2$	Tc-99m	αVβ3 integrin	OVCAR-3 human ovarian carcinoma	107
DOTA-E- c(RGDfK)$_2$	In-111	αVβ3 integrin	OVCAR-3 human ovarian carcinoma	107
DOTA-E{E- c(RGDfK)$_2$}$_2$	Cu-64	αVβ3 integrin	U87MG human glioblastoma	137
FB-PEG-c(RGDyK)	F-18	αVβ3 integrin	U87MG human glioblastoma	140
DOTA-PEG-c(RGDyK)	Cu-64	αVβ3 integrin	U87MG human glioblastoma	141
DOTA-PEG-E-c(RGDyK)$_2$	Cu-64	αVβ3 integrin	Human female lung adenocarcinoma	201
HPMA copolymer-RGD4C	Tc-99m	αVβ3 integrin	DU145 & PC-3 human prostate carcinoma	23, 178
HPMA copolymer-RGDfK	In-111	αVβ3 integrin	Murine lewis lung carcinoma	181
CGS27023A	F-18	MMP	Murine ehrlich breast tumor	85, 87
SAV03M	F-18	MMP	Murine ehrlich breast tumor	87
D-Tyr-HWGF	I-125	MMP-2, MMP-9	Murine lewis lung carcinoma	83
DOTA-HWGF	Cu-64	MMP-2, MMP-9	MDA-MB-435 human mammary carcinoma	90, 91
Magnetic resonance imaging				
Anti-αVβ3 antibody (LM609)-liposome	Gadolinium	αVβ3 integrin	Rabbit V2 carcinoma	126
Anti-αVβ3 antibody (DM101)-nanoparticle	Gadolinium	αVβ3 integrin	Rabbit corneal micropocket model	127
αVβ3 peptidomimetic antagonist-nanoparticle	Gadolinium	αVβ3 integrin	Rabbit V2 carcinoma	110
Optical imaging				
Cy5.5-c(RGDyK)	Cy5.5(NIR dye)	αVβ3 integrin	U87MG human glioblastoma	111
Cy5.5- E- c(RGDyK)$_2$	Cy5.5	αVβ3 integrin	U87MG human glioblastoma	129
Cy5.5-E{E- c(RGDfK)$_2$}$_2$	Cy5.5	αVβ3 integrin	U87MG human glioblastoma	129
Cy5.5-GPLGVRGK-PEG-PLL	Cy5.5	MMP-2	HT1080 human fibrosarcoma	94
NIRQ820-GVPLSLTMGC-Cy5.5	Cy5.5	MMP-7	HT1080 human fibrosarcoma	97

containing an Asn-Gly-Arg (NGR) motif.[54] NGR peptides have been used to deliver chemotherapeutic agents[64] and pro-apoptotic peptides[65] to tumor cells. In a murine breast carcinoma xenograft model, a NGR-doxorubicin conjugate significantly decreased metastasis to lymph nodes and substantially increased survival compared to free doxorubicin. In addition, the conjugates were less toxic to the liver and heart as compared to the free drug.[64]

Aminopeptidase A (APA) is a homodimeric membrane, spanning cell surface protein that is upregulated in the pericytes of tumor blood vessels.[66] Phage display studies have identified an APA binding peptide (CPRECESIC) that inhibits enzymatic activity of APA, suppresses both migration and proliferation of endothelial cells, and inhibits tumor growth.[67]

9.3.3 MATRIX METALLOPROTEINASES

MMPs are a family of over 21 zinc-dependent neutral endopeptidases that play an important role in tumor angiogenesis, tissue remodeling, and cell migration.[68–70] The two most important MMPs in cancer progression are MMP-2 (gelatinase A) and MMP-9 (gelatinase B).[71,72] In cancer, levels of these MMPs are abnormally elevated, enabling cancer cells to degrade the extracellular matrix (ECM), invade the vascular basement membrane, and metastasize to distant sites.[70]

MMP targeting for tumor imaging or therapy can be achieved using small molecule inhibitors of MMPs, peptide sequences that target MMP, tissue inhibitors of MMPs, and MMP peptide substrates that are cleaved at the tumor site. For example, cyclic peptides containing a HWGF motif can specifically inhibit MMP-2 and MMP-9 activities and prevent both tumor growth and metastasis.[73] Several small molecule MMP inhibitors such as the sulfonamide–hydroxamate compound CGS 27023A are currently in clinical trials.[69,74–76]

The protease functionality of MMP provides additional opportunities useful in enzyme prodrug therapy (Table 9.3). For example, drugs such as doxorubicin, auristatins, and duocarmycin may be conjugated to an acetyl-PLGL peptide sequence, MMP substrate, such that when incubated with MMP-2 or MMP-9, cleavage at the GL bond liberates a leucyl drug. Kline et al. showed that such a strategy produced no other cleavage intermediates and that proteolysis products were equipotent with the parent drugs.[77]

MMP-substrate-doxorubicin conjugates are found to be preferentially metabolized in murine fibrosarcoma xenografts relative to plasma and cardiac tissues.[78] There was a 10-fold increase in

TABLE 9.3
Conjugates for Angiogenesis Targeted Drug Delivery

Compound	Drug	Target	Tumor Model	Reference
NGR-doxorubicin	Doxorubicin	Aminopeptidase N	MDA-MB-435 human mammary carcinoma	64
mPEG-GPLGV-doxorubicin	Doxorubicin	MMP-2, MMP-9	Murine lewis lung carcinoma	79
Albumin- GPLGIAGQ-doxorubicin	Doxorubicin	MMP-2, MMP-9	A375 human melanoma	82
PLGL-doxorubicin	Doxorubicin	MMP-2, MMP-9	HT1080 human fibrosarcoma	78
RGD4C-doxorubicin	Doxorubicin	$\alpha_V\beta_3$ integrin	MDA-MB-435 human mammary carcinoma	64
RGD4C-doxorubicin	Doxorubicin	$\alpha_V\beta_3$ integrin	MH134 murine hepatoma	132
RGD4C-AFK-doxorubicin	Doxorubicin	$\alpha_V\beta_3$ integrin	HT1080 human fibrosarcoma	130, 131
E-c(RGDyK)$_2$-paclitaxel	Paclitaxel	$\alpha_V\beta_3$ integrin	MDA-MB-435 human mammary carcinoma	202
PTK787-albumin-PEG-c(RGDfK)	PTK787 (kinase inhibitor)	$\alpha_V\beta_3$ integrin	Human umbilical vein endothelial cells	21
c(RGDfC)-PEG-liposomes	5-FU	$\alpha_V\beta_3$ integrin	Murine B16F10 melanoma	185
RGD-PEG-liposomes	Doxorubicin	$\alpha_V\beta_3$ integrin	Murine B16F10 melanoma	203
c(RGDfC)-PEG-nanoparticles	Doxorubicin	$\alpha_V\beta_3$ integrin	Cl-66 murine metastatic mammary tumor carcinoma	186

doxorubicin tumor-to-heart ratio and a greater effectiveness than free doxorubicin at reducing tumor growth. In particular, the conjugate cured 8 of 10 mice at levels below the maximum tolerated dose, whereas doxorubicin cured 2 of 20 mice at its maximum tolerated dose. Furthermore, mice treated with the conjugate had no detectable change in body weight or circulating reticulocytes.[78]

Other studies have focused on molecular modifications to modify in vivo biodistribution kinetics. For example, MMP substrate peptide–doxorubicin conjugates have been coupled to polyethylene glycol (PEG) to increase blood circulation time and to reduce non-specific cytotoxicity.[79] The anti-cancer drug conjugate, mPEG–GPLGV–DOX, was synthesized by conjugating doxorubicin with GPLGV peptide (a substrate for MMP-2 and MMP-9) and PEG methyl ether (mPEG). In vivo experiments showed that a 50 mg/kg dose of mPEG–GPLGV–DOX had similar therapeutic effectiveness as a 10 mg/kg dose of doxorubicin, but it was associated with lower toxicity and prolonged life span relative to free drug.[79]

An interesting strategy to deliver drug therapy by EPR was investigated using a maleimide–doxorubicin conjugate of an octapeptide MMP substrate (GPLGIAGQ).[80] The water-soluble substrate was designed to form an in situ complex with an albumin thiol (cysteine-34).[81] The conjugate was efficiently cleaved by activated MMP-2 and MMP-9, liberating a doxorubicin tetrapeptide that showed antiproliferative activity in vitro in a murine renal cell carcinoma. When tested in a murine model of a human melanoma xenograft, the doxorubicin–octapeptide conjugate showed complete binding with albumin within 5 min and a 16 h stability half-life.[82] The maximum tolerated dose of the conjugate was 4-fold greater than free doxorubicin. Furthermore, the conjugate showed superior anti-tumor effects as compared to free doxorubicin with duration of remission for up to 40 days and almost a 4-fold reduction in tumor size as compared to free doxorubicin at the end of the 50 day trial period.

Although imaging MMP expression is a relatively new area of research, a number of radiolabeled imaging agents have been tested in animal models (Table 9.2).[19,20] I-125 labeled cyclic decapeptides containing HWGF have shown relatively disappointing results in mice bearing Lewis lung carcinoma.[83] There was low to moderate uptake in tumor, significant tumor washout overtime, high concentrations in the liver and kidney, and low metabolic stability of the iodo-tyrosine complex.[83] More promising results have been obtained with ^{18}F and ^{11}C labeled MMP inhibitors developed for positron emission tomography (PET) imaging.[84–86] In vitro studies showed no loss of MMP inhibitory activity of the radiolabeled compound as compared to the parent compound.[84,87] Dynamic microPET images of tumor-bearing mice showed substantially higher tumor uptake than other background organs.[88,89] The cyclic HWGF peptide has been conjugated with a 1,4,7,10-tetraazacylcododecane-N,N',N'',N'''-tetraacetic acid (DOTA) chelate to enable ^{64}Cu PET imaging.[90,91] Studies of the ^{64}Cu labeled peptide in a murine breast cancer model showed improved pharmacokinetics, tumor accumulation, and metabolic stability relative to a comparable ^{125}I radiolabeled peptide.

^{111}In-diethylenetriaminepentaacetic acid (DTPA)-N-TIMP-2, a radiolabeled high affinity endogenous tissue inhibitor of MMP, has been found to have similar activity as unmodified N-TIMP-2.[92] Unfortunately, ^{111}In-DTPA-N-TIMP-2 yielded disappointing results with insignificant tumor uptake in clinical studies in patients with Kaposi's sarcoma.[93]

Optical imaging methods to assess MMP activity has been achieved using a NIR fluorescence probe attached to a MMP-2 substrate.[94–96] The MMP-2 imaging probe consists of three elements: a quenched NIR fluorochrome (Cy5.5), a MMP-2 peptide substrate (GPLGVRGK), and a PEG-poly-L-lysine (PEG–PLL) graft copolymer.[94] The conjugate is designed such that the peptides are cleaved by MMP-2 at the tumor site, resulting in several hundred percent increases in the NIR fluorescence, allowing visualization of the MMP-2 in human fibrosarcoma tumor. Next MMP-2 inhibition was imaged after treatment with prinomastat where the treated tumor showed significantly lower NIR signal as compared to the untreated tumors.[94] A similar NIR fluorescence-based probe has been designed for MMP-7 that is over expressed in colon, breast, and pancreatic cancer.[97] In tumor extract containing MMP-7 in vitro, the fluorescence signal from the probe was enhanced 7-fold after enzymatic degradation. This probe could also be used to image tumor-associated enzyme activity and also inhibition of MMP-7 in vivo.

9.3.4 INTEGRINS

Integrins are a diverse group of heterodimeric cell surface receptors for ECM molecules.[98,99] This group consists of at least 25 integrins through the pairing 18 α and 8 β subunits. They mediate cell adhesion,[100] help in cell migration during metastasis,[101] and regulate cell growth.[102] $\alpha_V\beta_3$, $\alpha_V\beta_5$, and $\alpha_5\beta_1$ integrins are upregulated in angiogenic endothelial cells and are essential for angiogenesis.[103–106]

High affinity $\alpha_V\beta_3$ (and $\alpha_V\beta_5$) selective ligands containing the tripeptide Arg-Gly-Asp (RGD) have been identified by phage display studies[53] and have been used to deliver chemotherapeutic agents,[64] pro-apoptotic peptides,[65] and radiotherapy[107,108] to tumor tissues. RGD peptides and RGD mimetics[109] have been used to detect cancer when coupled to image contrast agents designed for MRI,[110] gamma scintigraphy,[19] PET,[19,51] and NIR fluorescence imaging.[111]

A large number of radiolabeled tracers based on the RGD tripeptide sequence have been developed (Table 9.2).[19,51,52] Structure activity relationship investigations of the cyclic pentapeptide, cyclo-(Arg^1-Gly^2-Asp^3-D-Phe^4-Val^5),[112–114] revealed that a hydrophobic amino acid in position 4 is essential for high $\alpha_V\beta_3$ affinity and that the amino acid in position 5 could be modified without loss of activity.[115] A tyrosine residue at position 4 or 5 provides a site to radio-iodinate the peptide, and lysine residue introduced at position 5 enables ^{18}F labeling.[116,117] Besides radiohalogenation, cyclic RGD peptides can be labeled with radio-metals by introducing a lysine residue at the terminal end of the peptide and conjugating that with a chelator such as DOTA,[118] DTPA,[119] hydrazinonicotinamide (HYNIC),[107] or N-ω-bis(2-pyridylmethyl)-L-lysine (DPK).[23] Other radio-metals such as yttrium-90 (^{90}Y) may provide a means to deliver molecularly targeted radiotherapy. For example, in a murine ovarian carcinoma xenograft model, a single injection of a ^{90}Y labeled, dimeric-RGD peptide-DOTA conjugate at its maximum tolerated dose (37 MBq) caused significant tumor growth delay and increased median survival time.[107]

Unfortunately, many of these tracers are lipophilic and are both concentrated and excreted by the liver.[117] Non-tumor bearing organs that retain the tracers may receive an unacceptably high radiation dose. To improve their pharmacokinetics, glucose or a galactose-based sugar amino acid has been introduced into the pentapeptide architecture. The resulting iodo-gluco-RGD[120] or [^{18}F]-galacto-RGD[121,122] shows reduced concentration in the liver and increased accumulation in tumor. In other attempts to improve pharmacokinetics, cyclic-RGD peptides have been conjugated to tetrapeptides of D-amino acids.[123] The pharmacokinetics and tumor accumulation of the resulting ^{18}F labeled conjugates containing D-asparatic acid are found to be comparable to [^{18}F]-galacto-RGD.[123]

Non-invasive imaging of angiogenesis markers should be important, both in therapy planning and in monitoring anti-angiogenic therapy. The first human images of $\alpha_V\beta_3$ expression have been obtained with [^{18}F]-galacto-RGD in patients with malignant melanoma and sarcoma.[124] These biodistribution studies demonstrated highly favorable pharmacokinetic profiles with specific receptor binding, rapid blood clearance primarily through the kidney, increasing tumor-to-background (T/B) ratios with time, and almost 4-fold higher tumor distribution volume as compared to muscle.[125] Tumor $\alpha_V\beta_3$ expression was imaged with good contrast in most parts of the body with the exception of the region near the urogenital tract, the tracer excretion route.

Detection of tumor angiogenesis by $\alpha_V\beta_3$ targeted MRI has been achieved using liposomes or nanoparticles encapsulating gadolinium ion that was derivatized with anti-$\alpha_V\beta_3$ antibody LM609,[126] antibody DM101[127] or RGD mimetics.[110,128] This approach provided detailed images of the tumor and delineation of the tumor from surrounding tissues, and it highlighted angiogenic vasculature. This enhanced detection of tumors may be helpful in directing therapeutic interventions.

Several optical angiogenesis imaging methods using cyclic RGD peptides have been described. One such method utilized a NIR dye Cy5.5 conjugated to RGD to visualize glioblastoma xenograft in nude mice.[111] The tumor could be visualized with high contrast to the background organs as early as 30 min post-injection. This study showed the feasibility of in vivo optical imaging of $\alpha_V\beta_3$ expressions. In a subsequent study, RGD mono-, di- and tetramer conjugated to Cy5.5 were compared in a glioblastoma xenograft model.[129] The tumor xenograft could be visualized with all three probes 30 min post-injection. The tumor uptake was in the order tetramer > dimer > monomer; however, the increase in the T/B ratios for the multimers was modest. For example, the dimer and the tetramer showed lower tumor-to-kidney ratios than the monomer because of higher net positive charge. Therefore, only modest enhancement of therapeutic index is possible by multimerization of RGD peptides and further optimization of these probes is needed.

RGD-targeted prodrugs have been tested as a means to improve chemotherapy toxicity profiles (Table 9.3). One strategy utilizes an enzymatically cleavable tripeptide sequence (D-Ala-Phe-Lys) recognized by the tumor-associated protease, plasmin, as a linker between RGD4C and doxorubicin.[130] Upon incubation with plasmin in vitro, doxorubicin is released and regains its cytotoxicity for endothelial and fibrosarcoma cells. Toxicity studies in BALB/c and nude mice showed significantly higher weight loss and mortality for free doxorubicin relative to an equimolar dose of the prodrug.[131] Furthermore, in vivo evaluation in mouse breast cancer and human adenocarcinoma models showed that prodrug anti-tumor efficacy was associated with strong inhibition of angiogenesis.[131] A similar RGD4C-doxorubicin conjugate was tested in an $\alpha_V\beta_3$ negative murine hepatoma model and was found to have superior anti-tumor effectiveness as compared to free doxorubicin.[132]

9.3.4.1 RGD Multimers

As experience with RGD-targeted compounds has evolved, efforts to enhance effectiveness have focused on improving tumor uptake and reducing non-specific localization in normal tissues. To enhance tumor targeting, RGD multimers have been studied as means to increase receptor binding affinity through polyvalency. A dimeric RGD peptide labeled with ^{111}In,[107] [99m]Tc,[133,134] [64]Cu,[135] and [18]F[136] have all shown higher receptor binding affinity and significantly higher tumor accumulation than the RGD monomer. Similarly, tetrameric RGD peptides have shown almost 3-fold and 10-fold higher integrin affinity than the dimer and monomer, respectively.[129,137] Biodistribution studies in M21 melanoma xenograft-bearing mice indicated that tumor uptake and T/B ratios increased as monomer < dimer < tetramer.[138] However, more recent biodistribution studies in human glioblastoma-bearing nude mice revealed that the higher tumor uptake of the tetramer as compared to dimer was offset by increased liver and kidney uptake.[129,137] This significantly lowered the T/B ratios of the multimers, substantially reducing the therapeutic index.

To address the unfavorable pharmacokinetics of these cyclic peptides, investigators have studied their conjugation to hydrophilic polymers such as PEG. A comparative biodistribution study showed that free peptide had a more rapid tumor washout than a pegylated cyclic RGD conjugate. In contrast, the PEG–RGD conjugate showed a gradual increase in both tumor accumulation and tumor-to-blood ratios.[139] Similarly, studies of [18]F[140] and [64]Cu[141] radiolabeled pegylated RGD have yielded high quality microPET images of glioblastoma xenografts with higher T/B ratios than the non-pegylated peptide.

As might be expected, studies combining multivalency and pegylation have shown significantly improved tumor retention and tumor-to-normal tissue distribution ratios.[142] The tumor uptake of pegylated RGD multimers increased by about 3-fold from monomer to tetramer. Tetramer tumor-to-liver ratios were 20- and 3-fold higher than the monomer and dimer, respectively.[142]

9.4 POLYMERIC CONJUGATES FOR ANGIOGENESIS TARGETING

Polymeric conjugates have been used for targeted delivery of drugs to tumor sites[143–146] because the attachment of drugs to water soluble polymers increases their aqueous solubility, reduces side effect, and overcomes multi-drug resistance; the large size of the conjugates increases the blood half-life and significantly alters the drug properties and pharmacokinetics; the conjugates can be tailor-made (i.e., side-chain content, molecular weight, charge, etc.) for specific targeting and delivery needs; they can be designed to passively (EPR) or actively target tumor sites; and site-specific drug release can be achieved by designing biodegradable spacers that can be enzymatically cleaved or that are pH sensitive. These advantages have led to the development of a wide range of polymer-anti-cancer drug conjugates, some of which are currently in clinical trials.[145]

N-(2-hydroxypropyl) methacrylamide (HPMA) copolymers have shown promise as drug carriers.[144,145,147,148] HPMA copolymers have been employed to modify the in vivo biodistribution of chemotherapeutic agents and enzymes. The advantages of HPMA copolymers over other water-soluble polymers is that they can be tailor-made with simple chemical modifications to regulate drug and targeting moiety content for biorecognition, internalization, or subcellular trafficking depending on specific therapeutic needs (Figure 9.2).[144,149,150] The overall molecular weight of HPMA copolymers is determined by the polymerization conditions, particularly the concentration of initiator and chain transfer agents.[151] Various side chain moieties (isotope chelators, targeting moieties, and drugs) (Figure 9.2) may be directly linked to the polymer chain via a biodegradable or non-biodegradable spacer.[143,144]

Polymer-based delivery systems have been used as carriers for passive and active targeting of drugs in the treatment of various diseases and as novel imaging agents.[145,152] Without a specific targeting ligand, moderate-sized (> 30 kD) polymers can passively (via EPR) accumulate in tumor tissues.[22,153] The EPR effect has been used to deliver macromolecular bioactive agents to solid tumors, including anti-angiogenic drugs.[154,155]

There are a number of important differences between small molecular weight drugs and polymeric conjugates of these small molecules. The advantage of polymer-based delivery

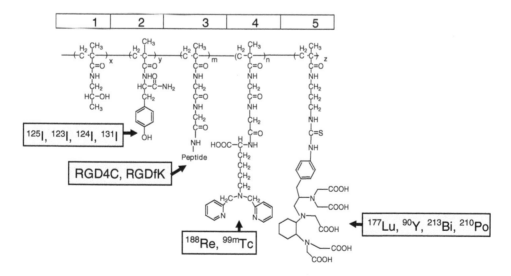

FIGURE 9.2 Structure of water-soluble HPMA copolymers for angiogenesis targeted delivery of radio-nuclides for imaging and therapy. The conjugate can contain side-chains of different chelating comonomers for a wide range of radionuclides for use as a single agent for both diagnosis and therapy. 1-HPMA, 2-MA-Tyr, 3-MA-GG-RGD4C, 4-MA-GG-DPK, and 5-APMA-CHX-A″-DTPA (see list of abbreviations for details). (From Mitra, A., et al., *Nucl. Med Biol.*, 33, 43, 2006. With permission.)

systems stem from decreased extravasation in normal tissues because of the large molecular weight of the conjugates.[156] Low extravasation of polymer conjugates in normal tissues generally results in reduced systemic toxicity. In this regard, the predominant liver[117] and kidney[120] uptake of small RGD peptides has been identified as a significant disadvantage of targeting tumor angiogenesis with small peptides.

The polymeric backbone may also provide a platform to support multivalent targeting ligands (Figure 9.1). As previously discussed, conjugation of a targeting peptide onto a polymer backbone significantly enhances the tumor to normal tissue uptake in comparison to the peptide itself. This is likely attributable to an increased target affinity because of the multivalency of the targeting moiety on the polymer backbone,[157] a combination of active targeting and passive EPR effect of the macromolecular conjugate,[158] and a decreased extravasation in normal tissues as a result of the large molecular weight of the conjugates.[156]

The strategy of targeting the tumor angiogenic endothelium using polymeric conjugates is particularly attractive where angiogenesis inhibitors show substantial toxicity at the effective dose, e.g., the fumagillin analogue TNP-470 shows extreme neurotoxicity at this level.[159] Conjugation of TNP-470 to a water-soluble polymer via an enzymatically degradable spacer allows passive localization and drug release at the tumor site and prevents blood–brain barrier penetration.[154]

9.4.1 POLYMER-BASED ANTI-ANGIOGENIC GENE THERAPY

Anti-angiogenic gene therapy strategies against several human tumors have shown encouraging results in animal models.[160,161] A number of strategies to deliver therapeutic genes to angiogenic endothelial cells using RGD guided viral[162,163] and non-viral systems (cationic polymers,[164] cationic lipids,[165] and cationic peptides[166]) have been investigated. The cationic polymer poly-ethylenimine (PEI) has been extensively used as a non viral gene carrier.[167] PEI has also been investigated as an RGD guided transfection system for angiogenic endothelial cells (Table 9.4). Broadly, these systems can be divided into two classes: those that are direct conjugates of PEI–RGD[168,169] and those where RGD is conjugated onto PEI via a hydrophilic PEG spacer (PEI–PEG–RGD).[164,168,170–172] Because of their reduced non-specific interaction with normal tissues, the charge-shielded pegylated systems have shown more efficient targeting and gene transfection.[171,172]

The PEI–PEG–RGD conjugate architecture has been used to deliver siRNA as a means to inhibit VEGFR-2 expression.[170] Specific tumor accumulation was observed in a murine neuroblastoma model, and there was lower non-specific uptake in lung and liver as compared to aqueous siRNA and non-targeted conjugate. With a 40 μg treatment dose repeated every three days, there

---

TABLE 9.4
Conjugates for Angiogenesis Targeted Gene Therapy

Compound	Gene	Target	Tumor Model	Reference
c(RGD)-PEG-PEI	siRNA	$\alpha_V\beta_3$ integrin	N2A neuroblastoma	170
RGD4C-PEG-PEI	sFlt-1	$\alpha_V\beta_3$ integrin	Human skin capillary endothelial cells	164
Pronectin F$^{+\circledR}$	LacZ	$\alpha_V\beta_3$ integrin	Murine meth-AR-1 fibrosarcoma	173
RGDC-PEG-PEI	Luciferase	$\alpha_V\beta_3$ integrin	Mewo human melanoma	168
RGD-PEI	Plasmid DNA	$\alpha_V\beta_3$ integrin	Human cervical carcinoma	169
RGD-PEG-PEI	DNA	$\alpha_V\beta_3$ integrin	Not reported	172
RGD-lipid nanoparticles	Raf-1	$\alpha_V\beta_3$ integrin	CT26 human colon carcinoma	165
RGD4C-PEG-cholesterol liposome	CAT	$\alpha_V\beta_3$ integrin	PO2 human colon carcinoma	204

was significant inhibition of tumor angiogenesis, reduced tumor growth rate, and a marked reduction of peritumoral vascularization.[170] In another study, Kim et al. reported an anti-angiogenesis strategy to block VEGF receptor function and thereby inhibit endothelial cell proliferation.[164] The authors conjugated PEI–PEG–RGD and a sFlt-1 gene that encoded a soluble fragment of Flt-1 (VEGF receptor antagonist) and found that this non-viral gene delivery system enabled stable expression of sFlt-1 by endothelial cells. The expressed soluble Flt-1 fragment, coupled to exogenous VEGF, blocked Flt-1 receptor binding and inhibited cultured endothelial cell proliferation.[164] Using another approach, Hosseinkhani and Tabata investigated the feasibility of tumor targeting using a non-viral gene carrier synthesized from a genetically engineered silk-like polymer containing repeated RGD sequences(Pronectin F+).[173] When cationized, Pronectin F+-plasmid DNA complexes with or without pegylation were intravenously injected into mice carrying a subcutaneous Meth-AR-1 fibrosarcoma mass, and the level of gene expression in the tumor was significantly higher for the PEG–DNA complex relative to non-pegylated complexes and free plasmid DNA.

Although anti-angiogenic gene therapy is still in its infancy, the studies beginning to appear in the literature suggest the prospects of development of new polymer-based gene delivery systems with improved transfection efficiency and reduced toxicity. It appears that many of the targeting and biodistribution lessons learned in studies of polymer-based drug delivery to sites of angiogenesis may be applied to enhance the success of polymer based gene delivery as well.

9.4.2 POLYMER-BASED NUCLEAR IMAGING AND RADIOTHERAPY

With the appropriate delivery system, radioisotopes have a significant advantage over other therapy agents, namely, the emission of energy that can kill at a distance from the point of radioisotope localization. This diameter of effectiveness helps to overcome the problem of tumor heterogeneity because, unlike other molecular therapy (cell toxins, chemotherapy, etc.), not all tumor cells need to take up the radioisotope to eradicate a tumor. There are also physical characteristics (type of particle emission, emission energy, half-life) of different radioisotopes that may be selected to enhance therapeutic effectiveness.[174] For example, different isotopes deliver beta particulate ionization over millimeters (^{131}I) to centimeters (^{90}Y). Long-lived isotopes such as ^{131}I that remain within the tumor target may provide extended radiation exposure and high radiation dose, especially if there is progressive renal clearance and high target to non-target ratios.

RGD peptides labeled with therapeutically relevant isotopes such as β-particle emitters have been investigated as potential angiogenesis targeted radiotherapy (Table 9.5). The chelation conditions for ^{90}Y and lutetium-177 (^{177}Lu) labeled RGD have revealed that time, temperature, pH, presence of trace metal contaminants, and stoichiometric ratio of chelator to isotope all have significant effects on the rate of chelation and radiolabeling efficiency.[175] A major challenge in development of therapeutic radiopharmaceuticals is radiolytic degradation of radiolabeled products because of production of free radicals in the presence of a large amount of high energy β-particles.[176] A study on the stability of ^{90}Y labeled, dimeric RGD peptide showed that presence

---

TABLE 9.5
Conjugates for Angiogenesis Targeted Radiotherapy

Compound	Radiolabel	Target	Tumor Model	Reference
DOTA-E- c(RGDfK)$_2$	Y-90	$\alpha_V\beta_3$ integrin	OVCAR-3 human ovarian carcinoma	107, 205
DTPA- Tyr3-octreotate-c(RGDyD)	In-111	$\alpha_V\beta_3$ integrin	CA20948 & AR42J rat pancreatic tumor	206,207
HPMA copolymer-RGD4C	Y-90	$\alpha_V\beta_3$ integrin	DU145 human prostate carcinoma	108

of free radical scavengers such as gentisic acid (GA) and ascorbic acid (AA), together with storage at low temperature ($-78°C$), stabilized the labeled compound for at least two half-lives of ^{90}Y, even at a high specific activity.[177]

To develop an HPMA copolymer, RGD-based targeted angiogenesis imaging and therapy agent, the biodistribution and tumor accumulation of HPMA copolymer-RGD4C conjugate as compared to a control conjugate containing the tripeptide Arg-Gly-Glu (HPMA copolymer-RGE4C conjugate) has been studied.[178] Figure 9.3 shows the scintigraphic image of SCID mice bearing prostate tumor xenografts 24 h post-injection of the conjugates. The HPMA copolymer-RGD4C conjugate shows greater tumor accumulation than the control. The tumor uptake of HPMA copolymer-RGD4C conjugate ($4.6 \pm 1.8\%$ injected dose/g) at 24 h post-injection was significantly higher than the control ($1.2 \pm 0.1\%$). A time-dependent biodistribution study showed sustained tumor accumulation of HPMA copolymer-RGD4C conjugate, efficient background clearance, and increasing T/B ratios over time. High T/B increases both tumor detection and therapeutic ratio.[178] Another advantage of using a polymeric conjugate of RGD was highlighted by comparing the biodistributions of HPMA copolymer-RGD4C conjugates and free RGD4C.[23] Organ distribution data in two murine xenograft human prostate models indicated higher tumor accumulation and lower extravasation in normal tissues for HPMA copolymer conjugates compared to free RGD4C peptides (Figure 9.4). Also, T/B was significantly higher for the macromolecular conjugate. These conjugates may be particularly advantageous for cancer radiotherapy because the combination of polyvalent interaction and EPR effect would help to retain the conjugate in the tumor, enhancing the radiation dose.[108,158]

The use of a water-soluble polymer (HPMA)-based conjugate of RGD peptide and the acyclic chelator cyclohexyl-diethylenetriamine pentaacetic acid ($\text{CHX-A}''\text{-DTPA}$) for angiogenesis directed (^{90}Y) radiotherapy has been studied.[108] After intravenous injection in prostate carcinoma

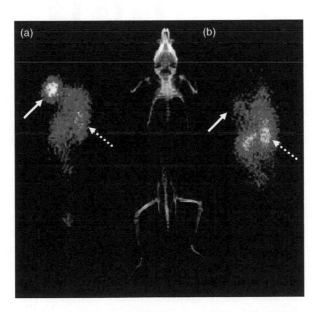

FIGURE 9.3 (See color insert following page 522.) Scintigraphic images of human prostate tumor-bearing SCID mice 24 h post-intravenous injection of ^{99m}Tc labeled HPMA copolymer conjugates. HPMA copolymer-RGD4C conjugate shows higher localization in tumor as compared to the control, HPMA copolymer-RGE4C conjugate (solid arrow). Additional higher activity was found in the kidney (broken arrow). The mouse radiograph (center) shows anatomic correlation of tumor and other organs. Figure legends: (a) HPMA copolymer-RGD4C conjugates; (b) HPMA copolymer-RGE4C conjugates. (From Mitra, A., et al., *J. Control. Release*, 102, 191, 2005. With permission.)

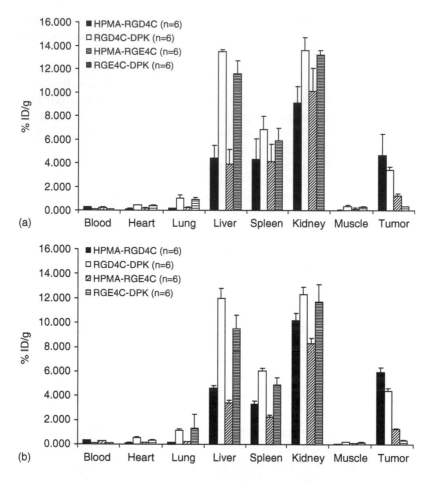

FIGURE 9.4 Residual radioactivity in % injected dose per gram of organ tissue (%ID/g) 24 h post-intravenous injection of [99mTc] labeled HPMA copolymer conjugates and free peptides in (a) DU145 and (b) PC-3 prostate tumor xenograft-bearing SCID mice. The HPMA copolymers showed significantly higher tumor accumulation and reduced background localization than the peptides. The organ data are expressed as mean $\pm$ SD (number of animals/group is shown). (From Line, B. R., et al., *J. Nucl. Med.*, 46, 1152, 2005. With permission.)

xenograft bearing SCID mice, the tumor accumulation of the conjugate peaked at 72 h post-injection, whereas the accumulation in other major organs significantly decreased during that period. A single injection of the [90Y] labeled conjugate at dose levels of 100 and 250 Ci caused significant reduction of tumor volume as compared to the untreated control that was evident from day 7 post-injection (Figure 9.5, top panel). Tumor histopathology showed cellular apoptosis in the treated groups, whereas no signs of acute toxicity were observed in the kidney, liver, and spleen (Figure 9.5, bottom panel).

The RGD4C peptide has a doubly cyclized structure containing two disulphide bonds that affords high binding affinity and specificity for the $\alpha_V\beta_3$ integrin.[178,179] However, the solution instability of the RGD4C disulphide bonds is a potential disadvantage that can lead to significant reduction in $\alpha_V\beta_3$ binding affinity.[180] To overcome this problem, highly stable RGD peptides having monocyclic structure with head-to-tail cyclization containing a D-amino acid, e.g., RGDfK or RGDyK, have been synthesized.[117] These have high affinity for $\alpha_V\beta_3$ and have been used for delivery of imaging agents and therapeutics to tumor angiogenesis.[19,21] The long term (up to 192 h) biodistribution and tumor targeting properties of multivalent HPMA copolymer

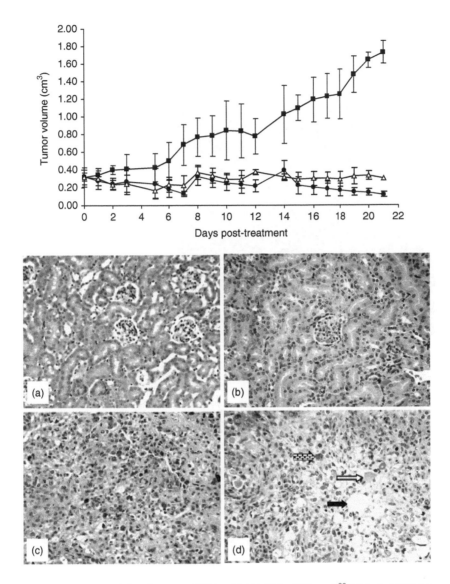

FIGURE 9.5 (See color insert following page 522.) TOP PANEL: Effect of ^{90}Y labeled HPMA copolymer-RGD4C conjugate treatment on human prostate tumor (DU145) growth in SCID mice. Animal groups treated with single dose of 100 µCi ($p < 0.03$) and 250 µCi ($p < 0.01$) ^{90}Y-HPMA-RGD4C conjugate showed significant tumor growth reduction as compared to the untreated controls by day seven post-treatment. Data are presented as mean of tumor volume $(cm^3) \pm SD$ ($n = 6$ mice per group). Figure legends: untreated control (closed square), 250 µCi of ^{90}Y-HPMA copolymer-RGD4C conjugate (closed circle), and 100 µCi of ^{90}Y-HPMA copolymer-RGD4C conjugate (open triangle). BOTTOM PANEL: Histological analysis of kidney and tumor samples at 21 days post-treatment following injection of 250 µCi ^{90}Y labeled HPMA copolymer-RGD4C conjugate or untreated control. Kidney samples taken from (a) control group and (b) 250 µCi treatment group were similar and showed no radiation induced toxicity. There was a complete lack of evidence of any tubular epithelial injury. Normal glomerular and proximal tubular anatomy was evident in both control and ^{90}Y treated specimens. (c) The tumor sections from the control animals showed high grade epithelial malignancy typical of DU145 xenografts. (d) Tumor samples from 250 µCi treatment animals showed large cellular drop out areas (black arrow), higher numbers of eosinophilic cytoplasmic hyaline globular bodies (thanatosomes, open arrow), and pronounced nuclear atypia (hatched arrow) indicative of treatment effect/induced cell damage (From Mitra, A., et al., *Nucl. Med. Biol.*, 33, 43, 2006. With permission.)

conjugates of RGD4C and RGDfK peptides have been studied.[181] Scintigraphic images and necropsy organ counts (Figure 9.6) showed that tumor accumulation of both HPMA-RGD4C and HPMA-RGDfK conjugates increased over time with peak accumulations at $4.9 \pm 0.9\%$ (96 h p.i.) and $5.0 \pm 1.2\%$ (48 h p.i.) ID/g, respectively. In contrast, the background organ distribution

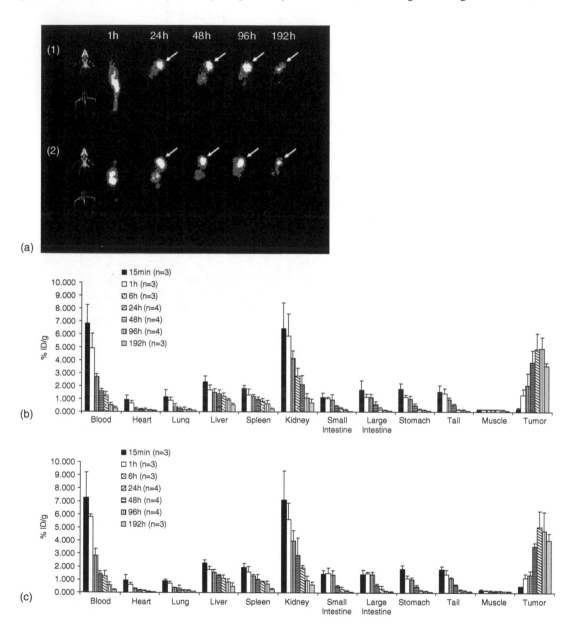

FIGURE 9.6 (See color insert following page 522.) (a) Scintigraphic images of Lewis lung carcinoma-bearing mice up to 192 h post-intravenous injection of ^{111}In labeled copolymers. (1) HPMA copolymer-RGD4C conjugate and (2) HPMA copolymer-RGDfK conjugate showed marked localization in tumor at 24 h p.i. and thereafter (solid arrow). The mouse radiograph shows anatomic correlation of tumor and other organs. Biodistribution of (b) ^{111}In-HPMA copolymer-RGD4C conjugate and (c) ^{111}In-HPMA copolymer-RGDfK conjugate. The organ activities, expressed as % injected dose per gram of organ tissue (%ID/g), showed persistent tumor localization and clearance of activity from the background organs (From Mitra, A., et al., *J. Control. Release*, 114, 175, 2006. With permission.)

rapidly cleared over time, resulting in significant increase in T/B ratios. The radioactive dose to organs as indicated by the area under curve was highest for the tumor. The polymer conjugates of RGD4C or RGDfK provides a means to enhance tumor uptake, decrease background accumulation, and enable selective delivery of therapeutic or diagnostic agents to tumor sites. Because both HPMA-RGD4C and HPMA-RGDfK conjugates have similar tumor targeting abilities and pharmacokinetics, RGDfK can be a suitable ligand because of higher solution stability and easier synthetic manipulation than the 12 amino-acid RGD4C.

These studies with HPMA copolymer-RGD conjugates demonstrate the feasibility and advantages of using multivalent polymer-peptide conjugates. These conjugates show prolonged retention at the tumor site and enhanced T/B ratios. This increased contrast at the tumor site will be necessary for development of a clinically relevant diagnostic agent for therapy planning. Further, the improved therapeutic index will be ideal for angiogenesis directed chemo- or radiotherapy.

9.5 SUMMARY AND FUTURE DIRECTIONS

There is a tremendous interest in targeted drug delivery to tumor vasculature given the genetic stability and accessibility of angiogenesis markers and the importance of angiogenesis in tumor growth and metastasis. A number of ligands targeting angiogenic markers have been identified, but the major focus has been on targeting $\alpha_V\beta_3$ integrins using RGD peptides.[21] Many reports of conjugating RGD ligands to a diverse group of macromolecular carrier systems have been published. These include water-soluble polymers (e.g., PEG,[139] N-(2-hydroxypropyl) methacrylamide,[178] and polyethylenimine[164]), proteins,[182,183] liposomes,[184,185] and nanoparticles.[165,186]

With respect to future development, a broad range of angiogenesis targeting polymer conjugates can be envisioned. As additional vascular targets are identified by in vitro and in vivo methods, angiogenesis targeting may become a critically important strategy for fighting neoplastic as well as non-neoplastic diseases. The first successful human trials for imaging $\alpha_V\beta_3$ integrins using [^{18}F]-galacto-RGD[124,125] and the results of pre-clinical studies using macromolecular conjugates of RGD argue the validity of this concept.

It is expected that the biokinetics of copolymer-conjugates will be continuously improved by tailoring the molecular weight[187–190] and/or electronegative charge[190,191] of the conjugates. This flexibility in the design and synthesis of HPMA and other copolymer conjugates is likely to be important in modifying the blood half-life or reducing non-specific accumulation in normal organs, factors that should increase the conjugate therapeutic index. These may incorporate either enzymatically hydrolysable (e.g., tetrapeptide GFLG)[143] or pH sensitive (e.g., hydrazone linker)[192] linkers for intracellular drug release. In addition, they will likely demonstrate targeting ligand multivalency to increase the tumor uptake and therapeutic efficacy.

It is also expected that there will be expanded use of therapeutic radionuclides to destroy angiogenic vasculature and surrounding tumor cells. Because radioactive emissions can kill at a distance from the point of radioisotope localization, they have a diameter of effectiveness that may overcome the problem of tumor heterogeneity that has plagued other molecular therapies. In short, the ever increasing interest in polymer-based therapeutics promises continued progress in angiogenesis targeted, tumor imaging, and therapy.

9.6 LIST OF ABBREVIATIONS

AFK	Ala-Phe-Lys
APA	Aminopeptidase A
APMA-CHX-A''-DTPA	N-methacryloylaminopropyl-2-amino-3-(isothiourea-phenyl)-propyl-cyclohexane-1,2-diamine-N,N-N',N',N'',N''-pentaacetic acid

APN	Aminopeptidase N
BBB	Blood–brain barrier
CAT	Chloramphenicol acetyl transferase
CDTRL	Cys-Asp-Thr-Arg-Leu
CGKRK	Cys-Gly-Lys-Arg-Lys
CHX-A″-DTPA	Cyclohexyl–diethylenetriamine pentaacetic acid
CPRECESIC	Cys-Pro-Arg-Glu-Cys-Glu-Ser-Ile-Cys
CRGRRST	Cys-Arg-Gly-Arg-Arg-Ser-Thr
CSRPRRSEC	Cys-Ser-Arg-Pro-Arg-Arg-Ser-Glu-Cys
DOTA	1,4,7,10-tetra-azacylcododecane-N', N'', N''', N''''-tetraacetic acid
DOX	Doxorubicin
DPK	N-ω-bis(2-pyridylmethyl)-L-lysine
DTPA	Diethylenetriaminepentaacetic acid
ECM	Extracellular matrix
EPR	Enhanced permeability and retention
FB	Fluorobenzoyl
FDA	Food and drug administration
FDG	2-Fluoro-2-deoxyglucose
5-Fu	5-Fluorouracil
GFLG	Gly-Phe-Leu-Gly
GPLGV	Gly-Pro-Leu-Gly-Val
GPLGIAGQ	Gly-Pro-Leu-Gly-Ile-Ala-Gly-Gln
GPLGVRGK	Gly-Pro-Leu-Gly-Val-Arg-Gly-Lys
GVPLSLTMGC	Gly-Val-Pro-Leu-Ser-Leu-Thr-Met-Gly-Cys
HPMA	N-(2-hydroxypropyl) methacrylamide
HWGF	His-Trp-Gly-Phe
HYNIC	Hydrazinonicotinamide
MA-GG-DPK	N-methacryloylglycylglycyl-(N-ω-bis(2-pyridylmethyl)-L-lysine)
MA-GG-RGD4C	N-methacryloylglycylglycyl-RGD4C
MA-Tyr	N-methacryloyltyrosinamide
MBq	Megabecquerel
mCi	Millicurie
MMP	Matrix metalloproteinases
MRI	Magnetic resonance imaging
NGR	Asn-Gly-Arg
NIR	Near-infrared
PEG	Polyethylene glycol
PEI	Polyethylenimine
PET	Positron emission tomography
PLGL	Pro-Leu-Gly-Leu
PLL	Poly-L-lysine
RGD	Arg-Gly-Asp
RGE	Arg-Gly-Glu
SCID	Severe combined immunodeficient
SMSIARL	Ser-Met-Ser-Ile-Ala-Arg-Leu
T/B	Tumor-to-background
TIMP	Tissue inhibitors of matrix metalloproteinases
VEGF	Vascular endothelial growth factor

ACKNOWLEDGMENT

Support from the American Russian Cancer Alliance and National Institutes of Health (1 R01 EB0207171) is acknowledged.

REFERENCES

1. Ferrara, N. and Kerbel, R. S., Angiogenesis as a therapeutic target, *Nature*, 438, 967, 2005.
2. Carmeliet, P., Angiogenesis in health and disease, *Nat. Med.*, 9, 653, 2003.
3. Carmeliet, P. and Jain, R. K., Angiogenesis in cancer and other diseases, *Nature*, 407, 249, 2000.
4. Carmeliet, P., Mechanisms of angiogenesis and arteriogenesis, *Nat. Med.*, 6, 389, 2000.
5. Thorpe, P. E., Vascular targeting agents as cancer therapeutics, *Clin. Cancer Res.*, 10, 415, 2004.
6. Siemann, D. W. et al., Differentiation and definition of vascular-targeted therapies, *Clin. Cancer Res.*, 11, 416, 2005.
7. Kerbel, R. and Folkman, J., Clinical translation of angiogenesis inhibitors, *Nat. Rev. Cancer*, 2, 727, 2002.
8. Ellis, L. M., Synopsis of angiogenesis inhibitors in oncology, *Oncology (Huntingt)*, 16, 14, 2002.
9. Risau, W., Mechanisms of angiogenesis, *Nature*, 386, 671, 1997.
10. Gasparini, G. et al., Angiogenic inhibitors: A new therapeutic strategy in oncology, *Nat. Clin. Pract. Oncol.*, 2, 562, 2005.
11. Folkman, J., Angiogenesis inhibitors: A new class of drugs, *Cancer Biol. Ther.*, 2, S127, 2003.
12. Ranieri, G. and Gasparini, G., Angiogenesis and angiogenesis inhibitors: A new potential anticancer therapeutic strategy, *Curr. Drug Targets Immune Endocr. Metabol. Disord.*, 1, 241, 2001.
13. Ferrara, N. et al., Discovery and development of bevacizumab, an anti-VEGF antibody for treating cancer, *Nat. Rev. Drug Discov.*, 3, 391, 2004.
14. Tozer, G. M., Kanthou, C., and Baguley, B. C., Disrupting tumour blood vessels, *Nat. Rev. Cancer*, 5, 423, 2005.
15. Narazaki, M. and Tosato, G., Targeting coagulation to the tumor microvasculature: Perspectives and therapeutic implications from preclinical studies, *J. Natl. Cancer Inst.*, 97, 705, 2005.
16. Ching, L. M. et al., Induction of intratumoral tumor necrosis factor (TNF) synthesis and hemorrhagic necrosis by 5,6-dimethylxanthenone-4-acetic acid (DMXAA) in TNF knockout mice, *Cancer Res.*, 59, 3304, 1999.
17. Tozer, G. M. et al., The biology of the combretastatins as tumour vascular targeting agents, *Int. J. Exp. Pathol.*, 83, 21, 2002.
18. Pedley, R. B. et al., Eradication of colorectal xenografts by combined radioimmunotherapy and combretastatin a-4 3-O-phosphate, *Cancer Res.*, 61, 4716, 2001.
19. Haubner, R. and Wester, H. J., Radiolabeled tracers for imaging of tumor angiogenesis and evaluation of anti-angiogenic therapies, *Curr. Pharm. Des.*, 10, 1439, 2004.
20. Li, W. P. and Anderson, C. J., Imaging matrix metalloproteinase expression in tumors, *Q. J. Nucl. Med.*, 47, 201, 2003.
21. Temming, K. et al., RGD-based strategies for selective delivery of therapeutics and imaging agents to the tumour vasculature, *Drug Resist. Updat.*, 8 (6), 381–402, 2005.
22. Maeda, H. et al., Tumor vascular permeability and the EPR effect in macromolecular therapeutics: A review, *J. Control. Release*, 65, 271, 2000.
23. Line, B. R. et al., Targeting tumor angiogenesis: Comparison of peptide and polymer-peptide conjugates, *J. Nucl. Med.*, 46, 1552, 2005.
24. Mahadevan, V. and Hart, I. R., Metastasis and angiogenesis, *Acta Oncol.*, 29, 97, 1990.
25. Hanahan, D. and Folkman, J., Patterns and emerging mechanisms of the angiogenic switch during tumorigenesis, *Cell*, 86, 353, 1996.
26. Folkman, J., Tumor angiogenesis: Therapeutic implications, *N. Engl. J. Med.*, 285, 1182, 1971.
27. Folkman, J., What is the evidence that tumors are angiogenesis dependent?, *J. Natl. Cancer Inst.*, 82, 4, 1990.
28. Weidner, N., Tumoural vascularity as a prognostic factor in cancer patients: The evidence continues to grow, *J. Pathol.*, 184, 119, 1998.

29. Weidner, N., Tumour vascularity and proliferation: Clear evidence of a close relationship, *J. Pathol.*, 189, 297, 1999.

30. Bergers, G. and Benjamin, L. E., Tumorigenesis and the angiogenic switch, *Nat. Rev. Cancer*, 3, 401, 2003.

31. Hanahan, D. and Weinberg, R. A., The hallmarks of cancer, *Cell*, 100, 57, 2000.

32. Folkman, J., Endogenous angiogenesis inhibitors, *APMIS*, 112, 496, 2004.

33. Nyberg, P., Xie, L., and Kalluri, R., Endogenous inhibitors of angiogenesis, *Cancer Res.*, 65, 3967, 2005.

34. Sund, M. et al., Function of endogenous inhibitors of angiogenesis as endothelium-specific tumor suppressors, *Proc. Natl. Acad. Sci. U.S.A.*, 102, 2934, 2005.

35. Relf, M. et al., Expression of the angiogenic factors vascular endothelial cell growth factor, acidic and basic fibroblast growth factor, tumor growth factor beta-1, platelet-derived endothelial cell growth factor, placenta growth factor, and pleiotrophin in human primary breast cancer and its relation to angiogenesis, *Cancer Res.*, 57, 963, 1997.

36. Yancopoulos, G. D. et al., Vascular-specific growth factors and blood vessel formation, *Nature*, 407, 242, 2000.

37. Hanahan, D. et al., Transgenic mouse models of tumour angiogenesis: The angiogenic switch, its molecular controls, and prospects for preclinical therapeutic models, *Eur. J Cancer*, 32A, 2386, 1996.

38. Harris, A. L., Hypoxia—a key regulatory factor in tumour growth, *Nat. Rev. Cancer*, 2, 38, 2002.

39. Semenza, G. L., HIF-1 and tumor progression: Pathophysiology and therapeutics, *Trends Mol. Med.*, 8, S62, 2002.

40. Jain, R. K., Munn, L. L., and Fukumura, D., Dissecting tumour pathophysiology using intravital microscopy, *Nat. Rev. Cancer*, 2, 266, 2002.

41. Morikawa, S. et al., Abnormalities in pericytes on blood vessels and endothelial sprouts in tumors, *Am. J. Pathol.*, 160, 985, 2002.

42. Jain, R. K., Molecular regulation of vessel maturation, *Nat. Med.*, 9, 685, 2003.

43. Baluk, P., Hashizume, H., and McDonald, D. M., Cellular abnormalities of blood vessels as targets in cancer, *Curr. Opin. Genet. Dev.*, 15, 102, 2005.

44. Ruoslahti, E., Specialization of tumour vasculature, *Nat. Rev. Cancer*, 2, 83, 2002.

45. Brekken, R. A. and Thorpe, P. E., Vascular endothelial growth factor and vascular targeting of solid tumors, *Anticancer Res.*, 21, 4221, 2001.

46. Burrows, F. J. and Thorpe, P. E., Vascular targeting—a new approach to the therapy of solid tumors, *Pharmacol. Ther.*, 64, 155, 1994.

47. Neri, D. and Bicknell, R., Tumour vascular targeting, *Nat. Rev. Cancer*, 5, 436, 2005.

48. Pasqualini, R. and Ruoslahti, E., Organ targeting in vivo using phage display peptide libraries, *Nature*, 380, 364, 1996.

49. Arap, W. et al., Steps toward mapping the human vasculature by phage display, *Nat. Med.*, 8, 121, 2002.

50. Ruoslahti, E., Vascular zip codes in angiogenesis and metastasis, *Biochem. Soc. Trans.*, 32, 397, 2004.

51. Haubner, R. H. et al., Radiotracer-based strategies to image angiogenesis, *Q. J. Nucl. Med.*, 47, 189, 2003.

52. McQuade, P. and Knight, L. C., Radiopharmaceuticals for targeting the angiogenesis marker alpha(v)beta(3), *Q. J. Nucl. Med.*, 47, 209, 2003.

53. Pasqualini, R., Koivunen, E., and Ruoslahti, E., Alpha v integrins as receptors for tumor targeting by circulating ligands, *Nat. Biotechnol.*, 15, 542, 1997.

54. Pasqualini, R. et al., Aminopeptidase N is a receptor for tumor-homing peptides and a target for inhibiting angiogenesis, *Cancer Res.*, 60, 722, 2000.

55. Isberg, R. R. and Tran, V. N., Binding and internalization of microorganisms by integrin receptors, *Trends Microbiol.*, 2, 10, 1994.

56. Hart, S. L. et al., Cell binding and internalization by filamentous phage displaying a cyclic Arg-Gly-Asp-containing peptide, *J. Biol. Chem.*, 269, 12468, 1994.

57. Ferrara, N., Gerber, H. P., and LeCouter, J., The biology of VEGF and its receptors, *Nat. Med.*, 9, 669, 2003.

58. Ferrara, N., Vascular endothelial growth factor: Basic science and clinical progress, *Endocr. Rev.*, 25, 581, 2004.

59. Ferrara, N., Vascular endothelial growth factor as a target for anticancer therapy, *Oncologist*, 9 (Suppl 1), 2, 2004.

60. Ahmed, S. I., Thomas, A. L., and Steward, W. P., Vascular endothelial growth factor (VEGF) inhibition by small molecules, *J. Chemother.*, 16 (Suppl 4), 59, 2004.

61. Sato, Y., Aminopeptidases and angiogenesis, *Endothelium*, 10, 287, 2003.

62. Chang, Y. W. et al., CD13 (aminopeptidase N) can associate with tumor-associated antigen L6 and enhance the motility of human lung cancer cells, *Int. J. Cancer*, 116, 243, 2005.

63. Ishii, K. et al., Aminopeptidase N regulated by zinc in human prostate participates in tumor cell invasion, *Int. J. Cancer*, 92, 49, 2001.

64. Arap, W., Pasqualini, R., and Ruoslahti, E., Cancer treatment by targeted drug delivery to tumor vasculature in a mouse model, *Science*, 279, 377, 1998.

65. Ellerby, H. M. et al., Anti-cancer activity of targeted pro-apoptotic peptides, *Nat. Med.*, 5, 1032, 1999.

66. Schlingemann, R. O. et al., Aminopeptidase a is a constituent of activated pericytes in angiogenesis, *J. Pathol.*, 179, 436, 1996.

67. Marchio, S. et al., Aminopeptidase A is a functional target in angiogenic blood vessels, *Cancer Cell*, 5, 151, 2004.

68. Vihinen, P. and Kahari, V. M., Matrix metalloproteinases in cancer: Prognostic markers and therapeutic targets, *Int. J. Cancer*, 99, 157, 2002.

69. Vihinen, P., Ala-aho, R., and Kahari, V. M., Matrix metalloproteinases as therapeutic targets in cancer, *Curr. Cancer Drug Targets*, 5, 203, 2005.

70. Egeblad, M. and Werb, Z., New functions for the matrix metalloproteinases in cancer progression, *Nat. Rev. Cancer*, 2, 161, 2002.

71. Nguyen, M., Arkell, J., and Jackson, C. J., Human endothelial gelatinases and angiogenesis, *Int. J. Biochem. Cell Biol.*, 33, 960, 2001.

72. Himelstein, B. P. et al., Metalloproteinases in tumor progression: The contribution of MMP-9, *Invas. Metast.*, 14, 246, 1994.

73. Koivunen, E. et al., Tumor targeting with a selective gelatinase inhibitor, *Nat. Biotechnol.*, 17, 768, 1999.

74. Hidalgo, M. and Eckhardt, S. G., Development of matrix metalloproteinase inhibitors in cancer therapy, *J. Natl. Cancer Inst.*, 93, 178, 2001.

75. Levitt, N. C. et al., Phase I and pharmacological study of the oral matrix metalloproteinase inhibitor, MMI270 (CGS27023A), in patients with advanced solid cancer, *Clin. Cancer Res.*, 7, 1912, 2001.

76. Eatock, M. et al., A dose-finding and pharmacokinetic study of the matrix metalloproteinase inhibitor MMI270 (previously termed CGS27023A) with 5-FU and folinic acid, *Cancer Chemother. Pharmacol.*, 55, 39, 2005.

77. Kline, T., Torgov, M. Y., Mendelsohn, B. A., Cerveny, C. G., and Senter, P. D., Novel antitumor prodrugs designed for activation by matrix metalloproteinases-2 and -9, *Mol. Pharm.*, 1, 9, 2004.

78. Albright, C. F. et al., Matrix metalloproteinase-activated doxorubicin prodrugs inhibit HT1080 xenograft growth better than doxorubicin with less toxicity, *Mol. Cancer Ther.*, 4, 751, 2005.

79. Bae, M. et al., Metalloprotease-specific poly(ethylene glycol) methyl ether–peptide–doxorubicin conjugate for targeting anticancer drug delivery based on angiogenesis, *Drugs Exp. Clin. Res.*, 29, 15, 2003.

80. Kratz, F. et al., Development and in vitro efficacy of novel MMP2 and MMP9 specific doxorubicin albumin conjugates, *Bioorg. Med. Chem. Lett.*, 11, 2001, 2001.

81. Kratz, F. et al., Probing the cysteine-34 position of endogenous serum albumin with thiol-binding doxorubicin derivatives. Improved efficacy of an acid-sensitive doxorubicin derivative with specific albumin-binding properties compared to that of the parent compound, *J. Med. Chem.*, 45, 5523, 2002.

82. Mansour, A. M. et al., A new approach for the treatment of malignant melanoma: Enhanced antitumor efficacy of an albumin-binding doxorubicin prodrug that is cleaved by matrix metalloproteinase 2, *Cancer Res.*, 63, 4062, 2003.

83. Kuhnast, B. et al., Targeting of gelatinase activity with a radiolabeled cyclic HWGF peptide, *Nucl. Med. Biol.*, 31, 337, 2004.

84. Zheng, Q. H. et al., Synthesis and preliminary biological evaluation of MMP inhibitor radiotracers [^{11}C]methyl-halo-CGS 27023A analogs, new potential PET breast cancer imaging agents, *Nucl. Med. Biol.*, 29, 761, 2002.

85. Furumoto, S., Iwata, R., and Ido, T., Design and synthesis of fluorine-18 labeled matrix metalloproteinase inhibitors for cancer imaging, *J. Labelled Compd. Rad.*, 45, 975, 2002.

86. Fei, X., Hutchins, G. D. et al., Synthesis of radiolabeled biphenylsulfonamide matrix metalloproteinase inhibitors as new potential PET cancer imaging agents, *Bioorg. Med. Chem. Lett.*, 13, 2217, 2003.

87. Furumoto, S. et al., Development of a new ^{18}F-labeled matrix metalloproteinase-2 inhibitor for cancer imaging by PET, *J. Nucl. Med.*, 43 (Suppl), 364, 2002.

88. Furumoto, S. et al., Tumor detection using ^{18}F-labeled matrix metalloproteinase-2 inhibitor, *Nucl. Med. Biol.*, 30, 119, 2003.

89. Zheng, Q. H. et al., Synthesis, biodistribution and micro-PET imaging of a potential cancer biomarker carbon-11 labeled MMP inhibitor ([2]R)-2-[[4-(6-fluorohex-1-ynyl)phenyl]sulfonylamino]-3-methylbutyric acid [^{11}C]methyl ester, *Nucl. Med. Biol.*, 30, 753, 2003.

90. Li, W. P. et al., ^{64}Cu-DOTA-CTTHWGFTLC: A selective gelatinase inhibitor for tumor imaging, *J. Nucl. Med.*, 43 (Suppl), 227, 2002.

91. Li, W. P. et al., In vitro and in vivo evaluation of a radiolabeled gelatinase inhibitor for microPET imaging of metastatic breast cancer, *Mol. Imaging Biol.*, 4 (Suppl), 23, 2002.

92. Giersing, B. K. et al., Synthesis and characterization of ^{111}In-DTPA-N-TIMP-2: A radiopharmaceutical for imaging matrix metalloproteinase expression, *Bioconjug. Chem.*, 12, 964, 2001.

93. Kulasegaram, R. et al., In vivo evaluation of ^{111}In-DTPA-N-TIMP-2 in Kaposi sarcoma associated with HIV infection, *Eur. J. Nucl. Med.*, 28, 756, 2001.

94. Bremer, C., Tung, C. H., and Weissleder, R., In vivo molecular target assessment of matrix metalloproteinase inhibition, *Nat. Med.*, 7, 743, 2001.

95. Bremer, C. et al., Optical imaging of matrix metalloproteinase-2 activity in tumors: Feasibility study in a mouse model, *Radiology*, 221, 523, 2001.

96. Bremer, C., Tung, C. H., and Weissleder, R., Molecular imaging of MMP expression and therapeutic MMP inhibition, *Acad. Radiol.*, 9 (Suppl 2), S314, 2002.

97. Pham, W. et al., Developing a peptide-based near-infrared molecular probe for protease sensing, *Bioconjug. Chem.*, 15, 1403, 2004.

98. Cox, D. et al., The pharmacology of the integrins, *Med. Res. Rev.*, 14, 195, 1994.

99. van Der, F. A. and Sonnenberg, A., Function and interactions of integrins, *Cell Tissue Res.*, 305, 285, 2001.

100. Sepulveda, J. L., Gkretsi, V., and Wu, C., Assembly and signaling of adhesion complexes, *Curr. Top. Dev. Biol.*, 68, 183, 2005.

101. Hood, J. D. and Cheresh, D. A., Role of integrins in cell invasion and migration, *Nat. Rev. Cancer*, 2, 91, 2002.

102. Giancotti, F. G. and Ruoslahti, E., Integrin signaling, *Science*, 285, 1028, 1999.

103. Brooks, P. C., Clark, R. A., and Cheresh, D. A., Requirement of vascular integrin alpha v beta 3 for angiogenesis, *Science*, 264, 569, 1994.

104. Friedlander, M. et al., Definition of two angiogenic pathways by distinct alpha v integrins, *Science*, 270, 1500, 1995.

105. Eliceiri, B. P. and Cheresh, D. A., Role of alpha v integrins during angiogenesis, *Cancer J.*, 6 (Suppl 3), S245, 2000.

106. Stupack, D. G. and Cheresh, D. A., Integrins and angiogenesis, *Curr. Top. Dev. Biol.*, 64, 207, 2004.

107. Janssen, M. L. et al., Tumor targeting with radiolabeled alpha(v)beta(3) integrin binding peptides in a nude mouse model, *Cancer Res.*, 62, 6146, 2002.

108. Mitra, A. et al., Polymer-peptide conjugates for angiogenesis targeted tumor radiotherapy, *Nucl. Med. Biol.*, 33, 43, 2006.

109. Goodman, S. L. et al., Nanomolar small molecule inhibitors for alphav(beta)6, alphav(beta)5, and alphav(beta)3 integrins, *J. Med. Chem.*, 45, 1045, 2002.

110. Winter, P. M. et al., Molecular imaging of angiogenesis in nascent Vx-2 rabbit tumors using a novel alpha(nu)beta3-targeted nanoparticle and 1.5 tesla magnetic resonance imaging, *Cancer Res.*, 63, 5838, 2003.

111. Chen, X., Conti, P. S., and Moats, R. A., In vivo near-infrared fluorescence imaging of integrin alphavbeta3 in brain tumor xenografts, *Cancer Res.*, 64, 8009, 2004.

112. Haubner, R., Finsinger, D., and Kessler, H., Stereoisomeric peptide libraries and peptidomimetics for designing selective inhibitors of the alphav beta3 integrin for a new cancer therapy, *Angew. Chem. Int. Ed. Engl.*, 36, 1374, 1997.

113. Aumailley, M. et al., Arg-Gly-Asp constrained within cyclic pentapeptides. Strong and selective inhibitors of cell adhesion to vitronectin and laminin fragment P1, *FEBS Lett.*, 291, 50, 1991.

114. Gurrath, M. et al., Conformation/activity studies of rationally designed potent anti-adhesive RGD peptides, *Eur. J. Biochem.*, 210, 911, 1992.

115. Haubner, R. et al., Structural and functional aspects of RGD containing cyclic pentapeptides as highly potent and selective avb3 antagonists, *J. Am. Chem. Soc.*, 118, 7461, 1996.

116. Haubner, R. et al., Comparison of tumor uptake and biokinetics of I-125 and F-18 labeled RGD-peptides, *J. Labelled Compd. Rad.*, 42, S36, 1999.

117. Haubner, R. et al., Radiolabeled alpha(v)beta3 integrin antagonists: A new class of tracers for tumor targeting, *J. Nucl. Med.*, 40, 1061, 1999.

118. Chen, X. et al., MicroPET and autoradiographic imaging of breast cancer alpha v-integrin expression using [18]F- and ^{64}Cu-labeled RGD peptide, *Bioconjug. Chem.*, 15, 41, 2004.

119. Wang, W. et al., Convenient solid-phase synthesis of diethylenetriaminepenta–acetic acid (DTPA)-conjugated cyclic RGD peptide analogues, *Cancer Biother. Radiopharm.*, 20, 547, 2005.

120. Haubner, R. et al., Glycosylated RGD-containing peptides: Tracer for tumor targeting and angiogenesis imaging with improved biokinetics, *J. Nucl. Med.*, 42, 326, 2001.

121. Haubner, R. et al., Noninvasive imaging of alpha(v)beta3 integrin expression using ^{18}F-labeled RGD-containing glycopeptide and positron emission tomography, *Cancer Res.*, 61, 1781, 2001.

122. Haubner, R. et al., [^{18}F]Galacto-RGD: Synthesis, radiolabeling, metabolic stability, and radiation dose estimates, *Bioconjug. Chem.*, 15, 61, 2004.

123. Haubner, R. et al., [F-18]-RGD-peptides conjugated with hydrophilic tetrapeptides for the non-invasive determination of the avb3 integrin, *J. Nucl. Med.*, 43 (Suppl), 89P, 2002.

124. Haubner, R. et al., Noninvasive visualization of the activated alphavbeta3 integrin in cancer patients by positron emission tomography and [(18)F]Galacto-RGD, *PLoS Med.*, 2, e70, 2005.

125. Beer, A. J. et al., Biodistribution and pharmacokinetics of the alphavbeta3-selective tracer ^{18}F-galacto-RGD in cancer patients, *J. Nucl. Med.*, 46, 1333, 2005.

126. Sipkins, D. A. et al., Detection of tumor angiogenesis in vivo by alphaVbeta3-targeted magnetic resonance imaging, *Nat. Med.*, 4, 623, 1998.

127. Anderson, S. A. et al., Magnetic resonance contrast enhancement of neovasculature with alpha(v)-beta(3)-targeted nanoparticles, *Magn. Reson. Med.*, 44, 433, 2000.

128. Winter, P. M. et al., Molecular imaging of angiogenesis in early-stage atherosclerosis with alpha(v)-beta3-integrin-targeted nanoparticles, *Circulation*, 108, 2270, 2003.

129. Cheng, Z. et al., Near-infrared fluorescent RGD peptides for optical imaging of integrin alphavbeta3 expression in living mice, *Bioconjug. Chem.*, 16, 1433, 2005.

130. de Groot, F. M. et al., Design, synthesis, and biological evaluation of a dual tumor-specific motive containing integrin-targeted plasmin-cleavable doxorubicin prodrug, *Mol. Cancer Ther.*, 1, 901, 2002.

131. Devy, L. et al., Plasmin-activated doxorubicin prodrugs containing a spacer reduce tumor growth and angiogenesis without systemic toxicity, *FASEB J.*, 18, 565, 2004.

132. Kim, J. W. and Lee, H. S., Tumor targeting by doxorubicin-RGD-4C peptide conjugate in an orthotopic mouse hepatoma model, *Int. J. Mol. Med.*, 14, 529, 2004.

133. Janssen, M. et al., Comparison of a monomeric and dimeric radiolabeled RGD-peptide for tumor targeting, *Cancer Biother. Radiopharm.*, 17, 641, 2002.

134. Liu, S. et al., ^{99m}Tc-labeling of a hydrazinonicotinamide-conjugated vitronectin receptor antagonist useful for imaging tumors, *Bioconjug. Chem.*, 12, 624, 2001.

135. Chen, X. et al., MicroPET imaging of breast cancer alphav-integrin expression with ^{64}Cu-labeled dimeric RGD peptides, *Mol. Imaging Biol.*, 6, 350, 2004.

136. Chen, X. et al., Micro-PET imaging of alphavbeta3-integrin expression with ^{18}F-labeled dimeric RGD peptide, *Mol. Imaging*, 3, 96, 2004.

137. Wu, Y. et al., MicroPET imaging of glioma integrin {alpha}v{beta}3 expression using (64)Cu-labeled tetrameric RGD peptide, *J. Nucl. Med.*, 46, 1707, 2005.

138. Poethko, T. et al., Two-step methodology for high-yield routine radiohalogenation of peptides: (18)F-labeled RGD and octreotide analogs, *J. Nucl. Med.*, 45, 892, 2004.

139. Chen, X. et al., Pharmacokinetics and tumor retention of ^{125}I-labeled RGD peptide are improved by PEGylation, *Nucl. Med. Biol.*, 31, 11, 2004.

140. Chen, X. et al., MicroPET imaging of brain tumor angiogenesis with ^{18}F-labeled PEGylated RGD peptide, *Eur. J. Nucl. Med. Mol. Imaging*, 31, 1081, 2004.

141. Chen, X. et al., Pegylated Arg-Gly-Asp peptide: [64]Cu labeling and PET imaging of brain tumor alphavbeta3-integrin expression, *J. Nucl. Med.*, 45, 1776, 2004.

142. Poethko, T. et al., Improved tumor uptake, tumor retention and tumor/background ratios of pegylated RGD-multimers, *J. Nucl. Med.*, 44, 46P, 2003.

143. Putnam, D. and Kopecek, J., Polymer conjugates with anticancer activity, *Adv. Polym. Sci.*, 122, 56, 1995.

144. Kopecek, J. et al., HPMA copolymer-anticancer drug conjugates: Design, activity, and mechanism of action, *Eur. J. Pharm. Biopharm.*, 50, 61, 2000.

145. Duncan, R., The dawning era of polymer therapeutics, *Nat. Rev. Drug Discov.*, 2, 347, 2003.

146. Greenwald, R. B. et al., Effective drug delivery by PEGylated drug conjugates, *Adv. Drug Deliv. Rev.*, 55, 217, 2003.

147. Kopecek, J. et al., Water soluble polymers in tumor targeted delivery, *J. Control. Release*, 74, 147, 2001.

148. Lu, Z. R. et al., Design of novel bioconjugates for targeted drug delivery, *J. Control. Release*, 78, 165, 2002.

149. Duncan, R. et al., Fate of N-(2-hydroxypropyl)methacrylamide copolymers with pendent galacto-samine residues after intravenous administration to rats, *Biochim. Biophys. Acta*, 880, 62, 1986.

150. Kopecek, J. and Duncan, R., Poly[N-(2-Hydroxypropyl)-methacrylamide] macromolecules as drug carrier systems, In *Polymers in Controlled Drug Delivery*, Illum, L. and Davis, S. S. Eds., Butter-worth-Heinemann, Oxford, UK, 1987.

151. Lu, Z. R. et al., Functionalized semitelechelic poly[N-(2-hydroxypropyl)methacrylamide] for protein modification, *Bioconjug. Chem.*, 9, 793, 1998.

152. Mitra, A., et al., Nanocarriers for nuclear imaging and radiotherapy of cancer, *Curr. Pharm. Des.*, in press, 2006.

153. Matsumura, Y. and Maeda, H., A new concept for macromolecular therapeutics in cancer chemotherapy: Mechanism of tumoritropic accumulation of proteins and the antitumor agent smancs, *Cancer Res.*, 46, 6387, 1986.

154. Satchi-Fainaro, R. et al., Targeting angiogenesis with a conjugate of HPMA copolymer and TNP-470, *Nat. Med.*, 10, 255, 2004.

155. Satchi-Fainaro, R. et al., Inhibition of vessel permeability by TNP-470 and its polymer conjugate, caplostatin, *Cancer Cell*, 7, 251, 2005.

156. Shiah, J. J. et al., Biodistribution of free and N-(2-hydroxypropyl)methacrylamide copolymer-bound mesochlorin e(6) and adriamycin in nude mice bearing human ovarian carcinoma OVCAR-3 xeno-grafts, *J. Control. Release*, 61, 145, 1999.

157. Mammen, M. et al., Polyvalent interactions in biological systems: Implications for design and use of multivalent ligands and inhibitors, *Angew. Chem. Int. Ed.*, 37, 2754, 1998.

158. Wester, H. J. and Kessler, H., Molecular targeting with peptides or Peptide-polymer conjugates: Just a question of size?, *J. Nucl. Med.*, 46, 1940, 2005.

159. Bhargava, P. et al., A Phase I and pharmacokinetic study of TNP-470 administered weekly to patients with advanced cancer, *Clin. Cancer Res.*, 5, 1989, 1999.

160. Dickson, P. V., Nathwani, A. C., and Davidoff, A. M., Delivery of antiangiogenic agents for cancer gene therapy, *Technol. Cancer Res. Treat.*, 4, 331, 2005.

161. Tandle, A., Blazer, D. G. III, and Libutti, S. K., Antiangiogenic gene therapy of cancer: Recent developments, *J. Transl. Med.*, 2, 22, 2004.

162. Witlox, A. M. et al., Conditionally replicative adenovirus with tropism expanded towards integrins inhibits osteosarcoma tumor growth in vitro and in vivo, *Clin. Cancer Res.*, 10, 61, 2004.

163. Okada, Y. et al., Optimization of antitumor efficacy and safety of in vivo cytokine gene therapy using RGD fiber-mutant adenovirus vector for preexisting murine melanoma, *Biochim. Biophys. Acta*, 1670, 172, 2004.

164. Kim, W. J. et al., Soluble Flt-1 gene delivery using PEI-g-PEG-RGD conjugate for anti-angiogenesis, *J. Control. Release*, 106, 224, 2005.

165. Hood, J. D. et al., Tumor regression by targeted gene delivery to the neovasculature, *Science*, 296, 2404, 2002.

166. Leng, Q. et al., Highly branched HK peptides are effective carriers of siRNA, *J. Gene Med.*, 7, 977, 2005.

167. Pack, D. W. et al., Design and development of polymers for gene delivery, *Nat. Rev. Drug Discov.*, 4, 581, 2005.

168. Kunath, K. et al., Integrin targeting using RGD-PEI conjugates for in vitro gene transfer, *J. Gene Med.*, 5, 588, 2003.

169. Erbacher, P., Remy, J. S., and Behr, J. P., Gene transfer with synthetic virus-like particles via the integrin-mediated endocytosis pathway, *Gene Ther.*, 6, 138, 1999.

170. Schiffelers, R. M. et al., Cancer siRNA therapy by tumor selective delivery with ligand-targeted sterically stabilized nanoparticles, *Nucl. Acids Res.*, 32, e149, 2004.

171. Suh, W. et al., An angiogenic, endothelial-cell-targeted polymeric gene carrier, *Mol. Ther.*, 6, 664, 2002.

172. Woodle, M. C. et al., Sterically stabilized polyplex: Ligand-mediated activity, *J. Control. Release*, 74, 309, 2001.

173. Hosseinkhani, H. and Tabata, Y., PEGylation enhances tumor targeting of plasmid DNA by an artificial cationized protein with repeated RGD sequences, Pronectin, *J. Control. Release*, 97, 157, 2004.

174. Flynn, A. A. et al., A model-based approach for the optimization of radioimmunotherapy through antibody design and radionuclide selection, *Cancer*, 94, 1249, 2002.

175. Liu, S. et al., (90)Y and (177)Lu labeling of a DOTA-conjugated vitronectin receptor antagonist useful for tumor therapy, *Bioconjug. Chem.*, 12, 559, 2001.

176. Garrison, W. M., Reaction mechanisms in radiolysis of peptides, polypeptides and proteins, *Chem. Rev.*, 87, 381, 1987.

177. Liu, S. and Edwards, D. S., Stabilization of (90)y-labeled DOTA-biomolecule conjugates using gentisic acid and ascorbic acid, *Bioconjug. Chem.*, 12, 554, 2001.

178. Mitra, A. et al., Targeting tumor angiogenic vasculature using polymer-RGD conjugates, *J. Control. Release*, 102, 191, 2005.

179. Koivunen, E., Wang, B., and Ruoslahti, E., Phage libraries displaying cyclic peptides with different ring sizes: Ligand specificities of the RGD-directed integrins, *Biotechnology (NY)*, 13, 265, 1995.

180. Assa-Munt, N. et al., Solution structures and integrin binding activities of an RGD peptide with two isomers, *Biochemistry*, 40, 2373, 2001.

181. Mitra, A., et al., Polymeric conjugates of mono- and bi-cyclic alphaVbeta3 binding peptides for tumor targeting, *J. Control. Release*, 114, 175, 2006.

182. Kok, R. J. et al., Preparation and functional evaluation of RGD-modified proteins as alpha(v)beta(3) integrin directed therapeutics, *Bioconjug. Chem.*, 13, 128, 2002.

183. Schraa, A. J. et al., Targeting of RGD-modified proteins to tumor vasculature: A pharmacokinetic and cellular distribution study, *Int. J. Cancer*, 102, 469, 2002.

184. Xiong, X. B. et al., Intracellular delivery of doxorubicin with RGD-modified sterically stabilized liposomes for an improved antitumor efficacy: In vitro and in vivo, *J. Pharm. Sci.*, 94, 1782, 2005.

185. Dubey, P. K. et al., Liposomes modified with cyclic RGD peptide for tumor targeting, *J. Drug Target*, 12, 257, 2004.

186. Bibby, D. C. et al., Pharmacokinetics and biodistribution of RGD-targeted doxorubicin-loaded nanoparticles in tumor-bearing mice, *Int. J. Pharm.*, 293, 281, 2005.

187. Seymour, L. W. et al., Effect of molecular weight (Mw) of N-(2-hydroxypropyl)methacrylamide copolymers on body distribution and rate of excretion after subcutaneous, intraperitoneal, and intravenous administration to rats, *J. Biomed. Mater. Res.*, 21, 1341, 1987.

188. Seymour, L. W. et al., Influence of molecular weight on passive tumour accumulation of a soluble macromolecular drug carrier, *Eur. J. Cancer*, 31A, 766, 1995.

189. Kissel, M. et al., Synthetic macromolecular drug carriers: Biodistribution of poly[(*N*-2-hydroxypropyl)methacrylamide] copolymers and their accumulation in solid rat tumors, *PDA J. Pharm. Sci. Technol.*, 55, 191, 2001.

190. Lammers, T. et al., Effect of physicochemical modification on the biodistribution and tumor accumulation of HPMA copolymers, *J. Control. Release*, 110, 103, 2005.

191. Mitra, A. et al., Technetium-99m-Labeled *N*-(2-hydroxypropyl) methacrylamide copolymers: Synthesis, characterization, and in vivo biodistribution, *Pharm. Res.*, 21, 1153, 2004.

192. Ulbrich, K. et al., HPMA copolymers with pH-controlled release of doxorubicin: In vitro cytotoxicity and in vivo antitumor activity, *J. Control. Release*, 87, 33, 2003.

193. Tucker, G. C., Alpha v integrin inhibitors and cancer therapy, *Curr. Opin. Investig. Drugs*, 4, 722, 2003.

194. Porkka, K. et al., A fragment of the HMGN2 protein homes to the nuclei of tumor cells and tumor endothelial cells in vivo, *Proc. Natl. Acad. Sci. U.S.A.*, 99, 7444, 2002.

195. Christian, S. et al., Nucleolin expressed at the cell surface is a marker of endothelial cells in angiogenic blood vessels, *J. Cell Biol.*, 163, 871, 2003.

196. Hoffman, J. A. et al., Progressive vascular changes in a transgenic mouse model of squamous cell carcinoma, *Cancer Cell*, 4, 383, 2003.

197. Arap, W. et al., Targeting the prostate for destruction through a vascular address, *Proc. Natl. Acad. Sci. U.S.A.*, 99, 1527, 2002.

198. Joyce, J. A. et al., Stage-specific vascular markers revealed by phage display in a mouse model of pancreatic islet tumorigenesis, *Cancer Cell*, 4, 393, 2003.

199. Ogawa, M. et al., Direct electrophilic radiofluorination of a cyclic RGD peptide for in vivo alpha(v)-beta3 integrin related tumor imaging, *Nucl. Med. Biol.*, 30, 1, 2003.

200. Su, Z. F. et al., In vitro and in vivo evaluation of a Technetium-99m-labeled cyclic RGD peptide as a specific marker of alpha(V)beta(3) integrin for tumor imaging, *Bioconjug. Chem.*, 13, 561, 2002.

201. Chen, X. et al., Integrin alpha v beta 3-targeted imaging of lung cancer, *Neoplasia*, 7, 271, 2005.

202. Chen, X. et al., Synthesis and biological evaluation of dimeric RGD peptide–paclitaxel conjugate as a model for integrin-targeted drug delivery, *J. Med. Chem.*, 48, 1098, 2005.

203. Xiong, X. B. et al., Enhanced intracellular uptake of sterically stabilized liposomal Doxorubicin in vitro resulting in improved antitumor activity in vivo, *Pharm. Res.*, 22, 933, 2005.

204. Thompson, B. et al., Neutral postgrafted colloidal particles for gene delivery, *Bioconjug. Chem.*, 16, 608, 2005.

205. Janssen, M. et al., Improved tumor targeting of radiolabeled RGD peptides using rapid dose fractionation, *Cancer Biother. Radiopharm.*, 19, 399, 2004.

206. Bernard, B. et al., Radiolabeled RGD-DTPA-Tyr3-octreotate for receptor-targeted radionuclide therapy, *Cancer Biother. Radiopharm.*, 19, 173, 2004.

207. Capello, A. et al., Increased cell death after therapy with an Arg-Gly-Asp-linked somatostatin analog, *J. Nucl. Med.*, 45, 1716, 2004.

10 Poly(L-Glutamic Acid): Efficient Carrier of Cancer Therapeutics and Diagnostics

Guodong Zhang, Edward F. Jackson, Sidney Wallace, and Chun Li

CONTENTS

10.1 Introduction 185
10.2 Synthesis and Properties of PG 186
    10.2.1 Chemistry 186
    10.2.2 Biodegradation and Biodistribution 187
10.3 PG-Anti-Cancer Drug Conjugates 187
    10.3.1 Anthracyclines 187
    10.3.2 Antimetabolites 188
    10.3.3 DNA-Binding Drugs 188
10.4 PG–TXL and PG–CPT: From the Laboratory to the Clinic 189
    10.4.1 PG–TXL 189
    10.4.2 PG–CPT 192
10.5 Combination of PG–TXL with Radiotherapy 193
10.6 PG as a Carrier of Diagnostic Agents 193
    10.6.1 Magnetic Resonance Imaging 193
    10.6.2 Near-Infrared Fluorescence Optical Imaging 195
10.7 Conclusions 195
Acknowledgments 196
References 196

10.1 INTRODUCTION

Most anti-cancer chemotherapeutic drugs used clinically are limited by a relatively low therapeutic index, owing to toxic side effects.[1-3] Over the past several decades, two strategies of improving the therapeutic efficacy of anti-cancer agents have emerged. The first approach is the design and development of agents that can modulate the molecular processes and pathways specifically associated with tumor progression. The success of this approach is shown by the successful introduction of a new breed of molecularly targeted anti-cancer agents such as imatinib mesylate (Gleevec), gefitinib (Iressa), trastuzumab (Herceptin), and cetuximab (C225, Erbitux). Alternatively, existing anti-cancer agents can be made more effective by using delivery systems that bring more drug molecules to the tumor site when compared with conventional formulation while reducing exposure of normal tissues to the drug. In this context, there have been important milestones with the use of polymer–drug conjugates in their own right. In particular, poly(L-glutamic acid) (PG)–paclitaxel (TXL) (CT2103, XYOTAX) has advanced to phase III clinical trials, and PG–camptothecin (CPT) (CT2106) has been tested in phase II clinical trials. The chemistry and applications of PG and its conjugates with various chemotherapeutic agents were previously reviewed.[1] In this chapter, the applications of PG in the delivery of both chemotherapeutic and diagnostic agents will be updated. The results of clinical studies of XYOTAX and CT2106 will also be summarized. Previous chapters in this book discuss, in depth, the rationale for the use of polymeric conjugates as cancer therapeutics.

10.2 SYNTHESIS AND PROPERTIES OF PG

PG anti-cancer drug conjugates consist of three parts: the PG backbone, the anti-cancer drug, and spacers that link the drug molecules with the PG backbone. Both the properties of PG (i.e., the molecular weight and its distribution) and the selection of spacers can directly affect the pharmacokinetics at both the whole-body and cellular level. For this reason, the synthetic chemistry of the preparation of PG will first be discussed. A summary of the physicochemical properties of PG conjugates will follow.

10.2.1 CHEMISTRY

PG is usually prepared by removing the benzyl protecting group of poly(γ-benzyl-L-glutamate) that is attained by polymerization of γ-benzyl-L-glutamate N-carboxyanhydride (NCA) monomer.[4]

Direct phosgenation of γ-benzyl-L-glutamate produces the corresponding NCA monomer.[5,6] In this reaction, γ-benzyl-L-glutamate is suspended in a dry inert solvent such as ethyl acetate, dioxane, or tetrahydrofuran, and it is allowed to heterogeneously react with the cyclizing reagent that is normally triphosgene. Researchers have used both the protic and aprotic initiators in the polymerization of γ-benzyl-L-glutamate NCA monomer.[7] Protic initiators such as primary amines are acylated by attack on the 5-position of the NCA, whereas aprotic initiators such as tertiary amines and alkoxides act as general bases. The ring-opening polymerization of NCA initiated by amines usually produces polymers with relatively broad molecular weight distributions. To circumvent this problem, Deming[8] developed the zero-valence nickel catalyst bipyNi(COD) (bipy: 2,2′-bipyridyl; COD: 1,5-cyclooctadiene) to initiate polymerization of NCAs. The active sites derived with this initiator are less accessible to side reactions when compared with that derived with amine initiators because of steric and electronic effects, resulting in polymers with a narrow molecular weight distribution (Mw/Mn, 1.05–1.15).

Impurities in NCAs can have a detrimental effect on the reproducibility of polyamino acid synthesis. NCAs are usually subjected to repeat recrystallization to eliminate trace amounts of impurities. A recent report described an efficient method of removal of hydrogen chloride and the hydrochloride salt of unreacted starting amino acids.[9] This method consists of extraction of NCA solution in ethyl acetate with water and an aqueous alkali solution at 0°C. Using highly purified monomer NCAs, Aliferis et al.[10] were able to achieve living polymerization initiated with primary amine initiators. This technique produced poly(γ-benzyl-L-glutamate) with an average molecular weight as high as 1.66×10^5 Da and with a relatively narrow molecular weight distribution (Mw/Mn, 1.40).

The benzyl protecting group in poly(γ-benzyl-L-glutamate) is removed by treatment with hydrogen bromide. Researchers have also used alkaline hydrolysis of poly(L-methyl glutamate) to prepare PG; however, racemization occurred during this process.[11] To avoid using harsh conditions for the removal of protecting groups, investigators have prepared glutamic acid NCAs having ester protecting groups that are labile under mild acidic conditions. For example, PG can be readily obtained from poly(γ-piperonyl-L-glutamate) by treating the polymer with trifluoroacetic acid.[12]

PG-based block copolymers can be prepared using macroinitiators or polymer coupling reactions. The block copolymer PG–vinylsulfone–polyethylene glycol (PEG) was prepared by directly reacting a heterofunctional PEG, vinylsulfone–PEG–N-hydroxysuccinimide (NHS), with an amino group at one end of the PG chain.[13] Also, Nishiyama et al.[14] used the macroinitiator PEG–NH$_2$ to initiate polymerization of γ-benzyl-L-glutamate NCA to produce a PEG–PG block copolymer.

Another polymer derived from poly(γ-benzyl-L-glutamate) is poly(hydroxyethylglutamate) (PHEG) that is also a suitable candidate drug carrier.[15] PHEG is a water-soluble, neutral polymer that can be readily prepared from poly(γ-benzyl glutamate) via aminolysis with 2-aminoethanol using 2-hydroxypyridine as a catalyst.[16]

The linkage of polymer–drug conjugates is important to the in vivo release of a drug that should be stable during circulation but should also allow drug release at an appropriate rate at the tumor tissue.[17] Generally, two kinds of linkage are involved in anti-cancer drug delivery by using polymer–drug conjugates: enzymatic hydrolysis and nonenzymatic hydrolyzable linkage. Many proteolytic enzymes such as cysteine proteinase cathepsins are expressed on the surface of metastatic tumor cells.[18,19] The linker chemistry can be designed in such a way that the active agents would be released from the polymer–drug conjugates only upon exposure to the proteinases in the tumors. Alternatively, hydrolytically labile, pH-sensitive linkages (i.e., hydrozone, ester, acetals, etc.) have exhibited their utility for tumorotropic or lysosomotropic delivery because the microenvironments of solid tumors are known to be acidic.[20,21] PG–TXL conjugate releases TXL through both mechanisms: backbone degradation mediated through proteolytic enzymes and side chain hydrolysis of the ester bond formed between glutamic acid and TXL.

10.2.2 BIODEGRADATION AND BIODISTRIBUTION

The enzymatic degradability of PG–drug conjugates is influenced by their structure, composition, and charge as well as the physicochemical properties of the drug attached to PG.[22–25] PG is more susceptible to lysosomal degradation than are poly(aspartic acid) and poly(D-glutamic acid).[26] Cysteine proteases, particularly cathepsin B, play key roles in the lysosomal degradation of PG.[22,27] Although researchers have identified oligomeric glutamic acids as the primary degradation products of PG,[28] more recent results demonstrated that monomeric L-glutamic acid is produced in the lysosomal degradation of PG.[29] Degradation of the PG backbone may not necessarily lead to the release of free drug in every case.

The biodistribution of PG and its drug conjugates depends on the molecular weight of PG. Polymers with a molecular weight lower than the renal clearance threshold are rapidly removed from the blood circulation by glomerular filtration. McCormick-Thomson and Duncan[27] found that the biodistribution of copolymers of glutamic acid and other hydrophobic amino acids was markedly influenced by the composition of the copolymers. These results suggest that the biodistribution of PG–drug conjugates is a function of the drugs used, the degree of modification, and the molecular weight of PG.

10.3 PG-ANTI-CANCER DRUG CONJUGATES

Since doxorubicin (Adriamycin [ADR]) was conjugated with PG in the 1980s,[30] investigators have studied various anti-cancer drugs such as anthracyclines, antimetabolites, DNA-binding drugs, TXL, and CPT. In particular, researchers have evaluated PG–TXL in phase III clinical trials and PG–CPT in phase II clinical trials.

10.3.1 ANTHRACYCLINES

The anthracycline group has been one of the most extensively studied groups of drugs used in polymer–drug conjugate delivery systems, especially in the 1980s. Based on the assumption that greater selectivity toward tumors over normal tissues can be achieved if PG–ADR conjugates only degrade and release ADR after being endocytosed by tumor cells, Van Heeswijk and colleagues[30] investigated three different PG conjugates containing ADR where the amine group in ADR was bound to the side chain carboxyl groups of high-molecular-weight PG either directly (with an amide bond) or indirectly through GlyGly and GlyGlyGlyLeu spacers, respectively. They investigated the degradability of the conjugates mediated by lysosomal enzymes and the subsequent release of ADR or ADR-peptide products with low molecular weights using reverse-phase high-performance liquid chromatography. The total amount of ADR released after 77 h of incubation was 3.6% for PG-Gly-GlyGlyLeu-ADR; 1.0% for PG-GlyGly-ADR; and 0.5% for PG–ADR, suggesting the importance of introducing an enzymatically degradable spacer between PG and the drug. In vitro, these conjugates exhibited reduced cytotoxicity against L1210 leukemia cells when compared with free ADR. In vivo, animals treated with the polymer conjugate containing the tetrapeptide spacer showed a similar mean survival duration as compared to those treated with free ADR, whereas the conjugate without a peptide spacer was completely inactive.[28,31]

Some have used hydrolytically labile ester bonds[32] and hydrozone bonds[33] to couple doxorubicin or daunorubicin (Dau) with PG. In these studies, the ester bond was formed by a reaction of 14-bromo-daunorubicin with the carboxylic group of PG via a nucleophilic substitution reaction in an alkaline aqueous medium. With intravenous administration of these conjugates into mice bearing MS-2 sarcoma or Gross' leukemia, these investigators found that the drug conjugates' potency and efficacy were correlated with the molecular weight of the carrier. For example, anti-tumor activity improved when the molecular weight increased from 14,000 to 60,000 Da at an equivalent doxorubicin dose of 30 mg/kg. These results can be attributed to the increased circulation times of conjugates with high molecular weights. Condensing the methylketone in Dau with hydrazide-derived PG yielded PG-hydrazone–Dau conjugates. The acid-sensitive conjugates were less cytotoxic than free Dau was in vitro against mouse lymphoma cells. However, these conjugates showed significant anti-tumor activity when intravenously injected into mice bearing intraperitoneally inoculated Yac lymphoma.[33]

10.3.2 ANTIMETABOLITES

To enhance the efficacy of 1-β-D-arabinofuranosylcytosine (ara-C) in simple dosage schedules, researchers synthesized two ara-C conjugates with PG via an amide bond: one where ara-C is directly coupled with the N-4 of ara-C to the carboxyl groups of PG and one where ara-C is linked with PG via the aminoalkylphosphoryl side chain introduced at the C-5$'$ of ara-C.[34] Both conjugates exhibited markedly decreased cytotoxicity in L1210 murine leukemia cells when compared with free ara-C. However, the anti-tumor activity of both conjugates was greater than or equal to that of free ara-C in mice bearing L1210 tumors after a single intraperitoneal injection of the conjugate. The authors suggested that both slow cleavage of free ara-C from the conjugates and protection of ara-C from deactivation by cytidine deaminase contributed to the enhanced anti-tumor activity of PG–ara-C conjugates in vivo.

10.3.3 DNA-BINDING DRUGS

Investigators have conjugated several DNA-binding drugs, including cyclophosphamide,[35] L-phenylalanine mustard (melphalan),[36] mitomycin C (MMC),[37–39] and cis-dichlorodiammineplatinum (II) (cisplatin [CDDP])[40,41] with PG. Moromoto et al.[36] prepared PG–melphalan by coupling the amine group of melphalan with the side-chain carboxyl group of PG in the presence of a water-soluble carbodiimide.[36] They investigated the anti-tumor activity of this conjugate in rats bearing sarcoma induced by subcutaneous inoculation of sarcoma cells. The authors used PG-^{3}H-phenyl-alanine as a model compound to demonstrate that free drug could be released from the conjugate and that the conjugate had a tendency to be absorbed through lymphatic routes when compared with free ^{3}H-phenylalanine after subcutaneous administration.[36] Roos et al.[37] conjugated MMC with PG by using the aziridine amine of MMC.[37,38] They found that the release rate, in vitro cytotoxicity, and in vivo anti-tumor activity of PG–MMC conjugates were influenced by the extent of MMC substitution. Furthermore, Seymour et al.[42] conjugated MMC with PHEG using an oligopeptide spacer and investigated the effect of this oligopeptide structure on the rate of drug release to optimize the MMC–oligopeptide–PHEG conjugates. The rate of MMC release from the conjugates was affected by the structure of the oligopeptide spacer; spacers bearing a terminal Gly had the fastest release of MMC when compared with other peptide spacers. Another study showed a correlation between the in vitro cytotoxicity of PG–MMC conjugates against B16F10 melanoma and C26 colorectal carcinoma cells and their hydrolytic stability.[38] Significant in vivo activity of the conjugates in mice bearing P338 leukemia or C26 colorectal carcinoma seems to result from both the stability of the conjugates in the blood and the rapid drug release, owing to combined chemical and enzymatic hydrolysis.

The clinical use of CDDP is limited by significantly toxic side effects such as acute nephrotoxicity and chronic neurotoxicity and a low therapeutic index. To reduce its toxicity and to maintain prolonged drug activity, investigators complexed CDDP with PG.[40,41] They found that PG–CDDP complex (60 mol CDDP/mol PG; molecular weight, 40,000 Da) was more thermodynamically stable and has reduced systemic toxicity when compared with CDDP. PG–CDDP was effective in suppressing the growth of OVCAR-3 human ovarian carcinoma in athymic mice and showed a broader therapeutic dose range (80% survival at 3–12 mg/kg) when compared with the narrow, inconsistent effective dose range of free CDDP (80% survival at 1.0–2.5 mg/kg), suggesting that the therapeutic index of CDDP can be improved by complexing it with PG.

10.4 PG–TXL AND PG–CPT: FROM THE LABORATORY TO THE CLINIC

10.4.1 PG–TXL

Taxanes are a class of the widely used and clinically active cytotoxic agents that exert their action by promoting tubulin polymerization and microtubule assembly.[43] Because of its poor aqueous solubility, use of TXL requires Cremophor and ethanol mixture as a vehicle, and it must be infused over 3–24 h. The conjugate PG–TXL where TXL is attached to PG via ester bonds has demonstrated significantly enhanced anti-tumor efficacy and improved safety when compared with TXL in preclinical studies.[44]

The release of TXL from PG–TXL and the metabolism of PG–TXL in both in vitro and in vivo models have been investigated.[45] These studies showed that when PG–TXL was incubated in buffered saline or plasma (mouse or human) for 24 h at 37°C, less than 14% of the bound TXL was released, suggesting that PG–TXL is relatively resistant to plasma esterase.

Tumor uptake of TXL and PG–TXL was compared using [^{3}H]TXL and PG–[^{3}H]TXL.[46] When free [^{3}H]TXL was incubated with MDA-MB453 cells, [^{3}H]TXL was taken up rapidly by the cells and was followed by a rapid drug efflux process. In contrast, when the tumor cells were incubated with PG–[^{3}H]TXL, a slower uptake and more persistent retention of radioactivity was observed. These data suggest that a fraction of PG–[^{3}H]TXL was taken up by tumor cells, possibly through pinocytosis, and that [^{3}H]TXL was more readily pumped out of the cells than were the more hydrophilic PG–[^{3}H]TXL and its degradation products. A recent report confirmed monoglutamyl-2′-TXL and diglutamyl-2′-TXL as the major intracellular metabolites of PG–TXL.[47] Hydrolysis of these metabolites led to release of free TXL. Specific enzyme inhibitors such as CA-074 methyl ester, a cell-permeable irreversible inhibitor of cathepsin B, and EST, a cell-permeable irreversible inhibitor of cysteine protease, decreased the formation of monoglutamate TXL and free TXL in a tumor cell line that had been incubated with PG–TXL. All of these findings are consistent with the results of proteolysis of the PG backbone through the action of cellular dipeptidases. Another experiment showed that the metabolism of PG–TXL in non tumor-bearing cathepsin B homozygous knockout mice is reduced, but not eliminated; this seems to confirm that in addition to cathepsin B, other cysteine proteases on the surface of tumor cells play important roles in the release of TXL from PG–TXL.[47]

PG–TXL's anti-tumor activity was first assessed in a variety of syngeneic and xenogeneic tumor models. For example, the maximum tolerated dose of PG–TXL after a single intravenous injection in rats and mice was 60 and 160 mg/kg, respectively.[44] In comparison, the maximum tolerated dose of TXL in rats and mice was 20 and 60 mg/kg, respectively. Therefore, PG–TXL's use represented a twofold and threefold improvement in toxicity in rats and mice, respectively. To determine if PG–TXL has a broad spectrum of anti-tumor activity, its therapeutic activity was evaluated against four syngeneic murine tumor cell lines (MCa-4 breast carcinoma, MCa-35 breast carcinoma, HCa-1 hepatocarcinoma, and FSa-II sarcoma) intramuscularly inoculated into C3Hf/Kam mice.[48] The anti-tumor and antimetastatic activities of PG–TXL were investigated using intraperitoneal injection of the SKOV3ip1 human ovarian tumor cell line in nude mice[49] and human MDA-MB-435-Lung2 breast tumors grown in the mammary fat pads of nude mice.[48] Treatment with PG–TXL exhibited significantly better anti-tumor activity than did treatment with TXL alone in all of the tumor models.

The pharmacokinetic profile and tissue distribution of PG–TXL were examined to verify the enhanced permeability and retention effect of macromolecules.[45] It was found that PG–[^{3}H]TXL has a much longer half-life in plasma (317 min) than [^{3}H]TXL does (29 min). Consequently, the area under the tissue-concentration-time curve (AUC) in tumors was five times greater when mice were injected with PG–TXL than when they were injected with TXL. Therefore, enhanced tumor uptake and sustained release of TXL from PG–TXL in tumor tissue seem to be among one of the major factors contributing to PG–TXL's markedly improved anti-tumor activity in vivo.

In the XYOTAX formulation used in clinical studies, the median molecular weight of PG–TXL is 48,000 Da, and the content of TXL is about 37% by weight, equivalent to approximately one TXL molecule for every 11 glutamic acid units in each PG polymer chain.[47] This formulation eliminated the use of Cremophor and alcohol, and it allows infusion of TXL over 30 min. In initial clinical trials in the United Kingdom, the investigators gave PG–TXL (CT-2103) to cancer patients in a 30-min infusion every three weeks at doses ranging from 30 to 720 mg/m^2.[50] They detected CT-2103 in all the patients' plasma and observed a long plasma half-life of up to 185 h. Importantly, the peak plasma concentration of released free TXL was less than 0.1 μM 24 h after administration of CT-2103 at doses up to 480 mg/m^2 (176 mg/m^2 TXL equivalent). As shown in Figure 10.1, the plasma concentrations of PG–TXL biphasically declined. The distribution phase was prolonged, and the apparent monoexponential terminal phase associated with drug elimination appeared approximately 48 h after administration of PG–TXL. The plasma concentration of PG–TXL in the terminal phase declined slowly, and the drug could be detected in plasma three weeks after administration at a dose of 200 mg/m^2. In comparison, the plasma concentration of free TXL paralleled with the PG–TXL concentration. Importantly, this study found that the AUC of free TXL was about 1–2% of the AUC of PG–TXL that supported the in vivo stability of PG–TXL in the plasma and the slow, prolonged release of the active moiety.[51] The steady-state volume of distribution ranged from 1.3 to 5.9 l/m^2 and was low that suggested that the distribution of PG–TXL was restricted mainly to the plasma and other extracellular body fluids.[47]

More than 400 patients have received PG–TXL in phase I and II trials.[52,53] Drug-related events that were reported in 10–20% of the patients included thrombocytopenia, diarrhea, leukopenia, myalgia, arthralgia, and anemia. Compared with conventional TXL-based treatment, PG–TXL showed three safety-related advantages. First, alopecia was rare, and complete hair loss was not observed. Second, nausea and vomiting were uncommon. Third, hypersensitivity reactions were rarely observed, and those that did occur were usually mild to moderate; therefore, routinely used prophylactic premedications were not required. The incidence of significant hypersensitivity reactions was less than 1% with no grade 4 hypersensitivity reactions.

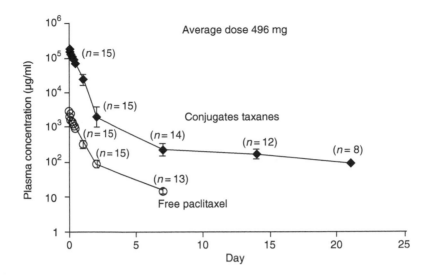

FIGURE 10.1 Plasma pharmacokinetics of PG–TXL and free TXL in cancer patients. The data are from four phase I dose-escalation studies using 1-, 2-, and 3-week schedules. In all of the studies, PG–TXL was administrated in a short intravenous infusion. (From Singer, J. W., Shaffer, S., and Baker, B., *Anti-cancer Drugs*, 16, 243–254, 2005. With permission.)

Phase II trials CT-2103, including the multicenter open-label studies (CTI-1071 and CTI-1069), have been completed.[47,53] The former study aimed to determine the rate of response and time to disease progression in a heterogeneous population of patients with advanced epithelial ovarian cancer who received CT-2103.[47,53] Ninety-nine patients registered in this trial received PG–TXL at a dose of 175 mg/m^2 every three weeks. The toxic effects were mild in this heavily pretreated population and consisted of the following: grade 3 neuropathy ($n = 15$), grade 3 neutropenia ($n = 10$), and grade 4 neutropenia ($n = 4$). Of 18 patients who received one or two prior chemotherapy regimens and had CDDP-sensitive disease, five (28%) had a response, and six (33%) had stable disease (SD). The cancer was difficult to treat in 21 patients with platinum-resistant disease who underwent pretreatment, yet the investigators observed responses in two patients (10%) and SD in four patients (19%). In CTI-1069,[47] the researchers aimed to evaluate the efficacy and tolerability of PG–TXL in patients with nonsmall cell lung cancer who were 70 years of age or older and had an Eastern Cooperative Oncology Group (ECOG) performance status of 2 (PS2). Thirty patients were registered in this trial, and they received PG–TXL at a dose of 175 mg/m^2 ($n = 28$) or 235 mg/m^2 ($n = 2$) every three weeks. Two patients had a partial response, whereas 16 patients had SD lasting at least 10 weeks. Among the 28 patients who received PG–TXL at 175 mg/m^2, the median survival duration was 8.1 months in those with an ECOG performance status of 0 (PS0) or 1 (PS1) and 5.4 months in PS2 patients. Both of the patients who received PG–TXL at 235 mg/m^2 died within 30 days after treatment; one died of neutropenia, whereas the other died of septic shock and renal failure.

More than 700 patients with lung cancer have participated in phase III trials PG–TXL(XYOTAX) that include two phase III trials of XYOTAX as first-line treatment in PS2 patients (STELLAR 3 and STELLAR 4) and one phase III trial of XYOTAX as second-line treatment in PS0, PS1, and PS2 patients (STELLAR 2).[47] The total number of patients in the STELLAR 2, STELLAR 3, and STELLAR 4 trials was 850, 400, and 477, respectively. Cell Therapeutics Inc. (CTI) announced the results of these trials in March and May 2005. Although none of the three trials met their primary end points, they did demonstrate similar efficacy, reduced side effects, and more convenient administration of XYOTAX when compared with the control drugs (TXL, docetaxel, gemcitabine, and vinorelbine).

In addition, clinical data from a pooled analysis of CTI's STELLAR 3 and 4 trials showed that in the 198 women treated on those trials, superior survival was observed in those who received XYOTAX ($p = 0.03$). The most notable impact was among women less than 55 years old and presumably pre-menopausal who were treated with XYOTAX compared to standard chemotherapy (median survival 10.0 vs. 5.3 months, hazard ratio = 0.51, log rank $p = 0.038$) while a survival trend ($p = 0.134$) was observed in women 55 years of age and older (post-menopausal) (http://www.ctiseattle.com). The favorable anti-tumor activity among women less than 55 years of age stimulated the initiation of the PIONEER 1 clinical trial where about 600 PS2 chemotherapy-naïve women with advanced stage NSCLC will be recruited. Each study arm of approximately 300 patients will be randomized to receive either XYOTAX at 175 mg/m^2 or TXL at 175 mg/m^2 once every three weeks. The primary endpoint is superior overall survival with several secondary endpoints including disease control, response rate in patients with measurable disease, time to disease progression, and disease-related symptoms.

CTI also reported the preliminary results of a phase II study of XYOTAX in combination with carboplatin for first-line induction and single-agent maintenance therapy for advanced-stage III/IV ovarian cancer in May 2005. Among the 82 patients in this study, 98% had a major tumor response, 85% had a complete response, and 12% had a partial response during induction of the therapy. At a dose of 175 mg/m^2, XYOTAX and carboplatin (AUC = 6) had grade 3/4 side effects, including thrombocytopenia (55%), neuropathy (23%), febrile neutropenia (19%), nausea (15%), anemia (11%), and vomiting (7%). The investigators found no grade 4 neuropathy, and only 4% patients needed dose delay because of neutropenia. CTI and the Gynecologic Oncology Group are collaborating on phase III studies for which they expect to enroll about 1550 patients to receive treatment with carboplatin and TXL initially followed by XYOTAX for consolidation in half of the patients.

10.4.2 PG–CPT

The CPTs are a family of synthetic and semisynthetic analogues of 20(S)-CPT that exhibit a broad range of anti-cancer activity by inhibiting topoisomerase-1 activity. Two properties of CPT compounds limit their therapeutic efficacy in humans: instability of the lactone form because of preferential binding of the carboxylate to serum albumin and a lack of aqueous solubility. PG is an effective solubilizing carrier of CPT and serves to protect the E-ring lactone structure in CPT. The PG–CPT conjugate was prepared by directly coupling the hydroxy group at the C20(S)-position of CPT with the carboxylic acid of PG.[54] When given intravenously in four doses every four days at an equivalent CPT dose of 40 mg/kg, PG–CPT delayed the growth of established H322 human lung tumors subcutaneously grown in nude mice. In mice that received intratracheal inoculation of H322 cells, the same treatment prolonged the median survival duration in mice by fourfold when compared with that in untreated control mice. These results showed that PG was as efficient carrier of CPT.

Studies have systematically investigated the structural effects of the anti-tumor efficacy of CPT, including linkers between PG and CPT, the point of attachment of PG on the CPT molecule, the polymer molecular weight, and drug loading.[55–57] First, coupling through the 20(S)-hydroxy group of CPT with or without a glycine linker yielded the most active conjugates because this site is located in close proximity to the lactone ring; therefore, the E-ring lactone is better protected by the linked PG chains. Second, increasing the molecular weight of PG from 33 to 50 kDa improved the anti-tumor efficacy of PG–CPT, probably because of an increased plasma half-life and reduced renal clearance. Third, based on the CPT-equivalent dosing levels, the investigators compared various linkers and found that PG-Gly-CPT and PG–(4-O-butyryl)–CPT seemed had the highest anti-tumor activity. Moreover, the linkers also affected the maximum drug payload; for example, only 15% (by weight) of CPT loading could be achieved for direct conjugation of PG–CPT by the ester linkage because of steric hindrance. On the other hand, up to 50% CPT loading could be achieved for the case with a glycine as linker. Fourth, the researchers observed improved anti-tumor efficacy of PG-Gly-CPT against HT-29 colon cancer and NCI-H460 lung carcinoma with increased CPT loading. However, increasing the loading of CPT to 47% resulted in significantly reduced solubility. Therefore, they recommended further investigation of PG-Gly-CPT with loading of CPT at 30–35%.

10.5 COMBINATION OF PG–TXL WITH RADIOTHERAPY

Combining chemotherapy and radiotherapy has significantly improved response and survival rates in patients with many solid tumors. Many chemotherapeutic agents can increase the radiosensitivity of tumors, potentiating the tumor response to radiation-caused damage. It was hypothesized that combining radiotherapy and chemotherapy using a polymer–drug conjugate may lead to a stronger radiosensitizing effect than using the drug alone. Irradiation can, in turn, potentiate the tumor response to polymer–drug conjugates by increasing tumor vascular permeability and the uptake of these conjugates into solid tumors. To test this hypothesis, PG–TXL was used as a model polymer–drug conjugate.[58] Administration of PG–TXL delayed the growth of OCa-1 syngeneic murine ovarian tumors in C3Hf/Kam mice. However, when PG–TXL was given in combination with tumor irradiation, significantly enhanced anti-tumor activity was observed. Using tumor growth delay as an end point, enhancement factors ranging from 1.36 to 4.4 were observed; these values depended on the doses of PG–TXL and radiation delivered. It was found that complete tumor regression occurred with the use of increased radiation doses (>10 Gy) and PG–TXL doses (>80 mg/kg equivalent TXL).[59] Similar results were observed in a mammary MCa-4 carcinoma model.[60] In contrast, it was found that combined radiotherapy and TXL treatment yielded an enhancement factor of less than 1.0 in MCa-4 tumors, indicating that conjugation of TXL with PG is necessary to improve that radiosensitization effect of TXL. When the treatment end point was tumor cure, enhancement factors as high as 8.4 and 7.2 were observed after fractionated and single-dose radiotherapy, respectively.[61,62]

To determine if prior irradiation affects tumor uptake of PG–TXL, [³H]PG–TXL was injected into mice with OCa-1 ovarian tumors 24 h after local irradiation at 15 Gy.[59] The uptake of [³H]PG–TXL in irradiated tumors was 28–38% higher than that in nonirradiated tumors at different times after injection of [³H]PG–TXL, suggesting that irradiation increased the accumulation of PG–TXL in the tumors. Therefore, the super-synergistic effect of combined radiotherapy and PG–TXL-based chemotherapy is partly ascribed to the enhanced permeability and retention effect of macromolecules caused by irradiation.

10.6 PG AS A CARRIER OF DIAGNOSTIC AGENTS

10.6.1 MAGNETIC RESONANCE IMAGING

Noninvasive imaging of intravascular compartments is of critical importance in the clinic. Many diseases such as infections, ulcers, cardiovascular diseases, and solid tumors involve gross hemorrhaging, abnormal vascular growth, and/or vascular occlusion.[63,64] Magnetic resonance imaging (MRI) with blood-pool contrast agents can be used to perform minimally invasive angiography, assess angiogenesis, quantify and measure the spacing of blood vessels, and measure blood volume and flow.[63,64] MRI blood-pool contrast agents are often paramagnetic gadolinium (Gd) chelates of high-molecular-weight polymers that are largely retained within the intravascular space during MRI.[65] Recent studies showed that tumor vascular permeability measurements obtained using polymeric contrast agents enhanced dynamic MRI that correlated with tumor microvessel density counts, suggesting that MRI can be used to characterize tumor angiogenic activity.[66,67]

Furthermore, MRI with polymeric contrast agents may permit more accurate grading of tumor invasiveness than those with smaller molecular weight contrast agents, such as Gd–diethylenetria-minepentaacetic acid (DTPA).[68,69]

Investigators have synthesized various polymeric contrast agents for MRI, including human serum albumin,[70] polylysine,[71,72] dextran,[73,74] dendrimers,[75–78] polyamide,[74,79] and grafted copo-lymers,[80] and they evaluated them as blood-pool imaging agents. Although most of these agents fulfill the criteria for long blood circulation time and high MRI relaxivity, their safety remains unestablished. The clinical applications of many current macromolecular contrast agents are limited by slow excretion from the body and the potential toxicity of free Gd^{3+} ions released by the metabolism of the contrast agents.[71–73,81,82] For albumin-based products, the possibility of immunogenic responses makes the use of serum proteins less attractive. Dendrimer-based blood-pool agents have the advantage of extremely narrow molecular weight distributions; however, these agents are not biodegradable.

Ideally, polymeric contrast agents are degraded and cleared from the body after completion of MRI. Based on its good biocompatibility and biodegradability, the metal chelator DTPA was conjugated with PG, and the physicochemical and imaging properties of the resulting polymeric contrast agents were evaluated.[83] One of these agents, PG–Bz–DTPA–Gd, was synthesized from PG and monofunctional *p*-aminobenzyl-DTPA (penta-*tert*-butyl ester). It was found that PG–Bz–DTPA–Gd was readily degraded upon exposure to an aqueous buffered solution containing cath-epsin B. The T_1 relaxivity of PG–Bz–DTPA–Gd at 1.5 T was four times greater than that of small-molecular-weight Gd–DTPA. Figure 10.2 compares the parametric AUC magnetic resonance images for signal intensity integrated over 90 s and 10 min after intravenous injection of gado-pentetate dimeglumine (Magnevist; Gd–DTPA) and PG–Bz–DTPA–Gd. Magnevist (743 Da) is a contrast agent that is clinically used. Whereas Magnevist rapidly diffused into the extravascular fluid space over 90 s, PG–Bz–DTPA–Gd was largely retained in the blood vessels for up to 10 min. Indeed, contrast enhancement of the vascular compartments was still visible 2 h after injection of PG–Bz–DTPA–Gd.[83]

In an effort to increase the rate of polymer degradation, Lu et al.[84] prepared PG–cystamine–[Gd(III)-1,4,7,10-tetraazacyclododecane-1,4,7,10-tetraacetic acid (DOTA)] (molecular weight of PG, 50,000 Da) where the metal chelator DOTA was conjugated with PG using a disulfide bond. They showed that glutathione and other endogenous sulfhydryl-containing biomolecules could exchange their free SH group with an S–S bond in PG–cystamine–[Gd(III)–DOTA], resulting in

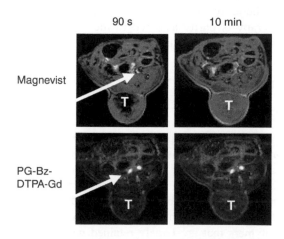

FIGURE 10.2 (See color insert following page 522.) Comparison of parametric AUC magnetic resonance images obtained at 90 s and 10 min after intravenous injection of Magnevist and PG–Bz–DTPA–Gd. Arrow: Blood vessel.

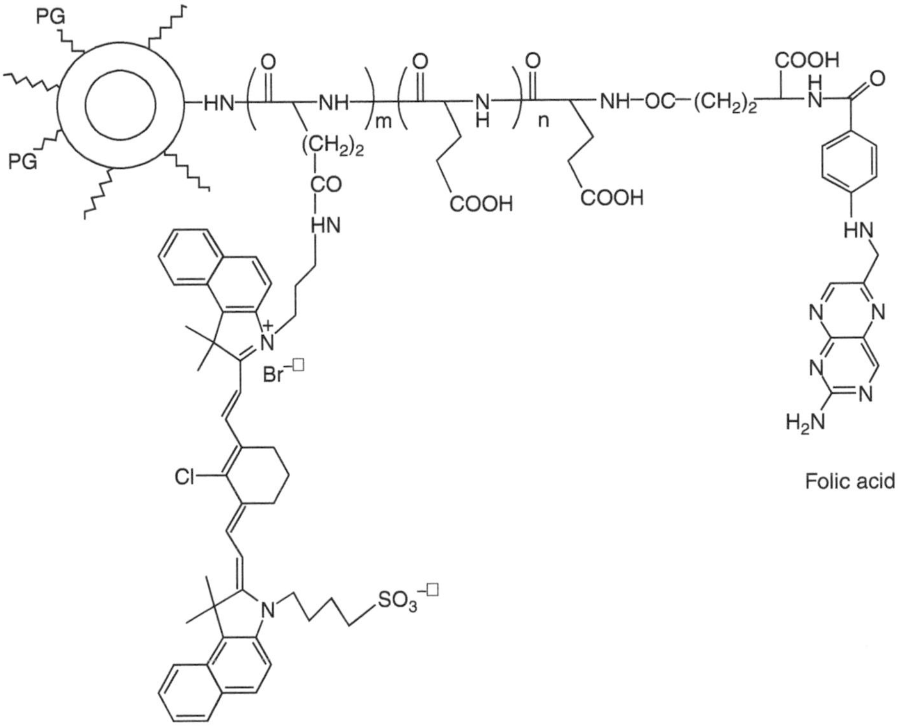

FIGURE 10.3 Structure of PAMAM$_{16}$–PG–(ICG)–folate containing folic acid molecules at the termini of the polymer and dye molecules (ICG-NH$_2$) at the side chains of the polymer. (From Tansey, W., Ke, S., Cao, X. Y. et al., *J. Control. Rel.*, 94, 39–51, 2004. With permission.)

rapid release of Gd(III)–DOTA from the polymer and subsequent clearance of Gd-containing species from the body. Use of PG–cystamine–[Gd(III)–DOTA] produced significant blood-pool contrast enhancement on MRI scans of the heart and blood vessels in nude mice bearing OVCAR-3 human ovarian carcinoma xenograft when compared with use of small-molecular-weight contrast agents.

10.6.2 NEAR-INFRARED FLUORESCENCE OPTICAL IMAGING

Near-infrared (NIR) optical imaging has unique advantages for diagnostic imaging of solid tumors. Specifically, NIR imaging is a potentially safe, noninvasive method of detecting solid tumors as NIR light (650- to 900-nm wavelengths) can penetrate several centimeters into tissue.[58,85] The NIR dye indocyanine green (ICG), a model diagnostic agent, has been conjugated with the side carboxylic acid of branched PG having a PAMAM core and the terminal folic acid group as a targeting moiety (Figure 10.3).[86] The resulting conjugate, PAMAM$_{16}$–PG–(ICG)–folate, exhibited selective binding to KB cells (overexpressing folate receptors) but not to SK-Br3 cells that cannot express folate receptors. Additionally, this selective binding was partially blocked by free folic acid.

10.7 CONCLUSIONS

Owing to its favorable physicochemical properties, PG has been shown to be an excellent polymeric carrier of both diagnostic and therapeutic agents. PG–TXL has become the first PG-based polymeric agent to be tested in clinical trials. The newer generation of PG-based polymers will have to meet a number of challenges, including the development of novel polymers with modulated rates of degradation, versatile conjugation chemistry allowing site-specific attachment of targeting moieties, and polymerization methods that allow accurate control of polymer molecular weights and molecular weight distributions.

ACKNOWLEDGMENTS

This work was supported in part by the National Cancer Institute (R29 CA74819), the Department of Defense (BC960384), and the John S. Dunn Foundation. We thank Donald R. Norwood for editing the manuscript.

REFERENCES

1. Li, C., Poly(L-glutamic acid)-anticancer drug conjugates, *Adv. Drug Deliv. Rev.*, 54, 695–713, 2002.
2. Duncan, R., The dawning era of polymer therapeutics, *Nat. Rev. Drug Discov.*, 2, 347–360, 2003.
3. Allen, T. M. and Cullis, P. R., Drug delivery systems: Entering the mainstream, *Science*, 303, 1818–1822, 2004.
4. Idelson, M. E. and Blout, E. R., High molecular weight poly(L-glutamic acid): Preparation and physical rotation changes, *J. Am. Chem. Soc.*, 80, 4631–4634, 1958.
5. Fuller, W. D., Verlander, M. S., and Goodman, M., A procedure for the facile synthesis of amino-acid *N*-carboxyanhydrides, *Biopolymers*, 15, 1869–1871, 1976.
6. Katakai, R. and Lisuka, Y., An improved method for the synthesis of *N*-carboxy amino acid anhydrides using trichloromethyl chloroformate, *J. Org. Chem.*, 50, 715–716, 1985.
7. Block, H., *Poly(g-benzyl-L-glutamate) and Other Glutamic Acid Containing Polymers*, Gordon and Breach, New York, 1983.
8. Deming, T. J., Facile synthesis of block copolypeptides of defined architecture, *Nature*, 390, 386–389, 1997.
9. Poche, D., Moore, M., and Bowles, J., An unconventional method for purifying the *N*-carboxyanhydride derivatives of *g*-alkyl-L-glutamates, *Synth. Commun.*, 29, 843–854, 1999.
10. Aliferis, T., Iatrou, H., and Hadjichristidis, N., Living polypeptides, *Biomacromolecules*, 5, 1653–1656, 2004.
11. Hanby, W. E., Waley, S. G., and Watson, J., Synthetic polypeptides. Part II. Polyglutamic acid, *J. Chem. Soc.*, 3239–3249, 1950.
12. Palacios, P., Bussat, P., and Bichon, D., Novel solid-phase synthesis of thiol-terminated-poly(alpha-amino acid)–drug conjugate, *J. Biochem. Biophys. Methods*, 23, 67–72, 1991.
13. Vega, J., Ke, S., Fan, Z. et al., Targeting doxorubicin to epidermal growth factor receptors by site-specific conjugation of C225 to poly(L-glutamic acid) through a polyethylene glycol spacer, *Pharm. Res.*, 20, 826–832, 2003.
14. Nishiyama, N., Okazaki, S., Cabral, H. et al., Novel cisplatin-incorporated polymeric micelles can eradicate solid tumors in mice, *Cancer Res.*, 63, 8977–8983, 2003.
15. De Marre, A. and Schacht, E., Preparation of 4-nitrophenol carborate esters of poly[*N*-5-(2-hydroxyethyl)-L-glutamine] and coupling with bioactive agents, *Markromol. Chem.*, 193, 3023–3030, 1992.
16. Romberg, B., Metselaar, J. M., deVringer, T. et al., Enzymatic degradation of liposome-grafted poly(hydroxyethyl L-glutamine), *Bioconjug. Chem.*, 16, 767–774, 2005.
17. D'Souza, A. J. and Topp, E. M., Release from polymeric prodrugs: Linkages and their degradation, *J. Pharm. Sci.*, 93, 1962–1979, 2004.
18. Thomssen, C., Schmit, M., Goretzki, L. et al., Prognostic value of the cysteine proteases cathepsins B and cathepsin L in human breast cancer, *Clin. Cancer Res.*, 1, 741–746, 1995.
19. Mai, J., Waisman, D. M., and Sloane, B. F., Cell surface complex of cathepsin B/annexin II tetramer in malignant progression, *Biochim. Biophys. Acta*, 1477, 215–230, 2000.
20. Kornguth, S. E., Kalinke, T., Robins, H. I., Cohen, J. D., and Turski, P., Preferential binding of radiolabeled poly-L-lysines to C6 and U87 MG glioblastomas compared with endothelial cells in vitro, *Cancer Res.*, 49, 6390–6395, 1989.

21. Hoste, K., De Winne, K., and Schacht, E., Polymeric prodrugs, *Int. J. Pharm.*, 277, 119–131, 2004.
22. Chiu, H. C., Kopeckova, P., Deshmane, S. S., and Kopecek, J., Lysosomal degradability of poly(alpha-amino acids), *J. Biomed. Mater. Res.*, 34, 381–392, 1997.
23. Tsutsumi, A., Perly, B., Forchioni, A., and Charahaty, C., A magnetic resonance study of the segmental motion and local conformations of poly(L-glutamic acid) in aqueous solutions, *Macromolecules*, 11, 977–986, 1978.
24. Dolnik, V., Novotny, M., and Chmelik, J., Electromigration behavior of poly(L-glutamate) conformers in concentrated polyacrylamide gels, *Biopolymers*, 33, 1299–1306, 1993.
25. Miller, W. G., Degradation of synthetic polypeptides II. degradation of poly-L-glutamic acid by proteolytic enzymes in 0.20 M sodium chloride, *J. Am. Chem. Soc.*, 86, 3913–3918, 1964.
26. Kishore, B. K., Kallay, Z., Lambricht, P., Laurent, G., and Tulkens, P. M., Mechanism of protection afforded by polyaspartic acid against gentamicin-induced phospholipidosis I. Polyaspartic acid binds gentamicin and displaces it from negatively charged phospholipid layers in vitro, *J. Pharmacol. Exp. Ther.*, 255, 867–874, 1990.
27. McCormick-Thomson, L. A. and Duncan, R., Poly(amino acid) copolymers as a potential soluble drug delivery system. I: Pinocytotic uptake and lysosomal degradation measured in vitro, *J. Bioact. Compat. Polym.*, 4, 242–251, 1989.
28. Hoes, C. J. T., Potman, W., Van Heeswijk, W. A. R. et al., Optimization of macromolecular prodrugs of the antitumor antibiotic Adriamycin, *J. Control. Rel.*, 2, 205–213, 1985.
29. Singer, J. W., Nudelman, E., De Vries, P., Shaffer, S., Metabolism of poly-L-glutamic acid (PG) paclitaxel (CT-2103): Intratumoral peptide cleavage followed by esterolysis, In *5th International Symposium on Polymer Therapeutics: From Laboratory to Clinical Practice*, Cardiff, UK, 2002.
30. Van Heeswijk, W. A. R., Hoes, C. J. T., Stoffer, T. et al., The synthesis and characterization of polypeptide–adriamycin conjugates and its complexes with Adriamycin. Part 1, *J. Control. Rel.*, 1, 301–315, 1985.
31. Hoes, C. J. T., Grootoonk, J., Duncan, R. et al., Biological properties of adriamycin bound to biodegradable polymeric carriers, *J. Control. Rel.*, 23, 37–54, 1993.
32. Zunino, F., Pratesi, G., and Micheloni, A., Poly(carboxylic acid) polymers as carriers for anthracyclines, *J. Control. Rel.*, 10, 65–73, 1989.
33. Hurwitz, E., Wilchek, M., and Pitha, J., Soluble macromolecules as carriers for daunorubicin, *J. Appl. Biochem.*, 2, 25–35, 1980.
34. Kato, Y., Saito, M., Fukushima, H., Takeda, Y., and Hara, T., Antitumor activity of 1-beta-D-arabinofuranosylcytosine conjugated with polyglutamic acid and its derivative, *Cancer Res.*, 44, 25–30, 1984.
35. Batz, H. G., Ringsdorf, H., and Ritter, H., Pharmacologically active polymers. 7. Cyclophosphamide- and steroid hormone-containing polymers as potential anticancer compounds, *Makromol. Chem.*, 175, 2229–2239, 1974.
36. Moromoto, Y., Sugibayashi, K., Sugihara, S. et al., Antitumor agent poly(amino acid) conjugates as a drug carrier in chemotherapy, *J. Pharm. Dyn.*, 7, 688–698, 1984.
37. Roos, C. F., Matsumoto, S., Takarura, Y., Hashida, M., and Sezaki, H., Physicochemical and antitumor characteristics of some polyamino acid prodrugs of mitomycin C, *Int. J. Pharm.*, 22, 75–87, 1984.
38. Soyez, H., Schacht, E., and Vanderkerken, S., The crucial role of spacer groups in macromolecular prodrug design, *Adv. Drug Deliv. Rev.*, 21, 81–106, 1996.
39. de Marre, A., Seymour, L. W., and Schacht, E., Evaluation of the hydrolytic and enzymatic stability of macromolecular mitomycin C derivatives, *J. Control. Rel.*, 31, 89–97, 1994.
40. Avichezer, D., Schechter, B., and Arnon, R., Functional polymers in drug delivery: Carrier-supported CDDP (cisplatin) complexes of polycarboxylates-effects on human ovarian carcinoma, *React. Funct. Polym.*, 36, 59–69, 1998.
41. Schlechter, B., Neumann, A., Wilchek, M., and Arnon, R., Soluble polymers as carriers of cis-platinum, *J. Control. Rel.*, 10, 75–87, 1989.
42. Seymour, L. W., Soyez, H., De Marre, A., Shoaibi, M. A., and Schacht, E. H., Polymeric prodrugs of mitomycin C designed for tumour tropism and sustained activation, *Anti-cancer Drug Des.*, 11, 351–365, 1996.
43. Horwitz, S. B., Taxol (Paclitaxel): Mechanisms of action, *Ann. Oncol.*, 5, S3–S6, 1994.

44. Li, C., Yu, D. F., Newman, R. A. et al., Complete regression of well-established tumors using a novel water-soluble poly(L-glutamic acid)–paclitaxel conjugate, *Cancer Res.*, 58, 2404–2409, 1998.

45. Li, C., Newman, R. A., Wu, Q. P. et al., Biodistribution of paclitaxel and poly(L-glutamic acid)–paclitaxel conjugate in mice with ovarian OCa-1 tumor, *Cancer Chemother. Pharmacol.*, 46, 416–422, 2000.

46. Oldham, E. A., Li, C., Ke, S., Wallace, S., and Huang, P., Comparison of action of paclitaxel and poly(L-glutamic acid)–paclitaxel conjugate in human breast cancer cells, *Int. J. Oncol.*, 16, 125–132, 2000.

47. Singer, J. W., Shaffer, S., Baker, B. et al., Paclitaxel poliglumex (XYOTAX; CT-2103): An intracellularly targeted taxane, *Anticancer Drugs*, 16, 243–254, 2005.

48. Li, C., Price, J. E., Milas, L. et al., Antitumor activity of poly(L–glutamic acid)–paclitaxel on syngeneic and xenografted tumors, *Clin. Cancer Res.*, 5, 891–897, 1999.

49. Auzenne, E., Donato, N. J., Li, C. et al., Superior therapeutic profile of poly-L-glutamic acid–paclitaxel copolymer compared with taxol in xenogeneic compartmental models of human ovarian carcinoma, *Clin. Cancer Res.*, 8, 573–581, 2002.

50. Todd, R., Sludden, J., Boddy, A. V. et al., Phase I and pharmacological study of CT-2103, a poly(L-glutamic acid)–paclitaxel conjugate, *Proc. Am. Soc. Clin. Oncol.*, 20, 2001 (abstr 439).

51. Bernareggi, A., Oldham, F., Barone, C., Tumor-targeted taxane: PK evidence for an enhanced permeability and retention effect with paclitaxel poliglumex (CT-2103), *Presented at the European Society for Medical Oncology (ESMO), October 29–November 2, 2004, Vienna, Austria.*

52. Markman, M., Improving the toxicity profile of chemotherapy for advanced ovarian cancer: A potential role for CT-2103, *J. Exp. Ther. Oncol.*, 4, 131–136, 2004.

53. Sabbatini, P., Aghajanian, C., Dizon, D. et al., Phase II study of CT-2103 in patients with recurrent epithelial ovarian, fallopian tube, or primary peritoneal carcinoma, *J. Clin. Oncol.*, 22, 4523–4531, 2004.

54. Zou, Y., Wu, Q. P., Tansey, W. et al., Effectiveness of water soluble poly(L-glutamic acid)–camptothecin conjugate against resistant human lung cancer xenografted in nude mice, *Int. J. Oncol.*, 18, 331–336, 2001.

55. Singer, J. W., De Vries, P., Bhatt, R. et al., Conjugation of camptothecins to poly(L-glutamic acid), *Ann. N Y Acad. Sci.*, 922, 136–150, 2000.

56. Singer, J. W., Bhatt, R., Tulinsky, J. et al., Water-soluble poly(L-glutamic acid)-Gly-camptothecin conjugates enhance camptothecin stability and efficacy in vivo, *J. Control. Rel.*, 74, 243–247, 2001.

57. Bhatt, R., de Vries, P., Tulinsky, J. et al., Synthesis and in vivo antitumor activity of poly(L-glutamic acid) conjugates of 20S-camptothecin, *J. Med. Chem.*, 46, 190–193, 2003.

58. Hawrysz, D. J. and Sevick-Muraca, E. M., Developments toward diagnostic breast cancer imaging using near-infrared optical measurements and fluorescent contrast agents, *Neoplasia*, 2, 388–417, 2000.

59. Li, C., Ke, S., Wu, Q. P. et al., Tumor irradiation enhances the tumor-specific distribution of poly(L-glutamic acid)–conjugated paclitaxel and its antitumor efficacy, *Clin. Cancer Res.*, 6, 2829–2834, 2000.

60. Ke, S., Milas, L., Charnsangavej, C., Wallace, S., and Li, C., Potentiation of radioresponse by polymer–drug conjugates, *J. Control. Rel.*, 74, 237–242, 2001.

61. Milas, L., Mason, K. A., Hunter, N., Li, C., and Wallace, S., Poly(L-glutamic acid)–paclitaxel conjugate is a potent enhancer of tumor radiocurability, *Int. J. Radiat. Oncol. Biol. Phys.*, 55, 707–712, 2003.

62. Tishler, R. B., Polymer–conjugated paclitaxel as a radiosensitizing agent-a big step forward for combined modality therapy?, *Int. J. Radiat. Oncol. Biol. Phys.*, 55, 563–564, 2003.

63. Bogdanov, A. A., Lewin, M., and Weissleder, R., Approaches and agents for imaging the vascular system, *Adv. Drug Deliv. Rev.*, 37, 279–293, 1993.

64. Kroft, L. J. and de Roos, A., Blood pool contrast agents for cardiovascular MR imaging, *J. Magn. Reson. Imaging*, 10, 395–403, 1999.

65. Brasch, R. C., New directions in the development of MR imaging contrast media, *Radiology*, 183, 1–11, 1992.

66. Su, M. Y., Muhler, A., Lao, X., and Nalcioglu, O., Tumor characterization with dynamic contrast-enhanced MRI using MR contrast agents of various molecular weights, *Magn. Reson. Med.*, 39, 259–269, 1998.

67. van Dijke, C. F., Brasch, R. C., Roberts, T. P. et al., Mammary carcinoma model: Correlation of macromolecular contrast-enhanced MR imaging characterizations of tumor microvasculature and histologic capillary density, *Radiology*, 198, 813–818, 1996.

68. Gossmann, A., Okuhata, Y., Shames, D. M. et al., Prostate cancer tumor grade differentiation with dynamic contrast-enhanced MR imaging in the rat: Comparison of macromolecular and small-molecular contrast media-preliminary experience, *Radiology*, 213, 265–272, 1999.

69. Daldrup, H., Shames, D. M., Wendland, M. et al., Correlation of dynamic contrast-enhanced magnetic resonance imaging with histologic tumor grade: Comparison of macromolecular and small-molecular contrast media, *Pediatr. Radiol.*, 28, 67–78, 1998.

70. van Bemmel, C. M., Wink, O., Verdonck, B., Viergever, M. A., and Niessen, W. J., Blood pool contrast-enhanced MRA: Improved arterial visualization in the steady state, *IEEE Trans. Med. Imaging*, 22, 645–652, 2003.

71. Schuhmann-Giampieri, G., Schmitt-Willich, H., Frenzel, T., Press, W. R., and Weinmann, H. J., In vivo and in vitro evaluation of Gd–DTPA–polylysine as a macromolecular contrast agent for magnetic resonance imaging, *Invest. Radiol.*, 26, 969–974, 1991.

72. Bogdanov, J., Alexei, A., Weissleder, R., and Brady, T. J., Long-circulating blood pool imaging agents, *Adv. Drug Deliv. Rev.*, 16, 335–348, 1995.

73. Wang, S. C., Wikstrom, M. G., White, D. L. et al., Evaluation of Gd–DTPA–labeled dextran as an intravascular MR contrast agent: Imaging characteristics in normal rat tissues, *Radiology*, 175, 483–488, 1990.

74. Rebizak, R., Schaefer, M., and Dellacherie, E., Polymeric conjugates of Gd(3 +)-diethylenetriaminepentaacetic acid and dextran. 1. Synthesis, characterization, and paramagnetic properties, *Bioconjug. Chem.*, 8, 605–610, 1997.

75. Wiener, E. C., Brechbiel, M. W., Brothers, H. et al., Dendrimer-based metal chelates: A new class of magnetic resonance imaging contrast agents, *Magn. Reson. Med.*, 31, 1–8, 1994.

76. Adam, G., Neuerburg, J., Spuntrup, E. et al., Gd–DTPA–cascade-polymer: Potential blood pool contrast agent for MR imaging, *J. Magn. Reson. Imaging*, 4, 462–466, 1994.

77. Kobayashi, H. M. and Brechbiel, W., Dendrimer-based macromolecular MRI contrast agents: Characteristics and application, *Mol. Imaging*, 2, 1–10, 2003.

78. Kim, Y. H., Choi, B. I., Cho, W. H. et al., Dynamic contrast-enhanced MR imaging of VX2 carcinomas after X-irradiation in rabbits: Comparison of gadopentetate dimeglumine and a macromolecular contrast agent, *Invest. Radiol.*, 38, 539–549, 2003.

79. Unger, E. C., Shen, D., Wu, G. et al., Gadolinium-containing copolymeric chelates—a new potential MR contrast agent, *Magma*, 8, 154–162, 1999.

80. Gupta, H., Wilkinson, R. A., Bogdanov, A. A., Callahan, R. J., and Weissleder, R., Inflammation: Imaging with methoxy poly(ethylene glycol)–poly-L-lysine–DTPA, a long-circulating graft copolymer, *Radiology*, 197, 665–669, 1995.

81. Rebizak, R., Schaefer, M., and Dellacherie, E., Polymeric conjugates of Gd(3 +)-diethylenetriaminepentaacetic acid and dextran. 2. Influence of spacer arm length and conjugate molecular mass on the paramagnetic properties and some biological parameters, *Bioconjug. Chem.*, 9, 94–99, 1998.

82. Franano, F. N., Edwards, W. B., Welch, M. J. et al., Biodistribution and metabolism of targeted and nontargeted protein-chelate–gadolinium complexes: Evidence for gadolinium dissociation in vitro and in vivo, *Magn. Reson. Imaging*, 13, 201–214, 1995.

83. Wen, X., Jackson, E. F., Price, R. E. et al., Synthesis and characterization of poly(L-glutamic acid) gadolinium chelate: A new biodegradable MRI contrast agent, *Bioconjug. Chem.*, 15, 1408–1415, 2004.

84. Lu, Z. R., Wang, X., Parker, D. L., Goodrich, K. C., and Buswell, H. R., Poly(L-glutamic acid) Gd(III)–DOTA conjugate with a degradable spacer for magnetic resonance imaging, *Bioconjug. Chem.*, 14, 715–719, 2003.

85. Ke, S., Wen, X., Gurfinkel, M. et al., Near-infrared optical imaging of epidermal growth factor receptor in breast cancer xenografts, *Cancer Res.*, 63, 7870–7875, 2003.

86. Tansey, W., Ke, S., Cao, X. Y. et al., Synthesis and characterization of branched poly(L-glutamic acid) as a biodegradable drug carrier, *J. Control. Rel.*, 94, 39–51, 2004.

11 Noninvasive Visualization of In Vivo Drug Delivery of Paramagnetic Polymer Conjugates with MRI

Zheng-Rong Lu, Yanli Wang, Furong Ye, Anagha Vaidya, and Eun-Kee Jeong

CONTENTS

11.1 Introduction 201
11.2 Magnetic Resonance Imaging 202
    11.2.1 Principles of MRI 202
    11.2.2 Contrast-Enhanced MRI 203
11.3 MR Imaging of Paramagnetic HPMA Copolymer Conjugates 204
    11.3.1 MR Imaging of a Paramagnetic HPMA Conjugate in an Animal Tumor Model 204
    11.3.2 MR Imaging of a Paramagnetic HPMA Conjugate for Bone Delivery 206
11.4 MR Imaging of Paramagnetic Poly(L-Glutamic Acid) Conjugates 206
    11.4.1 MR Imaging of PGA–(Gd-DOTA) Conjugate in an Animal Tumor Model 206
    11.4.2 MR Imaging of PGA–Mce$_6$–(Gd-DOTA) Conjugate in an Animal Tumor Model 208
11.5 Summary.. 210
References 210

11.1 INTRODUCTION

The conjugation of therapeutics to water-soluble biomedical polymers increases the aqueous solubility of hydrophobic drugs, prolongs in vivo drug retention time, reduces systemic toxicity, and enhances therapeutic efficacy.[1,2] Polymer drug conjugates can preferentially accumulate in solid tumor tissues due to the hyperpermeability of tumor blood vessels or the so-called "enhanced permeability and retention (EPR) effect."[3,4] Polymer drug conjugates can also down-regulate or overcome multiple drug resistance.[5,6] Because of these unique properties, polymer drug conjugates exhibit higher therapeutic efficacy than the corresponding therapeutics alone. Several polymer drug conjugates are currently used in clinical cancer treatment, and more are in the pipeline of clinical development. For example, styrene–maleic acid copolymer neocarzinostatin conjugate (SMANCS) is used for liver cancer treatment in Japan;[7] pegylated adenosine deaminase[8] and asparaginase[9,10] are used for enzyme replacement therapy for immunodeficiency and for the treatment of acute lymphoblastic leukemia, respectively. Many other polymer drug conjugates, including pegylated interferon α,[11] PEG–camptothecin conjugate,[12] poly(L-glutamic acid) paclitaxel conjugate,[13] poly[N-(2-hydroxypropyl)methacrylamide] (PHPMA)-cis-platinate conjugate,[14] and PHPMA–doxorubicin conjugate[15,16] are in various phases of clinical trials.

The in vivo behavior, including pharmacokinetics, biodistribution, drug delivery efficiency and therapeutic efficacy, of polymer drug conjugates has been traditionally evaluated by using blood and urine sampling, biopsy-based methods, and symptom-based observations. These methods are sometimes invasive and cannot accurately provide real time information on the interaction of polymers with various organs and tissues, the targeting and delivery efficiency of the drug delivery system, and therapeutic response. A large number of animals are also required in preclinical development. Sometimes, the data obtained by conventional biopsies may be misleading, which might be one of the causes of the efficacy discrepancy between preclinical development in animal models and human clinical trials and the failure of some clinical trials due to improper selection of drug candidates.

In vivo drug delivery by polymer drug conjugates involves circulation of the conjugates in the blood; interaction with the major organs including the liver, heart, lungs, spleen and kidneys, etc.; transport to target tissue; and uptake by target cells and drug release. Direct and continuous evaluation of drug delivery efficiency and tissue interaction of polymeric drug conjugates is critical to develop more efficacious drug delivery systems. Recent advancements in biomedical imaging technology have provided the essential tools for noninvasive and continuous in vivo evaluation for drug delivery. Nuclear medicine, including positron emission tomography (PET) and single photon emission computed tomography (SPECT), is a clinical imaging modality with high detection sensitivity (ca. 10^{-9} M). Magnetic resonance imaging (MRI) is a noninvasive clinical imaging modality with high spatial resolution. Both imaging modalities are able to noninvasively visualize in vivo drug delivery with polymeric drug conjugates. The polymer conjugates can be labeled with imaging probes and noninvasively and continuously monitored with the imaging modalities. The number of experimental animals can be dramatically reduced in the preclinical development of the conjugates. Biomedical imaging will provide real-time information of the in vivo behavior of polymeric drug conjugates. In this chapter, we will focus on noninvasive visualization of in vivo drug delivery with paramagnetic polymeric conjugates and contrast-enhanced MRI.

11.2 MAGNETIC RESONANCE IMAGING

11.2.1 PRINCIPLES OF MRI

MRI is a clinical imaging modality that produces three-dimensional anatomic images with high spatial resolution and provides in vivo physiological properties including physiochemical information, flow diffusion, and motion in the tissues. The physics of MRI is complicated and has been detailed in a number of monographs.[17,18] The fundamental principles of MRI are briefly reviewed in this section. A normal human adult has approximately 60% of the body weight as water. Water protons have a magnetic moment and the orientations of the magnetic moments are random in the absence of an external magnetic field. When a patient is placed in a MRI scanner, the proton magnetic moments align either along or against the static magnetic field (B_0) of the scanner and create a net magnetization pointing in the direction of the main magnetic field of the scanner. The magnitude of the magnetization is proportional to the external magnetic field strength as well as the amount of protons.

Nuclear magnetic resonance (NMR) measures the change of the magnetization by applying radiofrequency (RF) pulses. When an RF pulse is applied to create an oscillating electromagnetic field (B_1) perpendicular to the main field, the longitudinal magnetization is tipped away from the static magnetic field and deviates from the equilibrium state (in longitudinal space). A perfect $90°$ RF pulse flips the total longitudinal magnetization onto the transverse plane. Immediately after excitation with the RF pulse, the transverse magnetization then undergoes the decay process. The decaying transverse magnetization induces an NMR signal as an electromotive force (emf) on an RF coil. Simultaneously, the nuclear spin system approaches toward the equilibrium state with a characteristic recovery time T_1, which depends upon the physical and chemical environment of the molecules.

Longitudinal T_1 relaxation involves the return of protons from the high energy state back to equilibrium by dissipating their excess energy to their surroundings. The process is also called *spin-lattice relaxation*. Transverse (T_2) relaxation involves the transfer of the spin angular momentum among the protons via the interactions such as dipole–dipole interaction. The process is called *spin–spin relaxation*. In biological systems, the T_2 relaxation time is typically shorter than T_1 relaxation time. Because the main magnetic field of MRI scanners is not perfectly homogenous, the inhomogeneity of the magnetic field also causes dephasing of individual magnetizations of protons, resulting in more rapid loss of transverse magnetization. The process is called T_2^* *relaxation*.

In MR imaging, the position dependent frequency and phase are encoded into the transverse magnetization, and the measured signals are Fourier transformed to construct the spatial map (image). MR images are created based on the proton density, or T_1 and T_2 relaxation rates, and flow and diffusion properties in different tissues. These parameters vary between different tissues and create image contrast among the tissues. Tissues with a short T_1 relaxation time give a strong MR signal, resulting in bright images in T_1-weighted MRI. Tissues with a short T_2 relaxation time produce a weak MR signal, resulting in dark images in T_2-weighted MRI.

11.2.2 Contrast-Enhanced MRI

In many cases, the differences between the MR signal intensities of normal and diseased tissues are not large enough to produce obvious contrast for definitive diagnosis. Contrast agents are used to enhance the image contrast between normal and diseased tissues to improve diagnostic sensitivity and specificity.[19,20] MRI contrast agents are either paramagnetic metal chelates or ultrasmall superparamagnetic iron oxides. These agents interact with surrounding water molecules and increase the relaxation rates ($1/T_1$ and $1/T_2$) of water protons. The change of relaxation rate is linearly proportional to the concentration of contrast agents, [M]:

$$1/T_{i,\text{obs}} = 1/T_{i,\text{d}} + r_i[\text{M}], \tag{11.1}$$

where $i = 1,2$. In Equation 11.1, $1/T_{i,\text{obs}}$ is the observed relaxation rate with a contrast agent, $1/T_{i,\text{d}}$ the relaxation rate without the contrast agent, and r_i the relaxivity of the contrast agent. Relaxivity is a measurement of the efficiency of a contrast agent to enhance the relaxation rate of water protons or the efficiency to generate image contrast enhancement. Because the uptake of contrast agents varies between different tissues, MR image contrast is enhanced in the tissue with a high contrast agent concentration compared to that with a low concentration.

MRI contrast agents approved for clinical applications are mainly stable paramagnetic chelates, e.g., Gd(III) chelates[21–24] and Mn(II) chelates,[25] and ultrasmall superparamagnetic iron oxide (USPIO).[26] Gd(III) chelates of DTPA, DOTA, or their derivatives are commonly used for contrast-enhanced MRI. Both Gd(III) and Mn(II) chelates are mainly used as T_1 contrast agents, whereas USPIO is used as a T_2 contrast agent. These agents significantly enhance image contrast between normal tissue and diseased tissue and improve the diagnostic accuracy. Currently, contrast agents are used in approximately 30% of clinical MRI examinations. Macromolecular Gd(III) complexes have also been developed as MRI contrast agents.[27,28] These agents have a prolonged blood circulation time and preferentially accumulate in solid tumors due to the EPR effect. Macro-molecular Gd(III) complexes have demonstrated superior contrast enhancement in MR angiography and cancer imaging in animal models, but are not yet approved for clinical applications.

Contrast-enhanced MRI is a useful method for noninvasive in vivo evaluation of polymeric drug delivery systems after the systems are labeled with MRI contrast agents. The real-time pharmacokinetics and biodistribution of the labeled drug delivery systems can be continuously visualized by contrast-enhanced MRI. It has a potential to provide accurate four-dimensional information of in vivo properties of the drug delivery systems.

11.3 MR IMAGING OF PARAMAGNETIC HPMA COPOLYMER CONJUGATES

N-(2-Hydroxypropyl)methacrylamide (HPMA) copolymer is a biocompatible water-soluble poly-meric carrier for therapeutics.[1] The conjugation of anti-cancer drugs to HPMA copolymers results in many advantageous features over small molecular therapeutics, including improved water solu-bility and bioavailability, preferential accumulation of the conjugates in solid tumors, reduced nonspecific toxicity, and down-regulation of multi-drug resistance.[1,29] A number of HPMA copo-lymer drug conjugates have been systematically investigated both in vitro and in vivo. Several HPMA copolymer-anti-cancer drug conjugates are now in various phases of clinical development for cancer treatment.[30] The incorporation of MRI contrast agents into HPMA copolymer conjugates allows direct visualization of real-time drug delivery of the conjugates in animal models with contrast-enhanced MRI. The noninvasive imaging reveals some interesting features of drug delivery with the conjugates that cannot be obtained with conventional surgery based evaluations.

11.3.1 MR IMAGING OF A PARAMAGNETIC HPMA CONJUGATE IN AN ANIMAL TUMOR MODEL

HPMA copolymers have been labeled with radioactive probes including ^{131}I and ^{99m}Tc for noninvasive visualization of in vivo drug delivery of the conjugates with γ-scintigraphy.[31–33] Nuclear medicine is highly sensitive for detecting labeled conjugates, but its poor spatial resolution cannot provide detailed biodistribution of conjugates in tissues and organs. In comparison, contrast-enhanced MRI provides clear visualization of the biodistribution of paramagnetically labeled conjugates with high spatial resolution.

HPMA copolymers were labeled with an MRI contrast agent, Gd-DOTA, and investigated in an animal tumor model with contrast-enhanced MRI.[34] The structure of a labeled conjugate, poly[HPMA-co-(MA-Gly-Gly-1,6-hexanediamine-(Gd-DOTA))], is shown in Figure 11.1. The molecular weight of the polymer conjugate was 25.6 kDa (PD = 1.50) and the Gd content was 0.33 mmol Gd/g polymer. The T_1 relaxivity of the conjugate was 6.0 mM^{-1} s^{-1} per attached Gd(III) chelate at 3 T. Figure 11.2 shows the dynamic, contrast-enhanced, three-dimensional maximum intensity projection (MIP) (Figure 11.2a) and 2D coronal MR images (Figure 11.2b) of a mouse bearing a human prostate carcinoma DU-145 xenograft and a Kaposi's sarcoma xeno-graft. The conjugate was administrated intravenously at a Gd equivalent dose of 0.1 mmol/kg body weight. The pharmacokinetics and biodistribution of the conjugate were clearly revealed in the contrast-enhanced MR images.

Strong MRI signal was observed in the heart, liver, kidneys and vasculature in the 3D images at the initial stage after the injection of the conjugates. The signal intensity decreased over a 60-min period; meanwhile, the signal intensity in the urinary bladder increased gradually, indicating that the conjugate was excreted via renal filtration. Significant contrast enhancement was still visible in the heart at 60 min. After 22 h, most of the conjugate was cleared from the body except the liver, as shown in Figure 11.2. The signal intensity in the liver was stronger at 22 h than in the precontrast image, suggesting significant interaction of HPMA copolymers with the liver.

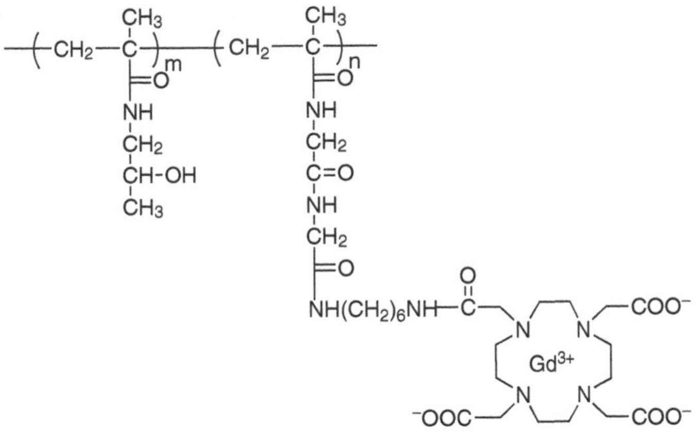

FIGURE 11.1 The structure of poly[HPMA-*co*-(MA-Gly-Gly-1,6-hexanediamine-(Gd-DOTA))].

The coronal images cross-sectioning the tumor tissues showed heterogeneous uptake of HPMA copolymers in two different tumors (Figure 11.2b). The prostate carcinoma had a smaller size, but stronger MR signal than the Kaposi's sarcoma. For Kaposi's sarcoma, the contrast enhancement was mainly observed in the periphery of the tumor tissue and to a lower extent in the inner tumor tissue. A plausible explanation is that the interstitium of the large tumor was possibly necrotic, which would limit the access of the conjugate. The results suggested that tumor uptake of HPMA copolymers may vary from tumor to tumor and with different tumor sizes. It might affect the efficacy of the polymer conjugates and more detailed studies are needed to understand the possible correlations.

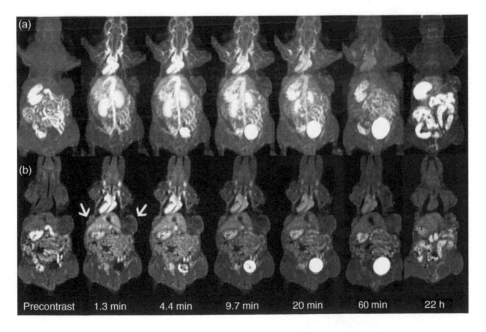

FIGURE 11.2 Three-dimensional maximum intensity projection (MIP) (a) and 2D coronal MR images (b) of the male nude mouse bearing prostate carcinoma DU-145 (left) and human Kaposi's sarcoma (SLK) tumor (right) at different time points. The arrow points to the tumor. Imaging parameters were TR = 7.8 ms, TE = 2.7 ms, 25° flip angle, and 0.4-mm coronal slice thickness.

11.3.2 MR Imaging of a Paramagnetic HPMA Conjugate for Bone Delivery

Kopecek and coworkers labeled HPMA copolymers (55 kDa) with Gd-DOTA to noninvasively evaluate macromolecular accumulation in arthritic joints in a rat model. The copolymers were labeled with fluorescein in addition to the paramagnetic metal chelate. The structure of the conjugate is shown in Figure 11.3. After intravenous administration of the conjugate in a rat model with adjuvant-induced arthritis, the biodistribution of copolymers was followed by contrast-enhanced MRI at various time points post-injection.[35]

Contrast-enhanced dynamic MRI clearly revealed the pharmacokinetic properties and biodistribution of the conjugate with high anatomic resolution in the adjuvant-induced arthritic rats (Figure 11.4). The copolymers had a relatively high molecular weight and a considerable amount of the copolymers was still circulating in the blood 3 h post-injection. Selective accumulation of HPMA copolymers was gradually shown in the arthritic joints of the adjuvant-induced arthritis rats. The copolymers were also cleared from the accumulation sites over time as shown by contrast-enhanced MRI. The uptake and retention of the MR contrast agent labeled polymer correlated well with the histopathological features of inflammation and local tissue damages. No significant contrast enhancement was observed in the hind-limb joints of healthy rats in a control study.

11.4 MR IMAGING OF PARAMAGNETIC POLY(L-GLUTAMIC ACID) CONJUGATES

Poly(L-glutamic acid) (PGA) is a negatively charged, biocompatible polymer that has been used as a carrier for drug delivery.[36] It is a homopolymer of L-glutamic acid linked through peptide bonds and the pendant γ-carboxyl groups are suitable to load therapeutic agents via chemical conjugation. Several poly(L-glutamic acid)-anti-cancer drug conjugates have been developed for cancer treatment.[37–39] The conjugation of a MRI contrast agent to PGA enables noninvasive investigation of its pharmacokinetics and in vivo drug delivery with contrast-enhanced MRI.

11.4.1 MR Imaging of PGA–(Gd-DOTA) Conjugate in an Animal Tumor Model

Poly(L-glutamic acid)–1,6-hexanediamine–(Gd-DOTA) conjugate was prepared using a synthetic procedure similar to that of PGA-cystamine-Gd-DOTA conjugate synthesis[40] to noninvasively

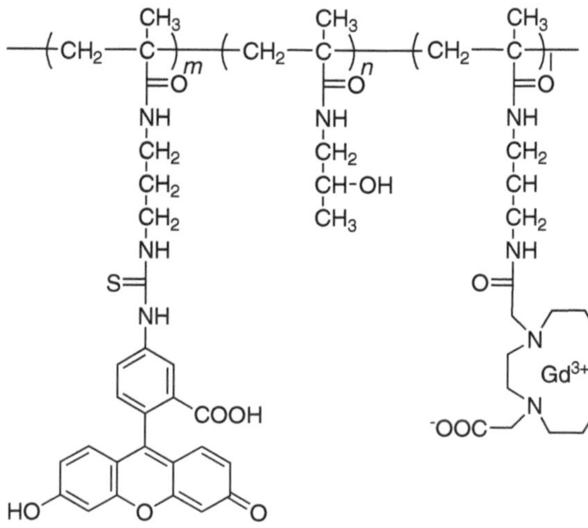

FIGURE 11.3 The structure of HPMA-MAFITC copolymers labeled with Gd(III)-DOTA.

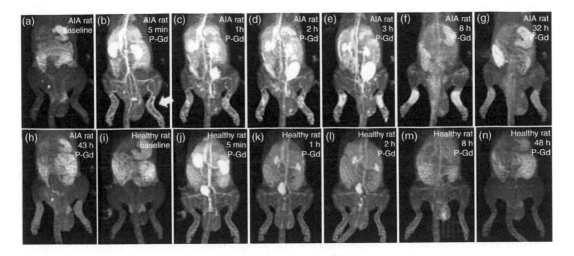

FIGURE 11.4 The MR images of rats taken at different time points. The acquired images were post-processed using the maximum intensity projection (MIP) algorithm. P-Gd(III)-DOTA is abbreviated as P-Gd. (A-H) AIA rat images at baseline (a) and 5 min (b), 1 h (c), 2 h (d), 3 h (e), 8 h (f), 32 h (g), 43 h (h) post-injection of P-Gd(III)-DOTA. (i–n) Healthy rat images at baseline (i) and 5 min (j), 1 h (k), 2 h (l), 8 h (m), 48 h (n) post-injection of P-Gd(III)-DOTA. Arrow points to the diseased joint (Adapted from Wang, D. etal., Pharm. Res., 21, 1741, 2004. With permission.)

study delivery efficiency of the conjugate. The structure of the conjugate is shown in Figure 11.5. PGA–1,6-hexanediamine–(Gd-DOTA) conjugates, H a high molecular weight (50 kDa, PD = 1.12) conjugate and a low molecular weight conjugate (20 kDa, PD = 1.08) were investigated by contrast-enhanced MRI to study the size effect of conjugates on the efficiency of in vivo drug delivery. The conjugates with different molecular weights had similar Gd content (41.7% for 50 kDa conjugate and 40.2 mol% for 20 kDa conjugate) and T_1 relaxivity (9.44 and 9.20 mM^{-1} s^{-1} for 50 and 20 kDa conjugates, respectively). The conjugates were intravenously administered into female nu/nu athymic mice bearing human breast carcinoma MB-231 xenografts at a dose of 0.07 mmol-Gd/kg. Dynamic contrast-enhanced MRI clearly revealed the size effect of the conjugates on their pharmacokinetics and in vivo tumor delivery efficiency in the animal model.

Figure 11.6 shows the T_1-weighted coronal MR images of mice before and at various time points after injection with the conjugates. Strong contrast enhancement was observed in the heart at the initial stage post-injection and then gradually faded away for both conjugates. The signal intensity decreased more rapidly for the low molecular weight conjugate than the high molecular weight conjugate, validating the concept that the blood circulation of the conjugates is prolonged

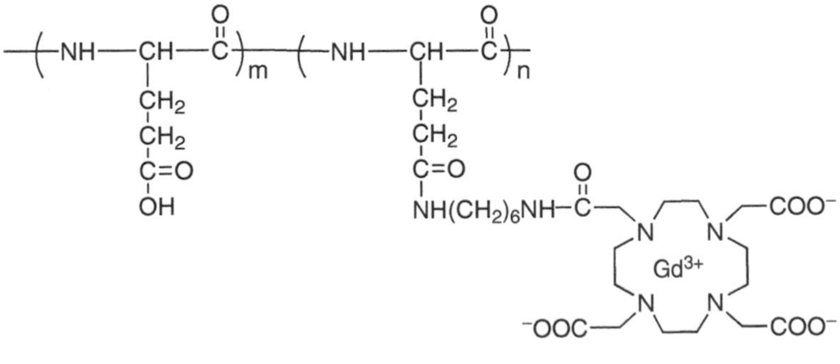

FIGURE 11.5 The structure of poly(L-glutamic acid)–1,6-hexanediamine–(Gd-DOTA) conjugate.

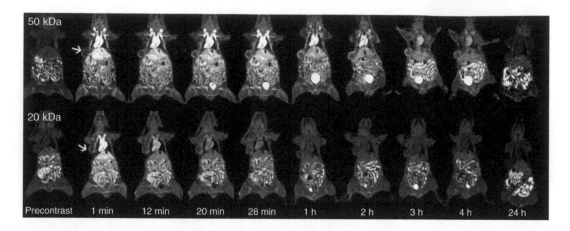

FIGURE 11.6 Coronal MR slice images of mice bearing MB-231 breast cancer tumors using PGA–1,6-hexa-nediamine–(Gd-DOTA) conjugates with different molecular weights. The images were taken at different time points using a 3D FLASH pulse sequence (1.74-ms TE, 4.3-ms TR, 25° RF tip angle, 120-mm FOV, and 1.6-mm coronal slice thickness).

with increase in molecular weights. The low-molecular-weight conjugate (20 kDa) was cleared from the blood circulation within 30 min after-injection. The dynamic changes in signal intensity in the liver correlated well to that in the heart. The enhancement in the liver returned to the background level 24 h after the injection, indicating low liver uptake of the PGA conjugates.

The contrast enhancement pattern in the tumor tissue was similar to that of HPMA conjugates in Kaposi's sarcoma. Strong signal was observed at tumor periphery and little in the inner tumor tissue. The dynamic MR images showed that the intensity and duration of tumor enhancement strongly depended on the size of the conjugates, indicating size-dependent tumor accumulation. Rapid clearance of the low molecular weight conjugate (20 kDa) from the blood circulation resulted in low and short accumulation in the tumor tissue. The high molecular weight conjugate (50 kDa) had a long blood circulation time, resulting in an increased contrast enhancement for at least 4 h. The enhancement also gradually extended into inner tumor tissue, indicating diffusion of the conjugate further into the tumor. The results clearly showed that drug delivery with polymeric conjugate into tumor tissue was a slow process and H conjugates with high molecular weight and long blood circulation time was more effective for tumor delivery than the conjugate with low molecular weight. Contrast-enhanced MRI clearly revealed the size effect of the conjugates on their pharmacokinetics and in vivo drug delivery efficiency.

11.4.2 MR IMAGING OF PGA–MCE$_6$–(Gd-DOTA) CONJUGATE IN AN ANIMAL TUMOR MODEL

Most anti-cancer drugs are lipophilic and the conjugation of lipophilic drugs to a polymeric carrier may also modify the in vivo behavior of the polymers. Mesochlorin e$_6$ (Mce$_6$) is a lipophilic photosentizer for photodynamic therapy. [41,42] A paramagnetic poly(L-glutamic acid) Mce$_6$ conjugate was prepared to investigate the pharmacokinetics, biodistribution and drug delivery efficiency of the PGA–Mce$_6$ conjugate with contrast-enhanced MRI. [43] The structure of the conjugate is shown in Figure 11.7. A PGA–1,6-hexanediamine–(Gd-DOTA) conjugate was also prepared from the same poly(L-glutamic acid) as a control. Mce$_6$ content was 2.5 mol% in PGA–Mce$_6$–1,6-hexanediamine–(Gd-DOTA) and Gd content was 20 and 29 mol% for PGA–Mce$_6$–1,6-hexanediamine–(Gd-DOTA) and PGA–1,6-hexanediamine–(Gd-DOTA), respectively. The T_1 relaxivity was 8.46 and 8.33 mM^{-1} s^{-1} for PGA–Mce$_6$–1,6-hexanediamine–(Gd-DOTA) and PGA–1,6-hexanediamine–(Gd-DOTA), respectively. The small amount

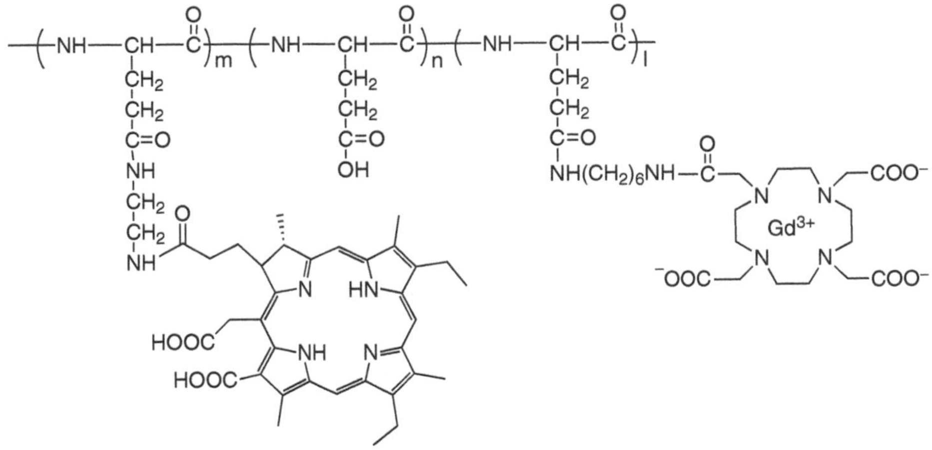

FIGURE 11.7 The structure of PGA–Mce$_6$–1,6-hexanediamine–(Gd-DOTA).

of Mce$_6$ in PGA–Mce$_6$–1,6-hexanediamine–(Gd-DOTA) significantly reduced the hydrodynamic volume and molecular weight distribution of the conjugate. The weight (M_w) and number (M_n) average molecular weights decreased from 101 and 49 kDa of PGA–1,6-hexadiamine–(Gd-DOTA) to 49 and 34 KDa for PGA–Mce$_6$–1,6-hexadiamine–(Gd-DOTA). The lipophilic Mce$_6$ may alter the conformation of the polymers, resulting in a smaller hydrodynamic volume for PGA–Mce$_6$–1,6-hexa-diamine–(Gd-DOTA).

A contrast-enhanced MRI study demonstrated that PGA–Mce$_6$–1,6-hexanediamine–(Gd-DOTA) had significantly different pharmacokinetics and biodistribution from PGA–1,6-hexanediamine–(Gd-DOTA) in female nu/nu athymic mice bearing human breast carcinoma MB-231 xenografts. Figure 11.8 shows coronal dynamic MR images of mice injected with PGA–Mce$_6$–1,6-hexanedia-mine–(Gd-DOTA) and PGA–1,6-hexanediamine–(Gd-DOTA) at a dose of 0.07 mmol-Gd/kg. The conjugates demonstrated different dynamic contrast enhancement patterns in the heart and liver. PGA–Mce$_6$–1,6-hexadiamine–(Gd-DOTA) had a strong contrast enhancement in the liver and relatively weak enhancement in the heart, whereas PGA–1,6-hexadiamine–(Gd-DOTA) exhibited

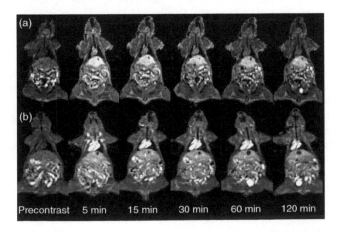

FIGURE 11.8 Coronal contrast-enhanced MR slice images for PGA–Mce$_6$–(Gd-DOTA) conjugate and PGA–(Gd-DOTA) conjugate through the heart (b) before and at 5, 15, 30, 60, and 120 min post-injection.

strong enhancement in the heart and relatively weak enhancement in the liver. The conjugation of Mce_6 significantly altered the pharmacokinetics and biodistribution of the poly(L-glutamic acid). The results indicated that PGA–Mce_6–1,6-hexadiamine–(Gd-DOTA) had higher liver accumulation than PGA–1,6-hexadiamine–(Gd-DOTA). Consequently, it cleared from the blood circulation more rapidly than PGA–1,6-hexadiamine–(Gd-DOTA), which could also be attributed to the larger hydrodynamic volume of PGA–1,6-hexadiamine–(Gd-DOTA).

11.5 SUMMARY

Contrast-enhanced MRI is effective for noninvasive visualization of the real-time pharmacokinetics and biodistribution of paramagnetically labeled polymer drug conjugates after intravenous administration. The technique provides three-dimensional anatomic images of soft tissues with high spatial resolution. The circulation and accumulation of the paramagnetic polymeric conjugates in organs or tissues result in bright signal for T_1-weighted images. Dynamic contrast-enhanced MRI clearly reveals circulation of the conjugates in the blood, interaction with the major organs, including the liver, heart, kidneys, etc., and accumulation and clearance in the target tissues. It also provides information on the dynamic changes of the distribution patterns of polymer conjugates in solid tumor tissues, which cannot be obtained with conventional pharmacokinetic methods. Such a detailed map of the delivery system deposition within the tissue and its correlation with the local pathologic features may help to optimize the structure of the polymeric drug conjugate to achieve high drug delivery efficiency. One limitation of contrast-enhanced MRI is the accurate quantification of the concentrations of the conjugates in the tissues and organs. Several technical factors control the accurate measurement of concentrations of contrast agents with conventional contrast-enhanced MRI. For quantitative study of in vivo drug delivery with paramagnetic conjugates, MR T_1 mapping can be used because T_1 relaxation rate ($1/T_1$) is linearly proportional to the concentration of MRI contrast agents. Data acquisition and algorithms to process the data for T_1 mapping are more complicated than for conventional contrast-enhanced MRI.

REFERENCES

1. Kopecek, J. et al., HPMA copolymer-anticancer drug conjugates: design, activity and mechanism of action, *Eur. J. Pharm. Biopharm.*, 50, 61, 2000.
2. Kopecek, J. et al., Water soluble polymers in tumor targeted delivery, *J. Control. Rel.*, 74, 147, 2001.
3. Hobbs, S. K. et al., Regulation of transport pathways in tumor vessels: role of tumor type and microenvironment, *Proc. Natl Acad. Sci. USA*, 95, 4607, 1998.
4. Maeda, H., Sawa, T., and Konno, T., Mechanism of tumor-targeted delivery of macromolecular drugs, including the EPR effect in solid tumor and clinical overview of the prototype polymeric drug SMANCS, *J. Control. Rel.*, 74, 47, 2001.
5. Minko, T., Kopeckova, P., and Kopecek, J., Chronic exposure to HPMA copolymer-bound adriamycin does not induce multidrug resistance in a human ovarian carcinoma cell line, *J. Controlled Rel.*, 59, 133, 1999.
6. Minko, T. et al., HPMA copolymer bound adriamycin overcomes MDR1 gene encoded resistance in a human ovarian carcinoma cell line, *J. Control. Rel.*, 54, 223, 1998.
7. Abe, S. and Otsuki, M., Styrene maleic acid neocarzinostatin treatment for hepatocellular carcinoma, *Curr. Med. Chem. Anti-Cancer Agents*, 2, 715, 2002.
8. Brewerton, L. J., Fung, E., and Snyder, F. F., Polyethylene glycol-conjugated adenosine phosphorylase: development of alternative enzyme therapy for adenosine deaminase deficiency, *Biochim. Biophys. Acta*, 1637, 171, 2003.
9. Pinheiro, J. P. et al., Drug monitoring of PEG-asparaginase treatment in childhood acute lymphoblastic leukemia and non-Hodgkin's lymphoma, *Leuk. Lymphoma*, 43, 1911, 2002.
10. Ettinger, L. J. et al., An open-label, multicenter study of polyethylene glycol-L-asparaginase for the treatment of acute lymphoblastic leukemia, *Cancer*, 75, 1176, 1995.

11. Motzer, R. J. et al., Phase II trial of branched peginterferon-alpha 2a (40 kDa) for patients with advanced renal cell carcinoma, *Ann. Oncol.*, 13, 1799, 2002.
12. Rowinsky, E. K. et al., A phase I and pharmacokinetic study of pegylated camptothecin as a 1-hour infusion every 3 weeks in patients with advanced solid malignancies, *J. Clin. Oncol.*, 21, 148, 2003.
13. Auzenne, E. et al., Superior therapeutic profile of poly-L-glutamic acid-paclitaxel copolymer compared with Taxol in xenogeneic compartmental models of human ovarian carcinoma, *Clin. Cancer Res.*, 8, 573, 2002.
14. Gianasi, E. et al., HPMA copolymer platinates as novel antitumor agents: in vitro properties, pharmacokinetics and antitumor activity in vivo, *Eur. J. Cancer*, 35, 994, 1999.
15. Thomson, A. H. et al., Population pharmacokinetics in phase I drug development: a phase I study of PK1 in patients with solid tumors, *Br. J. Cancer*, 81, 99, 1999.
16. Seymour, L. W. et al., Hepatic drug targeting: phase I evaluation of polymer-bound doxorubicin, *J. Clin. Oncol.*, 20, 1668, 2002.
17. Liang, Z. P. and Lauterbur, P. C., *Principles of Magnetic Resonance Imaging*, IEEE Press, New York, 1999.
18. Vlaardingerbroek, M. T. and den Boer, J. A., *Magnetic Resonance Imaging*, 3rd ed., Springer, New York, 2003.
19. Lauffer, R. B., Paramagnetic metal complexes as water proton relaxation agents for NMR imaging: theory and design, *Chem. Rev.*, 87, 901, 1987.
20. Merbach, A. E. and Toth, E., *The Chemistry of Contrast Agents in Medical Magnetic Resonance Imaging*, John Wiley & Sons, Inc., England, 2001 chap. 1–2
21. Weinmann, H. J. et al., Characteristics of gadolinium-DTPA complex: a potential NMR contrast agent, *Am. J. Roentgenol.*, 142, 619, 1984.
22. Kaplan, G. D., Aisen, A. M., Aravapalli, S. R. et al., Preliminary clinical trial of gadodiamide injection: a new nonionic gadolinium contrast agent for MR imaging, *J. Magn. Reson. Imaging*, 1, 57, 1991.
23. Magerstadt, M. et al., Gd(DOTA): An alternative to Gd(DTPA) as a $T_{1,2}$ relaxation agent for NMR imaging or spectroscopy, *Magn. Reson. Med.*, 3, 808, 1986.
24. Tweedle, M. F., The proHance story: the making of a novel MRI contrast agent, *Eur. Radiol.*, 7(Suppl 5), 225, 1997.
25. Youk, J. H., Lee, J. M., and Kim, C. S., MRI for detection of hepatocellular carcinoma: comparison of mangafodipir trisodium and gadopentetate dimeglumine contrast agents, *Am. J. Roentgenol.*, 183, 1049, 2004.
26. Weissleder, R. et al., Ultrasmall superparamagnetic iron oxide: an intravenous contrast agent for assessing lymph nodes with MR imaging, *Radiology*, 175, 494, 1990.
27. Wang, S. C. et al., Evaluation of Gd-DTPA-labeled Dextran as an intravascular MR contrast agent: imaging characteristics in normal rat tissues, *Radiology*, 175, 483, 1990.
28. Bryant, L. H. et al., Synthesis and relaxometry of high-generation ($G = 5$, 7, 9 and 10) PAMAM dendrimer-DOTA-gadolinium chelates, *J. Magn. Reson. Imaging*, 9, 348, 1999.
29. Lu, Z. R. et al., Design of novel bioconjugates for targeted drug delivery, *J. Control. Rel.*, 78, 165, 2002.
30. Duncan, R., The dawning era of polymer therapeutics, *Nat. Rev. Drug Discov.*, 2, 347, 2003.
31. Pimm, M. V. et al., Gamma scintigraphy of the biodistribution of ^{123}I labeled N-(2-hydroxypropyl)-methacrylamide copolymer-doxorubicin conjugates in mice with transplanted melanoma and mammary carcinoma, *J. Drug Target*, 3, 375, 1996.
32. Kissel, M. et al., Synthetic macromolecular drug carriers: biodistribution of poly [(N-2-hydroxypropyl)methacrylamide] copolymers and their accumulation in solid rat tumors, *J. Pharm. Sci. Tech.*, 55, 191, 2001.
33. Mitra, A. et al., Technetium-99m-Labeled N-(2-hydroxypropyl) methacrylamide copolymers: synthesis, characterization, and in vivo biodistribution, *Pharm. Res.*, 21, 1153, 2004.
34. Wang, Y.L., and Lu, Z.R. Unpublished data 2005.
35. Wang, D. et al., The arhrotropism of macromolecules in adjuvant-induced arthritis rat model: a preliminary study, *Pharm. Res.*, 21, 1741, 2004.
36. Li, C., Poly (L-glutamic acid)-anticancer drug conjugates, *Adv. Drug Deliv. Rev.*, 54, 695, 2002.

37. Hurwitz, E., Wilchek, M., and Pitha, J., Soluble macromolecules as carriers for daunorubicin, *J. Appl. Biochem.*, 2, 25, 1980.

38. Li, C. et al., Complete regression of well-established tumors using novel water-soluble poly (L-glutamic acid)-paclitaxel conjugates, *Cancer Res.*, 58, 2404, 1998.

39. Zou, Y. et al., Effectiveness of water soluble poly(L-glutamic acid)–camptothecin conjugate against resistant human lung cancer xenografted in nude mice, *Int. J. Oncol.*, 18, 331, 2001.

40. Lu, Z. R. et al., Poly(L-glutamic acid) Gd(III)-DOTA conjugate with a degradable spacer for magnetic resonance imaging, *Bioconjugate Chem.*, 14, 715, 2003.

41. Lu, Z. R., Kopeckova, P., and Kopecek, J., Polymerizable Fab' fragment for targeting of anticancer drugs, *Nature Biotechnol.*, 17, 1101, 1999.

42. Lu, Z. R. et al., Polymerizable Fab' antibody fragments targeted photodynamic cancer therapy in nude mice, *STP Pharm. Sci.*, 13, 69, 2003.

43. Vaidya, A. et al. Non-invasive in vivo imaging of a polymeric drug conjugate using contrast enhanced MRI. In *Transactions of 32nd Annual Meeting of the Controlled Release Society*, 2005.

Section 3

Polymeric Nanoparticles

12 Polymeric Nanoparticles for Tumor-Targeted Drug Delivery

Tania Betancourt, Amber Doiron,
and Lisa Brannon-Peppas

CONTENTS

12.1   Introduction 215
12.2   Targeting to Cancer..216
12.3   Passive Targeting and the EPR Effect 217
12.4   Targeting to Angiogenesis 218
    12.4.1   Targeting Using Vascular Endothelial Growth Factor Receptors.....................218
    12.4.2   Targeting Using Integrins 218
        12.4.2.1   Integrins as Targets for Imaging 220
12.5   Targeting Using Folate Receptors 220
    12.5.1   Antibodies and Folate Receptors 221
    12.5.2   Folate-Targeted Nanoparticles for Gene Delivery 222
12.6   Approaches for Cancer Targeting to Specific Cancer Types...........................223
    12.6.1   Prostate Cancer..224
12.7   Targeted Nanoparticles and Imaging of Cancer 225
12.8   Other Targets for Cancer 225
12.9   Avidin and Biotin Targeting 226
12.10 Conclusions 226
References..226

12.1 INTRODUCTION

Cancer is a disease that affects millions of people across the globe every year. The World Health Organization estimated that more than 10 million people developed a malignant tumor and more than 6.5 million people died from this disease during the year 2000.[1] In the United States, cancer is the second cause of deaths from disease after heart disease, accounting for more than half a million deaths every year. According to the American Cancer Society (ACS) cancer statistics, the overall cost for cancer for the United States in 2004 was $189.8 billion: $69.4 billion for direct medical costs, $16.9 billion for indirect morbidity costs, and $103.5 billion for indirect mortality costs.[2] Furthermore, while mortality rates of other major chronic diseases, such as heart and cerebrovascular disease, decreased significantly in the past half-century, cancer mortality rates have remained approximately constant.[2] This is a troubling fact because it suggests that recent detection and treatment options have not been able to improve mortality rates substantially.

Research in the past decade has focused on using unique characteristics of cancer cells and the vasculature surrounding those cells to deliver imaging agents, chemotherapeutic drugs, gene therapy, and other active agents directly and selectively to cancerous tissues. Many of these new formulations (described elsewhere in this volume) are liposomes, prodrugs, polymer conjugates, micelles, and dendritic systems. This chapter will concentrate on polymeric nanoparticles that have been studied as targeted systems for treatment and detection of cancer.

12.2 TARGETING TO CANCER

Nanoparticles may be targeted to the growing vasculature serving the growing cancer or to the cancer cells themselves. Targeted delivery utilizes unique phenotypic features of diseased tissues and cells in order to concentrate the drug at the location where it is needed. Targeted delivery can be divided into passive and active targeting. Passive targeting tries to minimize nonspecific interactions between the drug carrier and nontarget sites in the body by detailing the physiochemical properties of the aberrant tissue such as size, morphology, hydrophilicity, and surface charge.[3] When targeting tumor tissue, the enhanced permeability and retention effect (EPR) is an example of passive targeting approach; it allows passage of drug carriers ranging in size from 10 to 500 nm through the highly permeable blood vessels that supply growing tumors and leads to entrapment of large molecules as a result of deficient lymphatic drainage.[3–5] In fact, it has been reported that the intra-cellular openings in vascular endothelium of tumor blood vessels can be up to 2 μm in diameter and that the vessel leakiness in tumor vasculature can be up to an order of magnitude higher than that of normal blood vessels.[5] Active targeting utilizes biologically specific interactions including antigen–antibody and ligand–receptor binding and may seek drug uptake by receptor-mediated endocytosis through association of the drug or drug carrier with such antigen or ligand.[3] Receptor-mediated endocytosis commonly occurs through clathrin-coated vesicles and is carried out in mammalian cells continuously for the uptake of nutrients and for modulation of signal transduction through the up- or down-regulation of signaling receptors.[6] Targeted delivery avoids the need for high systemic drug levels for the drug to be effective and consequently offers a more economic alternative for treatment. Not only is it useful for therapeutic purposes; it is also beneficial in diagnosis. Recent research has pointed to its ability to concentrate imaging or contrast agents for the detection of malignancies and for monitoring the effects of therapeutic agents.[7,8] To date, most systems for targeted delivery have utilized drug conjugates, liposomes or micelles.[9–11] Targeting of particulate systems has focused more often on passive targeting based on size than on active targeting. But systems that combine both methods, starting with passive targeting through EPR and enhancing the targeting through specific interactions are beginning to show great promise.

While it is challenging to deliver a drug or imaging agent-containing nanoparticle directly and selectively to a cancerous cell or tissue, the additional challenges of then having that particle and/or its contents being transported into the targeted cell have often been overlooked. Couvreur presented some of these challenges at the 11th International Symposium on Recent Advances in Drug Delivery Systems and also summarized that presentation in a recent publication.[12] The tumor resistance can be due to the deliverance of nanoparticles to the tumor as well as to resistance to the active agent being delivered. In this article, many different pathways are described and the enhancement of drug delivery due to the presence of nanoparticles, especially polycyanoacrylate nanoparticles, is evaluated and summarized. The enhancement of drug permeability into cells due to interactions with biodegradation byproducts as well as the effect of nanoparticle surface charge is discussed.

In this chapter, we will first review recent work on targeting to the two most promising targets for cancer: Angiogenesis and folate receptors. We will then describe other potential targets for cancer imaging and therapy with nanoparticles, including antibodies strategies using biotin.

12.3 PASSIVE TARGETING AND THE EPR EFFECT

Although active targeting may be achieved through targeting to folate receptors, angiogenesis, or targets more specific to the cancer being treated, some enhancement of treatment to cancers on account of the EPR effect cannot be ignored and should be utilized until more specific targeting systems can be developed. Preparation of nanoparticle systems that can avoid uptake by the reticuloendothelial system (RES) is essential, and such particles are often referred to as "stealth" nanoparticles.

The most effective polymer used as a coating on nanoparticles to avoid detection by the RES is poly(ethylene glycol).[13] The latter can be achieved by adsorption of the PEG-containing polymer onto the surface of nanoparticles, direct conjugation of the PEG to the nanoparticles, or inclusion of PEG in the polymeric backbone that makes up the nanoparticles. The PEG itself may then also be modified to give targeting capabilities and to avoid uptake by the RES. The PEG works to mask the surface of the nanoparticles by reducing the plasma and protein adsorption to the particles, reducing the complement activation and hence the recognition of the PEG as a foreign substance in the blood stream. The longer circulation time afforded PEGylated nanoparticles allows them time to be targeted, whether passively or actively, to cancerous tissues.

Extended circulation time and enhanced tumor targeting were seen for poly(ethylene oxide)-modified poly(epsilon-caprolactone) nanoparticles in mice with tumors of MDA-MB-231, a human breast carcinoma.[14] These particles were loaded with [^{3}H]-tamoxifen. The amount of labeled tamoxifen at the tumor site, 6 h after injection, for those particles with PEO modification was at least twice that of particles without modification and four times that of labeled tamoxifen injection. The amount of labeled tamoxifen found in the blood stream at 6 h after injection was also at least twice that of the injection or unmodified nanoparticle formulations.

The stability and circulation of PLGA–mPEG nanoparticles containing cisplatin was investigated. It was found that while the mPEG content affected the drug release rate, the drug loading level had no effect on the drug release rate for in vitro studies.[15] The release was more that 60% completed within the first 12 h in all cases. Data for blood levels was only presented for 3 h, so it is hard to draw any valid conclusions from this information.

Nanoparticles of PLA and PEG–PPG–PEG were prepared containing irinotecan, a prodrug of an analogue of camptothecin.[16] Although little characterization beyond the average particles size (231 nm) was presented here and no in vitro studies were described, the in vivo studies are quite interesting. In this study, there was a modest increase in survival time in mice with M5076 tumors (early liver metastatic stage) after a single injection and more pronounced increases in survival time after either two or three repeat injections. It is noteworthy that the greatest survival times (20% survival at 45 days at the end of the study) were seen with two injections at days three and five after the implantation of the tumor.

Polycyanoacrylate nanoparticles have long been studied by Couvreur and collaborators as biodegradable nanoparticles for a variety of applications and therapies which are not limited to treatment of cancer.[17] A recent work describes the effectiveness of these nanoparticles as delivery systems for brain tumor targeting. Here they studied uncoated and PEG-coated nanoparticles and found that both types of particles showed accumulation in a well-established 9L gliosarcoma in rat studies. The PEG-coated particles showed the highest accumulation, with a tumor-to-brain ratio of 11.

Poly(butyl cyanoacrylate) nanoparticles containing doxorubicin were prepared with no surface modifications; the in vivo distribution of ^{99m}Tc labeled nanoparticles was evaluated in mice inoculated with Dalton's lymphoma tumor cells.[18] The nanoparticles were administered by subcutaneous injection and the concentration in a number of organs was followed for 48 hours and compared with that for ^{99m}Tc labeled doxorubicin alone. When compared with the amount of doxorubicin alone, the amount of radioactivity in the tumor was higher at all times tested, with a 13-fold increase seen at 48 h.

While the majority of work on targeted nanoparticles has been carried out with polylactides, polyglycolides and polycyanoacrylates, those are not the only materials that can be used in nanoparticles. Some recent work with radiolabeled gelatin nanoparticles modified with poly(ethylene glycol) showed that adding the PEG to the surface of these nanoparticles increased the circulation time in mice with double the amount of PEG-modified nanoparticles in the blood stream 3 h after injection compared to the amount of control nanoparticles.[19] In addition, there was a four-fold increase in the amount of PEG-modified gelatin nanoparticles found in the tumor 4 h after injection and later when compared to the number of nonmodified gelatin nanoparticles.

12.4 TARGETING TO ANGIOGENESIS

A recently explored and potentially promising target of cancer drug and gene nanoparticle therapy is tumor angiogenesis. It is now well established that tumor growth is dependent on new capillary infiltration from surrounding, preexisting vasculature.[20,21] This is an important control point in cancer as much research has proven that tumors cannot effectively grow past a small size or metastasize without blood supply.[22–24] Except for the cases of menstruation, wound healing, and tissue regeneration, capillaries do not increase in size or number under normal physiological conditions. Tumor growth is an exception to this physiological rule.

Tumors are typically unable to affect angiogenesis when they are small and surrounded by healthy tissue. However, at the point in growth where nutrients, oxygen, and growth factors can no longer reach the cancer cells, blood flow is required to allow further growth of the tumor. After what is occasionally a substantial time period, the tumor may abruptly induce angiogenesis into the tissue.[23] Because understanding this step in the progression of cancer is thought to be of great importance, much research has been and continues to be focused on pinpointing the progression of cancer and on targeting the event for therapy.

Much research has focused on targeted therapy of either chemotherapeutic agents to sites of tumor angiogenesis or of angiogenesis-inhibiting drugs to tumors with the goal of directly combatting the proliferation of newly forming capillaries in the tumor. Angiogenesis is a complex, multi-component process that involves many cell types, cytokines, growth factors and receptors, proteases, and adhesion molecules.[25] As a result, there are many potential targets for anti-angiogenic or chemotherapeutic therapy. Some recent advances in targeting approaches for nanoparticle drug delivery are discussed below.

12.4.1 TARGETING USING VASCULAR ENDOTHELIAL GROWTH FACTOR RECEPTORS

Vascular endothelial growth factor (VEGF) is particularly important in the process of angiogenesis and has been shown to greatly affect tumor growth in animal models.[26–28] The VEGF receptors have been used as a means to target the vascular bed in many instances. VEGF receptor-2 (VEGFR-2) has recently been used to target nanoparticles to tumor vascular beds by Li et al in mice with K1735-M2 tumors.[29] A succinyl-dextran-polymerized nanoparticle conjugated to rat anti-mouse VEGFR-2 antibody and radioisotope ^{90}Y caused a significant tumor growth delay compared to conventional radiolabeled antibody and other controls. Additionally, anti-CD31 staining showed a decreased vessel density and damage to tumor vessels after treatment with the anti-VEGFR-2-^{90}Y nanoparticles.

12.4.2 TARGETING USING INTEGRINS

The integrins represent another important cell surface molecule group for angiogenesis targeting because some integrins, such as $\alpha_v\beta_3$ and $\alpha_v\beta_5$, are upregulated on the endothelial cell surface of neovasculature.[30,31] The $\alpha_v\beta_3$ integrin is expressed on numerous tumor cell types; it is highly expressed on neovascular endothelial cells. Hood and collaborators used 40-nm diameter cationic-lipid-based nanoparticles coupled to an organic $\alpha_v\beta_3$ ligand that was shown to be specific for $\alpha_v\beta_3$ in cell studies. These nanoparticles, which contained a luciferase reporter plasmid, were injected into mice with the $\alpha_v\beta_3$-negative cell line M21-L. The nanoparticles targeted to the neovasculature within the tumor (but not to the tumor cells themselves) with no expression elsewhere in the mice as detected by luciferase expression. In order to test therapeutic efficacy, NPs were conjugated to a mutant Raf gene that blocks angiogenesis. Systemic injection in the M21-L tumor-expressing mice rapidly induced apoptosis of endothelial cells within the tumor. Tumor regression was seen within 10 days.[32]

In the same study with VEGFR-2 targeting discussed above, Li and collaborators also demonstrated targeting of the radioisotope ^{90}Y using nanoparticles targeted to the integrin $\alpha_v\beta_3$ with a small molecule integrin agonist.[29] In mice with K1735-M2 tumors, $\alpha_v\beta_3$-targeted ^{90}Y-nanoparticles significantly delayed tumor growth compared to untreated tumors. TUNEL staining of tumor sections showed widespread apoptosis in tumors treated with these targeted nanoparticles. The authors have postulated that this targeted nanoparticle radiotherapy has the potential to be used to treat a variety of solid tumors. They have also postulated that the use of nanoparticles increases efficacy due to the high payload delivered by the carriers.

Additionally, PEGylated polyethyleneimine (PEI) nanoplexes with a cyclic disulfide bond constrained Arg–Gly–Asp (RGD) peptide ligand at the distal end have been used to target integrin-expressing tumor neovasculature to deliver siRNA. Integrins are receptors for extracellular matrix components that contain a tripeptide RGD sequence. Therefore, RGD containing peptide sequences can target to cell surface integrins that are upregulated on neovasculature. The siRNA used inhibited angiogenesis by inhibiting VEGFR-2 expression. Intravenous injection of nanoparticles into nude mice with N2A tumors showed tumor uptake of the siRNA, inhibition of protein synthesis in the tumor, and inhibition of angiogenesis and tumor growth.[33] This study demonstrates tumor selective delivery through both the targeting ligand and gene pathway by using siRNA.

In another $\alpha_v\beta_3$ targeting approach using an RGD peptide, Kopelman has created multifunctional nanoparticles of 30–60 nm for the treatment of gliomas.[34] The nanoparticles are able to kill cancer cells by bombarding them with externally released reactive oxygen species created by photodynamic agents activated by laser light. The particles also contain superparamagnetic iron oxide and enhance imaging by magnetic resonance. The photodynamic sensitizer and MRI contrast agents are entrapped within a polyacrylamide core, the surface of which is coated with PEG chains and targeting RGD moieties. The particles containing photodynamic agents were shown to produce sufficient singlet oxygen to kill cells in vitro. Additionally, these nanoplatforms were injected into an in vivo rat intracerebral 9L tumor model, and diffusion MRI was performed at various times to evaluate the tumor diffusion, tumor growth, and tumor load. The gliomas treated with the nanoparticles and irradiated with laser light caused regional necrosis and significant shrinkage of tumor mass, a shrinkage that lasted for 12 days. The authors postulate that the light activated release of reactive oxygen from photosensitizer-containing nanoparticles is a viable approach for brain tumor treatment. Also, the incorporation of MRI contrast agents allows for monitoring of treatment and tumor progression in vivo.

Carbohydrate based nanoparticles have also been used to target drugs to neovasculature via an $\alpha_v\beta_3$/cyclic RGD peptide interaction. Inulin multi-methacrylate formed the core of the nanoparticles and was attached to the RGD targeting moiety with a PEG linker. Doxorubicin was loaded in the nanoparticles via covalent and noncovalent linkages. The pharmacokinetics and biodistribution of the doxorubicin loaded nanoparticles were studied over five days in female Balb/cJ mice with metastatic mammary tumor clone-66. A bi-exponential fix with a terminal half-life of 5.99 h was observed; decreasing drug concentrations with time in the heart, lungs, kidney, and plasma was also observed. Conversely, increasing drug accumulation was observed in the liver, spleen, and in the tumor where there was also the presence of high levels of doxorubicin metabolite. The presence of the high metabolite levels in the tumorsuggests nothing more than tumor-specific nanoparticle degradation and release of drug.[35]

12.4.2.1 Integrins as Targets for Imaging

In an effort to image angiogenesis, Mulder and collaborators have created MR-detectable and fluorescent liposomes that are targeted to $\alpha_v\beta_3$ integrin on neovasculature.[36] The MR contrast agent Gd-DTPA-bis(stearylamide) was incorporated into PEGylated liposomes covalently coupled to RGD peptide. It was observed that the liposomes effectively brought about an increase in signal on T_1-weighted images. Injection of these lipidic nanoparticles in nude mice with subcutaneously implanted LS174 human colon adenocarcinoma led to the ability to specifically image the vascular endothelial cells in the tumor. Ex vivo fluorescence confirmed that RGD liposomes specifically interacted with tumor endothelium associated with neovasculature.

In a study reported by Lanza, paramagnetic molecular imaging of angiogenesis was accomplished in vivo with $\alpha_v\beta_3$-targeted lipid encapsulated perfluorocarbon nanoparticles of about 250 nm.[37] A Vx-2 carcinoma tumor in New Zealand rabbits was imaged with a 1.5-T MRI system with either nontargeted or $\alpha_v\beta_3$-targeted nanoparticles. An eight-fold greater contrast enhancement was achieved with the targeted nanoparticles. In the second part of the study, paclitaxel loaded nanoparticles were targeted to tissue factor (TF) proteins on vascular smooth muscle cells with a specific TF antibody. The TF-targeted paclitaxel particles inhibited cell proliferation while delivery of nanoparticles to targeted cells was confirmed with fluorine spectroscopy.

12.5 TARGETING USING FOLATE RECEPTORS

During the past few decades, there has been great interest in the utilization of folate receptors for the targeted delivery of therapeutic and imaging agents. A number of delivery systems have been utilized for this purpose, including drug conjugates,[38–40] liposomes,[41] micelles,[42] viral vectors,[43] and nanoparticles.

Folate (vitamin B-9) is essential for the synthesis of nucleotides and amino acids. Two main groups of molecules are responsible for transport of folate molecules in vivo. Most cells in the body express a folate anion transporter with micro-molar affinities for folates that participates in the transport of coenzyme 5-methyltetrahydrofolate, the physiologic circulating reduced form of folate. Folate receptors (FR), by contrast, are members of the glycosylphosphatidylinositol (GPI)-linked membrane glycoprotein family and have high affinity for folic acid, an oxidized form of folate, and 5-methyltetrahydrofolate, with binding affinities being in the nanomolar range ($K_D < 1 \times 10^{-9}$ M) for the α isoform of FR.[44–46] It has been observed with few exceptions that only cells involved in pathologic conditions, including cancer cells, express the high affinity folate receptors. These receptors are able to transport folic acid, folate-bound molecules, and even particles through receptor-mediated endocytosis.[47,48]

FR are known to be overexpressed in various epithelial cancer cells, such as those of ovarian, mammary gland, colon, lung, prostate, and brain epithelial cancers, and in leukemic cells.[49–61] Folate receptor overexpression has been correlated to poor prognosis. In addition, metastasized cancer cells have been found to overexpress the folate receptor to a larger degree than localized tumor cells.[62] This finding is of great importance. The only nonpathological tissues where FR is expressed are choroid plexus, placenta, lungs, thyroid, and kidney.[46,63] FR expression is limited to the apical (luminal) side of polarized epithelial cells, except for the cells of the proximal tubules in the kidney. As a consequence, FR is practically inaccessible to blood-borne folate-linked systems.[44,45] These characteristics make folate receptors very advantageous for targeted delivery of nanoparticles with high payloads of therapeutic agents, imaging agents, and even genes for the treatment, detection, and monitoring of cancer. What is more, the macromolecular size of nanoparticles will prevent gromerular filtration and the consequent exposure of kidney tissue to folate-targeted nanoparticles.

12.5.1 Antibodies and Folate Receptors

The most common targeting moieties utilized for FR targeting include monoclonal antibodies to FR and folic acid.[46] To date, monoclonal antibodies to FR have not been utilized for targeted delivery of nanoparticles although successful experiences have been reported for radiopharmaceuticals.[64] The benefits of targeting with folic acid are many: it is small in size, has high stability, lacks immunogenicity, and costs very little.[62] Folic acid conjugation at its γ-carboxyl group is necessary for maintaining binding affinity to FR. The structure of folic acid is shown in Figure 12.1. Cellular uptake of drug carriers bound to folic acid is believed to be mediated by receptor mediated endocytosis although this process is not currently completely understood. A number of hypotheses have been proposed for this process, including clathrin and caveolar pathways.[63] It has been shown that folate-bound molecules are able to escape endosomes after receptor-mediated endocytosis because the process of endosomal acidification results in a conformational change in the receptor that facilitates folate ligand release. Consequently, folate-bound molecules provides a great opportunity for the delivery of pH-sensitive biopharmaceuticals.[45,62]

In most drug delivery systems investigated thus far, folic acid is incorporated to the drug delivery system through conjugation to a poly(ethylene glycol) (PEG) spacer utilizing well-known dicyclohexylcarbodiimide/N-hydroxysuccinimide (DCC/NHS) mediated chemistry. Such design aims to minimize steric hindrance for optimal folate recognition. This conjugation technique, however, can activate both the γ- and the α-carboxylic acids of folic acid. It should be said, though, that the γ group is more reactive and is responsible for most of the linkages.

Numerous attempts at nanoparticle targeting to the folate receptor have been reported. Many groups have formulated nanospheres of amphiphilic block copolymers including PEG. Park and collaborators reported in vitro results of the preparation and evaluation of methoxy poly(ethylene glycol) (PEG)-poly(ε-caprolactone) (PCL) block copolymer nanospheres loaded with paclitaxel in which folic acid was conjugated to a modified amino-terminated PCL with a carbodiimide-mediated reaction.[65] Dialysis was used to create these nanospheres that ranged in diameter from 50 to 120 nm, depending on the ratio of the block copolymers. Paclitaxel loading efficiencies of up to 55% were reported with this system, thus significantly increasing the effective solubility of this agent in aqueous systems like the body. Because the folate moiety was conjugated to the hydrophobic end of the block copolymer, it is expected that upon nanosphere formation it will be localized in the inner core of the particles. However, XPS characterization demonstrated the presence of nitrogen-containing molecules at the surface which could only be attributed to the folate linker. Although the targeting effectiveness of these particles was not determined, cytotoxicity studies revealed that encapsulation of paclitaxel into the nanospheres reduced its

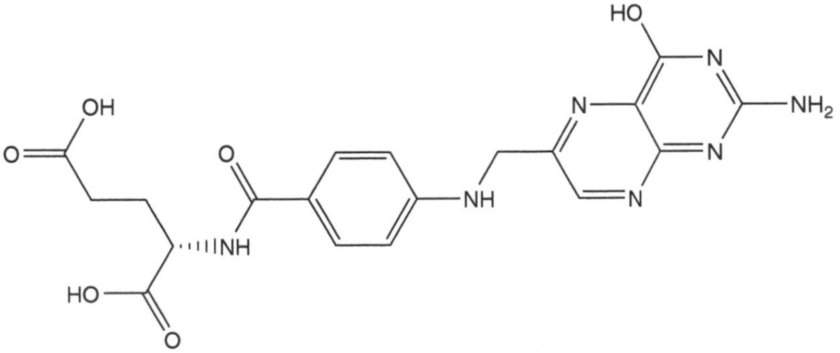

FIGURE 12.1 Chemical structure of folic acid. Conjugation to folic acid for folate targeting is commonly done at the γ-carboxyl group.

cytotoxicity. Consequently, encapsulation offered a safe alternative to direct administration of this chemotherapeutic agent. Further studies should evaluate whether conjugation of folic acid to hydrophobic end of the block copolymer results in sufficient surface availability of this targeting agent.

Another approach to folate targeting of nanoparticles has been the modification of surface properties of polymeric nanoparticles with polymer conjugates including the targeting agent. For example, Kim reported on the modification of anionic poly(lactic-co-glycolic acid) (PLGA) nanoparticles prepared by emulsification and solvent evaporation with a cationic poly(L-lysine)-poly(ethylene glycol)-folate (PLL–PEG–FOL) copolymer.[66] In this design, the polycation PLL block attaches to the PLGA nanoparticle through ionic interactions. Surface coating is achieved by simple incubation in an aqueous solution containing the PLL–PEG–FOL copolymer. The PEG-folate end of the copolymer is oriented toward the outer aqueous phase for better interaction between folate and the targeted cell membrane receptor. XPS characterization demonstrated the presence of nitrogen from folic acid on the surface of the coated nanoparticles. In vitro cellular studies with FITC-labeled nanoparticles revealed an increase in uptake of coated nanoparticles with increased conjugate-to-nanoparticle ratio in KB cells. Since a decrease in uptake was seen upon addition of free folic acid in the medium, the transport of nanoparticles into the cells was attributed to endocytosis mediated by the folate receptor.

Dendritic polymer systems of polyamidoamine (PAMAM) with folic acid as the targeting agents and drug and imaging agent (methotrexate or tritium and fluorescein or 6-carboxytetramethylrhodamine) were prepared and tested in mouse models.[67] It was found that targeted systems, as opposed to nontargeted systems, slowed the rate of tumor growth and even showed a complete cure in one mouse. This study was conducted with twice-weekly tail vein injections. The biodistribution studies showed that, for a single targeted injection of nanoparticles, a very high amount of the nanoparticles accumulated within the tumor by day 1; this level remained high through at least day 4.

Nanometric particles prepared from drug-polymer conjugates have also been reported. In one notable study, Yoo and collaborators reported on the formulation of doxorubicin-PEG-folate nanoaggregates encapsulating additional doxorubicin in their core.[11] These aggregates are formed spontaneously when an organic phase containing the copolymer and solubilized doxorubicin is dispersed into an aqueous phase containing triethylamine. The basic aqueous environment results in deprotonation of doxorubicin, causing it to form aggregates with the hydrophobic copolymer. The average aggregate diameter was approximately 200 nm. In vitro cellular studies showed increased uptake of the nano aggregates in cells expressing the folate receptor when folic acid was absent from cell media and increased cytotoxicity (anti-tumor efficacy) of the aggregates compared to the free drug in cells expressing the FR. In vivo studies in mouse xenografts (KB cells) showed that the nanoaggregates had superior therapeutic efficacy than both doxorubicin aggregates without the folic acid ligand and free doxorubicin in solution.

Nanoparticles of temperature-responsive hydrogels conjugated to folic acid have also been studied with the purpose of delivering chemotherapeutic agents. Nayak reported on poly(N-isopropylacrylamide) (pNIPAM) nanoparticles that exhibit lower critical solution temperature (LCST) behavior.[68] Fluorescent agents were included in the core of these nanoparticles to facilitate with tracking; amine comonomers were incorporated into the other pNIPAM shell for conjugation to folic acid. These nanoparticles swell when their temperature falls under their LCST. Indeed, the nanoparticle size increased from ∼50 nm at 37°C to ∼135 nm at 25°C. In vitro cell uptake studies showed a 10-fold increase in the intake of nanoparticles conjugated to folic acid compared to those without the targeting agent in KB cells (FR+).

12.5.2 FOLATE-TARGETED NANOPARTICLES FOR GENE DELIVERY

The use of folate-targeted nanoparticles for gene delivery has also been studied. In gene delivery, tissue targeting is very important for the efficacy and safety of treatment. To date, the transfection efficiency offered by synthetic gene delivery systems is very low compared to that of viral vectors. This is also true for nanoparticle-based systems. Extensive research is being done in the area to improve transfection efficiencies by designing better delivery systems with improved targeting ability, DNA stabilization, and cellular interaction.

Mansouri reported on the formulation of nanoparticles through the complex coacervation of chitosan-folic acid conjugates with DNA.[69] Chitosan is a biocompatible polycation that joins to form a complex with DNA through electrostatic interactions and provides protection against nuclease degradation. Complex coacervation is an optimal preparation technique for the encapsulation of DNA because it avoids the use of organic solvents and high-energy ultrasonication. Results revealed that the integrity of plasmid DNA in the particles was maintained and that conjugation of chitosan to folic acid did not interfere with the electrostatic interaction between chitosan and DNA. As expected, the ratio of chitosan to DNA, and consequent charge ratio, had an effect on particle size and zeta potential. Nanoparticles smaller than 200 nm were obtained with a chitosan amino group to DNA phosphate group ratio of more than 2. The use of these nanoparticles consequently offers a promising alternative for nonviral gene therapy for the treatment of cancer and other diseases in which folate receptors are overexpressed.

Nanoparticles with a poly(L-lactic acid) (PLL) core and a polyethyleneimide (PEI) surface conjugated to folate were utilized for delivery of plasmid DNA.[70] Folic acid was conjugated to the N-terminal amino group of PEI. PEI is a polycation that has been used in the past for DNA condensation and delivery because it protects DNA from degradation through an endosomal escape mechanism. Here nanoparticles were prepared through the self-assembly of the amphiphilic folate-PEI–PLL copolymer with DNA in an aqueous medium. Nanoparticles of approximately 100–150 nm in diameter and spherical shape were produced. In vitro luciferase transfection studies revealed that this system actually resulted in lower luciferase expression than PEI–DNA complexes.

In a separate report, folate-polyethyleneglycol distearoylphophatidylethanolamine conjugate (f-PEG–DSPE), 3([N-(N',N'-dimethylaminoethane)-carbamoyl] cholesterol, and Tween 80 were used to complex with DNA into cationic nanoparticles of 100–200 nm in diameter with a modified ethanol injection method.[71] The formulation was carried out by dissolving the lipids in ethanol and then removing the solvent through evaporation in the presence of water. The folate moiety, which is conjugated to the PEG end of one of the lipid conjugates, naturally localizes at the surface of the nanoparticles because of PEG migration toward the water phase. Tween 80, a nonionic surfactant, and PEG were incorporated with the purpose of improving the in vivo stability of the cationic nanoparticles through steric hindrance. The size of the nanoparticles with higher PEG content was maintained in the presence of serum. This suggests that these nanoparticles are better able to maintain their structural integrity in the presence of anionic competitors present in blood. Folate targeting enhanced association and transfection efficacy of nanoparticles complexed with a luciferase-encoding plasmid on FR(+) KB cells. The association and efficacy were reduced when folic acid was present in the medium, thus revealing the involvement of the folate receptor in the transport of the plasmid DNA into the cells.

A possible limitation of folate targeting is the noted variability of FR expression levels not only between patients, but also within a single tumor.[44] In addition, it has been reported that expression of FR in cancerous cell lines is not representative of those one sees in vivo.[44] Consequently, screening protocols for FR expression will need to be utilized clinically in order to determine if folate-targeted therapies are appropriate.

12.6 APPROACHES FOR CANCER TARGETING TO SPECIFIC CANCER TYPES

In a recent review, Kim and Nie describe passive and active targeting methods and then go into detail on a number of active targeting techniques.[72] The active target combinations they mention are lectin–carbohydrate, ligand–receptor and antibody–antigen. One limitation of using these targeting strategies is that the lectin–carbohydrate targeting systems are usually targeted to whole organs, making them inappropriate for targeting a cancerous part of a particular organ or tissue. New antibody systems show a great deal of promise. Unfortunately, they also have potentially harmful side effects such as advanced gastric adenocarcinoma. The latter arises when one attempts to target breast cancer on account of the fact that antigen-positive normal cells to the antibody BR96 in gastric mucosa, small intestine, and pancreas. Some of the aspects of angiogenic targeting mentioned here have already been covered elsewhere in this chapter. Specific targeting systems that are in use as cancer therapeutics are shown in Table 12.1.[8,72]

12.6.1 PROSTATE CANCER

Recently, aptamers have been used to target nanoparticulate systems to prostate-specific membrane antigen, a known prostate cancer tumor marker. A model drug, rhodamine-labeled dextran, was encapsulated in PEGylated poly(lactic acid) nanoparticles, which were subsequently surface modified with a prostate specific RNA aptamer (A10). Binding of the aptamer nanoparticles to LNCaP cells expressing prostate specific membrane antigen in vitro was significantly enhanced when compared to a control of nontargeted particles. Additionally, very low binding was seen on nonprostate specific membrane antigen expressing cells (PC3). The nanoparticles were shown to both target and be taken up by the prostate cancer epithelial cells. This evidence points to the conclusion that this novel aptamer-based, targeted nanoparticle delivery approach can be effective.[73]

Gao and collaborators report the use of quantum dots (QD) for in vivo targeting of prostate cancer and imaging of tumors.[7] The core-shell CdSe–ZnS quantum dots contain tri-n-octylphosphine oxide (TOPO) that binds to a covering of high molecular weight ABC triblock copolymer of polybutylacrylate, polyethylacrylate, polymethacrylic acid, and an 8-carbon alkyl side chain. The complex is functionalized with PEG molecules and monoclonal antibodies to prostate-specific membrane antigen. Specific binding was shown for prostate cancer lines whereas low binding was seen to normal cells. The QD-antibody formulations were studied in vivo in a mouse model human prostate cancer. The nanoparticles were shown to target to the tumor both by passive and active antibody targeting. Sensitive and multicolor fluorescence imaging of cancer cells in vivo was

TABLE 12.1
Targeting Systems Utilizing Antibodies Currently in Use to Treat Cancer

Mechanism	Antibody Target	Trade Name
Agonist activity	CD40, CD137	Various
Antagonist activity	CTLA4	MDX-010
Angiogenesis inhibition	VEGF	Avastin™
Antibody-dependent cell-mediated cytotoxicity	CD20	Rituxan®, HuMax-CD20, Zevalin®
Inhibition of binding of extracellular growth signals	HER-2/neu	Herceptin
Receptor blockage	EGF receptor	HuMax-EGFr
Toxin-mediated killing	CD33	Mytotarg®
Disruption signaling	HER-2/neu	Pertuzumab (2C4)
Complement-dependent cytotoxicity	CD20	Rituxan®, HuMax-CD20
Blockage ligand binding	EGF receptor	Erbutix™
Antibody-dependent lysis of leukemic cells following cell binding	CD52	Campath®
Inhibits phosphorilation of tyrosine kinases	EGF receptor	Iressa

accomplished in this study. Nonspecific uptake was seen in the liver and spleen with little or no uptake in other organs.

12.7 TARGETED NANOPARTICLES AND IMAGING OF CANCER

Hofmann and collaborators have evaluated the ability of super paramagnetic iron oxide nanoparticles (SPION) to interact with human melanoma cells in such a way that these particles could be selectively targeted to tumor cells ands then imaged using MRI.[74] They varied the coating placed on these particles and found that when comparing particles coated with poly(vinyl alcohol) (PVA), a vinyl alcohol/vinyl amine copolymer (amine-SPION), PVA with randomly distributed carboxylic groups or PVA with randomly distributed thiol groups, human cells in culture would interact strongly only with the amine-SPION particles. Furthermore, these particles showed the lowest cytotoxicity.

This human study involved intravenous administration of Combidex (Advanced Magnetics, Inc., ferumoxtran-10, a molecular imaging agent of iron oxide nanoparticles and a dense packing of dextran derivatives) to 18 men ages 21–46 with diagnosed testicular cancer.[75] The Combidex was imaged using MRI. From this study, it seems evident that those lymph nodes with a higher signal were classified as being malignant. However, based on the information from Advanced Magnetics, these particles should accumulate selectively in noncancerous lymph node tissue. The particles are still experimental and rightly so.

Although treatment of cancer with targeted nanoparticles is an important goal, more accurate imaging of cancer is needed to allow for the optimal treatment for each patient. Towards that end, a considerable amount of research is underway with various imaging techniques to establish more accurate determination of the presence and extent of cancer growth and metastases. Because magnetic resonance imaging (MRI) is a widely used imaging technique, much work is currently being done to develop targeted imaging agents for MRI. Some of these involve paramagnetic and superparamagnetic iron oxides due to their ability to affect water relaxation times T_1 and T_2. Gasco and collaborators have prepared solid lipid nanoparticles containing Endorem, superparamagnetic iron oxide nanoparticles (Guebert and Advanced Magnetics), using either a multiple emulsion technique or an oil in water emulsion technique.[76] Although the loading rates achieved were less than 1 wt% iron, it was possible to detect and image in vivo in rats. Incorporation of the Endorem in the SLN allowed passage across the blood brain barrier, passage which was not possible with Endorem alone.

Often development of nanoparticle systems is a two-pronged approach involving both drug and imaging agent. If a targeting system is successful, one should be able to enhance the imaging of a cancer and then kill it with the same system. One such study involved the preparation of glycol–chitosan nanoaggregates to which either fluorescein isothiocyanate (FITC) or doxorubicin (Dox) was conjugated.[12] Based on a single tail-vein injection of FITC-conjugates, levels remained high for eight days and gradually increased in the tumors of rats with II45 mesothelioma cells, in the kidney, and to a lesser extent in the spleen. Meanwhile, levels decreased in other organs. The liver showed some accumulation at day 3 but was significantly lower at day 8 relative to days 1 and 3. The performance of these systems, as evidenced by a decrease in tumor volume, was excellent with a consistent decrease in tumor volume after day 13 when a tail-vein injection of Dox-nanoaggregates is given at days 13 and 19.

12.8 OTHER TARGETS FOR CANCER

A new approach to targeting is the use of lectins, which are plant proteins that specifically recognize cell surface carbohydrates. The latter function as selective cancer-cell-targeting agents. PLGA nanoparticles of mean diameter 331 nm incorporating isopropyl myristate were used to deliver paclitaxel to malignant A549 and H1299 and normal CCL-186 pulmonary cells in vitro by means of wheat germ agglutinin lectin as the targeting molecule. The in vitro cytotoxicity against A549 and

H1299 cells was significantly increased with wheat germ agglutinin-conjugated PLGA nanoparticles containing IPM and paclitaxel compared to controls.[77] In a subsequent study, Mo and Lim evaluated in vivo efficacy of the same nanoparticles in a SCID mouse model injected with an A549 tumor nodule.[78] One injection of wheat germ agglutinin-conjugated PLGA nanoparticles containing IPM and a paclitaxel dose of 10 mg/kg inhibited tumor growth without appreciable weight loss. Tumor doubling was increased to 25 days compared to 11 days for conventionally formulated paclitaxel.

12.9 AVIDIN AND BIOTIN TARGETING

Although it is by far the most widely utilized polymer for surface modification of nanoparticles, PEG is not the only compound that can be included at the surface of nanoparticles. Nor is it necessary to achieve active targeting. Saltzman has recently reported a method for incorporating avidin-fatty acid conjugates into the surface of PLGA nanoparticles.[79] This method resulted in avidin at the surface of the nanoparticles that remained active for weeks. The ability of these nanoparticles to target to biotin was verified by targeting of the nanoparticles to biotinylated agarose beads. Not only could this system be used for targeted delivery; it can also be utilized to selectively modify surfaces for tissue engineering.

In another such example, Hunziker has prepared biotin-functionalized (poly(2-methyl-oxazoline)-*b*-poly(dimethylsiloxane)-*b*-poly(2-methyl-oxazoline) triblock copolymers.[80] The biotinylated targeting agents were added using streptavidin as a coupling agent. Uptake of these "nanocontainers" was seen in the presence of the target receptor, the macrophage scavenger receptor SRA1, but not in the absence of this target.

12.10 CONCLUSIONS

The amount of research in targeted, polymeric nanoparticles for cancer imaging and therapy has increased dramatically in the past 5–10 years. Seeing actual products using targeted therapies has no doubt fueled that work. In the next decade, we will certainly see products, whether with polymeric nanoparticles or some other type of delivery system, using folate receptors and carrying imaging agents. All of these technologies, driven by the fields of fundamental immunology, biochemistry, polymer chemistry, and biomedical engineering, are bringing us closer to the time when cancer may be treated on an individual basis. One patient's diagnosis and treatment will be unique to her condition and will be the most effective treatment possible for her. Until other scientists determine how to stop cancer from occurring, those mentioned in this chapter and many more besides them are doing their best to eliminate cancer.

REFERENCES

1. Ferlay, J. et al., *GLOBOCAN 2002: Cancer Incidence, Mortality and Prevalence Worldwide*, IARC CancerBase No. 5, Version 2.0., IARC Press, Lyon, 2004.
2. Society, A. C., *Cancer Facts and Figures 2005*, American Cancer Society, Atlanta, 2005.
3. Yokoyama, M. and Okano, T., Targetable drug carriers: Present status and a future perspective, *Advanced Drug Delivery Reviews*, 21, 77–80, 1996.
4. Torchilin, V. P., Strategies and means for drug targeting: An overview, In *Biomedical Aspects of Drug Targeting*, Muzykantov, V. and Torchilin, V., Eds., Kluwer Academic, Norwell, pp. 3–26, 2002.
5. McDonald, D. M. and Baluk, P., Significance of blood vessel leakiness in cancer, *Cancer Research*, 62, 5381–5385, 2002.
6. Conner, S. D. and Schmid, S. L., Regulated portals of entry into the cell, *Nature*, 422, 37–44, 2003.
7. Gao, X. et al., In vivo cancer targeting and imaging with semiconductor quantum dots, *Nature Biotechnology*, 22(8), 969–976, 2004.

8. Brannon-Peppas, L. and Blanchette, J. O., Nanoparticle and targeted systems for cancer therapy, *Advanced Drug Delivery Reviews*, 56, 1649–1659, 2004.

9. Luo, Y. et al., Targeted delivery of doxorubicin by HPMA copolymer-hyaluronan bioconjugates, *Pharmaceutical Research*, 19, 396–402, 2002.

10. Mastrobattista, E. et al., Targeted liposomes for delivery of protein-based drugs into the cytoplasm of tumor cells, *Journal of Liposome Research*, 12, 57–65, 2002.

11. Yoo, H. S. and Park, T. G., Folate receptor targeted biodegradable polymeric doxorubicin micelles, *Journal of Controlled Release*, 96, 273–283, 2004.

12. Vauthier, C. et al., Drug delivery to resistant tumors: The potential of poly(alkly cyanoacrylate) nanoparticles, *Journal of Controlled Release*, 93, 151–160, 2003.

13. Peracchia, M. T., Stealth nanoparticles for intravenous administration, *S.T.P. Pharma*, 13(3), 155–161, 2003.

14. Shenoy, D. B. and Amiji, M. M., Poly(ethylene oxide)-modified poly(e-caprolactone) nanoparticles for targeted delivery of tamoxifen in breast cancer, *International Jounal of Pharmaceutics*, 293, 261–270, 2005.

15. Avgoutsakis, K. et al., PLGA–mPEG nanoparticles of cisplatin: In vitro nanoparticles degradation, in vitro drug release and in vivo drug residence in blood properties, *Journal of Controlled Release*, 79, 125–135, 2002.

16. Machida, Y. et al., Efficacy of nanoparticles containing irinotecan prepared using poly(DL-lactic acid) and poly(ethylene glycol)-poly(propylene glycol)-poly(ethylene glycol) against M5076 tomur in the early liver metastatic stage, *S.T.P. Pharma*, 13(4), 225–230, 2003.

17. Brigger, I. et al., Poly(ethylene glycol)-coated hexadecylcyanoacrylate nanospheres display a combined effect for brain tumor targeting, *Journal of Pharmacology and Experimental Therapeutics*, 303, 928–936, 2002.

18. Reddy, L. H., Sharma, R. K., and Murthy, R. S. R., Enhanced tumour uptake of doxorubicin loaded poly(butyl cyanoacrylate) nanoparticles in mice bearing Dalton's lymphoma tumour, *Journal of Drug Targeting*, 12(7), 443–451, 2004.

19. Kaul, G. and Amiji, M., Biodistribution and targeting potential of poly(ethylene glycol)-modified gelatin naoparticlea in subcutaneous murine tumor model, *Journal of Drug Targeting*, 12(9–10), 585–591, 2004.

20. Folkman, J., Angiogenic-dependent diseases, *Seminars in Oncology*, 28(6), 536–542, 2001.

21. Lyden, D. et al., Impaired recruitment of bone-marrow-derived endothelial and hematopoietic presursor cells blocks tumor angiogenesis and growth, *Nature Medicine*, 7(11), 1194–1201, 2001.

22. Folkman, J., Vascularization of tumors, *Scientific American*, 234(5), 59–76, 1976.

23. Folkman, J., Fundamental concepts of the angiogenic process, *Current Molecular Medicine*, 3, 643–651, 2003.

24. Hanahan, D. and Folkman, J., Patterns and emerging mechanisms of the angiogenic switch during tumorgenesis, *Cell*, 86, 353–364, 1996.

25. Carmeliet, P. and Jain, R. K., Angiogenesis in cancer and other diseases, *Nature*, 407, 249–257, 2000.

26. Leenders, W. P. J., Targetting VEGF in anti-angiogenic and anti-tumour therapy: Where are we now?, *International Journal of Experimental Pathology*, 79, 339–346, 1998.

27. Claffey, K. P. et al., Expression of vascular permeasbility factor/vascular endothelial growth factor by melanoma cells increases tumor growth, angiogenesis, and experimental metastasis, *Cancer Research*, 56, 172–181, 1996.

28. Potgens, A. J. G. et al., Analysis of the tumor vasculature and metastatic behavior of xenografts of human melanoma cells lines transfected with vascular permeability factor, *American Journal of Pathology*, 148(4), 1203–1217, 1996.

29. Li, L. et al., A novel antiangiogenesis therapy using an integrin antagonist or anti-FLK-1 antibody coated with ^{90}Y-labeled nanoparticles, *International Journal of Radiation Oncology Biology Physics*, 58(4), 1215–1227, 2004.

30. Brooks, P. C., Clark, R. A. F., and Cheresh, D. A., Requirement of vascular integrin $\alpha_v\beta_3$ for angiogenesis, *Science*, 264, 569–571, 1994.

31. Varner, J. A. and Cheresh, D. A., Integrins and cancer, *Current Opinion in Cell Biology*, 8, 724–730, 1996.

32. Hood, J. D. et al., Tumor regression by targeted gene delivery to the neovasculature, *Science*, 296, 2404–2407, 2002.

33. Schiffelers, R. M. et al., Cancer siRNA therapy by tumor selective delivery with ligand-targeted sterically stabilized nanoparticle, *Nucleic Acids Research*, 32(19), e149, 2004.

34. Kopelman, R. et al., Multifunctional nanoparticle platforms for in vivo MRI enhancement and photo-dynamic therapy of a rat brain cancer, *Journal of Magnetism and Magnetic Materials*, 293, 404–410, 2005.

35. Bibby, B. C. et al., Pharmacokinetics and biodistribution of RGD-targeted doxorubicin-loaded nano-particles in tumor-bearing mice, *International Jounal of Pharmaceutics*, 293, 281–290, 2005.

36. Mulder, W. J. M. et al., MR molecular imaging and fluorescence microscopy for identification of activated tumor endothelium using a bimodal lipidic nanoparticle, *The FASEB Journal*, 19, 2008–2010, 2005.

37. Lanza, G. M., Magnetic resonance molecular imaging and targeted drug delivery with site-specific nanoparticles. In *BioMEMS and Biomedical Nanotech World*, Columbus, OH, 2002.

38. Leamon, C. P. et al., Synthesis and biological evaluation of EC72: A new folate-targeted chemo-therapeutic, *Bioconjugate Chemistry*, 16, 803–811, 2005.

39. Lee, J. W. et al., Synthesis and evaluation of taxol-folic acid conjugates as targeted antineoplastics, *Bioorganic Medicinal Chemistry*, 10, 2397–2414, 2002.

40. Paranjpe, P. V. et al., Tumor-targeted bioconjugate based delivery of camptothecin: Design, sythesis and in vitro evaluation, *Journal of Controlled Release*, 100, 275–292, 2004.

41. Gabizon, A. A., Liposome circulation time and tumor targeting: Implications for cancer chemotherapy, *Advanced Drug Delivery Reviews*, 16, 285–294, 1995.

42. Lee, D. H., Kim, D. I., and Bae, Y. H., Doxorubicin loaded pH-sensitive micelle targeting acidic extracellular pH of human ovarian A2780 tumor in mice, *Journal of Drug Targeting.*, 13(7), 391–397, 2005.

43. Douglas, J. T. et al., Targeted gene delivery by tropism-modified adenoviral vectors, *Nature Bio-technology*, 14, 1574–1578, 1996.

44. Elnakat, H. and Ratnam, M., Distribution, functionality and gene regulatio of folate receptor isoforms: Implications in targeted therapy, *Advanced Drug Delivery Reviews*, 56, 1067–1084, 2004.

45. Leamon, C. P. and Reddy, J. A., Folate-targeted chemotherapy, *Advanced Drug Delivery Reviews*, 56, 1127–1141, 2004.

46. Sudimack, J. and Lee, R. J., Targeted drug delivery via the folate receptor, *Advanced Drug Delivery Reviews*, 41, 147–162, 2000.

47. Antony, A., The biological chemistry of folate receptors, *Blood*, 79, 2807–2820, 1992.

48. Kamen, B. and Capdevila, A., Receptor-mediated folate accumulation is regulated by the cellular folate content, *Proceedings of the National Academy of Sciences of the United State of America*, 88, 5983–5987, 1986.

49. Mattes, M. et al., Patterns of antigen distribution in human carcinomas, *Cancer Research*, 50, 880s–884s, 1990.

50. Coney, L. et al., Distribution of the folate receptor GP38 in normal and malignant cell lines and tissues, *Cancer Research*, 52, 3396–3401, 1991.

51. Weitman, S. et al., Distribution of the folate receptor GP38 in normal and malignant cell lines and tissues, *Cancer Research*, 52, 3396–3401, 1992.

52. Ross, J., Chaadhuri, P., and Ratnam, M., Differential regulation of folate receptor isoforms in normal and malignant cell lines. Physiologic and clinical implications, *Cancer*, 73, 2432–2443, 1994.

53. Weitman, S. et al., Cellular localization of thr folate receptor: Potential role in drig toxicity and folate homeostasis, *Cancer Research*, 52, 6708–6711, 1992.

54. Weitman, S., Frazier, K., and Kamen, B., The folate receptor in central nervous system malignancies of childhood, *Journal of Neurooncology*, 21, 107–112, 1994.

55. Garin-Chesa, P. et al., Trophoblast and ovarian cancer antigen LK26. Sensitivity and specificity in immunopathology and molecular identification as a folate-binding protein, *American Journal of Pathology*, 142, 557–567, 1993.

56. Toffoli, G. et al., Overexpression of folate binding protein in ovaian cancers, *International Journal of Cancer*, 74, 193–198, 1997.

57. Holm, J. et al., High-affinity folate binding in human choroid plexus. Characterization of radioligand binding, immunoreactivity, molecular heterogeneity and hydrophobic domain of the binding protein, *Biochemical Journal*, 280, 267–271, 1991.

58. Shen, F. et al., Identification of a novel folate receptor, a truncated receptor and receptor type B in hematopoietic cells: cDNA cloning, expression, immunoreactivity, and tissue specificity, *Biochemistry*, 33, 1209–1215, 1994.

59. Sadasivan, E. et al., Characterization of multiple forms of folate-binding protein from himan leukemia cells, *Biochimica et Biophysica Acta*, 882, 311–321, 1986.

60. Sadasivan, E. et al., Purification, properties, and immunological characterization of folate-binding proteins from human leukemia cells, *Biochimica et Biophysica Acta*, 925, 36–47, 1987.

61. Morikawa, J. et al., Identification of signature genes by microarray for acute myeloid leukemia without maturation and acute promyelocytic leukemia with t(15;17)(q22;q12)(PML/RARalpha), *International Journal of Oncology*, 23, 617–625, 2003.

62. Hilgenbrink, A. R. and Low, P. S., Folate receptor-mediated drug targeting: From therapeutics to diagnostics, *Journal of Pharmaceutical Sciences*, 94(10), 2135–2146, 2005.

63. Sabharanjak, S. and Mayor, S., Folate receptor endocytosis and trafficking, *Advanced Drug Delivery Reviews*, 56, 1099–1109, 2004.

64. Coliva, A. et al., 90Y Labeling of monoclonal antibody MOv18 and preclinical validation for radio-immunotherapy of human ovarian carcinomas, *Cancer Immunology, Immunotherapy*, 54, 1200–1213, 2005.

65. Park, E. K., Lee, S. B., and Lee, Y. M., Preparation and characterization of methoxy poly(ethylene glycol)/poly(ε-caprolactone) amphiphilic block copolymeric nanospheres for tumor-specific folate-mediated targeting of anticancer drugs, *Biomaterials*, 26, 1053–1061, 2005.

66. Kim, S. H. et al., Target-specific cellular uptake of PLGA nanoparticles coated with poly(L-lysine)-poly(ethylene glycol)-folate conjugate, *Langmuir*, 21, 8852–8857, 2005.

67. Kukowska-Latallo, J. F. et al., Nanoparticle targeting of anticancer drug improves therapeutic response in animal model of human epithelial cancer, *Cancer Research*, 65(12), 5317–5324, 2005.

68. Nayak, S. et al., Folate-mediated cell targeting and cytotoxicity using thermoresponsive microgels, *Journal of the American Chemistry Society*, 126, 10258–10259, 2004.

69. Mansouri, S. et al., Characterization of folate–chitosan–DNA nanoparticles for gene therapy, *Biomaterials*, 27(9), 2060–2065, 2006.

70. Wang, C.-H. and Hsiue, G.-H., Polymer-DNA hybrid nanoparticles based on folate-polyethylene-imine-block-poly(L-lactide), *Bioconjugate Chemistry*, 16, 391–396, 2005.

71. Hattori, Y. and Maitani, Y., Enhanced in vitro DNA transfection efficiency by novel folate-linked nanoparticles in human prostate cancer and oral cancer, *Journal of Controlled Release*, 97, 173–183, 2004.

72. Kim, G. J. and Nie, S., Targeted cancer nanotherapy, *Nanotoday*, August, 28–33, 2005.

73. Farokhzad, O. C. et al., Nanoparticles-aptamer bioconjugates: A new approach for targeting prostate cancer cells, *Cancer Research*, 64, 7668–7672, 2004.

74. Petri-Fink, A. et al., Development of functionalized superparamagnetic iron oxide nanoparticles for interaction with human cancel cells, *Biomaterials*, 26, 2685–2694, 2005.

75. Harisinghani, M. G. et al., A pilot study of lymphotrophic nanoparticle-enhanced magnetic resonance imaging technique in early stage testicular cancer: A new method for noninvasive lymph node evaluation, *Urology*, 66, 1066–1071, 2005.

76. Peira, E. et al., In vitro and in vivo study of solid lipid nanoparticles loaded with superparamagnetic iron oxide, *Journal of Drug Targeting*, 11(1), 19–24, 2003.

77. Mo, Y. and Lim, L.-Y., Preparation and in vitro anticancer activity of wheat germ agglutinin (WGA)-conjugated PLGA nanoparticles loaded with paclitaxel and isopropyl myristate, *Journal of Controlled Release*, 107, 30–42, 2005.

78. Mo, Y. and Lim, L., Paclitaxel-loaded PLGA nanoparticles: Potentiation of anticancer activity by surface conjugation with wheat germ agglutinin, *Journal of Controlled Release*, 108, 244–262, 2005.

79. Fahmy, T. M. et al., Surface modification of biodegradable polyesters with fatty acid conjugates for improved drug targeting, *Biomaterials*, 26, 5727–5736, 2005.

80. Broz, P. et al., Cell targeting by a generic receptor-targeted polymer nanocontainer platform, *Journal of Controlled Release*, 102, 475–488, 2005.

13 Long-Circulating Polymeric Nanoparticles for Drug and Gene Delivery to Tumors

Sushma Kommareddy, Dinesh B. Shenoy, and Mansoor M. Amiji

CONTENTS

13.1 Introduction 231
13.2 Polymeric Nanoparticles 232
13.3 Nanoparticles for Drug Delivery 233
13.4 Nanoparticles for Gene Delivery 233
13.5 Long-Circulating Nanoparticles 234
13.6 Illustrative Examples of Polymeric Nanoparticles for Drug Delivery 235
    13.6.1 Polyethylene Oxide-Modified Poly(β-Amino Ester) Nanoparticles 235
    13.6.2 Poly(Ethylene Oxide)-Modified Poly(ε-Caprolactone) Nanoparticles 236
13.7 Illustrative Examples of Polymeric Nanoparticles for Gene Delivery 237
    13.7.1 Polyethylene Glycol-Modified Gelatin Nanoparticles....................................... 237
    13.7.2 Polyethylene Glycol-Modified Thiolated-Gelatin Nanoparticles 238
13.8 Concluding Remarks 238
Acknowledgments 239
References 239

13.1 INTRODUCTION

During the past few decades, there have been extensive efforts in the treatment and cure of cancer that is still the second leading cause of death next to the cardiovascular diseases. With advances in research techniques, there has been a surge of therapeutic agents in the form of proteins and nucleic acids as a source of new chemical entities for the treatment of cancer.[1–7] However, effective delivery of these novel agents to the target tissue has always been a problem.

The therapeutic agents used in cancer treatment are generally administered in the systemic circulation. The drug carrier, therefore, must overcome physiological barriers to reach the tumor cell in sufficient concentrations and to reside for the necessary duration to exert the pharmacological effect. These barriers include transport of the drugs within the blood vasculature and transport from the vasculature into the surrounding tumor tissues and through the interstitial spaces within the tumor. Solid tumors are characterized by vasculature that is heterogeneous in size and distribution, having a central avascular/necrotic region and vascularized peripheral region with discontinuous endothelium in the microvessels. Depending on the anatomic region of the tumor, the pore size of the endothelial junctions is found to vary from 100 to 780 nm with a mean of approximately 400 nm.[8–11] Tumor vasculature is also characterized by a lack of lymphatic drainage.[12–14] In addition to the complexity of tumor physiology, the development of multi-drug resistance (MDR) in tumor cells exacerbates the problem of achieving selective toxicity in tumor tissues.

Typically, a tumor consists of neoplastic cells, stromal cells, and the extracellular matrix and is associated with blood vessels, nerves, and immune cells that form an integral part of the stroma.[15,16] The tumor also contains leukocytes, fibroblasts, and other extracellular elements. The majority of the infiltrated leukocytes are macrophages, more popularly referred to as tumor-associated macrophages (TAM). They have increased phagocytic activity and play an active role in tumor progression. The presence of these activated macrophages in large numbers might lead to enhanced clearance of a drug-carrying vector from the tumor microenvironment.[15–18]

13.2 POLYMERIC NANOPARTICLES

Sub-micronic particles prepared from polymers are becoming increasingly popular for the delivery of drugs or genes to tumor tissues. When compared to all other colloidal delivery systems, nanoparticles have better stability in plasma and higher encapsulation efficiency, and they are amenable to large-scale preparation. These nanoparticles can be used to increase the solubility of hydrophobic drugs, lower the toxicity of drugs with a high therapeutic index, and enhance the stability of the payload, in the case of DNA, by protecting it from degradation by extracellular enzymes. Furthermore, these particles permit controlled release of the payload at the target site at relatively low doses.[19] In addition, there is a wide range of polymeric materials whose physicochemical properties can be tailored to achieve polymeric nanoparticles of the required nature.

Passive targeting of these particles at the tumor sites is generally achieved by the enhanced vascular permeability and lack of lymphatic drainage, together termed the *enhanced permeability and retention effect* (EPR) (Figure 13.1).[12–14] The accumulation of polymeric nanoparticles carrying drugs is dependent on the physical chemistry of the polymer, including molecular weight, surface charge, nature of the polymer, etc. The EPR is one of the main reasons for the success of polymeric nanoparticles in targeting tumors. Site-specific or active targeting of these particles can be achieved by conjugating with targeting moieties or ligands that are specific to the tumor cells. These targeting moieties present on the polymer backbone are used to exploit the differences between tumor cells and normal cells through receptor-mediated endocytosis. Active targeting, in particular, can be used to overcome the obstacle of tumor metastasis and to increase the specificity and efficacy of the polymeric carrier system. Transferrin, folate, epidermal growth factor, and argenine–glycine–aspartic acid (RGD) tripeptide are some of the moieties that are used for targeting tumor cells.[7,20–26]

The major drawback to the use of nanoparticles in cancer therapy is their initial burst release of the drug upon administration into systemic circulation that is followed by slow controlled release of the encapsulated drug. The polymers used in the preparation of nanoparticles generally lack the ability to encapsulate both hydrophilic and hydrophobic drugs in the same polymer system (i.e., the hydrophilic drugs have to be encapsulated in a hydrophilic polymer). Some of the synthetic polymers have cytotoxicity issues associated with them. However, this difficulty could be overcome by using biosynthetic polymers that are biodegradable and modified to have physicochemical properties similar to those of the synthetic polymers. The use of polymeric nanoparticles for systemic delivery must be limited to non-cationic or surface-modified polymers to prevent adsorption of plasma proteins onto the surface of the nanoparticles. Furthermore, some of the naturally occurring polymers have immunogenic reactions when injected into blood; such reactions could be prevented through the use of non-immunogenic biopolymers now available.

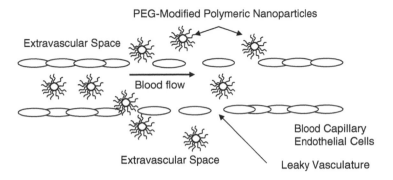

FIGURE 13.1 Schematic of enhanced permeability and retention effect. Passive targeting to the tumor cells is achieved by the extravasation of the polymeric nanoparticles through the leaky vasculature of the blood capillaries in the tumor. The leaky vasculature, along with the lack of lymphatic drainage, is called the EPR effect and is used to passively target nanocarriers to the tumor.

13.3 NANOPARTICLES FOR DRUG DELIVERY

The majority of the chemotherapeutic agents used in cancer therapy are low molecular weight drugs with a high volume of distribution, leading to the presence of toxic compounds throughout the body. These drugs must be given in high doses in order to reach therapeutic levels in tumor tissues. In the process, healthy tissues are exposed to cytotoxic drugs, resulting in side effects such as bone marrow suppression, alopecia, anorexia, etc. The low molecular weight anti-cancer agents are also rapidly cleared from the body. Continuous treatment of tumor cells with these drugs would exacerbate the problem of MDR that is one of the main reasons for the failure of chemotherapy.[19]

Numerous efforts in improving cancer chemotherapy have focused on increasing the therapeutic index and reducing the non-selective cytotoxicity of the drugs in use. Macromolecular carriers such as polymer conjugates, liposomes, microspheres, and nanoparticles are used for this purpose. Of these, the nanoparticles have been promising, mainly because of their advantages over other novel drug delivery vehicles such as their safety, enhanced stability, possibility of tailoring and surface modification, and industrial scale-up.[27] Nanoparticles prepared from polymer–drug conjugates have altered pharmacokinetic distribution and increased pharmacological activity at tolerable doses. This is due mainly to the controlled release of the drug and reduced renal clearance of the low molecular weight drugs.

The concept of targeting drugs at the site of action was first described as "magic bullets" by Paul Ehrlich.[28] Later, Ringsdorf presented a model of the polymer–drug conjugate that could be used to improve chemotherapy drugs for cancer.[29] The conjugation or encapsulation of these low molecular weight drugs within polymeric nanoparticles has distinct advantages: Increased plasma half-life of low molecular weight drugs; increased solubility of hydrophobic drugs; and controlled release of the drug at lower doses. The targeting of these nanoparticles to the tumor tissues is achieved either passively by the EPR effect or actively by surface conjugating a tumor-specific ligand.[30]

13.4 NANOPARTICLES FOR GENE DELIVERY

Due to the complexity and heterogeneity in cancer, gene therapy can provide a unique approach for treatment. Research on gene therapy focuses on the successful in vivo transfer of genes to the target tissue for the sustained expression of the genes of dysfunction. Various vectors, both viral and non-viral in origin, are used for this purpose.[31–34] The limitations associated with viral vectors, namely issues of integration with the host genome, self-replication, recombination potential, and immunogenicity, have encouraged researchers to seek alternatives such as nanoparticles, liposomes, and other cationic complexes.[35–37] Despite the low transfection efficiency associated with these non-viral vectors, they have the advantages of safety and high encapsulation efficiency. The DNA complexes of cationic liposomes and polymers have high transfection efficiency in vitro; however, owing to their toxicity and rapid clearance, these formulations have limited efficiency in vivo. The cationic liposomes and polymers, being positively charged, form aggregates with the negatively charged serum proteins and opsonins, resulting in enhanced phagocytosis and clearance from the blood circulation.[38]

In addition to cationic polymers, several other natural and synthetic polymers have been used in preparing DNA-encapsulated nanoparticles. Polymeric nanoparticles are increasingly popular, owing to such advantages as their small size, ease of production, and administration. Several polymers have been investigated as vectors for gene delivery applications. There has been a fair amount of success in reducing immunogenicity and cytotoxicity with the concomitant enhancement of the efficiency of transfection with these polymers. Besides protecting the DNA from degradation, the nanoparticles enhance the targeting of tumor tissues.[38]

13.5 LONG-CIRCULATING NANOPARTICLES

Long-circulating nanoparticles can be created through surface modification of conventional nanoparticles with water-soluble polymers such as polyethylene glycol (PEG) or polyethylene oxide (PEO) (Figure 13.2). The hydrophilic nature of these surface modifiers minimizes the interactions between the nanoparticles and plasma proteins (opsonins), resulting in reduced uptake by the reticulo-endothelial system (RES). The major outcome of modification with PEG or other hydrophilic flexible polymers is a significant increase in circulation time, the advantages of which include maintenance of optimal therapeutic concentration of the drug in the blood after a single administration of the drug carrier, increased probability of extravasation and retention of the colloidal carrier in areas of discontinuous endothelium, and enhancement in targetability of the system by use of a target-specific ligand.[39]

The protective action (Stealth® property) of PEG is mainly due to the formation of a dense, hydrophilic cloud of long polyethylene chains on the surface of the colloidal particle that reduces the hydrophobic interactions with the RES. The tethered or chemically anchored PEG chains can

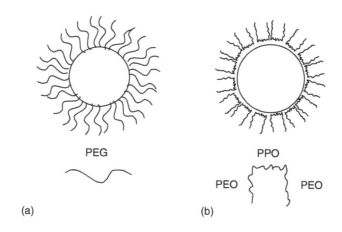

FIGURE 13.2 Schematic of long-circulating polymeric nanoparticles. The polymeric nanoparticles are made long-circulating by surface modification. The nanoparticles prepared from a hydrophilic polymer are modified using poly(ethylene glycol) (a), and the nanoparticles prepared from hydrophobic polymers are modified using Pluronic®, a triblock co-polymer of poly(ethylene oxide) and poly(propylene oxide) (b).

undergo spatial conformations, thereby preventing the opsonization of particles by the RES of the liver and spleen and improving the circulation time of molecules and particles in the blood. The greater the flexibility of the polymer, the greater the total number of possible conformations and transitions from one conformation to another.[40–43] Water molecules form a structured shell through hydrogen bonding to the ether oxygens of PEG. The tightly bound water around PEG chains forms a hydrated film around the particle and prevents protein interactions.[44] Furthermore, PEGylation may also increase the hydrodynamic size of the particles, decreasing their clearance through the kidneys, renal filtration being dependent on molecular mass and volume. This would ultimately result in an increase in the circulation half-life of the particles.[45,46]

The size, molecular weight, and shape of the PEG fraction and the linkage used to connect it to the entity of interest determine the consequences of PEGylation in relation to protein adsorption and pharmacokinetics such as volume of distribution, circulation time, and renal clearance. When formulated into colloidal particles, the PEG density on the colloidal surface can be changed by using PEG of appropriate molecular weight (PEG chain length) and molar ratio (the grafting efficiency). Longer PEG chains offer greater steric influence around the colloidal entity, similar to increased grafting density with shorter PEG chains. Longer PEG chains may also collapse onto the nanoparticle surface, providing a hydrophilic shield.[40]

Besides PEG, other hydrophilic polymers, including polyvinyl alcohol, polyacryl amide, polyvinyl pyrrolidone, poly-[N-(2-hydroxypropyl)methacrylamide], polysorbate-80, and block co-polymers such as poloxomer (Pluronic®) and poloxamine (Tetronic®), are also being used to modify the physicochemical properties of the colloidal carriers.[32,47,48]

The polymeric nanoparticles modified with PEG or PEO are mainly used to passively target tumors through the EPR effect. The surface-modified long-circulating polymeric nanoparticles are used to deliver both genes and drugs to the tumor tissues. Some of the polymeric nanoparticulate systems that were developed by this group will be discussed in the sections that follow.

13.6 ILLUSTRATIVE EXAMPLES OF POLYMERIC NANOPARTICLES FOR DRUG DELIVERY

13.6.1 POLYETHYLENE OXIDE-MODIFIED POLY(β-AMINO ESTER) NANOPARTICLES

A representative biodegradable, hydrophobic poly(β-amino ester) (PBAE) with pHsensitive solubility properties was synthesized by conjugate addition of 4,4′-trimethylenedipiperidine with 1,4-butanediol diacrylate, developed in Professor Robert Langer's lab at Massachusetts Institute of Technology. The paclitaxel-loaded nanoparticles (150–200 nm) prepared from PBAE were modified with Pluronic® F-108 (poloxamer 407), a triblock copolymer of polyethylene oxide/polypropylene oxide/polyethylene oxide (PEO/PPO/PEO). The PPO segment of the triblock polymer attaches to the hydrophobic surface of the nanoparticles, and the hydrophilic PEO segment contributes to the stealth properties of the polymeric nanoparticles. The pH-sensitive nature of the particles prepared from PBAE has already been shown by in vitro release studies carried out in the presence of buffers of pH ranging from 5.0 to 7.4 and was found to rapidly degrade in a medium of pH less than 6.5. Therefore, these nanoparticles were expected to readily release their contents within the acidic tumor microenvironment and in the endosomes and lysosomes of the cells upon internalization. This was confirmed by the in vitro cellular uptake of the PEO–PBAE nanoparticles encapsulated with tritiated [³H]-paclitaxel by human breast adenocarcinoma cells (MDA-MB231).[49,50]

The biodistribution of these PEO-modified PBAE nanoparticles was carried out by encapsulating a lipophilic form of the radionuclide indium-111 (¹¹¹indium oxine). Following tail vein injection in nude mice bearing a human ovarian xenograft, the radiolabeled PEO-modified PBAE nanoparticles were found to accumulate in the highly perfused organs such as the liver, spleen, and lungs with greater entrapment in the microvasculature of the lungs during the initial time points. The increasing concentrations in the kidney also indicate that the nanoparticles, once internalized, were disintegrated and eliminated through the kidney. The plasma half-life of the unmodified nanoparticles was reported to be one to ten minutes. By virtue of surface modification with PEO, the PBAE nanoparticles were shown to have improved circulation times, resulting in a mean residence time in the systemic circulation of 21 h. The paclitaxel-encapsulated PEO–PBAE nanocarriers were found to deliver the drug efficiently to solid tumors, resulting in a 5.2-fold and 23-fold higher concentration of the drug at one hour and five hours post-administration relative to the solution form of the drug.[40] From the tumor accumulation of the paclitaxel-loaded (^{3}H-labeled) nanoparticles, it is evident that the pH-sensitive PEO–PBAE nanoparticle formulations can deliver significantly higher concentrations of the drug into the tumor than the solution form.

13.6.2 POLY(ETHYLENE OXIDE)-MODIFIED POLY(ε-CAPROLACTONE) NANOPARTICLES

Poly(ε-caprolactone) (PCL) is another biodegradable polymer that has been used to encapsulate hydrophobic drugs. Using this polymer, nanoparticles were prepared by solvent displacement in an acetone-water system in the presence of Pluronic®. The solvent displacement technique used for the preparation of PCL nanoparticles facilitates instant adsorption of PPO–PEO groups when the organic solution of the polymer is introduced into aqueous solution containing the stabilizer.[51] In addition, it also favors the encapsulation of hydrophobic drugs such as tamoxifen that could be dissolved along with the polymer in the organic phase, resulting in a high entrapment efficiency of greater than 90% at loading levels of 20% of the weight of the drug. The intracellular uptake of these nanoparticles in MCF-7 estrogen receptor-positive breast cancer cells and MDA-MB231 human breast adenocarcinoma cells was monitored at different time points using tritiated ^{3}H]-tamoxifen. The results showed that the cell uptake followed saturable kinetics with most of the nanoparticles being internalized within the first 30 min of incubation.[52]

The in vivo disposition of these PEO-modified PCL nanoparticles was completed in mice bearing MDA-MB231 xenograft breast cancer tumors as it is a well-characterized and simpler model compared to MCF-7 that requires estrogen priming for growth. This study was used to compare the biodistribution profiles of the nanoparticles modified by the Pluronic® F-68 containing 30 residues of propylene oxide (PO) and 76 residues of ethylene oxide (EO) with those of Pluronic® F-108 that has 56 PO residues and 122 EO residues. The nanoparticulate formulations encapsulated with radiolabled tamoxifen were used. As expected, upon intravenous administration, the Pluronic®-modified nanoparticles had higher tumor concentrations in comparision to the tamoxifen solution and drug-loaded, unmodified nanoparticles. At early time points (one hour), the nanoparticles modified with Pluronic® F-108 had greater concentration in the tumor with no significant difference in the concentration of the Pluronic®-modified formulations at six hours post-injection. A similar trend has been observed in plasma concentration-time profiles with PEO–PCL formulations circulating for a longer time than the controls.[40]

In a parallel study, the PEO-modified PCL nanoparticles were radiolabeled by a similar procedure specific to PEO–PBAE nanoparticles. The nanoparticles, encapsulated with ^{3}H] tritium-labeled paclitaxel, were used to understand the change in concentration and localization of the drug in ovarian tumors (SKOV3). From the biodistribution studies, it was shown that the modification of PCL nanoparticles with PEO had extended the mean residence time to up to 25 h. Hydrophobic drugs such as paclitaxel were found to have high plasma concentrations as a result of their protein binding capacity; however, they were cleared from the blood within 24 h. The circulation time of such drugs has been enhanced by encapsulating them in PEO–PCL nanoparticles that, in turn, has resulted in higher concentrations of the drug in the tumors. The PEO–PCL nanoparticles have resulted in an 8.7-fold increase in drug concentration at five-hour time points when compared to the solution form of the drug.[40]

In another study, PEO-modified PCL nanoparticles were used for combination therapy of ceramide with paclitaxel in order to overcome MDR, particularly in cases of breast and ovarian cancers. Several MDR specimens of cancer were found to exhibit elevated levels of the enzyme glucosylceramide synthase (GCS) (also called ceramide glucosyl transferase) that is responsible for the inactivation of ceramide, a messenger in apoptotic signaling to its non-functional moiety glucosylceramide.[53–55] These findings suggest the importance of the role of ceramide in the mediation of a cytotoxic response and its function as an apoptotic messenger in the signaling pathway.

To potentially overcome MDR in ovarian cancer cell lines, C6-ceramide has been encapsulated along with paclitaxel into PEO-modified PCL nanoparticles. Upon treatment of the cells with paclitaxel, the MDR cell line SKOV3/TR exhibited $65.65 \pm 2.16\%$ viability at 1 µM dose; the sensitive cell line SKOV3 showed $16.37 \pm 0.41\%$ viability at 100 nM dose. Co-treatment of these cells along with 20 µM C6-ceramide in addition to paclitaxel (1 µM in the case of a resistant cell line and 100 nM in a sensitive cell line) resulted in a cell viability of $2.69 \pm 0.51\%$ with the resistant cells and $7.38 \pm 1.25\%$ with the sensitive cell lines, indicating a significant increase in cell death when compared to the paclitaxel treatment alone. Furthermore, the co-encapsulation of these drugs within PEO–PCL nanoparticles resulted in enhanced cell kill compared to the drugs alone. A 10 nM dose of paclitaxel, delivered in combination with ceramide in PEO–PCL nanoparticles, resulted in $63.98 \pm 4.9\%$ viability, and the free drugs in solution at these doses did not provoke any cell kill in the resistant cell line. The use of these drug-loaded nanoparticles resulted in a 100-fold increase in chemosensitivity of the MDR cells. These results demonstrate the clinical use of PEO–PCL nanoparticles in overcoming MDR by combination therapy.

13.7 ILLUSTRATIVE EXAMPLES OF POLYMERIC NANOPARTICLES FOR GENE DELIVERY

13.7.1 POLYETHYLENE GLYCOL-MODIFIED GELATIN NANOPARTICLES

To develop safe and effective systemically administered non-viral gene therapy vectors for solid tumors, PEGylated gelatin was synthesized and fabricated into nanoparticles. The PEG-modified gelatin was synthesized by reacting Type-B gelatin with PEG-epoxide that was further confirmed by electron spectroscopy for chemical analysis (ESCA), the results of which had shown an increase in the presence of an ether carbon (–C–O–) peak in the PEGylated gelatin nanoparticles. The experiments performed using these polymeric nanoparticulate carriers proved their ability to encapsulate DNA, improved transfection efficiency in vitro, and enhanced the biodistribution profile when the carriers were injected into mice bearing tumors.[56–58]

The in vitro cytotoxicity assays indicate that both gelatin and PEGylated gelatin are non-toxic to the cells. The cell uptake and trafficking studies of the control (gelatin) and PEGylated gelatin nanoparticles encapsulated with electron-dense gold nanoparticles confirmed that the particles were internalized by an endocytic pathway and remained stable during the vesicular transport. Further, the transfection studies carried out using the DNA (pEGFP-N1) encapsulated nanoparticles were internalized in NIH-3T3 fibroblast cells within the first six hours of incubation. Green fluorescent protein expression was observed after 12 h of nanoparticle incubation and remained stable for up to 96 hours. Flow cytometry results showed that the DNA transfection efficiency of PEGylated gelatin nanoparticles was better than that of gelatin.[57]

In order to determine the biodistribution profile, nanoparticles were radiolabeled with iodine-125 [^{125}I] and intravenously injected into mice bearing Lewis lung carcinoma (LLC) tumors. From the radioactivity of the plasma and other organs collected, it was evident that the majority of the PEGylated gelatin nanoparticles remained in the blood pool or were taken up by the tumor mass and liver. A two-fold increase in concentration of the PEGylated nanoparticles was observed in plasma even at three-hour time points, and about 4–5% of the recovered dose of PEGylated gelatin nanoparticles was found to be present in the tumor mass for up to 12 h. Upon non-compartmental pharmacokinetic analysis of the plasma and tumor concentration-time profiles, it was observed that PEGylated gelatin nanoparticles had greater mean residence time in plasma and a more than six-fold increase in half-life in the tumor. The results of this study showed the stealth nature of the PEGylated gelatin nanoparticles in avoiding uptake by RES, allowing the nanoparticles to circulate longer in plasma.[59]

The in vivo transfection efficiency of plasmid DNA-encapsulated (pCMV-β encoding β-galactosidase) PEGylated gelatin nanoparticles was evaluated by injecting these nanoparticles intravenously into LLC tumor-bearing female C57BL/6J mice. Following systemic administration, the animals were sacrificed, and the transgene expression in different organs was quantitatively determined by using o-nitrophenyl-β-d-galactopyranosidase (ONPG), a clear substrate of β-galactosidase that is converted to a yellow-colored product with an absorbance maximum at 420 nm, and qualitatively by X-gal® tissue staining. When compared to the gelatin nanoparticles, the PEGylated gelatin nanoparticles were found to efficiently transfect the tumor cells with the β-galactosidase expression increasing at time points up to 96 h post-transfection with the absorbance value at 420 nm increasing from 0.6 to 0.85 starting from 12 h post-transfection. The results of these studies clearly indicate that the long circulating, biocompatible, and biodegradable nanoparticles of PEGylated gelatin would be ideal for gene delivery applications to solid tumors.[60]

13.7.2 POLYETHYLENE GLYCOL-MODIFIED THIOLATED-GELATIN NANOPARTICLES

Another study synthesized thiolated gelatin and formulated it into nanoparticles that could rapidly release their payload in the presence of a high concentration of glutathione. The thiolated gelatins with different degrees of thiolation were characterized by Ellman's assay to determine degree of thiolation and by in vitro cytotoxicity assay. Upon carrying out the qualitative and quantitative transfection studies, the nanoparticles of thiolated gelatin, made with 20 mg of aminothiolane per gram of gelatin (SHGel-20) and encapsulated with plasmid DNA (pEGFP-N1), were found to have higher transfection efficiency in mouse fibroblast (NIH-3T3) cells in vitro (Figure 13.3).[61]

These nanoparticles, prepared by the desolvation method, were post-PEGylated with polyethylene glycol-succinimidyl glutarate (PEG-SG) of molecular weight 2,000 Da. The radiolabeled ([111]Indium) long-circulating (PEG-modified) thiolated-gelatin nanoparticles were injected intravenously into mice bearing human breast cancer (MDA-MB435) xenografts. The biodistribution and pharmacokinetics of these particles were compared to those of nanoparticles prepared from thiolated-gelatin, gelatin, and PEG-modified gelatin. The unmodified gelatin and thiolated gelatin nanoparticles were shown to have a plasma half-life of approximately three hours. Upon modification with PEG, the nanoparticles were found to have prolonged circulation times extending up to 10.7 h in the case of PEG-gelatin and 15.3 h with PEG-thiolated-gelatin nanoparticles. The successful surface hydrophilization of the gelatin and thiolated gelatin nanoparticles resulted in enhanced circulation times with almost 7% of the dose remaining in plasma at 12 h. The results also showed that the thiolated-gelatin nanoparticles (both non-modified and PEG-modified) had greater accumulations in the tumor than the nanoparticles prepared from non-thiolated gelatin. These thiolated-gelatin nanoparticles that are an improvement over gelatin nanoparticles can be made long-circulating by surface modification with PEG, and they can be used for rapid delivery of the payload the tumor tissues. The mechanisms of long-circulation through steric hindrance when in circulation and preferential accumulation in the tumor mass or inside the cells as a result of the cleavage of the disulfide bonds (arising from thiolation) in the highly reductive tumor-microenvironment can be exploited for the delivery of drugs and genes in clinical applications.[40,61]

13.8 CONCLUDING REMARKS

There are multiple factors affecting the delivery of drugs and genes to tumors. Factors such as blood flow, angiogenesis, microvessel density, interstitial pressure, macrophage activity, extracellular and intracellular components, and, most importantly, the physicochemical properties of the drug carrier

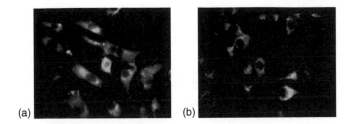

FIGURE 13.3 Transfection images of mouse fibroblast (NIH3T3) cells treated with thiolated-gelatin nanoparticles encapsulated with plasmid DNA (pEGFP-N1). The images show fluorescence of the expressed green fluorescent protein at 6 h (a) and 96 h post-transfection (b).

play an important role in the transport of drugs and macromolecules to tumors. This lab is particularly interested in enhancing the transport of drugs and genes to tumors by altering the physicochemical properties of the polymeric carrier used to encapsulate the drugs and genes. As shown in the illustrative examples, the biodistribution properties of the polymeric carriers could be modified using hydrophilic polymers such as PEG and PEO. This therapeutic strategy could be used to alter the passive/active targeting ability of the drug and gene carriers. However, the delivery of these newer agents is still a challenge, highlighting the necessity of additional research in this area.

ACKNOWLEDGMENTS

These studies have been supported by grants R01-CA095522 and R01-CA119617 from the National Cancer Institute, National Institutes of Health. Our research work on poly(beta-amino ester) nanoparticles is done in collaboration with Professor Robert Langer's lab at MIT (Cambridge, MA). We deeply appreciate the on-going, highly productive collaborations with his group and especially the assistance of Dr. Steven Little, Dr. Daniel Anderson, and Mr. David Nguyen.

REFERENCES

1. Boehm, T., Folkman, J., Browder, T., and O'Reilly, M. S., Antiangiogenic therapy of experimental cancer does not induce acquired drug resistance, *Nature*, 390(6658), 404–407, 1997.
2. Kerbel, R. S., A cancer therapy resistant to resistance, *Nature*, 390(6658), 335–336, 1997.
3. Kirsch, M., Strasser, J., Allende, R., Bello, L., Zhang, J., and Black, P. M., Angiostatin suppresses malignant glioma growth in vivo, *Cancer Research*, 58(20), 4654–4659, 1998.
4. Ganjavi, H., Gee, M., Narendran, A., Freedman, M. H., and Malkin, D., Adenovirus-mediated p53 gene therapy in pediatric soft-tissue sarcoma cell lines: Sensitization to cisplatin and doxorubicin, *Cancer Gene Therapy*, 12(4), 397–406, 2005.
5. Liu, Y., Huang, H., Saxena, A., and Xiang, J., Intratumoral coinjection of two adenoviral vectors expressing functional interleukin-18 and inducible protein-10, respectively, synergizes to facilitate regression of established tumors, *Cancer Gene Therapy*, 9(6), 533–542, 2002.
6. Van Meir, E. G., Polverini, P. J., Chazin, V. R., Su Huang, H. J., de Tribolet, N., and Cavenee, W. K., Release of an inhibitor of angiogenesis upon induction of wild type p53 expression in glioblastoma cells, *Nature Genetics*, 8(2), 171–176, 1994.
7. Schiffelers, R. M., Ansari, A., Xu, J., Zhou, Q., Tang, Q., Storm, G., Molema, G., Lu, P. Y., Scaria, P. V., Woodle, M. C. Cancer siRNA therapy by tumor selective delivery with ligand-targeted sterically stabilized nanoparticle *Nucleic Acids Research*, 32(19), e149, 2004 (http://nar.oxfordjournal.org/cgi/content/abstract/32/19/e149).

8. Yuan, F., Salehi, H. A., Boucher, Y., Vasthare, U. S., Tuma, R. F., and Jain, R. K., Vascular permeability and microcirculation of gliomas and mammary carcinomas transplanted in rat and mouse cranial windows, *Cancer Research*, 54(17), 4564–4568, 1994.

9. Yuan, F., Leunig, M., Huang, S. K., Berk, D. A., Papahadjopoulos, D., and Jain, R. K., Microvascular permeability and interstitial penetration of sterically stabilized (stealth) liposomes in a human tumor xenograft, *Cancer Research*, 54(13), 3352–3356, 1994.

10. Jain, R. K., Physiological barriers to delivery of monoclonal antibodies and other macromolecules in tumors, *Cancer Research*, 50(Suppl. 3), 814s–819s, 1990.

11. Jang, S. H., Wientjes, M. G., Lu, D., and Au, J. L., Drug delivery and transport to solid tumors, *Pharmaceutical Research*, 20(9), 1337–1350, 2003.

12. Maeda, H., SMANCS and polymer-conjugated macromolecular drugs: Advantages in cancer chemotherapy, *Advanced Drug Delivery Reviews*, 46(1–3), 169–185, 2001.

13. Maeda, H. and Matsumura, Y., Tumoritropic and lymphotropic principles of macromolecular drugs, *Critical Reviews in Therapeutic Drug Carrier Systems*, 6(3), 193–210, 1989.

14. Matsumura, Y. and Maeda, H., A new concept for macromolecular therapeutics in cancer chemotherapy: Mechanism of tumoritropic accumulation of proteins and the anti-tumor agent smancs, *Cancer Research*, 46(12 Pt 1), 6387–6392, 1986.

15. Balkwill, F. and Mantovani, A., Inflammation and cancer: Back to Virchow?, *Lancet*, 357(9255), 539–545, 2001.

16. Coussens, L. M. and Werb, Z., Inflammation and cancer, *Nature*, 420(6917), 860–867, 2002.

17. Lewis, C. and Murdoch, C., Macrophage responses to hypoxia: Implications for tumor progression and anti-cancer therapies, *American Journal of Pathology*, 167(3), 627–635, 2005.

18. Mantovani, A., Sozzani, S., Locati, M., Allavena, P., and Sica, A., Macrophage polarization: Tumor-associated macrophages as a paradigm for polarized M2 mononuclear phagocytes, *Trends in Immunology*, 23(11), 549–555, 2002.

19. Luo, D., Haverstick, K., Belcheva, N., Han, E., and Saltzman, M., Poly(ethylene glycol)-conjugated PAMAM dendrimer for biocompatible, high-efficiency DNA delivery, *Macromolecules*, 35, 3456–3462, 2002.

20. Kircheis, R., Blessing, T., Brunner, S., Wightman, L., and Wagner, E., Tumor targeting with surface-shielded ligand—polycation DNA complexes, *Journal of Controlled Release*, 72(1–3), 165–170, 2001.

21. Kim, S. H., Jeong, J. H., Cho, K. C., Kim, S. W., and Park, T. G., Target-specific gene silencing by siRNA plasmid DNA complexed with folate-modified poly(ethylenimine), *Journal of Controlled Release*, 104, 223–232, 2005.

22. Suh, W., Han, S. O., Yu, L., and Kim, S. W., An angiogenic, endothelial-cell-targeted polymeric gene carrier, *Molecular Therapy*, 6(5), 664–672, 2002.

23. Kursa, M., Walker, G. F., Roessler, V., Ogris, M., Roedl, W., Kircheis, R., and Wagner, E., Novel shielded transferrin-polyethylene glycol-polyethylenimine/DNA complexes for systemic tumor-targeted gene transfer, *Bioconjugate Chemistry*, 14(1), 222–231, 2003.

24. Lee, H., Kim, T. H., and Park, T. G., A receptor-mediated gene delivery system using streptavidin and biotin-derivatized, pegylated epidermal growth factor, *Journal of Controlled Release*, 83(1), 109–119, 2002.

25. Merdan, T., Callahan, J., Petersen, H., Kunath, K., Bakowsky, U., Kopeckova, P., Kissel, T., and Kopecek, J., Pegylated polyethylenimine-Fab' antibody fragment conjugates for targeted gene delivery to human ovarian carcinoma cells, *Bioconjugate Chemistry*, 14(5), 989–996, 2003.

26. Natarajan, A., Xiong, C. Y., Albrecht, H., DeNardo, G. L., and DeNardo, S. J., Characterization of site-specific ScFv PEGylation for tumor-targeting pharmaceuticals, *Bioconjugate Chemistry*, 16(1), 113–121, 2005.

27. Brannon-Peppas, L. and Blanchette, J. O., Nanoparticle and targeted systems for cancer therapy, *Advanced Drug Delivery Reviews*, 56(11), 1649–1659, 2004.

28. Ehrlich, P., *Studies in Immunity*, Wiley, New York, 1906.

29. Ringsdorf, H., Structure and properties of pharmacologically active polymers, *Journal of Polymer Science Polymer Symposium*, 51, 135–153, 1975.

30. Stastny, M., Strohalm, J., Plocova, D., Ulbrich, K., and Rihova, B., A possibility to overcome p-glycoprotein (PGP)-mediated multidrug resistance by antibody-targeted drugs conjugated to N-(2-hydroxypropyl)methacrylamide (HPMA) copolymer carrier, *European Journal of Cancer*, 35, 459–466, 1999.

31. El-Aneed, A., An overview of current delivery systems in cancer gene therapy, *Journal of Controlled Release*, 94(1), 1–14, 2004.

32. Fenske, D. B., MacLachlan, I., and Cullis, P. R., Long-circulating vectors for the systemic delivery of genes, *Current Opinion in Molecular Therapeutics*, 3(2), 153–158, 2001.

33. Han, S., Mahato, R. I., Sung, Y. K., and Kim, S. W., Development of biomaterials for gene therapy, *Moleular Therapeutics*, 2(4), 302–317, 2000.

34. McCormick, F., Cancer gene therapy: Fringe or cutting edge?, *Nature Reviews Cancer*, 1(2), 130–141, 2001.

35. Lehrman, S., Virus treatment questioned after gene therapy death, *Nature*, 401(6753), 517–518, 1999.

36. Marshall, E., Gene therapy. Second child in French trial is found to have leukemia, *Science*, 299(5605), 320, 2003.

37. Thomas, C. E., Ehrhardt, A., and Kay, M. A., Progress and problems with the use of viral vectors for gene therapy, *Nature Reviews Genetics*, 4(5), 346–358, 2003.

38. Kommareddy, S., Tiwari, S. B., and Amiji, M. M., Long-circulating polymeric nanovectors for tumor-selective gene delivery, *Technology in Cancer Research and Treatment*, 4(6), 615–626, 2005.

39. Torchilin, V. P., Drug targeting, *European Journal of Pharmaceutial Sciences*, 11(Suppl. 2), S81–S91, 2000.

40. Kommareddy, S., Shenoy, D. B., and Amiji, M., Nanoparticulate carriers of gelatin and gelatin derivatives, In *Biological and Pharmaceutical Nanomaterials*, Kumar, C., Ed., Wiley-VCH, Weinheim, Germany, pp. 330–347, 2005.

41. Torchilin, V. P., Polymer-coated long-circulating microparticulate pharmaceuticals, *Journal of Microencapsulation*, 15(1), 1–19, 1998.

42. Torchilin, V. P. and Papisov, M. I., Why do polyethylene glycol-coated liposomes circulate so long?, *Journal of Liposome Research*, 4(1), 725–739, 1994.

43. Zalipsky, S., Functionalized Poly(ethylene glycol) for preparation of biologically revalant conjugates, *Bioconjugate Chemistry*, 6, 150–165, 1995.

44. Gref, R., Domb, A. J., Quellec, P., Blunk, T., Müller, R. H., Verbavatz, J. M., and Langer, R., The controlled intravenous delivery of drugs using PEG-coated sterically stabilized nanospheres, *Advanced Drug Delivery Reviews*, 16(2–3), 215–233, 1995.

45. Delgado, C., Francis, G. E., and Fisher, D., The uses and properties of PEG-linked proteins, *Critical Reviews in Therapeutic Drug Carrier Systems*, 9(3–4), 249–304, 1992.

46. Mehvar, R., Dextrans for targeted and sustained delivery of therapeutic and imaging agents, *Journal of Controlled Release*, 69(1), 1–25, 2000.

47. Torchilin, V. P., How do polymers prolong circulation time of liposomes?, *Journal of Liposome Research*, 6(1), 99–116, 1996.

48. Oupicky, D., Howard, K. A., Konak, C., Dash, P. R., Ulbrich, K., and Seymour, L. W., Steric stabilization of poly-L-Lysine/DNA complexes by the covalent attachment of semitelechelic poly[N-(2-hydroxypropyl)methacrylamide], *Bioconjugate Chemistry*, 11(4), 492–501, 2000.

49. Lynn, D. M., Amiji, M. M., and Langer, R., pH-responsive polymer microspheres: Rapid release of encapsulated material within the range of intracellular pH, *Angewandte Chemie (International ed. in English)*, 40(9), 1707–1710, 2001.

50. Potineni, A., Lynn, D. M., Langer, R., and Amiji, M. M., Poly(ethylene oxide)-modified poly(beta-amino ester) nanoparticles as a pH-sensitive biodegradable system for paclitaxel delivery, *Journal of Controlled Release*, 86(2–3), 223–234, 2003.

51. Scholes, P. D., Coombes, A. G., Illum, L., Davis, S. S., Watts, J. F., Ustariz, C., Vert, M., and Davies, M. C., Detection and determination of surface levels of poloxamer and PVA surfactant on biodegradable nanospheres using SSIMS and XPS, *Journal of Controlled Release*, 59(3), 261–278, 1999.

52. Chawla, J. S. and Amiji, M. M., Cellular uptake and concentrations of tamoxifen upon administration in poly(epsilon-caprolactone) nanoparticles, *AAPS PharmSci*, 5(1), E3, 2003 (http://www.aapsj.org/view.asp?art=ps050103).

53. Lavie, Y., Cao, H., Bursten, S. L., Giuliano, A. E., and Cabot, M. C., Accumulation of glucosylcer-amides in multidrug-resistant cancer cells, *Journal of Biological Chemistry*, 271, 19530–19536, 1996.

54. Liu, Y. Y., Han, T. Y., Giuliano, A. E., and Cabot, M. C., Ceramide glycosylation potentiates cellular multidrug resistance, *FASEB Journal*, 15(3), 719–730, 2001.

55. Morjani, H., Aouali, N., Belhoussine, R., Veldman, R. J., Levade, T., and Manfait, M., Elevation of glucosylceramide in multidrug-resistant cancer cells and accumulation in cytoplasmic droplets, *International Journal of Cancer*, 94, 157–165, 2001.

56. Kaul, G. and Amiji, M., Long-circulating poly(ethylene glycol)-modified gelatin nanoparticles for intracellular delivery, *Pharmaceutical Research*, 19(7), 1062–1068, 2002.

57. Kaul, G. and Amiji, M., Cellular interactions and in vitro DNA transfection studies with poly(ethylene glycol)-modified gelatin nanoparticles, *Journal of Pharmaceutical Sciences*, 94(1), 184–198, 2004.

58. Kaul, G., Lee-Parsons, C., and Amiji, M., Poly(ethylene glycol)-modified gelatin nanoparticles for intracellular delivery, *Pharmaceutical Engineers*, 23(5), 1–5, 2003.

59. Kaul, G. and Amiji, M., Biodistribution and targeting potential of poly(ethylene glycol)-modified gelatin nanoparticles in subcutaneous murine tumor model, *Journal of Drug Targeting*, 12(9–10), 585–591, 2004.

60. Kaul, G. and Amiji, M., Tumor-targeted gene delivery using poly(ethylene glycol)-modified gelatin nanoparticles: In vitro and in vivo studies, *Pharmaceutical Research*, 22(6), 951–961, 2005.

61. Kommareddy, S. and Amiji, M., Preparation and evaluation of thiol-modified gelatin nanoparticles for intracellular DNA delivery in response to glutathione, *Bioconjugate Chemistry*, 16(6), 1423–1432, 2005.

14 Biodegradable PLGA/PLA Nanoparticles for Anti-Cancer Therapy

Sanjeeb K. Sahoo and Vinod Labhasetwar

CONTENTS

14.1 Introduction 243
14.2 General Principles of Drug Targeting to Cancer 244
    14.2.1 Passive Targeting 244
    14.2.2 Active Targeting... 244
14.3 PLGA as a Polymer for Nanoparticles 244
14.4 Application of PLGA/PLA Nanoparticles as Drug Delivery Vehicles
    to Cancer Tissues 245
14.5 PLGA Nanoparticles for Gene Delivery to Cancer.. 247
14.6 PLGA Nanoparticles for Photodynamic Therapy.. 248
14.7 Concluding Remarks 248
Acknowledgments 248
References.. 249

14.1 INTRODUCTION

Despite significant efforts in the field of oncology, cancer remains one of the leading causes of death in industrialized countries.[1] Of the various options available, surgery and radiation therapy are commonly used to treat localized tumors whereas chemotherapy is primarily used in the management and elimination of hematological malignancies and metastasized tumors.[2] The major limitation of the current cancer chemotherapeutics is their high pharmacokinetic volume of distribution and rapid elimination rates, requiring more frequent and high dose administration of these agents.[3] Therefore, cancer chemotherapeutics cause unacceptable damage to normal tissue when used at doses required to eradicate cancer cells. Furthermore, multidrug resistance in tumor cells exacerbates the problem of cancer chemotherapy.[4–7] Selective delivery of anti-cancer drugs to the tumor tissue offers a formidable solution to the problem. This could significantly enhance the therapeutic efficacy of these drugs and diminish their toxicity.[8,9] The overall goal of drug delivery strategies is to selectively attack cancer cells while saving normal tissue from drug toxicity.[10,11] The other issue with cancer chemotherapeutics is their systemic delivery, as most of these agents are poorly water soluble. Hence they require adjuvant or excipients to dissolve, which are often associated with serious side effects.[12] Therefore, drug carrier systems which would address the above problem of delivery as well as target drugs selectively to the tumor would be a major advancement in cancer chemotherapy. Many macromolecular carriers, including soluble synthetic and natural polymers, implants, liposomes, microspheres, dendrimers, nanoparticles, etc., have been tested for selective delivery of drugs to the tumor tissue.[13]

14.2 GENERAL PRINCIPLES OF DRUG TARGETING TO CANCER

14.2.1 PASSIVE TARGETING

Passive targeting refers to the accumulation of drug or drug-carrier system at a particular site due to physicochemical or pharmacological factors. Permeability of the tumor vasculature increases to the point where particulate carriers such as nanoparticles can extravasate from blood circulation and localize in the tumor tissue.[14,15] This occurs because as tumors grow and begin to outstrip the available supply of oxygen and nutrients, they release cytokines and other signaling molecules that recruit new blood vessels to the tumor, a process known as angiogenesis.[16] Angiogenic blood vessels, unlike the tight blood vessels in most normal tissues, have gaps as large as 600–800 nm between adjacent endothelial cells. Drug carriers in the nanometer size range can extravasate through these gaps into the tumor interstitial space.[17,18] Because tumors have impaired lymphatic drainage, the carriers concentrate in the tumor, resulting in higher drug concentration in the tumor tissue (10-fold or higher) than that can be achieved with the same dose of free drug. This is commonly referred to as *enhanced permeability and retention*, or the EPR effect.

14.2.2 ACTIVE TARGETING

Active targeting to the tumor can be achieved by molecular recognition of cancer cells either via ligand–receptor or antibody–antigen interactions. Active targeting may also lead to receptor-mediated cell internalization of drug carrier system. Nanoparticles and other polymer drug-conjugates offer numerous opportunities for targeting tumors through surface modifications which allow specific biochemical interactions with the proteins/receptors expressed on target cells.[19,20] For active and passive targeting of drug carrier systems, it is essential to avoid their uptake by the reticuloendothelial system (RES) so that they remain in the blood circulation and extravasate in the tumor vasculature. Particles with more hydrophobic surfaces are preferentially taken up by the liver, followed by the spleen and lungs.[21,22] Size of nanoparticles as well as their surface characteristics are the key parameters that can alter the biodistribution of nanoparticles. Particles smaller than 100 nm and coated with hydrophilic polymers such as amphiphilic polymeric compounds which are made of polyethylene oxide such as poloxamers, poloxamines, or polyethylene glycol (PEG) are being investigated to avoid their uptake by the RES. To improve the efficacy of targeting cancer chemotherapeutics to the tumor, a combination of passive and active targeting strategy is being investigated where long-circulating drug carriers are conjugated to tumor cell-specific antibody or peptides.[23] In addition to the above approach, direct intratumoral injection of the carrier system is feasible if the tumor is localized and can be accessed for administration of a carrier system.[24]

14.3 PLGA AS A POLYMER FOR NANOPARTICLES

A number of different polymers, both synthetic and natural, have been utilized in formulating biodegradable nanoparticles. Synthetic polymers have the advantage of sustaining the release of the encapsulated therapeutic agent over a period of days to several weeks as compared to natural polymers which, in general, have a relatively short duration of drug release. The polymers used for the formulation of nanoparticles include synthetic polymers such as polylactide–polyglycolide copolymers, polyacrylates, and polycaprolactones, or natural polymers such as albumin, gelatin, alginate, collagen, and chitosan. Of these polymers, polylactides (PLA) and poly (D,L-lactide-*co*-glycolide) (PLGA) have been most extensively investigated for drug delivery applications.[19] PLGA/PLA-based polymers have a number of advantages over other polymers used in drug and gene delivery, such as their biodegradability, biocompatibility, and approval by the FDA for human use.[19,25] PLGA/PLA polymers degrade in the body through hydrolytic cleavage of the ester linkage to lactic and glycolic acid, although there are reports of involvement of enzymes in their biodegradation.[19,25] These monomers are easily metabolized in the body via Krebs' cycle and eliminated as carbon dioxide and water. Biodegradation products of PLGA and PLA polymers are formed at a very slow rate, and they therefore do not affect normal cell function.[26] Furthermore, these polymers have been tested for toxicity and safety in extensive animal studies and are currently used in humans for resorbable sutures, bone implants and screws, contraceptive implants,[27,28] and also as graft materials for artificial organs and supporting scaffolds in tissue engineering research.[28–30] Long-term biocompatibility of these polymers was demonstrated by the absence of any untoward effects on intravascular administration of nanoparticles formulated using these polymers to the arterial tissue in pig and rat models of restenosis.[31]

14.4 APPLICATION OF PLGA/PLA NANOPARTICLES AS DRUG DELIVERY VEHICLES TO CANCER TISSUES

There are several studies regarding PLGA/PLA nanoparticles or some modification of these polymers for delivery of anti-cancer agents and other therapeutic agents.[19] We have recently demonstrated increased efficacy of transferrin conjugated paclitaxel-loaded PLGA nanoparticles (Figure 14.1a and b) both in vitro and in an animal model of prostate carcinoma.[24] Transferrin receptors are over-expressed in most cancer cells by two to tenfold more than in normal cells. We have demonstrated that transferring-conjugated nanoparticles have enhanced cellular uptake (Figure 14.1c) and retention than unconjugated nanoparticles.[4] A single-dose intratumoral injection of transferrin conjugated nanoparticles in animal models of prostate carcinoma demonstrated complete tumor regression and higher survival rate than animals that received either drug in solution or unconjugated nanoparticles.[24] The IC_{50} for paclitaxel with transferrin conjugated nanoparticles was fivefold lower than that with unconjugated nanoparticles or with drug in solution in PC3 [24] (Figure 14.1d) and in MCF-7 cells.[4] Kim et al. have demonstrated enhanced intracellular delivery of PLGA nanoparticles, which were surface-coated with cationic di-block copolymer, poly(L-lysine)–poly(ethylene glycol)–folate (PLL–PEG–FOL), in KB cells that overexpress folate receptors.[32] In another study, paclitaxel-loaded PLGA nanoparticles, which were conjugated to wheat germ agglutinin (WGA), demonstrated greater anti-proliferation activity in A549 and H1299 cells as compared to the conventional paclitaxel formulations. This enhanced activity of WGA-conjugated nanoparticles was attributed to greater intracellular accumulation of drug via WGA-receptor-mediated endocytosis of conjugated nanoparticles.[33] Cegnar et al. have developed cystatin-loaded PLGA nanoparticles with the strategy of inhibiting the tumor-associated activity of intracellular cysteine proteases cathepsins B and L. In an in vitro study, cystatin-loaded PLGA nanoparticles demonstrated 160-fold greater cytotoxic effect in MCF-10A neoT cells than free cystatin.[34] Similarly, interferon-alpha (IFN-alpha) loaded PLGA nanoparticles are being developed to improve the therapeutic efficacy of IFN-alpha while reducing its dose-related side effects.[35] In studies by Yoo et al., doxorubicin was chemically conjugated to a terminal end group of PLGA by an ester linkage and the doxorubicin–PLGA conjugate was formulated into nanoparticles. Nanoparticles containing the conjugate exhibited sustained doxorubicin release profiles over a one-month period, whereas those containing unconjugated doxorubicin showed a rapid drug release within five days. The conjugated doxorubicin nanoparticles demonstrated increased drug uptake in HepG2 cell line and exhibited slightly lower IC_{50} value compared to that of free doxorubicin.

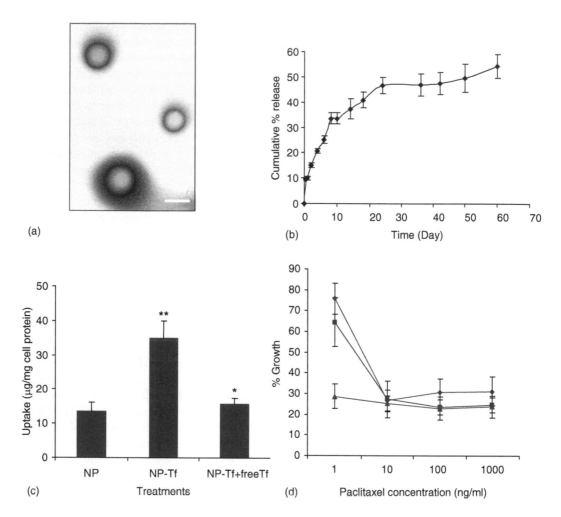

(a)

(b)

(c)

(d)

FIGURE 14.1 (a) Transmission electron micrographs of paclitaxel-loaded PLA nanoparticles (NPs) (bar = 100 nm). (b) Cumulative release of paclitaxel (Tx) from NPs in vitro. Tx loading in NPs is 5.4% (w/w). Data as mean ($\pm$s.e.m. ($n = 3$). (c) Uptake of unconjugated (NPs) and conjugated (NPs–Tf) in PC3 cells. A suspension of NPs (100 μg/mL) was incubated with PC3 cells (5×10^4 cells) for 1 h, cells were washed, and the NP levels in cells were determined by HPLC. To determine the competitive inhibition of uptake of Tf-conjugated NPs, excess dose of free Tf (50 μg) was added in the medium prior to incubation with NPs–Tf. Data as mean $\pm$ s.e.m. ($n = 6$).* $p < .05$ NPs–Tf+ free Tf vs. NPs, ** $p < .005$ NPs–Tf vs. NPs. NPs contain a fluorescent marker, 6-coumarin, to quantify their uptake. (d) Dose-dependent cytotoxicity of paclitaxel (Tx) in PC3 cells. Different concentration of Tx either as solution ($\blacklozenge$) or encapsulated in NPs (Tx–NPs ($\blacksquare$) or Tx–NPs–Tf, ($\blacktriangle$) was added to wells with NPs (without drug) or medium acting as respective controls. The medium was changed at two days and then on every alternate day with no further dose of the drug added. The extent of growth inhibition was measured at five days. The percentage of survival was determined by standardizing the absorbance of controls to 100% ($n = 6$). Data as mean $\pm$s.e.m. * $p < .005$ Tx–NPs–Tf vs. Tx–sol and Tx–NPs. (Data reproduced from Sahoo, S. K. and Labhasetwar, V., *Mol. Pharm.*, 2, 373, 2005; Sahoo, S. K., Ma, W., and Labhasetwar, V., *Int. J. Cancer*, 112, 335, 2004. With permission.)

The in vivo anti-tumor activity assay showed that a single injection of these nanoparticles had comparable activity to that of free doxorubicin when administered by daily injection.[36] Thus different strategies are being investigated using biodegradable PLGA-based nanoparticles to deliver anti-cancer drugs more effectively, both at the cellular level and also to the tumor tissue.

14.5 PLGA NANOPARTICLES FOR GENE DELIVERY TO CANCER

Gene therapy to cancer represents one of the most rapidly developing areas in pre-clinical and clinical cancer research. Crucial to the success of any gene therapy strategy is the efficiency with which the gene is delivered.[37] This in turn is dependent upon the type of delivery vector used. Many vectors have been developed based on either recombinant viruses or non-viral vectors. Research utilizing viral vectors has progressed much more rapidly than that with non-viral vectors. This is reflected in the fact that approximately 85% of current clinical protocols involve viral vectors. However, these vectors are still limited in many ways, particularly in relation to issues of safety, immunogenicity, limitations on the size of the gene that can be delivered, specificity, production problems, toxicity, cost and others.[38,39]

Although various polymeric systems are under investigation,[40] our efforts are focused on investigating biodegradable nanoparticles as a gene delivery system in cancer therapy.[19,41–43] These nanoparticles are formulated using PLGA and PLA polymers, with plasmid DNA (pDNA) entrapped into the nanoparticle matrix. The main advantage of PLGA or PLA-based nanoparticles for gene delivery is their non-toxic nature. Furthermore, the slow release of the encapsulated DNA from nanoparticles is expected to provide sustained gene delivery (Figure 14.2a). Blends of PLGA and polyoxyethylene derivatives such as poloxamer and poloxamine are also being used to modulate DNA release from nanoparticles.[44] In our study we demonstrated anti-proliferative activity of wild-type (wt) p53 gene-loaded nanoparticles in breast cancer cell line.[45] Cells transfected with wt-p53 DNA-loaded nanoparticles demonstrated sustained and significantly greater anti-proliferative effect than those treated with naked wt-p53 DNA (Figure 14.2b) or wt-p53 DNA complexed with a commercially available transfecting agent (Lipofectamine). This effect was attributed to sustained nanoparticles-mediated gene expression. This was evident from sustained p53 mRNA levels observed in cells transfected with nanoparticles compared with levels in cells which were transfected with naked wt-p53 DNA.

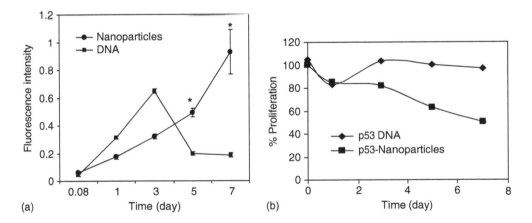

FIGURE 14.2 (a) Quantitative determination of intracellular DNA levels. Cells transfected with YOYO-labeled DNA-loaded nanoparticles demonstrated sustained and increase in intracellular DNA levels as opposed to transient DNA levels in the cells transfected with naked DNA. Data represented as mean ± s.e.m., $n = 6$, * $p < .001$. (b) Anti-proliferative activity of wt-p53 DNA-loaded nanoparticles (NPs) and naked wt-p53 DNA (DNA) in MDAMB-435S cells. Cells (2500 cells/well) grown in a 96-well plate were incubated with DNA-loaded NPs or equivalent amount of naked DNA Cell growth was followed using a standard MTS assay. NPs demonstrated increase in anti-proliferative activity with incubation time. Data represented as mean ± s.e.m., $n = 6$, * $p < .01$. (Data reproduced from Prabha, S. and Labhasetwar, V., *Mol. Pharm.*, 1, 211, 2004. With permission.)

Recently, He et al. have formulated thymidine kinase gene (TK gene)-loaded nanoparticles and investigated the expression of TK gene in hepatocarcinoma cells.[46] The expression of DNA encapsulated in nanoparticles was significantly greater than that with naked DNA. In another study, Cohen et al. have demonstrated significantly greater gene transfection (1–2 orders of magnitude) with PLGA nanoparticles than that with a liposomal formulation following their intramuscular injection in rats. Furthermore, the nanoparticle-mediated gene transfection was seen up to 28 days in the above study. Recently Kumar et al. have prepared cationic PLGA nanoparticles modified with cationic chitosan and demonstrated gene expression in A549 epithelial cells.[47]

14.6 PLGA NANOPARTICLES FOR PHOTODYNAMIC THERAPY

Photodynamic therapy (PDT) is a promising new treatment modality[48] which involves administration of a photosensitizing drug.[49] Upon illumination at a suitable wavelength, it induces a photochemical reaction resulting in generation of reactive oxygen species (ROS) that can kill tumor cells directly as well as the tumor-associated vasculature, leading to tumor infarction. Targeting is essential in PDT as singlet oxygen is extremely reactive. Several strategies such as chemical conjugation with various water-soluble polymers and encapsulation into colloidal carriers have been considered for the parenteral administration of photosensitizers. These colloidal carriers include liposomes, emulsions, polymeric micelles and nanoparticles. PLGA nanoparticles were evaluated for PDT using *meso*-tetra(4-hydroxyphenyl) porphyrin as photosensitizer.[50] Vargas et al. have encapsulated porphyrin in PLGA nanoparticles and demonstrated enhanced photodynamic activity against mammary tumor cells than free drug.[51] Colloidal carrier associated with photosensitizer showed enhanced photodynamic efficiency and selectivity of tumor targeting as compared with dye administered in homogenous aqueous solution.[52,53] Recently, Saxena et al. have formulated PLGA nanoparticles loaded with indocyanine green (ICG) and determined their biodistribution in mice. The results demonstrated that the nanoparticle formulation significantly increased the ICG concentration and circulation time in plasma as well as the ICG uptake, accumulation and retention in various organs as compared to ICG solution.[54] Such a formulation can be explored in tumor-diagnosis as well as for PDT.

14.7 CONCLUDING REMARKS

It is essential to understand the molecular mechanisms involved in nanoparticle-mediated drug or gene delivery to explore their therapeutic potentials for cancer therapy. Also, the important questions we need to address are: What are the barriers in targeted cancer drug therapy? What is the efficiency of drug targeting with carrier systems? Can it provide a therapeutic dose of the drug in the tumor tissue? How long should the drug effect in the tumor tissue be sustained to regress it completely? Is a combination of drugs, especially those that work by different pathways, more effective than single-drug therapy? These and other questions are critical as we move from a conceptual stage to reality. With better understanding of molecular targets, discovery of more potent drugs and simultaneous developments in nanotechnology it seems that an effective cancer therapy is forthcoming.

ACKNOWLEDGMENTS

Grant support from the National Institutes of Health (R01 EB003975) is gratefully acknowledged. The authors also thank Ms. Elaine Payne and Ms. DeAnna Loibl for providing administrative support.

REFERENCES

1. Greish, K. et al., Macromolecular therapeutics: Advantages and prospects with special emphasis on solid tumour targeting, *Clin. Pharmacokinet.*, 42, 1089, 2003.
2. Luo, Y. and Prestwich, G. D., Cancer-targeted polymeric drugs, *Curr. Cancer Drug Targets*, 2, 209, 2002.
3. Fang, J., Sawa, T., and Maeda, H., Factors and mechanism of "EPR" effect and the enhanced anti-tumor effects of macromolecular drugs including SMANCS, *Adv. Exp. Med. Biol.*, 519, 29, 2003.
4. Sahoo, S. K. and Labhasetwar, V., Enhanced antiproliferative activity of transferrin-conjugated paclitaxel-loaded nanoparticles is mediated via sustained intracellular drug retention, *Mol. Pharm.*, 2, 373, 2005.
5. Saeki, T. et al., Drug resistance in chemotherapy for breast cancer, *Cancer Chemother. Pharmacol.*, 56(Suppl 7), 84, 2005.
6. Kabanov, A. V., Batrakova, E. V., and Miller, D. W., Pluronic block copolymers as modulators of drug efflux transporter activity in the blood–brain barrier, *Adv. Drug Deliv. Rev.*, 55, 151, 2003.
7. Kabanov, A. V., Batrakova, E. V., and Alakhov, V. Y., Pluronic block copolymers for overcoming drug resistance in cancer, *Adv. Drug Deliv. Rev.*, 54, 759, 2002.
8. Jain, K. K., Nanotechnology-based drug delivery for cancer, *Technol. Cancer Res. Treat.*, 4, 407, 2005.
9. Singh, K. K., Nanotechnology in cancer detection and treatment, *Technol. Cancer Res. Treat.*, 4, 583, 2005.
10. Meisheid, A. M., Targeted therapies in the treatment of cancer, *J. Contin. Educ. Nurs.*, 36, 193, 2005.
11. Reddy, L. H., Drug delivery to tumours: Recent strategies, *J. Pharm. Pharmacol.*, 57, 1231, 2005.
12. Adams, J. D. et al., Taxol: a history of pharmaceutical development and current pharmaceutical concerns, *J. Natl. Cancer Inst. Monogr.*, 15, 141, 1993.
13. Sahoo, S. K. and Labhasetwar, V., Nanotech approaches to drug delivery and imaging, *Drug Discov. Today*, 8, 1112, 2003.
14. Maeda, H. et al., Tumor vascular permeability and the EPR effect in macromolecular therapeutics: a review, *J. Control. Release*, 65, 271, 2000.
15. Maeda, H., Sawa, T., and Konno, T., Mechanism of tumor-targeted delivery of macromolecular drugs, including the EPR effect in solid tumor and clinical overview of the prototype polymeric drug SMANCS, *J. Control. Release*, 74, 47, 2001.
16. Folkman, J. and Shing, Y., Angiogenesis, *J. Biol. Chem.*, 267, 10931, 1992.
17. Jain, R. K., Integrative pathophysiology of solid tumors: Role in detection and treatment, *Cancer J. Sci. Am.*, 4(Suppl 1), S48, 1998.
18. Jain, R. K., Delivery of molecular and cellular medicine to solid tumors, *J. Control. Release*, 53, 49, 1998.
19. Panyam, J. and Labhasetwar, V., Biodegradable nanoparticles for drug and gene delivery to cells and tissue, *Adv. Drug Deliv. Rev.*, 55, 329, 2003.
20. Minko, T. et al., Molecular targeting of drug delivery systems to cancer, *Curr. Drug Targets*, 5, 389, 2004.
21. Gref, R. et al., Biodegradable long-circulating polymeric nanospheres, *Science.*, 263, 1600, 1994.
22. Gref, R. et al., Poly(ethylene glycol)-coated nanospheres: Potential carriers for intravenous drug administration, *Pharm. Biotechnol.*, 10, 167, 1997.
23. Vasir, J. K. and Labhasetwar, V., Targeted drug delivery in cancer therapy, *Technol. Cancer Res. Treat.*, 4, 363, 2005.
24. Sahoo, S. K., Ma, W., and Labhasetwar, V., Efficacy of transferrin-conjugated paclitaxel-loaded nanoparticles in a murine model of prostate cancer, *Int. J. Cancer*, 112, 335, 2004.
25. Jain, R. A., The manufacturing techniques of various drug loaded biodegradable poly(lactide-*co*-glycolide) (PLGA) devices, *Biomaterials*, 21, 2475, 2000.
26. Shive, M. S. and Anderson, J. M., Biodegradation and biocompatibility of PLA and PLGA microspheres, *Adv. Drug Deliv. Rev.*, 28, 5, 1997.
27. Hanafusa, S. et al., Biodegradable plate fixation of rabbit femoral shaft osteotomies. A comparative study, *Clin. Orthop. Relat. Res.*, 262, 1995.

28. Matsusue, Y. et al., Tissue reaction of bioabsorbable ultra high strength poly (L-lactide) rod. A long-term study in rabbits, *Clin. Orthop. Relat. Res.*, 246, 1995.

29. Langer, R., Tissue engineering: a new field and its challenges, *Pharm. Res.*, 14, 840, 1997.

30. Mooney, D. J. et al., Long-term engraftment of hepatocytes transplanted on biodegradable polymer sponges, *J. Biomed. Mater. Res.*, 37, 413, 1997.

31. Guzman, L. A. et al., Local intraluminal infusion of biodegradable polymeric nanoparticles. A novel approach for prolonged drug delivery after balloon angioplasty, *Circulation*, 94, 1441, 1996.

32. Kim, S. H. et al., Target-specific cellular uptake of PLGA nanoparticles coated with poly(L-lysine)–poly(ethylene glycol)–folate conjugate, *Langmuir*, 21, 8852, 2005.

33. Mo, Y. and Lim, L. Y., Paclitaxel-loaded PLGA nanoparticles: Potentiation of anticancer activity by surface conjugation with wheat germ agglutinin, *J. Control. Release*, 108, 244, 2005.

34. Cegnar, M. et al., Poly(lactide-*co*-glycolide) nanoparticles as a carrier system for delivering cysteine protease inhibitor cystatin into tumor cells, *Exp. Cell Res.*, 301, 223, 2004.

35. Sanchez, A. et al., Biodegradable micro- and nanoparticles as long-term delivery vehicles for interferon-alpha, *Eur. J. Pharm. Sci.*, 18, 221, 2003.

36. Yoo, H. S. et al., In vitro and in vivo anti-tumor activities of nanoparticles based on doxorubicin–PLGA conjugates, *J. Control. Release*, 68, 419, 2000.

37. Yamamoto, M. and Curiel, D. T., Cancer gene therapy, *Technol. Cancer Res. Treat.*, 4, 315, 2005.

38. Seth, P., Vector-mediated cancer gene therapy: an overview, *Cancer Biol. Ther.*, 4, 512, 2005.

39. Maitland, N. J., Stanbridge, L. J., and Dussupt, V., Targeting gene therapy for prostate cancer, *Curr. Pharm. Des.*, 10, 531, 2004.

40. Vasir, J. K. and Labhasetwar, V., Polymeric nanoparticles for gene delivery, *Expt. Opin. Drug Del.*, 3, 325, 2006.

41. Panyam, J. et al., Rapid endo-lysosomal escape of poly(DL-lactide-*co*-glycolide) nanoparticles: Implications for drug and gene delivery, *FASEB J.*, 16, 1217, 2002.

42. Prabha, S. et al., Size-dependency of nanoparticle-mediated gene transfection: Studies with fractionated nanoparticles, *Int. J. Pharm.*, 244, 105, 2002.

43. Prabha, S. and Labhasetwar, V., Critical determinants in PLGA/PLA nanoparticle-mediated gene expression, *Pharm. Res.*, 21, 354, 2004.

44. Csaba, N. et al., PLGA: Poloxamer and PLGA:Poloxamine blend nanoparticles: New carriers for gene delivery, *Biomacromolecules*, 6, 271, 2005.

45. Prabha, S. and Labhasetwar, V., Nanoparticle-mediated wild-type p53 gene delivery results in sustained antiproliferative activity in breast cancer cells, *Mol. Pharm.*, 1, 211, 2004.

46. He, Q. et al., Preparation and characteristics of DNA-nanoparticles targeting to hepatocarcinoma cells, *World J. Gastroenterol.*, 10, 660, 2004.

47. Kumar, M. N. et al., Cationic poly(lactide-*co*-glycolide) nanoparticles as efficient in vivo gene transfection agents, *J Nanosci. Nanotechnol.*, 4, 990, 2004.

48. Dolmans, D. E., Fukumura, D., and Jain, R. K., Photodynamic therapy for cancer, *Nat. Rev. Cancer*, 3, 380, 2003.

49. Huang, Z., A review of progress in clinical photodynamic therapy, *Technol. Cancer Res. Treat.*, 4, 283, 2005.

50. Konan, Y. N. et al., Encapsulation of p-THPP into nanoparticles: Cellular uptake, subcellular localization and effect of serum on photodynamic activity, *Photochem. Photobiol.*, 77, 638, 2003.

51. Vargas, A. et al., Improved photodynamic activity of porphyrin loaded into nanoparticles: an in vivo evaluation using chick embryos, *Int. J. Pharm.*, 286, 131, 2004.

52. Konan, Y. N. et al., Enhanced photodynamic activity of meso-tetra(4-hydroxyphenyl)porphyrin by incorporation into sub-200 nm nanoparticles, *Eur. J. Pharm. Sci.*, 18, 241, 2003.

53. Konan, Y. N. et al., Preparation and characterization of sterile sub-200 nm meso-tetra (4-hydroxyl-phenyl)porphyrin-loaded nanoparticles for photodynamic therapy, *Eur. J. Pharm. Biopharm.*, 55, 115, 2003.

54. Saxena, V., Sadoqi, M., and Shao, J., Polymeric nanoparticulate delivery system for Indocyanine green: Biodistribution in healthy mice, *Int. J. Pharm.*, 308, 200, 2006.

15 Poly(Alkyl Cyanoacrylate) Nanoparticles for Delivery of Anti-Cancer Drugs

R. S. R. Murthy and L. Harivardhan Reddy

CONTENTS

15.1 Cancer Therapy 252
15.2 Nanoparticles as Drug Delivery Systems 252
    15.2.1 Polymeric Nanoparticles 253
        15.2.1.1 Nanoparticles Prepared by the Polymerization Process 253
            15.2.1.1.1 Dispersion Polymerization 253
            15.2.1.1.1 Emulsion Polymerization 254
        15.2.1.2 Poly(Alkyl Cyanoacrylate) Nanoparticles 254
        15.2.1.3 Poly(Alkyl Cyanoacrylate) Nanocapsules 256
    15.2.2 Loading of Drugs to Poly(Alkyl Cyanoacrylate) Nanoparticles 256
    15.2.3 Characterization of Nanoparticles 258
        15.2.3.1 Physicochemical Characterization 258
        15.2.3.2 Degradation of Poly(Alkyl Cyanoacrylate) Nanoparticles 260
        15.2.3.3 Influence of Enzymes on the Stability
                of Poly(Alkyl Cyanoacrylate) Nanoparticles 260
        15.2.3.4 Drug Release from Poly(Alkyl Cyanoacrylate) Nanoparticles 261
    15.2.4 In Vivo Distribution of Poly(Alkyl Cyanoacrylate) Nanoparticles 261
        15.2.4.1 Surface Modification to Alter Biodistribution 264
    15.2.5 Toxicity of Polyalkyl Cyanoacrylate Nanoparticles 264
        15.2.5.1 Toxicity of Drugs Associated with Poly(Alkyl Cyanoacrylate)
                Nanoparticles 266
    15.2.6 Drug Delivery in Cancer with Poly(Alkyl Cyanoacrylate) Nanoparticles 266
        15.2.6.1 Targeting to Cancer Cells and Tissues 266
        15.2.6.2 Drug Delivery to Resistant Tumor Cells 268
        15.2.6.3 Drug Delivery to Hepatocellular Carcinomas 272
        15.2.6.4 Drug Delivery to Brain Tumors 274
        15.2.6.5 Drug Delivery to Breast Cancer 275
        15.2.6.6 Drug Delivery to Lymphatic Carcinomas 276
        15.2.6.7 Drug Delivery to Gastric Carcinomas 276
        15.2.6.8 Chemoembolization Using Alkylcyanoacrylates 277
        15.2.6.9 Delivery of Peptide Anti-Cancer Drugs 278
15.3 Conclusions 280
References 280

15.1 CANCER THERAPY

Cancer is defined by two characteristics: uncontrolled cell growth and the ability to invade, metastasize, and spread to distant sites. Despite several modes of therapy—such as chemotherapy, immunotherapy, and radiotherapy—the therapy of cancer remains a challenge. Systemic chemotherapy is the widely adopted treatment for the disseminated malignant disease, while recent progress in drug therapy has resulted in curative chemotherapeutic regimens for several tumors. However, chronic administration of anti-cancer drugs leads to severe systemic toxicity. To improve the therapy with these anti-cancer agents and reduce this associated toxicity, drug delivery systems have been introduced to deliver the drugs directly to the site of interest.[1] Normal cells are also susceptible to the cytotoxic effects of chemotherapeutic agents and exhibit a dose response effect, but the response curve is shifted relative to that of malignant cells. Proliferative normal tissues such as bone marrow and gastrointestinal mucosa are generally the most susceptible to chemotherapy-induced toxicity. Because chemotherapy has been widely adopted for the treatment of cancer, several chemotherapeutic agents with potential anti-cancer activity have been introduced.

Several drug delivery systems facilitate effective chemotherapy with the anti-cancer agents. These systems include liposomes, microparticles, supramolecular biovectors, polymeric micelles, and nanoparticles. Although research on targeted therapeutic systems for cancer therapy has not significantly contributed to their commercialization for human application, the introduction into the market of products like doxorubicin long-circulating liposomes (Doxil) and Polifeprosan-20 with carmustine (BCNU, Gliadel Wafer) for cancer therapy has renewed interest in the field of targeted drug delivery to cancers.[2]

15.2 NANOPARTICLES AS DRUG DELIVERY SYSTEMS

The challenge of modern drug therapy is the optimization of the pharmacological action of the drugs coupled with the reduction of their toxic effects in vivo. The prime objectives in the design of drug delivery systems are the controlled delivery of the drug to its site of action at a therapeutically optimal rate and dosage to avoid toxicity and improve the drug effectiveness and therapeutic index.

Among the most promising systems to achieve this goal are the colloidal drug delivery systems, which include liposomes, niosomes, microemulsions, and nanoparticles. Nanoparticles are considered promising colloidal drug carriers, as they overcome the technological limitations and stability problems associated with liposomes, niosomes, and microemulsions. For instance, camptothecin-based drugs, because of their poor solubility and labile lactone ring, pose challenges for drug delivery. Williams et al.[3] developed a nanoparticle delivery system for a camptotheca alkaloid (SN-38) that is stable in human serum albumin; high lactone concentrations were observed even after 3 h and showed prolonged in vivo half-life of the active (lactone) form.

Nanoparticles are solid colloidal particles ranging in size from 10 to 1000 nm. Depending on the process used for their preparation, two types of nanoparticles are obtained: nanospheres and nanocapsules (Figure 15.1). Nanospheres possess a rigid matrix structure that incorporates the drug, whereas the nanocapsule contains an oily core incorporating the drug surrounded by a membrane structure.[4] These are prepared from either synthetic or natural macromolecules in which the drug is dissolved, entrapped, encapsulated, or adsorbed on the surface. Generally, the definition of nanoparticles includes not only the particles described by Birrenbach and Speiser[5] by the term "Nanopellets," but also nanocapsules and polymer lattices such as the "molecular scale drug entrapment" products described by Rhodes et al.[6] and Boylan and Banker.[7] Nanoparticles possess a matrix structure that incorporates the pharmacologically active substance and facilitates the controlled release of the active agent.

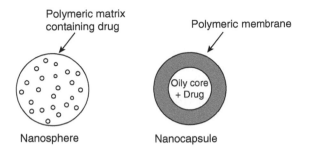

FIGURE 15.1 Representation of nanosphere and nanocapsule.

15.2.1 Polymeric Nanoparticles

In recent years, biodegradable polymeric nanoparticles have attracted attention as potential drug carriers because of their applications in the controlled delivery of drugs, their ability to target organs and tissues, their function as carriers for DNA in gene therapy, and their ability to perorally deliver proteins, peptides, and genes.[8,9] Langer[8] stated that the construction of fully biomimetic matrices containing all of the required chemical and topographical factors has not yet been achieved. Scientists are focusing on the design of new polymeric materials in which both topographical features and chemical cues act synergistically to qualify as drug-delivery carriers. Polymeric nanoparticles can be prepared either by the polymerization of monomers or from the preformed polymers. This chapter will focus on the former process.

15.2.1.1 Nanoparticles Prepared by the Polymerization Process

Historically, the first methods used to produce nanoparticles were derived from the field of latex engineering developed by polymer chemists. These methods were based on the in situ polymerization of monomers in various media. In the early 1970s, Birrenbach and Speiser, the pioneers in the field, produced the first polymerized nanoparticles for pharmaceutical use. Two types of polymerization processes were adopted for the preparation of polymeric nanoparticles: dispersion polymerization (DP) and emulsion polymerization (EP).

15.2.1.1.1 Dispersion Polymerization

A DP involves a monomer, an initiator, a solvent in which the newly formed polymer is insoluble, and a polymeric stabilizer. The polymer is formed in the continuous phase and precipitates into a new particle phase stabilized by the polymeric stabilizer. The nucleation is directly induced in the aqueous monomer solution. Small particles are formed by the aggregation of growing polymer chains precipitating from the continuous phase when these chains exceed a critical chain length. When enough stabilizer covers the particles, coalescence of these precursor particles with themselves and with their aggregates results in the formation of stable colloidal particles. In this technique, the monomer is dispersed or dissolved in an aqueous medium from which the formed polymer precipitates.

Essentially, the initiation of a monomer for polymerization requires an initiator that can generate ions or radicals to start the polymerization process. If the nucleation of the monomer is due to ions, then the mechanism is called "ionic polymerization." Depending on the type of ions produced, the ionic polymerization may be anionic or cationic. If the radical nucleates the monomer, then the mechanism is known as "radical polymerization."[10]

15.2.1.1.2 Emulsion Polymerization

Emulsion polymerization is a method for producing a latex and a polymer that exhibit any desired morphology, composition, sequence distribution, surface groups, molecular weight distribution, particle concentration, or other characteristics. In this technique, the monomer is emulsified in a nonsolvent-containing surfactant, which leads to the formation of monomer-swollen micelles and stabilized monomer droplets. The polymerization is performed in the presence of an initiator. The initiator generates either radicals or ions depending on the type of initiator, and these radicals or ions nucleate the monomeric units and start the polymerization process. The monomer-swollen micelles exhibit sizes in the nanometer range and thus have a much larger surface area than the monomer droplets. It has been assumed that, once formed, the free reactive monomers would more probably initiate the reaction within the micelles itself. In this case, the monomer droplets would act as monomer reservoirs. Being slightly soluble in the surrounding phase, the monomer molecules reach the micelles by diffusion from the monomer droplets, thus allowing the polymerization to be continued in the micelles.[11,12] Generally, the reaction continues until the monomer completely disappears. The particles obtained by EP are usually smaller (100–300 nm) than the original stabilized monomer droplets in the continuous phase. EP may be performed using either organic or aqueous media as the continuous phase.

Emulsion Polymerization in an Organic Continuous Phase. EP in the organic continuous phase was the first process reported for the preparation of nanoparticles.[5,13] In this process, the water-soluble monomers are polymerized. Acrylamide and the crosslinker N,N'-bisacrylamide were the prototype monomers to be polymerized using chemical initiators such as N,N,N',N'-tetramethyl-ethylenediamine and potassium peroxdisulphate or by light irradiation by γ- or UV-radiations. However, the high toxicity of monomers and the need for high amounts of organic solvents and surfactants limits the importance of this technique.

The cyanoacrylate monomers are relatively less toxic than acrylamide, and their polymers are biodegradable. Poly(alkyl cyanoacrylate) (PACA) nanoparticles were prepared by EP in the continuous organic phase. In this case, the monomer was added to the organic phase, creating nanoparticles with a shell-like structure (nanocapsules), as well as solid particles.[14] The mechanism of this nanocapsule formation was explained as following. The drug dissolved in aqueous phase was solubilized in the organic phase (containing surfactants) such as *iso*-octane, cyclohexane–chloroform, isopropyl myristate–butanol, and hexane. This result in a microemulsion with water-swollen micelles containing the drug. The monomer diffuse into the swollen micelles, and the OH$^-$ ions initiated the polymerization. The polymerization is so rapid that only an impermeable wall could be formed at the organic/water interface, preventing the diffusion of further monomers into the particles. The high amounts of organic solvents and surfactants required for this process, however, has greatly limited its application.

Emulsion Polymerization in an Aqueous Continuous Phase. This technique is widely used for the preparation of nanoparticles by the polymerization of a wide variety of monomers, including alkylcyanoacrylates. Employing very low quantities of surfactants, it is only used to stabilize the newly formed polymer particles. Apart from this, the polymerization of alkylcyanoacrylates has also been carried out in the absence of surfactants,[15] in aqueous media containing dextran or hydroxylpropyl-β-cyclodextrin (HPCD). Poly(ethyl cyanoacrylate) and poly(isobutyl cyanoacrylate) (PIBCA) nanospheres containing metaclopramide have been prepared using this technique. However, the resulting drug loading was only 9.2% and 14.8%, respectively.

15.2.1.2 Poly(Alkyl Cyanoacrylate) Nanoparticles

The first report on the synthesis of PACA nanoparticles appeared in 1979.[16] PACA nanoparticles are biodegradable and hence are eliminated rapidly from the body.[17] The cyanoacrylate monomers are polymerized in the aqueous medium by the anionic polymerization;[16,18] the mechanism is

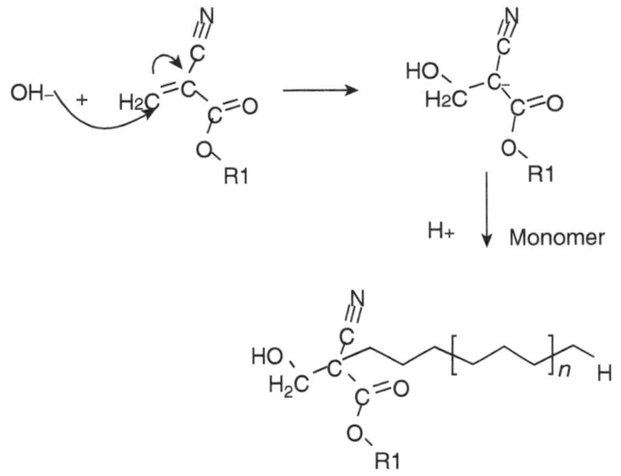

FIGURE 15.2 Mechanism of alkyl cyanoacrylate polymerization. (From Behan, N., Birkinshaw, C., and Clarke, N., *Biomaterials*, 22, 1335, 2001. With permission.)

depicted in Figure 15.2. The cyanoacrylate monomer is initiated by OH^- ions, and the polymerization was performed at a low pH to control the reaction, which otherwise would have led to rapid polymerization resulting in the precipitation of large polymer aggregates. The stabilizers used were found to have a significant influence on the size of the formed polymer particles. High concentrations of poloxamer 188 (above 2%) reduced the particle size of polyisobutyl cyanoacrylate nanoparticles from around 200 nm without an emulsifier, to as low as 31–56 nm.[19] Nanoparticles prepared without stabilizers or prepared using polysorbates were found to have monomodal molecular weight distribution, with mean molecular weights of 1000–4000 Da, whereas the stabilizers such as dextrans or poloxamers led to a distinctive bimodal distribution of the molecular weights, with peaks at 1000–4000 and 20,000–40,000 Da. This bimodal distribution is indicative of two separate polymerization reactions. The first reaction probably occurs in the aqueous phase where the termination of polymerization by H^+ ions is rapid, leading to small molecular weights. The second reaction is possibly the polymerization of captured growing unterminated polymer molecules within the primary particles. Due to the low concentration of H^+ ions in this environment, the termination frequency is reduced and hence the molecular weights would be much higher. Due to the initiation of cyanoacrylate polymerization by bases, the basic drugs also can act as initiators. However, because of the complexity of cyanoacrylate polymerization, the resulting particle size in such multi-component systems is difficult to predict.

Poly(butyl cyanoacrylate) (PBCA) nanoparticles of *n*-butyl cyanoacrylate containing methotrexate were prepared by DP and EP.[20] DP nanoparticles were prepared using dextran as a stabilizer; the EP nanoparticles were stabilized by poloxamer 188. A high zeta potential was observed for nanoparticles prepared by the DP method, whereas the incorporation of methotrexate resulted in a decrease in zeta potential. The DP nanoparticles exhibited a high release of methotrexate, suggesting the channelizing effect of dextran chains incorporated into nanoparticles during polymerization. When tested in two different release media such as 0.1 mol L^{-1} HCl and pH 7.4 phosphate buffer, a significant difference ($p < 0.01$) in release rates was found for DP and EP nanoparticles. Drug release from both the nanoparticles followed Fickian diffusion in 0.1-mol L^{-1} HCl, whereas the mechanism was found anomalous in pH 7.4 phosphate buffer. A similar study conducted on PBCA nanoparticles loaded with doxorubicin hydrochloride by incorporation and adsorption techniques[21] showed rapid drug release in 0.001-N HCl from both DP and EP nanoparticles; the release kinetics followed the Higuchi equation.

15.2.1.3 Poly(Alkyl Cyanoacrylate) Nanocapsules

Nanocapsules are spherical structures formed by an envelope surrounding a liquid central cavity. The technique for the preparation of PIBCA nanocapsules was first reported by Fallaouh et al.[22] The nanocapsules were obtained by injecting the organic phase, consisting of oil, with the dissolved drug, isobutylcyanoacrylate, and ethanol into the aqueous phase containing Pluronic F68 (a nonionic surfactant) under magnetic stirring. The colloidal suspension was concentrated by evaporation under a vacuum and filtered to obtain nanocapsules that possessed a shell-like structure with a wall 3-nm thick. This thickness was confirmed by transmission electron microscopy (TEM) after negative staining with phosphotungstic acid. The pH of the polymerization medium played an important role in obtaining nanocapsules. Nanocapsules with uniform size and low polydispersity were obtained only between pH 4 and 10. The concentration of alcohol above 15% only led to the formation of isolated nanocapsules. The mechanism of nanocapsule formation is probably the interfacial polymerization described by Florence et al.[23] for the manufacture of PACA microcapsules.

A different method was adopted by Chouinard et al.[24] for the preparation of poly(isohexyl cyanoacrylate) (PIHCA) nanocapsules. The monomer was treated with sulfur dioxide and dissolved in a solution containing Miglyol 810 (caprylic/capric triglyceride) in ethanol. This solution was added slowly, in a 1:2 ratio, into an aqueous phase containing a solution of 0.5% poloxamer 407 and 10 mM phosphate buffer under stirring. The nanocapsules were then purified by centrifugation and washing with distilled water. In this method, the sulfur dioxide acts as an inhibitor, preventing the polymerization of the cyanoacrylates by the ethanol. With the addition of the oily phase to the aqueous phase, the sulfur dioxide and ethanol migrate to the external phase, resulting in the formation of a very fine emulsion. In this case, the nanocapsule size was mainly influenced by Miglyol concentration. Freeze-fracture electron microscopy studies showed that polymeric walls of different thicknesses and densities can be prepared, which could be significant for the design of nanocapsules with tailored drug-release kinetics.

Palumbo et al.[25] adopted a similar technique for the preparation of polyethyl-2-cyanoacrylate nanocapsules containing idebenone (an antioxidant). In this method, the acetonic solution of Miglyol 812, idebenone, and monomer was added to 100 mL of aqueous phase (pH 7) containing Tween 80. The presence of nonionic surfactant allowed the polymerization of the ethylcyanoacrylate at the oil–water interface, thus encapsulating Miglyol 812 droplets. The immediate polymerization triggered the formation of drug-loaded nanocapsules, with the suspensions concentrated under a vacuum to remove any acetone. Tween 80 reduced the hydrodynamic size of the emulsion droplets as a function of its concentration, resulting in smaller-sized nanocapsules. These nanocapsules exhibited a negative charge that was determined by zeta-potential measurements. The freeze-fracture electron microscopy revealed the presence of internal oil droplets surrounded by the polymeric shell of the nanocapsule.

15.2.2 Loading of Drugs to Poly(Alkyl Cyanoacrylate) Nanoparticles

The drugs were loaded to PACA nanoparticles, either by incorporation during the polymerization process or by absorption onto the surface of the preformed particles. The former case can lead to the covalent coupling of the drug to the polymer,[26] or to the formation of a solid solution or solid dispersion of the drug in the polymer network. The addition of the drug to empty nanoparticles may lead to covalent coupling[27] of the drug; alternatively, the drug may be bound by sorption. The sorption can also lead either to the diffusion of the drug into the polymer network and the formation of a solid solution,[28] or to surface adsorption of the drug.[29] In the majority of cases, the type of drug loading will determine the carrier capacity. The carrier capacity is defined as the percentage of drug associated with a given amount of nanoparticles for a given initial concentration of drug without considering the characteristics of the adsorption isotherm.[30,31] Several drugs such as betaxolol chlorhydrate,[32] hematoporphyrin,[33] and primaquine[34] were adsorbed onto the preformed PACA nanoparticles, whereas drugs such as doxorubicin[35] and actinomycin D were incorporated into the nanoparticles during the polymerization process. The incorporation of rose bengal into the poly (butyl-2-cyanoacrylate) nanoparticles during the polymerization led to much greater carrier capacity than did the simple adsorption method.[29] The rose bengal incorporated into nanoparticles during polymerization is expected to be either dissolved, dispersed in, or adsorbed on the polymer matrix and remaining to be adsorbed on the particle surface. In contrast, after the adsorption process, the drug solely occupies the external surface of the nanoparticles, including minor pores and pits. Adsorption of hematoporphyrin onto PIBCA nanoparticles resulted in poor carrier capacity of the nanoparticles, because the drug was adsorbed mainly at the surface of the nanoparticles.[33] A similar report by Beck et al.[28] describes the preparation of PBCA nanoparticles loaded with mitoxantrone using both the incorporation and adsorption methods. The proportion of mitoxantrone bound to the particles was analyzed to be about 15% of the initial drug concentration with the incorporation method and about 8% with the adsorption method.

The type of drug binding onto the nanoparticles may result in different release mechanisms and rates.[36] Whether the drug is present in the form of a solid solution or a solid dispersion, its release characteristics mainly depend on the degradation rate of the polymer.[30] Drug desorption (tested in phosphate buffered medium pH 5.9) resulted essentially from a shift in the equilibrium conditions between free and adsorbed drug following the dilution of the nanoparticle suspension, modification of the pH conditions, or enzymatic hydrolysis by the added esterases. The release of hematoporphyrin was very rapid at physiological temperature and pH, and hence would not be useful for in vivo applications.[33] Finally, the surface charge of the particles and binding type between the drug and nanoparticles are the important parameters that determine the rate of desorption of the drug in the nanoparticles.[32] Recently, Poupaert and Couvreur[37] developed a computational derived structural model of doxorubicin interacting with oligomeric PACA in nanoparticles. The oligomeric PACAs are highly lipophilic entities and scavenge the amphiphilic doxorubicin during the polymerization process by extraction of protonated species from the aqueous environment to an increasingly lipophilic phase embodied by the growing PACAs. Interesting, hydrogen bonds were established between the N and H function of doxorubicin and the cyano groups of alkylcyanoacrylate. Therefore, the cohesion in this assembly comes from a blend of dipole-charge interaction, H-bonds, and hydrophobic forces. Upon hydrolytic erosion of the nanoparticles involving mainly the hydrolysis of ester groups to carboxylate, the assembly of doxorubicin–PACA tends to loosen the cohesion as the hydrophobic forces decline due to water intrusion and the increasing contribution of repulsive forces between anionic carboxylates. This creates, for example, a high local gradient of doxorubicin at the cell membrane surface, permitting MDR efflux to be overwhelmed. Very recently, heparin–PIBCA copolymer nanoparticles were developed as a carrier for hemoglobin,[38] which was indicated for treatment in thrombosis oxygen-deprived pathologies. In water, these copolymers spontaneously form nanoparticles with a ciliated surface of heparin. The heparin bound to the nanoparticle preserved its antithrombotic activity. The bound hemoglobin also maintained its capacity to bind ligands.

The integration of drugs with polymerization additives was observed in the case of 5-fluorouracil (5-FU)-loaded PBCA nanoparticles.[39] 5-FU in acidic solution (pH 2–3) may interfere in the initiation process through its amino groups via the formation of zwitterions. The proposed zwitterionic mechanism of initiation was supported by the molecular weight profiles of the polymer, as determined by gel permeation chromatography, and the covalent linkage of the cytostatic to the main polymer chain. ^{1}H NMR analyses demonstrated that a significant fraction of 5-FU was covalently bonded to the PBCA chains through its amino groups, preferentially through one of the two nitrogen atoms.

Stella et al.[40] developed a new concept to target the folate-binding protein by designing poly (ethylene glycol) (PEG)-coated biodegradable nanoparticles coupled to folic acid. This system is the soluble form of the folate receptor that is over expressed on the surface of many tumoral cells.

The copolymer poly[aminopoly(ethylene glycol)cyanoacrylate-*co*-hexadecyl cyanoacrylate] [poly(H$_2$NPEGCA-*co*-HDCA)] was synthesized, and their nanoparticles were prepared using the nanoprecipitation technique. The nanoparticles were then conjugated to the activated folic acid via PEG terminal amino groups and purified from unreacted products. The specific interaction between the conjugate folate-nanoparticles and the folate-binding protein was confirmed by surface plasmon resonance. This interaction did not occur with the nonconjugated nanoparticles used as a control.

15.2.3 CHARACTERIZATION OF NANOPARTICLES

15.2.3.1 Physicochemical Characterization

The methods used for the physicochemical characterization of nanoparticles are listed in Table 15.1. Particle size is the prominent feature of nanoparticles; the fastest and most routine methods of size analysis are photon correlation spectroscopy (PCS) and laser diffractometry (LD). The former method is useful for the determination of smaller particles,[41] whereas the latter method is useful for the determination of larger particles. PCS determines the hydrodynamic diameter of the nanoparticles via Brownian motion. The electron microscopy methods also allow the exact particle determination. Scanning electron microscopy requires the coating of the dry sample with a conductive material such as gold. The gold coating usually results in an estimate of particle sizes that is slightly greater than the normal. TEM, with or without staining, is a relatively easier method to determine particle size. Some nanoparticle materials are not electron dense and thus cannot be stained; others melt and sinter when irradiated by the electron beam of the microscope and thus cannot be visualized by this method. Such nanoparticles can be visualized in TEM after freeze fracture or freeze substitution.[42] This method is optimal because it also allows for the fracturing of the particles and consequently an observation of their interior. Unfortunately, the method is time-consuming and cannot be used for the routine determinations. Recently, new types of high-resolution microscopes such as the atomic-force microscope, the laser-force microscope, and the scanning-tunneling microscope were developed.[43–46] These microscopes are especially useful for the investigation of nanoparticle surfaces.

The molecular weight determination of nanoparticles can be performed after the particles have been dissolved in an appropriate solvent and analyzed by gel-permeation chromatography. Accurate results can be obtained only if the polymer standards have a similar structure and similar properties as the test material. The limitation of this technique, however, is that the

TABLE 15.1
List of Characterization Techniques for Nanoparticles

Parameter	Method of Characterization
Particle size	Photon correlation spectroscopy, laser diffraction, transmission electron microscopy, scanning electron microscopy
Molecular weight	Gel permeation chromatography
Density	Helium compression pycnometer
Crystallinity	X-ray diffractometry, differential scanning calorimetry
Surface charge	Electrophoresis, laser Doppler anemometry, amplitude-weighted phase structuration
Hydrophobicity	Hydrophobic interaction chromatography, contact angle measurements
Surface characteristics	Scanning electron microscopy, atomic force microscopy
Surface element analysis	X-ray photoelectron spectroscopy for chemical analysis (ESCA)

molecular weight of cross-linked polymers and nanoparticles from natural macromolecules cannot be determined.

Density measurements can be performed by the helium compression pycnometry and by density gradient centrifugation. A comparison of these two methods may offer information about the internal structure of nanoparticles.

Further information about the nanoparticle structure and crystallinity may be obtained using x-ray diffraction and thermoanalytical methods such as differential scanning calorimetry, differential thermal analysis, thermogravimetry, and thermal optical analysis.[47] The charge on nanoparticle surfaces is mainly determined by electrophoretic mobility, laser Doppler anemometry, and amplitude-weighted phase structuration.

Hydrophobicity of the nanoparticles can be determined by two major methods: contact angle measurements and hydrophobic interaction chromatography. The former can be performed only on flat surfaces and not directly on hydrated nanoparticles in their dispersion media. For this reason, hydrophobic interaction chromatography is the better method, wherein a differentiation between nanoparticles with different surface properties can be obtained by loading the particles on columns with alkyl-sepharose and eluting them with a Triton X-100 gradient.[48]

Kreuter[41] performed the physicochemical characterization of polymethyl, polyethyl, and PBCA nanoparticles. Because it gives only the mean diameter and the multimodal size distributions cannot be measured, determination of nanoparticle size by Photon Correlation Spectrometry is very susceptible to errors caused by the presence of bigger particles in the dispersion. Another disadvantage is that this method measures the Brownian motion of the particles, and the particle diameter is then calculated from the measurements via the diffusion coefficient.[41] Thus, the particle size determined by this method can be affected by the surrounding medium, including the adsorbed surfactants or hydration layers, and may therefore be different from the sizes obtained by electron microscopy. In case of characterization using electron microscopy, the individual particles can be analyzed and measured. According to Kopf et al.,[26] the surfactants present after the separation of the dispersion medium from the nanoparticles will coat the nanoparticles and lead to the disappearance of individual nanoparticle structures. This coating significantly reduces the particle surface area, and polycyanoacrylate nanoparticles could thus not be obtained surfactant-free in a nonaggregated form. The nanoparticles were x-ray amorphous, and their diffractograms indicated no sign of crystallinity.[20,21] The surface charge of colloidal particles expressed by their electrophoreitc mobility may have an important influence on their body distribution behavior in humans and animals.[49,50] The PACA nanoparticles possess negative charge. Charge measurements in serum yielded a more pronounced decrease caused by the adsorption of the serum contents. Another parameter that is of great importance for the fate of nanoparticles in the blood is the hydrophilicity/hydrophobicity of the particle surface. The latter has a profound influence on the interaction of the particles with the blood components.[49,51,52] The particles' hydrophilicity/hydrophobicity can be determined by measuring the water contact angle.[53] This was done by centrifugation, followed by washing with water and drying the poly(cyanoacrylate) nanoparticles. The dried material was dissolved in acetone and casted as a film with subsequent solvent evaporation.[41] The water droplet contact angles for polymethyl, polyethyl, and PBCA nanoparticles were reported as 55.6°, 64.7°, and 68.9°, respectively.[41]

Muller et al.[48] studied the physicochemical properties of alkyl cyanoacrylates of different chain lengths such as methyl, ethyl, isobutyl, and isohexyl cyanoacrylate. Zeta potential of the nanoparticles was determined as an alternative means of evaluating particle surface charge. Interaction of nanoparticles with serum proteins was assessed by the zeta-potential measurements. The nanoparticles were incubated with poloxamer 407 (Pluronic F-108), and the hydrophobicity determination by contact angle measurement revealed that the coating layer was very thin. The layer thickness was higher in isohexyl cyanoacrylate particles, indicating that they had the most hydrophobic nature of all the polymers used.

The molecular weight differences between polycyanoacrylate nanoparticles with different side-chain lengths as well as the molecular weight differences of polycyanoacrylate nanoparticles produced at different hydrochloric acid concentrations or surfactant concentrations appear to not be very pronounced.[54] Nevertheless, an increase in side-chain length led to the increasing molecular weights, whereas increasing the hydrochloric acid concentrations reduced the molecular weights.

15.2.3.2 Degradation of Poly(Alkyl Cyanoacrylate) Nanoparticles

The type of degradation (bulk or surface degradation) of polymethyl-, polyethyl-, and poly(isohexyl cyanoacrylate) nanoparticles was determined using photon correlation spectroscopic measurements.[48] In the case of surface degradation, the particle size should show an immediate increase while the polydispersity index remaining constant. If bulk degradation dominates, a lag period should occur, preceding the decrease in mean size due to disintegration of the particles. This process of disintegration would lead to a more heterogeneous distribution of particle sizes and consequently to an increase in the polydispersity index. Incubation of polyethyl cyanoacrylate nanoparticles with in 10^{-4} N NaOH led to an immediate, continuous decrease in particle size, as well as an unchanged polydispersity index indicating the predominant surface degradation. Incubation of PIHCA nanoparticles in 10^{-4} N NaOH led to no detectable size decrease; polydispersity was also unchanged, indicating the degradation by surface erosion at a much slower rate than poly(ethyl cyanoacrylate) nanoparticles. Coating the particles with poloxamers did not accelerate the particle degradation. At low electrolyte concentration, the size and dispersity of PIHCA nanoparticles remained unchanged during incubation in 10^{-4} N NaOH. At high ionic strength, the size increase of nanoparticles as a result of flocculation was larger than the size decrease due to degradation. In vitro degradation was also determined by turbidimetric measurements.[55] Poly(-methyl cyanoacrylate) and poly(ethyl cyanoacrylate) nanoparticles underwent the fastest degradation. A high electrolyte concentration in the medium led to the formation of larger aggregates accompanied by an increase in absorption of dispersion. The slowly degrading polymers PIBCA and PIHCA probably release a low concentration of degradation products over a prolonged period of time, indicating their low toxicity.

The stability studies on PBCA nanoparticles reveal that an acidic medium protects the nanoparticles against decomposition.[56,57] Higher temperatures promoted the degradation of nanoparticles. When incubated in human serum in vitro, the nanoparticles showed no particle agglomeration, suggesting their better tolerance in vivo.

15.2.3.3 Influence of Enzymes on the Stability of Poly(Alkyl Cyanoacrylate) Nanoparticles

The peroral route is the most convenient way of delivering drugs, leading to better patient compliance, especially during long-term treatment. However, many drugs, including proteins and peptides, are unstable in the gastrointestinal tract (GIT) or are insufficiently absorbed. Earlier studies have shown that cyanoacrylate nanoparticles[58,59] and nanocapsules[60,61] improve the absorption of several drugs like vincamin or insulin from the GIT. Maincent et al.[58] demonstrated the efficiency of PBCA nanoparticles as a drug delivery system for the GIT by improving the absorption of vincamin by more than 60% compared to an equimolar solution. Investigation of the parameters that influence the stability of nanoparticles in the GIT is much more difficult because of the complexity of the GIT. The pH changes from 2.0 in the stomach to 7–8 in the small intestine. Moreover, in different parts of the GIT there are different enzymes in differing concentrations. Degradation studies with cyanoacrylate was performed as early as 1972.[62] Lenaerts et al.[63] incubated PIBCA nanoparticles in rat liver microsomes. As these microsomes contain a mixture of enzymes, it was difficult to determine which enzyme took part in the degradation process. Scherer et al.[56] studied the influence of different enzymes in vitro on the stability of PBCA nanoparticles. The study used pepsin, amylase, and esterase enzymes, added in a buffer solution that contained dispersed nanoparticles. Amylase was selected because dextran is used as a stabilizer in nanoparticle preparation and was shown to be partially incorporated into the polymer chains.[64] Esterase was selected because hydrolysis of the ester side chain is the main route of degradation of PACA nanoparticles.[65] The study showed significant degradation of PBCA nanoparticles at 37°C and was dependent on the pH of the buffer solution. The addition of pepsin and amylase to the buffer solution was found to have no influence on the stability of nanoparticles. The results upon addition of amylase revealed that dextran either was not incorporated to a significant degree or that its incorporation did not alter the regular nonenzymatic or esterase-dependant degradation pathway and velocity of this polymer in the form of nanoparticles. In contrast, the esterase significantly influenced the stability of the nanoparticles. Esterase demonstrated the highest activity of nanoparticle degradation at neutral pH values; as well, the degradation of PBCA nanoparticles was proportional to the amount of esterase added. These observations suggest that the degradation of PBCA nanoparticles does not occur before the particles reach the small intestine, as esterase is released by the pancreas into the lumen of the small intestine, the major site for absorption of several drugs.

15.2.3.4 Drug Release from Poly(Alkyl Cyanoacrylate) Nanoparticles

Various methods were used to study the in vitro drug release of nanoparticles. They include side-by-side diffusion cells with artificial or biological membranes, the dialysis-bag diffusion technique, the reverse dialysis sac technique, ultrafiltration, ultracentrifugation, and the centrifugal ultrafiltration techniques. Despite the development of these methods, some technical difficulties remain,[66] including the separation of the nanoparticles from the release media when assessing the in vitro drug release from nanoparticles. Leavy and Benita[67] used a reverse dialysis method, in which the nanoparticles were added directly into the dissolution medium. The same technique was adopted by Calvo et al.[68] for the evaluation of nanoparticles, nanocapsules, and nanoemulsions. However, this method is not very sensitive for studying the rapid release formulations, although relatively speaking the dialysis bag technique is the best for assessing in vitro drug release. Because the nanoparticle dispersion is placed in the dialysis bag and immersed into the release medium, this technique avoids the problem of nanoparticle separation.[69,70] Considering the hindrance to drug diffusion caused by the dialysis bag, the release of the drug in solution form can be studied through the dialysis bag, and any diffusional barrier can be compensated for.

15.2.4 In Vivo Distribution of Poly(Alkyl Cyanoacrylate) Nanoparticles

Colloidal drug carriers such as liposomes and nanoparticles are able to modify the distribution of an associated substance. They can therefore be used to improve the therapeutic index of drugs by increasing their efficacy or reducing their toxicity. The polymer nanoparticles, particularly PACA nanoparticles, have attracted considerable attention as potential drug delivery systems, not only for their enhancement of therapeutic efficacy but also because they reduce the toxicity associated with a variety of drugs. Careful design of these delivery systems with respect to the target and route of administration may provide a solution to some of the delivery problems posed by many anti-cancer drugs, including doxorubicin and amphotericin B,[71] and new classes of active molecules such as peptides, proteins, genes, and oligonucleotides.

The use of colloidal particulate carrier systems with controlled particle size (25 nm to 1.7 μm in diameter) is significant in such applications where diminished uptake by mononuclear phagocytes and specific targeting of carriers to particular tissues or cells is of great importance. Studies conducted on the tissue distribution of PBCA nanoparticles with a diameter of 127 nm, loaded with 1-(2-chloroethyl)-3-(1-oxyl-2,2,6,6-tetramethyl piperidinyl)-1-nitrosourea (spin-labeled nitrosourea, SLCNU) suspensions injected intraperitoneally (i.p.) into Lewis lung carcinoma-bearing mice showed a relatively low accumulation of nanoparticles in the liver and spleen; the nanoparticles were mainly found in the lungs, kidneys, and heart.[72] The highest content of the particles was observed in the lungs of tumor-bearing experimental animals damaged by metastases, suggesting the possible usefulness of SLCNU–PBCA nanoparticles for targeting the lung metastases.

Reddy and Murthy[73] compared the pharmacokinetics and tissue distribution of doxorubicin (Dox) solution with Dox-loaded PBCA nanoparticles synthesized by DP and EP techniques after intravenous (i.v.) and intraperitoneal (i.p.) injection in healthy rats. The elimination half-life ($T_{1/2}$) and mean residence time (MRT) in blood was significantly higher than the free Dox after i.v. injection. After i.p. injection, DP nanoparticles quickly appeared in the blood and underwent rapid distribution to the organs of RES, whereas the EP nanoparticles were absorbed slowly into the blood and remained in the circulatory system for a longer time. The bioavailability (F) of DP (∼1.9-fold) and EP nanoparticles (∼2.12-fold) was higher compared to that of the Dox solution after i.p. injection. The distribution to the heart of both types of nanoparticles was low after i.v. and i.p. injection, compared to Dox. This was further proved by experiments using nanoparticles radiolabeled with ^{99m}Tc to study their distribution in Dalton's lymphoma implanted in the thigh region of mice, when administered subcutaneously below the tumor region.[74] A significantly high tumor uptake of ^{99m}Tc-EP nanoparticles (13-fold higher at 48 h post-injection) ($P < 0.001$) was found compared to the ^{99m}Tc-Dox solution.

Blank [^{14}C]-poly(hexyl cyanoacrylate) nanoparticles with diameters of 200–300 nm injected intravenously into nude mice bearing a human osteosarcoma showed distribution to the liver, spleen, lung, heart, kidney, GI tract, gonads, brain, and muscle, as well as in serum and transplanted tumor fragments when investigated by liquid scintillation counting.[75] The peak levels in all organs with the exception of tumors and the spleen were reached within 24 h; the highest levels of radioactivity were found after 7 days in the tumor and spleen. The tissue distribution of naked and either normal immunoglobulin G or monoclonal antibody (antitumor osteogenic sarcoma)-coated poly(hexyl-2-cyanoacrylate) nanoparticles in mice bearing human tumor xenografts showed similar results: the nanoparticles were deposited mainly in the liver and spleen, and no significant uptake was found in the tumors.[76]

The PACA nanoparticles exhibited rapid extravascular distribution after intravenous administration. Intravenous injection of isobutyl-2-cyanoacrylate nanoparticles resulted in high concentrations in the organs of the reticuloendothelial system such as the liver, spleen, and bone marrow. Similar observations were made by Kante et al.[77] in the case of PBCA nanoparticles injected intravenously in rats. In the case of mice with subcutaneously grafted Lewis lung carcinoma cells, the nanoparticles exhibited higher lung concentrations, indicating the use of nanoparticles in the treatment of lung metastasis.[17] However, the results were somewhat different for polybutyl-2-cyanoacrylate nanoparticles injected intravenously into rabbits.[78] The lung accumulation of nanoparticles was very low in that case, indicating that the nanoparticles (126 nm) were not mechanically filtered through the capillary bed of the lungs. The difference in lung concentrations compared to the observations of Grislain et al.[17] can be explained by the difference in particle size (254 nm). Coating with azidothymidine-loaded poly(hexyl cyanoacrylate) (PHCA) nanoparticles with polysorbate 80 led to higher brain levels of azidothymidine after intravenous injection in rats, indicating the facilitation of drug passage across the blood–brain barrier (BBB).[79] Physicochemical studies have shown that the coating of nanoparticles with block copolymers such as poloxamers and poloxamines induces a steric repulsion effect, minimizing the adhesion of particles to the surface of macrophages which in turn results in decreased phagocytic uptake and significantly higher levels in blood and non-RES organs, the brain, intestine, kidneys, etc. Borchard et al.[80] showed that coating the nanoparticles with surfactants resulted in uptake by brain endothelial cells, and that polysorbate 80 was the most effective surfactant for this purpose. Further investigations unequivocally demonstrated that polysorbate 80-coated PBCA nanoparticles loaded with enkaphalin analogue dalargin enabled the delivery of this drug across the BBB, achieving a significant pharmacological effect. Gulyaev et al.[81] reported significantly higher brain concentrations of intravenously injected polysorbate 80-coated Dox–PBCA nanoparticles in rats. Three probable mechanisms were proposed for the enhanced brain transport of polysorbate 80 coated nanoparticles: (1) endocytosis, (2) the opening of the tight endothelial junctions between the brain endothelial cells,[81] and (3) the inhibition of the *P-glycoprotein* efflux pump responsible for multi-drug resistance, which is a major obstacle to cancer chemotherapy.[82,83]

Isobutyl and isohexylcyanoacrylate nanoparticles were used by Ghanem et al.[84] as drug carriers, particularly for some anti-cancer drugs. Body distribution as well as pharmacokinetics in animals and partially in humans using a radiolabeled (^{111}In and ^{99m}Tc) free drug and its nanoparticles showed 60–75% accumulation in the reticuloendothelial system.

Skidan et al.[85] studied the antitumor efficacy of Dox loaded into polysorbate 80-coated nanoparticles to the brain. Pharmacokinetics were studied in rats after the intravenous injection of four Dox preparations: (1) solution in saline, (2) solution in polysorbate 80, intravenous (3) bound to nanoparticles, and (4) bound to nanoparticles coated with polysorbate 80. The results showed no significant difference in the body distribution between the two solution formulations, whereas the two nanoparticle formulations showed significantly decreased drug concentrations in the heart. High brain concentrations of Dox (>6 µg/g) were achieved with the nanoparticles overcoated with polysorbate 80 between 2 and 4 h, whereas the concentrations observed with the other three formulations were always negligible. Antitumor efficacy of Dox associated with polysorbate 80-coated nanoparticles as assessed in rats with intracranially implanted 101/8 glioblastoma resulted in a significant 3.6-fold increase in survival, whereas administration of free Dox or blank nanoparticles did not modify the mortality when compared to the control group.

Following the subcutaneous injections of metoclopramide solution (5 mg/kg) and three different metoclopramide nanosphere suspensions (10 mg/kg) on two phases in wistar rats, there was a rapid absorption, distribution, and elimination of the drug solution. The maximum drug concentration was observed after 30 min of subcutaneous injection of all the tested nanosphere formulations.[15] Polyethylcyanoacrylate-hydroxypropyl cyclodextrin showed the highest concentration, followed by polyisobutylcyanoacrylate–dextran and polyethylcyanoacrylate–dextran. The area under the curves of all nanoparticle formulations was 4.8- to 1.88-times higher than that obtained for the solution form.

Earlier studies showed that polystyrene and other biodegradable microspheres delivered to the intestine are preferentially absorbed by the M-cells of the Peyer's patches.[86,87] Intragastric administration of azidothymidine-loaded PIHCA nanoparticles resulted in high concentration in the intestinal mucosa as well as in the Peyer's patches.[88] Once localized in the mucous or intestinal tissue, the azidothymidine could be slowly released by the enzymatic action of esterases. Very low levels of azidothymidine bound to nanoparticles were detected in the lymph, suggesting an efficient capture of particles through the Peyer's patches resulting in an intracellular localization of the drug. As a result, some immune cells would carry azidothymidine to the mesenteric lymph nodes, the location of an important viral burden and a major replication site of HIV.

PACA nanoparticles are proved to be suitable delivery systems in comparison to other colloidal systems, including liposomes. Reszka et al.[89] observed that the mitanxantrone-loaded PBCA nanoparticles showed the highest tumor concentrations in B-16 melanoma-bearing mice after intravenous injection. Four different formulations—mitanxantrone solution, mitanxantrone-loaded negatively charged liposomes (small unilamellar vesicles), ^{14}C-PBCA nanoparticles, and poloxamine 1508-coated ^{14}C-PBCA nanoparticles—were studied for the biodistribution and tumor accumulation in the tumor-bearing mice. The liposomal formulation showed prolonged circulation in the blood and higher accumulation in the liver and spleen after 24 h of injection, compared to the nanoparticle formulations. Despite the longer circulation in the blood, however, the liposomal formulation could not result in high concentrations in the tumors, compared to the nanoparticle formulations, which showed relatively lower circulation time in the blood, indicating the efficacy of PBCA nanoparticles in delivering mitoxantrone to tumors. Similar results involving the higher efficacy of mitoxantrone-loaded PBCA nanoparticles in B-16 melanoma was reported by Beck et al.[28]

15.2.4.1 Surface Modification to Alter Biodistribution

The potential of injectable polymeric drug carriers (nanoparticles) is compromised by their accumulation in the tissues of the mononuclear phagocytic system (MPS). Thus, the therapeutic efficacy of the drugs associated with the nanoparticles is limited to the treatment of several liver diseases.[90,91] Long-circulating nanoparticles can be obtained by their surface modification with dysopsonic polymers such as PEG.[92,93] Indeed, these hydrophilic and flexible polymers can prevent the opsonin–nanoparticle interaction, which is the first step of the recognition by the immune system. In the tumor-bearing animal models, these long-circulating nanoparticles would be able to extravasate thorough the endothelium, allowing drugs to concentrate in the tumors.[94] Recently, it was shown that the synthesis of amphiphilic PEGylated polymers allows the direct preparation of PEG-coated nanoparticles, ensuring the stability of the PEG coating layer because the PEG chains remain chemically linked to the nanoparticle core.[95–97] Otherwise, a PEG coating layer simply adsorbed onto preformed nanoparticles is desorbed in vivo.[78] Peracchia et al.[98] synthesized a novel MePEGcyanoacrylate–hexadecylcyanoacrylate (PEG–PHDCA) copolymer, and the presence of a PEG layer on the nanoparticle surface was confirmed by surface chemical analysis.[99] Poly(alkyl cyanoacrylate) nanoparticles are rapidly biodegradable and hence could be able to release the associated drug into the bloodstream in a controlled manner before liver capture could occur. For solid tumors, extravasation could be followed by a rapid degradation of the polymer, leading to a rapid drug release.[100] The cytotoxicity of PHDCA polymer was decreased after its PEGylation in the mouse macrophage cell line J774.[101] This is due to the lower extent of the particle/cell interaction, because of the steric repulsion effect of PEG chains. Biodistribution studies with the above MePEGcyanoacrylate–hexadecylcyanoacrylate copolymer nanoparticles revealed that the nanoparticles exhibited long-circulating properties.[102] No effect of the degree of PEGylation on the plasma retention was observed because increasing the copolymer from 1:5 to 1:2 did not improve the nanoparticle circulation time. The 1:5 PEG–PHDCA copolymer probably already exhibited a PEG content that was able to determine a brush PEG configuration[103] and, thus, an optimal PEG density at the particle surface for the steric repulsion to occur efficiently. Together, the nanoparticles showed low liver accumulation and high spleen uptake, representing the possibility of targeting drugs to this tissue. More recently, a rapidly biodegradable copolymer poly(methoxyethylene glycol-cyanoacrylate-*co-n*-hexadecylcyanoacrylate) has been developed for the preparation of stealth nanoparticles.[102,104] Similarly, Ping et al.[105] synthesized poly(methoxyethyleneglycol cyanoacrylate-*co-n*-hexadecyl cyanoacrylate) (PEGylated PHDCA) nanoparticles loaded with an antitumor agent salvicine. The nanoparticles showed a significant initial burst of salvicine followed by a sustained release in 7.4 pH phosphate-buffered saline, thus suggesting its usefulness in tumor therapy.

15.2.5 Toxicity of Polyalkyl Cyanoacrylate Nanoparticles

Toxicity of PACA nanoparticles varies depending on the alkyl chain length in the polymer. Incubation of hepatocytes with 1% poly(methyl cyanoacrylate) nanoparticles resulted in complete perforation of cell membranes.[106,107] However, PIBCA nanoparticles showed no signs of toxicity at the same concentration. This toxicity is probably due to the presence of nanoparticle degradation products in the cytoplasm following their phagocytosis. The low toxicity of isobutyl product is due to its slower degradation than the methyl product. The toxicity and safety data on PBCA nanoparticles is somewhat contradictory. Kante et al.[106] found LD_{50} of 230 mg/kg of PBCA nanoparticles injected intravenously in mice. Olivier et al.[108] observed a 30–40% mortality rate in mice after the injection of 166 mg/kg of PBCA nanoparticles; some animals died even at low doses. Meanwhile, Simeonova et al.,[109] in an investigation of the immunomodulating properties of PBCA nanoparticles in mice, did not report any signs of toxicity at doses as high as 200 mg/kg. A much higher LD_{50} of 585 mg/kg was found for nanoparticles made of a similar polymer PHCA.[110] Gelperina et al.[111] reported no mortality in rats even after administering up to 400 mg/kg of empty PBCA nanoparticles.

The implantation into the tissue of monomers, which polymerize within the body in orthopedic surgery and percutaneous minimally invasive radiological procedures, was performed earlier. The polymerization was mostly exothermic, and is associated with the toxicity to the tissue.[112,113] The use of n-butylcyanoacrylate in the endovascular treatment of arteriovenous malformations has been well documented.[113,114] Implantation of n-butyl-2-cyanoacrylate (NBCA) has resulted in a good occlusive effect on the parenchyma of highly vascular tumors, with extensive intralesional necrosis. The antitumor effect of the NBCA at the implant surface on adjacent tumors appeared limited.[115] Histopathological examination of a surgically removed malignant pelvic para-ganglioma after a percutaneous injection of NBCA, three days after the NBCA injection, revealed variable degrees of tumoral necrosis.[116] Areas surrounding small volumes of NBCA at the periphery of the tumor showed practically no necrosis, whereas regions inside the tumor plastified with NBCA presented greater degrees of necrosis, especially toward the center of the tumoral mass (Figure 15.3). However, no necrosis was found at a macroscopic level outside the implant, indicating that the NBCA allowed for the destruction of tumor tissue within the heavily plastified areas. The carcinogenic potential of cyanoacrylate adhesive and isobutyl-2-cyanoacrylate were also reported respectively by Matsumoto[117] and Samson and Marshal.[118] Induction of malignant fibrous histiocytoma in female Fisher rats by implantation of foreign bodies was studied by Hatanaka et al.[119] Five kinds of foreign bodies (silicone, cellulose, polyvinyl chloride, and zirconia and alkyl-alpha cyanoacrylate) were implanted into the subcutaneous tissue of female Fisher rats. The tumors developed in almost all the cases was composed of a mixture of cells that resembled fibroblast,

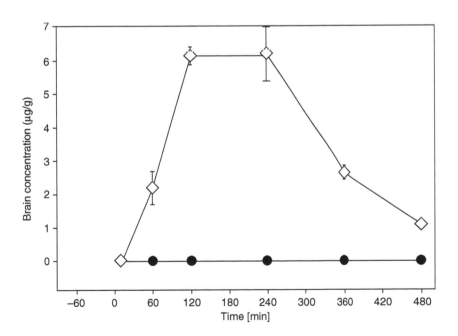

FIGURE 15.3 Uptake by rat brain of i.v. injected doxorubicin loaded to polysorbate 80-coated nanoparticles. Doxorubicin levels in rat brain (μg/g) versus time (min). ● Doxorubicin (5 mg/kg) in saline. ◇ Doxorubicin (5 mg/kg) bound to nanoparticles overcoated with 1% polysorbate 80. (From Gulyaev, A. E. *et al.*, *Pharm. Res.*, 16, 1564, 1999. With permission.)

Golgi apparatus, histiocytes, myofibroblasts, and immature mesenchymal cells. A rat malignant fibrous histiocytoma transplanted into the subcutaneous tissue of the syngeneic female Fisher rats grew and metastasized to the lungs. Canter et al.[120] reported the consecutive pathological findings of a patient who underwent surgery for facial hemangioma after percutaneous injection of NBCA for devascularization of a lesion. The patient underwent additional surgery one month and six months after the initial operation for the removal of the residual NBCA cast from the injection site. Acute inflammatory findings after injection of NBCA and the development of a chronic granulomatous foreign-body reaction support the histological findings of experimental animal studies and postmortem examinations of humans. Though the alkylcyanoacrylates of smaller alkyl chain lengths exhibited considerable toxicity, the alkylcyanoacrylates of larger chain lengths such as isohexylcyanoacrylate nanoparticles resulted in negligible toxicity[121] and drove them to their clinical trials.

15.2.5.1 Toxicity of Drugs Associated with Poly(Alkyl Cyanoacrylate) Nanoparticles

The study of the acute toxicity of doxorubicin associated with polysorbate 80-coated nanoparticles in healthy rats and rats with intracranially implanted 101/8 glioblastoma was reported by Gelperina et al.[111] Single intravenous administration of empty PBCA nanoparticles in the dose range of 100–400 mg/kg did not cause mortality within the period of observation and also did not affect body weight or the weight of the internal organs. Association of doxorubicin with PBCA nanoparticles did not produce significant changes of quantitative parameters of acute toxicity of the antitumor agent. Efficacy and toxicity of mitoxantrone-loaded PACA nanoparticles were compared with a drug solution and with a mitoxantrone-liposome formulation.[89] Furthermore, the influence of an additional coating surfactant, poloxamine 1508, which has been shown to change the body distribution of other polymeric nanoparticles, was investigated. PACA nanoparticles led to a significant reduction in tumor volume of B-16 melanoma in mice compared to liposomes. However, neither nanoparticles nor liposomes had overcome the mitoxantrone-associated leucocytopenia.

In a study by Gibaud et al.[122] involving the determination of the myelosuppressive effects of free and PACA-bound doxorubicin in mice in vivo, the alkyl cyanoacrylate nanoparticle-bound doxorubicin showed the highest myelosuppressive effect. The total and differential counts of blood, bone marrow, and spleen cells, as well as the number of granulocyte progenitors (CFU-GM), was determined by culture after the intravenous injection of 11 mg/kg body weight of doxorubicin, either free or bound to PIBCA or PIHCA nanoparticles. Doxorubcin bound to PIHCA nanoparticles showed the highest and longest myelosuppressive effects, which correlated well with a high concentration of the drug in the bone marrow and spleen. Furthermore, the PIHCA nanoparticles induced the release of colony-stimulating factors, which might account for the observed increase in the toxic effects of doxorubicin on bone marrow progenitors.

15.2.6 Drug Delivery in Cancer with Poly(Alkyl Cyanoacrylate) Nanoparticles

15.2.6.1 Targeting to Cancer Cells and Tissues

Increase in the antitumoral activity of many anti-cancer drugs was also observed when loaded in polyisohexylcyanoacrylate nanoparticles. A study[123] conducted with doxorubicin-loaded polyiso-hexylcyanoacrylate nanoparticles administered i.v. to the C57BL/6 mice consisting of metastases induced by i.v. inoculation of reticulosarcoma M5076 cell suspension showed a dramatic increase in the antitumoral activity of the drug. Doxorubicin measurements in healthy hepatic or neoplastic tissue, carried out with histological examinations using TEM, demonstrated that the hepatic tissue is an efficient reservoir of the drug when it was injected in association with nanoparticles. Accumulation of doxorubicin-associated nanoparticles suggests that the Kupffer cells created a gradient of drug concentration for a massive and prolonged diffusion of the free drug toward the neoplastic tissue. The negatively charged mitoxantrone PBCA nanoparticles (DHAQ-PBCA-NP) prepared by

EP method[124] following intravenous administration were concentrated mainly in liver tumors. The accumulation of DHAQ-PBCA-NP was higher in the liver tumors than in the liver tissue. Yang et al.[125] compared antitumor activity of mitoxantrone (DHAQ) and mitoxantrone-polybutyl cyanoacrylate-nanosphere (DHAQ-PBCA-NS) against experimental liver tumor H22 in mice. The results of antitumor activity determination of both drugs with different treatment schedules showed DHAQ-PBCA-NS to present higher activity than DHAQ; DHAQ-PBCA-NS is possessed of liver-targeting property.

Soma et al.[126] demonstrated that the Doxorubicin-loaded PACA nanoparticles are more efficient than free drugs in mice bearing hepatic metastasis of the M5076 tumor. High phagocytic activity of Kupffer cells in the liver played the role of drug reservoir after nanoparticle phagocytosis. The study also assessed the role of macrophages in mediating the cytotoxicity of doxorubicin-loaded nanoparticles on M5076 cells. After the phagocytosis of the doxorubicin-loaded nanoparticles, J774.A1 cells were able to release the active drug, allowing it to exert its cytotoxicity against M5076 cells. Intracellular accumulation and DNA binding of doxorubicin encapsulated in PIHCA nanospheres and of daunorubicin bound to polyglutamic acid (DGA) in comparison with free Dox and daunorubicin was studied by Bogush et al.[127] using methods involving interaction between an anthracycline and cellular DNA to understand how the drug reaches its nuclear targets. The results showed that the intracellular accumulation and DNA binding of Dox-loaded nanospheres and DGA are reduced by 30–40% in comparison with those obtained for free doxorubicin and daunorubicin, respectively.

Studies by Bennis et al.[128] on the cytotoxicity and accumulation of doxorubicin encapsulated in PIHCA nanospheres in a model of doxorubicin-resistant rat glioblastoma variants revealed that the cytotoxicity differed with the degree of resistance to this drug. The nanoparticle associated Dox was more cytotoxic than the free Dox, whereas co-administration of drug-free nanoparticles with free Dox did not significantly modify the cytotoxicity of Dox. However, Simeonova and Antcheva[129] reported that the cytotoxic activities of drug-free PBCA nanoparticles free of vinblastine, vinblastine-loaded nanoparticles (by incorporation and adsorption processes), and a mixture of vinblastine-free and drug-free PBCA nanoparticles, when compared in vitro on human erythroleukemic K-562 cells, showed enhanced cytotoxicity when vinblastine was either adsorbed on PBCA or mixed with them rather than free. When vinblastine was incorporated into the polymer matrix of nanoparticles, a lag period and a postponed cytotoxic effect on K-562 cells was observed. Preclinical studies[121] in 21 patients with refractory solid tumors also confirmed an increase in doxorubicin cytotoxicity and a decrease in cardiotoxicity when incorporated into biodegradable PIHCA by modifying tissue distribution. In clinical studies[130] conducted on 21 patients with Stages III–IV maxillary tumors, antitumor drugs deposited with a cyanoacrylate adhesive solution showed immediate and late results of therapy. This evidence of deposition of the drug and its regulated elimination from the composition helps create a high concentration of the agent at the operation site where the tumor elements remained after surgery.

Pan et al.[131] demonstrated the higher efficacy of 5-FU when loaded into PBCA nanocapsules with a diameter of 338 nm, compared to the free drug in S180, HepS, and EAC tumors.

The in vitro and in vivo antihepatoma effects of the liver-targeted drug delivery system lyophilized aclacinomycin-A loaded polyisobutylcyanoacrylate nanoparticle (ACM-IBC-NP) was found to be significantly concentration dependent.[132] The antihepatoma activity of lyophilized ACM-IBC-NP was higher than that of the aclacinomycin-A. Similarly, actinomycin-D absorbed into polymethylcyanoacrylate nanoparticles showed increased efficiency against an experimental tumor in comparison to the free drug.[133]

The antitumor efficacy of a medical-grade cyanoacrylate adhesive MK-8-based composition with prospidin has been studied by Chissov et al.,[134] and the kinetics of the drug elimination have been followed up in a rat model. The findings showed a prolongation of the relapse-free period and a longer survival of the experimental animals. Clinical trials of the composition carried out in 70 patients with esophageal carcinomas showed local disseminated forms in 52 patients. Application of the adhesive composition has not increased the rate of postoperative complications; in some cases, the life span of patients after palliative surgery was considerably longer. In a study by Pérez et al.[135] using an N-butyl-2-cyano acrylate-based tissue adhesive, Tisuacryl, as a nonsuture method for closing wounds in oral surgery involving apicectomy, molar extractions, and mucogingival grafting, the adhesive was well tolerated by the tissue and permitted immediate hemostasis and normal healing of incisions.

Association of 5-FU to PBCA nanoparticles also showed enhanced efficacy against Crocker sarcoma S 180 and a higher toxicity of the drug, as measured by induced leukopenia, body weight loss, and premature death.[31] The efficacy was further increased by an increase in the polymer-to-drug ratio. The nanoparticles yielded a prolonged persistence of the 5-FU in all organs examined, including the tumor.

15.2.6.2 Drug Delivery to Resistant Tumor Cells

Several cancers show resistance to conventional chemotherapy. Resistance to cancer chemotherapy involves both altered drug activity at the designated target and modified intra-tumor pharmacokinetics (e.g., uptake and metabolism). The membrane transporter *P-glycoprotein* plays a major role in pharmacokinetic resistance by preventing sufficient intracellular accumulation of several anti-cancer agents. Inhibiting the *P-glycoprotein* has great potential to restore chemotherapeutic effectiveness.[136] PACA nanoparticles were found to overcome multi-drug resistance (MDR) phenomena at both the cellular and noncellular level.[137,138]

Doxorubicin is a well known *P-glycoprotein* substrate. The resistant cells treated with doxorubicin-loaded PACA nanoparticles showed much higher sensitivity to the drug relative to the free doxorubicin.[139] Dox and Dox-loaded PIHCA nanoparticles were tested in two human tumor cell lines, K562 and MCF7, at increasing concentrations. The cell lines were more resistant to the free Dox than to the Dox-loaded nanoparticles. However, the MCF7 sublines showed a higher level of resistance to Dox-loaded nanoparticles than that of the free Dox. Different levels of overexpression of several genes involved in drug resistance (MDR1, MRP1, BCRP, and TOP2alpha) occurred in the resistant variants. MDR1 gene overexpression was consistently higher in free-Dox selected cells than in Dox-loaded nanoparticle-selected cells; this was the reverse for the BCRP gene. Overexpression of the MRP1 and TOP2alpha genes was also observed in the selected variants. The results indicate that several mechanisms may be involved in the acquisition of drug resistance and that drug encapsulation markedly alters or delays these processes.[140]

Further experiments, such as uptake studies in the presence of cytochalasin B or efflux studies, indicated a possible mechanism of nanoparticle–cell interaction. The degradation of the drug carrier was also shown to play a key role in the mechanism of action. The formation of an ion pair by the polycyanoacrylic acid resulting from the nanoparticle degradation with doxorubicin[141] has been considered as one of the possibilities (Figure 15.4). It was hypothesized that the nanoparticles adhere to the cell surface, followed by the release of doxorubicin and nanoparticle degradation products, which combine as an ion pair able to cross the cell membrane without being recognized by the *P-glycoprotein* (Figure 15.5).[142,143] The ion pair was considered to form between the positively charged doxorubicin and the negatively charged polycyanoacrylic acid, resulting in the charge masking of the doxorubicin and thereby making it more lipophilic and able to cross the cell membrane. The formation of an ion pair of 2-cyano-2-butyl hexanoic acid (a PACA-like molecule) with the doxorubicin has been assessed using ion pair reverse-phase high-performance liquid chromatography. As the PACA possess high molecular weight ($\sim$ 1000) and their nanoparticles possess large polydispersity, a PACA-like molecule, i.e., 2-cyano-2-butyl hexanoic acid (molecular weight of 197.28 g/mol) was used to investigate the formation of the ion pair. The ion pair formation was found between the charged amino group of doxorubicin and the carboxyl group of 2-cyano-2-butyl hexanoic acid, leading to the overall decrease in charge of the drug and the increase in drug lipophilicity (Figure 15.6). The experimental results were justified based on two

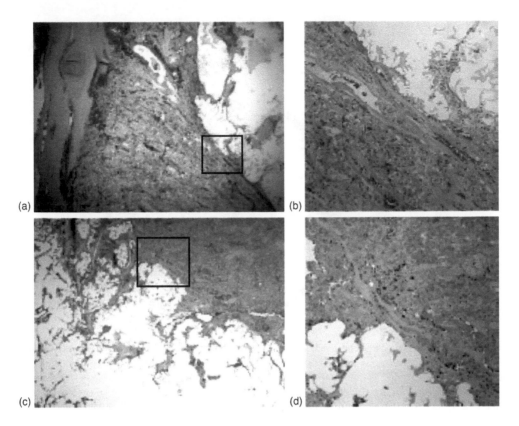

FIGURE 15.4 Histological section of resected paraganglioma after percutaneous NBCA injection. (a) Section close to the tumor surface showing NBCA (void) and viable tumor around the implant (H&E, original magnification X25). (b) High-power view of (a). Note absence of necrosis around the implant (H&E, original magnification X100). (c) Section in a deeper zone of the tumor. Note extensive necrosis (H&E, original magnification X25). (d) High-power view of (c). Note an acellular necrotic area (H&E, original magnification X100). (From San Millan Ruiz, D. et al., *Bone*, 25(Suppl. 2), 85S, 1999. With permission.)

assumptions: (1) the 2-cyano-2-butyl hexanoic acid and PACAs behave similarly in terms of acidic properties, and (2) the organic and aqueous buffer mixtures used in the chromatographic procedure could be correlated with the physiological medium.

In another experiment,[142] C6 rat glioblastoma cells—the sensitive, resistant C6 0.001 (continuously grown in 0.001 μg doxorubicin/mL medium) and the C6 0.5 cells (continuously grown in 0.5 μg doxorubicin/ml medium)—and the drug/formulation were separated by a polycarbonate membrane (pore size 0.2 μm) (Figure 15.7). The C6 sensitive cells showed the same cytotoxicity in the presence or absence of the membrane. While the doxorubicin-loaded PIHCA nanoparticles were found to overcome the resistance of C6 0.001 and C6 0.5 compared to the free drug and free drug + blank nanoparticles when in direct contact with the drug, a 4-fold and 5-fold resistance, respectively, was maintained when separated by the membrane. This clearly indicates the necessity of contact of the drug-loaded nanoparticles with the cell membranes to overcome the MDR in resistant cells. To confirm the possible interference of the nanoparticle degradation products with *P-glycoprotein*, the inhibition study of tritiated azidopine binding to the *P-glycoprotein* enriched membranes of C6 0.5 cells was performed. The blank nanoparticles could not inhibit the azidopine binding even after spontaneous degradation, whereas both the free doxorubicin and its nanoparticle formulation decreased the binding, indicating that the degradation products of nanoparticles do not have a direct role in *P-glycoprotein* (P-gp) inhibition and that it is the combination of both the

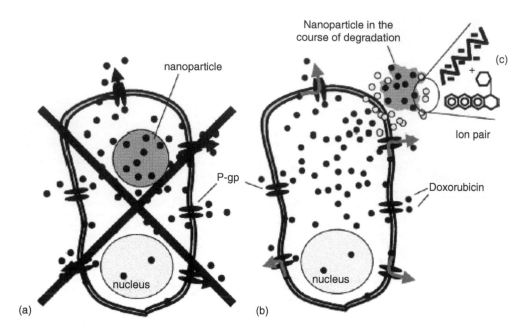

FIGURE 15.5 Hypothesis about the mechanism of action of PACA nanoparticles to overcome MDR at the cellular level. Drug-loaded nanoparticles are not endocytosed by the resistant cells (a) but adhere to the cell surface where they degrade and simultaneously release degradation products and the drug (b). The degradation products and the drug form ion pairs (c) that can penetrate the cells without being recognized by the P-gp and, by this means, increase the intracellular concentration of the anti-cancer drug in the resistant cells. (From Vauthier, C. *et al., J. Control. Release*, 93, 151, 2003. With permission.)

doxorubicin and the PACA in the nanoparticle formulation that plays a role in *P-glycoprotein* inhibition.

Among the various strategies implemented for bypassing the MDR in cancer cells (such as the use of ribozymes,[144] oligonucleotides,[145] and chemosensitizing compounds), the co-administration of chemosensitizing compounds with the anti-cancer agents has been widely investigated.[146] Of such compounds, cyclosporine A, a potent immunosuppressive agent, was shown to prevent MDR in resistant cells in culture[147] and exhibited promising clinical results in refractory leukemia.[148] Cyclosporine A is believed to bind directly to *P-gp*, thereby inhibiting the efflux pump of the cytotoxic agents, resulting in their higher intracellular accumulation. Co-encapsulation of cyclosporine A and doxorubicin into polyisobutyl cyanoacrylate nanoparticles has resulted in a significant improvement in the reversion of MDR in P388-resistant cells compared to the mixed solution of doxorubicin and cyclosporine A and the nanoparticle-loaded doxorubicin incubated with free cyclosporine A.[149] The nanoparticles target the loaded drugs close to the cell membrane, providing the higher drug concentrations at the cell surface,[150] and this could be the reason for the higher efficacy of drugs loaded into the nanoparticles compared to those in solution form.

Co-administration of cyclosporine A with anti-cancer agents was shown to overcome the multi-drug resistance in resistant cell lines.[147] To understand the contribution of macrophages in delivering the doxorubicin-loaded nanoparticles to the tumor cells, a co-culture of sensitive or resistant P388 tumor cells and macrophage monocyte cells J774.A1 were incubated with the free doxorubicin, doxorubicin-loaded isobutylcyanoacrylate nanoparticles, and a combined cyclosporine A and doxorubicin-loaded nanoparticle formulation.[151] When incubated directly with tumor cells,

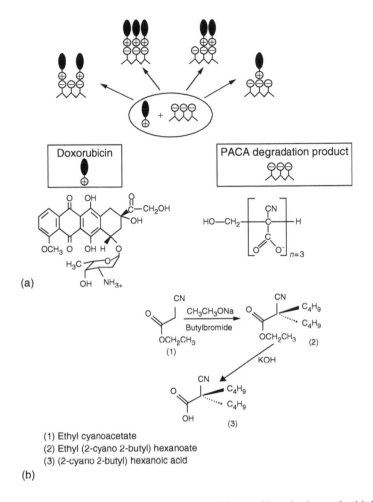

FIGURE 15.6 2-Cyano-2-butyl hexanoic acid and doxorubicin. (a) Hypothesis on the biological action of PACA ion pair formation with doxorubicin. (b) Structure of the hypothetic ion pair formation and synthesis of the 2-cyano-2-butyl hexanoic acid (From Pepin, X. *et al., J. Chromatogr. B*, 702, 181, 1997. With permission.)

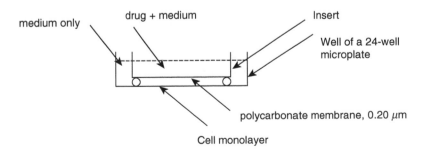

FIGURE 15.7 Schematic representation of the experimental device used for separating cell and drug compartments by a polycarbonate membrane (pore size 0.20 km). (From Hu, Y.-P. *et al., Cancer Chemother. Pharmacol.*, 37, 556, 1996. With permission.)

both the doxorubicin and the doxorubicin-loaded nanoparticle formulation were equally cytotoxic. However, in co-culture, the doxorubicin-loaded nanoparticle formulation was more effective than the free doxorubicin. In co-culture, whereas the nanoparticle-loaded doxorubicin enhanced the cytotoxicity compared to the free drug, it has partly overcome the multidrug resistance in resistant cells. The addition of cyclosporine A improved the cytotoxicity of doxorubicin irrespective of the formulation, but it could not overcome the MDR. However, the efficacy of cyclosporine A in overcoming the MDR was higher in the case of the combined nanoparticle formulation with the doxorubicin, compared to the effect seen with the mixed nanoparticles of cyclosporine A and the doxorubicin prepared separately. Such a combinatorial nanoparticulate system would have both enhanced efficacy against the tumor and be able to overcome the adverse effects associated with the administration in the free form.

Doxorubicin-14-O-hemiadipate (a 14-O-hemiadipate of doxorubicin) (H-Dox) was shown to partially circumvent the P-gp in MCF7/R human breast carcinoma and P388/R murine leukemia cell lines.[152] The comparative studies of the accumulation, retention, and intracellular localization of H-Dox and Dox in Dox-sensitive murine leukemia cell line P388/S and its resistant variant, Pgp-positive drug-resistant P388/R subline, in the presence or absence of cyclosporine A, showed a 38-fold decrease in Dox accumulation but only a 5-fold decrease in H-Dox accumulation, indicating a greater than 7-fold increase in H-Dox buildup in resistant cells. Coincubation with cyclosporine A resulted in a 54-fold increase in Dox accumulation and only a 5-fold increase in H-Dox uptake in P388/R cells, restoring the doxorubicin levels in P388/R to 100% of that found in P388/S cells. Once internalized by the resistant cells, H-Dox was retained better than Dox, regardless of the presence or absence of CsA. The H-Dox was localized in the nuclei of the P388/R cells but not the Dox.[153] Coadministration of chemosensitizers like the cyclosporine analogue SDZ PSC 833 [(3'-keto-Bmt1)-(Val2)-cyclosporin] (PSC 833) and anti-cancer drugs has been shown to possess powerful chemosensitization properties in vitro, in addition to being intrinsically nontoxic.[154]

Complete reversion of drug resistance in vitro was demonstrated by Cuvier et al.[137] with Doxorubicin-loaded PACA nanoparticles in terms of cell growth inhibition comparable with that obtained with sensitive cells exposed to free Dox. In vivo experiments showed significantly prolonged survival of P388-Adr-resistant leukemia bearing mice that had previously received P388-Adr-R cells by i.p. injections of Dox-NS, whereas free Dox injection was ineffective against this rapidly growing tumor. Intracellular concentration and the cytoplasmic and nuclear distribution of Dox as measured by laser microspectrofluorometry (LMSF) showed a similar accumulation and distribution of Dox, regardless of the form of Dox delivery in the sensitive cells, whereas Dox-NS accumulated the same amount in resistant cells.

Use of folate receptor-specific drug conjugates and their nanoparticles/liposomes forms is another strategy for bypassing multi-drug resistant efflux pumps. Folic acid, attached to polyethyleneglycol-derivatized, distearoyl-phosphatidylethanolamine, was used by Goren et al.[155] to target liposomes to folate receptor (FR)-overexpressing tumor cells in vitro. Additional experiments with DOX-loaded, folate-targeted liposomes (FTLs) indicate that liposomal DOX is rapidly internalized, released in the cytoplasmic compartment, and, shortly thereafter, detected in the nucleus, the entire process lasting 1–2 h. FR-mediated cell uptake of targeted liposomal DOX into a multi-drug resistant subline of M109-HiFR cells (M109R-HiFR) was unaffected by *P-glycoprotein*-mediated drug efflux, in sharp contrast to the uptake of free DOX.

15.2.6.3 Drug Delivery to Hepatocellular Carcinomas

Hepatocellular carcinomas (HCCs) are generally identified clinically at an advanced stage, usually in combination with cirrhosis. Though optimal, surgical resection is associated with a high rate of recurrence. Approaches to prevent recurrence, like chemoembolization after surgery, have not proven to be beneficial. Both doxorubicin and cisplatin are frequently used in the treatment of these carcinomas, but the overall response rates are low and neither approach seems to prolong the survival. Newly emerging agents with promising results include ^{90}Y microspheres, antiangiogenesis agents, inhibitors of growth factors and their receptors, and K vitamins.[156]

HCC is known to be chemoresistant to anti-cancer drugs due to the expression of multi-drug resistant (MDR) transporters. Doxorubicin loaded into the PACA nanoparticles enhanced the efficiency of treatment in the M5076 murine hepatic tumor metastasis in mice.[91] After intravenous injection, the doxorubicin-loaded PBCA nanoparticles resulted in higher hepatic concentrations in tumor-bearing mice. Histology of the mouse liver after the intravenous injection of the nanoparticle formulation revealed greater accumulation of nanoparticles in the Kupfer cells but not in the tumor cells.[123] This was supposed to be the result of macrophage uptake and subsequent capture by the Kupfer cells, of the nanoparticle formulation owing to the hydrophobicity of the nanoparticles, and then the drug release close to the tumor cells in the liver. To better understand the role of macrophages in delivering the nanoparticle-bound doxorubicin to the tumor cells, a co-culture system was developed consisting of two compartments separated by a porous polyester membrane (Transwell clear insert).[126] The results were compared with the control experiments involving direct incubation of drug/formulations with the tumor cells. The M5076 cells were seeded into the lower compartment, whereas the J774.A1 macrophage monocyte cells were placed on the top of the membrane. The doxorubicin and doxorubicin-loaded isobutylcyanoacrylate nanoparticle formulation were added to the upper compartment containing macrophages. The doxorubicin and nanoparticle formulation showed a 5-fold increase in IC_{50} of M5076 tumor cells in the co-culture system. In some groups of experiments, the recombinant mouse interferon-γ (IFN-γ) was added to the macrophage compartment before the addition of the formulation to activate the macrophages. This would result in the release of cytotoxic factors such as IFN-α and nitric oxide.[157] Doxorubicin and nanoparticle formulation were found to be more cytotoxic when the macrophages were activated by IFN-γ. In fact, this would probably be a result of a synergistic effect of the cytotoxic factors, especially nitric oxide released by the activated macrophages and the nanoparticle formulation because the role of antitumor factors in the inhibition of tumors both in vitro and in vivo has been well documented.[158,159] This co-culture system gives a clear indication of the role that macrophages play after intravenous injection in enhancing the cytotoxicity of the doxorubicin-loaded PACA nanoparticle formulation and its efficacy in liver tumor.

Recently, Barraud et al.[160] compared the antitumor efficacy of Dox-loaded PIHCA nanoparticles with free doxorubicin, both in vitro and in vivo, in HCC-bearing transgenic mice overexpressing the mdr1 and mdr3 genes. The 50% inhibition concentration (IC_{50}) of doxorubicin in vitro on different human hepatoma cell lines decreased with the nanoparticle formulation, compared to that of free Dox. Dox-loaded nanoparticles showed an enhanced MDR reversal in these cells.

Mitoxantrone-loaded PBCA nanoparticles showed higher tumor concentration after intravenous injection in the B-16 melanoma-bearing mice.[89] These nanoparticles have also shown higher efficacy in the treatment of B-16 melanoma.[28] The efficacy of mitoxantrone-loaded PBCA nanoparticles after intravenous administration was also tested in heterotopic and orthotopically transplanted human HCC-bearing mice.[161] The efficacy of the mitoxantrone-loaded nanoparticles was compared with that of free mitoxantrone and doxorubicin. In orthotopically transplanted mice, all three preparations produced a similar effect. Microscopic examination revealed that the mitoxantrone-loaded nanoparticles showed enhanced antitumor activity in heterotopically transplanted mice, which was supported by a low tumor cell proliferation. The nanoparticles also did not show any signs of toxicity of the heart, spleen, lung, and kidney, indicating their effectiveness in tumor therapy and that the fact that they had overcome the adverse effects associated with the anti-cancer agents.

15.2.6.4 Drug Delivery to Brain Tumors

Brain tumors are one of the most lethal forms of cancer. Many primary brain tumors are most resistant to chemotherapy, probably due to the presence of a tight BBB.[162] They are extremely difficult to treat due to a lack of therapeutic strategies capable of overcoming barriers for effective delivery of drugs to the brain. In fact, the vasculature of gliomas possess some special features: open endothelial gaps (inter-endothelial junctions and transendothelial channels having diameter of about 0.3 μm), fenestrations (5.5 nm maximum width), cytoplasmic vesicles such as caveolae (50–70 nm diameter), and vesicular vacuolar organelles (108 ± 32 nm diameter). All of these features are due to the secretion of a vascular endothelial growth factor (VEGF) that causes the loss of the barrier function of BBB, which is essential for delivering drugs to the brain.[163] Nanoparticles are considered suitable for delivering drugs to the brain, but they should possess the following ideal characteristics:[164] (1) a nontoxic, biodegradable, and biocompatible nature; (2) a particle diameter less than 100 nm; (3) exhibit physical stability in the blood (no aggregation); (4) avoidance of the MPS (no opsonization) prolonged blood circulation time; (5) a BBB-targeted brain delivery (receptor-mediated transcytosis across brain capillary endothelial cells); (6) a scalable and cost-effective manufacturing process; (7) be amenable to small molecules, peptides, proteins, or nucleic acids; (8) exhibit minimal nanoparticle excipient-induced drug alteration (chemical degradation/alteration, protein denaturation); and (9) show possible modulation of drug release profiles.

Extensive efforts have been made to develop novel strategies to overcome the obstacles for brain tumor drug delivery.[165] PACA nanoparticles enabled the delivery of a number of drugs, including doxorubicin, loperamide, tubocurarine, the NMDA receptor antagonist MRZ 2/576, and the peptides dalargin and kytorphin across the BBB after coating with surfactants.[166] However, only the surfactants polysorbate (Tween) 20, 40, 60, and 80, as well as some poloxamers (Pluronic F68), can induce this uptake. The mechanism for the delivery across the BBB most likely is endocytosis via the LDL receptor by the endothelial cells lining the brain blood capillaries after the injection of nanoparticles into the blood stream. This endocytotic uptake seems to be mediated by the adsorption of apolipoprotein B and/or E from the blood. After this process, the nanoparticle-associated drug may be released into the cells that diffuse into the internal parts of brain or the particles may be transcytosed.

Unfortunately, the conventional approaches resulted in only sub-therapeutic concentrations in the brain, due to the short plasma half-life of the compounds as well as the carrier system. Surface coating the carrier system with polyethylene glycol led to the enhancement of plasma residence time, due to the formation of a steric barrier preventing the interaction of the carrier with the plasma proteins (opsonins) and thus preventing the opsonization and further uptake by the MPS.[92,101]

PEG-coated-polyalkylcyanoacrylate nanoparticles consisting of the amphiphilic copolymer poly(PEGCA-co-hexadecyl cyanoacrylate) showed longer blood circulation time and better brain penetration than the nanoparticles coated with polysorbate 80 after intravenous injection in mice and rats. Unlike the polysorbate 80-coated nanoparticles, the penetration of poly(PEGCA-co-hexadecyl cyanoacrylate) nanoparticles into the brain occurred without any modification of the BBB. The brain permeability of the nanoparticles further increased in pathological conditions, where the permeability of the BBB itself is modified. This has been demonstrated with rats bearing an experimental allergic encephalomyelitis after intravenous injection.[167]

PEG-coated hexadecylcyanoacrylate nanospheres synthesized by Brigger et al.[168] showed a significant accumulation after intravenous injection in gliosarcoma in rats. The radioactive [^{14}C]-poly(hexadecylcyanoacrylate) ([^{14}C]-PHDCA) and [^{14}C]-poly(MePEG2000cyanoacrylate-co-hexadecylcyanoacrylate) 1:4 (PEG–PHDCA) ([^{14}C]-PEG–PHDCA) nanospheres were prepared using the nanoprecipitation technique. This procedure involves the dissolution of the radioactive polymer in a warm acetone, followed by the addition of the acetonic solution into the aqueous phase, resulting in the precipitation of nanospheres. The [^{14}C]-PHDCA and [^{14}C]-PEG–PHDCA nanospheres possessed a mean diameter of about 150 nm. Intravenous injection of the

[^{14}C]-PEG–PHDCA nanospheres into the rats inoculated intracerebrally with 9L gliosarcoma cells resulted in a long circulation time in the blood. Both [^{14}C]-PHDCA and [^{14}C]-PEG–PHDCA nanospheres were able to selectively extravasate the BBB and accumulated more in the gliosarcoma than in the peritumoral brain. However, this effect was significantly higher (3.1-times greater accumulation ($p < 0.05$)) for [^{14}C]-PEG–PHDCA nanospheres than for the [^{14}C]-PHDCA nanospheres. Intravenous injection of a hydrophilic tracer [^{3}H]-sucrose to the tumor-bearing rats (seven days after the inoculation of tumor cells) to assess the permeability of the BBB showed that the 9L gliosarcoma and, to a lesser extent, the peritumoral brain region were hyperpermeable to sucrose, indicating a selective disruption of the BBB at the pathological site and slight edematous surrounding the brain. The permeability of the BBB was not observed at the intracerebral injection site of the control rats, indicating that the BBB alteration was only due to the presence of cancerous cells.

15.2.6.5 Drug Delivery to Breast Cancer

Resistance to cancer chemotherapy involves both altered drug activity at the designated target and modified intra-tumor pharmacokinetics (e.g., uptake and metabolism). Two proteins in particular—*P-glycoprotein* (P-gp) and MDR-associated protein (MRP2)—are responsible for MDR associated with a variety of cancers.[169] BCRP (breast cancer resistant protein) is another type of protein that appears to play a major role in the MDR phenotype of a specific human breast cancer.[170] Cancer cells overexpressing these proteins are resistant to several chemotherapeutic agents, including the anthracyclines, and show a reduced nuclear accumulation of anthracyclines.[171] The ability of PACA nanoparticles to overcome the MDR of cancer cells has been well documented.[137,172] MRP1 is considered to be a transporter of some conjugated organic anions and drug-glutathione conjugates.[173] When delivered through PACA nanoparticles, doxorubicin forms an ion pair with the polycyanoacrylic acid and penetrates the cell membrane in resistant cells.[141] The efficacy of doxorubicin-loaded PIHCA nanoparticles was evaluated in resistant human breast adenocarcinoma cells, i.e., MCF7/VP and MCF7/doxorubicin overexpressing MRP1 and *P-glycoprotein*, respectively.[171] The MCF7/VP and MCF7/doxorubicin were 34 and 47 times resistant to doxorubicin alone, and 2.5 and 7.7 times resistant to nanoparticle formulation, respectively, compared to MCF7 cells. The nuclear accumulation of doxorubicin at 6 h after treatment was higher in MCF7 cells (133 ± 22 and 120 ± 41 µM when administered as free doxorubicin and nanoparticle formulation, respectively), followed by MCF7/VP (97 ± 19 and 107 ± 15 µM administered as free doxorubicin and nanoparticle formulation, respectively), and MCF7/doxorubicin (59 ± 12 and 66 ± 12 µM administered as free doxorubicin and nanoparticle formulation, respectively). At 12 h post incubation, the nuclear accumulation of doxorubicin in MCF7 cells did not increase significantly, whereas the accumulation was significantly higher in MCF7/VP cells (76 ± 10 and 118 ± 30 µM for free doxorubicin and nanoparticle formulation, respectively). At 18 h post incubation, the nuclear accumulation still increased in MCF7/VP cells (102 ± 20 and 158 ± 16 µM for free doxorubicin and nanoparticle formulation, respectively). However, the nuclear accumulation for MCF7/doxorubicin cells did not vary much with time. The confocal microscopic studies indicated the localization of doxorubicin in the nucleus of MCF7 cells, whereas the localization was observed in the cytoplasm alone in MCF7/doxorubicin cells after treatment with free doxorubicin and nanoparticle formulation. In contrast, for MCF7/VP cells the localization was in both the nucleus and the cytoplasm in cells treated with nanoparticle formulation; it was only in the cytoplasm after doxorubicin treatment.

It was hypothesized that after hydrolysis of the lateral ester bonds, the ion pair passively diffuse through the lipid bilayer of the cell membrane and some amount of the polycyanoacrylic acid (3–4 carboxylic group) could interact with glutathione and create a polycyanoacrylic acid–glutathione conjugate that would inhibit the efflux of MRP1–doxorubicin. The ion pair formation between doxorubicin and polyisohexylxyanoacrylate nanoparticles could also prevent the formation of the doxorubicin–glutathione–S-conjugate and hence the MRP1-mediated transport of the drug.[171] Thus, the doxorubicin-loaded PACA nanoparticles overcome the MDR in cells over-expressing *P-glycoprotein* and MRP1 accompanied by the nuclear accumulation of doxorubicin.

In support of the above results, Kisara et al.[174] observed that the two compounds buthionine sulfoximine (BSO) and cepharanthine (CE), which decrease the glutathione content in the MDR cells, showed greater reversal of multi-drug resistance to Dox in MDR cells. Treatment of the MDR cells with BSO resulted in an enhancement of the cytotoxic effect of Dox by 1.8-fold, whereas CE caused a greater reversal of drug resistance. BSO treatment also resulted in the decrease in GSH content of MDR cells compared to that of the sensitive ones. The combination of BSO with CE caused further potentiation of the antiproliferative effect of Dox in MDR cells.

15.2.6.6 Drug Delivery to Lymphatic Carcinomas

The lymphatic system is the physiological system that maintains the body water balance. It also controls certain immunological responses. At times, this system acts as a medium for the mestastais of cancer cells. Due to the peculiarity of the anatomy of the lymphatic interstitium, the achievement of drug localization into the lymphatic system is limited. Much effort has been made to achieve lymphatic targeting of drugs using colloidal carriers such as biodegradable nanoparticulate systems, including nanospheres, emulsions, and liposomes.[175] The major purpose of lymphatic targeting is (1) to provide an effective anti-cancer chemotherapy that will prevent the metastasis of tumor cells by accumulating the drug in the regional lymph node, and (2) to enhance the localization of diagnostic agents in the regional lymph node to visualize the lymphatic vessels before surgery. In the early days of lymphatic-targeting research, some success was achieved in delivering the emulsions and liposomes to the lymphatic system after intramuscular[176] and subcutaneous routes of administration.[177] For effective penetration into the lymphatic interstitium, the carrier systems need to possess a smaller size, hydrophobicity, and a negative surface charge.[178] PIBCA nanocapsules showed enhanced lymphatic targeting and regional lymph node retention of 12-(9-anthroxy) stearic acid, a lipophilic model compound.[175] The route of administration of the delivery system has a greater contribution in the delivery of nanoparticles to the lymphatic tumor in mice.[179] In vivo studies conducted by Reddy et al.[74] on mice bearing Dalton's Lymphoma tumor have showed that ^{99m}Tc-Dox-loaded PBCA nanoparticles exhibited enhanced accumulation in the lymph node, compared with the ^{99m}Tc-Dox, after subcutaneous injection. A significantly high tumor uptake of DPBC nanoparticles ($\sim$13-fold higher at 48 h post injection) ($P < 0.001$) was found, compared to free doxorubicin. The accumulation of the nanoparticles in the tumor increased with time, indicating their slow penetration from the injection site into the tumor. Such a nanoparticulate system would be advantageous in the lymphoma treatment, offering drug targeting and the prolonged availability of the drug to the tumor.

15.2.6.7 Drug Delivery to Gastric Carcinomas

Gastric carcinoma falls into the category of the difficult-to-treat tumors. Though the PACA nano-particles are considered most promising in the tumor-treatment applications, their use in the delivery of anti-cancer agents to gastric tumors remains limited.

Magnetically targeted systems can be classified in the new generation of drug carriers. The advantage of these carriers is their ability to minimize the uptake by the reticuloendothelial system.[180] Intra-arterial administration of these systems in the form of magnetic albumin micro-spheres[181] and magnetic liposomes[182] has been shown to treat the tumors successfully under the influence of a strong magnetic field. Recently, Gao et al.[183] prepared aclacinomycin-loaded magnetic iron oxide–PBCA nanoparticles using the interfacial polymerization technique. The technique involved dispersion of the drug in diluted hydrochloric acid into magnetic fluid, and the successive addition of the mixture to a hexane containing span 80 and polysorbate 80.

The buytlcyanoacrylate monomer was added into the above dispersion under stirring, and after 6 h the newly formed magnetic nanoparticles were separated using a magnet, washed with methanol and water, lyophilized, and sterilized by ^{60}Co irradiation (15 kGy). These magnetic particles were 210 nm in diameter, and they revealed a core-shell structure in which the iron oxide core was covered by the polymer coat. The aclacinomycin-loaded nanoparticles were evaluated in mice implanted with gastric carcinoma near the right fore feet. Administration of drug-loaded magnetic nanoparticles by intravenous injection into the tumor-bearing mice implanted with a magnet in the tumor resulted in a higher tumor-inhibition rate than the free aclacinomycin. The nanoparticle formulation greatly reduced the tumor mass compared to the free aclacinomycin and the drug-free carrier. The drug-loaded magnetic nanoparticles showed similar activity to that of the drug-free magnetic nanoparticles in vitro in the absence of a magnetic field in the gastric cancer cell line MKN-45. They did, however, show enhanced antitumor activity in vivo in tumor-bearing mice.

15.2.6.8 Chemoembolization Using Alkylcyanoacrylates

HCC is the most common liver tumor, with heterogeneity in the tumor behavior and the underlying liver disease. Recent combinations such as cisplatin, interferon, Adriamycin, and 5-FU are extremely toxic and yield response rates of only 20%, with no survival advantage compared to supportive care alone.[184] Higher concentrations of cancer chemotherapeutic agents can be delivered directly to the HCC via the hepatic arterial route. Considering that this route is the major vascular supply of these tumors, an even larger number of papers have reported the experience of hepatic artery chemotherapy or hepatic artery chemoembolization (TACE) with single agents, or with a dizzying combination of agents, and at doses not replicated by any two institutions. Loewe et al.[185] evaluated the potential of transarterial permanent embolization with the use of a mixture of cyanoacrylate and lipiodol for the treatment of unresectable primary HCC. Loewe et al.[186] used NBCA for hepatic artery embolization for the treatment of small-bowel neuroendocrine metastases to the liver. The results revealed that the permanent embolization of hepatic arteries as part of a multimodality treatment protocol is beneficial in long-term follow-up for patients with metastasized small-bowel neuroendocrine tumors. The use of cyanoacrylate is safe and effective as an embolic agent.

Transarterial embolization (TAE) with the use of microspheres and Lipiodol and cyanoacrylate for unresectable HCC is a feasible treatment modality. A retrospective analysis of 46 patients with histologically confirmed HCC was made who were treated with TAE of the hepatic arteries.[187] To induce permanent embolization, the microspheres (Embosphere; 100–700 µm) and a mixture of ethiodized oil (Lipiodol Ultrafluide) with cyanoacrylate (Glubran) was administrated. No patient died during embolization or within the first 24 h. Severe procedure-related complications were observed in 2 patients. At the time of the analysis, 38 of 46 patients were alive. The 180-, 360-, 520-, and 700-day cumulative survival rates for the total study population were 80.6%, 70.7%, 70.7%, and 47.1%, respectively, with a median survival of 666 days.

A procedure for effective and promising preoperative embolization of carotid body tumors was reported by Harman et al.[188] Ultrasound-guided direct percutaneous injection of n-butyl cyanoacrylate was given, and angiographic road map assistance was used for protection of parent arteries during the injection. After embolization, complete devascularization of the tumor was achieved without complications. The tumor was removed surgically with minimal blood loss. Transcatheter arterial embolization (TAE) of splanchnic arterial branches to allow continuous application of repeat hepatic arterial infusion chemotherapy (HAIC) was assessed.[189] One hundred and twenty-eight patients with unresectable advanced liver cancer were implanted with a percutaneous port catheter system and TAE of splanchnic arteries with coils and/or NBCA. The recanalization rate between coil-embolized and NBCA- or NBCA-coil-embolized arteries, and frequency of heterogeneously poor distribution was compared between patients with single arteries and those with multiple hepatic arteries. The arteries once embolized with coils alone spontaneously recanalized at a significantly higher rate than those with NBCA. A hepatic artery embolization study carried out by Loewe et al.[186] using NBCA and ethiodized oil for the treatment of small-bowel neuro-endocrine metastases to the liver for the treatment of liver metastases from neuroendocrine small-bowel tumors also concluded that the use of cyanoacrylate as an embolic agent is safe and effective.

The potential of transarterial permanent embolization with the use of a mixture of cyanoa-crylate and lipiodol for treatment of unresectable primary HCC was also assessed by Loewe et al.[185] The study included 36 patients with histologically proven HCC who were treated with transarterial embolization of the hepatic arteries. The study indicated that TAE with use of cyanoacrylate and lipiodol for unresectable HCC is a feasible treatment modality. A similar study conducted by Berghammer et al.[190] confirmed the safety of the procedure with minimum side effects. Thus, it constitutes a valuable therapeutic option for patients with Okuda stage I and II HCC.

A right gastric artery (RGA) embolization study to prevent acute gastric mucosal lesions caused by an influx of anti-cancer agents into the RGA in patients undergoing repeat HAIC was conducted on 217 patients with malignant hepatic tumors[191] using metallic coils and/or a mixture of n-butyl cyanoacrylate (n-BCA). RGA embolization was technically successful in the majority of patients (93%), with the lowest incidence of major complications. The clinical experience of Nadalini et al.[192] using isobutyl-2-cyanoacrylate vesical for embolization of the hypogastric arteries in cases of serious hemorrhage of the bladder and prostate is also reported. The effect was immediate results in the majority of cases, a decidedly positive outcome, especially considering the serious conditions of certain neoplastic patients. Isobutyl-2-cyanoacrylate suspension in Lipiodol was also used to treat percutaneous transcatheter embolization of the renal artery in clear cell carcinoma[193] in nine patients. In most of the patients, the procedure was found to be palliative, with no compli-cations attributed to the glue or to oil emboli. Preoperative embolization with isobutyl-2-cyanoacrylate by means of an intra-arterial catheter during selective angiography was adopted by Carmignani et al.[194] in two cases of carcinoma of the kidney.

Traditional preoperative embolization via a transarterial approach has proved beneficial, but it is often limited by complex vascular anatomy and unfavorable locations.[195] Paragangliomas, or glomus tumors, are the neoplasms of the head and neck. They are remarkably vascular, so surgical resection can be complicated by rapid and dramatic blood loss.[196] These tumors can develop in the middle ear (glomus tympanicum), the jugular foramen of the skull base (glomus jugulare), or the head and neck area (glomus caroticum, glomus vagale). Surgical removal of these tumors is also often associated with a significant intraoperative bleeding rate because of their vascular nature.[197–199] Direct percutaneous injection of n-butyl cyanoacrylate resulted in the effective devasculariza-tion of craniofacial tumors[200] and the embolization of oral tumors.[201] However, the technique involved additional risks and was not widely adopted. In a study in human patients by Abud et al.,[196] the presurgical devascularization was achieved by placing the diagnostic catheter in the common carotid artery to guide the puncture and perform the control angiography during and after the injection of the cyanoacrylate. The percutaneous devascularization of head and neck paragangliomas through the intralesional injection of cyanoacrylate resulted in effec-tive devascularization. It could be a safe and effective technique for managing such clinical lesions.

15.2.6.9 Delivery of Peptide Anti-Cancer Drugs

Ras mutations exist in almost 20–30% of human tumors, and the ras oncogenes possess single-point mutations in the sequence coding for the active site of RAS protein at codons 12, 13b, and 61.[202] These point mutations are good targets for the antisense oligonucleotides and can suppress the translation and targeted mutant mRNA. The mutated ras genes are involved in the cell proliferation and tumorigenicity. Antisense oligonucleotides (ODN) are molecules that are able to inhibit onco-gene expression, being therefore potentially active for the treatment of viral infections or cancer.

However, because of their poor stability in biological media and their weak intracellular penetration, colloidal drug carriers such as nanoparticles were employed for the delivery of oligonucleotides.[203] Association with nanoparticles protects the ODN against degradation and enables them to penetrate more easily into different types of cells. Thus, they were shown to improve the efficiency in inhibiting the proliferation of cells expressing the point-mutated Ha-ras gene. In vivo, the PACA nanoparticles were able to efficiently distribute the ODNs to the liver. As the ODNs have no affinity toward the PACA structure, the ion pair technique was adopted, using cationic surfactants such as cetyltrimethyl ammonium bromide or diethylaminoethyl dextran adsorbed onto the particle surface.[204,205] Spontaneous interpolymer complexation between cationic polyelectrolytes and DNA is known and is largely the result of cooperative electrostatic forces.[206]

ODNs adsorbed onto the PIHCA nanoparticles inhibited the Ha-Ras dependant tumor growth in mice. The ODN-nanoparticle formulation was tested in two cell lines, HBL100rasl and HBL100neo.[202] HBL100rasl is a clone obtained from the human mammary cell line. It expresses normal Ha-ras and Ha-ras carrying the $G \rightarrow U$ point mutation in codon 12 coding for Val instead of Gly at position 12. HBL100neo is a clone transformed only with the pSV2 vector, and it expresses only normal Ha-ras. Three 12-mer oligonucleotides—(1) an antisense oligonucleotide (AS-Val, 5′-CACCGACGGCGC-3′) directed against and centered at the point mutation in codon 12 of the Ha-ras mRNA, (2) an antisense oligonucleotide (AS-Gly, 5′-CACCGCCGGCGC-3′) targeted to the equivalent sequence of the normal Ha-ras mRNA, and (3) the 5′/3′ inverted sequence of AS-Val that contains the same bases as the antisense sequence but in reverse orientation (INV-Val, 5′-CGCCGGAGCCAC-3′). The inhibitory effect of AS-Val adsorbed on the nanoparticles was found to be at a concentration 100-times lower than that for the free AS-Val when tested on HBL100rasl. Cellular uptake studies using ^{32}P-labeling of free ODN and ODN adsorbed nanoparticle formulation revealed that the ODNs adsorbed onto the nanoparticles, incorporated at a lower rate than the free ODN but remained intact in the cells even after 72 h post incubation; the free ODN, in contrast, degraded after 3 h. Upon administration into the nude mice treated with HBL100rasl cells, the AS-Val nanoparticle formulation showed more potential tumor growth inhibition compared to the INV-Val nanoparticle formulation, indicating the sequence-dependant effect of ODN on the tumor-inhibition properties.

The cationic surfactants used for the preparation of ODN–PACA nanoparticles may cause toxicity. In addition, the esterase-induced erosion of PACAs causes rapid desorption of ODNs in serum.[207] To overcome these drawbacks, the PACA nanocapsules with aqueous core have been designed.[208,209] These nanocapsules enhanced the protection and serum stability of the ODNs.[209] Intratumoral injection of PACA nanoparticles containing Phosphorothioate oligonucleotides directed against EWS Fli-1 chimeric RNA led to a significant tumor growth inhibition compared to the free ODN in mice-bearing Ewing sarcoma.

Peptides like Leu-enkephalin dalargin and the Met-enkephalin kyotorphin normally do not cross the BBB when given systemically. Transport of these neuropeptides across the BBB was achieved by adsorption onto the surface of PBCA nanoparticles and successive coating of the nanoparticles with polysorbate (Tween) 80.[210]

Birrenbach and Speiser[5] described the polymerization process for the preparation of hydrophilic micelles containing solubilized drug molecules including labile proteins in a colloidal aqueous system of dissolved monomers. The hydrocarbon medium constituted the outer phase. After secondary solubilization with the aid of selected surfactants, the polymerization of micelles under different conditions was performed. The entrapped tagged material (human 125I-immunoglobulin G) showed a stable fixation in the nanoparticles during long-term in vitro liberation trials. Nanoparticles were spherical in shape, 80 nm in diameter, and embedded with antigenic material (tetanus toxoid and human immunoglobulin G) for parenteral use. These preparations showed intact biological activity and high antibody production in animals.

15.3 CONCLUSIONS

Drug delivery to tumors has recently been the area of significant interest in the field of drug delivery. Nanoparticles, due to their many inherent advantages, gained an important place in this area of research. Though the alkylcyanoacrylates were initially used as tissue adhesives, the discovery of the simple anionic polymerization of alkylcyanoacrylates has renewed interest in their applicability to drug-delivery research. Poly(alkyl cyanoacrylates) with a higher alkyl chain length are relatively less toxic than those with a lower alkyl chain length. Several interesting studies were reported on the usefulness of poly(alkyl cyanoacrylates) for delivering anti-cancer drugs to different types of cancers, including breast cancer and gastric cancer. Interesting results were observed in the case of delivery of doxorubicin-loaded poly(alkyl cyanoacrylate) nanoparticles for the treatment of liver tumors. Though the initial results on the delivery of doxorubicin-loaded poly-butylcyanoacrylate nanoparticles to lymphoma are encouraging, more experimental evidence is needed to evaluate their efficacy on tumors. The PEG modification of alkylcyanoacrylates has improved the delivery of their nanoparticles to the brain. Cellular resistance to multiple drugs represents a major problem in cancer chemotherapy. This drug resistance may appear clinically either as a lack of tumor size reduction or as the occurrence of clinical relapse after an initial positive response to treatment with antitumor agents. The resistance mechanism can have different origins, either directly linked to specific mechanisms developed by the tumor tissue or connected to the more general problem of distribution of a drug toward its targeted tissue. The role of *P-glyco-protein* in decreasing the uptake of anti-cancer drugs in multi-drug resistant (MDR) tumors through efflux is well studied. Reports on the importance of nanoparticles in reversion of multi-drug resistance mediated by *P-glycoprotein* in MDR tumors reveal the potential of using PBCA nano-particles in MDR tumors. A commercial application of such a strategy is on the way: doxorubicin-loaded poly(alkyl cyanoacrylate) nanoparticles have reached phase-II clinical trials for the treatment of resistant tumors. Several reports have elucidated the mechanism of nanoparticle uptake by the cells, but the intracellular trafficking and fate of these nanoparticles in the cell remain unclear. Of late, poly(alkyl cyanoacrylate) nanoparticles have also been considered as a potential system for delivering proteins and peptide drugs to tumors.

REFERENCES

1. Fenton, R. G. and Longo, D. I., Cell biology of cancer, In *Harrison's Internal Medicine*, Fauci, A. S., Braunwald, E., Isselbacher, K. J., Wilson, J. D., Martin, J. B., Kasper, D. L., Hauser, S. L., and Longo, D. I., Eds., vol. 1, 14th ed., McGraw Hill, New York, p. 505, 1998.
2. Reddy, L. H., Drug delivery to tumors: Recent strategies, *J. Pharm. Pharmacol.*, 57, 1231, 2005.
3. Williams, J. et al., Nanoparticle drug delivery system for intravenous delivery of topoisomerase inhibitors, *J. Control. Release*, 91, 167, 2003.
4. De Jaeghere, F., Doelker, E., and Gurny, R., Nanoparticles, In *Encyclopedia of Controlled Drug Delivery*, Mathiowitz, E., Ed., Wiley, New York, p. 641, 1999.
5. Birrenbach, B. and Speiser, P. P., Polymerized micelles and their use as adjuvants in immunology, *J. Pharm. Sci.*, 65, 1763, 1976.
6. Rhodes, C. T., Wai, K., and Banker, G. S., Molecular scale drug entrapment as a precise method of controlled drug release. II. Facilitated drug entrapment to polymeric colloidal dispersions, *J. Pharm. Sci.*, 59, 1578, 1970.
7. Boylan, J. C. and Banker, G. S., Molecular scale drug entrapment as a precise method of controlled drug release. IV. Entrapment of anionic drugs by polymer gelation, *J. Pharm. Sci.*, 62, 1177, 1973.
8. Langer, R., Biomaterials in drug delivery and tissue engineering: one laboratory's experience, *Acc. Chem. Res.*, 33, 94, 2000.
9. Lanza, R. P., Langer, R., and Chick, W. L., *Principles of Tissue Engineering*, Academic Press, Austin, TX, p. 405, 1997.

10. Gilbert, R. G., *Emulsion Polymerization: A Mechanistic Approach*, Academic Press, London, p. 30, 1995.

11. Bovey, F. A. et al., *Emulsion Polymerization*, vol. 1, Interscience Publishers, New York, p. 15, 1955.

12. Ballard, M. J., Napper, D. H., and Gilbert, R. G., Kinetics of emulsion polymerization of methyl methacrylate. Polymer chemistry special issue, *J. Polym. Sci.*, 22, 3225, 1984.

13. Ljungstedt, I., Ekman, B., and Sjoholm, I., Detection and separation of lymphocytes with specific surface receptors by using microparticles, *Biochem. J.*, 170, 161, 1978.

14. Krause, H. J., Schwarz, A., and Rohdewald, P., Interfacial polymerization, a useful method for the preparation of polymethylcyanoacrylate nanoparticles, *Drug Dev. Ind. Pharm.*, 12, 527, 1986.

15. Radwan, M. A., Preparation and in vivo evaluation of parenteral metoclopramide-loaded poly(alkylcyanoacrylate) nanospheres in rats, *J. Microencapsul.*, 18, 467, 2001.

16. Couvreur, P. et al., Polycyanoacrylate nanocapsules as potential lysosomotropic carriers: Preparation, morphological and sorptive properties, *J. Pharm. Pharmacol.*, 31, 331, 1979.

17. Grislain, L. et al., Pharmacokinetics and biodistribution of a biodegradable drug-carrier, *Int. J. Pharm.*, 15, 335, 1983.

18. Behan, N., Birkinshaw, C., and Clarke, N., Poly *n*-butyl cyanoacrylate nanoparticles: a mechanistic study of polymerization and particle formation, *Biomaterials*, 22, 1335, 2001.

19. Seijo, B. et al., Design of nanoparticles of less than 50 nm diameter: Preparation, characterization and drug loading, *Int. J. Pharm.*, 62, 1, 1990.

20. Reddy, L. H. and Murthy, R. S. R., Influence of polymerization technique and experimental variables on the particle properties and release kinetics of methotrexate from poly(butylcyanoacrylate) nanoparticles, *Acta Pharm.*, 54, 103, 2004.

21. Reddy, L. H. and Murthy, R. S. R., Study of influence of polymerization factors on formation of poly(butyl cyanoacrylate) nanoparticles and in vitro drug release kinetics, *ARS Pharm.*, 45, 211, 2004.

22. Fallaouh, N. A. K. et al., Development of a new process for the manufacture of polyisobutylcyanoacrylate nanocapsules, *Int. J. Pharm.*, 28, 125, 1986.

23. Florence, A. T., Whateley, T. L., and Wood, D. A., Potentially biodegradable microcapsules with poly-2-cyanoacrylate membranes, *J. Pharm. Pharmacol.*, 31, 422, 1979.

24. Chouinard, F. et al., Preparation and purification of polyisohexylcyanoacrylate nanocapsules, *Int. J. Pharm.*, 72, 211, 1991.

25. Palumbo, M. et al., Improved antioxidant effect of idebenone-loaded polyethyl-2-cyanoacrylate nanocapsules tested on human fibroblasts, *Pharm. Res.*, 19, 71, 2002.

26. Kopf, H. et al., Studium der Mizellpolymerisation in Gegenwart niedermolekularer Arzneistoffe. I. Herstellung und Isolierung der Nanopartikel, Restmonomerenbestimmung, physikalischchemische Daten, *Pharm. Ind.*, 38, 281, 1976.

27. Laakso, T. and Stjarnkvistjoholm, I., Biodegradable microspheres. VI. Lysosomal release of covalently bound antiparasitic drugs from starch microparticles, *J. Pharm. Sci.*, 76, 134, 1987.

28. Beck, P. et al., Influence of polybutyl cyanoacrylate nanoparticles and liposomes on the efficacy and toxicity of the anticancer drug mitoxantrone in murine tumor models, *J. Microencapsul.*, 10, 101, 1993.

29. Illum, L. et al., Evaluation of carrier capacity and release characteristic for poly(butyl-2-cyanoacrylate) nanoparticles, *Int. J. Pharm.*, 30, 17, 1986.

30. Couvreur, P. et al., Adsorption of antineoplastic drugs to polyalkyl cyanoacrylate nanoparticles and their release in calf serum, *J. Pharm. Sci.*, 68, 1521, 1979.

31. Kreuter, J. and Hartman, H. R., Comparative study on the cytostatic effects and tissue distribution of 5-fluorouracil in a free from and bound to polybutylcyanocrylate nanoparticles in sarcoma 180-bearing mice, *Oncology*, 40, 363, 1983.

32. Marchal-Heussler, I. et al., Antiglaucomatous activity of betaxolol chlorhydrate sorbed onto different isobutyl cyanoacrylate nanoparticle preparations, *Int. J. Pharm.*, 58, 115, 1990.

33. Brasseur, N., Brault, D., and Couvreur, P., Adsorption of hematoporphyrin onto polyalkyl cyanoacrylate nanoparticles: carrier capacity and drug release, *Int. J. Pharm.*, 70, 129, 1991.

34. Gaspar, R., Preat, V., and Roland, M., Nanoparticles of polyisohexyl cyanoacrylate (PIHCA) as carriers of primaquine: formulation, physico-chemical characterization and acute toxicity, *Int. J. Pharm.*, 68, 111, 1991.

35. Verdun, C. et al., Development of a nanoparticle controlled-release formulation for human use, *J. Control. Release*, 3, 205, 1986.

36. Harmia, T., Speiser, P., and Kreuter, J., Optimization of pilocarpoine loading onto nanoparticles by sorption procedures, *Int. J. Pharm.*, 33, 45, 1986.

37. Poupaert, J. H. and Couvreur, P. A., Computationally derived structural model of doxorubicin interacting with oligomeric polyalkylcyanoacrylate in nanoparticles, *J. Control. Release*, 92, 19, 2003.

38. Chauvierre, C. et al., Heparin coated poly(alkylcyanoacrylate) nanoparticles coupled to hemoglobin: A new oxygen carrier, *Biomaterials*, 25, 3081, 2004.

39. Simeonova, M. et al., Study on the role of 5-fluorouracil in the polymerization of butylcyanoacrylate during the formation of nanoparticles, *J. Drug Target.*, 12, 49, 2004.

40. Stella, B. et al., Design of folic acid-conjugated nanoparticles for drug targeting, *J. Pharm. Sci.*, 89, 1452, 2000.

41. Kreuter, J., Physicochemical characterization of polyacrylic nanoparticles, *Int. J. Pharm.*, 14, 43, 1983.

42. Kreuter, J. et al., Polybutyl cyanoacrylate nanoparticles for the delivery of [^{75}SE]norcholesterol, *Int. J. Pharm.*, 16, 105, 1983.

43. Magenheim, B. and Benita, S., Nanoparticle characterization: A comprehensive physicochemical approach, *STP Pharma Sci.*, 1, 121, 1991.

44. Edstrom, R. D. et al., Viewing molecules with scanning tunneling and atomic force microscopy, *FASEB J.*, 4, 3144, 1990.

45. Zasadzinski, J. A. N., Scanning tunneling microscopy with applications to biological surfaces, *Biotechniques*, 7, 174, 1989.

46. Fuchs, H. and Laschinski, R., Surface investigations with a combined scanning-electron scanning tunneling microscope, *Scanning*, 12, 126, 1990.

47. Gedde, U. W., Thermal analysis of polymers, *Drug Dev. Ind. Pharm.*, 16, 2465, 1990.

48. Muller, R. H. et al., Alkyl cyanoacrylate drug carriers. I. Physicochemical characterization of nanoparticles with different chain length, *Int. J. Pharm.*, 84, 1, 1992.

49. Andreade, J. D. et al., Contact angles at the solid–water interface, *J. Colloid Interface Sci.*, 72, 488, 1979.

50. Wilkins, D. J. and Myers, P. A., Studies on the relationship between the electrophoretic properties of colloids and their blood clearance and organ distribution in the rat, *Br. J. Exp. Pathol.*, 47, 568, 1966.

51. Baszkin, A. and Lyman, D. J., The interaction of plasma proteins with polymers. I. Relationship between polymer surface energy and protein adsorption/desorption, *J. Biomed. Mater. Res.*, 14, 393, 1980.

52. Lindsay, R. M. et al., Blood surface interactions, *Trans. Am. Soc. Artif. Intern. Organs*, 26, 603, 1980.

53. Andreade, J. D. et al., Surface characterization of poly(hydroxyethyl methacrylate) and related polymers. I. Contact angle methods in water, *J. Polym. Sci. Polym. Symp.*, 66, 131, 1979.

54. El-Egakey, M. A., Bentele, V., and Kreuter, J., Molecular weights of polycyanoacrylate nanoparticles, *Int. J. Pharm.*, 13, 349, 1983.

55. Muller, R. H. et al., In vitro model for the degradation of alkylcyanoacrylate nanoparticles, *Biomaterials*, 11, 590, 1990.

56. Scherer, D., Robinson, J. R., and Kreuter, J., Influence of enzymes on the stability of polybutylcyanoacrylate nanoparticles, *Int. J. Pharm.*, 101, 165, 1994.

57. Sommerfeld, P., Schroeder, U., and Sabel, B. A., Long-term stability of PBCA nanoparticle suspensions suggests clinical usefulness, *Int. J. Pharm.*, 155, 201, 1997.

58. Maincent, P. et al., Disposition kinetics and oral bioavailability of vincamine-loaded polyalkyl cyanoacrylate nanoparticles, *J. Pharm. Sci.*, 75, 955, 1986.

59. Kreuter, J., Peroral administration of nanoparticles, *Adv. Drug Deliv. Rev.*, 7, 71, 1991.

60. Damge, C. et al., New approach for oral administration of insulin with polyalkylcayanoacrylate nanocapsules as drug carrier, *Diabetes*, 37, 246, 1988.

61. Damge, C. et al., Nanocapsules as carriers for oral peptide delivery, *J. Control. Release*, 13, 233, 1990.

62. Wade, C. and Leonard, F., Degradation of poly-(methyl-2-cyanoacrylates), *J. Biomed. Mater. Res.*, 6, 215, 1972.
63. Lenaerts, V. et al., Degradation of poly(isobutylcyanoacrylate) nanoparticles, *Biomaterials*, 5, 65, 1984.
64. Douglas, S. J., Illum, L., and Davis, S. S., Particle size and size distribution of Poly(butyl-2-cyano-acrylate) nanoparticles. II. Influence of stabilizers, *J. Colloid Interface Sci.*, 103, 154, 1985.
65. Couvreur, P., Design of biodegradable polyalkylcyanoacrylate nanoparticles as a drug carrier, In *Microspheres and Drug Therapy*, Davis, S. S., Illum, L., McVie, J. G., and Tomlinson, E., Eds., Elsevier, Amsterdam, p. 103, 1984.
66. Washington, C., Drug release from monodisperse systems: A critical review, *Int. J. Pharm.*, 58, 1, 1990.
67. Leavy, M. V. and Benita, S., Drug release from submicron O/W emulsion: a new in vitro kinetic evaluation model, *Int. J. Pharm.*, 66, 29, 1990.
68. Calvo, P., Vila-Jato, J. L., and Alonso, M. J., Comparative in vitro evaluation of several colloidal systems, nanoparticles, nanocapsules and nanoemulsions as ocular drug carriers, *J. Pharm. Sci.*, 85, 530, 1996.
69. El-Samaligly, M. S., Rohdewald, P., and Mohmoud, H. A., Polyalkyl cyanoacrylate nanocapsules, *J. Pharm. Pharmacol.*, 38, 216, 1986.
70. LeRay, A. M. et al., End chain radiolabeling and in vitro stability studies of radiolabeled poly (hydroxyl) nanoparticles, *J. Pharm. Sci.*, 83, 845, 1994.
71. Barratt, G., Colloid drug carriers: achievements and perspectives, *Cell. Mol. Life Sci.*, 60, 21, 2003.
72. Simeonova, M. et al., Tissue distribution of polybutylcyanoacrylate nanoparticles loaded with spin-labelled nitrosourea in Lewis lung carcinoma-bearing mice, *Acta Physiol. Pharmacol. Bulg.*, 20, 77, 1994.
73. Reddy, L. H. and Murthy, R. S. R., Pharmacokinetics and bio-distribution of doxorubicin loaded poly(butyl) cyanoacrylate nanoparticles synthesized by two different techniques, *Biomed. Pap.*, 148, 1611, 2004.
74. Reddy, L. H., Sharma, R. K., and Murthy, R. S.R, Enhanced tumor uptake of doxorubicin loaded poly(butyl cyanoacrylate) nanoparticles in mice bearing dalton's lymphoma tumor, *J. Drug Target.*, 12, 443, 2004.
75. Gipps, E. M. et al., Distribution of polyhexyl cyanoacrylate nanoparticles in nude mice bearing human osteosarcoma, *J. Pharm. Sci.*, 75, 256, 1986.
76. Illum, L. et al., Tissue distribution of poly(hexyl 2-cyanoacrylate) nanoparticles coated with mono-clonal antibodies in mice bearing human tumor xenografts, *J. Pharmacol. Exp. Ther.*, 230, 733, 1984.
77. Kante, B. et al., Tissue distribution of [3H] actinomycin D adsorbed on polybutylcyanoacrylate nanoparticles, *Int. J. Pharm.*, 7, 45, 1980.
78. Douglas, S. J., Davis, S. S., and Illum, L., Biodistribution of poly(butyl-2-cyanoacrylate) nanopar-ticles in rabbits, *Int. J. Pharm.*, 34, 145, 1986.
79. Lobenberg, R. et al., Body distribution of azidothymidine bound to hexylcyanoacrylate nanoparticles after i.v. injection to rats, *J. Control. Release*, 50, 21, 1998.
80. Borchard, G. et al., Uptake of surfactant-coated poly(methylmethacrylate)-nanoparticles by bovine brain microvessel endothelial cell monolayers, *Int. J. Pharm.*, 110, 29, 1994.
81. Gulyaev, A. E. et al., Significant transport of doxorubicin into the brain with polysorbate 80-coated nanoparticles, *Pharm. Res.*, 16, 1564, 1999.
82. Woodcock, D. M. et al., Reversal of multidrug resistance by surfactants, *Br. J. Cancer*, 66, 62, 1992.
83. Nerurkar, M. M., Burto, P. S., and Borchardt, R. T., The use of surfactants to enhance the per-meability of peptides through Caco-2 cells by inhibition of an apically polarized efflux system, *Pharm. Res.*, 13, 528, 1996.
84. Ghanem, G. E. et al., Labelled polycyanoacrylate nanoparticles for human in vivo use, *Appl. Radiat. Isot.*, 44, 1219, 1993.
85. Skidan, I. N. et al., Modification of pharmacokinetics and antitumor efficacy of doxorubicin bound to polysorbate 80-coated polybutylcyanoacrylate nanoparticles. In Third Central European Conference on Human Tumor Markers, *Biomarkers Environ. J.*, 4, 39, 2001.

86. Pappo, J. and Ermak, T. H., Uptake and translocation of fluorescent latex particles by rabbit Peyer's patch follicle epithelium: a quantitative model for M cell uptake, *Clin. Exp. Immunol.*, 76, 144, 1989.

87. Eldridge, J. H. et al., Controlled release in the gut-associated lymphoid tissues. 1. Orally administered biodegradable microspheres target the Peyer's patches, *J. Control. Release*, 11, 205, 1990.

88. Dembri, A. et al., Targeting of 3′-Azido3′-deoxythymidine (AZT)-loaded poly(isohexylcyanoacrylate) nanospheres to the gastrointestinal mucosa and associated lymphoid tissues, *Pharm. Res.*, 18, 467, 2001.

89. Reszka, R. et al., Body distribution of free, liposomal and nanoparticle-associated mitoxantrone in B16-melanoma-bearing mice, *J. Pharmacol. Exp. Ther.*, 280, 232, 1997.

90. Fattal, E. et al., Treatment of experimental salmonellosis in mice with ampicillin-bound nanoparticles, *Antimicrob. Agents Chemother.*, 33, 1540, 1989.

91. Chiannilkulchai, N. et al., Doxorubicin-loaded nanoparticles: increased efficacy in murine hepatic metastases, *Selective Cancer Ther.*, 5, 1, 1989.

92. Storm, G. et al., Surface modification of nanoparticles to oppose uptake by the mononuclear phagocyte system, *Adv. Drug Deliv. Rev.*, 17, 31, 1995.

93. Gref, R. et al., In *Microparticulate Systems for the Delivery of Proteins and Vaccines*, Chen, S. and Bernstein, H., Eds., Marcel Dekker, New York, p. 279, 1996.

94. Gabizon, A., Selective tumor localization and improvement of therapeutic index of anthracyclines encapsulated in long circulating liposomes, *Cancer Res.*, 52, 891, 1992.

95. Gref, R., Biodegradable long-circulating polymeric nanospheres, *Science*, 263, 1600, 1994.

96. Bazile, D. et al., Stealth MePEG-PLA nanoparticles avoid uptake by the mononuclear phagocyte system, *J. Pharm. Sci.*, 84, 493, 1995.

97. Peracchia, M. T. et al., Development of sterically stabilized poly(isobutyl 2-cyanoacrylate) nanoparticles by chemical coupling of poly(ethylene glycol), *J. Biomed. Mater. Res.*, 34, 846, 1997.

98. Peracchia, M. T. et al., Synthesis of novel MePEGcyanoacrylate–hexadecylcyanoacrylate amphiphilic copolymer for nanoparticle technology, *Macromolecules*, 30, 846, 1997.

99. Peracchia, M. T. et al., Biodegradable PEGylated nanoparticles from a novel MePEGcyanoacrylate–hexadecylcyanoacrylate amphiphilic copolymer, *Pharm. Res.*, 15, 548, 1998.

100. Couvreur, P. et al., Tissue distribution of antitumor drugs associated with polyalkylcyanoacrylate nanoparticles, *J. Pharm. Sci.*, 69, 199, 1980.

101. Peracchia, M. T., Stealth® PEGylated polycyanoacrylate nanoparticles for intravenous administration and splenic targeting, *J. Control. Release*, 60, 121, 1999.

102. Peracchia, M. T. et al., Visualization of in vitro protein rejecting properties of PEGylated stealth polycyanoacrylate nanoparticles, *Biomaterials*, 20, 1269, 1999.

103. deGennes, P. G., Confirmation of polymers attached to an interface, *Macromolecules*, 13, 1069, 1980.

104. Peracchia, M. T. et al., Complement consumption by poly(ethylene glycol) in different confirmations chemically coupled to poly(isobutyl-2-cyanoacrylate) nanoparticles, *Life Sci.*, 61, 749, 1997.

105. Ping, L. Y. et al., PEGylated polycyanoacrylate nanoparticles as salvicine carriers: Synthesis, preparation, and in vitro characterization, *Acta Pharmacol. Sin.*, 22, 645, 2001.

106. Kante, B. et al., Toxicity of polalkylcyanoacrylate nanoparticles. I. Free nanoparticles, *J. Pharm. Sci.*, 71, 790, 1982.

107. Couvreur, P. et al., Toxicity of polyalkylcyanoacrylate nanoparticles. II. Doxorubicin-loaded nanoparticles, *J. Pharm. Sci.*, 71, 786, 1982.

108. Olivier, J. C. et al., Indirect evidence that drug brain targeting using polysorbate 80-coated polybutylcyanoacrylate nanoparticles is related to toxicity, *Pharm. Res.*, 16, 1836, 1999.

109. Simeonova, M., Chorbadjiev, K., and Antcheva, M., Study of the effect of polybutylcyanoacrylate nanoparticles and their metabolites on the primary immune response in mice to sheep red blood cells, *Biomaterials*, 19, 2187, 1998.

110. Couvreur, P. et al., Biodegrdable polymeric nanoparticles as drug carrier for antitumor agents, In *Polymeric Nanoparticles and Microspheres*, Guiot, P. and Couvreur, P., Eds., CRC Press, Boca Raton, FL, pp. 27–93, 1986.

111. Gelperina, S. E. et al., Toxicological studies of doxorubicin bound to polysorbate 80-coated poly (butyl cyanoacrylate) nanoparticles in healthy rats and rats with intracranial glioblastoma, *Toxicol. Lett.*, 126, 131, 2002.

112. Jefferiss, C. D., Lee, A. J., and Ling, R. S., Thermal aspects of self-curing polymethylmethacrylate, *J. Bone Joint Surg. Br.*, 57, 511, 1975.

113. Vinters, H. V. et al., The histotoxicity of cyanoacrylates: A selective review, *Neuroradiology*, 27, 279, 1985.

114. Vinters, H. V., Lundie, M. J., and Kaufmann, J. C., Long-term pathological follow-up of cerebral arteriovenous malformations treated by embolization with bucrylate, *N. Engl. J. Med.*, 314, 477, 1986.

115. Tranbahuy, P. et al., Direct intratumoral embolization of juvenile angiofibroma, *Am. J. Otolaryngol.*, 15, 429, 1994.

116. San Millan Ruiz, D. et al., Pathology findings with acrylic implants, *Bone*, 25(Suppl. 2), 85S, 1999.

117. Matsumoto, T., Carcinogenesis and cyanoacrylate adhesives, *J. Am. Med. Assoc.*, 202, 1057, 1967.

118. Samson, D. and Marshall, D., Carcinogenic potential of isobutyl-2-cyanoacrylate, *J. Neurosurg.*, 65, 571, 1986.

119. Hatanaka, S. et al., Induction of malignant fibrous histiocytoma in female Fisher rats by implantation of cyanoacrylate, zirconia, polyvinyl chloride or silicone, *In Vivo*, 7, 111, 1993.

120. Canter, H. I. et al., Tissue response to *N*-butyl-2-cyanoacrylate after percutaneous injection into cutaneous vascular lesions, *Ann. Plast. Surg.*, 49, 520, 2002.

121. Kattan, J. et al., Phase I clinical trial and pharmacokinetic evaluation of doxorubicin carried by polyisohexylcyanoacrylate nanoparticles, *Invest. New Drugs*, 10, 191, 1992.

122. Gibaud, S. et al., Increased bone marrow toxicity of doxorubicin bound to nanoparticles, *Eur. J. Cancer*, 30, 820, 1994.

123. Chiannilkulchai, N. et al., Hepatic tissue distribution of doxorubicin-loaded nanoparticles after i.v. administration in reticulosarcoma M5076 metastasis-bearing mice, *Cancer Chemother. Pharmacol.*, 26, 122, 1990.

124. Zhang, Z. R., Liao, G. T., and Hou, S. X., Study on mitoxantrone polycyanoacrylate nanospheres, *Yao Xue Xue Bao*, 29, 544, 1994.

125. Yang, Y. X. et al., Antitumor activity of mitoxantrone-nanosphere against murine liver tumor H22, *Sichuan Da Xue Xue Bao Yi Xue Ban*, 35, 68, 2004.

126. Soma, C. E. et al., Investigation of the role of macrophages on the cytotoxicity of doxorubicin and doxorubicin-loaded nanoparticles on M5076 cells in vitro, *J. Control. Release*, 68, 283, 2000.

127. Bogush, T. et al., Direct evaluation of intracellular accumulation of free and polymer-bound anthracyclines, *Cancer Chemother. Pharmacol.*, 35, 501, 1995.

128. Bennis, S. et al., Enhanced cytotoxicity of doxorubicin encapsulated in polyisohexylcyanoacrylate nanospheres against multidrug-resistant tumor cells in culture, *Eur. J. Cancer*, 30A, 89, 1994.

129. Simeonova, M. and Antcheva, M., In vitro study of cytotoxic activity of vinblastine in a free form and associated with nanoparticles, *Acta Physiol. Pharmacol. Bulg.*, 20, 31, 1994.

130. Rusakov, I. G. et al., Use of cytostatics in polymer-drug complexes in the treatment of malignant neoplasms of the maxilla, *Sov. Med.*, 1, 23, 1990.

131. Pan, W. S. and Hu, J., Studies of 5-fluorouracil nanocapsules, *Yao Xue Xue Bao*, 26, 280, 1991.

132. Jiang, X. H. et al., The antihepatoma effect of lyophilized aclacinomycin A polyisobutylcyanoacrylate nanoparticles in vitro and in vivo, *Yao Xue Xue Bao*, 30, 179, 1995.

133. Brasseur, F. et al., Actinomycin D absorbed on polymethylcyanoacrylate nanoparticles: increased efficiency against an experimental tumor, *Eur. J. Cancer*, 16, 1441, 1980.

134. Chissov, V. I. et al., Immediate results of the use of prospidin in a cyanoacrylate glue compound in cancer of the esophagus, *Sov. Med.*, 7, 28, 1989.

135. Pérez, M. et al., Use of *N*-butyl-2-cyanoacrylate in oral surgery: biological and clinical evaluation, *Artif. Organs*, 24, 241, 2000.

136. Walker, J., Martin, C., and Callaghan, R., Inhibition of *P-glycoprotein* function by XR9576 in a solid tumor model can restore anticancer drug efficacy, *Eur. J. Cancer*, 40, 594, 2004.

137. Cuvier, C. et al., Doxorubicin-loaded nanospheres bypass tumor cell multidrug resistance, *Biochem. Pharmacol.*, 44, 509, 1992.

138. Vauthier, C. et al., Poly(alkylcyanoacrylates) as biodegradable materials for biomedical applications, *Adv. Drug Deliv. Rev.*, 55, 519, 2003.

139. Colin de Verdiere, A. et al., Uptake of doxorubicin from loaded nanoparticles in multidrug-resistant leukemic murine cells, *Cancer Chemother. Pharmacol.*, 3, 504, 1994.

140. Laurand, A. et al., Quantification of the expression of multidrug resistance-related genes in human tumor cell lines grown with free doxorubicin or doxorubicin encapsulated in polyisohexylcyanoacrylate nanospheres, *Anticancer Res.*, 24, 3781, 2004.

141. Pepin, X. et al., On the use of ion-pair chromatography to elucidate doxorubicin release mechanism from polyalkyl cyanoacrylate nanoparticles at the cellular level, *J. Chromatogr. B*, 702, 181, 1997.

142. Hu, Y.-P. et al., On the mechanism of action of doxorubicin encapsulation in nanospheres for the reversal of multidrug resistance, *Cancer Chemother. Pharmacol.*, 37, 556, 1996.

143. Vauthier, C. et al., Drug delivery to resistant tumors: the potential of polyalkyl cyanoacrylate nanoparticles, *J. Control. Release*, 93, 151, 2003.

144. Kobayashi, H., Takamura, Y., and Miyachi, H., Novel approaches to reversing anticancer drug resistance using gene-specific therapeutics, *Hum. Cell.*, 14, 172, 2001.

145. Juliano, R. L. et al., Antisense pharmacodynamics: critical issues in the transport and delivery of antisense oligonucleotides, *Pharm. Res.*, 16, 494, 1999.

146. Maia, R. C., Carrico, M. K., Klumb, C. E., Noronha, H., Coelho, A. M., Vasconcelos, F. C., and Ruimanek, V. M., Clinical approach to circumvention, of multidrug resistance in refractory leukaemic patients: Association of cyclosporine A with etoposide, *J. Exp. Clin. Cancer Res.*, 16, 419, 1997.

147. Toffoli, G. et al., Reversal activity of cyclosporine A and its metabolites M1 M17 and M21 in multidrug resistant cells, *Int. J. Cancer*, 71, 900, 1997.

148. Den Boer, M. L. et al., The modulating effect of PSC 833 cyclosporine A, verapamil and genistein on in vitro cytotoxicity and intracellular content of daunorubicin in childhood acute lymphoblastic leukemia, *Leukemia*, 12, 912, 1998.

149. Soma, C. E. et al., Reversion of multidrug resistance by co-encapsulation of doxorubicin and cyclosporine A in polyalkyl cyanoacrylate nanoparticles, *Biomaterials*, 21, 1, 2000.

150. Colin de Verdiere, A. et al., Reversion of multidrug resistance with polyalkyl cyanoacrylate nanoparticles; towards a mechanism of action, *Br. J. Cancer*, 76, 198, 1997.

151. Soma, C. E. et al., Ability of doxorubicin-loaded nanoparticles to overcome multidrug resistance of tumor cells after their capture by macrophages, *Pharm. Res.*, 16, 1710, 1999.

152. Povarov, L. S. et al., Partial circumvention of *P-glycoprotein* (Pgp)-associated resistance to doxorubicin (Dox) in MCF7/R human breast carcinoma and P388/R murine leukemia cell lines by doxorubicin-14-*O*-hemiadipate (H-Dox), *Russ. J. Bioorg. Chem.*, 21, 797, 1995.

153. Leontieva, O. V., Preobrazhenskaya, M. N., and Bernacki, R. J., Partial circumvention of *P-glycoprotein*-mediated multidrug resistance by doxorubicin-14-*O*-hemiadipate, *Invest. New Drugs*, 20, 35, 2002.

154. Krishna, R. and Mayer, L. D., Liposomal doxorubicin circumvents PSC 833-free drug interactions, resulting in effective therapy of multidrug-resistant solid tumors, *Cancer Res.*, 57, 5246, 1997.

155. Goren, D. et al., Nuclear delivery of doxorubicin via folate-targeted liposomes with bypass of multidrug-resistance efflux pumps, *Clin. Cancer Res.*, 6, 1949, 2000.

156. Carr, B. I., Hepatocellular carcinoma: Current management and future trends, *Gastroenterology*, 127, S218, 2004.

157. Talmadge, J. E. and Hart, I. R., Inhibited growth of a reticulum cell sarcoma (M5076) Induced in vitro and in vivo by macrophage-activating agents, *Cancer Res.*, 44, 2446, 1984.

158. Talmadge, J. E. et al., Immunomodulatory and immunotherapeutic properties of recombinant g-interferon and recombinant tumor necrosis factor in mice, *Cancer Res.*, 47, 2563, 1987.

159. Iigo, M., Shimizu, I., and Sagawa, K., Synergistic antitumor effects of carboplatin and interferons on hepatic metastases of colon carcinoma 26 and M5076 reticulum cell sarcoma, *Jpn. J. Cancer Res.*, 84, 794, 1993.

160. Barraud, L. et al., Increase of doxorubicin sensitivity by doxorubicin-loading into nanoparticles for hepatocellular carcinoma cells in vitro and in vivo, *J. Hepatol.*, 42, 736, 2005.

161. Zhi-Rong, Z. et al., Study on the anticarcinogenic effect and acute toxicity of liver-targeting mitoxantrone nanoparticles, *World J. Gastroenterol.*, 5, 511, 1999.

162. Tamai, I. and Tsuji, A., Drug delivery through the blood–brain barrier, *Adv. Drug. Deliv. Rev.*, 19, 401, 1996.

163. Vajkoczy, P. and Menger, M. D., Vascular microenvironment in gliomas, *J. Neurooncol.*, 50, 99, 2000.

164. Olivier, J.-C., Drug transport to brain with targeted nanoparticles. *NeuroRx* (The American Society for Experimental NeuroTherapeutics, Inc.), 2, 108, 2005.

165. Rautioa, J. and Chikhale, P. J., Drug delivery systems for brain tumor therapy, *Curr. Pharm. Des.*, 10, 1341, 2004.

166. Kreuter, J., Influence of the surface properties on nanoparticle-mediated transport of drugs to the brain, *J. Nanosci. Nanotechnol.*, 4, 484, 2004.

167. Calvo, P. et al., Quantification and localization of PEGylated polycyanoacrylate nanoparticles in brain and spinal cord during experimental allergic encephalomyelitis in the rat, *Eur. J. Neurosci.*, 15, 1317, 2002.

168. Brigger, I. et al., Poly(ethylene glycol)-coated hexadecylcyanoacrylate nanospheres display a combined effect for brain tumor targeting, *J. Pharmacol. Exp. Ther.*, 303, 928, 2002.

169. Stouch, T. R. and Gudmundsson, O., Progress in understanding the structure–activity relationships of *P-glycoprotein*, *Adv. Drug Deliv. Rev.*, 54, 315, 2002.

170. Doyle, L. A. et al., A multidrug resistance transporter from human MCF-7 breast cancer cells, *Proc. Natl. Acad. Sci. USA*, 95, 15665, 1998.

171. Aouali, N. et al., Enhanced cytotoxicity and nuclear accumulation of doxorubicin-loaded nano-spheres in human breast cancer MCF7 cells expressing MRP1, *Int. J. Oncol.*, 23, 1195, 2003.

172. Henry-Toulme, N., Grouselle, M., and Ramaseilles, C., Multidrug resistance bypass in cells exposed to doxorubicin loaded nanospheres absence of endocytosis, *Biochem. Pharmacol.*, 50, 1135, 1995.

173. Marbeuf-Gueye, C. et al., Kinetics of anthracycline efflux from multidrug resistance protein-expres-sing cancer cell compared with *P-glycoprotein*-expressing cancer cells, *Mol. Pharmacol.*, 53, 141, 1998.

174. Kisara, S. et al., Combined effects of buthionine sulfoximine and cepharanthine on cytotoxic activity of doxorubicin to multidrug-resistant cells, *Oncol. Res.*, 7, 191, 1995.

175. Nishioka, Y. and Yoshino, H., Lymphatic targeting with nanoparticulate system, *Adv. Drug Deliv. Rev.*, 47, 55, 2001.

176. Hashida, M., Egawa, M., Muranishi, S., and Sezaki, H., Role of intra-muscular administration of water-in-oil emulsions as a method for increasing the delivery of anticancer agents to regional lymphatics, *J. Pharmacokinet. Biopharm.*, 5, 223, 1997.

177. Parker, R. J., Priester, E. R., and Sieber, S. M., Comparison of lymphatic uptake, metabolism, excretion and biodistribution of free and liposome entrapped [^{14}C]cytosine β-D-arabinofuranoside following intraperitoneal administration to rats, *Drug Metab. Dispos.*, 10, 40, 1982.

178. Porter, C. J., Drug delivery to the lymphatic system, *Crit. Rev. Ther. Drug Carrier Syst.*, 14, 333, 1997.

179. Reddy, L. H. et al., Influence of administration route on tumor uptake and biodistribution of etopo-side loaded solid lipid nanoparticles in Dalton's lymphoma tumor bearing mice, *J. Control. Release*, 105, 185, 2005.

180. Gupta, P. K. and Hung, C. T., Magnetically controlled targeted micro-carrier systems, *Life Sci.*, 44, 175, 1989.

181. Widder, K. J. et al., Selective targeting of magnetic albumin microspheres containing low-dose doxorubicin: total remission in Yoshida sarcomabearing rats, *Eur. J. Cancer Clin. Oncol.*, 19, 135, 1983.

182. Rudge, S. et al., Adsorption and desorption of chemotherapeutic drugs from a magnetically targeted carrier (MTC), *J. Control. Release*, 74, 335, 2001.

183. Gao, H. et al., Preparation of magnetic polybutyl cyanoacrylate nanospheres encapsulated with aclacinomycin A and its effect on gastric tumor, *World J. Gastroenterol.*, 10, 2010, 2004.

184. Yeo, W. et al., A phase II study of doxorubicin (A) versus cisplatin (P)/interferon γ-2b (I)/doxorubi-cin (A)/fluorouracil (F) combination chemotherapy (PIAF) for inoperable hepatocellular carcinoma (HCC), *Proc. Am. Soc. Clin. Oncol.*, 23, 4026, 2004.

185. Loewe, C. et al., Arterial embolization of unresectable hepatocellular carcinoma with use of cyano-acrylate and lipiodol, *J. Vasc. Interv. Radiol.*, 13, 61, 2002.

186. Loewe, C. et al., Permanent transarterial embolization of neuroendocrine metastases of the liver using cyanoacrylate and lipiodol: Assessment of mid- and long-term results, *Am. J. Roentgenol.*, 180, 1379, 2003.

187. Rand, T. et al., Arterial embolization of unresectable hepatocellular carcinoma with use of microspheres, lipiodol, and cyanoacrylate, *Cardiovasc. Intervent. Radiol.*, 28, 313, 2005.

188. Harman, M., Etlik, O., and Unal, O., Direct percutaneous embolization of a carotid body tumor with *n*-butyl cyanoacrylate: An alternative method to endovascular embolization, *Acta Radiol.*, 45, 646, 2004.

189. Yamagami, T. et al., Value of transcatheter arterial embolization with coils and *n*-butyl cyanoacrylate for long-term hepatic arterial infusion chemotherapy, *Radiology*, 230, 792, 2004.

190. Berghammer, P. et al., Arterial hepatic embolization of unresectable hepatocellular carcinoma using a cyanoacrylate/lipiodol mixture, *Cardiovasc. Intervent. Radiol.*, 21, 214, 1998.

191. Inaba, Y. et al., Right gastric artery embolization to prevent acute gastric mucosal lesions in patients undergoing repeat hepatic arterial infusion chemotherapy, *J. Vasc. Interv. Radiol.*, 12, 957, 2001.

192. Nadalini, V. F. et al., Therapeutic occlusion of the hypogastric arteries with isobutyl-2-cyanoacrylate in vesical and prostatic cancer, *Radiol. Med. (Torino)*, 67, 61, 1981.

193. Papo, J., Baratz, M., and Merimsky, E., Infarction of renal tumors using isobutyl-2 cyanoacrylate and lipiodol, *AJR Am. J. Roentgenol.*, 137, 781, 1981.

194. Carmignani, G., Belgrano, E., and Puppo, P., First clinical results with isobutyl-2-cyanoacrylate in the arterial embolisation of cancer of the kidney, *Radiol. Med. (Torino)*, 64, 991, 1978.

195. Abud, D. G. et al., Intratumoral injection of cyanoacrylate glue in head and neck paragangliomas, *Am. J. Neuroradiol.*, 25, 1457, 2004.

196. Derdeyn, C. P., Direct puncture embolization for paragangliomas: promising results but preliminary data, *Am. J. Neuroradiol.*, 25, 1453, 2004.

197. Patetsios, P. et al., Management of carotid body paragangliomas and review of a 30-year experience, *Ann. Vasc. Surg.*, 16, 331, 2002.

198. Muhm, M. et al., Diagnostic and therapeutic approaches to carotid body tumors. Review of 24 patients, *Arch. Surg.*, 132, 279, 1997.

199. Marchesi, M. et al., Surgical treatment of paragangliomas of the neck, *Int. Surg.*, 82, 394, 1997.

200. Casasco, A. et al., Devascularization of craniofacial tumors by percutaneous tumor puncture, *Am. J. Neuroradiol.*, 15, 1233, 1994.

201. Pierot, L. et al., Embolization by direct puncture of hypervascularized oral tumors, *Ann. Otolaryngol. Chir. Cervicofac.*, 111, 403, 1994.

202. Schwab, G. et al., Antisense oligonucleotides adsorbed to polyalkylcyanoacrylate nanoparticles specifically inhibit mutated Ha-ras-mediated cell proliferation and tumorigenicity in nude mice, *Proc. Natl. Acad. Sci. USA*, 91, 10460, 1994.

203. Lambert, G., Fattal, E., and Couvreur, P., Nanoparticulate systems for the delivery of antisense oligonucleotides, *Adv. Drug Deliv. Rev.*, 47, 99, 2001.

204. Chavany, C. et al., Polyalkylcyanoacrylate nanoparticles as polymeric carriers for antisense oligonucleotides, *Pharm. Res.*, 9, 441, 1992.

205. Zimmer, A., Antisense oligonucleotide delivery with poly-hexylcyanoacrylate nanoparticles as carriers, *Methods: A Companion to Methods in Enzymol.*, 18, 286, 1999.

206. Kircheis, R., Tumor targeting with surface-shielded ligand–polycation DNA complexes, *J. Control. Release*, 72, 165, 2001.

207. Nakada, Y. et al., Pharmacokinetics and biodistribution of oligonucleotide adsorbed onto poly(isobutylcyanoacrylate) nanoparticles after intravenous administration in mice, *Pharm. Res.*, 13, 38, 1996.

208. Lambert, G. et al., Effect of polyisobutylcyanoacrylate nanoparticles and lipofectin loaded with oligonucleotides on cell viability and PKC alpha neosynthesis in HepG2 cells, *Biochimie*, 80, 969, 1998.

209. Lambert, G. et al., Polyisobutylcyanoacrylate nanocapsules containing an aqueous core for the delivery of oligonucleotides, *Int. J. Pharm.*, 214, 13, 2001.

210. Schroeder, U. et al., Nanoparticle technology for delivery of drugs across the blood–brain barrier, *J. Pharm. Sci.*, 87, 1305, 1998.

16 Aptamers and Cancer Nanotechnology

Omid C. Farokhzad, Sangyong Jon, and Robert Langer

CONTENTS

16.1 Targeted Drug Delivery: An Introduction 289
16.2 Nucleic Acid Ligands (Aptamers): From Discovery to Practice 291
    16.2.1 Physicochemical Properties of Aptamers 292
    16.2.2 Methods for Isolation of Aptamers 293
    16.2.3 Examples of Aptamers for Targeted Delivery 294
        16.2.3.1 Alpha-v Beta-3 294
        16.2.3.2 Human Epidermal Growth Factor-3 295
        16.2.3.3 Prostate Specific Membrane Antigen 295
        16.2.3.4 Nucleolin 295
        16.2.3.5 Sialyl Lewis X 295
        16.2.3.6 Cytotoxic T-Cell Antigen-4 296
        16.2.3.7 Fibrinogen-Like Domain of Tenascin-C 296
        16.2.3.8 Pigpen 296
16.3 Aptamer–Nanoparticle Conjugates for Targeted Cancer Therapy 296
    16.3.1 Properties of Nanoparticles for Conjugation to Aptamers 297
        16.3.1.1 Size of Nanoparticles 297
        16.3.1.2 Polymers for Synthesis of Nanoparticles 298
        16.3.1.3 Charge of Nanoparticles 298
        16.3.1.4 Surface Modification of Nanoparticles 299
    16.3.2 Conjugation Strategies for Nanoparticle–Aptamer Conjugates 300
16.4 Aptamer–Drug Conjugates for Targeted Cancer Therapy 301
16.5 Aptamer–Nanoparticle Conjugates for Protein Detection 304
    16.5.1 Aptamer–QD Conjugates for the Detection of Proteins 304
    16.5.2 Aptamer–Gold-Nanoparticle Conjugates for the Detection of Proteins 305
16.6 Conclusion 306
Acknowledgments 306
References 306

16.1 TARGETED DRUG DELIVERY: AN INTRODUCTION

With advances in nanotechnology, it is becoming increasingly possible to combine specialized delivery vehicles and targeting approaches to develop highly selective and effective cancer therapeutic and diagnostic modalities.[1–4] In the case of therapeutic vehicles, drug targeting is expected to reduce the undesirable and sometimes life-threatening side effects common in anti-cancer therapy. Drug targeting can also enhance the efficacy of drugs because the drug is delivered directly to cancer cells, making it possible to deliver a cytotoxic dose of the drug in a controlled manner.

Cancer targeting may be achieved passively by continuously concentrating drug encapsulated nanoparticles in the tumor interstitial space due to the enhanced permeability and retention (EPR) effect.[5,6] This process occurs because of a differentially large quantity of nanoparticles extravasting out of tumor microvasculature, leading to an accumulation of drugs in the tumor interstitium. Multiple factors influence the EPR, including active angiogenesis and high vascularity, defective vascular architecture, impaired lymphatic drainage, and extensive production of vascular mediators, such as bradykinin, nitric oxide, vascular endothelial growth factor (VEGF), prosta-glandins, collagenase, and peroxynitrite. The ease of vehicle extravasation is a function of the maximum size of the transvascular transport pathways in tumor microvasculature and is determined mainly through the size of open interendothelial gap junctions and trans-endothelial channels. The pore cutoff size of these transport pathways has been estimated between 400 and 600 nm, and extravasation of liposomes into tumors in vivo suggests a cutoff size in the range of 400 nm.[8] As a general rule, particle extravasation is inversely proportional to size, and small particles (fewer than 200 nm size) should be most effective for extravasating the tumor microvasculature.[8–10] Passive tumor targeting has several limitations, including the inability to achieve a sufficiently high level of drug concentration at the tumor site, resulting in lack of drug efficacy.[11] Additionally, the lack of targeting efficiency contributes to undesirable systemic adverse effects (for reviews, see).[4,12]

Active tumor targeting may be achieved by both local and systemic administration of specially designed vehicles that recognize biophysical characteristics that are unique to the cancer cells. Most commonly, this represents binding of vehicles to target antigens using ligands that recognize tumor-specific or tumor-associated antigens. Some of these therapeutic conjugates are now under clinical development or are in clinical practice today. In the case of local drug delivery, such as through the injection of delivery vehicles within an organ, it is possible to achieve a desired effect within a subset of cells, as opposed to a generalized effect on all the cells of the targeted organ. In the case of cancer, the cytotoxic effects of a therapeutic agent would be directed to cancer cells while minimizing harm to noncancerous cells within and outside of the targeted organ. For example, suicide-targeted gene delivery has been demonstrated to be effective in killing prostate cancer but not healthy muscle cells in xenograft mouse models of prostate cancer.[13] Similar approaches have been used to deliver docetaxel via intratumoral injection of targeted controlled release polymer vehicles to prostate cancer cells in nude mice.[14] This approach is particularly useful for primary tumors that have not yet metastasized, such as localized breast or prostate cancer.

For metastatic cancers, the drug delivery vehicle would ideally be administered systemically because the location, abundance, and size of tumor metastasis within the body limits its visual-ization or accessibility, thus making local delivery approaches impractical. Despite the obvious advantages of systemic delivery, this approach is more challenging, and many physiological and biochemical barriers must be overcome for vehicles to reach the tumor and ultimately be capable of delivering therapeutically effective concentrations of the drugs directly to the cancer cells. One major challenge is the early systemic clearance of vehicles by the mononuclear phagocytic cells present in the liver, spleen, lung, and bone marrow.[15–18] This is in part due to the large percentage of cardiac output directed to these organs, and in part due to the dense population of macrophages and monocytes that home in these organs. Additionally, a variety of factors play an important role in the ultimate success of systemic targeted approaches, including those factors related to the biochemical and physical characteristics of the nanoparticles such as: (1) the chemical properties of the controlled release polymer system and the encapsulated drug; (2) the size of the nanoparticles; (3) surface charge and surface hydrophilicity of nanoparticles; and (4) the chemical nature of the targeting molecules.[19]

Several classes of molecules have been utilized for targeting applications, including various forms of antibody-based molecules,[20] such as chimeric human–murine antibodies, humanized antibodies, single chain Fv generated from murine hybridoma or phage display, and minibodies.[58] Multivalent antibody-based targeting structures, such as multivalent minibodies, single-chain dimers and dibodies, and multispecific binding proteins, including bispecific antibodies and antibody-based fusion proteins, have all been evaluated. More recently, other classes of ligands, such as carbohydrates[21] and nucleic acid ligands, also called aptamers,[22,23] have been used as escort molecules for targeted delivery applications. The concept of nucleic acid molecules acting as ligands was first described in the 1980s when it was shown that some viruses encode small structured RNA that binds to viral and host proteins with high affinity and specificity. These RNA nucleic acid ligands had evolved over time to enhance the survival and propagation of the viruses. Subsequently, it was shown that these naturally occurring RNA ligands can inhibit the viral replication and have therapeutic benefits.[24] Finally, methods were developed to perform in vitro evolution and to isolate novel nucleic acid ligands that bind to a myriad of important molecules for diverse applications in research and clinical practice.[25,26]

16.2 NUCLEIC ACID LIGANDS (APTAMERS): FROM DISCOVERY TO PRACTICE

The feasibility of using antibodies for targeted therapy, particularly for oncologic diseases, has been demonstrated repeatedly in the literature. Rituximab (Rituxan™) was the first therapeutic based on monocloncal antibodies to receive FDA approval in 1997 for treating patients with relapsed or refractory low-grade or follicular, CD20 positive, B-cell non-Hodgkin's lymphoma.[27] A wide variety of antibody-based drugs are now under clinical development or are in clinical practice today. For example, denileukin diftitox (Ontak™) is an FDA-approved immunotoxin for the treatment of cutaneous T-cell lymphoma.[28] Many other radioimmunoconjugates or chemoimmunoconjugates directed against cell surface antigens are currently in various stages of clinical and pre-clinical development. Despite the recent success of monoclonal antibodies as targeting moieties, the use of antibodies for drug targeting may have a number of potential disadvantages. First, antibodies are large molecules (~ 20 nm for intact antibodies)[29,30] and their use in developing nanoscale therapeutic and diagnostic tools may result in an increase in vehicle size without added advantage. Second, antibodies may be immunogenic. Despite the current engineering approaches to yield improved humanized antibodies, this problem is not universally solved. Third, the biological development of monoclonal antibodies can be difficult and unpredictable. For example, the target antigen may not be well tolerated by the animal used to produce the antibodies, or the target molecules may be inherently less immunogenic, making it difficult to raise antibodies against such targets (although, this problem is overcome with the use of phage display libraries).[31,32] Fourth, the production of antibodies involves a biological process that can result in batch-to-batch variability in their performance, particularly when production is scaled up. The ideal targeting molecule for the delivery of nanoscale therapeutic and diagnostic systems should, like monoclonal antibodies, bind with high affinity and specificity to a target antigen but overcome or ameliorate some of the problems associated with the use and production of monoclonal antibodies.

Nucleic acid aptamers are a novel class of ligands[25,26] that are small, nonimmunogenic, easy to isolate, characterize and modify, and exhibit high specificity and affinity for their target antigen. Aptamers derive their name from the Latin word *aptus*, meaning "to fit." In the short time since Jack Szostak and Larry Gold independently described the groundbreaking methodology for in vitro evolution of aptamers, this class of ligands has emerged as an important arsenal in research and clinical medicine.[22,33] Considering the many favorable characteristics of aptamers, which have resulted in their rapid progress into clinical practice,[33] researchers have begun to exploit this class of molecules for targeted delivery of controlled release polymer drug delivery vehicles. Recently, we described the first proof-of-concept drug delivery vehicles utilizing aptamers for targeted delivery of drug encapsulated nanoparticles[23] and have gone on to the show efficacy of similarly designed nanoparticles against prostate cancer tumors in vivo (Farokhzad et al., submitted).[14]

16.2.1 PHYSICOCHEMICAL PROPERTIES OF APTAMERS

Nucleic acid aptamers are single-stranded DNA, RNA, or unnatural oligonucleotides that fold into unique structures capable of binding to specific targets with high affinity and specificity. Aptamers are selected in vitro from a pool of ($\sim 10^{14}$–10^{15}) random oligonucleotides.[34] These molecules have a molecular weight in the 10–20 kD range, which is one order of magnitude lower than that of antibodies (150 kD).[35] The small size of aptamers (~ 5 nm for 30–60 base pair of aptamer)[36] is desirable when utilizing them as targeting molecules for the delivery of nanoscale drug delivery vehicles, as they do not substantially increase the overall size of the vehicle. When compared to small molecule ligands, aptamers have a larger surface area, offering more points of chemical contact with their targets. Aptamers fold through intramolecular interaction into tertiary conformations with specific binding pockets. These molecules bind with high specificity and affinity to a variety of target antigens with an equilibrium dissociation constant (K_d) in the 10 pM–10 μM range.[37] Unlike antibodies, their synthesis and large scale production is an entirely chemical process that does not rely on biological systems. This can translate into lower manufacturing cost and decrease batch-to-batch variability during production, which is a significant advantage for the commercialization of this class of molecules. Furthermore, due to their small size, aptamers exhibit superior tissue penetration,[22] and because of their similarity to endogenous molecules, aptamers are believed to be less immunogenic than antibodies.[38]

Unlike anti-sense oligonucleotides, which are single-stranded nucleic acids that require cellular uptake and affect the synthesis of a targeted protein by hybridizing to the mRNAs that encodes it, aptamers may inhibit a protein's function by directly binding to their targets, which may include other nucleic acids, proteins, peptides, and small molecules. Aptamers can be described by a sequence of approximately 15–60 nucleotides (A, U, T, C, and G). The conformation of the aptamer confers specificity for a target molecule by interacting with multiple domains or binding pockets. Small changes in the target molecule can foil interactions, and thus aptamers can distinguish between closely related but nonidentical targets. For example, specific RNAs were identified that have a high affinity for the bronchodilator theophylline (1,3-dimethylxanthine) yet exhibit a more than 10,000 times weaker binding affinity to caffeine (1,3,7-trimethylxanthine), which differs form theophylline only by the substitution of a methyl group at the nitrogen atom N7 position.[39] Based on their unique molecular recognition properties, aptamers have found great utility for applications in areas such as in vitro and in vivo diagnostics, analytical techniques, imaging, and therapeutics.[35,40,41]

Aptamers are highly stable and may tolerate a wide range of temperature, pH (~ 4–9), and organic solvents without loss of activity, but these molecules are susceptible to nuclease degradation and renal clearance in vivo. Therefore, their pharmacokinetic properties must be enhanced prior to in vivo applications. Several approaches have been adopted to optimize the properties of aptamers, such as: (1) capping their terminal ends; (2) substituting naturally occurring nucleotides with unnatural nucleotides that are poor substrates for nuclease degradation (i.e., 2'-F, 2'-OCH$_3$, or 2'-NH$_2$-modified nucleotides); (3) substituting naturally occurring nucleotides with hydrocarbon linkers; and (4) use of L-enantiomers of nucleotides to generate mirror image aptamers commonly referred to as spiegelmers.[42–45] Aptamers can also be stabilized using locked nucleic acid modifications to reduce conformational flexibility,[46] circularized, linked together in pairs, or clustered onto a substrate. Alternatively, a nuclease resistant aptamer may be selected de novo using a pool of oligonucleotides with 2'-F- or 2'-OCH$_3$-modified nucleotides. By combining some of these strategies, an aptamer's half-life can be prolonged from several minutes to many hours.[35]

To prolong the rate of clearance of aptamers, their size may be increased by conjugation with polymers such as polyethylene glycol (PEG).[47]

16.2.2 METHODS FOR ISOLATION OF APTAMERS

In vitro selection,[25] also called *systematic evolution of ligands by exponential enrichment* (SELEX),[26,48] is a protocol to isolate rare functional oligonucleotides (i.e., aptamers) from a pool of random oligonucleotides (Figure 16.1). Similar to phage display or other strategies used to isolate ligands from random libraries, SELEX is essentially an iterative selection and amplification protocol to isolate single-stranded nucleic acid ligands that bind to their target with high affinity and specificity. The complexity of the starting library is determined in part by the number of random nucleotides in the pool. For example, by using a library with 40 random nucleotides, a pool of 10^{24} distinct nucleotides can be generated. Practically speaking, the number of ligands in the starting pool for in vitro selection is closer to 10^{15}, representing 1 nmol of the library.

In the initial step, a library of random nucleotides flanked by fixed nucleotides is generated by solid phase oligonucleotide synthesis. The random nucleotides serve to add complexity to the pool while the fixed sequences are utilized for polymerase chain reaction (PCR) amplification. In the

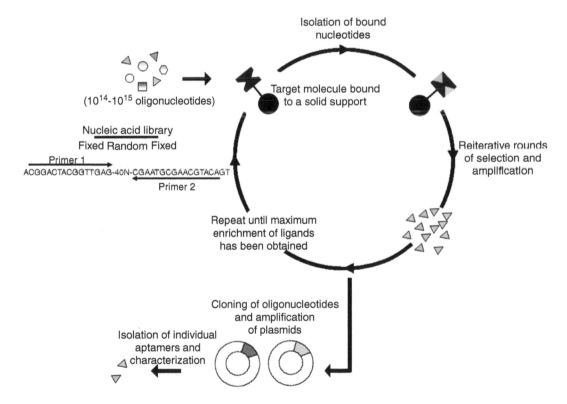

FIGURE 16.1 Schematic representation of SELEX. An oligonucleotide library is synthesized containing random sequences that are flanked by fixed sequences that facilitate PCR amplification. Target molecules are incubated with this pool of oligonucleotides, and bound and unbound oligonucleotides are partitioned. Bound oligonucleotides are isolated, and iterative rounds of selection and amplification are performed with increased stringency to isolate aptamers with high specificity and affinity for the target molecule. Oligonucleotide ligands representing the aptamers are subsequently cloned in plasmids, amplified, and sequenced. The net result of this enrichment process is a small number of highly specific aptamers that are isolated from a large library of random oligonucleotides.

case of isolating DNA aptamers, the oligonucleotide pool is incubated with the target of interest, and the bound fragments are partitioned and directly amplified by PCR system. The resulting pool is used in a follow-up round of selection and amplification, and the process is repeated until the affinity for the target antigen plateaus. Typically, this will be achieved in six to ten rounds of SELEX. After the last round of SELEX, aptamers are cloned in plasmids, amplified, sequenced, and their binding constants are determined (Figure 16.1). These aptamers may be subject to additional modification such as size minimization to truncate the nucleotides not necessary for binding characteristics and nuclease stabilization by replacing naturally occurring nucleotides with modified nucleotides (i.e., $2'$-F pyrimidines, $2'$-OCH$_3$ nucleotides) that are poor substrates for endo- and exonuclease degradation.

In contrast to the isolation of DNA aptamers, which require single-step amplification after portioning, the selection of RNA aptamers involves additional steps, including transcription of an RNA pool from the starting DNA library, reverse transcription of the partitioned RNA pool to generate a cDNA fragment and subsequent amplification of DNA, and transcription into RNA for the next round of selection.[34] The advantage of RNA SELEX, however, is that unnatural nucleotides such as $2'$-F pyrimidines and $2'$-OCH$_3$ nucleotides may be used in the transcription of the RNA pool because these modified bases are utilized by RNA polymerase as substrate. Furthermore, mutant RNA polymerases have also been described that are capable of improved incorporation of modified bases during transcription.[49] The resulting modified RNA pool can be used for isolation of nuclease-stable RNA aptamers. Recently, a fully $2'$-OCH$_3$-modified anti-VEGF aptamer was selected, and when conjugated to 40 kDa PEG, it demonstrated a circulating half-life of 23 h.[50] Conversely, a DNA polymerase that can incorporate unnatural bases such as $2'$-F and $2'$-OCH$_3$ has not been described, and, consequently, DNA aptamers must be nuclease stabilized after the SELEX procedure.

16.2.3 Examples of Aptamers for Targeted Delivery

Since their original description in 1990, aptamers have been isolated to a wide variety of targets, including intracellular proteins, transmembrane proteins, soluble proteins, carbohydrates, and small molecule drugs. As of the submission of this chapter, 624 articles related to aptamers have been published, and more than 200 aptamers have been isolated (comprehensively listed in the Aptamer Database),[51] and one aptamer, macugen (pegaptanib sodium), against the VEGF$_{165}$ protein was recently approved by the FDA in December, 2004 for the treatment of neovascular macular degeneration,[52,53] underscoring the rapid progress of aptamers from its original conception to clinical application. In choosing aptamers for targeting cancer cells, the aptamer must be directed towards receptors that are preferentially or exclusively expressed on the plasma membrane of cancer cells. Alternatively, they may be delivered to extracellular matrix molecules that are expressed preferentially in tumors. To date, several tumor-specific or tumor-associated aptamers have been isolated (reviewed by Pestourie et al.).[54] A partial listing of these aptamers, as well as their size and binding characteristics, are outlined below.

16.2.3.1 Alpha-v Beta-3

Alpha-v beta-3 ($\alpha v \beta 3$) integrin is a transmembrane glycoprotein that mediates numerous processes, including cell migration, cell growth, tumor growth and metastasis, angiogenesis, and vascular healing. The expression of this protein on the endothelial cells of tumor neovasculature is dramatically increased, making $\alpha v \beta 3$ an interesting protein for targeting approaches.[55] An 85-base-pair $2'$-flouropyrimidine RNA aptamer (Apt-$\alpha v \beta 3$) has been shown to bind to the $\alpha v \beta 3$ protein on the surface of endothelial cells in vitro and inhibit endothelial cell proliferation and survival.[56] The future in vivo use of the Apt-$\alpha v \beta 3$ may require post-SELEX optimization, including size minimization to facilitate large-scale synthesis, and enhance the pharmacokinetic properties of this molecule. This aptamer may be utilized for targeted cancer diagnostic and therapeutic applications.

16.2.3.2 Human Epidermal Growth Factor-3

Human epidermal growth factor-3 (HER-3) is a receptor tyrosine kinase that is over-expressed in several cancers. Over-expression of HER-3 is also associated with drug resistance in many HER-2 over-expressed tumors, making HER-3 a candidate target for drug delivery. A30 is an RNA aptamer that was isolated against the extracellular domain of the HER-3 and is shown to inhibit heregluin dependent tyrosine phosphorylation of HER-2 and heregluin-induced growth response of MCF-7 cells at $K_i = 10$ and 1 nM, respectively.[57] The future use of A30 for in vivo application may require post-SELEX optimization of this aptamer, including nuclease stabilization and size minimization.

16.2.3.3 Prostate Specific Membrane Antigen

Prostate specific membrane antigen (PSMA) encodes a folate carboxypeptidase. Its expression is tightly restricted to prostate acinar epithelium and is increased in prostatic intraepithelial neoplasia, prostatic adenocarcinoma, and in tumor-associated neovasculature. Two $2'$-flouropyrimidine RNA aptamers, named xPSM-A9 and xPSM-A10, against the extracellular domain of the PSMA have been isolated and characterized.[58] The aptamer xPSM-A9 inhibits the enzymatic function of the PSMA noncompetitively with a $K_i = 2.1$ nmol, and aptamer xPSM-A10 inhibits the enzymatic function of PSMA competitively with a $K_i = 11.9$ nmol. Aptamer xPSM-A10 has also been truncated from 71 nucleotides to its current size of 56 nucleotides ($\sim 18k$D).

16.2.3.4 Nucleolin

Nucleolin was originally described as a nuclear and cytoplasmic protein; however, a number of recent studies have shown that it can also be expressed at the cell surface.[59,60] Nucleolin is involved in the organization of the nuclear chromatin, rDNA transcription, packaging of the pre-RNA, ribosome assembly, nucleocytoplasmic transport, cytokinesis, nucleogenesis, and apoptosis. AS-1411 (formerly AGRO100) is an aptamer capable of making G-quadruplexes that bind to nucleolin on the cell surface[61] and interact with the nuclear factor kappa B (NFκB) essential modulator (NEMO) inside the cell.[62] The cytosolic localization of AS-1411 after binding to cell surface nucleolin may be exploited for the intracellular delivery of nanoparticles to cancer cells. The use of AS-1411 as a therapeutic modality has also shown promise for the treatment of cancer in humans, and Antisoma of United Kingdom is evaluating this aptamer in phase-I clinical trials.[63] The therapeutic benefit of AS-1411 is presumably attributed to the disruption of the NFκB signaling inside the cells.[64]

16.2.3.5 Sialyl Lewis X

Sialyl Lewis X (sLe^x) is a tetra saccharide glycoconjugate of transmembrane proteins that acts as a ligand for the selectin proteins during cell adhesion and inflammation. The sLe^x is also abnormally overexpressed on the surface of cancer cells and may play a role in cancer cell metastasis. An RNA aptamer referred to as clone 5 has been isolated and shown to bind to the sLe^x with sub-nanomolar affinity and block the sLe^x/selectin mediated adhesion of HL60 monocytic cell line in vitro.[65] The future use of clone 5 for in vivo targeted delivery applications may require post-SELEX optimization of this aptamer, including nuclease stabilization and size minimization.

16.2.3.6 Cytotoxic T-Cell Antigen-4

Cytotoxic T-cell antigen-4 (CTLA-4) is a transmembrane protein that is expressed on the surface of activated T-cells and functions to attenuate the T-cell response by raising the threshold needed for T-cell activation. D60 ($K_d = 33$ nM) is a size-minimized, 35-base-pair, 2′-fluoropyrimidine modified nuclease stable RNA aptamer that was selected against CTLA-4 and shown to inhibit CTLA-4 function in vitro and enhance tumor immunity in mice.[66]

16.2.3.7 Fibrinogen-Like Domain of Tenascin-C

Tenascin-C is an extracellular matrix protein that is over-expressed during tissue remodeling processes, such as fetal development, and wound healing, as well as tumor growth. TTA1 is an aptamer ($K_d = 5$ nM) that was isolated against the fibrinogen-like domain of Tenascin-C,[67] and may potentially be useful for cancer diagnostic and targeted therapeutic applications.

16.2.3.8 Pigpen

Pigpen is an endothelial protein of the Ewing's sarcoma family that parallels the transition from quiescent to angiogenic phenotypes in vitro. Using YPEN-1 endothelial cells and N9 micorglial cells, respectively, in a selection and counter-selection strategy in SELEX, a DNA aptamer named III.1 was isolated that preferentially bound to YPEN-1 cells.[68] The III.1 was also shown to selectively bind to the microvessels of experimental rat glioblastoma. The isolation and characterization of the III.1 target identified pigpen as the target antigen. The use of III.1 aptamer for targeting the microvasculature of tumors is a potentially powerful means of delivering drugs to the site of the cancer.

16.3 APTAMER–NANOPARTICLE CONJUGATES FOR TARGETED CANCER THERAPY

Virtually every branch of medicine has been dramatically impacted by controlled drug delivery strategies during the past four decades,[3,69–72] including cardiology,[73] ophthalmology,[74] endocrinology,[75] oncology,[76] immunology,[77] and orthopedics.[78] Drugs can be released in a controlled manner from within a material through surface or bulk erosion of the material, diffusion of the drug from the interior of the material, or swelling followed by diffusion or triggered by the environment or other external events,[2] such as changes in pH,[79] light,[80] temperature,[81] or the presence of an analyte, such as glucose.[82] In general, controlled-release polymer systems deliver drugs in the optimum dosage for long periods, thus increasing the efficacy of the drug, maximizing patient compliance and enhancing the ability to use highly toxic, poorly soluble, or relatively unstable drugs.

The conjugation of aptamers to drug encapsulated nanoparticles results in targeted delivery vehicles for therapeutic application. These may include delivery of small molecule drugs, protein based drugs, nucleic acid drugs (anti-sense oligoneucleotide, RNAi or gene therapy), and targeted delivery of agents for neutron capture therapy or photodynamic therapy. Aptamers may also be bound to imaging agents to facilitate diagnosis and identification of tumor metastases. For example, it may be useful to bind aptamers to optical imaging agents including fluorophores[68] and quantum dots (nanocrystals)[83] or MRI imaging agents such as magnetic nanoparticles[84,85] for detection of small foci of cancer metastasis. Multiplex systems comprising drug laden nanoparticle aptamer conjugates, together with imaging agents, represent a prospective avenue to future research.

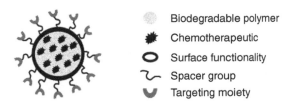

FIGURE 16.2 Schematic representation of targeted drug delivery vehicle composed of polymeric nanoparticles that are surface-modified with targeting agents.

16.3.1 PROPERTIES OF NANOPARTICLES FOR CONJUGATION TO APTAMERS

Nanoparticles are a particularly attractive drug delivery vehicle for cancer therapeutics because they can be synthesized to recognize tumor-specific antigens and deliver drugs in a controlled manner.[4,23] The design of targeted drug delivery nanoparticles combines drug encapsulated materials, such as biodegradable polymers, with a targeting moiety (Figure 16.2). Ideally, biodegradable targeted nanoparticles should be designed with the following parameters: (1) small size (preferably between 10 and 200 nm); (2) high drug loading and entrapment efficiency; (3) low rate of aggregation; (4) slow rate of clearance from the bloodstream; and (5) optimized targeting to the desired tissue with minimized uptake by other tissues. The following sections will discuss the various parameters that must be considered for engineering of nanoparticles for targeted drug delivery applications, including the development of nanoparticle–aptamer bioconjugates. This will include discussion of nanoparticle biomaterial, size, charge, and surface modification schemes to achieve the desired design parameters. It is important to note that a detailed review is beyond the scope of this chapter, and the reader is referred to the following reviews for further information: see references.[86–90]

16.3.1.1 Size of Nanoparticles

The biodistribution pattern of nanoparticles, active nanoparticle targeting to tumor antigens, and passive nanoparticle targeting by EPR[91] are all greatly effected by the size of the nanoparticle. Passive targeting of the nanoparticle occurs because microvasculatures of tumors are more "leaky," thus permitting selective permeation of nanoparticles into the desired tumor tissue. This phenomenon has been exploited to target liposomes, therapeutic and diagnostic nanoparticles, and drug-polymer conjugates to cancer tissue (reviewed by Maeda).[91] Smaller particles (fewer than 150 nm) are better suited for permeating through the leaky microvasculature of the tumor cells, and their more pronounced surface curvature may also reduce interaction with the receptors on the surface of macrophages and subsequent clearance of the particles.[12] Biodistribution studies using liposomes have shown that although particles that are larger than 200 nm are largely taken up by the spleen, those less than 70 nm are also efficiently cleared by the liver.[92] Taken together, the optimal nanoparticle size should be experimentally determined for each formulation because the interplay of various parameters (polymer system, encapsulated drug, surface charge, surface modification) makes it difficult to extrapolate the ideal nanoparticle size from seemingly similar studies. In our laboratory, we have tuned the size of polymeric nanoparticles made either by emulsions or nanoprecipitation to study the best formulation for a specific application. Generally, one can make use of the interaction of different solvents used for making the emulsions or tune the size of the nanoparticle by adjusting solvent ratios and polymer concentrations in solution. Our biodistribution studies using various sizes of poly(lactic-*co*-glycolic acid) (PLGA)–PEG nanoparticle–aptamer bioconjugates, has suggested a linear relationship with regards to uptake by liver and spleen such that smaller particles (~ 80 nm) are modestly better at avoiding uptake by these organs (unpublished results).

16.3.1.2 Polymers for Synthesis of Nanoparticles

Controlled release biodegradable nanoparticles for clinical applications can be made from a wide variety of polymers, including poly(lactic acid) (PLA),[93] poly(glycolic acid) (PGA), PLGA,[94] poly(orthoesters),[95] poly(caprolactone),[96] poly(butyl cyanoacrylate),[97] polyanhydrides,[98] and poly-N-isopropylacrylamide.[99] Although many fabrication methods exist, polymeric nanoparticles are frequently made using an oil-in-water emulsion,[18,100] which involves dissolving a polymer and drug in an organic solvent, such as methylene chloride, ethyl acetate, or acetone. The organic phase is mixed with an aqueous phase by vortexing and sonicating and then evaporated, forcing the polymer to precipitate as nanoparticles in the aqueous phase. The particles are then recovered by centrifugation and lyophilization.

One of the considerations with respect to the material used for drug delivery is its ability to encapsulate drugs as well as degrade over the appropriate times. This subject has been an active area of investigation by our group and other investigators in academic and industry laboratories for several decades. The result has been an increasing arsenal of polymers with distinct encapsulation and release characteristics for a myriad of research, industrial, and clinical applications.[101,102] PGA and PLA are common biocompatible polymers that are used for many biomedical applications. PGA is hydrophilic because it lacks a methyl group and is more susceptible to hydrolysis, making this polymer easily degradable. Alternatively, PLA is relatively more stable in the body.[103] Through these unique properties, polymers such as PLGA have been derived that are made from both glycolic acid and lactic acid components. The ability to change the ratio of these two components of the polymer can then be used to dramatically alter the rate of degradation. Therefore, by choosing the desired polymer system for the synthesis of nanoparticles the rate of degradation and subsequent release of the molecule may be tuned for the intended application.

16.3.1.3 Charge of Nanoparticles

Nanoparticle charge has been shown to be important for regulating its pharmacokinetic properties. For example, it has been shown that anionic and cationic liposomes activate the complement system through distinct pathways, suggesting that particle charge may impact particle opsonization and phagocytosis.[104] Cationic charge on liposomes has also been shown to reduce their circulating half-life in blood and to affect their biodistribution between the tumor microvasculature and interstitium without impacting overall tumor uptake.[105] Nanoparticles could be synthesized with charged surfaces either by using charged polymers, such as poly-L-lysine, polyethylenimine (PEI), or polysaccharides, or through surface modification approaches. For example, the layer-by-layer deposition of ionic polymers have been used to change surface properties of nanoparticles, such as quantum dots, by depositing ionic polymers of interest on the charged nanoparticle surfaces.[106] Furthermore, surface charge of nanoparticles has been shown to regulate their biodistribution.[107] For example, increasing the charge of cationic pegylated liposomes decreases their accumulation in the spleen and blood while increasing their uptake by the liver and an increasing in the accumulation of liposomes in tumor vessels.[105] These experiments suggest that optimizing surface physicochemical properties of nanoparticles to better match the biochemical and physiological features of tumors may enhance the intratumoral delivery of nanoparticles for systemic therapeutic approaches.

For conjugation of the negatively charged aptamers to nanoparticles, the surface charge of the nanoparticle may be important. For example, we believe that direct immobilization of aptamers on cationic nanoparticles made from PEI may result in formation of aptamer–PEI complex that render the aptamer ineffective as a targeting molecule (unpublished observation). Therefore, neutral polymers such as PLA, PLGA or those with a more negative charge, such as polyanhyrides, may be most suitable for conjugation to aptamers. We have used PLA–PEG block copolymers to generate aptamer–nanoparticles bioconjugates.[23,108] One approach that may facilitate the use of a wider array of biomaterials for aptamer targeted drug delivery is through methods of "masking" the surface charge of the particles. For example, the addition of neutrally charged hydrophilic layer of PEG on the surface of the nanoparticles may facilitate the use of positively charged materials for the synthesis of nanoparticles. These cationic nanoparticles are particularly useful for gene delivery applications and thus may enable efficient targeted gene delivery using aptamers.

16.3.1.4 Surface Modification of Nanoparticles

Nanoparticle surface modification may also be used to engineer their interaction with the surrounding tissue. These interactions could be positive (i.e., targeting molecules) or negative (i.e., nonadhesive coatings). The surface modification of nanoparticles is particularly important because intravenously applied nanoparticles may be captured by macrophages before ever reaching the target site. Therefore, surface modifying particles to render them invisible to macrophages is essential to making long-circulating nanoparticles.[18,109] The ability to control the biodistribution of nanoparticles is particularly important for drug carrying nanoparticles because the delivery of drugs to the normal tissues can lead to toxicity.[16,17]

Hydrophilic polymers such as PEG,[18,109] polysaccharides,[110,111] and small molecules[112] can be conjugated on the surface of nanoparticles to engineer particles with desirable biodistribution characteristics. For example, to enhance the rate of circulation within the blood and minimize uptake by nondesired cell types, nanoparticles may be coated with polymers such as PEG.[18,109] Various molecular weights and types of PEG (linear or branched) have been used to coat nanoparticles.[113] PEG coatings are also useful for minimizing nanoparticle aggregation and can be used to prevent clogging of small vasculature and improve size-based targeting. More recently, novel approaches aimed at conjugating small molecules on nanoparticles using high-throughput methods have yielded nanoparticle libraries that could be subsequently analyzed for targeted delivery.[112] The use of similar high-throughput approaches has significant potential in optimizing nanoparticle properties for cancer therapy.

Surface modification of nanoparticles can be achieved in a multistep approach by first generating nanoparticles and subsequently modifying the surface of particles to achieve the desired characteristics. Alternatively, amphiphilic polymers may be covalently linked prior to generating nanoparticles to simultaneously control the surface chemistry as well as encapsulate drugs and

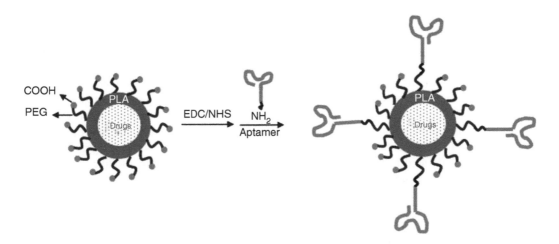

FIGURE 16.3 A schematic outlining a conjugation reaction between aptamers and polymer nanoparticles containing an encapsulated drug. Through incorporating a COOH-terminated, PEG-functionalized surface on the nanoparticle, NH_2-modified aptamers can be easily conjugated using simple aqueous chemistry.

eliminate the need for subsequent chemical modifications once the particle has been synthesized. This method may provide a more stable coating and better nanoparticle protection in contact with blood. For example, PLA, poly(caprolactone), and poly(cyanoacrylate) polymers, have been chemically conjugated to PEG polymers.[18,114,115] We have synthesized nanoparticles from amphiphilic copolymers composed of lipophilic PLGA and hydrophilic PEG (Figure 16.3) polymers where the PEG migrates to the surface of the nanoparticles in the presence of an aqueous solution.[18] A similar approach has also been used to generate pegylated PLA nanoparticles using PLA–PEG block copolymers.[23,108] These particles may be used to extend the nanoparticle residence times in circulation and enhance accumulation in tumor tissue through "passive targeting" and EPR effect.

In the case of engineering nanoparticles for active targeting, the polymer and its coating should have functional groups for the attachment of targeting moieties (which may be bound directly to the nanoparticle surface or though a spacer group). The targeting molecules can enhance the molecular interaction of the nanoparticles with a subset of cells or tissue.

16.3.2 CONJUGATION STRATEGIES FOR NANOPARTICLE–APTAMER CONJUGATES

Covalent conjugation of aptamers to substrates or drug delivery vehicles can be achieved most commonly through succinimidyl ester–amine chemistry that results in a stable amide linkage[43,43] or through maleimide–thiol chemistry.[23,108,116–120] Potential noncovalent strategies include affinity interactions (i.e., streptavidin–biotin) and metal coordination (i.e., between a polyhistidine tag at the end of the aptamer and Ni^{+2} chelates with immobilized nitrilotriacetic acid on the surface of the polymer particles). These covalent and noncovalent strategies have been used to immobilize a wide range of biomolecules, including proteins, enzymes, peptides, and nucleic acids to delivery vehicles.

We believe that covalently linked bioconjugates may result in enhanced stability in physiologic salt and pH while avoiding the unnecessary addition of biological components (i.e., streptavidin), thus minimizing immunologic reactions and potential toxicity. For covalent conjugation, the aptamer is typically modified to carry a terminal primary amine or thiol group that is in turn conjugated, respectively, to activated carboxylic acid N-hydroxysuccinimide (NHS) ester or maleimide functional groups present on the surface of drug delivery vehicles. These reactions are carried out under aqueous conditions with a product yield of 80–90%.[121] One potential difficulty with maleimide–thiol chemistry is the oxidation of the thiol group attached to aptamers during storage (formation of S–S bond between two thiol-modified aptamers), resulting in dimers of aptamers that are not able to participate in the conjugation reaction with the malimide group on particles. This problem can be partially alleviated by using a reducing agent, such as *tris*(2-carboxyethyl)phosphine (TCEP), β-mercaptoethanol, or dithiothreitol (DTT), during the conjugation reaction. A potential advantage of using NHS–amine chemistry is that the unreacted carboxylic acid groups on the particle surface make the particle surface charge (zeta potential) slightly negative, thus reducing nonspecific interaction between the negatively charged aptamers and the negative particle surface. Recently, controlled-release nanoparticles generated from PLA–PEG block copolymer with a terminal carboxylic acid group attached to the PEG were conjugated with primary-amine-terminated aptamers.[23,103] In this case, the hydrophilic PEG group facilitated the presentation of the carboxylic acid on the particle surface for conversion to activated carboxylic acid NHS ester and conjugation to the primary-amine-modified aptamers.

The conjugation of aptamers to nanoparticles can be qualitatively confirmed by fluorescent microscopy or flow cytometry through the use of fluorescent probes, such as fluorescein isothiocyanate (FITC), that are conjugated directly to the aptamers or indirectly to complementary oligonucleotides that hybridize to the aptamers.[23] Alternatively, analytical approaches, such as x-ray photoemission spectroscopy (XPS), may be used for characterization of the nanoparticle surface to confirm the extent of conjugation. The presence of a hydrocarbon spacer group between the nanoparticle surface and the aptamer should improve the probability of interaction between the aptamer and its target. Furthermore, a consistent density of the aptamer on the surface of nanoparticles can potentially be achieved by utilizing an excess molar amount of aptamer relative to the reactive group on the nanoparticle surface during the conjugation reactions. However, the optimal density of targeting molecule on nanoparticle surface may need to be experimentally determined.[122]

We have used the covalent conjugation approach to demonstrate a proof-of-concept for nanoparticle–aptamer bioconjugates that target the PSMA on the surface of prostate cancer cells and are taken up by cells that express the PSMA protein specifically and efficiently.[23] We have also shown using a microfluidic system that these nanoparticle–aptamer conjugates are capable of binding to their target cells under flow conditions suggesting their suitability for in vivo targeted drug-delivery applications.[108] Most recently, we have demonstrated the in vivo efficacy of docetaxel-encapsulated nanoparticle–aptamer conjugates using a xenograft prostate cancer nude mouse model.[14] These approaches have paved the way for the use of aptamers for targeted delivery of drug encapsulated nanoparticles to a myriad of human cancers.

16.4 APTAMER–DRUG CONJUGATES FOR TARGETED CANCER THERAPY

The clinical utility of radioimmunoconjugates and chemoimmunoconjugates in cancer treatment is well documented with several examples of each in clinical practice or under development at this time.[123–125] More recently aptamer conjugates for cancer therapeutic and diagnostic applications,[14,23,108,116,118–120,126] biosensors,[127–129] flow cytometry,[130] ELISAs,[131,132] and capillary electrophoresis for separation technology have been discribed.[40,133] As noted in Section 16.2, aptamers have distinct advantages over antibodies, and, for the purpose of targeted delivery of toxins, their greatest advantage lies in their relatively smaller size and remarkable stability. The smaller size of aptamers, as compared to antibodies, allows for a more efficient tumor penetration of aptamers, translating to a more efficient delivery of a cytotoxic payload to tumor cells.[119] Our research group is interested in the use of aptamer–toxin conjugates for therapeutic applications,[126] and we believe that aptamer–toxin conjugates may result in future novel therapeutic applications in oncology and other areas of medicine.

For the preparation of the aptamer–drug conjugates, conventional conjugation strategies that have been used for the preparation of immunoconjugates may also be employed with some modifications.[125] For targeted radiotherapy, beta-emitting radioisotopes, such as ^{131}I, ^{90}Y, or ^{67}Cu, or alpha-emitting radioisotopes such as ^{213}Bi or ^{211}At, may be used to develop radioaptamerconjugates. These radioisotopes may be linked directly to aptamers or immobilized through metal chelators attached to aptamers. For targeted chemotherapy, drug molecules (i.e., doxorubicin, calicheamicin, or auristatin) may be covalently coupled to one or more sites on the aptamer using simple chemistry (i.e., formation of amide, disulfide, or hydrazone bond).[134,135] The development of aptamer conjugates is facilitated by site-selective functionalization of aptamers at their $3'$- and/or $5'$-terminus, making it possible to reproducibly attach an exact number of drugs in a site-specific manner.[23,108,136] Aptamers may also be functionalized within the oligonucleotide chain using modified nucleotides with desired functional groups; however, this latter approach may negatively impact the aptamer conformation and function. In the case of conjugation of toxins to antibodies, the reaction is most commonly not site-directed, and this nonspecific conjugation approach may negatively impact the binding characteristics of antibodies.[136] Furthermore, unlike antibodies, the desired linkers or functional groups on aptamers for conjugation on other applications can be easily introduced during the chemical synthesis of these molecules. We have schematically demonstrated several proposed conjugation strategies between aptamers and drugs (Figure 16.4). Typical functional groups modifiable at the $5'$- or $3'$-ends of aptamers are mostly nucleophiles, including amine ($-NH_2$), sulfhydryl ($-SH$), and hydrazide ($-C(O)NHNH_2$). Appropriate functional groups on the drugs for immobilization on aptamers may be carboxylic acid (for amide bond formation with amine-modified aptamers), maleimide or unsaturated carbonyl (for thioether formation with sulfhydryl groups on aptamers), activated disulfide (for example, S-pyridyl disulfide) for disulfide

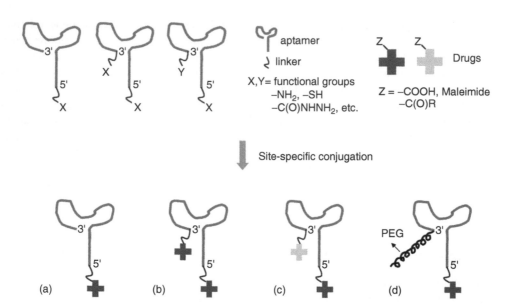

FIGURE 16.4 Available conjugation strategies between end-functionalized aptamers and drug molecules. Panels (a) through (d) demonstrate a series of single- and double-conjugation strategies.

formation with sulfhydryl groups on aptamers, and ketone for hydrazone formation with hydrazide on aptamers.[136]

The covalent bonds formed through the aforementioned strategies are all cleavable in vivo, allowing for the active drug to be released for therapeutic efficacy. The choice of conjugation strategy may allow for a spatial control of this release and improved clinical efficacy. In addition, by altering the size of aptamer–drug complex, the pharmacokinetics properties of the conjugate may be controlled, allowing for temporal control of the drug circulation and clearance. For example, a longer circulation time and reduced renal clearance may be achieved by linking the aptamer–drug conjugates to PEG polymers to increase the size of the complex above the threshold for renal clearance.[137] This conjugate form may result in improved drug delivery to cancers as a result of increased systemic circulation time. Intracellular versus extracellular drug release also have unique sets of design considerations.[138] In the case of the disulfide linkage approach, the aptamer–drug conjugate is expected to be stable in plasma during circulation with subsequent intracellular cleavage of disulfide bond and drug release due to high intracellular concentration of glutathione (GSH).[139] Hydrazone linkage also has favorable properties for conjugation of drugs to aptamers, allowing for a pH sensitive release of the drug in a relatively more acidic environment of the tumor tissues.[134,140,141] Therefore, disulfide and hydrazone linkages may be considered when developing aptamer conjugates for cancer therapeutics.

We have recently developed a novel strategy for generating aptamer–drug conjugates by intercalating drugs between nucleic acid bases of the aptamer (Figure 16.5).[126] Using doxorubicin (dox) as a model drug and the A10 PSMA aptamer as model aptamer, we have developed a stable aptamer–dox conjugate through physical interactions. The resulting physical conjugate demonstrated the following characteristics: (1) high discrimination between cancer cells expressing target antigen and cells lacking target antigen; and (2) efficient delivery of anti-cancer drug to target cancer cells.[126] Specifically, we developed aptamer–dox conjugates and demonstrated the stability of this system in vitro. Using LNCaP prostate cancer cell lines that express the PSMA antigen and PC3 prostate cancer cell lines that do not express the PSMA protein, we demonstrated a differential uptake of the aptamer–dox conjugate by LNCaP cells, translating to a more efficient cellular toxicity and anti-cancer efficacy in vitro (Figure 16.6). These results suggest that physical conjugate strategies such as intercalation may serve as a novel approach for developing aptamer–drug therapeutics.

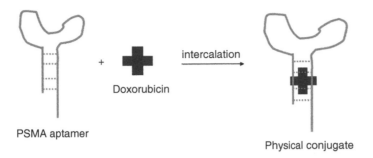

FIGURE 16.5 A schematic diagram outlining the formation of physical conjugate between aptamers and drugs capable of intercalating into base pairs of aptamers. Doxorubicin, an anti-cancer drug, can be intercalated into a PSMA aptamer to form a 1:1 complex.

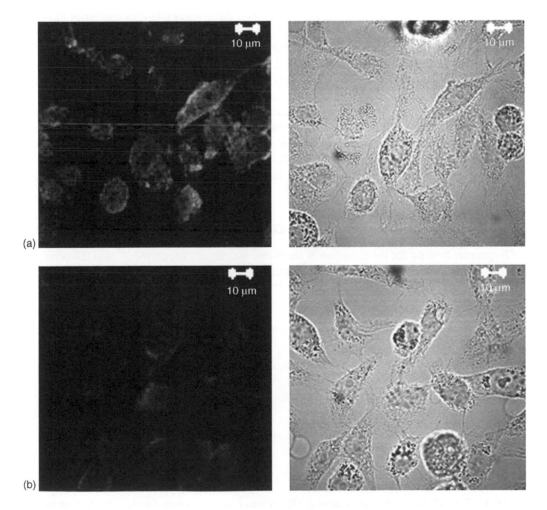

FIGURE 16.6 The confocal laser scanning microscopy images of LNCaP (a) and PC3 cells (b) after treatments of the aptamer–doxorubicin physical conjugate (1.5 μ*M*) for 2 h.

In addition to small molecule and radioisotopes, bio therapeutics such as small interfering RNAs (siRNA)[142–144] and protein-based toxins may also be conjugated to aptamers. For example, siRNAs can be coupled to aptamers through disulfide bond, allowing the release of siRNAs in the active form after cytosolic uptake by the targeted cells. A conjugate of aptamer–siRNA may represent a novel class of therapeutics with widespread application in medicine.[116]

16.5 APTAMER–NANOPARTICLE CONJUGATES FOR PROTEIN DETECTION

Quantum dots (QDs) are nanometer-sized semiconductor crystals with tunable fluorescence that can be developed as sensitive biosensors or probes for in vivo imaging and diagnostic applications.[145–147] Unlike conventional small molecule fluorophores and biological fluorophores, such as the green fluorescence protein (GFP), QDs possess several favorable characteristics, including: (1) greatly improved photostability (lack of photobleaching) that enable real time imaging of biomarkers over an extended period of time; (2) a broad excitation wavelength and a narrow emission wavelength enabling excitation of multiple QDs simultaneously using a single laser wavelength; (3) bright fluorescence with high quantum yield; and (4) ability to functionalize the surface of QDs with various molecules ranging from small molecules to antibodies and aptamers for desired targeted applications. Taken together, these favorable characteristics have made QDs an emerging class of imaging probes with broad applications in medicine.

Metallic nanoparticles, such as gold-nanoparticles (Gd-NPs), are capable of absorbing or scattering light and are now used in many biological applications, particularly visualization of cellular and tissue components by electron microscopy.[148,149] More recently, it has been reported that antibody–Gd-NP conjugates[150,151] can detect proteins[152] or viruses[153] in solution through surface plasmon resonance effect, which results in color change depending on the status of aggregation of the particles. This simple phenomenon together with the fact that Gd-NPs are biocompatible makes Gd-NPs useful detection probes of various in vivo applications.[154]

The surface functionalization of QDs and Gd-NPs with specific targeting molecules may result in the development of targeted imaging modalities with widespread applications in research and medicine.

16.5.1 APTAMER–QD CONJUGATES FOR THE DETECTION OF PROTEINS

A QD–aptamer beacon for the detection of thrombin was recently reported,[155] and the principle of this detection system is schematically illustrated (Figure 16.7). Briefly, QDs were surface

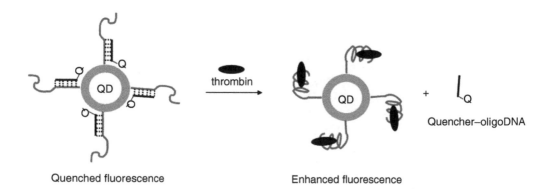

FIGURE 16.7 A schematic diagram outlining QD–aptamer beacon system for the detection of thrombin. Target protein replaces the bound quencher–oligoDNA by forming stable complex with aptamers on a QD, resulting in recovery of a native QD's fluorescence.

functionalized with an aptamer known to bind to the thrombin protein to develop a thrombin-detecting QD–aptamer beacon. A second oligonucleotide, having complementary sequences to the thrombin aptamer and bearing a quencher, was synthesized and utilized to quench the system in the absence of thrombin. When the quencher oligonucleotide was incubated with the QD–aptamer conjugate in the absence of thrombin, the hybridization of the quencher oligonucleotide and the thrombin aptamer proximated the quencher molecules to the surface of the QD, resulting in quenching effect through fluorescence resonance energy transfer (FRET).[156,157] In the presence of thrombin (1 μM), the quencher oligonucleotide is displaced, resulting in restoration of the fluorescence of the QD–aptamer conjugate. Using this QD–aptamer beacon system, a 1 μM concentration of thrombin could be detected specifically. We postulate that multiplexed detection of a panel of protein targets may be performed in parallel by formation of QD–aptamer conjugates from different aptamers, and QDs possessing different emitting fluorescence profile.[158,159] In addition, considering a growing number of tumor-specific or tumor-associated aptamers that are available today, it may be increasingly possible to develop more specific and more efficient cancer diagnostic probes for a myriad of important cancers. We have recently developed a QD–aptamer conjugate using the A10 PSMA aptamer and have demonstrated the differential binding and uptake of these vehicles by cells that express the PSMA antigen (unpublished results).

16.5.2 Aptamer–Gold-Nanoparticle Conjugates for the Detection of Proteins

A highly specific sensing system for the detection of platelet-derived growth factor (PDGF) has been developed using PDGF-specific, aptamer–conjugated gold-nanoparticles.[160] The principle of the detection using the conjugate is illustrated in Figure 16.8. The binding of soluble PDGF protein to Gd-NP–aptamer conjugates results in a change in the color of the solution due to the formation of aggregates between the Gd-NP conjugates. In the absence of PDGF, the conjugates are dispersed separately in solution (red), but in the presence of PDGF the Gd-NP aggregate closely to an inter-particle distances less than the average diameter of each nanoparticle, causing a dipole–dipole coupling between particles as well as increased scattering, resulting in a measurable color change (purple).[161–163] Using this detection system, Gd-NP–aptamer conjugate could detect nanomolar concentrations of PDGF in solution. Because Gd-NPs are biocompatible, easy to develop, and easy to functionalize with aptamers, it is expected that bioconjugates utilizing aptamers and Gd-NPs as target molecular probes may result in novel cancer diagnostic modalities in the future.

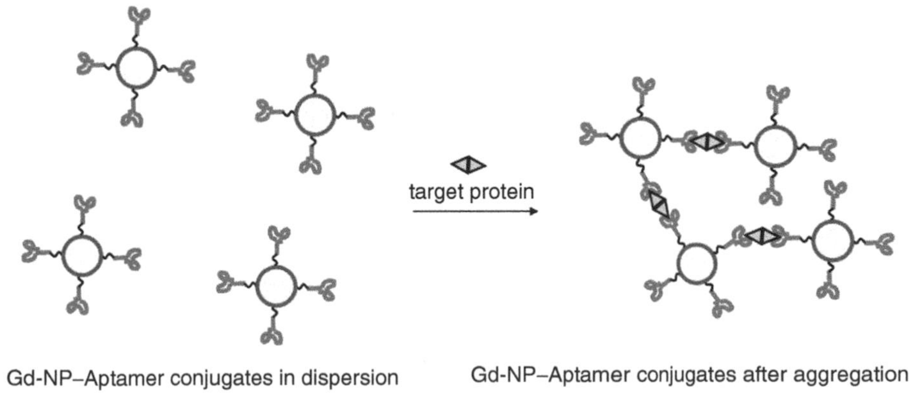

Gd-NP–Aptamer conjugates in dispersion   Gd-NP–Aptamer conjugates after aggregation

FIGURE 16.8 A schematic representation of the aggregation of gold-nanoparticle–aptamer conjugates in the presence of target proteins. Aggregation causes color change from red to purple.

16.6 CONCLUSION

Bioconjugates comprising nanoparticles and aptamers represent a potentially powerful tool for developing novel diagnostic and therapeutic modalities for cancer detection and treatment. As drug delivery vehicles for cancer therapy, nanoparticle–aptamer bioconjugates can be designed to target and be taken up by cancer cells for targeted delivery and controlled release of chemotherapeutic drugs over an extended time directly at the site of tumors. The successful achievement of this goal requires the isolation of aptamers that bind to the extracellular domain of antigens expressed exclusively or preferentially on the plasma membrane of cancer cells or on the extra cellular matrices of tumor tissue. In addition, nanoparticles would have to be designed with the optimized properties that facilitate targeting and delivery of the drugs to the desired tissues while avoiding uptake by the mononuclear phagocytic system in the body.

The targeted delivery of chemotherapeutic drugs for cancer therapy may minimize their side effects and enhance their cytotoxicity to cancer cells, resulting in a better clinical outcome. We anticipate that the combination of controlled release technology and targeted approaches may represent a viable approach for achieving this goal. One major clinical advantage of targeted drug-encapsulated nanoparticle conjugates over drugs that are directly linked to a targeting moiety is that large amounts of chemotherapeutic drug may be delivered to cancer cells per each delivery and biorecognition event. Another advantage would be the ability to simultaneously deliver two or more chemotherapeutic drugs and release each in a predetermined manner, thus resulting in effective combination chemotherapy, which is common for the management of many cancers. Antibodies and peptides have been widely used for the targeted delivery of drug encapsulated nanoparticles; however, the translation of these vehicles into clinical practice has lagged behind our advances in the laboratory. This has been largely due to the immunogenicity and nonspecific uptake of nanoparticle–antibody bioconjugates by nontargeted cells, resulting in toxicity or poor efficacy of these conjugates. Nanoparticle–aptamer bioconjugates face some of the same challenges of nonspecific uptake after systemic administration and thus must be engineered with surface physicochemical characteristics to avoid toxicity to nontargeted cells. We believe that optimal particle size and surface properties to sufficiently decrease the rate of nonspecific particle uptake while achieving successful targeting must be determined experimentally on a case-by-case basis, as this also depends on the polymer system, the drug being encapsulated and the tumor microenvironment including its vascularity. However, the advantage of nanoparticle–aptamer bioconjugates over their antibody conjugate counterparts lies in the ease of aptamer isolation, lack of immunogenicity, and a decreased batch-to-batch variability attributable to the chemical nature of aptamer synthesis and production, which can facilitate the translation of optimally engineered nanoparticle–aptamer bioconjugates into clinical practice.

ACKNOWLEDGMENTS

The authors thank Jack Szostak, Benjamin Teply, Jeffrey Karp, Jianjun Cheng, Chris Cannizarro, Etgar Levy-Nissenbaum, and Andrej Luptak for helpful discussions and review of this manuscript. This work was supported by NIH/NCI CA119349, NIH/NIBIB EB003647, and by the David Koch Cancer Research Fund.

REFERENCES

1. Peppas, N. A., Intelligent therapeutics: biomimetic systems and nanotechnology in drug delivery, *Adv. Drug Deliv. Rev.*, 56(11), 1529–1531, 2004.
2. Langer, R. and Tirrell, D. A., Designing materials for biology and medicine, *Nature*, 428(6982), 487–492, 2004.
3. Langer, R., Drug delivery and targeting, *Nature*, 392(6679), 5–10, 1998.
4. Ferrari, M., Cancer nanotechnology: opportunities and challenges, *Nat. Rev. Cancer*, 5(3), 161–171, 2005.

5. Matsumura, Y. and Maeda, H., A new concept for macromolecular therapeutics in cancer chemotherapy: mechanism of tumoritropic accumulation of proteins and the antitumor agent smancs, *Cancer Res.*, 46(12 Pt 1), 6387–6392, 1986.

6. Maeda, H. and Matsumura, Y., Tumoritropic and lymphotropic principles of macromolecular drugs, *Crit. Rev. Ther. Drug Carrier Syst.*, 6(3), 193–210, 1989.

7. Maeda, H., The enhanced permeability and retention (EPR) effect in tumor vasculature: the key role of tumor-selective macromolecular drug targeting, *Adv. Enzyme Regul.*, 41, 189–207, 2001.

8. Yuan, F., Dellian, M., Fukumura, D., Leunig, M., Berk, D. A., Torchilin, V. P., and Jain, R. K., Vascular permeability in a human tumor xenograft: molecular size dependence and cutoff size, *Cancer Res.*, 55(17), 3752–3756, 1995.

9. Kong, G., Braun, R. D., and Dewhirst, M. W., Hyperthermia enables tumor-specific nanoparticle delivery: effect of particle size, *Cancer Res.*, 60(16), 4440–4445, 2000.

10. Yuan, F., Leunig, M., Huang, S. K., Berk, D. A., Papahadjopoulos, D., and Jain, R. K., Microvascular permeability and interstitial penetration of sterically stabilized (stealth) liposomes in a human tumor xenograft, *Cancer Res.*, 54(13), 3352–3356, 1994.

11. Brigger, I., Morizet, J., Laudani, L., Aubert, G., Appel, M., Velasco, V., Terrier-Lacombe, M. J. et al., Negative preclinical results with stealth nanospheres-encapsulated Doxorubicin in an orthotopic murine brain tumor model, *J. Control Release*, 100(1), 29–40, 2004.

12. Brigger, I., Dubernet, C., and Couvreur, P., Nanoparticles in cancer therapy and diagnosis, *Adv. Drug Deliv. Rev.*, 54(5), 631–651, 2002.

13. Anderson, D. G., Peng, W., Akinc, A., Hossain, N., Kohn, A., Padera, R., Langer, R., and Sawicki, J. A., A polymer library approach to suicide gene therapy for cancer, *Proc. Natl Acad. Sci. U.S.A.*, 101(45), 16028–16033, 2004.

14. Farokhzad, O. C., Cheng, J., Teply, B. A., Sherifi, I., Jon, S., Kantoff, P. W., Richie, J. P., and Langer, R., Targeted nanoparticle-aptamer bioconjugates for cancer chemotherapy in vivo, *Proc. Natl Acad. Sci. U.S.A.*, 103(16), 6315–6320, 2006.

15. Brannon-Peppas, L. and Blanchette, J. O., Nanoparticle and targeted systems for cancer therapy, *Adv. Drug Del. Rev.*, 56(11), 1649–1659, 2004.

16. Demoy, M., Gibaud, S., Andreux, J. P., Weingarten, C., Gouritin, B., and Couvreur, P., Splenic trapping of nanoparticles: complementary approaches for in situ studies, *Pharm. Res.*, 14(4), 463–468, 1997.

17. Gibaud, S., Andreux, J. P., Weingarten, C., Renard, M., and Couvreur, P., Increased bone marrow toxicity of doxorubicin bound to nanoparticles, *Eur. J. Cancer*, 30A(6), 820–826, 1994.

18. Gref, R., Minamitake, Y., Peracchia, M. T., Trubetskoy, V., Torchilin, V., and Langer, R., Biodegradable long-circulating polymeric nanospheres, *Science*, 263(5153), 1600–1603, 1994.

19. Monsky, W. L., Fukumura, D., Gohongi, T., Ancukiewcz, M., Weich, H. A., Torchilin, V. P., Yuan, F., and Jain, R. K., Augmentation of transvascular transport of macromolecules and nanoparticles in tumors using vascular endothelial growth factor, *Cancer Res.*, 59(16), 4129–4135, 1999.

20. Guillemard, V. and Saragovi, H. U., Novel approaches for targeted cancer therapy, *Curr. Cancer Drug Targets*, 4(4), 313–326, 2004.

21. Zhang, X. Q., Wang, X. L., Zhang, P. C., Liu, Z. L., Zhuo, R. X., Mao, H. Q., and Leong, K. W., Galactosylated ternary DNA/polyphosphoramidate nanoparticles mediate high gene transfection efficiency in hepatocytes, *J. Control Release*, 102(3), 749–763, 2005.

22. Hicke, B. J. and Stephens, A. W., Escort aptamers: a delivery service for diagnosis and therapy, *J. Clin. Invest.*, 106(8), 923–938, 2000.

23. Farokhzad, O. C., Jon, S., Khademhosseini, A., Tran, T. N., Lavan, D. A., and Langer, R., Nanoparticle-aptamer bioconjugates: a new approach for targeting prostate cancer cells, *Cancer Res.*, 64(21), 7668–7672, 2004.

24. Sullenger, B. A., Gallardo, H. F., Ungers, G. E., and Gilboa, E., Overexpression of TAR sequences renders cells resistant to human immunodeficiency virus replication, *Cell*, 63(3), 601–608, 1990.

25. Ellington, A. D. and Szostak, J. W., In vitro selection of RNA molecules that bind specific ligands, *Nature*, 346(6287), 818–822, 1990.

26. Tuerk, C. and Gold, L., Systematic evolution of ligands by exponential enrichment: RNA ligands to bacteriophage T4 DNA polymerase, *Science*, 249(4968), 505–510, 1990.

27. James, J. S. and Dubs, G., FDA approves new kind of lymphoma treatment. Food and drug administration, *AIDS Treat. News*, No 284 2–3, 1997.

28. Foss, F. M., DAB(389)IL-2 (ONTAK): a novel fusion toxin therapy for lymphoma, *Clin. Lymphoma*, 1(2), 110–116, discussion 117, 2000.

29. Harris, L. J., Larson, S. B., Hasel, K. W., Day, J., Greenwood, A., and McPherson, A., The three-dimensional structure of an intact monoclonal antibody for canine lymphoma, *Nature*, 360(6402), 369–372, 1992.

30. Harris, L. J., Skaletsky, E., and McPherson, A., Crystallographic structure of an intact IgG1 monoclonal antibody, *J. Mol. Biol.*, 275(5), 861–872, 1998.

31. Marks, J. D., Selection of internalizing antibodies for drug delivery, *Methods Mol. Biol.*, 248, 201–208, 2004.

32. Marks, J. D., Ouwehand, W. H., Bye, J. M., Finnern, R., Gorick, B. D., Voak, D., Thorpe, S. J., Hughes-Jones, N. C., and Winter, G., Human antibody fragments specific for human blood group antigens from a phage display library, *Biotechnology (N Y)*, 11(10), 1145–1149, 1993.

33. Doggrell, S. A., Pegaptanib: the first antiangiogenic agent approved for neovascular macular degeneration, *Expert Opin. Pharmacother.*, 6(8), 1421–1423, 2005.

34. Fitzwater, T. and Polisky, B., A SELEX primer, *Methods Enzymol.*, 267, 275–301, 1996.

35. White, R. R., Sullenger, B. A., and Rusconi, C. P., Developing aptamers into therapeutics, *J. Clin. Invest.*, 106(8), 929–934, 2000.

36. Huang, D. B., Vu, D., Cassiday, L. A., Zimmerman, J. M., Maher, L. J., and Ghosh, G., Crystal structure of NF-kappaB (p50)2 complexed to a high-affinity RNA aptamer, *Proc. Natl Acad. Sci. U.S.A.*, 100(16), 9268–9273, 2003.

37. Hermann, T. and Patel, D. J., Adaptive recognition by nucleic acid aptamers, *Science*, 287(5454), 820–825, 2000.

38. Drolet, D. W., Nelson, J., Tucker, C. E., Zack, P. M., Nixon, K., Bolin, R., Judkins, M. B. et al., Pharmacokinetics and safety of an anti-vascular endothelial growth factor aptamer (NX1838) following injection into the vitreous humor of rhesus monkeys, *Pharm. Res.*, 17(12), 1503–1510, 2000.

39. Jenison, R. D., Gill, S. C., Pardi, A., and Polisky, B., High-resolution molecular discrimination by RNA, *Science*, 263(5152), 1425–1429, 1994.

40. Tombelli, S., Minunni, M., and Mascini, M., Analytical applications of aptamers, *Biosens. Bioelectron.*, 20(12), 2424–2434, 2005.

41. Charlton, J., Sennello, J., and Smith, D., In vivo imaging of inflammation using an aptamer inhibitor of human neutrophil elastase, *Chem. Biol.*, 4(11), 809–816, 1997.

42. Beigelman, L., McSwiggen, J. A., Draper, K. G., Gonzalez, C., Jensen, K., Karpeisky, A. M., Modak, A. S. et al., Chemical modification of hammerhead ribozymes. Catalytic activity and nuclease resistance, *J. Biol. Chem.*, 270(43), 25702–25708, 1995.

43. Aurup, H., Williams, D. M., and Eckstein, F., 2′-Fluoro- and 2′-amino-2′-deoxynucleoside 5′-triphosphates as substrates for T7 RNA polymerase, *Biochemistry*, 31(40), 9636–9641, 1992.

44. Pieken, W. A., Olsen, D. B., Benseler, F., Aurup, H., and Eckstein, F., Kinetic characterization of ribonuclease-resistant 2′-modified hammerhead ribozymes, *Science*, 253(5017), 314–317, 1991.

45. Eulberg, D. and Klussmann, S., Spiegelmers: biostable aptamers, *Chem. biochem.*, 4(10), 979–983, 2003.

46. Schmidt, K. S., Borkowski, S., Kurreck, J., Stephens, A. W., Bald, R., Hecht, M., Friebe, M., Dinkelborg, L., and Erdmann, V. A., Application of locked nucleic acids to improve aptamer in vivo stability and targeting function, *Nucleic Acids Res.*, 32(19), 5757–5765, 2004.

47. Tucker, C. E., Chen, L. S., Judkins, M. B., Farmer, J. A., Gill, S. C., and Drolet, D. W., Detection and plasma pharmacokinetics of an anti-vascular endothelial growth factor oligonucleotide-aptamer (NX1838) in rhesus monkeys, *J Chromatogr. B Biomed. Sci. Appl.*, 732(1), 203–212, 1999.

48. Gold, L. and Tuerk, L., *Methods of Producing Nucleic Acid Ligands*, U.S.A., 1997.

49. Chelliserrykattil, J. and Ellington, A. D., Evolution of a T7 RNA polymerase variant that transcribes 2′-O-methyl RNA, *Nat. Biotechnol.*, 22(9), 1155–1160, 2004.

50. Burmeister, P. E., Lewis, S. D., Silva, R. F., Preiss, J. R., Horwitz, L. R., Pendergrast, P. S., McCauley, T. G. et al., Direct in vitro selection of a 2′-O-methyl aptamer to VEGF, *Chem. Biol.*, 12(1), 25–33, 2005.

51. Lee, J. F., Hesselberth, J. R., Meyers, L. A., and Ellington, A. D., Aptamer database, *Nucleic Acids Res.*, 32, D95–D100, 2004.

52. Gragoudas, E. S., Adamis, A. P., Cunningham, E. T., Feinsod, M., and Guyer, D. R., Pegaptanib for neovascular age-related macular degeneration, *N Engl. J. Med.*, 351(27), 2805–2816, 2004.

53. Vinores, S. A., Technology evaluation: pegaptanib. Eyetech/Pfizer, *Curr. Opin. Mol. Ther.*, 5(6), 673–679, 2003.

54. Pestourie, C., Tavitian, B., and Duconge, F., Aptamers against extracellular targets for in vivo applications, *Biochimie*, 87(9-10), 921–930, 2005.

55. McNeel, D. G., Eickhoff, J., Lee, F. T., King, D. M., Alberti, D., Thomas, J. P., Friedl, A. et al., Phase I trial of a monoclonal antibody specific for alphavbeta3 integrin (MEDI-522) in patients with advanced malignancies, including an assessment of effect on tumor perfusion, *Clin. Cancer Res.*, 11(21), 7851–7860, 2005.

56. Mi, J., Zhang, X., Giangrande, P. H., McNamara, J. O., Nimjee, S. M., Sarraf-Yazdi, S., Sullenger, B. A., and Clary, B. M., Targeted inhibition of alphavbeta3 integrin with an RNA aptamer impairs endothelial cell growth and survival, *Biochem. Biophys. Res. Commun.*, 338(2), 956–963, 2005.

57. Chen, C. H., Chernis, G. A., Hoang, V. Q., and Landgraf, R., Inhibition of heregulin signaling by an aptamer that preferentially binds to the oligomeric form of human epidermal growth factor receptor-3, *Proc. Natl Acad. Sci. U.S.A.*, 100(16), 9226–9231, 2003.

58. Lupold, S. E., Hicke, B. J., Lin, Y., and Coffey, D. S., Identification and characterization of nuclease-stabilized RNA molecules that bind human prostate cancer cells via the prostate-specific membrane antigen, *Cancer Res.*, 62(14), 4029–4033, 2002.

59. Deng, J. S., Ballou, B., and Hofmeister, J. K., Internalization of anti-nucleolin antibody into viable HEp-2 cells, *Mol. Biol. Rep.*, 23(3-4), 191–195, 1996.

60. Christian, S., Pilch, J., Akerman, M. E., Porkka, K., Laakkonen, P., and Ruoslahti, E., Nucleolin expressed at the cell surface is a marker of endothelial cells in angiogenic blood vessels, *J. Cell Biol.*, 163(4), 871–878, 2003.

61. Dapic, V., Abdomerovic, V., Marrington, R., Peberdy, J., Rodger, A., Trent, J. O., and Bates, P. J., Biophysical and biological properties of quadruplex oligodeoxyribonucleotides, *Nucleic Acids Res.*, 31(8), 2097–2107, 2003.

62. Barnhart, K. M., Laber, D. A., Bates, P. J., Trent, J. O., and Miller, D. M., In *AGRO100: The translation from lab to clinic of a tumor-targeted nucleic acid aptamer. Proceedings from the American Society for Clinical Oncology Meeting*, American Society for Clinical Oncology, New Orleans, LA, 2004.

63. Laber, D. A., Sharma, V. R., Bhupalam, L., Taft, B., Hendler, F. J., and Barnhart, K. M. In *Update on the First Phase I Study of AGRO100 in Advanced Cancer, American Society of Clinical Oncology Annual Meeting*, Abstract 3064, 2005.

64. Girvan, A. C., Barve, S. S., Thongboonkerd, V., Klein, J. B., and Pierce, W. M., In *Inhibition of NFkB signaling by an oligonucleotide aptamer*, Proceedings of the American Association of Cancer Research, 2004.

65. Jeong, S., Eom, T., Kim, S., Lee, S., and Yu, J., In vitro selection of the RNA aptamer against the Sialyl Lewis X and its inhibition of the cell adhesion, *Biochem. Biophys. Res. Commun.*, 281(1), 237–243, 2001.

66. Santulli-Marotto, S., Nair, S. K., Rusconi, C., Sullenger, B., and Gilboa, E., Multivalent RNA aptamers that inhibit CTLA-4 and enhance tumor immunity, *Cancer Res.*, 63(21), 7483–7489, 2003.

67. Hicke, B. J., Marion, C., Chang, Y. F., Gould, T., Lynott, C. K., Parma, D., Schmidt, P. G., and Warren, S., Tenascin-C aptamers are generated using tumor cells and purified protein, *J. Biol. Chem.*, 276(52), 48644–48654, 2001.

68. Blank, M., Weinschenk, T., Priemer, M., and Schluesener, H., Systematic evolution of a DNA aptamer binding to rat brain tumor microvessels. selective targeting of endothelial regulatory protein pigpen, *J. Biol. Chem.*, 276(19), 16464–16468, 2001.

69. Santini, J. T., Cima, M. J., and Langer, R., A controlled-release microchip, *Nature*, 397(6717), 335–338, 1999.

70. Edwards, D. A., Hanes, J., Caponetti, G., Hrkach, J., Ben-Jebria, A., Eskew, M. L., Mintzes, J., Deaver, D., Lotan, N., and Langer, R., Large porous particles for pulmonary drug delivery, *Science*, 276(5320), 1868–1871, 1997.

71. LaVan, D. A., McGuire, T., and Langer, R., Small-scale systems for in vivo drug delivery, *Nature Biotechnol.*, 21(10), 1184–1191, 2003.

72. Elisseeff, J., McIntosh, W., Fu, K., Blunk, B. T., and Langer, R., Controlled-release of IGF-I and TGF-beta1 in a photopolymerizing hydrogel for cartilage tissue engineering, *J. Orthop. Res.*, 19(6), 1098–1104, 2001.

73. Tanabe, K., Regar, E., Lee, C. H., Hoye, A., van der Giessen, W. J., and Serruys, P. W., Local drug delivery using coated stents: new developments and future perspectives, *Curr. Pharm. Des.*, 10(4), 357–367, 2004.

74. Ebrahim, S., Peyman, G. A., and Lee, P. J., Applications of liposomes in ophthalmology, *Surv. Ophthalmol.*, 50(2), 167–182, 2005.

75. Haak, T., New developments in the treatment of type 1 diabetes mellitus, *Exp. Clin. Endocrinol. Diabetes*, 107(suppl. 3), S108–S113, 1999.

76. Guerin, C., Olivi, A., Weingart, J. D., Lawson, H. C., and Brem, H., Recent advances in brain tumor therapy: local intracerebral drug delivery by polymers, *Invest. New Drugs*, 22(1), 27–37, 2004.

77. Jiang, W., Gupta, R. K., Deshpande, M. C., and Schwendeman, S. P., Biodegradable poly(lactic-co-glycolic acid) microparticles for injectable delivery of vaccine antigens, *Adv. Drug Deliv. Rev.*, 57(3), 391–410, 2005.

78. Takahira, N., Itoman, M., Higashi, K., Uchiyama, K., Miyabe, M., and Naruse, K., Treatment outcome of two-stage revision total hip arthroplasty for infected hip arthroplasty using antibiotic-impregnated cement spacer, *J. Orthop. Sci.*, 8(1), 26–31, 2003.

79. Little, S. R., Lynn, D. M., Ge, Q., Anderson, D. G., Puram, S. V., Chen, J., Eisen, H. N., and Langer, R., Poly-beta amino ester-containing microparticles enhance the activity of nonviral genetic vaccines, *Proc. Natl Acad. Sci. U.S.A.*, 101(26), 9534–9539, 2004.

80. Lendlein, A., Jiang, H., Junger, O., and Langer, R., Light-induced shape-memory polymers, *Nature*, 434(7035), 879–882, 2005.

81. Lendlein, A. and Langer, R., Biodegradable, elastic shape-memory polymers for potential biomedical applications, *Science*, 296(5573), 1673–1676, 2002.

82. Zion, T. C. and Ying, J. Y., In *Glucose responsive nanoparticle for crontrolled insulin delivery.*, Materials Research Society Fall Meeting, Boston, Dec 1–5, 2003.

83. Chan, W. C. and Nie, S., Quantum dot bioconjugates for ultrasensitive nonisotopic detection, *Science*, 281(5385), 2016–2018, 1998.

84. Harisinghani, M. G., Barentsz, J., Hahn, P. F., Deserno, W. M., Tabatabaei, S., van de Kaa, C. H., de la Rosette, J., and Weissleder, R., Noninvasive detection of clinically occult lymph-node metastases in prostate cancer, *N. Engl. J. Med.*, 348(25), 2491–2499, 2003.

85. Bonnemain, B., Superparamagnetic agents in magnetic resonance imaging: physicochemical characteristics and clinical applications. A review., *J. Drug Target*, 6(3), 167–174, 1998.

86. Bala, I., Hariharan, S., and Kumar, M. N., PLGA nanoparticles in drug delivery: the state of the art, *Crit. Rev. Ther. Drug Carrier Syst.*, 21(5), 387–422, 2004.

87. Drotleff, S., Lungwitz, U., Breunig, M., Dennis, A., Blunk, T., Tessmar, J., and Gopferich, A., Biomimetic polymers in pharmaceutical and biomedical sciences, *Eur. J. Pharm. Biopharm.*, 58(2), 385–407, 2004.

88. Moghimi, S. M., Hunter, A. C., and Murray, J. C., Nanomedicine: current status and future prospects, *Faseb J.*, 19(3), 311–330, 2005.

89. Panyam, J. and Labhasetwar, V., Biodegradable nanoparticles for drug and gene delivery to cells and tissue, *Adv. Drug Deliv. Rev.*, 55(3), 329–347, 2003.

90. Ravi Kumar, M., Hellermann, G., Lockey, R. F., and Mohapatra, S. S., Nanoparticle-mediated gene delivery: state of the art, *Expert Opin. Biol. Ther.*, 4(8), 1213–1224, 2004.

91. Maeda, H., Wu, J., Sawa, T., Matsumura, Y., and Hori, K., Tumor vascular permeability and the EPR effect in macromolecular therapeutics: a review, *J. Control Release*, 65(1-2), 271–284, 2000.

92. Liu, D., Mori, A., and Huang, L., Role of liposome size and RES blockade in controlling biodistribution and tumor uptake of GM1-containing liposomes, *Biochim. Biophys. Acta*, 1104(1), 95–101, 1992.

93. Alonso, M. J., Gupta, R. K., Min, C., Siber, G. R., and Langer, R., Biodegradable microspheres as controlled-release tetanus toxoid delivery systems, *Vaccine*, 12(4), 299–306, 1994.

94. Davda, J. and Labhasetwar, V., Characterization of nanoparticle uptake by endothelial cells, *Int. J. Pharm.*, 233(1-2), 51–59, 2002.

95. Deng, J. S., Li, L., Tian, Y., Ginsburg, E., Widman, M., and Myers, A., In vitro characterization of polyorthoester microparticles containing bupivacaine, *Pharm. Dev. Technol.*, 8(1), 31–38, 2003.

96. Molpeceres, J., Chacon, M., Guzman, M., Berges, L., and del Rosario Aberturas, M., A polycapro-lactone nanoparticle formulation of cyclosporin-A improves the prediction of area under the curve using a limited sampling strategy, *Int. J. Pharm.*, 187(1), 101–113, 1999.

97. Sommerfeld, P., Sabel, B. A., and Schroeder, U., Long-term stability of PBCA nanoparticle suspensions, *J. Microencapsul.*, 17(1), 69–79, 2000.

98. Gao, J., Niklason, L., Zhao, X. M., and Langer, R., Surface modification of polyanhydride microspheres, *J. Pharm. Sci.*, 87(2), 246–248, 1998.

99. Huang, G., Gao, J., Hu, Z., St John, J. V., Ponder, B. C., and Moro, D., Controlled drug release from hydrogel nanoparticle networks, *J. Control Release*, 94(2-3), 303–311, 2004.

100. Win, K. Y. and Feng, S. S., Effects of particle size and surface coating on cellular uptake of polymeric nanoparticles for oral delivery of anticancer drugs, *Biomaterials*, 26(15), 2713–2722, 2005.

101. Langer, R., Drug delivery. Drugs on target, *Science*, 293(5527), 58–59, 2001.

102. Langer, R. and Peppas, N. A., Advances in biomaterials, drug delivery, and bionanotechnology, *AICHE J.*, 49(12), 2990–3006, 2003.

103. Matsusue, Y., Hanafusa, S., Yamamuro, T., Shikinami, Y., and Ikada, Y., Tissue reaction of bioabsorbable ultra high strength poly (L-lactide) rod. A long-term study in rabbits, *Clin. Orthop. Relat. Res.*, 317, 246–253, 1995.

104. Chonn, A., Cullis, P. R., and Devine, D. V., The role of surface charge in the activation of the classical and alternative pathways of complement by liposomes, *J. Immunol.*, 146(12), 4234–4241, 1991.

105. Campbell, R. B., Fukumura, D., Brown, E. B., Mazzola, L. M., Izumi, Y., Jain, R. K., Torchilin, V. P., and Munn, L. L., Cationic charge determines the distribution of liposomes between the vascular and extravascular compartments of tumors, *Cancer Res.*, 62(23), 6831–6836, 2002.

106. Jaffar, S., Nam, K. T., Khademhosseini, A., Xing, J., Langer, R., and Belcher, A. M., Layer-by-layer surface modification and patterned electrodeposition of quantum dots, *Nano Letters*, 4(8), 1421–1425, 2004.

107. Wang, Y., Ameer, G. A., Sheppard, B. J., and Langer, R., A tough biodegradable elastomer, *Nat. Biotechnol.*, 20(6), 602–606, 2002.

108. Farokhzad, O. C., Khademhosseini, A., Jon, S., Hermmann, A., Cheng, J., Chin, C., Kiselyuk, A., Teply, B., Eng, G., and Langer, R., Microfluidic system for studying the interaction of nanoparticles and microparticles with cells, *Anal. Chem.*, 77(17), 5453–5459, 2005.

109. Gref, R., Minamitake, Y., Peracchia, M. T., Domb, A., Trubetskoy, V., Torchilin, V., and Langer, R., Poly(ethylene glycol)-coated nanospheres: potential carriers for intravenous drug administration, *Pharm. Biotechnol.*, 10, 167–198, 1997.

110. Lemarchand, C., Gref, R., and Couvreur, P., Polysaccharide-decorated nanoparticles, *Eur. J. Pharm. Biopharm.*, 58(2), 327–341, 2004.

111. Lemarchand, C., Gref, R., Passirani, C., Garcion, E., Petri, B., Muller, R., Costantini, D., and Couvreur, P., Influence of polysaccharide coating on the interactions of nanoparticles with biological systems, *Biomaterials*, 27(1), 108–118, 2006.

112. Weissleder, R., Kelly, K., Sun, E. Y., Shtatland, T., and Josephson, L., Cell-specific targeting of nanoparticles by multivalent attachment of small molecules, *Nat. Biotechnol.*, 23(11), 1418–1423, 2005.

113. Mosqueira, V. C., Legrand, P., Morgat, J. L., Vert, M., Mysiakine, E., Gref, R., Devissaguet, J. P., and Barratt, G., Biodistribution of long-circulating PEG-grafted nanocapsules in mice: effects of PEG chain length and density, *Pharm. Res.*, 18(10), 1411–1419, 2001.

114. Lode, J., Fichtner, I., Kreuter, J., Berndt, A., Diederichs, J. E., and Reszka, R., Influence of surface-modifying surfactants on the pharmacokinetic behavior of 14C-poly (methylmethacrylate) nanoparticles in experimental tumor models, *Pharm. Res.*, 18(11), 1613–1619, 2001.

115. Bazile, D., Prud'homme, C., Bassoullet, M. T., Marlard, M., Spenlehauer, G., and Veillard, M., Stealth Me.PEG-PLA nanoparticles avoid uptake by the mononuclear phagocytes system, *J. Pharm. Sci.*, 84(4), 493–498, 1995.

116. McNamara, J. O., Andrechek, E. R., Wang, Y., Viles, K. D., Rempel, R. E., Gilboa, E., Sullenger, B. A., and Giangrande, P. H., Cell type-specific delivery of siRNAs with aptamer-siRNA chimeras, *Nat. Biotechnol.*, 24(8), 1005–1015, 2006.

117. Farokhzad, O. C., Karp, J. M., and Langer, R., Nanoparticle-aptamer bioconjugates for cancer targeting, *Expert Opin. Drug Deliv.*, 3(3), 311–324, 2006.

118. Chu, T. C., Shieh, F., Lavery, L. A., Levy, M., Richards-Kortum, R., Korgel, B. A., and Ellington, A. D., Labeling tumor cells with fluorescent nanocrystal-aptamer bioconjugates, *Biosens. Bioelectron.*, 21(10), 1859–1866, 2006.

119. Chu, T. C., Marks, J. W., Lavery, L. A., Faulkner, S., Rosenblum, M. G., Ellington, A. D., and Levy, M., Aptamer:toxin conjugates that specifically target prostate tumor cells, *Cancer Res.*, 66(12), 5989–5992, 2006.

120. Cheng, J., Teply, B. A., Sherifi, I., Sung, J., Luther, G., Gu, F. X., Levy-Nissenbaum, E., Radovic-Moreno, A. F., Langer, R., and Farokhzad, O.C., Formulation of functionalized PLGA-PEG nanoparticles for in vivo targeted drug delivery, *Biomaterials*, 2006. In press.

121. Sehgal, D. and Vijay, I. K., A method for the high efficiency of water-soluble carbodiimide-mediated amidation, *Anal. Biochem.*, 218(1), 87–91, 1994.

122. Takae, S., Akiyama, Y., Otsuka, H., Nakamura, T., Nagasaki, Y., and Kataoka, K., Ligand Density Effect on Biorecognition by PEGylated Gold Nanoparticles: Regulated Interaction of RCA(120) Lectin with Lactose Installed to the Distal End of Tethered PEG Strands on Gold Surface, *Biomacromolecules*, 6(2), 818–824, 2005.

123. Polakis, P., Arming antibodies for cancer therapy, *Curr. Opin. Pharmacol.*, 5(4), 382–387, 2005.

124. Waldmann, T. A., Immunotherapy: past, present and future, *Nat. Med.*, 9(3), 269–277, 2003.

125. Wu, A. M. and Senter, P. D., Arming antibodies: prospects and challenges for immunoconjugates, *Nat. Biotechnol.*, 23(9), 1137–1146, 2005.

126. Bagalkot, V., Farokhzad, O. C., Langer, R., and Jon, S., Aptamer-Doxorubicin physical conjugate as a novel trageted drug delivery platform, *Angew Chem. Int. Ed. Engl.*, 2006, In press.

127. Nutiu, R. and Li, Y., Structure-switching signaling aptamers: transducing molecular recognition into fluorescence signaling, *Chemistry*, 10(8), 1868–1876, 2004.

128. Srinivasan, J., Cload, S. T., Hamaguchi, N., Kurz, J., Keene, S., Kurz, M., Boomer, R. M. et al., ADP-specific sensors enable universal assay of protein kinase activity, *Chem. Biol.*, 11(4), 499–508, 2004.

129. Stojanovic, M. N. and Kolpashchikov, D. M., Modular aptameric sensors, *J. Am. Chem. Soc.*, 126(30), 9266–9270, 2004.

130. Yang, X., Li, X., Prow, T. W., Reece, L. M., Bassett, S. E., Luxon, B. A., Herzog, N. K. et al., Immunofluorescence assay and flow-cytometry selection of bead-bound aptamers, *Nucleic Acids Res.*, 31(10), e54, 2003.

131. Baldrich, E., Restrepo, A., and O'Sullivan, C. K., Aptasensor development: elucidation of critical parameters for optimal aptamer performance, *Anal. Chem.*, 76(23), 7053–7063, 2004.

132. Yan, X. R., Gao, X. W., Yao, L. H., and Zhang, Z. Q., [Novel methods to detect cytokines by enzyme-linked oligonucleotide assay], *Sheng Wu Gong Cheng Xue Bao*, 20(5), 679–682, 2004.

133. Schou, C. and Heegaard, N. H., Recent applications of affinity interactions in capillary electrophoresis, *Electrophoresis*, 27(1), 44–59, 2006.

134. Dillman, R. O., Johnson, D. E., Shawler, D. L., and Koziol, J. A., Superiority of an acid-labile daunorubicin-monoclonal antibody immunoconjugate compared to free drug, *Cancer Res.*, 48(21), 6097–6102, 1988.

135. Jaracz, S., Chen, J., Kuznetsova, L. V., and Ojima, I., Recent advances in tumor-targeting anticancer drug conjugates, *Bioorg. Med. Chem.*, 13(17), 5043–5054, 2005.

136. Hermanson, G. T., *Bioconjugate Techniques*, Academic Press, San Diego, 1996.

137. Moghimi, S. M., Hunter, A. C., and Murray, J. C., Long-circulating and target-specific nanoparticles: theory to practice, *Pharmacol. Rev.*, 53(2), 283–318, 2001.

138. Payne, G., Progress in immunoconjugate cancer therapeutics, *Cancer Cell*, 3(3), 207–212, 2003.

139. Saito, G., Swanson, J. A., and Lee, K. D., Drug delivery strategy utilizing conjugation via reversible disulfide linkages: role and site of cellular reducing activities, *Adv. Drug Deliv. Rev.*, 55(2), 199–215, 2003.

140. Trail, P. A., Willner, D., Bianchi, A. B., Henderson, A. J., TrailSmith, M. D., Girit, E., Lasch, S., Hellstrom, I., and Hellstrom, K. E., Enhanced antitumor activity of paclitaxel in combination with the anticarcinoma immunoconjugate BR96-doxorubicin, *Clin. Cancer Res.*, 5(11), 3632–3638, 1999.

141. Mosure, K. W., Henderson, A. J., Klunk, L. J., and Knipe, J. O., Disposition of conjugate-bound and free doxorubicin in tumor-bearing mice following administration of a BR96-doxorubicin immunoconjugate (BMS 182248), *Cancer Chemother. Pharmacol.*, 40(3), 251–258, 1997.

142. Lu, P. Y., Xie, F. Y., and Woodle, M. C., siRNA-mediated antitumorigenesis for drug target validation and therapeutics, *Curr. Opin. Mol. Ther.*, 5(3), 225–234, 2003.

143. Cheng, J. C., Moore, T. B., and Sakamoto, K. M., RNA interference and human disease, *Mol. Genet. Metab.*, 80(1-2), 121–128, 2003.

144. Cheng, J. C. and Sakamoto, K. M., The emerging role of RNA interference in the design of novel therapeutics in oncology, *Cell Cycle*, 3(11), 1398–1401, 2004.

145. Smith, A. M., Gao, X., and Nie, S., Quantum dot nanocrystals for in vivo molecular and cellular imaging, *Photochem. Photobiol.*, 80(3), 377–385, 2004.

146. Gao, X., Cui, Y., Levenson, R. M., Chung, L. W., and Nie, S., In vivo cancer targeting and imaging with semiconductor quantum dots, *Nat. Biotechnol.*, 22(8), 969–976, 2004.

147. Gao, X., Yang, L., Petros, J. A., Marshall, F. F., Simons, J. W., and Nie, S., In vivo molecular and cellular imaging with quantum dots, *Curr. Opin. Biotechnol.*, 16(1), 63–72, 2005.

148. Hayat, M., *Colloidal Gold: Principles, Methods and Applications*, Academic Press, San Diego, 1989.

149. West, J. L. and Halas, N. J., Engineered nanomaterials for biophotonics applications: improving sensing, imaging, and therapeutics, *Annu. Rev. Biomed. Eng.*, 5, 285–292, 2003.

150. Loo, C., Hirsch, L., Lee, M. H., Chang, E., West, J., Halas, N., and Drezek, R., Gold nanoshell bioconjugates for molecular imaging in living cells, *Opt. Lett.*, 30(9), 1012–1014, 2005.

151. Loo, C., Lowery, A., Halas, N., West, J., and Drezek, R., Immunotargeted nanoshells for integrated cancer imaging and therapy, *Nano Lett.*, 5(4), 709–711, 2005.

152. Xu, S., Ji, X., Xu, W., Li, X., Wang, L., Bai, Y., Zhao, B., and Ozaki, Y., Immunoassay using probe-labelling immunogold nanoparticles with silver staining enhancement via surface-enhanced Raman scattering, *Analyst*, 129(1), 63–68, 2004.

153. Driskell, J. D., Kwarta, K. M., Lipert, R. J., Porter, M. D., Neill, J. D., and Ridpath, J. F., Low-level detection of viral pathogens by a surface-enhanced Raman scattering based immunoassay, *Anal. Chem.*, 77(19), 6147–6154, 2005.

154. Copland, J. A., Eghtedari, M., Popov, V. L., Kotov, N., Mamedova, N., Motamedi, M., and Oraevsky, A. A., Bioconjugated gold nanoparticles as a molecular based contrast agent: implications for imaging of deep tumors using optoacoustic tomography, *Mol. Imaging Biol.*, 6(5), 341–349, 2004.

155. Levy, M., Cater, S. F., and Ellington, A. D., Quantum-dot aptamer beacons for the detection of proteins, *Chem. biochem.*, 6(12), 2163–2166, 2005.

156. Meyer, T. and Teruel, M. N., Fluorescence imaging of signaling networks, *Trends Cell Biol.*, 13(2), 101–106, 2003.

157. Mank, M., Reiff, D. F., Heim, N., Friedrich, M. W., Borst, A., and Griesbeck, O., A FRET-Based Calcium Biosensor with Fast Signal Kinetics and High Fluorescence Change, *Biophys. J.*, 90(5), 1790–1796, 2006.

158. Chan, W. C., Maxwell, D. J., Gao, X., Bailey, R. E., Han, M., and Nie, S., Luminescent quantum dots for multiplexed biological detection and imaging, *Curr. Opin. Biotechnol.*, 13(1), 40–46, 2002.

159. Han, M., Gao, X., Su, J. Z., and Nie, S., Quantum-dot-tagged microbeads for multiplexed optical coding of biomolecules, *Nat. Biotechnol.*, 19(7), 631–635, 2001.

160. Huang, C. C., Huang, Y. F., Cao, Z., Tan, W., and Chang, H. T., Aptamer-modified gold nanoparticles for colorimetric determination of platelet-derived growth factors and their receptors, *Anal. Chem.*, 77(17), 5735–5741, 2005.

161. Elghanian, R., Storhoff, J. J., Mucic, R. C., Letsinger, R. L., and Mirkin, C. A., Selective colorimetric detection of polynucleotides based on the distance-dependent optical properties of gold nanoparticles, *Science*, 277(5329), 1078–1081, 1997.

162. Storhoff, J. J., Marla, S. S., Bao, P., Hagenow, S., Mehta, H., Lucas, A., Garimella, V. et al., Gold nanoparticle-based detection of genomic DNA targets on microarrays using a novel optical detection system, *Biosens. Bioelectron.*, 19(8), 875–883, 2004.

163. Storhoff, J. J., Lucas, A. D., Garimella, V., Bao, Y. P., and Muller, U. R., Homogeneous detection of unamplified genomic DNA sequences based on colorimetric scatter of gold nanoparticle probes, *Nat. Biotechnol.*, 22(7), 883–887, 2004.

Section 4

Polymeric Micelles

17 Polymeric Micelles for Formulation of Anti-Cancer Drugs

*Helen Lee, Patrick Lim Soo, Jubo Liu,
Mark Butler, and Christine Allen*

CONTENTS

17.1 Introduction 318
17.2 General Properties of Polymeric Micelles for Drug Formulation 319
    17.2.1 Building Blocks for Polymeric Micelles 319
    17.2.2 Physico-Chemical Properties of Polymeric Micelles 321
        17.2.2.1 Size and Size Distribution 321
        17.2.2.2 Morphology 321
        17.2.2.3 Stability 322
    17.2.3 Preparation of Block Copolymer Micelle Formulations 323
    17.2.4 Properties of Micelle Formulations 324
        17.2.4.1 Drug Partitioning..324
        17.2.4.2 Factors Affecting Drug Loading 324
        17.2.4.3 Factors Influencing Drug Release from Polymeric Micelles ...325
17.3 Physical Characteristics of Polymeric Micelles for Clinical
    Applications in Cancer Therapy 326
    17.3.1 Doxorubicin 327
        17.3.1.1 NK911 327
        17.3.1.2 SP1049C 328
    17.3.2 Paclitaxel 329
        17.3.2.1 Genexol-PM 329
        17.3.2.2 NK105 330
17.4 Interactions of Polymeric Micelles with Cancer Cells......................................330
    17.4.1 Drug Diffusion...331
    17.4.2 Cellular Internalization..332
    17.4.3 In Vitro Cytotoxicity 334
    17.4.4 Influence or Effect on Multi-Drug Resistance.......................................335
17.5 In Vivo Biological Performance of Block Copolymer Micelles 336
    17.5.1 Fate of Polymeric Micelles In Vivo 337
    17.5.2 In Vivo Drug Release Profiles 338
    17.5.3 Toxicity, Pharmacokinetics, and Anti-Cancer Efficacy of Drugs
        Formulated in Block Copolymer Micelles 339
        17.5.3.1 Doxorubicin Formulations (SP1049C and NK911) 339

            17.5.3.2   Paclitaxel Formulations (Genexol-PM and NK105) 339
17.6   Evolution of Polymeric Micelles 340
        17.6.1   Triggered Stimulation of Drug Activation or Release 340
                 17.6.1.1   Ultrasonic Irradiation 340
                 17.6.1.2   Thermosensitive Systems 341
                 17.6.1.3   Photodynamic Therapy 342
        17.6.2   Active Targeting...342
                 17.6.2.1   Ligand Coupling...343
                 17.6.2.2   pH-Responsive Systems 344
17.7   Future Perspectives..345
Acknowledgments 345
References 345

17.1 INTRODUCTION

The use of conventional formulations for the systemic administration of many small molecule anti-cancer agents often results in an unsatisfactory therapeutic effect because of the narrow therapeutic window of these drugs. The low therapeutic index of these drugs is mostly attributed to the dose-limiting toxicities that are associated with them and/or the excipients used in the formulation. Therefore, colloidal carriers that are composed of biocompatible and biodegradable materials with sustained drug release have been developed for formulation and delivery. Nanosized colloidal carriers have been shown to allow for increased accumulation of drug at tumors via the enhanced permeation and retention (EPR) effect.[1] As a result of the altered pharmacokinetics, the exposure of chemotherapeutic drugs to healthy tissues is reduced. Colloidal carriers including liposomes, nanoparticles, and polymeric micelles have been explored extensively for the delivery of a wide range of drugs as anti-cancer therapy.[1,2]

Polymeric micelles are nanosized assemblies of amphiphilic block copolymers that are suitable for the delivery of hydrophobic and amphiphilic agents. In an aqueous medium, micelles consist of a hydrophilic shell that minimizes clearance by the mononuclear phagocytic system (MPS) and a hydrophobic core that functions as a reservoir for hydrophobic drugs. In the past two decades, four polymeric micelle formulations loaded with chemotherapeutic drugs (i.e., NK911, SP1049C, Genexol-PM, and NK105) have entered clinical trial development. The results from the clinical studies have indicated that the polymeric micelle formulations reduce the toxicity associated with conventional formulations of these drugs that, in turn, results in a higher therapeutic index. Many preclinical fundamental studies have evaluated the relationships between the composition of the copolymers and the physico-chemical properties of the micelles. The properties of the micelles such as polymer–drug compatibility, thermodynamic and kinetic stability, and the drug release profiles have been shown to influence the in vivo performance and therapeutic effectiveness of the micelle-formulated drugs. These studies serve as guidelines for the optimization of polymeric micelles for clinical applications.

This chapter is divided into six major sections and provides a discussion of the various aspects relating to use of polymeric micelles for formulation of anti-cancer drugs. Section 17.2 contains information on the physico-chemical properties of micelles (e.g., composition, size and size distribution, morphology, stability), micelle preparation techniques as well as drug loading and release properties of micelles. Section 17.3 highlights the preclinical development of several polymeric micelle formulations that are currently under clinical trial evaluation for use as cancer therapy. Specifically, the optimization of the physico-chemical characteristics of NK911, SP1049C, Genexol-PM, and NK105 are discussed. In Section 17.4, the interaction between polymeric micelles and cancer cells, including cellular internalization, in vitro cytotoxicity, and chemosensitization of multi-drug resistant (MDR) cancer cells is reviewed. In Section 17.5, the physico-chemical properties of polymeric micelles are related to their in vivo performance such as in vivo drug release, pharmacokinetics, toxicity profiles, and anti-cancer efficacy. Finally, in Section 17.6, the development of several advanced polymeric micelle systems that are capable of targeted delivery to tumors via external stimuli (e.g., ultrasound, heat, light) or active targeting mechanisms (e.g., ligand coupling, pH-sensitive) are introduced.

17.2 GENERAL PROPERTIES OF POLYMERIC MICELLES FOR DRUG FORMULATION

In order to fully exploit the benefits of polymeric micelles for drug delivery, the drug-loaded micelle system requires extensive optimization in terms of its physico-chemical characteristics. Factors that may influence the performance of drug-loaded micelle systems include the selection of the building blocks to form the micelles, physico-chemical properties, drug partitioning, drug loading, and drug release profile.

17.2.1 BUILDING BLOCKS FOR POLYMERIC MICELLES

Amphiphilic block copolymers are the typical building blocks used for the formation of micelles. The term *amphiphilic* refers to a molecule that consists of both hydrophobic and hydrophilic segments. In aqueous solution, these block copolymers can self-assemble to form micelles consisting of a hydrophobic core and a hydrophilic corona or shell.[3–6] The hydrophilic corona

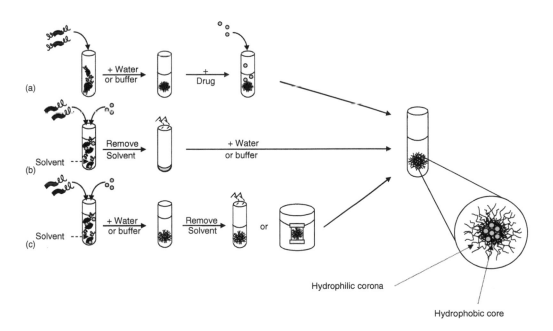

FIGURE 17.1 Techniques for preparation of drug-loaded block copolymer micelles. (a) Copolymers and drugs are directly dissolved in water and form drug-loaded micelles spontaneously using the direct dissolution method. (b) In the evaporation method, copolymers and drugs are dissolved in an organic solvent. The solvent is evaporated, and drug-loaded micelles are formed with the addition of an aqueous solution. (c) In the dialysis method, copolymers and drugs are dissolved in an organic solvent that is miscible with water. Subsequent addition of water induces formation of micelles with swollen cores. Finally, the solvent is removed by evaporation or dialysis to form intact drug-loaded micelles. The structure of a drug-loaded micelle is also shown above.

TABLE 17.1

Examples of Polymers That Are Commonly Used as Building Blocks for Block Copolymer Micelles in Drug Delivery

Hydrophilic segment (corona-forming block)

Poly(ethylene glycol) or poly(ethylene oxide)	PEG or PEO	$\left[CH_2\text{--}CH_2\text{--}O\right]_m$
Poly(*N*-vinyl-pyrrolidone)	PVP	$\left[CH_2\text{--}CH\right]_m$ (pyrrolidone ring)

Hydrophobic segment (core-forming block)

Poly(caprolactone)	PCL	$\left[\overset{O}{\overset{\|}{C}}\text{--}(CH_2)_5\text{--}O\right]_n$
Poly(D,L-lactide)	PDLLA	$\left[\overset{O}{\overset{\|}{C}}\text{--}\underset{CH_3}{CH}\text{--}O\right]_n$
Poly(glycolide)	PGA	$\left[\overset{O}{\overset{\|}{C}}\text{--}CH_2\text{--}O\right]_n$
Poly(D,L-lactide-*co*-glycolide)	PLGA	$\left[\overset{O}{\overset{\|}{C}}\text{--}CH_2\text{--}O\right]_x\left[\overset{O}{\overset{\|}{C}}\text{--}\underset{CH_3}{CH}\text{--}O\right]_y$
Poly(aspartic acid)	PAsp	$-NH\left[\overset{O}{\overset{\|}{C}}\text{--}\underset{CH_2\text{--}COOH}{CH}\text{--}NH\right]_x\left[\overset{O}{\overset{\|}{C}}\text{--}CH_2\text{--}\underset{COOH}{CH}\text{--}NH\right]_y$
Poly(β-benzyl-L-aspartate)	PBLA	$NH\left[\overset{O}{\overset{\|}{C}}\text{--}\underset{CH_2\text{--}COOCH_2\text{--}C_6H_5}{CH}\text{--}NH\right]_n$

Pluronic® copolymer (triblock)

$$\left[O\text{--}CH_2\text{--}CH_2\right]_{m/2}\left[O\text{--}\underset{CH_3}{CH}\text{--}CH_2\right]_n\left[O\text{--}CH_2\text{--}CH_2\right]_{m/2}$$

Poly(ethylene oxide)-*b*-poly(propylene oxide)-*b*-poly(ethylene oxide)

acts as the interface between the core and the exterior environment, and the hydrophobic core serves as the reservoir for the incorporation of lipophilic drugs (Figure 17.1).

In drug delivery, biodegradable and biocompatible copolymers are chosen as a result of their degradation into non-toxic oligomers or monomers that are eventually absorbed in the body and/or eliminated. Many of the amphiphilic block copolymers used in drug delivery contain either a polyester, polyether, or a poly(amino acid) derivative as the hydrophobic block. A number of these hydrophobic polymers that have been used as the core of a micelle are listed in Table 17.1. Polyesters such as poly(D,L-lactide) (PDLLA), poly(glycolide) (PGA), and poly(ε-caprolactone) (PCL) have been chosen as the core block because these polymers are biocompatible, biodegradable, and FDA approved for use in medical devices. Poly(propylene oxide) (PPO) is a polyether that has been used as the hydrophobic block in the well-studied Pluronic® family, the triblock copolymers of poly(ethylene glycol)-*block*-poly(propylene oxide)-*block*-poly(ethylene glycol) (PEG-*b*-PPO-*b*-PEG).[7] Finally, poly(amino acid)s such as poly(aspartic acid) (PAsp) and poly (β-benzyl-L-aspartate) (PBLA) have also been widely explored as materials for the hydrophobic block because they can be easily modified synthetically. The wide range of materials available for selection as the hydrophobic block of the copolymer enables the preparation of micelles' having a variety of microenvironments within their cores.

In contrast, the hydrophilic block is important for stabilizing the micelle in its aqueous environment. The criteria for the selection of the hydrophilic block is that the polymer should be both uncharged and water soluble (e.g., poly(ethylene glycol), polyacrylamide, polyhydroxyethylmethacrylate, poly(*N*-vinyl-pyrrolidone) (PVP), and poly(vinyl alcohol)).[8] Poly(ethylene glycol) (PEG) is most commonly used as the hydrophilic segment in micelles. The molecular weights of PEG used in block copolymers typically range between 1000 and 12,000 g/mol with a chain length that is typically equal to or greater than the length of the core-forming block.[4] PEG can also be referred to as poly(ethylene oxide) (PEO), a polymer with a molecular weight that is greater than 20,000 g/ mol, whereas PEG refers to polymers with molecular weights below this value.[9] Because of its high water solubility and large excluded volume, PEG can stabilize the micelles by sterically excluding other polymers from the surface of the micelles.[1,2,10] In addition, PEG can inhibit the surface adsorption of biological components such as proteins, improve the residence time of the micelles in the circulation, and limit the micelle clearance by the MPS.[10–13]

The selection of building blocks for the micelle system includes the type of polymers and the length of the core- and corona-forming polymer blocks; these will influence the physico-chemical properties of the micelles and also affect the overall therapeutic efficacy of the delivery system.

17.2.2 Physico-Chemical Properties of Polymeric Micelles

In order to design an effective micellar drug delivery system, several key physico-chemical properties of the micelles should be considered as a means to optimize performance. These include micelle size and size distribution, morphology, and stability.

17.2.2.1 Size and Size Distribution

Block copolymer micelles typically range in size from 10 to 100 nm. As a result of their small size and the steric stabilization provided by the hydrophilic corona, micelles are less susceptible to MPS clearance.[14,15] In addition, their small size allows them to extravasate through the leaky endothelium and accumulate at the tumor interstitium. Therefore, micelles can passively target tumors via the enhanced permeability and retention (EPR) effect.[1,16] Micelles can also pass through the pores of the renal filtration system as individual polymer chains following disassembly. A renal threshold limit of less than 50,000 g/mol has been reported for polymeric carriers,[17] so it is important to design a micelle system where the individual polymer chains have a molecular weight that is lower than this threshold limit. The small size of the micelles also allows for sterilization by filtration in order to remove bacteria prior to intravenous administration.

Filtration can also be used to reduce the size distribution of colloidal carriers that have broad distributions following initial preparation. Micelles tend to have a relatively narrow size distribution; however, they are prone to secondary aggregation because of insufficient coverage of the hydrophobic core by the hydrophilic chains. Dilution has been shown to be useful in breaking apart the secondary aggregates of primary micelles.[18]

17.2.2.2 Morphology

When the corona-forming block of a copolymer is smaller than the core-forming block, crew-cut micelles are produced.[3] These copolymers may be used to prepare a wide range of morphologies including spheres, rods, lamellae, vesicles, large compound micelles, etc.[19,20] Of particular interest

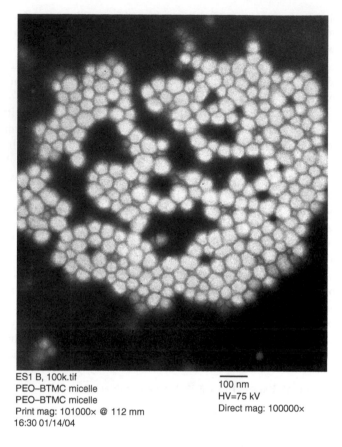

ES1 B, 100k.tif
PEO–BTMC micelle
PEO–BTMC micelle
Print mag: 101000× @ 112 mm
16:30 01/14/04

100 nm
HV=75 kV
Direct mag: 100000×

FIGURE 17.2 Transmission electron micrograph of block copolymer micelles formed from methoxy poly (ethylene glycol)-block-poly(5-benxyloxy-trimethylene carbonate) copolymers. (Reproduced with permission from *Biomacromolecules* 2004, 5, 1810–1817. Copyright 2004, American Chemical Society).

are block copolymer vesicles that may prove to be effective for the simultaneous delivery of both hydrophilic and hydrophobic agents.[21,22] However, most of the block copolymer micelles used in drug delivery consist of spherical, star-type micelles as they are formed from copolymers that have a hydrophilic block that is of equal or greater length than the hydrophobic core-forming block.[3] A sample transmission electron micrograph of a population of spherical micelles is shown in Figure 17.2.

17.2.2.3 Stability

The physical stability of block copolymer micelles is important because the early release of the drug into the bloodstream as a result of the disassembly of copolymer chains results in insufficient accumulation of the drug at the target site. It is critical that the copolymer micelles possess sufficient stability to remain intact in the circulation. The thermodynamic tendency for micelles to disassemble into individual chains is reflected by the critical micelle concentration (CMC) of the copolymer. Below the CMC, the copolymer is in the form of single polymer chains or unimers, and above the CMC, the copolymer exists as micelles in equilibrium with a small population of single chains. In this way, the CMC represents an important parameter as it determines the thermodynamic stability of the micelles during dilution.[23] Small molecule surfactants form micelles that tend to have higher CMC values that make them more susceptible to disassembly following dilution. In contrast, amphiphilic block copolymers tend to be more thermodynamically stable with lower CMC values in the range of 10^{-7}–10^{-6} M.[3,24,25]

The CMC has been found to be dependent on the nature and length of the hydrophobic block, the length of the hydrophilic block, and the total molecular weight of the copolymer.[26,27] Specifically, increasing the hydrophobicity of the core-forming block reduces the CMC and, in turn, improves the thermodynamic stability of the micelles. For example, Leroux et al. have demonstrated that increasing the hydrophobic PDLLA block from 27 to 55 mol% in a series of PVP-b-PDLLA-b-PVP copolymers results in a decrease in the CMC from 19.9 to 5.1 mg/L.[28] In contrast, when the hydrophilic block length is increased and the hydrophobic block is kept constant, the CMC increases as has been shown for the Pluronic® copolymers.[29]

Even at concentrations below the CMC, micelles can remain kinetically stable for extended periods of time depending on their glass transition temperature (T_g) and/or T_m if the core-forming block is semicrystalline. Micelles with a core-forming block that has a lower T_g (below 37°C) will tend to disassemble at a faster rate than those with a higher T_g as a result of the free motion of the core-forming polymer chains at temperatures above the T_g. For example, Kataoka et al. found a dramatic increase in the CMC of PEG-b-PDLLA micelles at a temperature above the T_g of the hydrophobic PDLLA polymer that suggested a decrease in the kinetic stability of the system.[30] Kang et al. improved the kinetic stability of PEG-b-poly(lactide) (PLA) block copolymer micelles by blending equimolar mixtures of PEG-b-poly(D-lactide) (PDLA) and PEG-b-poly(L-lactide) (PLLA) enantiomeric copolymers. They reported superior kinetic stability as a result of the increased van der Waals interactions between the PLA polymer chains that promoted a denser packing and a tighter conformation.[31] Finally, the presence of the hydrophobic drug in the core has also been shown to promote kinetic stability of the formulation.[32]

17.2.3 PREPARATION OF BLOCK COPOLYMER MICELLE FORMULATIONS

The selection of the appropriate preparative technique for micelle formation is largely based on the solubility of the copolymer and the drug in aqueous media. There are several methods that have been developed for the preparation of drug-loaded block copolymer micelles, including the direct dissolution method, evaporation method, and dialysis method as depicted in Figure 17.1.

The direct dissolution method is employed for block copolymers that are relatively soluble in water (e.g., Pluronic®). In this method, the copolymer and the drug are directly dissolved in water or an aqueous solution such as phosphate-buffered saline (PBS) with subsequent heating and/or stirring to induce micellization. Successful loading of doxorubicin, etoposide, nystatin, and haloperidol into Pluronic® micelles has been achieved using this method.[33,34]

For copolymers and drugs with limited aqueous solubility, the evaporation method can be used to prepare the micelle formulations. This method involves first dissolving the copolymer and the drug in a common solvent or a mixture of two miscible solvents. The mixture is then stirred, and the solvent is allowed to evaporate, yielding a copolymer–drug film that can be reconstituted in warm water or a buffer to form the drug-loaded micelles.[35–38]

The most commonly used method for the preparation of drug-loaded micelles from poorly water-soluble copolymers is the dialysis method. The copolymer and the drug are first dissolved in a common organic solvent that is miscible with water. Slow addition of water results in the formation of micelles with swollen cores. The solution is then dialyzed against a large volume of water to remove the organic solvent.[3,28,32,39] As the percentage of solvent decreases, the degree of swelling in the micelle core decreases to form intact drug-loaded micelles. Alternatively, the solvent can be removed by evaporation of the solvent from the polymer–drug mixture to form micelles (i.e., emulsification method).[38,40,41] The selection of the organic solvent and the ratio of water-to-solvent employed have been shown to greatly affect the physical properties and drug loading of the micelles.[39]

17.2.4 Properties of Micelle Formulations

The drug loading and drug release profiles of block copolymer micelle formulations will influence the in vivo therapeutic efficacy of the system. There is a balance between the amount of drug that can be loaded into the hydrophobic core and the amount of drug that can be released from the same core.[36] The drug loading and drug release from the micelles is highly dependent on the partitioning behavior of the drug into the micelle.

17.2.4.1 Drug Partitioning

Drug loading into block copolymer micelles is achieved by the partitioning of the lipophilic agents into the micelle core. The partition coefficient (K_v) refers to the extent to which a drug selectively partitions into a micelle and can be expressed as follows:

$$K_v = \frac{[\text{Drug}]_{\text{micelle}}}{[\text{Drug}]_{\text{aqueous}}} \qquad (17.1)$$

where $[\text{Drug}]_{\text{micelle}}$ and $[\text{Drug}]_{\text{aqueous}}$ represent the molar concentrations of the drug in the micelle phase and aqueous phase, respectively.[23,42] Many different experimental methods including surface tension,[23] UV–vis spectroscopy,[43,44] fluorescence,[33,42,45,46] solubility in model solvents,[47,48] and theoretical methods[49,50] have been used to study the partitioning behavior of molecules in block copolymer micelle solutions. The length of the hydrophobic block influences both the CMC and the partition coefficient. An increase in either the length or the hydrophobicity of the core-block has been shown to improve the partitioning behavior of hydrophobic agents into the micelle core.[51,52]

Hydrophobic molecules such as pyrene preferentially reside in the hydrophobic micelle core.[50] However, the micelle core is not the only site for drug solubilization. Various amphiphilic solubilizates can reside at the core–corona interface or in the corona itself.[45,47,53]

17.2.4.2 Factors Affecting Drug Loading

The partitioning behavior of drugs into the micelle core, core–corona interface and/or corona controls the amount of drug that can be solubilized in the micelles. However, for a block copolymer micelle system to act as a true carrier, it must retain the drug within the micelles for prolonged periods of time following administration. Also, it is highly unlikely that a single block copolymer micelle system can effectively deliver all types of drugs. Therefore, selecting a core block for the drug of interest is an important consideration in the design of an effective micelle delivery system. Micelle drug loading is influenced by several factors, including the hydrophobicity and length of the core-forming block, the length of the corona-forming block, the specific interactions between the drug and core, and the method of micelle preparation.

The hydrophobicity and length of the core-forming block have a great influence on the size of the micelles and the loading into the core of the micelle. For example, Lin et al. reported that increasing the degree of hydrophobicity of the core-forming block improved the drug loading into the micelles.[54] Also a larger amount of drug can be incorporated into micelles formed from copolymers having a larger core-forming block in comparison to a shorter hydrophobic block. For example, Shuai et al. demonstrated that for doxorubicin-loaded micelles formed from PEG-b-PCL copolymers with a constant PEG block length, increasing the PCL block length resulted in an increase in doxorubicin loading from 3.1% to 4.3% (wt./wt.). The hydrophilic block length can also influence drug loading. By increasing the corona-forming block, the CMC will also increase, and this may result in a decrease in the aggregation number (N_{agg}). For example, Gadelle et al.

showed that increasing the PEG block length caused an increase in the CMC, a decrease in the N_{agg}, and a decrease in the degree of solubilization of hydrophobic agent into the micelles.[55]

Specific interactions between the core-forming block and entrapped drug can also affect drug loading in the micelles. This is referred to as polymer–drug compatibility where the compatibility is defined as interaction between the two species that does not involve any chemical change to either the drug or the polymer. Typically, the interactions that are present between polymer and drug are hydrogen bonding, ionic, and pi–pi. Lee et al. showed that by increasing the number of functionalized carboxylic acid (COOH) groups from 0% to 19.5% (weight percent of repeat units), they were able to augment papaverine loading from 3.5% to 15% (wt./wt.) in PEG-b-PDLLA micelles. They attributed the increase in papaverine loading levels to hydrogen bonding interactions between the COOH groups of the copolymer and the unpaired electron groups of the oxygen and/or nitrogen atoms in the drug.[56] Other approaches include improving drug loading by enhancing the local microenvironment of the micelle core through the means of conjugation of a small amount of drug[16] or conjugation of side chains with similar structures to the drug.[57,58]

Finally, the method that is selected for preparation of the micelles has also been shown to have an effect on the overall degree of drug loading. Yokoyama et al. have shown that using different micelle preparation techniques led to distinctly different drug loading efficiencies for campthothecin (CPT) in a series of PEG-b-poly(aspartate ester) copolymers.[38,59] The use of the dialysis method resulted in a CPT loading efficiency of 19% and a cloudy formulation with large aggregates and drug precipitates. In contrast, when CPT was loaded using the evaporation method, a loading efficiency of 73% was obtained and also a clear solution was observed.[38] In addition to the selection of the preparation method, method-specific parameters (nature of organic solvent, solvent ratio) have also been proven to affect drug loading into the micelles. Yokoyama et al. showed that the use of dimethylformamide to prepare KRN 5500 loaded PEG-b-poly(16-β-cetyl-L-aspartate-co-β-benzyl-L-aspartate) micelles resulted in a higher drug loading level when compared to using dimethylsulfoxide.[60] The ability to achieve a high drug loading level is important as this allows for maximization of the drug to polymer ratio that, in turn, means that less copolymer will need to be administered per dose of formulation.

17.2.4.3 Factors Influencing Drug Release from Polymeric Micelles

A drug can be localized in potentially three different areas in a block copolymer micelle: in the core, at the core–corona interface, and at the corona.[53] The drug in the micelle core has a longer diffusion path relative to the drug that is located at the core–corona interface or the corona. The drug that is released from the core–corona interface or the corona is typically referred to as a burst release, an initial discharge of drug in a relatively short period of time (typically hours). For example, Bromberg et al. observed a burst release (5–11%) of β-estradiol from the poly(acrylic acid) (PAA) corona of Pluronic®-PAA micelles.[61] In most cases, after a burst release, a slower and more consistent release that typically lasts for a longer period of time (typically days to months) is observed. For example, Kim et al. demonstrated slow release of indomethacin (i.e., less than 30% released over a period of approximately 100 h) from Pluronic®/PCL micelles.[62]

The kinetics of drug release from micellar systems has been studied extensively by many groups.[36,45,61,63–65] Despite the extensive release studies, the mechanism of drug release from block copolymer micelles is not fully understood. However, many groups believe that diffusion and copolymer degradation are the two principal mechanisms of release provided that the micelles are stable under the release conditions investigated.[63] If the interaction between the polymer and the drug is strong and the rate of biodegradation is fast, then the degradation rate would govern the rate of drug release. However, the rate of drug release from micelles, as observed in vitro, usually exceeds the rate of copolymer degradation. As a result, polymer degradation can most often be ruled out as one of the main mechanisms of drug release. Therefore, diffusion of the drug may be considered the principal mechanism of release from micelles.

Drug release is influenced by three main factors: the characteristics of the drug (amount of drug, molecular weight of the drug, and molecular volume of the drug), the properties of the core-forming block (molecular weight of the core, nature of the core, and physical state of the core), and the degree of polymer–drug compatibility.

The amount of drug present in the micelle has been found to influence the release. Most often it has been found that an increase in the concentration of drug present results in a decrease in the rate of drug release.[36,65–67] For example, Gref et al. reported that higher concentrations of lidocaine in PEG-*b*-poly(D,L-lactide-*co*-glycolide) (PEG-*b*-PLGA) micelles resulted in a slower release rate for the drug.[66] Similarly, Lim Soo et al. observed that increasing the amount of estradiol in PEG-*b*-PCL micelles decreased the rate of drug release.[65]

The physical properties of the drug, including molecular weight and molecular volume have also been shown to influence their rate of release from micelles. An increase in the molecular weight of the drug forces a greater reorientation of the polymer chains for movement of the drug molecules through the polymer matrix. An increase in the volume of the drug will also require a greater effort to reorganize the polymer to accommodate its movement. As a result, larger or heavier molecular weight drugs will tend to have lower diffusion coefficients as opposed to smaller molecular weight agents.

The rate of diffusion of a drug in and out of the micelle is also greatly influenced by the properties of the micelle core. In general, increasing the molecular weight of the core-forming block will increase the size of the core provided that the hydrophilic block length is held constant. This increase in the molecular weight of the hydrophobic block and the size of the core has been found to decrease the overall rate of release rate of the entrapped agent.[63,68–71]

The nature of the core, specifically the hydrophobicity and/or polarity of the core, influences the release rate of the drug as it determines the permeability of the core to aqueous media. Micelles with a highly hydrophobic core will likely have less aqueous fluid in the core in comparison to micelles that have a more hydrophilic or polar core. Therefore, micelles prepared from more hydrophilic cores will likely have an accelerated rate of drug release. The physical state of the core will also affect the diffusion of drug from the micelles. Polymers that have a high T_g or bulky groups present on their backbone are limited in terms of their ability to reorient. As a result, these polymers will form micelle cores with a high microscopic viscosity, resulting in a low diffusion coefficient for the drug that is incorporated because of limited movement.[72] This is in contrast to polymers that are more rubbery or gel-like that will tend to form cores that have a low microscopic viscosity and a higher rate of diffusion for the incorporated drugs.

Finally, the degree of polymer–drug compatibility greatly influences the rate of drug release from the micelles. Drugs that are highly miscible with the core-forming block have a high degree of polymer–drug compatibility. It is expected that these drugs would be molecularly dissolved in the core and not exhibit any crystallinity such that the release would be relatively fast. However, it is necessary to consider that the actual effect of a high degree of miscibility on drug release depends on the strength and extent of the core–drug interactions.[36,45] An increase in the extent of interaction between a drug and the polymer will result in a decrease in the diffusion coefficient of the drug (slower release).[73] The ability to provide sustained release of the drug from micelle systems is particularly beneficial for their use in therapeutic applications.

17.3 PHYSICAL CHARACTERISTICS OF POLYMERIC MICELLES FOR CLINICAL APPLICATIONS IN CANCER THERAPY

In the past two decades, many groups have reported on the development and optimization of polymeric micelles for the delivery of anti-cancer drugs. In particular, much effort has focused on the formulation of doxorubicin and paclitaxel in micelles because of their dose-limiting toxicities in currently approved formulations and high potency against a wide variety of cancers. To

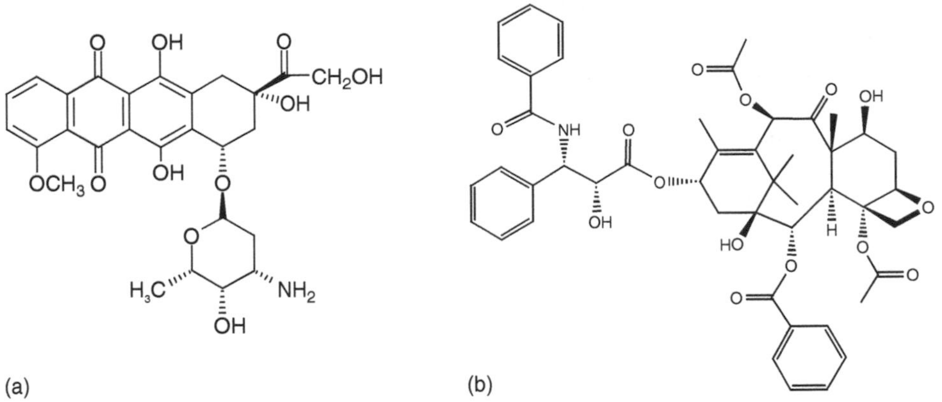

FIGURE 17.3 Structures of anti-cancer drugs, (a) doxorubicin and (b) paclitaxel, that have been formulated in polymeric micelle formulations for clinical trial development.

date, several doxorubicin- (i.e., NK911, SP1049C) and paclitaxel-loaded (i.e., Genexol-PM, NK105) micelle formulations have entered clinical trial development.

17.3.1 DOXORUBICIN

Doxorubicin (Adriamycin®), a member of the anthracycline antibiotics family, is currently employed for treatment of many types of cancer because of its broad anti-tumor spectrum. It is approved, either alone or in combination with other chemotherapeutic drugs, for treatment of breast, ovarian, bladder, lung, thyroid, and gastric cancer as well as malignant lymphoma, acute forms of leukemia, and sarcomas.[74] The cytotoxic effect of doxorubicin is said to result from a combination of cell damage mechanisms, some that have not been conclusively identified. One of the effects of doxorubicin that is known to contribute to its anti-tumor activity is the inhibition of cellular replication and DNA transcription as a result of an irreversible change in the DNA tertiary structure.[75] This non-cell type specific cytotoxic effect of doxorubicin is mainly attributed to its ability to intercalate between DNA base pairs with its planar structure (Figure 17.3a). Indeed, the planar moiety of doxorubicin has been shown to improve the stability of micelles when loaded in the micelle core.[76] Doxorubicin has been successfully loaded into micelles formed from PEG-*b*-poly(ε-caprolactone) (PEG-*b*-PCL),[41] PEG-*b*-poly(D,L-lactide-*co*-glycolide) (PEG-*b*-PLGA),[77,78] PEG-*b*-poly(propylene oxide)-*b*-PEG (PEG-*b*-PPO-*b*-PEG; Pluronic®),[26,79–81] and PEG-*b*-poly (aspartic acid) (PEG-*b*-PAsp).[16,32,82,83] Currently, two doxorubicin-loaded micelle formulations, NK911 (i.e., PEG-*b*-PAsp) and SP1049C (i.e., Pluronic®) are being investigated in clinical trials.

17.3.1.1 NK911

NK911 is a doxorubicin-loaded PEG-*b*-PAsp copolymer micelle formulation developed by Kataoka and co-workers. Doxorubicin is both physically entrapped in the micelle core and chemically conjugated to the aspartic acid side chains of the core-forming block via amide linkages. NK911 is currently undergoing phase II clinical trial evaluation for efficacy and toxicity profiles. The formulation entered phase I in 2001, and the results from this trial revealed that NK911 exhibits a longer circulation half-life, a larger area under the curve (AUC), and reduced toxicities in comparison to the conventional formulation of doxorubicin.[82] The prolonged circulation lifetime of NK911 is attributed to its composition and the monomodal size distribution of the micelles with a mean diameter of approximately 50 nm.[16,82]

The amphiphilic or polar nature of doxorubicin (log P value of 0.52) facilitates the partitioning of this drug from the micelle core into the external medium.[84] Therefore, chemical conjugation of doxorubicin to the core-forming block can enhance drug retention and retard drug release from the micelles. The initial development of the PEG-b-Asp micelles for drug delivery consisted of only conjugated doxorubicin to the PAsp core. Specifically, PEG-b-PAsp copolymers were synthesized from PEG-b-PBLA with the subsequent addition of doxorubicin to the carboxyl groups on the side chains of the core-forming block. As a result of the conjugation, a 37 mol% substitution of doxorubicin was achieved that is equivalent to 20 mg/mL doxorubicin.[85] The doxorubicin residues of the conjugates stabilize the micelle core through pi–pi stacking and hydrophobic interactions, providing the micelles with excellent kinetic stability in vivo.[76]

Subsequent studies revealed that the chemically conjugated doxorubicin exerts negligible anti-tumor activity in vitro and in vivo as a result of the lack of a cleavable linkage between the drug and the copolymer. Therefore, the next generation of the PEG-b-PAsp micelles was prepared to include physically encapsulated doxorubicin within the micelle core as well as the chemically conjugated drug. The amount of doxorubicin that could be physically entrapped within the core was found to be proportional to the amount of conjugated doxorubicin present. The enhanced compatibility between the physically entrapped drug and the micelle core is attributed to the presence of the chemically conjugated drug. In vivo studies confirmed that micelles with the highest amount of physically entrapped doxorubicin had the most potent anti-cancer activity.[32]

The formation of a dimerized form of the physically entrapped doxorubicin was detected within the micelles. Despite the fact that the dimer was found to have insignificant anti-tumor activity in vitro and in vivo, it plays a supplementary role to stabilize the entrapped doxorubicin. The addition of dimers to the micelles as part of the physically entrapped components can further prolong the circulation lifetime of the micelles in vivo.[86] However, the addition of doxorubicin dimers was also found to decrease the aqueous solubility of the drug-loaded micelles that compromised the stability of the micelles following lyophilization and long-term storage. Therefore, the fully optimized NK911 formulation was prepared such that it only contained doxorubicin monomer.[82]

17.3.1.2 SP1049C

SP1049C is a doxorubicin-loaded Pluronic® copolymer micelle formulation developed by Kabanov's group and Supratek Inc. (Montreal, Canada). In this formulation, doxorubicin is physically encapsulated in micelles assembled from two different types of Pluronic® copolymers, L61 and F127 in a 1:8 w/w ratio.[26] In October 2005, SP1049C was granted orphan drug status by the FDA for the treatment of oesophageal carcinoma based on results from the phase I study; meanwhile, its phase II clinical trial results are currently under final review. In contrast to NK911, the phase I clinical trial of SP1049C deliberately recruited patients with cancers that were refractory following conventional treatment and patients with previous anthracycline treatment.[81] This is due to the established chemosensitizing effect of SP1049C against multi-drug resistant (MDR) cancers, an effect that is attributed to its Pluronic® copolymer composition.

The Pluronic® copolymers are the class of copolymers that have been most widely evaluated and employed in pharmaceutical applications. Studies have indicated that Pluronic® copolymers having an hydrophile-lipophile balance (HLB) of less than 19 and a moderate molecular weight are most effective in overcoming MDR.[34] In a separate study, Pluronic® L61 that has an HLB value of 7 was found to be the most effective chemosensitizer among other Pluronic® copolymers.[26,34] Doxorubicin loaded in Pluronic® L61 micelles was found to be 2–3 orders of magnitude more toxic than conventional doxorubicin when evaluated in a range of drug-resistant cancer cells (e.g., MCF-7/ADR breast carcinoma, SKVLB ovarian carcinoma, KBV oral epithelial carcinoma, LoVo/Dox colon adenocarcinoma, P388-Dox murine leukemia cells, SP2/0Dox myeloma as well as others).[26] Furthermore, mechanistic studies have shown that it is the unimers rather than the micelles that are responsible for the reversal of MDR.[26,34] More details on the mechanisms and effects of Pluronic® as chemosensitizers are given in Section 17.4.

Despite the effectiveness of micelles formed from only Pluronic® L61, the hydrophobicity of the copolymer results in unfavorable stability issues. Phase separation and micelle aggregation leading to microembolic effects have been observed in preliminary in vivo toxicity studies of the L61 formulation. Kabanov and co-workers solved this problem by introducing a higher molecular weight and more hydrophilic Pluronic® copolymer (F127) into the L61 micelle formulation. By varying the ratio of the L61 to the F127 copolymer, it was found that at 0.25 w/w% L61 (i.e., effective dosing concentration of L61) and 2.0 w/w% F127, the Pluronic® micelles maintained their effective diameter of 22.4 nm with no aggregation.[26] In addition, the IC_{50} of doxorubicin loaded in the L61/F127 micelles was similar to that of the L61 formulation. Therefore, the optimized SP1049C formulation utilized for the clinical studies contains 2.0 mg/mL doxorubicin, 2.5 mg/mL L61, and 20 mg/mL F127 in 0.9% sodium chloride.[81]

17.3.2 PACLITAXEL

Similar to doxorubicin, paclitaxel is a chemotherapeutic drug that has been loaded into block copolymer micelles for investigation in clinical trials. Paclitaxel is an anti-neoplastic drug that was first extracted from the bark of Pacific yew trees (Figure 17.3b). The cytotoxic effect of paclitaxel is mainly attributed to its ability to promote the assembly as well as prevent the depolymerization of microtubules.[87] Stabilization of the microtubule networks inhibits normal interphase and mitotic functions that, in turn, impedes cell division. Paclitaxel is approved for treatment of ovarian, breast, and non-small cell lung cancers as well as AIDS-related Kaposi's sarcoma.[87] Paclitaxel is highly hydrophobic with a log P value of 3.96 and low aqueous solubility of 1 μg/mL.[88,89] The commercialized formulation of paclitaxel, known as Taxol®, contains the drug dissolved in a mixture of Cremophor EL and ethanol that increases its solubility by more than 6000-fold. However, administration of Cremophor EL leads to severe dose-limiting toxicities such as hypersensitivity reactions and neuropathy.[90] As a result, much effort has gone into the development of a novel formulation of paclitaxel using biocompatible and biodegradable polymeric micelles. Paclitaxel has been successfully loaded into PEG-b-poly(D,L-lactide) (PEG-b-PDLLA),[35,37,90,91] poly(N-vinyl-pyrrolidone)-b-PDLLA (PVP-b-PDLLA),[28,92] PEG-b-PCL,[93–95] and PEG-b-poly(δ-valerolactone) (PEG-b-PVL).[95] Genexol-PM (i.e., PEG-b-PDLLA) and NK105 (i.e., PEG-b-PAsp) represent two paclitaxel-loaded polymeric micelle formulations that have entered clinical trial development.

17.3.2.1 Genexol-PM

Genexol-PM, developed by Samyang Co. (Korea), is a paclitaxel formulation based on PEG-b-PDLLA micelles. In 2004, results from a phase II clinical trial that evaluated Genexol-PM and cisplatin as treatment for patients with advanced gastric cancer was reported.[96] In the first clinical trial, the AUC and C_{max} of Genexol-PM was found to be lower than that of Taxol®; however, the maximum tolerated dose (MTD) was increased from 135 to 390 mg/m².[90] Preclinical studies also revealed similar pharmacokinetic profiles for Genexol-PM and Taxol® in the B16 melanoma murine model. There are two explanations for the above observations. First, the liquid-like core of PDLLA acts as a solubilization site for paclitaxel as opposed to an actual drug carrier,[35] and second, the decrease in AUC and C_{max} occurs as a result of the shift in the drug accumulation to the tumor tissues that was evidenced in preclinical biodistribution studies.[97] Formulation development of paclitaxel-loaded PEG-b-PDLLA micelles was mostly reported by Burt and co-workers.[35,37,91]

Genexol-PM is prepared by the evaporation method that involves co-dissolving paclitaxel and PEG-b-PDLLA in acetonitrile with subsequent evaporation of the solvent to form a copolymer–drug matrix. The matrix is then heated to 60°C with the addition of water or buffer for self-assembly of paclitaxel-loaded micelles.[97] The evaporation method is much more suitable than the direct dissolution method because paclitaxel and PEG-*b*-PDLLA are relatively insoluble in aqueous media. Indeed, the method employed for preparation of the paclitaxel-loaded PEG-*b*-PDLLA micelles can influence the extent of drug loading and physical stability of the formulation. The choice of solvent strongly influences the extent to which paclitaxel can be solubilized in PEG-*b*-PDLLA micelles as reported by Zhang et al.[37] Among many of the solvents tested, only acetonitrile resulted in a clear micelle solution with no evidence of drug precipitation. It was suggested that acetonitrile enhances the miscibility between paclitaxel and the PDLLA block that, in turn, reduces the possibility of phase separation during the solvent evaporation process.[37]

Burt and co-workers also studied the use of PEG-*b*-poly(D,L-lactide-*co*-caprolactone) and PEG-*b*-poly(glycolide-*co*-caprolactone) copolymers for encapsulation of paclitaxel. The addition of hydrophobic caprolactone to PDLLA could potentially increase paclitaxel loading through the increased hydrophobicity of the core-forming block. However, these micelles were found to have poor physical stability and resulted in the precipitation of paclitaxel.[35] The PEG-*b*-PDLLA micelles can solubilize as much as 5% (w/v) paclitaxel and remain stable for up to 24 h. This is mainly attributed to the absence of crystalline paclitaxel in the paclitaxel/PEG-*b*-PDLLA matrix as revealed by x-ray diffraction studies. In fact, paclitaxel is molecularly dissolved in the PDLLA core that prevents drug precipitation and results in excellent solubilization of the drug.[37,91]

The amorphous state of the paclitaxel–PDLLA core can be problematic for in vivo applications. Studies have indicated that the presence of crystalline drug in the core can retard drug release.[69,98] Therefore, the amorphous core of the paclitaxel-loaded PEG-*b*-PDLLA micelles can lead to a much faster drug release profile. As demonstrated by Burt et al. following intravenous administration paclitaxel rapidly dissociates from the micellar components in the circulation.[35] Therefore, it is likely that Genexol-PM acts as a solubilizer for paclitaxel rather than as a true drug carrier.

17.3.2.2 NK105

In 2005, Kataoka and co-workers, in collaboration with NanoCarrier Co., Ltd. and Nippon Kayaku Co., Ltd., first reported on NK105, a paclitaxel-incorporated micelle formulation.[99] The phase I clinical trial of NK105 first began in April 2004. Similar to NK911, the building block of NK105 is PEG-*b*-PAsp copolymers but with a slight modification to the aspartate blocks such that the carboxyl residues are converted into 4-phenyl-1-butanolate by an esterification reaction. The modification increases the hydrophobicity of the core-forming block that results in enhanced entrapment and retention of paclitaxel in the micelles. The resulting drug-loaded micelles contain 23% (w/w) paclitaxel with a weight–average diameter of approximately 85 nm.

With the improved micelle core stability, NK105 is predicted to act as a true drug carrier rather than a solubilizer. As indicated by the in vivo preclinical studies, NK105 is extremely stable in vivo with a 90-fold increase in the AUC and a 4–6-fold increase in the terminal phase half-life.[99] In addition, the tumor AUC was found to be 25-times higher than that of Taxol®. The superior in vivo anti-tumor activity of NK105 in a human colorectal cancer xenograft model (HT-29) was a result of the enhanced drug accumulation in the tumor, allowing for more drug-loaded micelle–cell interactions. These results highlight the importance of polymer–drug compatibility and core stability in the design of effective micellar delivery systems. Therefore, the recent studies that focus on customizing copolymer compositions to accommodate the distinct properties of different drugs such as camptothecin,[38,100] KRN 5500,[60] and cisplatin[101,102] may lead to the development of true polymeric micelle carriers for delivery to tumor tissues.

17.4 INTERACTIONS OF POLYMERIC MICELLES WITH CANCER CELLS

Once the drug-loaded micelles have reached the tumor site, there are several mechanisms that have been proposed for the delivery of micelle-incorporated drugs into the cancer cells as

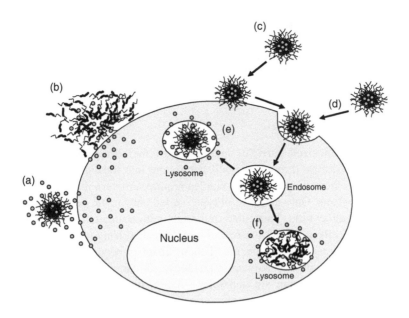

FIGURE 17.4 Proposed mechanisms for the delivery of micelle-formulated drugs into cancer cells. (a) Drug is released from the micelles and diffused into the cell as free drug. (b) The drug-loaded micelles diassemble in the extracellular space and release the drug; the released drug then diffuses across the cell membrane into the intracellular space. (c) Intact drug-loaded micelles are internalized via adsorptive pinocytosis after adsorption onto the cell surface or (d) by fluid-phase endocytosis. The internalized drug-loaded micelles eventually reach the lysosome where the drug can be released (e) by diffusion or (f) following disassembly or degradation of the micelles.

illustrated in Figure 17.4.[103–106] In general, drugs can enter the cells via diffusion as free molecules following release from the micelles (Figure 17.4a and b) or be internalized as drug molecules entrapped within the micelles (Figure 17.4c–f). In addition, certain block copolymers have also been found to act as chemosensitizers or modulators of drug efflux pumps.[26,34,107–111]

17.4.1 DRUG DIFFUSION

A free drug can diffuse across the cell membrane following disassembly of the micelles or release of the drug from intact micelles. The extent and rate of diffusion of the free molecules across the cell membrane will depend largely on the physico-chemical properties of the drug (water solubility, charge, molecular weight, and partition coefficient). Small hydrophobic molecules can easily permeate the membrane and enter the intracellular space. Studies have shown that the rate of cell uptake of a free drug (non-micelle-incorporated) is faster than that of drug incorporated in micelles.[103–105] In this way, the release profile of the drug from the micelles can have a significant influence on the rate of cellular uptake of the drug. As the drug is released from the micelles, there is an increase in the gradient between the concentrations of free drug on the exterior and interior of the cell that, in turn, drives diffusion of the drug into the intracellular compartment. A micelle system with a slow drug release profile will only have a limited pool of free drug available for diffusion into the cell. For example, Maysinger et al. demonstrated that for PEG-*b*-PCL micelles loaded with benzo[a]pyrene that has an apparent partition coefficient (K_v) of 690, the rate of cell uptake of the micelle formulated probe was similar to that of the rate of uptake of free benzo[a]pyrene. In contrast, for micelles loaded with DiI that has a K_v of 5800, a much lower degree of cell uptake of the micelle formulated probe was observed when compared to free DiI.[45,103] Therefore, at the site of interest, an increase in the rate of drug release from the micelles can result in a greater extent of intracellular drug accumulation if the drug can easily diffuse across the cell membrane. Micelle systems that respond to external stimuli as a means to promote drug release at the tumor have also been designed and are discussed further in Section 17.6. Another route for the entry of micelle-formulated drug into cells includes the internalization or endocytosis of the drug-loaded micelles.[103,104,106]

17.4.2 CELLULAR INTERNALIZATION

Macromolecules and nanosized particles including polymer–drug conjugates, liposomes, and micelles are unable to diffuse through the cell membrane into the intracellular compartment. In order to enter the cell, these drug delivery vehicles require an internalization mechanism such as phagocytosis or pinocytosis. Only certain cell types called phagocytic cells (macrophages, neutrophils) can undergo phagocytosis. Particulate materials with diameters of less than 1 μm can be internalized into these cells in membrane-bound vesicles called phagosomes.[112] Conversely, pinocytosis is known to be non-cell type specific. This mechanism occurs by invagination of the cell membrane that is similar to the mechanism in phagocytosis. However, the pinocytic vesicles that pinch off from the cell membrane are generally smaller than phagosomes and have diameters ranging from 100 to 200 nm.[112] The pinocytosis mechanism can be further categorized into fluid-phase and adsorptive.

Fluid-phase pinocytosis involves the engulfment of soluble materials and extracellular fluid during continual cell membrane ruffling. This is in contrast to adsorptive pinocytosis that involves the association of particulate materials with cell membrane components prior to internalization.[113]

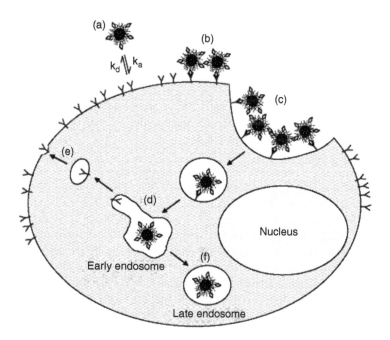

FIGURE 17.5 Internalization of ligand-conjugated micelles via receptor-mediated endocytosis. (a) The ligand-coupled micelles approach the cancer cells and (b) bind to the cell surface receptors depending on the association constant (k_a) and the dissociation constant (k_d) of the receptor–ligand complexes. (c) The micelles are then rapidly internalized into the cells via receptor-mediated endocytosis. (d) The internalized receptor and the ligand-conjugated micelles then undergo endosomal sorting in the early endosomes. In most cases, (e) the receptor is recycled to the cell surface, and (f) the micelles enter the late endosome that eventually fuses with the lysosome.

For example, micelles that are conjugated with specific ligands can bind to the corresponding receptors that are expressed on the cell surface; as a result, the micelles can be internalized via receptor-mediated endocytosis.[78,114,115] More details on receptor-mediated endocytosis are provided in Section 17.6 and Figure 17.5. In most cases, micelles are internalized into the cells through fluid-phase endocytosis.[77,116,117]

Fluid-phase endocytosis is expected to result in a lower degree of cell uptake in comparison to adsorptive pinocytosis or diffusion of free hydrophobic agents. Differences in the cell uptake of micelle-incorporated agents and non-micelle-incorporated agents have been demonstrated in various studies. For example, using a hydrophobic fluorescent probe as a tracker for micelles, the rate and amount of intracellular fluorescence was found to be much lower than incubation with free probes.[104] This has also been confirmed by other studies that relied on confocal microscopy as a qualitative tool.[103,105] Rapoport et al. also demonstrated that the incorporation of doxorubicin and ruboxyl into Pluronic® micelles resulted in a decrease in the extent of cell uptake of the agents in ovarian cancer cells.[118] In a drug accumulation study at copolymer concentrations above the CMC of the Pluronic® copolymers, the accumulation of the fluorescent probe R123 was substantially decreased in both non-multi- and multi-drug resistant cancer cells. It was suggested that the R123-loaded micelles were transported into the cells via a vesicular route and followed accumulation kinetics different from that of the non-micelle incorporated R123.[34]

Endocytosis is known to be an energy-, time-, and pH-dependent process.[113] Many pharmacological manipulations have been employed in order to demonstrate that micelles are internalized via an endocytotic pathway. For example, a low pH and incubation at 4°C resulted in the reduction of the internalization of micelles.[104] Also, pre-treatment of cells with metabolic inhibitors such as sodium azide and 2-deoxyglucose (DOG) that deplete the energy source for membrane ruffling resulted in a decrease in the extent of uptake of radiolabeled-drug loaded-PEG-b-PCL micelles by at least 80%. Another study conducted by Luo et al. reported similar findings, yet in this case, the micelles were tracked using fluorescently labeled-copolymer.[106] Furthermore, the pre-treatment of cells with Brefeldin A, known to alter the morphology of endosomes, was also found to reduce the extent of uptake of micelles. This suggests that an endosomal compartment may be involved in the internalization of the micelles that further supports the claim that micelle internalization does proceed by an endocytotic mechanism.[104]

In order to confirm that micelles have been internalized via the endocytotic pathway, different methods have been developed for the detection of micelles in the relevant organelles. For example, a pH-sensitive fluorescent probe has been used to confirm the presence of the micelles in acidic organelles using flow cytometry.[118,119] The subcellular localization of micelles can also be visualized using confocal microscopy by fluorescently labeling the micelles and counterstaining the intracellular organelles.[41,105,115] Hydrophobic fluorescent probes physically entrapped in micelles can also be used to track the micelles. The use of these probes to follow the pathway or fate of the micelles requires virtually no or minimal release of the probe during the time frame of the experiments. However, cell culture media often contain protein that can act to accelerate drug release.[120] Therefore, the accuracy associated with using physically entrapped probes to track micelles is questionable because of the different subcellular localization patterns of free and micelle-incorporated drugs.[41,105] Ideally, fluorescent probes that are chemically conjugated to the copolymers provide the most reliable tool for following the micelles in vitro. Although copolymers also exist in unimeric form, the concentration of unimers in a micelle solution does not exceed the CMC of the copolymer and represents only a minor population.[106] Therefore, it is reasonable to assume that the fluorescence detected corresponds to the micelle population provided that the conjugation of the probe to the copolymers is stable.

Internalized micelles have been shown to localize mainly in cytoplasmic compartments, in particular, the accumulation of micelles in acidic organelles (endosomes or lysosomes) has been observed.[105,118,119] Rapoport et al. have reported on the intracellular uptake and trafficking of Pluronic® micelles. Their results suggested that after the micelles were internalized in acidic organelles, the non-ionic Pluronic® copolymers induced permeabilization of the membranes, resulting in the release of the encapsulated drugs into the cytosol or into the nucleus via diffusion of the drug.[118] The nuclear pore complexes on the nucleus allow for passive diffusion of small molecules that have diameters of less than 9 nm, whereas macromolecules as large as 39 nm can be actively transported into the nucleus.[121] Theoretically, micelles that are greater than 9 nm cannot be translocated into the nucleus unless nuclear targeting signals are present on the particles. For example, in cancer cells that have been incubated with tetramethylrhodamine-5-carbonyl azide (TMRCA)-conjugated PEG-*b*-PCL micelles for 24 h, only cytoplasmic localization of the micelles was observed, and the nuclear compartment was devoid of micelles.[105] Similar results were obtained by Shuai et al. with doxorubicin-loaded PEG-*b*-PCL micelles that have a delayed release profile for doxorubicin from the micelles. The authors demonstrated that following incubation of cells with micelle formulated doxorubicin for 2 and 24 h, the drug was only localized in cytoplasmic compartments, whereas incubation of free drug with cells resulted in nuclear accumulation.[41] To date, the cellular internalization pathway and fate of micelles and micelle-encapsulated drugs as a function of time have not been clearly elucidated. However, the in vitro pathway and fate of micelles can have a significant impact on the cytotoxic effects of the formulated drug. The various anti-cancer agents that have been traditionally formulated in micelles have different sites of action within the cell. For example, paclitaxel and doxorubicin are required to bind to microtubules in the cytoplasm and DNA in the nucleus, respectively, in order to exert their cytotoxic effects. Therefore, if micelles could be engineered or designed to achieve a specific subcellular localization that matched the required localization of the drug, the cytotoxicity of the drug could be enhanced.

17.4.3 IN VITRO CYTOTOXICITY

The subcellular localization and the release profile of drug-loaded micelles can impact the cytotoxic effects of the drugs. In most cases, free drugs have been found to have lower IC_{50} values than drugs formulated in micelles. Such an increase in the IC_{50} of the micelle-formulated drugs is likely because of the delayed release of drugs from the micelle core.[41,95,122]

Shuai et al. reported delayed cell death in MCF-7 breast cancer cells when using a doxorubicin formulated PEG-*b*-PCL micelle system in comparison to free doxorubicin.[41] The confocal microscopy images showed that the micelle-incorporated doxorubicin resides in the cytoplasm, and free doxorubicin accumulates quickly in the cell nucleus. The delayed cell death could be due to the difference in the subcellular localization of free and micellized drug because doxorubicin can only exert its cytotoxic effect after reaching the nucleus.[41] Interestingly, as reported by Yoo and Park, doxorubicin-conjugated PEG-*b*-PLGA micelles, led to a 10-fold increase in the cytotoxicity of this drug. However, in this case, doxorubicin was found to localize in both the cytoplasm and the nucleus when delivered by micelles.[77] The discrepancy between the intracellular localization of doxorubicin formulated in the two distinct systems may be explained by their different drug release profiles. The PCL core is more hydrophobic than the PLGA core and can, therefore, provide better drug retention. Although doxorubicin was conjugated to the PLGA system that should result in a much slower release, the hydrolytic linkage is susceptible to cleavage and results in a faster release of the drug when compared to the PCL system. The cytotoxicity observed in the PEG-*b*-PLGA micelle system can be attributed to the doxorubicin that has been cleaved from the copolymer and released from the micelles because the conjugated drug was found to be non-cytotoxic.[16,32]

Furthermore, using a thermosensitive micelle system, Liu et al. obtained accumulation of doxorubicin in the cytoplasm and the nucleus following a temperature increase, and no nuclear accumulation of the drug was observed prior to the increase in temperature.[122] The delayed cytotoxicity of micelle-formulated drugs may seem to be a limitation in terms of the utilization of this technology as cancer therapy. However, the retention of drug in the micelles in vivo is required in order to ensure successful delivery of the drug to the tumor site. The relationship between drug retention in micelles and in vivo efficacy is discussed in Section 17.5.

Copolymer micelles have also been found to amplify the effect of chemotherapeutic drugs in cancer cells. In the case of multi-drug-resistant cancer cells, certain copolymers can also act as chemosensitizers and increase the intracellular accumulation of the drug. The Pluronic® copolymers are the family of materials that have been investigated most extensively in terms of their role as chemosensitizers or modulators of drug efflux.

17.4.4 Influence or Effect on Multi-Drug Resistance

Multi-drug resistance (MDR) is one of the most significant obstacles in chemotherapy and is often found in refractory cancers. MDR may be a result of the undesired efflux of a drug from the intracellular compartment. Transporter proteins that are overexpressed on the cell membrane of certain cancer cells are responsible for drug efflux. The drug efflux protein that has been studied most extensively is the ATP-dependent P-glycoprotein (P-gp).[26,123] This drug transporter is also expressed in healthy tissues such as the epithelial cells of the intestine and the endothelial cells in the blood–brain barrier. The normal physiological function of P-gp is to regulate molecular transport and provide protection or detoxification of the cell.[124] The normal physiological levels of drug efflux protein also function to limit the bioavailability of drugs delivered via the oral route as well as to limit drug delivery to the brain.

Kabanov's group has made significant contributions in the development of Pluronic® micelles as a tool for enhancing drug delivery both across the blood–brain barrier and into MDR tumors.[7,26,34,108,110,124–127] Their work in this area originated from the finding that doxorubicin formulated in Pluronic® micelles was more cytotoxic than free doxorubicin in drug-resistant cells. Pluronic® can increase the cytotoxicity of anthracyclines in drug-resistant cancer cells by 2–3 orders of magnitude. For instance, the IC_{50} of daunorubicin in the presence of Pluronic® copolymers was decreased by 700-fold in resistant ovarian cancer cells (SKVLB).[125] Additional results on the cytotoxicity of anti-cancer drugs formulated in Pluronic® micelles in resistant cancer cells are summarized elsewhere.[108,125] Further mechanistic studies have demonstrated that the ability of Pluronic® copolymers to inhibit drug efflux proteins depends on their composition.

Specific compositions of Pluronic® copolymers were evaluated in a screening study in order to demonstrate that the inhibitory effects of the copolymers on drug efflux were dependent on the length of the PEG and PPO segments. It was found that hydrophobic Pluronic® copolymers with values of HLB < 19 and with an intermediate length of the PPO segments such as L61 (PEG_2–PPO_{30}–PEG_2) and P85 (PEG_{26}–PPO_{40}–PEG_{26}) are the most effective at inhibiting P-gp.[34] SP1049C is a Pluronic®-based micelle formulation that has been selected for further preclinical development and has entered phase II clinical trial for the treatment of relapsed cancers. L61 was chosen as the base component for the SP1049C formulation with the addition of F172 (another high molecular weight hydrophilic Pluronic®) copolymers to stabilize the formulation as discussed in Section 17.3.[26]

In many of the studies, the P85 copolymer was fundamental in the investigation of the mechanisms of Pluronic® copolymers for overcoming drug resistance in cancer.[23,108,109,124] To date, several mechanisms have been hypothesized to account for the role that Pluronic® copolymers play as drug efflux modulators, and these have been reviewed in detail elsewhere.[26,123,125] It has been confirmed that the unimers, as opposed to the copolymer micelles, are responsible for inhibiting drug efflux. For example, accumulation and cytotoxicity of doxorubicin were demonstrated to increase with an increase in the Pluronic® concentration up to the CMC of the copolymer in drug-resistant human breast cancer cells (MCF-7/ADR) as well as head and neck cancer (KBv) cells. However, after reaching the CMC of the copolymers (the concentration when micelles are formed), both the intracellular doxorubicin content and the cytotoxicity were found to level off and/or decrease.[34] Similarly, the extent of intracellular accumulation of fluorescein, a multi-drug resistance-associated protein (MRP) substrate, in an MRP-expressing human pancreatic adenocarcinoma cell line (Panc-1) was found to be dependent on the concentration of Pluronic® present.[127]

The ability of Pluronic® unimers to inhibit drug efflux transporters is mainly attributed to ATP-depletion.[26,109] Use of certain metabolic inhibitors such as sodium azide can inhibit P-gp and result in increased drug accumulation by depleting the energy source for efflux across the cell membrane. Similar effects are obtained with the addition of Pluronic® copolymers to the cells as they result in an abolishment of the energy supply within the MDR cells.[34,109,125,127] For example, P85 and P105 were found to reduce the activity of the electron transport chain reactions in the mitochondria in HL-60 cells, decreasing the degree of oxidative metabolism, indicating that Pluronic® copolymers can interrupt the production of ATP.[128] In addition, cell exposure to P85 has been found to induce reversible transient ATP depletion in Jurkat lymphoma T-cells.[129] These effects are much more profound in MDR cells because the drug efflux mechanisms are ATP-driven.[109] It was believed that MDR cells consume energy at a much higher rate than non-MDR cells; therefore, their activity is more susceptible to inhibition via ATP-depletion. As a result, the Pluronic® copolymers were capable of chemosensitizing the MDR cells to a greater extent than non-MDR cells. All of the above suggest that Pluronic®-induced ATP depletion is the major mechanism that is responsible for inhibiting drug efflux in MDR cells.

Another proposed mechanism of drug efflux inhibition is membrane fluidization caused by Pluronic® copolymers. Non-ionic surfactants have been shown to increase membrane fluidity and contribute to the inhibition of P-gp efflux, resulting in the enhanced cytotoxicity of anthracyclines in MDR cells.[125,130] However, Pluronic® cannot increase the intracellular accumulation of P-gp substrate in MRP-expressing cells, indicating that membrane fluidization on its own is insufficient to inhibit drug efflux in drug-resistant cells. It was believed that the membrane fluidization effect works synergistically with the ATP-depletion effect induced by Pluronic® copolymers to chemosensitize MDR cells.[127]

Recently, Burt's group reported on the synthesis of a series of low molecular weight PEG-*b*-PCL for inhibition of P-gp efflux in human colon adenocarcinoma (Caco-2). Interestingly, at concentrations above the CMC, the PEG-*b*-PCL copolymers were found to increase the accumulation of P-gp substrate, and little activity was observed at concentrations below the CMC.[111] Nevertheless, the inhibition of P-gp efflux was attributed to the unimers, and the inhibition mechanisms of the PEG-*b*-PCL copolymers were said to be similar to that of the Pluronic® copolymers.[107] These low molecular weight PEG-*b*-PCL copolymers are a class of materials with potential applications in the delivery of drugs to MDR cancer cells or the formulation of drugs requiring improved bioavailability.

17.5 IN VIVO BIOLOGICAL PERFORMANCE OF BLOCK COPOLYMER MICELLES

In order to successfully reach and interact with the tumor cells, the micelle-formulated drug must first circumvent the MPS. As systemic drug delivery vehicles, block copolymer micelles may serve as either solubilizers and/or true drug carriers depending on their capability for drug retention and stability in vivo. Several of the micelle formulations that have been developed for hydrophobic drugs have been shown to provide a significant increase in their water solubility.[36] The use of biocompatible and biodegradable copolymers to form these micelles may result in formulations with reduced toxicity in comparison to formulations formed from some conventional surfactants or cosolvents. As a result, these copolymers may allow for higher doses of the drug to be administered, and they result in improved efficacy when compared to a conventional formulation of the drug.

Micelles that function merely as solubilizers increase the aqueous solubility of the drug, but they do not retain the drug for prolonged periods of time following systemic administration. In this way, the pharmacokinetics and biodistribution profile of the drug may remain largely unchanged. By contrast, micelles may function as true carriers that are able to retain the drug until reaching the

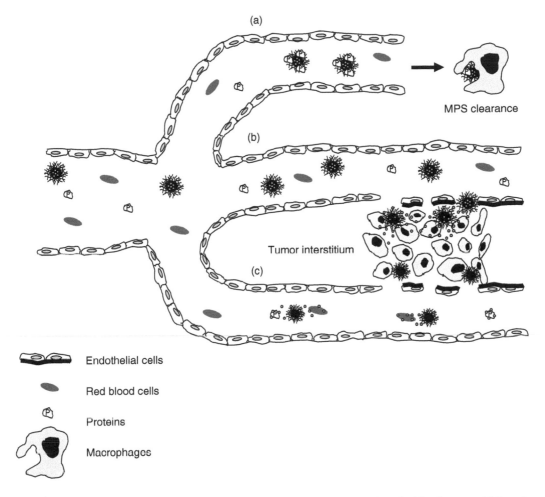

FIGURE 17.6 In vivo fate of drug-loaded micelles following administration into the bloodstream. (a) Proteins can be adsorbed onto the micelle surface, leading to eventual clearance by the mononuclear phagocytic system (MPS). (b) The drug-loaded micelles can also circulate in the bloodstream, accumulate in the tumor interstitium via the enhanced permeability and retention (EPR) effect, and subsequently release the encapsulated drug to the proximal tumor cells. (c) In the case where the micelle-encapsulated drug has a high affinity for blood components, the drugs may be released from the micelle core at an accelerated rate to bind to proteins that are present in the circulation prior to reaching the tumor.

target or desired site.[16,99,101] Micelles that function as carriers can result in significant improvements in the toxicity profile and/or therapeutic efficacy of the encapsulated agents.

The ability of a micelle system to function as a solubilizer or a true carrier depends on the following two factors: the stability of the micelles in vivo and the drug release profile in the presence of various blood components. The stability of block copolymer micelles in vivo depends on both thermodynamic and kinetic components as discussed in Section 17.2.

17.5.1 FATE OF POLYMERIC MICELLES IN VIVO

The MPS serves as one of the major mechanisms for the elimination of nanosized drug delivery systems such as liposomes, nanoparticles, and micelles.[131–133] Clearance by the MPS is first mediated by an array of blood components that interact with the delivery vehicle or other foreign bodies present within the systemic circulation, a process termed *opsonization*.[134–136] Following intravenous injection, the micelles quickly distribute into various tissues and organs by way of the blood and lymphatic system. The primary elimination route for micelles is postulated to include uptake by the liver and spleen followed by clearance through the kidneys via the excretion of urine[35,76,137–139] and/or hepatobiliary excretion in the feces.[131,140] In this process, various plasma proteins play an important role because they adsorb to the surface of a delivery vehicle within the first few minutes of exposure, especially if the surface is charged or hydrophobic (Figure 17.6a).[141] The protein adsorption could cause opsonization, leading to rapid clearance of both carrier and encapsulated drug by the MPS.[10,142,143] As a result, a reduction in the extent of protein adsorption to delivery vehicles such as micelles has been considered to be one of the key strategies to increase the circulation lifetime of the vehicles in vivo.

Various efforts to prolong the lifetime of micelles in vivo have been explored, including optimization of micelle size, size distribution, and surface properties. To this point, the micelles explored for applications in drug delivery have mostly been formed from copolymers that include PEG as the hydrophilic block. The presence of PEG at the surface of a delivery vehicle is known to impart stealth properties.[1,2,10] Good correlations have been found between the length of the PEG block and the extent of stabilization provided.[137] The degree of stabilization is largely determined by the thickness of the PEG layer. The PEG chains are present in a coil-like conformation on the surface of the micelles. The thickness of the surface layer (δ_h) may be calculated from the PEG block length using the following equation: $\delta_h = a^{0.84}$ where a is the number of PEG units.[144] In a study by Gref et al. in Balb/C mice, a significant increase in the circulation lifetime of PEG-*b*-PLGA nanospheres was observed when the length of the PEG block was increased from 0 to 20,000 g/mol, and the length of the PLGA block was held constant. The increased thickness of the protective PEG layer likely provides complete coverage of the more hydrophobic PLGA components. In the same study, the increase in the PEG length also resulted in a significant reduction in the extent of liver uptake of the particles that provided further evidence that PEG could effectively decrease the extent of opsonization and, therefore, limit MPS elimination.[66]

For materials to be eliminated by glomerular filtration through the kidneys, they must have a molecular weight of less than 50,000 g/mol.[17] The total molecular weight of a block copolymer micelle is typically on the order >200,000 g/mol, and the molecular weight of the individual copolymers ranges between 3000 and 25,000 g/mol.[145] Therefore, it may be postulated that only the copolymer single chains rather than intact micelles are eliminated by renal excretion.

17.5.2 In Vivo Drug Release Profiles

Micelles introduced into the circulation are exposed to a vast array of proteins, blood cells, and other blood components. A high affinity between the drug and major blood components commonly results in an acceleration of drug release from the micelles in vivo that, in turn, leads to a rapid elimination of the drug from the body (Figure 17.6c).[16] Studies have shown that the in vivo release profile of drug from micelles is largely determined by the compatibility between the drug and the core-forming block as well as the affinity between the drug and plasma proteins.[36,120,146,147] Ramaswamy et al. investigated the distribution of paclitaxel in vitro in human plasma samples following incubation of free drug or drug incorporated in PEG-*b*-PDLLA micelles with the plasma. In this study, similar plasma distribution profiles were obtained following incubation of free and micelle-formulated paclitaxel with lipoprotein deficient and lipoprotein rich fractions.[147] In this way, this study clearly illustrates the extent to which the in vivo retention of drugs in colloidal carriers may be determined by the drugs' affinity for plasma components. Therefore, the micro-environment within the delivery system must have the capability to compete with the various plasma components in order to achieve prolonged drug retention within the micelles in vivo.

17.5.3 Toxicity, Pharmacokinetics, and Anti-Cancer Efficacy of Drugs Formulated in Block Copolymer Micelles

As previously discussed, various micelle-based formulations have been demonstrated, in both preclinical and clinical studies, to decrease the toxicity and increase the efficacy of the encapsulated agents.[16,82,90,148] To date, two of the drugs that have been most commonly formulated in block copolymer micelles are doxorubicin and paclitaxel (Figure 17.3). A general description and overview of both the doxorubicin and paclitaxel formulations in clinical development has been given in Section 17.3. At present, there are two micelle formulations of doxorubicin (i.e., SP1049C and NK911) and two of paclitaxel (Genexol-PM and NK105) that are in clinical trial development.

17.5.3.1 Doxorubicin Formulations (SP1049C and NK911)

SP1049C includes doxorubicin formulated in L61 and F127 Pluronic® copolymers at a ratio of 1:8 (wt/wt).[26] For doxorubicin administered in the SP1049C formulation at a dose of 10 mg/kg, the AUC in healthy mice was found to be 14.6 µg h/mL while the maximum concentration (C_{max}) in plasma was 4.47 µg/g. For doxorubicin administered as free drug, the AUC is 7.1 µg h/mL and the C_{max} in plasma is 2.77 µg/g. In this way, the SP1049C formulation provides a 2.1-fold increase in the AUC.

NK911 is another micelle formulation of doxorubicin and includes both chemically and physically entrapped doxorubicin within PEG-b-PAsp micelles. In its phase I trial, the administration of the NK911 formulation at a dose of 50 mg/m^2 was shown to provide a significant increase in the half-life ($t_{1/2}$) and the AUC as well as a significant decrease in the steady state volume of distribution (V_{ss}). Specifically, for doxorubicin administered to humans in NK911, the $t_{1/2}$ values were reported to be $t_{1/2,\alpha} = 7.5$ min, $t_{1/2,\beta} = 2.8$ h, and $t_{1/2,\gamma} = 64.2$ h, and the AUC was found to be 3262.7 ng h/mL, and the V_{ss} was 14.9 L/kg. In comparison to free doxorubicin, the $t_{1/2}$ values were reported to be $t_{1/2,\alpha} = 2.4$ min, $t_{1/2,\beta} = 0.8$ h, and $t_{1/2,\gamma} = 25.8$ h, and the AUC was found to be 1620.3 ng h/mL, and the V_{ss} was 24 L/kg.[82] In this way, the encapsulation of doxorubicin in the NK911 formulation increases the AUC and plasma C_{max} for this drug by 36.4 and 28.6-fold, respectively. The change in the pharmacokinetic profile of the drug translates into changes in the efficacy and toxicity of the encapsulated agent. Specifically, the anti-tumor activity was evaluated in four tumor xenograft models in mice, namely, Colon 26, M5076, P388, and Lu-24. For the groups of animals receiving treatment with the micelle-formulated drug, a significant percentage of mice bearing colon 26 and MX-1 tumors were cured, whereas no cures were found for mice who received the same dose of free doxorubicin. Moreover, in all tumor models, treatment with the free drug resulted in more severe toxicity than that observed in animals receiving even the highest dose of NK911.

17.5.3.2 Paclitaxel Formulations (Genexol-PM and NK105)

Similar to doxorubicin, there are two promising formulations of paclitaxel that are under clinical trial evaluation: Genexol-PM and NK105. The main pharmacokinetic parameters of paclitaxel administered as Genexol-PM at a dose of 135 mg/m^2 are 5473 ng h/mL for the AUC and 12.7 h for the $t_{1/2\beta}$; whereas for paclitaxel administered as Taxol®, the AUC is 9307.5 ng h/mL, and the $t_{1/2} = 20.1$ h.[90,149] In this way, the Genexol-PM formulation does not improve the pharmacokinetic profile of paclitaxel in vivo; therefore, the micelles in this formulation function primarily as solubilizers. In fact, it has been found that the drug is released quite rapidly from the PEG-b-PDLLA micelles following administration. However, as a result of the low toxicity of the micelles in comparison to Cremophor EL (the primary excipient in Taxol®), the advantage of this formulation is the significant improvement in the maximum tolerated dose (i.e., MTD = 390 mg/m^2) when compared to the MTD of paclitaxel administered as Taxol® (i.e., 135–200 mg/m^2).[90,149] In this case, the increase in the MTD is attributed to the biocompatibility or non-toxicity of the

PEG-*b*-PDLLA copolymers. Therefore, an improvement in the anti-cancer efficacy may be expected because of the higher dose of paclitaxel that may be administered when given as the Genexol-PM formulation.

NK105 is a micelle formulation of paclitaxel wherein the micelles have been shown to perform as true carriers. The NK105 micelles are formed from copolymers with PEG as the hydrophilic block and modified PAsp as the core-forming block. The PAsp block is conjugated with 4-phenyl-1-butanol as a means to increase the compatibility between the micelle core and paclitaxel. Paclitaxel administered in the NK105 formulation has been shown to have a much longer circulation lifetime than paclitaxel administered as Taxol®. Specifically, the AUC for paclitaxel was increased by 50–86-fold and the $t_{1/2}$ was increased 4–6 times for the NK105 formulation in comparison to paclitaxel administered as Taxol®.[99] In addition, the therapeutic efficacy of NK105 and Taxol® formulations were investigated in an HT-29 colon cancer xenograft model in nude mice. The NK105 formulation was found to exhibit statistically superior anti-tumor activity in comparison to Taxol®. The anti-tumor activity of NK105 administered at a paclitaxel-equivalent dose of 25 mg/kg was comparable to that obtained following the administration of Taxol® at a paclitaxel dose of 100 mg/kg. In addition, all tumors disappeared in the NK105 treatment group following one dose of 100 mg/kg, and all mice remained tumor-free for the 33-day duration of the study. This significant improvement in the anti-cancer activity of paclitaxel following encapsulation in this micelle system may be attributed to the ability of the NK105 micelles to function as true drug carriers.

17.6 EVOLUTION OF POLYMERIC MICELLES

Recently, much effort has been devoted to increasing drug accumulation at diseased sites. Advances in synthetic chemistry and a more comprehensive understanding of biological systems have facilitated the development of smart micelle systems that are capable of targeting drugs to specific tissues or cellular organelles. Through chemical modification, conventional micelles may be designed to deliver to and/or release drugs at a specific site of interest by active targeting or external stimulation.

17.6.1 Triggered Stimulation of Drug Activation or Release

Following administration of drug-loaded micelles, external stimuli may be locally applied at the tumor in order to release or activate the drug. This type of drug delivery mechanism may ensure a more accurate temporal and spatial accumulation of drugs in the tumor as well as reduce toxicity to other tissues. In this way, localized drug delivery may be achieved in a non-invasive manner using external stimuli such as ultrasound, heat, and light.

17.6.1.1 Ultrasonic Irradiation

The application of ultrasonic irradiation at tumor sites has been found to be effective in enhancing the accumulation of drugs in tumor cells and has resulted in improved therapeutic efficacy. The use of ultrasonic irradiation for micelle-based chemotherapy can be optimized by varying many factors including the sonication frequency, the types of waves applied, the time between drug injection and the application of ultrasound as well as the physico-chemical properties of the micelles.[150,151] Micelle properties such as the hydrophobicity and state of the micelle core were found to be the most influential in increasing drug uptake in tumor cells upon ultrasonic irradiation.[79,128,152]

To date, several mechanisms have been proposed for enhancing tumor cell uptake of micelle-formulated drugs via ultrasonic irradiation. It had been suggested that ultrasound promotes extravasation of drug-loaded carriers because of increased vasculature permeability.[79] However, a recent in vivo study revealed that the application of ultrasound does not enhance extravasation of micelles. Rather, the enhanced drug accumulation in tumor cells requires initial accumulation of drug-loaded micelles in the tumor interstitium via the enhanced permeability and retention (EPR) effect.[150] This indicates that the ultrasonic effect occurs at the cellular level and could potentially play a role in micelle–cell interactions. In vitro studies have suggested that ultrasound does not increase the extent of endocytosis of intact micelles.[119] In fact, ultrasound has been shown to trigger drug release as well as enhance drug uptake by increasing cell membrane permeability.[119,151,152]

The application of ultrasound perturbs the drug-loaded micelles, shifting the equilibrium from micelle-encapsulated drug to free drug. In this way, the release of drugs from the micelles is triggered.[151,152] Furthermore, ultrasonication can cause reversible cell membrane disruption, creating temporary pores in the membrane, resulting in enhanced drug diffusion into the cells following drug release.[119] This mechanism is also effective in treating multi-drug resistant cancer cells because of the enhanced partitioning of the released drug into the intracellular compartment.[79,118] The percentage of drug released after ultrasonication was found to decrease with increasing hydrophobic interactions between the drug and the micelle core.[151] The use of Pluronic® copolymers to form micelles that have liquid-like cores facilitates drug release after ultrasonication. Nonetheless, following injection, these micelles are very unstable in vivo and may result in reduced drug accumulation at tumor sites. PEG-conjugated distearoylphosphatidylethanolamine (DSPE) has been added to Pluronic® micelles to improve micelle stability and was proven to be as effective as PEG-b-PBLA micelles that have solid-like cores in ultrasound-enhanced chemotherapy.[150] Interestingly, after inactivation of ultrasonication, the released drug is re-encapsulated in the micelles, and the intact drug-loaded micelles are capable of recirculating until they reach the tumor again. As a result, more than a 3-fold increase in drug uptake in tumor cells is observed in vivo, and a reduction in the accumulation of the drug in normal tissues is achieved when compared to non-sonicated tumors.[150]

17.6.1.2 Thermosensitive Systems

In addition to ultrasonication, micelles can also be designed to trigger the release of drugs at the tumors using heat as an external stimulus. Application of thermo-responsive micelles for the treatment of solid tumors involves the use of a thermosensitive polymer as the micelle-building block and the local administration of heat at the tumor site. Thermo-responsive copolymers including poly(N-isopropylacrylamide-b-D,L-lactide) (PIPAAm-b-PDLLA), poly(N-isopropylacrylamide-b-(butyl methacrylate)) (PIPAAm-b-PBMA), poly(N-isopropylacrylamide-co-N,N-dimethylacrylamide)-b-poly(D,L-lactide-co-glycolide) (P(IPAAm-co-DMAAm)-b-PLGA) have been used for the preparation of thermo-responsive micelles.[122,153–155] These micelles undergo reversible structural changes that facilitate drug release above the lower critical solution temperature (LCST) of the thermosensitive polymer. In the case of PIPAAm micelles, dehydration of the outer shells of the micelles also occurs above the LCST, resulting in the enhanced adsorption of drug-loaded micelles to cells via increased hydrophobic interactions.[155–157]

Okano's group developed a thermo-responsive, doxorubicin-loaded micelle system with an LCST of 32°C. By toggling the temperature below and above the LCST, the release of doxorubicin may be switched reversibly between an on and off state. At temperatures above the LCST, doxorubicin release is accelerated, whereas decreasing the temperature to levels below the LCST results in minimal drug release.[155] This effect is also influenced by the state of the micelle core in addition to the thermosensitivity of the PIPAAm outer shells. For instance, micelles that are formed from copolymers with liquid-like cores (e.g., $T_g = 20°C$ for PBMA) are more sensitive to thermal triggered drug release than micelles with solid-like cores (e.g., $T_g = 100°C$ for polystyrene (PS)) at temperatures above the LCST. As a result, the doxorubicin-loaded PPIAAm-b-PBMA micelles were three times more cytotoxic than the PPIAAm-b-PS micelles above the LCST mainly as a result of their distinct doxorubicin release profiles. At temperatures below the LCST, both systems exhibited almost no cytotoxicity.[153]

For in vivo delivery, the LCST of the thermo-responsive micelles should be several degrees above physiological temperature (i.e., LCST > 37°C).[157] The LCST can be manipulated by fine-tuning the hydrophilicity of the PIPAAm segment. The addition of hydrophilic groups to the PIPAAm polymer can elevate the LCST of the micelles formed.[122,155–158] For example, Liu et al. reported that by adding hydrophilic DMAAm to the PIPAAm segment, the P(IPAAm-co-DMAAm)-b-PLGA micelles have an LCST of approximately 39°C. As a result, the doxorubicin-loaded P(IPAAm-co-DMAAm)-b-PLGA micelles have been shown to be more potent in the 4T1 mouse breast cancer cells at 39.5°C (IC_{50} = 3.1 mg/L) than at 37°C (IC_{50} = 7.9 mg/L).[122] Therefore, heat-activated drug release in tumors (above 37°C) can reduce the possible toxicity associated with the drug in healthy tissues (at 37°C).

17.6.1.3 Photodynamic Therapy

In photodynamic therapy (PDT), light is used to activate the drug (a photosensitizer) at diseased sites. Initially, PDT was used for the treatment of superficial tumors such as skin cancer. Recent improvements in the technology have allowed for the treatment of deep tumors such as lung cancers.[159] The tumor area is exposed to light of a specific wavelength that activates the drug, leading to the destruction of tumor cells and proximal tumor vasculature via the production of reactive oxygen species or by other mechanisms.[160,161] In order for the PDT agents to be effective in destroying tumor cells, it is necessary that there is no aggregation of the drug. However, in aqueous media, the hydrophobic photosensitizers tend to aggregate, resulting in reduced potency. Pluronic® micelles were reported to be able to solubilize benzoporphyrin while avoiding drug aggregation in the micelle core, implying that the potency of the photosensitizer was maintained.[162] In this way, polymeric micelles become an attractive alternative for the solubilization and delivery of photosensitizers.

Kataoka's group has developed polyion complex (PIC) micelles based on anionic copolymers (PEG-b-poly(L-lysine) or PEG-b-PAsp) for the encapsulation of a polycationic photosensitizer. These micelles are stabilized by electrostatic interactions, and the micelles are pH-responsive, allowing intratumoral drug release or intracellular endosomal escape.[160,162,163] The photosensitizer-loaded PIC micelles showed no skin phototoxicity in rats upon exposure to broadband visible light, and severe skin phototoxicity was observed with the administration of Photofrin® (a currently approved PDT drug formulated in 0.9% sodium chloride).[164] The photosensitizer-loaded PIC micelles were also shown to exhibit enhanced efficacy and reduced toxicity in the Lewis lung carcinoma (LLC) cell line in comparison to free photosensitizer.[160]

It has also been reported that the subcellular localization of the photosensitizer influences its efficacy. Leroux's group has developed pH-sensitive aluminum chloride pthalocyanine (AlClPc)-loaded N-isopropylacrylamide copolymer micelles for treatment against EMT-6 tumors in vivo. The tumor accumulation of the micelle-formulated drug is lower than that of AlClPc dissolved in Cremophor EL. However, the copolymer micelle formulation has been shown to have enhanced therapeutic efficacy that is most likely attributed to an increase in the cytoplasmic accumulation of the drug.[165,166] In many cases, the accumulation of photosensitizer at the mitochondria has been found to be the most efficacious.[161,167] Therefore, the use of pH-responsive polymeric micelles for PDT could enhance drug accumulation in the mitochondria following endosomal/lysosomal escape of the micelles. In the future, the design of polymeric micelles for use in PDT will likely be aimed at improving tumor and subcellular localization by fine-tuning the copolymer composition in order to achieve active targeting.

17.6.2 Active Targeting

The phenomenon of active targeting of micellar delivery vehicles loaded with drugs implies that the micelles can accumulate at the targeted cells or tissues without external stimuli via specific sensing

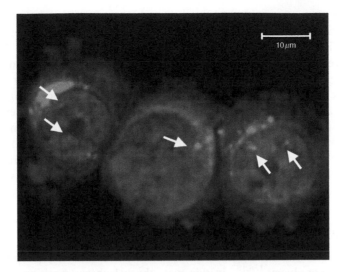

FIGURE 17.7 (See color insert following page 522.) Confocal fluorescence microscopy image of epidermal growth factor receptor (EGFR)-overexpressed breast cancer cells (MDA-MB-468) following incubation with EGF-conjugated poly(ethylene glycol)-*block*-poly(δ-valerolactone) micelles (as shown in red) for 2 h at 37°C. The cell nuclei are stained with Hoechst 33258 (blue). The EGF-conjugated micelles mainly localized at the perinuclear region as well as within the nucleus as shown by the white arrows.

mechanisms that are integrated into the carriers. The carriers can be engineered by means of ligand coupling or addition of pH-sensitive moieties according to the biological characteristics of the diseased site. In particular, many cancers are associated with over expression of certain cell surface receptors and acidosis at the tumor sites.

17.6.2.1 Ligand Coupling

It has been demonstrated that ligand-conjugated drug carriers can selectively and efficiently accumulate in target cells.[168–172] The addition of biomolecules to the surface of micelles enables the drug-loaded micelles to enter the cells via receptor-mediated endocytosis as shown in Figure 17.5. Through the highly specific binding affinity between the ligand–receptor complex, the attachment of ligands to micelles surface allows for targeted drug delivery to specific cells that express the corresponding cell surface receptors. In particular, many types of cancer cells express a higher-than-normal level of cell surface receptors that have binding affinities for growth-dependent molecules as a result of their rapid growth rate.[173] Ligands such as sugar moieties,[174–177] transferrin,[178–180] folate,[78,95,181] epidermal growth factor (EGF),[115,167,178] and antibody fragments[182] have been attached to the surface of copolymer micelles to target different types of cancers.

Kabanov's group was one of the first to apply the ligand coupling concept to micellar drug delivery systems. The authors demonstrated that by chemically conjugating α_2-glycoprotein, a ligand that targets brain glial cells, to FITC-containing Pluronic® micelles, the accumulation of FITC in brain tissues was increased and clearance of FITC by the lung was decreased in comparison to delivery using conventional Pluronic® micelles.[126,183] Recently, another ligand, folate, was conjugated to PEG-*b*-PLGA micelle system for targeting folate receptor over expressing cancers. The doxorubicin-loaded folate-PEG-*b*-PLGA micelles were found to enhance the intracellular accumulation of doxorubicin and resulted in a lower IC_{50} in folate receptor overexpressing KB epidermal carcinoma cells. In vivo studies revealed that treatment with doxorubicin-loaded folate-PEG-*b*-PLGA micelles resulted in a high suppression of tumor growth.[78]

In cases where nuclear targeting is desired, ligands such as EGF and the HIV-1 TAT peptide that have nuclear translocation properties, may be employed.[6,115,184] Zeng et al. reported on an EGF-conjugated PEG-*b*-poly(δ-valerolactone) (PEG-*b*-PVL) micelle system that targets EGFR-over expressing breast cancers. The EGF-PEG-*b*-PVL micelles were shown to mainly localize in the perinuclear region and in the nucleus of MDA-MB-468 breast cancer cells as shown in Figure 17.7.[115] Nuclear targeting is critical for the delivery of anti-cancer drugs and oligonucleotides whose site of action is located in the nucleus. Wagner's group has developed EGF- and transferrin-conjugated PEG-*b*-poly(ethylenimine) (PEG-*b*-PEI) micelle systems for gene delivery to tumor tissues.[178,180,185,186] Molecules such as oligonucleotides are more susceptible to degradation in the endosomal or lysosomal compartments; therefore, pH-responsive micelles that are capable of destabilizing the endosomal or lysosomal membranes could potentially improve the therapeutic effect of these molecules.

17.6.2.2 pH-Responsive Systems

The pH where drug release is triggered can be adjusted by varying the pH-sensitive moiety incorporated in the copolymer.[158,187] In this way, the pH phase transition of a micelle system can be designed to suit targets that have different pH values. For instance, micelles that are stable at pH~2 and are disrupted at pH>7.2 may be suitable for oral delivery of hydrophobic drugs. As an example, a 10-fold increase in the bioavailability of fenofibrate formulated in pH-responsive micelles was reported.[188] In the case of tumor targeting, micelles with a phase transition of approximately pH 7.0 could be beneficial as tumor tissues are known to be slightly more acidic (mean pH 7.0) than normal tissues (pH 7.4).[187,189,190] If subcellular targeting is desired, cytoplasmic accumulation of drugs can be enhanced by utilizing micelles that are responsive to pH values ranging from 5.0 to 6.0 (endosomal/lysosomal pH).[187,191,192]

pH-responsive micelles have also been developed by conjugating drugs to the hydrophobic blocks of copolymers using pH-sensitive linkages. For instance, doxorubicin can be conjugated to PEG-*b*-P(Asp) copolymers via a hydrazone linkage. The micelles formed from this copolymer were found to be extracellularly stable, and they provided selective release of doxorubicin in the endosomes where the pH ranges from 5.0 to 6.0. The released drug was found to first localize in the cytoplasm and eventually in the nucleus. The pH-responsive doxorubicin micelles were also shown to be capable of infiltrating avascular multicellular tumor spheroids in a uniform manner. Furthermore, the micelles were effective in suppressing tumor growth in a murine C26 colon tumor model and increased the MTD by more than 2.5 times because of their secure drug encapsulation.[191,192]

Another approach to develop pH-responsive micelles is to disrupt micelle formation in the acidic environment by introducing titrable groups into the copolymer.[187] Leroux's group introduced a hydrophilic titrable monomer (methacrylic acid or MAA) into the poly(*N*-isopropylacrylamide)-*co*-octadecyl acrylate (poly(NIPA)-*co*-ODA) copolymer to form pH-responsive micelles for PDT.[158] These micelles were found to undergo a pH phase transition and structural changes in the micelle core at a pH of approximately 5.8. The enhanced cytotoxicity of the AlClPc-loaded poly(NIPA)-*co*-ODA-*co*-MAA micelles were found to be highly dependent on the pH-gradient between the cytoplasm and the endosomal/lysosomal compartments.[158]

Bae and co-workers have developed several pH-responsive micelle systems by using poly (L-histidine) (polyHis) as the pH-sensitive moiety. The CMC of the PEG-*b*-polyHis copolymer was found to increase as the pH of the media was decreased with an abrupt increase in the CMC occurring at a pH of 7.2. Therefore, with the appropriate alteration in pH, these micelles become thermodynamically unstable and dissociate, resulting in the release of their contents.[193] Doxorubicin-loaded PEG-*b*-polyHis micelles were shown to result in a significant increase in the in vitro intracellular accumulation of drug, in vivo tumor localization of drug, and tumor growth inhibition in the A2780 ovarian tumor model via extracellular pH-dependent targeting of the tumor.[194] The addition of polyHis to PEG-*b*-PDLLA further reduces the pH phase transition of the micelles to the range of 6.6–7.2. These micelles were further modified with ligand coupling to achieve a double-targeting effect.[181,190] For example, Lee et al. prepared polyHis-*b*-PEG-*b*-PDLLA mixed micelles with biotin as the targeting ligand. Biotin is protected by the PEG shell at pH > 7.0 and is exposed on the surface of micelles when the pH is decreased to 6.5–7.0 (extracellular tumor pH). As the pH is further reduced to less than 6.5 (endosomal pH), the micelles destabilize, triggering drug release and disrupting endosomal membranes via the proton-sponge effect.[190] In the future, it is anticipated that combinations of sensing mechanisms will be used to improve the effectiveness of micellar-drug targeting to the site of interest.

17.7 FUTURE PERSPECTIVES

The field of block copolymer micelles for drug delivery is rapidly expanding and has great potential in cancer therapeutics. Although there are four promising polymeric micelle formulations currently in clinical trial development, many preclinical fundamental studies are on-going in order to fully exploit the use of block copolymer micelles for drug delivery. The advances in synthetic polymer science continue to provide new biodegradable and biocompatible polymers that can be tailored and designed for the challenges that anti-cancer drug delivery continues to face. By enhancing the compatibility between the copolymer and the drug, block copolymer micelles have emerged as true carriers rather than solubilizers for pharmaceutical agents. The mechanisms of loading and release of drugs from micelles are being elucidated and will provide a greater understanding of the relationships between the physico-chemical properties of micelles and their invivo performance. In the future, multiple targeting mechanisms may be integrated into one micelle system to fully exploit the use of polymeric micelle as delivery vehicles for anti-cancer drugs.

ACKNOWLEDGMENTS

The authors are grateful to NSERC for funding their research in the area of block copolymer micelles for applications in drug delivery. In addition, the authors thank Lorne F. Lambier for a scholarship to H. Lee, CIHR/Rx&D for a career award to C. Allen, a post-doctoral fellowship to P. Lim Soo, NSERC for a scholarship to J. Liu, and OGS for a scholarship to M. Butler.

REFERENCES

1. Moghimi, S. M., Hunter, A. C., and Murray, J. C., Long-circulating and target-specific nanoparticles: Theory to practice, *Pharmacological Reviews*, 53, 283–318, 2001.
2. Torchilin, V. P., Recent advances with liposomes as pharmaceutical carriers, *Nature Reviews Drug Discovery*, 4, 145–160, 2005.
3. Allen, C., Maysinger, D., and Eisenberg, A., Nano-engineering block copolymer aggregates for drug delivery, *Colloids and Surfaces B*, 16, 3–27, 1999.
4. Kwon, G. S., Diblock copolymer nanoparticles for drug delivery, *Critical Reviews in Therapeutic Drug Carrier Systems*, 15, 481–512, 1998.
5. Kataoka, K., Kwon, G. S., Yokoyama, M., Okano, T., and Sakurai, Y., Block-copolymer micelles as vehicles for drug delivery, *Journal of Controlled Release*, 24, 119–132, 1993.
6. Gaucher, G., Dufresne, M. H., Sant, V. P., Kang, N., Maysinger, D., and Leroux, J. C., Block copolymer micelles: Preparation, characterization and application in drug delivery, *Journal of Controlled Release*, 109, 169–188, 2005.
7. Kabanov, A. V., Batrakova, E. V., and Alakhov, V. Y., Pluronic (R) block copolymers as novel polymer therapeutics for drug and gene delivery, *Journal of Controlled Release*, 82, 189–212, 2002.
8. Elbert, D. L. and Hubbell, J. A., Surface treatments of polymers for biocompatibility, *Annual Review of Material Science*, 26, 365–394, 1996.

9. Harris, J. M., Ed., Poly(ethylene glycol) chemistry: Biotechnical and biomedical applications, In *Topics in Applied Chemistry*, Plenum, New York; London, 385, 1992.

10. Allen, C., Dos Santos, N., Gallagher, R., Chiu, G. N. C., Shu, Y., Li, W. M., Johnstone, S. A., Janoff, A. S., Mayer, L. D., Webb, M. S., and Bally, M. B., Controlling the physical behavior and biological performance of liposome formulations through use of surface grafted poly(ethylene glycol), *Bioscience Representatives*, 22, 225–250, 2002.

11. Gref, R., Domb, A., Quellec, P., Blunk, T., Muller, R. H., Verbavatz, J. M., and Langer, R., The controlled intravenous delivery of drugs using peg-coated sterically stabilized nanospheres, *Advanced Drug Delivery Reviews*, 16, 215–233, 1995.

12. Hunter, R. J. and White, L. R., *Foundations of colloid science*, Oxford Science Publications, Clarendon Press, Oxford [Oxfordshire]; Oxford University Press, New York, p. 2 v. (1089), 1987.

13. Müller, R. H., *Colloidal Carriers for Controlled Drug Delivery and Targeting: Modification, Characterization and In Vivo Distribution*, Wissenschaftliche Verlagsgesellschaft, Stuttgart; CRC Press; Sole distributor for North America CRC Press Inc., Boca Raton, FL, p. 379, 1991.

14. Kato, Y., Watanabe, K., Nakakura, M., Hosokawa, T., Hayakawa, E., and Ito, K., Blood clearance and tissue distribution of various formulations of alpha-tocopherol injection after intravenous administration, *Chemical and Pharmaceutical Bulletin (Tokyo)*, 41, 599–604, 1993.

15. Oku, N. and Namba, Y., Long-circulating liposomes, *Critical Reviews in Therapeutic Drug Carrier Systems*, 11, 231–270, 1994.

16. Nakanishi, T., Fukushima, S., Okamoto, K., Suzuki, M., Matsumura, Y., Yokoyama, M., Okano, T., Sakurai, Y., and Kataoka, K., Development of the polymer micelle carrier system for doxorubicin, *Journal of Controlled Release*, 74, 295–302, 2001.

17. Seymour, L. W., Duncan, R., Strohalm, J., and Kopecek, J., Effect of molecular weight (M_w) of N-(2-hydroxypropyl)methacrylamide copolymers on body distribution and rate of excretion after subcutaneous, intraperitoneal, and intravenous administration to rats, *Journal of Biomedical Materials Research*, 21, 1341–1358, 1987.

18. Allen, C., Yu, Y., Maysinger, D., and Eisenberg, A., Polycaprolactone-*b*-poly(ethylene oxide) block copolymer micelles as a novel drug delivery vehicle for neurotrophic agents FK506 and L-685,818, *Bioconjugate Chemistry*, 9, 564–572, 1998.

19. Zhang, L. F. and Eisenberg, A., Formation of crew-cut aggregates of various morphologies from amphiphilic block copolymers in solution, *Polymers for Advanced Technologies*, 9, 677–699, 1998.

20. Choucair, A. and Eisenberg, A., Control of amphiphilic block copolymer morphologies using solution conditions, *European Physical Journal E*, 10, 37–44, 2003.

21. Discher, D. E. and Eisenberg, A., Polymer vesicles, *Science*, 297, 967–973, 2002.

22. Lim Soo, P. and Eisenberg, A., Preparation of block copolymer vesicles in solution, *Journal of Polymer Science Part B-Polymer Physics*, 42, 923–938, 2004.

23. Kabanov, A. V., Nazarova, I. R., Astafieva, I. V., Batrakova, E. V., Alakhov, V. Y., Yaroslavov, A. A., and Kabanov, V. A., Micelle formation and solubilization of fluorescent-probes in poly(oxyethylene-*b*-oxypropylene-*b*-oxyethylene) solutions, *Macromolecules*, 28, 2303–2314, 1995.

24. Wilhelm, M., Zhao, C. L., Wang, Y. C., Xu, R. L., Winnik, M. A., Mura, J. L., Riess, G., and Croucher, M. D., Polymer micelle formation. 3. Poly(styrene-ethylene oxide) block copolymer micelle formation in water—a fluorescence probe study, *Macromolecules*, 24, 1033–1040, 1991.

25. Liu, T. B., Kim, K., Hsiao, B. S., and Chu, B., Regular and irregular micelles formed by A LEL triblock copolymer in aqueous solution, *Polymer*, 45, 7989–7993, 2004.

26. Alakhov, V., Klinski, E., Li, S. M., Pietrzynski, G., Venne, A., Batrakova, E., Bronitch, T., and Kabanov, A., Block copolymer-based formulation of doxorubicin. From cell screen to clinical trials, *Colloids and Surfaces B: Biointerfaces*, 16, 113–134, 1999.

27. Lee, H., Zeng, F. Q., Dunne, M., and Allen, C., Methoxy poly(ethylene glycol)-*block*-poly(delta-valerolactone) copolymer micelles for formulation of hydrophobic drugs, *Biomacromolecules*, 6, 3119–3128, 2005.

28. Kang, N. and Leroux, J. C., Triblock and star-block copolymers of N-(2-hydroxypropyl) methacrylamide or N-vinyl-2-pyrrolidone and D,L-lactide: Synthesis and self-assembling properties in water, *Polymer*, 45, 8967–8980, 2004.

29. Alexandridis, P., Holzwarth, J. F., and Hatton, T. A., Micellization of poly(ethylene oxide)–poly(-propylene oxide)–poly(ethylene oxide) triblock copolymers in aqueous-solutions—thermodynamics of copolymer association, *Macromolecules*, 27, 2414–2425, 1994.
30. Yamamoto, Y., Yasugi, K., Harada, A., Nagasaki, Y., and Kataoka, K., Temperature-related change in the properties relevant to drug delivery of poly(ethylene glycol)–poly(D,L-lactide) block copolymer micelles in aqueous milieu, *Journal of Controlled Release*, 82, 359–371, 2002.
31. Kang, N., Perron, M. E., Prud'homme, R. E., Zhang, Y. B., Gaucher, G., and Leroux, J. C., Stereocomplex block copolymer micelles: Core-shell nanostructures with enhanced stability, *Nano Letters*, 5, 315–319, 2005.
32. Yokoyama, M., Fukushima, S., Uehara, R., Okamoto, K., Kataoka, K., Sakurai, Y., and Okano, T., Characterization of physical entrapment and chemical conjugation of adriamycin in polymeric micelles and their design for in vivo delivery to a solid tumor, *Journal of Controlled Release*, 50, 79–92, 1998.
33. Croy, S. R. and Kwon, G. S., The effects of pluronic block copolymers on the aggregation state of nystatin, *Journal of Controlled Release*, 95, 161–171, 2004.
34. Batrakova, E., Lee, S., Li, S., Venne, A., Alakhov, V., and Kabanov, A., Fundamental relationships between the composition of pluronic block copolymers and their hypersensitization effect in MDR cancer cells, *Pharmaceutical Research*, 16, 1373–1379, 1999.
35. Burt, H. M., Zhang, X. C., Toleikis, P., Embree, L., and Hunter, W. L., Development of copolymers of poly(D,L-lactide) and methoxypolyethylene glycol as micellar carriers of paclitaxel, *Colloids and Surfaces B: Biointerfaces*, 16, 161–171, 1999.
36. Liu, J., Xiao, Y., and Allen, C., Polymer–drug compatibility: A guide to the development of delivery systems for the anticancer agent, ellipticine, *Journal of Pharmaceutical Sciences*, 93, 132–143, 2004.
37. Zhang, X. C., Jackson, J. K., and Burt, H. M., Development of amphiphilic diblock copolymers as micellar carriers of taxol, *International Journal of Pharmaceutics*, 132, 195–206, 1996.
38. Yokoyama, M., Opanasopit, P., Okano, T., Kawano, K., and Maitani, Y., Polymer design and incorporation methods for polymeric micelle carrier system containing water-insoluble anticancer agent camptothecin, *Journal of Drug Targeting*, 12, 373–384, 2004.
39. Kohori, F., Yokoyama, M., Sakai, K., and Okano, T., Process design for efficient and controlled drug incorporation into polymeric micelle carrier systems, *Journal of Controlled Release*, 78, 155–163, 2002.
40. Jette, K. K., Law, D., Schmitt, E. A., and Kwon, G. S., Preparation and drug loading of poly(ethylene glycol)-*block*-poly(epsilon-caprolactone) micelles through the evaporation of a cosolvent azeotrope, *Pharmaceutical Research*, 21, 1184–1191, 2004.
41. Shuai, X., Ai, H., Nasongkla, N., Kim, S., and Gao, J., Micellar carriers based on block copolymers of poly(epsilon-caprolactone) and poly(ethylene glycol) for doxorubicin delivery, *Journal of Controlled Release*, 98, 415–426, 2004.
42. Zhao, J. X., Allen, C., and Eisenberg, A., Partitioning of pyrene between "crew cut" block copolymer micelles and H2O/DMF solvent mixtures, *Macromolecules*, 30, 7143–7150, 1997.
43. Sharma, P. K. and Bhatia, S. R., Effect of anti-inflammatories on pluronic (R) F127: Micellar assembly, gelation and partitioning, *International Journal of Pharmaceutics*, 278, 361–377, 2004.
44. Barreiro-Iglesias, R., Bromberg, L., Temchenko, M., Hatton, T. A., Concheiro, A., and Alvarez-Lorenzo, C., Solubilization and stabilization of camptothecin in micellar solutions of pluronic-*g*-poly(acrylic acid) copolymers, *Journal of Controlled Release*, 97, 537–549, 2004.
45. Lim Soo, P., Luo, L., Maysinger, D., and Eisenberg, A., Incorporation and release of hydrophobic probes in biocampatible polycaprolactone-*block*-poly(ethylene oxide) micelles: Implications for drug delivery, *Langmuir*, 18, 9996–10004, 2002.
46. Letchford, K., Zastre, J., Liggins, R., and Burt, H., Synthesis and micellar characterization of short block length methoxy poly(ethylene glycol)-*block*-poly(caprolactone) diblock copolymers, *Colloids and Surfaces B: Biointerfaces*, 35, 81–91, 2004.
47. Choucair, A. and Eisenberg, A., Interfacial solubilization of model amphiphilic molecules in block copolymer micelles, *Journal of the American Chemical Society*, 125, 11993–12000, 2003.

48. Goldenberg, M. S., Bruno, L. A., and Rennwantz, E. L., Determination of solubilization sites and efficiency of water-insoluble agents in ethylene oxide-containing nonionic micelles, *Journal of Colloid and Interface Science*, 158, 351–363, 1993.

49. Nagarajan, R., Solubilization of "guest" molecules into polymeric aggregates, *Polymers for Advanced Technologies*, 12, 23–43, 2001.

50. Nagarajan, R. and Ganesh, K., Comparison of solubilization of hydrocarbons in (PEO–PPO) diblock versus (PEO–PPO–PEO) triblock copolymer micelles, *Journal of Colloid and Interface Science*, 184, 489–499, 1996.

51. Kozlov, M. Y., Melik-Nubarov, N. S., Batrakova, E. V., and Kabanov, A. V., Relationship between pluronic block copolymer structure, critical micellization concentration and partitioning coefficients of low molecular mass solutes, *Macromolecules*, 33, 3305–3313, 2000.

52. Xing, L. and Mattice, W. L., Strong solubilization of small molecules by triblock-copolymer micelles in selective solvents, *Macromolecules*, 30, 1711–1717, 1997.

53. Teng, Y., Morrison, M. E., Munk, P., Webber, S. E., and Prochazka, K., Release kinetics studies of aromatic molecules into water from block polymer micelles, *Macromolecules*, 31, 3578–3587, 1998.

54. Lin, W. J., Juang, L. W., and Lin, C. C., Stability and release performance of a series of pegylated copolymeric micelles, *Pharmaceutical Research*, 20, 668–673, 2003.

55. Gadelle, F., Koros, W. J., and Schechter, R. S., Solubilization of aromatic solutes in block copolymers, *Macromolecules*, 28, 4883–4892, 1995.

56. Lee, J., Cho, E. C., and Cho, K., Incorporation and release behavior of hydrophobic drug in functionalized poly(D,L-lactide)-*block*-poly(ethylene oxide) micelles, *Journal of Controlled Release*, 94, 323–335, 2004.

57. Lavasanifar, A., Samuel, J., and Kwon, G. S., Micelles self-assembled from poly(ethylene oxide)-*block*-poly(*N*-hexyl stearate L-aspartamide) by a solvent evaporation method: Effect on the solubilization and haemolytic activity of amphotericin B, *Journal of Controlled Release*, 77, 155–160, 2001.

58. Adams, M. L. and Kwon, G. S., Relative aggregation state and hemolytic activity of amphotericin B encapsulated by poly(ethylene oxide)-*block*-poly(*N*-hexyl-L-aspartamide)-acyl conjugate micelles: Effects of acyl chain length, *Journal of Controlled Release*, 87, 23–32, 2003.

59. Oh, K. T., Bronich, T. K., and Kabanov, A. V., Micellar formulations for drug delivery based on mixtures of hydrophobic and hydrophilic pluronic((R)) block copolymers, *Journal of Controlled Release*, 94, 411–422, 2004.

60. Yokoyama, M., Satoh, A., Sakurai, Y., Okano, T., Matsumura, Y., Kakizoe, T., and Kataoka, K., Incorporation of water-insoluble anticancer drug into polymeric micelles and control of their particle size, *Journal of Controlled Release*, 55, 219–229, 1998.

61. Bromberg, L. and Magner, E., Release of hydrophobic compounds from micellar solutions of hydrophobically modified polyelectrolytes, *Langmuir*, 15, 6792–6798, 1999.

62. Kim, S. Y., Ha, J. C., and Lee, Y. M., Poly(ethylene oxide)–poly(propylene oxide)–poly(ethylene oxide)/poly(epsilon-caprolactone) (PCL) amphiphilic block copolymeric nanospheres. II. Thermo-responsive drug release behaviors, *Journal of Controlled Release*, 65, 345–358, 2000.

63. Kim, S. Y., Kim, J. H., Kim, D., An, J. H., Lee, D. S., and Kim, S. C., Drug-releasing kinetics of MPEG/PLLA block copolymer micelles with different PLLA block lengths, *Journal of Applied Polymer Science*, 82, 2599–2605, 2001.

64. Cho, Y. W., Lee, J., Lee, S. C., Huh, K. M., and Park, K., Hydrotropic agents for study of in vitro paclitaxel release from polymeric micelles, *Journal of Controlled Release*, 97, 249–257, 2004.

65. Lim Soo, P., Lovric, J., Davidson, P., Maysinger, D., and Eisenberg, A., Polycaprolactone-*block*-poly(ethylene oxide) micelles: A nanodelivery system for 17beta-estradiol, *Molecular Pharmacology*, 2, 519–527, 2005.

66. Gref, R., Minamitake, Y., Peracchia, M. T., Trubetskoy, V., Torchilin, V., and Langer, R., Biodegradable long-circulating polymeric nanospheres, *Science*, 263, 1600–1603, 1994.

67. Nah, J. W., Jeong, Y. I., and Cho, C. S., Norfloxacin release from polymeric micelle of poly(gamma-benzyl L-glutamate) poly(ethylene oxide) poly(gamma-benzylL-glutamate) block copolymer, *Bulletin of the Korean Chemical Society*, 19, 962–967, 1998.

68. Ryu, J., Jeong, Y. I., Kim, I. S., Lee, J. H., Nah, J. W., and Kim, S. H., Clonazepam release from core-shell type nanoparticles of poly(epsilon-caprolactone)/poly(ethylene glycol)/poly(epsilon-caprolactone) triblock copolymers, *International Journal of Pharmaceutics*, 200, 231–242, 2000.

69. Kim, S. Y., Shin, I. L. G., Lee, Y. M., Cho, C. S., and Sung, Y. K., Methoxy poly(ethylene glycol) and epsilon-caprolactone amphiphilic block copolymeric micelle containing indomethacin. II. Micelle formation and drug release behaviours, *Journal of Controlled Release*, 51, 13–22, 1998.

70. Nah, J. W., Jeong, Y. I., Cho, C. S., and Kim, S. I., Drug-delivery system based on core-shell-type nanoparticles composed of poly(gamma-benzyl-L-glutamate) and poly(ethylene oxide), *Journal of Applied Polymer Science*, 75, 1115–1126, 2000.

71. Leroux, J. C., Allemann, E., DeJaeghere, F., Doelker, E., and Gurny, R., Biodegradable nanoparticles—from sustained release formulations to improved site specific drug delivery, *Journal of Controlled Release*, 39, 339–350, 1996.

72. Kwon, G. S., Naito, M., Kataoka, K., Yokoyama, M., Sakurai, Y., and Okano, T., Block copolymer micelles as vehicles for hydrophobic drugs, *Colloids and Surfaces B: Biointerfaces*, 2, 429–434, 1994.

73. Saltzman, W. M., Drug delivery: Engineering principles for drug delivery, *Topics in Chemical Engineering*, Oxford University Press, New York, 372, 2001.

74. Doxorubicin, U.S. National Library of Medicine and National Institutes of Health http://www.nlm.nih.gov/medlineplus/druginfo/medmaster/a682221.html.

75. Lipp, H.-P. and Bokemeyer, C., Anticancer drug toxicity, In *Anthracyclines and Other Intercalating Agents*, Lipp, H. P., Ed., Marcel Dekker Inc., New York, pp. 81–113, 1999.

76. Kwon, G. S., Yokoyama, M., Okano, T., Sakurai, Y., and Kataoka, K., Biodistribution of micelle-forming polymer–drug conjugates, *Pharmaceutical Research*, 10, 970–974, 1993.

77. Yoo, H. S. and Park, T. G., Biodegradable polymeric micelles composed of doxorubicin conjugated PLGA–PEG block copolymer, *Journal of Controlled Release*, 70, 63–70, 2001.

78. Yoo, H. S. and Park, T. G., Folate receptor targeted biodegradable polymeric doxorubicin micelles, *Journal of Controlled Release*, 96, 273–283, 2004.

79. Rapoport, N., Pitt, W. G., Sun, H., and Nelson, J. L., Drug delivery in polymeric micelles: From in vitro to in vivo, *Journal of Controlled Release*, 91, 85–95, 2003.

80. Pruitt, J. D. and Pitt, W. G., Sequestration and ultrasound-induced release of doxorubicin from stabilized pluronic P105 micelles, *Drug Delivery*, 9, 253–258, 2002.

81. Danson, S., Ferry, D., Alakhov, V., Margison, J., Kerr, D., Jowle, D., Brampton, M., Halbert, G., and Ranson, M., Phase I dose escalation and pharmacokinetic study of pluronic polymer-bound doxorubicin (SP1049C) in patients with advanced cancer, *British Journal of Cancer*, 90, 2085–2091, 2004.

82. Matsumura, Y., Hamaguchi, T., Ura, T., Muro, K., Yamada, Y., Shimada, Y., Shirao, K., Okusaka, T., Ueno, H., Ikeda, M., and Watanabe, N., Phase I clinical trial and pharmacokinetic evaluation of NK911, a micelle-encapsulated doxorubicin, *British Journal of Cancer*, 91, 1775–1781, 2004.

83. Kwon, G., Suwa, S., Yokoyama, M., Okano, T., Sakurai, Y., and Kataoka, K., Enhanced tumor accumulation and prolonged circulation times of micelle-forming poly(ethylene oxide-aspartate) block copolymer–adriamycin conjugates, *Journal of Controlled Release*, 29, 17–23, 1994.

84. Dollery, C. T., *Therapeutic Drugs*, Churchill Livingstone, Edinburgh, 1999.

85. Yokoyama, M., Miyauchi, M., Yamada, N., Okano, T., Sakurai, Y., Kataoka, K., and Inoue, S., Polymer micelles as novel drug carrier—adriamycin-conjugated poly(ethylene glycol) poly(aspartic acid) block copolymer, *Journal of Controlled Release*, 11, 269–278, 1990.

86. Fukushima, S., Machida, M., Akutsu, T., Skimizu, K., Tanaka, S., Okamoto, K., and Mashiba, H., Roles of adriamycin and adriamycin dimer in antitumor activity of the polymeric micelle carrier system, *Colloids and Surfaces B: Biointerfaces*, 16, 227–236, 1999.

87. Paclitaxel, RxList Inc., 2005 http://www.rxlist.com/cgi/generic/paclitaxel.htm.

88. Liggins, R. T., Hunter, W. L., and Burt, H. M., Solid-state characterization of paclitaxel, *Journal of Pharmaceutical Sciences*, 86, 1458–1463, 1997.

89. Estimated by CAChe 6.1 (computer-aided molecular design modeling software).

90. Kim, T. Y., Kim, D. W., Chung, J. Y., Shin, S. G., Kim, S. C., Heo, D. S., Kim, N. K., and Bang, Y. J., Phase I and pharmacokinetic study of genexol-PM, a cremophor-free, polymeric micelle-formulated paclitaxel, in patients with advanced malignancies, *Clinical Cancer Research*, 10, 3708–3716, 2004.

91. Liggins, R. T. and Burt, H. M., Polyether–polyester diblock copolymers for the preparation of paclitaxel loaded polymeric micelle formulations, *Advanced Drug Delivery Reviews*, 54, 191–202, 2002.

92. Le Garrec, D., Gori, S., Luo, L., Lessard, D., Smith, D. C., Yessine, M. A., Ranger, M., and Leroux, J. C., Poly(*N*-vinylpyrrolidone)-*block*-poly(D,L-lactide) as a new polymeric solubilizer for hydrophobic anticancer drugs: In vitro and in vivo evaluation, *Journal of Controlled Release*, 99, 83–101, 2004.

93. Shuai, X., Merdan, T., Schaper, A. K., Xi, F., and Kissel, T., Core-cross-linked polymeric micelles as paclitaxel carriers, *Bioconjugate Chemistry*, 15, 441–448, 2004.

94. Kim, S. Y. and Lee, Y. M., Taxol-loaded block copolymer nanospheres composed of methoxy poly(ethylene glycol) and poly(epsilon-caprolactone) as novel anticancer drug carriers, *Biomaterials*, 22, 1697–1704, 2001.

95. Park, E. K., Lee, S. B., and Lee, Y. M., Preparation and characterization of methoxy poly(ethylene glycol)/poly(epsilon-caprolactone) amphiphilic block copolymeric nanospheres for tumor-specific folate-mediated targeting of anticancer drugs, *Biomaterials*, 26, 1053–1061, 2005.

96. Park, S. R., Oh, D. Y., Kim, D. W., Kim, T. Y., Heo, D. S., Bang, Y. J., Kim, N. K., Kang, W.-K., Kim, H.-T., Im, S.-A., Kim, J.-H., and Kim, H.-K., A multi-center, late phase II clinical trial of genexol (paclitaxel) and cisplatin for patients with advanced gastric cancer, *Oncology Reports*, 12, 1059–1064, 2004.

97. Kim, S. C., Kim, D. W., Shim, Y. H., Bang, J. S., Oh, H. S., Wan Kim, S., and Seo, M. H., In vivo evaluation of polymeric micellar paclitaxel formulation: Toxicity and efficacy, *Journal of Controlled Release*, 72, 191–202, 2001.

98. Yu, J. J., Jeong, Y. I., Shim, Y. H., and Lim, G. T., Preparation of core-shell type nanoparticles of diblock copolymers of poly(L-lactide)/poly(ethylene glycol) and their characterization in vitro, *Journal of Applied Polymer Science*, 85, 2625–2634, 2002.

99. Hamaguchi, T., Matsumura, Y., Suzuki, M., Shimizu, K., Goda, R., Nakamura, I., Nakatomi, I., Yokoyama, M., Kataoka, K., and Kakizoe, T., NK105, a paclitaxel-incorporating micellar nanoparticle formulation, can extend in vivo antitumour activity and reduce the neurotoxicity of paclitaxel, *British Journal of Cancer*, 92, 1240–1246, 2005.

100. Watanabe, M., Kawano, K., Yokoyama, M., Opanasopit, P., Okano, T., and Maitani, Y., Preparation of camptothecin-loaded polymeric micelles and evaluation of their incorporation and circulation stability, *International Journal of Pharmaceutics*, 308, 183–189, 2006.

101. Nishiyama, N. and Kataoka, K., Preparation and characterization of size-controlled polymeric micelle containing *cis*-dichlorodiammineplatinum(II) in the core, *Journal of Controlled Release*, 74, 83–94, 2001.

102. Uchino, H., Matsumura, Y., Negishi, T., Koizumi, F., Hayashi, T., Honda, T., Nishiyama, N., Kataoka, K., Naito, S., and Kakizoe, T., Cisplatin-incorporating polymeric micelles (NC-6004) can reduce nephrotoxicity and neurotoxicity of cisplatin in rats, *British Journal of Cancer*, 93, 678–687, 2005.

103. Maysinger, D., Berezovska, O., Savic, R., Lim Soo, P., and Eisenberg, A., Block copolymers modify the internalization of micelle-incorporated probes into neural cells, *Biochimica et Biophysica Acta*, 1539, 205–217, 2001.

104. Allen, C., Yu, Y., Eisenberg, A., and Maysinger, D., Cellular internalization of PCL(20)-*b*-PEO(44) block copolymer micelles, *Biochimica et Biophysica Acta*, 1421, 32–38, 1999.

105. Savic, R., Luo, L., Eisenberg, A., and Maysinger, D., Micellar nanocontainers distribute to defined cytoplasmic organelles, *Science*, 300, 615–618, 2003.

106. Luo, L., Tam, J., Maysinger, D., and Eisenberg, A., Cellular internalization of poly(ethylene oxide)-*b*-poly(epsilon-caprolactone) diblock copolymer micelles, *Bioconjugate Chemistry*, 13, 1259–1265, 2002.

107. Zastre, J., Jackson, J., and Burt, H., Evidence for modulation of P-glycoprotein-mediated efflux by methoxypolyethylene glycol-*block*-polycaprolactone amphiphilic diblock copolymers, *Pharmaceutical Research*, 21, 1489–1497, 2004.

108. Alakhov, V., Moskaleva, E., Batrakova, E. V., and Kabanov, A. V., Hypersensitization of multidrug resistant human ovarian carcinoma cells by pluronic P85 block copolymer, *Bioconjugate Chemistry*, 7, 209–216, 1996.

109. Batrakova, E. V., Li, S., Elmquist, W. F., Miller, D. W., Alakhov, V. Y., and Kabanov, A. V., Mechanism of sensitization of MDR cancer cells by pluronic block copolymers: Selective energy depletion, *British Journal of Cancer*, 85, 1987–1997, 2001.

110. Batrakova, E. V., Han, H. Y., Alakhov, V., Miller, D. W., and Kabanov, A. V., Effects of pluronic block copolymers on drug absorption in Caco-2 cell monolayers, *Pharmaceutical Research*, 15, 850–855, 1998.

111. Zastre, J., Jackson, J., Bajwa, M., Liggins, R., Iqbal, F., and Burt, H., Enhanced cellular accumulation of a P-glycoprotein substrate, rhodamine-123, by Caco-2 cells using low molecular weight methoxypolyethylene glycol-*block*-polycaprolactone diblock copolymers, *European Journal of Pharmaceutics and Biopharmaceutics*, 54, 299–309, 2002.

112. Robinson, J. R. and Lee, V. H. L., Eds., Controlled drug delivery: Fundamentals and applications, *Drugs and the Pharmaceutical Sciences*, Vol. 29, Marcel Dekker, New York, pp. xix, 716, 1987.

113. Pastan, I. H. and Willingham, M. C., Eds., *Endocytosis*, Plenum Press, New York, pp. xviii, 326, 1985.

114. Torchilin, V. P., Targeted polymeric micelles for delivery of poorly soluble drugs, *Cellular and Molecular Life Sciences*, 61, 2549–2559, 2004.

115. Zeng, F., Lee, H., and Allen, C., Epidermal growth factor-conjugated poly(ethylene glycol)-*block*-poly(δ-valerolactone) copolymer micelles for targeted delivery of chemotherapeutics, Bioconjugate Chemistry, 2006, in Press.

116. Kakizawa, Y. and Kataoka, K., Block copolymer micelles for delivery of gene and related compounds, *Advanced Drug Delivery Reviews*, 54, 203–222, 2002.

117. Muniruzzaman, M., Marin, A., Luo, Y., Prestwich, G. D., Pitt, W. G., Husseini, G., and Rapoport, N. Y., Intracellular uptake of pluronic copolymer: Effects of the aggregation state, *Colloids and Surfaces B: Biointerfaces*, 25, 233–241, 2002.

118. Rapoport, N., Marin, A., Luo, Y., Prestwich, G. D., and Muniruzzaman, M. D., Intracellular uptake and trafficking of pluronic micelles in drug-sensitive and MDR cells: Effect on the intracellular drug localization, *Journal of Pharmaceutical Science*, 91, 157–170, 2002.

119. Husseini, G. A., Runyan, C. M., and Pitt, W. G., Investigating the mechanism of acoustically activated uptake of drugs from pluronic micelles, *BMC Cancer*, 2, 20, 2002.

120. Liu, J., Zeng, F., and Allen, C., Influence of serum protein on polycarbonate-based copolymer micelles as a delivery system for a hydrophobic anti-cancer agent, *Journal of Controlled Release*, 103, 481–497, 2005.

121. Pante, N. and Kann, M., Nuclear pore complex is able to transport macromolecules with diameters of about 39 nm, *Molecular Biology of the Cell*, 13, 425–434, 2002.

122. Liu, S. Q., Tong, Y. W., and Yang, Y. Y., Incorporation and in vitro release of doxorubicin in thermally sensitive micelles made from poly(*N*-isopropylacrylamide-*co*-*N*,*N*-dimethylacrylamide)-*b*-poly(D,L-lactide-*co*-glycolide) with varying compositions, *Biomaterials*, 26, 5064–5074, 2005.

123. Kabanov, A. V., Batrakova, E. V., and Miller, D. W., Pluronic block copolymers as modulators of drug efflux transporter activity in the blood–brain barrier, *Advanced Drug Delivery Reviews*, 55, 151–164, 2003.

124. Batrakova, E. V., Li, S., Miller, D. W., and Kabanov, A. V., Pluronic P85 increases permeability of a broad spectrum of drugs in polarized BBMEC and Caco-2 cell monolayers, *Pharmaceutical Research*, 16, 1366–1372, 1999.

125. Kabanov, A. V., Batrakova, E. V., and Alakhov, V. Y., Pluronic block copolymers for overcoming drug resistance in cancer, *Advanced Drug Delivery Reviews*, 54, 759–779, 2002.

126. Kabanov, A. V., Batrakova, E. V., Meliknubarov, N. S., Fedoseev, N. A., Dorodnich, T. Y., Alakhov, V. Y., Chekhonin, V. P., Nazarova, I. R., and Kabanov, V. A., A new class of drug carriers—micelles of poly(oxyethylene)–poly(oxypropylene) block copolymers as microcontainers for drug targeting from blood in brain, *Journal of Controlled Release*, 22, 141–157, 1992.

127. Miller, D. W., Batrakova, E. V., and Kabanov, A. V., Inhibition of multidrug resistance-associated protein (MRP) functional activity with pluronic block copolymers, *Pharmaceutical Research*, 16, 396–401, 1999.

128. Rapoport, N., Marin, A. P., and Timoshin, A. A., Effect of a polymeric surfactant on electron transport in HL-60 cells, *Archives of Biochemistry and Biophysics*, 384, 100–108, 2000.

129. Slepnev, V. I., Kuznetsova, L. E., Gubin, A. N., Batrakova, E. V., Alakhov, V., and Kabanov, A. V., Micelles of poly(oxyethylene)–poly(oxypropylene) block copolymer (pluronic) as a tool for low-molecular compound delivery into a cell: Phosphorylation of intracellular proteins with micelle incorporated [gamma-32P]ATP, *Biochemistry International*, 26, 587–595, 1992.

130. Regev, R., Assaraf, Y. G., and Eytan, G. D., Membrane fluidization by ether, other anesthetics, and certain agents abolishes P-glycoprotein ATPase activity and modulates efflux from multidrug-resistant cells, *European Journal of Biochemistry*, 259, 18–24, 1999.

131. Yamamoto, Y., Nagasaki, Y., Kato, Y., Sugiyama, Y., and Kataoka, K., Long-circulating poly(-ethylene glycol)–poly(D,L-lactide) block copolymer micelles with modulated surface charge, *Journal of Controlled Release*, 77, 27–38, 2001.

132. Poznansky, M. J. and Juliano, R. L., Biological approaches to the controlled delivery of drugs: A critical review, *Pharmacological Reviews*, 36, 277–336, 1984.

133. Drummond, D. C., Meyer, O., Hong, K., Kirpotin, D. B., and Papahadjopoulos, D., Optimizing liposomes for delivery of chemotherapeutic agents to solid tumors, *Pharmacological Reviews*, 51, 691–743, 1999.

134. Patel, H. M., Serum opsonins and liposomes: Their interaction and opsonophagocytosis, *Critical Reviews in Therapeutic Drug Carrier Systems*, 9, 39–90, 1992.

135. Moghimi, S. M. and Patel, H. M., Tissue specific opsonins for phagocytic cells and their different affinity for cholesterol-rich liposomes, *FEBS Letters*, 233, 143–147, 1988.

136. Ulrich, F. and Zilversmit, D. B., Release from alveolar macrophages of an inhibitor of phagocytosis, *American Journal of Physiology*, 218, 1118–1127, 1970.

137. Dunn, S. E., Brindley, A., Davis, S. S., Davies, M. C., and Illum, L., Polystyrene–poly(ethylene glycol) (PS–PEG2000) particles as model systems for site specific drug delivery. 2. The effect of PEG surface density on the in vitro cell interaction and in vivo biodistribution, *Pharmaceutical Research*, 11, 1016–1022, 1994.

138. Batrakova, E. V., Li, S., Li, Y., Alakhov, V. Y., Elmquist, W. F., and Kabanov, A. V., Distribution kinetics of a micelle-forming block copolymer pluronic P85, *Journal of Controlled Release*, 100, 389–397, 2004.

139. Grindel, J. M., Jaworski, T., Piraner, O., Emanuele, R. M., and Balasubramanian, M., Distribution, metabolism, and excretion of a novel surface-active agent, purified poloxamer 188, in rats, dogs, and humans, *Journal of Pharmaceutical Science*, 91, 1936–1947, 2002.

140. Novakova, K., Laznicek, M., Rypacek, F., and Machova, L., I-125-labeled PLA/PEO block copolymer: Biodistribution studies in rats, *Journal of Bioactive and Compatible Polymers*, 17, 285–296, 2002.

141. Ishihara, K., Nomura, H., Mihara, T., Kurita, K., Iwasaki, Y., and Nakabayashi, N., Why do phospholipid polymers reduce protein adsorption? *Journal of Biomedical Materials Research*, 39, 323–330, 1998.

142. Devine, D. V. and Marjan, J. M., The role of immunoproteins in the survival of liposomes in the circulation, *Critical Reviews in Therapeutic Drug Carrier Systems*, 14, 105–131, 1997.

143. Patel, H. M. and Moghimi, S. M., Serum-mediated recognition of liposomes by phagocytic cells of the reticuloendothelial system—the concept of tissue specificity, *Advanced Drug Delivery Reviews*, 32, 45–60, 1998.

144. Tan, J. S., Butterfield, D. E., Voycheck, C. L., Caldwell, K. D., and Li, J. T., Surface modification of nanoparticles by PEO/PPO block copolymers to minimize interactions with blood components and prolong blood circulation in rats, *Biomaterials*, 14, 823–833, 1993.

145. Allen, C., Maysinger, D., and Eisenberg, A., Nano-engineering block copolymer aggregates for drug delivery, *Colloids and Surfaces B: Biointerfaces*, 16, 3–27, 1999.

146. Seki, J., Sonoke, S., Saheki, A., Koike, T., Fukui, H., Doi, M., and Mayumi, T., Lipid transfer protein transports compounds from lipid nanoparticles to plasma lipoproteins, *International Journal of Pharmaceutics*, 275, 239–248, 2004.

147. Ramaswamy, M., Zhang, X., Burt, H. M., and Wasan, K. M., Human plasma distribution of free paclitaxel and paclitaxel associated with diblock copolymers, *Journal of Pharmaceutical Science*, 86, 460–464, 1997.

148. Mizumura, Y., Matsumura, Y., Hamaguchi, T., Nishiyama, N., Kataoka, K., Kawaguchi, T., Hrushesky, W. J., Moriyasu, F., and Kakizoe, T., Cisplatin-incorporated polymeric micelles eliminate nephrotoxicity, while maintaining antitumor activity, *Japan Journal of Cancer Research*, 92, 328–336, 2001.

149. Gianni, L., Kearns, C. M., Giani, A., Capri, G., Vigano, L., Lacatelli, A., Bonadonna, G., and Egorin, M. J., Nonlinear pharmacokinetics and metabolism of paclitaxel and its pharmacokinetic/pharmacodynamic relationships in humans, *Journal of Clinical Oncology*, 13, 180–190, 1995.

150. Gao, Z. G., Fain, H. D., and Rapoport, N., Controlled and targeted tumor chemotherapy by micellar-encapsulated drug and ultrasound, *Journal of Controlled Release*, 102, 203–222, 2005.

151. Husseini, G. A., Myrup, G. D., Pitt, W. G., Christensen, D. A., and Rapoport, N. Y., Factors affecting acoustically triggered release of drugs from polymeric micelles, *Journal of Controlled Release*, 69, 43–52, 2000.

152. Marin, A., Muniruzzaman, M., and Rapoport, N., Mechanism of the ultrasonic activation of micellar drug delivery, *Journal of Controlled Release*, 75, 69–81, 2001.

153. Chung, J. E., Yokoyama, M., and Okano, T., Inner core segment design for drug delivery control of thermo-responsive polymeric micelles, *Journal of Controlled Release*, 65, 93–103, 2000.

154. Kohori, F., Sakai, K., Aoyagi, T., Yokoyama, M., Sakurai, Y., and Okano, T., Preparation and characterization of thermally responsive block copolymer micelles comprising poly(N-isopropylacrylamide-b-DL-lactide), *Journal of Controlled Release*, 55, 87–98, 1998.

155. Chung, J. E., Yokoyama, M., Yamato, M., Aoyagi, T., Sakurai, Y., and Okano, T., Thermo-responsive drug delivery from polymeric micelles constructed using block copolymers of poly(N-isopropylacrylamide) and poly(butylmethacrylate), *Journal of Controlled Release*, 62, 115–127, 1999.

156. Chung, J. E., Yokoyama, M., Aoyagi, T., Sakurai, Y., and Okano, T., Effect of molecular architecture of hydrophobically modified poly(N-isopropylacrylamide) on the formation of thermoresponsive core-shell micellar drug carriers, *Journal of Controlled Release*, 53, 119–130, 1998.

157. Cammas, S., Suzuki, K., Sonc, C., Sakurai, Y., Kataoka, K., and Okano, T., Thermo-responsive polymer nanoparticles with a core-shell micelle structure as site-specific drug carriers, *Journal of Controlled Release*, 48, 157–164, 1997.

158. Taillefer, J., Jones, M. C., Brasseur, N., van Lier, J. E., and Leroux, J. C., Preparation and characterization of pH-responsive polymeric micelles for the delivery of photosensitizing anticancer drugs, *Journal of Pharmaceutical Science*, 89, 52–62, 2000.

159. Houlton, S., Blocking the way forward, 2003 http://www.users.globalnet.uk/~sarahx/articles/mcblocks.htm.

160. Zhang, G. D., Harada, A., Nishiyama, N., Jiang, D. L., Koyama, H., Aida, T., and Kataoka, K., Polyion complex micelles entrapping cationic dendrimer porphyrin: Effective photosensitizer for photodynamic therapy of cancer, *Journal of Controlled Release*, 93, 141–150, 2003.

161. Dolmans, D. E., Fukumura, D., and Jain, R. K., Photodynamic therapy for cancer, *Nature Reviews Cancer*, 3, 380–387, 2003.

162. van Nostrum, C. F., Polymeric micelles to deliver photosensitizers for photodynamic therapy, *Advanced Drug Delivery Reviews*, 56, 9–16, 2004.

163. Jang, W. D., Nishiyama, N., Zhang, G. D., Harada, A., Jiang, D. L., Kawauchi, S., Morimoto, Y., Kikuchi, M., Koyama, H., Aida, T., and Kataoka, K., Supramolecular nanocarrier of anionic dendrimer porphyrins with cationic block copolymers modified with polyethylene glycol to enhance intracellular photodynamic efficacy, *Angewandte Chemie International Edition in English*, 44, 419–423, 2005.

164. Ideta, R., Tasaka, F., Jang, W. D., Nishiyama, N., Zhang, G. D., Harada, A., Yanagi, Y., Tamaki, Y., Aida, T., and Kataoka, K., Nanotechnology-based photodynamic therapy for neovascular disease using a supramolecular nanocarrier loaded with a dendritic photosensitizer, *Nano Letters*, 5, 2426–2431, 2005.

165. Taillefer, J., Brasseur, N., van Lier, J. E., Lenaerts, V., Le Garrec, D., and Leroux, J. C., In-vitro and in-vivo evaluation of pH-responsive polymeric micelles in a photodynamic cancer therapy model, *Journal of Pharmacy and Pharmacology*, 53, 155–166, 2001.

166. Le Garrec, D., Taillefer, J., Van Lier, J. E., Lenaerts, V., and Leroux, J. C., Optimizing pH-responsive polymeric micelles for drug delivery in a cancer photodynamic therapy model, *Journal of Drug Targeting*, 10, 429–437, 2002.

167. Woodburn, K. W., Vardaxis, N. J., Hill, J. S., Kaye, A. H., Reiss, J. A., and Phillips, D. R., Evaluation of porphyrin characteristics required for photodynamic therapy, *Photochemistry and Photobiology*, 55, 697–704, 1992.

168. Mamot, C., Drummond, D. C., Greiser, U., Hong, K., Kirpotin, D. B., Marks, J. D., and Park, J. W., Epidermal growth factor receptor (EGFR)-targeted immunoliposomes mediate specific and efficient drug delivery to EGFR- and EGFRvIII-overexpressing tumor cells, *Cancer Research*, 63, 3154–3161, 2003.

169. Torchilin, V. P., Lukyanov, A. N., Gao, Z. G., and Papahadjopoulos-Sternberg, B., Immunomicelles: Targeted pharmaceutical carriers for poorly soluble drugs, *Proceedings of the National Academy of Sciences of the United States of America*, 100, 6039–6044, 2003.

170. Torchilin, V. P., Fluorescence microscopy to follow the targeting of liposomes and micelles to cells and their intracellular fate, *Advanced Drug Delivery Reviews*, 57, 95–109, 2005.

171. Farokhzad, O. C., Jon, S. Y., Khademhosseini, A., Tran, T. N. T., LaVan, D. A., and Langer, R., Nanopartide-aptamer bioconjugates: A new approach for targeting prostate cancer cells, *Cancer Research*, 64, 7668–7672, 2004.

172. Lee, H., Jang, I. H., Ryu, S. H., and Park, T. G., N-terminal site-specific mono-PEGylation of epidermal growth factor, *Pharmaceutical Research*, 20, 818–825, 2003.

173. Mendelsohn, J. and Baselga, J., The EGF receptor family as targets for cancer therapy, *Oncogene*, 19, 6550–6565, 2000.

174. Wakebayashi, D., Nishiyama, N., Yamasaki, Y., Itaka, K., Kanayama, N., Harada, A., Nagasaki, Y., and Kataoka, K., Lactose-conjugated polyion complex micelles incorporating plasmid DNA as a targetable gene vector system: Their preparation and gene transfecting efficiency against cultured HepG2 cells, *Journal of Controlled Release*, 95, 653–664, 2004.

175. Yasugi, K., Nakamura, T., Nagasaki, Y., Kato, M., and Kataoka, K., Sugar-installed polymer micelles: Synthesis and micellization of poly(ethylene glycol)–poly(D,L-lactide) block copolymers having sugar groups at the PEG chain end, *Macromolecules*, 32, 8024–8032, 1999.

176. Jule, E., Nagasaki, Y., and Kataoka, K., Lactose-installed poly(ethylene glycol)–poly(D,L-lactide) block copolymer micelles exhibit fast-rate binding and high affinity toward a protein bed simulating a cell surface. A surface plasmon resonance study, *Bioconjugate Chemistry*, 14, 177–186, 2003.

177. Lim, D. W., Yeom, Y. I., and Park, T. G., Poly(DMAEMA-NVP)-*b*-PEG-galactose as gene delivery vector for hepatocytes, *Bioconjugate Chemistry*, 11, 688–695, 2000.

178. Kursa, M., Walker, G. F., Roessler, V., Ogris, M., Roedl, W., Kircheis, R., and Wagner, E., Novel shielded transferrin–polyethylene glycol–polyethylenimine/DNA complexes for systemic tumor-targeted gene transfer, *Bioconjugate Chemistry*, 14, 222–231, 2003.

179. Vinogradov, S., Batrakova, E., Li, S., and Kabanov, A., Polyion complex micelles with protein-modified corona for receptor-mediated delivery of oligonucleotides into cells, *Bioconjugate Chemistry*, 10, 851–860, 1999.

180. Ogris, M., Walker, G., Blessing, T., Kircheis, R., Wolschek, M., and Wagner, E., Tumor-targeted gene therapy: Strategies for the preparation of ligand-polyethylene glycol-polyethylenimine/DNA complexes, *Journal of Controlled Release*, 91, 173–181, 2003.

181. Lee, E. S., Na, K., and Bae, Y. H., Polymeric micelle for tumor pH and folate-mediated targeting, *Journal of Controlled Release*, 91, 103–113, 2003.

182. Merdan, T., Callahan, J., Peterson, H., Bakowsky, U., Kopeckova, P., Kissel, T., and Kopecek, J., Pegylated polyethylenimine-Fab′ antibody fragment conjugates for targeted gene delivery to human ovarian carcinoma cells, *Bioconjugate Chemistry*, 14, 989–996, 2003.

183. Kabanov, A. V. and Alakhov, V. Y., Amphiphilic Block Copolymers: Self-Assemble and Applications In *Micelles of Amphiphilic Block Copolymers as Vehicles for Drug Delivery*, Alexandris, P. and Lindman, B., Eds., Elsevier, The Netherlands, 1997.

184. Lin, S. Y., Makino, K., Xia, W. Y., Matin, A., Wen, Y., Kwong, K. Y., Bourguignon, L., and Hung, M. C., Nuclear localization of EGF receptor and its potential new role as a transcription factor, *Nature Cell Biology*, 3, 802–808, 2001.

185. Kircheis, R., Kichler, A., Wallner, G., Kursa, M., Ogris, M., Felzmann, T., Buchberger, M., and Wagner, E., Coupling of cell-binding ligands to polyethylenimine for targeted gene delivery, *Gene Therapy*, 4, 409–418, 1997.

186. Blessing, T., Kursa, M., Holzhauser, R., Kircheis, R., and Wagner, E., Different strategies for formation of PEGylated EGF-conjugated PEI/DNA complexes for targeted gene delivery, *Bioconjugate Chemistry*, 12, 529–537, 2001.

187. Gillies, E. R. and Frechet, J. M. J., Development of acid-sensitive copolymer micelles for drug delivery, *Pure and Applied Chemistry*, 76, 1295–1307, 2004.

188. Sant, V. P., Smith, D., and Leroux, J. C., Enhancement of oral bioavailability of poorly water-soluble drugs by poly(ethylene glycol)-*block*-poly(alkyl acrylate-*co*-methacrylic acid) self-assemblies, *Journal of Controlled Release*, 104, 289–300, 2005.

189. Tannock, I. F. and Rotin, D., Acid pH in tumors and its potential for therapeutic exploitation, *Cancer Research*, 49, 4373–4384, 1989.

190. Lee, E. S., Na, K., and Bae, Y. H., Super pH-sensitive multifunctional polymeric micelle, *Nano Letters*, 5, 325–329, 2005.

191. Bae, Y., Fukushima, S., Harada, A., and Kataoka, K., Design of environment-sensitive supramolecular assemblies for intracellular drug delivery: Polymeric micelles that are responsive to intracellular pH change, *Angewandte Chemie International Edition in English*, 42, 4640–4643, 2003.

192. Bae, Y., Nishiyama, N., Fukushima, S., Koyama, H., Yasuhiro, M., and Kataoka, K., Preparation and biological characterization of polymeric micelle drug carriers with intracellular pH-triggered drug release property: Tumor permeability, controlled subcellular drug distribution, and enhanced in vivo antitumor efficacy, *Bioconjugate Chemistry*, 16, 122–130, 2005.

193. Lee, E. S., Shin, H. J., Na, K., and Bae, Y. H., Poly(L-histidine)–PEG block copolymer micelles and pH-induced destabilization, *Journal of Controlled Release*, 90, 363–374, 2003.

194. Gao, Z. G., Lee, D. H., Kim, D. I., and Bae, Y. H., Doxorubicin loaded pH-sensitive micelle targeting acidic extracellular pH of human ovarian A2780 tumor in mice, *Journal of Drug Targeting*, 13, 391–397, 2005.

18 PEO-Modified Poly(L-Amino Acid) Micelles for Drug Delivery

Xiao-Bing Xiong, Hamidreza Montazeri Aliabadi, and Afsaneh Lavasanifar

CONTENTS

18.1 Introduction 358
18.2 Synthesis of PEO-*b*-PLAA Block Copolymers 358
18.3 Micellization of PEO-*b*-PLAA Block Copolymers 361
    18.3.1 Dialysis Method 362
    18.3.2 Solvent Evaporation Method 362
    18.3.3 Co-Solvent Evaporation Method 362
18.4 Rational Design and Functional Properties of PEO-*b*-PLAA Micelles in
Drug Delivery 363
    18.4.1 The PEO Shell 363
    18.4.2 The PLAA Core 363
    18.4.3 Micellar Dimensions 363
    18.4.4 Micellar Stability 367
    18.4.5 Drug Incorporation and Release Properties 368
18.5 PEO-*b*-PLAA Micelles for the Delivery of Anti-Cancer Drugs 369
    18.5.1 Cyclophosphamide 369
    18.5.2 Doxorubicin 369
    18.5.3 Paclitaxel 370
    18.5.4 Cisplatin and Its Derivative 370
    18.5.5 Methotrexate 371
    18.5.6 KRN5500 371
    18.5.7 Camptothecin 372
18.6 PEO-*b*-PLAA Micelles for the Delivery of Other Therapeutic Agents ... 372
    18.6.1 Amphotericin B 372
    18.6.2 Indomethacin 373
18.7 PEO-*b*-PLAA Micelles for Active Drug Targeting 373
18.8 PEO-*b*-PLAA Micelles for Gene Delivery 374
References 376

18.1 INTRODUCTION

Polymeric micelles are core/shell structures formed through the self-assembly of amphiphilic block copolymers. The nanoscopic dimension as well as unique properties offered by separated core and shell domains in the structure of polymeric micelles has made them one of the most promising carriers for passive or active drug targeting in cancer. The nanoscopic size of polymeric micelles makes the carrier unrecognizable by the phagocytic cells of the reticuloendothelial system (RES), elongating their blood circulation, and facilitating the carrier's extravasation from tumor vasculature. The small size of polymeric micelles is also expected to ease penetration of the carrier within the tumor tissue and further internalization of polymeric micelles into the tumor cells. Hydrophobic core of polymeric micelles provides an excellent host for the incorporation and stabilization of anti-cancer agents that are mostly hydrophobic. Steric effect induced by the dense hydrophilic brush on the micellar surface protects the carrier against attachment of proteins on the micellar surface, avoiding early uptake and clearance of the carrier by RES leading to prolonged circulation time and higher accumulation of the carrier and the encapsulated drug in selective tissues that have leaky vasculature (e.g., tumor or inflammation sites). Polymeric micelles have been the focus of several reviews in recent years.[1–11]

Among different micelle-forming block copolymers developed to date, those with poly(ethylene oxide) (PEO) as the shell-forming block and poly(L-amino acid)s (PLAA)s and poly(ester)s as the core-forming block are in the front line of drug development. The placement is owed to the biocompatibility of the PEO and biodegradability of PLAA and poly(ester) structures. The primary advantage of PEO-*b*-PLAA block copolymers over PEO-*b*-poly(esters) is the chemical flexibility of the PLAA structure that makes nano-engineering of the carrier a feasible approach.[4] To date, research on PEO-*b*-PLAA for drug delivery has been mainly conducted on amino acids with functional side groups in their chemical structure, including L-aspartic acid, L-glutamic acid, L-lysine, and L-histidine. The presence of free functional side groups on the PLAA block provide sites for the attachment of drugs, drug compatible moieties, or charged therapeutics such as DNA. Moreover, a systemic alteration in the structure of the core-forming block also may be used to better control the extent of drug loading, release, or activation.

Chemical modifications in the structure of the PLAA block have led to the development of optimal PEO-*b*-PLAA-based polymeric micellar formulation for the delivery of a number of potent therapeutic agents, such as doxorubicin (DOX),[12–23] paclitaxel (PTX),[24] cisplatin (CDDP),[25–32] amphotericin B (AmB),[33–41] etc. To this point, three PEO-*b*-PLAA-based polymeric micellar formulations (all for the solubilization and delivery of anti-cancer agents) have successfully passed the phase of bench-top development and advanced to the stage of clinical evaluations (Table 18.1).[23,24,32] This chapter provides an overview on the preparation; micellar properties and biological performance of PEO-modified PLAA- based polymeric micellar formulations. Emphasis has been placed on the application of PEO-*b*-PLAA- micelles for the targeted delivery of anti-cancer agents. At the end, new advancements in the field, including a second generation of PEO-*b*-PLAA micelles (polymeric micelles for active targeting) and development of PEO-*b*-PLAA micelles for gene delivery are briefly discussed.

18.2 SYNTHESIS OF PEO-*b*-PLAA BLOCK COPOLYMERS

The traditional method for the synthesis of PLAA relies on the ring-opening polymerization (ROP) of α-amino acid-*N*-carboxyanhydrides (NCAs) (Figure 18.1).[42] NCAs can be pre-prepared from α-amino acids using a solution of phosgene in THF by the Fuchs–Farthing method.[43,44] NCA polymerization can be initiated with different nucleophiles or bases, the most common being primary amines and alkoxide anions.[45–48] This synthetic method is appealing to the pharmaceutical industry because it involves simple reagents, and it is economical; high molecular weight polymers can be prepared in both good yield and large quantity by this method.

TABLE 18.1
Most Common PEO-Modified PLAA Micelles Investigated for Drug Delivery

Drug Incorporation Method	PLAA Block	Incorporated Drug	Latest Reported Phase of Progress	References
Chemical conjugation[a]	P(Lys)	Cyclophosphamid sulfide	Development	96
	P(Asp)	Doxorubicin	Development	12, 18, 22, 98, 100
	PHEA	Methotrexate	Development	33, 87, 90
Physical encapsulation[b]	PBLA	Doxorubicin	Development	13, 94, 95
	P(Asp)-DOX	Doxorubicin	Clinical trials phase II	12, 17, 21, 23, 100
	PBLG	Doxorubicin	Development	133
	PPBA	Paclitaxel	Clinical trials phase I	24
	PBL-C16(Asp)	KRN-5500	Development	65, 110
	PBLA	Camptothecin	Development	113
	P(n-butyl-L-Asp)	Camptothecin	Development	113
	P(lauryl-L-Asp)	Camptothecin	Development	113
	P(methylnaphtyl-L-Asp)	Camptothecin	Development	113
	PBLA	Amphotericin B	Development	40, 41
	PHSA	Amphotericin B	Development	33–39
	PBLA	Indomethacin	Development	64
Electrostatic complexation[c]	P(Asp)	Cisplatin	Development	27, 30, 31
	P(Glu)	Cisplatin	Clinical trials phase I	29, 32

[a] Most common chemical structures shown in Figure 18.5.
[b] Most common chemical structures shown in Figure 18.6.
[c] Most common chemical structures shown in Figure 18.7.

For the synthesis of PEO-*b*-PLAA block copolymers, α-methoxy-ω-amino PEO has been used as the initiator. Synthesis of PEO-*b*-poly(β-benzyl-L-aspartate) (PEO-*b*-PBLA) and PEO-*b*-poly[ε-(benzyloxycarbonyl)-L-lysine] (PEO-*b*-PBLL) from polymerization of β-benzyl-L-aspartate-*N*-carboxyanhydride (BLA–NCA) and *N*-carboxyanhydride of ε-(benzyloxycarbonyl)-L-lysine (BLL–NCA), respectively, using α-methoxy-ω-amino PEO as initiator has been reported.[49] When primary amines are used as initiators, the NCA polymerization may proceed by two different mechanisms: amine mechanism and activated monomer (AM) mechanism (Figure 18.2A and Figure 18.2B).[48,50] The amine mechanism is a nucleophilic ring-opening chain-growth process where the polymer linearly grows with monomer conversion (Figure 18.2A). In the AM mechanism, NCA will be deprotonated, forming a nucleophile that initiates chain growth (Figure 18.2B). Polymerization by the AM mechanism may lead to side reactions. Besides, the initiator will not be part of the final product. In a given polymerization process, the system can

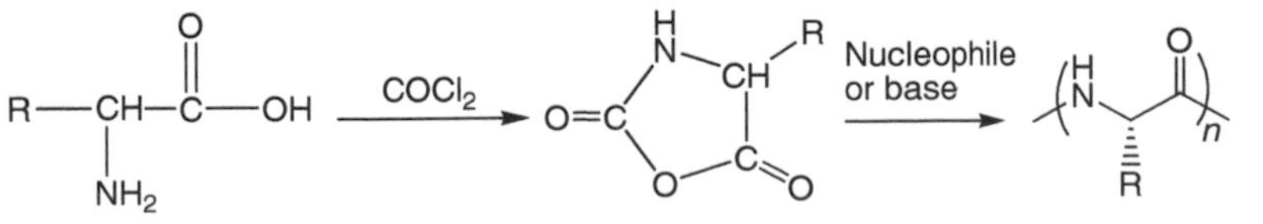

FIGURE 18.1 General scheme for the synthesis of PEO-*b*-PLAA-based block copolymers by ring-opening polymerization of α-amino acid-*N*-carboxyanhydrides (NCAs).

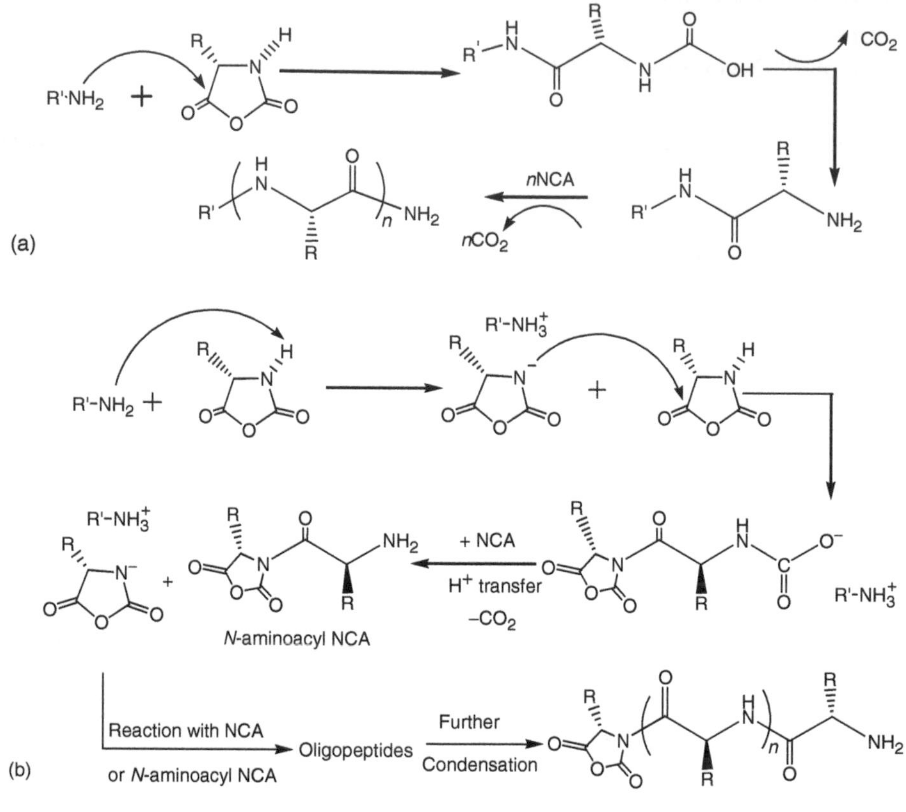

FIGURE 18.2 (a) Amine mechanism and (b) activated monomer (AM) mechanism of NCA polymerization. (From Deming, T. J., *Advanced Drug Delivery Reviews*, 54, 8, 2002. With permission.)

alternate between the amine and AM mechanisms. Because a propagation step for one mechanism is a side reaction for the other and vice versa, block copolymers prepared from the NCA method using amine initiators have structures different from those predicted by monomer feed compositions and, most likely, have considerable homopolymer contamination.

To avoid side reactions, transition metal initiators have been developed[51–55] that use transition metal complexes as the end groups to control the addition of each NCA monomer to the polymer chain ends. However, the presence of chain-transfer reactions prevents the preparation of high molecular weight PLAAs by this process.

Living polymerization is a relatively novel method to prepare PLAA that has been developed to overcome the mentioned limitations of existing methods (Figure 18.3).[56–58] In this method, the transition metal initiator activates the monomers and forms covalent active species that permit the formation of polypeptides via the living polymerization of NCAs. The metals react identically with NCA monomers to form metallacyclic complexes by oxidative addition across the anhydride bond of NCA. The AB diblock, PEO-*b*-poly(L-lysine) (PEO-*b*-P(L-Lys)), ABA triblock, poly(γ-benzyl-L-glutamate)-*b*-PEO-*b*-poly(γ-benzyl-L-glutamate) (PBLG-*b*-PEO-*b*-PBLG) copolymers, and diblock copolymers of poly(methyl acrylate)-*b*-(PBLG) of high molecular weights have been synthesized by living polymerization mechanism.[50,56,59]

Alternatively, PEO can be coupled to PLAA after the polymerization of PLAA. For instance, poly(L-histidine) (P(L-His)) has been synthesized by base-initiated ROP of protected NCA of L-His and coupled to carboxylated PEO to form PEO-*b*-P(L-His) via an amide linkage using dicyclohexyl carbodiimide (DCC) and *N*-hydroxysuccinimide (NHS).[60]

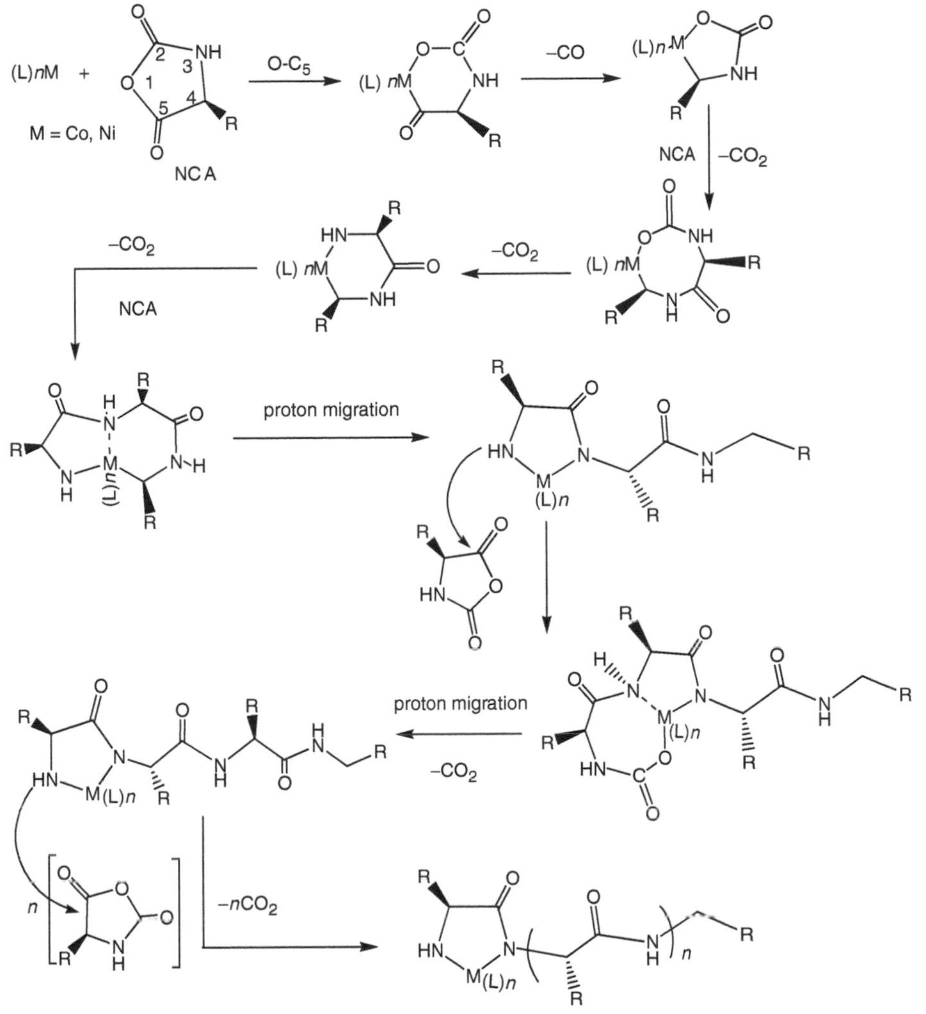

FIGURE 18.3 Living polymerization of NCA.

PEO-*co*-P(L-aspartic acid) (PEO-*co*-P(L-Asp))-bearing amine groups on the P(L-Asp) block have been synthesized by the melt polycondensation of *N*-(benzyloxycarbonyl)-L-aspartic acid anhydride (*N*-CBZ-L-ASP anhydride) and low molecular weight PEO. The product was an alternating copolymer having reactive amine groups on the P(L-Asp) residue. The backbone was linked by ester bond that is more biodegradable than the amide bond between PEO and P(L-Asp).[61]

Preparation of micelle-forming PEO-*b*-PLAA-based block copolymers with poly(leucine), poly(tyrosine), poly(phenylalanine) (PPA), and a composition peptide, i.e., poly(phenylalanine-*co*-leucine-*co*-tyrosin-*co*-tryptophan) P(FLYW), as the core forming block through solid phase peptide synthesis has also been reported.[62] Methoxy PEO-*b*-PLAA dendrimers have also been prepared by the liquid phase peptide synthesis and used in drug delivery.[63]

18.3 MICELLIZATION OF PEO-*b*-PLAA BLOCK COPOLYMERS

Assembly of micelle-forming PEO-*b*-PLAA block copolymers through one of the following methods has been reported.

18.3.1 DIALYSIS METHOD

In this method, block copolymer is dissolved in a water miscible organic solvent first. Then, this solution is dialyzed against water. The semi-permeable membrane keeps the micelles inside the dialysis bag, but it allows removal of organic solvent. Gradual replacement of the organic solvent with water, i.e., the non-solvent for the core-forming block, triggers self-association of block copolymers (Figure 18.4A).[13,35,40,64,65]

18.3.2 SOLVENT EVAPORATION METHOD

In this process, the polymer is dissolved in a volatile organic solvent. The organic solvent is then removed completely by evaporation (usually under reduced pressure), leading to the formation of polymer films. Next, this film is reconstituted in an aqueous phase by vigorous shaking.[36,66] The solvent evaporation method of micellization cannot be utilized for block copolymers having large hydrophobic segments because the polymer film cannot be reconstituted easily in an aqueous phase for those structures (Figure 18.4B).

18.3.3 CO-SOLVENT EVAPORATION METHOD

In this approach, polymer is dissolved in a volatile water miscible organic solvent such as methanol. Then self-assembly is triggered by the gradual addition of aqueous phase (non-solvent for the core-forming block) to the organic phase. This step is followed by the evaporation of the organic co-solvent (Figure 18.4C).[33,34]

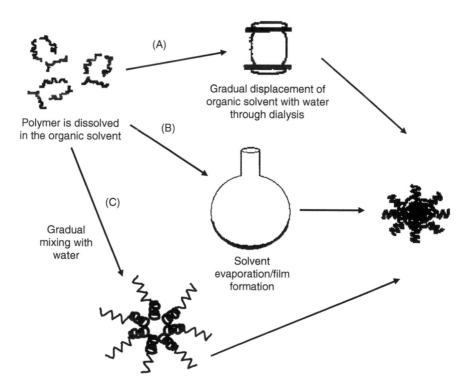

FIGURE 18.4 Micellization of PEO-*b*-PLAA-based block copolymers through (A) dialysis, (B) solvent evaporation, and (C) co-solvent evaporation methods.

18.4 RATIONAL DESIGN AND FUNCTIONAL PROPERTIES OF PEO-*b*-PLAA MICELLES IN DRUG DELIVERY

18.4.1 THE PEO SHELL

PEO has a safe history of application as a pharmaceutical ingredient for the design of non-immuno-genic peptides, carriers, or biomedical surfaces.[67,68] In PEO-*b*-PLAA, the presence of hydrophilic PEO chains introduces amphiphilicity, leading to the self-assembly of block copolymer and forma-tion of association colloids. The PEO shell is expected to induce steric repulsive forces and to stabilize the colloidal interface against aggregation or absorption of proteins to the carrier as well.[10,69–71] As a result, the colloidal carrier will stay within the appropriate size range and surface properties that can avoid uptake by RES. A major barrier against targeted tumor delivery by colloidal carriers is the non-specific uptake and clearance of the carrier from the blood stream by RES. Avoiding RES will prolong the circulation of polymeric micelles in blood, providing more chance for the extravasation of the carrier at sites with leaky vasculature (e.g., solid tumors).

The extent of steric stabilization is found to be dependent on the length and density of the PEO chain on the surface of colloidal carriers.[20,72–74] Elongation of the PEO chain or an increase in the aggregation number of PEO-*b*-PLAA micelles may lead to the reduced chance of micellar aggre-gation and longer circulation time in blood and a higher probability of passive targeting to the tumor site for polymeric micelles.[10] The same parameter may reduce the adherence and interaction of the polymeric micelles with target cells.[75]

18.4.2 THE PLAA CORE

Chemical flexibility of the PLAA structure as the core-forming block is the primary advantage of the PEO-*b*-PLAA-based polymeric micelles. Existence of several functional groups on a PLAA block provides a number of sites for the attachment of drugs, drug compatible moieties, or charged therapeutics to a single polymeric backbone in micelle-forming PEO-*b*-PLAA block copolymers. This may lead to a lower dose of administration for drugs that are chemically conjugated to the PLAA block of PEO-*b*-PLAA. On the other hand, the diversity of functional groups in a PLAA chain (amino, hydroxyl, and carboxylic groups) allows conjugation of different chemical entities to the polymeric backbone and provides an opportunity for the design of polyion complex and/or pH responsive polymeric micelles. Finally, through changes in the chemical structure of the PLAA block, it will also be possible to fine tune the structure of PEO-*b*-PLAA micelles to achieve optimal properties for drug targeting.[4] From a pharmaceutical point of view, PEO-*b*-PLAA micelles are considered advantageous over PEO-*b*-poly(ester)s because they can easily be reconstituted after freeze–drying and do not need lyoprotection.

The most effort for the production of PEO-*b*-PLAA micellar delivery systems has been focused on the application of P(L-Asp), poly(L-glutamic acid) (P(L-Glu)), and P(L-Lys)-based polymers (Table 18.1, Figure 18.5 through 18.7). This is owed to the presence of free carboxyl and amino groups that makes these structures useful for the chemical conjugation and electrostatic complexa-tion of drugs and DNA to the PLAA block.[76]

A major concern for the application of PLAA-based pharmaceuticals is the possibility of immu-nogenic reactions and long-term accumulation and toxicity by these agents. Unfortunately, the information on the biocompatibility and biodegradability of PLAAs is limited.[77,78] Nevertheless, the results of ongoing clinical trials in Japan on PEO–PLAA-based micelles for the delivery of DOX, PTX, and CDDP is expected to provide more information on the safety of these structures.

18.4.3 MICELLAR DIMENSIONS

PEO-*b*-PLAA micelles developed to date are mostly in a diameter range of 10–100 nm.[79] At this specific size range, PEO-*b*-PLAA micelles are expected to be big enough to avoid glomerular

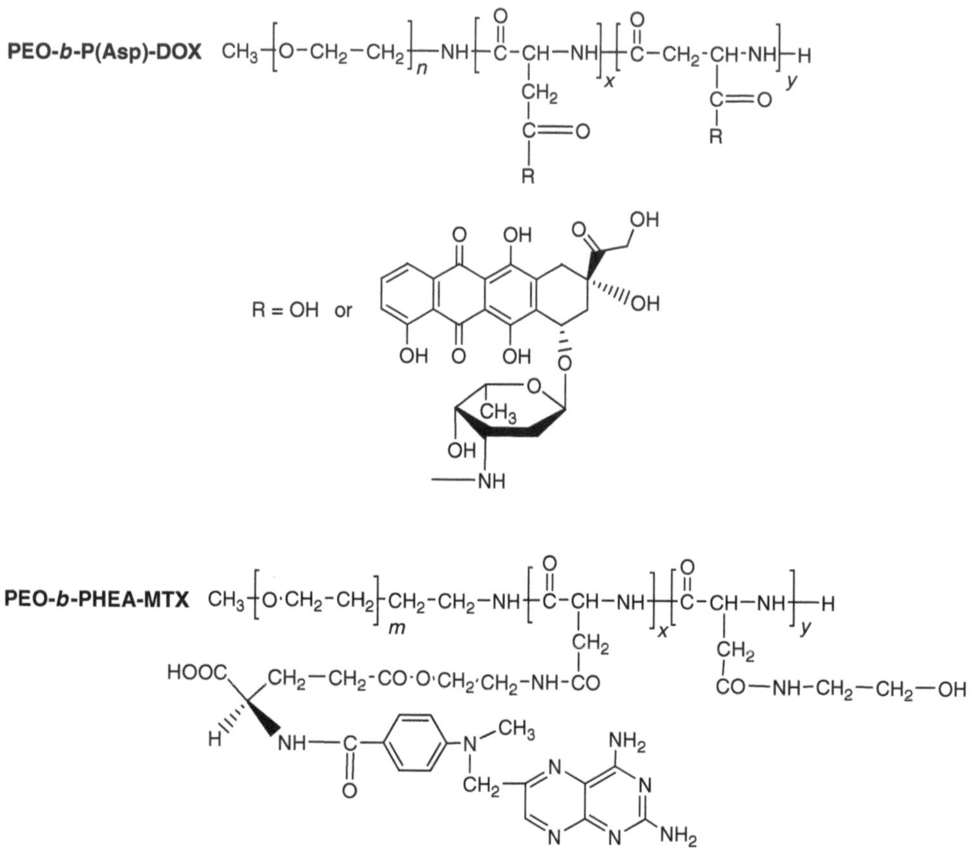

FIGURE 18.5 Chemical structure of some micelle-forming PEO-*b*-PLAA drug conjugates designed for drug delivery.

filtration and elimination by kidney and sufficiently small to escape RES clearance.[80,81] As a result, polymeric micelles may stay in circulation in the blood for prolonged periods, decreasing the clearance (CL), and volume of distribution (Vd), but increasing the area under the plasma concentration versus time curve (AUC) of the encapsulated drug if the drug stays within the carrier in biological system. This change in the pharmacokinetic profile may allow for a reduction in the dose of administration and/or decreased toxicity for the incorporated drug.

Carriers of less than 200 nm with prolonged blood circulation properties are shown to passively accumulate in solid tumors.[82] The new blood vessels in the tumor, formed to provide access to nutrients and oxygen, are usually poorly made and have large gaps (150–730 nm) in their endothelium. This allows the extravasation of nanoparticles of the appropriate size range to the extravascular space surrounding the tumor cells.[83,84] Because the lymphatic system that drains fluids out of other organs is absent in tumors, the permeated nanocarrier is expected to become trapped in the tumor. This phenomenon, known as the enhanced permeation and retention (EPR) effect, is believed to be the reason for the passive accumulation of colloidal carriers (e.g., liposomes, polymerosomes, and polymeric micelles) in solid tumors.[85,86] The polymeric micelles' smaller size compared to other colloidal carriers is expected to ease further penetration of the delivery system inside the dense environment of solid tumors. Other advantages associated with the small dimensions of polymeric micelles include the ease of sterilization via filtration and safety of administration.

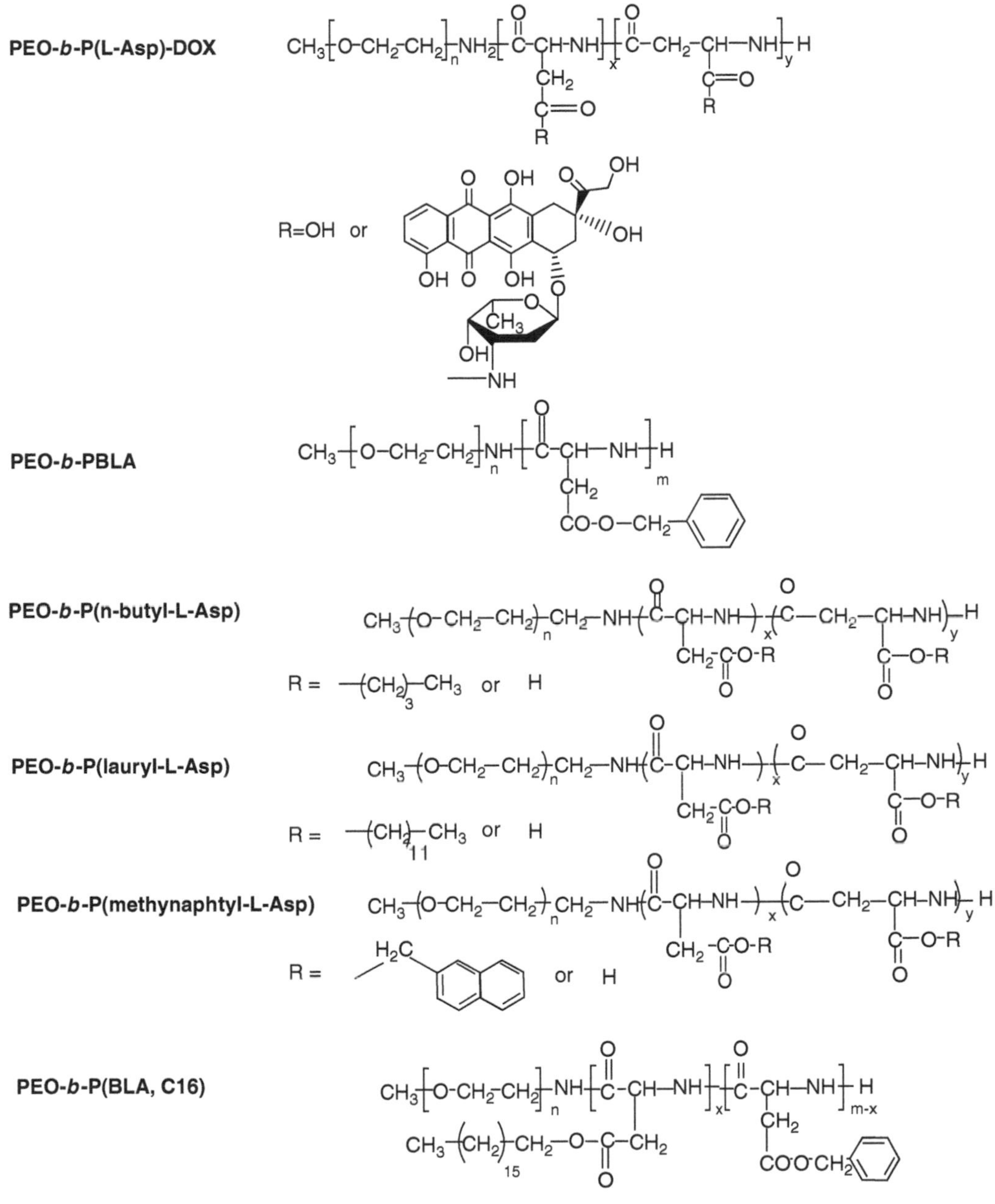

FIGURE 18.6. Chemical structure of most commonly used PEO-b-PLAA-based polymers for physical encapsulation of drugs.

The downside of polymeric micelles' smaller size is the restricted space for drug encapsulation in the carrier. Second, this provides a limitation in the polymeric micelles' ability for a sustained drug release because of a large total surface area, and finally, it allows for possible carrier accumulation in normal tissues bearing leaky vasculature such as the liver. Chemical engineering of the core structure in PEO-b-PLAA micelles may be used to modify the final dimensions of the carrier to an optimum scale so the polymeric micellar carrier can avoid each of these problems. Elongation of

PEO-*b*-PHSA

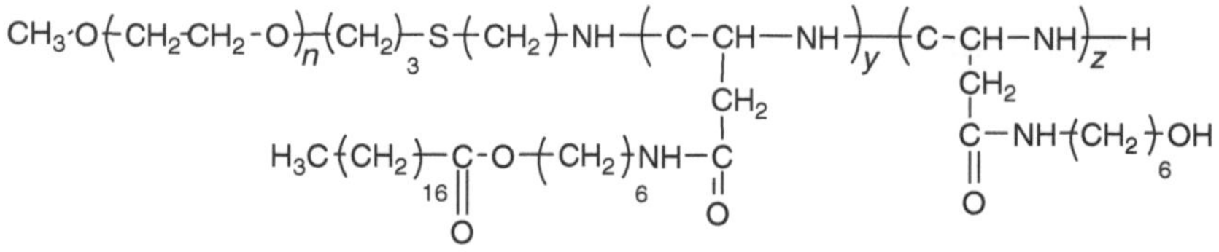

PEO-*b*-PBLG

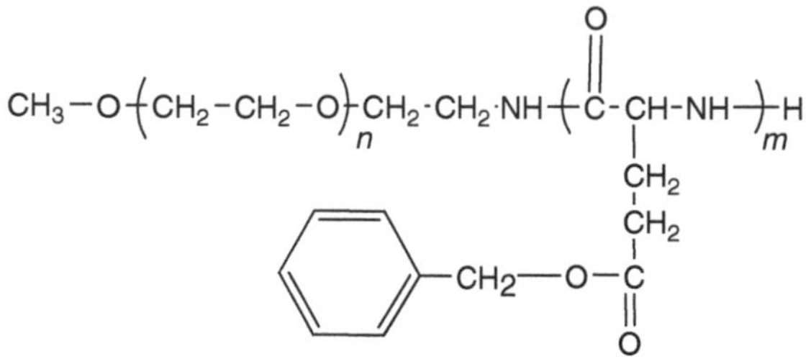

PEO-*b*-PBLG star block copolymer

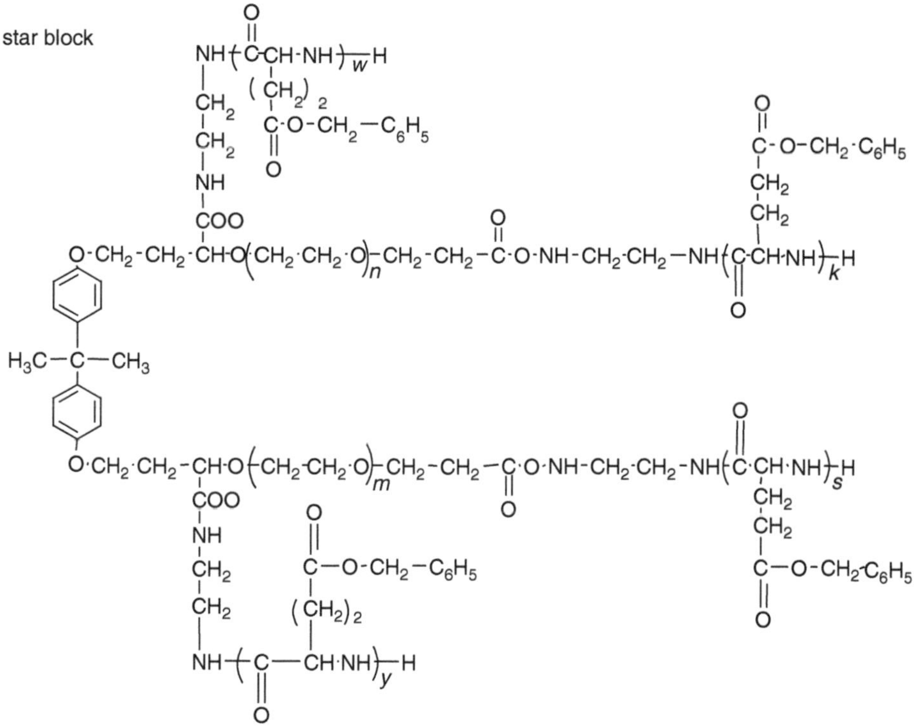

PEO-*b*-PPA

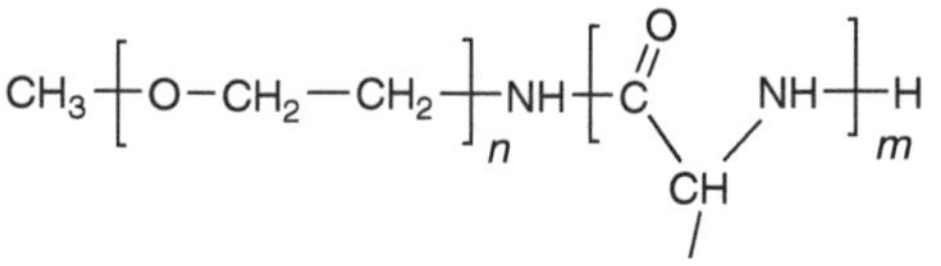

Figure 18.6 (*continued*)

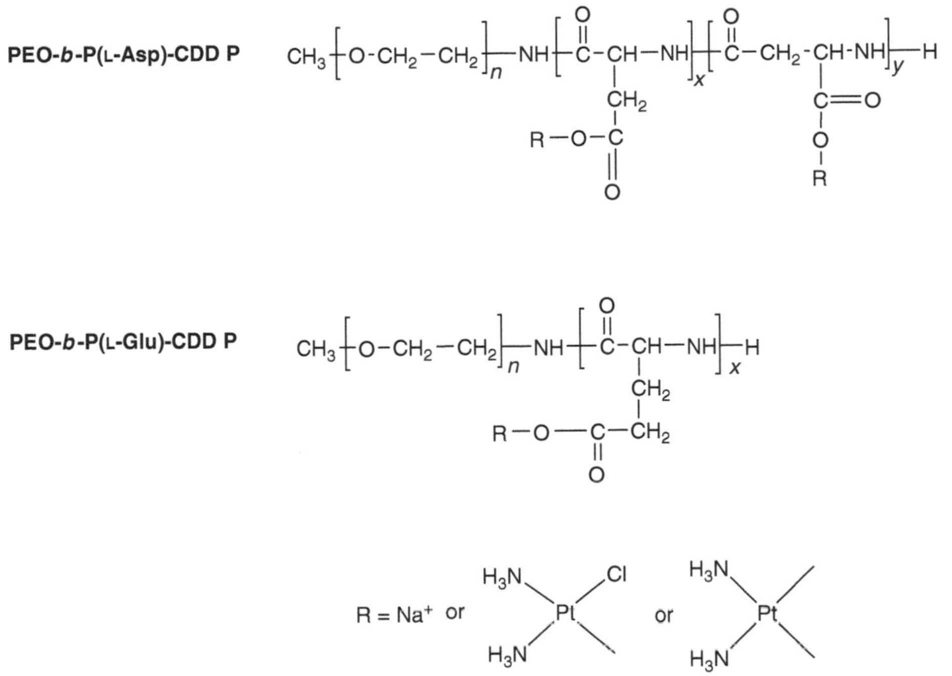

FIGURE 18.7 PEO-*b*-PLAA-based polymers used for the formation of polyion complex micelles.

the PLAA block and an increase in the level of substitution or the length of the side chain have shown an increasing effect on the final size of the prepared polymeric micelle.[34,37,87]

18.4.4 MICELLAR STABILITY

Thermodynamically and kinetically stable polymeric micellar structures may retain their integrity in a biological environment for longer periods and, more effectively, avoid uptake by RES and elimination through the kidney and possibly change the normal organ distribution of the encapsulated drug the same way. PEO-*b*-PLAA block copolymers are reported to be thermodynamically stable and characterized by very low critical micellar concentrations (CMC) in μM range.[37,87,88] Therefore, even after intravenous application and dilution in blood, PEO-*b*-PLAA micelles may still stay above CMC and in a micellar form. The chemical structure of the PLAA core may be modified toward more hydrophobicity to push the CMC to even lower values.[35,37]

Formation of hydrophobic, electrostatic, and hydrogen bonds in the core of PEO-*b*-PLAA micelles after assembly of block copolymers makes the micellar structure kinetically stable. As a result, polymeric micelles may not necessarily exist in equilibrium with polymeric unimers above CMC, and their dissociation will be very slow below CMC. The degree of kinetic stability is suggested to be high, especially for block copolymer structure with stiff or bulky core-forming blocks. Cross-linking of the core structure may enhance the kinetic stability of polymeric micelles as well.[89]

Currently, there has been no report on the development of experimental tools that can assess directly the in vivo stability of micellar structures. Instead, micellar stability is assessed indirectly through in vitro drug release and micellar dissociation studies under diluting condition of gel permeation chromatography (GPC). For instance, the stability of methotrexate (MTX) esters of PEO-*block*-poly(2-hydroxyethyl-L-aspartamide) (PEO-*b*-PHEA) (Figure 18.5) was investigated by GPC, *in vitro*. At high drug conjugation levels, MTX conjugate micelles were shown to be quite stable and entirely elute as micelles during GPC. Likewise, the level of stearic acid substitution in micelles of PEO-*block*-poly(*N*-hexyl stearate-L-aspartamide) (PEO-*b*-PHSA) (Figure 18.6) was shown to influence micellar stability during in vitro studies.[87] A comparison between the pharmacokinetics and biodistribution of polymeric micellar carrier-encapsulated drug and free drug (or commercially available solubilized form of the drug) in animal models has provided an indirect measure for the evaluation of the carrier's in vivo stability. For example, the presence of DOX dimers in PEO-*b*-P(Asp)-based formulation of DOX has been found to improve the pharmacokinetic and biodistribution profile of DOX for passive targeting. This effect has been attributed to the stabilization of the micellar formulation by DOX dimmers.[17] A similar effect is observed for CDDP polymeric micellar formulations when the structure of the core-forming block is made more hydrophobic by changing PLAA structure from P(L-Asp) to P(L-Glu) (Figure 18.7).[29]

18.4.5 DRUG INCORPORATION AND RELEASE PROPERTIES

Incorporation of drugs by PEO-*b*-PLAA micelles has been accomplished through three different means: chemical conjugation of drug to the PLAA block and formation of micelle-forming polymer/drug conjugates; physical encapsulation of drugs in PEO-*b*-PLAA micelles; and formation of an electrostatic complex between the PLAA core and charged drug followed by the self-assembly of the polyion complex (Table 18.1, Figure 18.5 through 18.7).

Drug cleavage from the polymeric backbone by non-specific hydrolysis or enzymes is a prerequisite for the activity of drug-block copolymer conjugates. If the micellar structure can withstand dissociation in a biological environment, then the hydrophobic core of polymeric micelles may protect the degradable bonds between the polymeric backbone and drug from exposure to water and hydrolysis. In this case, the micelle-forming block copolymer drug conjugate may prevent the premature release of the incorporated drug in blood circulation and provide a delayed mode of release for the cytotoxic agent that is only triggered after accumulation and uptake of the carrier by tumor cells. The other possible scenario will be the excessive stability of the polymeric pro-drug that may lead to the inactivity of the final product. Incorporation of hydrophilic moieties, formation of pH sensitive bonds between polymer and drug, and changes in the molecular weight of the core-forming block may be used to modify micellar stability and drug release properties of micelle-forming PEO-*b*-PLAA-based drug conjugates.[87,90,91]

Generally, physical encapsulation of drugs within polymeric micelles is a more attractive approach because many drug molecules do not bear reactive functional groups for chemical conjugation. In this system, formation of hydrophobic interactions or hydrogen bonds between the micelle-forming block copolymer and physically encapsulated drug provides basis for drug solubilization and stabilization within the micellar carrier. Unlike polymer–drug conjugates that may provide a delayed release for the incorporated drug, physical encapsulation of hydrophobic drugs in polymeric micelles usually leads to a sustained mode of drug release from the carrier. It is possible that the drug content of physically loaded polymeric micelles is prematurely released in the blood circulation, i.e., before the carrier reaches its target. In this case, the released drug and carrier will follow separate paths in the biological system, and the colloidal carrier will be ineffective in terms of drug targeting. To lower the rate of micellar dissociation, drug diffusion, and the overall rate of drug release from the micellar carrier, the PLAA core of PEO-*b*-PLAA micelles may be tailored either to bear more drug-compatible moieties, to become glassy under physiological condition (37°C), or to become cross-linked.[16,33,38,39,65,89,92–94] Finally, the method of drug incorporation in polymeric micelles also may be modified to improve the extent of drug loading, localization, or physical state of the loaded drug providing other means for controlling the rate of drug release from polymeric micelles.[36,66,95]

Polyion complex micelles can incorporate charged therapeutics through electrostatic interactions between oppositely charged polymer/drug combinations. Neutralization of the charge on the core-forming segment of the block copolymer will trigger self-assembly of the polyion complex and further stabilization of the complex within the hydrophobic environment of the micellar core. The extent of drug release from polyion complex micelles is dependent on the rate of drug exchange with free ions and proteins in the physiological media. The presence of ions and proteins at high concentrations in the biological environment has been shown to lead to a relatively fast rate of drug release and micellar instability for polyion complex micelles in vivo. In this context, chemical manipulation of the PLAA core may be used to increase the hydrophobicity and rigidity of the micellar core, restricting the penetration of free ions to the micellar core in polyion complex micelles.[26,29] In addition, the micellar core may be cross-linked to avoid its dissociation.[89,92,93] This may lead to a sustained or even delayed mode of drug release from the carrier.

18.5 PEO-*b*-PLAA MICELLES FOR THE DELIVERY OF ANTI-CANCER DRUGS

18.5.1 Cyclophosphamide

The first attempt for the design of PEO-*b*-PLAA-based micellar systems for targeted delivery of anti-cancer drugs was made by Ringsdorf et al. in the 1980s.[96] In this study, cyclophosphamide (CP) sulfide and palmitic acid were attached to the P(L-Lys) block of a PEO-*b*-P(L-Lys). Palmitic acid was used as a hydrophobic moiety to induce the required amphiphilicity for micelle-formation to the block copolymer. CP is an alkylating agent that is transformed to active alkylating metabolites via hepatic and intracellular enzymes. The polymeric micellar formulation was found to be efficient in the stabilization of active CP metabolite, and it caused a five-fold increase in the life span of L1210 tumor-bearing mice even at a reduced CP-equivalent dose.

18.5.2 Doxorubicin

The PEO-*b*-PLAA-based micellar formulation of DOX was developed by Kataoka et al. in the late 1980s and early 1990s and then entered clinical trials in 2001 in Japan. In early studies, micelle-forming PEO-*b*-P(L-Asp) conjugates of DOX (Figure 18.5) were synthesized and tried in different in vitro and in vivo cancer models for their pharmacokinetics, biodistribution, toxicity, and efficacy.[18,22,97–100] The results of these studies pointed to a higher blood and tumor levels for PEO-*b*-P(L-Asp)-DOX micelles compared to free drug, especially for micelles having longer PEO chain lengths. A tumor/heart concentration ratio of 12, 8.1, and 0.9 was demonstrated for PEO-*b*-P(L-Asp)-DOX with PEO chains of 12000, 5000 gmol^{-1}, and free DOX, respectively, 24 h after intravenous injection. Cardiomyopathy is the most important and dose-limiting adverse effect of DOX. The maximum tolerable dose (MTD) of DOX was increased (almost 20-fold) by its PEO-*b*-P(L-Asp) conjugate that allowed administration of higher drug levels. However, compared to free drug, higher levels of conjugated DOX were required for an equal anti-tumor activity in vitro and in vivo. The formulation was found to contain some un-conjugated, physically entrapped DOX that was, in fact, responsible for the anti-tumor activity of this system.

In further studies, PEO-*b*-P(L-Asp)-DOX micelles were utilized as micellar nanocontainers for the physical encapsulation of DOX.[12,16,17] To delay the release of DOX unimers from the formulation in a biological environment, DOX dimers were co-incorporated with DOX unimer within the micellar core. Compared to free DOX, within 24 h, DOX encapsulated in PEO-*b*-P(L-Asp)-DOX micelles showed 28.9-fold higher AUC in plasma, 3.4-fold higher tumor AUC, less toxicity, and superior in vivo activity in solid and hematological cancers in mice. After 24 h, the pharmacokinetics of encapsulated and free DOX became similar, pointing to drug release from the micellar carrier.

Because lyophilized micelles that contained DOX dimers were not reconstitutable in water after long periods of storage, the dimer form of DOX has been removed from this formulation for clinical trials.[21,23] The physically encapsulated DOX in PEO-*b*-P(L-Asp)-DOX micelles, i.e., NK911, showed similar side effects to free DOX in phase I clinical trails. In pharmacokinetic evaluations in humans, NK911 exhibited 2.5-fold increase in half-life, 2.2-fold decrease in CL, 1.6-fold decrease in Vd, and 2-fold increase in plasma AUC compared to free DOX. The lower degree of change in pharmacokinetic parameters for the NK911 compared to liposomal formulation of DOX pointed to the lower stability of polymeric micellar formulation. However, infusion-related symptoms observed after administration of liposomal DOX were not seen for NK911. This formulation has entered phase II clinical trials for the treatment of metastatic pancreatic cancer.

Another PEO-*b*-PLAA-based micellar structure with benzyloxy group as the core-forming block has also been used for physical encapsulation of DOX by the same research group.[13,94,95] PEO-*b*-PBLA (Figure 18.6) efficiently encapsulated DOX and sustained its rate of release (50% drug release within 72 h). The formulation was not as effective as PEO-*b*-P(L-Asp)-DOX carrier of DOX in animal models in reducing the toxicity of DOX. MTD of DOX was raised 2.3-fold by PEO-*b*-PBLA in C-26-bearing mice (compared to a 20-fold increase in MTD observed for the PEO-*b*-P(L-Asp)-DOX carrier).

18.5.3 PACLITAXEL

PTX is a poorly water-soluble drug (water solubility of 0.4 mg/mL). For clinical use, PTX is solubilized using a significant amount of Cremophor EL in Taxol® formulation (Bristol–Myers Squibb). Several attempts have been made to develop safe solubilizing agents for this important chemotherapeutic agent.[1] However, the problems of low solubility, drug leakage, and precipitation upon dilution in biological system for PTX have not been completely resolved.

The hydrophobic nature of PTX and the absence of safe carriers that can reduce the toxicity and enhance the efficacy of injectable PTX have driven a tremendous amount of attention to the application of polymeric micellar carriers for this drug.

The only polymeric micellar formulation showing benefit in drug targeting for PTX is its PEO-*b*-PLAA-based micellar formulation, namely NK105. Recently, Hamaguchi et al. reported on the development of a PEO-*b*-PLAA-based polymeric micellar formulation for PTX delivery.[24] The formulation consists of physically encapsulated PTX in micelles of PEO-*b*-poly(4-phenyl-1-buta-noate)L-aspartamide (PEO-*b*-PPBA). In preclinical studies in animal models, NK105 has shown 86-fold increase in its AUC in plasma, 86-fold decrease in CL, and 15-fold decrease in steady-state volume of distribution (Vdss) compared to Taxol® after intravenous injection.[24] This has resulted in a 25-fold increase in drug AUC in tumor and stronger anti-tumor activity in C-26 tumor-bearing mice models. At a PTX equivalent dose of 100 mg/kg, a single administration of NK105 resulted in the disappearance of tumors, and all mice remained tumor-free thereafter. This formulation is currently in phase I clinical trials in Japan.

PTX polymeric micellar formulations based on PEO-*b*-poly(ester) block copolymers have also entered clinical evaluations in Canada and South Korea.[101] Although PEO-*b*-poly(ester) micelles were remarkably successful in raising the solubility of PTX, they have mostly failed in retaining their drug content in biological environment, thereby regarded as ineffective in terms of drug targeting.[102–107]

18.5.4 CISPLATIN AND ITS DERIVATIVE

CDDP is another anti-cancer agent that has been tried for encapsulation in polymeric micelles. Polyion complexes of positively charged CDDP and negatively charged free carboxyl group of the PEO-*b*-P(L-Asp) block copolymer were formed and assembled to micellar structures by Kataoka et al. (Figure 18.7).[26–31,76] Incorporation of CDDP in polymeric micelles has led to modest changes in the pharmacokinetic and biodistribution properties of CDDP in Lewis Lung carcinoma mice. Compared to free drug, 5.2- and 4.6-fold increases in the plasma and tumor AUC (calculated based on total platinum content) were observed for CDDP-loaded micelles. The low degree of change in pharmacokinetics of CDDP was attributed to the instability of the micellar formulation.

In further studies, to restrict the ion exchange rate with the micellar core and stabilize the polymeric micellar formulation, the P(L-Asp) core of PEO-b-PLAA micelles was replaced with the more hydrophobic structure of P(L-Glu).[29] This change resulted in an increase in platinum levels in plasma and tumor for CDDP. In anti-tumor activity studies, four out of ten C-26-bearing mice receiving the PEO-b-P(L-Glu) formulation of CDDP showed complete cure, whereas, with free drug, no mice showed complete tumor regression. Recently, this formulation (named NC-6004) has concluded the stage of preclinical assessments showing 65-fold increase in plasma AUC, 9-fold decrease in total CL, and decreased nephro and neurotoxicity in rats. NC-6004 was also found to be equally effective as free CDDP against MKN-45 (human gastric cancer) tumor xenografts in mice. This formulation has recently moved to clinical trial in Japan.[32]

PEO-b-P(L-Glu) micelles have also been tried as carriers for a less toxic but more hydrophobic analog of CDDP, i.e., dichloro(1,2-diaminocyclohexane)platinum(II) (DACHPt) with hopes to increase the stability of polyion complex micellar formulation. In pharmacokinetic studies, a slightly higher platinum plasma levels for this system was observed compared to CDDP formulation, but tumor accumulation was similar for both drugs.[108]

18.5.5 METHOTREXATE

Kwon et al. first reported on the conjugation of MTX to a PEO-b-P(L-Asp) derivative in 1999.[90] In this study, MTX was attached through an ester bond to PEO-b-PHEA, obtained by aminolysis of PEO-b-PBLA with ethanolamine. The yield of the reaction was greater than 90%, and MTX content was 9–20 wt%. The PEO-b-PHEA–MTX conjugates (Figure 18.5) showed self-assembly in an aqueous phase using a dialysis method that resulted in micelles with an average size of 14 nm and a narrow size distribution. For the in vitro release, a sample in a dialysis bag was placed into 0.1 M phosphate buffer at varied pH (2.0–9.9). Free MTX was quickly released from the dialysis bag within 10 h. The in vitro release of MTX from self-assembled PEO-b-PHEA–MTX conjugate was slowest at pH 5 and somewhat faster at pH 2.2 and 7 (less than 20% of the drug was released over 260 h). At pH 10, release of MTX was rapid and almost complete in 80 h.

In 2000, the same authors continued their study on this micellar formulation by changing the level of MTX substitution on the polymeric chain.[87] They obtained three different MTX substitution levels of 3.2, 9.0, and 19.4 wt% and showed a decrease in the CMC of PEO-b-PHEA–MTX with an increase in MTX substitution degree. The average diameter of micelles also increased from 11.7 to 27.1 nm as the level of MTX substitution was raised from 3.2 to 19.4 wt%. The results of in vitro release experiments in PBS pH 7.4 showed a slower release rate at higher MTX substitution levels. For PEO-b-PHEA–MTX with 3.2, 9.0, and 19.4% of MTX substitution, 21, 10, and 5% of drug was released after 20 days, respectively.

18.5.6 KRN5500

KRN5500 is a new anti-cancer drug currently in clinical trials in Japan and the USA. It is a semi-synthetic, water-insoluble analog of spicamycin derived from *Streptomyces alanosinicus*. KRN5500 is an inactive prodrug that is metabolized to 4-*N*-glycylspicamycin amino nucleoside (SAN-Gly) by a cytosomal enzyme. Fatty acid chains present in the chemical structure of KRN5500 are lost in SAN-Gly. As a result, SAN-Gly does not cross cellular membranes as easily as KRN5500 and is found to be 1000-fold less cytotoxic than KRN5500 in vitro.[109]

Development of a polymeric micellar formulation for KRN5500 was reported by Yokoyama et al.[65,110] To encapsulate KRN5500 that contains an aliphatic residue in its chemical structure, the aromatic group of the core-forming block in PEO-b-PBLA has been replaced with cetyl esters (Figure 18.6). Polymeric micellar KRN5500 and free drug were found to be similar in terms of anti-tumor activity against HT-29 (human colonic cancer) and MKN-45 xenografts in a mouse model, but polymeric micellar formulation was less toxic. The vascular damage with fibrin clot or an increase in the plasma levels of BUN seen after intravenous administration of free drug was not observed for the polymeric micellar formulation. In a bleomycin-induced lung injury rat model, free KRN5500 at a dose of 3 mg/kg caused extensive pulmonary pathological changes with widespread hemorrhage; however, after administration of polymeric micellar formulation at a similar dose, no pathological change was observed, and the lung resembled the control group.

18.5.7 CAMPTOTHECIN

Camptothecin (CPT) is a naturally occurring cytotoxic alkaloid isolated from *Camptotheca accuminata*. It exists in two forms: biologically active lactone form and inactive carboxylate form.[111,112] The lactone form of CPT exhibits poor aqueous solubility; therefore, it is a good drug candidate for incorporation in polymeric micelles.

Camptothecin has been physically encapsulated in micelles of PEO-*b*-P(L-Asp) having benzyl, *n*-butyl, lauryl, and methylnaphtyl attached to the core-forming block through dialysis, emulsion, and solvent evaporation methods.[113] The presence of aromatic structures, e.g., benzyl or methylnaphtyl groups, and application of a solvent evaporation method of physical encapsulation were shown to be more efficient, resulting in the stabilization of encapsulated CPT in the micellar structures. The optimized polymeric micellar formulation of CPT was shown to stabilize the lactone form of CPT even in the presence of a stimulated physiological condition. In the presence of serum, 72% of CPT structure was protected by polymeric micelles, but free drug lost 80% of its active form.

18.6 PEO-*b*-PLAA MICELLES FOR THE DELIVERY OF OTHER THERPEUTIC AGENTS

18.6.1 AMPHOTERICIN B

AmB is the most potent anti-fungal agent used in systemic mycosis. It is an amphiphilic drug with very low water solubility (aqueous solubility of 1 μg/mL). For clinical use, it is solubilized with the aid of a low molecular weight surfactant, sodium deoxycholate, in Fungizone®. It is also available as three lipid-based formulations: a lipid complex (Ablect®), a complex with cholesteryl sulfate (Amphotec®), and a conventional liposome (Ambisome®).

Micelles of PEO-*b*-PHSA (Figure 18.6) having aliphatic structures in their core were prepared and used for the solubilization of AmB through a solvent evaporation method.[33–36,38,39] In this system, the level of aliphatic chain substitution was found to affect the encapsulation efficiency and release rate of AmB where higher levels of aliphatic side chains in the micellar core led to higher AmB encapsulation and a lower rate of drug release. As a result of a sustained mode of drug release, hemolytic activity of AmB was reduced in its polymeric micellar formulation in comparison to free AmB or Fungizone®. The anti-fungal efficacy of polymeric micellar formulation was found to be comparable to that of free AmB. In further studies, the hemolytic activity of AmB incorporated in PEO-*b*-PLAA micelles having different fatty acid chain lengths as the core-forming block (at a range between 2 and 18 carbons) but similar substitution levels (86–95%) were compared. The results pointed to an advantage for stearic acid (saturated fatty acid with an 18 carbon chain length) in terms of reduced AmB hemolytic activity. In vivo efficacy studies of this formulation in a neutropenic murine model of disseminated candidiasis indicated a similar efficacy for PEO-*b*-PHSA formulation of AmB and Fingizone®. The physical encapsulation of AmB in PEO-*b*-PBLA at an alkaline pH has also been reported.[40,41]

18.6.2 INDOMETHACIN

Indomethacin is a non-steroidal anti-inflammatory drug (NSAID), showing a low solubility in water (35 µg/mL). A properly designed polymeric micellar carrier may increase the solubility of indomethacin, control its renal side effects, and increase accumulation of this drug in inflamed areas. PEO-*b*-PBLA micelles have been used to solubilize indomethacin, and they showed a sustained mode of release for the encapsulated drug. The release rate of drug from PEO-*b*-PBLA micelles was found to be dependent on the ionization state of indomethacin where maximum control on drug release was observed at pH values below the pK_a of indomethacin (4.5). At this condition, indomethacin was unionized and favored the non-polar environment of PBLA core in polymeric micelles.[64]

18.7 PEO-*b*-PLAA MICELLES FOR ACTIVE DRUG TARGETING

The clinical advantage of polymeric micelles in terms of drug targeting may be further improved if the carrier can take advantage of differences between cancer and normal tissue at a cellular level. This may be achieved through the attachment of targeting ligands that recognize cancerous tissue to the surface of polymeric micelles, rendering them effective in active drug targeting. PEO-*b*-PLAA micelles are excellent candidates for active drug targeting because chemical modifications on both core- and shell-forming blocks are possible in PEO-*b*-PLAA-based carriers. Targeting ligands can be attached at the end of the PEO chains, and the anti-cancer drugs can be physically or chemically entrapped in the core. To date, various ligands such as different sugars, transferrin, folate residues, and peptides have been attached to polymeric micelles for active targeting.[114–117]

Park et al. developed a folate-mediated (FOL-mediated) intracellular delivery system that can transport therapeutic proteins or other bioactive macromolecules into specific cells that over express FOL receptors.[118,119] The di-block copolymer conjugate, FOL–PEO-*b*-P(L-Lys) (Figure 18.8), was physically complexed with fluorescein isothiocyanate conjugated bovine serum albumin (FITC–BSA) in an aqueous phase by ionic interactions. Cellular uptake of FOL–PEO-*b*-P(L-Lys)/FITC–BSA complexes was greatly enhanced against a FOL receptor over expressing cell line (KB cells) compared to a FOL receptor deficient cell line (A549 cells). The presence of an excess amount of free FOL in the medium inhibited the intracellular delivery of FOL–PEO-*b*-P(L-Lys)/FITC–BSA complexes. Therefore, the enhanced cellular uptake of FITC–BSA by KB cells was attributed to FOL receptor-mediated endocytosis of the complexes having FOL moieties on the surface. The FOL–PEO-*b*-P(L-Lys) di-block copolymer can potentially be applied for intracellular delivery of a wide range of other active agents that carry negative charge.

In a separate research, the novel pH-sensitive, FOL-modified polymeric mixed micelles composed of FOL–PEO-*b*-P(L-His) and PEO-*b*-poly(L-lactic acid) block copolymers were prepared. The anti-cancer drug, DOX, was physically encapsulated in the micelles.[120] The polymeric micellar carrier was shown to be more efficient for the delivery of DOX to tumor cells in vitro, demonstrating the potential for solid tumor treatment through combining targetability and pH sensitivity.

Attachment of C225, i.e., the antibody against epidermal growth factor (EGF) receptors, to the PEO terminus of a PEO-*b*-P(L-Glu)-DOX has been reported (Figure 18.9).[121] The polymeric immuno-conjugate C225–PEO-*b*-P(L-Glu)-DOX selectively bound to human vulvar squamous carcinoma (A431) cells that over express EGF receptors. Receptor-mediated uptake of C225–PEO-*b*-P(L-Glu)-DOX occurred rapidly (within 5 min), but non-specific uptake of PEO-*b*-P(L-Glu)-DOX required 24 h. Binding of C225–PEO-*b*-P(L-Glu)-DOX to A431 cells was blocked by pretreatment with C225 antibody. The results indicate that C225–PEO-*b*-P(L-Glu)-DOX was more potent than free DOX in inhibiting the growth of A431 cells after a 6-h exposure period.

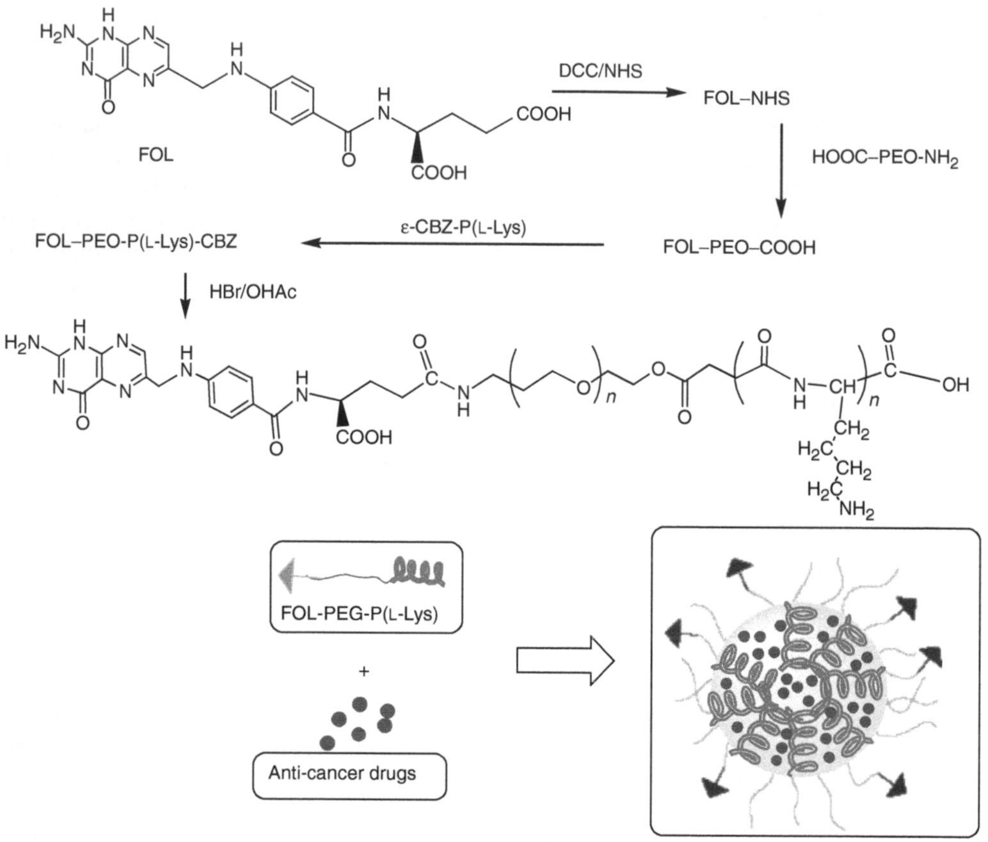

FIGURE 18.8 Synthesis of FOL–PEO-*b*-P(L-Lys) and formation of polyion complex micelles with charged anti-cancer drugs or genes.

18.8 PEO-*b*-PLAA MICELLES FOR GENE DELIVERY

Polymer-based systems have attracted great interest for gene delivery because compared to viral vectors, they are simple to prepare, rather stable, easy to modify, and relatively safe. For the purpose of gene delivery, the polymeric carriers should bear positive charges that can interact with the negative charges of phosphate groups on DNA, resulting in condensation of DNA by the carrier.[122] The polymer/DNA complexes are usually prepared in the presence of an excess amount of cationic polymer, leading to a net positive charge for the complex. The net positive charge of the complex will increase the chance of its interaction with negatively charged cell membranes and facilitates cellular uptake of the polymer/DNA complex via endocytosis. P(L-Lys) and its derivatives have widely been investigated for gene delivery because of its high charge density. However, the high charge density of P(L-Lys) also results in its cytotoxicity and its rapid clearance from circulation after intravenous injection. PEO has been attached to P(L-Lys) to shield the high charge density of the P(L-Lys), reducing its toxicity and elimination from the body. PEO-P(L-Lys) copolymers are capable of association and micelle formation. To date, two types of PEO-P(L-Lys) copolymers have been used as gene carriers: AB type block copolymers, i.e., PEO-*b*-P(L-Lys) and comb-shaped PEO grafted P(L-Lys), i.e., PEO-*g*-P(L-Lys) (Figure 18.10).

The PEO-*b*-P(L-Lys) was synthesized and investigated as a gene carrier by Seymour et al. It formed carriers with an average diameter of around 100 nm.[123] The cytotoxicity of the PEO-*b*-P(L-Lys)/DNA complex was found to be significantly lower than P(L-Lys)/DNA complex. PEO-*b*-

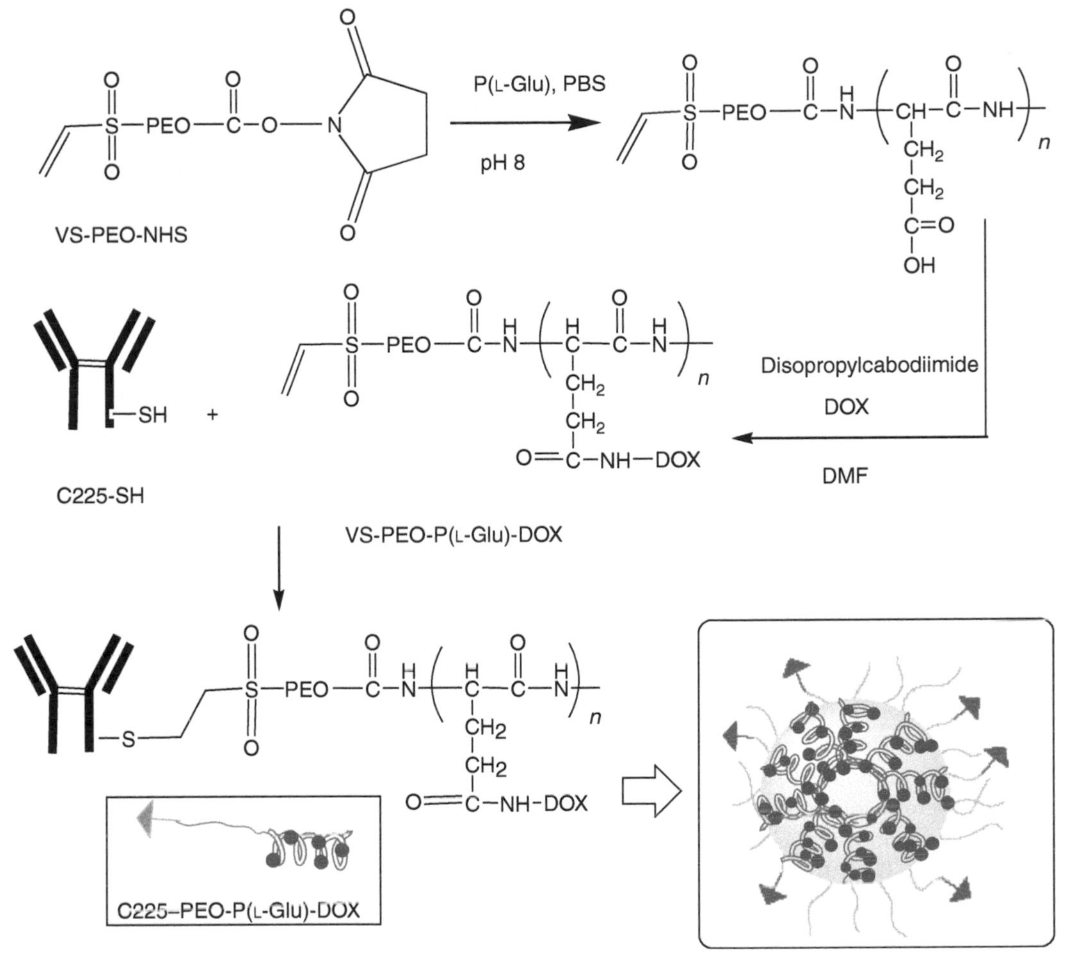

FIGURE 18.9 Synthesis of C225–PEO-*b*-P(L-Glu)-DOX and micelles with chemically loaded anti-cancer drugs.

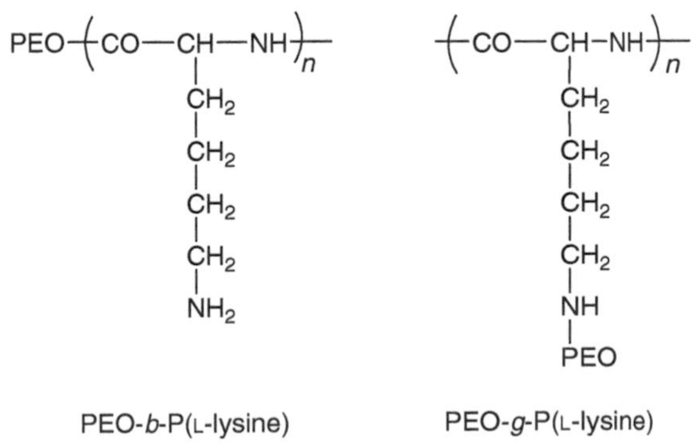

FIGURE 18.10 Chemical structure of PEO-*b*-P(L-Lys)[123–127] and PEO-*g*-P(L-Lys)[129] used for gene delivery.

P(L-Lys)s were further studied by Kataoka et al. for DNA delivery.[124–127] Complexation of DNA and PEO-*b*-P(L-Lys) in their study led to the formation of polyion complex micelles with an average diameter of 48.5 nm. The polyion complex micelles showed higher transfection efficiency in human hepatoma HepG2 cells at a 4:1 positive to negative charge ratio when compared to P(L-Lys) of the same molecular weight. Southern blotting assay showed that naked plasmid DNA was degraded in the blood within 5 min after intravenous injection. On the contrary, when plasmid DNA was administered as complex with PEO-*b*-P(L-Lys) micelles, the supercoiled DNA was detected in the blood for 30 min. Methoxy PEO-*b*-P(L-Lys) dendrimers have also been studied as gene carriers by Choi et al., forming a spherical polymer/DNA complex at a 2:1 positive to negative charge ratio.[63] Multiblock copolymers of PEO-*b*-P(L-Lys)-*g*-P(L-His) with ester bonds between PEO and P(L-Lys) have also been investigated as gene carriers.[128]

The application of PEO-*g*-P(L-Lys) as gene carriers was evaluated in vitro and in vivo by Kim et al. PEO-*g*-P(L-Lys) showed a slightly lower DNA condensing effect when compared to P(L-Lys), but it had a 5–30-fold increase in transfection efficiency in HepG2 cells (a human liver carcinoma cell line).[129] PEO-*g*-P(L-Lys) demonstrated low cytotoxicity, early gene expression, and maintenance of gene expression level for up to 96 h. PEO-*g*-P(L-Lys) was then evaluated as a carrier of the anti-sense glutamic acid decarboxylase (GAD) plasmid to the pancreas for the prevention of type 1 diabetes in mice.[130] The anti-sense mRNA was expressed in the pancreas of mouse for more than 3 days when delivered as PEO-*g*-P(L-Lys) complex.

Attachment of galactose or lactose to PEO-*b*-P(L-Lys) nanocarriers has been accomplished to target the asialoglycoprotein receptor of hepatocytes. Lactose-PEO-*b*-P(L-Lys) showed ten times higher transfection efficiency in HepG2 and lower cytotoxicity when compared to P(L-Lys)/DNA complex.[131]

A synthetic peptide based on apolipoprotein B100 with arterial wall-binding effects was selected and introduced to the end of PEO-*g*-P(L-Lys) for targeting the polymer/DNA complex to the aorta endothelial cells.[132] The arterial wall binding (AWBP) conjugate of PEO-*g*-P(L-Lys) condensed plasmid DNA and formed a spherical shape complex with a size of 100 nm. In gene expression studies, the transfection efficiency of the AWBP–PEO-*g*-P(L-Lys)/DNA complex to bovine aorta endothelial cells and smooth muscle cells was 150–180 times higher than that of P(L-Ly) or PEO-*g*-P(L-Lys) complexes with DNA. Free AWBP decreased the transfection efficiency of the AWBP–PEO-*g*-P(L-Lys), suggesting a targeted gene delivery by the carrier to the aorta endothelial cells via receptor-mediated endocytosis.

REFERENCES

1. Aliabadi, H. M. and Lavasanifar, A., Polymeric micelles for drug delivery, *Expert Opinion on Drug Delivery*, 3(1), 139–162, 2006.
2. Gaucher, G., Dufresne, M. H., Sant, V. P., Kang, N., Maysinger, D., and Leroux, J., Block copolymer micelles: Preparation, characterization and application in drug delivery, *Journal of Controlled Release*, 109(1–3), 169–188, 2005.
3. Kwon, G. S., Polymeric micelles for delivery of poorly water-soluble compounds, *Critical Reviews in Therapeutic Drug Carrier Systems*, 20(5), 357–403, 2003.
4. Lavasanifar, A., Samuel, J., and Kwon, G. S., Poly(ethylene oxide)-*block*-poly(L-amino acid) micelles for drug delivery, *Advanced Drug Delivery Reviews*, 54(2), 169–190, 2002.
5. Allen, C., Eisenberg, A., and Maysinger, D., Copolymer drug carriers: Conjugates, micelles and microspheres, *STP Pharma Sciences*, 9(1), 139–151, 1999.
6. Allen, C., Maysinger, D., and Eisenberg, A., Nano-engineering *block* copolymer aggregates for drug delivery, *Colloids and Surfaces B: Biointerfaces*, 16(1–4), 3–27, 1999.
7. Kataoka, K., Harada, A., and Nagasaki, Y., Block copolymer micelles for drug delivery: Design, characterization and biological significance, *Advanced Drug Delivery Reviews*, 47(1), 113–131, 2001.

8. Le Garrec, D., Ranger, M., and Leroux, J. C., Micelles in anticancer drug delivery, *American Journal of Drug Delivery*, 2(1), 15–42, 2004.

9. Jones, M. C. and Leroux, J. C., Polymeric micelles—A new generation of colloidal drug carriers, *European Journal of Pharmaceutics and Biopharmaceutics*, 48(2), 101–111, 1999.

10. Kwon, G. S. and Kataoka, K., Block copolymer micelles as long-circulating drug vehicles, *Advanced Drug Delivery Reviews*, 16(2–3), 295–309, 1995.

11. Torchilin, V. P., Structure and design of polymeric surfactant-based drug delivery systems, *Journal of Controlled Release*, 73(2–3), 137–172, 2001.

12. Yokoyama, M., Okano, T., Sakurai, Y., and Kataoka, K., Improved synthesis of adriamycin-conjugated poly(ethylene oxide)-poly(aspartic acid) block copolymer and formation of unimodal micellar structure with controlled amount of physically entrapped adriamycin, *Journal of Controlled Release*, 32(3), 269–277, 1994.

13. Kwon, G. S., Naito, M., Yokoyama, M., Okano, T., Sakurai, Y., and Kataoka, K., Physical entrapment of adriamycin in AB block copolymer micelles, *Pharmaceutical Research*, 12(2), 192–195, 1995.

14. Kwon, G., Naito, M., Yokoyama, M., Okano, T., Sakurai, Y., and Kataoka, K., Block copolymer micelles for drug delivery: Loading and release of doxorubicin, *Journal of Controlled Release*, 48(2–3), 195–201, 1997.

15. Kataoka, K., Matsumoto, T., Yokoyama, M., Okano, T., Sakurai, Y., Fukushima, S., Okamoto, K., and Kwon, G. S., Doxorubicin-loaded poly(ethylene glycol)-poly(beta-benzyl-L-aspartate) copolymer micelles: Their pharmaceutical characteristics and biological significance, *Journal of Controlled Release*, 64(1–3), 143–153, 2000.

16. Yokoyama, M., Fukushima, S., Uehara, R., Okamoto, K., Kataoka, K., Sakurai, Y., and Okano, T., Characterization of physical entrapment and chemical conjugation of adriamycin in polymeric micelles and their design for in vivo delivery to a solid tumor, *Journal of Controlled Release*, 50(1–3), 79–92, 1998.

17. Yokoyama, M., Okano, T., Sakurai, Y., Fukushima, S., Okamoto, K., and Kataoka, K., Selective delivery of adriamycin to a solid tumor using a polymeric micelle carrier system, *Journal of Drug Targeting*, 7(3), 171–186, 1999.

18. Kwon, G. S., Yokoyama, M., Okano, T., Sakurai, Y., and Kataoka, K., Biodistribution of micelle-forming polymer-drug conjugates, *Pharmaceutical Research*, 10(7), 970–974, 1993.

19. Yokoyama, M., Miyauchi, M., Yamada, N., Okanao, T., Sakurai, Y., Kataoka, K., and Inoue, S., Polymer micelles as novel drug carrier: Adriamycin-conjugated poly(ethylene glycol)–poly(aspartic acid) block copolymer, *Journal of Controlled Release*, 11(1–3), 269–278, 1990.

20. Kwon, G., Suwa, S., Yokayama, M., Okano, T., Sakurai, Y., and Kataoka, K., Enhanced tumor accumulation and prolonged circulation times for micelle-forming poly(ethylene oxide-aspartate) block copolymer–adriamycin conjugates, *Journal of Controlled Release*, 29, 17–23, 1994.

21. Nakanishi, T., Fukushima, S., Okamoto, K., Suzuki, M., Matsumura, Y., Yokoyama, M., Okano, T., Sakurai, Y., and Kataoka, K., Development of the polymer micelle carrier system for doxorubicin, *Journal of Controlled Release*, 74(1–3), 295–302, 2001.

22. Yokoyama, M., Okano, T., Sakurai, Y., Ekimoto, H., Shibazaki, C., and Kataoka, K., Toxicity and antitumor activity against solid tumors of micelle-forming polymeric anticancer drug and its extremely long circulation in blood, *Cancer Research*, 51(12), 3229–3236, 1991.

23. Matsumura, Y., Hamaguchi, T., Ura, T., Muro, K., Yamada, Y., Shimada, Y., Shirao, K., Okusaka, T., Ueno, H., Ikeda, M., and Watanabe, N., Phase I clinical trial and pharmacokinetic evaluation of NK911, a micelle-encapsulated doxorubicin, *British Journal of Cancer*, 91(10), 1775–1781, 2004.

24. Hamaguchi, T., Matsumura, Y., Suzuki, M., Shimizu, K., Goda, R., Nakamura, I., Nakatomi, I., Yokoyama, M., Kataoka, K., and Kakizoe, T., NK105, a paclitaxel-incorporating micellar nanoparticle formulation, can extend in vivo antitumour activity and reduce the neurotoxicity of paclitaxel, *British Journal of Cancer*, 92, 1240–1246, 2005.

25. Bogdanov, A. A., Martin, C., Bogdanova, A. V., Brady, T. J., and Weissleder, R., An adduct of *cis*-diamminedichloroplatinum(II) and poly(ethylene glycol)poly(L-lysine)-succinate: Synthesis and cytotoxic properties, *Bioconjugate Chemistry*, 7(1), 144–149, 1996.

26. Nishiyama, N. and Kataoka, K., Preparation and characterization of size-controlled polymeric micelle containing *cis*-dichlorodiammineplatinum(II) in the core, *Journal of Controlled Release*, 74(1–3), 83–94, 2001.

27. Nishiyama, N., Kato, Y., Sugiyama, Y., and Kataoka, K., Cisplatin-loaded polymer–metal complex micelle with time-modulated decaying property as a novel drug delivery system, *Pharmaceutical Research*, 18(7), 1035–1041, 2001.

28. Nishiyama, N., Koizumi, F., Okazaki, S., Matsumura, Y., Nishio, K., and Kataoka, K., Differential gene expression profile between PC-14 cells treated with free cisplatin and cisplatin-incorporated polymeric micelles, *Bioconjugate Chemistry*, 14(2), 449–457, 2003.

29. Nishiyama, N., Okazaki, S., Cabral, H., Miyamoto, M., Kato, Y., Sugiyama, Y., Nishio, K., Matsumura, Y., and Kataoka, K., Novel cisplatin-incorporated polymeric micelles can eradicate solid tumors in mice, *Cancer Research*, 63(24), 8977–8983, 2003.

30. Nishiyama, N., Yokoyama, M., Aoyaga, T., Okano, T., Sakurai, Y., and Kataoka, K., Preparation and characterization of self-assembled polymer–metal complex micelle from *cis*-dichlorodiammineplatinium (II) and poly(ethylene glycole)–poly(aspartic acid) block copolymer in an aqueous medium, *Langmuir*, 15, 377–383, 1999.

31. Mizumura, Y., Matsumura, Y., Hamaguchi, T., Nishiyama, N., Kataoka, K., Kawaguchi, T., Hrushesky, W. J., Moriyasuand, F., and Kakizoe, T., Cisplatin-incorporated polymeric micelles eliminate nephrotoxicity, while maintaining antitumor activity, *Japanese Journal of Cancer Research*, 92(3), 328–336, 2001.

32. Uchino, H., Matsumura, Y., Negishi, T., Koizumi, F., Hayashi, T., Honda, T., Nishiyama, N., Kataoka, K., Naito, S., and Kakizoe, T., Cisplatin-incorporating polymeric micelles (NC-6004) can reduce nephrotoxicity and neurotoxicity of cisplatin in rats, *British Journal of Cancer*, 93(6), 678–687, 2005.

33. Adams, M. L., Andes, D. R., and Kwon, G. S., Amphotericin B encapsulated in micelles based on poly(ethylene oxide)-*block*-poly(L-amino acid) derivatives exerts reduced in vitro hemolysis but maintains potent in vivo antifungal activity, *Biomacromolecules*, 4(3), 750–757, 2003.

34. Adams, M. L. and Kwon, G. S., Relative aggregation state and hemolytic activity of amphotericin B encapsulated by poly(ethylene oxide)-*block*-poly(*N*-hexyl-L-aspartamide)-acyl conjugate micelles: Effects of acyl chain length, *Journal of Controlled Release*, 87(1–3), 23–32, 2003.

35. Lavasanifar, A., Samuel, J., and Kwon, G. S., Micelles of poly(ethylene oxide)-*block*-poly(*N*-alkyl stearate L-aspartamide): Synthetic analogues of lipoproteins for drug delivery, *Journal of Biomedical Materials Research*, 52(4), 831–835, 2000.

36. Lavasanifar, A., Samuel, J., and Kwon, G. S., Micelles self-assembled from poly(ethylene oxide)-*block*-poly(*N*-hexyl stearate L-aspartamide) by a solvent evaporation method: Effect on the solubilization and haemolytic activity of amphotericin B, *Journal of Controlled Release*, 77(1–2), 155–160, 2001.

37. Lavasanifar, A., Samuel, J., and Kwon, G. S., The effect of alkyl core structure on micellar properties of poly(ethylene oxide)-*block*-poly(L-aspartamide) derivatives, *Colloids and Surfaces B: Biointerfaces*, 22(2), 115–126, 2001.

38. Lavasanifar, A., Samuel, J., and Kwon, G. S., The effect of fatty acid substitution on the in vitro release of amphotericin B from micelles composed of poly(ethylene oxide)-*block*-poly(*N*-hexyl stearate-L-aspartamide), *Journal of Controlled Release*, 79(1–3), 165–172, 2002.

39. Lavasanifar, A., Samuel, J., Sattari, S., and Kwon, G. S., Block copolymer micelles for the encapsulation and delivery of amphotericin B, *Pharmaceutical Research*, 19(4), 418–422, 2002.

40. Yu, B. G., Okano, T., Kataoka, K., and Kwon, G. S., Polymeric micelles for drug delivery: Solubilization and haemolytic activity of amphotericin B, *Journal of Controlled Release*, 53(1–3), 131–136, 1998.

41. Yu, B. G., Okano, T., Kataoka, K., and Kwon, G. S., In vitro dissociation of antifungal efficacy and toxicity for amphotericin B-loaded poly(ethylene oxide)-*block*-poly(beta-benzyl-L-aspartate) micelles, *Journal of Controlled Release*, 56(1–3), 285–291, 1998.

42. Smeenk, J. M., Lowik, D. W. P. M., and van Hest, J. C. M., Peptide-containing block copolymers: Synthesis and potential applications of bio-mimetic materials, *Current Organic Chemistry*, 9(12), 1115–1125, 2005.

43. Fuller, W. D., Verlander, M. S., and Goodman, M., A procedure for the facile synthesis of amino-acid *N*-carboxyanhydrides, *Biopolymers*, 15(9), 1869–1871, 1976.

44. Li, L. Y., Sun, P. C., Yao, Y., Chen, T. H., Li, B. H., Jin, Q. H., and Ding, D. T., Synthesis of amphiphilic diblock copolymer poly(L-alanine)-b-poly(hydroxyethyl glutamine) and its self-assembly in water, *Chemical Journal of Chinese Universities—Chinese*, 26(8), 1548–1551, 2005.

45. Harwood, H. J., Comments concerning the mechanism of strong base initiated NCA polymerization. Abstract in *Abstracts of papers of the American Chemical Society*, 187(APR),72-Poly, 1984.

46. Kricheldorf, H. R. and Mulhaupt, R., Mechanism of the NCA polymerization. 7. Primary and secondary amine-initiated polymerization of beta-amino acid NCAs, *Makromolekulare Chemie-Macromolecular Chemistry and Physics*, 180(6), 1419–1433, 1979.

47. Kricheldorf, H. R., Von Lossow, C., and Schwarz, G., Primary amine and solvent-induced polymerizations of L- or D,L-phenylalanine N-carboxyanhydride, *Macromolecular Chemistry and Physics*, 206(2), 282–290, 2005.

48. Sekiguchi, H., Mechanism of N-carboxy-alpha-amino acid anhydride (NCA) polymerization, *Pure and Applied Chemistry*, 53(9), 1689–1714, 1981.

49. Harada, A. and Kataoka, K., Formation of polyion complex micelles in an aqueous milieu from a pair of oppositely-charged block-copolymers with poly(ethylene glycol) segments, *Macromolecules*, 28(15), 5294–5299, 1995.

50. Deming, T. J., Methodologies for preparation of synthetic block copolypeptides: Materials with future promise in drug delivery, *Advanced Drug Delivery Reviews*, 54(8), 1145–1155, 2002.

51. Deming, T. J., Transition metal-amine initiators for preparation of well-defined poly(gamma-benzyl L-glutamate), *Journal of the American Chemical Society*, 119(11), 2759–2760, 1997.

52. Freireic, S., Gertner, D., and Zilkha, A., Polymerization of N-carboxy anhydrides by organotin catalysts, *European Polymer Journal*, 10(5), 439–443, 1974.

53. Tsuruta, T., Matsuura, K., and Inoue, S., Copolymerization of propylene oxide with N-carboxy-DL-alanine anhydride by organometallic systems, *Die Makromolekulare Chemie*, 83, 289–291, 1965.

54. Yamashit, S. and Tani, H., Polymerization of gamma-benzyl L-glutamate N-carboxyanhydride with metal acetate-tri-normal-butylphosphine catalyst system, *Macromolecules*, 7(4), 406–409, 1974.

55. Yamashit, S., Waki, K., Yamawaki, N., and Tani, H., Stereoselective polymerization of alpha-amino-acid N-carboxyanhydrides with nickel DL-2-methylbutyrate-tri-normal-butylphosphine catalyst system, *Macromolecules*, 7(4), 410–415, 1974.

56. Brzezinska, K. R. and Deming, T. J., Synthesis of AB diblock copolymers by atom-transfer radical polymerization (ATRP) and living polymerization of alpha-amino acid-N-carboxyanhydrides, *Macromolecular Bioscience*, 4(6), 566–569, 2004.

57. Dvorak, M. and Rypacek, F., Preparation and polymerization of N-carboxyanhydrides of alpha-amino-acids, *Chemicke Listy*, 89(7), 423–436, 1995.

58. Vayaboury, W., Giani, O., Cottet, H., Deratani, A., and Schue, F., Living polymerization of alpha-amino acid N-carboxyanhydrides (NCA) upon decreasing the reaction temperature, *Macromolecular Rapid Communications*, 25(13), 1221–1224, 2004.

59. Deming, T. J., Living polymerization of alpha-amino acid-N-carboxyanhydrides, *Journal of Polymer Science Part A—Polymer Chemistry*, 38(17), 3011–3018, 2000.

60. Lee, E. S., Shin, H. J., Na, K., and Bae, Y. H., Poly(L-histidine)–PEG block copolymer micelles and pH-induced destabilization, *Journal of Controlled Release*, 90(3), 363–374, 2003.

61. Won, C. Y., Chu, C. C., and Lee, J. D., Synthesis and characterization of biodegradable poly(L-aspartic acid-co-PEG), *Journal of Polymer Science Part A—Polymer Chemistry*, 36(16), 2949–2959, 1998.

62. Van Domeselaar, G. H., Kwon, G. S., Andrew, L. C., and Wishart, D. S., Application of solid phase peptide synthesis to engineering PEO-peptide block copolymers for drug delivery, *Colloids and Surfaces B: Biointerfaces*, 30(4), 323–334, 2003.

63. Choi, J. S., Lee, E. J., Choi, Y. H., Jeong, Y. J., and Park, J. S., Poly(ethylene glycol)-*block*-poly(L-lysine) dendrimer: Novel linear polymer/dendrimer block copolymer forming a spherical water-soluble polyionic complex with DNA, *Bioconjugate Chemistry*, 10(1), 62–65, 1999.

64. La, S. B., Okano, T., and Kataoka, K., Preparation and characterization of the micelle-forming polymeric drug indomethacin-incorporated poly(ethylene oxide)–poly(beta-benzyl L-aspartate) block copolymer micelles, *Journal of Pharmaceutical Sciences*, 85(1), 85–90, 1996.

65. Yokoyama, M., Satoh, A., Sakurai, Y., Okano, T., Matsumura, Y., Kakizoe, T., and Kataoka, K., Incorporation of water-insoluble anticancer drug into polymeric micelles and control of their particle size, *Journal of Controlled Release*, 55(2–3), 219–229, 1998.

66. Opanasopit, P., Yokoyama, M., Watanabe, M., Kawano, K., Maitani, Y., and Okano, T., Influence of serum and albumins from different species on stability of camptothecin-loaded micelles, *Journal of Controlled Release*, 104(2), 313–321, 2005.

67. Working, P., Newman, M., Johnson, J., and Cornacoff, J., Safety of poly(ethylene glycol) and poly(ethylene glycol) derivatives, In *Poly(ethylene glycol) Chemistry and Biological Applications*, Harris, J. and Zalipsky, S., Eds., American Chemical Society, Washington, DC, pp. 44–57, 1997.

68. Zalipsky, S. and Harris, J., Introduction to chemistry and biological applications of poly(ethylene glycol), In *Poly(ethylene glycol) Chemistry and Biological Applications*, Harris, J. and Zalipsky, S., Eds., American Chemical Society, Washington, DC, pp. 1–13, 1997.

69. Stolnik, S., Illum, L., and Davis, S. S., Long circulating microparticulate drug carriers, *Advanced Drug Delivery Reviews*, 16, 195–214, 1995.

70. Zhang, F., Kang, E., Neoh, K., and Huang, W., Modification of gold surface by grafting of poly(ethylene glycol) for reduction in protein adsorption and platelet adhesion, *Journal of Biomaterial Science Polymer Edition*, 12, 515–531, 2001.

71. Trubetskoy, V. and Torchilin, V., Use of polyoxyethylene-lipid conjugates as long-circulating carriers for delivery of therapeutic and diagnostic agents, *Advanced Drug Delivery Reviews*, 16, 311–320, 1995.

72. Dunn, S., Brindley, A., Davis, S., Davies, M., and Illum, L., Polystyrene-poly(ethylene glycol) (PS-PEG) particles as model systems for site specific drug delivery. 2. The effect of PEG surface density on the in vitro cell interaction and in vivo biodistribution, *Pharmaceutical Research*, 11, 1016–1022, 1994.

73. Mosqueira, V. C., Legrand, P., Morgat, J. L., Vert, M., Mysiakine, E., Gref, R., Devissaguet, J. P., and Barratt, G., Biodistribution of long-circulating PEG-grafted nanocapsules in mice: Effects of PEG chain length and density, *Pharmaceutical Research*, 18(10), 1411–1419, 2001.

74. Gref, R., Luck, M., Quellec, P., Marchand, M., Dellacherie, E., Harnisch, S., Blunk, T., and Muller, R. H., 'Stealth' corona-core nanoparticles surface modified by polyethylene glycol (PEG): Influences of the corona (PEG chain length and surface density) and of the core composition on phagocytic uptake and plasma protein adsorption, *Colloids and Surfaces B: Biointerfaces*, 18(3–4), 301–313, 2000.

75. Mahmud, A. and Lavasanifar, A., The effect of block copolymer structure on the internalization of polymeric micelles by human breast cancer cells, *Colloids and Surfaces B: Biointerfaces*, 45(2), 82–89, 2005.

76. Nishiyama, N., Bae, Y., Miyata, K., Fukushima, S., and Kataoka, K., Smart polymeric micelles for gene and drug delivery, *Drug Discovery Today: Technologies*, 2(1), 21–26, 2005.

77. McCormick-Thomson, L., Sgouras, D., and Duncan, R., Poly(amino acid) copolymers as potential soluble drug delivery system. 2. Body distribution and preliminary biocompatibility testing in vitro and in vivo, *Journal of Bioactive and Compatible Polymers*, 4, 252–268, 1989.

78. Chiu, H., Kopeckova, P., and Deshmane, A. S. K. J., Lysosomal degradability of poly(a-amino acids), *Journal of Biomedical Materials Research*, 34, 381–392, 1997.

79. Kwon, G. S., Diblock copolymer nanoparticles for drug delivery, *Critical Reviews in Therapeutic Drug Carrier Systems*, 15(5), 481–512, 1998.

80. Litzinger, D. C., Buiting, A. M., van Rooijen, N., and Huang, L., Effect of liposome size on the circulation time and intraorgan distribution of amphipathic poly(ethylene glycol)-containing liposomes, *Biochimica et Biophysica Acta*, 1190, 99–107, 1994.

81. Papahadjopoulos, D. and Gabizon, A., Liposomes designed to avoid the reticuloendothelial system, *Progress in Clinical and Biological Research*, 343, 85–93, 1990.

82. Charrois, G. J. R. and Allen, T. M., Rate of biodistribution of STEALTH liposomes to tumor and skin: Influence of liposome diameter and implications for toxicity and therapeutic activity, *Biochimica et Biophysica Acta*, 1609(1), 102–108, 2003.

83. Jain, R. K., Delivery of molecular and cellular medicine to solid tumors, *Advanced Drug Delivery Reviews*, 46(1–3), 149–168, 2001.

84. Dass, C., Tumour angiogenesis, vascular biology and enhanced drug delivery, *Journal of Drug Targeting*, 12(5), 245–255, 2004.

85. Duncan, R., The dawning era of polymer therapeutics, *Nature Reviews*, 2, 347–360, 2003.

86. Maeda, H., Wu, J., Sawa, T., Matsumura, Y., and Hori, K., Tumor vascular permeability and the EPR effect in macromolecular therapeutics: A review, *Journal of Controlled Release*, 65(1–2), 271–284, 2000.

87. Li, Y. and Kwon, G. S., Methotrexate esters of poly(ethylene oxide)-*block*-poly(2-hydroxyethyl-L-aspartamide). Part I: Effects of the level of methotrexate conjugation on the stability of micelles and on drug release, *Pharmaceutical Research*, 17(5), 607–611, 2000.

88. Kwon, G., Naito, M., Yokoyama, M., Okano, T., Sakurai, Y., and Kataoka, K., Micelles based on AB block copolymers of poly(ethylene oxide) and poly(beta-benzyl L-aspartate), *Langmuir*, 9, 945–949, 1993.

89. Kakizawa, Y., Harada, A., and Kataoka, K., Glutathione-sensitive stabilization of block copolymer micelles composed of antisense DNA and thiolated poly(ethylene glycol)-*block*-poly(L-lysine): A potential carrier for systemic delivery of antisense DNA, *Biomacromolecules*, 2(2), 491–497, 2001.

90. Li, Y. and Kwon, G. S., Micelle-like structures of poly(ethylene oxide)-block-poly(2-hydroxyethyl aspartamide)-methotrexate conjugates, *Colloids and Surfaces B: Biointerfaces*, 16, 217–226, 1999.

91. Yokoyama, M., Kwon, G. S., Okano, T., Sakurai, Y., Naito, M., and Kataoka, K., Influencing factors on in vitro micelle stability of adriamycin–block copolymer conjugates, *Journal of Controlled Release*, 28(1–3), 59–65, 1994.

92. Jaturanpinyo, M., Harada, A., Yuan, X., and Kataoka, K., Preparation of bionanoreactor based on core–shell structured polyion complex micelles entrapping trypsin in the core cross-linked with glutaraldehyde, *Bioconjugate Chemistry*, 15(2), 344–348, 2004.

93. Miyata, K., Kakizawa, Y., Nishiyama, N., Harada, A., Yamasaki, Y., Koyama, H., and Kataoka, K., Block catiomer polyplexes with regulated densities of charge and disulfide cross-linking directed to enhance gene expression, *Journal of the American Chemical Society*, 126(8), 2355–2361, 2004.

94. Kataoka, K., Matsumoto, T., Yokoyama, M., Okano, T., Sakurai, Y., Fukushima, S., Okamoto, K., and Kwon, G. S., Doxorubicin-loaded poly(ethylene glycol)–poly(beta-benzyl-L-aspartate) copolymer micelles: Their pharmaceutical characteristics and biological significance, *Journal of Controlled Release*, 64(1–3), 143–153, 2000.

95. Kwon, G., Naito, M., Yokoyama, M., Okano, T., Sakurai, Y., and Kataoka, K., Block copolymer micelles for drug delivery: Loading and release of doxorubicin, *Journal of Controlled Release*, 48, 195–201, 1997.

96. Ringsdorf, H., Dorn, K., and Hoerpel, G., Polymeric antitumor agents on molecular and cellular level, In *Bioactive Polymer Systems: An Overview*, Gebelein, C. G. and Carraher, C. E., Eds., Plenum Press, New York, pp. 531–585, 1985.

97. Kwon, G., Suwa, S., Yokoyama, M., Okano, T., Sakurai, Y., and Kataoka, K., Enhanced tumor accumulation and prolonged circulation times of micelle-forming poly(ethylene oxide-aspartate) block copolymer–adriamycin conjugates, *Journal of Controlled Release*, 29, 17–23, 1994.

98. Yokoyama, M., Inoue, S., Kataoka, K., Yui, N., and Sakurai, Y., Preparation of adriamycin-conjugated poly(ethylene glycol)–poly(aspartic acid) block copolymer: A new type of polymeric anticancer agent, *Makromolekulare Chemie*, 8, 431–435, 1987.

99. Yokoyama, M., Kwon, G. S., Okano, T., Sakurai, Y., Seto, T., and Kataoka, K., Preparation of micelle forming polymer–drug conjugates, *Bioconjugate Chemistry*, 3, 295–301, 1992.

100. Yokoyama, M., Miyauchi, M., Yamada, N., Okano, T., Sakurai, Y., and Kataoka, K., Characterization and anticancer activity of the micelle-forming polymeric anticancer drug adriamycin-conjugated poly(ethylene glycol)–poly(aspartic acid) block copolymer, *Cancer Research*, 50, 1693–1700, 1990.

101. Kim, T. Y., Kim, D. W., Chung, J. Y., Shin, S. G., Kim, S. C., Heo, D. S., Kim, N. K., and Bang, Y. J., Phase I and pharmacokinetic study of Genexol-PM, a cremophor-free, polymeric micelle-formulated paclitaxel, in patients with advanced malignancies, *Clinical Cancer Research*, 10(11), 3708–3716, 2004.

102. Kim, S. C., Kim, D. W., Shim, Y. H., Bang, J. S., Oh, H. S., Wan Kim, S., and Seo, M. H., In vivo evaluation of polymeric micellar paclitaxel formulation: Toxicity and efficacy, *Journal of Controlled Release*, 72(1–3), 191–202, 2001.

103. Le Garrec, D., Gori, S., Luo, L., Lessard, D., Smith, D. C., Yessine, M. A., Ranger, M., and Leroux, J. C., Poly(*N*-vinylpyrrolidone)-*block*-poly(D,L-lactide) as a new polymeric solubilizer for hydrophobic anticancer drugs: In vitro and in vivo evaluation, *Journal of Controlled Release*, 99(1), 83–101, 2004.

104. Leung, S. Y., Jackson, J., Miyake, H., Burt, H., and Gleave, M. E., Polymeric micellar paclitaxel phosphorylates Bcl-2 and induces apoptotic regression of androgen-independent LNCaP prostate tumors, *Prostate*, 44(2), 156–163, 2000.

105. Zhang, X., Burt, H. M., Mangold, G., Dexter, D., Von Hoff, D., Mayer, L., and Hunter, W. L., Antitumor efficacy and biodistribution of intravenous polymeric micellar paclitaxel, *Anti-Cancer Drugs*, 8(7), 696–701, 1997.

106. Zhang, X., Burt, H. M., Von Hoff, D., Dexter, D., Mangold, G., Degen, D., Oktaba, A. M., and Hunter, W. L., An investigation of the antitumour activity and biodistribution of polymeric micellar paclitaxel, *Cancer Chemotherapy and Pharmacology*, 40(1), 81–86, 1997.

107. Zhang, X., Jackson, J. K., and Burt, H. M., Development of amphiphilic diblock copolymers as micellar carriers of taxol, *International Journal of Pharmaceutics*, 132(1–2), 195–206, 1996.

108. Cabral, H., Nishiyama, N., Okazaki, S., Koyama, H., and Kataoka, K., Preparation and biological properties of dichloro(1,2-diaminocyclohexane)platinum(II) (DACHPt)-loaded polymeric micelles, *Journal of Controlled Release*, 101(1–3), 223–232, 2005.

109. Sakai, T., Kawai, H., Kamishohara, M., Odagawa, A., Suzuki, A., Uchida, T., Kawasaki, T., Tsuruo, T., and Otake, N., Structure–antitumor activity relationship of semi-synthetic spicamycin derivatives, *Journal of Antibiotics (Tokyo)*, 48(12), 1467–1480, 1995.

110. Matsumura, Y., Yokoyama, M., Kataoka, K., Okano, T., Sakurai, Y., Kawaguchi, T., and Kakizoe, T., Reduction of the side effects of an antitumor agent, KRN5500, by incorporation of the drug into polymeric micelles, *Japanese Journal of Cancer Research*, 90(1), 122–128, 1999.

111. Barreiro-Iglesias, R., Bromberg, L., Temchenko, M., Hatton, T. A., Concheiro, A., and Alvarez-Lorenzo, C., Solubilization and stabilization of camptothecin in micellar solutions of pluronic-*g*-poly(acrylic acid) copolymers, *Journal of Controlled Release*, 97(3), 537–549, 2004.

112. Mi, Z. and Burke, T. G., Differential interactions of camptothecin lactone and carboxylate forms with human blood components, *Biochemistry*, 33(34), 10325–10336, 1994.

113. Yokoyama, M., Opanasopit, P., Okano, T., Kawano, K., and Maitani, Y., Polymer design and incorporation methods for polymeric micelle carrier system containing water-insoluble anticancer agent camptothecin, *Journal of Drug Targeting*, 12(6), 373–384, 2004.

114. Leamon, C. P., Weigl, D., and Hendren, R. W., Folate copolymer-mediated transfection of cultured cells, *Bioconjugate Chemistry*, 10(6), 947–957, 1999.

115. Nagasaki, Y., Yasugi, K., Yamamoto, Y., Harada, A., and Kataoka, K., Sugar-installed polymeric micelle for a vehicle of an active targeting drug delivery system, Abstract in *Abstracts of Papers of the American Chemical Society*, publication no. 221, U434–U434, 2001.

116. Nasongkla, N., Shuai, X., Ai, H., Weinberg, B. D., Pink, J., Boothman, D. A., and Gao, J. M., cRGD-functionalized polymer micelles for targeted doxorubicin delivery, *Angewandte Chemie-International Edition*, 43(46), 6323–6327, 2004.

117. Vinogradov, S., Batrakova, E., Li, S., and Kabanov, A., Polyion complex micelles with protein-modified corona for receptor-mediated delivery of oligonucleotides into cells, *Bioconjugate Chemistry*, 10(5), 851–860, 1999.

118. Kim, S. H., Jeong, J. H., Joe, C. O., and Park, T. G., Folate receptor mediated intracellular protein delivery using PLL–PEG–FOL conjugate, *Journal of Controlled Release*, 103(3), 625–634, 2005.

119. Kim, S. H., Jeong, J. H., Chun, K. W., and Park, T. G., Target-specific cellular uptake of PLGA nanoparticles coated with poly(L-lysine)-poly(ethylene glycol)–folate conjugate, *Langmuir*, 21(19), 8852–8857, 2005.

120. Lee, E. S., Na, K., and Bae, Y. H., Polymeric micelle for tumor pH and folate-mediated targeting, *Journal of Controlled Release*, 91(1–2), 103–113, 2003.

121. Vega, J., Ke, S., Fan, Z., Wallace, S., Charsangavej, C., and Li, C., Targeting doxorubicin to epidermal growth factor receptors by site-specific conjugation of C225 to poly(L-glutamic acid) through a polyethylene glycol spacer, *Pharmaceutical Research*, 20(5), 826–832, 2003.

122. Kakizawa, Y. and Kataoka, K., Block copolymer micelles for delivery of gene and related compounds, *Advanced Drug Delivery Reviews*, 54, 203–222, 2002.

123. Wolfert, M. A., Schacht, E. H., Toncheva, V., Ulbrich, K., Nazarova, O., and Seymour, L. W., Characterization of vectors for gene therapy formed by self-assembly of DNA with synthetic block co-polymers, *Human Gene Theraphy*, 7(17), 2123–2133, 1996.

124. Katayose, S. and Kataoka, K., Remarkable increase in nuclease resistance of plasmid DNA through supramolecular assembly with poly(ethylene glycol)–poly(L-lysine) block copolymer, *Journal of Pharmaceutical Sciences*, 87(2), 160–163, 1998.

125. Katayose, S. and Kataoka, K., Water-soluble polyion complex associates of DNA and poly(ethylene glycol)–poly(L-lysine) block copolymer, *Bioconjugate Chemistry*, 8(5), 702–707, 1997.

126. Harada-Shiba, M., Yamauchi, K., Harada, A., Takamisawa, I., Shimokado, K., and Kataoka, K., Polyion complex micelles as vectors in gene therapy—pharmacokinetics and in vivo gene transfer, *Gene Theraphy*, 9(6), 407–414, 2002.

127. Harada, A., Togawa, H., and Kataoka, K., Physicochemical properties and nuclease resistance of antisense-oligodeoxynucleotides entrapped in the core of polyion complex micelles composed of poly(ethylene glycol)–poly(L-lysine) block copolymers, *European Journal of Pharmaceutical Sciences*, 13(1), 35–42, 2001.

128. Bikram, M., Ahn, C. H., Chae, S. Y., Lee, M. Y., Yockman, J. W., and Kim, S. W., Biodegradable poly(ethylene glycol)-*co*-poly(L-lysine)-*g*-histidine multiblock copolymers for nonviral gene delivery, *Macromolecules*, 37(5), 1903–1916, 2004.

129. Choi, Y. H., Liu, F., Kim, J. S., Choi, Y. K., Park, J. S., and Kim, S. W., Polyethylene glycol-grafted poly-L-lysine as polymeric gene carrier, *Journal of Controlled Release*, 54(1), 39–48, 1998.

130. Lee, M., Han, S. O., Ko, K. S., Koh, J. J., Park, J. S., Yoon, J. W., and Kim, S. W., Repression of GAD autoantigen expression in pancreas beta-cells by delivery of antisense plasmid/PEG-*g*-PLL complex, *Molecular Therapy*, 4(4), 339–346, 2001.

131. Choi, Y. H., Liu, F., Park, J. S., and Kim, S. W., Lactose-poly(ethylene glycol)-grafted poly-L-lysine as hepatoma cell-targeted gene carrier, *Bioconjugate Chemistry*, 9(6), 708–718, 1998.

132. Nah, J. W., Yu, L., Han, S. O., Ahn, C. H., and Kim, S. W., Artery wall binding peptide–poly(-ethylene glycol)-grafted-poly(L-lysine)-based gene delivery to artery wall cells, *Journal of Controlled Release*, 78(1-3), 273–284, 2002.

133. Jeong, Y. I., Nah, J. W., Lee, H. C., Kim, S. H., and Cho, C. S., Adriamycin release from flower-type polymeric micelle based on star-block copolymer composed of poly(gamma-benzyl L-glutamate) as the hydrophobic part and poly(ethylene oxide) as the hydrophilic part, *International Journal of Pharmaceutics*, 188(1), 49–58, 1999.

19 Hydrotropic Polymer Micelles for Cancer Therapeutics

Sang Cheon Lee, Kang Moo Huh, Tooru Ooya, and Kinam Park

CONTENTS

19.1   Introduction 385
19.2   General Review of Polymer Micelle Drug Carriers 387
      19.2.1   Preparative Methods of Polymer Micelles Based on Block Copolymers 387
      19.2.2   Characterization and Properties of the Polymeric Micelles 388
      19.2.3   Key Parameters for Morphology and Stabilization
              of the Polymeric Micelles 389
      19.2.4   Drug Loading, Solubilization, and Drug Release 390
19.3   Polymer Micelles for Cancer Chemotherapy 391
      19.3.1   Polymer Micelles as Carriers of Anti-Cancer Drugs 391
      19.3.2   Polymer Micelles for Solubilization of Poorly Soluble Anti-Cancer Drugs 391
      19.3.3   Targeting Systems Using Polymer Micelles 392
      19.3.4   Other Applications of Polymer Micelles for Cancer Therapy 393
19.4   Hydrotropic Polymer Micelles 393
      19.4.1   Hydrotropy and Hydrotropic Agents 393
      19.4.2   Hydrotropic Agents in Pharmaceutics 394
      19.4.3   Mechanistic Studies and Structure–Property Relationship of Hydrotropic
              Solubilization 395
      19.4.4   Design Strategy of Hydrotropic Polymer Micelle Systems 395
      19.4.5   Hydrotropic Polymer Micelles as Carriers for Anti-Cancer Drugs 400
19.5   Conclusions and Future Perspectives.. 404
References 404

19.1 INTRODUCTION

Recently, polymer micelles derived from amphiphilic block copolymers in an aqueous phase have attracted attention as a promising formulation for poorly water-soluble drugs.[1–4] Block copolymer micelles are nanosized particles with a typical core-shell structure.[5] The core solubilizes the hydrophobic drugs, while the corona allows the suspension of micelles in an aqueous medium. The use of block copolymer micelles as drug-carrying vehicles was proposed by Ringsdorf's group in the 1980s.[6] The rationale for incorporating low-molecular-weight drugs into micelles is to overcome the problems of drug formulations such as toxic side effects, poor pharmacokinetics, and limited solubility in water. Hydrophobic drugs can be solubilized into hydrophobic core structures of polymeric micelles, and solubilized at higher concentrations than their intrinsic water-solubility.[7–10] The chemical composition of polymeric micelles is attractive because it can be tailored to have desirable physicochemical properties for drug solubilization. Therefore, various amphiphilic block copolymers have been synthesized and investigated for micelle formulations of poorly soluble drugs.[11–14] In most polymeric micelles, hydrophobic drugs can be incorporated into the hydrophobic core of micelles by hydrophobic interaction and other additional interactions, such as the metal–ligand coordination bond, receptor–ligand interaction, and the electrostatic interaction.[15–17] It is believed that the more compatible drugs are with the cores of the micelles, the greater their ability to be dissolved.

As drug carriers, polymeric micelles have been widely investigated for a diverse class of anti-cancer agents including paclitaxel, adriamycin, and methotrexate.[3,12,18,19] However, several limitations, for example, limited drug-solubilizing ability, poor stability in water after drug loading, and the lower stability with the higher loading content, are known to limit successful clinical applications.[20,21]

Hydrotropy has many advantages in enhancing the water solubility of poorly soluble drugs. Hydrotropes (or hydrotropic agents) self-associate to form noncovalent assemblies of nonpolar microdomains to solubilize poorly water-soluble solutes.[22] The high concentration of hydrotropes (greater than 1 M) is a key factor in enhancing water-solubility of poorly soluble solutes.[23,24] Hydrotropes often exhibit a higher selectivity in solubilizing guest hydrophobic molecules than in surfactant micelles. Thus, identifying hydrotrope structures that effectively solubilize a specific drug molecule is important. Recently, Park et al. have examined a number of hydrotropes for solubilization of paclitaxel, a model poorly soluble anti-cancer agent.[25] Through screening the effective structures of the hydrotropes for paclitaxel solubilization, N,N-diethylnicotinamide (DENA) and N-picolylnicotinamide (PNA) were found to be the best hydrotropes. They increased the water-solubility of paclitaxel by 3–5 orders of magnitude over its normal solubility in pure water (about 0.3 μg/mL). They also examined polymers and hydrogels, based on DENA and PNA hydrotropes, to develop new polymeric solubilizing systems maintaining the benefits of hydrotropy.[26] The hydrotropic property of the hydrotropes was maintained in their polymeric forms, and the highly localized concentration of the hydrotrope in polymers and hydrogels was found to be a main contributor to effective solubilization of paclitaxel. However, upon dilution in aqueous media, drugs solubilized in hydrotropic polymers precipitate due to the low physical stability of the formulations. Thus, a need exists for the hydrotropic formulations with a high stability.

To overcome the limitations of current polymeric micelles, hydrotropic polymer micelle systems that have had core components designed with a high solubilizing capacity for poorly soluble anti-cancer agents, like paclitaxel, show enhanced long-term stability even at high drug loading.[27] The key in the polymer system designs is to introduce the identified structure of hydrotropes for a specific drug to the drug-solubilizing micellar cores. The benefit of using self-assembled structures is the congestion of hydrotropic moieties with high local concentration at the micellar inner core. To date, many polymeric micelles have shown limited solubilizing capacity for paclitaxel, and, in most cases, the maximum content of paclitaxel loaded in micelles was around 20 wt%.[20,21] In addition to the limited loading capacity, the stability of the drug-loaded polymeric micelles is poor. Stability decreases as the drug loading increases. The poor colloidal stability of existing polymeric micelles is normally caused by the enhanced hydrophobicity of micelles after solubilization of paclitaxel, leading to aggregation of micelles. Because the hydrotropic moiety is relatively hydrophilic, it retains the hydrophilicity of the colloidal micelles even after loading of large amounts of paclitaxel.

This chapter gives a general review on recent developments of polymer micelles for cancer chemotherapeutics, the basic concept of hydrotropic solubilization, and a systemic rationale of hydrotropic polymer micelle systems. In particular, it focuses on the design strategy and unique properties of hydrotropic micelle formulations for the delivery of poorly soluble anti-cancer drugs.

19.2 GENERAL REVIEW OF POLYMER MICELLE DRUG CARRIERS

Micelles are defined as colloidal dispersions, including a category of a dispersed system that consists of the particulate matter or the dispersed phase in a continuous phase or dispersion medium.[28,29] Micellization phenomena have been studied widely in the field of drug delivery. Amphiphilic small molecules have been utilized for preparing micelles, but amphiphilic macromolecules are better than the small molecules because of their lower critical micelle concentration (CMC) and higher stability. Polymeric micelles represent a separate class of micelles. Polymeric micelles consist of amphiphilic block copolymers which include multi-block copolymers, AB-type di-block copolymers, and ABA-type tri-block copolymers with a hydrophobic core (A or B segment) and a hydrophilic shell (A or B segment). Representative block copolymers are summarized in Table 19.1.[5,9,12] The amphiphilic block copolymer forms spherical micelles in aqueous environments with a hydrophilic shell and hydrophobic core parts (Figure 19.1). In general, the core part represents a molten-liquid globule and a swollen, hydrophilic corona. These micelles have a high solubilization capacity of poorly soluble drugs, and good potential to minimize drug degradation, loss, and harmful side effects. In this chapter, polymeric micelles for drug delivery applications are briefly summarized in terms of methods/mechanism of micelle formation, influencing factors of stabilization, and solubilization of poorly soluble drugs.

19.2.1 PREPARATIVE METHODS OF POLYMER MICELLES BASED ON BLOCK COPOLYMERS

There are two ways to prepare polymeric micelles: the direct dissolution method and the dialysis method (Figure 19.2).[1] One can select a method depending on the block copolymer's solubility in an aqueous medium. As for direct dissolution, the block copolymer is simply dissolved in water or buffers at a concentration above its CMC. If needed, the temperature is elevated during the micelle formation. A good example of this method is poly(ethylene oxide)-*block*-poly(propylene oxide)-*block*-poly(ethylene oxide) (PEO–PPO–PEO; Pluronics®).[30,31] If block copolymers are not easily water soluble, the dialysis method is useful to prepare polymeric micelles. In this case, block copolymer is dissolved in a water-miscible organic solvent such as dimethylformamide (DMF), dimethylsulfoxide (DMSO), and acetonitrile. The mixed solution is then dialyzed against distilled water. During the dialysis, the organic solvent is removed by exchanging the outer medium (distilled water). The pore size of the semi-permeable membrane (usually SpectraPor®) should be carefully selected, based on the expected size of polymeric micelles. The choice of organic solvent is important because the solvent effects the size and size-population distribution of polymeric micelles. La et al. reported that the use of DMSO as the organic solvent gave rise to PEO-*b*-poly(β-benzyl L-aspartate) (PEO–PBLA) micelles, of which size was only 17 nm.[32] However, only

---

TABLE 19.1
Representative Block Copolymers for Poorly Soluble Drugs

Block Copolymers

Pluronics® (PEO–PPO–PEO)
Methoxy-PEG-*b*-poly(D,L-lactide) (PEG–PLA)
PEG-*b*-poly(lactic acid-*co*-glycolic acid)-*b*-PEG
Methoxy-PEG-*b*-polycaprolactone (PEG–PCL)
Poly(β-maleic acid)-*b*-poly(β-alkylmaleic acid alkyl ester)
Poly(*N*-isopropyl acrylamide)-*b*-PEG
Poly(aspartic acid)-*b*-PEG
PEG-*b*-phosphatidyl ethanolamine

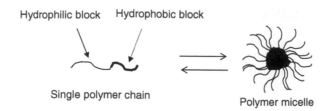

Hydrophilic block   Hydrophobic block

Single polymer chain

Polymer micelle

FIGURE 19.1 General scheme of micelle formation using amphiphilic block copolymers.

6% of the PEO–PBLA formed the micelle. When dimethylacetamide (DMAc) was used as the organic solvent, the size was 19 nm with a high yield. Therefore, one should examine the effect of organic solvents on the size and size distribution, as well as the composition, of block copolymers.

19.2.2 CHARACTERIZATION AND PROPERTIES OF THE POLYMERIC MICELLES

The micellization of di- or tri-block copolymers is usually characterized by the following methods:

1. Dye-solubilization methods using 1,6-diphenyl-1,3,5-hexatriene (DPH) and pyrene (Figure 19.3)[11,33–37]
2. Static and dynamic light-scattering measurements.[32,38,39]

The dye-solubilization method is useful to determine CMC, local viscosities, and polarity of the micellar core.[40] For example, DPH is preferentially partitioned into the hydrophobic core of micelles with the formation of micelles, which causes an increase in the absorbance of the dyes. When the concentration of the block copolymers changes, CMC can be determined as the cross-point of the extrapolation of the change in absorbance over a wide concentration range. Pyrene has

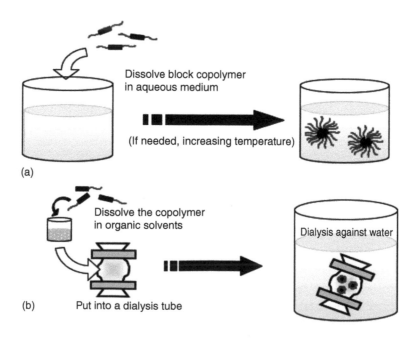

FIGURE 19.2 Preparation of micelles by direct dissolution (a) and dialysis methods (b).

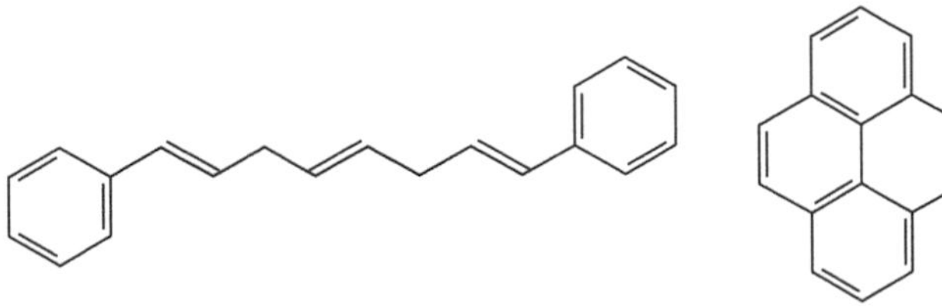

1,6-Diphenyl-1,3,5-Hexatriene (DPH)   Pyrene

FIGURE 19.3 Chemical structures of DPH and pyrene.

been well-studied as a fluorescent probe to determine CMC, as well as DPH.[4,41] Below CMC, a small amount of pyrene is solubilized in water (the saturated concentration is as low as 6×10^{-7} M). In the presence of polymeric micelles, pyrene is preferentially solubilized in the nonpolar micelle core. By increasing the concentration of the block copolymer in the presence of pyrene, an increase in emission intensity can be seen, at which point pyrene becomes associated with the micelle core. When the concentration of the block copolymers changes, CMC can be determined as the point from which to extrapolate the change in absorbance in a wide range of the concentration.

Weight-average molecular weight (M_W) of micelles is calculated by the results of static light-scattering (SLS) measurements using a Debye plot[39]:

$$K(C - \text{CMC})/(R - R_{\text{CMC}}) = 1/M_W + 2A_2(C - \text{CMC})$$

where K indicates $4\pi^2(\mathrm{d}n/\mathrm{d}c)^2/N_A\lambda^4$, R and R_{CMC} the excess Rayleigh ratios at concentration C and CMC, n is the refractive index of solution at CMC, $\mathrm{d}n/\mathrm{d}c$ is the refractive index increment, N_A is Avogadro's number, λ is the wavelength of the laser light, and A_2 is the second virial coefficient. Hydrodynamic radii (R_H) of micelles can be determined by dynamic light-scattering (DLS) measurements using the Stokes–Einstein equation[39,41]:

$$R_H = k_B T/(3\pi\eta D)$$
$$D = \Gamma/K^2$$

where k_B is the Boltzmann constant, T is the absolute temperature, η is the solvent viscosity, D is the diffusion coefficient obtained from the average characteristic line width (Γ), and K^2 is the magnitude of the scattering vector. Because polymeric micelles are a kind of spherical assembly, the diffusion coefficients should be independent of the detection angle, due to the undetectable rotational motions.

19.2.3 KEY PARAMETERS FOR MORPHOLOGY AND STABILIZATION OF THE POLYMERIC MICELLES

As for AB-type di-block copolymers, one can imagine that the di-block copolymer can form spherical micelles in an aqueous solution. If the hydrophilic block is too long, the di-block copolymers exist in water as a unimer or a macromolecular micelle. On the other hand, if the hydrophobic block is too long, it can show nonmicellar morphology, such as rods and lamellar structures. Eisenberg and his group reported that different morphologies are formed at equilibrium, near-equilibrium, and nonequilibrium conditions.[42] When the length of the hydrophobic segment is significantly shorter than that of the hydrophilic part, the block copolymer usually forms nonspherical micelles. The formation of "crew-cut" aggregates has been suggested by a force-balance effect between the hydrophobic part's degree of stretching, the interfacial energy of the micelle core with water, and the interaction between hydrophilic chains as a shell.[42–44] Copolymer compositions, concentrations, and organic solvents used for the micelle preparation significantly affect the morphology.

From the pharmaceutical point of view, stability of polymeric micelles in the blood stream is important because administration of micelles into the body always accompanies dilution, which causes dissociation of the micelles.[1] The stability of polymeric micelles both in vitro and in vivo depends on their CMC values. However, the CMC values have been used as a thermodynamic stability: Equilibrium between unimers and micelles can be shifted below or above the CMC. On the other hand, kinetic stability, which represents the actual rate of micelle dissociation below the CMC, depends on the size of the hydrophobic part, the physical state of the micelle core, and the hydrophilic/hydrophobic ratio. The hydrophobic block plays an especially critical role. For example, an increase in the length of the hydrophobic block at a given length of the hydrophilic block causes a significant decrease in the CMC value, and an increase in the stability of the micelle. On the other hand, an increase in the length of hydrophilic block leads to only a small rise of CMC value. Additionally, CMC values of di-block copolymers are generally lower than those of tri-block copolymer of the same molecular weight and hydrophilic/hydrophobic ratio. Thus, the molecular design of block copolymers for polymeric micelles should be optimized for pharmaceutical applications. To increase stability, unimolecular micelles that mimic polymeric micelles regarding their morphological properties have been proposed since 1991.[45,46] These micelles are intrinsically stable upon dilution, because their formation is independent of polymer concentration. Both dendrimers and star polymers have been designed as unimolecular micelles.[45,47–49]

19.2.4 DRUG LOADING, SOLUBILIZATION, AND DRUG RELEASE

Drug loading methods mostly depend on the methods of micelle formation.[50,51] In the case of micelles prepared by the direct dissolution in water, a stock solution of drugs in some organic solvents is prepared in an empty vial. Then the organic solvent is allowed to evaporate, followed by adding an aliquot of water containing the block copolymer. The oil-in-water emulsion method is another method: the drug, in a nonpolar solvent such as chloroform, is added dropwise to the water containing the block copolymer. In the case of the dialysis method, the drug is mixed with the block copolymer in common organic solvents, and the organic solvents in dialysis bags are exchanged with a large amount of water to induce assembly of micelles. Upon micellization, the poorly soluble drugs are trapped in the hydrophobic inner cores of micelles.

It is known that the solubilization of poorly soluble drugs in the micelle core strongly depends on the types and efficacy of the interactions between the solubilized drug and the hydrophobic micelle core.[1] Drug-loading capacity depends on the compatibility between the loaded drug and the hydrophobic core. Drug characteristics such as polarity, hydrophobicity, and charge often affect compatibility; consequently, the structure and the length of hydrophobic block should be carefully examined to determine maximum compatibility with any poorly soluble drugs. A good parameter that has been used to assess compatibility is the Flory–Huggins interaction parameter, χ_{sp}. This parameter is described by the following equation[1]:

$$\chi_{sp} = (\delta_s - \delta_p)^2 V_s / RT,$$

where χ_{sp} is the interaction parameter between the solubilized drug (s) and the hydrophobic block part (p), δ is the Scatchard–Hildebrand solubility parameter of the hydrophobic block part, and V_s is the molar volume of the solubilized drug. If the χ_{sp} shows low value, the compatibility between drug and hydrophobic block part is great. By using this parameter, Gadelle et al. suggested the following mechanism for solubilization of aromatic solutes in PEO-b-PPO-b-PEO[28]:

1. The addition of apolar solutes (drugs) promotes aggregation of the block copolymer molecules.

2. The micelle core contains some water molecules.
3. Solubilization is initially a replacement process in which water molecules are displaced from the micelle core by the drug.

Drug-release kinetics generally depend upon the rate of the drug's diffusion from the micelles. There are several factors that affect the release rate: Polymer–drug interactions, localization of the drug within the micelle, the physical state of the micelle core, the length of the hydrophobic part of the block copolymer, the molecular volume of the drug, and the physical state of the drug in the micelle.[1] For instance, if the drug is located in the micelle core as a crystal, it may act as a reinforcing filter. If the magnitude of interaction between the copolymer and the surface of crystallite is strong, the drug crystal may cause the glass-transition temperature to increase. Thus, solubilization of the drug, which depends on the characteristics of both the drug and the block copolymer, is important in designing polymeric micelles for poorly soluble drugs, such as anti-cancer drugs.

19.3 POLYMER MICELLES FOR CANCER CHEMOTHERAPY

19.3.1 POLYMER MICELLES AS CARRIERS OF ANTI-CANCER DRUGS

Significant progress in cancer therapy has been made with the development of new anti-cancer agents and related technologies. However, the major challenge remains in delivery technologies that can effectively, selectively deliver anti-cancer agents to tumor sites, and avoiding systemic toxicity and adverse effects to normal tissues or cells. Recently, polymer micelles have attracted attention as a novel drug carrier system, in particular for anti-cancer agents, due to its various, promising properties. These nanoscaled delivery systems have been developed to demonstrate a series of required properties as drug carriers, such as biocompatibility, stability both in vitro and in vivo, targeting, and drug loading capacity. Recent research into the development of polymer micelle delivery system has focused on two issues, drug solubilization and targeted delivery system, to enhance the efficacy of the delivered drug.[52–55]

19.3.2 POLYMER MICELLES FOR SOLUBILIZATION OF POORLY SOLUBLE ANTI-CANCER DRUGS

Because many anti-cancer drugs and drug candidates are water-insoluble or poorly water-soluble, applicability of solubilizing systems to these drugs is indispensable. For instance, paclitaxel is one of the most effective anti-cancer agents, and currently formulated with use of surfactants, including Cremophor EL, due to its extremely poor water solubility.[56] Despite good efficacy of the formulation, its clinical use is limited by serious side effects resulting from use of Cremophor EL.[57–59] Diverse approaches—such as chemical and physical modification, use of a cosolvent, and emulsification—have been explored for increasing the aqueous solubility of poorly soluble drugs.[60–66] Although several systems have shown high solubilizing effects, they have been limited by their stability and toxicity. Thus, development of nontoxic and stable effective solubilizing systems for poorly soluble drugs is important for enhancing drug efficacy with high bioavailability.

Over the past decade polymer micelles have been extensively investigated as a potent drug carrier that can effectively solubilize poorly soluble anti-cancer drugs. Because polymer micelles have been made with biocompatible polymers such as poly(ethylene oxide), poly(D,L-lactide), etc., they are much less toxic than other solubilizing agents such as cosolvents or surfactants.[20,58] Additionally, recent advances in polymer synthesis make it possible to sophisticate the design of polymer composition and structure so that the resulting micelle shows high solubilizing capacity with prolonged stability.

Presently a number of polymer micelle systems with different compositions and structures have been tried to solubilize and deliver various anti-canter drugs. Several examples of the polymer micelles used for solubilization of poorly water-soluble anti-cancer drugs are summarized in Table 19.2.[2–14,19,20,42,52,54] Drug-loading tests have shown that, in most polymer micelle systems, not only molecular design of block copolymers, but also the drug incorporation method are crucial factors to increase the solubilizing capacity.[1,14]

19.3.3 TARGETING SYSTEMS USING POLYMER MICELLES

Unique core-shell structure of polymer micelle systems with a nanoscaled size may ensure prolonged circulation times that are favorable for passive drug targeting.[10] Furthermore, polymer micelles can be selectively accumulated to tumor sites by the enhanced permeability and retention (EPR) effect, which makes them useful as a tumor-targeting system. An alternative, passive-targeting strategy is to make smart micelles using stimuli-responsive amphiphilic block copolymers that can sense and respond to the unique tumor environment.[54,67] Kataoka et al. have developed a novel intracellular pH-sensitive polymeric micelle drug carrier that controls the systemic, local, and subcellular distributions of pharmacologically-active drugs using amphiphilic block copolymers, poly(ethylene glycol)–poly(aspartate hydrazone adriamycin).[68] Because the anti-cancer drug, adriamycin, is conjugated to the hydrophobic segments through acid-sensitive hydrazone links, this polymer micelle can stably preserve drugs under physiological conditions (pH 7.4), but selectively release the drug by sensing the intracellular pH decrease in endosomes and lysosomes (pH 5–6).

The active targeting is usually achieved by attaching specific-targeting ligand molecules, such as monoclonal antibodies, to the micelle surface, which provides a preferential accumulation in the tumor sites or cells. Such ligand and other targeting moieties are introduced in other literatures.[69] When linked with tumor-targeting ligands, polymer micelles can be used to target tumor sites with

TABLE 19.2
Use of Polymer Micelle Systems for Solubilization of Poorly Soluble Anti-Cancer Agents

Polymer Composition[a]	Loaded Drug	Loading Content (wt%)[b]
PEG–PDLLA	Paclitaxel	5–25
	Methotrexate	3.7–2.8
PEG–PDLLACL	Paclitaxel	15–25
PEG–PGACL	Paclitaxel	8–11
PEtOz–PCL	Paclitaxel	0.5–7.6
PEG–PCL	Paclitaxel	4.1–20.8
PEG–pHPMAmDL	Paclitaxel	22
PVP-b-PDLLA	Paclitaxel	5
	Docetaxel	4
	Teniposide	10
	Etoposide	20
PEG-poly(aspartate)	Camptothecin	13
PEG–P(Asp(ADR)	Doxorubicin Ellipticine	8.4–7.8
PEG–PBTMC	Doxorubicin	10
Poly(NIPAAm-co-DMAAm)-PLGA		2–8

[a] Calculated by: (weight of loaded drug)×100/(weight of drug-loaded micelle).
[b] PDLLA, Poly(D,L-lactide); PDLLACL, Poly(D,L-lactide-co-caprolactone); PGACL, Poly(glycolide-co-caprolactone); PEtOz, poly(2-ethyl-2-oxazoline); pHPMAmDL, poly(N-(2-hydroxypropyl) methacrylamide lactate); PVP, poly(N-vinyl-pyrrolidone); P(Asp(ADR), adriamycin-conjugated poly(aspartic acid) block copolymer; PBTMC, poly(5-benzyloxytrimethylene carbonate); NIPAAm, N-isopropylacrylamide; DMAAm, N,N-dimethylacrylamide; PLGA, poly(D,L-lactide-co-glycolide).

high affinity and specificity. Recently a multifunctional polymer micelle system has been proposed to endow enhanced tumor selectivity and endosome-disruption property on the carrier.[68] The polymer micelle is designed to expose the cell interacting ligand (biotin) only under slightly acidic environmental conditions of various solid tumors, and show pH-dependent dissociation, causing the enhanced release of the loaded drug from the carrier in early endosomal pH.

19.3.4 OTHER APPLICATIONS OF POLYMER MICELLES FOR CANCER THERAPY

Polymer micelles have supramolecular functionalities that are not available from either individual polymer chains or bulk polymer solids. Based on supramolecular architecture, polymer micelles have much more surface area and a separate core that can be available for chemically or physically introducing other active agents, including optical, radioisotopic, magnetic diagnostic, and photosensitive agents.[70,71] Although the delivery of chemotherapeutic agents by polymer micelle systems is relatively well established, the application for delivering imaging and photosensitive agents remains in the early stages of research.[47,70,71] The polymer design and targeting strategies can be also utilized in delivery of other active agents for cancer treatments. Recently, photodynamic therapy (PDT) has been considered as a promising method of treating cancer and other diseases.[72] PDT involves the systemic administration of photosensitizers followed by a local application of light, introducing photochemical reactions that generate cytotoxical substances that produce necrosis of a cancer cells.[70] Polymer micelles can be a good candidate for delivery of photosensitizers in PDT because of their high solubilizing capacity for hydrophobic drugs and potential targeting property. The delivery of photosensitizers by polymer micelles may allow selective accumulation to solid tumor tissue and resultant higher photocytotoxicity with reduced side effects.[47,72] Also, hydrophobic photosensitizers that are poorly soluble or tend to aggregate in aqueous environments can be effectively solubilized in polymer micelle system, showing high photodynamic efficacy by overcoming problems from intrinsic poor solubility.[70]

19.4 HYDROTROPIC POLYMER MICELLES

19.4.1 HYDROTROPY AND HYDROTROPIC AGENTS

Hydrotropy is a collective molecular phenomenon describing a solubilization process whereby the presence of large amounts of a second solute, called hydrotropes, results in an increased aqueous solubility of a poorly soluble compound.[22,73] Hydrotropic agents are a diverse class of substances with wide industrial usage, including solubilizing agents in drug formulations, agents in the separation of isomeric mixtures, catalysts in heterogeneous phase chemical reactions, and fillers in cleaning formulations and cosmetics. Some examples of hydrotropic agents are nicotinamide and its derivatives, anionic benzoate, benzosulfonate, neutral phenol such as catechol and resorcinol, aliphatic glycolsulfate, and amino acid L-proline.[22,74] Figure 19.4 shows a variety of compounds that are classified as hydrotropic agents. Hydrotropic agents are characterized by a short, bulky, and compact moiety such as an aromatic ring, whereas surfactants have long hydrocarbon chains. In general hydrotropic agents have a shorter hydrophobic segment, leading to a higher water solubility, than that of surfactants. Other examples of hydrotropic materials are sodium gentisate, sodium glycinate, sodium toluate, sodium naphtoate, sodium ibuprofen, pheniramine, lysine, tryptophan, isoniazid, and urea.[22,23] At concentrations higher than the minimal hydrotrope concentration, hydrotropic agents self-associate and form noncovalent assemblies of lowered polarity (i.e., nonpolar microdomains) to solubilize hydrophobic solutes. The self-aggregation of hydrotropic agents is different from surfactant self-assemblies (i.e., micelles), in that hydrotropes form planar or open-layer structures instead of forming compact spheroid assemblies.

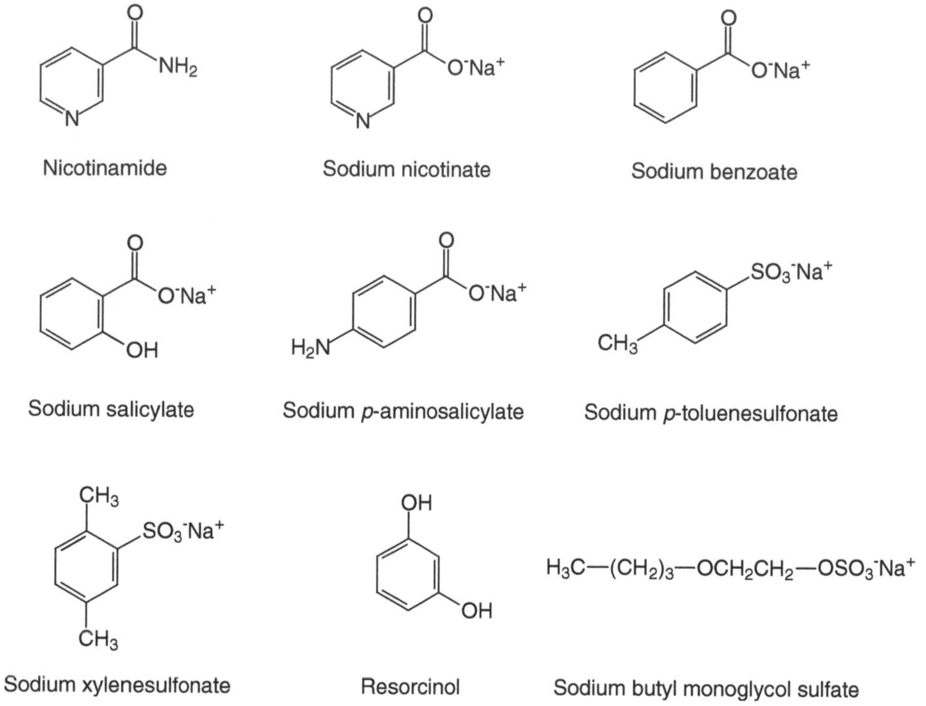

FIGURE 19.4 Chemical structures of hydrotropes used in the literature.

19.4.2 HYDROTROPIC AGENTS IN PHARMACEUTICS

Delivery of poorly water-soluble drugs by oral route has been limited by poor absorption from the gastrointestinal tract (GI).[75] One of the governing factors in the transport of drugs across the intestinal barrier is the carrier-mediated efflux mechanism, which is referred to as multidrug resistance (MDR). This phenomenon is mediated by the increased expression of energy-dependent drug efflux proteins such as P-glycoproteins, multidrug-resistance-associated proteins, and lung-resistance proteins. Since the excretion of the absorbed drugs by this mechanism is the saturable process, the solubility enhancement of poorly water-soluble drugs might overcome the barrier. At high concentration of drugs, the efflux protein will be saturated, thereby resulting in the increase of the absorbed amount of drugs and bioavailability. The hydrotrope approach is a promising new method with great potential for poorly soluble drugs. Using hydrotropic agents is one of the easiest ways of increasing water-solubility of poorly soluble drugs because it only requires mixing the drugs with hydrotropes in water.[24,25] Hydrotropic agents are commonly used to enhance the water-solubility of poorly soluble drugs, and in many instances the water-solubility of these drugs is increased by orders of magnitude. The use of hydrotropic agents offers numerous benefits over other solubilization methods such as micellar solubilization, cosolvency, and salting-in.[74] To date, various hydrotropic agents have been utilized to enhance the aqueous solubility of many hydrophobic drugs including paclitaxel, ketoprofen, diazepam, allopurinol, indomethacin, and griseofulvin. In most cases, the high concentration of hydrotropes is key to increasing water solubility of poorly soluble drugs. Thus, the approach to increase the local concentration of hydrotropes may provide an opportunity to develop new hydrotrope-based solubilizing systems. Despite the advantages of hydrotropes, application of low molecular weight hydrotropes in drug delivery has not been practical because it may result in absorption of a significant amount of hydrotropes themselves into the body, along with the drug. Recently polymeric forms of hydrotropes, such as hydrotropic polymers and hydrogels, were developed to overcome the drug formulations based on low molecular weight hydrotropes.[26] The polymeric hydrotropes were found to maintain the hydrotropic activity of its low molecular weight hydrotropes. Their solubilizing ability for paclitaxel was well examined, and the solubilization mechanism correlated with the assembly of polymeric hydrotropes. However, the low physical stability of drug formulations is also reported to limit the practical applications of these formulations.

19.4.3 Mechanistic Studies and Structure–Property Relationship of Hydrotropic Solubilization

The term hydrotropy does not mean a specific mechanism, but implies a collective solubilization phenomenon that is incompletely understood. There have been various theoretical and experimental studies aimed at explaining hydrotropic solubilization. Most of proposed mechanisms can be classified into following two schemes. The first one involves the complex formation between the hydrotrope and the solute. As a representative example, nicotinamide—a nontoxic vitamin B_3—has been shown to enhance the solubilities of a wide variety of hydrophobic drugs through complexation. Using nicotinamide and its derivatives, such as N-methylnicotinamide and N,N-diethylnicotinamide (DENA), Rasool et al. described the aromaticity of pyridine ring, which might promote the stacking of molecules through its planarity, as a most significant contributor in complexation, because the abilities of aromatic amide ligands to enhance the aqueous solubilities of tested drugs was higher than those of the aliphatic amide ligands.[76] The other mechanism for hydrotropic solubilization is self-association of the hydrotrope in an aqueous phase. This view is supported by experimental data, proving that some hydrotropes, including nicotinamide and aromatic sulfonates, associate in aqueous solutions. Using nicotinamide-riboflavin system, Kildsig et al. showed that the self-association of nicotinamide contributed to the solubility increase of riboflavin, rather than complexation between two species.[77]

Each hydrotropic agent is effective in increasing the water solubility of selected hydrophobic drugs, and no hydrotropic agents were found that are universally effective. Therefore, the structure–activity relationship between selected pairs of the hydrotropic agent and the drug is another important concern to discuss. Park et al. reported on the intensive studies of the elucidation of structure–activity relationship for the hydrotropic solubilization of paclitaxel using not only nicotinamide and its analogues as aromatic amides, but also various ureas as aliphatic amides.[25] They showed that nicotinamide enhanced the aqueous solubility of paclitaxel to a greater extent than the aliphatic analogues of nicotinamide such as nipecotamide and N,N-dimethylacetamide. In addition, the aqueous solubility of paclitaxel was found to be strongly dependent on the alkyl substituent on the amide nitrogen of nicotinamide. DENA of 3.5 M enhanced the aqueous solubility of paclitaxel up to 39.07 mg/mL, whereas N,N-dimethylnicotinamide showed paclitaxel solubility of 1.77 mg/mL at the same concentration. They synthesized new hydrotropic agents based on the structure of nicotinamide by varying the substituent to the amide nitrogen with a pyridine ring or an ally group. The aqueous solution of N-picolylnicotinamide (PNA) (3.5 M), having another pyridine ring as a substituent on the amide nitrogen, was highly effective in the enhancement of paclitaxel solubility (29.44 mg/mL), compared with the hydrotropic effect of nicotinamide on the paclitaxel (0.69 mg/mL). Besides, N-allylnicotinamide highly contributed to the enhancement of paclitaxel solubility (14.18 mg/mL) compared to that of N,N-dimethylnicotianmide (1.77 mg/mL). The most interesting find is that the threshold concentration where the association occurred is consistent with the threshold concentration where paclitaxel solubility began to increase.

19.4.4 Design Strategy of Hydrotropic Polymer Micelle Systems

Hydrotropic polymer micelles were firstly developed based on the self-assemblies of amphiphilic block copolymers containing the hydrophilic block, and the hydrophobic block containing pendent hydrotropic moieties. The main strategy for hydrotropic systems is to maximize the local concentration of hydrotropic structure because the drug solubilizing capacity increases when increasing the local concentration of hydrotropes.[26] The hydrophobic block possesses hydrotropes assembled in the limited volume in micellar core region. Considering several hundreds of aggregation numbers of micelles, the local concentration of hydrotropes would be maximized as compared to other polymeric hydrotropes. In addition to the enhanced solubilizing ability for poorly soluble drugs, the intrinsic hydrophilicity of hydrotropic structure can retain the colloidal stability by compensation of enhanced hydrophobicity of drug-loaded polymer micelles. Using atom-transfer radical polymerization (ATRP) of modified hydrotropes, Park et al. started the synthesis of amphiphilic block copolymers consisting of PEG and poly(2-(4-(vinylbenzyloxyl)-N,N-diethylnicotinamide) (PDENA) or poly(2-(4-(vinylbenzyloxyl)-N-picolylnicotinamide) (P(2-VBOPNA)) as shown in Figure 19.5.[27]

PEG-*b*-PDENA copolymers associated to form polymeric micelles with the hydrotrope-rich core and the PEG outer shell in aqueous media. The mean hydrodynamic diameters were in the range of 30–100 nm. In these polymer systems, the solubilization of paclitaxel is based on a synergistic effect of the unique micelle characteristics and hydrotropic activity. Hydrotropic micelles demonstrated not only higher loading capacity (up to 37 wt% of paclitaxel) but also enhanced physical stability in aqueous media.[27] The enhanced stability is due to the intrinsic hydrophilicity of hydrotropic moieties in the micellar core. The drug loading into the hydrotropic polymeric micelle core is mainly based on the attractive interactions between the hydrotropic moiety and paclitaxel.

The structure of hydrotropic polymer micelles can be varied by introducing a diverse class of hydrotropic structures into the block copolymers through ATRP.[28,29,78–80] Table 19.3 lists some of the block and graft copolymers consisting of poly(ethylene glycol) (PEG) and the hydrotropic polymers that have been synthesized based on the molecular structures of identified hydrotropic agents for paclitaxel, such as DENA and PNA.

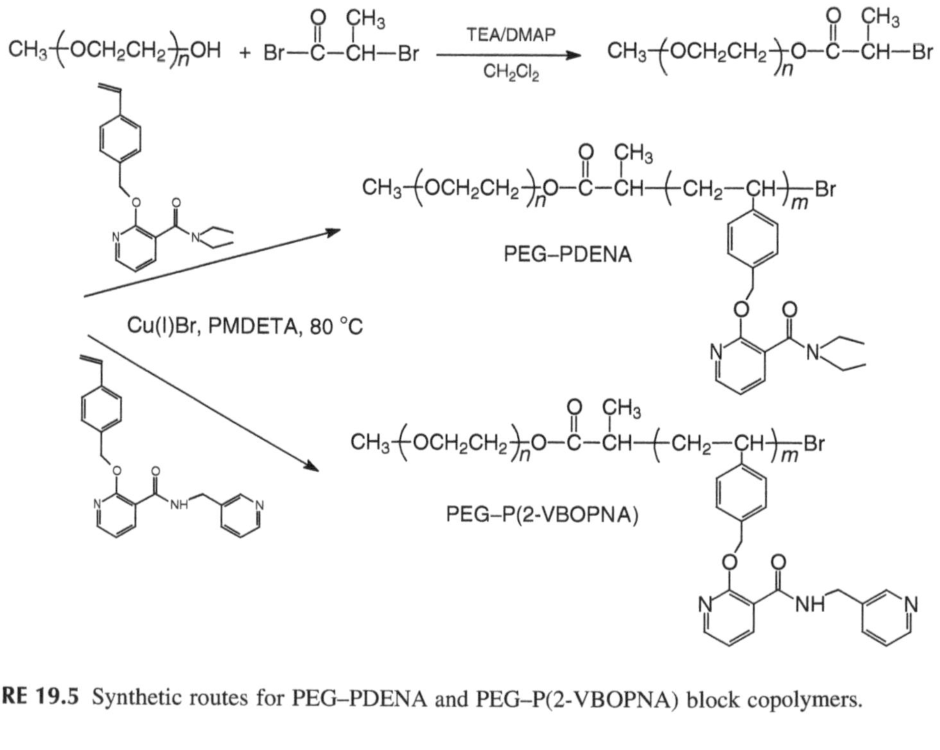

FIGURE 19.5 Synthetic routes for PEG–PDENA and PEG–P(2-VBOPNA) block copolymers.

The main components necessary for the synthesis of the block copolymers are PEG modified with bromine or chlorine, and the modified hydrotropic agent with polymerizable vinyl, acryl, and styryl groups. Table 19.4 lists the useful hydrotropic agents modified with double bond functionality. The relevant combination of monomers listed in Table 19.4 can lead to a diverse class of the copolymers capable of making the hydrotropic micelles.

In addition to the structural variation, a new property such as stimuli-responsibility can be endowed with the hydrotropic polymer micelles. One example is stimuli-responsive hydrotropic polymer micelles which were derived from PEG–P(2-VBOPNA) copolymers. In this system, PEG is freely water-soluble, independent of pH, whereas P(2-VBOPNA) is known to be soluble in water only when the hydrotropic PNA groups are protonated. P(2-VBOPNA) is soluble in low pH but loses its water-solubility above a critical pH (pH 2.5). Thus, the PEG–P(2-VBOPNA) copolymers, as expected, dissolved molecularly as unimers at low pH but become amphiphilic above pH 2.5. The combination of PEG and P(2-VBOPNA) in a block copolymer provides an opportunity to undergo pH-responsive micellization by controlling solution pH. Figure 19.6 shows pH-dependency of scattering intensity of PEG–P(2-VBOPNA) by dynamic light scattering.

As solution pH increases, scattering intensity increases abruptly above pH 2.0, accompanied by the formation of micelles around pH 3.0 of which sizes were about 35 nm. As pH of the aqueous solution of the block copolymers is increased from 2 to 3, the degree of protonation of PNA groups decreased, and thus the P(2-VBOPNA) block becomes progressively more hydrophobic, as expected. ^{1}H NMR analysis is also a good tool to confirm the pH-responsive micellization of PEG–P(2-VBOPNA) (Figure 19.7).

At pH 1, the block copolymers are fully solvated, and thus the signals assigned for the each block are clearly visible. On the other hand, at pH 7.4, the resonance peaks of P(2-VBOPNA) disappear while the signals from PEG are still evident, which indicates the poor solvation and/or limited molecular motion of P(2-VBOPNA) blocks. This result strongly supports the formation of micelles consisting of P(2-VBOPNA) as a hydrophobic inner core and PEG as a hydrated outer shell. Using this unique pH-sensitive micelle system, it may be suggested that PEG-*b*-P(2-VBOPNA) allows a unique drug-loading method into polymeric micelles, where drug molecules are initially solubilized exclusively by the interaction with hydrotropic P(2-VBOPNA) block at low pH, and then drugs are possibly encapsulated into inner core of micelles upon pH-responsive micellization. The use of hydrotropic block copolymers showing pH-responsive micellization does not need organic solvents during the drug loading process.

The ideal oral delivery systems for anti-cancer drugs need the two following physicochemical properties to maximize oral bioavailability of drugs: the fast drug release during the transit of GI, and the prolonged retention time in the GI tract. The former was known to be the characteristic of hydrotropic polymer micelles. Thus, if the design of hydrotropic polymer micelles showing mucoadhesion property is possible, the idealized delivery systems can be developed based on hydrotropic polymer micelles. The study on the degradation kinetics of paclitaxel in aqueous

---

TABLE 19.3

Exemplary Copolymers of PEG and Hydrotropic Polymers Synthesized from Modified Hydrotropic Agents

Poly(ethylene glycol)-*block*-poly(2-(4-vinylbenzyloxy)-*N*,*N*-diethylnicotinamide)

Poly(ethylene glycol)-*block*-poly(2-(4-vinylbenzyloxy)-*N*-picolylnicotinamide)

Poly(ethylene glycol)-*block*-poly(2-(4-vinylbenzyloxy)-nicotinamide)

Poly(oligoethylene glycol methacrylate-*co*-poly(2-(4-vinylbenzyloxy)-*N*,*N*-diethylnicotinamide)

Poly(oligoethylene glycol methacrylate-*co*-poly(2-(4-vinylbenzyloxy)-*N*-picolylnicotinamide)

Poly(oligoethylene glycol methacrylate-*co*-poly(2-(4-vinylbenzyloxy)-nicotinamide)

TABLE 19.4
Modified Hydrotropic Monomers That Can Be Used in the Synthesis of Hydrotropic Polymer Micelles

| 6-acryloyl-N-picolylnicotinamide | 6-(2-(acryloyl) ethoxyethoxy-ethoxy)-N-picolylnicotinamide | 6-(4-vinylbenzyl oxy)-N-picolyl nicotinamide | 2-(2-(acryloyl) ethoxyethoxyethoxy)-N, N-diethylnicotinamide | 2-(4-vinylbenzyloxy)-N, N-diethyl nicotinamide | 2-acryloyl-N, N-diethylnicotinamide |

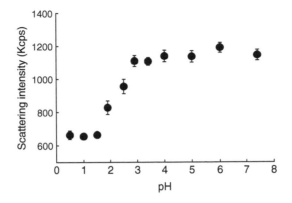

FIGURE 19.6 Variation of scattering intensity (●) as a function of pH for PEG_{5000}-$P(2\text{-VBOPNA})_{2070}$ at 5 mg/mL ($n = 3$).

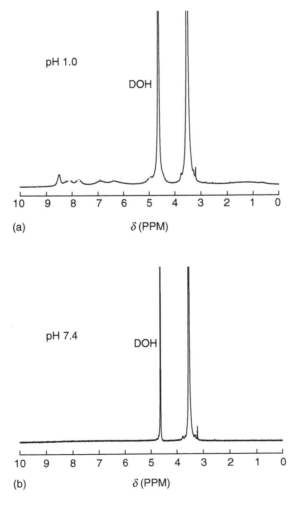

FIGURE 19.7 ^1H NMR spectra of PEG_{5000}-$P(2\text{-VBOPNA})_{2070}$ at 20 mg/mL in D_2O: (a) pH 1.0; (b) pH 7.4.

solutions at different pH values shows that a maximum stability of paclitaxel occurred in the pH 3–5 region.[61] This provides a perfect opportunity to prepare mucoadhesive hydrotropic polymer micelle formulation for oral paclitaxel delivery. One of the best mucoadhesive polymers is poly(acrylic acid) (PAA), which is highly adhesion to biological mucosa at a pH lower than 5. PAA at neutral pH is highly charged and highly water-soluble, and can provide a hydrophilic segment of the block copolymers. Upon contact with the stomach's low pH, the polymer becomes highly mucoadhesive. Alternatively, PAA can be treated at pH 5 to provide the mucoadhesive property. Because of the highly hydrophilic properties of PAA, it can replace the PEO segment of the block copolymers. Thus, the design of the block copolymer consisting of PAA and the hydrotropic block may provide the useful oral formulations of poorly soluble anti-cancer agents.

19.4.5 HYDROTROPIC POLYMER MICELLES AS CARRIERS FOR ANTI-CANCER DRUGS

Hydrotropic micelles were first examined for their ability to solubilize paclitaxel, a good example of an anti-cancer drug with extremely low water solubility.[27] Due to the high potential of paclitaxel, many formulations have been developed to increase water solubility. Although the polymeric micelles based on the amphiphilic block copolymers have shown high potential as drug solubilizing systems, most polymeric micelles have shown limited solubilizing capacity for paclitaxel, and, in most cases, maximum contents of paclitaxel loaded into micelles was around 20 wt%.[9,20,25] Besides, a simple polymer design may not predict whether the resulting polymer micelles show high solubilizing capacity. A more serious limitation is the poor stability of paclitaxel-solubilized polymeric micelles in water, which becomes lower as the content of paclitaxel increases.[20] Of the properties of solubilizing systems, a high solubilizing capacity and a good physical stability are the two most important factors in determining whether the drug delivery systems are clinically useful. The goal of using hydrotropic polymer micelles is to develop a hydrotropic polymeric micelle that has a high solubilizing capacity for poorly water-soluble drugs, as well as a good long-term stability. Because an identified hydrotrope for a specific drug is introduced as a core component in a highly localized way, paclitaxel solubilization may be presented by the synergistic effect of both hydrotropic and micellar solubilization. The poor colloidal stability of existing micelles is normally caused by the enhanced hydrophobicity of micelles after solubilization of paclitaxel. Hydrotropic moieties, characterized by a strong hydrophilic nature, are expected to allow good stability for paclitaxel-loaded micelles in water. Huh et al. described the superiority of the hydrotropic polymer micelles relative to current polymeric micelles, and how the hydrotropic polymer micelles are different from existing systems.[27] These questions can be answered by comparing their drug loading and physical stability. The loading capacity of the normal polymer micelles for poorly soluble drugs is decided by various factors, such as the length of the core-forming polymer, and compatibility between drugs and the core-forming polymers.[1] Of these factors, the compatibility is the most significant in determining the solubilizing capacity. One parameter that has been used to assess the compatibility between solubilizates and the polymer is Flory–Huggins interaction parameter as described in Section 19.2.4.[1] This value is dependent on both the selected drug and the polymer. Due to the uniqueness of each drug, no single core-forming block can maximize the solubilization level for all drugs. Therefore, the first priority is to find or synthesize the right structure for effective solubilization of selected drugs. However, the number of biocompatible polymers is limited, and an added difficulty of the synthesis step is the screening of a large number of the polymer structures for effective solubilization of the selected drug. On the other hand, the hydrotropic approach is simpler and less labor-intensive, even though a screening process is also required. Many hydrotropes can be identified for a selected drug by a simple mixing procedure, and the broad range of the chemical structure can be readily screened.[25] The key concept of the hydrotropic polymer micelles is based on the hydrotrope containing core-forming polymers. Thus, the systematic design affords more efficient systems for solubilizing poorly soluble anti-cancer drugs. Another advantage of this approach is the enhanced physical stability of the formulations. This property is one measure that distinguishes hydrotropic polymer micelles from normal polymer micelles. Conventional polymer micelles have a polymer with a strong hydrophobic nature, specifically the drug solubilization has been expected only from the hydrophobic interaction between drugs and the inner core. Often, polymer micelles with drugs cannot overcome the enhanced hydrophobicity and the secondary aggregation between micelles, resulting in precipitation. From this point of view, the hydrotropic polymer micelles provide the formulations with good stability, even at a high loading of poorly soluble anti-cancer drugs, due to the hydrophilic nature of the hydrotropes residing in the micellar core domains. Of course, the same approach can be used for solubilization of other poorly soluble drugs. Huh et al. reported results on the solubilizing (loading) effect of the hydrotropic polymer micelle for paclitaxel by a dialysis method. The results are illustrated in Figure 19.8.

The hydrotropic polymeric micelles solubilized paclitaxel at a level of 18.4–37.4 wt%, depending on the organic solvents used in the dialysis and the initial feed weight ratio of paclitaxel to the block copolymer. The loading content normally increases to certain feed-weight ratios of paclitaxel to the block copolymer. However, as the amount of paclitaxel was further increased, precipitates of unloaded paclitaxel formed during dialysis, resulting in the decreased loading contents. The maximum loading was observed with initial feed weight ratio of 1:0.5. When acetonitrile and the feed weight ratio of 1:0.5 were used the loading content was as high as 37.4 wt%, which was not possible with existing polymeric micelle systems. The maximum loading content of paclitaxel in a control micelle of PEG_{2000}–$PDLLA_{2000}$ was estimated to be 27.6 wt%, which is close to the literature value. Loading capacity of PEG–PDENA micelles for paclitaxel was enhanced by increasing the block length of PDENA.

Poor physical stability of drug-loaded micelles, which causes serious problems in drug formulation, has been often addressed. There is often a compromise between loading content and stability in polymeric micelle systems. Drug-loaded micelles may break up more easily in aqueous media, resulting in drug precipitation. The physical stability of polymer micelles was investigated by several methods and compared. PEG–PLA micelles that have been extensively used to solubilize paclitaxel demonstrated a high loading capacity, but showed drug precipitation after 24 h. The initial transparent micelle solution became translucent, or turbid, after 24 h. Similar results have been reported in the literature.[20,21,25] On the other hand, PEG–PDENA micelle solutions were observed to be stable for several weeks without drug precipitation at the same or higher loading contents of paclitaxel.

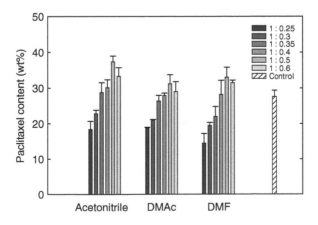

FIGURE 19.8 Paclitaxel lading contents in hydrotropic polymer micelles and comparison with PEG_{2000}–$PDLLA_{2000}$ micelles.

The loading content of paclitaxel in polymer micelles was reported as a function of time. As shown in Figure 19.9, the paclitaxel content in PEG–PDENA micelles was maintained for more than 30 days.[27]

In contrast, the PEG–PLA micelle showed a dramatic decrease in paclitaxel content after three days, which resulted from drug precipitation. The stability of drug-loaded PEG–PDENA micelles was further confirmed by observing no change in micelle size and scattering intensity using dynamic light-scattering measurements.

The paclitaxel release profiles of PEG–PDENA and PEG–PLA micelles were compared as shown in Figure 19.10.[27]

For PEG–PDENA micelles, the micelles of 25.9 wt% loading released the most drug within 48 h, whereas the micelles of 31.3 wt% drug loading showed a much faster drug release that was complete within 24 h. It was found that the faster release was from the micelles with higher drug loading. The micelles with higher drug loading lead to a relatively lower polymer concentration. Thus, there is less polymer–drug interaction in the micelles, thereby resulting in faster release kinetics. The in vitro release data showed that paclitaxel was released faster from the hydrotropic polymer micelles than from PEG to PLA micelles. The PEG–PLA micelles required almost 72 h to release all the loaded paclitaxel. As indicated by the higher CMC values of PEG–PDENA micelles, the hydrotropic polymeric micelles have less hydrophobic cores that may release the loaded paclitaxel faster than more hydrophobic PLA cores. On the other hand, the release from paclitaxel bulk powder was negligible, even after 74 h. (Figure 19.10).

The cytotoxicity of anti-cancer agents loaded in polymer micelle formulations is one measure for determining whether the formulation is effective in delivering the drugs into the tumor cells. The anti-tumor cytotoxicities of hydrotropic polymeric micelles and other control micelle systems on various cell lines, as measured by ED_{50}, are shown in Table 19.5. The resulting cytotoxicity of hydrotropic polymer micelles and control micelles clearly show the superior cytotoxic properties of hydrotropic micelles. The ED_{50} values of hydrotropic micelles are much lower than PEG–PLA and PEG–PPA micelles. The most widely used polymeric micelles is PEG–PLA micelles, with the maximum paclitaxel loading capacity of 24 wt%. Although the hydrotropic polymer micelles have higher paclitaxel, loading up to 37 wt%, hydrotropic polymer micelles with only 20 and 25 wt% paclitaxel loading were used to compare their efficacy with that of PEG–PLA micelles. At equivalent paclitaxel loading, i.e., 25 wt% loading, hydrotropic polymer micelles were

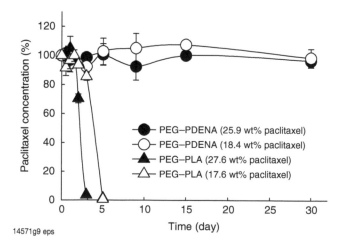

14571g9 eps

FIGURE 19.9 Changes in paclitaxel concentration of polymer micelles in distilled water. (From Huh, K.M., Lee, S.C., Cho, Y.W., Lee, J., Jeong, J.H., and Park, K., *Journal of Controlled Release*, 101, 59–68, 2005. With permission.)

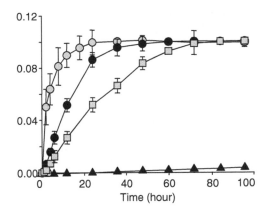

FIGURE 19.10 Release kinetics from polymer micelles and paclitaxel powders in 0.8 M sodium salicylate at 37°C. The total amount of loaded paclitaxel was 0.1 mg. (From Huh, K. M., Lee, S. C., Cho, Y. W., Lee, J., Jeong and Park, K., *Journal of Controlled Release*, 101, 59–68, 2005. With permission.)

substantially more effective. In the case of the MDA231 cell line, hydrotropic polymer micelles were more than two orders of magnitude more effective. The data clearly indicate that hydrotropic polymer micelles are not only more stable in aqueous solution, but also more effective for paclitaxel delivery to the cells.

As mentioned in Section 19.4.2, the enhanced drug solubility can improve oral bioavailability of drugs by suppressing the MDR effect. The excellent solubilizing ability of hydrotropic polymer micelles for poorly soluble drugs can overcome this barrier, thereby enhancing drug absorption in the GI tract. As a preliminary study, PEG–PDENA micelles were examined for the bioavailability of loaded paclitaxel using an in vivo, chronically-catheterized rat model. Our initial study showed that oral delivery of paclitaxel using PEG–PDENA micelles could result in the blood paclitaxel

TABLE 19.5

ED_{50} (µg/mL) of Paclitaxel and Paclitaxel (PTX)-Loaded Polymeric Micelles on Various Tumor Cell Lines

	Cancer Cell Lines			
Samples[a]	HT-29	MDA231	MCF-7	SKOV-3
Doxorubicin (positive control)	0.044	0.050	0.773	0.611
Paclitaxel	0.003	0.033	0.043	0.006
PTX-loaded HTM (25 wt% loading)	0.005	0.002	0.002	0.001
PTX-loaded HTM (20 wt% loading)	0.006	0.004	< 0.001	0.008
PTX-loaded PLA-PEG (24 wt% loading)	0.014	0.305	< 0.001	0.015
PTX-loaded PEG–PPA	0.012	1.077	0.002	1.543
HTM alone	4.221	5.650	5.767	0.048
PEG–PLA alone	4.672	8.435	5.533	—
PEG–PPA alone	0.277	4.881	6.194	—

[a] PTX, paclitaxel; HTM, hydrotropic micelle; PLA, poly(lactic acid); PEG, poly(ethylene glycol); PPA, poly(phenylalanine).

concentrations that are clinically significant. To increase oral bioavailability, hydrotropic polymer micelles with other interesting functions can be developed. The systematic design of hydrotropic polymer micelles for its faster drug releasing properties, as well as the prolonged retention during through the GI tract, can provide an opportunity to develop ideal oral delivery systems for poorly soluble anti-cancer drugs.

19.5 CONCLUSIONS AND FUTURE PERSPECTIVES

Hydrotropic polymer micelle formulation presents an alternative and promising approach in formulation of poorly soluble drugs. Based on synergistic effect of the micellar characteristics and hydrotropic activity, the hydrotropic polymer micelles exhibit a high drug loading capacity with enhanced long-term stability in aqueous media. These unique properties make it highly attractive for applications of controlled delivery of poorly soluble anti-cancer drugs. The hydrotropic polymer micelles can be applied to many poorly soluble drugs and drug candidates by introducing the identified hydrotropic structures for specific drug molecules. The hydrotropic polymer micelle is in the early stages of the development, but due to its unique properties, the systematic design may generate effective oral formulations applicable to many poorly soluble anti-cancer agents. The high versatility of the hydrotropic polymer micelle in creating a broad range of drug formulations gives it the potential for a broad range of formulations in pharmaceutical and biomedical applications.

REFERENCES

1. Allen, C., Maysinger, D., and Eisenberg, A., Nanoengineering blocks copolymer aggregates for drug delivery, *Colloids and Surfaces B: Biointerfaces*, 16, 3–27, 1999.
2. Lee, S. C., Kim, C., Kwon, I. C., Chung, H., and Jeong, S. Y., Polymeric micelles of poly(2-ethyl-2-oxazoline)-block-poly(ε-caprolactone) copolymer as a carrier for paclitaxel, *Journal of Controlled Release*, 89, 437–446, 2003.
3. Yokoyama, M., Okano, T., Sakurai, Y., Fukushima, S., Okamoto, K., and Kataoka, K., Selective delivery of adriamycin to a solid tumor using a polymeric micelle carrier system, *Journal of Drug Targeting*, 7, 171–186, 1999.
4. Lee, S. C., Chang, Y., Yoon, J.-S., Kim, C., Kwon, I. C., Kim, Y.-H., and Jeong, S. Y., Synthesis and micellar characterization of amphiphilic diblock copolymers based on poly(2-ethyl-2-oxazoline) and aliphatic polyesters, *Macromolecules*, 32, 1847–1852, 1999.
5. Kwon, G., Naito, M., Yokoyama, M., Okano, T., Sakurai, Y., and Kataoka, K., Micelles based on AB block copolymers of poly(ethylene oxide) and poly(beta-benzyl L-aspartate), *Langmuir*, 9, 945–949, 1993.
6. Bader, H., Ringsdorf, H., and Schmidt, B., Watersoluble polymers in medicine, *Die Angewandte Makromolekulare Chemie*, 123/124, 457–485, 1984.
7. Jeong, Y.-I., Cheon, J.-B., Kim, S.-H., Nah, J.-W., Lee, Y.-M., Sung, Y.-K., Akaike, T., and Cho, C. S., Clonazepam release from core-shell type nanoparticles in vitro, *Journal of Controlled Release*, 51, 169–178, 1998.
8. Kim, I. S. and Kim, S. H., Development of a polymeric nanoparticulate drug delivery system: in vitro characterization of nanoparticles based on sugar-containing conjugates, *International Journal of Pharmaceutics*, 245, 67–73, 2002.
9. Kim, S. Y. and Lee, Y. M., Taxol-loaded block copolymer nanospheres composed of methoxy poly(ethylene glycol) and poly(ε-caprolactone) as novel anticancer drug carriers, *Biomaterials*, 22, 1697–1704, 2001.
10. Kwon, G. S. and Okano, T., Polymeric micelles as new drug carriers, *Advanced Drug Delivery Review*, 21, 107–116, 1996.
11. Creutz, S., van Stam, J., De Schryver, F. C., and Jerome, R., Dynamics of poly((dimethylamino)alkyl methacrylate-block-sodium methacrylate) micelles. Influence of hydrophobicity and molecular architecture on the exchange rate of copolymer molecules, *Macromolecules*, 31, 681–689, 1998.

12. Zhang, Y., Jin, T., and Zhuo, R., Methotrexate-loaded biodegradable polymeric micelles: Preparation, physicochemical properties and in vitro drug release, *Colloids and Surfaces B. Biointerfaces.* 44, 104–109, 2005.

13. Yokoyama, M., Fukushima, S., Uehara, R., Okamoto, K., Kataoka, K., Sakurai, Y., and Okano, T., Characterization of physical entrapment and chemical conjugation of adriamycin in polymeric micelles and their design for in vivo delivery to a solid tumor, *Journal of Controlled Release*, 50, 79–92, 1998.

14. Yokoyama, M., Opanasopit, P., Okano, T., Kawano, K., and Maitani, Y., Polymer design and incorporation methods for polymeric micelle carrier system containing water-insoluble anti-cancer agent camptothecin, *Journal of Drug Targeting*, 12, 373–384, 2004.

15. You, L.-C., Lu, F.-Z., Li, Z.-C., Zhang, W., and Li, F. -M, Glucose-sensitive aggregates formed by poly(ethylene oxide)-block-poly(2-glucosyloxyethyl acrylate) with concanavalin A in dilute aqueous medium, *Macromolecules*, 36, 1–4, 2003.

16. Harada, A. and Kataoka, K., Formation of polyion complex micelles in an aqueous milieu from a pair of oppositrly-charged block copolymers with poly(ethylene glycol) segments, *Macromolecules*, 28, 5294–5299, 1995.

17. Harada, A. and Kataoka, K., Novel polyion complex micelles entrapping enzyme molecules in the core: Preparation of narrowly-distributed micelles from lysozyme and poly(ethylene glycol)–poly (aspartic acid) block copolymers in aqueous medium, *Macromolecules*, 31, 288–294, 1998.

18. Suh, H., Jeong, B., Rathi, R., and Kim, S. W., Regulation of smooth muscle cell proliferation using paclitaxel-loaded poly(ethylene oxide)–poly(lactide/glycolide) nanospheres, *Journal of Biomedical Research Materials*, 42, 331–338, 1998.

19. Soga, O., van Nostrum, C. F.., Fens, M., Rijcken, C. J. F., Schiffelers, R. M., Storm, G., and Hennink, W. E., Thermosensitive and biodegradable polymeric micelles for paclitaxel delivery, *Journal of Controlled Release*, 103, 341–353, 2005.

20. Burt, H. M., Zhang, X., Toleikis, P., Embree, L., and Hunter, W. L., Development of copolymers of poly(D,L-lactide) and methoxypolyethylene glycol as micellar carriers of paclitaxel, *Colloids and Surfaces B: Biointerfaces*, 16, 161–171, 1999.

21. Zhang, X., Jackson, J. K., and Burt, H. M., Development of amphiphilic diblock copolymers as micellar carriers of taxol, *International Journal of Pharmaceutics*, 132, 195–206, 1996.

22. Balasubramanian, D., Srinivas, V., Gaikar, V. G., and Sharma, M. M., Aggregation behavior of hydrotropic compounds in aqueous solution, *Journal of Physical Chemistry*, 93, 3865–3870, 1989.

23. Srinivas, V. and Balasubramanian, D., Proline is a protein-compatible hydrotrope, *Langmuir*, 11, 2830–2833, 1995.

24. Srinivas, V. and Balasubramanian, D., When does the switch from hydrotropy to micellar behavior occur?, *Langmuir*, 14, 6658–6661, 1998.

25. Lee, J., Lee, S. C., Acharya, G., Chang, C.-J., and Park, K., Hydrotropic solubilization of paclitaxel: Analysis of chemical structures for hydrotropic property, *Pharmaceutical Research*, 20, 1022–1030, 2003.

26. Lee, S. C., Acharya, G., Lee, J., and Park, K., Hydrotropic polymers: Synthesis and characterization of polymers containing picolylnicotinamide moieties, *Macromolecules*, 36, 2248–2255, 2003.

27. Huh, K. M., Lee, S. C., Cho, Y. W., Lee, J., Jeong, J. H., and Park, K., Hydrotropic polymer micelle system for delivery of paclitaxel, *Journal of Controlled Release*, 101, 59–68, 2005.

28. Gadelle, F., Koros, W. J., and Schechter, R. S., Solubilization of aromatic solutes in block copolymers, *Macromolecules*, 28, 4883–4892, 1995.

29. Gan, Z., Jim, T. F., Li, M., Yuer, Z., Wang, S., and Wu, C., Enzymatic biodegradation of poly (ethylene oxide-*b*-ε-caprolactone) diblock copolymer and its potential biomedical applications, *Macromolecules*, 32, 590–594, 1999.

30. Alexandridis, P., Holzwarth, J. F., and Hatton, T. A., Micellization of poly(ethylene oxide)–poly (propylene oxide)–poly(ethylene oxide) triblock copolymers in aqueous solutions: Thermodynamics of copolymer association, *Macromolecules*, 27, 2414–2425, 1994.

31. Kabanov, A. V., Slepnev, V. I., Kuzetsova, L. E., Batrakova, E. V., Alakhov, V. Y., Melik-Nubarov, N. S., Sveshinikov, P. G., and Kabanov, V. A., Pluronic micelles as tool for low-molecular compound vector delivery into a cell: Effect of Staphylococcus aureus enterotoxin B on cell loading with micelle incorporated fluorescent dye, *Biochemistry International*, 26, 1035–1042, 1992.

32. La, S. B., Okano, T., and Kataoka, K., Preparation and characterization of the micelle-forming polymeric drug indomethacin-incorporated poly(ethylene oxide)–poly(benzyl-2′-aspartate) block copolymer micelles, *Journal of Pharmaceutical Sciences*, 85, 85–90, 1996.

33. Kalyanasundaram, K. and Thomas, J. K., Environmental effects on vibronic band intensities in pyrene monomer fluorescence and their application in studies of micellar systems, *Journal of the American Chemical Society*, 99, 2039–2044, 1977.

34. Jeong, B., Bae, Y. H., and Kim, S. W., Biodegradable thermosensitive micelles of PEG–PLGA–PEG triblock copolymers, *Colloids and Surfaces B: Biointerfaces*, 16, 185–193, 1999.

35. Huh, K. M., Lee, K. Y., Kwon, I. C., Kim, Y.-H., Kim, C., and Jeong, S. Y., Synthesis of triarmed poly(ethylene oxide)–deoxycholic acid conjugate and its micellar characteristics, *Langmuir*, 16, 10566–10568, 2000.

36. Kabanov, A. V., Nazarova, I. R., Astafieva, I. V., Batrakova, E. V., Alakhov, V. Y., Yaroslavov, A. A., and Kabanov, V. A., Micelle formation and solubilization of fluorescent probes in poly(oxyethylene-*b*-oxypropylene-*b*-oxyethylene) solutions, *Macromolecules*, 28, 2303–2314, 1995.

37. Lee, K. Y., Jo, W. H., Kwon, I. C., Kim, Y.-H., and Jeong, S. Y., Structural determination and interior polarity of self-aggregates prepared from deoxycholic acid-modified chitosan in water, *Macromolecules*, 31, 378–383, 1998.

38. Kim, C., Lee, S. C., Kwon, I. C., Chung, H., and Jeong, S. Y., Complexation of poly(2-ethyl-2-oxazoline)-block-poly(ε-caprolactone) micelles with multifunctional carboxylic acids, *Macromolecules*, 35, 193–200, 2002.

39. Xu, R., Winnik, M. A., Hallett, F. R., Riess, G., and Croucher, M. D., Light-scattering study of the association behavior of styrene–ethylene oxide block copolymers in aqueous solution, *Macromolecules*, 24, 87–93, 1991.

40. Ringsdorf, H., Venzmer, J., and Winnik, F. M., Fluorescence studies of hydrophobically modified poly(*N*-isopropylacrylamide), *Macromolecules*, 24, 1678–1686, 1991.

41. Wilhelm, M., Zhao, C.-L., Wang, Y., Xu, R., and Winnik, M. A., Poly(styrene–ethylene oxide) block copolymer micelle formation in water: a fluorescence probe study, *Macromolecules*, 24, 1033–1040, 1991.

42. Zhang, L. and Eisenberg, A., Multiple morphologies and characteristics of "crew-cut" micelle-like aggregates of polystyrene-*b*-poly(acrylic acid) diblock copolymers in aqueous solutions, *Journal of the American Chemical Society*, 118, 3168–3181, 1996.

43. Zhang, L. and Eisenberg, A., Multiple morphologies of "Crew-Cut" aggregates of polystyrene-*b*-poly(acrylic acid) block copolymers, *Science*, 268, 1728–1731, 1995.

44. Zhang, L., Yu, K., and Eisenberg, A., Ion-induced morphological changes in "Crew-Cut" aggregates of amphiphilic block copolymers, *Science*, 272, 1777–1779, 1996.

45. Heise, A., Hedrick, J. L., Frank, C. W., and Miller, R. D., Starlike block copolymers with amphiphilic arms as models for unimolecular micelles, *Journal of the American Chemical Society*, 121, 8647–8648, 1999.

46. Newkome, G. R., Moorefield, C. N., Baker, G. R., Saunders, M. J., and Grossman, S. H., Unimolecular micelles, *Angewandte Chemie International Edition in English*, 30, 1178–1180, 1991.

47. Zhang, G.-D., Harada, A., Nishiyama, N., Jeang, D.-L., Koyama, H., Aida, T., and Kataoka, K., Polyion complex micelles entrapping cationic dendrimer porphyrin: Effective photosensitizer for photodynamic therapy of cancer, *Journal of Controlled Release*, 93, 141–150, 2003.

48. Jones, M.-C., Ranger, M., and Leroux, J.-C., pH-Sensitive unimolecular polymeric micelles: Synthesis of a novel drug carrier, *Bioconjugate Chemistry*, 14, 774–781, 2003.

49. Liu, H., Farrell, S., and Uhrich, K., Drug release characteristics of unimolecular polymeric micelles, *Journal of Controlled Release*, 68, 167–174, 2000.

50. Nagasaki, Y., Okada, T., Scholz, C., Iijima, M., Kato, M., and Kataoka, K., The reactive polymeric micelles based on an aldehyde-ended poly(ethylene glycol)/poly(lactide) block copolymer, *Macromolecules*, 31, 1473–1479, 1998.

51. Torchilin, V. P., Structure and design of polymeric surfactant-based drug delivery systems, *Journal of Controlled Release*, 73, 137–172, 2001.

52. Le Garrec, D., Gori, S., Luo, L., Lessard, D., Smith, D. C., Yessine, M.-A., Ranger, M., and Leroux, J.-C., Poly(*N*-vinylpyrrolidone)-block-poly(D,L-lactide) as a new polymeric solubilizer for hydrophobic anticancer drugs: in vitro and in vivo evaluation, *Journal of Controlled Release*, 99, 83–101, 2004.

53. Liggins, R. T. and Burt, H. M., Polyether–polyester diblock copolymers for the preparation of paclitaxel loaded polymeric micelle formulations, *Advanced Drug Delivery Reviews*, 54, 191–202, 2002.

54. Liu, S. Q., Tong, Y. W., and Yang, Y. Y., Incorporation and in vitro release of doxorubicin in thermally sensitive micelles made from poly(*N*-isopropylacrylamide-*co-N,N*-dimethylacrylamide)-*b*-poly(D,L-lactide-*co*-glycolide) with varying compositions, *Biomaterials*, 26, 5064–5074, 2005.

55. Lukyanov, A. N. and Torchilin, V. P., Micelles from lipid derivatives of water-soluble polymers as delivery systems for poorly soluble drugs, *Advanced Drug Delivery Reviews*, 56, 1273–1289, 2004.

56. U.S. National Institutes of Health. National Cancer Institute, 1986. Developmental Therapeutics Program. *NCI Investigational Drugs: Pharmaceutical Data*, pp. 70–72, (http://dtp.nci.nih.gov/docs/idrugs/chembook.html/).

57. Adams, J. D., Flora, K. P., Goldspiel, B. R., Wilson, J. W., and Arbuck, S. G., Taxol: a history of pharmaceutical development and current pharmaceutical concerns, *Journal of the National Cancer Institute Monogram*, 15, 141–147, 1993.

58. Gref, R., Minamitake, Y., Peracchia, M. T., Trubetskoy, V., Torchilin, V., and Langer, R., Biodegradable long-circulating polymeric nanospheres, *Science*, 263, 1600–1603, 1994.

59. Mazzo, D. J., Nguyen-Huu, J.-J., Pagniez, S., and Denis, P., Compatibility of docetaxel and paclitaxel in intravenous solutions with poyvinyl chloride infusion materials, *American Journal of Health-System Pharmacy*, 54, 566–569, 1997.

60. Alkan-Onyuksel, H., Ramakrishnan, S., Chai, H.-B., and Pezzuto, J. M., A mixed micellar formulation suitable for the parenteral administration of taxol, *Pharmaceutical Research*, 11, 206–212, 1994.

61. Dordunoo, S. K. and Burt, H. M., Solubility and stability of taxol: Effects of buffer and cyclodextrins, *International Journal of Pharmaceutics*, 133, 191–201, 1996.

62. Muller, R. H. and Bohm, B., Nanosuspensions, In *Emulsions and Nanosuspensions for the Formulation of Poorly Soluble Drugs*, Muller, Rainer H., Benita, S., and Bohm, B., Eds., Medpharm Scientific Publishers, Stuttgart, pp. 149–174, 1998.

63. Rubino, J. T. and Yalkowsky, S. H., Cosolvency and cosolvent polarity, *Pharmaceutical Research*, 4, 220–230, 1987.

64. Serajuddin, A. T. M., Solid dispersion of poorly water-soluble drugs: Early promises, subsequent problems, and recent breakthroughs, *Journal of Pharmaceutical Science.*, 88, 1058–1066, 1999.

65. Sharma, A. and Straubinger, R. M., Novel taxol formulations: Preparation and characterization of taxol-containing liposomes, *Pharmaceutical Research*, 11, 889–896, 1994.

66. Tarr, B. D., Sambandan, T. G., and Yalkowsky, S. H., A new parenteral emulsion for the administration of taxol, *Pharmaceutical Research*, 4, 162–165, 1987.

67. Nishiyama, N., Bae, Y., Miyata, K., Fukumura, S., and Kataoka, K., *Drug Discovery Today: Technologies*, 2, 21–26, 2005.

68. Bae, Y., Nishiyama, N., Fukushima, S., Koyama, H., Yasuhiro, M., and Kataoka, K., Preparation and biological characterization of polymeric micelle drug carriers with intracellular pH-triggered drug release property: Tumor permeability, controlled subcellular drug distribution, and enhanced in vivo antitumor efficacy, *Bioconjugate Chemistry*, 16, 122–130, 2005.

69. Jaracz, S., Chen, J., Kuznetsova, L. V., and Ojima, I., Recent advances in tumor–targeting anticancer drug conjugates, *Bioorganic & Medicinal Chemistry*, 13, 5043–5054, 2005.

70. van Nostrum, C. F., Polymeric micelles to deliver photosensitizers for photodynamic therapy, *Advanced Drug Delivery Reviews*, 56, 9–16, 2004.

71. Torchilin, V. P., Polymeric micelles in diagnostic imaging, *Colloids and Surfaces B: Biointerfaces*, 16, 305–319, 1999.

72. Dolmans, D. and Fukumura, D., Photodynamic therapy for cancer, *Nature Reviews Cancer*, 3, 380–387, 2003.

73. Yalkowsky, S. H., *Solubility and Solubilization in Aqueous Media*, American Chemical Society, Washington, DC, 1999.

74. Gandhi, N. N., Kumar, M. D., and Sathyamurthy, N., Effect of hydrotropes on solubility and mass-transfer coefficient of butyl acetate, *Journal of Chemical & Engineering Data*, 43, 695–699, 1998.

75. Lobenberg, R., Amidon, G. L., and Vierira, M., Solubility as a limiting factor to drug absorption, In *Oral Drug Absorption: Prediction and Assessment*, Dressman, J. H. and Lennernas, H., Eds., Marcel Dekker, New York, pp. 137–153, 2000.

76. Rasool, A. A., Hussain, A. A., and Dittert, L. W., Solubility enhancement of some water-insoluble drugs in the presence of nicotinamide and related compounds, *Journal of Pharmaceutical Science*, 80, 387–393, 1991.

77. Coffman, R. E. and Kildsig, D. O., Hydrotropic solubilization-Mechanistic studies, *Pharmacal Research*, 13, 1460–1463, 1996.

78. Xia, J., Zhang, X., and Matyjaszewski, K., Atom transfer radical polymerization of 4-vinylpyridine, *Macromolecules*, 32, 3531–3533, 1999.

79. Matyjaszewski, K. and Xia, J., Atom transfer radical polymerization, *Chemical Reviews*, 101, 2921–2990, 2001.

80. Patten, T. E., Xia, J., Abernathy, T., and Matyjaszewski, K., Polymers with very low polydispersities from atom transfer radical polymerization, *Science*, 272, 866–868, 1996.

20 Tumor-Targeted Delivery of Sparingly-Soluble Anti-Cancer Drugs with Polymeric Lipid-Core Immunomicelles

Vladimir P. Torchilin

CONTENTS

20.1 Introduction 409
20.2 Micelles, Micellization, and Drug Delivery 410
20.3 Tumor Targeting of Polymeric Micelles 411
References.. 417

20.1 INTRODUCTION

Various drug delivery and drug targeting systems-based soluble polymers, nanoparticles made of natural and synthetic polymers, nanocapsules, lipoproteins, liposomes, micelles, and many other pharmaceutical carriers are currently widely used to minimize drug degradation and loss upon administration, prevent harmful or undesirable side-effects, and increase drug bioavailability and the fraction of the drug accumulated in the required (pathological) zone.[1,2] Many of those drug carriers are long-circulating[3,4] because prolonged circulation maintains the required therapeutic level of pharmaceuticals in the blood for extended time intervals, and it allows for slow accumulation of high-molecular-weight drugs or drug-containing nanocarriers in pathological sites with affected and leaky vasculature via the enhanced permeability and retention effect (EPR).[5,6] It also provides better targeting effects for specific ligand-modified drugs and drug carriers.[7]

The development of biocompatible, biodegradable, long-circulating, and specifically targeted drug carriers for sparingly soluble pharmaceuticals that possess high loading capacities is attracting a lot of attention because the therapeutic application of hydrophobic, poorly water-soluble agents might be associated with serious problems as low water-solubility results in poor absorption and low bioavailability upon oral administration.[8] The aggregation of poorly soluble drugs upon intravenous administration might lead to an embolism[9] and increased local toxicity at the sites of aggregate deposition.[10] However, the hydrophobicity (low water solubility) allows a drug molecule to better penetrate a cell membrane when the molecular target for the drug is intracellularly located,[11,12] and this is especially important for certain anti-cancer agents, many of which are bulky polycyclic compounds.[13] To overcome some drugs' poor solubility and to increase their bioavailability, certain clinically acceptable organic solvents are used in pharmaceutical formulations such as ethanol or polyethoxylated castor oil (Cremophor EL)[10] as well as drug carriers such as liposomes[14] and cyclodextrins.[15] In addition, a lot of interest is currently expressed toward using various micelle-forming substances in formulations of insoluble drugs.[10]

20.2 MICELLES, MICELLIZATION, AND DRUG DELIVERY

Micelles represent dispersed systems, consisting of particulate matter or dispersed phase that are distributed within a continuous phase or dispersion medium and, under certain conditions (concentration and temperature), are spontaneously formed by amphiphilic or surface-active agents (surfactants) that are molecules that consist of two clearly distinct moieties differing in their affinity toward a given solvent.[16] They typically have particle sizes within a 5 to 50–100 nm range. The concentration of a monomeric amphiphile where micelles (aggregates, including several dozens of amphiphilic molecules and usually spherical in shape) appear is called the critical micelle (or micellization) concentration (CMC). The lower the CMC value of a given amphiphilic polymer, the more stable micelles are even at low net concentration of an amphiphile in the medium. This is important from the practical point of view because upon dilution with the large volume of the blood, only micelles with low CMC value still exist, whereas micelles with high CMC value may dissociate into uinimers, and their content may precipitate in the blood. In aqueous media, hydrophobic fragments of amphiphilic molecules form the core of a micelle that can solubilize poorly soluble pharmaceuticals,[17] whereas polar molecules will be adsorbed on the micelle surface, and substances with intermediate polarity/solubility will be distributed in various intermediate positions.

Micellization of biologically active substances is a wide-spread phenomenon that is important for pharmacological action. For example, the increase in the bioavailability of a lipophilic drug upon oral administration is attributed to drug solubilization in the gut by naturally occurring biliary lipid/fatty acid-containing mixed micelles produced by organism upon the digestion of the dietary fat. Surfactant micelles are also widely used as adjuvants and drug carrier systems, and in many areas of pharmaceutical technology and controlled drug delivery research, micelles as drug carriers provide a set of clear advantages.[18–22] The solubilization of sparingly soluble drugs by micelle-forming surfactants results in an increased water solubility of sparingly soluble drug and its improved bioavailability, extended blood half-life upon intravenous administration, reduction of toxicity and other adverse effects, enhanced permeability across the physiological barriers, and substantial changes in drug biodistribution. In addition, being in a micellar container, the drug is well-protected from possible inactivation under the effect of biological surroundings and does not provoke toxic side effects on non-target tissues.

Polymeric micelles, especially popular in drug delivery research, are formed from block-copolymers consisting of hydrophilic and hydrophobic monomer units with the length of a hydrophilic block exceeding, to some extent, that of a hydrophobic one. These micelles contain the core of the hydrophobic blocks stabilized by the corona of hydrophilic polymeric chains. If the length of a hydrophilic block is too high, copolymers exist in water as unimers (individual molecules), whereas molecules with very long hydrophobic block forms structure with non-micellar morphology.[23] The core compartment of the pharmaceutical polymeric micelle should demonstrate high loading capacity, controlled release profile for the incorporated drug, and good compatibility between the core-forming block and incorporated drug. The micelle corona should provide an effective steric protection for the micelle. It should also determine for the micelle hydrophilicity, charge, the length and surface density of hydrophilic blocks, and the presence of reactive groups suitable for further micelle derivatization such as an attachment of targeting moieties. The listed properties control key biological characteristics of a micellar carrier including pharmacokinetics, biodistribution, biocompatibility, longevity, surface adsorption of biomacromolecules, adhesion to biosurfaces, and targetability.[1,24–30]

PEG still remains the hydrophilic block of choice, although some other hydrophilic polymers may be used to make corona blocks,[31,32] whereas a variety of polymers was used to build hydrophobic core-forming blocks including propylene oxide,[26,33] L-lysine,[34,35] aspartic acid,[36,37] β-benzoyl-L-aspartate,[38,39] caprolactone,[40,41] D,L-lactic acid,[27,42] and many others. Phospholipid residues can also be successfully used as hydrophobic core-forming groups.[43] The use of lipid moieties as hydrophobic blocks capping hydrophilic polymer (such as PEG) chains can provide additional advantages for particle stability when compared with conventional amphiphilic polymer micelles because of the existence of two fatty acid acyls that may considerably contribute to an increase in the hydrophobic interactions in the micelle's core. Although diacyllipid–PEG conjugates have been introduced into the area of controlled drug delivery as polymeric surface modifiers for liposomes,[44] the diacyllipid–PEG molecule itself represents a typical amphiphilic copolymer with a bulky hydrophilic (PEG) portion and a very short, but extremely hydrophobic, diacyllipid part. Similar to other PEG-containing amphiphilic block-copolymers, diacyllipid–PEG conjugates were found to form micelles in an aqueous environment.[45] A series of PEG–phosphatidylethanolamine (PEG–PE) conjugates was synthesized using egg PE and N-hydroxysuccinimide esters of methoxy–PEG succinates with different MW of PEG fragments.[44] HPLC-based gel permeation chromatography showed that these polymers formed micelles of different sizes in water. The stability of the polymeric micelles was confirmed by the fact that no dissociation into individual polymeric chains was found following the chromatography of the serially diluted samples of PEG–PE up to a polymer concentration of ca. 1 µl/ml corresponds to a micromolar CMC value that is at least 100-fold lower than those of conventional detergents.[46] Such low CMC values indicate that micelles prepared from these polymers will maintain their integrity even upon strong dilution (for example, in the blood during a therapeutic application). High stability of polymeric micelles allows also for good retention of encapsulated drugs in the solubilized form upon parenteral administration. PEG–PE micelles also can efficiently incorporate a broad variety of sparingly soluble and amphiphilic substances including paclitaxel, tamoxifen, camptothecin, porphyrine, vitamin K3, and others.[47–49]

Among other factors influencing the efficacy of drug loading into the micelle are the sizes of both core-forming and corona-forming blocks.[50] In the first case, the larger the hydrophobic block, the bigger core size and the ability of micelle to entrap hydrophobic drugs. In the second case, the increase in the length of the hydrophilic block results in the increase of the CMC value, i.e., at a given concentration of the amphiphilic polymer in solution the smaller fraction of this polymer will be present in the micellar form and the quantity of the micelle-associated drug drops. Incorporation into polymeric micelles is beneficial for various sparingly soluble drugs. Therefore, doxorubicin incorporated into Pluronic® micelles demonstrated superior properties as compared to free drugs in the experimental treatment of murine tumors (leukemia P388, myeloma, and Lewis lung carcinoma) and human tumors (breast carcinoma MCF-7) in mice.[51] In addition, the reduction of the side effects of the drug was observed in many cases. The toxicity decrease (smaller vascular damage and liver focal necrosis) was found for the polymeric micelle-incorporated anti-tumor drug KRN5500.[52]

20.3 TUMOR TARGETING OF POLYMERIC MICELLES

Making micelles capable of specific accumulation in desired areas in the body (tumors) can further increase the efficiency of micelle-encapsulated pharmaceuticals. The simplest way to achieve this is to rely on previously mentioned preferential accumulation of drug-loaded micelles in areas with compromised vasculature (tumors and infarcts) via the EPR effect that is based on the spontaneous penetration of long-circulating macromolecules, molecular aggregates, and particulate drug carriers into the interstitium through the leaky blood vessels in certain pathological sites in the body.

It has been shown that the EPR effect is typical for solid tumors and infarcts.[5,6] Micelles prepared from PEG–PE conjugates with different length of the PEG block demonstrated much higher accumulation in tumors compared to non-target tissue (muscle) in experimental Lewis lung carcinoma (LLC: tumor with a relatively low vascular permeability)[53,54] in mice. The peak uptake is achieved around five hours post-injection, and the maximal total tumor uptake of the injected dose within the observation period (AUC) was found for micelles made PEG–PE with the largest size of the PEG block (5,000 Da). This was explained by the fact that these micelles had little extravasation into the normal tissue compared to micelles prepared from the shorter PEG–PE conjugates. Micelles prepared from PEG–PE conjugates with shorter PEG, however, might be more efficient carriers of poorly soluble drugs because they have a greater hydrophobic-to-hydrophilic phase ratio and can be loaded with drugs more efficiently on a weight-to-weight basis. Similar results were also obtained in another murine tumor model, EL4 T cell lymphoma (EL4);[55] see Figure 20.1.

It is important that tumor diffusion and accumulation parameters of nanosized pharmaceutical carriers strongly depend on the cutoff size of tumor blood vessel wall, and the cutoff size varies for different tumors.[56–58] Because tumor vasculature permeability depends on the particular type of the tumor,[57] the use of micelles as drug carriers could be specifically justified for tumors whose vasculature has the low cutoff size (below 200 nm). The use of PEG–PE micelles for the effective delivery of a model protein drug (soybean trypsin inhibitor or STI, MW 21.5 kDa) to subcutaneously established LLC in mice was reported.[54]

An interesting targeting mechanism is based on the fact that many pathological processes in various tissues and organs are accompanied with local temperature increase and/or acidosis.[59,60] Micelles made capable of disintegration under the increased temperature or decreased pH values in pathological sites (such as tumors) can combine the EPR effect-mediated accumulation with stimuli-responsiveness. It was shown that micelles made of thermo- or pH-sensitive components such as poly(N-isopropylacrylamide) and its copolymers with poly(D,L-lactide) and other blocks acquire the ability to disintegrate in target areas, releasing the micelle-incorporated drug.[61–65] pH-responsive polymeric micelles loaded with phtalocyanine seem to be promising carriers for the photodynamic cancer therapy,[66] whereas doxorubicin-loaded polymeric micelles containing acid-cleavable linkages provided an enhanced intracellular drug delivery into tumor cells and, therefore, higher efficiency.[67]

As with many other delivery systems, the drug delivery potential of polymeric micelles may be further enhanced by attaching targeting ligands to their surface. The attachment of various specific ligand to the water-exposed termini of hydrophilic blocks has been used to improve the targeting of micelles and micelle-incorporated drugs and DNA.[26,68,69] Among those ligands, various sugar moieties can be named as well as[70] transferrin,[71] and folate residues[72] because many target cells,

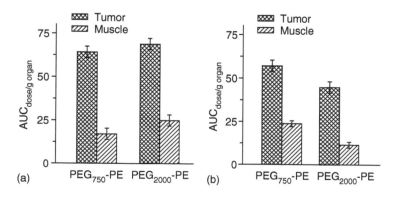

FIGURE 20.1 Accumulation of PEG–PE micelles in tumors in mice compared to muscle accumulation. (a) LLC tumor; (b) EL4 tumor.

especially cancer cells, over express appropriate receptors (such as transferrin and folate receptors) on their surface. Transferrin-modified micelles based on PEG and polyethyleneimine with sizes between 70 and 100 are expected to target tumors with over expressed transferrin receptors.[71] Similar targeting approaches were successfully tested with folate-modified micelles.[73] Poly(L-histidine)/PEG and poly(L-lactic acid)/PEG block copolymer micelles carrying folate residue on their surfaces were shown to be efficient for the delivery of adriamycin to tumor cells in vitro, demonstrating potential for solid tumor treatment and combined targetability and pH-sensitivity.[74]

Among all specific ligands, monoclonal antibodies, primarily of the IgG class, and their Fab fragments provide the broadest opportunities in terms of diversity of targets and specificity of interaction. Several successful attempts to covalently attach an antibody to a surfactant or polymeric micelles (i.e., to prepare immunomicelles) have been already described.[22,26,71,75] By adapting the coupling technique developed for attaching specific ligands to liposomes utilizing PEG–PE with the free PEG terminus activated with p-nitrophenylcarbonyl (pNP) group,[76] PEG–PE-based immunomicelles have been prepared modified with monoclonal antibodies (see the principal reaction scheme in Figure 20.2). The micelle-attached protein was quantified using fluorescent labels or by SDS-PAGE,[75,77] and it was calculated that 10–20 antibody molecules could be attached to a single micelle. Antibodies attached to the micelle corona preserve their specific binding ability, and immunomicelles specifically recognize their target substrates as was confirmed by ELISA experiments while the micelles maintain essentially the same size as non-modified ones.[75]

One has to keep in mind, however, that whole antibody attachment to the micelles (as well as to any other drug carriers) might provoke faster clearance from the circulation as a result of uptake by Fc receptor-bearing Kupffer cells. To test if the antibody attachment affects the blood clearance of PEG–PE micelles, the comparison was made of the clearance characteristics of plain and antibody-modified micelles in mice; it was found that PEG–PE micelle modification with an antibody had a very small effect on their blood clearance.[75]

To specifically enhance the tumor accumulation of PEG–PE-based micelles, they have been modified with tumor-specific monoclonal antibodies.[75,77] Non-pathogenic monoclonal antinuclear autoantibodies with the nucleosome-restricted specificity (monoclonal antibody 2C5, mAb 2C5 being among them) that recognize the surfaces of numerous tumor, but not normal, cells via tumor cell surface-bound nucleosomes[78,79] were used in these experiments. Because these antibodies bind a broad variety of cancer cells, they may serve as specific ligands for the delivery of drugs and drug carriers into tumors. The data shown in Figure 20.3a indicate that fluorescence (rhodamine)-labeled 2C5-immunomicelles effectively bind to the surface of several unrelated human and murine tumor cells lines: human BT20 (breast adenocarcinoma), murine LLC (Lewis lung carcinoma), and EL4 (T lymphoma) cells. The incubation of antibody-free micelles with the same cells results in virtually no micelle-to-cell association. Drug(paclitaxel)-loaded 2C5-immunomicelles also effectively bound various tumors cells (Figure 20.3b). Such specific recognition of cancer cells by drug-loaded mAb 2C5-immunomicelles results in dramatically improved cancer cell killing by such micelles. Figure 20.3c presents the results of in vitro experiments with human breast cancer MCF-7 cells that showed a clearly superior efficiency of paclitaxel-loaded 2C5-immunomicelles compared to paclitaxel-loaded plain or modified with a non-specific IgG micelles or free drug.

In vivo experiments with LLC tumors-bearing mice revealed a dramatically enhanced tumor uptake of paclitaxel-loaded radiolabeled 2C5-immunomicelles compared to non-targeted micelles (Table 20.1). For tumor accumulation experiments in vivo, LLC tumors have been grown in mice subcutaneously injected with 20,000 LLC cells in 50 µL of 10 mM HBS into the left rear flank. When tumor diameters reached 3–7 mm (8–12 days post inoculation), the mice were injected with ^{111}In-labeled plain or mAb 2C5-bearing paclitaxel-loaded PEG–PE micelles via the tail vein. At 30 min or 2 h post injection, mice were sacrificed; tumors and muscle samples were collected and analyzed for the presence of ^{111}In radioactivity to estimate micelle accumulation in the tumor. To estimate the actual tumor accumulation of the drug, samples of tumor tissues from mice receiving ca. 100 µg paclitaxel per animal in plain PEG–PE micelles, in 2C5-immunomicelles, and the same

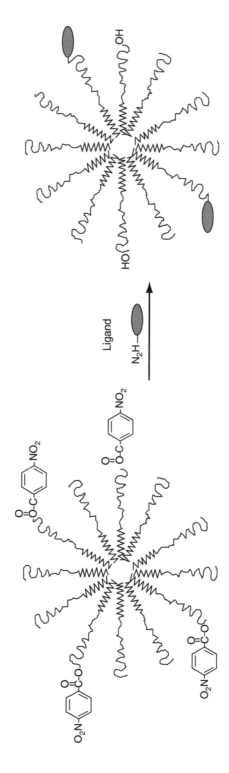

FIGURE 20.2 Schematic pattern of antibody (or other amino-containing ligand) attachment to activated PEG–PE micelles via the pNP group.

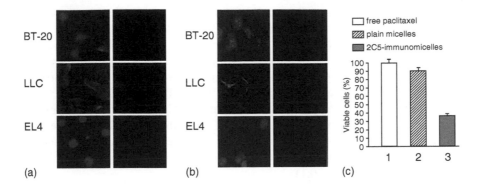

FIGURE 20.3 Specific properties of PEG–PE-based 2C5-immunomicelles in vitro. (a) Enhanced binding of rhodamine-labeled 2C5-immunomicelles with various cancer cells. (b) Enhanced binding of rhodamine-labeled drug(paclitaxel)-loaded 2C5-immunomicelles with various cancer cells. (c) Increased cytotoxicity of paclitaxel-loaded 2C5-immunomicelles toward cancer cells (with MCF7 cells as an example).

quantity of paclitaxel in Cremophor EL/ethanol/saline mixture (free paclitaxel) were analyzed 30 min and 2 h post injection by HPLC. The results presented in the data in the Table 20.1 clearly demonstrate that paclitaxel-loaded mAb 2C5-immunomicelles accumulate in tumors significantly better than drug-loaded plain micelles or free drugs and bring into tumors more drug (paclitaxel) than any other preparation. An enhanced accumulation of 2C5-targeted micelles over plain micelles in the tumor (up to 30%) was observed both at 30 min and at 2 h post injection indicating the specific recognition and tumor binding of 2C5-targeted immunomicelles. These data suggest the possibility that drug-loaded immunomicelles may also be better internalized by tumor cells similar to antibody-targeted liposome,[80] delivering more drug inside tumor cells that might be achieved in case of simple EPR effect-mediated tumor accumulation. By analyzing the absolute quantity of tumor-accumulated paclitaxel (HPLC[81]) delivered by different drug formulations, it was shown that 2C5-immunomicelle was bringing into tumors substantially higher quantities of paclitaxel than in the case of paclitaxel-loaded non-targeted micelles or free drug formulation (Taxol®)(see Table 20.1). The difference in tumor accumulation of paclitaxel in immunomicelles compared to other drug preparations was larger at 2 h post injection than at 30 min post injection, explained by the accumulation of different preparations in different tumor compartments. Non-targeted paclitaxel formulations accumulate in the interstitial space of the tumor via the EPR effect

TABLE 20.1
Paclitaxel-Loaded PEG–PE-Based Micelles In Vivo

	Paclitaxel in Pain Micelles	Paclitaxel in 2C5-Micelles	Free Paclitaxel
Tumor accumulation of micelles (% dose/g of tumor)	4.15±0.36	6±0.10	n/a
Tumor accumulation of paclitaxel (ng/g of tumor)	200±30	850±40	130±25
Inhibition of tumor growth (%)	35±5	74±34	23±4

Accumulation measured 2 h post injection. Tumor growth inhibition measured over 10 days. Immunomicelles show the highest accumulation of all preparations, bring the maximum quantity of the drug into tumor, and allow for maximum inhibition of tumor growth.

and eventually drug could be cleared from the site after gradual micelle degradation. Contrary to that, paclitaxel-loaded 2C5-immunomicelles are internalized by cancer cells and keep the drug inside the tumor in a way similar to what was observed with drug-loaded anti-her2 immunoliposomes.[80] The internalization by tumor cells may be highly therapeutically significant for many anti-tumor agents. For example, a much higher tumor regression was observed with a carrier capable of intracellular drug delivery for an equal doxorubicin dose delivered to the tumor.[80] It results in a higher therapeutic efficiency of paclitaxel-loaded mAb 2C5-micelles. The average weight of excised tumors in the group treated with paclitaxel in mAb 2C5-immunomicelles was approx. 0.7 g compared to 1.6 g and 1.4 g in groups treated with the same quantity of paclitaxel as free drug or in plain PEG–PE micelles, respectively, ($P<.05$ in both cases). The weight of untreated tumors was around 2.0 g (Table 20.1).

Recently, a poorly soluble photodynamic therapy (PDT) agent, *meso*-tetratphenylporphine (TPP), was effectively solubilized using non-targeted and tumor-targeted polymeric micelles prepared with PEG–PE.[82] Encapsulation of TPP into PEG–PE-based micelles and 2C5-immunomicelles (bearing an anti-cancer monoclonal 2C5 antibody) resulted in significantly improved anti-cancer effects of the drug at PDT conditions against murine (LLC, B16) and human (MCF-7, BT20) cancer cells in vitro. For this purpose, the cells were incubated for 6 h or 18 h with the TPP or TPP-loaded PEG–PE micelles/immunomicelles and then light-irradiated for 30 min. The phototoxic effect depended on the TPP concentration and the light dose (the duration of light-irradiation) (see some data in Figure 20.4). An increased level of apoptosis was shown in the PDT-treated cultures. The attachment of the anti-cancer 2C5 antibodies to TPP-loaded micelles provided the maximum level of cell killing at a given time. Therefore, drug-loaded immunomicelles may also represent a useful formulation of the photosensitizer for practical PDT.

Certain polymeric micelles possess a very high stability and ability to carry a variety of poorly soluble pharmaceuticals, and they can be used as targeted drug delivery systems into tumors. The tumor targeting can be achieved via the EPR effect by making micelles of stimuli-responsive amphiphilic block-copolymers or by attaching specific targeting ligand molecules (such as folate of transferrin) to the micelle surface. Tumor-specific immunomicelles can be prepared by coupling monoclonal antibody (such as antinucleosomal monoclonal antibody) molecules to pNP groups on the water-exposed terimini of the micelle corona-forming blocks. The micelle-coupled antibodies preserve their specific activity, and immunomicelles prepared with the use of the cancer-specific mAb 2C5 specifically bind to different cancer cells in vitro, demonstrating increased accumulation

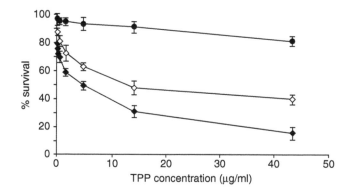

FIGURE 20.4 Improved killing of cancer cells with the drug loaded into tumor-specific 2C5-immunomicelles under the PDT conditions. Phototoxicity of free TPP and TPP encapsulated in PEG–PE micelles and 2C5-PEG–PE-immunomicelles after light irradiation towards BT20 (human breast adenocarcinoma) cells; all preparations were pre-incubated with cells for 18 h): ● free TPP; ◇ TPP in PEG–PE micelles; ◆ TPP in 2C5-PEG–PE-immunomicelles.

in experimental tumors in vivo. Being loaded with poorly soluble anti-cancer drugs (such as paclitaxel, camptothecin,[83] or photodynamic therapy agents),[82] mAb 2C5-immunomicelles demonstrate significantly increased cytotoxicity toward tumor cells in vitro and in vivo.

REFERENCES

1. Müller, R. H., *Colloidal Carriers for Controlled Drug Delivery and Targeting*, Wissenschaftliche Verlagsgesellschaft/CRC Press, Stuttgart, Germany/Boca Raton, 1991.
2. Cohen, 2. S. and Bernstein, H., Eds., *Microparticulate Systems for the Delivery of Proteins and Vaccines*, Marcel Dekker, Inc., New York, 1996.
3. Lasic, D. D. and Martin, F., Eds., *Stealth Liposomes*, CRC Press, Boca Raton, 1995.
4. Torchilin, V. P. and Trubetskoy, V. S., Which polymers can make nanoparticulate drug carriers long-circulating?, *Adv. Drug Deliv. Rev.*, 16, 141, 1995.
5. Palmer, T. N. et al., The mechanism of liposome accumulation in infarction, *Biochim. Biophys. Acta*, 797, 363, 1984.
6. Maeda, H. et al., Tumor vascular permeability and the EPR effect in macromolecular therapeutics: A review, *J. Control. Rel.*, 65, 271, 2000.
7. Torchilin, V. P., Polymer-coated long-circulating microparticular pharmaceuticals, *J. Microencapsul.*, 15, 1, 1998.
8. Lipinski, C. A. et al., Experimental and computational approaches to estimate solubility and permeability in drug discovery and development settings, *Adv. Drug Deliv. Rev.*, 46, 3, 2001.
9. Fernandez, A. M. et al., *N*-Succinyl-(beta-alanyl-L-leucyl-L-alanyl-L-leucyl)doxorubicin: An extracellularly tumor-activated prodrug devoid of intravenous acute toxicity, *J. Med. Chem.*, 44, 3750, 2001.
10. Yalkowsky, S. H., Ed,. *Techniques of Solubilization of Drugs*, Marcel Dekker, New York, 1981.
11. Yokogawa, K. et al., Relationships in the structure-tissue distribution of basic drugs in the rabbit, *Pharm. Res.*, 7, 691, 1990.
12. Hageluken, A. et al., Lipophilic beta-adrenoceptor antagonists and local anesthetics are effective direct activators of G-proteins, *Biochem. Pharmacol.*, 47, 1789, 1994.
13. Shabner, B. A. and Collings, J. M., Eds., *Cancer Chemotherapy: Principles and Practice*, J.B. Lippincott Co., Philadelphia, 1990.
14. Lasic, D. D. and Papahadjopoulos, D., *Medical Applications of Liposomes*, Elsevier, New York, 1998.
15. Thompson, D. and Chaubal, M. V., Cyclodextrins (CDS)—excipients by definition, drug delivery systems by function (part I: Injectable applications), *Drug Deliv. Technol.*, 2, 34, 2000.
16. Mittal, K. L. and Lindman, B., Eds., *Surfactants in Solution*, Vol. 1–3, Plenum Press, New York, 1991.
17. Lasic, D. D., Mixed micelles in drug delivery, *Nature*, 355, 279, 1992.
18. Kwon, G. S. and Kataoka, K., Block copolymer micelles as long-circulating drug vehicles, *Adv. Drug Deliv. Rev.*, 16, 295, 1995.
19. Kwon, G. S., Diblock copolymer nanoparticles for drug delivery, *Crit. Rev. Ther. Drug Carrier Syst.*, 15, 481, 1998.
20. Kabanov, A. V., Batrakova, E. V., and Alakhov, V. Y., Pluronic block copolymers as novel polymer therapeutics for drug and gene delivery, *J. Control. Rel.*, 82, 189, 2002.
21. Jones, M. and Leroux, J., Polymeric micelles-a new generation of colloidal drug carriers, *Eur. J. Pharm. Biopharm.*, 48, 101, 1999.
22. Torchilin, V. P., Structure and design of polymeric surfactant-based drug delivery systems, *J. Control. Rel.*, 73, 137, 2001.
23. Zhang, L. and Eisenberg, A., Multiple morphologies of "crew-cut" aggregates of polystyrene-*b*-poly(acrylic acid) block copolymers, *Science*, 268, 1728, 1995.
24. Gref, R. et al., Biodegradable long-circulating polymeric nanospheres, *Science*, 263, 1600, 1994.
25. Gref, R. et al., The controlled intravenous delivery of drugs using PEG-coated sterically stabilized nanospheres, *Adv. Drug Deliv. Rev.*, 16, 215, 1995.
26. Kabanov, A. V. et al., The neuroleptic activity of haloperidol increases after its solubilization in surfactant micelles, *FEBS Lett.*, 258, 343, 1989.

27. Hagan, S. A. et al., Polylactide-poly(ethelene glycol) copolymers as drug delivery systems 1. Caracterization of water dispersible micelle-forming systems, *Langmuir*, 12, 2153, 1996.

28. Inoue, T. et al., An AB block copolymers of oligo(methyl methacrylate) and poly(acrylic acid) for micellar delivery of hydrophobic drugs, *J. Control. Rel.*, 51, 221, 1998.

29. Hunter, R. J., *Foundations of Colloid Science*, Vol. 1, Oxford University Press, New York, 1991.

30. Kuntz, R. M. and Saltzman, W. M., Polymeric controlled delivery for immunization, *Trends Biotech.*, 15, 364, 1997.

31. Torchilin, V. P. et al., Amphiphilic vinyl polymers effectively prolong liposome circulation time in vivo, *Biochim. Biophys. Acta*, 1195, 181, 1994.

32. Torchilin, V. P. et al., New sunthetic amphiphilic polymers for steric protection of liposomes in vivo, *J. Pharm. Sci.*, 84, 1049, 1995.

33. Miller, D. W. et al., Interactions of pluronic block copolymers with brain microvessel endothelial cells: Evidence of two potential pathways for drug absorption, *Bioconjug. Chem.*, 8, 649, 1997.

34. Katayose, S. and Kataoka, K., Remarkable increase in nuclease resistance of plasmid DNA through supramolecular assembly with poly(ethylene glycol)–poly(L-lysine) block copolymer, *J. Pharm. Sci.*, 87, 160, 1998.

35. Trubetskoy, V. S. et al., Block copolymer of polyethylene glycol and polylysine as a carrier of organic iodine: Design of a long circulating particulate contrast medium for x-ray computed tomography, *J. Drug Target.*, 4, 381, 1997.

36. Yokoyama, M. et al., Characterization and anticancer activity of the micelle-forming polymeric anticancer drug adriamycin-conjugated poly(ethylene glycol)–poly(aspartic acid) block copolymer, *Cancer Res.*, 50, 1693, 1990.

37. Harada, A. and Kataoka, K., Novel polyion complex micelles entrapping enzyme molecules in the core. Preparation of narrowly-distributed micelles from lysozyme and poly(ethylene glycol)–poly (aspartic acid) block copolymer in aqueous medium, *Macromolecules*, 31, 288, 1998.

38. La, S. B., Okano, T., and Kataoka, K., Preparation and characterization of the micelle-forming polymeric drug indomethacin-incorporated poly(ethylene oxide)–poly(-benzyl L-aspartat) block copolimer micelles, *J. Pharm. Sci.*, 85, 85, 1996.

39. Kwon, G. S. et al., Block copolymer micelles for drug delivery: Loading and release of doxorubicin, *J. Control. Rel.*, 48, 195, 1997.

40. Kim, S. Y. et al., Metoxy poly(ethylene glucol) and ε-caprolactone amphiphilic block copolymeric micelle containing indomethacin, II. Micelle formation and drug release behaviors, *J. Control. Rel.*, 51, 13, 1998.

41. Allen, C. et al., Polycaprolactone-*b*-poly(ethylene oxide) block copolymer micelles as a novel drug delivery vehicle for neurotrophic agents FK506 and L-685,818, *Bioconjug. Chem.*, 9, 564, 1998.

42. Ramaswamy, M. et al., Human plasma distribution of free paclitaxel and paclitaxel associated with diblock copolymers, *J. Pharm. Sci.*, 86, 460, 1997.

43. Trubetskoy, V. S. and Torchilin, T. P., Use of polyoxyethylene–lipid conjugates as long-circulating carriers for delivery of therapeutic and diagnostic agents, *Adv. Drug Deliv. Rev.*, 16, 311, 1995.

44. Klibanov, A. L. et al., Amphipathic polyethyleneglycols effectively prolong the circulation time of liposomes, *FEBS Lett.*, 268, 235, 1990.

45. Lasic, D. D. et al., Valentincic, phase behavior of "stealth–lipid" decithin mixtures, *Period. Biol.*, 93, 287, 1991.

46. Ray, R. et al., *Handbook of Pharmaceutical Excipients*, APhA Publications, Washington, 2003.

47. Trubetskoy, V. S. and Torchilin, V. P., Polyethylene glycol based micelles as carriers of therapeutic and diagnostic agents, *S.T.P. Pharma Sci.*, 6, 79, 1996.

48. Gao, Z. et al., Diacyl-polymer micelles as nanocarriers for poorly soluble anticancer drugs, *Nano Lett.*, 2, 979, 2002.

49. Wang, J. et al., Preparation and in vitro synergistic anticancer effect of Vitamin K# and 1,8-diazabicyclo[5,4,0]undec-7-ene in poly(ethylene glycol)-diacyllipid micelles, *Int. J. Pharm.*, 272, 129, 2004.

50. Allen, C., Maysinger, D., and Eisenberg, A., Nano-engineering block copolymer aggregates for drug delivery, *Coll. Surf. B: Biointerf.*, 16, 1, 1999.

51. Alakhov, V. Yu. and Kabanov, A. V., Block copolymeric biotransport carriers as versatile vehicles for drug delivery, *Expert. Opin. Invest. Drugs*, 7, 1453, 1998.

52. Matsumura, Y. et al., Reduction of the side effects of an antitumor agent, KRN5500, by incorporation of the drug into polymeric micelles, *Jpn. J. Cancer Res.*, 90, 122, 1999.

53. Wang, J., Mongayt, D., and Torchilin, V. P., Polymeric micelles for delivery of poorly soluble drugs: Preparation and anticancer activity in vitro of paclitaxel incorporated into mixed micelles based on poly(ethylene glycol)–lipid conjugate and positively charged lipids, *J. Drug Target.*, 13, 73, 2005.

54. Weissig, V., Whiteman, K. R., and Torchilin, V. P., Accumulation of liposomal- and micellar-bound protein in solid tumor, *Pharm. Res.*, 15, 1552, 1998.

55. Lukyanov, A. N. et al., Polyethylene glycol-diacyllipid micelles demonstrate increased accumulation in subcutaneous tumors in mice, *Pharm. Res.*, 19, 1424, 2002.

56. Yuan, F. et al., Vascular permeability in a human tumor xenograft: Molecular size dependence and cutoff size, *Cancer Res.*, 55, 3752, 1995.

57. Hobbs, S. K. et al., Regulation of transport pathways in tumor vessels: Role of tumor type and microenvironment, *Proc. Natl Acad. Sci. USA*, 95, 4607, 1998.

58. Monsky, W. L. et al., Augmentation of transvascular transport of macromolecules and nanoparticles in tumors using vascular endothelial growth factor, *Cancer Res.*, 59, 4129, 1999.

59. Helmlinger, G. et al., Interstitial pH and pO2 gradients in solid tumors in vivo: High-resolution measurements reveal a lack of correlation, *Nat. Med.*, 3, 177, 1997.

60. Tannock, I. T. and Rotin, D., Acid pH in tumors and its potential for therapeutic exploitation, *Cancer Res.*, 49, 4373, 1989.

61. Kwon, G. S. and Okano, T., Soluble self-assembled block copolymers for drug delivery, *Pharm. Res.*, 16, 597, 1999.

62. Cammas, S. et al., Thermoresponsive polymer nanoparticles with a core-shell micelle structure as site specific drug carriers, *J. Control. Rel.*, 48, 157, 1997.

63. Chung, J. E. et al., Effect of molecular architecture of hydrophobically modified poly(*N*-isopropylacrylamide) on the formation of thermoresponsive core shell micellar drug carriers, *J. Control. Rel.*, 53, 119, 1998.

64. Kohori, F. et al., Preparation and characterization of thermally responsive block copolymer micelles comprising poly(*N*-isopropylacrylamide-*b*-DL-lactide), *J. Control. Rel.*, 55, 87, 1998.

65. Meyer, O., Papahadjopoulos, D., and Leroux, J. C., Copolymers of *N*-isopropylacrylamide can trigger pH sensitivity to stable liposomes, *FEBS Lett.*, 41, 61, 1998.

66. Le Garrec, D. et al., Optimizing pH-responsive polymeric micelles for drug delivery in a cancer photodynamic therapy model, *J Drug Target.*, 10, 429, 2002.

67. Yoo, H. S., Lee, E. A., and Park, T. G., Doxorubicin-conjugated biodegradable polymeric micelles having acid-cleavable linkages, *J Control. Rel.*, 82, 17, 2002.

68. Bae, Y. et al., Multifunctional polymeric micelles with folate-mediated cancer cell targeting and pH-triggered drug releasing properties for active intracellular drug delivery, *Mol. Biosyst.*, 1, 242–250, 2005.

69. Yokoyama, M. et al., Stabilization of disulfide linkage in drug-polymer-immunoglobulin conjugate by microenvironmental control, *Biochem. Biophys. Res. Commun.*, 164, 1234–1239, 1989.

70. Nagasaki, Y. et al., Sugar-installed block copolymer micelles: Their preparation and specific interaction with lectin molecules, *Biomacromolecules*, 2, 1067, 2001.

71. Vinogradov, S. et al., Polyion complex micelles with protein-modified corona for receptor-mediated delivery of oligonucleotides into cells, *Bioconjug. Chem.*, 10, 851, 1999.

72. Leamon, C. P., Weigl, D., and Hendren, R. W., Folate copolymer-mediated transfection of cultured cells, *Bioconjug. Chem.*, 10, 947, 1999.

73. Leamon, C. P. and Low, P. S., Folate-mediated targeting: From diagnostics to drug and gene delivery, *Drug Discov. Today*, 6, 44, 2001.

74. Lee, E. S., Na, K., and Bae, Y. H., Polymeric micelle for tumor pH and folate-mediated targeting, *J Control. Rel.*, 91, 103, 2003.

75. Torchilin, V. P. et al., Immunomicelles: Targeted pharmaceutical carriers for poorly soluble drugs, *Proc. Natl Acad. Sci. USA*, 100, 6039, 2003.

76. Torchilin, V. P. et al., *p*-Nitrophenylcarbonyl–PEG–PE–liposomes: Fast and simple attachment of specific ligands, including monoclonal antibodies, to distal ends of PEG chains via *p*-nitrophenyl-carbonyl groups, *Biochim. Biophys. Acta*, 1511, 397, 2001.

77. Gao, Z. et al., PEG–PE/phosphatidylcholine mixed immunomicelles specifically deliver encapsulated taxol to tumor cells of different origin and promote their efficient killing, *J. Drug Target.*, 11, 87, 2003.

78. Iakoubov, L. Z. and Torchilin, V. P., A novel class of antitumor antibodies: Nucleosome-restricted antinuclear autoantibodies (ANA) from healthy aged nonautoimmune mice, *Oncol. Res.*, 9, 439, 1997.

79. Torchilin, V. P., Iakoubov, L. Z., and Estrov, Z., Therapeutic potential of antinuclear autoantibodies in cancer, *Cancer Ther.*, 1, 179, 2003.

80. Park, J. W. et al., Tumor targeting using anti-her2 immunoliposomes, *J. Control. Rel.*, 74, 95, 2001.

81. Sharma, A., Conway, W. D., and Straubinger, R. M., Reversed-phase high-performance liquid chromatographic determination of taxol in mouse plasma, *J. Chromatogr. B. Biomed. Appl.*, 655, 315, 1994.

82. Roby, A., Erdogan, S., and Torchilin, V. P., Solubilization of poorly soluble PDT agent, mesa-tetraphenylporphin, in plain or immunotargeted PEG–PE micelles results in dramatically improved cancer cell killing in vitro. *Eur. J. Pharm. Biopharm.*, 62(3):235–240,2006.

83. Mu, L., Elbayoumi, T. A., and Torchilin, V. P., Mixed micelles made of poly(ethylene glycol)–phosphatidylethanolamine conjugate and *d*-alpha-tocopheryl polyethylene glycol 1000 succinate as pharmaceutical nanocarriers for camptothecin, *Int. J. Pharm.*, 306(1–2): 142–149, 2005.

21 Combined Cancer Therapy by Micellar-Encapsulated Drugs and Ultrasound

Natalya Rapoport

CONTENTS

21.1 Introduction 421
21.2 Polymeric Micelles as Drug Carriers 422
    21.2.1 Advantages and Shortcomings of Polymeric Micelles as Drug Carriers 422
    21.2.2 Polymeric Micelles in Clinical Trials 423
21.3 Ultrasound in Drug Delivery 425
    21.3.1 Introduction to Ultrasound in Medicine 425
    21.3.2 Biological Effects of Ultrasound 425
        21.3.2.1 Cellular Level 425
        21.3.2.2 Systemic Level 427
    21.3.3 Ultrasound-Induced Drug Release from Polymeric Micelles 428
    21.3.4 Ultrasound-Enhanced Intracellular Drug Uptake 428
21.4 In Vivo Evaluation of the Drug/Micelle/Ultrasound Tumor-Targeting Modality 430
    21.4.1 Ultrasound-Induced Drug Release from Micelles In Vivo 430
    21.4.2 Biodistribution of Polymeric Micelles and Micellar-Encapsulated Drugs 430
    21.4.3 Effect of the Time of Ultrasound Application 432
    21.4.4 Inhibition of Tumor Growth Using Micelle/Ultrasound Drug Delivery 432
        21.4.4.1 Drug-Sensitive Tumors 432
        21.4.4.2 Drug Resistant Tumors 433
        21.4.4.3 Poorly Vascularized Tumors 435
    21.4.5 Mechanisms of Micelle/Ultrasound Bioeffects 436
21.5 Combining Ultrasound Imaging and Treatment 437
    21.5.1 Dual-Modality Contrast Agents/Drug Carrier Systems 437
21.6 Conclusions 438
References 438

21.1 INTRODUCTION

Chemotherapy is often complicated by toxic side effects of anti-cancer drugs. Also, effective chemotherapy regimens are frequently hindered by the resistance of tumor cells to one or more drugs [cross-resistance or multidrug resistance (MDR)]. The MDR is either inherent or acquired in the process of chemotherapy. Developing new tumor-localized chemotherapeutic modalities for treatment of drug-sensitive and MDR tumors is a major goal of current research in academia and industry. Towards this end, nanoparticle-based drug delivery combined with a localized drug release from carrier in the tumor volume is a promising approach to targeted chemotherapy. This chapter reviews the state of the art in drug targeting to tumors using polymeric micelles as drug carriers and tumor-localized ultrasound as a means to trigger drug release from micelles in the tumor volume. Ultrasound enhances the intracellular drug uptake and sensitizes MDR tumors to the action of conventional drugs.

21.2 POLYMERIC MICELLES AS DRUG CARRIERS

The polymeric micelles are formed by hydrophobic–hydrophilic block copolymers at concentrations above the critical micelle concentration (CMC). Below the CMC, block copolymer molecules exist in a solution in the form of individual molecules (unimers). At concentrations above the CMC, copolymer molecules self-assemble into dense micelles with hydrophobic cores and hydrophilic corona. The amphiphilic character of block copolymer micelles, their size (approximately 10–150 nm), and surface properties provide for a high drug loading capacity and long circulation time in the vascular system. This makes them attractive drug carriers.[1,2]

21.2.1 ADVANTAGES AND SHORTCOMINGS OF POLYMERIC MICELLES AS DRUG CARRIERS

Drug encapsulation in micelles decreases systemic concentration of drug and diminishes intracellular drug uptake by normal cells, thus reducing unwanted drug interactions with healthy tissues. The important advantage offered by polymeric micelles is a so-called "enhanced penetration and retention (EPR) effect"[3] that provides for a selective accumulation of micellar-encapsulated drugs in tumor interstitium; in turn, this offers prospects for controlled drug release in the tumor volume.

The simplicity of micelle formation by self-assembly of amphiphilic block copolymer molecules and drug encapsulation by physical mixing rather than chemical conjugation are extremely attractive features of polymeric micelles.

Another important advantage offered by polymeric micelles is related to solubilization of drugs with poor aqueous solubility. Most anti-cancer drugs are lipophilic; thus, they have low aqueous solubility; conventional solubilizing agents, currently in use for the formulation of low-solubility drugs, are usually very toxic. Use of polymeric micelles as solubilizing agents results in dramatically increased aqueous solubility, and substantially decreased systemic toxicity of clinical formulations of paclitaxel (PTX), doxorubicin (DOX), and many other anti-cancer drugs.[4–18]

Various polymeric micellar systems have been designed to optimize important parameters of drug performance, such as aqueous solubility, on-demand release, and biological distribution. In this context, the most thoroughly studied are the micelles that comprise of poly(ethylene oxide) (PEO) as a hydrophilic block (or blocks). Examples of these include PEO-b-poly(propylene oxide)-b-PEO copolymers (Pluronic®), PEO-b-polyesters, and PEO-b-poly(amino acid)s.

The application of biodegradable, pH-sensitive micelles like those of poly(ethylene glycol)-co-poly(L-lactide) (PEG–PLLA) or poly(ethylene glycol)-co-poly(caprolactone) (PEG–PCL), may be especially attractive. If internalized by tumor cells via endocytosis, the micelles end up in the acidic environment of endosomes and lysosomes, where their accelerated degradation will release the encapsulated drug.

In vitro drug release from micellar systems is usually assessed either by dialysis through semipermeable membranes or by gel-permeation chromatography. Unfortunately, the information provided by these techniques is not directly pertinent to the in vivo behavior of drug-loaded polymeric micelles. The systemic injection of micellar formulations is associated with substantial dilutions and sink conditions. Because micelle formation is thermodynamically driven, when diluted below the CMC, micelles dissociate into individual molecules (unimers), thus releasing their drug load. To prevent premature micelle dissociation and drug release upon injection, micellar systems should have low CMC, glassy cores, or cross-linked cores. This may be achieved by induction of strong hydrophobic interactions or hydrogen bonds in the micelle cores.[19–21]

21.2.2 Polymeric Micelles in Clinical Trials

Many micellar drug delivery systems showed promising results in in vitro and animal experiments, but showed disappointing behavior in clinical trials.

The first clinical trials of micellar drug delivery systems began in this decade. Physically encapsulated DOX in PEO_{5000}-b-$P(Asp)_{4000}$-DOX micelles (NK911 formulation) entered clinical trials in Japan in 2001.[22] In animal studies, NK911 demonstrated about 30-fold higher area-under-the-curve (AUC) in plasma, higher tumor drug levels, lower toxicity, and higher therapeutic activity. However, in clinical trials, NK911 showed the same spectrum of side effects as free DOX, despite somewhat more favorable pharmacokinetic parameters. The discrepancy between the expected and observed clinical results was most likely caused by a premature drug release from micelles due to a very strong dilution of infused micellar solutions in the human circulatory system (volume: ~ 5 L) when compared to that of mice (volume: ~ 2 mL). A drop-like infusion of micellar drug delivery systems appears to have no prospect due to the dilution problem. Only bolus injections of reasonably stable micellar systems could provide for micelle preservation and drug retention in circulation.

Aqueous Pluronic® solutions are another polymeric system suggested for DOX delivery, especially for drug resistant cancers. Pluronic® copolymers in unimeric concentrations were shown to hyper-sensitize drug resistance carcinoma cells and enhance response of resistant tumors to chemotherapeutic agents.[23–31] The reason behind this phenomenon is related to an increased drug uptake caused by deactivation of drug efflux pumps; the latter process is associated with Pluronic®-induced energy depletion in resistant tumor cells.[32] In addition, favorable changes in the intracellular trafficking of DOX occur in the MDR cells under Pluronic® action.[30,33,34]

Pluronic® formulation of DOX, known as SP1049C, was thoroughly investigated in animal studies by Alakhov et al.[35] Although this formulation is probably partly micellar, it is not expected to retain micellar structure upon intravenous injections. In vivo, DOX and Pluronic® unimers mostlikely move along independent routes. Not unexpectedly, the toxic level of DOX in SP1049C formulation was shown to be similar to that of free DOX. The results of pharmacokinetic and biodistribution experiments on SP1049C in Lewis lung carcinoma bearing C57BL/6 mice showed a very moderate increase in plasma AUC. In anti-tumor efficacy studies using hematological and solid animal tumor models, SP1049C was shown to be more effective than free DOX at an equal dose of 5 mg/kg.[30]

SP1049C entered preclinical and clinical trials in Canada in 1999.[36] The toxicity profiles of free DOX and Pluronic® formulation were identical, except for the histopathological changes in skin and thymus, that were less pronounced with SP1049C. In 2004, Danson et al. reported on the results of phase-I clinical trials of SP1049C.[36] In this study, SP1049C was given to patients with histologically proven cancer refractory to conventional treatments. The dose-limiting toxicity (DLT) in humans, i.e., myelosuppression, was reached at 90 mg/m^2 of SP1049C; therefore, a dose of 70 mg/m^2 was recommended for future trials. The pharmacokinetic profile of SP1049C was similar to that of a clinical DOX formulation, except for a slower terminal clearance stage. SP1049C induced temporal partial response in several patients with advanced solid tumors. The results were considered promising, and the phase-II trials were initiated; their first results were recently reported by Valle et al.[37] In phase-II trials, SP1049C was evaluated in patients with less severe forms of cancer. Patients received 75 mg/m^2 of SP1049C (30 min intravenous infusion), four times a week, for up to six cycles. The authors reported partial responses in some patients after 4–6 treatment cycles. However, data showed appearance of hematological and nonhematological (especially cardiac) toxicities, presumably because DOX was not encapsulated and, therefore, not prevented from the attack on the heart muscle.

PTX is another potential drug candidate for the delivery in polymeric micelles. PTX is a highly hydrophobic molecule; for clinical use, PTX is solubilized with the aid of a significant amount of Cremophor EL and ethyl alcohol.[38] These formulations (Taxol® or PTX Injections) cause significant systemic toxicity. Considerable efforts of various groups have been dedicated to developing micellar formulations of PTX.[4–6,9,39] However, except for the Japanese formulation NK105 and Genexol® (see below), no other formulations have progressed to clinical trials due to a rapid loss of drug from micellar carriers, resulting in the absence of advantages over clinical formulations.

In 2001, Kim et al. reported on PTX-loaded PEO-b-poly(D,L-lactide) (PEO$_{2000}$-b-PDLLA$_{1750}$) micelles (Genexol®-PM) prepared by a solvent evaporation technique.[39] The maximum tolerated doses (MTDs) of Genexol®-PM and Taxol® for the intravenous (i.v.) administration in nude mice were 60 and 20 mg/kg, respectively. This allowed a threefold higher administration dose of drug. In animal studies, Genexol®-PM showed a better anti-tumor effect in ovarian and breast cancer mice models than clinical Taxol®.

However, despite an increase in the administered dose, plasma AUC was 20-fold lower for the micellar formulation, which was attributed to a fast release of drug from the micelles. Another factor involved in the fast elimination of micellar PTX from the circulation could be related to the uptake by the cells of the reticulo-endothelial system (RES). In all organs except plasma, PTX biodistribution was comparable for a clinical formulation or Genexol-PM. In phase-I human trials, micellar-encapsulated PTX allowed administration of a higher drug dose, but resulted in a lower plasma AUC and a shorter PTX half life.[40]

Another type of polymeric micelles suggested for PTX delivery comprise PDLLA as a core-forming block and poly(N-vinylpyrrolidone) (PVP) as a shell-forming block.[41] The in vivo evaluation of this system showed results similar to those for Genexol®-PM in regards to a reduced plasma AUC.[42] For the micellar formulation, the AUC was also reduced in liver, kidney, spleen, and heart, but was not significantly changed in the tumor. For the polymeric micellar formulation, MTD was not reached even at 100 mg/kg, whereas the MTD of Taxol® was 20 mg/kg. This allowed a threefold increase of the therapeutic dose of PTX, resulting in a significant increase of the anti-tumor activity of the polymeric system compared to Taxol®.

Polymeric micellar PTX formulation NK105 reported by Hamaguchi et al.[43] comprises micelle-forming block copolymer poly(ethelene oxide)-b-poly(4-phenyl-1-butanoate)-L-apartamide (PEO-b-PPBA). This formulation demonstrated stealth and targeting properties. NK105 manifested a 86-fold increase in PTX plasma AUC compared to Taxol®, accompanied by a 25-fold increase for drug AUC in tumor, and a higher anti-tumor activity in C-26 tumor-bearing mice. At a PTX dose of 100 mg/kg, tumors were completely resolved after a single administration of NK105. This formulation is currently in phase-I clinical trials in Japan.

In summary, only a few polymeric micellar formulations have demonstrated success in clinical trials. Polymeric micelles as drug carriers are either too stable, therefore incapable of providing adequate drug release at the tumor site, or too unstable, thus prematurely releasing their drug payload.

To overcome these complications, we have developed Pluronic®-based micellar systems that withstand the destabilizing effects of the biological environment and effectively target tumors while retaining their drug load.[44] Various routes of micelle stabilization were tested for Pluronic micelles;[19] the optimal stability, low toxicity, and long circulation time of micelle-encapsulated DOX was achieved by forming mixed micelles of Pluronic® and PEO-diacyl phospholipids[44] or by introducing a small concentration of oil in the micelle core to increase core hydrophobicity.[19] The on-demand release of DOX from micelles and an effective intracellular drug uptake by the tumor cells was achieved by local ultrasound treatment of the tumor at a particular time after the drug injection.[44]

Important advantages of ultrasound are that it is noninvasive, can penetrate deep into the interior of the body, and can be focused and carefully controlled. Therapeutic tumor treatment by ultrasound requires tumor imaging by diagnostic ultrasound prior to therapy. During the last decade, combining diagnostic and therapeutic ultrasound in the same instrument has attracted ever growing attention. Developing dual-modality systems that combine properties of ultrasound contrast agents and drug carriers is a timely task.

21.3 ULTRASOUND IN DRUG DELIVERY

21.3.1 INTRODUCTION TO ULTRASOUND IN MEDICINE

Ultrasound consists of pressure waves generated by piezoelectric transducers, with frequencies at or greater than 20 kHz. Like optical or audio waves, ultrasonic waves can be focused, reflected, refracted, and propagated through a medium, thus allowing the waves to be directed to and focused on a particular volume of tissue in a spatially and temporally controlled manner. Ultrasound technology allows for a high degree of spatial and dynamic control due to a favorable range of energy penetration characteristics in soft tissue and the ability to shape energy deposition patterns.

In medicine, ultrasound is used as either a diagnostic or a therapeutic modality. The main advantage of ultrasound is its noninvasive nature; the transducer is placed in contact with a water-based gel or water layer on the skin, and no insertion or surgery is required.

In current therapeutic applications, ultrasound is used mostly for tissue thermoablation. Progress in imaging techniques has allowed ablating large tumor masses by high intensity focused ultrasound (HIFU). HIFU instruments noninvasively deliver high intensity ultrasound energy to predetermined volumes of the body.

However, the ablative techniques require delivering high acoustic energy to a well delineated volume in the body over a substantial time period (up to several hours), which is associated with a potential risk of ablating surrounding tissues due to a patient's breathing or motion. Motion correction is a serious concern in HIFU applications. In addition, some locations in the body are screened by bones and not accessible for HIFU treatment.

In our lab, we have been developing a different mode of application of therapeutic ultrasound. In this modality, tumor cell killing is based on the ultrasound-enhanced cytotoxic action of tumor-targeted chemotherapeutic drugs rather than the mechanical or thermal action of ultrasound.[45–56] This technology requires order of magnitude lower ultrasound energy, reducing the risk of coagulative necrosis of healthy tissues associated with HIFU.

21.3.2 BIOLOGICAL EFFECTS OF ULTRASOUND

21.3.2.1 Cellular Level

The effect of ultrasound on the cellular level is usually attributed to formation of plasma membrane defects (sonoporation) induced by cavitating microbubbles; the formation and collapse of microbubbles is called inertial cavitation. This process has been investigated using a voltage clamp technique.[57–59] It was shown that sonoporation was substantially enhanced by ultrasound contrast agents that nucleate inertial cavitation.[60] The degree of membrane damage increased with increasing duty cycle of pulsed ultrasound.[59] Some membrane defects are resealed when ultrasound is turned off (this is called transient sonoporation); a key role in membrane resealing is played by Ca^{2+}.[58]

In our experiments, ultrasound bioeffects on the cellular level were compared for the drug-sensitive and multidrug resistant (MDR) A2780 ovarian carcinoma cells in the presence or absence of DOX.[61] Focused 1.1-MHz ultrasound was applied at varying peak negative pressures and DOX concentrations. Cell viability, a degree of membrane damage, DOX uptake, and cell proliferation were measured. Cell incubation with the nucleic acid stain Sytox allowed discriminating between the cells experiencing various degrees of plasma membrane damage. Fluorescence of Sytox is negligible in the extracellular environment; when Sytox enters the cells and intercalates the DNA, its fluorescence increases by several orders of magnitude. For the present application it is important that Sytox does not penetrate into the cells with intact membranes; however, when cell membranes are damaged, Sytox enters the cells and intercalates the DNA, which results in a dramatic fluorescence increase.

Major results of our experiments on ultrasound bioeffects are listed below.

- Under the action of ultrasound, cells with various degrees of membrane damage, intermediate between viable and dead cells, were generated.
- At the same delivered ultrasound energy, higher ultrasound pressure caused more severe membrane damage.
- A proliferation rate of sonicated multidrug resistant ovarian carcinoma A2780/AD cells with compromised membranes was significantly lower than the proliferation rate of the intact cells (Figure 21.1).
- The MDR A2780/AD cells manifested a higher susceptibility to ultrasound than parental drug-sensitive A2780 cells (Figure 21.2).
- Introduction of DOX in the incubation medium increased cell susceptibility to the ultrasound action. For the same ultrasound energy, the cells sonicated in the presence of DOX manifested significantly higher membrane damage than the cells without DOX (Figure 21.3). For the same initial DOX concentration in the incubation medium, the effect of DOX was stronger at 37°C than at room temperature suggesting that an increased cell susceptibility to ultrasound was caused by DOX internalized by the cells. We hypothesize that the effect of DOX may be associated with a destabilizing action on the DNA molecule at the intercalation sites.

Very interesting data on the effect of ultrasound on the gene expression in multidrug resistant Hep2D/ADR hepatic carcinoma cells were recently obtained in Chongqing Medical University by Shao and Wu (Wu, F., personal communication). These researchers observed a noticeable suppression of the MDR1 gene expression in the cells sonicated by 0.8-MHz ultrasound; the *BCL2/BAX* expression ratio was also substantially reduced, resulting in an enhanced apoptosis. The last effect was significantly stronger in the MDR compared to parental cells.

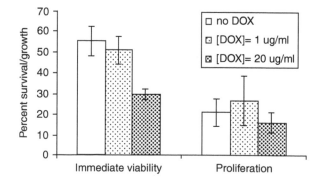

FIGURE 21.1 Immediate cell viability (left group of bars) and proliferation at 37°C for 72 h (right group of bars) for the A2780/AD cells sonicated (0.55 MPa, 15 s) in the presence or absence of various concentrations of DOX. (Adapted from Kamaev, P., Christensen, D. A., and Rapoport, N., unpublished results.)

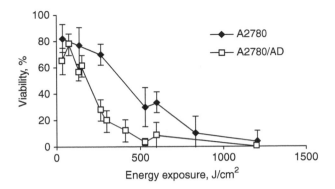

FIGURE 21.2 Immediate cell viability after ultrasound exposure decreases with increasing acoustic energy; cells were irradiated in suspensions by focused 1.1-MHz ultrasound. The MDR cells manifest higher susceptibility to the ultrasound action than their wild-type counterparts. (Adapted from Kamaev, P., and Rapoport, N., *Therapeutic Ultrasound: 5th International Symposium on Therapeutic Ultrasound*, Clement, G. T., McDannold, N. J., and Hynynen, K., Eds., American Institute of Physics, Melville, NY, pp. 543–547, 2006. With permission.)

These clinically important findings can be used for effective therapy of multidrug-resistant tumors.

21.3.2.2 Systemic Level

In living organisms, the bioeffects of ultrasound are not limited to sonoporation. On a systemic level, a very important role may be played by enhanced extravasation. Because of the adverse effects that this may have in diagnostic ultrasound, extensive work has been done on studying extravasation at various locations in the body and from various blood vessels, especially in the process of heart imaging. Extravasation was substantially enhanced by ultrasound contrast agents. Enhanced leakage of the die and erythrocytes from heart vessels was observed in several studies.[62–64] Erythrocyte extravasation under the action of contrast-enhanced ultrasound was also observed in the kidney; it was shown that the effect of pulsed ultrasound was stronger than that of continuous wave (CW) ultrasound.[65]

Although enhanced extravasation is an undesirable effect in ultrasound imaging, it may play a very positive role in drug delivery, especially for the local delivery of macromolecular drugs or

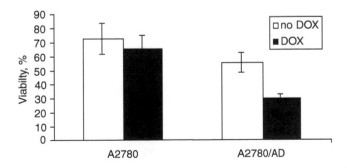

FIGURE 21.3 Effect of DOX on the immediate viability of drug-sensitive and MDR ovarian carcinoma cells sonicated in suspensions by focused ultrasound; DOX concentration in the media is 10 μg/mL for the MDR cells and 20 μg/mL for the sensitive cells. Ultrasound conditions: frequency 1.1 MHz, pressure 0.55 MPa, duration 15 s, duty cycle 33%, room temperature. (Adapted from Kamaev, P., and Rapoport, N., *Therapeutic Ultrasound: 5th International Symposium on Therapeutic Ultrasound*, Clement, G. T., McDannold, N. J., and Hynynen, K., Eds., American Institute of Physics, Melville, NY, pp. 543–547, 2006. With permission.)

drug-loaded nanoparticles. It was shown that ultrasound enhanced the extravasation of the macro-molecular MRI contrast agent.[66]

21.3.3 ULTRASOUND-INDUCED DRUG RELEASE FROM POLYMERIC MICELLES

Ultrasound was shown to release DOX from the unstabilized and stabilized Pluronic® micelles.[46,55] A degree of drug release depended on a number of ultrasound parameters—frequency, power density, pulse length, and inter-pulse intervals.[46] Examples of drug release profiles under CW or pulsed ultrasound are presented in Figure 21.4. Measurements were based on a decrease of DOX fluorescence when drug was transferred from the hydrophobic environment of micelle cores into the aqueous environment. Measurements were performed in situ in the absence of cells.

As indicated by fluorescence profiles, drug release under the action of ultrasound pulses was reversible and drug re-encapsulation proceeded during the inter-pulse intervals. Pulsed ultrasound induced alternating cycles of micelle perturbation and restoration. Kinetic parameters of the drug release and re-encapsulation indicated that both processes were controlled by motion of micelle-forming macromolecules rather than diffusion of relatively small drug molecules.[55]

Low-frequency ultrasound was more effective in drug release from micelles than high-frequency ultrasound (Figure 21.5). Low-frequency ultrasound can penetrate deeper into the tissue than high-frequency ultrasound, but it does not allow sharp focusing.[50] Therefore, in vivo, high-frequency ultrasound could be used for irradiating small and relatively superficial tumors, while low-frequency ultrasound could be used for larger and deeper located tumors.

21.3.4 ULTRASOUND-ENHANCED INTRACELLULAR DRUG UPTAKE

Without ultrasound, drug encapsulation in polymer micelles results in a significantly reduced intra-cellular uptake. The degree of drug shielding depends on drug/micelle interaction, and is higher for

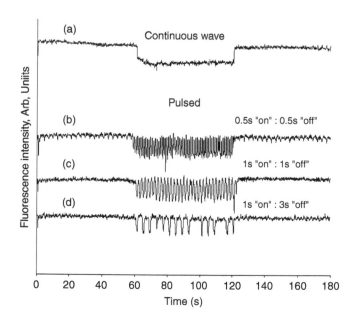

FIGURE 21.4 Doxorubicin (DOX) release profiles from 10% Pluronic micelles under continuous wave (CW) and pulsed 20 kHz ultrasound at a power density of 0.058 W/cm² at various durations of ultrasound pulses and inter-pulse intervals indicated in the figure. Measurements are based on the decrease of fluorescence intensity when DOX is transferred from the hydrophobic environment of micelles cores into the aqueous environment. (Adapted from Husseni, G. A. et al., *J. Control. Release*, 69, 43–52, 2000. With permission.)

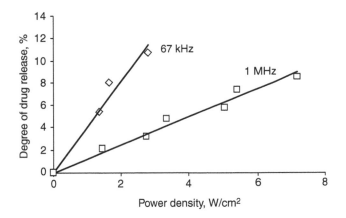

FIGURE 21.5 Effect of ultrasound frequency and power density on DOX release from 10% Pluronic P-105 micelles. (Adapted from Marin, A. et al., *J. Control. Release*, 84, 39–47, 2002.)

more lipophilic drugs. As an example, DOX uptake by ovarian carcinoma A2780 cells dropped by 50% after encapsulation in 10% Pluronic® micelles. For a more hydrophobic analog of DOX called Ruboxyl, micellar encapsulation resulted in a decrease of the intracellular uptake by 66%.

Drug shielding by micelles reduces systemic concentrations of toxic drugs, diminishing side effects. As an example, DOX encapsulation in PEO-*b*-poly(α-benzyl L-aspartate) (PEO–PBLA) micelles allowed administering high concentrations of DOX to tumor-bearing mice without toxic side effects.[67] The same results were manifested by micelle-encapsulated PTX[39–43] (see Section 21.2.2).

Ultrasound substantially enhanced the intracellular drug uptake from/with polymeric micelles.[45,50,56] Typical examples are shown in Figure 21.6.

Based on the results presented above, a new modality of tumor-targeted chemotherapy has been suggested. According to this technology, micelle-encapsulated drugs are injected intravenously to tumor-bearing subjects. Drug-loaded micelles are long-circulating (Figure 21.7), which provides for their gradual accumulation in the tumor interstitium via the EPR effect (see below). At the time corresponding to a maximal micelle accumulation, ultrasound is applied locally to the tumor. Under the ultrasound action, drug is released from micelles in the tumor volume; the intracellular uptake of drug by tumor cells is enhanced due to ultrasound-induced plasma membrane perturbation.

Important advantages of this technique are its noninvasive character, on-demand drug release, and a prospect for sensitization of drug-resistant tumors.

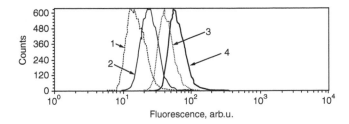

FIGURE 21.6 Fluorescence histogram of A2780 cells. (1) Control; (2)–(4): cells incubated for 30 min. with 20 µg/mL DOX encapsulated in PEG–PLLA micelles. (2) No ultrasound; (3) sonication for 30 s by CW 3-MHz ultrasound at 2.0 W/cm²; (4) sonication for 30 s by 1-MHz ultrasound at 3.4 W/cm² and 33% duty cycle. (From Gao, Z. and Rapoport, N., unpublished results.)

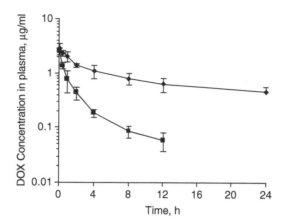

FIGURE 21.7 Pharmacokinetics of DOX dissolved in PBS (squares) or encapsulated in stabilized Pluronic P-105 micelles (diamonds); mean values plus/minus standard deviations are presented, $n = 4$. (From Gao, Z. and Rapoport, N., unpublished results.)

In the following section, the results of the in vivo evaluation of the new therapeutic modality are discussed.

21.4 IN VIVO EVALUATION OF THE DRUG/MICELLE/ULTRASOUND TUMOR-TARGETING MODALITY

21.4.1 ULTRASOUND-INDUCED DRUG RELEASE FROM MICELLES IN VIVO

Drug release from micelles in vivo was confirmed in experiments described in studies by Gao, et al.[44] Ovarian carcinoma A2780 cells (2×10^6) were inoculated in peritoneal cavities of athymic (nu/nu) mice. The next day, mice were treated intraperitoneally by 3 mg/kg DOX that was either dissolved in PBS (designated "free DOX" hereafter) or encapsulated in 10% Pluronic® P-105 micelles. Treatment was combined or not combined with the ultrasonic irradiation of the abdominal region of a mouse. Some groups of mice were treated by ultrasound without DOX. Tumor yields and animal survival rates were monitored. The results are presented in Table 21.1.[44]

Without DOX, no effect of ultrasound on the tumor staging was observed; the tumors were developed in 100% mice. Only a moderate effect of free DOX on the tumor development was observed without ultrasound, tumor yield being reduced by 30%. However, when DOX and ultrasound were combined, tumor yields were reduced by about 60% for both free and micellar-encapsulated DOX, which was accompanied by a statistically significant increase of the average survival rates of corresponding animal groups. No differences were observed between the effects of free or micellar-encapsulated DOX, suggesting that ultrasound did, indeed, trigger DOX release from micelles in vivo.

The absence of the ultrasound effect in the absence of DOX indicated that tumor cell death was associated with the ultrasound-enhanced cytotoxic action of DOX rather than mechanical or thermal action of ultrasound itself.[44]

21.4.2 BIODISTRIBUTION OF POLYMERIC MICELLES AND MICELLAR-ENCAPSULATED DRUGS

The effect of ultrasound on biodistribution of polymeric micelles was studied using fluorescently labeled Pluronic® P-105 micelles.[47] The uptake of micelles by various organ cells was measured by flow cytometry. A significant advantage of this technique is that it does not require organ

TABLE 21.1
Tumor Yields and Survival Rates of Mice Inoculated on Day 0 with i.p. Injections of A2780 Ovarian Carcinoma Cells

Group	Number of Animals in Group	Number of Internal Tumors	Tumor Yield (%)	Survival Rate (Days)
PBS	17	17	100	40.6 ± 8.2
PBS + US	5	5	100	39.6 ± 6.5
P-105 empty micelles	13	13	100	51.9 ± 21.6
P-105 empty micelles/US	13	13	100	47.2 ± 9.0
DOX/PBS[a]	10	7	70	61.5 ± 10[b]
DOX/PBS/US[a]	5	2	40	77.3 ± 24
DOX/micelle/US[a]	11	4	36	91 ± 27[b]

[a] The mice were treated on days 1, 4, 7, and 11; DOX at 3 mg/kg in various delivery systems was i.p. injected. One hour after the DOX injection, some groups of animals were sonicated in the abdominal region for 30 s by 1-MHz ultrasound at a power density of 1.2 W/cm^2. By day 25, all control mice developed tumors. The last three groups marked by "a" were treated on days 1, 26, and 35.

[b] $p = .01$.

Source: Gao, Z., Fain, H., and Rapoport, N., *J. Control. Release*, 102, 203–221, 2005.

homogenization, thus providing for direct measurements of carrier or drug concentrations inside the cells, i.e., at the site of action.

A 100-µL solution of fluorescently labeled stabilized Pluronic® micelles was injected intravenously through the tail vein to ovarian carcinoma bearing nu/nu mice. At various time points ranging from 2 to 12 h after the i.v. injection, mice were sonicated for 30 s by 1-MHz or 3-MHz ultrasound. Ten minutes after the ultrasound application, mice were sacrificed, tumor and other organs excised, dried with filter paper, weighed, and sliced in trypsin. The individual cells of various organs were fixed by 2.5% glutaraldehyde. Fluorescence of Pluronic® P-105 molecules internalized by the cells of various organs was measured by flow cytometry.

Without ultrasound, a multimodal distribution of tumor cell fluorescence was observed for the intravenous injections of the fluorescently labeled, stabilized Pluronic® micelles. A significant fraction of the cells manifested a relatively low fluorescence (Figure 21.8, regular line). Fluorescence intensity of other organ cells was at the level of the autofluorescence of corresponding cells. These data indicated low efficiency of the intracellular uptake of intact Pluronic® micelles. A short (30-s ultrasound exposure) tumor sonication strongly enhanced the intracellular uptake of Pluronic® molecules by the tumor cells (Figure 21.8, bold line), suggesting that before sonication, micelles were effectively accumulated in the tumor interstitium but were not effectively internalized by the tumor cells. Tumor sonication not only substantially enhanced micelle internalization, but also resulted in a much more uniform fluorescence distribution over the tumor cell population, suggesting an enhanced micelle diffusion over the tumor volume.[47]

The biodistribution of micellar-encapsulated DOX closely followed that of a micellar carrier.[44] For micellar-encapsulated DOX without ultrasound, DOX uptake by the tumor cells was relatively low. A local tumor sonication by 1-MHz ultrasound dramatically increased the intracellular uptake of drug by the tumor cells (Figure 21.9). A similar effect was observed for PEG–PBLA micelles (Figure 21.10).[44] Somewhat enhanced DOX uptake by the cells of other organs was observed for 1-MHz ultrasound (data not shown). This was associated with a deep penetration of 1-MHz ultrasound in the mouse body, resulting in ultrasound scattering at various internal interfaces and/or

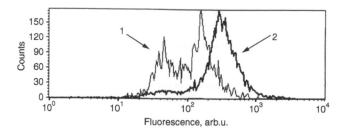

FIGURE 21.8 Fluorescence histogram of the tumor cells in (1) unsonicated, and (2) sonicated mouse; 5% stabilized Pluronic micelles comprising 0.1% fluorescently labeled Pluronic P-105 were injected intravenously to two subcutaneous A2780 tumor bearing mice; 4 h after the injection, the tumor of one mouse was sonicated for a total of 60 s by 1-MHz ultrasound at 3.4 W/cm^2 power density at 50% duty cycle (two 30-s sonications were performed, with a 30-s interval between the sonications for cooling the ultrasound probe); both mice were sacrificed 10 min after the completion of sonication. (Modified from Gao, Z., Fain, H., and Rapoport, N., *Mol. Pharmaceutics*, 1, 317–330, 2004.)

reflection from the skin/air interface at the site opposite to the transducer. This unwanted effect was eliminated when tumor was irradiated by a more localized 3-MHz ultrasound (Figure 21.10).

21.4.3 EFFECT OF THE TIME OF ULTRASOUND APPLICATION

Measuring micelle accumulation in the tumor cells for various times of ultrasound application after drug injection enabled the selection of optimal ultrasound application times. For the stabilized Pluronic® micelles, the optimal time of ultrasound application was found to be between four and eight hours after injection.[44,47] The results were interpreted as follows. At shorter times, a large number of micelles were still circulating in the vasculature while a lower concentration of micelles was accumulated in the tumor interstitial space. At longer times, some clearance of micelles from the tumor volume could already have occurred. The optimal time of ultrasound application is expected to depend on the pharmacokinetics of a particular delivery system.

21.4.4 INHIBITION OF TUMOR GROWTH USING MICELLE/ULTRASOUND DRUG DELIVERY

21.4.4.1 Drug-Sensitive Tumors

Targeting of micellar-encapsulated drugs to tumor cells and uniform drug distribution in the tumor volume result in a significant suppression of the growth of subcutaneous ovarian carcinoma tumors

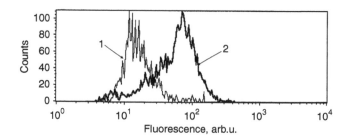

FIGURE 21.9 Fluorescence histograms of the tumor cells in (1) nonsonicated, and (2) sonicated mice injected i.v. with DOX (6 mg/kg) encapsulated in mixed micelles. Mice were sonicated for 30 s by 1-MHz CW ultrasound at a power density of 1.7 W/cm^2 applied 8 h after the drug injection; nonsonicated and sonicated mice were sacrificed 10 min later. (Modified from Gao, Z., Fain, H., and Rapoport, N., *J. Control. Release*, 102, 203–221, 2005.)

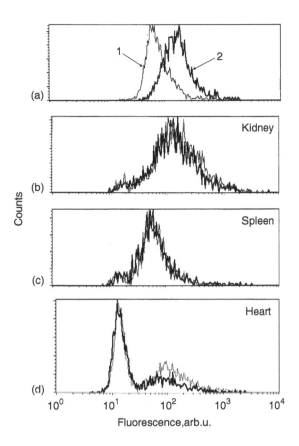

FIGURE 21.10 Fluorescence histograms of (a) tumor, (b) kidney, (c) spleen, and (d) heart cells of (1) nonsonicated (regular lines), and (2) sonicated (bold lines) mice injected with DOX (6 mg/kg) encapsulated in PEG–PBLA micelles. Mice were sonicated for 30 s by 3-MHz CW ultrasound 12 h after the drug injection at a power density of 1.8 W/cm². (Adapted from Gao, Z., Fain, H., and Rapoport, N., *J. Control. Release*, 102, 203–221, 2005.)

inoculated in nu/nu mice (Figure 21.11).[44] However, the success of the treatment depended dramatically on the stage of tumor progression at the start of the treatment. When the initial tumor volume at the start of the treatment exceeded 250 mm³, none of the treatment modalities was effective in suppressing tumor growth, which emphasizes a need for early tumor detection.

21.4.4.2 Drug Resistant Tumors

We have attained substantial success in treating multidrug resistant, poorly differentiated MCF-7/AD$_{mt}$ breast cancer tumors by micellar-encapsulated PTX combined with ultrasonic tumor irradiation. The MCF7/AD$_{mt}$ tumor cells expressed the *MDR1* gene[25] and were highly resistant to treatment by clinical PTX formulation, Taxol® (PTX injection). However, tumor growth was substantially suppressed when animals were treated by intravenous injections of a micellar PTX formulation in conjunction with local tumor sonication (Figure 21.12) (the micellar formulation of PTX, Genexol® was kindly donated by Samyang Pharmaceuticals Inc. (Korea)).

In vitro without ultrasound, the intracellular uptake of PTX from the micellar formulation was significantly lower than that with PTX injection, which can be advantageous when used in vivo for preventing unwanted drug interactions with healthy cells. Under ultrasound, drug uptake by the MCF7/AD$_{mt}$ cells from micellar PTX formulation was increased more than 20-fold, which resulted in the inhibition of cellular proliferation by nearly 90%.

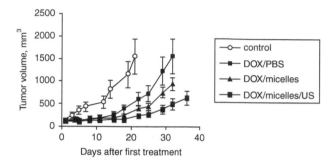

FIGURE 21.11 Growth curves of s.c. A2780 ovarian carcinoma tumors inoculated in female nu/nu mice; DOX (3 mg/kg) was either dissolved in PBS (positive control) or encapsulated in mixed micelles; drug injections were combined or not combined with tumor sonication. Treatments were initiated when the tumor volume reached 75–125 mm^3; three consecutive treatments were applied on days 1, 3, and 5. Ultrasound parameters: frequency, 1 MHz; power density, 3.4 W/cm^2; duty cycle 50%; duration, 30 s; time of application: 4 h after the i.v. injection of the drug. (Adapted from Gao, Z., Fain, H., and Rapoport, N., *J. Control. Release*, 102, 203–221, 2005.)

The in vivo results are shown in Figure 21.12. As could be expected, without ultrasound, clinical PTX and Genexol® both manifested low efficacy; in contrast, injections of Genexol® combined with ultrasonic tumor irradiation resulted in a complete tumor resolution. Most importantly, the tumors treated by Genexol®/ultrasound did not metastasize, whereas tumors treated without ultrasound manifested a high rate of metastasizing into lymph nodes and/or spleen.

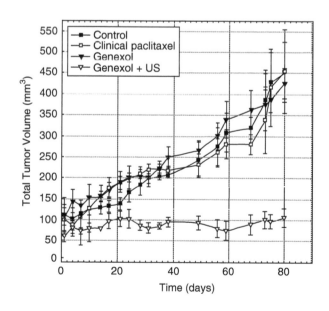

FIGURE 21.12 Growth curves of subcutaneous MCF7/AD$_{mt}$ tumors inoculated in female nu/nu mice; bolus injections of a clinical paclitaxel formulation Paclitaxel injections or a micellar formulation Genexol® provided by Samyang Pharmaceuticals, Inc. (Korea) were administered twice a week for two weeks intravenously through the tail vein at a dose rate of 13 mg/kg/day. (From Howard, B., Gao, Z., and Rapoport, N., *Am. J. Drug Deliv.*, in press.)

The aqueous base of the drug formulation, reduced systemic toxicity, potentials for tumor targeting, and on-demand drug release triggered by ultrasound are important additional advantages of micellar formulations of PTX.

21.4.4.3 Poorly Vascularized Tumors

A different type of drug resistance was observed for subcutaneously grown colon cancer HCT116 tumors. These tumors demonstrated very loose capillary networks and large necrotic areas filled with ascetic fluid. HCT116 colon cancer tumors manifested strong resistance to the intravenous drug injections independent on the delivery modality (Figure 21.13, upper set of curves). The resistance was caused by poor vascularization of the tumor volume, resulting in insufficient drug delivery to tumor cells.

Poor tumor vascularization represents a unique type of tumor resistance to chemotherapy. Upon intravenous drug injections, avascular tumor regions receive insufficient levels of chemotherapeutic agents; cancerous cells in these regions remain viable and promote tumor growth. This type of resistance has not been widely discussed in the literature.

In our laboratory, successful treatment of HCT116 tumors was achieved by direct intratumoral injections of micellar-encapsulated DOX (Figure 21.13, lower set of curves). The encapsulated drug was effectively retained in the tumor volume, with very low, if any, accumulation in the cells of other organs. In contrast, free DOX quickly leaked from the tumor tissue, and a short time after the intratumoral injection of free drug, DOX was already found in the heart cells and those of other organs, causing associated toxicity.

With the intratumoral injections of micellar encapsulated DOX, effective tumor suppression was observed even without ultrasound (Figure 21.13). We hypothesize that a high hydrostatic pressure generated by intratumoral injections forced drug-loaded micelles through cell membranes.

For therapy of poorly vascularized tumors, intratumoral injections of micellar-encapsulated drugs present a viable alternative to inefficient intravenous injections.

The advantages of intratumoral injections were also observed for highly vascularized A2780 tumors. The beneficial effect of ultrasound was revealed in drug biodistribution experiments. For the intravenous injections of micellar-encapsulated DOX, tumor sonication resulted in a

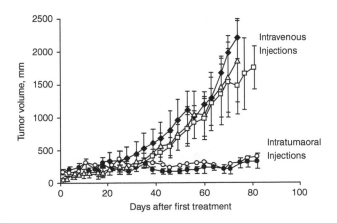

FIGURE 21.13 Growth curves of HCT116 tumors upon the intravenous (upper group of curves) or intratumoral (lower group of curves) injections of DOX at a dose rate of 3 mg/kg. Intravenous injections: filled diamonds—control, open triangles—DOX/PBS, open squares—DOX/mixed micelles/ultrasound; Intratumoral injections: open circles—DOX/mixed micelles, closed circles—DOX/mixed micelles/ultrasound. (From Gao, Z. and Rapoport, N., unpublished results.)

dramatically increased intracellular drug uptake; for the intratumoral injections of the same formulation, the main effect of ultrasound was related to a much more uniform drug distribution over the tumor cell population (Figure 21.14).

Intratumoral chemotherapy by nanoparticle encapsulated drugs could be advantageous for tumors with well-defined primary lesions such as breast, colorectal, prostate, and skin cancers. These procedures may be performed noninvasively or minimally invasively through intraluminal, laparoscopic or percutaneous means under the guidance of an appropriate imaging modality such as MRI, CT, or ultrasound. During the last decade, the interest in intratumoral chemotherapy has significantly increased.[68]

21.4.5 MECHANISMS OF MICELLE/ULTRASOUND BIOEFFECTS

Ultrasound irradiation may play multiple roles in tumor therapy. Ultrasound may (1) induce thermal effects, (2) enhance extravasation of drug-loaded micelles into the tumor interstitium, (3) enhance drug diffusion through the tumor interstitium, (4) trigger drug release from micelles in the tumor volume, and (5) enhance the intracellular uptake of both released and encapsulated drug at the irradiation site. Though the relative importance of these factors remains to be further explored, some conclusions can be drawn based on the results presented above.

Drug release from perturbed micelles under the action of ultrasound is undoubtedly an important factor in the ultrasound-enhanced drug uptake by the tumor cells from micellar carriers. Ultrasound can also enhance the intracellular uptake of intact drug-loaded micelles due to cell membrane perturbation. Another important component of the ultrasound effect is associated with the ultrasound-enhanced diffusion of free or micellar-encapsulated drug over tumor tissue.

In summary, drug encapsulation in polymeric micelles combined with local ultrasonic tumor irradiation results in effective tumor targeting and suppression of growth of drug-sensitive and multidrug resistant tumors. The effect is associated with the ultrasound-enhanced cytotoxic action of passively targeted or intratumorally injected micellar-encapsulated chemotherapeutic agents.

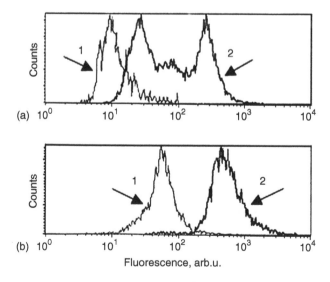

FIGURE 21.14 Fluorescence histograms of A2780 tumor cells in (a) unsonicated, and (b) sonicated mouse; DOX encapsulated in stabilized Pluronic micelles was injected at a dose of at 6 mg/kg (1) intravenously, and (2) intratumorally; sonicated tumors were irradiated for 30 s by 1-MHz ultrasound at 3.4 W/cm^2 power density and 50% duty cycle 4 h after the drug injection. (From Gao, Z. and Rapoport, N., unpublished results.)

This allows on-demand drug delivery and control of systemic, local, and subcellular drug distribution.

21.5 COMBINING ULTRASOUND IMAGING AND TREATMENT

Ultrasonic tumor treatment requires tumor imaging prior to therapy. Using ultrasound for both tumor imaging and treatment is especially attractive. Development of ultrasound contrast agents and concomitant use of harmonic ultrasound imaging has recently made possible imaging of liver,[69] breast,[70] and prostate tumors.[71] Ultrasound-stimulated acoustic emission of microbubbles was used for color Doppler imaging of liver metastases.[72]

Some dual ultrasonic imaging/treatment techniques are already in clinical practice, others are in the investigational stage. Examples include Sonablate® for prostate cancer treatment (Focus Surgery, Inc., Indianapolis, Indiana) and a miniature ultrasound-based dual-modality device for image guided soft tissue ablation (Guided Therapy Systems, Mesa, Arizona).

With dual-modality ultrasound instruments already on the market, developing dual-modality ultrasound imaging/drug carrier systemswould be very timely.

21.5.1 DUAL-MODALITY CONTRAST AGENTS/DRUG CARRIER SYSTEMS

Ultrasonically enhanced site-specific drug delivery using acoustically activated drug-loaded microbubbles was suggested by Ungar[73] and Ferrara;[74–77] these authors developed echogenic microbubbles stabilized by lipid bilayers. Drug was incorporated in the oil layer located on the microbubble surface. An effective ultrasound-triggered transfer of a solute from the modified lipospheres to the tumor cells has been demonstrated. Polymeric ultrasound contrast agents with targeting potentials have also been explored by the Wheatley group.[78–81]

Nano/microbubbles oscillate in the ultrasound field and can be cavitated for site-specific localized delivery of chemotherapeutic agents; they concentrate ultrasound energy and may be used for enhancing drug delivery from micelles and other nanoparticles. Nano/microemulsions and microbubbles are efficient reflectors of ultrasound energy, which may be used for contrast-enhanced ultrasound imaging of tumor neovasculature.

We have recently developed a new class of these agents that combine diagnostic and therapeutic properties and can be used for ultrasonically enhanced drug delivery and tumor imaging. Our microbubbles differ in a number of ways from the products on the market or those under development by other groups:

- The microbubbles are produced in situ upon injection of specially designed micro-emulsions.
- The microbubbles have strong walls produced by biodegradable diblock copolymers; the walls stabilize microbubbles for ultrasound imaging.
- The diblock copolymer not only forms microbubble walls but also forms polymeric micelles that effectively encapsulate chemotherapeutic agents and act as drug carriers.
- Upon injection, a localized ultrasonic irradiation of the tumor in the presence of microbubbles provides for effective intracellular drug uptake by the tumor cells, which results in inhibition of tumor growth.

Echogenic nano/microbubbles designed to interact specifically with neovascular epithelium may allow ultrasound imaging of small metastases. To ensure molecular targeting, the nano/microparticle surface should be decorated by site-specific ligands. In the future, we anticipate conjugating block copolymer molecules with vectors that would target micelles and nano/microbubbles to the receptors that are overexpressed on the endothelial lining of tumor neovasculature.

21.6 CONCLUSIONS

In animal models, drug encapsulation within polymeric micelles combined with ultrasonic tumor irradiation results in effective tumor targeting and inhibition of tumor growth. The effect is based on ultrasonically triggered localized drug release from micelles and enhanced intracellular drug uptake. This technique offers prospects for treating multidrug resistant tumors that fail conventional chemotherapy. Dual-modality drug delivery/ultrasound contrast agent systems used with dual-modality ultrasound imaging/treatment instruments are expected to allow precisely controlled, ultrasound-enhanced tumor chemotherapy in a clinical environment.

REFERENCES

1. Kwon, G. S. and Kataoka, K., Block copolymer micelles as long circulating drug vehicles, *Drug Deliv. Rev.*, 16, 295, 1995.
2. Aliabadi, H. and Lavasanifar, A., Polymeric micelles for drug delivery, *Expert Opin. Drug Deliv.*, 3, 139, 2006.
3. Matsumura, Y. and Maeda, H., A new concept for macromolecular therapeutics in cancer chemotherapy: Mechanism of tumoritropic accumulation of proteins and the antitumor agent smancs, *Cancer Res.*, 46, 6387, 1986.
4. Zhang, X., Burt, H. M., and Von Hoff, D., An investigation of the antitumor activity and biodistribution of polymeric micellar paclitaxel, *Cancer Chemother. Pharmacol.*, 40, 81, 1997.
5. Zhang, X., Jackson, J. K., and Burt, H. M., Development of amphiphilic diblock copolymers as micellear carriers of taxol, *Int. J. Pharm.*, 132, 195, 1996.
6. Zhang, X., Burt, H. M., and Mangold, G., Anti-tumor efficacy and biodistribution of intravenous polymeric micellar paclitaxel, *Anticancer Drug*, 8, 696, 1997.
7. Adams, M. L., Andes, D. R., and Kwon, G. S., Amphotericin B encapsulated in micelles based on poly(ethylene oxide)-*b*-poly(L-amino acid) derivatives exerts reduced in vitro hemolysis but maintains potent in vivo antifungal activity, *Biomacromolecules*, 4, 750, 2003.
8. Adams, M. L. and Kwon, G. S., Relative aggregation state and hemolytic activity of amphotericin B encapsulated by poly(ethylene oxide)-*b*-poly(N-hexyl-L-aspartamide)-acyl conjugate micelles: Effects of acyl chain length, *J. Control. Release*, 87, 23, 2003.
9. Leung, S., Jackson, J., Miyake, H., Burth, H., and Gleave, M. E., Polymeric micellar paclitaxel phosphorylates Bcl-2 and induces apopototic regression of androgen-independent LNCaP prostate tumors, *Prostate*, 44, 156, 2000.
10. Shin, I. G. et al., Methoxy poly(ethylene glycol)/epsilon-caprolactone amphiphilic block copolymeric micelle containing indomethacin. I. Preparation and characterization, *J. Control. Release*, 51, 1, 1998.
11. Kim, S. C. et al., In vivo evaluation of polymeric micellar paclitaxel formulation: Toxicity and efficacy, *J. Control. Release*, 72, 191, 2001.
12. Benahmed, A., Ranger, M., and Leroux, J. C., Novel polymeric micelles based on the amphiphilic diblock copolymer poly(N-vinyl-2-pyrrolidone)-*b*-poly(D,L-lactide), *Pharm. Res.*, 18, 323, 2001.
13. Yu, B. G. et al., Polymeric micelles for drug delivery: Solubilization and haemolytic activity of amphotericin B, *J. Control. Release*, 53, 131, 1998.
14. Lavasanifar, A., Samuel, J., and Kwon, G. S., Micelles of poly(ethylene oxide)-*b*-poly(N-alkyl stearate L-aspartamide): Synthetic analogues of lipoproteins for drug delivery, *J. Biomed. Mater. Res.*, 52, 831, 2000.
15. Piskin, E., Kaitian, X., Denkbas, E. B., and Kucukyavuz, Z., Novel PDLLA/PEG copolymer micelles as drug carriers, *J. Biomater. Sci. Polym. Ed.*, 7, 359, 1995.
16. Francis, M. F., Piredda, M., and Winnik, F. M., Solubilization of poorly water soluble drugs in micelles of hydrophobically modified hydroxypropylcellulose copolymers, *J. Control. Release*, 93, 59, 2003.
17. Jette, K. K. et al., Preparation and drug loading of poly(ethylene glycol)-*b*-poly(epsilon-caprolactone) micelles through the evaporation of a cosolvent azeotrope, *Pharm. Res.*, 21, 1184, 2004.
18. Nakanishi, T. et al., Development of the polymer micelle carrier system for doxorubicin, *J. Control. Release*, 74, 295, 2001.

19. Rapoport, N., Stabilization and acoustic activation of pluronic micelles for tumor-targeted drug delivery, *Colloids Surf. B Biointerfaces*, 3, 93, 1999.
20. Shuai, X. et al., Core-cross-linked polymeric micelles as paclitaxel carriers, *Bioconj. Chem.*, 15, 441, 2004.
21. Lee, J., Cho, E. C., and Cho, K., Incorporation and release behavior of hydrophobic drug in functionalized poly(D,L-lactide)-*b*-poly(ethylene oxide) micelles, *J. Control. Release*, 94, 323, 2004.
22. Matsumura, Y. et al., Phase I clinical trial and pharmacokinetic evaluation of NK911, a micelle-encapsulated doxorubicin, *Br. J. Cancer*, 91, 1775, 2004.
23. Alakhov, V. et al., Hypersensitization of multidrug resistant human ovarian carcinoma cells by pluronic P85 block copolymer, *Bioconjug. Chem.*, 7, 209, 1996.
24. Batrakova, E. V. et al., Anthracycline antibiotics non-covalently incorporated into the block copolymer micelles: In vivo evaluation of anti-cancer activity, *Br. J. Cancer*, 74, 1545, 1996.
25. Venne, A. et al., Hypersensitizing effect of pluronic L61 on cytotoxic activity, transport, and subcellular distribution of doxorubicin in multiple drug-resistant cells, *Cancer Res.*, 56, 3626, 1996.
26. Batrakova, E. V. et al., Effects of pluronic block copolymers on drug absorption in Caco-2 cell monolayers, *Pharm. Res.*, 850, 1998.
27. Batrakova, E. V. et al., Effects of pluronic P85 unimers and micelles on drug permeability in polarized BBMEC and Caco-2 cells, *Pharm. Res.*, 15, 1525, 1998.
28. Miller, D. W., Batrakova, E. V., and Kabanov, A. V., Inhibition of multidrug resistance-associated protein (MRP) functional activity with pluronic block copolymers, *Pharm. Res.*, 16, 396, 1999.
29. Batrakova, E. et al., Fundamental relationships between the composition of pluronic block copolymers and their hypersensitization effect in MDR cancer cells, *Pharm. Res.*, 16, 1373, 1999.
30. Alakhov, V. et al., Block copolymer-based formulation of doxorubicin. From cell screen to clinical trials, *Colloids Surf. B Biointerfaces*, 16, 113, 1999.
31. Batrakova, E. V. et al., Optimal structure requirements for pluronic block copolymers in modifying P-glycoprotein drug efflux transporter activity in bovine brain microvessel endothelial cells, *J. Pharmacol. Exp. Ther.*, 304, 845, 2003.
32. Kabanov, A. V., Batrakova, E. V., and Alakhov, V. Y., An essential relationship between ATP depletion and chemosensitizing activity of Pluronic block copolymers, *J. Control. Release*, 91, 75, 2003.
33. Venne, A. et al., Hypersensitizing effect of pluronic L61 on cytotoxic activity, transport, and subcellular distribution of doxorubicin in multiple drug-resistant cells, *Cancer Res.*, 56, 3626, 1996.
34. Rapoport, N. et al., Intracellular uptake and trafficking of Pluronic micelles in drug-sensitive and MDR cells: Effect on the intracellular drug localization, *J. Pharm. Sci.*, 91, 157, 2002.
35. Danson, S. et al., Phase I dose escalation and pharmacokinetic study of pluronic polymer-bound doxorubicin (SP1049C) in patients with advanced cancer, *Br. J. Cancer*, 90, 2085, 2004.
36. Danson, S. et al., Phase I dose escalation and pharmacokinetic study of pluronic polymer-bound doxorubicin (SP1049C) in patients with advanced cancer, *Br. J. Cancer*, 90, 2085, 2004.
37. Valle, J. W. et al., A phase II, window study of SP1049C as first-line therapy in inoperable metastatic adenocarcinoma of the esophagus, *J. Clin. Oncol.*, 22, 4195, 2004.
38. Gelderblom, H. et al., Cremophor EL: The drawbacks and advantages of vehicle selection for drug formulation, *Eur. J. Cancer*, 37, 1590, 2001.
39. Kim, S. C. et al., In vivo evaluation of polymeric micellar paclitaxel formulation: Toxicity and efficacy, *J. Control. Release*, 72, 191, 2001.
40. Kim, T. Y. et al., Phase I and pharmacokinetic study of Genexol-PM, a cremophor-free, polymeric micelle-formulated paclitaxel, in patients with advanced malignancies, *Clin. Cancer Res.*, 10, 3708, 2004.
41. Fournier, E. et al., A novel one-step drug-loading procedure for water-soluble amphiphilic nanocarriers, *Pharm. Res.*, 21, 962, 2004.
42. Le Garrec, D. et al., Poly(*N*-vinylpyrrolidone)-*b*-poly(D,L-lactide) as a new polymeric solubilizer for hydrophobic anticancer drugs: In vitro and in vivo evaluation, *J. Control. Release*, 99, 83, 2004.
43. Hamaguchi, T. et al., NK105, a paclitaxel-incorporating micellar nanoparticle formulation, can extend in vivo antitumour activity and reduce the neurotoxicity of paclitaxel, *Br. J. Cancer*, 92, 1240, 2005.
44. Gao, Z. G., Fain, H. D., and Rapoport, N., Controlled and targeted tumor chemotherapy by micellar-encapsulated drug and ultrasound, *J. Control. Release*, 102, 203, 2005.

45. Marin, A., Muniruzzaman, M., and Rapoport, N., Mechanism of the ultrasound activation of micellar drug delivery, *J. Control. Release*, 75, 69, 2001.
46. Husseini, G. A. et al., Factors affecting acoustically triggered release of drugs from polymeric micelles, *J. Control. Release*, 69, 43, 2000.
47. Gao, Z., Fain, H. D., and Rapoport, N., Ultrasound-enhanced tumor targeting of polymeric micellar drug carriers, *Mol. Pharm.*, 1, 317, 2004.
48. Rapoport, N., Tumor targeting by polymeric assemblies and ultrasound activation, In *Kentus Books MML Series*, Arshadi, R. and Kono, K., Eds., Vol. 8, Kentus Books, London, pp. 305–362, 2005.
49. Rapoport, N., Combined cancer therapy by micellar-encapsulated drug and ultrasound, *Int. J. Pharm.*, 277, 155, 2004.
50. Rapoport, N., Marin, A., and Christensen, D. A., Ultrasound-activated drug delivery, *Drug Deliv. Syst. Sci.*, 2, 37, 2002.
51. Rapoport, N., Factors affecting ultrasound interactions with polymeric micelles and viable cells, In *Carrier-Based Drug Delivery*, ACS Symposium Series, Swenson, S., Ed., Vol. 879, Chapter 12, American Chemical Society, Washington, DC, 2004.
52. Rapoport, N., Marin, A., Muniruzzaman, M., and Christensen, D., Controlled drug delivery to drug-sencitive and multidrug resistant cells: Effects of pluronic micelles and ultrasound, In *Advances in Controlled Drug Delivery*, ACS Symposium Book Series, Dinh, S. M. and Liu, P., Eds., American Chemical Society, Washington, DC, pp. 85–101, 2003.
53. Rapoport, N. et al., Drug delivery in polymeric micelles: From in vitro to in vivo, *J. Control. Release*, 91, 85, 2003.
54. Rapoport, N. Y. et al., Ultrasound-triggered drug targeting of tumors in vitro and in vivo, *Ultrasonics*, 42, 943, 2004.
55. Husseini, G. A. et al., Kinetics of ultrasonic release of Doxorubicin from Pluronic P-105 micelles, *Colloids Surf. B Biointerfaces*, 24, 253, 2002.
56. Marin, A. et al., Drug delivery in pluronic micelles: Effect of high-frequency ultrasound on drug release from micelles and intracellular uptake, *J. Control. Release*, 84, 39, 2002.
57. Deng, C. X. et al., Ultrasound-induced cell membrane porosity, *Ultrasound Med. Biol.*, 30, 519, 2004.
58. Honda, H. et al., Role of intracellular calcium ions and reactive oxygen species in apoptosis induced by ultrasound, *Ultrasound Med. Biol.*, 30, 683, 2004.
59. Pan, H. et al., Study of sonoporation dynamics affected by ultrasound duty cycle, *Ultrasound Med. Biol.*, 31, 849, 2005.
60. Miller, D. L. and Quddus, J., Sonoporation of monolayer cells by diagnostic ultrasound activation of contrast-agent gas bodies, *Ultrasound Med. Biol.*, 26, 661, 2000.
61. Kamaev, P. P. and Rapoport, N. Y., Effect of anticancer drug on the cell sensitivity to ultrasound *in vitro* and *in vivo*, AIP Conf. Proc., 829, 543, 2006.
62. Li, P., Armstrong, W. F., and Miller, D. L., Impact of myocardial contrast echocardiography on vascular permeability: Comparison of three different contrast agents, *Ultrasound Med. Biol.*, 30, 83, 2004.
63. Li, P. et al., Impact of myocardial contrast echocardiography on vascular permeability: An in vivo dose response study of delivery mode, pressure amplitude and contrast dose, *Ultrasound Med. Biol.*, 29, 1341, 2003.
64. Miller, D. L. et al., Histological characterization of microlesions induced by myocardial contrast echocardiography, *Echocardiography*, 22, 25, 2005.
65. Wible, J. H. et al., Microbubbles induce renal hemorrhage when exposed to diagnostic ultrasound in anesthetized rats, *Ultrasound Med. Biol.*, 28, 1535, 2002.
66. Bednarski, M. D. et al., In vivo target-specific delivery of macromolecular agents with MR-guided focused ultrasound, *Radiology*, 204, 263, 1997.
67. Kataoka, K. et al., Doxorubicin-loaded poly(ethylene glycol)-poly(beta-benzyl-L-aspartate) copolymer micelles: Their pharmaceutical characteristics and biological significance, *J. Control. Release*, 64, 143, 2000.
68. Goldberg, E. P. et al., Intratumoral cancer chemotherapy and immunotherapy: Opportunities for non-systemic preoperative drug delivery, *J. Pharm. Pharmacol.*, 54, 159, 2002.

69. Wilson, S. R. et al., Harmonic hepatic US with microbubble contrast agent: Initial experience showing improved characterization of hemangioma, hepatocellular carcinoma, and metastasis, *Radiology*, 215, 153, 2000.
70. Kedar, R. P. et al., Microbubble contrast agent for color Doppler US: Effect on breast masses. Work in progress, *Radiology*, 198, 679, 1996.
71. Halpern, E. J., Rosenberg, M., and Gomella, L. G., Prostate cancer: Contrast-enhanced ultrasound for detection, *Radiology*, 219, 219, 2001.
72. Bauer, A. et al., Ultrasonic imaging of organ perfusion with SH U 563A, *Acad. Radiol.*, 9(Suppl 1), S46, 2002.
73. Unger, E. C. et al., Therapeutic applications of lipid-coated microbubbles, *Adv. Drug Deliv. Rev.*, 56, 1291, 2004.
74. Shortencarier, M. J. et al., A method for radiation-force localized drug delivery using gas-filled liposheres, *IEEE Trans. Ultrason. Ferroelectr. Freq. Control*, 51, 822, 2004.
75. Bloch, S. H., Dayton, P. A., and Ferrara, K. W., Targeted imaging using ultrasound contrast agents. Progess and opportunities for clinical and research applications, *IEEE Eng. Med. Biol. Mag.*, 23, 18, 2004.
76. Bloch, S. H. et al., The effect of size on the acoustic response of polymer-shelled contrast agents, *Ultrasound Med. Biol.*, 31, 439, 2005.
77. Dayton, P. A. and Ferrara, K. W., Targeted imaging using ultrasound, *J. Magn. Reson. Imaging*, 16, 362, 2002.
78. Lathia, J. D., Leodore, L., and Wheatley, M. A., Polymeric contrast agent with targeting potential, *Ultrasonics*, 42, 763, 2004.
79. El-Sherif, D. M. et al., Ultrasound degradation of novel polymer contrast agents, *J. Biomed. Mater. Res. A*, 68, 71, 2004.
80. El-Sherif, D. M. and Wheatley, M. A., Development of a novel method for synthesis of a polymeric ultrasound contrast agent, *J. Biomed. Mater. Res. A*, 66, 347, 2003.
81. Basude, R. and Wheatley, M. A., Generation of ultraharmonics in surfactant based ultrasound contrast agents: Use and advantages, *Ultrasonics*, 39, 437, 2001.

22 Polymeric Micelles Targeting Tumor pH

Eun Seong Lee and You Han Bae

CONTENTS

22.1 Introduction 443
22.2 Acidic Tumor pH$_e$ 444
22.3 Multidrug Resistance and Cancer 444
22.4 PolyHis-PEG-Based pH-Sensitive Micelles: Preparation and Characteristics 445
    22.4.1 Synthesis... 445
    22.4.2 PolyHis/PEG Copolymer Micelles 446
    22.4.3 Mixed Micelles of PolyHis/PEG and PLLA/PEG............................ 448
22.5 Tumor pH$_e$ Targeting 451
22.6 Early Endosomal pH Targeting for Reversal of MDR 455
22.7 Conclusions 460
References... 461

22.1 INTRODUCTION

The dose–response curves for cell cycle-phase nonspecific anti-cancer drugs, such as doxorubicin, have been constructed from in vitro experimental results. The survival fraction of both wild-type sensitive and resistant cancer cells decreases in a sigmoidal pattern as a function of log of dose size (the survival fraction of the cells remains the same as the untreated control at low concentrations, starts to sharply decrease above a specific concentration and reaches a plateau at high concentrations). The inflection point of the curves is drug specific and is shifted to a lower concentration with an increasing duration of drug exposure. The inflection point for the resistant cells occurs at a higher concentration region and shows a more gradual decrease in viability with increasing drug concentration.[1,2] Overall, the experimental results and theoretical fitting of these results give us a simple implication that high dose exposure for a longer duration results in effective destruction.

Two classical nanosized anti-cancer drug carriers are liposomes and polymeric micelles. Each carrier system shows advantages as well as disadvantages over the other. For instance, liposomal carriers show continuous leakage of an entrapped drug with instability during storage and circulation. Polymeric micelles demonstrate much improved stability, but they may still dissociate upon dilution after injection and interact with blood components.[3,4]

Considering the dose–response relationship, it was suggested that future improvement in carrier design, providing a triggerable mechanism for drug release upon reaching the tumor sites, could be the most efficacious delivery strategy.[5] Various approaches for externally stimulated triggered release have been attempted, including hyperthermic conditions[6] used for thermosensitive liposomes or polymeric micelles, and sonication[7] used to trigger release from Pluronic® micelles.

In this chapter, polymeric micelles for the triggered release either by slightly acidic extracellular pH (pH_e) or by early endosomal pH, are introduced. Triggered release by tumor pH_e may allow localized high-dose therapy, which is effective for sensitive cells. Active intracellular translocation of the micellar carriers via specific interactions, combined with triggered release in early endosomes and endosomolytic activity of destabilized micelle components, has been proven to be effective for treating multidrug-resistant tumors.

22.2 ACIDIC TUMOR pH_e

The pH_e of normal tissues and blood is around pH 7.4, and the intracellular pH (pH_i) of normal cells is around pH 7.2. However, in most tumor cells the pH gradient is reversed ($pH_i > pH_e$). Particularly, pH_e is lower than in normal tissues.[8–11] Although there is a distribution of in vivo pH_e measurements of human patients having various solid tumors (adenocarcinoma, squamous cell carcinoma, soft tissue sarcoma, and malignant melanoma) in readily accessible areas (limbs, neck, or chest wall) using needle type microelectrodes, the mean pH value is reported to be 7.0 with a full range of 5.7–7.8,[11] the variation being dependent on tumor histology and tumor volume. Recent measurements of pH_e by noninvasive technologies, such as ^{19}F, ^{31}P, or ^1H probes in magnetic resonance spectroscopy in human-tumor xenografts in animals, further verified consistently low pH_e.[12,13] All measurements of pH_e of human and animal solid-tumors by either invasive or noninvasive method showed that more than 80% of all measured values fall below pH 7.2. This tumor pH can be further lowered by hyperglycemic conditions.[14]

The high rate of glycolysis in tumor cells, under either aerobic or anaerobic conditions, has been thought to be a major cause of low pH_e. The tumor cells synthesize ATP both by mitochondrial oxidative phosphorylation and glycolysis. Glycolysis produces two moles of lactic acid ($pK_a = 3.9$) and two moles of ATP upon consumption of one mole of glucose. The hydrolysis of generated ATP produces protons. Despite the high production rate of protons in tumor cells, their cytosolic pH, particularly of resistant cells, is maintained to be alkaline, creating a favorable condition for glycolysis. The mechanisms responsible for exporting protons out of the cells remain unknown. This, along with the help of inadequate blood supply, poor lymphatic drainage, and high interstitial pressure in tumor tissues, leads to low pH_e.[8–10] However, it is of interest to note that in one study, glycolysis-deficient variant cells (lacking lactate dehydrogenase) produced negligible quantities of lactic acid but still presented an acidic environment.[15] This finding has indicated the acidity may be a phenotype of tumor cells rather than a consequence of cell metabolic events. The acidic environment may benefit the cancer cell because it promotes invasiveness (metastasis) by destroying the extracellular matrix of surrounding normal tissues.

A few attempts for pH-triggered release using pH-sensitive liposomes have been described in the literature,[16,17] but a practical system that triggers the release at tumor pH_e is far off because of technical difficulties in endowing the desired pH sensitivity to the liposomes.

22.3 MULTIDRUG RESISTANCE AND CANCER

The resistance of cancer cells to multiple structurally unrelated chemotherapeutic drugs has been termed *multidrug resistance* (MDR). Clinical resistance of some cancers to cytotoxic drugs has been observed since the beginning of chemotherapy. The degree of resistance varies widely depending on the type of cancer. In past years, various mechanisms for MDR have been proposed, including the expression of cell surface multidrug efflux pumps (P-glycoprotein (Pgp)[18] and multidrug resistance-associated proteins (MRP1)[19]) altered glutathione metabolism,[20] reduced activity of topoisomerase II,[21] and various changes in cellular proteins[22–24] and mechanisms.[25]

Strong evidence shows that the expression of (Pgp) is associated with drug-resistance in cancer. Many attempts to circumvent Pgp-based MDR in cancer chemotherapy have utilized Pgp blocking agents (MDR modulators or reverters),[26–30] that can be co-administered with the anti-cancer drug. This approach is based on the premise that inhibiting Pgp function will result in increased accumulation of most anti-cancer drugs in the tumor cells and restore anti-tumor activity. However, co-administration of MDR modulators with anti-cancer drugs has often resulted in exacerbated toxicity of the anti-cancer drugs and very little improvement in chemosensitization of MDR cancers because of the blockade of Pgp excretory functions in healthy tissues, such as liver and kidney, which markedly reduces anti-cancer drug clearance.

Independently, acidic intracellular organelles also participate in resistance to chemotherapeutic drugs. Parental drug sensitive cells are characterized to have rather acidic and diffused pH profile inside the cells, while multidrug resistant cells develop more acidic organelles (recycling endosome, lysosome, and trans-Golgi network) compared to cytosolic and nucleoplasmic pH.[31–33]

Because most anti-cancer drugs are in an ionizable form, the pH of extracellular matrix and intracellular compartments is a critical factor in determining drug partitioning and distribution. For instance, weakly-basic drugs accumulate more in an acidic phase, where the ionized form predominates, because an ionized drug is less permeable through cellular or subcellular membranes. This is more pronounced in MDR cells.

It is claimed that the drug sequestration followed by transport to the plasma membrane and extrusion into the external medium is another major mechanism for MDR for weakly-acidic drugs.[31,34] It is interesting to note that transfection of P-gp into certain sensitive cells is often coupled with increased cytosolic pH. However, this is not the case for MCF-7 cell line. The transfection did not influence the internal pH but induced 20-fold increased resistance. When the same cell line was selected from an escalated doxorubicin treatment schedule, the cells presented the overexpression of P-gp together with typical MDR pH profile and increased the resistance 980-fold. These observations indicate that the activity of the efflux pump, combined with drug sequestration and exocytosis mechanism, seem to be highly influential factors in MDR.[35] Thus, this subcellular pH effect must be taken into account in the carrier-design to avoid the sequestration effect and to improve drug availability to the cytosol and nucleus.[34] Few investigators have addressed the drug sequestration issue in drug-carrier design.

This chapter summarizes our attempt for the triggered release of doxorubicin, either by tumor pH$_e$ or by early endosomal pH, from pH-sensitive polymeric micelles which are based on poly(L-histidine) core.

22.4 POLYHIS-PEG-BASED pH-SENSITIVE MICELLES: PREPARATION AND CHARACTERISTICS

22.4.1 SYNTHESIS

L-Histidine (His) was derivatized by introducing carbobenzoxy (CBZ) group to α amino group, and amino group in imidazole ring of N^α-CBZ-L-histidine was protected with dinitrophenyl (DNP) group. N^α-CBZ-N^{im}-DNP-L-His was then transformed to the N-carboxy anhydride (NCA) form by thionyl chloride. The ring-opening polymerization of N^α-CBZ-N^{im}-DNP-L-His NCA·HCl in dimethylformamide (DMF) was performed using different molar ratios of the monomer to initiator (isopropylamine or n-hexylamine) (M/I ratio).[36]

Monocarboxy-PEG was used to prepare an activated N-hydroxysuccinyl (NHS)–poly(ethylene glycol) (PEG). The coupling reaction between poly(N^{im}-DNP-L-histidine) and NHS–PEG (1:1 functional group ratio) was carried out in tetrahydrofuran (THF) for two days at room temperature. After the reaction, the block copolymer was precipitated in a mixed solvent (THF/n-hexane) and filtered. 2-Mercaptoethanol was added to the poly(N^{im}-DNP-L-histidine)-PEG diblock copolymer in DMF to deprotect the imidazole group (Scheme 22.1). Deprotection was complete within one day at room temperature. The polymer solution was added to diethyl ether (800 mL) at 0°C to precipitate polymer, while excess 2-mercaptoethanol and DNP-mercaptoethanol remained soluble. The polymer was filtered, washed with diethyl ether, and dried in vacuo for two days. For further purification, the polymer was dissolved in a minimum volume of 3 N HCl and stored in a 0°C for one day. After the precipitation of DNP was complete, the solution was filtered through a 0.45-μm filter and the product was precipitated by adding an excess amount of acetone to the solution, and then dried in vacuo for two days. Dialysis (Spectra/Por; molecular weight cut off (MWCO) 5000) was used to remove uncoupled polymers. Before subsequent experiments, the HCl salt of poly-histidine (1 mmol) in dimethylsulfoxide (DMSO) was converted to the free base by treatment with TEA (2 mmol) for 3 h at room temperature; the free base was recrystallized from a solution mixture of ethanol and DMSO (10:1).

It has been documented in literature that histidine residues in proteins significantly contribute to the protein buffering capacity[37] at physiological pH. The acid-base titration profiles of the polyHis homopolymer and polyHis/PEG block copolymers are presented in Figure 22.1. All of the polymer solutions exhibited a buffering pH region of pH 4–9. The titration curve confirmed that the poly-mers with a higher molecular-weight polyHis block had a higher buffering capacity (Figure 22.1a) in the physiological pH range of pH 5.5–8.0 due to the higher concentrations of imidazole group. PolyHis ($M_n \sim 5000$)/PEG ($M_n \sim 2000$) (PolyHis5K/PEG2K) and PolyHis ($M_n \sim 3000$)/PEG ($M_n \sim 2000$) (polyHis3K/PEG2K) had inflexion points around pH 7.0 (pK_b). PolyHis5K and Poly-His3K showed pK_b values around pH 6.5. The pK_b shift of the block copolymer compared to the polyHis homopolymer may be due to increased hydration by PEG.[38]

22.4.2 PolyHis/PEG Copolymer Micelles

The deprotonated polyHis at high pH is hydrophobic, whereas PEG is soluble in water at all pHs. This amphiphilicity is responsible for the formation of polymeric micelles. Lowering the solution pH below the pK_b can affect the micellar structure because protonation decreases the hydrophobi-city of polyhistidine.[36]

The polyHis/PEG block copolymer micelles were prepared by the dialysis of the copolymer solution in DMSO against a pH 8.0 medium. This was done in the presence of pyrene as

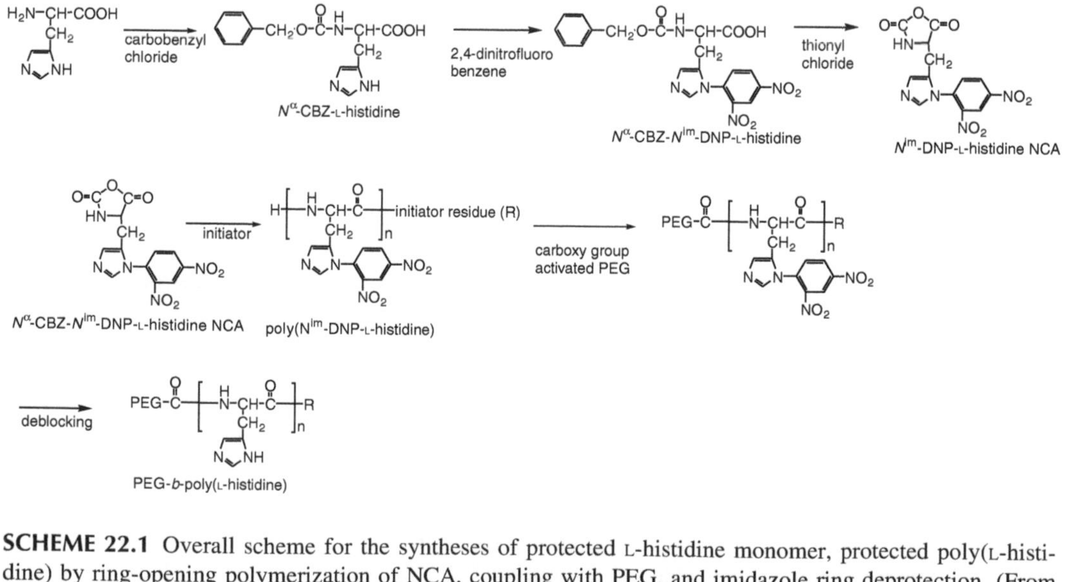

SCHEME 22.1 Overall scheme for the syntheses of protected L-histidine monomer, protected poly(L-histi-dine) by ring-opening polymerization of NCA, coupling with PEG, and imidazole ring deprotection. (From Lee, E. S., Shin, H. J., Na, K., and Bae, Y. H., *J. Control Release*, 90, 363, 2003. With permission.)

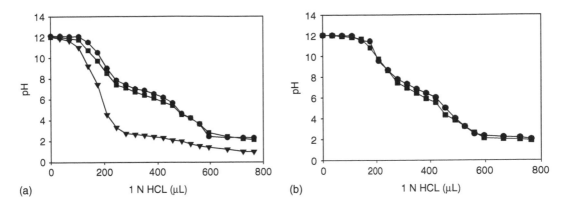

FIGURE 22.1 The pH profile of (a) polyHis5K/PEG2K (●), polyHis5K (■), NaCl (▼) and (b) polyHis3K/PEG2K (●), polyHis3K (■) by acid-base titration. (From Lee, E. S., Shin, H. J., Na, K., and Bae, Y. H., *J. Control Release*, 90, 363, 2003. With permission.)

a fluorescent probe so that the micelle formation could be monitored by fluorometry. Pyrene strongly fluoresces in a nonpolar environment, while in a polar environment it shows weak fluorescence intensity. The change of total emission intensity vs. polymer concentration indicates the formation of micelles or the change from micelles to unimers (the dissociatation of polymers from the micelles).[39] PolyHis3K/PEG2K micelles exhibited instability over time, as evidenced by increasing light transmittance, while the stability of polyHis5K/PEG2K micelles was maintained for two days (data not shown). The instability of polyHis3K/PEG2K was probably due to short polyHis block length and indicates that there may be a critical polyHis length for stability somewhere between M_n 3000 and 5000. Therefore, polyHis5K/PEG2K was used for further investigation.

The effect of the dialysis pH on critical micelle concentration (CMC) was examined and the results are presented in Figure 22.2. At a dialysis pH between 8.0 and 7.4, the CMC of polyHis5K/PEG2K micelle increased slightly with decreasing pH. However, the CMC was significantly elevated below pH 7.2. It is evident that the protonation of the imidazole group in the copolymer at lower pH causes a reduction in hydrophobicity, leading to an increase in CMC.

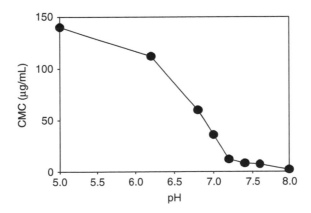

FIGURE 22.2 The pH effect on the CMC of polyHis5K/PEG2K polymeric micelles that were prepared at various pHs (pH 8.0–5.0). (From Lee, E. S., Shin, H. J., Na, K., and Bae, Y. H., *J. Control Release*, 90, 363, 2003. With permission.)

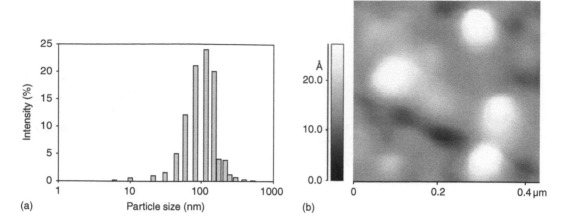

FIGURE 22.3 The particle size distribution of polyHis5K/PEG2K polymeric micelles measured by (a) zeta-sizer and (b) AFM. (From Lee, E. S., Shin, H. J., Na, K., and Bae, Y. H., *J. Control Release*, 90, 363, 2003. With permission.)

In addition, below pH 5.0 (typically pH 4.8), the CMC of polyHis5K/PEG2K micelle could not be detected. Taken together, at pH 8.0 the less protonated polyHis constitutes the hydrophobic core in the micellar structure, but in the range of pH 5.0–7.4 the polymer produced less-stable micelles.

The size of the micelles formed from polyHis5K/PEG2K measured by a zetasizer was around 110 nm, based on the intensity-average diameter with a unimodal distribution (Figure 22.3a), and the ratio of weight-average particle size to number-average particle size was 1.2. The relatively large size of polyHis5K/PEG2K micelles may be the result of dialysis conditions, which probably produced a secondary micelle structure that is a cluster of individual single micelles.[40] However, the polyHis5K/PEG2K micellar shape was regular and spherical, as visualized in the atomic force microscopy photograph (Figure 22.3b). The morphology of polymeric micelles was further evident from dynamic light scattering (DLS) measurements.

A noticeable increase in light-scattering intensity was observed after dialyzing the block copolymer solutions at pH 8.0, while the unimer solution in DMSO was practically transparent. The pH-dependent micelle stability was confirmed again by the measurement of transmittance of the micelle solution. Figure 22.4a shows the transmittance change of the polymeric micelle solution as a function of pH. As the pH of the micelle solution decreased from the dialysis conditions (pH 8, ionic strength 1.0), the transmittance increased. In particular, this increase was prominent from pH 7.4 to 6.0, reaching a plateau around pH 6.4. This result implies that polyHis5K/PEG2K micelles start dissociation starting at pH 7.4 and show a sharp transition between pH 7.4 and 7.0 followed by a gradual transition between pH 7.0 and 6.0.

Figure 22.4b shows the changes in the particle size distribution of the micelles as a function of pH. The instability of the micelles on reduction of pH resulted in release of polyHis5K/PEG2K unimers or unimicelles isolated from secondary micelles, resulting in a bimodal size distribution with existing secondary micelles. The intensity of the peak at size 2–30 nm gradually increased with micelle dissociation at a lower pH. Below pH 7.4, an abrupt increase in the intensity of small-sized particles was apparent, which is consistent with the transmittance change of the micelle solution at this pH (Figure 22.4a).

22.4.3 MIXED MICELLES OF POLYHIS/PEG AND PLLA/PEG

The PolyHis/PEG micelle was relatively unstable and began to gradually disintegrate at pH 7.4. The incorporation of a nonionizable block copolymer (Poly(L-lactic acid) (PLLA)/PEG) into the

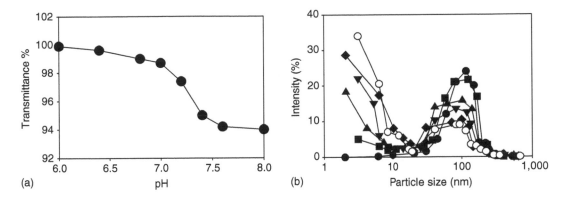

FIGURE 22.4 pH-dependent micelle destabilization. (a) The transmittance change of polyHis5K/PEG2K polymeric micelles (0.1 g/L) with pH. The polyHis5K/PEG2K polymeric micelles were prepared in NaOH–Na$_2$B$_4$O$_7$ buffer solution (pH 8.0) and exposed at each pH for 24 h. (b) The change of particle size distribution in polyHis5K/PEG2K micelles exposed to each pH solution for 24 h; pH 8.0 (●), 7.4 (■), 7.2 (▲), 6.8 (▼), 6.4 (◆) and 6.0 (○). (From Lee, E. S., Shin, H. J., Na, K., and Bae, Y. H., *J. Control Release*, 90, 363, 2003. With permission.)

micellar structure of polyHis/PEG improved micelle stability at pH 7.4 and shifted destabilization to lower pHs. In particular, the mixed micelles composed of polyHis/PEG (75 wt%) and PLLA/PEG (25 wt%) showed a unimodal size distribution with an average size of 70-nm diameter at pH 9.0. Other mixed micelles prepared from different contents of PLLA/PEG ranged 70–100 nm in diameter with unimodal size distributions at pH 9.0.[41]

Figure 22.5 shows the effect of PLLA/PEG content on the CMC of the mixed micelles. When the mixed micelle was prepared at pH 9.0 and tested at pH 9.0, the CMC of the mixed micelles was very slightly influenced by PLLA/PEG content in a range of 0–25 wt%. However, above 25 wt% PLLA/PEG the CMC gradually increased, reaching to the higher CMC value of the PLLA/PEG micelle, although the CMC change is not statistically significant (Figure 22.5a). This indicates that a small amount of PLLA/PEG does not disturb the core stability.

The presence of PLLA/PEG did significantly influence the pH-dependent CMC of the micelles. The CMC of polyHis/PEG micelle sharply increased from 2.7 to 140 μg/mL as pH decreased from 8 to 5, with a major transition occurring between pH 7.0 and 6.8 (Figure 22.5b). However, the mixed micelle having 25 wt% PLLA/PEG showed a small increase in CMC, from 2.6 to 3.5 μg/mL, and the transition occurred in a pH range of 7.5–6.8 (Figure 22.5c).

Figure 22.6 shows the pH-dependent stability of the mixed micelles monitored by relative light transmittance ($T/T_i\%$; the transmittance of the mixed micelle at each pH/transmittance at pH 9.0). The $T/T_i\%$ of polyHis/PEG micelle slightly increased with decreasing pH, especially below pH 7.6, which is attributed to micelle destabilization followed by dissociation. PLLA/PEG micelle, being pH insensitive, showed no change in turbidity and preserved its stability in the entire pH range tested.

It is interesting to note that the $T/T_i\%$ of the mixed micelles decreased rather than increased with decreasing pH. This result seems to be associated with ionization of polyHis/PEG, accompanied by separation and isolation of PLLA/PEG from the micelles, which in turn became segregated in water, and indirectly reflects the disintegration of the micelles by ionization of the imidazole groups along polyHis chains. The destabilizing pH was influenced by the amount of PLLA/PEG block copolymer in the mixed micelles. The addition of up to 10 wt% of PLLA/PEG to the micelle showed a slight shift in destabilizing pH. Although a 25 wt% addition considerably enhanced the micelle stability at pH 7.4 and destabilization only occurred below pH 7.0. At 40 wt% PLLA/PEG,

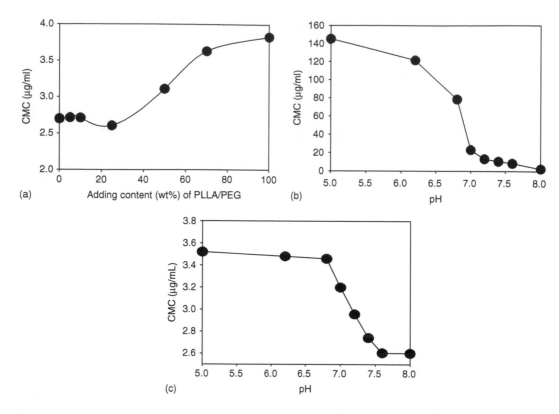

FIGURE 22.5 The CMC change of the mixed micelles prepared with (a) different contents of PLLA/PEG in NaOH–Na$_2$B$_4$O$_7$ pH 9.0 buffer, (b) pH-dependent CMC of polyHis/PEG micelle in two solutions (HCl or NaOH–Na$_2$B$_4$O$_7$ ionic strength = 0.1), and (c) pH-dependent CMC of the mixed micelle composed of 25 wt% PLLA/PEG and 75 wt% polyHis/PEG in the same solutions ($n = 3$). (From Lee, E. S., Shin, H. J., Na, K., and Bae, Y. H., *J. Control Release*, 90, 363, 2003. With permission.)

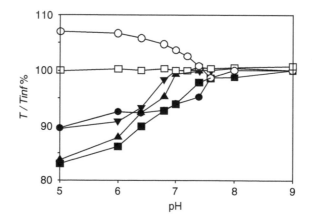

FIGURE 22.6 pH-dependent relative turbidity (T/T_i%) of the mixed micelles composed of polyHis/PEG and PLLA/PEG (PLLA/PEG content: 0 wt% (○); 5 wt% (●); 10 wt% (■); 25 wt% (▲); 40 wt% (▼); and 100 wt% (□)). T_i is the transmittance at 500 nm at pH 9.0 and T is the transmittance at a given pH after micelles were stabilized for 24 h. (From Lee, E. S., Na, K., and Bae, Y. H., *J. Control Release*, 91, 103, 2003. With permission.)

the destabilization pH was shifted a bit further downward with less release of PLLA/PEG due to its high content.

22.5 TUMOR pH$_e$ TARGETING

The pH-dependent micelle property prompted us to test the micelles with pH-induced drug release. When doxorubicin (DOX) was incorporated into the mixed micelles (0, 10, 25, and 40 wt% PLLA/PEG) during dialysis, the micelle sizes ranged from 50 to 70 nm as determined by dynamic light scattering. The drug loading efficiency was about 75–85% and the drug content in the micelles was 15–17 wt%.[41]

Figure 22.7a shows the cumulative DOX release from polyHis/PEG micelle. The DOX release pattern followed nearly first-order kinetics and reached a plateau in 24 h and 30 wt% of loaded DOX was released for 24 h at pH 8.0. Decreasing pH accelerated the release rate. The pH-dependent release patterns of the mixed micelles are presented in Figure 22.7b by plotting cumulative amount of DOX for 24 h vs. release medium pH. The mixed micelles containing 25 wt% PLLA/PEG showed a favorable pH-dependency such that 32 wt% of DOX was released at pH 7.0, 70 wt% of DOX at pH 6.8, and 82 wt% at pH 5.0. On the other hand, the mixed micelles containing 40 wt% of PLLA/PEG suppressed the release of DOX to 35 wt% at pH 6.8, and 64 wt% at pH 6.6. This indicates that the content of PLLA/PEG controlled pH-dependent release from the mixed micelles by destabilization; the transition pH coincided with the results shown in Figure 22.6. The results support the idea that the mixed micelles truly recognize the minute difference in pH and discriminate the tumor pH by destabilization and release of micelle content.

When the blank micelles were tested with human breast adenocarcinoma (MCF-7) cells, no noticeable cytotoxicity was observed up to 100 µg/mL of polymeric micelle for a 48-h culture, regardless of the culture medium pH. However, DOX-loaded micelles did present tumor cell toxicity in a pH-dependent manner.

DOX-loaded polyHis/PEG micelles demonstrated a certain degree of cytotoxic effect at pH 7.4 because of low micelle stability and DOX release. But the cell viability was much reduced at pH 6.8 due to enhanced DOX release (Figure 22.8). The mixed micelles with 10 wt% PLLA/PEG (Figure 22.8b) caused low cytotoxicity at pH 7.4, which increases dramatically below pH 7.2. The above results are distinguished from the cytotoxicity of free DOX, which is almost independent of pH. The mixed micelles having 25 wt% PLLA/PEG (Figure 22.8c) appeared more sensitive to

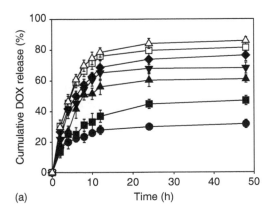
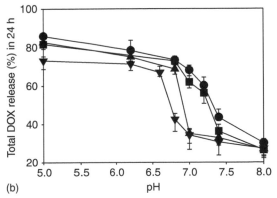

FIGURE 22.7 pH-dependent cumulative DOX release from the mixed micelles composed of polyHis/PEG and PLLA/PEG with varying PLLA/PEG content in the mixed micelles: (a) 0 wt%; pH 8.0 (●); 7.4 (■), 7.2 (▲), 7.0 (▼), 6.8 (◆), 6.2 (□), 5.0 (△). (b) 0 wt% (●), 10 wt% (■), 25 wt% (▲) and 40 wt% (▼) after 24 h. (From Lee, E. S, Na, K., and Bae, Y. H., *J. Control Release*, 91, 103, 2003. With permission.)

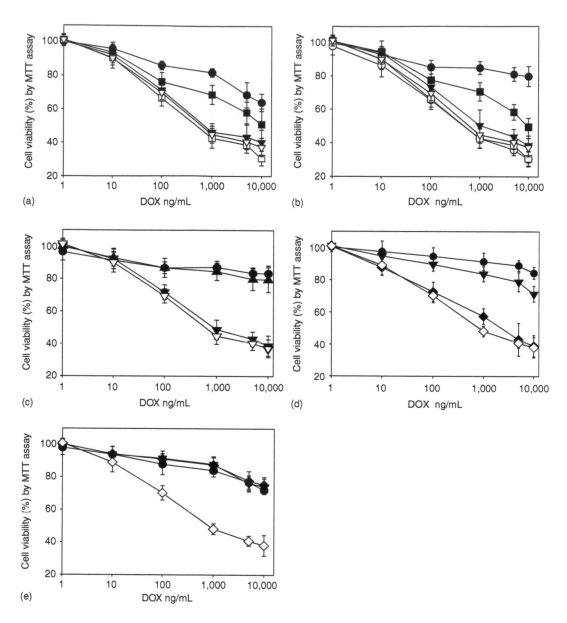

FIGURE 22.8 The cytotoxicity of DOX-loaded mixed micelles with different PLLA/PEG contents (a) 0 wt%, (b) 10 wt%, (c) 25 wt%, (d) 40 wt%, and (e) 100 wt% after 48 h incubation at varying pH (pH 7.4 (●); 7.2 (■); 7.0 (▲); 6.8 (▼); and 6.6 (◆)). Free-DOX cytotoxicity is presented as a reference at pH 7.4 (○), 7.2 (□), 7.0 (△), 6.8 (▽), 6.6 (▣). (From Lee, E. S., Na, K., and Bae, Y. H., *J. Control Release*, 91, 103, 2003. With permission.)

pH, differentiating especially between pH 7.0 and 6.8. The mixed micelles with 40 wt% PLLA/PEG micelles (Figure 22.8d) distinguished between pH 6.8 and 6.6 in a similar cytotoxicity study. Considering that a PLLA/PEG micelle (Figure 22.8e) has no pH-dependent cytotoxicity and neither do blank micelles, the pH-dependent cytotoxicity solely relies on the release of DOX at different pHs. The change in pH from 8.0 to 6.6 may influence cell surface charges, and cellular physiology and viability, noting that the low pH is a favorable environment for tumor cells but opposite for normal cells. The cell viability expressed in this study is relative to those at different pHs in the absence of DOX, and the direct pH effect on the cell viability was not monitored with the MCF-7 cell line. No noticeable difference in cell viability by pH with free DOX was observed.

In vitro results of the mixed micelles with 25 wt% PLLA/PEG (referred as PH-sensitive mixed micelle (PHSM) hereafter) showed the pH-dependent cytotoxicity of DOX against sensitive MCF-7 cells, and were comparable with free DOX and a conventional pH-insensitive PLLA/PEG micelle (referred to as pH-insensitive micelle (*PHIM*). The free DOX showed the highest destruction, which is not significantly influenced by pH variation between 6.8 and 7.4. PHIM showed the least cyto-toxicity at the two tested pH values. However, PHSM showed a distinct pH effect by switching the cytotoxicity from the level of PHIM at pH 7.0 (MCF-7 cell viability was 85% after the treatment of equivalent DOX 1 μg/mL) to a similar level of free DOX at pH 6.8 (MCF-7 cell viability was 36% after the treatment of equivalent DOX 1 μg/mL). This was attributed to pH-triggered release of DOX.

Figure 22.9a shows the in vivo results of the anti-cancer activity of the tested DOX formu-lations i.v. injected on days 0 and 3. The DOX loaded PHSM (equivalent DOX = 10 mg/kg) exhibited significant inhibition ($P < 0.05$ compared with free DOX or saline solution) of the growth of s.c. MCF-7 xenografts. The tumor volume of mice treated with the PHSM ($P < 0.05$ compared with free DOX) was approximately 4.5 and 3.6 times smaller than that of saline solution or free DOX treatment after 6 weeks, respectively. The triggered release of DOX by tumor pH_e ($<$ 7.0), after accumulation of the micelle in the tumor sites via the enhanced permeation and retention (EPR) mechanism,[42,43] may present a more effective modality in tumor chemotherapy, providing higher local concentrations of the drug at tumor sites (targeted high-dose cancer therapy), while providing a minimal release of the drug from micelles while circulating in the blood (pH 7.4). When compared with PHIM, the volume of tumors treated with the PHSM was approximately 3.0 times smaller (particle size = 50–70 nm, data now shown) than those whose properties are not influenced by pH. In addition, the destabilization of PHSM (converting to unimers) may help extravasation of additional micelles from the blood compartment and their penetration into tumors by reducing the physical barrier effect, which can be generated by stable particles, such as conventional liposomes and polymeric micelles. It was reported that drug carriers, such as PEG modified liposomes, are in an insoluble form and only reside in the vicinity of the leaky blood vessels after extravasation, rather than migrating into the deeper sites of tumors. This "road block effect" could obstruct the subsequent accumulation of the carriers at tumor sites.[44] It is interesting to note that only PHSM administration showed decreased tumor size for the initial first week. No obvious changes in mice

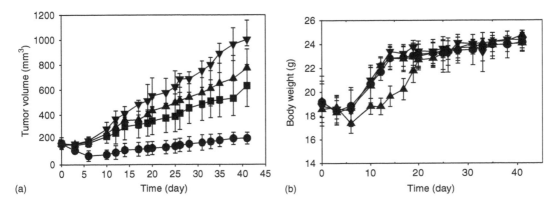

FIGURE 22.9 Tumor growth inhibition of s.c. human breast MCF-7 carcinoma xenografts in BALB/c nude mice. Mice were injected i.v. with 10 mg/kg DOX equivalent dose: PHSM (●), PHIM (■), free DOX (▲), and saline solution (▼). Two i.v. injections on day 0 and 3 were administered. (a) Tumor volume change and (b) body weight change ($n = 5$). Values are the means ± the standard deviations (SD). (From Lee, E. S., Na, K., and Bae, Y. H., *J. Control Release*, 103, 405, 2005.)

body weight in experimental groups, except for the free DOX group, were observed. Free DOX caused more weight-loss after i.v. administration (Figure 22.9b) than micelle formulations.

"Stealth" particles decorated with PEG are known to be invisible to macropharges and have prolonged half-lives in the blood compartment. This property endows the particles a higher probability to extravasate in pathological sites, like tumors or inflamed regions where a leaky vascular structure persists.[45] Recent investigations have reported that the size cut-off of tumor vasculature for the EPR effect grew to 200–700 nm.[46] Nevertheless, the nanoparticle penetration to tumor vasculature is heterogeneous in a solid tumor, and the vascular structure or cut off size differs from tumor to tumor and is influenced by tumor location.[46] A more recent study demonstrated that the optimal size of stealth liposomes was determined to be around 100 nm in an animal model.[47] This means that the size of micelle or liposome affects the in vivo bio-distribution, post injection. Smaller size causes more accumulation in the liver, while larger ones cause more accumulation in the spleen.[48]

We hypothesized that the average particle size of 50–70 nm of PHSM may meet the optimal size for extravasation or long-circulation suggested by several research groups.[42,44,47,48] A significant portion of DOX carried by the PHSM, however, was accumulated in the liver and spleen. In Figure 22.10a, DOX levels in the blood after i.v. administration of PHSM to the animals bearing s.c. MCF-7 carcinoma xenografts were 27% at 4 h, 8% at 12 h, 4% at 48 h in the blood and 18% at 4 h, 15% at 12 h, 3% at 48 h in the tumors and 14% at 4 h, 7% at 12 h, 1.5% at 48 h in the liver and 2% at 4 h, 1% at 12 h, barely detectable levels at 48 h in the spleen. The mononuclear phagocytes system (MPS), especially in the liver and spleen,[42,44,46,49] seems to be responsible for the accumulation in these organs even with small size of the particles.

As shown in Figure 22.8b, because of very short half-life ($t_{1/2}$, within few minutes) of free DOX in the blood,[50] DOX levels in the blood at 12–48 h were not detectable. DOX levels in the tumors were 1.5% at 4 h, 1% at 12 h, and barely detectable levels were found at 48 h. In contrast, high concentrations of DOX were found in the heart, liver and considerable DOX levels were shown in

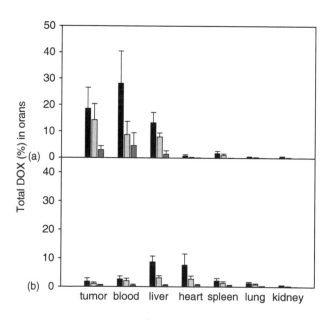

FIGURE 22.10 Total DOX (%) levels (total DOX remaining in organs to DOX amount first injected) in blood and tissues (tumor, liver, heart, spleen, lung, kidney) after i.v. administration (10 mg/kg DOX equivalent dose) of (a) PHSM micelles, and (b) free DOX. The mice (s.c. human breast MCF-7 carcinoma xenografts) were sacrificed at 4 (■), 12 (■), and 48 h (■) after i.v. administration. Values are the means ± the standard deviations ($n = 4$, SD). (From Lee, E. S., Na, K., and Bae, Y. H., *J. Control Release*, 103, 405, 2005.)

the lung, spleen and kidney. These results are in striking contrast to DOX biodistribution when carried by PHSM. Especially, PHSM is responsible for the enhanced tumor accumulation.[51] The combined mechanisms of pH_e-triggered release and EPR effect can be beneficial to treat solid tumors by minimizing the road block effect of the particles after extravasation, increasing local concentration for diffusion to the tumors and more active internalization. This, however, requires further investigation for verification.

22.6 EARLY ENDOSOMAL pH TARGETING FOR REVERSAL OF MDR

PolyHis has been known for endosomolytic activity and recently low MW polyHis-conjugated poly(L-lysine) has demonstrated improved gene transfection.[52] However, the difficulty in controlling its molecular weight and limited solubility of high MW polymer in organic solvents has restricted the fabrication and applications of polyHis-based drug carriers. The mixed micelles, who's surface was decorated with folate, consisting of polyHis/PEG-folate (Figure 22.11a) and PLLA/PEG-folate (Figure 22.11b), showed a similar release pattern of the micelles without folate (PHSM); cumulative DOX release for 24 h of 30–35 wt% in the pH range of 7.0–7.4, which is slightly more than 20–30 wt% initial burst release from conventional pH-insensitive micelles,[53] but about 70 wt% at pH 6.8 during the same time period. During this process, it was observed that the drug or polymer was distributed in the cytosol and the nucleus in a sensitive breast cancer cell line (MCF-7).[41]

From these results, we hypothesized that the folate receptor-mediated internalization route and the micelle destabilization reverse the drug resistance of solid tumors. For proof of concept, a MCF-7/DOX^R cell was selected after being stepwise exposed to 0.001–10 µg/mL of free DOX as reported.[54–58]

Figure 22.12 shows the decreased cytotoxicity of free DOX against MCF-7/DOX^R cells, exhibiting a typical drug resistance curve. The cytotoxicity of DOX against sensitive MCF-7 cells slightly decreased by lowering pH from 8.0 to 6.8 because of an increased degree of ionization of DOX, causing less diffusivity through the plasma membrane.[59] For the resistant cells, the pH effect was not noticed even at a high DOX concentration range. The IC_{50} value of free DOX increased from 470 ng/mL for wild MCF-7 cells to 25,800 ng/mL for MCF-7/DOX^R cells at pH 7.4.

Figure 22.13 visualizes different distributions of free DOX in MCF-7 (Figure 22.13a) and MCF-7/DOX^R (Figure 22.13b) cells. In the MCF-7/DOX^R cells, there was a limited DOX uptake in the cytosolic compartment, while the sensitive MCF-7 cells accumulated DOX in the cytosolic compartment and nucleus. Figure 22.14 shows the cytotoxic effects of free DOX or DOX-loaded

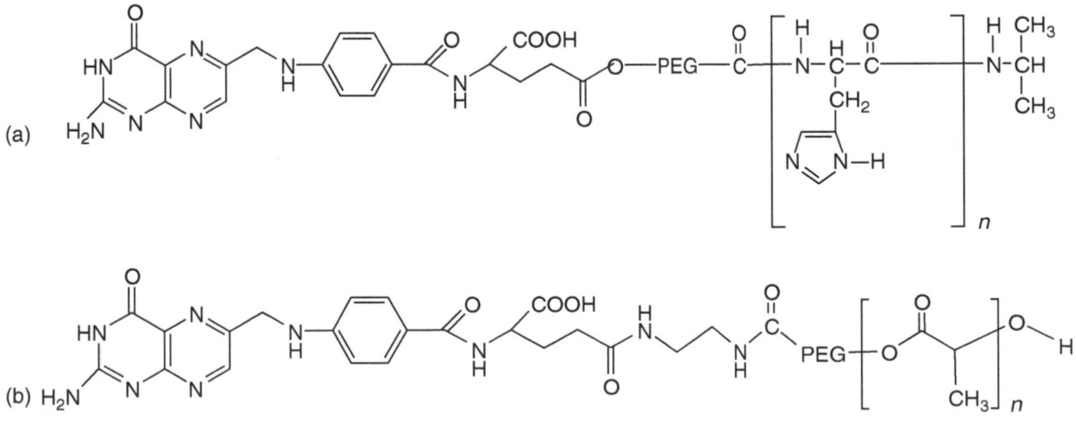

FIGURE 22.11 The chemical structures of (a) polyHis/PEG-folate and (b) PLLA/PEG-folate.

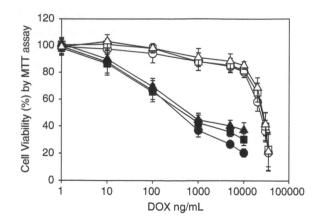

FIGURE 22.12 Characterization of micelles on MCF-7/DOX^R cells: The cytotoxicity of free DOX against MCF-7 (close circle) and MCF-7/DOX^R (open circle) cells at pH 8.0 (●), pH 7.4 (■) and pH 6.8 (▲) ($n = 7$). (From Lee, E. S., Na, K., and Bae, Y. H., *J. Control Release*, 103, 405, 2005.)

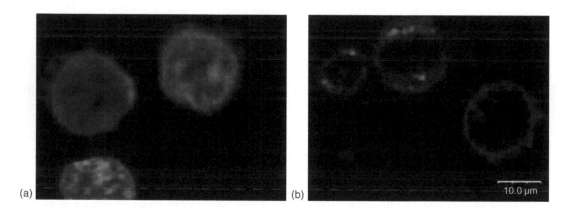

FIGURE 22.13 Confocal microscopy studies on (a) DOX-sensitive MCF-7 and (b) DOX-resistant MCF-7 (MCF-7/DOX^R) cells treated with free DOX at pH 6.8.

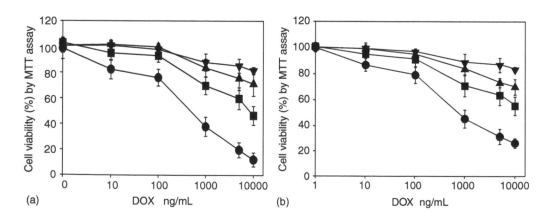

FIGURE 22.14 The cytotoxicity of DOX-loaded PHSM/f (●), polyHis/PEG-folate micelles (■), PLLA/PEG-folate micelles (▲), free DOX (▼) against MCF-7/DOX^R cells at (a) pH 7.4 and (b) pH 6.8 after 48 h incubation ($n = 7$). (From Lee, E. S., Na, K., and Bae, Y. H., *J. Control Release*, 103, 405, 2005.)

micelles against MCF-7/DOX^R cells at pH 7.4 and 6.8. The viability of MCF-7/DOX^R cells was not dramatically decreased as expected at even high concentrations of free DOX due to the influence of Pgp function. DOX-loaded PLLA/PEG-folate micelles (pH-insentive micelles as a control) showed limited cytotoxicity against MCF-7/DOX^R cells. DOX loaded polyHis/PEG-folate micelles demonstrated a similar degree of cytotoxicity against MDR cells as that of free DOX against the sensitive cells.[41] This effect is most likely due to enhanced DOX release and the endosomal escaping activity of polyHis/PEG-folate unimer released after internalization of the micelles and subsequent destabilization. This can be regarded as cytosolic delivery of DOX, disturbing drug sequestration mechanism of MDR.[41] However, it is shown that the viability of cells treated by PHSM with folate (PHSM/f) was about 10% at 10 µg/mL (DOX concentration), whereas for polyHis/PEG-folate micelles (without PLLA/PEG-folate) the value is about 40% at similar DOX concentration (Figure 22.14a). This enhanced effect suggests that there could be an additional mechanism of PHSM/f beside the functions of polyHis. Hoffman's group reported that the membrane disruption mechanisms of poly(2-ethylacrylic acid) (PEAA) is presumably related to their increase in hydrophobicity at low pH values.[60] Similar to that, PLLA/PEG-folate released from destabilized PHSM/f could interact with the endosomal membrane due to the exposed hydrophobic PLLA portion, disturbing membrane stability. There is no significant difference between two pHs tested (pH 7.4 (Figure 22.14a) and 6.8 (Figure 22.14b)). Figure 22.14b suggests the feasibility of MDR reversal by PHSM/f at acidic tumor pH$_e$.[61]

In the case of sensitive MCF-7 cells, PHSM/f showed significant cytotoxicity against cells. Figure 22.15 demonstrates this effect. The DOX-loaded micelles with folate were internalized via folate-receptor mediated endocytosis. The PHSM/f micelles induced greatly enhanced cytotoxicity (cell viability 16% at DOX 10 µg/mL at pH 8.0 (this pH was employed to ensure the micelle stability before internalization)) resulting from increased micelle concentration inside the cell due to folate receptor-mediated endocytosis and fusogenic effect of polyHis, while the PLLA/PEG-folate showed a slight increase in cytotoxicity. These observations clarify the role of polyHis in anti-cancer activity as two folds: (1) destabilization for drug release and (2) fusogenicity. Figure 22.15b indicates that, once internalization by folate receptors grew predominant, the pH effect on drug release became insignificant. This observation can be explained by the facts that (1) the mixed micelles were internalized into tumor cells via folate-receptor mediated endocytosis and underwent destabilization in a slightly acidic early endosomal compartment by protonation of an imidazole group and (2) DOX released from the mixed micelles under the influence of the extra-cellular pH, regardless of folate-receptor mediated endocytosis, has effective an drug portion for cell killing. Figure 22.15c shows the cell killing rates of the mixed micelles and free DOX. The free DOX diffuses only passively to cells. Unlike free DOX, PHSM/f facilitated faster cell killing in 0–4 h of incubation. This observation is consistent with the high number of folate-receptors on tumor cells. In addition, there is little difference in the cell killing rate between pH 6.8 and 7.4 (data not shown), thus explaining quick cytotoxic action of DOX-loaded mixed micelles against MCF-7 cells due to the active internalization.[58,59]

For further evidence of the functioning of PHSM/f, a confocal-microscopic study was conducted to reveal the distribution of DOX in the cells. Figure 22.16a shows the localized DOX in MCF-7/DOX^R cells treated with DOX-loaded PLLA/PEG-folate micelles. The PLLA/PEG-folate micelles were entrapped in endosome or multivesicular bodies resulting from folate receptor-mediated endocytosis. This could be responsible for the discrete DOX-intensity and biased DOX-intensity distribution. The entrapment seems to be a consequence of the absence of fusogenic activity. Unlike PLLA/PEG-folate micelles, PHSM/f incubation resulted in broad distribution of DOX in cytosol as well as nuclear compartments (Figure 22.16b).

To bypass one of the major MDR mechanisms, Pgp pumping, we employed folate receptor-mediated endocytosis, because MCF-7 cells over-express folate receptors.[62] Along with this entry mechanism, the further enhanced drug release at endosomal pH[36,41] and the disruption of the endosomal membrane[41,51,52,63] are expected to kill MDR cells.

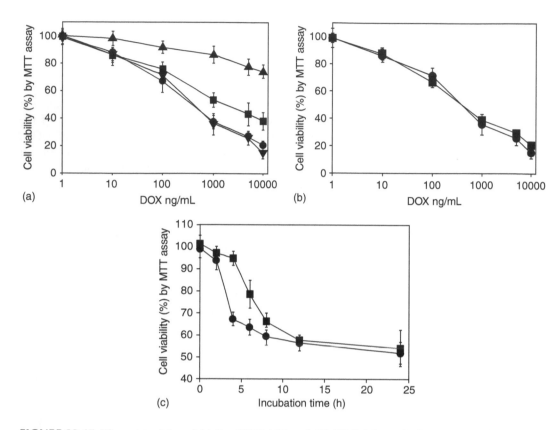

FIGURE 22.15 The cytotoxicity of (a) free DOX (●), polyHis/PEG-folate micelle (■), PLLA/PEG-folate micelle (▲) and PHSM/f (▼) at pH 8.0 and of (b) PHSM/f at pH 7.4 (●) and pH 6.8 (■) against sensitive MCF-7 cells after 48 h incubation. (c) The cytotoxicity against MCF-7 cells treated with PHSM/f (●) and free DOX (■) at pH 6.8. The DOX content in the micelles was adjusted to be equivalent to free DOX concentration (500 ng/mL) and MCF-7 cells were treated for a given incubation time. (From Lee, E. S., Na, K., and Bae, Y. H., *J. Control Release*, 91, 103, 2003. With permission.)

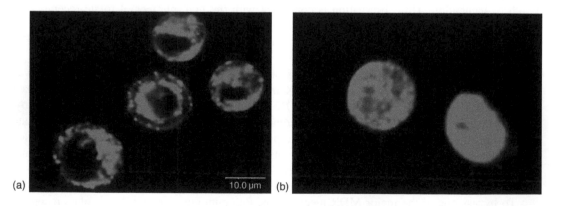

FIGURE 22.16 Confocal microscopy studies on MCF-7/DOX^R cells treated with (a) DOX-loaded PLLA/PEG-folate micelles and (b) DOX-loaded PHSM/f micelles at pH 6.8. The cells were incubated with the carriers for 30 min. (From Lee, E. S., Na, K., and Bae, Y. H., *J. Control Release*, 103, 405, 2005.)

In the in vitro studies with MCF-7/DOX^R cells,[51] the PHSM/f showed a significantly enhanced efficacy in tumor cell cytotoxicity, most likely due to enhanced internalization (folate receptor-mediated endocytosis), and subsequent pH-induced micelle destabilization and endosomal escape. However, the PHSM or PHIM with folate (PHIM/f) showed low efficacy in killing MCF-7/DOX^R cells, due to lack of entry mechanism and/or endosomal escape.

The MCF-7/DOX^R xenografts in nude mice were used for the investigation of in vivo efficacy. The tumor bearing animals were treated by multiple i.v. injections on days 0, 3, and 6 (Figure 22.17a). Neither complete tumor regression, nor toxicity-induced death was observed in all experimental groups, though, the tumor growth in mice treated by PHSM/f was inhibited and the size was reduced for two weeks after the third i.v. injection. The tumor volume in mice treated by PHSM/f ($P < 0.05$ compared with free DOX) was approximately 2.7-times smaller than those treated with free DOX or PHIM, and approximately 1.9-times smaller than those treated with PHSM after six weeks. These results indicate that the folate-mediated endocytosis pathway of PHSM/f enhances the regression of tumors and is more efficient for MDR tumor chemotherapy than PHSM. This being said, the tumor regression efficacy in PHIM/f was not significant because some folate conjugates may have recycled back to the surface[64] before drug unloading. In addition, slow or limited drug release from PHIM maybe associated with drug sequestration in acidic organelles in MDR cells.[31,33] The active internalization, enhanced drug release rate, and endosomal escaping activity justify the efficacy of PHSM/f. Free DOX administration exhibited more weight-loss in mice than the groups treated by any micelles (Figure 22.17b), signifying lower toxicity of DOX-loaded micelles.

As shown in Figure 22.18, s.c. MCF-7/DOX^R carcinoma xenografts showed different DOX levels in tumor as compared to s.c. MCF-7 carcinoma xenografts, particularly for PHSM. DOX levels in the tumors of s.c. MCF-7/DOX^R carcinoma xenografts in which the DOX levels were 21% at 4 h, 12% at 12 h, and 5% at 48 h for PHSM/f; 7% at 4 h, 5% at 12 h, and 2% at 48 h for PHSM; and 1% at 4 h and barely detectable levels at 12 and 48 h for free DOX. This suggests that the role of the Pgp pump in s.c. MCF-7/DOX^R carcinoma xenografts is significant in determining DOX levels in the resistant tumors. In the case of DOX delivered by PHSM into the resistant tumor, the released DOX cannot diffuse in the cells but is instead being washed away from the site. On the other hand, DOX carried by PHSM/f shows similar accumulation levels in the resistant tumor to that of sensitive tumor, resulting from active internalization by folate receptor-mediated endocytosis.

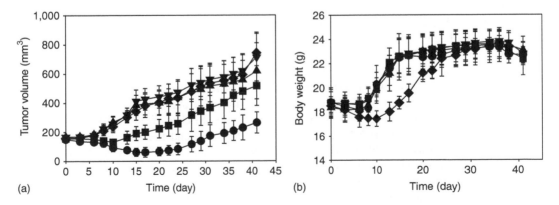

FIGURE 22.17 Tumor growth inhibition of s.c. human breast MCF-7/DOX^R carcinoma xenografts in BALB/c nude mice. Mice were injected i.v. with 10 mg/kg DOX equivalent dose: PHSM/f (●), PHSM (■), PHIM/f (▲), PHIM (▼), and free DOX (◆). Three i.v. injections on day 0, 3, and 6 were made. (a) Tumor volume change and (b) body-weight change ($n = 5$). Values are the means ± the standard deviations (SD). (From Lee, E. S., Na, K., and Bae, Y. H., *J. Control Release*, 103, 405, 2005.)

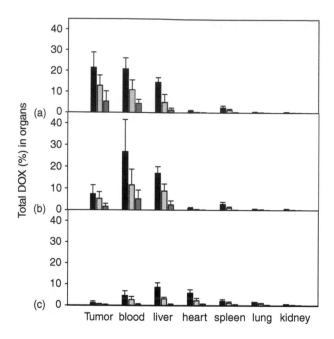

FIGURE 22.18 Total DOX (%) levels (total DOX remaining in organs to DOX content first injected) in blood and tissues (tumor, liver, heart, spleen, lung, and kidney) after i.v. administration (10 mg/kg DOX equivalent dose) of (a) PHSM/f micelles, (b) PHSM micelles, and (c) free DOX. Mice (s.c. human breast MCF-7/DOX[R] carcinoma xenografts) were sacrificed at 4 (■), 12 (■), and 48 h (■) after i.v. administration. Values are the means ± the standard deviations (*n* = 4, SD). (From Lee, E. S., Na, K., and Bae, Y. H., *J. Control Release*, 103, 405, 2005.)

This, combined with the endosomal escaping capacity of PHSM/f, resulted in a remarkable regression effect for resistant tumors, as shown in Figure 22.18.

22.7 CONCLUSIONS

Although important findings in scientific research and technological advances, such as long-circulating carries, EPR effect, tumor specific receptor-mediated targeting, and MDR modulator and reverters, have been achieved in the last few decades for effective solid tumor targeting, current chemotherapy is still facing major challenges to improving drug accumulation in tumor sites and reversing intrinsic or acquired drug-resistance of tumors.

An accelerated release rate of doxorubicin at tumor extracellular pH was provided by pH-sensitive polyHis/PEG micelles, but the micelle was not fully stable at pH 7.4. When a pH-insensitive block copolymer, PLLA–PEG, was blended with polyHis-PEG to form a mixed micelle, the stability of the micelles at pH 7.4 was significantly improved and the micelles' destabilization pH shifted from 7.2 to 6.6 with an increase in the amount of PLLA–PEG in the mixed micelle. The micelles were further modified with folate. These pH-sensitive micelles proved to be much more effective in regressing wild-type MCF-7 cells in vitro and in vivo studies with less weight-loss in animals, when compared with the pH-insensitive PLLA–PEG micelle. The folate effect was not apparent with the sensitive cells in the in vitro results. It seems that doxorubicin released either extracellularly or intracellularly is effective for cell-kill due to its high diffusivity through the plasma membrane.

The pH-sensitive micelle-with-folate showed a dramatic effect when tested with drug-resistant tumors in vitro and in vivo, demonstrating a similar cell-kill rate as that of sensitive cells.

The underlying mechanisms involve (1) fast internalization of the micelles to bypass the Pgp, (2) fast release rate of doxorubicin at early endosomes, and (3) quick disruption of the endosomal membrane by virtue of polyHis protonation and possible membrane interactions of freed polyHis-PEG. These mechanisms avoid the major MDR mechanisms of the Pgp pump and drug sequestration in acidic organelles and endosomal recycling, which resulted in uniform intracellular distributions of doxorubicin and polyHis-PEG, including the nucleus, In a resistant tumor xenograft model in mice, the tumor was regressed by iv administration of the folate-decorated pH-sensitive micelles with less weight-loss of the treated animal than control animals treated by pH-insensitive micelles and free doxorubicin.

REFERENCES

1. Levasseur, L. M., Slocum, H. K., Rustum, Y. M., and Greco, W. R., Modeling of the time-dependency of in vitro drug cytotoxicity and resistance, *Cancer Res.*, 58, 5749, 1998.
2. Gardner, S. N., A mechanistic, predictive model of dose–response curves for cell cycle phase-specific and -nonspecific drugs, *Cancer Res.*, 60, 1417, 2000.
3. Moghimi, S. M., Hunter, A. C., and Murray, J. C., Long-circulating and target-specific nanoparticles: Theory and practice, *Pharmacol. Rev.*, 53, 283, 2001.
4. Yokoyama, M., Novel passive targetable drug delivery with polymeric micelles, In *Biorelated Polymers and Gels*, Okano, T., Ed., Academic Press, San Diego, pp. 193–229, 1998.
5. Drummond, D. C., Meyer, O., Hong, K., Kirpotin, D. B., and Papahadjppoulos, D., Optimizing liposomes for delivery of chemotherapeutic agents to solid tumors, *Pharmacol. Rev.*, 51, 691, 1999.
6. Kong, G., Braun, R. D., and Dewhirst, M. W., Hyperthermia enables tumor-specific nanoparticle delivery: Effect of particle size, *Cancer Res.*, 60, 4440, 2000.
7. Munshi, N., Rapoport, N., and Pitt, W. G., Ultrasonic activated drug delivery from Pluronic P-105 micelles, *Cancer Lett.*, 118, 13, 1997.
8. Hobbs, S. K., Monsky, W. L., Yuan, F., Roberts, W. G., Griffith, L., Torchilin, V. P., and Jain, R. K., Regulation of transport pathways in tumor vessels: Role of tumor type and microenvironment, *Proc. Natl Acad. Sci. USA*, 95, 4607, 1998.
9. Tannock, I. F. and Rotin, D., Acid pH in tumors and its potential for therapeutic exploitation, *Cancer Res.*, 49, 4373, 1989.
10. Stubbs, M., McSheehy, R. M. J., Griffiths, J. R., and Bashford, L., Causes and consequences of tumour acidity and implications for treatment, *Opinion*, 6, 15, 2000.
11. Engin, K., Leeper, D. B., Cater, J. R., Thistlethwaite, A. J., Tupchong, L., and McFarlane, J. D., Extracellular pH distribution in human tumors, *Int. J. Hyperthermia*, 11, 211, 1995.
12. van Sluis, R., Bhujwalla, Z. M., Raghunand, N., Ballesteros, P., Alvarez, J., Cerdán, S., Galons, J. P., and Gillies, R. J., In vivo imaging of extracellular pH using ^{1}H MRSI, *Magn. Reson. Med.*, 41, 743, 1999.
13. Ojugo, A. S. E., McSheehy, P. M. J., Mcintyre, D. J. O., McCoy, C., Stubbs, M., Leach, M. O., Judson, I. R., and Griffiths, J. R., Measurement of the extraceullar pH of solid tumours in mice by magnetic resonace spectroscopy: A comparison of exogenous ^{19}F and ^{31}P probes, *NMR Biomed.*, 12, 495, 1999.
14. Leeper, D. B., Engin, K., Thistlethwaite, A. J., Hitchon, H. D., Dover, J. D., Li, D. J., and Tupchong, L., Humon tumor extracellular pH as a function of blood glucose concentration, *Int. J. Radiat. Oncol. Biol. Phys.*, 28, 935, 1994.
15. Yamagata, M., Hasuda, K., Stamato, T., and Tannock, I. F., The contribution of lactic acid to acidification of tumours: Studies of variant cells lacking lactate dehydrogenase, *Br. J. Cancer*, 77, 1726, 1998.
16. Gerasimov, O. V., Boomer, J. A., Qualls, M. M., and Thompson, D. H., Cytosolic drug delivery using pH- and light-sensitive liposomes, *Adv. Drug Deliv. Rev.*, 38, 317, 1999.
17. Drummond, D. C., Zignani, M., and Leroux, J. C., Current status of pH-sensitive liposomes in drug delivery, *Prog. Lipid Res.*, 39, 409, 2000.
18. Gottesman, M. M., Pastan, I., and Ambudkar, S. V., P-glycoprotein and multi drug resistance, *Curr. Opin. Genet. Dev.*, 6, 610, 1996.

19. Narasaki, F., Matsuo, I., Ikuno, N., Fukuda, M., Soda, H., and Oka, M., Multidrug resistance-associated protein (MRP) gene expression in human lung cancer, *Anticancer Res.*, 16, 2079, 1996.

20. Zaman, G. J., Lankelma, J., van Tellingen, Q., Beijnen, J., Dekker, H., Paulusma, C., Oude Elferink, R. P., Baas, F., and Borst, P., Role of gluthathione in the export of compounds from cells by the multidrug resistance-associated protein, *Proc. Natl Acad. Sci. USA*, 92, 7690, 1995.

21. De Jong, S., Zijlstra, J. G., de-Vries, E. G., and Mulder, N. H., Reduced DNA topoisomerase II activity and drug-induced DNA cleavage activity in an adriamycin-resistant human small cell lung carcinoma cell line, *Cancer Res.*, 50, 304, 1990.

22. Sugawara, I., Akiyama, S., Scheper, R. J., and Itoyama, S., Lung resistance protein (LRP) expression in human normal tissues in comparison with that of MDR1 and MRP, *Cancer Lett.*, 112, 23, 1997.

23. Pohl, G., Filipits, M., Suchomel, R. W., Stranzl, T., Depisch, D., and Pirker, R., Expression of the lung resistance protein (LRP) in primary breast cancer, *Anticancer Res.*, 19, 5051, 1999.

24. William, S. D. and Scheper, R. J., Lung resistance-related protein: Determining its role in multidrug resistance, *J. Natl Cancer Inst.*, 91, 1604, 1999.

25. Gonçlaves, A., Braguer, D., Kamath, K., Martello, L., Briand, C., Horwitz, S., Wilson, L., and Jordan, M. A., Resistance to taxol in lung cancer cells associated with increased microtubule dynamics, *Proc. Natl. Acad. Sci. USA*, 98, 11737, 2001.

26. Rogan, A. M., Hamilton, T. C., Young, R. C., Klecker, R. W., and Ozols, R. F., Reversal of adriamycin resistance by verapamil in human ovarian cancer, *Science*, 224, 994, 1984.

27. Germann, U. A., Shlyakhter, D., Mason, V. S., Zelle, R. E., Duffy, J. P., Gallulo, U., Armistead, D. M., Saunders, J. O., Boger, J., and Harding, M. W., Cellular and biochemical characterization of VX-710 as a chemosensitizer: Reversal of P-glycoprotein-mediated multidrug resistance in vitro, *Anticancer Drugs*, 8, 125, 1997.

28. Dale, I. L., Tuffley, W., Callaghan, R., Holmes, J. A., Martin, K., Luscombe, M., Mistry, P., Ryder, H., Stewart, A. J., Charlton, P., Twentyman, P. R., and Bevan, P., Reversal of P-glycoprotein-mediated multidrug resistance by XR9051, a novel diketopiperazine derivative, *Br. J. Cancer*, 78, 885, 1998.

29. Fisher, G. A., Lum, B. L., Hausdorff, J., and Sikic, B.I, Pharmacological considerations in the modulation of multidrug resistance, *Eur. J. Cancer*, 32A, 1082, 1996.

30. Mistry, P., Stewart, A. J., Dangerfield, W., Okiji, S., Liddle, C., Bootle, D., Plumb, J. A., Templeton, D., and Charlton, P., In vitro and in vivo reversal of P-glycoprotein-mediated multidrug resistance by a noel potent modulator, XR9576, *Cancer Res.*, 61, 749, 2001.

31. Simon, S. M., Role of organelle pH in tumor cell biology and drug resistance, *Drug Discov. Today*, 4, 32, 1999.

32. Busa, W. B. and Nuccitelli, R., Metabolic regulatoin via intracellular pH, *Am. J. Physiol.*, 246, R409, 1984.

33. Simon, S., Roy, D., and Schindler, M., Intracellular pH and the control of multidrug resistance, *Proc. Natl Acad. Sci. USA*, 91, 1128, 1994.

34. Belhoussine, R., Morjani, H., Millot, J. M., Sharonov, S., and Manfait, M., Confocal scanning microspectrofluorometry reveals specific anthracycline accumulation in cytoplasmic organelles of multidrug-resistant cancer cells, *J. Histochem. Cytochem.*, 46, 1369, 1998.

35. Tang, F., Horie, K., and Borchardt, R. T., Are MDCK cells transfected with the human MDR1 gene a good model of the human intestinal mucosa?, *Pharm. Res.*, 19, 765, 2002.

36. Lee, E. S., Shin, H. J., Na, K., and Bae, Y. H., Poly(L-histidine)-PEG block copolymer micelles and pH-induced destabilization, *J. Control Release*, 90, 363, 2003.

37. Patchornik, A., Berger, A., and Katchalski, E., Poly-L-histidine, *J. Am. Chem. Soc.*, 79, 5227, 1957.

38. Urry, D. W., Peng, S., Gowda, D. C., Parker, T. M., and Harris, R. D., Comparison of electrostatic- and hydrophobic-induced pKa shifts in polypentapeptides. The lysine residue, *Chem. Phys. Lett.*, 225, 97, 1994.

39. Marques, C. M., Bunchy micelles, *Langmuir*, 13, 1430, 1997.

40. La, S. B., Okano, T., and Kataoka, K., Preparation and characterization of the micelle-forming polymeric drug indomethacin-incorporated poly(ethylene oxide)-poly(beta-benzyl-L-aspartate) block copolymer micelle, *J. Pharm. Sci.*, 85, 85, 1996.

41. Lee, E. S., Na, K., and Bae, Y. H., Polymeric micelle for tumor pH and folate mediated targeting, *J. Control Release*, 91, 103, 2003.

42. Ning, Z., Da, D., Rudoll, T. L., Needham, D., Whorton, A. R., and Dewhirst, M. W., Increased microvascular permeability contributes to preferential accumulation of stealth liposomes in tumor tissue, *Cancer Res.*, 53, 3764, 1993.

43. Maeda, H., Wu, J., Sawa, T., Matsumura, Y., and Hori, K., Tumor vascular permeability and the EPR effect in macromolecular therapeutics: A review, *J. Control Release*, 65, 271, 2000.

44. Yuan, F., Leunig, M., Huang, S. K., Berk, D. A., Papahadjopoulos, D., and Jain, R. K., Microvascular permeability and interstitial penetration of sterically stabilized (stealth) liposomes in a human tumor xenograft, *Cancer Res.*, 54, 4564, 1994.

45. Oku, N., Namba, Y., and Okada, Y., Tumor accumulation of novel RES-avoiding liposomes, *Biochim. Biophys. Acta.*, 1126, 225, 1992.

46. Hobbs, S. K., Monsky, W. L., Yuan, F., Roberts, W. G., Griffith, L., Torchilin, V. P., and Jain, R. K., Regulation of transport pathways in tumor vessels: Role of tumor type and microenvironment, *Proc. Natl Acad. Sci. USA*, 95, 4607, 1998.

47. Charrois, G. J. R. and Allen, T. M., Rate of biodistribution of STEALTH® liposomes to tumor and skin: Influence diameter and implications for toxicity and therapeutic activity, *Biochim. Biophys. Acta.*, 1609, 102, 2003.

48. Takeuchi, H., Kojima, H., Toyoda, T., Yamamoto, H., Hinox, T., and Kawashima, Y., Prolonged circulation time of doxorubicin-loaded liposomes coated with a modified polyvinyl alcohol after intravenous injection in rats, *Eur. J. Pharm. Biopharm.*, 48, 123, 1999.

49. Goren, D., Horowitz, A. T., Tzemach, D., Tarshish, M., Zalipsky, S., and Gabizon, A., Nuclear delivery of doxorubicin via folate-targeted liposomes with bypass of multidrug-resistantce efflux pump, *Clin. Cancer Res.*, 6, 1949, 2000.

50. Shiah, J. G., Dvorak, M., Kopeckova, P., Sun, Y., Peterson, C. M., and Kopecek, J., Biodistribution and antitumour efficacy of long-circulating N-(2-hydroxypropyl) methacrylamide copolymer-doxorubicin conjugates in nude mice, *Eur. J. Cancer*, 37, 131, 2001.

51. Lee, E. S., Na, K., and Bae, Y. H., Doxorubicin loaded pH-sensitive polymeric micelles for reversal of resistant MCF-7 tumor, *J. Control Release*, 103, 405, 2005.

52. Benns, J. M., Choi, J. S., Mahato, R. I., Park, J. S., and Kim, S. W., pH-sensitive cationic polymer gene delivery vehicle: N-Ac-poly(L-histidine)-graft-poly(L-lysine) comb shaped polymer, *Bioconjug. Chem.*, 11, 637, 2000.

53. Na, K. and Bae, Y. H., Self-assembled hydrogel nanoparticles responsive to tumor extracellular pH from pullulan derivative/sulfonamide conjugate: Characterization, aggregation and adriamycin release in vitro, *Pharm. Res.*, 19, 681, 2002.

54. Gottesman, M. M. and Pastan, I., Biochemistry of multidrug resistance mediated by the multidrug transporter, *Ann. Rev. Biochem.*, 62, 385, 1993.

55. Thiebaut, F., Tsuruo, T., Hamada, H., Gottesman, M. M., Pastan, I., and Willingham, M. C., Cellular localization of the multidrug-resistance gene product P-glycoprotein in normal human tissues, *Proc. Natl Acad. Sci. USA*, 84, 7735, 1987.

56. Baldini, N., Scotlandi, K., Serra, M., Shikita, T., Zini, N., Ognibene, A., Santi, S., Ferrachini, R., and Maraldi, N. M., Nuclear immunolocalization of P-glycoprotein in multidrug-resistant cell lines showing similar mechanisms of doxorubicin distribution, *Eur. J. Cell Biol.*, 68, 226, 1995.

57. Gottesman, M. M., Pastan, I., and Ambudkar, S. V., P-glycoprotein and multidrug resistance, *Curr. Opin. Genet. Dev.*, 6, 610, 1996.

58. Mechetner, E., Kyshtoobayeva, A., Zonis, S., Kim, H., Stroup, R., Garcia, R., Parker, R. J., and Fruehauf, J. P., Levels of multidrug resistance (MDR1) P-glycoprotein expression by human breast cancer correlate with in vitro resistance to taxol and doxorubicin, *Clin. Cancer Res.*, 4, 389, 1998.

59. Mahoney, B. P., Raghunand, N., Baggett, B., and Gillies, R. J., Tumor acidity, ion trapping and chemotherapeutics. I. Acid pH affects the distribution of chemotherapeutic agents in vitro, *Biochem. Pharmacol.*, 66, 1207, 2003.

60. Murthy, N., Robichaud, J. R., Tirrell, D. A., Stayton, P. S., and Hoffman, A. S., The design and synthesis of polymers for eukaryotic membrane disruption, *J. Control Release*, 61, 137, 1999.

61. van Sluis, R., Bhujwalla, Z. M., Raghunand, N., Ballesteros, P., Alvarez, J., Cerdán, S., Galons, J. P., and Gillies, R. J., In vivo imaging of extracellular pH using ^{1}H MRSI, *Magn. Reso. Med.*, 41, 743, 1999.

62. Goren, D., Horowitz, A. T., Tzemach, D., Tarshish, M., Zalipsky, S., and Gabizon, A., Nuclear delivery of doxorubicin via folate-targeted liposomes with bypass of multidrug-resistantce efflux pump, *Clin. Cancer Res.*, 6, 1949, 2000.
63. Putnam, D., Gentry, C. A., Pack, D. W., and Langer, R., Polymer-based gene delivery with low cytotoxicity by a unique balance of side-chain termini, *Proc. Natl Acad. Sci.*, 98, 1200, 2001.
64. Sabharanjak, S. and Mayor, S., Folate receptor endocytosis and trafficking, *Adv. Drug Deliv. Rev.*, 56, 1099, 2004.

23 cRGD-Encoded, MRI-Visible Polymeric Micelles for Tumor-Targeted Drug Delivery

Jinming Gao, Norased Nasongkla, and Chalermchai Khemtong

CONTENTS

23.1 Introduction 465
23.2 Integrin $\alpha_v\beta_3$-Mediated Drug Targeting to Angiogenic Tumor Vasculature 466
23.3 cRGD-Encoded Micelles Facilitate Doxorubicin Delivery to
Tumor Endothelial Cells 467
23.4 cRGD-Encoded Micelles Increase Doxorubicin Toxicity over
Non-cRGD Micelles 468
23.5 Non-Invasive Imaging Methods for Pharmacokinetic Studies 469
23.6 SPIO as MR T2 Contrast Agent 470
23.7 Development of MRI-Visible Micelles 470
23.8 Non-Invasive Monitoring of Cell and Tumor Targeting by
cRGD-Encoded, SPIO-Loaded Micelles 472
23.9 Conclusions 473
Acknowledgments 474
References 474

23.1 INTRODUCTION

Remarkable progress has been made in recent years to establish polymer micelles as a novel multifunctional platform of nanoscale constructs for diagnostic and therapeutic applications.[1–5] Polymer micelles are composed of amphiphilic block copolymers that contain distinguished hydrophobic and hydrophilic segments. The distinct chemical nature of the two blocks results in thermodynamic phase separation in aqueous solution and formation of nanoscopic supramolecular core/shell structures. During the micellization process, the hydrophobic blocks associate to form the core region, whereas the hydrophilic blocks form the shell that separates the core from the aqueous medium. This unique architecture enables the micelle core to serve as a nanoscopic depot for therapeutic/contrast agents and the shell as biospecific surfaces for targeting applications (Figure 23.1).

Biocompatible and biodegradable block copolymers are currently used in micelle design. For the hydrophilic segment, poly(ethylene glycol) (PEG) blocks with a molecular weight of 2–15 kD are most often used. Meanwhile, the hydrophobic blocks include a variety of polymers such as

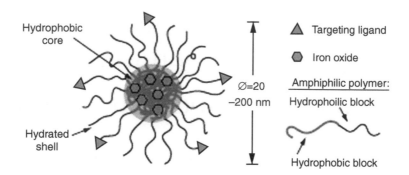

FIGURE 23.1 Micellar nanomedicine platform for targeted drug delivery.

poly(D, L-lactide), poly(ε-caprolactone), poly(β-benzoyl-L-aspartate), and poly(γ-benzyl-L-glutamate).[1,5] Different hydrophobic blocks can be used to provide different hydrophobic environments (e.g., aromatic vs. aliphatic) to maximize interactions with encapsulated agents.

Pioneering work from several leading groups has significantly advanced polymeric micelles as versatile carrier systems for the delivery of hydrophobic drugs. For instance, Kataoka's group has recently developed pH-sensitive micelles by attaching hydrophobic doxorubicin (DOX) to the polymer backbone via a hydrazone linkage. This acid-labile linkage undergoes cleavage at endosomal/lysosomal pH to release the conjugated DOX that leads to significantly improved anti-tumor efficacy.[6,7] A powerful means to improve the delivery efficiency of micelles is to introduce targeting ligands on the micelle surface to achieve an active targeting function. Among the yet limited literature reports dedicated to the active targeting of polymer-related micelle systems, recent work by Torchilin's group provides a landmark study in this direction.[8] Torchilin and coworkers have introduced two kinds of antibodies, mAb 2C5 and mAb 2G4, to the distal PEG end of phosphatidylethanolamine–poly(ethylene glycol) (PEG–PE) micelles. Not only can the immunomicelles with cancer-specific 2C5 antibody specifically bind to cancer cells in vitro, they can also increase drug accumulation in Lewis lung carcinoma tumors in mice. When compared to administration with either taxol or non-targeting micelles, immunomicelles demonstrate the highest efficacy of tumor growth inhibition. These advances bring out the unprecedented promise that polymeric micelles can serve as effective carriers for a large variety of poorly soluble drugs.

23.2 INTEGRIN $\alpha_v\beta_3$-MEDIATED DRUG TARGETING TO ANGIOGENIC TUMOR VASCULATURE

Since the original report by Cheresh and coworkers,[9] numerous studies have demonstrated that $\alpha_v\beta_3$ integrin is an important receptor affecting growth, local invasiveness, and metastatic potential of tumors.[10-12] In general, integrins are a family of heterodimeric membrane receptors that bind to cell-surface adhesion molecules and extracellular matrix proteins (ECM), and they play a critical role in tissue morphogenesis, repair, and remodeling. The function of integrins during angiogenesis has been most extensively studied with $\alpha_v\beta_3$ that is not readily detectable in quiescent vessels but becomes highly expressed in angiogenic vessels.[9] Recently, the crystal structures of the extracellular domains of $\alpha_v\beta_3$ and $\alpha_v\beta_3$/c(-RGDfV-) complex have been determined by Xiong et al.[13,14] These structures provide molecular insights and better understanding of $\alpha_v\beta_3$ receptor-ligand interactions and establish the structural basis for future development of more potent and specific $\alpha_v\beta_3$-binding ligands.[15]

Vascular targeting via $\alpha_v\beta_3$-dependent mechanisms offers significant opportunities for the development of cancer-targeted therapeutics and diagnostics. For therapeutic applications, Ruoslahti and coworkers have conjugated DOX to a cyclic RGD peptide (RGD-4C) as identified by a phage library.[16] The DOX-RGD-4C conjugates have demonstrated significantly improved anti-tumor efficacy in nude mice bearing human breast tumor xenografts.[17] Although quantitative pharmacokinetics of DOX-RGD-4C was not reported, it is expected that the renal clearance of the peptide-drug conjugate may be fast as a result of its small molecular weight. Recently, Ghanderhari and coworkers have shown that RGD-4C-functionalized hydroxypropylmethacrylamide (HPMA) polymer led to significantly increased tumor targeting and localization in severe combined immunodeficiency disease mice bearing DU145 or PC-3 prostate tumor xenografts over the RGE-4C-HPMA control.[18,19] This work opens up many exciting opportunities to broaden HPMA scaffold in targeted drug delivery applications. Recently, Cheresh and coworkers have reported the success of $\alpha_v\beta_3$-targeted gene therapy of cancer with cationic polymerized lipid-based nanoparticles. A mutant raf gene was delivered to tumor endothelial cells that led to tumor cell apoptosis and sustained regression of primary and metastatic melanomas in mice.[20] These data provide useful precedence for introducing $\alpha_v\beta_3$-targeted ligands on the nano delivery systems to enhance their targeting efficiency to tumor vasculature.

23.3 cRGD-ENCODED MICELLES FACILITATE DOXORUBICIN DELIVERY TO TUMOR ENDOTHELIAL CELLS

Previous work in this lab has demonstrated the feasibility of producing a cyclic pentapeptide c(Arg-Gly-Asp-d-Phe-Lys, cRGDFK—also referred to as cRGD herein)-encoded polymeric micelles and micelle targeting to $\alpha_v\beta_3$-overexpresing tumor endothelial cells (SLK cells).[21] Figure 23.2 illustrates the schematic diagram of synthesis of maleimide–terminated poly(ethylene glycol)–poly(ϵ-caprolactone) (MAL–PEG–PCL) copolymer and cRGD attachment on the micelle surface. Doxorubicin was also encapsulated inside the micelle core. The intrinsic fluorescent properties of doxorubicin ($\lambda_{ex} = 485$ nm, $\lambda_{em} = 595$ nm) permit the study of cell uptake by flow cytometry and confocal laser scanning microscopy.

Figure 23.3a shows that the percentage of cell uptake increased with increasing cRGD density on the micelle surface. With 5% cRGD surface density, a modest 3-fold increase of cell uptake was

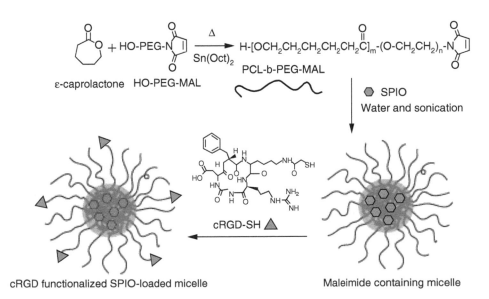

FIGURE 23.2 Synthesis of MAL–PEG–PCL copolymer and preparation of cRGD-encoded, DOX-loaded micelles. (From Nasongkla, N., Shuai, X., Ai, H., Weinberg, B. D., Pink, J., Boothman, D. A., and Gao, J., *Angew. Chem. Int. Ed. Engl.*, 43, 6323, 2004. With permission.)

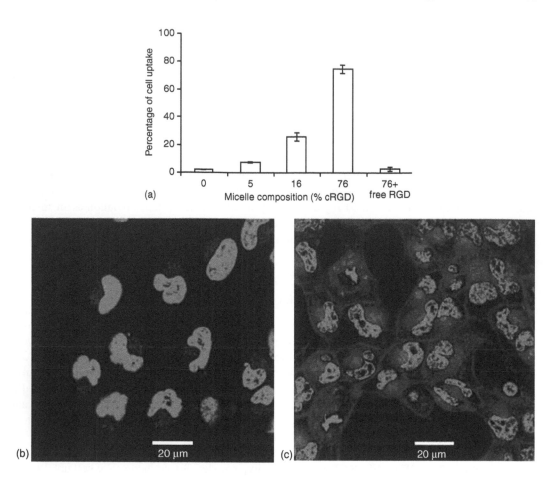

FIGURE 23.3 (See color insert following page 522.) (a) Percentage of micelle uptake in SLK tumor endothelial cells by flow cytometry as a function of cRGD density (0–76%) on the micelle surface. The last panel shows that the cell uptake of 76% cRGD-micelles is inhibited by the presence of free RGD ligands (9 mM) in solution. (b, c) Confocal laser scanning microscopy images of SLK cells treated with 0% (b) and 16% (c) cRGD-micelles after incubation for 2 h. The scale bars are 20 µm in both images. (From Nasongkla, N., Shuai, X., Ai, H., Weinberg, B. D., Pink, J., Boothman, D. A., and Gao, J., *Angew. Chem. Int. Ed. Engl.*, 43, 6323, 2004. With permission.)

observed. A 30-fold increase was observed by flow cytometry with 76% cRGD attachment. In the presence of excess free RGD ligands, the $\alpha_v\beta_3$-mediated cell uptake can be completely inhibited (last panel, Figure 23.3a). Confocal laser scanning microscopy further supports the cRGD-enhanced cell uptake (Figure 23.3b and c) and demonstrates that a majority of the micelles were taken up via $\alpha_v\beta_3$-mediated endocytosis and localized mainly in the cytoplasmic compartments in the cell (as shown by the red punctated dots). Cell nuclei were stained blue by Hoechst 33342 ($\lambda_{ex} = 352$ nm, $\lambda_{em} = 455$ nm) and overlaid with doxorubicin fluorescent images ($\lambda_{ex} = 485$ nm, $\lambda_{em} = 595$ nm).

23.4 cRGD-ENCODED MICELLES INCREASE DOXORUBICIN TOXICITY OVER NON-cRGD MICELLES

Figure 23.4 shows the growth inhibition of SLK cells by different micelle formulations. DOX concentration was kept the same at 1 µM for all samples. In these experiments, different DOX

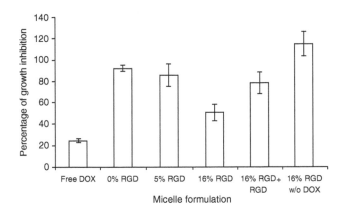

FIGURE 23.4 Cell growth in SLK cells for different compositions of SPIO-loaded micelles at 1 μM DOX concentration. The percentage of growth was calculated as the ratio of cell numbers for treated samples over untreated control. Error bars were obtained from triplicate samples.

samples were exposed to SLK cells for 1 h. After 1 h, the DOX-containing media were replaced with cell culture media and kept for an additional four days until the cells in the untreated control group reached confluence. Percentage of growth (PG) was calculated as the ratio of the number of cancer cells of the treated group over the untreated control. At 1 μM DOX dose, free drug demonstrated the lowest growth at 24.4 ± 2%. For micelle-delivered DOX, growth decreased with an increase of cRGD density on the micelle surface. The PG values for 0%, 5%, and 16% cRGD-micelles were 92 ± 8%, 85 ± 10%, and 51 ± 2.58%, respectively. The cRGD micelle toxicity was inhibited by the free RGD ligand co-incubated with the micelles (PG 78 ± 10%, Figure 23.4). In addition, cRGD-encoded, DOX-free micelles did not show much tumor growth inhibition (115 ± 11%) at the same micelle dose. These data demonstrate the success of cRGD-encoded micelles in enhancing the drug targeting efficiency and cytotoxic response in $\alpha_v\beta_3$-overexpresing SLK cells.

23.5 NON-INVASIVE IMAGING METHODS FOR PHARMACOKINETIC STUDIES

Conventional clinical pharmacokinetic studies require the collection of blood or tissue samples to assess the clearance rate, tissue distribution, and concentration vs. time relationships of chemotherapy. Such procedures are invasive and not compliant with patients especially when multiple samples need to be collected at different time points. In addition, the quantitative interpretation of these data is complicated by individual variability and uncertainty in the technical procedure (e.g., extraction of drugs from tissue may not be complete).

To circumvent these limitations, significant research efforts have focused on the development of non-invasive imaging techniques to characterize the drug pharmacokinetics in the target tissue in vivo.[22–25] Such imaging techniques include single photon emission computed tomography (SPECT),[26,27] positron-emission tomography (PET, dual photons),[26,27] and magnetic resonance imaging (MRI).[28] SPECT imaging requires the use of radionuclides (e.g., ^{111}In, ^{99m}Tc) that emit high-energy gamma photons (typically 60–600 keV) whereas PET detects positron annihilation radiation (511 keV). Although these two methods provide the most sensitive measurement of radiolabeled drugs (10^{-10}–10^{-12} mol), they have very low spatial resolutions. For example, the resolutions of small animal SPECT and PET instruments are approximately 2 and 5 mm, respectively.[26,27] Such resolution is not adequate to accurately assess drug distribution within tumor tissue, especially in a small animal tumor model (e.g., tumor xenograft in mice).

MRI is another imaging modality that can be used to study in situ pharmacokinetics.[28] MRI detects the magnetization changes of nuclei that have net angular momentum such as ^1H, ^{13}C, and ^{19}F. Examples of drugs studied by MRI include 5-fluorouracil (^{19}F), ^{13}C-labelled temozolomide, and ifosfamide (^{31}P). MRI has much higher resolution (~ 100 μm) than nuclear imaging methods. Its main limitation is low sensitivity, making it difficult to detect drugs at their therapeutic concentrations.[28] Use of exogenous MR contrast agents (e.g., SPIO) can dramatically increase the sensitivity of MR detection.

23.6 SPIO AS MR T2 CONTRAST AGENT

Superparamagnetic iron oxide (SPIO) nanoparticles have been extensively studied in the past decade for their use as MR contrast agents. Unlike the low molecular weight, paramagnetic metal chelates such as Gd-DTPA (T1 contrast agent) and SPIO nanoparticles are considered T2-negative contrast agents with substantially higher T2 and T1 relaxivity compared to T1 agents.[29] SPIO agents are composed of iron oxide nanocrystals with the general formula $Fe_2^{3+}O_3M^{2+}O$, where M^{2+} is a divalent metal ion such as iron, manganese, nickel, cobalt, or magnesium. When M^{2+} is ferrous iron (Fe^{2+}), SPIO becomes magnetite. In the absence of an external magnetic field, the magnetic domains inside SPIO are randomly oriented with no net magnetic field. An external magnetic field can cause the magnetic dipoles of the magnetic domains to reorient, leading to dramatically increased magnetic moments and significantly shortened relaxations in both T1 and T2/T2* relaxation processes.[29]

SPIO nanoparticles have been classified into four different categories based on their sizes and applications: large, standard, ultrasmall, and monocrystalline agents. Large SPIO agents (>200 nm) are mainly used for oral gastrointestinal lumen imaging. For example, AMI-121 (Advanced Magnetics) is a clinically approved T2 agent with a hydrodynamic diameter of 300 nm and T2 and T1 relaxivities of 72 and 3.2 Fe $mM^{-1} s^{-1}$, respectively.[30] Standard SPIO agents (50–200 nm) are easily sequestered by the RES in the liver and spleen upon intravenous injection and are primarily used for liver and spleen imaging. The iron oxide crystal (4.8–5.6 nm) in AMI-25 (Endorem® by Guerbet and Feridex® by Berlex Lab), a clinically approved agent, is coated with dextran to yield a final hydrodynamic diameter between 80 and 150 nm. The T2 and T1 relaxivities are 98.3 and 23.9 Fe $mM^{-1} s^{-1}$, respectively.[31] AMI-25 efficiently accumulates in the liver ($\sim 80\%$ of injected dose) and spleen (5–10%) within minutes of administration, and it has a short blood half-life (6 min).

Weissleder and coworkers have pioneered the development of ultrasmall (USPIO) and monocrystalline iron oxide nanoparticles (MION).[32–35] Two USPIO agents, AMI-227 and CLARISCAN, are currently in phase III clinical trials. AMI-227 is composed of a crystalline core of 4–6 nm surrounded by a dextran coating. Its hydrodynamic diameter is 20–40 nm, and it has T2 and T1 relaxivities of 44.1 and 21.6 Fe $mM^{-1} s^{-1}$, respectively.[36] Because of its smaller size, AMI-227 showed an increased blood half-life over 24 h.[37] Consequently, AMI-227 can be used as a blood pool agent for MR angiography. Eventually, these particles are taken up by the RES of lymph nodes and become particularly useful for MR lymphography.[38,39]

23.7 DEVELOPMENT OF MRI-VISIBLE MICELLES

Recently, a new class of MRI-visible nanoparticles was developed through the incorporation of hydrophobic SPIO nanoparticles inside polymer micelles.[40] High loading of SPIO was achieved inside the micelle core that led to a significant increase in T2 relaxivity. The resulting SPIO-loaded micelles have shown superb sensitivity under a clinical 1.5 T MR scanner.

First, monodisperse SPIO (Fe_3O_4) nanoparticles with different sizes were synthesized following published work by Sun and coworkers.[41] Transmission electron microscopy (TEM)

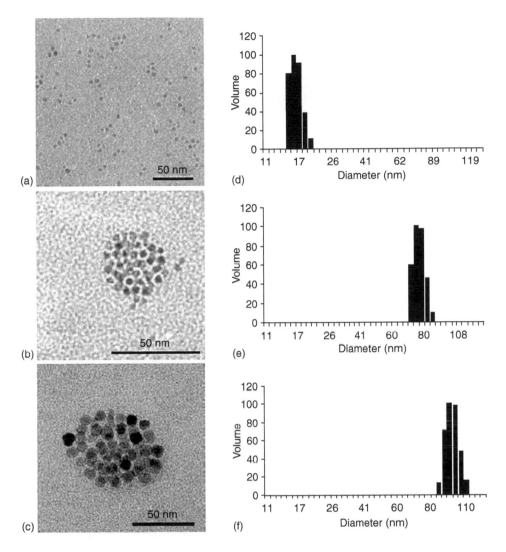

FIGURE 23.5 TEM images and DLS data of different SPIO-micelle formulations. (a, d) Single SPIO (4 nm in diameter) encapsulated by PEG5k–DSPE lipid; (b, e) SPIO (4 nm)-loaded PEG5k-b-PCL5k micelles; (c, f) SPIO (8 nm)-loaded PCL5k-b-PEG5k micelles. A cluster of SPIO particles are loaded inside each PCL5k-b-PEG5k micelle as shown in (b) and (c). (From Ai, H., Flask, C., Weinberg, B., Shuai, X., Pagel, M., Farrell, D., Duerk, J., and Gao, J., *Adv. Mat.*, 17, 1949, 2005. With permission.)

characterization of SPIO nanoparticles showed 4, 8, and 16 nm in diameter. The size distribution is uniform for all particle sizes. Distearoylphosphatidyl–ethanolamine–poly(ethylene glycol) (PEG5k–DSPE) lipid and amphiphilic diblock copolymer of PCL5k-b-PEG5k were used for the micelle formation. Interestingly, PEG5k–DSPE lipid micelles provide individual encapsulation of single 4 nm SPIO nanoparticles (Figure 23.5a). In comparison, PEG5k-b-PCL5k polymer micelles allow the incorporation of a cluster of 4 and 8 nm SPIO nanoparticles, respectively (Figure 23.5b and Figure 23.5c). Dynamic light scattering (DLS) was used to study the size distribution of SPIO-loaded PEG5k-b-PCL5k polymer micelles. DLS experiments were performed using the micelle solutions with 90° scattering angle at 25°C. Figure 23.5d shows a representative DLS diagram of PEG5k–DSPE micelles containing 4 nm SPIO nanoparticles. The mean hydrodynamic diameter was measured to be 17 ± 1 nm. The mean hydrodynamic diameters were 75 ± 4 and 97 ± 6 nm for

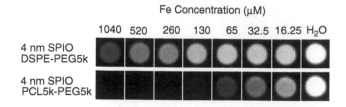

FIGURE 23.6 T2-weighted MRI scan (TR = 5000 ms, TE = 40 ms) of single 4 nm SPIO encapsulated by PEG5k-DSPE and a cluster of 4 nm SPIO loaded in PEG5k-PCL5k micelles. MRI intensity at the same Fe concentration is compared. (From Ai, H., Flask, C., Weinberg, B., Shuai, X., Pagel, M., Farrell, D., Duerk, J., and Gao, J., *Adv. Mat.*, 17, 1949, 2005. With permission.)

PEG-b-PCL micelles containing 4 and 8 nm SPIO particles (Figure 23.5e and Figure 23.5f), respectively.

The T1 and T2 relaxivities of different micelle particles were determined on a clinical 1.5 T scanner (Siemens Sonata). The T1 relaxivities for different micelle formulations were comparable and ranged from 1.3 to 2.9 Fe mM^{-1} s^{-1}. In contrast, the T2 relaxivities of micelles notably depended on SPIO clustering, SPIO diameter, and loading density. Figure 23.6 illustrates the MR signal intensity as a function of Fe concentration for two micelle formulations. The top row shows SPIO (4 nm) particles individually encapsulated by PEG5k–DSPE lipid (sample in Figure 23.5a), and the bottom row shows PEG-b-PCL micelles loaded with the same 4 nm SPIO particles (sample in Figure 23.5b). At the same Fe concentration, SPIO-loaded PEG-b-PCL micelles showed significantly increased contrast in these T2-weighted images. The T2 relaxivity increased from 25 for single SPIO micelles to 169 Fe mM^{-1} s^{-1} for SPIO-loaded polymeric micelles ($\sim$7 fold increase). Clustering of SPIO particles considerably increased the T2 relaxivity per Fe concentration. For polymeric micelles, further increase in SPIO diameter (e.g., from 4 to 8 and 16 nm) also led to significantly increased loading density and enhanced T2 relaxivity. The values of T2 relaxivity reached 318 and 471 Fe mM^{-1} s^{-1} for micelles loaded with 8 and 16 nm Fe$_3$O$_4$ particles, respectively.[40]

The MR sensitivity of detection was measured for the micelle formulations using a T2-weighted scan (TR = 5000 ms, TE = 40 ms) in an in vitro test tube experiment. The sensitivity limit is defined as the micelle concentration where the MR signal intensity decreases to 50% of the signal of pure water. For PEG-b-PCL micelles loaded with 16 nm SPIO particles, the highest sensitivity limit was observed at 5.2 μg/mL. In this micelle formulation, the micelle diameter is 110 ± 9 nm, and Fe$_3$O$_4$ loading is 54.2%. Because the molecular weight of typical micelles is on the order of 10^6 g/mol,[2] the above sensitivity limit corresponds to a micelle concentration of approximately 5 nM.

23.8 NON-INVASIVE MONITORING OF CELL AND TUMOR TARGETING BY cRGD-ENCODED, SPIO-LOADED MICELLES

Recently, cRGD-encoded, SPIO-loaded polymer micelles using similar procedures were successfully developed as shown in Figure 23.2. In this system, 8 nm SPIO nanoparticles were encapsulated (6.7 wt.%) into cRGD-encoded (50%) and non-cRGD (0%) PEG–PLA micelles. The mean hydrodynamic diameters of both micelle formulations are approximately 40 nm by dynamic light scattering measurement. In cell uptake experiments, tumor SLK endothelial cells (1.2×10^6 cells) were co-incubated with SPIO micelles at the final iron concentrations of 0, and 12.5 μg/mL in 10% fetal bovine serum cell culture medium for two hours. The medium was then removed, and the cells were gently washed twice with HBSS. Cells were detached by trypsin, and

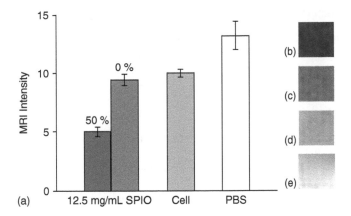

FIGURE 23.7 (a) MRI signal intensity of SLK cells treated with SPIO-loaded PEG–PLA micelles. SLK cells with micelle treatment and PBS buffer were used as control. Images (b) and (c) correspond to the cells incubated with 50% and 0% cRGD-micelles at 12.5 μg Fe/mL, respectively. (d) MR image of the vial containing SLK cells without any micelle incubation. (e) MR image of the vial containing PBS. MRI intensity is significantly darkened in cells incubated with cRGD-encoded micelles (b).

cell pellets were collected after centrifugation at 1200 rpm for 5 min. Cells were then fixed with 2% paraformaldehyde in phosphate-buffered saline (PBS) (200 μL) for two hours and concentrated by centrifugation again. Finally, each sample was dispersed in 20 μL of 2% agarose gel in PBS. A 384-well plate was used to contain the samples for MRI imaging. MR images were collected with conventional T2-weighted spin echo acquisition parameters (TR = 6,000 ms, TE = 90 ms).

Figure 23.7 shows the MR signal intensity of SLK cells incubated with cRGD-encoded and non-cRGD micelles. Three major conclusions can be drawn from these experiments. First, cRGD-encoded micelles led to significant reduction of MR signal amplitude over non-cRGD micelles as a result of the increased micelle uptake in SLK cells. Second, non-specific uptake of non-cRGD micelles is relatively small with minimal change in MR intensity over the Fe concentration range from 0 to 12.5 μg/mL. Third, MR detection is highly sensitive in imaging tumor cells (1.2×10^6 SLK cells at 12×10^6 cells/mL) with specific $\alpha_v\beta_3$-mediated uptake of Fe_3O_4 micelles. At 12.5 Fe μg/mL, MR amplitude decreased from 9.44 for non-cRGD micelles to 4.98 for cRGD-encoded micelles. The above data demonstrate the feasibility in producing biologically specific, highly sensitive MR molecular probes based on the micelle construct and design.

23.9 CONCLUSIONS

The long-term goal of this research is to establish polymer micelles as a safe and efficacious nanomedicine platform for targeted cancer therapy. The cRGD-encoded micelles have shown superb targeting efficiency in cell culture in vitro that establishes the micelle foundation for future in vivo evaluations in tumor-bearing mice. To facilitate the evaluation of tumor targeting efficiency of these micelles, MRI-visible micelles have been developed. Clustering of SPIO particles inside the micelle core led to a dramatic increase in T2 relaxivity. Together with high loading density of Fe_3O_4 inside polymer micelles, they led to an ultra sensitivity of detection by MRI per micelle basis (~ 5 nM). The much improved sensitivity of SPIO-loaded micelles may prove essential in detecting low concentrations of micelle nanoparticles for in vivo intratumoral tissue distribution studies by MRI.

ACKNOWLEDGMENTS

This work was supported by the National Institutes of Health grants R01 CA90696 and R21 EB005394 to Jinming Gao. Norased Nasongkla acknowledges the Royal Thai Government for a predoctoral fellowship support.

REFERENCES

1. Adams, M. L., Lavasanifar, A., and Kwon, G. S., Amphiphilic block copolymers for drug delivery, *J. Pharm. Sci.*, 92, 1343–1355, 2003.
2. Kwon, G. S., Polymeric micelles for delivery of poorly water-soluble compounds, *Crit. Rev. Ther. Drug. Carrier Syst.*, 20, 357–403, 2003.
3. Lukyanov, A. N. and Torchilin, V. P., Micelles from lipid derivatives of water-soluble polymers as delivery systems for poorly soluble drugs, *Adv. Drug Deliv. Rev.*, 56, 1273–1289, 2004.
4. Otsuka, H., Nagasaki, Y., and Kataoka, K., PEGylated nanoparticles for biological and pharmaceutical applications, *Adv. Drug Deliv. Rev.*, 55, 403–419, 2003.
5. Torchilin, V. P., Targeted polymeric micelles for delivery of poorly soluble drugs, *Cell. Mol. Life Sci.*, 61, 2549–2559, 2004.
6. Bae, Y., Fukushima, S., Harada, A., and Kataoka, K., Design of environment-sensitive supramolecular assemblies for intracellular drug delivery: Polymeric micelles that are responsive to intracellular pH change, *Angew. Chem. Int. Ed. Engl.*, 42, 4640–4643, 2003.
7. Bae, Y., Nishiyama, N., Fukushima, S., Koyama, H., Yasuhiro, M., and Kataoka, K., Preparation and biological characterization of polymeric micelle drug carriers with intracellular pH-triggered drug release property: Tumor permeability, controlled subcellular drug distribution, and enhanced in vivo antitumor efficacy, *Bioconjug. Chem.*, 16, 122–130, 2005.
8. Torchilin, V. P., Lukyanov, A. N., Gao, Z., and Papahadjopoulos-Sternberg, B., Immunomicelles: Targeted pharmaceutical carriers for poorly soluble drugs, *Proc. Natl. Acad. Sci. USA*, 100, 6039–6044, 2003.
9. Brooks, P. C., Clark, R. A., and Cheresh, D. A., Requirement of vascular integrin alpha v beta 3 for angiogenesis, *Science*, 264, 569–571, 1994.
10. Ruegg, C., Dormond, O., and Foletti, A., Suppression of tumor angiogenesis through the inhibition of integrin function and signaling in endothelial cells: Which side to target? *Endothelium*, 9, 151–160, 2002.
11. Teti, A., Migliaccio, S., and Baron, R., The role of the alphaVbeta3 integrin in the development of osteolytic bone metastases: A pharmacological target for alternative therapy? *Calcif. Tissue Int.*, 71, 293–299, 2002.
12. Varner, J. A. and Cheresh, D. A., Tumor angiogenesis and the role of vascular cell integrin alphav-beta3, *Important Adv. Oncol.*, 69–87, 1996.
13. Xiong, J. P., Stehle, T., Diefenbach, B., Zhang, R., Dunker, R., Scott, D. L., Joachimiak, A., Goodman, S. L., and Arnaout, M. A., Crystal structure of the extracellular segment of integrin alpha Vbeta3, *Science*, 294, 339–345, 2001.
14. Xiong, J. P., Stehle, T., Zhang, R., Joachimiak, A., Frech, M., Goodman, S. L., and Arnaout, M. A., Crystal structure of the extracellular segment of integrin alpha V beta3 in complex with an Arg–Gly–Asp ligand, *Science*, 296, 151–155, 2002.
15. Gottschalk, K. E. and Kessler, H., The structures of integrins and integrin-ligand complexes: Implications for drug design and signal transduction, *Angew. Chem. Int. Ed. Engl.*, 41, 3767–3774, 2002.
16. Pasqualini, R., Koivunen, E., and Ruoslahti, E., Alpha v integrins as receptors for tumor targeting by circulating ligands, *Nat. Biotechnol.*, 15, 542–546, 1997.
17. Arap, W., Pasqualini, R., and Ruoslahti, E., Cancer treatment by targeted drug delivery to tumor vasculature in a mouse model, *Science*, 279, 377–380, 1998.
18. Line, B. R., Mitra, A., Nan, A., and Ghandehari, H., Targeting tumor angiogenesis: Comparison of Peptide and polymer-Peptide conjugates, *J. Nucl. Med.*, 46, 1552–1560, 2005.
19. Mitra, A., Mulholland, J., Nan, A., McNeill, E., Ghandehari, H., and Line, B. R., Targeting tumor angiogenic vasculature using polymer-RGD conjugates, *J. Control. Release*, 102, 191–201, 2005.
20. Hood, J. D., Bednarski, M., Frausto, R., Guccione, S., Reisfeld, R. A., Xiang, R., and Cheresh, D. A., Tumor regression by targeted gene delivery to the neovasculature, *Science*, 296, 2404–2407, 2002.

21. Nasongkla, N., Shuai, X., Ai, H., Weinberg, B. D., Pink, J., Boothman, D. A., and Gao, J., cRGD-functionalized polymer micelles for targeted doxorubicin delivery, *Angew. Chem. Int. Ed. Engl.*, 43, 6323–6327, 2004.
22. Fischman, A. J., Alpert, N. M., and Rubin, R. H., Pharmacokinetic imaging: A noninvasive method for determining drug distribution and action, *Clin. Pharmacokinet.*, 41, 581–602, 2003.
23. Port, R. E. and Wolf, W., Noninvasive methods to study drug distribution, *Invest. New Drugs*, 21, 157–168, 2003.
24. Seddon, B. M. and Workman, P., The role of functional and molecular imaging in cancer drug discovery and development, *Br. J. Radiol.*, 76, S128–S138, 2003.
25. Singh, M. and Waluch, V., Physics and instrumentation for imaging in-vivo drug distribution, *Adv. Drug Deliv. Rev.*, 41, 7–20, 2000.
26. Bhatnagar, A., Hustinx, R., and Alavi, A., Nuclear imaging methods for non-invasive drug monitoring, *Adv. Drug Deliv. Rev.*, 41, 41–54, 2000.
27. Saleem, A., Aboagye, E. O., and Price, P. M., In vivo monitoring of drugs using radiotracer techniques, *Adv. Drug Deliv. Rev.*, 41, 21–39, 2000.
28. Griffiths, J. R. and Glickson, J. D., Monitoring pharmacokinetics of anticancer drugs: Non-invasive investigation using magnetic resonance spectroscopy, *Adv. Drug Deliv. Rev.*, 41, 75–89, 2000.
29. Wang, Y., Hussain, S., and Krestin, G., Superparamagnetic iron oxide contrast agents: Physicochemical characteristics and applications in MR imaging, *Eur. Radiol.*, 11, 2319–2331, 2001.
30. Hahn, P. F., Stark, D. D., Lewis, J. M., Saini, S., Elizondo, G., Weissleder, R., Fretz, C. J., and Ferrucci, J. T., First clinical trial of a new superparamagnetic iron oxide for use as an oral gastrointestinal contrast agent in MR imaging, *Radiology*, 175, 695–700, 1990.
31. Weissleder, R., Stark, D. D., Engelstad, B. L., Bacon, B. R., Compton, C. C., White, D. L., Jacobs, P., and Lewis, J., Superparamagnetic iron oxide: Pharmacokinetics and toxicity, *AJR Am. J. Roentgenol.*, 152, 167–173, 1989.
32. Moore, A., Weissleder, R., and Bogdanov, A. Jr., Uptake of dextran-coated monocrystalline iron oxides in tumor cells and macrophages, *J. Magn. Reson. Imaging*, 7, 1140–1145, 1997.
33. Schaffer, B. K., Linker, C., Papisov, M., Tsai, E., Nossiff, N., Shibata, T., Bogdanov, A. Jr., Brady, T. J., and Weissleder, R., MION-ASF: Biokinetics of an MR receptor agent, *Magn. Reson. Imaging*, 11, 411–417, 1993.
34. Weissleder, R., Elizondo, G., Wittenberg, J., Rabito, C. A., Bengele, H. H., and Josephson, L., Ultrasmall superparamagnetic iron oxide: Characterization of a new class of contrast agents for MR imaging, *Radiology*, 175, 489–493, 1990.
35. Weissleder, R., Lee, A. S., Fischman, A. J., Reimer, P., Shen, T., Wilkinson, R., Callahan, R. J., and Brady, T. J., Polyclonal human immunoglobulin G labeled with polymeric iron oxide: Antibody MR imaging, *Radiology*, 181, 245–249, 1991.
36. Jung, C. W. and Jacobs, P., Physical and chemical properties of superparamagnetic iron oxide MR contrast agents: Ferumoxides, ferumoxtran, ferumoxsil, *Magn. Reson. Imaging*, 13, 661–674, 1995.
37. McLachlan, S. J., Morris, M. R., Lucas, M. A., Fisco, R. A., Eakins, M. N., Fowler, D. R., Scheetz, R. B., and Olukotun, A. Y., Phase I clinical evaluation of a new iron oxide MR contrast agent, *J. Magn. Reson. Imaging*, 4, 301–307, 1994.
38. Anzai, Y., Blackwell, K. E., Hirschowitz, S. L., Rogers, J. W., Sato, Y., Yuh, W. T., Runge, V. M., Morris, M. R., McLachlan, S. J., and Lufkin, R. B., Initial clinical experience with dextran-coated superparamagnetic iron oxide for detection of lymph node metastases in patients with head and neck cancer, *Radiology*, 192, 709–715, 1994.
39. Anzai, Y., Brunberg, J. A., and Lufkin, R. B., Imaging of nodal metastases in the head and neck, *J. Magn. Reson. Imaging*, 7, 774–783, 1997.
40. Ai, H., Flask, C., Weinberg, B., Shuai, X., Pagel, M., Farrell, D., Duerk, J., and Gao, J., Magnetite-loaded polymeric micelles as novel magnetic resonance probes, *Adv. Mat.*, 17, 1949–1952, 2005.
41. Sun, S. and Zeng, H., Size-controlled synthesis of magnetite nanoparticles, *J. Am. Chem. Soc.*, 124, 8204–8205, 2002.

24 Targeted Antisense Oligonucleotide Micellar Delivery Systems

Ji Hoon Jeong, Sun Hwa Kim, and Tae Gwan Park

CONTENTS

24.1 Introduction 477
24.2 Micellar Antisense Oligonucleotide Delivery Systems 478
    24.2.1 Poly(D, L-Lactic-Co-Glycolic Acid)-Oligonucleotides Hybrid Polymeric Micelles 478
    24.2.2 Oligonucleotide Delivery Systems Based on Polyelectrolyte Complex Micelles 478
    24.2.3 In Vivo Applications 481
24.3 Ligand-Mediated Targeting of Antisense Oligonucleotides 483
    24.3.1 Ligand-Mediated ODN Targeting Systems Based on PEC Micelles 483
    24.3.2 In Vivo Applications 483
24.4 Conclusions 484
References 484

24.1 INTRODUCTION

Antisense oligonucleotides (antisense ODNs) are receiving increasing attention because of their high selectivity toward the sequence of a target mRNA.[1–4] The sequence-specific binding results in the degradation of the target mRNA, leading to blockage of the target protein expression. Owing to the features, antisense ODNs have been popularly employed for the modulation of specific gene expression. The clinical use of antisense ODN, however, is still limited as a result of its intrinsic properties: vulnerability to enzymatic digestion by humoral and cellular nucleases, and poor cellular uptake in a target tissue.[5–7] The susceptibility of ODN to enzymatic digestion has been partly improved by modifying the ODN backbone with either phosphorothioate linkage or amide linkage peptide nucleic acid (PNA).[8–10] The problem with poor cellular uptake could be addressed by employing cationic lipids and polymers.[11–15] The electrostatic interaction between a cationic lipid/polymer and negatively charged DNA or ODN can lead to the formation of polyelectrolyte complex (PEC) nanoaggregates that can greatly facilitate cellular uptake of the polyanions by an adsorption-induced internalization process called endocytosis. Despite their satisfactory performance in intracellular gene transfer in vitro, the use of PEC is still limited because of its rapid clearance and undesirable pharmacokinetic profiles upon intravenous administration.[16,17]

A series of recent developments for micellar gene delivery systems could provide a chance to overcome the potential problems concerning *in vivo* application of PEC systems. Self-assembling polymeric micelles have been considered as potential drug carriers for anti-cancer drugs, including nucleic acid-based therapeutic drugs such as antisense ODNs. Most polymeric micelles generally have a core-shell structure composed of hydrophobic parts as an internal core and hydrophilic parts as a surrounding corona. The hydrophilic corona could protect micelles from nonspecific adsorption of serum proteins in the blood stream, leading to an avoidance of opsonization-induced clearance of the micelles. The size of polymeric micelles, usually less than 100 nm, provides them with a chance to escape from renal exclusion and reticulo-endothelial system (RES). Because of the small size of the micelles, the enhanced permeability through loosened endothelial cell junctions in the vicinity of solid tumors can also be expected. A hybrid conjugate of ODN with hydrophobic biodegradable poly(D, L-lactic-*co*-glycolic acid) (PLGA) that exhibited micellar properties in aqueous solution was recently introduced.[6] PEC micelles are also a new class of gene delivery systems based on electrostatically interacted DNA/polymer complexes. The PEC micelles could be prepared from the interaction between ODN and cationic block copolymers, including poly(ethylene glycol) (PEG)-*b*-poly(L-lysine) di-block copolymer and PEG-grafted polyethylenimine (PEI).[18,19] Alternatively, the PEC micelles could be also formed by interacting PEG-conjugated ODN (ODN–PEG) and cationic polymers or peptides.[20] These systems could provide ODN with the enhanced solubility of nano-carriers and an improved stability against nuclease attacks. Introduction of cell binding ligands such as transferring and folate could also enhance the cellular uptake and improve the *in vivo* tissue distribution of the PEC micelles.

24.2 MICELLAR ANTISENSE OLIGONUCLEOTIDE DELIVERY SYSTEMS

24.2.1 POLY(D, L-LACTIC-CO-GLYCOLIC ACID)-OLIGONUCLEOTIDES HYBRID POLYMERIC MICELLES

Micelle-forming amphiphilic copolymers have been utilized as a solubilizer and carrier of various hydrophobic drugs, including several water insoluble anti-cancer drugs. An A–B type block copolymer of PEG-*b*-PLGA, containing doxorubicin conjugated to the terminal end of PLGA, could form micelles and remarkably improve the cellular uptake of doxorubicin.[21] Slow degradation of PLGA chains permit sustained release of the conjugated drug over a desired period. An alternative micelle-forming copolymer composed of PLGA and ODN (ODN/PLGA) was synthesized for intracellular ODN delivery.[6] Water-insoluble PLGA segments serve as a hydrophobic core and hydrophilic ODN as a surrounding corona, resulting in an amphiphilic structure similar to A–B type di-block copolymers. The ODN/PLGA could form spherical micelles with a size of ca. 80 nm. The micelles have a critical micelle concentration (cmc) ranging from 5 to 10 mg/L, depending on the analytical methods. When ODN/PLGA micelles were subjected to incubation in an aqueous solution, the micelles released ODN in a sustained manner by controlled degradation of conjugated PLGA chains. The micelles demonstrated much higher intracellular delivery efficiency in NIH3T3 mouse fibroblast cells when compared to naked ODN. The fluorescently labeled micelles were distributed over the cytoplasm of the cells, suggesting the micelles were taken up by the cells via endocytosis.

24.2.2 OLIGONUCLEOTIDE DELIVERY SYSTEMS BASED ON POLYELECTROLYTE COMPLEX MICELLES

Electrostatic interaction between oppositely charged polyelectrolytes (e.g., polycations and DNA) leads to the formation of nanoparticular polyelectrolyte complex (PEC) by charge neutralization. The nanoparticulates are usually unstable in aqueous milieu as a result of the high incidence of collision among particles and their high surface energy. Because they tend to aggregate one another, often resulting in precipitation, the complex formation between polycations and polyanions (ODN or plasmid DNA) should be nonstoichiometric; there the surface charge of the complexes should be negative (polyanions excess) or positive (polycation excess). The nonstoichiometric condition between polycation and polyanion can prevent the premature inter-particular aggregation by charge repulsion among the complexes during their formation.[22] However, the complexes with negatively charged surfaces are not appropriate for intracellular gene transfer because of the negatively charged characteristics of the cellular membrane. The positively charged particles are also undesirable for *in vivo* application because of their susceptibility to the nonspecific interaction of serum components such as negatively charged albumin that could often lead to inter-particular aggregation in the blood stream or opsonization-mediated clearance by reticuloendothelial cells. The introduction of hydrophilic PEG segments to the complexes' surfaces greatly increases their stability, not only enhancing the solubility of the complexes, but also efficiently preventing inter-particular aggregation by the steric effect of the flexible PEG chains.[23,24] Depending on the length (molecular weight) and density (the number of PEG present on the surface), the PEG chains around the PEC are often enough to shield the surface charge of the nonstoichiometric complexes to achieve charge neutrality of the complex surface.[25–28]

Block or graft copolymers consisting of a cationic polymer and PEG were reported to form PEC micelles by interacting with negatively charged ODN or plasmid DNA. When a poly(1-lysine)-PEG (PLL-*b*-PEG) di-block copolymer interacts with negatively charged ODN or plasmid DNA, self-assembled micelles can generate.[18] The charge neutralization between polycation and negatively charged polynucleotide could form a core-shell type nanoparticulate, entrapped and surrounded by a hydrophilic PEG segment. Similarly, the formation of self-associated micelles based on the interaction between counter ions could also be observed in a mixture of PEG-grafted PEI and alkyl sulfates.[19] The surrounded hydrophilic shell not only stabilizes the micelles, but also increase water solubility of the particles. The PEC micelles formed from PLL-*b*-PEG, and ODN showed a spherical shape with a diameter of about 50 nm as determined by dynamic light scattering. The formation of PEC micelles also greatly enhanced the stability of particles. Overnight incubation at room temperature and even storage under frozen conditions did not affect the size distribution of the micelles.[18] The addition of oppositely charged poly(aspartic acid) to PLL-*b*-PEG/DNA PEC micelles caused exchange reaction between poly(aspartic acid) and DNA, leading to a release of DNA in the medium.[29] This result suggested that the PEC micelle-based delivery system could stabilize the complex and increase its water solubility by efficiently compartmentalizing it within PEG shells while retaining its DNA releasing capability by exchange reaction with counter ions. The PEC micelles also efficiently protect the encapsulated ODN or plasmid DNA from enzymatic attack.[30,31] Environmental stimuli-sensitive PEC micelles were prepared using thiolated PLL-*b*-PEG. The micelles formed from thiolated PLL-*b*-PEG and ODN were dissociated to release ODN in the presence of reduced glutathione, suggesting that the ODN entrapped in PEC micelles could be released in a cytoplasmic environment for proper therapeutic activity of ODN.[32] Other polyamine-PEG conjugates including PEI–PEG and polyspermine (PSP)-PEG also showed PEC micelles forming capability when introduced to ODN.[33,34] Diameters of the PEC micelles from PEI–PEG and PSP–PEG were ca. 32 nm and ca. 12 nm as determined by dynamic light scattering and transmission electron microscopy (TEM). The micelles could be stably stored in a solution for several months. The micellar formulation could efficiently minimize the nonspecific interaction with a major component of serum proteins, namely, bovine serum albumin (BSA).

Recently, an alternative class of PEC micelles-based ODN delivery systems where PEC micelles could form from the interaction between PEG-conjugated oligonucleotide and polycation was introduced. Figure 24.1 shows the schematic diagram of the formation of PEC micelles. An A–B block-type conjugate, ODN–PEG, was synthesized.[20] An acid labile phosphoroamidate linkage was employed between ODN and PEG as a stimuli-sensitive linkage for the efficient release of ODN from the micelles in acidic endosomal pH. The ODN–PEG conjugate could self-associate to form PEC micelles, by interacting with several functional polycations such as fusogenic KALA peptide, PEI, and protamine (Figure 24.2). The charge-neutralized

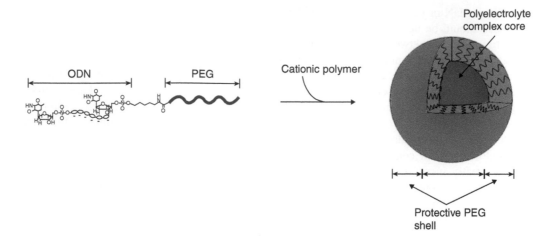

FIGURE 24.1 The schematic diagram of the formation of PEC micelles.

ODN/polycation complex was entrapped and compartmentalized by surrounding PEG corona to form a stable core-shell structure. The spontaneously formed PEC micelles showed a spherical shape with a diameter of ca. 70 nm and a narrow distribution.[35] The PEC micellar formulation demonstrated much higher intracellular transport efficiency when compared to naked ODN. When formulated with antisense ODN corresponding to c-myc or c-raf mRNA, the PEC micelles could remarkably inhibit the proliferation of smooth muscle cells (A7R5) or ovarian cancer cells (A2780), respectively.[35,36] The formation of PEC micelles between ODN–PEG conjugate and linear PEI provides ODN with enhanced nuclease resistance.[37] The PEC formulation also showed minimal adsorption of serum proteins, suggesting that the surrounding PEG shell could provide the nano-particulate with improved water solubility and with an efficient protection from the nonspecific or

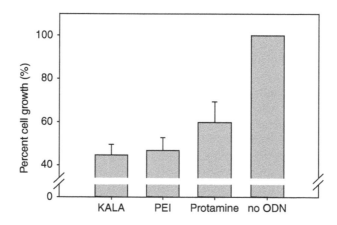

FIGURE 24.2 Influence of PEC micelles containing antisense *c-myb* on the proliferation of smooth muscle cells. Each PEC micelle was prepared by ODN–PEG conjugate with different polycations, KALA, PEI, and protamine. The size and surface zeta-potential values of the different micellar formulations were similar to those of the ODN–PEG/KALA formulation. The concentration of antisense ODN used in each formulation was 20 µg/ml. (From Jeong, J. H., Kim, S. W., and Park, T. G., *Bioconjug. Chem.*, 14, 2003. With permission.)

specific interaction with extracellular or intracellular proteins. These features would make PEC micellar formulation feasible for *in vivo* applications, especially for a systemic administration.

24.2.3 IN VIVO APPLICATIONS

There are two important factors that should be considered in designing polymeric gene carriers for *in vivo* application: pharmacokinetic behaviors in systemic circulation and cellular uptake and intracellular trafficking of the carriers. Systemically administered polymeric gene carriers could experience a number of barriers upon injection, including the attacks of extracellular nucleases and systemic clearance systems such as RES. The accumulation of the carriers in a desired tissue or organ could also be one of the barriers the carriers should overcome. Even after arriving at the appropriate destination, there are more cellular barriers, including cellular uptake, escape from endosomal compartment, and localization into the nucleus when it is required. A number of previous reports on micelle-based drug delivery systems demonstrated improved pharmacokinetics and cellular internalization.[21,38–42] The PEC micelles have a similar core-shell structure to polymeric micelles. The surrounding hydrophilic shell could minimize the nonspecific adsorption of serum proteins to avoid the surveillance of the systemic clearance systems, allowing prolonged circulation of the carrier. Also, the colloidal size of PEC micelles makes it feasible for them to be used for passive tumor targeting. Nanoparticulates, generally having less than 100 nm, could be expected to experience enhanced permeability and retention in the vicinity of a solid tumor.[43] In addition, the use of functional polymers that can facilitate the endosomal escape of ODN will be advantageous to improve intracellular trafficking. For example, PEI, one of the commonly used cationic gene carriers, has a high buffering capacity that may facilitate escape of the complexes from the endosomal compartment.[44]

In vivo anti-tumor activity of the PEC micellar formulation (ODN–PEG/PEI) containing antisense ODN that blocks the expression of c-raf gene was evaluated by intratumoral injection in nude mice with subcutaneously implanted solid tumor xenograft derived from human ovarian cancer.[18] Significant retardation of solid tumor growth was observed in PEC formulation with antisense c-raf ODN compared to the formulations with a scrambled ODN and PBS (Figure 24.3). When the PEC micelles containing antisense c-raf ODN were intravenously administered to the mice bearing a

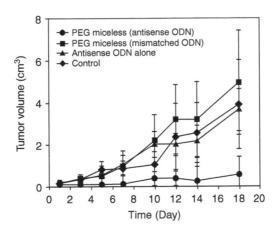

FIGURE 24.3 Effect of ODN–PEG/PEI PEC micelles containing antisense c-raf ODN on the growth of human ovarian cancer-derived tumor xenograft in nude mice. Each ODN formulation was administered intratumorally at 2.5 mg kg^{-1} injection^{-1}. (From Jeong, J. H., Kim, S. W., and Park, T. G., *J. Control. Rel.*, 93, 2003. With permission.)

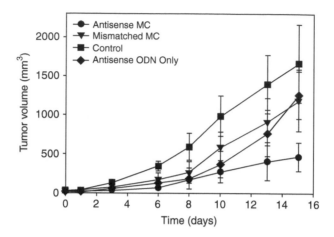

FIGURE 24.4 Effect of systemically administered ODN–PEG/PEI PEC micelles containing antisense c-raf ODN on the growth of human lung cancer-derived tumor xenograft in nude mice. Each ODN formulation was administered through tail vein at 2.5 mg kg^{-1} injection^{-1}. (From Jeong, J. H., Kim, S. H., Kim, S. W., and Park, T. G., *Bioconjug. Chem.*, 16(4), 2005. With permission.)

solid tumor derived from human lung cancer, the proliferation rate of the tumor was significantly reduced (Figure 24.4).[35] The PEC micellar formulation also exhibited much higher deposition of ODN in a solid tumor region than in a naked ODN after 12 h of intravenous injection through a tail vein (Figure 24.5). Taken together, the results suggested the hydrophilic PEG shell surrounding charge-neutralized polyelectrolyte core could successfully inhibit the nonspecific interaction of serum proteins in the blood stream. Also, through loosened vasculature near the rapidly growing tumors, the nanosized particles could be targeted to the solid tumor by the enhanced permeation and retention (EPR) effect.[43]

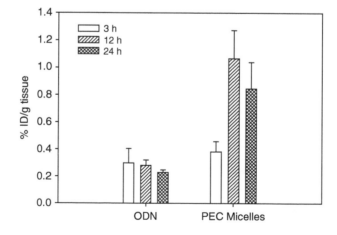

FIGURE 24.5 Accumulation of ODN–PEG/PEI PEC micelles in solid tumor (human lung cancer) after intravenous administration. (From Jeong, J. H., Kim, S. H., Kim, S. W., and Park, T. G., *Bioconjug. Chem.*, 16(4), 2005. With permission.)

24.3 LIGAND-MEDIATED TARGETING OF ANTISENSE OLIGONUCLEOTIDES

Nonviral carriers do not have an inherent selectivity of cell surface binding and internalization. The limitation can be addressed by conjugating various ligands targeting specific cells or tissure to polymeric gene carriers. The PEC micelles demonstrated promising potential as a carrier for systemic cancer treatment that could target solid tumors. In conjunction with the passive targeting property, it would be advantageous if the PEC micelles have cell- or tissue-specific ligands on the surface. The specific ligand could play an additional role for enhanced internalization of PEC micelles into the cells by receptor mediated endocytosis.

24.3.1 LIGAND-MEDIATED ODN TARGETING SYSTEMS BASED ON PEC MICELLES

A graft copolymer composed of PEI and PEG where the terminal end of PEG was biotinylated for further conjugation was synthesized and used for ligand-mediated targeting of cancer cells. PEC micelles could be generated by interacting biotinylated PEI-g-PEG and ODN. Avidin was attached to the biotin located at the end of the PEG segment and biotinylated trasferrin (Tf) was subsequently attached to the avidin to obtain a PEI–PEG-Tf conjugate.[45] Enhanced cellular uptake of PEI–PEG-Tf/ODN PEC micelles was shown in cancer cells. When formulated with antisense ODN targeting multi-drug efflux transporter protein, P-glycoprotein (P-gp), the PEI–PEG-Tf/ODN PEC micelles efficiently inhibit the expression of the P-gp, leading to intracellular accumulation of a P-gp specific probe (rhodamine 123) in multi-drug resistant (MDR) cancer cells.

Based on the same idea of PEG–ODN, a small interference lactosylated PEG-siRNA (Lac-PEG-siRNA) conjugate was recently synthesized.[46] An acid-labile β-thiopropionate linkage was introduced between siRNA and PEG to facilitate acid-triggered intracellular release of siRNA in the cytoplasm. The PEC micelles were prepared by interacting the PEG-siRNA conjugate with PLL, and they formed spherical shape nanoparticulates with an average diameter of. 117 nm. The lactose moiety at the end of the PEG segment could target the asialogycoprotein of hepatocytes to facilitate the receptor-mediated endocytosis of the PEC micelles. The PEC micelles demonstrated a significant RNA interference (RNAi) activity in human hepatoma cells (HuH-7), comparable to that of commercial cationic liposome-based carrier oligofectAMINE.

24.3.2 IN VIVO APPLICATIONS

PEC micelles demonstrated potential as a tumor targeting carrier for antisense ODN. Owing to their small size (generally less than 100 nm), the PEC micelles could be passively located in a solid tumor region through the loosened endothelial vasculature of the tumors by EPR effect. In addition to the passive targeting property, it would be more desirable if the carrier could also target a specific cell or tissue. Folate (FA) has been recognized as a potential targeting ligand for cancer cells. Expression of FA receptor is frequently up-regulated in a number of malignant tumors and highly restricted in normal cells and tissues.[47,48] Therefore, FA was coupled to several polymeric carriers for drug targeting to cancer cell.[38,49,50] FA has also been used for the modification of cationic polymers for tumor-targeted gene delivery. [51–53] Recently, ODN–PEG conjugate having FA at the end of PEG was synthesized for the systemic treatment of cancer based on PEC micelle delivery system.[54] The size of ODN–PEG–FA/PEI PEC micelles was ca. 90 nm with narrow distribution. Systemic administration of the micellar formulation to nude mice bearing human tumor xenograft derived from KB cells, FA-over-expressed the cells (KB cells), demonstrating the enhanced deposition of ODN in solid tumor region when compared to ODN–PEG/PEI PEC micelles. However, no significance in tumor deposition of ODN was observed between ODN–PEG–FA and ODN–PEG formulations in FA-receptor negative tumor xenograft model (A549 cells).

24.4 CONCLUSIONS

Polymeric gene carriers have been developed as substitutes for viral vectors that could cause a series of unexpected side effects such as immunogenecity and oncogenecity. Micelle-based gene carriers exhibited a number of advantageous features for systemic delivery, including colloidal stability in solution, reduced interaction with components of biological fluid, improved *in vivo* pharmacokinetics, and the possibility of passive and active targeting of cancers. The intracellular pharmacokinetic behavior of the nucleic acid-based therapeutics could also be improved by employing functional polymers with the existing formulation. Although there is still room for further improvement, the micellar gene delivery system could be considered a good example of a multi-functional drug delivery platform that has the potential to overcome a number of intrinsic barriers in the *in vivo* application of therapeutic genes.

REFERENCES

1. Simons, M., Edelman, E. R., DeKeyser, J. L., Langer, R., and Rosenberg, R. D., Antisense c-myb oligonucleotides inhibit intimal arterial smooth muscle cell accumulation *in vivo*, *Nature*, 359, 67, 1992.
2. Stein, C. A. and Cheng, Y. C., Antisense oligonucleotides as therapeutic agents—is the bullet really magical?, *Science*, 261, 1004, 1993.
3. Elbashir, S. M., Harborth, J., Lendeckel, W., Yalcin, A., Weber, K., and Tuschl, T., Duplexes of 21-nucleotide RNAs mediate RNA interference in cultured mammalian cells, *Nature*, 411, 494, 2001.
4. Elbashir, S. M., Lendeckel, W., and Tuschl, T., RNA interference is mediated by 21- and 22-nucleotide RNAs, *Genes Dev.*, 15, 188, 2001.
5. Wagner, R. W., Gene inhibition using antisense oligodeoxynucleotides, *Nature*, 372, 333, 1994.
6. Jeong, J. H. and Park, T. G., Novel polymer-DNA hybrid polymeric micelles composed of hydrophobic poly(D,L-lactic-co-glycolic acid) and hydrophilic oligonucleotides, *Bioconjug. Chem.*, 12, 917, 2001.
7. Braasch, D. A., Paroo, Z., Constantinescu, A., Ren, G., Oz, O. K., Mason, R. P., and Corey, D. R., Biodistribution of phosphodiester and phosphorothioate siRNA, *Bioorg. Med. Chem. Lett.*, 14, 1139, 2004.
8. Agrawal, S., Goodchild, J., Civeira, M. P., Thornton, A. H., Sarin, P. S., and Zamecnik, P. C., Oligodeoxynucleoside phosphoramidates and phosphorothioates as inhibitors of human immunodeficiency virus, *Proc. Natl. Acad. Sci. USA*, 85, 7079, 1988.
9. Nielsen, P. E., Egholm, M., Berg, R. H., and Buchardt, O., Peptide nucleic acids (PNAs): Potential antisense and anti-gene agents, *Anticancer Drug Des.*, 8, 53, 1993.
10. Egholm, M., Buchardt, O., Christensen, L., Behrens, C., Freier, S. M., Driver, D. A., Berg, R. H., Kim, S. K., Norden, B., and Nielsen, P. E., PNA hybridizes to complementary oligonucleotides obeying the Watson–Crick hydrogen-bonding rules, *Nature*, 365, 566, 1993.
11. Pagnan, G., Stuart, D. D., Pastorino, F., Raffaghello, L., Montaldo, P. G., Allen, T. M., Calabretta, B., and Ponzoni, M., Delivery of c-myb antisense oligodeoxynucleotides to human neuroblastoma cells via disialoganglioside GD(2)-targeted immunoliposomes: Antitumor effects, *J. Natl. Cancer Inst.*, 92, 253, 2000.
12. Juliano, R. L. and Akhtar, S., Liposomes as a drug delivery system for antisense oligonucleotides, *Antisense Res. Dev.*, 2, 165, 1992.
13. Barron, L. G., Meyer, K. B., and Szoka, F. C. Jr., Effects of complement depletion on the pharmacokinetics and gene delivery mediated by cationic lipid-DNA complexes, *Hum. Gene Ther.*, 9, 315, 1998.
14. Meyer, O., Kirpotin, D., Hong, K., Sternberg, B., Park, J. W., Woodle, M. C., and Papahadjopoulos, D., Cationic liposomes coated with polyethylene glycol as carriers for oligonucleotides, *J. Biol. Chem.*, 273, 15621, 1998.
15. Kim, J. S., Kim, B. I., Maruyama, A., Akaike, T., and Kim, S. W., A new non-viral DNA delivery vector: The terplex system, *J. Control. Rel.*, 53, 175, 1998.

16. Hashida, M., Takemura, S., Nishikawa, M., and Takakura, Y., Targeted delivery of plasmid DNA complexed with galactosylated poly(L-lysine), *J. Control. Rel.*, 53, 301, 1998.

17. Nishikawa, M., Takemura, S., Takakura, Y., and Hashida, M., Targeted delivery of plasmid DNA to hepatocytes *in vivo*: Optimization of the pharmacokinetics of plasmid DNA/galactosylated poly(L-lysine) complexes by controlling their physicochemical properties, *J. Pharmacol. Exp. Ther.*, 287, 408, 1998.

18. Kataoka, K., Togawa, H., Harada, A., Yasugi, K., Matsumoto, T., and Katayose, S., Spontaneous formation of polyion complex micelles with narrow distribution from antisense oligonucleotide and cationic block copolymer in physiological saline, *Macromolecules*, 29, 8556, 1996.

19. Bronich, T. K., Cherry, T., Vinogradov, S. V., Eisenberg, A., Kabanov, V. A., and Kabanov, A. V., Self-assembly in mixtures of poly(ethyleneoxide)-graft-poly(ethylenimine) and alkyl sulfates, *Langmuir*, 14, 6101, 1998.

20. Jeong, J. H., Kim, S. W., and Park, T. G., Novel intracellular delivery system of antisense oligonucleotide by self-assembled hybrid micelles composed of DNA/PEG conjugate and cationic fusogenic peptide, *Bioconjug. Chem.*, 14, 473, 2003.

21. Yoo, H. S. and Park, T. G., Biodegradable polymeric micelles composed of doxorubicin conjugated PLGA–PEG block copolymer, *J. Control. Rel.*, 70, 63, 2001.

22. Jeong, J. H., Song, S. H., Lim, D. W., Lee, H., and Park, T. G., DNA transfection using linear poly(ethylenimine) prepared by controlled acid hydrolysis of poly(2-ethyl-2-oxazoline), *J. Control. Rel.*, 73, 391, 2001.

23. Lee, H., Jeong, J. H., and Park, T. G., PEG grafted polylysine with fusogenic peptide for gene delivery: High transfection efficiency with low cytotoxicity, *J. Control. Rel.*, 79, 283, 2002.

24. Lee, H., Jeong, J. H., and Park, T. G., A new gene delivery formulation of polyethylenimine/DNA complexes coated with PEG conjugated fusogenic peptide, *J. Control. Rel.*, 76, 183, 2001.

25. Brus, C., Petersen, H., Aigner, A., Czubayko, F., and Kissel, T., Efficiency of polyethylenimines and polyethylenimine-graft-poly(ethylene glycol) block copolymers to protect oligonucleotides against enzymatic degradation, *Eur. J. Pharm. Biopharm.*, 57, 427, 2004.

26. Brus, C., Petersen, H., Aigner, A., Czubayko, F., and Kissel, T., Physicochemical and biological characterization of polyethylenimine-graft-poly(ethylene glycol) block copolymers as a delivery system for oligonucleotides and ribozymes, *Bioconjug. Chem.*, 15, 677, 2004.

27. Kunath, K., von Harpe, A., Petersen, H., Fischer, D., Voigt, K., Kissel, T., and Bickel, U., The structure of PEG-modified poly(ethylene imines) influences biodistribution and pharmacokinetics of their complexes with NF-kappaB decoy in mice, *Pharm. Res.*, 19, 810, 2002.

28. Jeong, J. H., Lee, M., Kim, W. J., Yockman, J. W., Park, T. G., Kim, Y. H., and Kim, S. W., Anti-GAD antibody targeted non-viral gene delivery to islet beta cells, *J. Control. Rel.*, 107, 562, 2005.

29. Katayose, S. and Kataoka, K., Water-soluble polyion complex associates of DNA and poly(ethylene glycol)-poly(1-lysine) block copolymer, *Bioconjug. Chem.*, 8, 702, 1997.

30. Katayose, S. and Kataoka, K., Remarkable increase in nuclease resistance of plasmid DNA through supramolecular assembly with poly(ethylene glycol)-poly(L-lysine) block copolymer, *J. Pharm. Sci.*, 87, 160, 1998.

31. Harada, A., Togawa, H., and Kataoka, K., Physicochemical properties and nuclease resistance of antisense–oligodeoxynucleotides entrapped in the core of polyion complex micelles composed of poly(ethylene glycol)-poly(L-lysine) block copolymers, *Eur. J. Pharm. Sci.*, 13, 35, 2001.

32. Kakizawa, Y., Harada, A., and Kataoka, K., Glutathione-sensitive stabilization of block copolymer micelles composed of antisense DNA and thiolated poly(ethylene glycol)-block-poly(L-lysine): A potential carrier for systemic delivery of antisense DNA, *Biomacromolecules*, 2, 491, 2001.

33. Vinogradov, S. V., Bronich, T. K., and Kabanov, A. V., Self-assembly of polyamine-poly(ethylene glycol) copolymers with phosphorothioate oligonucleotides, *Bioconjug. Chem.*, 9, 805, 1998.

34. Kabanov, V. A. and Kabanov, A. V., Interpolyelectrolyte and block ionomer complexes for gene delivery: Physico-chemical aspects, *Adv. Drug Deliv. Rev.*, 30, 49, 1998.

35. Jeong, J. H., Kim, S. H., Kim, S. W., and Park, T. G., Polyelectrolyte complex micelles composed of c-raf anti-sense oligodeoxynucleotide-poly(ethylene glycol) conjugate and poly(ethylenimine): Effect of systemic administration on tumor growth, *Bioconjug. Chem.*, 16(4), 1034–1037, 2005.

36. Jeong, J. H., Kim, S. W., and Park, T. G., A new antisense oligonucleotide delivery system based on self-assembled ODN–PEG hybrid conjugate micelles, *J. Control. Rel.*, 93, 183, 2003.

37. Oishi, M., Sasaki, S., Nagasaki, Y., and Kataoka, K., pH-responsive oligodeoxynucleotide (ODN)-poly(ethylene glycol) conjugate through acid-labile beta-thiopropionate linkage: Preparation and polyion complex micelle formation, *Biomacromolecules*, 4, 1426, 2003.

38. Yoo, H. S. and Park, T. G., Folate receptor targeted biodegradable polymeric doxorubicin micelles, *J. Control. Rel.*, 96, 273, 2004.

39. Yoo, H. S., Lee, E. A., and Park, T. G, Doxorubicin-conjugated biodegradable polymeric micelles having acid-cleavable linkages, *J. Control. Rel.*, 82, 17, 2002.

40. Harada-Shiba, M., Yamauchi, K., Harada, A., Takamisawa, I., Shimokado, K., and Kataoka, K., Polyion complex micelles as vectors in gene therapy–pharmacokinetics and *in vivo* gene transfer, *Gene Ther.*, 9, 407, 2002.

41. Kakizawa, Y. and Kataoka, K., Block copolymer micelles for delivery of gene and related compounds, *Adv. Drug Deliv. Rev.*, 54, 203, 2002.

42. Ideta, R., Yanagi, Y., Tamaki, Y., Tasaka, F., Harada, A., and Kataoka, K., Effective accumulation of polyion complex micelle to experimental choroidal neovascularization in rats, *FEBS Lett.*, 557, 21, 2004.

43. Maeda, H., Wu, J., Sawa, T., Matsumura, Y., and Hori, K., Tumor vascular permeability and the EPR effect in macromolecular therapeutics: A review, *J. Control. Rel.*, 65, 271, 2000.

44. Boussif, O., Lezoualc'h, F., Zanta, M. A., Mergny, M. D., Scherman, D., Demeneix, B., and Behr, J. P., A versatile vector for gene and oligonucleotide transfer into cells in culture and *in vivo*: Poly-ethylenimine, *Proc. Natl. Acad. Sci. USA*, 92, 7297, 1995.

45. Vinogradov, S., Batrakova, E., Li, S., and Kabanov, A., Polyion complex micelles with protein-modified corona for receptor-mediated delivery of oligonucleotides into cells, *Bioconjug. Chem.*, 10, 851, 1999.

46. Oishi, M., Nagasaki, Y., Itaka, K., Nishiyama, N., and Kataoka, K., Lactosylated poly(ethylene glycol)-siRNA conjugate through acid-labile beta-thiopropionate linkage to construct pH-sensitive polyion complex micelles achieving enhanced gene silencing in hepatoma cells, *J. Am. Chem. Soc.*, 127, 1624, 2005.

47. Lu, Y. and Low, P. S., Folate-mediated delivery of macromolecular anticancer therapeutic agents, *Adv. Drug Deliv. Rev.*, 54, 675, 2002.

48. Antony, A. C., The biological chemistry of folate receptors, *Blood*, 79, 2807, 1992.

49. Kim, S. H., Jeong, J. H., Chun, K. W., and Park, T.G, Target-specific cellular uptake of PLGA nanoparticles coated with poly(1-lysine)-poly(ethylene glycol)-folate conjugate, *Langmuir*, 21, 8852, 2005.

50. Yoo, H. S. and Park, T. G., Folate-receptor-targeted delivery of doxorubicin nano-aggregates stabilized by doxorubicin-PEG-folate conjugate, *J. Control. Rel.*, 100, 247, 2004.

51. Cho, K. C., Kim, S. H., Jeong, J. H., and Park, T. G., Folate receptor-mediated gene delivery using folate-poly(ethylene glycol)-poly(L-lysine) conjugate, *Macromol. Biosci.*, 5, 512, 2005.

52. Benns, J. M., Maheshwari, A., Furgeson, D. Y., Mahato, R. I., and Kim, S. W., Folate-PEG-folate-graft-polyethylenimine-based gene delivery, *J. Drug Target.*, 9, 123, 2001.

53. Kim, S. H., Jeong, J. H., Cho, K. C., Kim, S. W., and Park, T. G., Target-specific gene silencing by siRNA plasmid DNA complexed with folate-modified poly(ethylenimine), *J. Control. Rel.*, 104, 223, 2005.

54. Jeong, J. H., Kim, S. H., Kim, S. W., and Park, T. G., In vivo tumor targeting of ODN–PEG-folic acid/PEI polyelectrolyte complex micelles, *J. Biomater. Sci. Polym. Ed.*, 16, 1409, 2005.

Section 5

Dendritic Nanocarriers

25 Dendrimers as Drug and Gene Delivery Systems

Tae-il Kim and Jong-Sang Park

CONTENTS

25.1 Introduction 489
25.2 Drug Delivery 490
    25.2.1 Introduction 490
    25.2.2 Non-Covalent Encapsulation of Drugs 490
    25.2.3 Covalent Conjugation of Drugs 493
    25.2.4 Neutron Capture Therapy 494
        25.2.4.1 Boron Neutron Capture Therapy 494
        25.2.4.2 Gadolinium Neutron Capture Therapy 495
    25.2.5 Photodynamic Therapy 496
25.3 Gene Delivery 498
    25.3.1 Introduction 498
    25.3.2 PAMAM Dendrimers 498
    25.3.3 Polypropylenimine Dendrimers 501
    25.3.4 Other Dendrimers 502
25.4 Conclusion 504
Acknowledgments 504
References.. 504

25.1 INTRODUCTION

Dendrimers are highly branched, mono-disperse macromolecules, characterized by layers from a core focal point called generations and multivalent end group functionalities. The word *dendrimer* originated from the Greek *dendron*, meaning "tree" and *meros*, meaning "part."

Dendrimers are prepared by iterative synthetic steps. First, Vögtle and coworkers reported the synthesis of dendritic structures as cascade molecules.[1] A lot of dendrimers have been developed, and they have become a subject of intense research, including within electrochemistry, photophysics, and supramolecular chemistry.[2,3] Over the past decade, characteristics such as a highly defined structure, modifiable surface functionality, and an internal cavity of dendrimers have aroused scientists' interest in the biomedical field. In recent years, dendrimers have shown potential in fields, including drug delivery, gene delivery, magnetic resonance imaging (MRI), and anti-cancer therapeutics.[4–6]

The focus of this review is the history and the current trends of dendrimer application for drug and gene delivery systems.

25.2 DRUG DELIVERY

25.2.1 Introduction

The development of an efficient drug delivery system is very important to improve the pharmacological activity of drug molecules. Various types of drug delivery systems have been designed, including liposomes, polymeric micelles, etc.[7-9] Dendrimers have emerged as new alternatives and efficient tools for delivery of drug molecules. Figure 25.1 shows the molecular structures of various dendrimers used for delivery systems. Generally, bioactive drugs have hydrophobic moieties in their structure, resulting in low water-solubility that inhibits efficient delivery into cells. Dendrimers designed to be highly water-soluble and biocompatible have been shown to be able to improve drug properties such as solubility and plasma circulation time via various formulations and to deliver drugs efficiently. In comparison to linear polymeric carriers, the multivalent functionalities of dendrimers can be linked to drug molecules or ligands in a well-defined manner and can be used to increase the binding efficiency and affinity of therapeutic molecules to receptors via synergistic interaction. In addition, the low polydispersity of dendrimers can show a reproducible pharmacokinetic pattern.

25.2.2 Non-Covalent Encapsulation of Drugs

Drug delivery using unimolecular micelles has the advantage that the micellar structure is maintained and not disrupted at all concentrations, contrary to the behavior of conventional polymeric

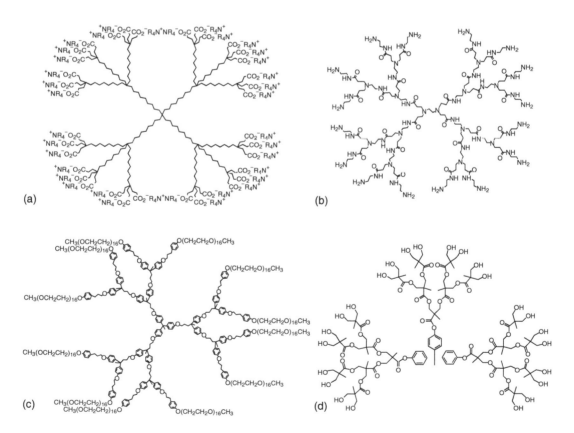

FIGURE 25.1 Structures of various dendrimers. (a) Unimolecular micelle, (b) poly(amidoamine) dendrimer, (c) polyaryl ether dendrimer, (d) polyester dendrimer based on 2,2-bis(hydroxymethyl)propionic acid.

micelles because all segments are covalently connected. Newkome et al. reported that a dendritic unimolecular micelle was shown to solubilize various hydrophobic guest molecules in the internal hydrophobic cavities.[10] However, the problem of these systems for drug delivery is their high hydrophobicity. So, water-soluble and biocompatible poly(ethylene glycol) (PEG) has been introduced to the drug delivery systems.

Initially, dendrimers were used for delivery systems mediated by non-covalent encapsulation of drugs as unimolecular micelles and dendritic boxes. Meijer and coworkers first reported the dendritic box based on the construction of a chiral shell of protected amino acids onto poly(propyleneimine) (PPI) dendrimers with 64 amine end groups (Figure 25.2).[11] They showed that the guest's and cavity's shapes determine the number of guests entrapped in the dendritic box; the architecture of the dendrimer is important, and shape-selective liberation of guests could be achieved by removing the shell in two steps.[12] Meijer and coworkers also developed a new host–guest system by synthesizing oligoethyleneoxy-modified PPI dendrimers, and they studied the interactions in aqueous media by UV/Vis titration.[13] PEGylated PPI dendrimers were also prepared as drug delivery systems by Pales and coworkers.[14] PEGylated PPI dendrimers were

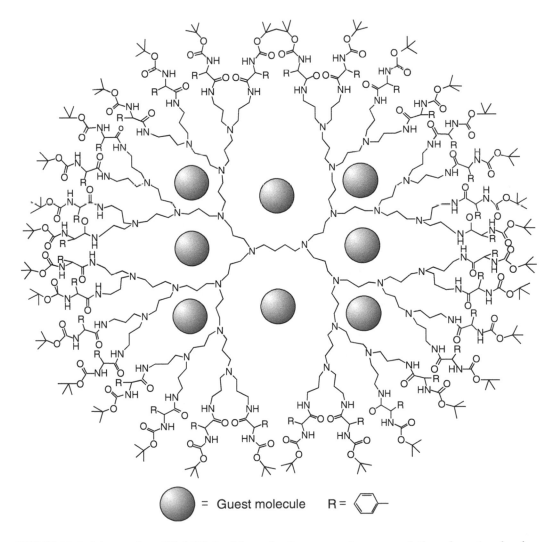

= Guest molecule   R = ⟨benzene⟩–

FIGURE 25.2 Scheme of modified PPI dendrimers for the non-covalent encapsulation of guest molecules.

shown to enhance water-solubility of encapsulated pyrene and betamethasones, securing their application as promising controlled release drug carriers. Encapsulated molecules were found to be solubilized in both regions of the interior of dendrimers and the PEG coat. The same group reported that the introduction of quaternary ammonium groups to PPI dendrimers not only enhanced the water solubility, but they also affected the protonation titration profile, leading to the release of the entrapped molecule within a narrower pH region.[15] Pyrene is released when the internal tertiary amines get protonated between pH 4 and 2, suggesting these dendrimers are potential candidates for pH-sensitive controlled-release drug delivery systems.

Poly(amidoamine) (PAMAM) dendrimer that was first developed by Tomalia et al.[16] also has been extensively studied for non-covalent encapsulation of drug molecules. PAMAM dendrimers are water-soluble and biocompatible, and they possess internal cavities hosting guest molecules. Twyman et al. modified the ester end groups of half-generation PAMAM dendrimers with Tris.[17] This highly water-soluble dendrimer could solubilize hydrophobic guest molecules at pH 7.0, but it lost its binding capability to guest molecules at pH 2.0, presumably because of the protonation of the internal tertiary amines.

To improve biocompatibility and water-solubility, PEG was grafted on the surface of PAMAM dendrimers. PEG-attached PAMAM dendrimers could encapsulate anti-cancer drugs, adriamycin, and methotrexate, and their ability to encapsulate these drugs increased with the increasing dendrimer generation and chain length of PEG grafts.[18] The dendrimers slowly released the drugs in low ionic strength solution; however, drugs were readily released in isotonic solutions. Another group explored the use of PEGylated PAMAM dendrimers for delivery of the anti-cancer drug 5-fluorouracil (5-FU) in vitro and in albino rats.[19] PEGylation of dendrimers was found to increase their drug-loading capacity and reduce the drug release rate and hemolytic toxicity, suggesting the use of PEGylated PAMAM dendrimers as prolonged delivery systems.

Haba et al. prepared PEG-modified PAMAM dendrimers having methacryloyl groups on the chain ends of the dendrimer for drug delivery.[20] After linking of the methacryloyl groups, they found that the dendrimer decreased its affinity against rose Bengal because of its hiding the dendrimer interior of the peripheral network and that the dendrimer having the peripheral network retained Rose Bengal dye molecules tightly in their interior. This result indicated that the linking of polymerizable groups attached to the periphery of the dendrimer could be an efficient approach for retaining guest molecules in their interior, and that they could be used as photo-sensitizers for photodynamic therapy (PDT) and as boron10-containing compounds for neutron capture therapy because these pharmaceutical molecules could exhibit their therapeutic effects, even if they are not released from their carriers.

Frechet and coworkers synthesized novel dendritic unimolecular micelles with a hydrophobic polyether dendrimer core surrounded by hydrophilic PEGs.[21] For the G-3 micelle, the loading of indomethacin was found to be of 11%, a value corresponding to approximately nine to ten drug molecules per micelle, and preliminary in vitro release tests showed that sustained release characteristics were achieved. Recently, linear dendritic block copolymers comprising PEG and either a polylysine or polyester dendrons were designed, and highly acid-sensitive cyclic acetals were attached to the dendrimer periphery for pH-responsive micelle systems.[22] At pH 7.4, the fluorescence of micelle-encapsulated Nile Red was constant, indicating it was retained in the micelle, whereas at pH 5, the fluorescence decreased, consistent with its release into aqueous solution. The rate of release was strongly correlated with the rate of acetal hydrolysis. These results suggest that the loading and controlled-release of drugs are dependent on the chemical structure of the dendrimer.

A novel triblock copolymer composed of citric acid dendritic blocks and PEG core was synthesized and applied as a drug delivery system.[23] The dendrimers were found to be able to trap various hydrophobic drugs in their suitable sites and to release them in a controlled manner in in vitro experiments.

25.2.3 COVALENT CONJUGATION OF DRUGS

Another approach for dendritic drug carriers is based on covalent conjugation of drug molecules to the dendrimer periphery using the multivalency of dendrimers. Although many drug molecules were conjugated to the dendrimers through non-degradable linkages, introducing degradable bonds between the drug and the dendrimer can control the release of drug molecules. Figure 25.3 shows the scheme of drug-conjugated dendrimers.

Dendrimers were first used for immunoconjugates. Chemical modification of the antibody with drug molecules often diminished its biological activity, but approaches to link the antibody to another macromolecule including the dendrimer were reported to increase drug loading while retaining the activity. Roberts et al. conjugated antibody to porphyrin to chelate radiolabeled copper ions using PAMAM dendrimer as a linker.[24] They found that 100% of the porphyrin bound to the heavy chain of the antibody, and dendrimer conjugates contributed 90% of the immunoactivity of unmodified antibody and showed a high radiolabeling level in contrast to the conjugates obtained through the random-coupling method. PAMAM dendrimers were also modified chemically by reaction with 1,4,7,10-tetraazacyclododecanetetraacetic acid (DOTA) and diethylenetriaminepentaacetic acid (DTPA) type bifunctional metal chelators and used as a linker to monoclonal antibody (moAb) 2E4.[25] The dendrimer–antibody conjugates were found to retain biological activity and to be easily radiolabeled, suggesting use of these dendrimer conjugates for moAb-based radiotherapy or imaging. Recently, Patri et al. conjugated anti-PSMA (prostate-specific membrane antigen) to PAMAM dendrimer labeled with FITC and capped with acetic anhydride to minimize the non-specific interaction with the cell surfaces for targeting prostate cancer.[26] This dendrimer conjugate showed great potential to target PSMA-positive LNCaP cell line with minimal loss of immunoreactivity. Meanwhile, PAMAM dendrimers were utilized to investigate the potential for cancer chemotherapy by conjugating with anti-cancer drugs. Cisplatin, a potent anti-cancer drug, was conjugated to PAMAM dendrimer, giving a dendrimer–platinate that was highly water-soluble and released platinum slowly in vitro.[27,28] This dendrimer conjugate showed more potent anti-tumor activity than cisplatin and selective accumulation in solid tumor tissue and was also less toxic than cisplatin. Zhuo et al. synthesized a water-soluble acetylated PAMAM dendrimer series with a core of 1,4,7,10-tetraazacyclododecane and linked it to 1-bromoacetyl-5-fluorouracil to form dendrimer–5-FU conjugates.[29] Hydrolysis of the conjugates in a phosphate buffer solution could release free 5-FU in a generation-dependent manner, showing the dendrimer conjugate seems to be a promising carrier for the controlled release of anti-tumor drugs. Wang et al. used PAMAM dendrimers for doxorubicin (DOX) delivery.[30] They conjugated semitelechelic poly[N-(2-hydroxypropyl)-methacrylamide] macromolecules (ST-PHPMA,) to the periphery of PAMAM dendrimer and introduced DOX to the dendrimer conjugate, star-like HPMA copolymer. It showed lower cytotoxicity than other brush-like polymer-DOX in A2780 human ovarian carcinoma cell line and a slow rate of in vitro DOX release, demonstrating that the differences of DOX internalization because of the structure of the polymers would result in different

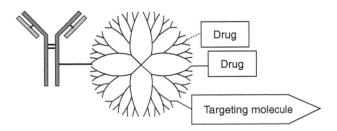

FIGURE 25.3 Schematic view of drug–dendrimer conjugate. A dotted line, degradable linkage; a black line, non-degradable linkage.

intracellular DOX concentrations with cytotoxicity changes. 5-Aminosalicylic acid (5-ASA) was also conjugated to PAMAM dendrimers using two different spacers containing azo-bond, p-amino-benzoic acid (PABA), and p-aminohippuric acid (PAH) for colonic delivery.[31] The dendrimer conjugates incubated in rat fecal contents showed colon-specific and prolonged release of 5-ASA through reductive cleavage of azo-bond bycolonic bacteria, suggesting the potential for colon-specific drug delivery carriers. In addition, Quintana et al. attached folic acid to acetamide-, hydroxyl-, or carboxyl-capped PAMAM dendrimers conjugated with methotrexate (MTX) to target tumor cells through the folate receptor.[32] These dendrimer conjugates showed enhanced cellular uptake, and this targeted delivery improved the cytotoxic response of the cells to MTX 100-fold over the free drug. Recently, Tansey et al. also conjugated folic acid and indocyanine green, a model diagnostic agent to biodegradable poly(L-glutamic acid) (PG) chain-terminated PAMAM dendrimer for tumor targeting.[33] The PG–dendrimer conjugates showed a more prolonged degradation pattern than linear PG chain in the presence of lysosomal enzyme. After conjugation of PEG, the dendrimer conjugates were found to show reduced non-specific interaction and to bind selectively to tumor cells expressing folate receptors.

Folic acids have already been conjugated to acid-labile hydrazide-terminated polyaryl ether dendrimers to target tumor cells by Frechet and coworkers.[34] The conjugates are soluble in aqueous medium above pH 7.4. The anti-tumor drug methotrexate was also attached to the dendrimers. The same group also prepared polyaryl ether dendrimer–PEG conjugates as model drug carriers.[35] Cholesterol and two amino acid derivatives were attached to the dendrimer via carbonate, ester, and carbamate linkages.

Greenwald and coworkers synthesized novel PEG–poly(aspartic) acid dendron conjugates and attached an anti-tumor agent, cytosine arabinoside (ara-C), to the dendron conjugates.[36] These conjugates showed greater tumor inhibition than free drugs and increased aqueous solubility, resulting in high loading and controlled release of drugs. Another study of conjugation of ara-C to PEG-dendrons composed of aminoadipic acid was accomplished by Schiavon et al.[37] The conjugates also presented prolonged blood residence time and higher loading of drugs.

Polyester dendrimers based on the monomer unit 2,2-bis(hydroxymethyl)propanoic acid were also developed as drug carriers both in vitro and in vivo.[38] These dendrimers were found to be both water soluble and non-toxic. A 3-arm PEG–polyester dendrimer hybrid that conjugated with DOX via a hydrazone linkage susceptible to acidic hydrolysis showed pH-dependent drug release in vitro and a degree of anti-cancer activity in cancer cell lines. In addition, the little accumulation of the dendrimer conjugate found in any organ examined, including the liver, heart, and lungs, suggested that it is highly water soluble and biocompatible.

Recently, a novel approach to developing covalently conjugated dendritic drug carriers was presented by two different groups.[39,40] These cascade-release or self-immolative conjugates were able to simultaneously release all drug molecules connected to the dendritic termini by a single trigger, resulting in whole dendrimer collapse. Moreover, Shabat and coworkers demonstrated self-immolative dendrimers as a platform for multi-prodrugs, activated by a single enzymatic clea-vage.[41] The bioactivation of the heterodendritic prodrugs of DOX and campothecin (CPT) was evaluated using the Molt-3 leukemia cell line, and the conjugates showed a mild to significant increase in toxicity in comparison with the classical monomeric prodrugs.

25.2.4 Neutron Capture Therapy

25.2.4.1 Boron Neutron Capture Therapy

Boron neutron capture therapy (BNCT) is a cancer-therapeutic application based on a nuclear capture reaction of a lethal $^{10}B(n,\alpha)^7Li^{3+}$ reaction.[42] If ^{10}B can be delivered at a concentration of at least 10^9 atoms per cell to tumor tissue, subsequent irradiation with thermal or epithermal neutrons produces highly energetic α particles and Li^{3+} ions that damage the tumor cells in nuclear fission reaction. To deliver high levels of boron in tumor tissue, it is proposed to use dendrimers as boron carriers for their well-defined molecular structure and multivalency.

Initially, Barth et al. utilized PAMAM dendrimers in order to boronate moAb IB16-6 that targeted murine B16 melanoma using isocyanato polyhedral borane ($Na(CH_3)_3NB_{10}H_8NCO$).[43] The boronated dendrimers were found to localize in the liver and spleen in in vivo distribution study, leaving a question of if the properties of boronated dendrimers can be modified to reduce their hepatic and splenic localization. Calssonn and coworkers attached epithermal growth factor (EGF) to boronated PAMAM dendrimers in order to target EGF receptor (EGFR) that is over expressed in brain tumors.[44] The boronated dendrimers-EGF were bound to malignant glioma cell membrane and endocytosed, resulting in the accumulation of boron in lysosome. Although these boronated dendrimers were localized in the liver and accumulated a little in the tumor after intravenous injection into rats having intracerebral C6 glioma transfected with EGFR gene, direct intratumoral injection could selectively deliver them to EGFR-positive gliomas.[45] Recently, use of another moAb that is directed against EGFR and EGFRvII, cetuximab (IMC-C225), was investigated for boronated PAMAM dendrimer-mediated BNCT.[46] There were ~1100 boron atoms per molecule of cetuximab with only a slight reduction of K_a. Localization study of these conjugates in F98 rats bearing intracerebral implants of either EGFR-transfected or wild-type gliomas by intratumoral injection showed specific molecular targeting of EGFR and accumulation of boron.

To reduce the hepatic uptake of boronated dendrimers, PEG was introduced to the dendrimers.[47] Moreover, folic acid was also conjugated to the dendrimers to target the folate receptors in cancer cells. Among all the prepared combinations, boronated dendrimers with 1–1.5 PEG2000 units exhibited the lowest hepatic uptake in C57BL/6 mice. Interestingly, dendrimers containing ~13 decaborate clusters, ~1 PEG2000 unit, and ~1 PEG800 unit with folic acid attached to the distal end showed receptor-dependent uptake in contrast to dendrimers containing ~15 decaborate clusters and ~1 PEG2000 unit with folic acid attached to the distal end in in vitro studies using folate receptor (+) KB cells. Biodistribution studies with former dendrimer in C57BL/6 mice bearing folate receptor (+) murine 24JK-FBP sarcomas resulted in selective tumor uptake (6.0% ID/g tumor), and they also showed an increase in uptake of the dendrimers by the liver and kidney, warranting further effort in the optimization of biodistribution profiles of these boronated dendrimers.

Polylysine dendrons also were examined for BNCT. Borane moieties were attached to the periphery of the dendrons and peptide spacers linked the dendrons to targeting molecules such as antibodies, a fluorescent moiety, and a PEG chain (Figure 25.4).[48] This offered a promising perspective for application in BNCT.

25.2.4.2 Gadolinium Neutron Capture Therapy

Although the vast majority of NCT research has involved various aspects of boron chemistry for BNCT, [157]Gd has also been suggested from theoretical and experimental studies with limited success.[49–51] To some extent, Gd complexes have been clinically employed as an MRI contrasting agent.[52] The potential to trace the in vivo pharmacokinetics of a Gd-NCT agent directly by MRI would be useful for planning the neutron irradiation and to predict the therapeutic effect.

Kobayashi et al. synthesized avidin-G6-(1B4M-Gd)$_{254}$(Av-G6Gd) from PAMAM dendrimer G6, biotin, avidin, and 2-(p-isothiocyanatobenzyl)-6-methyl-diethylenetriaminepentaacetic acid (1B4M).[53] A sufficient amount (162 ppm) of Av-G6Gd was accumulated and internalized into the SHIN3 cells both in vitro and in vivo to kill the cell using [157/155]Gd with external irradiation with an appropriate neutron beam while monitoring with MRI, suggesting that Av-G6Gd may be a promising agent for Gd neutron capture therapy of peritoneal carcinomatosis, and it may have the potential to permit monitoring of its pharmacokinetic progress with MRI.

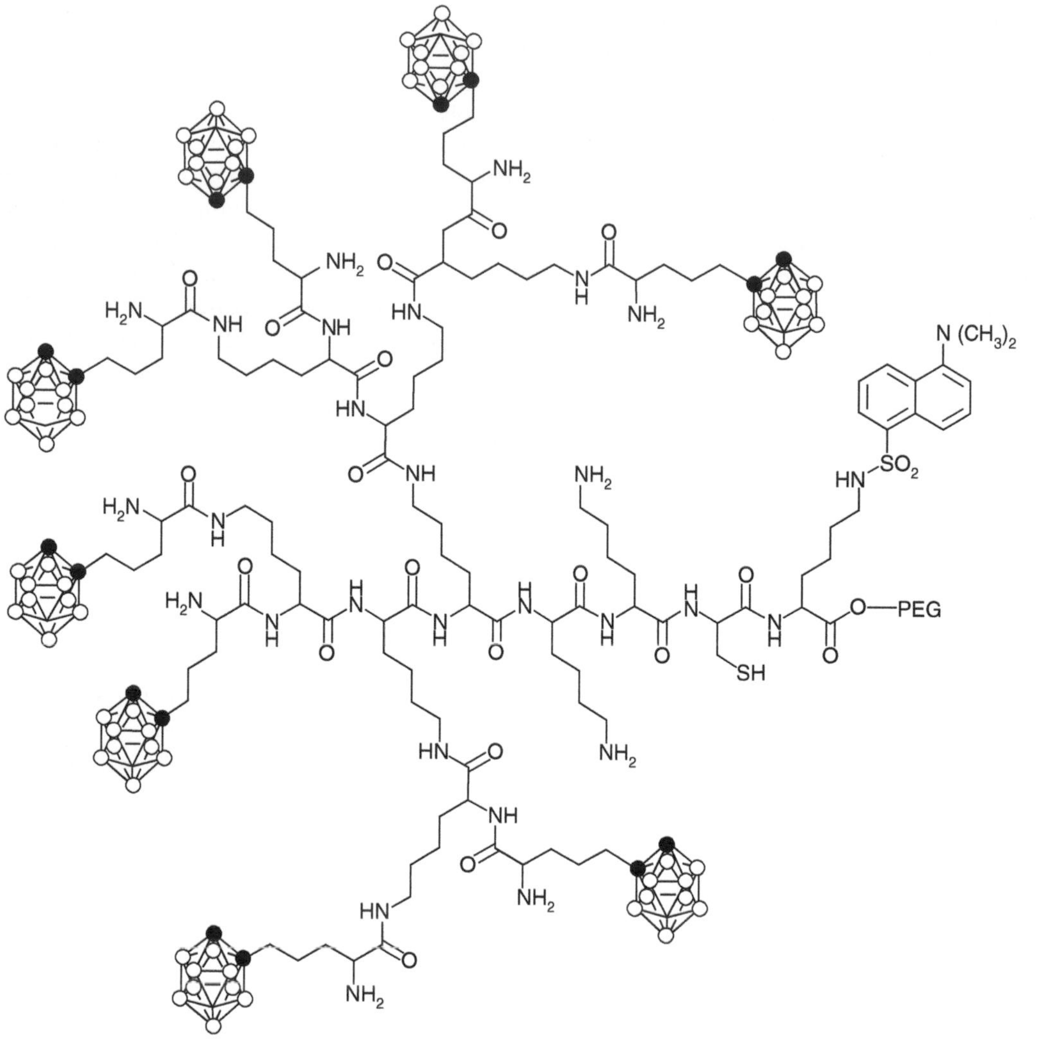

FIGURE 25.4 Polylysine dendrimer conjugate used in boron neutron capture therapy (BNCT).

25.2.5 PHOTODYNAMIC THERAPY

PDT is a technique for inducing tissue damage with light, following administration of a light-activated photosensitizing drug that can be selectively concentrated in malignant or diseased tissue.[54] Activation of the photosensitizer results in the generation of reactive oxygen species (ROS), primarily singlet oxygen that oxidizes intracellular substrates such as lipids and amino acid residues, leading to cell death. The most widely studied application of PDT to date is in cancer therapy. However, the use of conventional photosensitizer systems bring skin phototoxicity, damage to normal tissue because of poor tumor selectivity, poor water solubility, and difficulties in treating solid tumors. In order to improve the disadvantages, dendrimers would be manufactured as promising photosensitizer carriers through modification of peripheral functionality.

Battah et al. synthesized dendrimers with 5-aminolevulinic acid (ALA) moieties conjugated to the periphery by ester bonds that are cleavable by esterase inside the cells for PDT.[55] ALA is a natural precursor of an effective photosensitizer protophorphyrin IX (PpIX), and its administration can increased the cellular concentration of PpIX.[56] These ALA dendrimers showed increased production of PpIX and higher cytotoxicity after irradiation relative to free ALA in tumorigenic keratinocyte PAM 212 cell line.

Another approach of designing dendritic photosensitizers for PDT involved aryl ether dendrimer porphyrins (DP) with either 32 quaternary ammonium groups or carboxylic acid moieties on their periphery (Figure 25.5).[57] Hydrophilic DPs were found to be localized in membrane-limited organelles but hydrophobic PIX diffused through the cytoplasm, except the nucleus. DPs showed remarkably higher $^1O_2$-induced cytotoxicity against Lewis lung carcinoma (LLC) cells and far lower dark toxicity than PIX, demonstrating their highly selective photosensitizing effect in combination with a reduced systemic toxicity.

Zhang et al. entrapped DP with 32 amine groups on the periphery, $[NH_2CH_2CH_2NHCO]_{32}DPZn$ with polyion micells composed of PEG and poly(aspartic) acid for effective PDT.[58] Compared to $[NH_2CH_2CH_2NHCO]_{32}DPZn$, a relatively low cellular uptake of $[NH_2CH_2CH_2NHCO]_{32}DPZn$

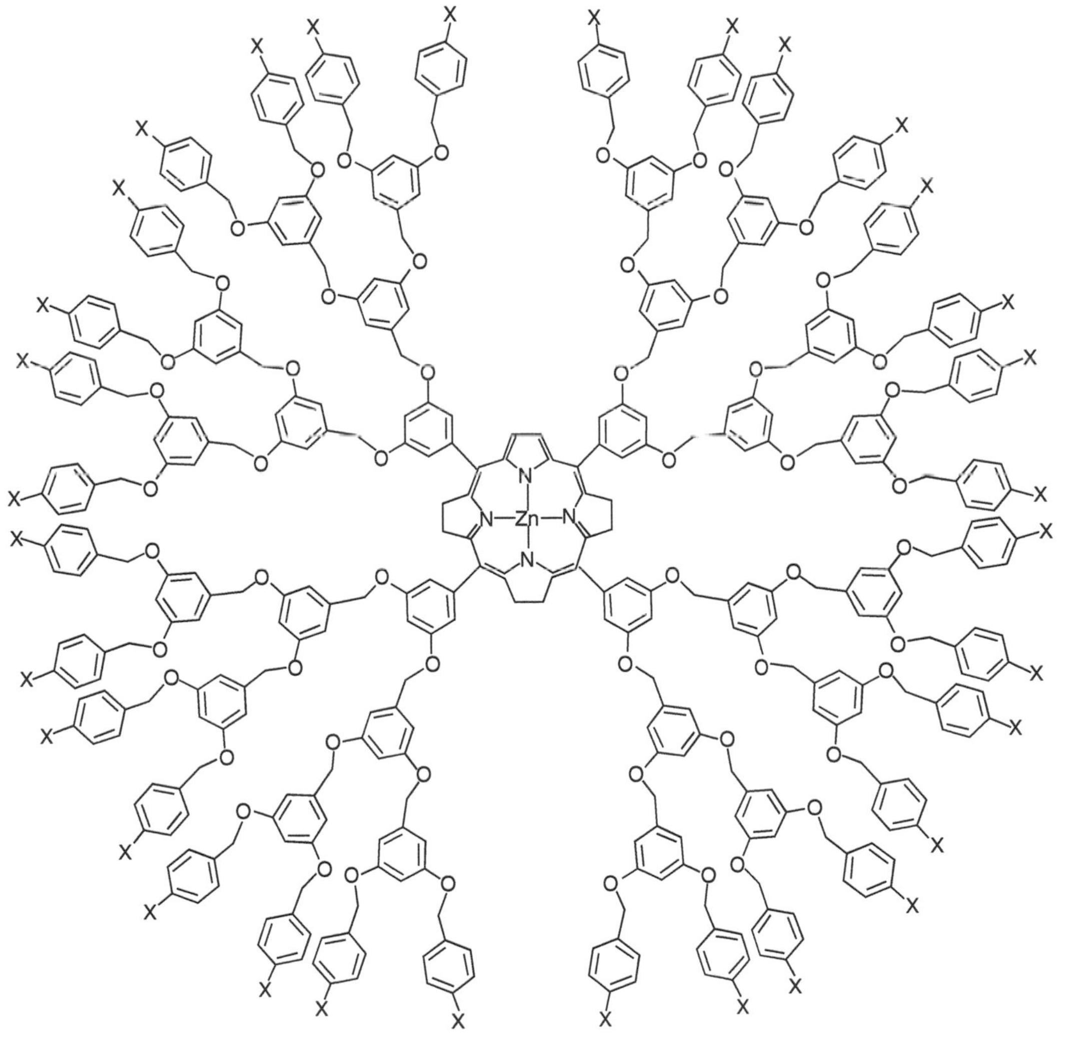

$X = CONH(CH_2)_2N^+Me_3Cl^-$ or COO^-H^+

FIGURE 25.5 Porphyrin-based dendrimer for photodynamic therapy.

incorporated in the PIC micelle was observed, yet the latter exhibited enhanced photodynamic efficacy on the LLC cell line. Also, this system was found to reduce the dark toxicity of the cationic dendrimer porphyrin, probably as a result of the biocompatible PEG shell of the micelles.

Another approach for deeper tissue penetration was tried using two-photon absorption (TPA) with near-infrared lasers. Porphyrin sensitizers have low TPA cross-sections, limiting their usefulness in TPA applications. Recently, Dichtel et al. attached donor chromophores capable of efficient TPA to a central porphyrin acceptor for an enhanced TPA cross-section.[59] This dendrimer was found to enhance the effective TPA cross-section of a porphyrin acceptor, allowing more efficient generation of singlet oxygen using 780 nm light. This result showed a potential use of PDT to tumors below the skin surface.

25.3 GENE DELIVERY

25.3.1 INTRODUCTION

Cationic dendrimers have been reported to deliver genetic materials into cells by forming complexes with negatively charged genetic materials through electrostatic interaction. The complexes bound to cell membranes were known to be internalized into cells by endocytosis (Figure 25.6). It has been postulated that the branched cationic polymers such as polyethyleneimine or PAMAM dendrimer, having a high buffer capacity lead to polymeric swelling in acidic endosomes, disrupting the membrane barrier and resulting in release of the complexes.[60] However, the uptake and the transfection mechanism of cationic polymers are not yet fully understood. Recently, two different mechanisms, caveolae and clathrin-mediated endocytosis, were also reported to involve the uptake mechanism of the polyplexes.[61]

Several dendrimers have been synthesized and used for gene delivery carriers from a decade ago. The history is not long, but the intense research on developing dendritic gene delivery carriers has been carried out because of many characteristic advantages of dendrimers such as monodisperse size of dendrimers and controllable multivalent end-group functionalities.

25.3.2 PAMAM DENDRIMERS

It is only for a decade that the potential of dendrimers as gene delivery carriers has been investigated. The first applied dendrimer for gene delivery was PAMAM dendrimer that was reported by Haensler and Szoka.[62] They showed that the dendrimers mediate high efficiency transfection of a variety of suspension- and adherent-cultured mammalian cells, and dendrimer-mediated

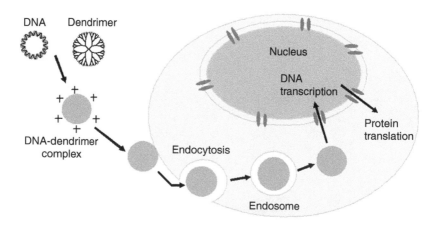

FIGURE 25.6 Scheme of dendrimer-mediated transfection.

transfection is a function both of the dendrimer/DNA ratio and the diameter of the dendrimer. When GALA, a water soluble, membrane-destabilizing peptide, is covalently attached to the dendrimer via a disulfide linkage, transfection efficiency is increased by 2–3 orders of magnitude. Also, they concluded that the high transfection efficiency of the dendrimers may be caused by the low pK_a of the amines in the polymer that permit the dendrimer to buffer the pH change in the endosomal compartment. Kukowska-Latallo et al. also reported a high degree of transfection of PAMAM dendrimers with minimal cytotoxicity in various cell lines.[63] The addition of DEAE-dextran increased the transfection efficiency, suggesting that dextran alters the nature of dendrimer/DNA complexes. The most efficient transfection was achieved with a lysosomotropic agent such as chloroquine, indicating that endosomal localization is a rate-determining step in transfection. Mazda and coworkers investigated the transfection of the Epstein–Barr virus (EBV)-based plasmid vector/PAMAM dendrimer complexes to improve the efficiency in hepatocellular carcinoma cells (HCC) and cholangiocarcinoma cells.[64,65] They found highly efficient suicide gene transfer into various HCC cells by EBV-based plasmid vectors/PAMAM dendrimer complexes in vitro, suggesting a possible application of this non-viral vector system to gene therapy of HCC, and they reported that EBV-based plasmid vectors/PAMAM dendrimer complexes equipped with the carcinoembryonic antigen (CEA) promoter showed high transfection in cholangiocarcinoma cells, providing an efficient non-viral method for the targeted gene therapy of CEA-producing malignancies.

Thermally degraded PAMAM dendrimers with heterodisperse structure (fractured PAMAM) showed a dramatically enhanced transfection efficiency in in vitro experiments.[66] Transfection activity was found to be related both to the initial size of the dendrimer and to its degree of degradation. The increased transfection after the heating process was thought to be principally due to the increase in flexibility that enabled the fractured dendrimer to be compact when complexed with DNA and swell when released from DNA. The dendrimers are presently sold commercially by Qiagen as SuperFect™.

Bielinska et al. investigated the ability of PAMAM dendrimers to function as an effective delivery system for antisense oligonucleotides (ODNs) and antisense expression plasmids for the targeted modulation of gene expression.[67] Transfections of ODNs or antisense cDNA plasmids using PAMAM dendrimers resulted in a specific and dose-dependent inhibition of luciferase expression. Binding of the phosphodiester oligonucleotides to dendrimers also extended their intracellular survival. Delong et al. also evaluated PAMAM dendrimer (generation 3, Mr 6909) as a potential delivery vehicle for oligonucleotides.[68] Use of dendrimers resulted in a 50-fold enhancement in cell uptake of oligonucleotide as determined by flow cytometry, enhanced cytosolic, and nuclear availability as shown by confocal microscopy. Interestingly, fluorescent dye, Oregon green 488-conjugated dendrimer, was proved to be a much better delivery agent for antisense compounds than unmodified dendrimer.[69] This result suggested that coupling of relatively hydrophobic small molecules to PAMAM dendrimers may provide a useful means of enhancing their capabilities as delivery agents for nucleic acids.

The efficiency of cell transfection by PAMAM dendrimers was shown to be greatly increased when anionic oligomers (oligonucleotide or dextran sulfate) were mixed with plasmids before addition of the dendrimer.[70] This enhancement of gene transfer by anionic oligomers depended on their size, structure, and charge. This result may be due to modification of the interaction between plasmids and dendrimers induced by the oligonucleotide that effect leads to a less condensed structure of the polyplexes, thereby facilitating their penetration into cells.

The transfection efficiency of PAMAM dendrimers was also increased by addition of substituted β-cyclodextrins (β-CD).[71] In vitro CAT expression was increased approximately 200-fold when dendrimer/DNA/β-CD formulations were applied on the surface of collagen membranes, indicating that β-CDs affected the physicochemical properties of dendrimer/DNA complexes. Besides, PAMAM dendrimers conjugated with α-cyclodextrins (α-CDs) were investigated for their potential as a gene delivery system.[72] The gene transfer activity of α-CD dendrimer conjugate

(G3) was higher than that of dendrimers (G2, G3, and G4) and of the α-CD conjugate (G2), and the cytotoxicity of the α-CD conjugate (G3) was lower than that of the α-CD conjugate (G4). This greatest gene transfer activity could be attributed to the synergistic interplay of the proton sponge effect of the dendrimer (G3) and the tentative membrane-disrupting effect of α-CD on endosomal membranes. Recently, the transfection efficiency of α-CD dendrimer conjugates was improved by surface modification of dendrimers with mannose.[73] The mannosylated conjugates were found to have a much higher gene transfer activity than unmodified α-CD dendrimer in various cell lines, which was independent of the expression of cell surface mannose receptors. Also, the conjugates provided higher gene transfer activity than dendrimer and α-CD dendrimer conjugate in kidney 12 h after intravenous injection in mice.

Arginine conjugation to the periphery of PAMAM dendrimer was reported to greatly increase the transfection efficiency.[74] Arginine residues of HIV-Tat peptide are thought to perform an important role in the intracellular translocation of molecular delivery.[75] The arginine-conjugated PAMAM dendrimer (PAMAM-Arg) showed enhanced gene expression in HepG2 cells, Neuro 2A cells, and primary rat vascular smooth muscle cells in comparison with native PAMAM dendrimer. It was presumed that the increased gene expression might be attributed to the difference in either the cell-penetrating activity during uptake, the nuclear localization efficiency after entry into the cytosol of arginine residues, or the synchronous function of both effects. Recently, Kang et al. conjugated Tat peptide to PAMAM dendrimer G5 for cellular delivery of antisense and siRNA oligonucleotides designed to inhibit MDR1 gene expression in vitro.[76] However, the dendrimer conjugate/oligonucleotide complexes weakly inhibited the gene expression, showing that conjugation of the dendrimer with the Tat cell-penetrating peptide failed to further enhance the effectiveness of the dendrimer.

Lee et al. synthesized internally quaternized PAMAM-OH dendrimer (QPAMAM-OH) by methylation of internal tertiary amines and investigated the gene delivery potency.[77] Although the dendrimer did not show high transfection efficiency in in vitro experiments, it could form a complex with plasmid DNA unlike with unmodified PAMAM-OH dendrimer, and its cytotoxicty was revealed to be very low in comparison with amine-terminated PAMAM dendrimer.

PAMAM dendrimers were also applied for ex vivo or in vivo gene delivery. Hudde et al. investigated the efficiency of activated polyamidoamine dendrimers (SuperFect™) to transfect rabbit and human corneas in ex vivo culture.[78] Whole-thickness rabbit or human corneas were transfected ex vivo with complexes of dendrimers and plasmids containing lacZ or TNFR-Ig genes. Six to ten percent of the corneal endothelial cells expressed the marker gene, and corneas transfected with TNFR-Ig plasmid were found to express TNFR-Ig protein. The fifth generation of intact PAMAM dendrimers was also investigated for its ability to enhance gene transfer and expression in a clinically relevant murine vascularized heart transplantation model.[79] The complexes of the plasmid pMP6A-beta-gal, encoding beta-galactosidase (beta-Gal), and dendrimer were perfused via the coronary arteries during donor graft harvesting. The grafts infused with pMP6A-beta-gal/dendrimer complexes showed beta-Gal expression in myocytes from 7 to 14 days. The complexes containing 20 µg of DNA and 260 µg of dendrimer (1:20 charge ratio) in a total volume of 200 µL resulted in the highest gene expression in the grafts. Prolonged incubation (cold ischemic time) up to 2 h and pretreatment with serotonin were found to further enhance gene expression.

Kukowska-Latallo et al. evaluated in vivo gene transfer and expression by intravascular and endobronchial routes using DNA complexed with G9 PAMAM dendrimer.[80] After intravenous administration of the dendrimer/pCF1CAT plasmid complexes, CAT expression was never observed in organs other than the lung. Repeated intravascular doses, administered four times at 4-day intervals, maintained expression at 15–25% of peak concentrations achieved after the initial dose. However, in contrast with vascular delivery, intranasal delivery of the complex was found to give lower levels of CAT expression than that was observed with naked plasmid DNA. In situ localization of CAT enzymatic activity suggested that vascular administration seemed to achieve expression in the lung parenchyma, mainly within the alveoli, whereas endobronchial administration primarily targeted bronchial epithelium, showing that intravenously administered G9 dendrimer was an effective vector for pulmonary gene transfer and that transgene expression could be prolonged by repeated administration. PAMAM G4 dendrimer was also investigated for in vivo tumor targeting and transfection.[81] Biotinylated and ^{111}In labeled 20-mer antisense multi-amino-linked oligonucleotide–avidin conjugate (^{111}In-oligo–Av) and the complexes of the oligonucleotide with PAMAM G4 (^{111}In-oligo/G4), biotinylated PAMAM G4–avidin (^{111}In-oligo/G4–Av) were found to significantly enhance the tumor delivery of ^{111}In-oligo compared with delivery without carrier after i.p. injection.

25.3.3 POLYPROPYLENIMINE DENDRIMERS

Polypropylenimine (PPI) dendrimer is also a commercially available cationic dendrimer, but much less interest has been taken in its application to gene delivery systems compared to PAMAM dendrimer. Since Buhleier et al. synthesized low molecular weight PPI dendrimer in 1978,[1] synthesis of the dendrimer has been improved through the Michael addition of amines to acrylonitrile, followed by the reduction of the nitriles to primary amines.[82] Diaminobutane (DAB) and diaminoethane cores are most commonly used for PPI dendrimer synthesis.

The potential of PPI dendrimers as gene delivery carriers was first compared with other cationic polymers including branched PEI (25 kDa), linear PEI (22 kDa), and PAMAM dendrimer and others in Cos-7 cell line.[83] Although PPI dendrimer G5 (Astramol™-64) showed very low transfection efficiency, MTT assay demonstrated that its cytotoxicity was relatively low compared with other polymers in treating polyplexes or free polycations. However, PPI dendrimers (generations 1–5) with DAB core were re-evaluated as efficient gene delivery systems by Uchegbu and coworkers.[84] They reported that DNA binding increased with dendrimer generation through molecular modeling, and experimental data and cytotoxicity followed the trend DAB 64 (terminal amine number) > DAB 32 > DAB 16 > DOTAP > DAB 4 > DAB 8, whereas transfection efficacy followed the trend DAB 8 = DOTAP = DAB 16 > DAB 4 > DAB 32 = DAB 64. They concluded that the generation 2 polypropylenimine dendrimer (DAB 8) with a low level of cytoxicity gave optimum gene transfer activity.

PPI dendrimers were also chemically modified for improved physiological properties. Park and coworkers investigated ternary complex of PPI dendrimer linked with DABs to its periphery (PPI–DAB), DNA, and cucurbituril (CB) for a non-covalent strategy in developing a gene delivery carrier.[85] CB is a large-cage compound composed of glycoluril units interconnected with methylene bridges, and it is able to form stable pseudorotaxanes with strings derived from DAB moeities of PPI–DAB through multiple non-covalent interactions.[86,87] The DNA binding capacities of PPI, PPI–DAB, and PPI–DAB/CB decreased with increasing generation and increased in the order, PPI > PPI–DAB ≥ PPI–DAB/CB. Fluorescence experiment using DAB-tethered acridine indicated that there was no change in the CB beads threaded on PPI–DAB after the addition of DNA to the PPI–DAB/CB binary complex. Importantly, in vitro transfection results showed that CB had no negative influence on the transfection efficiency of polymer/DNA binary complex, and the ternary complex was a highly active form of gene delivery vector, suggesting that this promising system can be applied to other non-covalent molecular recognitions, and it would be possible to create a really functional gene delivery carrier if the functionalized-CB were synthesized. Recently, another approach to developing gene delivery carriers with PPI dendrimers involved methyl quaternary ammonium derivatives of PPI dendrimers with varying generation.[88] Quaternization of DAB-cored PPI dendrimer (DAB 8) with eight surface amines (Q 8) proved to be critical in enhancing DNA binding and resulted in improved colloidal stability and vector tolerability on i.v. injection of the polyplex.

Quaternization was also found to improve in vitro cell biocompatiblity of DAB 16 and DAB 32 polyplexes. Interestingly, the intravenous administration of DAB 16 and Q8 polyplexes resulted in liver-targeted gene expression as opposed to the lung-targeted gene expression obtained with the control polymer, Exgen 500. This intrinsic targeting ability without the need for targeting ligands may be exploited for the targeted delivery of genes in the treatment of liver diseases. Also, Dufes et al. synthesized a vector system based on PPI dendrimer and investigated its transfection in tumors after i.v. injection.[89] The systemic tumor necrosis factor alpha (TNFalpha) gene delivery by the dendrimer was found to be efficient and non-toxic in the treatment of various established carcinomas. Interestingly, these dendrimers and other common polymeric transfection agents also exhibited plasmid-independent anti-tumor activity, ranging from pronounced growth retardation to complete tumor regression, indicating that the combination of the pharmacological active dendrimer and gene would be more potent.

25.3.4 OTHER DENDRIMERS

Dendritic poly(L-lysine)s (DPKs) of several generations were synthesized and investigated as gene transfection reagents.[90] The DPKs of the fifth and sixth generation showed efficient and serum-tolerant gene transfection ability into several cell lines without significant cytotoxicity. Several uptake inhibitor experiments suggested that the uptake of the DNA complex of DPK is mediated by the endocytosis pathway, especially the macropinocytotic process. In addition, terminal amino acids of DPK G6 were replaced by arginines and histidines to investigate the effect of substituting terminal cationic groups on the gene delivery.[91] Arginine-conjugated polylysine dendrimer (KGR6) showed higher transfection efficiency than unmodified KG6. In contrast, histidine-conjugated polylysine dendrimer (KGH6) showed high transfection efficiency only after polyplex formation under acidic conditions because of imdazole groups with a low pK_a value of histidine residues. pH-dependent complex formation and transfection of KGH6 showed the potential for a functional gene delivery system.

Choi et al. synthesized PEG–dendrimer conjugates PEG–poly(L-lysine) dendrimers and poly (L-lysine) dendrimer–PEG–poly(L-lysine) dendrimers (PLLD–PEG–PLLD) and characterized their self-assembly with plasmid DNA.[92,93] PEG was used as a core and poly(L-lysine) dendrimers were extended via a divergent peptide synthesis method. The dendrimers could form nanosized and water-soluble complexes with plasmid DNA in a generation-dependent manner and showed low cytotoxicity, but their transfection was not investigated. Thereafter, Kim ct al. prepared triblock copolymers and PAMAM–PEG–PAMAM dendrimers for gene delivery (Figure 25.7).[94] The copolymers showed the generation-dependent formation of nanosized complexes with plasmid DNA as PLLD–PEG–PLLD and high water-solubility because of their PEG cores. The G5 copolymer was found to possess low cytotoxicity and a much enhanced transfection ability in comparison to PLLD–PEG–PLLD in HepG2 cells and 293 cells.

Joester et al. developed amphiphilic dendrimers with rigid tolane cores for gene delivery (Figure 25.8).[95] The dendrimers showed hydrophobicity, nominal charge-dependent DNA binding affinity, low cytotocity, and very high transfection efficiency superior to commercial transfection agents. However, the significant reduction of transfection efficiency with serum is thought to need research into the interaction of the dendrimer with anionic serum proteins.

A series of water-soluble dendrimers with a phosphoramidothioate backbone (P-dendrimers) were prepared for gene delivery.[96] The polymers showed rather moderate cytotoxicity toward HeLa, HEK 293, and HUVEC cells, and they efficiently delivered oligodeoxyribonucleotide into HeLa cells in serum-containing medium.

Hussain et al. developed another gene delivery system based on anionic dendrimer by directly conjugating ODNs to pentaerythritol-based phosphoroamidite dendrimer.[97] The cellular uptake of the ODN–dendrimer conjugates was up to 4-fold greater than for naked ODN in cancer cells. The ODN–dendrimer conjugates were found to effectively inhibit cancer cell growth, showing a marked knockdown in EGFR protein expression.

Recently, Nishiyama et al. reported the application of photochemical internalization (PCI)-mediated gene delivery in vivo using dendrimeric photosentisizer.[98] PCI is that the cytoplasmic

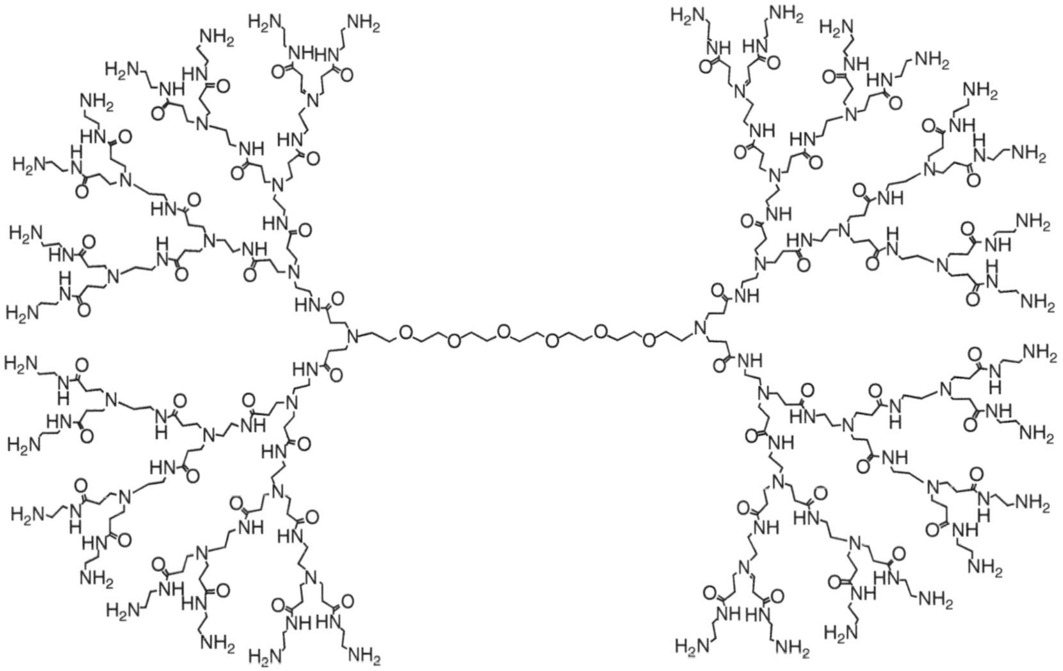

FIGURE 25.7 Molecular structure of PAMAM–PEG–PAMAM G4.

delivery of macromolecular compounds is enhanced by the photochemical disruption of the endosomal membrane in using light and a hydrophilic photosensitizer.[99] They prepared ternary complexes having cores of pDNA/cationic peptide polyplexes enveloped with anionic dendrimer phthalocyanine (DPc). The complexes showed greatly enhanced transfection efficiency and low

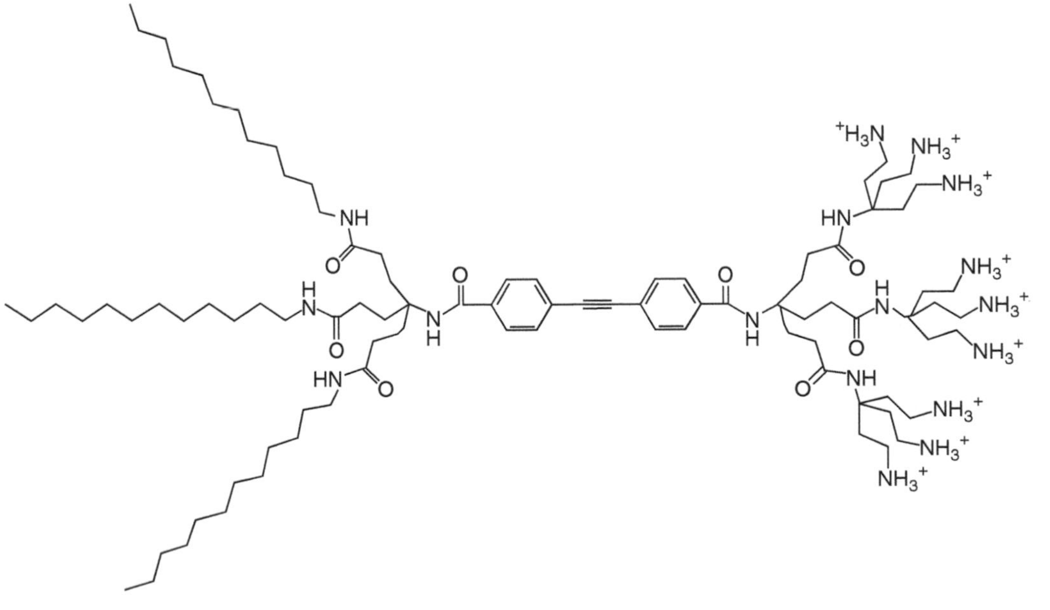

FIGURE 25.8 Molecular structure of amphiphilic dendrimer with tolane core.

photocytotoxicity in HeLa cells after photoirradiation, and they achieved significant gene expression in conjunctival tissue of rat eyes in contrast to ExGen500 that is one of the most efficient vectors.

25.4 CONCLUSION

In the past decade, various dendrimers have been developed and investigated for drug and gene delivery systems because of their unique structural properties such as multivalent and controllable functionalities on the periphery and highly branched and globular architectures as shown in this review. It is apparent that the application of dendrimers to drug and gene delivery systems has achieved considerably successful results, and much of the work reported in medical and pharmaceutical journals has demonstrated the growing interest in dendrimers.

Although it is not presented in this chapter, the application of dendrimers has been extended in other medical fields such as diagnostics, antiviral, or antibacterial therapy. PAMAM dendrimer-based Gd^{III} chelates were synthesized for use as a MRI contrast agent.[100,101] MRI is a powerful technique in medical diagnostics and used to visualize organs and blood vessels. The dendrimer chelates showed excellent MRI images and long blood circulation times in in vivo experiments. Also, chemically modified polylysine dendrimers and other dendrimers were synthesized for useful antiviral or antibacterial drugs.[102–105] In addition, numerous dendrimeric glycans or glycosylated dendrimers were shown to efficiently interact with natural carbohydrates receptors and were focused in their potential as microbial anti-adhesins, microbial toxin antagonists, and anti-cancer drugs.[106,107]

However, research about stable and safe biodistribution and efficient tissue localization as well as the biological function and fate of dendrimers is not well-established. Therefore, it is thought to be very important to explore the physiological safety and pharmacokinetics of developed dendrimers over a vaster range. It is also expected that a more convenient methodology to synthesizing novel dendrimers giving high water-solubility, good biocompatibility, and perfect structure will be designed for efficient application of dendrimers to biological systems.

ACKNOWLEDGMENTS

This work was supported by the National R&D Program Grant of The Ministry of Commerce, Industry and Energy (M10422010006-04N2201-00610), and the Gene Therapy Project of The Ministry of Science and Technology (M1053403004-05N3403-00410). The authors also wish to acknowledge helpful assistance received from Jung-un Baek.

REFERENCES

1. Buhleier, E., Wehner, W., and Vögtle, F., Cascade- and nonskid-chain-like" syntheses of molecular cavity topologies, *Synthesis*, 155–158, 1978.
2. Zeng, F. and Zimmerman, S. C., Dendrimers in supramolecular chemistry: From molecular recognition to self-assembly, *Chem. Rev.*, 97, 1681–1712, 1977.
3. Fischer, M. and Vögtle, F., Dendrimers: From design to application—a progress report, *Angew. Chem. Int. Ed.*, 38, 884–905, 1999.
4. Stiriba, S. E., Frey, H., and Haag, R., Dendritic polymers in biomedical applications: From potential to clinical use in diagnostics and therapy, *Angew. Chem. Int. Ed.*, 41, 1329–1334, 2002.
5. Patri, A. K., Majoros, I. J., and Baker, J. R. Jr., Dendritic polymer macromolecular carriers for drug delivery, *Curr. Opin. Chem. Biol.*, 6, 466–471, 2002.
6. Gilles, E. R. and Frechet, J. M. J., Dendrimers and dendritic polymers in drug delivery, *Drug Discov. Today*, 10, 35–43, 2005.

7. Blume, G. and Cevc, G., Liposomes for sustained drug release in vivo, *Biochim. Biophys. Acta*, 1029, 97, 1990.

8. Lian, T. and Ho, R. J. Y., Trends and developments in liposome drug delivery systems, *J. Pharm. Sci.*, 90, 680, 2001.

9. Rosler, A., Vandermeulen, G. W. M., and Klok, H. A., Advanced drug delivery devices via self-assembly of amphiphilic block copolymers, *Adv. Drug Deliv. Rev.*, 53, 95–108, 2001.

10. Newkome, G. R. et al., Unimolecular micelles, *Angew. Chem. Int. Ed. Engl.*, 30, 1178–1180, 1991.

11. Jansen, J. F. G. A., de Brabander-van den Berg, E. M. M., and Meijer, E. W., Encapsulation of guest molecules into a dendritic box, *Science*, 266, 1226–1229, 1994.

12. Jansen, J. F. G. A., Meijer, E. W., and de Brabander-van den Berg, E. M. M., The dendritic box: Shape-selective liberation of encapsulated guests, *J. Am. Chem. Soc.*, 117, 4417–4418, 1995.

13. Baars, M. W. P. L. et al., The localization of guests in water-soluble oligoethyleneoxy-modified poly(propylene imine) dendrimers, *Angew. Chem. Int. Ed.*, 39, 1285–1288, 2000.

14. Sideratou, Z., Tsiourvas, D., and Paleos, C. M., Solubilization and release properties of PEGylated diaminobutane poly(propylene imine) dendrimers, *J. Colloid Interface Sci.*, 242, 272–276, 2001.

15. Sideratou, A., Tsiourvas, D., and Paleos, C. M., Quaternized poly(propylene imine) dendrimers as novel pH-sensitive controlled-release systems, *Langmuir*, 16, 1766–1769, 2000.

16. Tomalia, D. A. et al., A new class of polymers: Starbust-dendritic macromolecules, *Polym. J.*, 17, 117–132, 1985.

17. Twyman, L. J. et al., The synthesis of water soluble dendrimers, and their application as possible drug delivery systems, *Tetrahedron Lett.*, 40, 1743–1746, 1999.

18. Kojima, C. et al., Synthesis of polyamidoamine dendrimers having poly(ethylene glycol) grafts and their ability to encapsulate anticancer drugs, *Bioconjug. Chem.*, 11, 910–917, 2000.

19. Bhadra, D. et al., A PEGylated dendritic nanoparticle carrier of fluorouracil, *Int. J. Pharm.*, 257, 111–124, 2003.

20. Haba, Y. et al., Synthesis of biocompatible dendrimers with a peripheral network formed by linking of polymerizable groups, *Polymer*, 46, 1813–1820, 2005.

21. Liu, M., Kono, K., and Frechet, J. M. J., Water-soluble unimolecular micelles: Their potential as drug delivery agents, *J. Control. Release*, 65, 121–131, 2000.

22. Gillies, E. R., Jonsson, T. B., and Frechet, J. M. J., Stimuli-responsive supramolecular assemblies of linear-dendritic copolymers, *J. Am. Chem. Soc.*, 126, 11936–11943, 2004.

23. Namazi, H. and Adeli, M., Dendrimers of citric acid and poly(ethylene glycol) as the new drug-delivery agents, *Biomaterials*, 26, 1175–1183, 2005.

24. Roberts, J. C. et al., Using starbust dendrimers as linker molecules to radiolabel antibodies, *Bioconjug. Chem.*, 1, 305–308, 1990.

25. Wu, C. et al., Metal–chelate–dendrimer–antibody constructs for use in radioimmunotherapy and imaging, *Bioorg. Med. Chem. Lett.*, 4, 449–454, 1994.

26. Patri, A. K. et al., Antibody–dendrimer conjugates for targeted prostate cancer therapy, *Polym. Mater. Sci. Eng.*, 86, 130, 2002.

27. Duncan, R. and Malik, N., Dendrimers: Biocompatibility and potential for delivery of anticancer agents, *Proc. Int. Symp. Control. Release Bioact. Mater.*, 23, 105–106, 1996.

28. Malik, N., Evagorou, E. G., and Duncan, R., Dendrimer–platinate: A novel approach to cancer chemotherapy, *Anticancer Drugs*, 10, 767–776, 1999.

29. Zhuo, R. X., Du, B., and Lu, Z. R., In vitro release of 5-fluorouracil with cyclic core dendritic polymer, *J. Control. Release*, 57, 249–257, 1999.

30. Wang, D. et al., Synthesis of starlike *N*-(2-hydroxypropyl)methacrylamide copolymers: Potential drug carriers, *Biomacromolecules*, 1, 313–319, 2000.

31. Wiwattanapatapee, R., Lomlim, L., and Saramunee, K., Dendrimers conjugates for colonic delivery of 5-aminosalicylic acid, *J. Control. Release*, 88, 1–9, 2003.

32. Quintana, A. et al., Design and function of a dendrimer-based therapeutic nanodevice targeted to tumor cells through the folate receptor, *Pharm. Res.*, 19, 1310–1316, 2002.

33. Tansey, W. et al., Synthesis and characterization of branched poly(L-glutamic acid) as a biodegradable drug carrier, *J. Control. Release*, 94, 39–51, 2004.

34. Kono, K., Liu, M., and Fréchet, J. M. J., Design of dendritic macromolecules containing folate or methotrexate residues, *Bioconjug. Chem.*, 10, 1115–1121, 1999.

35. Liu, M., Kono, K., and Fréchet, J. M. J., Water-soluble dendrimer–polyethylene glycol starlike conjugates as potential drug carriers, *J. Polym. Sci. A*, 37, 3492–3503, 1999.

36. Choe, Y. H. et al., Anticancer drug delivery systems: Multi-loaded N^4-acyl poly(ethylene glycol) prodrugs of ara-C. II. Efficacy in ascites and solid tumors, *J. Control. Release*, 79, 55–70, 2002.

37. Schiavon, O. et al., PEG–ara-C conjugates for controlled release, *Eur. J. Med. Chem.*, 39, 123–133, 2004.

38. Padilla De Jesu's, O. L. et al., Polyester dendritic systems for drug delivery applications: In vitro and in vivo evaluation, *Bioconjug. Chem.*, 13, 453–461, 2002.

39. de Groot, F. M. H. et al., "Cascade-release dendrimers" liberation all end groups upon a single triggering event in the dendritic core, *Angew. Chem. Int. Ed.*, 42, 4490–4494, 2003.

40. Amir, R. J. et al., Self-immolative dendrimers, *Angew. Chem. Int. Ed.*, 42, 4494–4499, 2003.

41. Shamis, M., Lode, H. N., and Shabat, D., Bioactivation of self-immolative dendritic prodrugs by catalytic antibody 38C2, *J. Am. Chem. Soc.*, 126, 1726–1731, 2004.

42. Hawthorne, M. F., The role of chemistry in the development of boron neutron capture therapy of cancer, *Angew. Chem. Int. Ed. Engl.*, 32, 950–984, 1993.

43. Barth, R. F. et al., Boronated starbust dendrimer–monoclonal antibody immunoconjugates: Evaluation as a potential delivery system for neutron capture therapy, *Bioconjug. Chem.*, 5, 58–66, 1994.

44. Capala, J. et al., Boronated epidermal growth factor as a potential targeting agent for boron neutron capture therapy of brain tumors, *Bioconjug. Chem.*, 7, 7–15, 1996.

45. Yang, W. et al., Intratumoral delivery of boronated epidermal growth factor for neutron capture therapy of brain tumors, *Cancer Res.*, 57, 4333–4339, 1997.

46. Wu, G. et al., Site-specific conjugation of boron-containing dendrimers to anti-EGF receptor monoclonal antibody cetuximab(IMC-C225) and its evaluation as a potential delivery agent for neutron capture therapy, *Bioconjug. Chem.*, 15, 185–194, 2004.

47. Shukla, S. et al., Synthesis and biological evaluation of folate receptor-targeted boronated PAMAM dendrimers as potential agents for neutron capture therapy, *Bioconjug. Chem.*, 14, 158–167, 2003.

48. Qualmann, B. et al., Synthesis of boron-rich lysine dendrimers as protein labels in electron microscopy, *Angew. Chem. Int. Ed. Engl.*, 35, 909–911, 1996.

49. Matsumoto, T., Transport calculations of depth–dose distributions for gadolinium neutron capture therapy, *Phys. Med. Biol.*, 37, 155–162, 1992.

50. Khokhlov, V. F. et al., Neutron capture therapy with gadopentetate dimeglumine: Experiments on tumor-bearing rats, *Acad. Radiol.*, 2, 392–398, 1995.

51. Hofmann, B. et al., Gadolinium neutron capture therapy (GdNCT) of melanoma cells and solid tumors with the magnetic resonance imaging contrast agent Gadobutrol, *Invest. Radiol.*, 34, 126–133, 1999.

52. Caravan, P. et al., Gadolinium(III) chelates as MRI contrast agents: Structure, dynamics, and applications, *Chem. Rev.*, 99, 2293–2352, 1999.

53. Kobayashi, H. et al., Avidin-dendrimer-(1B4M-Gd)$_{254}$: A tumor-targeting therapeutic agent for gadolinium neutron capture therapy of intraperitoneal disseminated tumor which can be monitored by MRI, *Bioconjug. Chem.*, 12, 587–593, 2001.

54. Dougherty, T. J. et al., Photodynamic therapy, *J. Natl. Cancer Inst.*, 90, 889–905, 1998.

55. Battah, S. H. et al., Synthesis and biological studies of 5-aminolevulinic acid-containing dendrimers for photodynamic therapy, *Bioconjug. Chem.*, 12, 980–988, 2001.

56. Peng, Q. et al., 5-Aminolevulinic acid-based photodynamic therapy: Principles and experimental research, *Photochem. Photobiol.*, 65, 235–251, 1997.

57. Nishiyama, N. et al., Light-harvesting ionic dendrimer porphyrins as new photosensitizers for photodynamic therapy, *Bioconjug. Chem.*, 14, 58–66, 2003.

58. Zhang, G. D. et al., Polyion complex micelles entrapping cationic dendrimer porphyrin: Effective photosensitizer for photodynamic therapy of cancer, *J. Control. Release*, 93, 141–150, 2003.

59. Dichtel, W. R. et al., Singlet oxygen generation via two-photon excited FRET, *J. Am. Chem. Soc.*, 126, 5380–5381, 2004.

60. Boussif, O. et al., A versatile vector for gene and oligonucleotide transfer into cells in culture and in vivo: Polyethylenimine, *Proc. Natl Acad. Sci.*, 92, 7297–7301, 1995.

61. Rejman, J., Bragonzi, A., and Conese, M., Role of clathrin- and caveolae-mediated endocytosis in gene transfer mediated by lipo- and polyplexes, *Mol. Ther.*, 12, 468–474, 2005.

62. Haensler, J. and Szoka, F. C. Jr., Polyamidoamine cascade polymers mediate efficient transfection of cells in culture, *Bioconjug. Chem.*, 4, 372–379, 1993.

63. Kukowska-Latallo, J. F. et al., Efficient transfer of genetic material into mammalian cells using starburst polyamidoamine dendrimers, *Proc. Natl Acad. Sci. USA*, 93, 4897–4902, 1996.

64. Harada, Y. et al., Highly efficient suicide gene expression in hepatocellular carcinoma cells by Epstein–Barr virus-based plasmid vectors combined with polyamidoamine dendrimer, *Cancer Gene Ther.*, 7, 27–36, 2000.

65. Tanaka, S. et al., Targeted killing of carcinoembryonic antigen (CEA)-producing cholangiocarcinoma cells by polyamidoamine dendrimer-mediated transfer of an Epstein–Barr virus (EBV)-based plasmid vector carrying the CEA promoter, *Cancer Gene Ther.*, 7, 1241–1250, 2000.

66. Tang, M. X., Redemann, C. T., and Szoka, F. C. Jr., In vitro gene delivery by degraded polyamidoamine dendrimers, *Bioconjug. Chem.*, 7, 703–714, 1996.

67. Bielinska, A. et al., Regulation of in vitro gene expression using antisense oligonucleotides or antisense expression plasmids transfected using starburst PAMAM dendrimers, *Nucleic Acids Res.*, 24, 2176–2182, 1996.

68. Delong, R. et al., *J. Pharm. Sci.*, 86, 762–764, 1997.

69. Yoo, H. and Juliano, R. L., Enhanced delivery of antisense oligonucleotides with fluorophore-conjugated PAMAM dendrimers, *Nucleic Acids Res.*, 28, 4225–4231, 2000.

70. Maksimenko, A. V. et al., Optimisation of dendrimer-mediated gene transfer by anionic oligomers, *J. Gene Med.*, 5, 61–71, 2003.

71. Roessler, B. J. et al., Substituted β-cyclodextrins interact with PAMAM dendrimer–DNA complexes and modify transfection efficiency, *Biochem. Biophys. Res. Commun.*, 283, 124–129, 2001.

72. Kihara, F. et al., Effects of structure of polyamidoamine dendrimer on gene transfer efficiency of the dendrimer conjugate with α-cyclodextrin, *Bioconjug. Chem.*, 13, 1211–1214, 2002.

73. Wada, K. et al., Improvement of gene delivery mediated by mannosylated dendrimer/α-cyclodextrin conjugates, *J. Control. Release*, 104, 397–413, 2005.

74. Choi, J. S. et al., Enhanced transfection efficiency of PAMAM dendrimer by surface modification with L-arginine, *J. Control. Release*, 99, 445–456, 2004.

75. Tung, C. H. and Weissleder, R., Arginine containing peptides as delivery vectors, *Adv. Drug Deliv. Rev.*, 55, 281–294, 2003.

76. Kang, H. et al., Tat-conjugated PAMAM dendrimers as delivery agents for antisense and siRNA oligonucleotides, *Pharm. Res.*, 22, 2099–2106, 2005.

77. Lee, J. H. et al., Polyplexes assembled with internally quaternized PAMAM-OH dendrimer and plasmid DNA have a neutral surface and gene delivery potency, *Bioconjug. Chem.*, 14, 1214–1221, 2003.

78. Hudde, T. et al., Activated polyamidoamine dendrimers, a non-viral vector for gene transfer to the corneal endothelium, *Gene Ther.*, 6, 939–943, 1999.

79. Wang, Y. et al., DNA/dendrimer complexes mediate gene transfer into murine cardiac transplants ex vivo, *Mol. Ther.*, 2, 602–608, 2000.

80. Kukowska-Latallo, J. F. et al., Intravascular and endobronchial DNA delivery to murine lung tissue using a novel, nonviral vector, *Hum. Gene Ther.*, 11, 1385–1395, 2000.

81. Sato, N. et al., Tumor targeting and imaging of intraperitoneal tumors by use of antisense oligo-DNA complexed with dendrimers and/or avidin in mice, *Clin. Cancer Res.*, 7, 3606–3612, 2001.

82. de Brabander-van den Berg, E. M. M. and Meijer, E. W., Poly(propylene imine) dendrimers: Large-scale synthesis by hetereogeneously catalyzed hydrogenations, *Angew. Chem. Int. Ed. Engl.*, 32, 1308–1311, 1993.

83. Gebhart, C. L. and Kabanov, A. V., Evaluation of polyplexes as gene transfer agents, *J. Control. Release*, 73, 401–416, 2001.

84. Zinselmeyer, B. H. et al., The lower-generation polypropylenimine dendrimers are effective gene-transfer agents, *Pharm. Res.*, 19, 960–967, 2002.

85. Lim, Y. b. et al., Self-assembled ternary complex of cationic dendrimer, cucurbituril, and DNA: Noncovalent strategy in developing a gene delivery carrier, *Bioconjug. Chem.*, 13, 1181–1185, 2002.

86. Whang, D. et al., Self-assembly of a polyrotaxane containing a cyclic "bead" in every structural unit in the solid state: Cucurbituril molecules threaded on a one-dimensional coordination polymer, *J. Am. Chem. Soc.*, 118, 11333–11334, 1996.

87. Lee, J. W. et al., Novel pseudorotaxane-terminated dendrimers: Supramolecular modification of dendrimer periphery, *Angew. Chem. Int. Ed.*, 40, 746–749, 2001.

88. Schatzlein, A. G. et al., Preferential liver gene expression with polypropylenimine dendrimers, *J. Control. Release*, 101, 247–258, 2005.

89. Dufes, C. et al., Synthetic anticancer gene medicine exploits intrinsic antitumor activity of cationic vector tocure established tumors, *Cancer Res.*, 65, 8079–8084, 2005.

90. Ohsaki, M. et al., In vitro gene transfection using dendritic poly(L-lysine), *Bioconjug. Chem.*, 13, 510–517, 2002.

91. Okuda, T. et al., Characters of dendritic poly(L-lysine) analogues with the terminal lysines replaced with arginines and histidines as gene carriers in vitro, *Biomaterials*, 25, 537–544, 2004.

92. Choi, J. S. et al., Poly(ethylene glycol)-*block*-poly(L-lysine) dendrimer: Novel linear polymer/dendrimer block copolymer forming a spherical water-soluble polyionic complex with DNA, *Bioconjug. Chem.*, 10, 62–65, 1999.

93. Choi, J. S. et al., Synthesis of a barbell-like triblock copolymer, poly(L-lysine) dendrimer-*block*-poly(ethylene glycol)-*block*-poly(L-lysine) dendrimer, and its self-assembly with plasmid DNA, *J. Am. Chem. Soc.*, 122, 474–480, 2000.

94. Kim, T. I. et al., PAMAM–PEG–PAMAM: Novel triblock copolymer as a biocompatible and efficient gene delivery carrier, *Biomacromolecules*, 5, 2487–2492, 2004.

95. Joester, D. et al., Amphiphilic dendrimers: Novel self-assembling vectors for efficient gene delivery, *Angew. Chem. Int. Ed.*, 42, 1486–1490, 2003.

96. Maszewska, M. et al., Water-soluble polycationic dendrimers with a phosphoramidothioate backbone: Preliminary studies of cytotoxicity and oligonucleotide/plasmid delivery in human cell culture, *Oligonuleotides*, 13, 193–205, 2003.

97. Hussain, M. et al., A novel anionic dendrimer for improved cellular delivery of antisense oligonucleotides, *J. Control. Release*, 99, 139–155, 2004.

98. Nishiyama, N. et al., Light-induced gene transfer from packaged DNA enveloped in a dendrimeric photosensitizer, *Nat. Mater.*, 4, 934–941, 2005.

99. Høgest, A. et al., Photochemical internalization in drug and gene delivery, *Adv. Drug Deliv. Rev.*, 56, 95–115, 2004.

100. Wiener, E. C. et al., Dendrimer-based metal chelates: A new class of magnetic resonance imaging contrast agents, *Magn. Reson. Med.*, 31, 1–8, 1994.

101. Bryant, L. H. Jr. et al., Synthesis and relaxometry of high-generation ($G = 5, 7, 9,$ and 10) PAMAM dendrimer–DOTA–gadolinium chelates, *J. Magn. Reson. Imaging*, 9, 348–352, 1999.

102. Bourne, N. et al., Dendrimers, a new class of candidate topical microbicides with activity against herpes simplex virus infection, *Antimicrob. Agents Chemother.*, 44, 2471–2474, 2000.

103. Witvrouw, M. et al., Potent anti-HIV (type 1 and type 2) activity of polyoxometalates: Structure–activity relationship and mechanism of action, *J. Med. Chem.*, 43, 778–783, 2000.

104. Landers, J. J. et al., Prevention of influenza pneumonitis by sialic acid-conjugated dendritic polymers, *J. Infect. Dis.*, 186, 1222–1230, 2002.

105. Chen, C. Z. S. and Cooper, S. L., Recent advances in antimicrobial dendrimers, *Adv. Mater.*, 12, 843–846, 2000.

106. Veprek, P. and Jezek, J., Peptide and glycopeptide dendrimers. Part II, *J. Pept. Sci.*, 5, 203–220, 1999.

107. Andre, S. et al., Wedgelike glycodendrimers as inhibitors of binding of mammalian galectins to glycoproteins, lactose maxiclusters, and cell surface glycoconjugates, *Chembiochem*, 2, 822–830, 2001.

26 Dendritic Nanostructures for Cancer Therapy

*Ashootosh V. Ambade, Elamprakash N. Savariar,
and S. (Thai) Thayumanavan*

CONTENTS

26.1 Introduction 509
26.2 Dendrimer–Drug Conjugates 511
26.3 Encapsulation of the Drug in a Dendrimer.. 514
26.4 Polymeric Drug Carriers Based on Dendrimers 517
26.5 Summary... 518
References... 518

26.1 INTRODUCTION

Nanomaterials have emerged from academic curiosity and are being applied in many diverse areas such as information technology and biotechnology.[1–4] Synthetic nanomaterials based on macromolecules are relevant in biomedical applications because their size scale is comparable with that of biomacromolecules.[5] Dendrimers provide a prominent example of such synthetic macromolecules that have attracted great attention in recent years because of their potential in drug delivery among other promising applications.[6–12] Dendrimers are highly branched polymers of a few nanometer dimensions that possess a globular shape at high molecular weights.[13,14] These are synthesized by repetitive organic transformations and, as a result, are almost monodisperse. An interesting structural feature of dendrimers is the spatially isolated core in the dendritic interior that is reminiscent of site-isolation in biomolecules such as proteins. The branches emanating from this core provide cavities in the interior that can be exploited for guest encapsulation. Every layer of branches that is added to the core during the synthesis is termed as a generation. The highly branched nature of dendrimers leads to large number of peripheral groups. These surface functionalities can be modified by attachment of drugs or targeting ligands to meet specific applications that often involve controlled interaction with cell walls and other biological targets. A representative structure is shown in Figure 26.1 that illustrates the typical features—core, intermediate layers, and peripheral groups—of a dendrimer. The dendritic structure in terms of type of repeat units, size of interior cavities, and type of peripheral groups can be tuned by using either convergent or divergent synthetic strategy.[15,16] It is then obvious that such versatile molecules are being intensely pursued in cancer therapeutics.[17,18]

A major problem with anti-cancer agents is that they indiscriminately attack healthy cells as well as cancerous cells. These harmful side effects can be reduced by developing a drug delivery vehicle that is specific to tumor cells. This can be achieved by employing a strategy called active

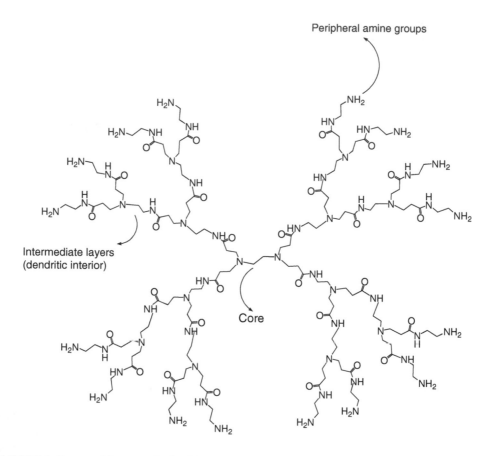

FIGURE 26.1 Structural features of a dendrimer are illustrated with a structure of amine-terminated PAMAM dendrimer.

targeting wherein functionalities that respond to over-expressed receptors on tumor cells are attached to the drug carrier. Dendrimers provide an opportunity in this direction in the form of a large number of tunable peripheral groups where a variety of functionalities can be incorporated (Figure 26.2). Another strategy is the passive targeting that relies on the well-known enhanced permeation and retention (EPR) effect. This effect, first described by Maeda et al., [19] takes advantage of poorly formed (leaky) vasculature of solid tumors that allows accumulation of polymer–drug

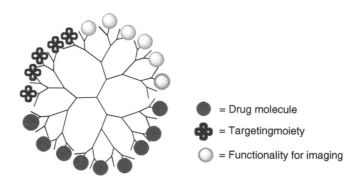

FIGURE 26.2 A dendrimer with multiple functionalities attached to it is illustrated by a cartoon.

conjugates ranging in size between 10 and 500 nm within tumors than of free drugs. The polymeric molecules are retained following the accumulation due to their larger size whereas free drug molecules are easily eliminated from the cells. Dendrimers have an advantage for this strategy as well because drugs can be physically encapsulated into the interior of nanometer-sized dendrimers, thereby increasing the uptake of drugs into tumors. Before describing the various ways in which dendritic structure has been utilized for drug delivery in cancer therapy, it is useful to discuss the techniques that are employed for this purpose.

Photodynamic therapy (PDT), also known as photoradiation therapy or photochemotherapy, involves the use of a light-activated drug called photosensitizer that can be administered orally or intraperitoneally, or it can be applied to the skin.[20,21] When light of a specific wavelength that is absorbed by the dye is shined on the targeted tissue, the photosensitizer is excited from ground singlet state to excited singlet state. It then undergoes intersystem crossing to a longer-lived excited triplet state that reacts with the ground triplet state of molecular oxygen present in the tissues. This energy transfer creates highly-reactive excited singlet-state oxygen species. The singlet oxygen reacts destructively with nearby biomolecules and induces cell death either by apoptosis or necrosis. PDT, however, has limitations because it is applicable only in areas where light can reach.

Neutron capture therapy (NCT) is another promising technique for treatment of cancer.[22] It is based on the nuclear reaction that occurs when a stable isotope, for example, boron ^{10}B, is irradiated with low energy or thermal neutrons to produce high-energy alpha particles (^{4}He nuclei) and ^{7}Li ions. The combined path length of these two particles is about 12 µm that is close to the diameter of a cell. Therefore, if the tumor cells can be supplied with a high concentration of ^{10}B atoms, the effect of the radiation will be limited to tumor cells, and the normal cells may be spared. This is particularly important in cancer therapy where prevention of indiscriminate destruction of healthy cells is a challenge. ^{157}Gd is another isotope of interest because of its large neutron capture cross section. Success of the above two therapies depends on localizing sufficient concentration of the photosensitizer or radioactive material in the tumor cells while having negligible accumulation in the healthy cells.

Magnetic resonance imaging (MRI) is one of the most widely used imaging techniques that has greatly improved medical diagnosis.[23] It is based on the relaxation properties of hydrogen nuclei in water in body tissues when placed in a uniform magnetic field. In clinical practice, for detection of cancer in particular, it uses the fact that the relaxation times of protons in cancerous and normal tissues markedly differ. One of the advantages of MRI over other imaging techniques is that it is a non-radiative technique and harmless to the patient.

This chapter will focus on the structural aspects of dendritic molecules that have been used to design efficient and multifunctional drug delivery vehicles in cancer therapy with the help of selected illustrative examples. There are only few cases where these macromolecules have been studied in vivo, i.e., in animal models, and the biological studies will be discussed only cursorily.

26.2 DENDRIMER–DRUG CONJUGATES

The term *dendrimer–drug conjugate* is mainly applied when a drug molecule is covalently linked to a dendrimer. The drug can be covalently attached to a dendrimer on the periphery or at the core and quite rarely at the branching points, i.e., in the interior layers. Attachment of a drug to multiple peripheral groups in a dendrimer greatly increases the effective concentration of a drug once it is released at the targeted site. This is particularly advantageous for the use of prodrugs. In the case of PDT, however, it is desirable to isolate the drug molecule to prevent fluorescence quenching because of aggregation, and this can be achieved by attaching the photosensitizer to the core of a dendrimer. A dendrimer–drug conjugate can provide certain advantages over conventional polymeric drug delivery vehicles because dendrimers are monodisperse, structurally defined macromolecules with a controlled molecular weight and size.

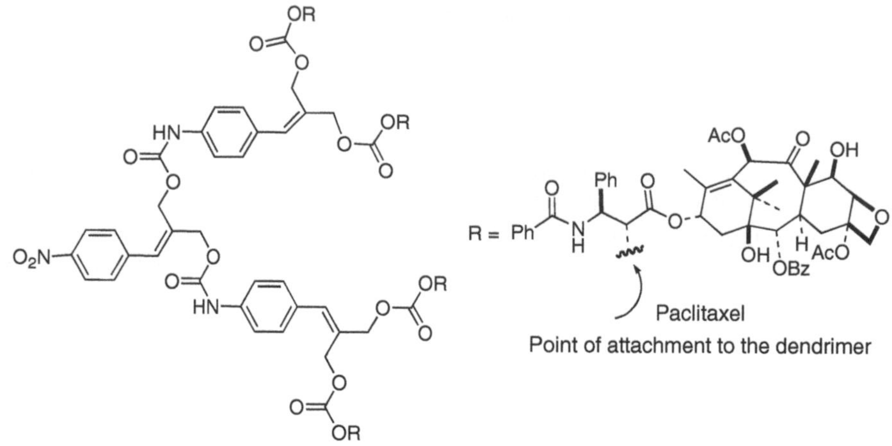

FIGURE 26.3 Paclitaxel–dendrimer conjugate with a cascade-release mechanism. The NO_2 group at the focal point is converted to NH_2 under reducing conditions triggering the cascade-release.

When a drug is attached to peripheral groups of a dendrimer, the linker between drug and dendrimer is of crucial importance because the drugs need to be efficiently released in active form at the targeted site. Advantage is taken of the acidic or reducing environment near the tumor cells by attaching the drugs to dendritic periphery via acid-labile or disulphide (susceptible to reduction) linker (see Section 26.4 for examples). In this case, each drug is separately cleaved over a long period of time. An elegant design was reported in this direction wherein a single triggering event at the core results in disassembly of the dendrimer and release of all the paclitaxel molecules attached to the periphery.[24] The drug release in these dendrimers incorporating a masked 4-aminocinnamyl alcohol linker is activated by reduction of nitro to amino group under mild reducing conditions (Figure 26.3).

Poly(amidoamine) (PAMAM) dendrimers have been widely used for preparation of dendrimer–drug conjugates because of easily modifiable terminal amine groups, commercial availability of higher generation dendrimers, and biocompatibility.[25–27] For example, methylprednisolone (MP) was attached to PAMAM periphery through a glutaric acid linker.[28] Because the MP–glutaric acid compound was attached to a preformed dendrimer, only 12 drug molecules (32 wt%) could be conjugated to a dendrimer that has 32 terminal moieties as a result of steric hindrance. However, considering that MP is a bulky steroidal drug, this was a significant increase over earlier payloads.

Monoclonal antibodies (mAb) are useful for targeted delivery because these can specifically bind to certain receptors over-expressed on tumor cells. Antibody–PAMAM conjugates have been prepared by attaching J591 anti-PSMA antibody to a fluorescein-labeled dendrimer for targeted delivery in prostate cancer therapy. It was based on the fact that prostate-specific membrane antigen (PSMA) is over-expressed on tumor cells than on normal tissue.[29] Attachment of multiple functionalities for targeting (folic acid) and imaging (fluorescein) along with an anti-cancer drug (methotrexate) to a dendrimer periphery has also been demonstrated with PAMAM dendrimers[30] and is elaborated below with examples on MRI and NCT. Similarly, different PAMAM dendrimers attached with fluorescein and folic acid were covalently attached to oligonucleotides and assembled via hybridization to afford multifunctional dendrimer clusters as illustrated by a cartoon in Figure 26.4.[31] On a different note, carbohydrate moieties have been attached to PAMAM dendrimers to study multivalent protein–carbohydrate interactions. This is particularly relevant to cancer metastasis that involves these interactions in cellular recognition processes.[32–34]

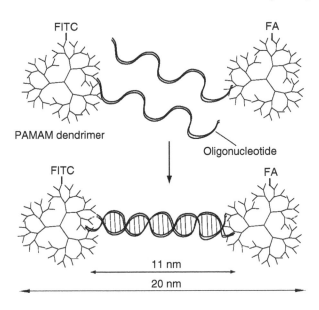

FIGURE 26.4 PAMAM clusters with multiple functionalities prepared via DNA hybridization.

PAMAM dendrimers have also been used for preparation of MRI contrast agents.[35] Gd(III) complexes were attached to G6-PAMAM dendrimers along with avidin to afford a targeted delivery vehicle for NCT.[36] [157]Gd provides two advantages. First, it has a large neutron capture cross section that can be exploited in NCT, and second, Gd(III) complexes can be used as MRI contrast agents. In fact, dendrimers bearing Gd(III) complexes on their periphery have undergone clinical trials as MRI contrast agents.[37,38] Therefore, the Gd-dendrimer conjugate described above serves a dual purpose, therapeutic as well as imaging. The Gd-dendrimer complex was also used as a contrast agent for dynamic MRI to evaluate the perfusion of an antibody viz. herceptin into tumor tissue because it has similar size (~ 10 nm).[39] Similarly, folic acid-conjugated[40] and epidermal growth factor (EGF)-conjugated[41] boronated PAMAM dendrimers have been prepared as targeting agents in boron NCT. Slow clearance of macromolecular MRI agents is an issue that limits their scope, particularly in the investigation of tumor angiogenesis. To address this problem, the avidin chase system was employed where biotinylated dendrimeric Gd-complexes were rapidly cleared (~ 2 min) from blood pool to the liver on injection of avidin.[42] Dendrimeric Gd(III) chelates with high relaxivity and fast water exchange have been developed to improve upon the currently available gadolinium complexes that have slow water exchange rates.[43] In rare examples of interior functionalized dendrimers, boron clusters have been incorporated in the intermediate layers of water-soluble dendrimers for application in NCT.[44,45]

Porphyrins and phthalocyanines are two important macrocycles for use as photosensitizers in PDT because of large π-conjugation domains and high singlet oxygen quantum yields.[21] However, large π-conjugation and hydrophobicity in these molecules also lead to a tendency to aggregate that reduces their solubility in water and precludes use in biological systems. The aggregation also causes self-quenching and decreases the efficiency of singlet oxygen formation. Incorporation of a porphyrin or phthalocyanine chromophore at the core of a benzyl ether dendrimer has been shown to provide the site isolation and to necessarily prevent the aggregation and self-quenching. Hydrophilic groups such as oligoethylene glycol were attached to the periphery of these dendrimers, making them useful in biological applications.[46] Oligoethylene glycols are important in biological applications because they provide solubility in water, reduce hepatic uptake, and enhance biodistribution. It is to be noted that benzyl ether dendrimers are another class of dendrimers, apart from PAMAM dendrimers, that can be readily synthesized and have been widely studied for many applications. Attachment of various functionalities and drugs on periphery and in the interior can also be carried out with benzyl ether dendrimers using recently developed synthetic strategies.[47–50]

One of the current goals in PDT is to design porphyrin sensitizers that can absorb in the near infra-red (NIR) region. This is highly desirable because skin is more transparent to light of this wavelength. However, porphyrins have low two-photon absorption (TPA) cross-sections that limits their use in singlet oxygen generation. Direct chemical modification of the macrocycles has been attempted to enhance the TPA cross-section and water solubility.[21] An alternate strategy using dendrimers was recently developed where donor chromophores with a high TPA cross-section were covalently attached to the periphery of a benzyl ether dendrimer, containing a porphyrin moiety at the core.[51] It was observed that emission from porphyrin moiety in the dendritic derivative was 17 times higher than from a porphyrin without the dendritic donor substituents. These dendrimers were then decorated with water-soluble triethyleneglycol moieties to further expand the scope of their application.[52] In another study, benzyl ether dendrimers with porphyrin unit at the core were made water-soluble by attachment of quaternary ammonium or carboxylic groups (Figure 26.5), and their efficiency as photosensitizers was evaluated.[53] To further enhance water-solubility and cellular uptake of the dendrimer porphyrins with charged periphery, poly-ion complex (PIC) micelles composed of polyethylene glycol–poly(L-lysine) block copolymers were developed that could entrap the negatively charged dendrimers through electrostatic association (Figure 26.6).[54] These polymeric micelle-based nanocarriers showed enhanced cellular uptake, higher photocytotoxicity, and similar singlet oxygen generation compared to free dendrimer porphyrins. The higher cellular uptake was due to shielding of negatively charged dendrimer periphery inside a charge-neutral micelle. The PIC micelles carrying anionic dendrimers were further tested in vivo for treatment of age-related macular degeneration (AMD) caused by choroidal neovascularization (CNV).[55] Experimental CNV lesions were created in rats by laser photocoagulation, and PIC micelles were injected after seven days. The dendrimer-loaded micelles were found to accumulate preferentially in CNV lesions and were retained even after 24 h, whereas free dendrimers were cleared within 24 h. When cationic dendrimer porphyrins were encapsulated in the polyethylene glycol–poly(aspartic acid) block copolymers, the cellular uptake was lower, and the photodynamic efficacy was higher compared to free dendrimers.[56] In all these PIC micelles, the dark systemic toxicity (cytotoxicity of dendrimers under dark conditions) was found to be reduced probably as a result of a biocompatible poly(ethylene glycol) (PEG) shell.

In a different approach to PDT, 5-aminolevulinic acid (ALA), a natural precursor of protoporphyrin IX, was attached to the periphery of first and second generation dendrimers via ester linkages to obtain dendrimer–ALA prodrugs (Figure 26.7).[57] ALA can be intracellularly released by non-specific esterases leading to higher intracellular ALA levels and higher protoporphyrin IX concentration. Also, the dendrimer–ALA conjugate is more lipophilic than only ALA and can easily cross the cell membranes as was evident by higher uptake of the second-generation dendrimer studied using tumorigenic cell line PAM 212 in vitro. These dendrimers also showed negligible dark systemic toxicity.

26.3 ENCAPSULATION OF THE DRUG IN A DENDRIMER

Most of the anti-cancer drugs are hydrophobic, and their entrapment in a hydrophilic dendrimer is a useful strategy to enhance their solubility for drug delivery. This kind of drug carriers composed of dendrimeric micelles have been a subject of a recent review[12] and is discussed here only briefly.

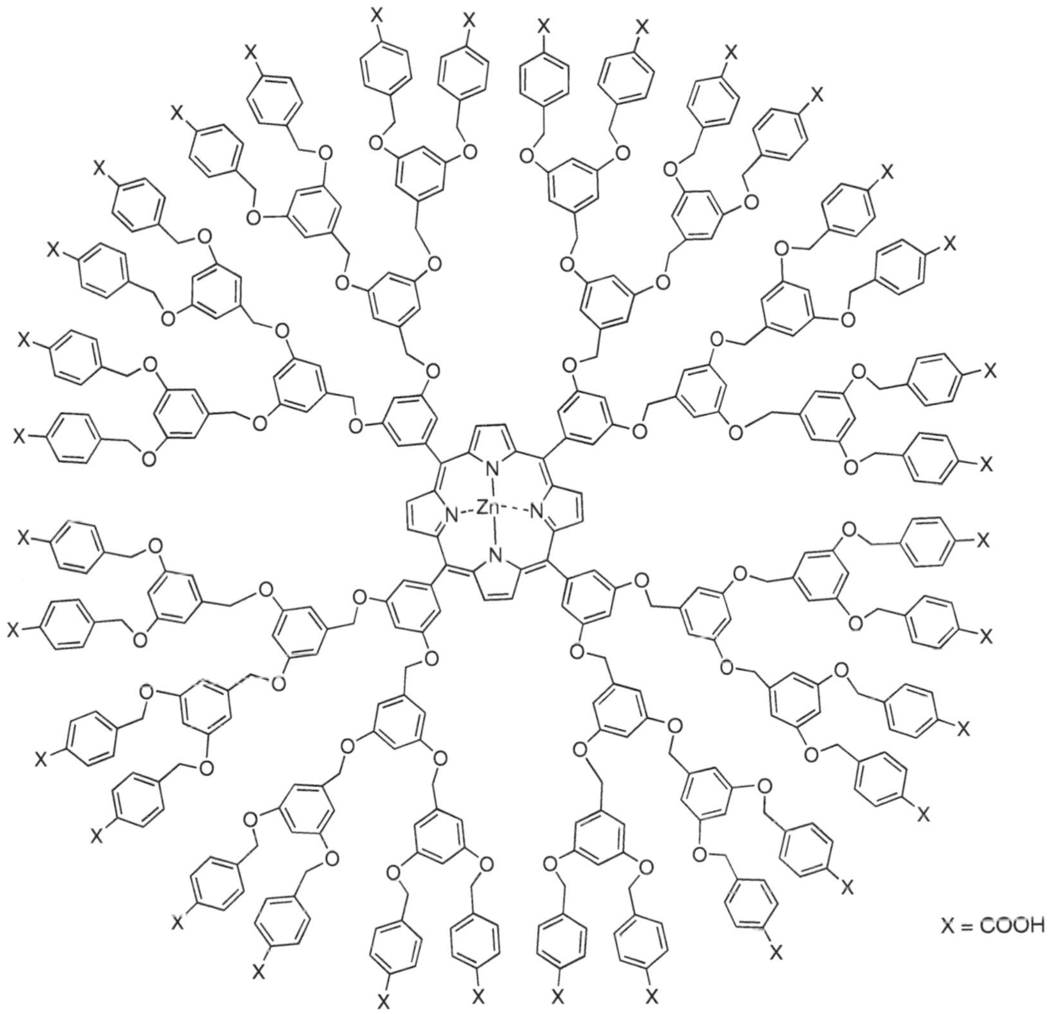

FIGURE 26.5 A water-soluble benzyl ether dendrimer with porphyrin moiety at the core for PDT.

Physical encapsulation prevents the drug from being prematurely degraded. The periphery of the dendrimer is rendered hydrophilic, either by attachment of oligoethylene glycol chains or charged functional groups such as carboxylates, and the hydrophobic interior is suitable for sequestration of hydrophobic molecules.[58–61] The drug loading capacity in a dendrimer by encapsulation depends upon its generation and size; however, higher payloads can be obtained only by attaching the drug molecules to the peripheral groups. Several examples of modified PAMAM dendrimers used for physical encapsulation of drugs have been reported.[62,63]

Melamine-based dendrimers modified on the surface with cationic, anionic, and oligoethylene glycol groups have been used to encapsulate methotrexate and 6-mercaptopurine, and their hepatotoxicity has been studied in vivo.[64,65] A synthetic strategy has been developed for these dendrimers as well to incorporate multiple functionalities on the periphery. A hydrophobic anticancer drug, 10-hydroxycamptothecin, was encapsulated in hydrophilic poly(glycerol succinic acid) (PGLSA) dendrimers.[66] Cytotoxicity assays of these drug-loaded dendrimers on human breast cancer cells showed that the drug retained its activity upon encapsulation.

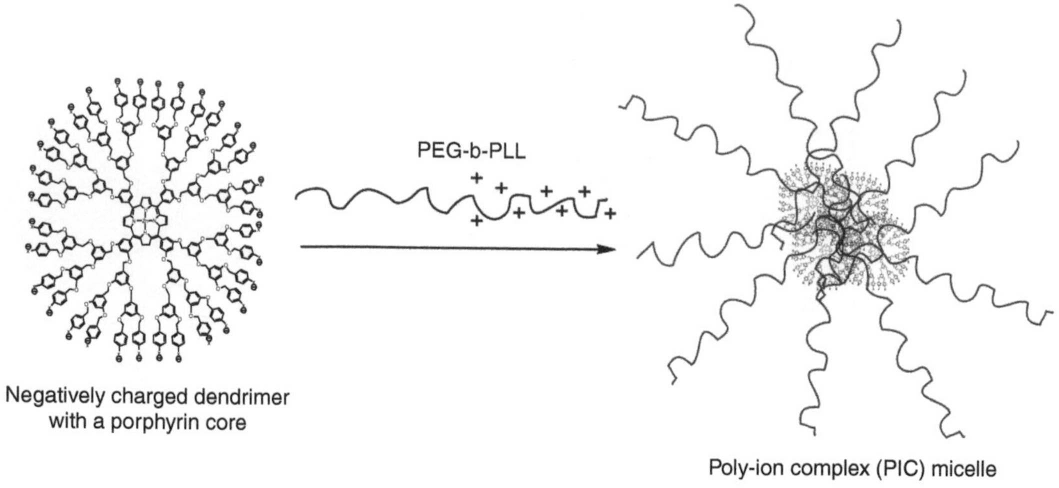

FIGURE 26.6 A cartoon showing poly-ion complex (PIC) micelles encapsulating dendrimer–porphyrins.

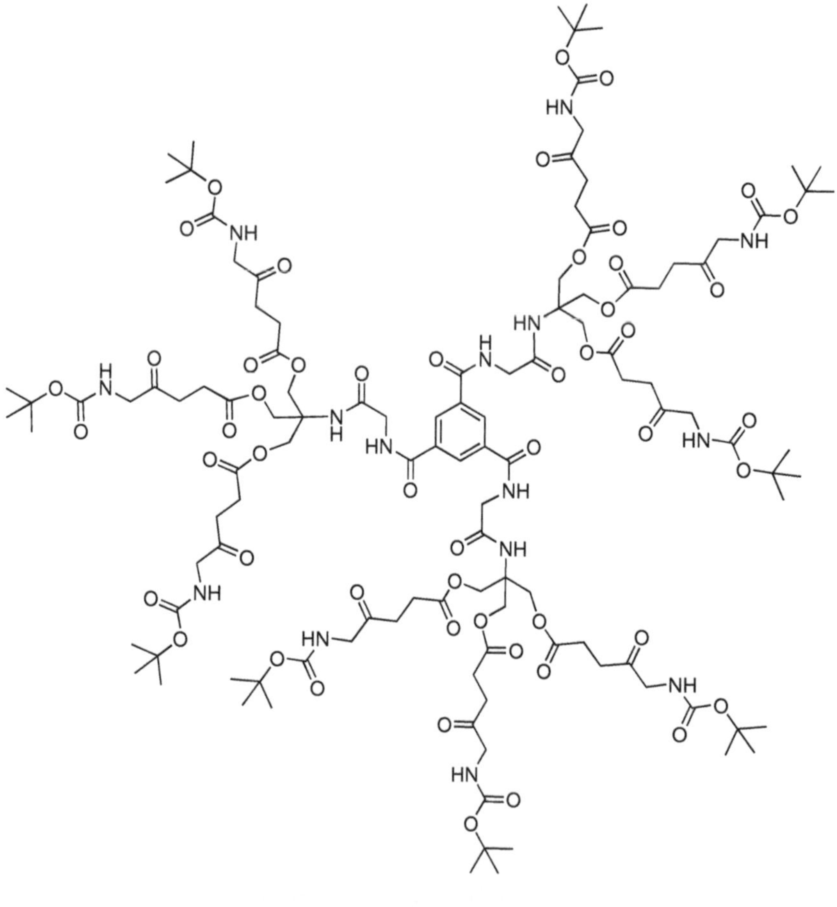

FIGURE 26.7 Boc-protected ALA–dendrimer conjugate for PDT.

26.4 POLYMERIC DRUG CARRIERS BASED ON DENDRIMERS

Dendrimers have been explored as building blocks in the construction of macromolecular scaffolds for drug delivery to circumvent the following problems: rapid release of drugs physically encapsulated in a dendrimer, and shorter circulation time of dendrimers themselves. The dendrimer component in linear–dendritic hybrid polymers provides the large number of functionalities required for high payload, whereas the linear component affords high molecular weight unattainable in dendrimers. Such a combination of desirable properties in these hybrid polymers is expected to provide higher plasma half-life (circulation time) and enhanced cellular uptake because of EPR effect and high potency. With this premise, dendritic aliphatic polyesters to which doxorubicin (DOX) was attached via an acid-labile hydrazone linker were connected to a three-arm polyethylene glycol star polymer to obtain high molecular weight hybrid polymers (Figure 26.8).[67] This polymer–drug conjugate showed less cytotoxicity (human breast cancer cell lines MDA-MD-231 and MDA-MD-435 using sulforhodamine B assay) than the free drug and longer circulation time (tested using ^{125}I radiolabeling) than only dendrimers. These dendritic aliphatic polyesters were also used to prepare linear–dendritic bow-tie hybrids by attaching PEG chains to dendrimer periphery (illustrated by a schematic in Figure 26.9) and evaluated for biodistribution.[68] Bow-tie polymers with molecular weights up to 1,60,000 g/mol were synthesized, and those with higher molecular weight (>40,000 g/mol) and more number of linear arms were studied. When tested on mice bearing subcutaneous B16F10 tumors, these polymers were found to accumulate mainly in

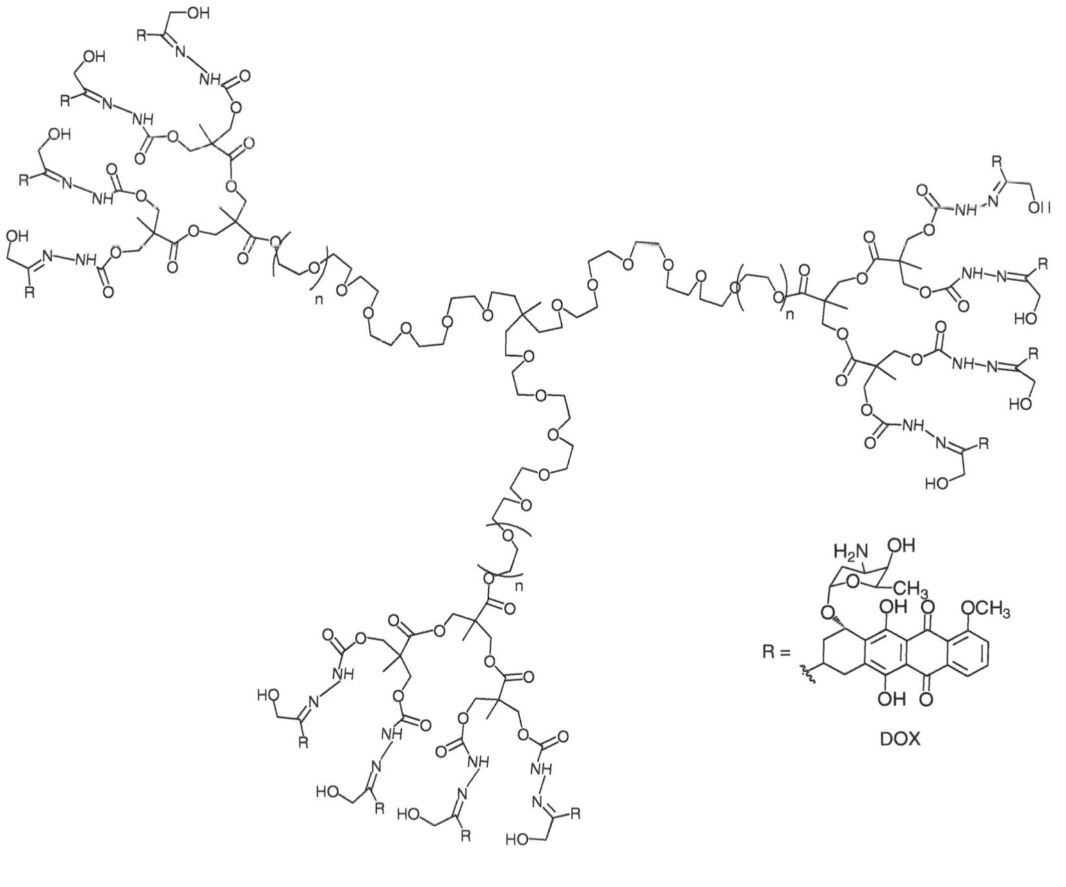

FIGURE 26.8 A linear–dendritic hybrid star polymer with doxorubicin molecules attached to the periphery via acid-labile hydrazone linkages.

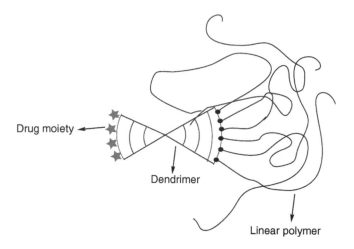

FIGURE 26.9 Structure of a dendritic–linear bow-tie hybrid polymer is illustrated with a cartoon.

tumors with little concentration in the liver, kidney, and lungs. Star-like polymers containing DOX have also been synthesized with PAMAM dendrimer as the core. For example, starlike poly-N-(2-hydroxypropyl) methacrylamide (HPMA) was synthesized by attachment of the copolymers to PAMAM dendrimers of generation 2, 3, and 4.[69] DOX was attached to the methacrylamide backbone via a peptide linker (GlyPheLeuGly) because this linker has been shown to get cleaved by an enzyme Cathepsin B that is over-expressed in malignant cells.[70] The effect of molecular architecture on release of DOX in vitro was clearly evident by the slower release of DOX from the star-like polymer compared to that from the linear polymer.

26.5 SUMMARY

Dendrimers have clearly emerged as important tools in drug delivery systems. Although tedious synthesis of dendrimers hampers their widespread use in biological applications, studies on commercially available dendrimers have clearly demonstrated their potential in drug delivery. Toward realizing their full potential, however, a combination of useful structural features of dendrimers with those of well-established biocompatible linear polymers is necessary. This will provide the key to address important issues involving nanoscale medical devices in cancer therapy.

REFERENCES

1. Hedrick, J. L. et al., Application of complex macromolecular architectures for advanced micro-electronic materials, *Chem. Eur. J.*, 8, 3308, 2002.
2. Son, S. J. et al., Magnetic nanotubes for magnetic-field-assisted bioseparation, biointeraction, and drug delivery, *J. Am. Chem. Soc.*, 127, 7316, 2005.
3. Weissleder, R. et al., Cell-specific targeting of nanoparticles by multivalent attachment of small molecules, *Nat. Biotechnol.*, 23, 1418, 2005.
4. Katz, E. and Willner, I., Nanobiotechnology: Integrated nanoparticle–biomolecule hybrid systems: Synthesis, properties, and applications, *Angew. Chem. Int. Ed.*, 43, 6042, 2004.
5. Duncan, R., The dawning era of polymer therapeutics, *Nat. Rev. Drug Discov.*, 2, 347, 2003.
6. Gillies, E. R. and Frechet, J. M. J., Dendrimers and dendritic polymers in drug delivery, *Drug Discov. Today*, 10, 35, 2005.
7. Kim, Y. and Zimmerman, S. C., Applications of dendrimers in bio-organic chemistry, *Curr. Opin. Chem. Biol.*, 2, 733, 1998.

8. Boas, U. and Heegaard, P. M. H., Dendrimers in drug research, *Chem. Soc. Rev.*, 33, 43, 2004.

9. Tully, D. C. and Frechet, J. M. J., Dendrimers at surfaces and interfaces: Chemistry and applications, *Chem. Commun.*, 14, 1229, 2001.

10. Patri, A. K., Majoros, I. J., and Baker, J. R., Dendritic polymer macromolecular carriers for drug delivery, *Curr. Opin. Chem. Biol.*, 6, 466, 2002.

11. Haag, R. and Vogtle, F., Highly branched macromolecules at the interface of chemistry, biology, physics, and medicine, *Angew. Chem. Int. Ed.*, 43, 272, 2004.

12. Ambade, A. V., Savariar, E. N., and Thayumanavan, S., Dendrimeric micelles for controlled drug release and targeted delivery, *Mol. Pharm.*, 2, 264, 2005.

13. Bosman, A. W., Janssen, H. M., and Meijer, E. W., About dendrimers: Structure, physical properties, and applications, *Chem. Rev.*, 99, 1665, 1999.

14. Basu, S., Sandanaraj, B. S., and Thayumanavan, S., Molecular recognition in dendrimers, In *Encyclopedia of Polymer Science and Technology*, 4th ed., Mark, H. F., Ed., Wiley, New York, 2004.

15. Grayson, S. M. and Frechet, J. M. J., Convergent dendrons and dendrimers: From synthesis to applications, *Chem. Rev.*, 101, 3819, 2001.

16. Newkome, G. R., Moorefield, C. N., and Vogtle, F., Eds., *Dendrimers and Dendrons: Concepts, Syntheses, Applications*, 2nd ed., Wiley-VCH, Weinheim, 2001.

17. Abu-Rmaileh, R., Attwood, D., and D'Emanuele, A., Dendrimers in cancer therapy, *Drug Deliv. Syst. Sci.*, 3, 65, 2003.

18. Sampathkumar, S. -G. and Yarema, K. J., Targeting cancer cells with dendrimers, *Chem. Biol.*, 12, 5, 2005.

19. Matsumura, Y. and Maeda, H., A new concept for macromolecular therapeutics in cancer chemotherapy: Mechanism of tumoritropic accumulation of proteins and antitumor agent smancs, *Cancer Res.*, 46, 6387, 1986.

20. Henderson, B. W. and Dougherty, T. J., Eds., *Photodynamic Therapy: Basic Principles and Clinical Applications*, Marcel Dekker, New York, 1992.

21. Lunardi, C. N. and Tedesco, A. C., Synergic photosensitizers: A new trend in photodynamic therapy, *Curr. Org. Chem.*, 9, 813, 2005.

22. Barth, R. F. et al., Boron neutron capture therapy of cancer: Current status and future prospects, *Clin. Cancer Res.*, 11, 3987, 2005.

23. Matthews, S. E., Pouton, C. W., and Threadgill, M. D., Macromolecular systems for chemotherapy and magnetic resonance imaging, *Adv. Drug Deliv. Rev.*, 18, 219, 1996.

24. De Groot, F. M. H. et al., Cascade-release dendrimers liberate all end groups upon a single triggering event in the dendritic core, *Angew. Chem. Int. Ed.*, 42, 4490, 2003.

25. Esfand, R. and Tomalia, D. A., Poly(amidoamine) (PAMAM) dendrimers: From biomimicry to drug delivery and biomedical applications, *Drug Discov. Today*, 6, 427, 2001.

26. Malik, N. et al., Dendrimers: Relationship between structure and biocompatibility in vitro, and preliminary studies on the biodistribution of ^{125}I labeled polyamidoamine dendrimers in vivo, *J. Control. Release*, 65, 133, 2000.

27. Shi, X., Majoros, I. J., and Baker, J. R., Capillary electrophoresis of poly(amidoamine) dendrimers: From simple derivatives to complex multifunctional medical nanodevices, *Mol. Pharm.*, 2, 278, 2005.

28. Khandare, J. et al., Synthesis, cellular transport and activity of polyamidoamine dendrimer-methyl-prednisolone conjugates, *Bioconjug. Chem.*, 16, 330, 2005.

29. Patri, A. K. et al., Synthesis and in vitro testing of J591 antibody–dendrimer conjugate for targeted prostate cancer therapy, *Bioconjug. Chem.*, 15, 1174, 2004.

30. Majoros, I. J. et al., Poly(amidoamine) dendrimer-based multifunctional engineered nanodevice for cancer therapy, *J. Med. Chem.*, 48, 5892, 2005.

31. Choi, Y. et al., Synthesis and functional evaluation of DNA-assembled polyamidoamine dendrimer clusters for cancer cell-specific targeting, *Chem. Biol.*, 12, 35, 2005.

32. Bezouska, K., Design, functional evaluation and biomedical applications of carbohydrate dendrimers (glycodendrimers), *Rev. Mol. Biotechnol.*, 90, 269, 2002.

33. Woller, E. K. et al., Altering the strength of lectin binding interactions and controlling the amount of lectin clustering using mannose/hydroxyl-functionalized dendrimers, *J. Am. Chem. Soc.*, 125, 8820, 2003.

34. Page, D. and Roy, R., Synthesis and biological properties of mannosylated starburst poly(amidoamine) dendrimers, *Bioconjug. Chem.*, 8, 714, 1997.

35. Venditto, V. J. et al., PAMAM dendrimer based macromolecules as improved contrast agents, *Mol. Pharm.*, 2, 302, 2005.

36. Kobayashi, H. et al., Avidin–dendrimer–(1B4M-Gd)$_{254}$: A tumor-targeting therapeutic agent for gadolinium neutron capture therapy of intraperitoneal disseminated tumor which can be monitored by MRI, *Bioconjug. Chem.*, 12, 587, 2001.

37. Wiener, E. and Narayanan, V. V., Magnetic resonance imaging contrast agents: Theory and the role of dendrimers, *Adv. Dendr. Macromol.*, 5, 129, 2002.

38. Stiriba, S. E., Frey, H., and Haag, R., Dendritic polymers in biomedical applications: From potential to clinical use in diagnostics and therapy, *Angew. Chem. Int. Ed.*, 41, 1329, 2002.

39. Kobayashi, H. et al., Rapid accumulation and internalization of radiolabeled herceptin in an inflammatory breast cancer xenograft with vasculogenic mimicry predicted by the contrast-enhanced dynamic MRI with the macromolecular contrast agent G6-(1B4M-Gd)$_{256}$, *Cancer Res.*, 62, 860, 2002.

40. Shukla, S. et al., Synthesis and biological evaluation of folate receptor-targeted boronated PAMAM dendrimers as potential agents for neutron capture therapy, *Bioconjug. Chem.*, 14, 158, 2003.

41. Capala, J. et al., Boronated epidermal growth factor as a potential targeting agent for boron neutron capture therapy of brain tumors, *Bioconjug. Chem.*, 7, 7, 1996.

42. Kobayashi, H. et al., Activated clearance of a biotinylated macromolecular MRI contrast agent from the blood pool using an avidin chase, *Bioconjug. Chem.*, 14, 1044, 2003.

43. Pierre, V. C., Botta, M., and Raymond, K. N., Dendrimeric gadolinium chelate with fast water exchange and high relaxivity at high magnetic field strength, *J. Am. Chem. Soc.*, 127, 504, 2005.

44. Newkome, G. R. et al., Chemistry of micelles. 37. Internal chemical transformations in a precursor of a unimolecular micelle: Boron supercluster via site-specific addition of $B_{10}H_{14}$ to cascade molecules, *Angew. Chem. Int. Ed. Engl.*, 33, 666, 1994.

45. Parrott, M. C. et al., Synthesis and properties of carborane-functionalized aliphatic polyester dendrimers, *J. Am. Chem. Soc.*, 127, 12081, 2005.

46. Brewis, M. et al., Phthalocyanine-centred aryl ether dendrimers with oligo(ethyleneoxy) surface groups, *Tetrahedron Lett.*, 42, 813, 2001.

47. Freeman, A. W., Chrisstoffels, L. A. J., and Frechet, J. M. J., A Simple method for controlling dendritic architecture and diversity: A parallel monomer combination approach, *J. Org. Chem.*, 65, 7612, 2000.

48. Sivanandan, K., Sandanaraj, B. S., and Thayumanavan, S., Sequences in dendrons and dendrimers, *J. Org. Chem.*, 69, 2937, 2004.

49. Vutukuri, D., Sivanandan, K., and Thayumanavan, S., Synthesis of dendrimers with multifunctional periphery using an ABB' monomer, *Chem. Commun.*, 21, 796–797, 2003.

50. Sivanandan, K., Vutukuri, D., and Thayumanavan, S., Functional group diversity in dendrimers, *Org. Lett.*, 4, 3751, 2002.

51. Dichtel, W. R. et al., Singlet oxygen generation via two-photon excited FRET, *J. Am. Chem. Soc.*, 126, 5380, 2004.

52. Oar, M. A. et al., Photosensitization of singlet oxygen via two-photon excited fluorescence resonance energy transfer in a water-soluble dendrimer, *Chem. Mater.*, 17, 2267, 2005.

53. Nishiyama, N. et al., Light-harvesting ionic dendrimer porphyrins as new photosensitizers for photodynamic therapy, *Bioconjug. Chem.*, 14, 58, 2003.

54. Jang, W. -D. et al., Supramolecular nanocarrier of anionic dendrimer porphyrins with cationic block copolymers modified with polyethylene glycol to enhance intracellular photodynamic efficacy, *Angew. Chem. Int. Ed.*, 44, 419, 2005.

55. Ideta, R. et al., Nanotechnology-based photodynamic therapy for neovascular disease using a supramolecular nanocarrier loaded with a dendritic photosensitizer, *Nano Lett.*, 5(12), 2426–2431, 2005.

56. Zhang, G. -D. et al., Polyion complex micelles entrapping cationic dendrimer porphyrin: Effective photosensitizer for photodynamic therapy of cancer, *J. Control Release*, 93, 141, 2003.

57. Battah, S. H. et al., Synthesis and biological studies of 5-aminolevulinic acid-containing dendrimers for photodynamic therapy, *Bioconjug. Chem.*, 12, 980, 2001.

58. Vutukuri, D. R., Basu, S., and Thayumanavan, S., Dendrimers with both polar and apolar nanocontainer characteristics, *J. Am. Chem. Soc.*, 126, 15636, 2004.

59. Gopidas, K. R., Whitesell, J. K., and Fox, M. A., Metal-core-organic shell dendrimers as unimolecular micelles, *J. Am. Chem. Soc.*, 125, 14168, 2003.
60. Hawker, C. J., Wooley, K. L., and Frechet, J. M. J., Unimolecular micelles and globular amphiphiles: Dendritic macromolecules as novel recyclable solubilization agents, *J. Chem. Soc. Perkin Trans.*, 1, 1287, 1993.
61. Aathimanikandan, S. V., Savariar, E. N., and Thayumanavan, S., Temperature-sensitive dendritic micelles, *J. Am. Chem. Soc.*, 127, 14922, 2005.
62. Beezer, A. E. et al., Dendrimers as potential drug carriers; encapsulation of acidic hydrophobes within water soluble PAMAM derivatives, *Tetrahedron*, 59, 3873, 2003.
63. Kojima, C. et al., Synthesis of polyamidoamine dendrimers having poly(ethylene glycol) grafts and their ability to encapsulate anticancer drugs, *Bioconjug. Chem.*, 11, 910, 2000.
64. Neerman, M. F. et al., Reduction of drug toxicity using dendrimers based on melamine, *Mol. Pharm.*, 1, 390, 2004.
65. Zhang, W. et al., Orthogonal, convergent syntheses of dendrimers based on melamine with one or two unique surface sites for manipulation, *J. Am. Chem. Soc.*, 123, 8914, 2001.
66. Morgan, M. T. et al., Dendritic molecular capsules for hydrophobic compounds, *J. Am. Chem. Soc.*, 125, 15485, 2003.
67. Padilla De Jesus, O. L. et al., Polyester dendritic systems for drug delivery applications: In vitro and in vivo evaluation, *Bioconjug. Chem.*, 13, 453, 2002.
68. Gillies, E. R. et al., Biological evaluation of polyester dendrimer: Poly(ethylene oxide) "bow-tie" hybrids with tunable molecular weight and architecture, *Mol. Pharm.*, 2, 129, 2005.
69. Wang, D. et al., Synthesis of starlike N-(2-hydroxypropyl) methacrylamide copolymers: Potential drug carriers, *Biomacromolecules*, 1, 313, 2000.
70. Fuchs, S. et al., Fluorescent dendrimers with a peptide cathepsin B cleavage site for drug delivery applications, *Chem. Commun.*, 14, 1830–1832, 2005.

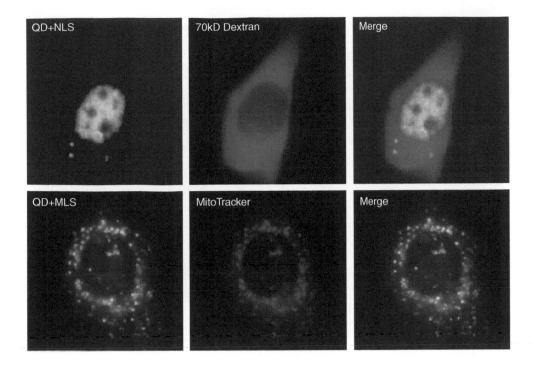

FIGURE 5.2

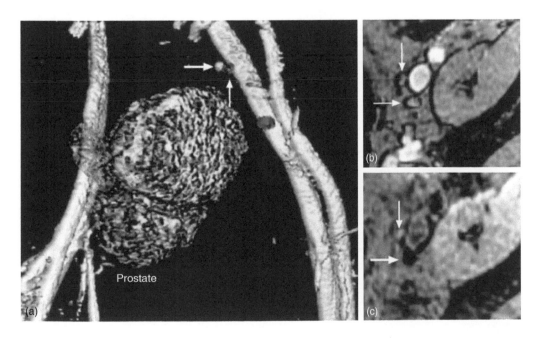

FIGURE 5.3

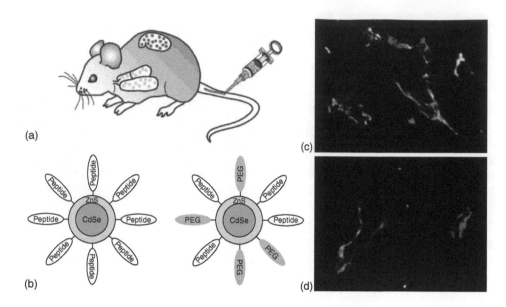

FIGURE 5.4

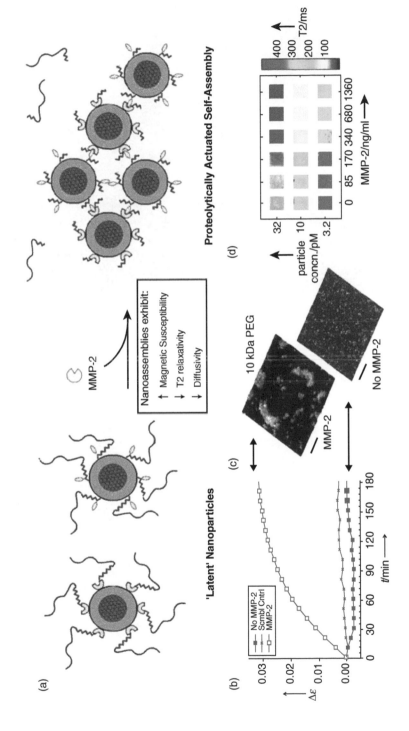

FIGURE 5.5

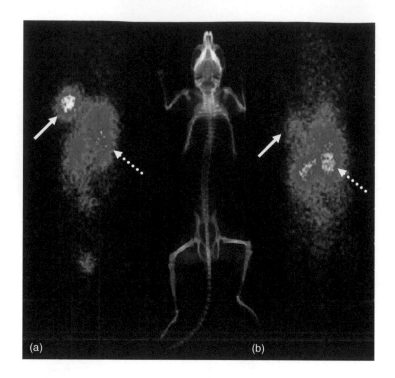

FIGURE 9.3

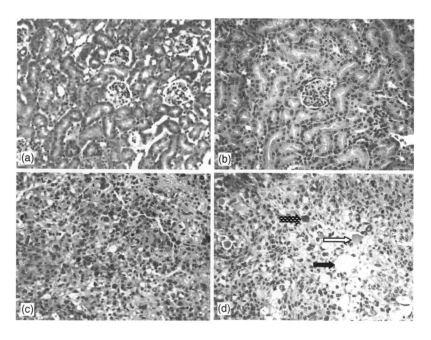

FIGURE 9.5

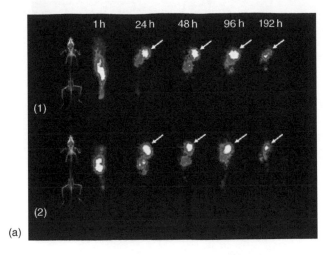

FIGURE 9.6

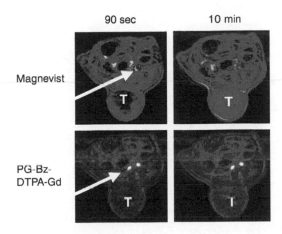

FIGURE 10.2

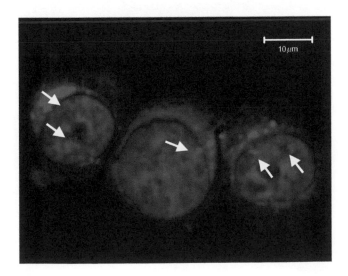

FIGURE 17.7

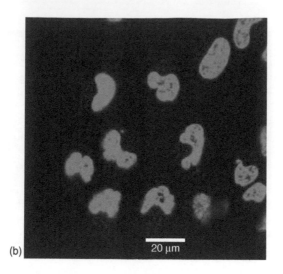

FIGURE 23.3

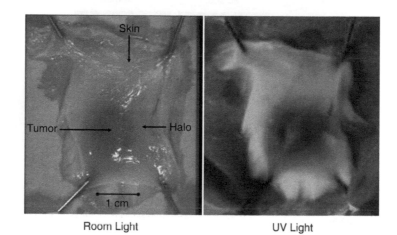

FIGURE 29.2

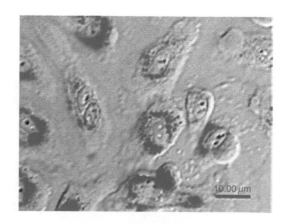

FIGURE 30.1

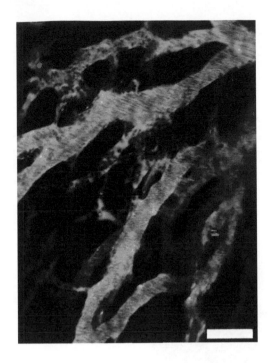

FIGURE 30.2

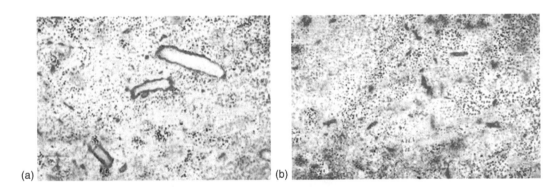

FIGURE 32.10

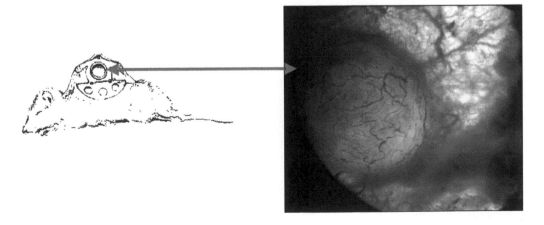

FIGURE 34.6

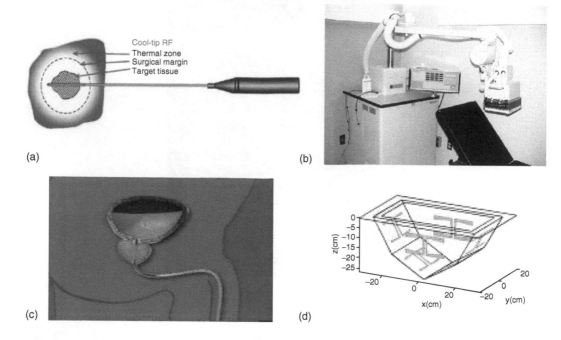

(a)

(b)

Cool-tip RF
Thermal zone
Surgical margin
Target tissue

(c)

(d)

FIGURE 34.30

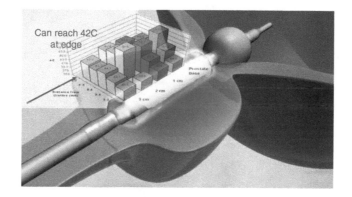

FIGURE 34.31

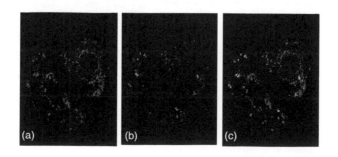

FIGURE 38.3

27 PEGylated Dendritic Nanoparticulate Carriers of Anti-Cancer Drugs

D. Bhadra, S. Bhadra, and N. K. Jain

CONTENTS

27.1   Introduction 523
27.2   Problems in Cancer Chemotherapy 524
27.3   Nanoparticles in Cancer Chemotherapy 525
     27.3.1   Challenges of Nanotechnology 527
27.4   Poly(Ethylene Glycol) Conjugation 527
     27.4.1   Derivatization and Activation of Poly(Ethylene Glycol) 528
     27.4.2   Properties of Poly(Ethylene Glycol) Conjugates 528
27.5   Dendrimer as Drug Delivery System: Selection, Applicability, and Rationality 529
     27.5.1   Differences Between Hyperbranched Structures and Dendrimers 532
     27.5.2   Classification of Dendrimers 532
     27.5.3   Features of Dendrimers as Nanodrugs 533
27.6   Applications of PEGylated Dendrimers 534
     27.6.1   Enhanced Biocompatibility for Drug Delivery 535
     27.6.2   Micellar Means of Enhancing Solubility 537
     27.6.3   Enhancing Pharmacokinetic Properties 538
     27.6.4   Dendrimer Use as Transfection Reagents 539
     27.6.5   Stimuli Responsive Carriers 540
     27.6.6   Alteration of Toxicity 541
     27.6.7   Effective Photodynamic Therapy 543
     27.6.8   Dendritic Gels 543
     27.6.9   Ligand-Based Drug Delivery 543
     27.6.10 Increasingly Effective Boron Neutron Capture Therapy 544
     27.6.11 Improved MRI Contrast Agents 544
27.7   Conclusions 545
References 545

27.1 INTRODUCTION

Cancer is a leading cause of death. Irrespective of etiology, cancer is basically a disease of cells characterized by loss of normal cellular growth, maturation, and multiplication that lead to disturbance of homeostasis. The main features of cancer are (Barar 2000) excessive cell growth, invasiveness, undifferentiated cells or tissues, the ability to metastasize or spread to newer sites and establish new growths, a type of acquired heredity where the progeny of cancer cells also retain cancerous properties, and a shift of cellular metabolism, leading to increased production of macro-molecules from nucleosides and amino acids that causes an increased metabolism of carbohydrates for cellular energy, ultimately resulting in the host's illness. This illness is attributed to pressure effects as a result of local tumor growth, destruction of the organs involved by the primary growth or its metastases, and systemic effects as a result of the new growths.

A single cancerous cell surrounded by healthy tissue will replicate at a rate higher than the other cells, placing a strain on the nutrient supply and elimination of metabolic waste products. Tumor cells will displace healthy cells until the tumor reaches a diffusion-limited maximal size. Although tumor cells will typically not initiate apoptosis in a low nutrient environment, they do require the normal building blocks of cell function like oxygen, glucose, and amino acids. The vasculature that was designed to supply the now-extinct healthy tissue could not place as high a demand for nutrients because of its slower growth rate. Tumor cells will continue dividing because they do so without regard to nutrient supply, but many tumor cells will perish because the amount of nutrients is insufficient. The tumor cells at the outer edge of a mass have the best access to nutrients, and cells on the inside die creating a necrotic core within tumors that rely on diffusion to deliver nutrients and eliminate waste products. In essence, a steady state tumor size forms as the rate of proliferation is equal to the rate of cell death until a better connection with the circulatory system is created. This diffusion-limited maximal size of most tumors is around 2 mm^3 (Grossfeld, Carrol, and Lindeman 2002). To grow beyond this size, the tumor must recruit the formation of blood vessels to provide the necessary nutrients to fuel its continued expansion. It is thought that there could be numerous tumors at this diffusion-limited maximal size throughout the body. Until the tumor can gain that access to the circulation, it will remain at this size, and the process can take years.

27.2 PROBLEMS IN CANCER CHEMOTHERAPY

In the past, chemotherapy was considered as a last resort after more successful treatments like surgery and radiotherapy had failed. The main problem of cancer chemotherapy is the lack of highly selective drugs, and the rapidly dividing normal cells of the bone marrow, gut, lymphoid tissue, spermatogenic cells, fetus, and hair follicles are also killed. Most of the antineoplastic drugs act on the processes such as DNA synthesis, transcription, or the mitotic phase, and they are labeled as cell cycle phase-specific drugs (also known as phase-dependent drugs). The phase-specific drugs do not act on G_0 phase. In contrast, there are certain drugs that kill the cells during all or most phases of the cycle, and they are labeled as cell cycle phase-nonspecific drugs (also known as phase-independent drugs). The phase-specific drugs have proven to be effective in hematological malignancies and tumors with high rate of proliferation or high growth fraction (Barar 2000).

The various obstacles for drug delivery to tumors include differences in cellular morphology, tissue immunogenecity, uncontrolled rate of growth, capacity of metastasize, and poor response to chemotherapeutic agents. Furthermore, there are variations in blood flow and vessel permeability within different regions of some tissues (Reynolds 1996). Various drug delivery systems are employed for delivery of chemotherapeutic agents to neoplastic cells. This helps in targeting drugs directly into cells and preventing drug interaction with normal tissues and alleviating side effects to normal cells. Because of a lack of highly selective drugs for tumor cells and tissues and decreased penetration of drug into neoplastic cells, a number of novel drug delivery systems, prodrugs, and other chemically modified forms of drugs having altered tissue distributions were brought in use and recently reviewed for targeting to cells located in various parts of body (Brigger, Dubernet, and Couvreur 2002).

There are various means to effectively fight the tumors such as achieving targeting by avoiding reticuloendothelial system (RES); targeted delivery through enhanced permeability and retention (EPR) effects; tumor-specific targeting; targeting through angiogenesis; targeting tumor vasculaturel; etc. (Brannon-Peppas and Blanchette 2004). Particles with longer circulation times and greater ability to target to the site of interest should be 100 nm or less in diameter and have a hydrophilic surface in order to reduce clearance by macrophages (Storm et al. 1995). Coatings of hydrophilic polymers can create a cloud of chains at the particle surface that will repel plasma proteins, and work in this area began by adsorbing surfactants to the nanoparticles surface.

Other routes include forming the particles from branched or block copolymers with hydrophilic and hydrophobic domains. One potential advantage in treatment of advance stage cancerous tissue is the inherent leaky vasculature that allows for greater accumulation of colloidal systems. The check the defective vascular architecture, created as a result of the rapid vascularization necessary to serve fast-growing cancers, coupled with poor lymphatic drainage allows an enhanced permeation and retention effect (EPR effect) (Sledge and Miller 2003). The ability to target treatment to very specific cancer cells also uses a cancer's own structure in that many cancers over express particular antigens, even on their surface. This makes them ideal targets for drug delivery as long as the targets for a particular cancer cell type can be identified with confidence and are not expressed in significant quantities anywhere else in the body. Tumor-activated prodrug therapy uses the approach that a drug conjugated to a tumor-specific molecule will remain inactive until it reaches the tumor (Chari 1998). These systems would ideally be dependent on interactions with cells found specifically on the surface of cancerous cells and not the surface of healthy cells. Most linkers are usually peptidase cleavable or acid labile but may not be stable enough in vivo to give desirable clinical outcomes.

Limitations also exist because of the lower potency of some drugs after being linked to targeting moieties when the targeting portion is not cleaved correctly or at all. For example, adriamycin-conjugated poly(ethylene glycol) linker with enzymatically cleavable peptide sequences (alanyl-valine, alanyl-proline, and glycyl-proline) or using monoclonal antibodies has shown a greater selectivity to cleavage at tumor cells (Suzawa et al. 2002). A number of targeted cancer treatments using antibodies for specific cancer types have been approved by the U.S. Food and Drug Administration like rituximab, trastuzumab, gemtuzumabozogamicin, alemtuzumab, ibritumomab tiuxetan, gefitinib, etc. (Abou-Jawde et al. 2003).

27.3 NANOPARTICLES IN CANCER CHEMOTHERAPY

Nanotechnology applied to cancer treatment may offer several promising advantages over conventional drugs. Nanoscale devices are two orders of magnitude smaller than tumor cells, making it possible for them to directly interact with intracellular organelles and proteins. Because of their molecule-like size, nanoscale tools may be capable of early disease detection using minimal amounts of tissue, even down to a single malignant cell (NCI 2005). These tools may not only prevent disease by monitoring genetic damage, but also treat cells in vivo while minimizing interference with healthy tissue. By combining different kinds of nanoscale tools on a single device, it may be possible to run multiple diagnostic tests simultaneously (www.math.uci.edu/~cristini/publications/nanochap.pdf). In particular, it is hoped that cancer drug therapy involving nanotechnology will be more effective in targeting malignant cells and sparing healthy tissue. In this regard, the role of nanoparticles loaded with chemotherapeutic drugs has been receiving much attention. Research and development in this area is expected to dramatically increase in importance in the coming years. In general, nanoscale drug delivery systems (Figure 27.1) for chemotherapy can be divided into two categories: polymer- and lipid-based (Langer 2000; Sahoo and Labhasetwar, 2003).

Polymers are usually larger than lipid molecules, and they form a solid phase such as polymeric nanoparticles, films, pellets, dendrimers; whereas, lipids form a liquid (or liquid crystalline phase) such as liposomes, cubersomes, micelles, and other emulsions (Feng and Chien 2003). Whereas

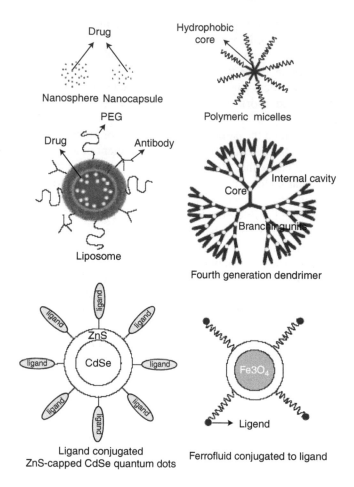

FIGURE 27.1 Representation of different nanoparticulate carriers used for drug delivery. (Adapted from Sahoo, S. K. and Labhasetwar, V., *Drug Discov. Today*, 8, 2003, www.drugdiscoverytoday.com. With permission.)

polymer-based systems are considered biologically more stable than lipid-based systems, the latter are generally more biocompatible. Polymer-based systems might possess good drug targeting ability because their uptake may be different for cells in different tissues (Maruyama 2000). In fact, Feng and Chien (2003) have suggested that a combination of polymer- and lipid-based systems could integrate their advantages while avoiding their respective disadvantages. An example of such a nanoparticle would be a liposomes-in-microspheres (LIM) system where drugs are first loaded into liposomes and then encapsulated into polymeric microspheres. This way, both hydrophobic and hydrophilic drugs can be delivered in one nanoparticle. The bioactivity of peptides and proteins would be preserved in the liposomes whose stability is protected by the polymeric matrix (Feng and Chien 2003).

Chemotherapy using nanoparticles has been studied in clinical trials for several years, and numerous studies have been published in this regard (Kreuter 1994). Two liposomally delivered drugs are currently on the market: daunorubicin and doxorubicin (Massing and Fuxius 2000). These encapsulated drugs can be formulated to maximize their half-life in the circulation. For example, a stealth version of liposomal doxorubicin coated with polyethylene glycol to reduce its uptake by the RES can extend its half-life in blood for up to 50–60 h (CCO Formulary 2003).

27.3.1 CHALLENGES OF NANOTECHNOLOGY

The difficulties faced by nanotechnology in the service of clinical medicine are numerous. These difficulties should be kept in mind while considering chemotherapeutic treatment involving nanotechnology and the potential role of biocomputation. First, there are basic physical issues that matter on a small scale. Because matter behaves differently on the nanolevel than it does at micro- and macro levels, most of the science at the nanoscale has been devoted to basic research, designed to expand understanding of how matter behaves on this length scale (www.math.uci.edu/~cristini/publications/nanochap.pdf). Because nanomaterials have large surface areas relative to their volumes, phenomena such as friction are more critical than they are in larger systems. The small size of nanoparticles may result in significant delay or speed-up in their intended actions. They may accumulate at unintended sites in the body. They may provoke unexpected immune system reactions. Cells may adapt to the nanoparticles, modifying the body's behavior in unforeseen ways (www.math.uci.edu/~cristini/publications/nanochap.pdf).

The efficacy of nanoparticles may be adversely affected by their interaction with the cellular environment. For instance, the RES may clear nanoscale devices, even stealth versions, too rapidly for them to be effective because of the tendency of the RES to phagocytose nanoparticles (Kreuter 1994). Nanoparticles can be taken up by dendritic cells (Elamanchili et al. 2004) and by macrophages (Cui, Hsu, and Mumper 2003). RES accumulation of nanoparticles could potentially lead to a compromise of the immune system. On the other hand, larger nanoparticles may accumulate in larger organs, leading to toxicity. Perhaps the biggest issue of all is that the physically compromised tumor vasculature may prevent most of the nanodevices from reaching the target cells by vascular transport or diffusion. Alterations in the tumor vasculature may adversely affect the convection of the nanodevices in the blood stream (Brigger, Dubernet, and Couvreur 2002). Local cell density and other stromal features may hamper drug or nanodevice diffusion through tumoral tissue.

27.4 POLY(ETHYLENE GLYCOL) CONJUGATION

Poly(ethylene glycol) is a non-toxic water-soluble polymer that resists recognition by the immune system. The term PEG is used for poly(ethylene glycol) that is often referred for polymer chains with molecular weight (MWs) below 20,000 Da, and poly(ethylene oxide) (PEO) refers to higher MW polymers (Harris et al. 1984; Peppas et al. 1999). These are homogeneous polymers of $M_w/M_n < 1.1$ represented by the formula $HO(CH_2CH_2O)_nH$ and is called as poly(oxy-1, 2-ethanediyl), or α-hydro-ω-hydroxy-poly(ethylene glycol). It is an additional polymer of ethylene oxide and water, where n represents the average number of oxy-ethylene groups. These are generally soluble in water, alcohol, acetone, and chloroform, but miscible with glycols and practically insoluble in ether. These are also known as Macrogols (Reynolds 1996) that are generally of higher molecular weights such as Macrogols of 20,000 Da. The pH of 5% solution is about 4–7.5.

The aqueous solutions of PEGs can be sterilized by filtration, autoclaving, etc. and are needed to be stored in airtight containers. The PEGs are relatively less toxic, but toxicity, if any appears to be greatest with macrogols of lower molecular weights. On topical administration, it may cause stinging, especially to mucous membrane. Such administration of PEGs may be associated with hypersensitivity reactions such as urticaria, and local gastrointestinal discomfort, bloating and nausea, abdominal cramps, vomiting and irritation, may result on its use in bowel cleansing preparation. Macrogols demonstrate oxidizing ability, leading to incompatibility with activity of bacitracin or benzyl-penicillin that may get reduced in macrogol bases. Estimated acceptable daily intake may be upto 10-mg/kg body weight (Reynolds 1996). Its presence in aqueous solution has no deleterious effect on protein conformation or activity of enzymes. It exhibits rapid clearance from the body, and it has been approved for a wide range of biomedical applications. PEG may transfer its properties to other molecules when it is covalently bound to that molecule. This can result in modification of number of properties of molecules as illustrated later on (Zalipsky 1995).

PEG has polyether backbone that is inert in biological environment as well as in most chemical reaction conditions as in chemical modifications and/or conjugation reactions. The chemical derivatisation of end groups of PEG is an essential first step in preparation of bioconjugates. The various properties are altered because of polyethylene glycol conjugation that is discussed in the following sections. PEG conjugation was tried and found most applicable in case of many nanoparticulate systems, using activated PEGs with appropriate spacers.

27.4.1 Derivatization and Activation of Poly(Ethylene Glycol)

Poly(ethylene glycol) has two equivalent hydroxyl groups, so this could act as potential cross-linking agent for any systems to which it is attached. These hydroxyl groups can be attached to various bio-active species by covalent coupling using number of simple PEG analogues. Some important stable analogues are bromo, amino, aminoethyl, carboxymethyl, succinimidosuccinate, tosylate, mesylate, aldehyde, octadecylamine, monopalmitate, stearoyloxy derivative of PEG, or monomethoxy-PEG (MPEG) (Buckmann and Morr 1981; Harris et al. 1984). These were prepared for reactions with sensitive bio-active systems under mild conditions.

One very important derivative that has been used in number of derivatisation reactions that has one hydroxyl group blocked is monomethyl-ether of polyethylene-glycol or monomethoxy-polyethylene-glycol (MPEG), i.e., considered equivalent in terms of reaction for formation of PEG derivatives. This was generally brought in use for conjugation to bioactive species. It is generally used when multiple chains of polymers had to be linked to the intended substrates. Because of structural simplicity and possession of only one-derivatizable end group, the use of MPEG minimizes cross-linking possibilities and leads to improved homogeneity of conjugate. Therefore, it usually is a starting material of choice for the covalent modification of proteins, biomaterials, particulates, lipids, drugs, etc. (Zalipsky 1995).

$$MeOH + H(OCH_2CH_2)_nOH \rightarrow MeO(CH_2CH_2O)_nH$$

Commercially available MPEGs are often contaminated with significant amounts (equivalent to 25%) of dihydroxy-terminated polymers and may be purified by chromatography and ethyl ether precipitation. The use of these polymers as starting materials of choice for covalent modification of proteins, biomaterials, and particles is dependent for not forming cross-linked conjugates (Harris et al. 1984).

In most cases, commercially available MPEGs of 2000–5000 MWs are used for preparation of number of reagents. MPEG-linked and based electrophiles were synthesized and used for linking with number of available attachment sites, e.g., amino acids and other nucleophilic groups. These are called activated PEGs (Zalipsky 1995). The intermediate linkers are called spacers, linking to bioactive groups, especially proteins and enzymes. These generally require mild conditions of reactions and specific properties of reactants for binding.

In order of reaction and based on ease of introduction, these can roughly be arranged as follows: arylating agents > acylating agents > alkylating agents (Zalipsky 1995). For example, tresylates, succinimidyl succinate derivatives, succinimidyl ester of carboxymethyl PEG, oxy-carboxylamino acid derivative of PEG, acetaldehyde, and hydrazide derivatives were mainly used for PEGylation of proteins and enzymes.

27.4.2 Properties of Poly(Ethylene Glycol) Conjugates

This group reviewed various properties affected by PEGylation of carriers (Bhadra et al. 2002). PEGylation has pronounced effects on various parameters of drug delivery systems (Katre 1993; Zalipsky 1995). As far as bio-distribution and pharmacokinetic, PEG conjugation increases blood circulation half-life, by reducing the tissue distribution, RES, liver, spleen, and macrophage uptake of the carriers. (Veronese et al. 1989; Papahadjopoulus et al. 1991; Phillips et al. 1999). PEG coatings alter solubility of the systems (Choi et al. 1998;Kim et al. 1999). It reduces the toxicity by decreasing the release and contact of active constituent, decreasing immunogenicity, cell-adherence, thrombogenicity, and protein adsorption. PEGylation can also lead to stabilization of delivery systems. It can also improve utilization as drug a delivery system by increasing drug loading per system, improving drug targeting as in case of liposomes by diffusion controlling mechanisms, and promoting its use as a micellar delivery system. This process can sustain and control drug delivery safely and appropriately by decreasing burst effects and by decreasing drug leakage and increasing stability of system. It can improve transfection capability (Choi et al. 1998). This can increase tumor uptake (Papahadjopoulus et al. 1991; Ishida et al. 1999) of drug and delivery systems.

Manipulation of the upper layers (corona) of nanoparticles or dendrimers confers advantageous properties to such particles, e.g., increased solubility and biocompatibility (McNeil 2005). Attaching hydrophilic polymers to the surface such as PEG greatly increases the hydration (i.e., solubility) of the nanoparticles and can protect attached proteins from enzymatic degradation when used for in vivo applications (Bhadra et al. 2002; Harris and Chess 2003). Nanoparticles with hydrophilic polymers such as PEG attached to their surface can act as a platform for lipophilic molecules, and they can overcome the solubility barrier. Insoluble compounds can be attached, adsorbed, or otherwise encapsulated in the hydrated nanoparticles. The surface addition of PEG and other hydrophilic polymers also increases the in vivo compatibility of nanoparticles. When intravascularly injected, uncoated nanoparticles are rapidly cleared from the bloodstream by the reticuloendothelial system (RES) (Brigger, Dubernet, and Couvreur 2002). Nanoparticles coated with hydrophilic polymers have prolonged half-lives, believed to result from decreased opsonization and subsequent clearance by macrophages (Moghimi and Szebeni 2003).

Cancer therapy has benefited from the use of liposomal doxorubicin, a formulation that again increases the therapeutic index of the active agent through a combination of passive tumor targeting and reduced toxicity (Gabizon and Martin 1997). In this case, coating the liposome with PEG significantly decreases uptake by macrophages and allows the liposomes to concentrate in tumors by escaping from the leaky vasculature surrounding solid tumors (Dvorak et al. 1988) through EPR effect (Maeda 2001). The surface chemistry of the nanoshells can be modified with PEG to increase biocompatibility and with sulfide-based linkers to allow the particles to be functionalized with targeting ligands.

27.5 DENDRIMER AS DRUG DELIVERY SYSTEM: SELECTION, APPLICABILITY, AND RATIONALITY

Dendrimers are a relatively new class of polymers with structures that depart rather dramatically from traditional linear polymers that are built from so-called AB monomers (Newkome, Moorefield, and Vogtle 1996). Because they are built from AB_n, monomers (where n is usually 2 or 3), dendrimers are highly branched and have three distinct structural features: a core, multiple peripheral (end-) groups, and branching units that link the two. The peripheral groups and branching units are together called dendrons, or informally, wedges. Branching units used to date have contained virtually every type of functional group, ranging from those comprising pure hydrocarbons or aromatic groups to modified carbohydrates and nucleic acids. Dendrimers are iteratively synthesized so that the number of intervening branching units between the core and one end-group (i.e., the number of layers), called the generation number, is determined by the number of synthetic cycles. In the commercially available poly(amidoamine) (PAMAM) dendrimers, ammonia represents the zeroth generation (Tomalia, Naylor, and Goddard 1990). As the generation number increases from one to seven, the number of peripheral groups follows the geometric series 3, 6,

12,…,192. These peripheral groups largely control the solubility of the compound so that even with highly hydrophobic internal units, a dendrimer will dissolve in water if the end groups are sufficiently hydrophilic. As will be discussed here, this feature and the high local concentration of end groups enable a number of applications in bio-organic chemistry. Dendrimers represent a new class of highly branched polymers whose interior cavities and multiple peripheral groups facilitate potential applications in biomedicine and bio-organic chemistry (Kim and Zimmerman 1998). Major advances in the past years were made in the synthesis and study of new carbohydrate, nucleic acid, and peptide dendrimers as well as in the use of dendrimers as magnetic resonance imaging contrast agents, agents for cellular delivery of nucleic acids, and scaffolds for biomimetic systems.

Researchers are now trying to synthesize dendrimer clusters for targeted therapy. Scientists at the University of Michigan have devised a method to easily create multipurpose molecules for use in targeted anti-cancer therapy (NCI 2005). By attaching single-stranded DNA linkers to dendrimers, the researchers could bridge together individual dendrimer subunits with specific functions and create a multifunctional dendrimer cluster. This bridging technique could potentially lead the way to developing customized therapies for individual patients. Dendrimers are promising candidates for use as the base of anti-cancer drugs to which functional molecules can be attached.

Ideally, an anti-cancer therapeutic agent would have multiple functional groups such as a targeting molecule to bind a cell, a radioactive compound to kill the cell, and a fluorescent probe so the process can be imaged. However, there are problems synthesizing these compounds, and different drugs would have to be designed for each different tumor with this approach. Dr. James Baker, Jr. and colleagues improved compound synthesis through a building block approach in one NCI-funded study. They created two single-function dendrimers; one with a molecule to bind folate receptors and the other with a fluorescein molecule for imaging. They then added complimentary 34-base stretches of single-stranded DNA to each, and the two dendrimers conjugated through base pairing. Using fluorescence imaging, they observed that the dendrimer cluster specifically bound to a cancer cell line over expressing the folate receptor. Compared with traditional chemistry, DNA-linked dendrimers could provide a more effective way to mix and match different therapeutic combinations to better treat individual patients.

Dendrimers represent one nanostructured material that may soon find its way into medical therapeutics. Starburst dendrimers are tree-shaped synthetic molecules with a regular branching structure emanating outward from a core that forms nanometer by nanometer with the number of synthetic steps or generations dictating the exact size of the particles, typically a few nanometers in spheroidal diameter (Freitas 2005). The peripheral layer can be made to form a dense field of molecular groups that serve as hooks for attaching other useful molecules such as DNA that can enter cells while avoiding triggering an immune response unlike viral vectors commonly employed today for transfection. Upon encountering a living cell, dendrimers of a certain size trigger a process called endocytosis where the cell's outermost membrane deforms into a tiny bubble or vesicle. The vesicle encloses the dendrimer that is then admitted into the cell's interior. Once inside, the DNA is released and migrates to the nucleus where it becomes part of the cell's genome. The technique has been tested on a variety of mammalian cell types (Kukowska-Latallo et al. 2000) and in animal models (Vincent et al. 2003) though clinical human trials of dendrimer gene therapy remain to be done.

Glycodendrimer nanodecoys have also been used to trap and deactivate some strains of influenza virus particles (Reuter et al. 1999; Landers et al. 2002). James Baker's group at the University of Michigan is extending this work to the synthesis of multi-component nanodevices called tecto-dendrimers built up from a number of single-molecule dendrimer components (http://nano.med.u-mich.edu/projects/Dendrimers.html; Quintana et al. 2002). Tecto-dendrimers have a single core dendrimer surrounded by additional dendrimer modules of different types, each type designed to perform a function necessary to a smart therapeutic nanodevice (Figure 27.2). Baker's group has

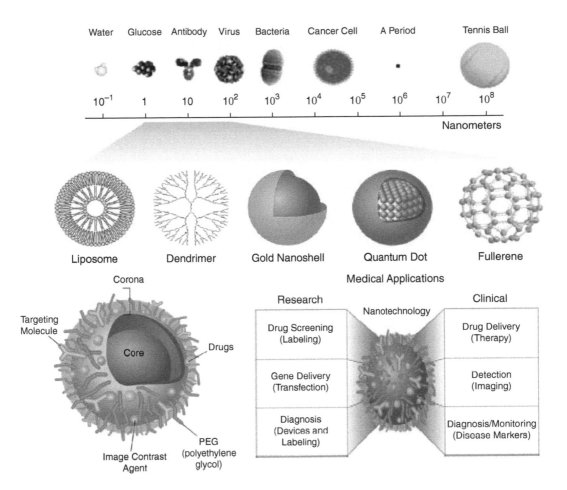

FIGURE 27.2 Morphological and size-based comparison of various novel drug delivery carriers such as liposome, dendrimer, gold nanoshell, quantum dot, fullerene, and nanometric scale of representation of various similar systems. (Adapted from McNeil, S. E., *J. Leukoc. Biol.*, 78, 1–10, 2005. With permission.)

built a library of dendrimeric components from which a combinatorially large number of nanodevices can be synthesized.

The initial library contains components that will perform the following tasks: diseased cell recognition, diagnosis of disease state, drug delivery, report location, and report outcome of therapy. By using this modular architecture, an array of smart therapeutic nanodevices can be created with little effort. For instance, once apoptosis-reporting, contrast-enhancing, and chemotherapeutic-releasing dendrimer modules are made and attached to the core dendrimer, it should be possible to make large quantities of this tecto-dendrimer as a starting material. This framework structure can be customized to fight a particular cancer simply by substituting any one of many possible distinct cancer recognition or targeting dendrimers, creating a nanodevice customized to destroy a specific cancer type and no other while also sparing the healthy normal cells. These nanodevices were synthesized using an ethylenediamine core with folic acid, fluorescein, and methotrexate covalently attached to the surface to provide targeting, imaging, and intracellular drug delivery capabilities (Freitas 1999). The "targeted delivery improved the cytotoxic response of the cells to methotrexate 100-fold over free drug" (Quintana et al. 2002).

At least a half dozen cancer cell types have already been associated with at least one unique protein that targeting dendrimers could use to identify the cell as cancerous, and as the genomic revolution progresses, it is likely that proteins unique to each kind of cancer will be identified, allowing Baker to design a recognition dendrimer for each type of cancer (http://nano.med.umich.edu/projects/Dendrimers.html). The same cell-surface protein recognition-targeting strategy could be applied against virus-infected cells and parasites. Molecular modeling has been used to determine optimal dendrimer surface modifications for the function of tecto-dendrimer nanodevices and to suggest surface modifications that improve targeting (Quintana et al. 2002).

NASA and the National Cancer Institute have funded Baker's lab to produce dendrimer-based nanodevices that can detect and report cellular damage because of radiation exposure in astronauts on long-term space missions (Sparks 2002). By mid-2002, the lab had built a dendrimeric nanodevice to detect and report the intracellular presence of caspase-3, one of the first enzymes released during cellular suicide or apoptosis (programmed cell death) that is one sign of a radiation-damaged cell. The device includes one component that identifies the dendrimer as a blood sugar so that the nanodevice is readily absorbed into a white blood cell and a second component using fluorescence resonance energy transfer (FRET) that employs two closely bonded molecules. Before apoptosis, the FRET system stays bound together, and the white cell interior remains dark upon illumination. Once apoptosis begins and caspase-3 is released, the bond is quickly broken, and the white blood cell is awash in fluorescent light. If a retinal scanning device measuring the level of fluorescence inside an astronaut's body reads above a certain baseline, counteracting drugs can be taken.

27.5.1 DIFFERENCES BETWEEN HYPERBRANCHED STRUCTURES AND DENDRIMERS

Dagani (1996) discussed and compared hyperbranched polymers and dendrimers. Hyperbranched polymers have highly branched structure, but unlike dendrimers, their structure is neither regular nor highly symmetrical. The dendrimers were obtained by careful stepwise growth of successive layers or generations. Hyperbranched polymers are obtained in a single step by polycondensation of A_2B monomers that contain two reactive groups of type A and one of type B. Functional groups A and B are selected in a way that they react with each other to form a covalent bond. Structure of hyperbranched polymers is thought to be intermediate between those of linear polymers and dendrimers. Because of steric constraints and statistical nature of coupling steps, hyperbranched polymers have irregular structures as some of the A groups remain unreacted, leading to incorporation of linear segments (i.e., monomers units coupled at two rather than three points) (Frechet 1994). Voit, Turner, and Mourey (1993) showed that viscosity-MW behavior of the hyperbranched polymers does not show a maximum as observed in the case of dendrimers. Instead, they obey the Mark–Hownick–Sakunada relation. Although their viscosities are anomalously lower when compared to linear polymers of comparable size. Analogous to dendrimers, A_2B hyperbranched polymers had essentially same number of reactive groups as they have monomeric units. However, unlike dendrimers, these functional groups are not all located at chain ends (Frechet 1994).

27.5.2 CLASSIFICATION OF DENDRIMERS

The dendrimers that are most studied and reported up to higher generations (Bosman, Janssen, and Meijer 1999) can be classified chemically viz. peptide dendrimers that were patented and described upto higher generations by Denkewalter in the early 1980s. Those can well be characterized by size-exclusion chromatography only. PAMAM dendrimers are another class of dendritic structures that have been thoroughly investigated and had received wide spread attention. These were divergently constructed, e.g., PAMAM described by Tomalia and arborol systems of Newkome. Polypropyleneimine dendrimers were produced by Vogtle by divergent synthesis. It was also produced by heterogeneously catalyzed hydrogenation by De Brabander-van-den Berg and Meijer (1993) and is also studied widely. Frechet produced Poly-ether dendrimers by convergent procedures (Wooley et al. 1994). These also include Poly(aryl ether) dendrimers. Moore produced Phenyl acetylene dendrimers by convergent approach. Tetrathiofulvalene $(TTF)_{21}$-glycol dendrimers are another class of dendrimers that are intra-dendrimer aggregates of TTF cation radicals (Christensen et al. 1998). These are evident by spectrochemical studies of partially oxidized TTF units. These were prepared by convergent approaches. Pentaporphyrin is another starburst porphyrin polymer that is considered as a first generation dendrimers (Norsten and Branda 1998). These had hybrid properties of porphyrin and dendrimers where ether linkages are found for iterations. Aryl ester monodispersed dendrimers based on 1, 3, 5-benzene tricarboxylic acid were initially described by Miller, Kwock, and Neenan (1992) and prepared by convergent synthesis. This was possibly used for MWs standards as polymeric rheological modifier, molecular inclusion hosts, etc. Two types of dendrimeric nucleic acids have been studied, viz. those that are covalently connected, and those that self-assemble via complementary base-pairing schemes (Kim and Zimmerman 1998). An example of a covalent DNA dendrimer was reported, wherein a second-generation phosphodiester-based dendrimer derived from pentaerythritol held nine pentathymidy-late units on the end-groups and either a 15-mer or 26-mer at the core.

Additionally, many other types of interesting, valuable, aesthetically pleasing dendritic systems have been developed such as Dendrophanes that were first described by Diederich and coworkers. They called these phenyl methane-based cyclophane. These were designed as globular proteins. Metallodendrimers include Ruthenium-terpyridine complexed dendrimers by Newkome and coworkers, dendritic iron (II) complexes by Chow and coworkers, zinc-porphyrin dendrimers by Diederich and coworkers, etc. and they contain metal complexed in dendrimer structure. Polyamino phosphine containing dendrimers are phosphorous containing dendrimers. Mesogen functionalized carbosilane dendrimer involves functionalization by 36 mesogenic units attached through C-5 spacer to liquid crystalline dendrimer that form smectic-A phase in a temperature range of 117–130°C. Dendritic box was based on construction of a chiral shell of protected amino acids onto polypropyleneimine dendrimers with 64 amino end groups. These monodispersed dendritic containers of nanometric dimensions have physically entrapped or locked in guest molecules (Jansen et al. 1994; Jansen and Meijer 1995). Toyokuni et al. (1994) described an example of carbohydrate dendrimers for tumor-associated carbohydrate antigens. They linked it without the use of macromolecular carrier or an adjuvant, and they conjugated it with starburst PAMAM dendrimers to elicit antibody responses.

Dendrimeric systems can also be classified on the basis of physical characteristics of the system such as simple dendrimers having simple monomeric units; liquid crystalline dendrimers having mesogenic liquid crystalline substances; chiral dendrimers having optical activities; and micellar dendrimers having unimolecular micellar structures or water-soluble hyperbranched dendrimers of polyphenylene, etc. Hybrid dendrimers are combinations of dendritic and linear polymers in hybrid block or graft copolymers forms, whereas amphiphilic dendrimers are a class of globular dendri-mers having unsymmetrical, but highly controlled, distributions of chain end chemistry. These are oriented at interfaces forming interfacial liquid membranes for stabilizing aqueous organic emul-sion. These may be oriented under influence of external stimulus, e.g., electric field.

27.5.3 Features of Dendrimers as Nanodrugs

The various features of dendrimeric formulations in use include cost and ease of manufacture for cost effective (compared with small molecule drugs), and such formulations can be scaled-up for Current Good Manufacturing Practices (cGMP) guidelines manufacture. These are effective and active against a wide range of diseases and for their novel modes of action. Also at therapeutic doses, the dendrimers are safe and stable as solids, and they are thus available in a variety of Pharmaceutical formulations. The formulations are reproducible and defined as far as identity is concerned. They appear in most cases as white to off-white powder. Starpharma-like Pharma-ceuticals (www.starpharma.com/framemaster.htm) are creating value from such dendrimer-based nanotechnology. Their core business strategies include developing high-value dendrimer nano-drugs to address unmet market needs, partnering with pharmaceutical companies to create new opportunities and solutions to problems with the application of dendrimer nanotechnology, extending core skills and know-how through licensing and partnering with others, and investing in non-pharmaceutical applications of dendrimer technology. Rational drug design is being used by the biotechnology and pharmaceutical industry to design novel drugs based on known biological targets (e.g., receptors involved in the pathogenesis of disease). In this context, Starpharma's dendrimer technologies offer a unique ability to design functionalized, polyvalent dendrimers as defined species for specific biological targets.

The dendrimers are being eyed as drug-delivery agents, micelle mimics, and nanoscale building blocks. There are also added possibilities for grafting of active groups on surface of dendrimer by using cross-linkers. Also fatty acids, and polymers such as poly(ethyleneglycol) (PEG), etc. can be attached to dendrimers due to $-COOH$ and $-NH_2$ terminal active groups present on its surface. PEG coatings, also called as PEGylation, could also have many possible numerous advantages so could be selected for coating (Zeng and Zimmerman 1997). Gitsov and Fréchet (1996) published some preliminary results on macromolecules that change shape when the polarity of the solvent changes. These macromolecules consist of four long PEG chains extending out from a central carbon atom. Each of these hydrophilic PEG arms is terminated with a hydrophobic wedge-like dendritic group based on 3,5-dihydroxybenzyl alcohol. The arms are long enough and flexible enough to allow this hybrid star to assume several very different conformations in solution. In tetrahydrofuran, the hybrid star forms a unimolecular micelle that has a hydrophilic core of tightly packed PEG arms surrounded by a loose hydrophobic shell of dendritic wedges. In chlorinated solvents such as chloroform where both building blocks are readily soluble, the PEG arms (and their dendritic extremities) extend outward, making the core more accessible. In polar or aqueous media such as methanol, the hydrophilic PEG arms loop around the hydrophobic dendritic wedges, pushing the wedges into the star's core and leaving an outer PEG layer exposed to the outside world. The hybrid stars rearrange their two types of components in this way to minimize the overall free energy of the system.

A novel hybrid star prepared in Fréchet's lab at the University of California Berkeley changes its conformation when the polarity of the solvent medium changes. In THF (tetrahydrofuran), the hydrophilic PEG core is compact, whereas in chloroform, it is more extended and open. In methanol, the hydrophilic PEG arms envelope the hydrophobic dendritic wedges, leading to two possible conformations. In one, the dendritic wedges are tightly packed toward the middle of the micelle with the PEG arms forming loops in the surrounding medium. In the other arrangement, each dendritic moiety is individually wrapped in a PEG arm and somewhat extended into the medium. Freemantle (1999) also explored blossoming of dendrimers in a conference that explored design, synthesis, structure, and potential applications of highly branched macromolecules.

27.6 APPLICATIONS OF PEGYLATED DENDRIMERS

Dendrimers are synthetic, nanoscale structures (1–100 nm) types of polyvalent pharmaceuticals (www.starpharma.com/framemaster.htm) that can be tailored for many pharmaceutical applications. Specialized chemistry techniques allow for precise control of the physical and chemical properties of dendrimers. They are constructed in a series of controlled steps (see below) that increase the number of small branching molecules around a central core molecule. The final generation of molecules added to the growing structure makes up the polyvalent surface of the dendrimer. The core branching and surface molecules are chosen to give desired properties and functions. Dendrimers allow researchers, for the first time, to produce highly defined and biocompatible nanoscale objects built from the bottom up. This high definition enables unique functionality in life sciences applications. For readers interested in a historical account, terminology, nomenclature, or other background material, an excellent monograph by Newkome, Moorefield, and Vogtle (1996) is recommended. Zeng and Zimmerman (1997) broadly reviewed applications of dendrimers in bioorganic chemistry that have been reported in the past year, and, although somewhat different in focus, it updates earlier reviews published. An excellent review of dendrimers that act as biological mimics was also published by Smith and Diederich (1998). Beyond discussing a couple of results in the area of biomimetic dendrimers, this review highlighted various work on dendrimers whose monomer units are derived from carbohydrates, amino acids, or nucleotides and also several promising biomedical applications under active investigation.

27.6.1 ENHANCED BIOCOMPATIBILITY FOR DRUG DELIVERY

PAMAM dendrimers having poly(ethylene glycol) (PEG) grafts were designed as a novel drug carrier that possesses an interior for the encapsulation of drugs and a biocompatible surface. PEG monomethyl ether with the average MW of 550 or 2000 Da was combined to essentially every chain end of the dendrimer of the third or fourth generation via urethane bond (Kojima et al. 2000). The PEG-attached dendrimers encapsulating anti-cancer drugs adriamycin and methotrexate were prepared by extraction with chloroform from mixtures of the PEG-attached dendrimers and varying amounts of the drugs. Their ability to encapsulate these drugs increased with increasing dendrimer generation and chain length of PEG grafts. Among the PEG-attached dendrimers prepared, the highest ability was achieved by the dendrimer of the fourth generation having the PEG grafts with the average MW of 2,000 Da that could retain 6.5 adriamycin molecules or 26 methotrexate molecules/dendrimer molecule. The methotrexate-loaded PEG-attached dendrimers slowly released the drug in an aqueous solution of low ionic strength. However, in isotonic solutions, methotrexate and doxorubicin were readily released from the PEG-attached dendrimers.

Padilla et al. (2002) described the use of high MW polymers (> 20,000 Da) as soluble drug carriers to improve drug targeting and therapeutic efficacy. They evaluated various dendritic architectures composed of a polyester dendritic scaffold based on the monomer unit 2,2-bis(hydroxymethyl) propanoic acid for their suitability as drug carriers both in vitro and in vivo. These systems are both water-soluble and non-toxic. In addition, the potent anti-cancer drug doxorubicin was covalently bound via a hydrazone linkage to a high MW 3-arm poly(ethylene oxide) (PEO)-dendrimer hybrid. Drug release was a function of pH, and the release rate was more rapid at pH < 6. The cytotoxicity of the DOX-polymer conjugate measured on multiple cancer lines in vitro was reduced but not eliminated, indicating that some active doxorubicin was released from the drug polymer conjugate under physiological conditions. Furthermore, biodistribution experiments show little accumulation of the DOX-polymer conjugate in vital organs, and the serum half-life of doxorubicin attached to an appropriate high MW polymer has been significantly increased when compared to the free drug. Therefore, this new macromolecular system exhibits promising characteristics for the development of new polymeric drug carriers.

Kolhe et al. (2003) attached PEG grafts to the terminal amine groups of the dendrimer and enhanced biocompatibility of higher generation PAMAM dendrimer. Similarly, Bhadra et al. (2003) described the use of uncoated and PEGylated newer PAMAM dendrimers for delivery of an anti-cancer drug 5-fluorouracil. The 4.0 G PAMAM dendrimer was PEGylated using N-hydroxysuccinimide-activated carboxymethyl MPEG-5000. The PEGylation of the systems was found to have increased their drug-loading capacity, reduced their drug release rate, and hemolytic toxicity. TEM study revealed quite uniform surface of the systems. The systems were found suitable for prolonged delivery of an anti-cancer drug by in vitro and blood-level studies in albino rats without producing any significant hematological disturbances. PEG modification has been found to be suitable for modification of PAMAM dendrimers for reduction of drug leakage and hemolytic toxicity. This, in turn, could improve drug-loading capacity and stabilize such systems in the body. The study suggests use of such PEGylated dendrimeric systems as nanoparticulate depot type of system for drug administration (Figure 27.3).

The entrapment of 5-fluorouracil in PEG modification dendrimers significantly increased by 12 times because of more sealing of dendrimeric structure by PEG coating at the peripheral portions of

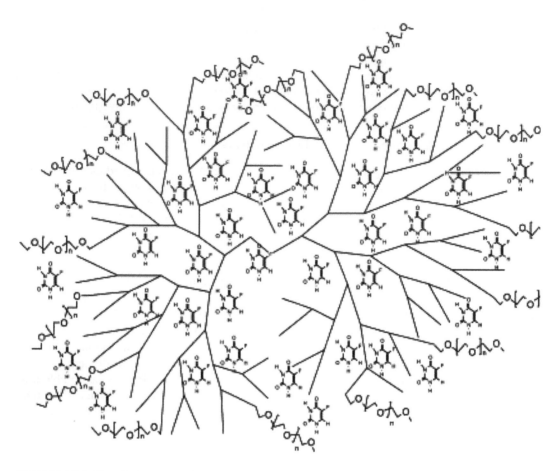

FIGURE 27.3 Chemical structural representation of PEG coated PAMAM loaded with 5-fluorouracil. (Adapted from Bhadra, et al., *Int. J. Pharm.*, 257, 111, 2003. With permission.)

dendrimers as coat that prevented drug release by enhancing complexation probably by steric and electronic effects of the additional functional groups made available by MPEG (Figure 27.4). This, in turn, also reduced the rate of drug release from such systems across dialysis membrane as observed by their release profile. The PAMAM dendrimers on PEGylation behaved similar to

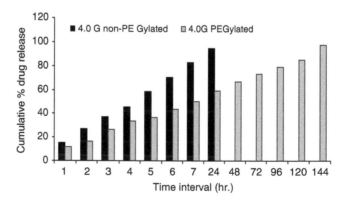

FIGURE 27.4 Comparative cumulative percentage release of 5-fluorouracil from PEG coated and uncoated PAMAM dendrimers. (Adapted from Bhadra, et al., *Int. J. Pharm.*, 257, 111, 2003. With permission.)

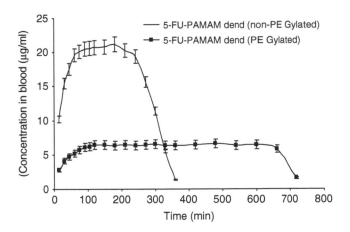

FIGURE 27.5 Blood level data of 5-fluorouracil from PEG coated and uncoated PAMAM dendrimers. (Adapted from Bhadra, et al., *Int. J. Pharm.*, 257, 111, 2003. With permission.)

spherical unimolecular polymer micelles with chains of hydrated methoxy-poly(ethyleneglycol) on its surface as coat. This increased the entrapment capacity of dendrimers and made them act as nanometric containers carrying drugs. Such PEGylated dendrimers can be suggested for sustaining delivery of the drug 5-fluorouracil as also observed by the blood level data where blood level was much prolonged and was detectable up to 12 h. The formulations were found to be following sustained release characteristics for 5-FU (Figure 27.5) as shown by the relative increase in MRT for both non-PEGylated and PEGylated drug-dendrimer complexes as compared to plain drug solution (6.024 and 13.31 times, respectively). The release rate, however, was found to have increased in vivo as compared to in vitro data, possibly because of the metabolism by the enzymes and hydrolysis in the body. The blood level of the drug was found to be lower in case of PEGylated systems than that of non-PEGylated as a result of slower release rates of the drug similar to the trend found in vitro for the drug release. This might also be attributed to better cellular penetration and adhesion properties of the PEG chains of PEGylated systems that might lead to entanglement of such systems within capillary fenestration from which such systems might release drugs acting as nanoparticulate depot type carrier present in blood circulation.

27.6.2 MICELLAR MEANS OF ENHANCING SOLUBILITY

Chapman and coworkers prepared amphiphilic copolymers derived from linear PEO and tBoC-terminated poly-$\alpha, \in$-lysine dendrimers. The use of PEO as a platform for the dendrimers synthesis greatly facilitated separation because product up to fourth generation could be precipitated from reaction mixtures by ether. Surface tension method was employed to determine Critical Micelle Concentration (CMC) and micelle formation behavior. The surface of the aggregates was suggested to be highly compact. The existence of transition concentration for dye solubility of dye orange OT by G4 hydramphiphiles in water supports the micellar behavior (Chapman et al. 1994).

Frechet and coworkers also synthesized another novel class of amphiphilic star copolymers. A four-armed PEG star was attached to four polyether dendrons derived from a pentaerythritol core. Results of SEC/VISC and ^{1}HNMR studies indicated the formation of unimolecular micelles in chloroform, tetrahydrofuran, or methanol, but with strikingly different structures. The star copolymers could self organizes into different micellar structures as a function of the environment. A potential application of these stimuli responsive copolymers involves their uses as solvent specific encapsulation agents (Gitsov and Frechet 1996).

Water-soluble dendritic unimolecular micelles as potential drug delivery agents had been explored by Liu et al. (2000) with the hydrophobic core surrounded by hydrophilic shell. These were prepared by coupling dendritic hypercores with PEG-mesylates. The monomeric core that was selected to build it was 4,4-bis(4'hydroxy phenyl) pentanol as this larger monomeric unit provides flexibility to the dendritic structures while contributing to the container capacity of the overall structure. Four generations of dendritic hypercores with 6, 12, 24, and 48 phenolic end groups were prepared. Subsequent coupling reaction with PEG-mesylates afforded four generations of dendritic unimolecular micelles (Gitsov and Fréchet 1993). The container property was demonstrated by pyrene solubilization in aqueous solution. In general, copolymers containing low generation dendrons, especially G1, tended to form unimolecular micelle, whereas G2 and G3 copolymers formed multi molecular micelles, presumably driven by hydrophobic effects and π–π interaction between dendritic blocks. Coexistence of two well-separated peaks in the SEC indicated a slow exchange process between these two species of different sizes.

Ooya, Lee, and Park (2003) developed new methods and pharmaceutical compositions to increase the aqueous solubility of paclitaxel (PTX), a poorly water-soluble drug. Graft and star-shaped graft polymers consisting of PEG400 graft chains increased the PTX solubility in water by three orders of magnitude. Polyglycerol dendrimers (dendriPGs) dissolved in water at high concentrations without significantly increasing the viscosity and at 80 wt% were found to increase the solubility of PTX 10,000-fold. The solubilized PTX was released from graft polymers, star-shaped graft polymers, and the dendriPGs into the surrounding aqueous solution. The release rate was a function of the star shape and the dendrimer generation. The availability of the new graft, star, and dendritic polymers having ethylene glycol units should permit development of novel delivery systems for other poorly water-soluble drugs.

27.6.3 Enhancing Pharmacokinetic Properties

New polyester dendrimer-PEO bow-tie hybrids were evaluated for their potential as drug delivery vehicles and to explore the effect of MW and architecture on their pharmacokinetic properties (Elizabeth et al. 2005). In vitro experiments showed that these polymers were non-toxic to MDA-MB-231 cells and that they degraded to lower MWs at the normal physiological pH of 7.4 and at the mildly acidic pH of 5.0 that may be encountered upon uptake of these molecules by endocytosis and subsequent trafficking to endosomes and lysosomes. Biodistribution studies showed that all carriers with MWs of 40,000 and greater had long plasma circulation times, whereas those with lower MWs were cleared more rapidly with significant quantities excreted in the urine. In general, it was found that the more branched [G-3] polymers exhibited increased circulation times and decreased renal clearance relative to the less branched polymers, a property likely attributable to their decreased flexibility and resulting difficulty in passing through glomerular pores. It was also demonstrated that the more branched structures provide increased levels of steric protection for the payload on the drug-carrying dendron at the core of the molecule. High levels of tumor accumulation were found for two [G-3] polymers in mice bearing subcutaneous B16F10 melanoma. Overall, the features of these new branched carriers include degradability, lack of toxicity, long circulation half-lives, and high levels of tumor accumulation make them very promising polymers for therapeutic applications. High MW polymers have shown promise in terms of improving the properties and the efficacy of low MW therapeutics. However, new systems that are highly biocompatible are biodegradable, have well-defined MW, and have multiple functional groups for drug attachment are still needed. The biological evaluation of a library of eight polyester dendrimer-PEO bow-tie hybrids is described here.

The group of evaluated polymers was designed to include a range of MWs (from 20,000 to 160,000) and architectures with the number of PEO arms ranging from two to eight. In vitro experiments revealed that the polymers were non-toxic to cells and were degraded to lower MW species at pH 7.4 and pH 5.0. Biodistribution studies with ^{125}I-radiolabeled polymers showed that the high MW carriers (>40,000) exhibited long circulation half-lives. Comparison of the renal clearances for the four-arm versus eight-arm polymers indicated that the more branched polymers were excreted more slowly into the urine, a result attributed to their decreased flexibility. Because of their essentially linear architecture that does not provide for good isolation of the iodinated phenolic moieties, the polymers with two arms were rapidly taken up by the liver. The biodistributions of two long-circulating high MW polymers in mice bearing subcutaneous B16F10 tumors were evaluated, and high levels of tumor accumulation were observed. These new carriers are promising for applications in drug delivery and are also useful for improving the understanding of the effect of polymer architecture on pharmacokinetic properties (Elizabeth et al. 2005).

27.6.4 DENDRIMER USE AS TRANSFECTION REAGENTS

The recent completion of the human genome sequence has provided promising tools for defining cancer-specific targets, including genes and their expression patterns. In addition, the development in proteomics and achievements of high-throughput screenings offer great potential for discovering effective drugs, including DNA drugs, for gene therapy. Therefore, cancer therapy is no longer limited by the identification of genes; it is limited by an inability to deliver them efficiently to cancer cells, and the bottleneck in cancer gene therapy will soon be shifting from gene discovery to gene delivery (Luo 2004). The goal is to provide adequate amounts of drugs that are close to targeted cells because the actual destination for all gene delivery is the cell nucleus. Only drugs that are close to cells can be internalized. Once inside, intracellular barriers are considerably different. For example, diffusion is no longer the bottleneck. Rather, cytosol survival and nuclear targeting are more important (for non-viral gene delivery). An important and difficult task for cancer gene therapy is increasing the expression level of the therapeutic gene (that is usually toxic) in targeted tumor cells, but at the same time, avoiding its expression in normal tissue (Luo and Saltzman 2000).

Methoxypoly(ethylene glycol)-*block*-poly(L-lysine) dendrimer was designed to form a water-soluble complex with plasmid DNA. The copolymer was synthesized by the liquid-phase peptide synthesis method (Choi et al. 1999). Agarose gel electrophoresis and DNase 1 protection assay proved that this linear polymer/dendrimer block copolymer spontaneously assembled with plasmid DNA, forming a water-soluble complex that increased the stability of the complexed DNA. Atomic force microscopy of the complex was evaluated at various charge ratios, showing that the copolymer/DNA complex was like a globular shape.

Choi et al. (2000) synthesized and used a barbell-like triblock copolymer, poly(L-lysine) dendrimer-*block*-poly(ethylene glycol)-*block*-poly(L-lysine) dendrimer, for delivery of plasmid DNA by its self-assembly with the structures. A barbell-like ABA-type triblock copolymer, poly(L-lysine) dendrimer-*block*-poly(ethylene glycol)-*block*-poly(L-lysine) dendrimer (PLLD–PEG–PLLD), was synthesized by the liquid-phase peptide synthesis method. The self-assembling complex formation of the third and fourth generation of the copolymer with plasmid DNA was studied. ^{1}H NMR and matrix-assisted laser desorption/ionization-time-of-flight mass spectrometry (MALDI-TOF MS) were used for the characterization of the synthesized copolymer. The self-assembling behavior of the third and fourth generations of the copolymer with plasmid DNA was investigated by electrophoretic mobility shift assay, DNase I protection assay, and ethidium bromide exclusion assay. They observed great differences in the self-assembling ability of the third and fourth generations of the polymer. This suggests that the number of positively charged amines per polymer molecule should be an important factor for the potential for self-assembling complex formation with DNA. Atomic force microscopy (AFM) and ζ potentials were used for evaluating the shape, size distribution, and surface charge of the complexes at various charge ratios. From AFM images, it was observed that the shape of the complex was nearly spherical, and its size was about 50–150 nm in diameter. The in vitro cytotoxicity of the copolymer was compared with that of poly(L-lysine), poly(D-lysine), and polyethylenimine.

Luo et al. (2002) demonstrated a simple and successful synthetic approach to devise a highly efficient DNA delivery system with low cytotoxicity and low cost. PAMAM dendrimer is a highly efficient DNA delivery agent when compared to other chemical transfection reagents. Partially degraded, high-generation dendrimers offer even higher efficiency, presumably because of enhanced flexibility of the otherwise rigid dendrimer chains. It was hypothesized that chemical modification of low generation dendrimer with biocompatible PEG chains would create a conjugate of PAMAM core with flexible PEG chains that mimics the fractured high-generation dendrimer and produces high transfection efficiency. Generation 5 PAMAM dendrimer was modified with 3400 MW PEG. The novel conjugate produced a 20-fold increase in transfection efficiency compared with partially degraded dendrimer controls. The cytotoxicity of PEGylated dendrimers was very low. This extremely efficient, highly biocompatible, low-cost DNA delivery system can be readily used in basic research laboratories and may find future clinical applications.

This system can be readily used in basic research laboratories and may be modified and applied in future clinical usage. Furthermore, similar approaches could be adopted with other DNA delivery agents that may create new opportunities for using DNA as a drug via a synthetic delivery route.

Mannisto et al. (2002) investigated the influence of shape, MW, and PEGylation of linear, grafted, dendritic, and branched poly(L-lysine) on their DNA delivery properties. DNA binding, condensation, complex size and morphology, cell uptake, and transfection efficiency were determined. Most poly(L-lysine) condense DNA linear polymers and are at the same time more efficient than most dendrimers. At low MWs of PLL, DNA binding and condensation were less efficient, particularly with dendrimers. PEGylation did not decrease DNA condensation of PLLs at less than 60% (fraction of MW) of PEG. PEGylation sterically stabilized the complexes but did not protect them from interaction with polyanionic chondroitin sulfate. Cell uptake of poly(L-lysine)/DNA complexes was high and PEGylation increased the transfection efficacy. However, overall transfection level of poly(L-lysine) is low possibly as a result of the inadequate escape of the complexes from endosomes or poor release of DNA from the complexes. Physicochemical and biological structure-property relationships of poly(L-lysine) were demonstrated, but no clear correlations between the tested physicochemical determinants (size of complexes, zeta-potentials, condensation of DNA, and the shape of complexes) and biological activities were seen. Intracellular factors and/or still unknown features of DNA complexation may ultimately determine transfection activity with the carriers.

Kim et al. (2004) synthesized and applied a novel triblock copolymer, PAMAM-*block*-PEG-*block*-PAMAM, as a gene carrier. PAMAM dendrimer is proven to be an efficient gene carrier itself, but it is associated with certain problems such as low water solubility and considerable cytotoxicity. Therefore, they introduced PEG to engineer a non-toxic and highly transfection-efficient polymeric gene carrier because PEG is known to convey water-solubility and biocompatibility to the conjugated copolymer. This copolymer could achieve self-assembly with plasmid DNA, forming compact nanosized particles with a narrow size distribution. Fulfilling expectations, the copolymer was found to form highly water-soluble polyplexes with plasmid DNA, showed little cytotoxicity despite its poor degradability, and finally achieved high transfection efficiency comparable to PEI. Consequently, they showed that an approach involving the introduction of PEG to create a tree-like cationic copolymer possesses a great potential for use in gene delivery systems. Logically, this approach could be extended to cancer treatment.

27.6.5 Stimuli Responsive Carriers

Gillies and Frechet (2002) designed and prepared a new polyester dendrimer, PEO hybrid systems, for drug delivery and related therapeutic applications. These systems consist of two covalently attached polyester dendrons where one dendron provides multiple functional handles for the attachment of therapeutically active moieties, whereas the other is used for attachment of solubilizing PEO chains. By varying the generation of the dendrons and the mass of the PEO chains, the MW, architecture, and drug loading can be readily controlled. The bow-tie shaped dendritic scaffold was synthesized using both convergent and divergent methods with orthogonal protecting groups on the periphery of the two dendrons. PEO was then attached to the periphery of one dendron using an efficient coupling procedure. A small library of eight carriers with MWs ranging from about 20 to 160 kDa were prepared and characterized by various techniques, confirming their well-defined structures.

Gillies, Jonsson, and Frechet (2004) used a pH-responsive micelle system, linear-dendritic block copolymers comprising PEO and either a polylysine or polyester dendron. These were prepared, and hydrophobic groups were attached to the dendrimer periphery by highly acid-sensitive cyclic acetals. These copolymers were designed to form stable micelles in aqueous solution at neutral pH and disintegrate into unimers at mildly acidic pH following loss of the hydrophobic groups upon acetal hydrolysis. Micelle formation was demonstrated by encapsulation of the fluorescent probe Nile Red, and the micelle sizes were determined by dynamic light scattering. The structure of the dendrimer block, its generation, and the synthetic method for linking the acetal groups to its periphery all had an influence on the CMC and the micelle size. The rate of hydrolysis of the acetals at the micelle core was measured for each system at pH 7.4 and pH 5, and it was found that all systems were stable at neutral pH but that they underwent significant hydrolysis at pH 5 over several hours. The rate of hydrolysis at pH 5 was dependent on the structure of the copolymer, most notably the hydrophobicity of the core-forming block. To demonstrate the potential of these systems for controlled release, the release of Nile Red as a model payload was examined. At pH 7.4, the fluorescence of micelle-encapsulated Nile Red was relatively constant, indicating it was retained in the micelle, and at pH 5, the fluorescence decreased, consistent with its release into the aqueous environment. The rate of release was strongly correlated with the rate of acetal hydrolysis and was controlled by the chemical structure of the copolymer. The mechanism of Nile Red release was investigated by monitoring the change in size of the micelles over time at acidic pH. Dynamic light scattering measurement showed a size decrease over time, eventually reaching the size of a unimer, providing evidence for the proposed micelle disintegration.

27.6.6 ALTERATION OF TOXICITY

In one of the studies Bhadra et al. (2003) found that PEGylation significantly decreased the hemolysis of red blood cells (RBCs) to below 5% in case of whole generation amine terminated charged dendrimers having about 15.3–17.3% hemolytic toxicity, similar to negligible in case of half generations of carboxylic acid terminated dendrimers. This was due to inhibition of interaction of RBCs with the charged quaternary ammonium ion as determined by interaction with RBCs. The hemolytic toxicity of the dendrimers was enough to preclude its use as a drug delivery system. The toxicity was due to the poly-cationic nature of the PAMAM dendrimers that was also responsible for their cytotoxicity (Malik et al. 2000), particularly in the case of whole generation amine terminated charged dendrimers, but not in the case of half generations of carboxylic acid terminated dendrimers.

The PEGylation of the dendrimers were found to have also considerably lower rate of hemolysis of the RBCs because of significantly reduced interaction of RBCs with the charged quaternary ammonium ion that is generally present on the amine terminated whole generations of dendrimers. Also, the hematological study was undertaken to assess the relative effects of the PEGylated and non-PEGylated systems as compared to the plain drug on various blood parameters. The blood parameters undergoing major changes as to normal values of blood levels are RBC count, WBC count, and differential lymphocytes count. The RBC count of non-PEGylated 5-FU-dendrimers was found to have decreased below normal values by about $1 \times 10^6/\mu l$ RBCs to $2 \times 10^6/\mu l$ RBCs as against similar PEGylated systems. The WBC count of non-PEGylated 5-FU-dendrimer complex increased by $3 \times 10^3/\mu l$ cells to $4 \times 10^3/\mu l$ cells as compared to normal values. However, for 5 FU-PEGylated dendrimer complexes, the increase was by $1 \times 10^3/\mu l$ cells to $1.5 \times 10^3/\mu l$ WBCs as compared to normal count in controlled group. The increase in WBC count is significant in case of non-PEGylated systems. Similarly, relatively greater increase in lymphocyte count was observed by non-PEGylated dendrimer-drug complexes that were by about $2 \times 10^3/\mu l$ cells. This was similar to blood toxicity and cytotoxicity effects of acrylates nanoparticulates (Malik et al. 2000), which are known to be stimulating the macrophage level and WBC count. This was found to be true in the case of these non-PEGylated drug-dendrimeric systems but not in case of PEGylated systems, conforming the trends of the PEGylated carriers (Veronese et al. 1989; Papahadjopoulus et al. 1991; Phillips et al. 1999) undergoing lesser phagocytic uptake.

Neerman et al. (2004) also showed that dendrimers based on melamine could reduce the organ toxicity of solubilized cancer drugs administered by intraperitoneal injection. Methotrexate and 6-mercaptopurine, both FDA approved anti-cancer drugs, are known hepatotoxins. The solubility of these molecules can be increased by mixing them with a dendrimer based on melamine. C3H mice were administered subchronic doses of methotrexate or 6-mercaptopurine with and without a solubilizing dendrimer. Forty-eight hours after dosing, the mice were sacrificed, and serum was collected for biochemical analyses. The levels of alanine transaminase, ALT, were used to probe liver damage. When the drugs are encapsulated by the dendrimer, a significant reduction in hepatotoxicity is observed; ALT levels from the rescued groups (drug + dendrimer) were 27% (methotrexate) and 36% (6-mercaptopurine) lower than those of animals treated with the drug alone.

Initial studies of a generation 3, cationic dendrimer bearing 24 primary amino groups revealed that this dendrimer was toxic both in vitro and in vivo. Extensive liver damage was observed when mice were challenged with subchronic doses of this dendrimer. Other groups hypothesized that surface group charge was the determining factor in predicting a dendrimer's toxicity as cationic surface groups will electrostatically adhere to cell surface, inducing bulk adhesion that leads to the cell's demise. Surface group modification of a generation 3, cationic dendrimer bearing 48 primary amino groups to possess anionic or neutral surface groups had a significant impact on both in vitro and in vivo toxicity. The introduction of neutral PEG groups afforded the most protection in vitro and in vivo. Shielding of the dendrimer's charged groups attenuates the bulk adhesion of these molecules, leading to a decrease in the toxicity of these dendrimers. It appeared that PEGylation afforded the most protection against toxicity of dendrimers based on melamine. As a result, the candidate vehicle was subjected to a series of analyses. Albumin interaction studies showed that this dendrimer interacted with albumin that could have a profound effect on the dendrimer's pharmacokinetic parameters. In vitro drug release profiles of the anti-cancer drugs methotrexate and paclitaxel revealed that methotrexate was rapidly released while paclitaxel showed a more controlled, sustained release. This discrepancy in drug release was attributed to disparities in the drug's hydrophobicity; the more hydrophobic drug, paclitaxel, formed a stronger association with the dendrimer's hydrophobic core. As a result, release was more controlled. In terms of the in vitro cytotoxicity of these dendrimer/drug formulations, free drug was more toxic than the dendrimer-associated drug. This was attributed to differences in the cellular uptake of free versus dendrimer-associated drug. To assess if a difference in cellular uptake existed between free and dendrimer associated substrates, cells exposed to free rhodamine B showed higher intracellular levels of dye than those cells exposed to dendrimer-associated rhodamine B. The preliminary in vivo biodistribution screen of the Cy5.5-labeled dendrimer showed that the dendrimer was still present in the body 24 h post injection while exhibiting a high degree of liver accumulation after such time (https://txspace.tamu.edu/dev-xml/bitstream/1969.1/5330/1/etd-tamu-2005A-TOXI-Neerman.pdf).

Chen et al. (2004) designed a small library of dendrimers prepared from a common precursor that is available in 5 g scale in five linear steps at 56% overall yield. The precursor is a generation three dendrimer that displays 48 peripheral sites by incorporating AB4 surface groups. Manipulation of these sites provided six dendrimers that vary in the chemistry of the surface group (amine, guanidine, carboxylate, sulfonate, phosphonate, and PEGylated) that were evaluated for hemolytic potential and cytotoxicity. Cationic dendrimers were found to be more cytotoxic and hemolytic than anionic or PEGylated dendrimers. The PEGylated dendrimer was evaluated for acute toxicity in vivo. No toxicity, neither mortality nor abnormal blood chemistry, based on blood urea nitrogen levels or alanine transaminase activity was observed in doses up to 2.56 g/kg i.p. and 1.28 g/kg intravenously.

27.6.7 Effective Photodynamic Therapy

Photosensitizers play a crucial role in the photodynamic therapy (PDT) of cancer. Zhang et al. (2003) observed that the use of PIC micelles as a delivery system reduced the dark toxicity of the cationic dendrimer porphyrin, probably because of the biocompatible PEG shell of the micelles. They compared relatively low cellular uptake of dendrimer porphyrin $[NH_2CH_2CH_2NHCO]_{32}$-DPZ_n incorporated in the PIC micelle as was observed, yet the latter exhibited enhanced photodynamic efficacy on the Lewis Lung Carcinoma (LLC) cell line. In this study, a third-generation aryl ether dendrimer porphyrin with 32 primary amine groups on the periphery, $[NH_2CH_2CH_2NHCO]_{32}DPZ_n$, and pH-sensitive, polyion complex micelles (PIC) composed of the porphyrin dendrimer and PEG-b-poly(aspartic acid) were evaluated as new photosensitizers (PSs) for PDT in the LLC cell line. The preliminary photophysical characteristics of $[NH_2CH_2$-$CH_2NHCO]_{32}DPZ_n$ and the corresponding micelles were investigated. Electrostatic assembly resulted in a red-shift of the Soret peak of the porphyrin core and the enhanced fluorescence.

Jang et al. (2005) described a supramolecular nanocarrier of anionic dendrimer porphyrins with cationic block copolymers modified with PEG to enhance intracellular photodynamic efficacy.

27.6.8 Dendritic Gels

Luman, Smeds, and Grinstaff (2003) developed a process of high-yield convergent synthesis of dendrons, dendrimers, and dendritic-linear hybrid macromolecules composed of succinic acid, glycerol, and PEG. This convergent synthesis relies on two orthogonal protecting groups, namely, the benzylidene acetal (bzld) for the protection of the 1,3-hydroxyls of glycerol and the tert-butyldiphenylsialyl (TBDPS) ester for protection of the carboxylic acid of succinic acid. These novel polyester dendritic macromolecules are entirely composed of building blocks known to be biocompatible or degradable in vivo to give natural metabolites. Derivatization of the dendritic periphery with a methacrylate affords a polymer that can be subsequently photo-cross-linked. The three-dimensional cross-linked gels formed by ultraviolet irradiation are optically transparent with mechanical properties dependent on the initial cross-linkable dendritic macromolecule.

27.6.9 Ligand-Based Drug Delivery

Architectural features of synthetic ligands were systematically varied to optimize inhibition of mast cell degranulation initiated by multivalent crossing of IgE-receptor complexes Baird et al. (2003). A series of ligands were generated by end-capping PEG polymers and amine-based dendrimers with the hapten 2,4-dinitrophenyl (DNP). These were used to explore the influence of polymeric backbone length, valency, and hapten presentation on binding to anti-DNP IgE and inhibition of stimulated activation of RBL cells. Monovalent MPEG(5000)-DNP (IC(50) = 50 nM), bivalent DNP-PEG(3350)-DNP (IC(50) = 8 nM), bismonovalent MPEG(5000)-DNP(2) (IC(50) = 20 nM), bisbivalent DNP(2)-PEG(3350)-DNP(2) (IC(50) = 3 nM), and DNP(4)-dendrimer ligands (IC(50) = 50 nM) all effectively inhibit cellular activation caused by multivalent antigen, DNP-bovine serum albumin. For different DNP ligands, evidence is available for more effective inhibition because of preferential formation of intra-IgE cross-links by bivalent ligands of sufficient length, self-association of monovalent ligands with longer tails, and higher probability of binding for bisvalent ligands. They also showed that larger DNP(16)-dendrimers of higher valency trigger degranulation by cross-linking

IgE-receptor complexes, whereas smaller DNP-dendrimers are inhibitory. Therefore, features of synthetic ligands can be manipulated to control receptor occupation, aggregation, and inhibition of the cellular response.

27.6.10 INCREASINGLY EFFECTIVE BORON NEUTRON CAPTURE THERAPY

A recent methodology in cancer treatment is the boron neutron-capture therapy (BNCT) (Hawthorne 1993). In this therapy, the generation of cytotoxic and energetic products from nuclear fission reactions of low-energy neutrons and ^{10}B nuclei is used to destroy malignant cells. An efficient agent for the therapy is water-soluble and has a high local density of boron clusters, requirements that are met for several synthesized boron-containing and water-soluble dendrimers. Qualmann et al. (1996a, 1996b) have, in addition, introduced antigen selectivity by coupling a lysine-based boronated dendrimer to antibody fragments (Fab¢). In these agents, the attachment of a poly(ethylene glycol)(PEG) moiety is necessary to keep the conjugates water-soluble. The covalent nature of the boronated Fab¢ fragments leads to a better stability of these conjugates as compared to, for example, borate-coated polystyrene beads.

Successful treatment of cancer by boron neutron capture therapy (BNCT) requires the selective delivery of ^{10}B to constituent cells within a tumor. The expression of the folate receptor is amplified in a variety of human tumors and potentially might serve as a molecular target for BNCT. Shukla et al. (2003) investigated the possibility of targeting the folate receptor on cancer cells using folic acid conjugates of boronated poly(ethylene glycol) (PEG) containing third generation PAMAM dendrimers to obtain ^{10}B concentrations necessary for BNCT by reducing the uptake of these conjugates by the reticuloendothelial system. First, they covalently attached 12–15 decaborate clusters to third generation PAMAM dendrimers. Varying quantities of PEG units with varying chain lengths were then linked to these boronated dendrimers to reduce hepatic uptake. Among all prepared combinations, boronated dendrimers with 1–1.5 PEG(2000) units exhibited the lowest hepatic uptake in C57BL/6 mice (7.2–7.7% injected dose (ID)/g liver). Therefore, two folate receptor-targeted boronated third generation PAMAM dendrimers were prepared: one containing approximately 15 decaborate clusters and approximately 1 PEG(2000) unit with folic acid attached to the distal end, and the other containing approximately 13 decaborate clusters with, approximately one PEG(800) unit with folic acid attached to the distal end. In vitro studies using folate receptor (+) KB cells demonstrated receptor-dependent uptake of the latter conjugate. Biodistribution studies with this conjugate in C57BL/6 mice bearing folate receptor (+) murine 24JK-FBP sarcomas resulted in selective tumor uptake (6.0% ID/g tumor), but also high hepatic (38.8% ID/g) and renal (62.8% ID/g) uptake, indicating that attachment of a second PEG unit and/or folic acid may adversely affect the pharmacodynamics of this conjugate.

27.6.11 IMPROVED MRI CONTRAST AGENTS

Macromolecules conjugated with polyethylene glycol (PEG) acquire more hydrophilicity, resulting in a longer half-life in circulation and lower immunogenicity. Kobayashi et al. (2001) synthesized two novel conjugates for MRI contrast agents from a generation-4 polyamidoamine dendrimer (G4D), 2-(p-isothiocyanatobenzyl)-6-methyl-diethylenetriaminepentaacetic acid (1B4M), and one or two PEG molecules with a MW of 20,000 Da (PEG(2)-G4D-(1B4M-Gd)(62) (MW: 96 kD), PEG(1)-G4D-(1B4M-Gd)(63) (MW: 77 kD). Their pharmacokinetics, excretion, and properties as vascular MRI contrast agents were evaluated and compared with those of G4D-(1B4M-Gd)(64) (MW: 57 kD). PEG(2)-G4D-(1B4M-Gd)(62) remained in the blood significantly longer and accumulated significantly less in the liver and kidney than the other two preparations ($P < 0.01$). Although the blood clearance was slower, PEG(2)-G4D-(1B4M-Gd)(62) was excreted more readily without renal retention than the other two preparations. The positive effects of PEG conjugation on a macromolecular MRI contrast agent were found to be prolonging retention in the circulation, increased excretion, and decreased accumulation in the organs.

27.7 CONCLUSIONS

Nanotechnology refers to the interactions of cellular and molecular components and engineered materials—typically clusters of atoms, molecules, and molecular fragments—at the most elemental level of biology. Such nanoscale objects—typically, though not exclusively, with dimensions smaller than 100 nm—can be useful by themselves or as part of larger devices, containing multiple nanoscale objects. At the nanoscale, the physical, chemical, and biological properties of materials fundamentally differ, and often nanotechnology can change the very foundations of cancer diagnosis, treatment, and prevention. The novel nanodevices are capable of many clinically important functions, including detecting cancer at its earliest stages, pinpointing its location within the body, delivering anti-cancer drugs specifically to malignant cells, and in killing malignant cells. As these nanodevices are evaluated in clinical trials, researchers envision that nanotechnology will serve as multifunctional tools that will not only be used with any number of diagnostic and therapeutic agents, but will also change the very foundations of cancer diagnosis, treatment, and prevention (NCI 2004).

Such nanoparticulate carriers also include PEG conjugated dendrimers. These are globular, hyperbranched polymers possessing a high concentration of surface functional groups and internal cavities. These unique features make them very useful in many biomedical applications, especially as carrier molecules. There are many such technological opportunities associated with the dendrimers. Starpharma-like companies used these dendrimers to build large active compounds that present a polyvalent array to receptors, and others platform opportunities for the delivery of small molecules as a toolbox for rational drug design and other nanotechnology applications in life sciences. There are many product opportunities in dendrimers and their use such as with VivaGel™, a topical microbicide gel for prevention of HIV and other STDs in women. They can act as novel chemotherapeutic agents, angiogenesis inhibitors, and they can be used for targeting a range of tropical and exotic diseases and as bio-defense agents in addition to their use for targeting a broad range of viral, respiratory, and cancer-like diseases in the future. Development of drug delivery systems with low toxicity and low cost is the key to a nanaotechnological approach to cancer treatment, including DNA-based concepts. Targeting the drugs to specific receptors exclusively expressed by tumor cells also appears to be an approach worth exploring. This biochemical approach is expected to pay rich dividends in the chemotherapy of cancer. Cytosol survival of anti-cancer agents and folate receptor-based approaches also hold promise. Carbon nanotubes offer yet another attractive possibility.

REFERENCES

Abou-Jawde, R. et al., An overview of targeted treatments in cancer, *Clin. Ther.*, 25, 2121, 2003.

Baird, E. J. et al., Highly effective poly(ethylene glycol) architectures for specific inhibition of immune receptor activation, *Biochemistry*, 42, 12739, 2003.

Barar, F. S. K., *Essential of Pharmacotherapeutics*, S. Chand and Company Ltd, New Delhi, 2000.

Bhadra, D. et al., Pegnology: A review of PEG-ylated systems, *Pharmazie*, 57, 5, 2002.

Bhadra, D. et al., A PEGylated dendritic nanoparticulate carrier of fluorouracil, *Int. J. Pharm.*, 257, 111, 2003.

Bosman, A. W., Janssen, H. M., and Meijer, E. W., About dendrimers: Structure, physical properties and applications, *Chem. Rev.*, 99, 1665, 1999.

Brannon-Peppas, L. and Blanchette, J. O., Nanoparticle and targeted systems for cancer therapy, *Adv. Drug Deliv. Rev.*, 56, 1649, 2004.

Brigger, I., Dubernet, C., and Couvreur, P., Nanoparticles in cancer therapy and diagnosis, *Adv. Drug Deliv. Rev.*, 54, 631, 2002.

Buckmann, A. F. and Morr, M., Functionalization of polyethylene glycol and *mono*-methoxy-polyethylene glycol, *Makromol. Chem.*, 182, 1379, 1981.

CCO Formulary, Liposomal Doxorubicin, Oct. 2003.

Chapman, T. M. et al., Hydramphiphiles: Novel, linear, dendritic block copolymeric surfactants, *J. Am. Chem. Soc.*, 116, 11195, 1994.

Chari, R. V. J., Targeted delivery of chemotherapeutics: Tumor activated prodrug therapy, *Adv. Drug Deliv. Rev.*, 31, 89, 1998.

Chen, H. T. et al., Cytotoxicity, hemolysis, and acute in vivo toxicity of dendrimers based on melamine, candidate vehicles for drug delivery, *J. Am. Chem. Soc.*, 126, 10044, 2004.

Choi, Y. H. et al., Polyethylene glycol grafted poly-L-lysine as polymeric gene carrier, *J. Controlled Release*, 54, 39, 1998.

Choi, J. S. et al., Poly(ethylene glycol)-*block*-poly(L-lysine) dendrimer: Novel linear polymer/dendrimer block copolymer forming a spherical water-soluble polyionic complex with DNA, *Bioconjugate Chem.*, 10, 62, 1999.

Choi, J. S. et al., Synthesis of a barbell-like triblock copolymer, poly(L-lysine) dendrimer-*block*-Poly(ethylene glycol)-*block*-Poly(L-lysine) dendrimer, and its self-assembly with plasmid DNA, *J. Am. Chem. Soc.*, 122, 474, 2000.

Christensen, C. A. et al., Synthesis and electrochemistry of a tetrathiafulvalene $(TTF)_{21}$-glycol dendrimer: Intradendrimer aggregation of TTF cation radicals, *Chem. Commun.*, 509, 1998.

Cui, Z., Hsu, C. H., and Mumper, R. J., Physical characterization and macrophage cell uptake of mannan-coated nanoparticles, *Drug Dev. Ind. Pharm.*, 29, 689, 2003.

Dagani, R., Chemists explore potential of dendritic macromolecules as functional materials, CandEN Washington, Chem. and Eng. News, June 3, 1996.

De Brabander-Van den Berg, E. M. M. and Meijer, E. W., Poly (propyleneimine) dendrimers: Large-scale synthesis by heterogeneously catalyzed hydrogenations, *Angew. Chem. Int. Ed. Engl.*, 32, 1308, 1993.

Dvorak, H. F. et al., Identification and characterization of the blood vessels of solid tumors that are leaky to circulating macromolecules, *Am. J. Pathol.*, 133, 95, 1988.

Elamanchili, P. et al., Characterization of poly(D,L lactic-*co*-glycolic acid) based nanoparticulate system for enhanced delivery of antigens to dendritic cells, *Vaccine*, 22, 2406, 2004.

Elizabeth, R. et al., Biological evaluation of polyester dendrimer: Poly(ethylene oxide) "bow-tie" hybrids with tunable molecular weight and architecture, *Mol. Pharmacol.*, 2, 129, 2005.

Feng, S. S. and Chien, S., Chemotherapeutic engineering: Application and further development of chemical engineering principles for chemotherapy of cancer and other diseases, *Chem. Eng. Sci.*, 58, 4087, 2003.

Frechet, J. M., Functional polymers and dendrimers: Reactivity, molecular architecture, and interfacial energy, *Science*, 263, 1710, 1994.

Freemantle, M., Blossoming of dendrimers, *Chem. & Engg. News London, Science/Technology*, 77, 27, 1999.

Freitas, R. A. Jr., Nanomedicine, In *Basic Capabilities*, Landes Bioscience, Georgetown, TX, I, (http://www.nanomedicine.com/NMI.htm), 1999.

Freitas, R. A., Current status of nanomedicine and medical nanorobotics, *J. Comput. Theor. Nanosci.*, 2, 1, 2005.

Gabizon, A. and Martin, F., Polyethylene glycol-coated (PEGylated) liposomal doxorubicin. Rationale for use in solid tumours, *Drugs*, 54, 15, 1997.

Gillies, E. R. and Frechet, J. M., Designing macromolecules for therapeutic applications: Polyester dendrimer-poly(ethylene oxide) "bow-tie" hybrids with tunable molecular weight and architecture, *J. Am. Chem. Soc.*, 124, 14137, 2002.

Gillies, E. R., Jonsson, T. B., and Frechet, J. M., Stimuli-responsive supramolecular assemblies of linear-dendritic copolymers, *J. Am. Chem. Soc.*, 126, 11936, 2004.

Gitsov, I. and Fréchet, J. M. J., Solution and solid state properties of hybrid linear-dendritic block copolymers, *Macromolecules*, 26, 6536, 1993.

Gitsov, I. and Fréchet, J. M. J., Stimuli responsive hybrid macromolecules: Novel amphiphilic star copolymers with dendritic groups at the periphery, *J. Am. Chem. Soc.*, 118, 3785, 1996.

Grossfeld, G. D., Carrol, P. R., and Lindeman, N., Thrombospondin-1 expression in patients with pathologic state T3 prostate cancer undergoing radical prostatectomy: Association with p53 alterations, tumor angiogenesis and tumor progression, *Urology*, 59, 97, 2002.

Harris, J. M. et al., Synthesis and characterisation of polyethylene glycol derivatives, *J. Polymer Sci. Polymer Chem.*, 22, 341, 1984.

Harris, J. M. and Chess, R. B., Effect of pegylation on pharmaceuticals, *Nat. Rev. Drug Discov.*, 2, 214, 2003.

Hawthorne, M. F., The role of chemistry in the development of boron neutron capture therapy of cancer, *Angew. Chem. Int. Ed. Engl.*, 32, 950, 1993.

http://nano.med.umich.edu/Dendrimers.html.

http://press2.nci.nih.gov/sciencebehind/nanotech.

http://www.cancer.gov/clinicaltrials/results/dose-dense0604.

Ishida, O. et al., Size dependent extravagation and interstitial localization of polyethylene glycol liposomes in solid tumour bearing mice, *Int. J. Pharm.*, 190, 49, 1999.

Jang, W. D. et al., Supramolecular nanocarrier of anionic dendrimer porphyrins with cationic block copolymers modified with polyethylene glycol to enhance intracellular photodynamic efficacy, *Angew. Chem. Int. Ed. Engl.*, 44, 419, 2005.

Jansen, J. F. G. A. and Meijer, E. W., The dendritic box: Shape-selective liberation of encapsulated guests, *J. Am. Chem. Soc.*, 117, 4417, 1995.

Jansen, J. F. G. A., Brabander Vanden Berg, E. M. M., and Meijer, E. W., *Science*, 266, 1226, 1994.

Katre, N. V., The conjugation of protein with poly ethylene glycol and other polymers altering properties of proteins to enhance their therapeutic potential, *Adv. Drug Deliv. Rev.*, 10, 91, 1993.

Kim, A. et al., Pharmacodynamics of insulin in polyethylene glycol coated liposomes, *Int. J. Pharmacol.*, 180, 75, 1999.

Kim, T. I. et al., PAMAM–PEG–PAMAM: Novel triblock copolymer as a biocompatible and efficient gene delivery carrier, *Biomacromolecules*, 5, 2487, 2004.

Kim, Y. and Zimmerman, S. C., Applications of dendrimers in bio-organic chemistry, *Curr. Opin. Chem. Biol.*, 2, 733, 1998. http://biomednet.com/elecref/1367593100200733.

Kobayashi, H. et al., Positive effects of polyethylene glycol conjugation to generation-4 polyamidoamine dendrimers as macromolecular MR contrast agents, *Magn. Reson. Med.*, 46(4), 781, 2001.

Kojima, C. et al., Synthesis of polyamidoamine dendrimers having poly(ethylene glycol) grafts and their ability to encapsulate anticancer drugs, *Bioconjugate Chem.*, 11(6), 910, 2000.

Kolhe, P. et al., Drug complexation, in vitro release and cellular entry of dendrimers and hyperbranched polymers, *Int. J. Pharm.*, 259, 143, 2003.

Kreuter, J., Nanoparticles, In *Colloidal Drug Delivery Systems*, Kreuter, J., Ed., Marcel Dekker, Inc., New York, 1994.

Kukowska-Latallo, J. F. et al., Intravascular and endobronchial DNA delivery to murine lung tissue using a novel, nonviral vector, *Hum. Gene Ther.*, 11, 1385, 2000.

Langer, R., Biomaterials in drug delivery and tissue engineering: One laboratory's experience, *Acc. Chem. Res.*, 33, 94, 2000.

Landers, J. J. et al., Prevention of influenza pneumonitis by sialic acid-conjugated dendritic polymers, *J. Infect. Dis.*, 186, 1222, 2002.

Liu, M., Kono, K., and Frechet, J. M., Water-soluble dendritic unimolecular micelles: their potential as drug delivery agents, *J. Control. Rel.*, 65, 121, 2000.

Luman, N. R., Smeds, K. A., and Grinstaff, M. W., The convergent synthesis of poly(glycerol-succinic acid) dendritic macromolecules, *Chemistry*, 9, 5618, 2003.

Luo, D., A new solution for improving gene delivery, *Trends Biotechnol.*, 22, 101, 2004.

Luo, D. and Saltzman, W. M., Synthetic DNA delivery systems, *Nat. Biotechnol.*, 18, 33, 2000.

Luo, D. et al., Poly(ethylene glycol)-conjugated PAMAM dendrimer for biocompatible, high-efficiency DNA delivery, *Macromolecules*, 35, 3456, 2002.

Maeda, H., The enhanced permeability and retention (EPR) effect in tumor vasculature: The key role of tumor-selective macromolecular drug targeting, *Adv. Enzyme Regul.*, 41, 189, 2001.

Malik, N. et al., Dendrimers: Relationship between structure and biocompatibility in vitro, and preliminary studies on the biodistribution of 125I-labelled polyamidoamine dendrimers in vivo, *J. Controlled Release*, 65, 133, 2000.

Mannisto, M. et al., Structure-activity relationships of poly(L-lysines): Effects of pegylation and molecular shape on physicochemical and biological properties in gene delivery, *J. Controlled Release*, 83(1), 169, 2002.

Maruyama, K., In vivo targeting by liposomes, *Biol. Pharm. Bull.*, 23, 791, 2000.

Massing, U. and Fuxius, S., Liposomal formulations of anticancer drugs: Selectivity and effectiveness, *Drug Resist. Updat.*, 3, 171, 2000.

McNeil, S. E., Nanotechnology for the biologist, *J. Leukoc. Biol.*, 78, 1–10, 2005.

Miller, T. M., Kwock, E. W. K., and Neenan, T. X., *Macromolecules*, 25, 3143–3148, 1992.

Moghimi, S. M. and Szebeni, J., Stealth liposomes and long circulating nanoparticles: Critical issues in pharmacokinetics, opsonization and protein-binding properties, *Prog. Lipid Res.*, 42, 463, 2003.

NCI Cancer Bulletin, National Cancer Institute, U.S. Department of Health and Human Services, National Institutes of Health, NIH Publication, Press Office No. (301) pp. 496–6641, Sep., (http://www.cancer.gov), 2004.

NCI Cancer Bulletin, National Cancer Institute, U.S. Department of Health and Human Services, National Institutes of Health, NIH Publication No. 05-5498, Feb., vol. 2, (http://www.cancer.gov), 2005.

Neerman, M. F. et al., Reduction of drug toxicity using dendrimers based on melamine, *Mol. Pharmacol.*, 1, 390, 2004.

Newkome, G. R., Moorefield, C. N., and Vogtle, F., *Dendritic Molecules: Concepts, Syntheses, Perspectives*, VCH, Weinheim, 1996.

Norsten, T. and Branda, N., A starburst porphyrin polymer: A first generation dendrimer, *Chem. Commun.*, 1257, 1998.

Ooya, T., Lee, J., and Park, K., Effects of ethylene glycol-based graft, star-shaped, and dendritic polymers on solubilization and controlled release of paclitaxel, *J. Controlled Release*, 93, 121, 2003.

Padilla, D. J. et al., Polyester dendritic systems for drug delivery applications: In vitro and in vivo evaluation, *Bioconjugate Chem.*, 13, 453, 2002.

Papahadjopoulos, D. et al., Sterically stabilized liposomes improvement in pharmacokinetics and anti tumour therapeutic efficacy, *Proc. Natl Acad. Sci.*, 88, 11460, 1991.

Peppas, N. A. et al., Polyethylene glycol containing hydrogels in drug delivery, *J. Controlled Release*, 62, 81, 1999.

Phillips, W. T. et al., Polyethylene glycol modified liposomes encapsulated hemoglobin: A long circulating red blood cell substitute, *J. Pharmacol. Exp. Ther.*, 288, 665, 1999.

Qualmann, B. et al., Synthesis of boron-rich lysine dendrimers as protein labels in electron microscopy, *Angew. Chem. Int. Ed. Engl.*, 35, 909, 1996.

Qualmann, B. et al., Electron spectroscopic imaging of antigens by reaction with boronated antibodies, *J. Microsc.*, 183, 69, 1996.

Quintana, A. et al., Design and function of a dendrimer-based therapeutic nanodevice targeted to tumor cells through the folate receptor, *Pharm. Res.*, 19, 1310, 2002.

Reuter, J. D. et al., Inhibition of viral adhesion and infection by sialic-acid-conjugated dendritic polymers, *Bioconjugate Chem.*, 10, 271, 1999.

Reynolds, J. E. F., *Martindale Extra Pharmacopoeia*, 31st ed., Royal Pharmacol Society, London, 1996.

Sahoo, S. K. and Labhasetwar, V., Nanotech approaches to drug delivery and imaging, *Drug Discov. Today*, 8, 2003, www.drugdiscoverytoday.com.

Shukla, S. et al., Synthesis and biological evaluation of folate receptor-targeted boronated PAMAM dendrimers as potential agents for neutron capture therapy, *Bioconjugate Chem.*, 14, 158, 2003.

Sledge, G. and Miller, K., Exploiting the hallmarks of cancer: The future conquest of breast cancer, *Eur. J. Cancer*, 39, 1668, 2003.

Smith, D. K. and Diederich, F. N., Functional dendrimers: Unique biological mimics, *Chem. Eur. J.*, 4, 1353, 1998.

Sparks H., How miniature radiation detectors will keep astronauts safe in deep space. http://www.space.com/businesstechnology/technology/radiation_nanobots_020717.html, 2002.

Storm, G., Belliot, S. O., Daemen, T., and Lasic, D., Surface modification of nanoparticles to oppose uptake by the mononuclear phagocyte system, *Adv. Drug Deliv. Rev.*, 17, 31, 1995.

Suzawa, T. et al., Enhanced tumor cell selectivity of adriamycin-monoclonal antibody conjugate via a poly(ethylene-glycol)-based cleavable linker, *J. Controlled Release*, 9, 229, 2002.

Tomalia, D. A., Naylor, A. M., and Goddard, W. A., Starburst dendrimers: Molecular-level control of size, shape, surface chemistry, topology, and flexibility from atoms to macroscopic matter, *Angew. Chem. Int. Ed. Engl.*, 29, 138, 1990.

Toyokuni, T., Hakomori, S. I., and Singhal, A. K., *Bioorg. Med. Chem.*, 2, 1119, 1994.

Veronese, F. M. et al., Preparation, physicochemical and pharmcokinetic characterisation of methoxy poly-ethylene glycol derivatized superoxide dismutase, *J. Controlled Release*, 10, 145, 1989.

Vincent, L. et al., Efficacy of dendrimer-mediated angiostatin and TIMP-2 gene delivery on inhibition of tumor growth and angiogenesis: In vitro and in vivo studies, *Int. J. Cancer*, 10, 419, 2003.

Voit, B., Turner, S. R., and Mourey, T., Hyperbranched polyesters: 1. All aromatic hyperbranched polyesters with phenol and acetate end groups; synthesis and characterization, *Macromolecules*, 26, 4617, 1993.

Wooley, K. L., Hawker, C. J., and Frechet, J. M. J., A branched-monomer approach for the rapid synthesis of dendrimers, *Angew. Chem. Int. Ed. Engl.*, 33, 82, 1994.

Zalipsky, S., Chemistry of polyethylene glycol conjugates with biologically active molecules, *Adv. Drug Deliv. Rev.*, 16, 157, 1995.

Zeng, F. and Zimmerman, S. C., Dendrimers in supramolecular chemistry: From molecular recognition to self-assembly, *Chem. Rev.*, 97, 1681, 1997.

Zhang, G. D. et al., Polyion complex micelles entrapping cationic dendrimer porphyrin: Effective photosensi-tizer for photodynamic therapy of cancer, *J. Controlled Release*, 93, 141, 2003.

28 Dendrimer Nanocomposites for Cancer Therapy

Lajos P. Balogh and Mohamed K. Khan

CONTENTS

28.1 Composite Nanodevices 552
    28.1.1 Introduction 552
    28.1.2 Dendrimers 555
    28.1.3 Toxicity 558
    28.1.4 Gold and Radioactive Gold 558
    28.1.5 Dendrimer Nanocomposites 559
        28.1.5.1 Synthesis of Gold/Dendrimer Nanocomposites 561
        28.1.5.2 Characterization of Dendrimer Hosts and Gold
                 Nanocomposites Used in Biodistribution Experiments 562
        28.1.5.3 Fabrication of Nanocomposite Devices for Biologic
                 Experiments 562
        28.1.5.4 Synthesis of Tritium Labeled Dendrimers 564
        28.1.5.5 Folate-Targeted Devices 564
        28.1.5.6 Gold Composite Nanodevice Synthesis 564
        28.1.5.7 Gold in the Nanocomposite Particles After Radiation
                 Polymerization 569
        28.1.5.8 Visualization of Gold/PAMAM Nanocomposite Uptake in
                 Tumor Cells 569
28.2 Biology 571
    28.2.1 Targeting and Nanodevice Delivery 572
        28.2.1.1 Quantitative In Vivo Biodistribution of Dendrimer and Gold
                 Composite Nanodevices (CND) 573
        28.2.1.2 Related Techniques in the Literature 573
        28.2.1.3 The Instrumental Neutron Activation Analysis (INAA) Method 574
        28.2.1.4 Determination of Gold Content Using INAA—Precision
                 and Sensitivity 575
        28.2.1.5 Biodistribution of Positive and Neutral Surface Generation Five
                 Dendrimers—Non-Specific Uptake 575
        28.2.1.6 Biodistribution of 5 nm Gold Nanocomposites Compared with
                 Dendrimer Templates 577
        28.2.1.7 Biologic Differences Between Dendrimers and Gold Composite
                 Nanodevices 577
        28.2.1.8 The Importance of Size on Nanocomposite Biodistribution 579
        28.2.1.9 Biodistribution Summary 580
    28.2.2 Toxicity Studies 580

28.2.2.1   Long-Term Toxicity Studies of Tritiated
           Dendrimer Nanoparticles 580
28.2.2.2   Long-Term Toxicity of {Au(0)} Gold Composite
           Nanodevices 582
28.2.2.3   Treatment of Tumors in Mouse Model Systems 582
Acknowledgments 585
Definitions 585
References 586

28.1 COMPOSITE NANODEVICES

28.1.1 INTRODUCTION

Every year, the United States alone reports more than half a million cancer-related deaths and approximately 1.3 million new cases.[3] Cancer is customarily treated with drugs (extracts, synthetic molecules, and/or biologic materials, e.g., proteins, antibodies, etc.). In an ideal case, the intervening medication should only act at the disease site and influence only those mechanisms that are causing the illness. Most of the unwanted side-effects are due to therapeutic materials that arrive at unwelcome sites (cells, tissues, organs) and influence normal mechanisms in a wrong way. Targeted drug delivery (without side effects) has been the ultimate goal of the treatment of diseases. Living objects have developed mechanisms to interact and deal with molecules and organic or inorganic materials (particles) introduced in the blood stream; there are filter organs and defense systems to eliminate the unwanted ones and keep those that are useful.

Most of the anti-cancer drugs are cytotoxic—they kill cells. To use these in a clinical setting, it is mandatory to minimize the side effects, i.e., eliminate cancer cells without compromising the normal operation of other cells. The ongoing revolution in nanoscience and nanotechnology offers novel ways to treat diseases. These man-made complex objects provide an opportunity to better understand how the body works and simultaneously offer new ways of imaging and treatment. Targeted nanodevices open up many new avenues for therapy and diagnosis of cancer.

Understanding biodistribution is a key to success for any medication. Biodistribution is determined by the interactions between the properties of the medicine and the living biologic system. Therefore, to ensure that a therapeutic material has only one concerted action at the molecular level, molecules/particles/devices with identical properties are needed. Mixtures of materials will have fractions with different properties, each with an individual biodistribution (of which only an envelope is observed). Different fractions with differing properties of biodistribution can lead to different actions. In simple terms, only identical particles will have a well-defined biodistribution.

Because of these reasons, such materials need to be prepared that are composed of identical molecules/objects, presenting identical properties. In other words, only identical nanodevices with well-defined properties will have a defined biodistribution, making homogeneity (or, at least, narrow distribution) a fundamental requirement for a successful and reproducible targeted delivery.

Nanoscience and nanotechnology encompass a broad range of areas with different degrees of maturity, namely synthesis and controlled formation of nanostructures with a variety of specific functionalities that are enabled by a widespread availability of tools to observe, characterize, and manipulate at the nanoscale level. This is a cyclic and synergistic process; improved characterization (new methods, higher spatial resolution, and higher sensitivity) leads to a better understanding of composition–structure–property relations that motivates better nanoparticle and nanostructure fabrication and allows better control of size and placement that, in turn, stimulates improved characterization.

Properties of nanosized particles depend not only on composition, but also on size (diameter, surface, volume), shape, structure, architecture, flexibility, conformations, etc. Materials composed of nanoparticles are characterized not only by their individual properties, but also by the distribution of those individual properties over a very large number of particles. To employ nanodevices (nanosized multifunctional particles) in therapy and imaging, it is necessary to clearly understand the relevant properties of the materials used, how the given biologic system works, and how nanodevices interact with the biologic system.

To ensure the existence of these properties, appropriate characterization methods must be used that provide information about relevant properties and their distributions, including nanoparticle composition, size, shape, surface, and interior properties. At the nanoscale (1–100 nm), the surface-to-volume ratios are very large, and surface forces (Coulomb interactions, hydrogen bonding, van der Waals forces) greatly determine the characteristics of these devices and control the interactions of these devices with biological systems.

Reproducibility of nanoparticle properties is necessary to observe a reproducible biodistribution that is key to successful therapy. (Presently, most knowledge of nanoparticles is based on demonstrations of concept.)

Properties and biologic behavior of nanodevices are not yet completely understood, and detailed chemical–material science–physics–biology studies are being performed to generate more materials with appropriate properties and a better understanding in the future. Empiric studies are necessary to uncover any unexpected properties of nanodevices.

Radiotherapy has been used for decades as standard cancer therapy in almost all forms of cancers with varying degrees of success. One of the difficult challenges with radiation therapy is the delivery of a lethal dose of radiation to the tumor while leaving the surrounding normal tissue unharmed. Radioisotopes are often practically insoluble in water or body fluids, and they are rarely used as elements. Typically radioisotopes are attached to an appropriate carrier molecule. Tumor-directed antibody (or peptide) radiation therapy has been used to achieve this goal with limited success. In this technique, the radiation emitter is attached to the targeting antibody or peptide, often by a cage-type carrier, that is linked to the targeting molecule. The caged carrier can carry only a limited number of radioactive atoms/molecules and usually requires a linker to prevent destruction of the targeting molecule. This creates a fundamental limitation for the dose that can be delivered per targeting molecule that may limit the ability to successfully treat solid tumors with radioactive antibody approaches. For any targeted dose delivery approach, it has to be demonstrated that the treatment will be effective if the dose is directly delivered to the tumor

Composites are physical mixtures or two or more components that display improved properties in one or more areas as compared to their individual bulk constituents. They are widely used in many diverse industries and form the basics of engineering plastics, structural adhesives, and matrices. Nanocomposites can be defined as having at least one component with at least one of their dimensions less than 100 nm. Therefore, nanocomposite properties will be strongly influenced by extensive interfacial interactions between the components at the interphase, and the rule of mixtures may fail to provide estimates of nanocomposite properties.

Guest–host nanodevices composed of therapeutic (or imaging) materials carried in micelles, liposomes, polymeric nanocarriers, dendrimers, nanoemulsions, nanocomposites, etc., are a well-studied class of multifunctional nanomaterials with several potential medical uses, including targeted drug-delivery, cancer imaging, and therapy. Typically, the common feature in these structures is that the carried material is not bound to the carrier with covalent bonds, but it is constrained by physical forces. Composition, 3-D structure, and surface chemistry of the carrier are critical parameters for targeted delivery. Charge and size of the nanodevices determine their in vivo biodistribution, and thereby, the efficacy for imaging and therapies.

Composite nanodevices possess chemical and physical (spectral) properties, both the inorganic and the organic components, and physical and biologic interactions of the nanoparticles with each other and with the environment (solubility, biocompatibility, etc.) are dominated by the size, shape, and contact surface of the host. As a consequence, in hybrid composite nanodevices, a specific nanoparticle property (e.g., radioactivity belonging to the inorganic guest) can be manipulated as if it were a property of the organic host (e.g., dendrimer).

Targeted composite nanodevices are constructed by covalently binding targeting moieties to the organic template that dominates the surface of the nanocomposite. The surface can be chemically modified with various substituents, including proteins to achieve the necessary organ/tissue specificity and create targeted composite nanodevices. The interaction of these nanodevices with the biologic environment (solubility, biocompatibility) is controlled by the contact surface of the host molecule, i.e., by their size and surface properties. The separation of overall properties from the surface chemistry (targeting, charge, and size) is a unique advantage in the optimization of these nanodevices because properties of the components can be individually selected and optimized.[4] Nearly monodisperse gold/dendrimer nanocomposites can be synthesized in various predetermined sizes, and their interaction with biological objects may be adjusted by modifying their surface properties. They can readily penetrate cells and are easy to observe by various methods.[5] Composite nanoparticles with either cationic, anionic, neutral, lipophilic, lipophobic, or mixed surfaces can be created by encapsulating passive or active therapeutic agent(s) that adsorb or emit radiations, respectively. Guests in radioactive nanocomposite devices may be metals or other compounds containing isotopes of medical importance permitting imaging labels, (e.g., ^{125}I, ^{3}H) or radiation therapy, for example, ^{198}Au that can deliver β-radiation to tumors.

Using targeted nanodevices allows one to deliver various functionalities via a single biohybrid nanoparticle, making these nanodevices a potentially attractive tool in radiation medicine. An analysis of tumor type should allow for the selection and rapid immobilization of the appropriate radioactive isotope in the pre-made template that can then be efficiently delivered to specifically kill tumor cells with minimal collateral damage. The charge, size (i.e., surface chemistry), composition, and 3-D structure of these nanoparticles are critical in determining their in vivo biodistribution, and therefore, the efficacy of nanodevice imaging and therapies.

Some of the fundamental problems that can be overcome by the use of targeted composite nanodevices are solubility and dose delivered. First, many radioisotopes are often practically insoluble in water or body fluids, but they may become readily available in nanocomposites, i.e., in different sizes, contents, and surfaces. Second, use of composite nanodevices permits at least a log-fold higher delivery of radioactivity than that available with current radioactive antibody therapies. As opposed to labeled antibodies, at least 1–1000 radioisotope nuclei can be encapsulated in one nanocomposite device 6. Therefore, radioactive nanocomposite particles will be able to deliver therapeutic dose amounts to tumor cells. (Recent data indicate that even one radioactive alpha-particle decay could trigger cell death if it occurs in the cell.)[7]

The specific advantage of the targeted dose delivery by composite nanodevices approach over intracellular drug delivery or simple diffusion of drug into cells is that cell surface delivery is sufficient and no internalization is needed. As the specific activity of one composite {^{198}Au} nanodevice is extremely small (in the range of 1E-15 Ci/nanoparticles, i.e., 3.7E-25 Bq/nanoparticle), single particles delivering β-radiation at the cell surface (low-level non-specific uptake by T-cells) are harmless, but concentrations above a certain critical level (i.e., when large number of particles are delivered to specific cells or tissues) can be therapeutic.

Practically, for any targeted delivery approach, it has to be demonstrated that the treatment is effective if the therapeutic material accumulates in the tumor. These conceptual experiments must be followed by biodistribution studies and the optimization of devices.

In a proof-of-concept research, {^{198}Au} nanodevices were tested, but conceptually, a number of other nuclides with the desired specific activity could be used (allowing the use of gamma, beta, alpha, or combinations of these radiations in the future). As gold is non-toxic and the hot and cold gold isotopes are chemically identical, biodistribution studies may conveniently be done with nanocomposites containing non-radioactive gold.

28.1.2 DENDRIMERS

The term *dendrimer* may refer to a mathematical structure, a chemical structure of a single molecule of a certain class of macromolecules, or to a certain material (usually a product of a specific synthesis procedure)[8,9] (Figure 28.1). A dendrimer material is typically a mixture of very similar polymer molecules.

A number of characteristic general properties of dendrimers can be concluded from the dendritic structure itself as a result of symmetry and maximal degree of branching. These characteristics are the function of the core structure, the connector length, the degree of branching, etc., and they appear in all the families (for example, the tendency of dendrons to move in a concerted fashion).[10–13]

Dendrimer molecules are symmetric molecules with possible molecular masses of up to several million daltons (i.e., dendritic structures built from atoms) that contain connectors and branching units composed of repetitive shell structures around a core following a predefined molecular motif. Dendrimer molecules always have specific chemistry and specific properties.

There are many ways to construct polymer molecules with dendritic architecture, depending on which part of the macromolecule is dendritic (and what different authors consider to be a dendrimer). To date, more than 200 various dendritic structures have been reported in the literature. The major synthesis strategies are the divergent route that leads to higher yields with lower dendritic purity.[14] Convergent strategies[15] usually result in more uniform molecules but with lower overall yields (a great number of other strategies have been also reported). Dendrimer molecules are well-defined and highly symmetrical molecules containing a large number of regularly spaced internal and external functional groups. (Dendritic molecules are built to maximize symmetric branching.) As a result, the interior of a dendrimer may considerably differ from its exterior and may be hydrophilic or hydrophobic, depending on the design and synthesis route.

Constructing a dendritic network from atoms necessarily invokes specific properties because the atoms themselves have specific properties. Dendrimer families contain a certain group of molecular motifs (PAMAM, PPI, polylysine dendrimers, polyether dendrimers, polyacetylene dendrimers, etc.). They are synthesized on different synthetic principles and use different routes; therefore, they have different chemistries. Dendrimer families may be very different.

Generations and their subclasses. Within any families there are low, high, and middle generation dendrimers. Molecules of low generation dendrimers (LGD) structurally are relatively small organic molecules with fully flexible loose networks and with a few interactive sites (e.g., the

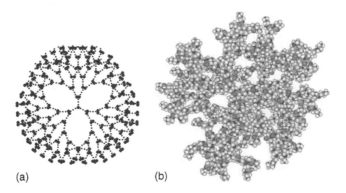

(a)   (b)

FIGURE 28.1 (a) Dendritic structure (three-functional core, bifunctional branching, persistent angles) and a computer model of an ideal PAMAM dendrimer molecule (b). (The left panel has been reproduced from Bielinska, A., Eichman, J. D., Lee, I., Baker, J. R., and Balogh, L., *J. Nanopart. Res.*, 4, 395–403, 2002. With permission.)

chemical accessibility of the functions are roughly equal). High generation dendrimers (HGD) are essentially highly charged organic nanoparticles where the accessibility of surface groups is much higher than of the interior. These organic nanoparticles have a well-defined generational size and a persistent surface, properties that may be quite different from its interior. Medium generation dendrimers (MGD) are transitional and may display properties of both groups.

A dendrimer material is typically a mixture of very similar polymer molecules as only perfect reactions (complete conversion with 100% yield) or perfect purification would lead to perfect structures in the consecutive synthesis steps. Consequently, practical dendrimer materials are oligomers or macromolecules with narrow polydispersity, and they always contain molecules with imperfect structures. However, the high level of synthetic control and effective purification between the synthesis steps allows the synthesis of defined polymers with very low polydispersities. Material properties are usually different for different generations; for example, in viscosity measurements, LGD materials can be described as draining networks, whereas HGD materials behave as ideal Newtonian fluids. There may be measurable differences between batches of individually manufactured or synthesized dendrimer materials; therefore, one must be very careful to generalize experimental results to the class of dendrimers. Dendritic properties of a given material can neither be assessed nor evaluated without taking into account the family, generation, composition, various properties, and their distribution (Figure 28.2).

Polyionic water-soluble dendrimers are those where tertiary nitrogen provides the branching. Poly(amidoamine) (PAMAM) dendrimers contain beta-alanine subunits and can be synthesized to be biofriendly.[16–19] They are well characterized and are also commercially available. PAMAM dendrimer molecules undergo changes in size, shape, and flexibility as a function of increasing generations (from a small branching molecule at generation 0–2 (LGD) through flexible macromolecules (MGD: G3, G4, G5–G6) to dense organic particles behaving as hard spheres at generations seven and above) (Figure 28.2). Branching structure of MGD and HGD PAMAMs can efficiently entrap therapeutic molecules.[20–22] They have been used as delivery vehicles for oligonucleotides and antisense oligonucleotides and as probes for oligonucleotide arrays.[23,24] PAMAM dendrimers are not broken down by enzymes in vivo and generally are removed from the bloodstream by the filter organs.[25–27]

Diversities that exist in dendrimer materials are of generational, skeletal, and/or substitutional origin. It is common that traces of various generations are present in materials (Figure 28.3a, peaks A–B–C in SEC-MALLS); the individual molecules may or may not be perfectly dendritic (see: skeletal diversity), and less than complete substitution of the end groups results in slightly different substitutions on the individual molecules (because of the random nature of chemical reactions, i.e.,

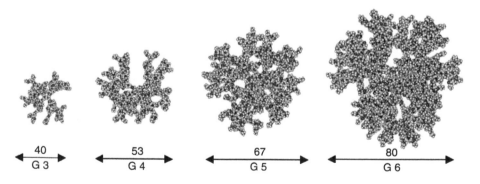

FIGURE 28.2 Size and shape of ethylenediamine core PAMAM dendrimers as a function of generation (fully protonated structures in water, i.e., maximally extended conformations). Dendritic polymers provide a multifunctional surface with a variable size and tunable compatibility. (Reproduced from Bielinska, A., Eichman, J. D., Lee, I., Baker, J. R., and Balogh, L., *J. Nanopart. Res.*, 4, 395–403, 2002. With permission.)

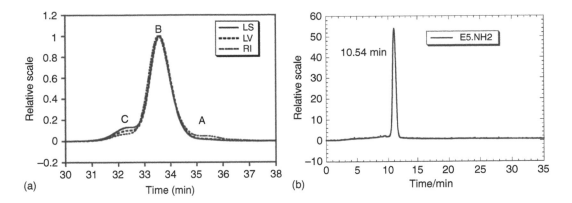

FIGURE 28.3 (a) Size exclusion chromatogram (SEC-MALLS) showing mass (Mn = 28,940) and polydispersity (PDI = 1.067) A: G4 trailing generation; B: G5 main component; C: dimers of G5 present in the material; and (b) a capillary electropherogram (CE) illustrating the almost identical charge/mass ratio for an actual ethylenediamine core primary amine-terminated generation five PAMAM dendrimer material.

substitutional diversity).[28] The most important properties of dendrimers are mass and charge (Figure 28.3). In case of HGD (e.g., G>6 in the case of PAMAMs) that are dense-packed, contribution to the surface may be solely due to terminal groups. In the more open structures of lower generations, other functional groups also contribute in an increasing fashion as the generational number decreases.

Dendrimers have a very rich synthetic chemistry describing how molecules with dendritic architectures can be synthesized and/or modified (to date, more than 5000 papers have been published on dendrimers).[8,9] Large dendritic structures may be synthesized by reacting dendrimers as building blocks (e.g., as core and shell reagents) with each other (Figure 28.4). These molecules have been termed tecto-dendrimers.[21,29,30]

Among other means, PAMAM dendrimers have been proposed as versatile vehicles of drug transport to specific organs and tissues because of their unique external and internal functionalities that can be altered according to need. These spherical macromolecules have modifiable surface functionalities offering a versatile mode to covalently attach drugs, diagnostic/imaging modules, and/or targeting moieties, and they allow the control of effective surface charge. However, the expensive starting materials, the required multistep synthesis, and subsequent characterization of the multifunctional nanodevices often present a serious challenge.[31] This complexity may be

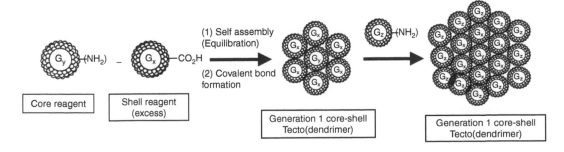

FIGURE 28.4 Scheme of tecto-(dendrimer) synthesis. The method is used to make large templates and supramolecular nanoparticles for multi-functional composite nanodevice fabrication.

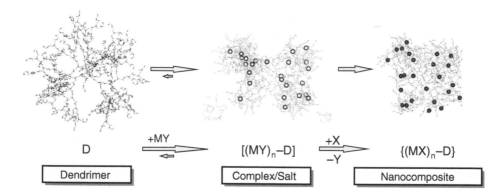

FIGURE 28.5 General scheme of dendrimer nanocomposite synthesis through complex formation followed by reactive encapsulation. (Reproduced from Bielinska, A., Eichman, J. D., Lee, I., Baker, J. R., and Balogh, L., *J. Nanopart. Res.*, 4, 395–403, 2002. With permission.)

considerably simplified using hybrid composite nanoparticles templated by monodisperse dendrimers (Figure 28.5).

Polyionic dendrimers are close to being optimal templates for these composite nanoparticles because the better the templates, the more homogenous the resulting composite particles become. PAMAM and PPI can interact with a wide variety of ions and in situ synthesized compounds.[32–38] These tertiary nitrogens act as focal points for pre-orientation either of cations to coordinatively bind through donor–acceptor interactions of the lone electron pair or of anions that can electrostatically bind to the positive charge of the protonated $-N=$ ligands. Multiple binding within the large molecule (a nanoscopic size pseudo-phase) also contributes to the unusually strong binding and high binding capacity.[39,40] Therefore, using PAMAM dendrimers as templates, the pre-orientation and dispersion of the inorganic components are controlled by the appropriate dendritic polymer host[32,34,35] and the size, size-distribution, shape, and surface of the resulting composite particle can be designed. Countless combinations are possible between metal cations or anions and dendrimers, but examples will be limited to aqueous solutions of generation (g4–g6) ethylenediamine core poly(amidoamine) (PAMAM) dendrimers and gold in the form of complex $[AuCl_4]^-$ (tetrachloroaurate) anions.[41,42]

28.1.3 Toxicity

Dendrimer biocompatibility and toxicity have been recently reviewed, and details can be found in the literature.[16] Briefly, in in vitro experiments, it was observed[43] that amine-terminated G5 PAMAM dendrimers decreased the integrity of the cell membrane and allowed the diffusion of cytosolic proteins out of the cell and dye molecules into the cell. Although neither G5 amine- nor acetamide-terminated PAMAM dendrimers were cytotoxic up to 500 nM concentration, the dose dependent release of the cytoplasmic proteins lactate dehydrogenase (LDH) and luciferase (Luc) indicated the presence of holes. The induction of permeability caused by the amine-terminated dendrimer was not permanent, and leaking of cytosolic enzymes returned to normal levels upon removal of the dendrimers. Diffusion of dendrimers through holes is sufficient to explain the non-specific uptake of positively charged PAMAM dendrimers into cells and is consistent with the lack of uptake of neutral surface PAMAM dendrimers.

28.1.4 Gold and Radioactive Gold

Noble metals, especially gold, have many applications in optical and biological sciences as a result of their stability. Metallic gold is usually produced by reduction using Na-citrate, citric acid, hydrazine, hydrogen, or electrochemical deposition, and the color of these particles is the function of size and shape of the metal domains formed.[44]

Gold-based therapeutic agents have been recently reviewed.[45] Gold nanoparticles[46] and nanoshells[47–49]have recently been successfully tested for photothermal therapy of cancer. Radioactive gold has been approved by FDA for radiation treatment and the use of ^{198}Au colloids for cancer therapy has already been attempted in the 1950s.[45,50–52] Radioactive ^{198}Au is used in interstitial brachytherapy for treatment of prostate cancer, initially in the form of gold colloids (micron-sized particles),[53] and later with seed technique.[54]

With biopsy results approaching a negative rate of 80%, and at five years, a cancer specific survival of 100% for Stages A and B1, 90% for Stage B2, and 76% for Stage C, this form of treatment offers an effective and well-tolerated alternative mode of therapy for patients with localized prostate cancer. Present evaluation of the accumulated Au(0) related biodistribution data is almost impossible because of the partial or full lack of reliable particle characterization describing size and/or surface properties [51,52] and literature thereof.

Use of radioactive gold cluster immunoconjugates (containing 11–33 Au atoms) was reported in the early 1990s, but because of the cumbersome preparation and insufficient delivered dose, the research was discontinued.[55] Recently, 30–34 nm multifunctional, TNF-conjugated gold nanoparticles were prepared and tested for targeting solid tumors in mice.[36,41,56,57]

28.1.5 DENDRIMER NANOCOMPOSITES

Dendrimer templated nanocomposites are synthesized by reactive encapsulation.[58,59] According to the general concept of DNCs,[32] precursors to a desired product are pre-organized by an appropriately selected dendrimer in the first step. This pre-organization is followed by in situ chemical reaction(s) or physical treatment (irradiation, etc.) that generates reaction products immobilized in the nanoscopic size dendrimer molecules (Figure 28.5). This procedure yields dispersed small domains of guest molecules that are integrated with the dendrimer molecule(s) without creating covalent bonds between the dendrimer and the topologically entrapped matter. Dendrimer templated nanocomposites may be constructed with flexible architectures and in variable compositions.

Structure of nanocomposite particles was found to be the function of the dendrimer structure and surface groups as well as the formation mechanism and the involved chemistry. Based on the locality of the guest atoms, three different types of single nanocomposite architectures have been identified such as internal (I), external (E), and mixed (M) type nanocomposites where the inorganic matter either entrapped in the interior, in the exterior, or in both part of the dendrimer, respectively. Dominant properties of the composite nanoparticles approach the dendrimer properties with decreasing M/D ratios (i.e., with the molar ratio of the inorganic reactants and the organic template). Stable internal structures form when medium and high generation polyionic dendrimers with a weak binding periphery are employed with proper chemistry and at relatively low metal/-dendrimer ratios. (In this case, the inorganic domains are located in the interior of the dendrimer templates.) Single composite nanoparticles may be used as building blocks in the synthesis of nanostructured materials and devices. Details of experimental procedures and analytical techniques can be found in previous studies.[41]

Dendrimer nanocomposites can be constructed in precise sizes and well-defined surfaces with various covalently attached targeting moieties and with variable content. DNCs display stability and robustness in addition to properties represented by their individual components as a result of the molecular level's mixing of uniform macromolecules and nanoparticles. The unique advantage of this approach is that host and guest components can be separately selected, developed, and optimized. As a result of various synthetic options, the interior and/or the exterior of the host can be cationic, anionic, or non-ionic, depending on their termini and interior functionalities and the pH.

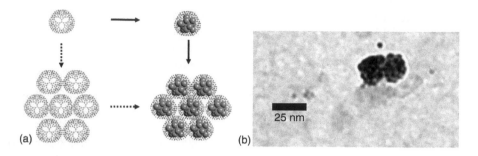

FIGURE 28.6 Scheme of assembling templates and synthesis of nanocomposites of well-defined sizes and surfaces (a) and TEM image of an {Au} nanocomposite single particle and a dimer of {Au-PAMAM}$_n$ multi-particle (b). (Reproduced from Bielinska, A., Eichman, J. D., Lee, I., Baker, J. R., and Balogh, L., *J. Nanopart. Res.*, 4, 395–403, 2002. With permission.)

Preparation of modular composite structures is possible either (a) by the synthesis of modular templates from simple dendrimer molecules (i.e., using tecto-dendrimers), or alternatively, (b) synthesizing of modular nanocomposites from single nanocomposite units. Variation of nanocomposite size is possible by using larger templates such as tecto-dendrimers or by combining existing nanocomposite particles. Alternatively, the size of nanoparticles may also be varied by reacting the surface of premade nanoparticles (Figure 28.6a and Figure 28.6b). Stable internal composite structures cannot be formed using linear macromolecules, only from dendrimers. This is the sub-structure that has great potential in combining the properties of organic and inorganic materials for biomedical applications. For these nanocomposite structures, the particle surface is supplied by the macromolecule; therefore, the solubility and surface compatibility of the nanocomposite particles is very similar to the template dendrimer macromolecules (Figure 28.7). These DNC structures are most frequently amorphous or nearly amorphous.[60]

Radioisotopes are often practically insoluble in water or body fluids; however, the nanocomposites are soluble in different sizes, contents, and surfaces. An analysis of tumor type should allow for the selection of a targeter, allowing rapid immobilization of the appropriate radioactive isotope in the pre-made template. Targeted composite nanoparticles can then be efficiently delivered to specifically kill tumor cells with minimal collateral damage.

As a basis for a future targeted [198]Au-based image-guided therapy, the effects of size and surface charge on the in vivo biodistribution, excretion, and toxicity of gold/dendrimer

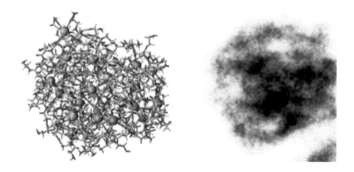

FIGURE 28.7 (a) Computer simulation of a dendrimer nanocomposite; (b) HRTEM of a gold nanocomposite particle containing 14 Au atoms. (Reproduced from Khan, M. K., Nigavekar, S. S., Minc, L. D., Kariapper, M. S., Nair, B. M., Lesniak, W. G., and Balogh, L. P., *Technol. Cancer Res. Treat.* 4(6), 603–613, 2005. With permission.)

nanocomposites in a mouse melanoma tumor model system were studied. The nanodevices tested include 5 and 22 nm primary amine surface, 5 nm carboxylate surface, and 11 nm carboxylate surface gold/dendrimer nanocomposites that display positive and negative surface charges at biological pH values. Gold nanocomposites offer the high electron density and radioactivity of the immobilized gold atoms, and the interactions of the composite particle with the surrounding biologic environment are determined by the surface. Biologic fate can be influenced by increasing or decreasing the number and/or strength of interacting forces on the nanparticle surface by changing the number and character of the dendrimer terminal groups.

28.1.5.1 Synthesis of Gold/Dendrimer Nanocomposites

To synthesize gold/dendrimer composite nanodevices in nearly uniform sizes and with close to identical properties, polyionic PAMAM dendrimer templates with low (4–10) M/D ratios were used. This method leads to well-controlled and uniform composite particles. The product is a synthetic dendrimer-dominated material that is water-soluble and stable. These hybrid nanoparticles have tunable properties; they can be made to be non-immunogenic, and they are similar in size to fundamental proteins present in the blood.[61]

Synthesis of {Au(0)} gold/dendrimer composite nanoparticles from dendrimer templates (Figure 28.5) involves two synthesis steps: (a) forming a PAMAM-tetrachloroaurate salt, and (b) the transformation of $[AuCl_4]^-$ through Au^{3+} ions to Au(0). Hydrolysis of $[AuCl_4]^-$ is a complex process[62] and easily occurs in aqueous solutions of PAMAMs as the nitrogens neutralize HCl, the hydrolysis byproduct. Reduction is usually performed either by adding external reducing agents (N_2H_4 or $NaBH_4$)[41,57] or utilizing the reductive nature of terminal groups of PAMAMs.[63]

In the two extreme cases, (depending on the reaction conditions) the dendrimers may either encapsulate the colloidal gold,[57] or alternatively, gold nanocrystals may grow in the interior of the PAMAM molecules.[6] Reduction of Au^{3+} ions by the PAMAM itself may be accelerated by applying UV-irradiation.[64] The usual size of dendrimer templated nanocomposite particles is $d = 3$–10 nm, depending on the size of the template, the gold/dendrimer ratio, and the type of the composite particle.[41,65] Smaller (LGD) PAMAMs provide larger, multiparticle composites because of their open structures, whereas for MGD and HGD (G5–G9) PAMAMs and at low metal/dendrimer ratios, the composite particles preserve the size of the template.[66,67]

Controlled fabrication of large nanocomposite particles requires large templates. Synthesis of positively charged large ($d > 10$ nm) PAMAM dendrimer structures, (e.g., generation 9 PAMAM with an Mn = 467,000 and $d = 11.4$ nm) the routinely used divergent route requires several months. The multi-step synthesis and purification (altogether 18 step for G9) results in expensive (presently \$16,500/g), materials.[68] Applying tecto-dendrimers[69–71] as templates is a more favorable approach, but tecto-PAMAMs can provide only methylester or carboxylate terminated (i.e., negatively charged) composite nanoparticles.

It has been found that neutron/gamma irradiation also induces elemental gold formation in aqueous solutions or gels (submitted for publication). A novel and simple procedure to fabricate {(^{198}Au(0)$_n$-PAMAM} + positively charged radioactive gold nanocomposites in $d = 10$–30 nm sizes directly from PAMAM tetrachloroaurate salts or $d = 5$ nm {Au(0)$_n$-PAMAM_E5.NH$_2$}$^{k+}$ nanocomposites by simultaneously utilizing both the neutron and gamma-radiation have been devised and developed. This method was used to fabricate positively charged radioactive gold nanocomposites in sizes between 10 nm and 30 nm, also to be used as dose delivery agents (Nano Letters, 2006, submitted).

Particle size, mass, and charge of dendrimer nanocomposites may be very similar to their templates. Shown in Figure 28.8 is the PAGE electropherogram of different generations of amine-terminated dendrimer templates and the corresponding dendrimer nanocomposites. These Au-nanocomposites were templated by polycationic (amine terminated) PAMAM dendrimers of different generations with the same molar ratio of nitrogen ligand/gold atom and were characterized

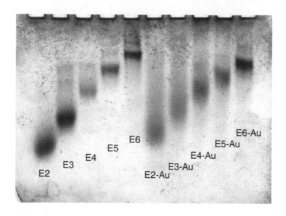

FIGURE 28.8 PAGE electropherograms of several polycationic PAMAM dendrimer templates and their corresponding gold-dendrimer nanocomposites. Observe the similarity between the movement patterns of the dendrimer templates and the nanocomposites, indicating similar charge/mass ratios for the templates and the gold nanocomposites (D nitrogen ligand/Au ratio = 4.5–5.2). (Reproduced from Khan, M. K., Nigavekar, S. S., Minc, L. D., Kariapper, M. S., Nair, B. M., Lesniak, W. G., and Balogh, L. P., *Technol. Cancer Res. Treat.* 4(6), 603–613, 2005. With permission.)

after a prolonged storage, i.e., in thermodynamic equilibrium. The amine-terminated gold-dendrimer nanocomposites and their respective dendrimer templates all display very similar migration patterns. (Au-nanocomposites display somewhat lower electrophoretic mobility probably because of the slightly higher mass of the composite particles.)[65]

In any case, using a good quality template is essential for success. Templates can ease the difficulty of defining size and shape, but the quality of the product will be determined by the template's precision. In short, the product cannot be better than the template and the procedure used. In an ideal case, one would have a perfect template and a subsequent perfect procedure to get a homogenous and uniformly sized nanomaterial of a predefined shape. Any imperfection in the template or the procedure results in a broadening distribution of composition and geometry (and, subsequently, of the properties) in the product.

28.1.5.2 Characterization of Dendrimer Hosts and Gold Nanocomposites Used in Biodistribution Experiments

Determination of fundamental properties (composition, size, and charge) is vital for reproducibility prior to initiating any biological experiment. A variety of analytical techniques to were applied to investigate the purity, properties, and structural characteristics of dendrimer hosts and host-guest nanocomposites, including polyacrylamide gel electrophoresis (PAGE), capillary electrophoresis (CE),[72] size exclusion chromatography (SEC), and HPLC,[73,74] UV–visible and fluorescence spectrophotometry, acid–base potentiometric titration, mass spectrometry (ESI-MS and MALDI-TOF techniques), and various NMR techniques.[65,75,76] (available at www.mrs.org).[77] Gold/dendrimer nanocomposites have been characterized by UV–visible and fluorescence spectrophotometry, NMR, DLS (dynamic light scattering), and TEM. Particle structure of the neutron-irradiated gold nanocomposites have also been studied (after isotope decay) by methods described above in order to evaluate the effect of direct activation on the stability of the polymer host.[4,42,61]

28.1.5.3 Fabrication of Nanocomposite Devices for Biologic Experiments

In these studies, non-targeted and folic acid targeted hosts were synthesized and characterized (i.e., $d = 5$ nm) ^{3}H labeled PAMAM dendrimers with positive and neutral surface charges in normal and

TABLE 28.1
Molecular Characteristics of Polyanionic PAMAM Dendrimer Templates

PAMAM	E3.SAH	E5.SAH	E5(E3.SAH)$_9$
Mw[a]	10,109	41,626	147,542
Mn[b]	6300	38,400	152,200
Mw[b]	9790	38960	182,100
Polydispersity[b]	1.022	1.045	1.196
Theoretical no. of COOH groups[a]	32	128	256
Practical no. of carboxylate groups[c]	30.6	113.2	296
Charge/mass (mol/g)[a,d]	-3.18×10^{-3}	-3.08×10^{-3}	-1.74×10^{-3}
Charge/mass (mol/g)[b,c]	-3.32×10^{-3}	-2.90×10^{-3}	-1.94×10^{-3}

[a] Theoretical molecular weight assuming complete conversions.

[b] As measured by SEC at pH 2.74.

[c] Measured by potentiometric titration.

[d] As measured by CE.

tumor tissues to examine the biodistribution and the toxicity. Then, gold nanocomposites with different surface charges and in different sizes were prepared and studied.

The combined analytical results and properties of the PAMAM dendrimers used as templates are presented in Table 28.1 and Table 28.2. Details of template synthesis and characterization are described in the literature.

Zeta potential measurements of $\{(Au^0)_9\text{-PAMAM\_E5.}(NH_2)_{110}\}$ in aqueous solutions confirmed that these nanoparticles are positively charged, indicating that the terminal amines can still be protonated after the formation of the nanocomposites. The net charge of amine-terminated dendrimers and their Au-nanocomposites was compared by PAGE and proved to be similar to their corresponding templates in the pH $= 8.3$ running buffer.[65]

TABLE 28.2
Molecular Characteristics of Polycationic PAMAM Dendrimer Templates

Dendrimer	E3.NH$_2$	E5.NH$_2$
Mn[a]	6909	28,826
Mn[b]	6648	27250[c]
Diameter [a] (nm)	3.6	5.4
Theoretical no. of NH$_2$ groups [a]	32	128
Total no. of N ligands [a]	62	254
Charge/Mw ratio[d] (C/g)	8.973×10^{-3}	8.811×10^{-3}
Average no. of –NH$_2$ groups[b]	26.3	120[c]
Total average no. of –N = ligands [b]	54	238[c]
Charge/Mw ratio[b] (C/g)	8.12×10^{-3}	8.73×10^{-3}

[a] Theoretical values.

[b] Practical values determined from titration and GPC measurements.

[c] Literature data.

[d] Assuming theoretical values and full protonation.

28.1.5.4 Synthesis of Tritium Labeled Dendrimers

Earlier biodistribution studies of PAMAM dendrimers have been carried out with full generation PAMAM dendrimers that were radiolabeled by partially quaternizing the primary amine termini using ^{14}C, containing methyl iodide or ^{125}I labels.[17,78] Because these radiolabeled macromolecules carry permanent positive charges, this labeling procedure results in a mixture of products where the labeled molecules are chemically different from the unmodified population of dendrimers. The iodine-labeling procedure may lead to data where the distribution of radiolabeled materials is different from the overall distribution of the nanoparticles.

Labeling of dendrimers by ^{3}H allows proper and accurate quantification of the PAMAM derivatives for the true in vivo biodistribution as a function of time because the labeled molecules are chemically identical with the unlabeled dendrimers. Regulation of surface charge with simultaneous labeling with tritium was achieved via acetylation by partially or fully reacting terminal primary amine groups of the PAMAM molecules. This synthetic step reduces the in vivo toxicity associated with the polyamine character (the terminal primary amine groups are positively charged at biologic pH values) and provides stable labels for the dendrimers. Increasing the degree of acetylation decreases the surface net positive charges; the fully acetylated generation five PAMAM displays about $+5$ mV zeta potential as opposed to $+40$ mV measured for the primary amine functional PAMAM_E5.NH$_2$.

The biodistribution of the fully acetylated dendrimer (NSD, Neutral Surface Dendrimer) and a partially acetylated one that has a partially positive surface charge have been compared. (The term of *positively charged* refers to the surface character in dilute aqueous solutions at biologic pH.)

28.1.5.5 Folate-Targeted Devices

Briefly, partially acetylated PAMAM was conjugated to folic acid as a targeting agent, and the remaining primary amines were capped using tritium labeled acetic anhydride. These conjugates were intravenously injected into immunodeficient mice bearing human KB tumors that over-express the folic acid receptor. In contrast to non-targeted polymer, folate-conjugated nanoparticles concentrated in the tumor and liver tissue over fours days after administration. The tumor tissue localization of the folate-targeted polymer could be attenuated by prior i.v. injection of free folic acid. Using fluorescein labels instead of tritium labels, confocal microscopy was also used to confirm the internalization of the conjugates into the tumor cells. PAMAM dendrimers substituted with folic acid and methotrexate increased its anti-tumor activity and markedly decreased its toxicity, allowing therapeutic responses not possible with a free drug.[79]

28.1.5.6 Gold Composite Nanodevice Synthesis

The {Au} gold/dendrimer composite nanodevices described below have been fabricated using four methods:

1. By reactive encapsulation of gold in commercial (primary amine terminated and carboxylate terminated) PAMAM templates;
2. By performing post-synthetic modification on the surface groups of the {Au} gold nanocomposites;
3. By reactive encapsulation of gold in carboxylate functional PAMAM tecto-dendrimer templates; and
4. By radiation polymerization of {Au} nanocomposites (Table 28.3).

Method 28A1 $\{(Au^0)_{9.08} - PAMAM\_E5.(NH_2)_{120}\}^+_{d=5\ nm}$, for short: $\{Au(0)\}^+_{d=5\ nm}$, i.e., $d = 5$ nm amine surface gold/dendrimer nanocomposite (positively charged at pH $= 7.4$).

TABLE 28.3
Summary of Nanodevices and Notations

Description	Full Annotation	Acronym
PSD	$PAMAM\_E5.(NH_2)_{44}(NHCOCH_3^*)_{66}$	$E5.(NH_2)(NHOAc^a)$
NSD	$PAMAM\_E5.(NHCOCH_3^*)_{110}$	$E5.OAc^a$
Targeted NSD	$PAMAM\_E5.(NHCOCH_3^*)_{104}(FA)_6$	$E5.OAc^a\text{-}FA$
Neutral surface 5 nm gold CND	$\{(Au^0)_{9.08} - PAMAM\_E5.(NHCOCH_3)_{120}\}^+_{d=5\ nm}$	$\{Au(0)\}^0_{d=5\ nm}$
Positive surface 5 nm gold CND	$\{(Au^0)_{9.08} - PAMAM\_E5.(NH_2)_{120}\}^+_{d=5\ nm}$	$\{Au(0)\}^+_{d=5\ nm}$
Negative surface 5 nm gold CND	$\{(Au^0)_{6.45} - PAMAM\_E4.5(COOH)_{60}\}^-_{d=5\ nm}$	$\{Au(0)\}^-_{d=5\ nm}$
Negative surface 11 nm gold CND	$\{(Au^0)_{60.09} - (PAMAM\_((E5)(E3.(COOH)_{296}))$ $\}^-_{d=11\ nm}$	$\{Au(0)\}^-_{d=11\ nm}$
Polymerized PCND[b]	$\{(Au^0)_{5.7} - E5.(NH_2)\}^+_{87,\ d=22.2\ nm}$ and radioactive $\{^{198}(Au^0) - E5.(NH_2)\}^+_{87,\ d=22.2\ nm}$	$\{(Au^0)_{5.8}\}^+_{d=22\ nm}$ $\{^{198}(Au^0)\}^+_{d=22\ nm}$
Polymerized PCND[b]	$\{(Au^0)_{5.7} - E5.(NH_2)_{209,\ d=29.7\ nm}^+\}$ and $\{^{198}(Au^0) - E5.(NH_2)\}^+_{209,\ d=29.7\ nm}$	$\{(Au^0)_{5.7} - E5.(NH_2)\}^+_{d=29.7\ nm}$ $\{^{198}(Au^0) - E5.(NH_2)\}^+_{d=29.7\ nm}$

PSD: Positive Surface Dendrimer; NSD: Neutral Surface Dendrimer; Ng: Negative; CND: Composite Nanodevice

[a] Denotes radioactivity.

[b] Numbers of primary particles have been calculated from volume, assuming full space-filling.

The $d = 5$ nm positive nanodevices (Figure 28.7) were fabricated from PAMAM dendrimers and HAuCl$_4$ as previously described.[5,36] Briefly, solution of HAuCl$_4$ in methanol was added dropwise to a methanol solution of PAMAM dendrimer (Mn = 26,000, functionality = 120), resulting in a yellow solution of the tetrachloroaurate salt of the PAMAM dendrimer with a gold/dendrimer ratio of Au/D = 9.08), i.e., [$PAMAM\_E5.(NH_2)_{120})(AuCl_4)_{9.08}$]. The resulting yellowish solid was redissolved in water that then resulted in hydrolysis of the complex and led to nanocomposite formation (composite gold content = 6.44%). (Positive composite nanodevice, PCND)(Figure 28.9).

Method 28A2 $\{(Au^0)_{6.45} - PAMAM\_E4.5(COOH)_{60}\}^-_{d=5\ nm}$, for short: $\{Au(0)\}^-_{d=5\ nm}$, $d = 5$ nm carboxylate surface gold/dendrimer nanocomposite (negatively charged at pH = 7.4)

Briefly, methanol solution of HAuCl4 was added drop-wise to an aqueous solution of generation 4.5 PAMAM dendrimer (Mn = 26,258, functionality = 60), resulting in a red solution of the correspondent dendrimer nanocomposite (Au/D = 6.45). (Composite gold content = 4.41%) (Negative composite nanodevice, NgCND).

Method 28B. Acetylation of the $\{Au(0)\}^+_{d=5\ nm}$ amine surface material, resulted in $\{(Au^0)_{9.08} - PAMAM\_E5.(NHOCCH_3)_{110}\}^0_{d=5\ nm}$, for short: $\{Au(0)\}^0_{d=5\ nm}$, an acetamide surface $d = 5$ nm gold/dendrimer nanocomposite that is nearly neutral at pH = 7.4 (neutral composite nanodevice, NCND).

Method 28C. The $\{(Au^0)_{60.09} - (PAMAM\_((E5)(E3.(COOH)_{296}))\}^-_{d=11\ nm}$, for short: $\{Au(0)\}^-_{d=11\ nm}$, i.e., the $d = 11$ nm carboxylate surface (negatively charged at pH = 7.4) gold/dendrimer nanocomposite was synthesized from PAMAM_E5(E3.COOH)$_9$, a succinamic acid-terminated tecto-dendrimer (see Table 28.1). Briefly, methanol solution of HAuCl$_4$ was added drop-wise to an aqueous solution of PAMAM_E5(E3.COOH)$_9$ tecto-dendrimer (Mn = 120,000, functionality = 296), resulting in a red solution of the corresponding dendrimer nanocomposite (Au/D = 60.09; composite gold content = 8.94%).

Method 28D. Positively charged $d > 10$ nm dendrimer nanocomposites were made from generation three and generation five templated nanocomposites by simultaneously exposing them to gamma and neutron radiation in a nuclear reactor above Tg, the glass transition temperature of

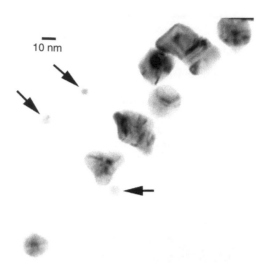

FIGURE 28.9 TEM of single $\{(Au(0)_n - E5\ NH_2\}$ particles (marked with arrows) and various nanocomposite aggregates formed during sample preparation. (Reproduced from Balogh, L., Bielinska, A., Eichman, J. D., Valluzzi, R., Lee, I., Baker, J. R., Lawrence, T. S., and Khan, M. K., *Chim. Oggi (Chem. Today)*, 20 (5), 35–40, 2002. With permission.)

the polymers, (for PAMAMs Tg = 20–35 C). Under these conditions, the neutrons activate the trapped gold metal clusters in the $\{(Au^0)_{5.76}\}$ primary composite nanoparticles, and the gamma-radiation cross links the organic network. Irradiation times were varied from 30 min to 3 h. Level of activation of ^{197}Au within the allowable degree of polymerization of the dendrimer host was controlled by setting the neutron irradiation time and limiting the gamma dose by led filters. Gold (^{197}Au) captures neutrons very efficiently because of its large cross-section. ^{198}Au ($t_{1/2} =$ 2.69 days) decays predominantly by beta-radiation (99%, 0.96 MeV) with a small gamma component (0.98%, 1.1 MeV) into a stable ^{198}Hg isotope. The build-up of activity for any given isotope is a function of several intrinsic factors, including isotopic abundance and neutron cross-section as well as a series of irradiation parameters, including neutron flux and energy spectrum. Therefore, although the general principles are valid for all the reactors, detailed experimental parameters of the procedure have to be independently worked out. Under these experimental conditions, efficiency was 50% of the target (theoretical) value (Figure 28.10).

There were no changes observed in nanocomposite particle size at low doses. At higher irradiation doses, the size of nanoparticles increased and the particle size distribution became somewhat wider.

The exact mechanism of cross linking is yet unknown. It is speculated that radical mechanisms that are responsible for beta-alanine cross linking (as well as local heat effects that are known to result in retro-Michael synthesis) may result in the rearrangement of the molecular structure.[80]

Data indicate that the product (similarly to silver/dendrimer nanocomposites)[76] is composed of primary composite nanoparticles and their aggregates (Table 28.4). Volume weighted averages are directly proportional to %w/w concentrations. Nevertheless, interaction with biologic objects (e.g., cells) is determined by the number of interacting objects, thereby necessitating determination of number averages as well. By number averages, each sample is predominantly composed of single particles (Figure 28.11).

The size of the cross linked primary particles depends on the irradiation conditions; brief exposure to neutrons and gamma rays does not change particle size (Figure 28.12, left side), but it results in low activities. Longer irradiation activates the gold and simultaneously melts the

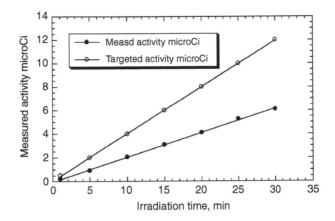

FIGURE 28.10 Comparison of targeted and measured specific activities of ^{198}Au as a function of irradiation time. (Reproduced from Balogh, L. P., Nigavekar, S. S., Cook, A. C., Minc, L., and Khan, M. K., *PharmaChem* 2(4), 94–99, 2003. With permission.)

dendrimers and cross links the primary particles into spherical and still soluble radioactive nano-composites. Figure 28.12 shows the empirical correlation between particle diameter and irradiation time when pre-made $\{(Au^0)_{6.53}\text{-PAMAM\_E5.NH}_2\}$ nanocomposites were treated in a reactor.

In TEM, large particle structures that are probably responsible for the observed intense light scattering have been identified (Figure 28.13b). Visualization of these loose aggregates offers a few mechanistic clues. First, size and size distribution are unaffected when samples are irradiated with low doses in solution (Figure 28.13a). At higher doses, larger {Au} spherical particles form with a few superclusters present that are composed of primary $\{Au\}_n$ clusters loosely attached to each other by their surface.

For approximate calculations, complete space-filling (100% packing) values for the cross linked primary particles and partial (50% packing) values for their aggregates were assumed. Based on volume averaged size data by DLS, a $d = 22.2 \pm 1.9$ nm polymerized particle

TABLE 28.4
Particle Size Distributions of Soluble $\{^{198}Au\}^+$ Nanocomposites After Irradiation of $\{(Au^0)_{6.53}\text{-E5.NH}_2\}$ (NICOMP Data)

Sample $\{(Au^0)_{6.53}$ -E5.NH$_2\}$. A, B, C	Volume Weighted			Number Weighted			Irradiation Time (Hr.)	Zeta Potential (mV)
	Mean Diam. (Nm)	Std. Dev. (Nm)	Present (%)	Mean Diam. (Nm)	Std. Dev. (Nm)	Present (%)		
$\{(Au^0)_{6.53}\}_{n=,}^{+23.3\ mV}{}_{d=14.0\pm1.9\ nm}$	14.0	1.9	99.4	13.4	1.4	100	1.0	+23.3
Aggregate	453	45.1	0.6	0	0	0		
$\{(Au^0)_{6.53}\}_{n=,}^{+22.7\ mV}{}_{d=16.6\pm1.7\ nm}$	16.6	1.7	94.6	15.8	1.9	99.9	1.5	+22.7
Aggregate	73.6	10.3	5.4	70.4	6.2	0.1		
$\{(Au^0)_{6.53}\}_{n=,}^{+22.7\ mV}{}_{d=21.1\pm2.9\ nm}$	21.1	2.9	91.2	20.2	2.2	99.9	2.0	+25.2
Aggregate	111.2	17.2	8.8	104.9	4.0	0.1		

Data were measured after loss of radioactivity.

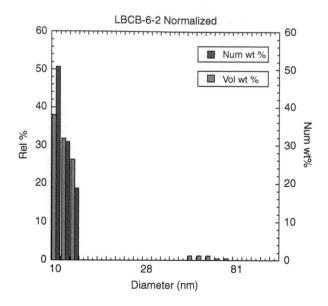

FIGURE 28.11 Comparison of number weighted and volume weighted distributions reveals that although this nanocomposite solution contains a few clusters, each sample is predominantly composed of particles of similar size.

$\{(Au^0)^+_{5.76}\}_{87,\ d=22.2\ nm}$ contains 87 primary ($d = 5$ nm) DNC units and 499 Au^0 atoms per particle. A $d = 29.7 \pm 2.9$ nm polymerized gold/dendrimer nanocomposite particle ($\{(Au^0)^+_{5.69}\}_{209,\ d=29.7\ nm}$) contains 209 primary ($d = 5$ nm) DNC units and 1123 Au atoms. The aggregate of this secondary nanoparticles (with a packing fraction of 50%) would be composed of $d = 22$ nm particles (each containing 489 Au atoms in this example) that would suggest the presence of 51.4 $\{(^{198}Au)_n\}$

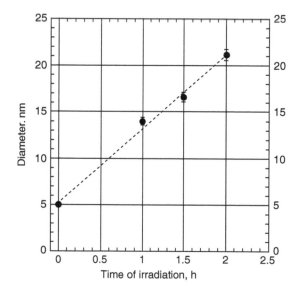

FIGURE 28.12 Average diameter of Peak 1 in aqueous solutions of gold nanocomposites as a function of irradiation time (in equivalent position). Extensively irradiated nanocomposite samples became insoluble in water.

TABLE 28.5
Particle Diameters of Radiation Polymerized Gold/Dendrimer Nanocomposites
(Au/D = 5.69) According to Dynamic Light Scattering Data with NICOMP Data Analysis

	Volume Weighted		Number Weighted		No. of Au atoms	
Sample	Mean Diameter (nm)	Percent (%)	Mean Diameter (nm)	Percent (%)	Per Host	Gold mass Fraction (%)
Primary particle: $\{(Au^0)_{5.76}\}_{87}$	22.2	97.9	21.7	99.9	504	99.3
Aggregate $(\{(Au^0)_{5.76}\}_{87})_{12.5}$	51.5	2.1	51.4	0.1	3113	0.7

composite nanoparticles per aggregate (on average), containing 31,002 gold atoms in total. Data are shown in Table 28.5 and Table 28.6. The resulting nanocomposites preserved a net positive charge although measured zeta potential values (Table 28.6) were approximately half of what was measured for $\{(Au)_{56.61}$-PAMAM_E5.NH$_2\}$ nanocomposites.[65] (The surface zeta potential for the latter nanocomposites was 41 mV as calibrated by $d = 50$ nm poly(styrene)sulfonate standards from Duke Scientific). Decrease of the surface positive charge indicates involvement of primary amine terminal groups in the polymerization reactions followed by their subsequent partial elimination from the host during radiation. The net positive zeta potential also supports the existence of the primary amine surface functions, explaining the formation of superclusters through surface attachment.

28.1.5.7 Gold in the Nanocomposite Particles After Radiation Polymerization

It is well known that the maximum of the surface plasmon resonances of noble metal nanocrystals in the UV–vis spectrum shifts toward longer wavelengths (from red to blue, i.e., toward lower energies) when particles aggregate.[81,82] However, the maxima in the UV–visible spectra of the {Au} nanocomposites are practically the same before and after radiation polymerization. The lack of shift indicates that the structural change is restricted to the host and the Au nanodomains remain isolated (Figure 28.14).

28.1.5.8 Visualization of Gold/PAMAM Nanocomposite Uptake in Tumor Cells

In this demonstrative experiment, amine surface gold PAMAM nanocomposites were used to visualize $\{Au(0)\}+$ in tumor tissue by TEM. One million B16F10 melanoma cells were

TABLE 28.6
NICOMP Size Distribution Data After Irradiation and Radioactive Decay for the Nano-
composite (Au/D = 5.76) Used in the Preliminary Therapeutic Experiments

	Volume Weighted		Number Weighted		No. of Au atoms	
Sample	Mean Diameter (nm)	Percent (%)	Mean Diameter (nm)	Percent (%)	Per particle	Mass Fraction (%)
Primary particle: $\{(Au^0)_{5.69}\}_{209}$	29.7	94.1	27.7	99.9	1123	97.5
Aggregate $(\{(Au^0)_{5.69}\}_{209})_{51.4}$	110.4	5.9	106.4	0.1	31,002	2.5

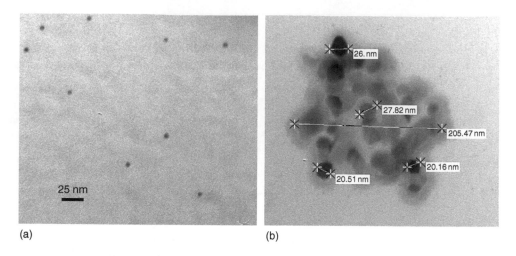

(a)   (b)

FIGURE 28.13 (a) The TEM image illustrates the unperturbed size distribution of the $\{Au_{5.7}\text{-}$ PAMAM_E5.NH$_2\}$ gold nanocomposite when irradiated with low doses. (b) Almost identical d = 22 polymerized $\{Au\}_n$ particles forming a loose $(\{Au\}_n)_m$ aggregate. (Reproduced from Balogh, L. P., Nigavekar, S. S., Cook, A. C., Minc, L., and Khan, M. K., *PharmaChem* 2(4), 94–99, 2003. With permission.)

subcutaneously injected on the dorsal surface of a C57Bl6/J mouse and tumor allowed to develop for two weeks. The mice were then injected with the nanocomposite solution. Tumor and other organs (liver and lung tissue) were isolated and ultra-thin sections examined on a Philips CM100 electron microscope. Figure 28.15 shows a TEM image of one of these sections with three tumor cells. The image indicates that 1.25 h after the direct injection only a few larger gold nanocomposite clusters are present in the interstitium. Most of the smaller clusters were found in the nucleus and cytoplasm. During the internalization of gold nanocomposites, diffusion through the cell wall

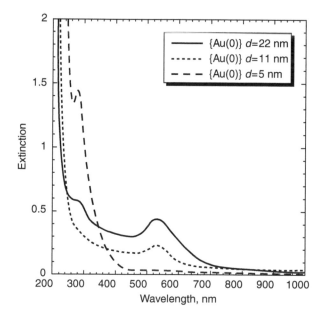

FIGURE 28.14 Comparison of UV–visible spectra of $d = 5$ nm (dashed line), $d = 11$ nm (dotted line), and $d = 22$ (continuous line) nm $\{Au(0)\}$ gold nanocomposites.

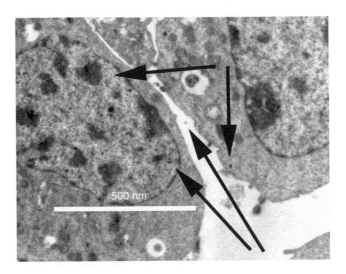

FIGURE 28.15 Electron microscope image of mouse tumor tissue 1.25 h after {Au} injection. Unstained specimen, black dots represent gold nanocomposite clusters, dark areas within the nucleus are preferred for {Au}. (Reproduced from Balogh, L., Bielinska, A., Eichman, J. D., Valluzzi, R., Lee, I., Baker, J. R., Lawrence, T. S., and Khan, M. K., *Chim. Oggi (Chem. Today)* 20(5), 35–40, 2002. With permission.)

appears to be the dominant mechanism although transfer through a nuclear pore has also been observed.

28.2 BIOLOGY

As previously shown, nanocomposite dimensions are in the size range of small cellular machinery. The interactions that devices this small will have with biologic systems is truly unknown, and mathematical or other guiding principles to determine how exactly these small devices will behave in complex biologic systems that have potential interactions at the intracellular, cellular, multi-cellular, organ, and multi-organ system level are yet unknown. Research has demonstrated that until understanding of this process is had, it will be critical to empirically test these interactions as simple assumptions about nanodevice behavior in biologic systems are currently often incorrect.

Interactions of nanoparticles with biologic objects should typically be surface interactions. In other words, they depend on the surface characteristics of both the biologic entity and the nano-particle. This has been borne out in several of this team's experiments and will be detailed below. The nanoparticle's size is important, and this may be due to cellular or macromolecular interactions that are affected by size such as the ability to move through membranes, pores, membrane junctions, or recognition by certain cells (such as macrophages) for uptake. Charge of the nanoparticle is important because the Coulomb interactions are the strongest physical forces between any two objects. Other three-dimensional changes within the nanodevices that can affect their surface size/charge would also then be expected to affect biologic interactions.

Biodistribution, or how these nanodevices move throughout complex biologic multi-organ systems, is a dynamic property and must be looked at as a function of time. Biodistribution can be affected by many factors, including site of injection or uptake, blood flow/rate, uptake resident time in the blood and various organs and organ systems, and clearance from the body (urine, feces, and sweat, for example) to name only a few of these factors. Nanodevices intravascularly injected will travel first to the heart then lungs, then back to the heart, then throughout the rest of the body. The heart and lung have first exposure to the nanodevices. Certain organ systems may have architecture (fenestrated structures for example) that may hasten or slow the time blood is exposed to the organ, the ability to enter organs, and the ability to stay in certain organs. Size/charge characteristics (particularly size) could have large influences on the route of clearance of these nanodevices. Detailed empiric biodistribution studies may eventually permit more detailed mathematical understanding of how different size/charge nanodevices will distribute over time in complex biologic systems.

The vascular system is a critical organ system that needs detailed examination as it is the first organ system seen by intravascularly injected nanodevices and the main route of distribution to the rest of the body. Of particular importance to the study of cancer (and potentially other diseases such as macular degeneration, rheumatoid arthritis, etc.) is the angiogenic microvasculature. When these pathologic conditions occur or when a wound is healing, the normal existing microvasculature is stimulated to build a new microvasculature via budding off of the existing microvasculature. This process of neovascularization is called angiogenesis. It has potentially important implications for potential uses of nanodevices in pathologic and some non-pathologic biologic conditions as will be discussed below. A detailed understanding of the in vivo biodistribution of nanodevices will eventually permit the design of nanodevices that can specifically or selectively target specific organs or tissues in order to improve medical therapy.

28.2.1 TARGETING AND NANODEVICE DELIVERY

Targeting of tumor cells may be performed by either selective or specific ways as described in related literature. Direct receptor specific targeting is the one most often thought about, and nanocomposites permit the optimization of surface properties (via modulation of charge and attachment of targeting moieties) to maximally permit this direct targeting. Targeting groups can be conjugated to the host dendrimer's surface[27] to allow the imaging agent to selectively bind to specific sites such as receptors on tumor cells to improve detection. For example, there are a number of cell surface receptors that are known to be over expressed on cancer cells, including epidermal growth factor receptor (EGFR) and the laminin receptor VLA-683. Nanocomposites could be constructed to specifically target these receptors.

Nanodevices can also be constructed to target exposed receptors on the angiogenic microvasculature. Over the last few years, data have emerged supporting the concept that the microvasculatures of tissues and/or organs also have antigenic surface differences. Using a phage display technique, investigators have isolated peptide fragments that specifically bind to different tissue or organ microvasculatures in mice.[84–86] As part of this work, they also isolated peptides that specifically bind to angiogenic tumor microvasculature.[87–89] One of these angiogenic microvascular targeting peptides isolated is a three amino acid peptide, arginine–glycine–aspartate (RGD). RGD has been used to target chemotherapy and other agents to the angiogenic tumor microvasculature.[87,90,91] Radio-labeled linear or cyclic forms of the RGD peptide itself have been used to target the tumor microvasculature, permitting imaging with PET or SPECT.[92–97] One could envision the use of RGD or other small molecules binding angiogenic microvasculature and present on the surface of the composite nanodevices to directly target the microvasculature. These studies are currently underway in this team's laboratories.

Non-target specific or non-selective targeting is another way to affect biodistribution of the nanodevices by not relying on the binding to a specific receptor but by taking advantage of the existing structural properties of complex biologic systems. It is well known that the angiogenic microvasculature is very different from the fully formed microvasculature. It is leakier than normal microvasculature, and there is also evidence that the tumor angiogenic microvasculature often lacks pericytes and has aberrant morphologies.[98–101] These differences between the tumor microvasculature and the normal microvasculature may result in enhanced permeability and retention effect (EPR), suggested first by Maeda.[102] The accumulation of macromolecules in the tumor was also found after i.v. injection of an albumin-dye complex (Mw = 69,000) as well as after injection into normal and tumor tissues. The complex was retained only by tumor tissue for prolonged periods. With liposomes, the most effective size preferably retained in the tumor was suggested to be around 50 nm. The Duncan group has also shown[103] that appropriately sized dendrimers and dendrimer-cis-platin complexes may become trapped in tumor vasculature as a result of the EPR effect.[104]

Using PAMAM dendrimers, it has also been demonstrated that polymers of different sizes have different permeability or cellular uptake, depending on the size (generation) and surface of the dendrimer (discussed in more detail below). For future therapy, these size and charge characteristics are exploited to specifically deliver nanoparticles through the tumor microvasculature and to the tumor or directly to the microvasculature itself using additional surface recognition to further improve avidity.

28.2.1.1 Quantitative In Vivo Biodistribution of Dendrimer and Gold Composite Nanodevices (CND)

As previously explained, composite nanodevices have several potential advantages over current technologies for the delivery or radiation or other therapies to tumors or the tumor microvasculature. In particular, even the smallest gold composite nanodevice can deliver greater than ten gold atoms per device to a target. This is a log fold more radiation than previously possible with the usual radioactive antibody techniques and offers a real potential to have greater than a log fold more radiation delivery per device than previously possible. The composite nanodevice architecture permits the simultaneous optimization of the carried inorganic component (gold or other inorganic material) and the dendrimer network (with or without receptor-specific targeting), providing the surface components of the fully formed nanodevice. To think of this clinical point of view, for example, targeting the nanodevice by the covalent linking of specific moieties to its surface and then carrying radioactive gold for imaging/therapy applications, the gold placement would not interfere with the targeter functioning on the surface. This optimization of the different compartments is a critical advantage of the composite nanodevices compared to other structures.

28.2.1.2 Related Techniques in the Literature

Illustrative of a few of these points is a study by Haifield et al. in the early 1990s.[55] This group carried out a partially successful attempt to increase the gold carrying content of antibodies by covalently attaching gold clusters using sulfhydryl linkages to antibody fragments, and it then carried out biodistribution studies of these devices in mice. They were able to get some of the gold clusters averaging about 11 gold atoms to bind to various antibody fragments. The characterization of these clusters was done with TEM, and much more detailed characterization using several other techniques would be needed to be certain of the actual complexes formed. The size of the final assembly was not presented. They were able to make an antibody fragment-Au-cluster device that retained about 80% of its original binding activity (because of the interference of the gold clusters with the antibody fragment binding site). To stabilize the clusters, 21 phenyl groups were used, and this hydrophobic layer around the Au clusters caused increased uptake arguably in the blood cells. There was also very high uptake (trapping of clusters) in the kidney, probably as a result of the size of the antibody-Au cluster constructs. In both cases, the Au cluster chemistry/architecture interfered with the targeting component of the devices. More recent work on colloidal gold preparations by Paciotti et al.[56] examined the in vivo biodistribution of a colloidal gold nanoparticle devices made of thiol–PEG (polyethylene glycol) and gold and recombinant human TNF (PT–cAu–TNF). The mechanism of the TNF linkage to the colloidal gold could not be delineated, but the preparations did appear to be 33 nm in size, and crude biodistribution measurements in a few organs showed relatively rapid clearance from the blood and deposition of the TNF in tumors. Much of the biodistribution of the nanoparticle was done by observing the color of organs as they turned more purple as they picked up the nanoparticle (possibly because of the formation of large gold nanoparticle aggregates that shifts the maximum of the plasmon peak toward blue). The data were not quantitative for whole particle determinations. The measurement of TNF in a few organs was quantitative. This nanoparticle appeared to have some interesting data supporting improved delivery to tumors compared to the non-PEGylated form of the nanodevice. It is unclear from the work, however, how this approach could be extended to permit other targeting with different agents or what improvements need to be made. There was no indication of how the Au–PEG–TNF components could be modified or optimized. Given the lack of knowledge of the actual chemistry holding the device together for the short few-hour time course of the experiment, it is unclear if this system can be modified or improved. These reports emphasize the advantages possible with the composite nanodevices, the need for much more detailed chemical characterization and biological understanding of how these devices distribute throughout the body, and how simple modifications may affect this biodistribution.[56,105]

Quantitative and detailed in vivo biodistribution data for dendrimer nanoparticles and, for the first time, composite nanodevices in tumor mouse model systems are presented below. Although there are many theories on how nanodevices might interact with complex animal systems, this team decided to empirically measure nanodevice biodistribution with several planned modifications of the nanodevices as a way to eventually begin to understand guiding principles of how nanodevices and, in particular, nanocomposites interact with organs/tissues. This data will eventually be used to develop a more detailed mathematical understanding of the process and to aid in the design of future nanodevices aimed at specific biologic understanding or therapy.

As part of this approach, these groups carried out the first detailed and quantitative biodistribution studies on dendrimer nanodevices in tumor model systems.[77] They then expanded these studies to examine the first detailed and quantitative biodistribution studies of nanocomposites.[4] Important initial principles of the interactions of nanocomposites were determined, and predictions were made about how to examine nanodevices in general in complex biologic systems.[4,77,79]

28.2.1.3 The Instrumental Neutron Activation Analysis (INAA) Method

Gold content of biologic samples by Au content of the nanodevices and tissue samples was determined by direct neutron irradiation. This approach was possible because (a) in the absence of oxygen, the gold/dendrimer nanocomposites are stable for a number of months (metal nanoclusters are effective oxygen-transfer catalysts), (b) PAMAMs do not break down in vivo because of enzymatic activity, and (c) the natural abundance of gold in mice and humans is close to zero.

In the INAA technique, samples are activated by irradiation with neutrons, usually within the high flux fields of a nuclear reactor. Specific isotopes become radioactive by neutron capture-type nuclear reactions. After removal from the reactor, samples are allowed to decay (cool) to permit unwanted short-lived activity (for example, from Na) to diminish. Gamma-ray spectra from activated samples are then measured using solid-state detectors.[24] Specific isotopes may be identified by their gamma rays of characteristic energy, and quantitative determinations are made by comparing peak areas in the spectrum to those of a standard reference material that has been irradiated and counted under identical conditions.

Advantages of this analysis method are that (a) determination of CND biodistribution is very precise, (b) the biologic experiments can be performed on their own timescale and schedule, (samples can be prepared, accumulated, stored, and analyzed upon reactor availability), (c) high sensitivity may be achieved for gold, and (d) there is no biohazard material present after irradiation. However, it is a disadvantage that the dried and irradiated tissues are not suited for further study (TEM, pathology, immunology, etc.) The build-up of activity for a given isotope is a function of several intrinsic factors, including isotopic abundance and neutron cross-section as well as a series of irradiation parameters, including neutron flux and energy spectrum. Because irradiation parameters can vary slightly over time within a range of operating conditions, quantification of the activated isotope is rarely approached through direct calculation. Rather, elemental concentrations are determined through comparison on a weight-ratio basis from activities (measured as gamma counts-per-second) generated in a standard reference material of known composition, assuming all other parameters are held constant, including irradiation time, flux, decay time, counting time, and detector geometry.

28.2.1.4 Determination of Gold Content Using INAA—Precision and Sensitivity

All tissue samples were lyophilized and analyzed either at the Ford Nuclear Reactor of the Phoenix Memorial Laboratory (University of Michigan) or at the Oregon State University Radiation Center (OSU-RC, Corvallis, OR). A series of six gel check standards prepared with known gold concentrations (in 7% gelatin to mimic the organic tissue samples) were prepared to monitor measurement precision and sensitivity. Four replicates of a gold standard (NBS-SRM-2128-1, 1% Au by weight) prepared by pipetting approximately 100 mg of the liquid standard on to filter paper placed in sample vials were similarly weighed and desiccated. Samples and standards were irradiated for 2 h in core-face location 112B with a nominal thermal neutron flux of $10E + 11$. Following an initial 2-day decay, a 15,000 s (live-time) count of gamma activity for Au-198 was recorded for each sample using a HPGe detector (33% relative efficiency) with a $4''$ counting geometry. Gold concentrations were determined through direct comparison on a weight ratio basis with the mean activity generated in four replicates of the Au standard. Sensitivities of <75 ppb were obtained, corresponding to 10 ng of Au (Figure 28.16). This method permits detailed and quantitative biodistribution measurements in mouse tumor model systems.

28.2.1.5 Biodistribution of Positive and Neutral Surface Generation Five Dendrimers—Non-Specific Uptake

In the first study of PAMAM dendrimers, the surface of PAMAM dendrimers were labeled with tritium[77] and studied the biodistribution of PAMAM_E5.$(NH_2)_{44}(NHCOCH_3^*)_{66}$ a tritiated generation five (5 nm) positive surface dendrimer and PAMAM_E5.$(NHCOCH_3^*)_{110}$ neutral surface PAMAM dendrimers (PSD and NSD, respectively) in a mouse melanoma tumor model and in a prostate cancer mouse model. The feasibility of dendrimer delivery and subsequent quantification was also tested. Systemic administration of NSDs and PSDs via tail vein injection were carried out in tumor bearing mice. At different time points, the mice were euthanized and their organs

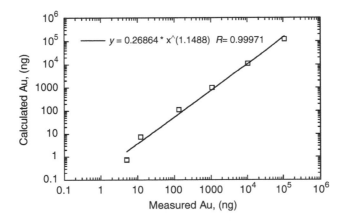

FIGURE 28.16 Comparison of Au measured by INAA and the theoretical (known) Au content measured in gel sample (gold concentrations were determined through direct comparison on a weight ratio basis with the mean activity generated in four replicates of the Au standard).

harvested, weighed, and processed. The in vivo distribution of dendrimers was then determined via liquid scintillation counting of recovered tritium from the organs.

Although neither particles localized selectively to tumor tissue, they did distribute to all organs tested and were recoverable within one hour post-injection. The vascular delivery of dendrimers could be homogenous in the organ (as seen in multiple samples of the liver analyzed). Both dendrimers rapidly cleared from blood, the % ID/g organ levels for PSD and NSD being 20.6 and 21.6 at 5 min, 14.3 and 9.9 at 1 h, 1.7 and 0.9 at 1 day, 0.3 and 0.1 at 4 days, and 0.2 and 0.1 at 7 days post-injection, respectively. These organic nanoparticles localized to all major organs and tumor tissue and were rapidly cleared from the blood. Following an initial rapid clearance during the first day post-injection, concentrations were maintained at a relatively stable level in every tissue with a very slow decline over time. Lungs, kidneys, liver, and heart exhibited the highest uptake within an hour. Levels in the spleen and tumor followed this, and the negligible amounts were found in the brain tissue. More detailed tables and graphs of both PSD and NSD biodistributions are available in this team's previous publication.[77] The PSD dendrimer nanoparticle biodistribution is shown in Figure 28.17. NSD biodistribution is very similar, but with lesser amounts throughout. In all organs and tissues tested (including tumor) the ratio of PSD:NSD was higher than 2:1, most likely because of increased retention of the positively charged dendrimers relative to neutral derivatives. It shows, in a quantitative manner, a global and differential effect of charge on organ and tissue biodistribution when the dendrimers are almost the same size ($d = 5$ nm).

In another experiment, the nanoparticle biodistribution was determined in another tumor model system (human DU145 prostate cancer xenograft), and comparison was made to the previously determined B16 melanoma tumor model system (see Figure 28.17). The results are very similar in both models, supporting the general validity of the findings to multiple tumor and mouse types.

In detailed excretion studies, it was shown that the dendrimers were excreted via both urine and feces (Figure 28.18). A total of 48.3% of NSD and 29.6% of PSD were excreted via urine over seven days. Within the first two hours, urinary NSD excretion was more than three-fold that of PSD, but at later sampling points, the daily excretion of both dendrimers was equivalent. This correlated with the increased retention of PSD seen in organs and tissues. Dendrimers were excreted via feces to a much lesser extent with a total of 5.41% of NSD and 2.96% of PSD recovered over seven days

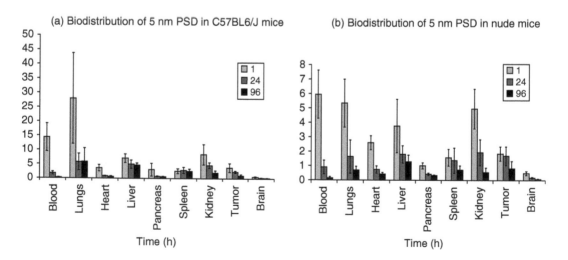

FIGURE 28.17 Organ biodistribution of positive surface 5 nm nanoparticle (PSD) depicted as percent injected dose recovered per gram of organ (% ID/g). Panel (a) shows biodistribution in C57BL6/J mice subcutaneously carrying B16 melanoma tumors on the dorsal surface. Panel (b) shows the same nanoparticle injected into nude (nu/nu) mice carrying the human prostate carcinoma line DU145. $n = 6$ for panel (a) and n = 3 for panel (b).

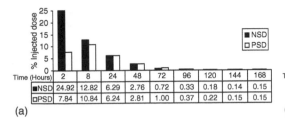
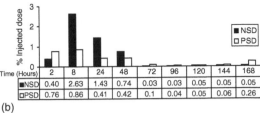

(a)   (b)

FIGURE 28.18 Excretion of positive surface dendrimers (PSD) and neutral surface dendrimers (NSD) via urine and feces (a and b, respectively) at time points indicated. The data obtained from B16F10 melanoma mouse model ($n = 5$) are presented as a percentage of injected dose excreted per mouse.

(Figure 28.18). The dendrimers were mainly excreted via urine with a significant amount of urinary excretion occurring within 24 h post-injection. This was the first quantitative demonstration of the manner in which dendrimers are excreted from the body (urine) and of how this occurs over time.

28.2.1.6 Biodistribution of 5 nm Gold Nanocomposites Compared with Dendrimer Templates

As the next step in this research, this team examined composite nanodevices of differing surface.[4] It compared the biodistribution of 5 nm composite nanodevices with those of their respective dendrimer templates of the same surface charges and size. It found that although charge affected biodistribution of the same size nanodevices, surprisingly, the gold/dendrimer composite nanodevices had a very different organ/tissue biodistribution pattern in mice compared to the dendrimer nanoparticles of similar size and surface charge.

28.2.1.7 Biologic Differences Between Dendrimers and Gold Composite Nanodevices

This comparison is most accurate and particularly striking in the case of the acetamide (neutral) surface nanodevices where the size and surface charge are the same. There is significant and consistent high-level accumulation over time (selectivity) in the liver and spleen of mice intravenously injected with the gold nanocomposites (Figure 28.19a). In contrast, no such organ specific accumulation (selectivity) was seen with the NSD (Figure 28.19b).

The overall (global) effects of surface charge are also different when the dendrimer nanoparticles are compared with the CNDs. With the dendrimers (Figure 28.19a and Figure 28.20a), and as discussed above, the partially positive surface macromolecules distributed into all organs at higher levels (about 2-fold higher levels) than that seen with the fully acetylated (neutral surface) dendrimer nanoparticles.[77] There were no significant organ-to-organ differences (or selectivity) seen. With the gold composite nanodevices, however, there are large organ-to-organ differences (selectivity) seen (Figure 28.19a and Figure 28.20b). There is significant spleen and liver accumulation (selectivity) seen with the $\{(Au^0)_{9.08} - PAMAM\_E5.(NHCOCH_3)_{120}\}^+_{d=5\ nm}$ neutral surface nanodevices over a prolonged time period. With the positive surface $\{(Au^0)_{9.08} - PAMAM\_E5.(NH_2)_{120}\}^+_{d=5\ nm}$ nanocomposite, there was significant accumulation noted in the kidney.

Finally, a further major difference noted was that the unmodified primary amine terminated PAMAM dendrimer was previously shown to be toxic in vivo (causing the death of mice) as discussed in the introduction, and this is exactly why partially acetylated PAMAMs were used for in vivo studies as PSD. With the gold/dendrimer composite nanodevices, however, unmodified hosts were used to safely prepare respective nanocomposites for these mouse tumor model studies. Although the exact reasons are yet unknown, dendrimers and dendrimer nanocomposites are

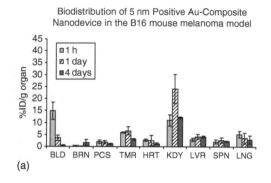

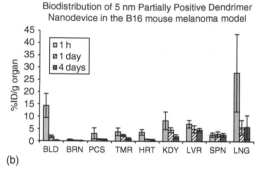

FIGURE 28.19 Organ/tissue biodistribution of neutral surface Au-composite nanodevice or (Au0)9-PAMAM_E5.(NHCOCH3)110} (a) and neutral surface dendrimer or PAMAM_E5.(NHCOCH3)110 (b) in tumor bearing mice. Abbreviations: BLD, blood; BRN, brain; PCS, pancreas, TMR, tumor; HRT, heart; KDY, kidney; LVR, liver; SPN, spleen; and LNG, lung. (The (b) has been reproduced from Nigavekar, S. S., Sung, L. Y., Llanes, M., El-Jawahri, A., Lawrence, T. S., Becker, C. W., Balogh, L., and Khan, M. K., *Pharm. Res.*, 21(3), 476–483, 2004. With permission.)

different enough to yield different biologic interactions (even though they are similar by charge and size in materials characterization), resulting in a different toxicity profile.

The surface-to-volume ratio of nanodevices is much greater than for micron objects, enabling the surface to govern the interactions of nanoparticles with proteins, carbohydrates, or other macromolecules within the organs, tissues, or cells. It was previously believed that the surface of the 5 nm positive or NSDs would largely govern the biodistribution of the gold/dendrimer composite nanodevices. Even though gold nanocomposites of primary amine terminated PAMAMs typically form mixed structures,[41] the fact that the $\{(Au^0)_{9.08} - \text{PAMAM\_E5.(NHCOCH}_3)_{120}\}^+_{d=5\ nm}$ and $\{(Au^0)_{9.08} - \text{PAMAM\_E5.(NH}_2)_{120}\}^+_{d=5\ nm}$ nanodevices have very different biodistribution within the mouse tumor model system in comparison to the dendrimer templates indicates that, from a biologic point of view, these devices interact differently with macromolecules present on or in organs/tissues. A classic three-dimensional conformational change (as can be seen with proteins) and perhaps affecting these surface interactions cannot account for the difference in biodistribution

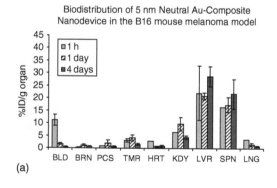

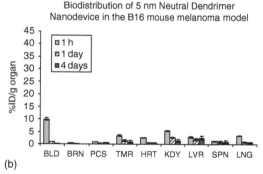

FIGURE 28.20 Organ/tissue biodistribution of positive surface Au-composite nanodevice or {(Au0)9-PAMAM_E5.(NH2)110 (a) and partially positive surface dendrimer or PAMAM_E5.(NH2)44 (NHCOCH3)66 (b) in tumor bearing mice. Abbreviations: BLD, blood; BRN, brain; PCS, pancreas; TMR, tumor; HRT, heart; KDY, kidney; LVR, liver; SPN, spleen; and LNG, lung. (The (b) has been reproduced from Nigavekar, S. S., Sung, L. Y., Llanes, M., El-Jawahri, A., Lawrence, T. S., Becker, C. W., Balogh, L., and Khan, M. K., *Pharm. Res.*, 21(3), 476–483, 2004. With permission.)

as all the studied particles are spherical. The increased rigidity and density of {Au(0)} composite nanodevices compared to the relative flexibility of the template dendrimers may explain some of this difference. Alternatively, perhaps another changed physiochemical property may partially be responsible for some differences in biodistribution. This aspect is presently under investigation.

The results again indicate that those working on nanoparticles for biologic and medical uses need to understand in detail how nanodevices truly behave in vivo and that new nanodevices will have to be rigorously tested for biodistribution, toxicity, and other critical measures needed for the use of these devices in biologic systems. As future multifunctional nanodevices are made, it will be critical to carry out detailed biologic testing of these devices with respect to changes in charge, size, complexation, and surface substitution to gain an in depth understanding of how they will behave in these systems. Studies on dendrimers and composites have shown that this rigorous testing is critical and will be needed for other nanomaterials as well.

28.2.1.8 The Importance of Size on Nanocomposite Biodistribution

This team conducted a series of further studies that are currently in submission for publication where the effect of size and charge on biodistribution of nanocomposites is further examined. It examined 5, 11, and 22 nm nanocomposites with varying surface charge (partially positive, negative, and neutral). A key finding of these studies was that there are different organs that are selectively uptake charged nanoparticles, depending on the size and charge tested (see Figure 28.21). Importantly, when surface charge was kept constant and the size of the nanodevice was varied as with the 5 and 11 nm carboxylate (negative) surface nanocomposites, different nanodevice biodistribution was seen with differential organ selective uptake noted.

Nanodevice levels in blood are higher for the $\{Au(0)\}^-_{d=5 \text{ nm}}$ gold nanocomposites than they are for the $\{Au(0)\}^-_{d=11 \text{ nm}}$ gold nanocomposites. Smaller size might preclude quick clearance from blood circulation, perhaps by uptake of the nanodevices into circulating blood cells. Nanodevice accumulation in the spleen appears to be determined by size as the $\{Au(0)\}^-_{d=11 \text{ nm}}$ negative surface gold nanocomposites show much higher levels of accumulation in the spleen than the 5 nm negative

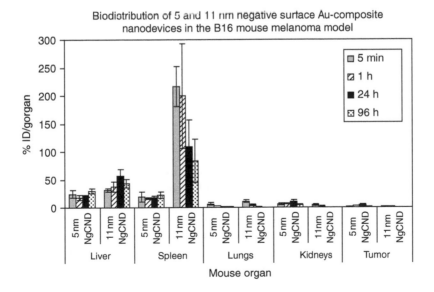

FIGURE 28.21 Biodistribution of 5 and 11 nm negative surface Au-composite nanodevices in C57BL6/J mice bearing B16F10 melanoma. Shown here are nanodevice accumulation in tumor as well as in organs that show the highest accumulation of nanodevice that include the liver, spleen, lungs, and kidneys. Nanodevice levels were tested at times 5 min and hours 1, 24, and 96 post-injection.

surface Au-composite nanodevice. Liver accumulation was also consistently higher in the 11 nm nanocomposites. These studies indicate that size can affect distribution of the nanodevices with similar surface charges. This will have to be taken into consideration if this nanodevice were to be used to deliver radiation or toxic therapeutics, especially in the absence of a tumor targeting moiety on the surface of the nanodevice. This is the first set of data demonstrating the critical importance of the size of the nanodevices on their biodistribution in complex organisms.

28.2.1.9 Biodistribution Summary

Detailed quantitative biodistribution studies demonstrated the critical importance of surface size and charge on the interactions of nanocomposites with complex biologic systems. This team was able to empirically determine how differing size and charge affects this biodistribution, showing that selective organ uptake can be achieved with modulation of these parameters. It also showed that biodistribution profiles of Au-composite nanodevices were different from profiles of dendrimers of the same size and charge. These studies also revealed that an important principle of nanodevice design is to be fully aware of these parameters, including the importance of flexibility and density. Importantly, this team also demonstrated that it is difficult to predict what size/charge modification would produce a given biodistribution. The behaviors appear to be complex at this time. Completion of its ongoing research will provide a deeper insight into the biodistribution of nanoparticles of differing size and charge and create an important background to examine targeted devices. Without this detailed type of baseline study, there would be no way to ensure true targeting is occurring as envisioned targeting may be more a result of selective uptake because of size, charge, and other properties, and they are not due to the targeting moiety on the nanodevice. Understanding nanodevice behavior in complex animal systems will help to design the best nanodevices for targeting and suggest ways to protect organs from undesired effects. The detailed studies above will also allow researchers to begin a more detailed mathematical and pharmacokinetic analysis to explain nanodevice size/charge effects on biodistribution and permit them to make educated predictions as to their behavior in biologic systems in the future.

28.2.2 TOXICITY STUDIES

28.2.2.1 Long-Term Toxicity Studies of Tritiated Dendrimer Nanoparticles

Initial long-term toxicity studies were carried out on both the dendrimer nanoparticles and on several of the composite nanodevices as indicated in Table 28.7. As in vitro toxicity data cannot yet be directly transposed to in vivo conditions, this team has carried out initial toxicity studies on the tritiated dendrimers. Because tumor-bearing mice would die as a result of their tumor-burden, long-term toxicity/biodistribution studies were carried out with normal C57BL/6J mice. PAMAM _E5.$(NH_2)_{44}(NHCOCH_3^*)_{66}$ generation 5 PAMAM dendrimer derivative with a partially positive surface (PSD) and a neutral surface PAMAM_E5.$(NHCOCH_3^*)_{110}$ dendrimer (NSD) were evaluated for in vivo toxicity in 6–8 week old male C57BL6/J mice. The mice were observed for a period of 75 days. There were five mice in each group with a control group that was given PBS. Clinical toxicity tables were kept on each set of mice, and daily notations were made of any signs of weakness, lethargy, dehydration, change in mobility, or reflex. Weights of mice-given dendrimers did not differ from weights of mice in the control group (see Figure 28.22), and none of the mice died during observation. The labeled dendrimers were detectable in all organs tested (see Table II in Nigavekar et al. 2004).[77] This demonstrated that the nanoparticles could distribute to all tested organs and remain there at relatively stable levels for at least 12 weeks. The mice appeared healthy at that time as well, and no long-term clinical toxicity or weight loss was observed. The mice weights increased over the first 15 days and then were stably maintained with slight increment over the period observed. This team's conclusion was that these dendrimers are non-toxic in this animal model.

TABLE 28.7

Male C57BL6/J, 6–8-Week-Old Mice Were Injected with Tritiated Dendrimer Nanoparticles (5 nm PSD and NSD) and Au-Composite Nanodevices (5 nm PCND, 11 nm NgCND, and 22 nm PCND) and The Were Monitored Over a Period of Time for Change in Weights

	Toxicity analysis				
Analysis of Mice Weights	5 (nm) PSD	5 nm NSD	5 nm PCND	11 nm NgCND	22 nm PCND
Time course of analysis (in days)	75	75	75	49	49
Analysis of clinical Toxicity	√	√	√	√	√

The mice were also monitored for signs of clinical toxicity such as weakness, lethargy, dehydration, change in mobility, or reflex. Death, as an end-point was noted as well.

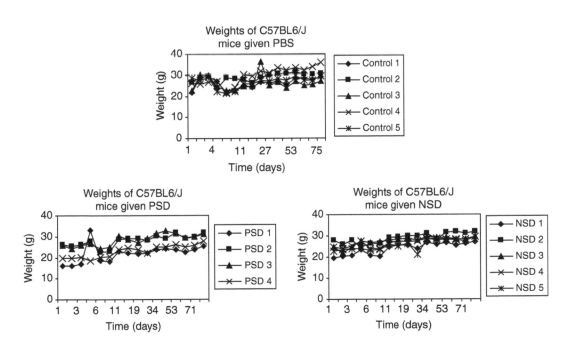

FIGURE 28.22 Long-term monitoring of weights of non-tumor bearing C57BL/6J mice given (a) PBS as control ($n = 5$), (b) PSD ($n = 4$) and (c) NSD ($n = 5$). Individual weights of each mouse have been plotted as a function of time (days). PBS, phosphate buffered saline; PSD, positive surface dendrimer; NSD, neutral surface dendrimer. The notation *PSD 1*, for example, means the positive surface dendrimer injected into mouse 1.

28.2.2.2 Long-Term Toxicity of {Au(0)} Gold Composite Nanodevices

Very limited data are available on the toxicity of noble metal dendrimer nanocomposites. Some preliminary data in cell culture systems provide some understanding of metal composite nanodevices in vitro.[5,76] It appears that toxicity is very similar to that of the dendrimer template, i.e., if the template is non-toxic, the noble metal composite nanoparticles are also non-toxic.[76]

To gain the insight into the in vivo toxicity of composite nanodevices, the team completed detailed in vivo toxicity analysis of $\{Au(0)\}^+_{d=5\ nm}$ the 5 nm positively charged CND, $\{Au(0)\}^-_{d=11\ nm}$ the 11 nm negatively charged CND, and $\{(Au^0)_{5.8}\}^+_{d=22\ nm}$ the 22 nm positively charged CND in healthy mice. The mice were injected with the nanodevices as described above for the tritiated dendrimers. In brief, the 5 nm PCND, the 11 nm NgCND, and the 22 nm PCND were all evaluated for toxicity in 6–8 week old male C57BL6/J mice. Mice given the 5 nm nanodevice were observed for 75 days, whereas those given the 11 nm and 22 nm devices were observed for 49 days. There were five mice in each group with a control group that was given PBS. Daily toxicity tables were kept on each group of mice, and any signs of weakness, lethargy, dehydration, change in mobility, or reflexes noted. There was no clinical toxicity found. Weights of mice that were given dendrimers did not differ from weights of mice in the control group (see Figure 28.23), and none of the mice died during observation.

28.2.2.3 Treatment of Tumors in Mouse Model Systems

Reactive encapsulation permits the entrapment of different radionuclides within composite nanoparticles of defined size and targeting properties. An advantage of this approach would be the development of an image guided therapy where the radioactivity delivered to a tumor can be increased either by increasing the particle size or by increasing the number or specific activity of the guest atoms without destroying the targeting ability of the nanocomposite device. Additionally,

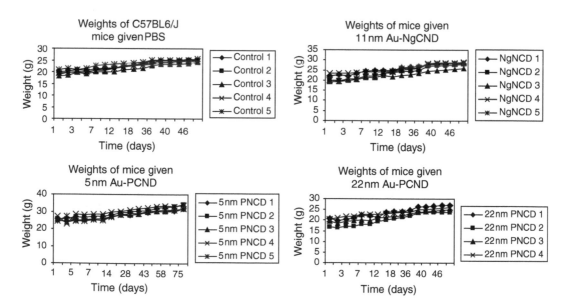

FIGURE 28.23 Long-term monitoring of weights of non-tumor bearing C57BL/6J mice given PBS ($n = 5$) as control ($n = 5$), 5 nm Au-PCND ($n = 5$), 22 nm Au-PCND ($n = 4$), and 11 nm Au-NgCND ($n = 5$). Individual weights of each mouse have been plotted as a function of time (days). Au-PCND = gold positive surface composite nanodevice; Au-NgCND = gold negative surface composite nanodevice; PBS = phosphate buffered saline. The notation *PCND 1*, for example, means that particular nanodevice injected into mouse 1.

varying the particular nature of the atoms (metals, isotopes) encapsulated in the CNDs will permit the application of different imaging techniques (e.g., film, SPECT, PET). Because these CNDs can encapsulate a great number $(2-1024)^{6,39}$ of radionuclides per macromolecule, they can deliver at least a log fold more radioactivity to tumors than is possible with current radioimmunotherapeutic technologies. This offers the added potential of imaging microscopic tumor burdens as seen in micrometastases or minimal residual disease states. In the future, this amplified signal and molecular targeting capability could greatly aid in advancing the field of molecular imaging.[106] Significantly, these composite nanodevices could be used for a combined imaging and therapy of the tumors.[6,39,106]

Many attempts have been made to target radionuclides to tumors[107–114] with the best known approaches being the use of radiolabeled monoclonal antibody therapy or the use of surface chelators bound to dendrimers. A detailed review and critique of these approaches is beyond the scope of this chapter, and several useful reviews exist.[115–118] In the past, radiolabeled antibodies for direct tumor cell targeting have received wide interest for the purpose of receptor imaging. In general, however, the penetration of these macromolecules into tumor tissue has been problematic with typically only 0.001%–0.01% of the injected dose being localized to each gram of solid human tumor tissue in man. Several factors account for this poor penetration: antibodies in the circulation must travel across the endothelial cell layer and often through dense fibrous stroma before encountering tumor cells; the dense packing of tumor cells and tight junctions between epithelial tumor cells hinders transport of the antibody within the tumor mass; the absence of lymphatics within the tumor contributes to the buildup of a high interstitial pressure that opposes the influx of molecules into the tumor core; antibodies entering the tumor become absorbed to the perivascular regions by the first tumor cells encountered, leaving none to reach tumor cells at sites farther from the blood vessels; and subpopulations of the rapidly mutating tumors can readily lose antigens targeted by the antibodies. Therefore, researchers have turned toward the development of small radiolabeled peptides that do not suffer from most of the aforementioned drawbacks.[107–118]

Tumor-directed antibody (or peptide) therapy has also not been able to deliver sufficient targeted dose because only one radioactive molecule can often be linked to a given antibody (or peptide), using a caged carrier to hold the radioactivity (e.g., DOTA).[119] The specific activity of the single radioactive moiety is also limited as it can damage the antibody if it is too high. This limits the detection of microscopic tumors, and higher doses leading to better detection have not been possible. This has also greatly limited any combination of imaging and therapy (where much higher dose delivery is needed). This derives at least partly from the fundamental problem that, to increase activity, one must increase substitution that tends to modify or obliterate the specificity of the antibody or targeting dendrimer. Finally, the ability to carry out different forms of imaging (PET or SPECT for example) is also limited by the chemistry necessary to place different isotopes onto the same targeting antibody. These fundamental problems can be overcome by the use of CNDs to deliver radioisotopes.[119]

Some of the fundamental problems that can be overcome by the use of targeted CNDs are those of solubility and dose delivery. First, many radioisotopes are often practically insoluble in water or body fluids, but they may become readily available in CNDs, i.e., in different sizes, contents, and surfaces. Second, use of CNDs permits at least a log-fold higher delivery of radioactivity than that available with current radioactive antibody therapies. As opposed to labeled antibodies that can carry only one radioisotope, 1–1000 nuclei can be encapsulated in one CND[6], allowing the delivery of therapeutic dose amounts to tumor cells.

To justify the targeted dose delivery by the any approach, it has to be first demonstrated that the treatment will be effective if the dose is delivered directly to the tumor (nanobrachytherapy). A proof-of-concept research of the therapeutic efficacy on {^{198}Au} radioactive composite nanodevices in tumors (conceptually, other nuclides with the desired specific activity could also be used in the future, allowing the use of gamma, beta, alpha, or combinations of these radiations) has been conducted. To take advantage of the flexibility of composite nanodevice systems, Systemically

Targeted RadioTherapy (StaRT) is being developed to target radioactive nanodevices either to tumors, to the tumor angiogenic microvasculature, or to both for the treatment of cancer. Before carrying out these targeted experiments, however, an idea of the level of doses that could be delivered by CNDs and if these doses were enough to slow or halt the growth of tumors was needed. In order to test this idea of nanobrachytherapy, specific doses of radioactive CNDs were directly injected into growing tumors, and the dose needed to significantly slow tumor growth was examined. Once this dose is known, calculating if targeted devices intravascularly injected could deliver these types of doses can be conducted (i.e., if it is feasible to proceed with targeted nanodevice delivery).

B16 melanoma carrying mice were randomly sorted into three groups of seven mice with approximately similar tumor volumes (~ 500 mm^3). The control group was intra-tumorally injected with PBS buffer, whereas the remaining two groups were intra-tumorally injected with 35 and 74 µCi of radioactive ^{198}Au carrying composite nanodevice, respectively. A second control experiment was carried out with the same nanodevice after the radioactivity of its guest component decayed over time. The now cold {Au(0)} nanocomposite (corresponding to the original 74 µCi sample) was injected intra-tumorally into one set of tumor bearing mice, and the second was again injected with PBS buffer as control. The tumor volumes of all mice were recorded every other day for eight days.

When compared to tumors injected with either cold nanocomposite or PBS (untreated), there was a statistically significant slowing of tumor growth observed in tumor injected with 74 µCi of radioactive gold nanocomposite ($p = 0.03$ in both cases). Although injection with the 35 µCi CND did not significantly slow tumor growth, there was a noticeable trend toward slow tumor growth (see Figure 28.24). It should be possible with the use of fractionated experiments (multiple injections over time) to be able to safely deliver these types of doses to tumors using targeted devices. With radioactive antibodies, intraperitoneal/intravenous injections of 300–800 µCi of various isotope/antibody combinations have been successfully utilized to slow tumor growth in mouse xenograft models.[120–123] It was also assumed that about 5%–10% of the 800 µCi injected dose (approximately 40–80 µCi) to the tumor could be targeted as supported by data on folate-surfaced targeted dendrimers[79] and the literature above on other targeted devices. With these calculations, the 35–74 µCi needed for tumor delivery from intravascular injections are feasible, especially if fractionated treatment is contemplated.

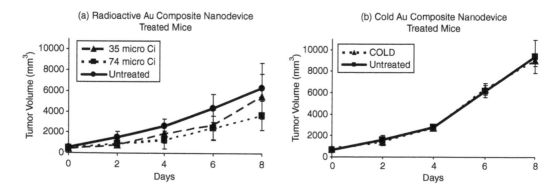

FIGURE 28.24 Treatment of melanoma tumors using radioactive Au-composite nanodevice. B16F10 melanoma tumors were grown in C57BL6/J mice to sizes of ~ 500 mm^3. (a) The mice were intra-tumorally injected with PBS buffer (untreated control) and 35 and 74 µCi of radioactive ^{198}Au carrying composite nanodevice (CND), respectively ($n = 7$ for all three groups). (b) Mice were intra-tumorally injected with PBS buffer (untreated control) and cold nanodevice corresponding to the original 74 µCi radioactive Au-CND ($n = 3$ for both groups). The tumor volumes of all mice were recorded every other day for eight days.

These experiments demonstrate for the first time that radiation therapy (nanobrachytherapy or systemic STaRT) with targeted composite nanodevices is feasible, and experiments are now being conducted to move this entire field forward with more detailed determination of the appropriate fractionation schemes for maximal tumor control using targeted composite nanodevices. Direct intratumoral injection gave a rough idea of the doses needed to do this, and they seemed reasonable given prior radioactive antibody data currently available in the literature. Importantly, for some tumors, direct injection of radioactive nanodevices may itself be a treatment.

ACKNOWLEDGMENTS

This research has been funded with federal funds in part from the U.S. Department of Energy under Award No. FG01-00NE22943 in part from the National Cancer Institute and National Institutes of Health, under Contract No. NO1-CO-97111 and RO1 CA104479 and in part by the Department of Defense Grant# DAMD17-03-1-0018. Silver nanocomposite research was sponsored by the Army Research Laboratory under the contract of DAAL-01-1996-02-0044. Thanks to D. Sorenson and C. Edwards of the Microscope and Imaging Laboratory at the University of Michigan for their technical support, to Leah Minc OSU-RC for INAA analysis, to Shraddha Nigavekar, Bindu Nair, Muhammed Taju Kariapper, Lok Yun Sung, Mikel Llanes, Kerstin May, Areej El-Jawari, Alla Kwitney, and Amber Warnat for their significant contributions on various components of the biologic components of these projects, and to Wojciech Lesniak, Kai Sun, X. Shi, and the University of Michigan Electron Microbeam Analysis Laboratory for their contributions to the synthesis and characterization components of these studies.

DEFINITIONS

Nanoscience The science of the phenomena related to nanoscale materials and/or devices. It refers to the multi-disciplinary study of materials, processes, and mechanisms that have distinctive properties (chemical, physical, and biological) from the molecular (i.e., <1 nm) to the sub-micrometer scales (i.e., >100 nm) because of quantum-size effects.

Nanotechnology A wide range of possible technologies that encompass the incorporation of engineered or manufactured devices or materials with critical length scales between approximately 1 nm and 100 nm, and applications that exploit specific properties and functions of nanosized materials and/or devices.

Particle Any object that has a persistent surface and an interior that may be different from the surface.

Nanodevice A surface modified multifunctional nanoparticle that is able to perform different functions.

Cluster Aggregations of atoms that are too large to be referred to as molecules and too small to resemble small pieces of crystals. Clusters generally do not have the same structure or atomic arrangement as a bulk solid and can change structure with the addition of just one or a few atoms. As the number of atoms (and their degree of freedom) increases, eventually a crystal-like structure may be established. Cluster science is devoted to understanding the changes in fundamental properties of materials, as a function of size, evolving from isolated atoms or small molecules to a bulk phase.[1,2]

Aggregates Dendrimers or nanoparticles held together by weak physical forces.

Nomenclature To describe the complex structure of these nanoscopic complexes and nanocomposites, the following convention is used to describe the composition of the synthesized nanomaterials: brackets denote complexes, and braces represent nanocomposite structures. Within the brackets or braces, the complexed or encapsulated components are listed (with an index of their average number per dendrimer molecule) followed by the family of dendrimers and the terms used for identification, i.e., core, generation, and surface. Naming of the materials will follow this pattern. For instance, PPI_DAB4.(NH$_2$)$_{16}$ describes a diaminobutane core poly(propyleneimine) dendrimer molecule with 16 primary amine terminal groups, and PAMAM_E5.(NH$_2$)$_{44}$(NHOAc)$_{66}$ denotes a generation five ethylenediamine core poly(amidoamine) dendrimer material with an average of 44 primary amine and 66 acetamide terminal groups (Please observe that the sum of these numbers are less than the theoretical 128, indicating that these are measured average numbers). All known properties can be listed by using this principle. Nanocomposites are also named according to the same pattern {(component#1)i(component#2)j...-FAMILY_Core.Generation.Terminal group}Therefore, the formula {(Au^0)$_{14.5}$-PAMAM_E4.NH$_2$} denotes a gold dendrimer nanocomposite where a generation 4 amine terminated ethylenediamine (EDA or E for short) core poly(amidoamine) (PAMAM) dendrimer contains 14.5 zerovalent gold atoms per dendrimer molecule on average. The above scheme provides a relatively simple, but consistent, way to identify materials with a complicated structure. Additional descriptors may be added to the superscripts/superscripts and parentheses can be embedded into each other, e.g.,

$$\{(Au^0)_{10}\text{-E5.NH}_2\}_{250\pm10}$$

The composite listed above denotes an aggregate particle composed of 250 ± 10 primary composite nanoparticles. When it is trivial, that dendrimer family is used; one may follow the shorter pattern and omit naming the family. Therefore, {(Au^0)$_{10}$-PAMAM_E5.NH$_2$} and {(Au^0)$_{10}$-E5.NH$_2$} may be considered to be equivalent names if only PAMAMs are used.

REFERENCES

1. Alivisatos, A. P. and Schultz, P. G., Organization of 'nanocrystal molecules' using DNA, *Nature*, 382 (6592), 609–611, 1996.
2. Schmid, G., Large clusters and colloids. Metals in the embryonic state, *Chem. Rev.*, 92 (8), 1709–1727, 1992.
3. Jemal, A., Tiwari, R. C., Murray, T., Ghafoor, A., Samuels, A., Ward, E., Feuer, E. J., and Thun, M. J., Cancer statistics, *CA Cancer J. Clin.*, 54 (1), 8–29, 2004.
4. Khan, M. K., Nigavekar, S. S., Minc, L. D., Kariapper, M. S., Nair, B. M., Lesniak, W. G., and Balogh, L. P., In vivo biodistribution of dendrimers and dendrimer nanocomposites—implications for cancer imaging and therapy, *Technol. Cancer Res. Treat.*, 4 (6), 603–613, 2005.
5. Bielinska, A., Eichman, J. D., Lee, I., Baker, J. R., and Balogh, L., Imaging {Au0-PAMAM} gold–dendrimer nanocomposites in cells, *J. Nanopart. Res.*, 4, 395–403, 2002.
6. Grohn, F., Bauer, B. J., Akpalu, Y. A., Jackson, C. L., and Amis, E. J., Dendrimer templates for the formation of gold nanoclusters, *Macromolecules*, 33 (16), 6042–6050, 2000.
7. McDevitt, M. R., Ma, D., Lai, L. T., Simon, J., Borchardt, P., Frank, R. K., Wu, K., Pellegrini, V., Curcio, M. J., and Miederer, M., et al., Tumor therapy with targeted atomic nanogenerators, *Science*, 294 (5546), 1537–1540, 2001.
8. Newkome, G. R., Moorefield, C. N., and Vögtle, F., Eds., *Dendritic Macromolecules: Concepts, Syntheses, Perspectives*, Wiley-VCH, New York, 1996.
9. Frechet, J. M. J. and Tomalia, D. A., Eds., *Dendrimers and Other Dendritic Polymers*, Wiley, New York, p. 647, 2001.
10. Mansfield, M. L. and Klushin, L. I., Monte Carlo studies of dendrimer macromolecules, *Macromolecules*, 26 (16), 4262–4268, 1993.
11. Karatasos, K., Adolf, D. B., and Davies, G. R., Statics and dynamics of model dendrimers as studied by molecular dynamics simulation, *J. Chem. Phys.*, 115, 5310–5318, 2001.

12. Murat, M. and Grest, G. S., Molecular dynamics study of dendrimer molecules in solvents of varying quality, *Macromolecules*, 29 (4), 1278–1285, 1996.

13. Welch, P. and Muthukumar, M., Tuning the density profile of dendritic polyelectrolytes, *Macromolecules*, 31 (17), 5892–5897, 1998.

14. Tomalia, D. A., Baker, H., Dewald, J., Hall, M., Kallos, G., Martin, S., Roeck, J., Ryder, J., and Smith, P., *Polym. J. (Tokyo)*, 17 (1), 117–132, 1985.

15. Grayson, S. M. and Fréchet, J. M. J., Convergent dendrons and dendrimers: From synthesis to applications, *Chem. Rev.*, 101 (12), 3819–3868, 2001.

16. Duncan, R. and Izzo, L., Dendrimer biocompatibility and toxicity, *Adv. Drug Deliv. Rev.*, 57 (15), 2215–2237, 2005.

17. Roberts, J. C., Bhalgat, M. K., and Zera, R. T., Preliminary biological evaluation of polyamidoamine (PAMAM) starburst dendrimers, *J. Biomed. Mater. Res.*, 30 (1), 53–65, 1996.

18. Tomalia, D. A., Dewald, J. R., Hall, M. J., Martin, S.J., and Smith, P. B., *First SPSJ International Polymer Conference*, Kyoto, Japan, p. 65, 1984.

19. Tomalia, D. A., Naylor, A. M., and Goddard, W. A., Starburst dendrimers: Molecular-level control of size, shape, surface chemistry, topology, and flexibility from atoms to macroscopic matter, *Angew. Chem. Int. Ed. Engl.*, 29 (2), 138–175, 1990.

20. D'Emanuele, A. and Attwood, D., Dendrimer–drug interactions, *Adv. Drug Deliv. Rev.*, 57 (15), 2147–2162, 2005.

21. Svenson, S. and Tomalia, D. A., Dendrimers in biomedical applications—reflections on the field, *Adv. Drug. Deliv. Rev.*, 57 (15), 2106–2129, 2005.

22. Tomalia, D. A. and Dvornic, P. R., In *Polymeric Materials Encyclopedia*, Salamone, J. C., Ed., Vol. 3, CRC Press, Boca Raton, FL, pp. 1814–1830, 1996.

23. Bielinska, A., Kukowska-Latallo, J. F., Johnson, J., Tomalia, D. A., and Baker, J. R. Jr., Regulation of in vitro gene expression using antisense oligonucleotides or antisense expression plasmids transfected using starburst PAMAM dendrimers, *Nucleic Acids Res.*, 24 (11), 2176–2182, 1996.

24. El-Sayed, M., Kiani, M. F., Naimark, M. D., Hikal, A. H., and Ghandehari, H., Extravasation of poly(amidoamine) (PAMAM) dendrimers across microvascular network endothelium, *Pharm. Res.*, 18 (1), 23–28, 2001.

25. Malik, N., Evagorou, E. G., and Duncan, R., Dendrimer-platinate: A novel approach to cancer chemotherapy, *Anticancer Drugs*, 10 (8), 767–776, 1999.

26. Raduchel, B., Schmitt-Willich, S., Ebert, J., Frenzel, T., Misselwitz, B., and Weinmann, H. J., *Synthesis and Characterization of Novel Dendrimer-Based Gadolinium Complexes as MRI Contrast Agents for the Vascular System*, American Chemical Society, Washington, DC, p. 278, 1998.

27. Wilbur, D. S., Pathare, P. M., Hamlin, D. K., Buhler, K. R., and Vessella, R. L., Biotin reagents for antibody pretargeting. 3. Synthesis, radioiodination, and evaluation of biotinylated starburst dendrimers, *Bioconjug. Chem.*, 9 (6), 813–825, 1998.

28. Shi, X., Banyai, I., Islam, M. T., Lesniak, W., Davis, D. Z., Baker, J. R. J., and Balogh, L. P., Generational, skeletal and substitutional diversities in generation one poly(amidoamine) dendrimers, *Polymer*, 46, 3022–3034, 2005.

29. Uppuluri, S., Swanson, D. R., Brothers, H. M., Piehler, L. T., Li, J., Meier, D. J., Hagnauer, G. L., and Tomalia, D. A., Tecto(dendrimer) core-shell molecules: Macromolecular tectonics for the systemic synthesis of larger controlled structure molecules, *Polym. Mater. Sci. Eng.*, 80, 55–56, 1999.

30. Uppuluri, S., Swanson, D. R., Brothers, H. M. I., Piehler, L. T., Li, J., Meier, D. J., H., G. L., Tomalia, D.A, *Core-Shell Tecto(Dendrimers): A New Class of Regio-Specifically Cross-Linked Polymers, Polymers for Advanced Technologies (PAT)*, Tokyo, Japan, August 31–September 5, pp. 94–95, 1999.

31. Quintana, A., Raczka, E., Piehler, L., Lee, I., Myc, A., Majoros, I., Patri, A. K., Thomas, T., Mule, J., and Baker, J. R. Jr., Design and function of a dendrimer-based therapeutic nanodevice targeted to tumor cells through the folate receptor, *Pharm. Res.*, 19 (9), 1310–1316, 2002.

32. Balogh, L., Swanson, D.R., Spindler, R., Tomalia., D.A., Formation and characterization of dendrimer-based water soluble inorganic nanocomposites, In *Proceedings of the American Chemical Society, Division of Polymeric Materials Science and Engineering*, Vol. 77, Los Vegas, VV, pp. 118–119, 1997.

33. Tan, N. B., Balogh, L., and Trevino, S., Structure of metallo-organic nanocomposites produced from dendrimer complexes, *Proc. ACS Polym. Mater.: Sci. Eng.*, 77, 120, 1997.

34. Balogh, L. and Tomalia, D. A., Dendrimer-templated nanocomposites. I. Synthesis of zero-valent copper nanoclusters, *J. Am. Chem. Soc.*, 120, 7355–7356, 1998.

35. Zhao, M., Sun, L., and Crooks, R. M., Preparation of Cu nanoclusters within dendrimer templates, *J. Am. Chem. Soc.*, 120 (19), 4877–4878, 1998.

36. Esumi, K., Suzuki, A., Aihara, N., Usui, K., and Torigoe, K., Preparation of gold colloids with UV irradiation using dendrimers as stabilizer, *Langmuir*, 14 (12), 3157–3159, 1998.

37. Larré, C., Donnadieu, B., Caminade, A. M., and Majoral, J. P., Regioselective gold complexation within the cascade structure of phosphorus-containing dendrimers, *Chem. Eur. J.*, 4 (10), 2031–2036, 1998.

38. Bosman, A. W., Janssen, H. M., and Meijer, E. W., About dendrimers: Structure, physical properties, and applications, *Chem. Rev.*, 99, 1665–1688, 1999.

39. Diallo, M., Balogh, L., Shafagati, A., Johnson, J. H. J., Goddard, W. A. I., and Tomalia, D. A., 2-Poly(amidoamine) dendrimers: A new class of high capacity chelating agents for Cu (II) ions, *Environ. Sci. Technol.*, 33 (5), 820–824, 1999.

40. Diallo, M. S., Christie, S., Swaminathan, P., Balogh, L., Shi, X., Um, W., Papelis, C., Goddard, W. A. III, and Johnson, J. H. Jr., Dendritic chelating agents. 1. Cu(II) binding to ethylene diamine core poly(amidoamine) dendrimers in aqueous solutions, *Langmuir*, 20 (7), 2640–2651, 2004.

41. Balogh, L., Valuzzi, R., Laverdure, K. S., Gido, S. P., Hagnauer, G. L., and Tomalia, D. A., Formation of silver and gold dendrimer nanocomposites, *J. Nanopart. Res.*, 1 (3), 353–368, 1999.

42. Balogh, L., Laverdure, K. S. Gido, S. P., Mott A. G., Miller, M. J., Ketchel, B. P., Tomalia, D. A. Dendrimer–metal nanocomposites, In *Organic/Inorganic Hybrid Materials*, pp. 69–75, 1999.

43. Hong, S., Bielinska, A. U., Mecke, A., Keszler, B., Beals, J. L., Shi, X., Balogh, L., Orr, B. G., Baker, J. R. Jr., and Banaszak Holl, M. M., Interaction of poly(amidoamine) dendrimers with supported lipid bilayers and cells: Hole formation and the relation to transport, *Bioconjug. Chem.*, 15 (4), 774–782, 2004.

44. Martin, C. R., *Science*, 266, 1961–1966, 1994.

45. Shaw, C. F. III, Gold-based therapeutic agents, *Chem. Rev.*, 99, 2589–2600, 1999.

46. El-Sayed, I. H., Huang, X., and El-Sayed, M. A, Selective laser photo-thermal therapy of epithelial carcinoma using anti-EGFR antibody conjugated gold nanoparticles, *Cancer Lett.*, 2005.

47. Hirsch, L. R., Stafford, R. J., Bankson, J. A., Sershen, S. R., Rivera, B., Price, R. E., Hazle, N. J., and West, J. L., Nanoshell-mediated near-infrared thermal therapy of tumors under magnetic resonance guidance, *Proc. Natl Acad. Sci. U.S.A.*, 100 (23), 13549–13554, 2003.

48. Loo, C., Lowery, A., Halas, N., West, J., and Drezek, R., Immunotargeted nanoshells for integrated cancer imaging and therapy, *Nano Lett.*, 5 (4), 709–711, 2005.

49. Loo, C., Lin, A., Hirsch, L., Lee, M., Barton, J., Halas, N., West, J., and Drezek, R., Nanoshell-enabled photonics-based imaging and therapy of cancer, *Technol. Cancer Res. Treat.*, 3, 33–40, 2004.

50. Holm, H. H. III, The history of interstitial brachytherapy of prostatic cancer, *Semin. Surg. Oncol.*, 13, 431–437, 1997.

51. Root, S. W., Andrews, G. A., Knieseley, R. M., and Tyor, M. P., The distribution and radiation effects of intravenously administered colloidal Au 198 in man, *Cancer*, 7, 856–866, 1954.

52. Rubin, P. and Levitt, S. H., The response of disseminated reticulum cell sarcoma to the intravenous injection of colloidal radioactive gold, *J. Nucl. Med.*, 5, 581–594, 1964.

53. Flocks, R. H., Kerr, H. D., Elkins, H. B., and Culp, A., Treatment of carcinoma of the prostate by interstitial radiation with radio-active gold: A preliminary report, *J. Urol.*, 68, 510–522, 1952.

54. Loening, S. A., Gold seed implantation in prostate brachytherapy, *Semin. Surg. Oncol.*, 13 (6), 419–424, 1997.

55. Hainfeld, J. F., Foley, C. J., Srivastava, S. C., Mausner, L. F., Feng, N. I., Meinken, G. E., and Steplewski, Z., Radioactive gold cluster immunoconjugates: Potential agents for cancer therapy, *Int. J. Radiat. Appl. Instrum. Part B, Nucl. Med. Biol. (Oxford)*, 17 (3), 287–294, 1990.

56. Paciotti, G. F., Myer, L., Weinreich, D., Goia, D., Pavel, N., McLaughlin, R. E., and Tamarkin, L., Colloidal gold: A novel nanoparticle vector for tumor directed drug delivery, *Drug Deliv.*, 11 (3), 169–183, 2004.

57. Garcia, M. E., Baker, L. A., and Crooks, R. M., Preparation and characterization of dendrimer–gold colloid nanocomposites, *Anal. Chem.*, 71 (1), 256–258, 1999.

58. Tomalia, D. A., Balogh, L. P., US Patent No. 6,664,315B2, 2003.

59. Balogh, L., Tomalia, D. A., and Hagnauer, G. L., Revolution of nanoscale proportions, *Chem. Innovation*, 30 (3), 19–26, 2000.

60. Ottaviani, M. F., Valluzzi, R., and Balogh, L., Internal structure of silver-poly(amidoamine) dendrimer complexes and nanocomposites, *Macromolecules*, 35 (13), 5105–5115, 2002.

61. Balogh, L., Bielinska, A., Eichman, J. D., Valluzzi, R., Lee, I., Baker, J. R., Lawrence, T. S., and Khan, M. K., Dendrimer nanocomposites in medicine, *Chim. Oggi (Chem. Today)*, 20 (5), 35–40, 2002.

62. Cotton F. A., Willinson G., *Advanced Inorganic Chemistry*, 4 ed., Wiley, New York, pp. 1052–1055, 1980.

63. Lee, W. I., Bae, Y., and Bard, A. J., Strong blue photoluminescence and ECL from OH-terminated PAMAM dendrimers in the absence of gold nanoparticles, *J. Am. Chem. Soc.*, 126, 8358–8359, 2004.

64. Esumi, K., Suzuki, A., Yamahira, A., and Torigoe, K., Role of poly(amidoamine) dendrimers for preparing nanoparticles of gold, platinum, and silver, *Langmuir*, 16 (6), 2604–2608, 2000.

65. Balogh, L., Ganser, T. R., and Shi, X. In *Characterization of Dendrimer–Gold Nanocomposite Materials*, Materials Research Society Symposium Proceedings, Warrendale, PA, 2005; Sanchez, C., Schubert, U., Laine, R. M., Chujo, Y., Materials Research Society, Warrendale, PA, EE13.33, pp. 1–6, 2005.

66. Esumi, K., Dendrimers for nanoparticle synthesis and dispersion stabilization, *Top. Curr. Chem.*, 227, 31–52, 2003.

67. Seo, Y.-S., Kim, K. S., Shin, K., White, H., Rafailovich, M., Sokolov, J., Lin, B., Kim, H. J., Zhang, C., and Balogh, L., Morphology of amphiphilic gold/dendrimer nanocomposite monolayers, *Langmuir*, 18 (15), 5927–5932, 2002.

68. Tomalia, D. A., Baker, H., Dewald, J., Hall, M., and Kallos, G., Dendritic macromolecules: Synthesis of starburst dendrimers, *Macromolecules*, 19, 2466–2468, 1986.

69. Uppuluri, S., Swanson, D. R., Piehler, L. T., Li, J., Hagnauer, G. L., and Tomalia, D. A., Core-shell tecto(dendrimers): I. Synthesis and characterization of saturated shell models, *Adv. Mater.*, 12 (11), 796–800, 2000.

70. Li, J., Swanson, D. R., Qin, D., Brothers, H. M., Piehler, L. T., Tomalia, D., and Meier, D. J., Characterizations of core-shell tecto-(dendrimer) molecules by tapping mode atomic force microscopy, *Langmuir*, 15 (21), 7347–7350, 1999.

71. Tomalia, D. A., Piehler, L. T., Durst, H. D., and Swanson, D. R., Partial shell-filled core-shell tecto(dendrimers): A strategy to surface differentiated nano-clefts and cusps, *Proc. Natl Acad. Sci. U.S.A.*, 99 (8), 5081–5087, 2002.

72. Shi, X., Patri, A. K., Lesniak, W., Islam, M. T., Zhang, C., Baker, J. R., and Balogh, L. P., Analysis of poly(amidoamine)-succinamic acid dendrimers by slab-gel electrophoresis and capillary zone electrophoresis, *Electrophoresis*, 26, 2960–2967, 2005.

73. Islam, M. T., Shi, X., Balogh, L. P., and Baker, J. R., HPLC separation of different generations of poly(amidoamine) dendrimers modified with various terminal groups, *Anal. Chem.*, 77, 2063–2070, 2005.

74. Shi, X., Lesniak, W., Islam, M. T., Muñiz, M. C., Balogh, L. P., and Baker, J. R., Comprehensive characterization of surface-functionalized poly(amidoamine) dendrimers with acetamide, hydroxyl, and carboxyl groups, *Colloids Surf., A: Physicochem. Eng. Aspects*, 272, 139–150, 2006.

75. Shi, X., Bányai, I., Islam, M. T., Lesniak, W., Davis, D. Z., Baker, J. R., and Balogh, L. P., Generational, skeletal and substitutional diversities in generation one poly(amidoamine) dendrimers, *Polymer*, 46 (9), 3022–3034, 2005.

76. Lesniak, W., Bielinska, A. U., Sun, K., Janczak, K. W., Shi, X., Baker, J. R., and Balogh, L. P., Silver/dendrimer nanocomposites as biomarkers: Fabrication, characterization, in vitro toxicity, and intracellular detection, *Nano Lett.*, 5 (11), 2123–2130, 2005.

77. Nigavekar, S. S., Sung, L. Y., Llanes, M., El-Jawahri, A., Lawrence, T. S., Becker, C. W., Balogh, L., and Khan, M. K., 3H dendrimer nanoparticle organ/tumor distribution, *Pharm. Res.*, 21 (3), 476–483, 2004.

78. Malik, N., Wiwattanapatapee, R., Klopsch, R., Lorenz, K., Frey, H., Weener, J. W., Meijer, E. W., Paulus, W., and Duncan, R., Dendrimers: Relationship between structure and biocompatibility in vitro, and preliminary studies on the biodistribution of 125I-labelled polyamidoamine dendrimers in vivo, *J. Control. Release*, 65 (1–2), 133–148, 2000.

79. Kukowska-Latallo, J. F., Candido, K. A., Cao, Z., Nigavekar, S. S., Majoros, I. J., Thomas, T. P., Balogh, L. P., Khan, M. K., and Baker, J. R. Jr., Nanoparticle targeting of anticancer drug improves therapeutic response in animal model of human epithelial cancer, *Cancer Res.*, 65 (12), 5317–5324, 2005.

80. Tang, M. X., Redemann, C. T., and Szoka, F. C., In vitro gene delivery by degraded polyamidoamine dendrimers, *Bioconjug. Chem.*, 7, 703–714, 1996.

81. Rosi, N. L. and Mirkin, C. A., Nanostructures in biodiagnostics, *Chem. Rev.*, 105, 1547–1562, 2005.

82. Burda, C., Chen, X., Narayanan, R., and El-Sayed, M. A., Chemistry and properties of nanocrystals of different shapes, *Chem. Rev.*, 105, 1025–1102, 2005.

83. Bonkhoff, H. and Remberger, K., Widespread distribution of nuclear androgen receptors in the basal cell layer of the normal and hyperplastic human prostate, *Virchows Arch. A Pathol. Anat. Histopathol.*, 422 (1), 35–38, 1993.

84. Pasqualini, R. and Ruoslahti, E., Organ targeting in vivo using phage display peptide libraries, *Nature*, 380 (6572), 364–366, 1996.

85. Rajotte, D. and Ruoslahti, E., Membrane dipeptidase is the receptor for a lung-targeting peptide identified by in vivo phage display, *J. Biol. Chem.*, 274 (17), 11593–11598, 1999.

86. Pasqualini, R., Koivunen, E., Kain, R., Lahdenranta, J., Sakamoto, M., Stryhn, A., Ashmun, R. A., Shapiro, L. H., Arap, W., and Ruoslahti, E., Aminopeptidase N is a receptor for tumor-homing peptides and a target for inhibiting angiogenesis, *Cancer Res.*, 60 (3), 722–727, 2000.

87. Arap, W., Pasqualini, R., and Ruoslahti, E., Cancer treatment by targeted drug delivery to tumor vasculature in a mouse model, *Science*, 279 (5349), 377–380, 1998.

88. Pasqualini, R., Koivunen, E., and Ruoslahti, E., Alpha v integrins as receptors for tumor targeting by circulating ligands, *Nat. Biotechnol.*, 15 (6), 542–546, 1997.

89. Burg, M. A., Pasqualini, R., Arap, W., Ruoslahti, E., and Stallcup, W. B., NG2 proteoglycan-binding peptides target tumor neovasculature, *Cancer Res.*, 59 (12), 2869–2874, 1999.

90. Ellerby, H. M., Arap, W., Ellerby, L. M., Kain, R., Andrusiak, R., Rio, G. D., and Krajewski, S., Anti-cancer activity of targeted pro-apoptotic peptides, *Nat. Med.*, 5 (9), 1032–1038, 1999. http://www.java/Propub/medicine/nm0999_1032.fulltextjava/Propub/medicine/nm0999_1032.abstract.

91. Dmitriev, I., Krasnykh, V., Miller, C. R., Wang, M., Kashentseva, E., Mikheeva, G., Belousova, N., and Curiel, D. T., An adenovirus vector with genetically modified fibers demonstrates expanded tropism via utilization of a coxsackievirus and adenovirus receptor-independent cell entry mechanism, *J. Virol.*, 72 (12), 9706–9713, 1998.

92. DeNardo, S. J., Burke, P. A., Leigh, B. R., O'Donnell, R. T., Miers, L. A., Kroger, L. A., and Goodman, S. L., Neovascular targeting with cyclic RGD peptide (cRGDf-ACHA) to enhance delivery of radioimmunotherapy, *Cancer Biother. Radiopharm.*, 15 (1), 71–79, 2000.

93. Janssen, M. L., Oyen, W. J., Dijkgraaf, I., Massuger, L. F., Frielink, C., Edwards, D. S., Rajopadhye, M., Boonstra, H., Corstens, F. H., and Boerman, O. C., Tumor targeting with radiolabeled alpha(v)-beta(3) integrin binding peptides in a nude mouse model, *Cancer Res.*, 62 (21), 6146–6151, 2002.

94. Ogawa, M., Hatano, K., Oishi, S., Kawasumi, Y., Fujii, N., Kawaguchi, M., Doi, R., Imamura, M., Yamamoto, M., Ajito, K. *et al.*, Direct electrophilic radiofluorination of a cyclic RGD peptide for in vivo alpha(v)beta(3) integrin related tumor imaging(1), *Nucl. Med. Biol.*, 30 (1), 1–9, 2003.

95. Haubner, R., Wester, H. J., Weber, W. A., Mang, C., Ziegler, S. I., Goodman, S. L., Senekowitsch-Schmidtke, R., Kessler, H., and Schwaiger, M., Noninvasive imaging of alpha(v)beta3 integrin expression using 18F-labeled RGD-containing glycopeptide and positron emission tomography, *Cancer Res.*, 61 (5), 1781–1785, 2001.

96. Sivolapenko, G. B., Skarlos, D., Pectasides, D., Stathopoulou, E., Milonakis, A., Sirmalis, G., Stuttle, A., Courtenay-Luck, N. S., Konstantinides, K., and Epenetos, A. A., Imaging of metastatic melanoma utilising a technetium-99m labelled RGD-containing synthetic peptide, *Eur. J. Nucl. Med.*, 25 (10), 1383–1389, 1998.

97. Liu, S., Edwards, D. S., Ziegler, M. C., Harris, A. R., Hemingway, S. J., and Barrett, J. A., 99mTc-labeling of a hydrazinonicotinamide-conjugated vitronectin receptor antagonist useful for imaging tumors, *Bioconjug. Chem.*, 12 (4), 624–629, 2001.

98. Folkman, J., Tumor Angiogenesis, In *Cancer Medicine*, Holland, J. F., Bast, R. C., Morton, D. L., Frei, E., Kufe, D. W., and Weichselbaum, R. R., Eds., Vol. 1, Williams and Wilkens, Baltimore, MD, pp. 181–204, 1996.

99. Jain, R. K., Determinants of tumor blood flow: A review, *Cancer Res.*, 48 (10), 2641–2658, 1998.

100. Ribatti, D., Vacca, A., and Dammacco, F., The role of the vascular phase in solid tumor growth: A historical review, *Neoplasia*, 1 (4), 293–302, 1999.

101. Schlingemann, R. O., Rietveld, F. J., Kwaspen, F., van de Kerkhof, P. C., de Waal, R. M., and Ruiter, D. J., Differential expression of markers for endothelial cells, pericytes, and basal lamina in the microvasculature of tumors and granulation tissue, *Am. J. Pathol.*, 138 (6), 1335–1347, 1991.

102. Matsumura, Y. and Maeda, H., A new concept for macromolecular therapeutics in cancer chemotherapy: Mechanism of tumoritropic accumulation of proteins and the antitumor agent smancs, *Cancer Res.*, 46 (12 Pt 1), 6387–6392, 1986.

103. Duncan, R., Malik, N., Richardson, S., and Ferruti, P., Polymer conjugates for anti-cancer agent and DNA delivery, *Polym. Prepr. (Am. Chem. Soc. Div., Polym. Chem.)*, 39, 180, 1998.

104. Muggia, F. M., Doxorubicin–polymer conjugates: Further demonstration of the concept of enhanced permeability and retention, *Clin. Cancer Res.*, 5 (1), 7–8, 1999.

105. Hainfeld, J. F., Foley, C. J., Srivastava, S. C., Mausner, L. F., Feng, N. I., Meinken, G. E., and Steplewski, Z., Radioactive gold cluster immunoconjugates: Potential agents for cancer therapy, *Int. J. Rad. Appl. Instrum. B*, 17 (3), 287–294, 1990.

106. Sharma, V., Luker, G. D., and Piwnica-Worms, D., Molecular imaging of gene expression and protein function in vivo with PET and SPECT, *J. Magn. Reson. Imaging*, 16 (4), 336–351, 2002.

107. Berning, D. E., Katti, K. V., Volkert, W. A., Higginbotham, C. J., and Ketring, A. R., 198Au-labeled hydroxymethyl phosphines as models for potential therapeutic pharmaceuticals, *Nucl. Med. Biol.*, 25 (6), 577–583, 1998.

108. Brechbiel, M. W., Gansow, O. A., Wu, C., Garmestani, K., Yordanov, A., Deal, K., and Chappell, L., In *Chelated Metal Ions for Therapeutic and Diagnostic Applications*, 215th ACS Natioial Meeting, Dallas, March 29–April 2, ACS, Washington, DC, Dallas, NUCL-106, 1998.

109. Wiener, E. C., Brechbiel, M. W., Brothers, H., Magin, R. L., Gansow, O. A., Tomalia, D. A., and Lauterbur, P. C., Dendrimer-based metal chelates: A new class of magnetic resonance imaging contrast agents, *Magn. Reson. Med.*, 31 (1), 1–8, 1994.

110. Wu, C., Brechbiel, M. W., Kozak, R. W., and Gansow, O. A., Metal–chelate–dendrimer–antibody constructs for use in radioimmunotherapy and imaging, *Bioorg. Med. Chem. Lett.*, 4 (3), 449–454, 1994.

111. Wiener, E. C., Konda, S., Shadron, A., Brechbiel, M., and Gansow, O., Targeting dendrimer-chelates to tumors and tumor cells expressing the high-affinity folate receptor, *Investig. Radiol.*, 32 (12), 748–754, 1997.

112. Brechbeil, M., *J. Nucl. Med.*, 35 (5 suppl. S), 62, 1994.

113. Barth, R. F., Soloway, A. H., and Brugger, R. M., Boron neutron capture therapy of brain tumors: Past history, current status, and future potential, *Cancer Investig.*, 14 (6), 534–550, 1996.

114. Margerum, L. D., Campion, B. K., Koo, M., Shargill, N., Lai, J., Marumoto, A., and Sontum, P. C., *J. Alloys Compd.*, 249 (1–2), 185–190, 1997.

115. Von Kleist, S., Ten years of tumor imaging with labelled antibodies, *In Vivo*, 7 (6B), 581–584, 1993.

116. Blankenberg, F. G. and Strauss, H. W., Nuclear medicine applications in molecular imaging, *J. Magn. Reson. Imaging*, 16 (4), 352–361, 2002.

117. Kwekkeboom, D., Krenning, E. P., and de Jong, M., Peptide receptor imaging and therapy, *J. Nucl. Med.*, 41 (10), 1704–1713, 2000.

118. Ang, E. S. and Sundram, F. X., Prospects for tumour imaging with radiolabelled antibodies, *Ann. Acad. Med. Singapore*, 22 (5), 776–784, 1993.

119. Deshpande, S. V., DeNardo, S. J., Kukis, D. L., Moi, M. K., McCall, M. J., DeNardo, G. L., and Meares, C. F., Yttrium-90-labeled monoclonal antibody for therapy: Labeling by a new macrocyclic bifunctional chelating agent, *J. Nucl. Med.*, 31 (4), 473–479, 1990.

120. Beaumier, P. L., Venkatesan, P., Vanderheyden, J. L., Burgua, W. D., Kunz, L. L., Fritzberg, A. R., Abrams, P. G., and Morgan, A. C. Jr., 186Re radioimmunotherapy of small cell lung carcinoma xenografts in nude mice, *Cancer Res.*, 51 (2), 676–681, 1991.

121. Chalandon, Y., Mach, J. P., Pelegrin, A., Folli, S., and Buchegger, F., Combined radioimmunotherapy and chemotherapy of human colon carcinoma grafted in nude mice, advantages and limitations, *Anticancer Res.*, 12 (4), 1131–1139, 1992.

122. Cheung, N. K., Landmeier, B., Neely, J., Nelson, A. D., Abramowsky, C., Ellery, S., Adams, R. B., and Miraldi, F., Complete tumor ablation with iodine 131-radiolabeled disialoganglioside GD2-specific monoclonal antibody against human neuroblastoma xenografted in nude mice, *J. Natl Cancer Inst.*, 77 (3), 739–745, 1986.

123. Esteban, J. M., Schlom, J., Mornex, F., and Colcher, D., Radioimmunotherapy of athymic mice bearing human colon carcinomas with monoclonal antibody B72.3: Histological and autoradiographic study of effects on tumors and normal organs, *Eur. J. Cancer Clin. Oncol.*, 23 (6), 643–655, 1987.

124. Balogh,, L. P., Nigavekar, S. S., Cook, A. C., Minc, L., and Khan, M. K., Development of dendrimer–gold radioactive nanocomposites to treat cancer microvasculature, *PharmaChem*, 2 (4), 94–99, 2003.

Section 6

Liposomes

29 Applications of Liposomal Drug Delivery Systems to Cancer Therapy

Alberto A. Gabizon

CONTENTS

29.1 Introduction 595
29.2 The Challenge of Cancer Therapy 596
29.3 The Rationale for the Use of Liposomal Drug Carriers 597
29.4 Differential Factors on the Pharmacokinetics of Low-Molecular-
Weight Drugs and Liposomes 597
29.5 Correlation Between Liposome Formulation and Liposome
Pharmacokinetics-Stealth Liposomes 598
29.6 Preclinical Observations with Liposomal Anthracyclines 600
29.7 Liposomal Anthracyclines in the Clinic—Doxil (PLD) 602
29.8 Development of Other Liposome-Entrapped Cytotoxic Agents 605
29.9 Future Avenues of Liposome Development—Receptor
Targeted Liposomes 606
29.10 Conclusions 607
References 607

29.1 INTRODUCTION

In the last two decades, major advances in the use of injectable drug delivery systems for the treatment of cancer have occurred. These include macromolecular conjugates, liposomes, and other nanoparticles. Poly(ethylene glycol) (PEG)-modified liposomal doxorubicin (Doxil, Caelyx) was the first liposomal anti-cancer drug to be approved by the Food and Drug Administration,[1] whereas paclitaxel albumin-bound particle suspension (ABI007, Abraxane) was recently approved for the treatment of metastatic breast cancer.[2] The ultimate goal behind the development of these sophisticated drug delivery systems is to improve the efficacy and decrease the side effects of new and old anti-cancer drugs. To achieve this goal, changes in the pharmacokinetics, biodistribution, and bioavailability profile leading to a positive impact on the drug pharmacodynamics are required.

This chapter will focus on injectable liposome-based drug delivery systems of anti-cancer drugs, the most widely used drug nanoparticle in cancer. Liposomes are vesicles with an aqueous interior surrounded by one or more concentric bilayers of phospholipids with a diameter ranging from a minimal diameter of ~ 30 nm to several microns.[3] However, for injectable clinical applications, practically all liposome formulations are in the submicron ultrafilterable range (<200 nm size) and can be considered as nanosize particulate systems. Liposomes are spontaneously formed when amphiphilic lipids such as phospholipids are dispersed in water. The ensuing structures are physically stable supramolecular assemblies, and, unlike polymerized particles, they are not covalently bound. Whether the drug is encapsulated in the aqueous core or in the surrounding bilayer of the liposome is dependent on the characteristics of the drug and the encapsulation process.[4] In general, water-soluble drugs are encapsulated within the central aqueous core, whereas lipid-soluble drugs are incorporated directly into the lipid membrane.

The current trend is to classify liposomes into a class of pharmaceutical devices in the nano-scale range engineered by physical and/or chemical means, and referred to as nanomedicines.[5] The field of nanomedicines is rapidly evolving and aims at increased sophistication in the design of nanosize devices and their interactions with cellular targets at the nanoscale level. In fact, lipo-somes are the first generation of nanomedicines approved for treatment of cancer (Doxil, described later in this chapter) and fungal infections (Ambisome, containing amphotericin B). Current lipo-some formulations involve a slow drug release system and a passive targeting process known as enhanced permeability and retention (EPR) that will be discussed later in this chapter.

29.2 THE CHALLENGE OF CANCER THERAPY

Current understanding of the molecular processes underlying the pathologic behavior of cancer cells has progressed enormously in the last decade.[6] Of particular relevance to cancer targeting is the fact that a number of receptors, mostly growth factor receptors, have been found to be over-expressed in tumor cells and to play an important role as catalysts of growth. Receptor profiling of tumors may offer a potential Achilles heel for targeting specific ligands or antibodies with or without delivery of a cytotoxic drug cargo.[7] In addition, the pathophysiology of tumor neovascu-lature and the interaction of tumor with stroma have been recognized as processes that play a major role in tumor development. Cancer is ultimately a disease caused by somatic gene mutations that result in the transformation of a normal cell into a malignant tumor cell. Eventually, the tumor cell phenotype progresses along three major steps:[8]

1. Increased proliferation rate and/or decreased apoptosis, causing an increase of tumor cell mass.
2. Invasion of surrounding tissues and switch on of angiogenesis. This is a critical step that differentiates in situ, non-invasive tumors with no metastatic potential from invasive tumors with metastatic and life-threatening potential.
3. Metastases, i.e., abnormal migration of tumor cells from the primary tumor site via blood vessels or lymphatics to distant organs with formation of secondary tumors. Most commonly, this is the process that causes death of the host because of disruption of the function of vital organs or systems (brain, lung, liver, kidney, bone marrow, coagu-lation, intestinal passage, and other).

Despite formidable advances in clinical imaging, the diagnosis of a tumor mass requires usually the presence of a nodule of ~ 10 mm diameter representing a cluster of 10^9 cells.* Because the lethal tumor burden is in the order of 10^{12} cells in most cancer patients, this implies that tumors have already gone through 3/4 of their doubling cell expansion process by the time of clinical diagnosis. As a result, significant heterogeneity and phenotypic diversity are already present in

---

* This does not apply to superficial skin tumors that can be recognized often when they are only 2–3 mm diameter and contain clusters of $\sim 10^7$ cells. Occasionally, modern imaging techniques (high resolution CT scan, MRI) can detect deep-seated lesions with suspected cancer features that are smaller than 10 mm.

most diagnosed cancers, posing a major therapeutic challenge as a result of the development of metastatic ability and drug resistance.

In addition to the lack of specificity of chemotherapeutic (cytotoxic) agents, a number of physiologic factors can seriously limit the efficiency of drug distribution from plasma to tumors and neutralize their effects. These include competition for drug uptake of well-perfused tissues such as liver and kidneys, rapid glomerular filtration and urinary excretion of low-molecular-weight drugs, protein binding with drug inactivation (e.g., cisplatin), and stability problems in biological fluids (e.g., hydrolysis of nitrosoureas, opening of lactone ring of camptothecin analogs).

29.3 THE RATIONALE FOR THE USE OF LIPOSOMAL DRUG CARRIERS

The rationale for the use of liposomes in cancer drug delivery is based on the following pharmacological principles:[1]

1. Slow drug release: drug bioavailability depends on drug release from liposomes. Entrapment of drug in liposomes will slow down drug release and reduce renal clearance to a variable extent. Slow release may range from a mere blunting of the peak plasma levels of free drug to a sustained release of drug, mimicking continuous infusion. These pharmacokinetic changes may have important pharmacodynamic consequences with regard to toxicity and efficacy of the liposome delivered agents.
2. Site avoidance of specific tissues: the biodistribution pattern of liposomes may lead to a relative reduction of drug concentration in tissues specifically sensitive to the delivered drug. This may have implications with regard to the therapeutic window of various cytotoxic drugs such as the cardiotoxic anthracyclines, provided that anti-tumor efficacy is not negatively affected.
3. Accumulation in tumors: prolongation of the circulation time of liposomes results in significant accumulation in tissues with increased vascular permeability. This is often the case of tumors, especially in those areas with active neoangiogenesis.[9] Tumor localization of long-circulating liposomes such as pegylated liposomes, sometimes referred to as Stealth or sterically-stabilized,[10] is a passive targeting effect, enabling substantial accumulation of liposome-encapsulated drug in the interstitial fluid at the tumor site,[11] a phenomenon sometimes referred as EPR effect.

29.4 DIFFERENTIAL FACTORS ON THE PHARMACOKINETICS OF LOW-MOLECULAR-WEIGHT DRUGS AND LIPOSOMES

There are a number of differential effects of physiologic factors on clearance and biodistribution of low-molecular-weight drugs and nanoparticles such as liposomes.

- Protein binding: low-molecular-weight drugs may be inactivated and/or irreversibly bound by plasma proteins, reducing the bioavailability toward cellular target molecules. This is the case of cisplatin, a widely used anti-cancer cytotoxic drug. In the case of nanoparticles, plasma proteins can adsorb to their surface, a process known as opsonization that results in tagging the particle for recognition and removal by macrophages. In addition, protein binding to the liposome surface may de-stabilize the bilayer and accelerate the leakage of liposome contents. PEG coating (PEGylation) of liposomes reduces opsonization and the effects associated with it.
- Reticulo-endothelial system (RES) clearance: it is unimportant for low-molecular-weight drugs, but it plays a major role in the clearance of nanoparticles, reducing the

TABLE 29.1
Parameters Affecting Delivery of Liposomal Drugs to Tumors

Tumor Factors	Liposome Factors
Blood flow	Long circulation time
Vascular permeability	Stability (drug retention)
Interstitial pressure	Small vesicle size
Phagocytic activity	Saturation of the RES
For a detailed discussion of these factors, see text.	

fraction available for distribution to tumor tissue. Kupffer cell macrophages lining the liver sinusoids remove opsonized liposomes and other nanoparticles from circulation and represent a major factor in the clearance of particulate carriers.

- Glomerular filtration: unless they become protein-bound, low-molecular-weight drugs can be filtered out by kidney glomeruli. In contrast, liposomes and other nanoparticles are non-filterable because their radius exceeds the glomerular filterable threshold size.
- Microvascular permeability: enhanced microvascular permeability with fenestrations in capillaries and post-capillary venules is critical for extravasation of nanoparticles from the blood stream to the interstitial fluid of target tissues. The presence of fenestrations is irrelevant for tissue delivery of small molecules.
- Extravascular transport: diffusion is the predominant mechanism of transport for small molecules. In contrast, convective transport plays a major role in the extravascular movement of nanoparticles for which diffusion rates are very slow.[12] Large tumors tend to develop high interstitial pressure that reduces the rate of convective transport. In fact, in an animal model, it has been shown that liposomes accumulate significantly less in larger tumors on a per gram tissue basis.[13] In agreement with this, large tumor size predicts poor response to liposome-delivered chemotherapy in ovarian cancer.[14]

Table 29.1 lists a number of tumor and liposome factors that play an important role in the delivery of liposomal drugs. On the tumor side, a rich blood flow and highly permeable microvascular bed will increase the probability of liposome deposition, whereas a high interstitial fluid pressure is likely to reduce the movement of molecules and particles into the tumor compartment. On the liposome side, avoiding drug leakage and prolonging the circulation time will result in more liposomes reaching the tumor vascular bed with an intact drug payload, and a small vesicle size will facilitate extravasation through the endothelium gaps or fenestrations. There are also data indicating that saturation of the RES will prolong circulation time and indirectly enhance liposome deposition in tumors.[15]

29.5 CORRELATION BETWEEN LIPOSOME FORMULATION AND LIPOSOME PHARMACOKINETICS-STEALTH LIPOSOMES

In 1971, Gregoriadis et al.[16] published the first research work where liposomes were used as drug carriers for medical applications. This initial study led to growing interest in liposomes, and many laboratories began examining liposome pharmacokinetics and biodistribution in animals as well as in vitro stability in serum. The early liposome work was mostly based on formulations composed of neutral egg lecithin (PC), often in combination with negatively or positively charged lipids. These liposomes were found to rapidly release a large fraction their encapsulated contents in circulation and were quickly removed from circulation by macrophages of the RES residing in liver and spleen.

Reformulation with high phase-transition temperature (Tm) lipids (distearoyl-PC, dipalmitoyl-PC, sphingomyelin) and an addition of cholesterol led to improved retention of liposome contents and prolongation of circulation time, especially when the vesicles were properly downsized to < 100 nm diameter. However, these relatively improved liposome formulations would still largely accumulate in the RES, and a greater improvement in circulation half-life appeared to be required for cancer targeting. Surface modifications of liposomes that could reduce the RES affinity were investigated based on the erythyrocyte paradigm whereby a layer of carbohydrate groups prolong circulation for nearly three months. A number of glycolipids such as monosialoganglioside (GM1), phosphatidyl-inositol, and cerebroside sulfate were included in the formulations and, indeed, extended liposome circulation time.[17,18] However, a major leap forward took place when the hydrophilic polymer PEG that was known to reduce immunogenicity and prolong circulation time when attached to enzymes and growth factors was introduced into liposomes in the early 1990s. PEG that is inexpensive as a result of easy synthesis and that could be prepared in high purity and large quantities had distinct advantages over the other glycolipid surface modifiers. Addition of a conjugate of PEG with a lipid anchor, distearoyl-phosphatidylethanolamine (PEG–DSPE), to the liposomal formulation was shown to significantly prolong liposome circulation time[10,19–22] and formed a pivotal element of the pharmaceutical development of the DOXIL formulation described thereafter in this chapter. Because of their ability to avoid/escape RES clearance mechanisms, PEG-coated liposomes have been coined Stealth* liposomes.[23] In general, the pharmacokinetic variables of the liposomes depend on the liposomes' physicochemical characteristics such as size, surface charge, membrane lipid packing, steric stabilization as conferred by PEG coating, dose, and route of administration.

It should be noted that the pharmacokinetic disposition of liposome-associated agents is dependent on the carrier until the drug is released from the carrier. Therefore, the pharmacology and pharmacokinetics of drugs delivered in liposomes are complex, and ideally, studies must be done to evaluate the disposition of the entrapped form of the drug and of the released active drug. A prerequisite for these types of studies is to develop the proper methodology to discriminate between both forms of the drug in plasma.

In parallel to the development of stable formulations with long circulation half-life, it was soon realized that a prolonged residence time in circulation was a critical pharmacokinetic factor for liposome deposition in tumors and that there was a strong correlation between liposome circulation time and tumor uptake.[18] A number of studies have addressed the mechanism of liposome accumulation in tumors. Microscopic observations with colloidal gold-labeled liposomes[24] and morphologic studies with fluorescent liposomes in the skin-fold chamber model[25] have demonstrated that liposomes extravasate into the tumor extracellular fluid through gaps in tumor microvessels and are found predominantly in the perivascular area with minimal uptake by tumor cells. Studies with ascitic tumors demonstrate a steady extravasation process of long circulating liposomes into the ascitic fluid with gradual release of drug followed by drug diffusion into the ascitic cellular compartment.[26,27] The process underlying the preferential tumor accumulation of liposomes as well as other macromolecular and particulate carriers is known as EPR effect.[28] This is a passive and non-specific process, resulting from increased microvascular permeability and defective lymphatic drainage in tumors creating an in situ depot of liposomes in the tumor interstitial fluid. Circulating liposomes cross the leaky tumor vasculature, moving from plasma into the interstitial fluid of tumor tissue following convective transport and diffusion processes. Although convective transport of plasma fluid occurs also in normal tissues, the continuous, non-fenestrated endothelium and basement membrane prevents the extravasation of liposomes. EPR is relatively a slow process where long-circulating liposomes possess a distinct advantage because of the repeated

---

* Stealth is a registered trademark of Alza Corp., Mountain View, CA.

passage through the tumor microvascular bed and their high concentration in plasma during an extended period of time.

For any intra-vascular drug carrier device to access the tumor cell compartment and interact with tumor cell receptors, it must first cross the vascular endothelium and diffuse into the interstitial fluid because, with few exceptions, tumor cells and their surface receptors are not directly exposed to the blood stream. Therefore, the EPR effect is not only important for the tumor accumulation of non-targeted liposomes but also for that of ligand-targeted liposomes. This has led to the assertion that the extravasation process is the rate-limiting step of liposome accumulation in tumors.[29] Experimental data with targeted and non-targeted liposomes have so far lent consistent support to this hypothesis.[30]

In most instances, delivery of drug to tumor cells depends on the release of drug from liposomes in the interstitial fluid because liposomes are seldom taken up by tumor cells unless they are tagged with specific ligands. The factors controlling this process and its kinetics are not well understood and may vary among tissues, depending on the liposome formulation in question. In the case of remote-loaded formulations, that is often the case of anthracyclines, a gradual loss of the liposome gradient retaining the drug in addition to disruption of the integrity of the liposome bilayer by phospholipases may be involved in the release process. Uptake by tumor-infiltrating macrophages could also contribute to liposomal drug release. In any case, once the drug is released from liposomes, it will freely diffuse through the interstitial tumor space and reach deep layers of tumor cells. This is an inherent advantage of this delivery system as opposed to covalently bound drug-carrier systems. It is also a critical factor for the success of the liposomal drug approach because most of the liposomes appear to remain in interstitial spaces immediately surrounding the blood vessels,[24,25] and therefore would not be able to interact with more than one layer of tumor cells. As a result, the ability of the liposome to effectively carry the anti-cancer agent to the tumor and the ability to release the drug at a satisfactory rate in the tumor extracellular fluid are equally important factors in determining the anti-tumor effect of liposomal-encapsulated anti-cancer agents.

The EPR effect has been confirmed in a variety of implanted tumor models. Its validity regarding human tumors and, particularly, cancer metastases is unclear as yet. One point of concern is the fact that interstitial fluid pressure increases in most tumors once they grow above a certain size threshold,[31] therefore hindering extravascular transport and liposome delivery. Unfortunately, there is a paucity of imaging studies in cancer patients with radiolabeled liposomes. One of the few studies with radiolabeled pegylated liposomes demonstrated significant liposome accumulation based on tumor imaging findings in 15/17 patients tested.[32] In another study where tumor metastases and normal muscle tissues of two breast cancer patients were examined for doxorubicin concentration after injection of pegylated liposomal doxorubicin (PLD), liposomal drug was found at 10-fold greater concentration in tumor as compared to muscle.[33] Another important piece of work in this area is the study of Northfelt et al.[34] that pointed to an enhanced deposition of drug in Kaposi's sarcoma (KS) skin lesions of patients receiving PLD as compared to normal skin of the same patients and to doxorubicin concentration in KS biopsies of patients receiving free doxorubicin.

29.6 PRECLINICAL OBSERVATIONS WITH LIPOSOMAL ANTHRACYCLINES

The drugs most frequently incorporated and evaluated in liposomal formulations are anthracyclines, including doxorubicin and daunorubicin. The choice of doxorubicin by many of the early research groups examining the role of liposomes as drug carriers in cancer chemotherapy stems from its broad spectrum of anti-tumor activity on the one hand, and its disturbing cumulative dose-limiting cardiac toxicity. Anthracyclines such as doxorubicin and daunorubicin cause acute toxic side effects, including bone marrow depression, alopecia, and stomatitis, and they are dose limited by a serious and mostly irreversible characteristic cardiomyopathy.[35] The first study describing the encapsulation of anthracyclines into liposomes appeared in 1979.[36] Work from various research groups followed soon after, supporting the general principle that liposomal formulations reduced the toxicity of anthracyclines in animal models.[37]

Based on the Stealth concept and using an elegant drug-loading mechanism, a formulation of PLD known as DOXIL (Caelyx in Europe) has been developed (Figure 29.1). The loading mechanism, coined remote (active) loading, is based on an ammonium sulfate gradient, leading to marked accumulation of doxorubicin inside the aqueous phase of the liposome ($\sim$ 15,000 doxorubicin molecules/vesicle). At high concentration, the entrapped drug forms a crystallinelike precipitate, contributing to the stability of the entrapment.[38,39] This loading technology provides substantial stability with negligible drug leakage in circulation while still enabling satisfactory rates of drug release in tissues and malignant effusions.[40]

Studies in animal tumor models with doxorubicin encapsulated in pegylated and other long-circulating liposomes have demonstrated increased anti-tumor activity of liposomal drug as compared to optimal doses of free drug in various rodent models of syngeneic and human tumors and increased accumulation of liposomal drug in various transplantable mouse and human tumors as compared to free drug with a delayed peak tumor concentration and slow tissue clearance after injection of liposomal drug.[40]

The most valuable pharmacokinetic advantage of the Stealth liposomal delivery system is the enhancement of tumor exposure to doxorubicin as a result of the accumulation of liposomes in tumors by the EPR effect (Figure 29.2) as demonstrated in animal models and in some forms of human cancer. When the tissue uptake of PLD was examined in a couple of syngeneic mouse tumor models, it was found that tumor drug uptake linearly correlated with dose, and liver drug uptake showed a saturation profile. In the case of free doxorubicin, liver uptake linearly increased with dose, and tumor uptake marginally increased with dose. As a result, the delta of tumor drug concentration in favor of PLD was substantially greater at high doses.[15] These results suggest

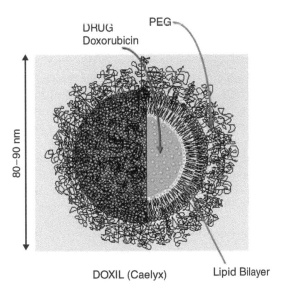

FIGURE 29.1 Schematic diagram of pegylated (Stealth) liposome encapsulating doxorubicin in the water phase (DOXIL, PLD). The PEG coating is critical for stabilizing and protecting the liposome surface from opsonization by plasma proteins, conferring prolonged circulation time. The gradient-driven loading of doxorubicin results in a high drug concentration and gelification in the liposome water phase, ensuring stable drug retention during circulation.

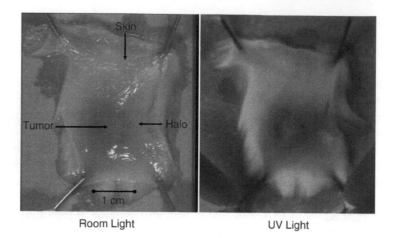

FIGURE 29.2 (**See color insert following page 522.**) Deposition of PLD in M109 carcinoma subcutaneously implanted in BALB/c mice by the EPR effect. The mouse was sacrificed 48 h after intravenous injection of 20 mg/kg PLD. Left panel: note the red-orange to bluish color in the tumor and peritumoral halo, suggesting accumulation of doxorubicin. Right panel: note the strong tumor fluorescence observed with UV light, confirming the presence of large amounts of doxorubicin. The drug levels measured in doxorubicin-equivalents were: tumor, 86 µg/g (i.e., ~35% of injected dose per gram); peritumoral halo, 40 µg/g; normal skin, 0.5 µg/g.

a passive process of liposomal uptake into tumor with non-saturable kinetics, whereas the liver uptake is mediated by receptor-mediated endocytosis and shows a saturable profile.

In preclinical therapeutic studies using a variety of rodent tumors and human xenografts in immunodeficient nude mice, PLD was more effective than free doxorubicin and other (non-pegy-lated) formulations of liposomal doxorubicin.[11] In a few instances, the activity of PLD preparations was matched, but not surpassed, by non-pegylated preparations of liposomal doxorubicin.[1] In many of these studies, the improved efficacy of PLD was obtained at milligram-equivalent doses to the MTD of free doxorubicin, indicating that there was a net therapeutic gain per milligram drug, independently of toxicity buffering. A study addressing this issue examined the activity of esca-lating doses of PLD and doxorubicin against implants of the mouse 3LL tumor and concluded that the activity of 1–2 mg/kg DOXIL was approximately equivalent to 9 mg/kg doxorubicin, i.e., a 6-fold enhancement in efficacy.[41] Similar observations were made in the M109 model pointing to a 4-fold advantage for PLD as compared to free doxorubicin, i.e., a dose of 2.5 mg/kg PLD was at least as effective as 10 mg/kg free doxorubicin.[15]

There is a large body of preclinical data on other liposome formulations of anti-cancer agents moving into clinical development or already approved for clinical use. A tentative list of these formulations is presented in Table 29.2. Obviously, in some cases, the added value of these formulations has not been or will not be sufficient to justify further development despite positive preclinical data.

29.7 LIPOSOMAL ANTHRACYCLINES IN THE CLINIC—DOXIL (PLD)

The anthracycline antibiotic doxorubicin has a broad spectrum of antineoplastic action and a correspondingly widespread degree of clinical use. In addition to its role in the treatment of breast cancer, doxorubicin is indicated in the treatment of various cancers of the lymphatic and hematopoyetic systems, gastric carcinoma, small-cell cancer of the lung, soft tissue, and bone sarcomas as well as cancer of the uterus, ovary, bladder, and thyroid. Unfortunately, toxicity

TABLE 29.2
Formulations of Liposomal Anti-Cancer Agents Clinically Tested[a]

Pegylated	Non-Pegylated	Regional Therapy	Non-Cytotoxic
DOXIL (PLD)[a]	Myocet[71]	DepoCyt (intrathecal)[72]	L-MTP-PE[73]
PLD with DSPC[74,b]	DaunoXome[75]	L-NDDP (intrapleural)[76]	Liposomal ATRA[77]
MCC-465 Immunoliposome[78]	Liposomal Annamycin[79]	Liposomal Camptothecin (aerosol)[80]	BLP25 Liposomal Vaccine[81]
SPI-077[63]	Liposomal Vincristine[57]	Liposomal IL2 (aerosol)[82]	Liposomal antisense ODN[83]
Lipoplatin[84]	Lurtotecan/OSI-211[56,85]	Liposomal E1A (intra-tumoral)[86]	
S-CKD602[87,c]	OSI-7904L (TS inhibitor)[88]		
	LE-Paclitaxel[58]		

[a] This list refers to formulations tested in the last 10 years and does not intend to be comprehensive given the difficulties involved in tracking all the relevant publications.

[b] This formulation is from Taiwan and differs from DOXIL in that DSPC is the main lipid component.

[c] Published information on this formulation of a topo I inhibitor is available only in abstract form. The reference given here refers to a preclinical study done with a similar formulation.

often limits the therapeutic activity of doxorubicin and may preclude adequate dosing. Other common complications of conventional anthracycline therapy include alopecia and dose-limiting myelosuppression. Most importantly, cardiotoxicity limits the cumulative dose of conventional anthracycline that can be given safely.[42]

Encapsulation of anthracyclines within liposomes significantly alters the drug pharmacokinetic profile with a tendency to selective enhancement of drug concentration in tumors.[43] In animal studies, these pharmacologic effects resulted in maintained or enhanced anthracycline efficacy and safety in a variety of experimental tumor types.[44] Improved safety and/or therapeutic profile of liposomal anthracycline therapy in KS, ovarian cancer, breast cancer, and multiple myeloma have been reported.[45] The relative lack of cardiotoxicity with liposomal anthracycline therapy is an important asset of the liposomal approach.[42] Therefore, in most instances where conventional anthracycline therapy is effective but the required course of treatment could lead to a high risk of toxicity, the use of liposomal anthracyclines should be preferred.

The most commonly used liposomal anthracycline formulation in the clinic is DOXIL (DOXIL®, Caelyx®, PLD). PLD embodies the pharmacokinetic advantages of stealth liposomes (Figure 29.1). The PLD formulation consists of doxorubicin encapsulated in pegylated liposomes formulated with a hydrogenated soybean PC and cholesterol.[40] PLD was granted initial market clearance in 1995 by the U.S. Food and Drug Administration (FDA) for use in treatment of AIDS-related KS. In 1999, PLD was granted U.S. market clearance for use in the treatment of recurrent metastatic carcinoma of the ovary based on a superior therapeutic index over topotecan, the comparator drug. In January 2003, the European Commission of the European Union granted centralized marketing authorization to PLD as monotherapy for metastatic breast cancer based on equivalent efficacy to doxorubicin and reduced cardiac toxicity. In addition, a phase III trial recently completed in the U.S. and Europe investigating the safety and efficacy of PLD as a replacement for doxorubicin in multiple myeloma found equal efficacy and reduced toxicity for PLD treatment, paving the way for PLD to become standard therapy in multiple myeloma.[46]

PLD was recognized 10 years ago as a liposomal doxorubicin formulation with unique pharmacokinetics and had a major change in the clinical toxicity profile. Clinical pharmacokinetic studies have indicated that PLD dramatically prolongs the circulation time of doxorubicin in agreement with preclinical studies. In 1994, the results of a pharmacokinetic study where

15 patients were sequentially given the same dose in drug-equivalents of free doxorubicin and PLD were published.[47] A dramatic reduction in the drug clearance and volume of distribution resulting in a 1000-fold increase in the area under the curve (AUC) was observed with the liposomal formulation. It was also found that practically all the drug circulating in plasma is in liposome-encapsulated form. Metabolites in plasma were undetectable or were present at very low levels. However, they were readily detected in urine 24 hours or later after injection, indicating that the drug had become bioavailable. The following drug distribution picture had emerged from this initial study and more recent ones:[40]

1. Drug circulates in plasma for prolonged periods of time (half-life in the range of ~50–80 hours) in liposome-encapsulated form. Despite its prolonged presence in blood, the drug is not bioavailable as long as it remains in the interior of a circulating liposome.
2. Most of the injected drug (>95%) is distributed to tissues in liposome-encapsulated form. Once in tissues, drug leakage and liposome breakdown with or without liposome internalization by cells gradually provides a pool of bioavailable drug.
3. The rate of metabolite production is slower than the rate of renal clearance of metabolites. As a result, metabolites do not accumulate in plasma but can be detected in urine.
4. A small fraction of injected drug (<5%) leaks from circulating liposomes and is handled as free drug with fast plasma clearance and rapid metabolism. This drug fraction is the source of small amounts of metabolites that can be sometimes detected in plasma.

In 1995, a phase I study of PLD in patients with solid tumors[48] provided clear evidence of a major change in the toxicity profile with muco-cutaneous toxicities as the major dose-limiting toxicities. In contrast, myelosuppression and alopecia were minor and cardiotoxicity was conspicuously absent. The maximal tolerated dose was established as 60 mg/m^2 with mucositis being the dose-limiting toxicity. It was also found that the optimal dosing interval for retreatment was four weeks, rather than the standard 3-week schedule of doxorubicin. The dose-schedule limiting toxicity was a form of skin toxicity known as hand-foot syndrome, also referred to as palmar-plantar erythrodysesthesia (PPE), that appears to be related to the long half-life of PLD. Therefore, it became well-established that the PLD formulation imparts a significant pharmacokinetic–pharmacodynamic change to the drug doxorubicin, unprecedented in magnitude for any intravenous drug delivery system. Later, data gathered from further clinical studies in metastatic breast cancer and recurrent ovarian cancer brought down the recommended dose of PLD from 60 to 40–50 mg/m^2 once every four weeks.[49] This dose reduction was needed mainly to prevent skin toxicity, resulting from successive courses of therapy.

KS is a multi-focal tumor affecting skin and sometimes mucosas well known for its very high microvascular permeability. Profuse extravasation of colloidal gold-labeled Stealth liposomes in a transgenic mouse model of KS has been shown.[50] In AIDS patients, KS is frequent and has an aggressive course. Therefore, this condition was chosen for initial clinical testing of PLD in phase II-III studies. Indeed, PLD as single agent therapy demonstrated a significantly greater efficacy and better safety than standard chemotherapy (combinations of bleomycin and vincristine with or without doxorubicin) and was also effective as second line chemotherapy in pretreated patients. Because of the extreme sensitivity of KS to chemotherapy, a low and relatively subtoxic dose of 20 mg/m^2 every three weeks is sufficient for effective treatment.[51]

Further to the successful application in various malignant conditions, there was another major development in the clinical research with PLD related to the prevention of the cardiac toxicity of doxorubicin. Evaluation of the cardiac function in patients receiving PLD revealed a major risk reduction of cardiotoxicity as compared to free doxorubicin historical data. A retrospective analysis of patients treated with large cumulative doses of PLD did not reveal any significant cardiac toxicity despite the fact that some of these patients were treated with three times as much as the maximal cumulative dose acceptable for free doxorubicin.[52] Two additional reports focusing on cardiac biopsies of PLD-treated patients at high cumulative doses demonstrated minimal pathology findings in line with the lack of clinical cardiotoxicity.[53,54] The cardiac safety of PLD was confirmed in a phase III study in metastatic breast cancer, showing a drastic reduction in cardiotoxicity in PLD-treated patients in comparison to free doxorubicin treatment.[55]

29.8 DEVELOPMENT OF OTHER LIPOSOME-ENTRAPPED CYTOTOXIC AGENTS

Besides anthracyclines, other cytotoxic compounds currently being developed in liposomes include: Topoisomerase I inhibitory properties such as camptothecin analogs,[56] mitotic spindle poisons such as Vinca alkaloids and taxanes,[57,58] and DNA-reactive agents such as cisplatin,[59] and mitomycin C (MMC).[60] For cell cycle phase-specific cytotoxic drugs of the first two groups (e.g., topotecan, vinorelbine), the anti-tumor activity tends to increase with liposome encapsulation as the time of exposure is extended by exploiting the liposomal slow release features. An additional advantage for liposome encapsulation of camptothecin analogs derives from their improved stability in the mildly acidic environment of the liposome water phase as opposed to the instability of their active lactone configuration when present in free form at pH 7.4 in plasma. For cisplatin, a major advantage of liposome encapsulation would be reduction of the nephrotoxicity. At this time, none of these compounds have yet been approved by a regulatory agency and none has gone through phase III studies.

It should be noted that the development of liposomal formulations of cytotoxic agents has often failed for different reasons. The formulation of a liposomal anti-cancer agent is a complex process with at least three distinct variables that may affect the outcome and the risk of failure: the design of the liposome carrier, the choice of the drug, and the method of drug encapsulation. One example illustrating the complexity of the relationship between formulation and pharmacological effects is provided by a product known as SPI-77, consisting of cisplatin encapsulated in pegylated liposomes. These liposomes have long-circulating characteristics, retain the drug in plasma exceedingly well, reduce drug toxicity, and produce a high and long-lasting accumulation of drug in implanted tumors.[61] However, because the Pt exposure was measured in tumor extracts, it is unclear whether the Pt measured was SPI-77 (i.e., liposomally encapsulated Pt), protein-bound Pt, or unbound Pt. The anti-tumor activity of SPI-77 in preclinical models was variable and, in some cases, inferior by comparison to free drug.[61] In clinical studies, SPI-77 was inactive and showed no dose-limiting toxicities even at doses greater than 2-fold the MTD of free cisplatin.[62] It turned out that cisplatin release from these liposomes was minimal in in vitro[61] as well as in vivo systems as indicated by the low occurrence of DNA adducts, resulting in a major reduction of bioavailability.[63] Indeed, microdialysis experiments in animal tumor models suggest that the amount of unbound Pt released by SPI-77 in tumor extracellular fluid is significantly lower when compared with free cisplatin.[64] The lack of activity of SPI-77 in early-phase clinical studies has led to the discontinuation of its further clinical development.

In some instances, a strategy based on chemically modifying the drug is needed to achieve stable encapsulation. It is a formidable challenge to develop liposome formulations that retain the drug payload in the blood stream during prolonged periods of circulation, thereby taking full advantage of the pharmacokinetic and biodistribution benefits of the pegylated liposomal carriers.[40] Passive encapsulation of MMC in liposomes and even encapsulation of cyclodextrin–MMC complexes in liposomes did not result in stable formulations as a result of the rapid leakage of MMC. Recently, an alternative approach to delivering MMC via long-circulating liposomes has been proposed based on the design of a lipophilic prodrug with excellent retention in liposomes, but gradually releasing active MMC upon exposure to appropriate environmental conditions.[60]

Attachment of drugs to bilayer-compatible lipids can ensure prolonged association with a lipo-some.[65,66] The MMC prodrug consists of a conjugate of MMC attached to 1,2-distearoyl glycerol lipid via a cleavable dithiobenzyl linker. Upon thiolytic cleavage of the disulfide-substituted benzyl urethane, MMC is released and becomes bioavailable. Intravenous administration of a pegylated liposomal formulation of the MMC prodrug in rats resulted in long circulation time and slow clearance consistent with Stealth pharmacokinetics, indicating stable prodrug retention in lipo-somes. The therapeutic index of this liposomal formulation was superior to that of free MMC in three animal tumor models, including a mouse tumor model with a multidrug-resistant phenotype where PLD was essentially inactive. The potential added value of this approach is 2-fold: effective treatment of multidrug- resistant tumors while significantly buffering the toxicity of MMC.

29.9 FUTURE AVENUES OF LIPOSOME DEVELOPMENT—RECEPTOR TARGETED LIPOSOMES

As liposomes remain one of the most attractive platforms for systemic drug delivery, an increased expansion and sophistication of these systems would be expected in the forthcoming future. The most logical improvement is the coupling of a ligand to the surface of the liposome that will target the vesicles to a specific cell-surface receptor followed by internalization and intracellular delivery of the liposome drug cargo. Examples in this direction are the targeting of PLD to HER2-expressing or folate-receptor expressing cancer cells using, respectively, a specific anti-HER2 scFv[67] or a folate conjugate anchored to the liposome surface.[30] Another example is the tumor vascular targeting of liposomes with endothelium-specific peptides associated to liposomes.[68] A major advantage of targeted liposomal nanocarriers over ligand-drug bioconjugates is the delivery-ampli-fying effect of the former that sometimes can provide to the target cell a ratio of 1,000 drug molecules per single ligand–receptor interaction.

One important technological and conceptual advance in this field is the ability to insert ligands onto preformed liposomes using lipophilic tails as anchors into the bilayer. Ligand post-insertion (as opposed to pre-insertion during liposome formation) has been demonstrated for antibody frag-ments[69] and small ligands such as folate.[30] This approach enables, on the one hand, the possibility of using end-product formulations without interfering with the pharmaceutical methods of prep-aration, and, on the other hand, the flexibility of ligand choice, depending on the target to be addressed. One encouraging observation, at least for folate-targeted liposomes, is that pegylated liposomes are able to retain during in vivo circulation and extravasation to malignant ascites the targeting ligand and the ability to bind to target cells, suggesting that these systems are stable enough for in vivo applications, requiring prolonged circulation before encountering the target.[70]

Except for rare instances, tumor cells are not directly exposed to the blood stream. Therefore, for an intra-vascular targeting device to access the targeted tumor cell receptor, it has first to cross the vascular endothelium and diffuse into the interstitial fluid. Experimental data with antibody-targeted liposomes and folate-targeted liposomes indicate that liposome deposition in tumors is similar for both targeted and non-targeted systems,[29,70] supporting the hypothesis that extravasation is the rate-limiting step of liposome accumulation in tumors. However, once liposomes have penetrated the tumor interstitial fluid, binding of targeted liposomes to the targeted receptor may occur, shifting the intra-tumor distribution from the extracellular compartment to the tumor cell compartment as recently shown for a mouse ascitic tumor.[70] Retrograde movement of liposomes into the blood stream, if any, will be reduced for liposomes with binding affinity to a tumor cell receptor. Therefore, the theoretical advantages of a targeted liposomal drug delivery system over a non-targeted one are primarily derived from a relative shift of liposome distribution from the extracellular fluid to the tumor cell compartment, and, possibly, from prolonged liposome retention in tumor. On the negative side, the main disadvantage of a targeted system to a cancer cell receptor is the difficulty of a large nanosize assembly such as liposomes to penetrate a solid tumor mass, confronted by the high interstitial fluid pressure that often characterizes tumor masses at clinically detectable size.[31] One such system targeted to the Her2 receptor has shown superior therapeutic efficacy over its non-targeted counterpart. With the maturation of the pharmaceutical technology required for the application of targeted liposomal drug delivery systems, the final word regarding their added value over non-targeted systems awaits clinical trials.

29.10 CONCLUSIONS

Despite huge research efforts, with liposomal drugs is cancer therapy anthracycline formulations, and specifically, the leading compound, PLD, are the only liposome products that have been able, so far, to fill a niche in the anti-cancer armamentarium. As this chapter has discussed, this is due, at least in part, to the complexity of the pharmaceutical technology involved that requires drug-specific solutions and has a direct impact on the pharmacology of the end compound. Many of the oncology opinion leaders regard liposome formulations, with liposomal in cancer therapy drugs as a cosmetic improvement of toxic drugs that will become obsolete in the near future once smart new drugs are developed, given the increasing knowledge of the molecular biology of cancer cells. However, a realistic scenario favored by the majority of clinicians predicts that the use of broad-spectrum cytotoxic drugs will remain as a major tool in cancer therapy for a few more decades to come. Moreover, the nanomedicine approach is perceived today as a key to future breakthroughs because of the complexity of living systems. These factors may reinvigorate the interest in particulate drug delivery systems such as liposomes and augur that the use of liposome carriers in cancer therapy will expand in the foreseeable future.

REFERENCES

1. Gabizon, A. A., Pegylated liposomal doxorubicin: Metamorphosis of an old drug into a new form of chemotherapy, *Cancer Invest.*, 19(4), 424–436, 2001.
2. Gradishar, W. J., Tjulandin, S., Davidson, N., Shaw, H., Desai, N., Bhar, P. et al., Phase III trial of nanoparticle albumin-bound paclitaxel compared with polyethylated castor oil-based paclitaxel in women with breast cancer, *J. Clin. Oncol.*, 23(31), 7794–7803, 2005.
3. Lasic, D. D. and Papahadjopoulos, D., *Medical Applications of Liposomes*, Elsevier, Amsterdam, 1998.
4. Barenholz, Y., Relevancy of drug loading to liposomal formulation therapeutic efficacy, *J. Liposome Res.*, 13(1), 1–8, 2003.
5. Moghimi, S. M., Hunter, A. C., and Murray, J. C., Nanomedicine: Current status and future prospects, *FASEB J.*, 19(3), 311–330, 2005.
6. Hanahan, D. and Weinberg, R. A., The hallmarks of cancer, *Cell*, 100(1), 57–70, 2000.
7. Van Den Bossche, B. and Van de Wiele, C., Receptor imaging in oncology by means of nuclear medicine: Current status, *J. Clin. Oncol.*, 22(17), 3593–3607, 2004.
8. Kastan, M. B., Molecular biology of cancer: The cell cycle, In *Cancer: Principles & Practice of Oncology*, DeVita, V. T., Hellman, S., and Rosenberg, S. A., Eds., 5th ed., Lippincott-Raven, Philadelphia, pp. 121–134, 1997.
9. Jain, R. K., Delivery of molecular medicine to solid tumors: Lessons from in vivo imaging of gene expression and function, *J. Control. Release*, 74(1-3), 7–25, 2001.
10. Papahadjopoulos, D., Allen, T. M., Gabizon, A., Mayhew, E., Matthay, K., Huang, S. K. et al., Sterically stabilized liposomes: Improvements in pharmacokinetics and antitumor therapeutic efficacy, *Proc. Natl Acad. Sci. USA*, 88(24), 11460–11464, 1991.
11. Gabizon, A. and Martin, F., Polyethylene glycol-coated (pegylated) liposomal doxorubicin. Rationale for use in solid tumours, *Drugs*, 54(Suppl. 4), 15–21, 1997.
12. Swabb, E. A., Wei, J., and Gullino, P. M., Diffusion and convection in normal and neoplastic tissues, *Cancer Res.*, 34(10), 2814–2822, 1974.

13. Harrington, K. J., Rowlinson-Busza, G., Syrigos, K. N., Abra, R. M., Uster, P. S., Peters, A. M. et al., Influence of tumour size on uptake of ^{111}In-DTPA-labelled pegylated liposomes in a human tumour xenograft model, *Br. J. Cancer*, 83(5), 684–688, 2000.

14. Safra, T., Groshen, S., Jeffers, S., Tsao-Wei, D. D., Zhou, L., Muderspach, L. et al., Treatment of patients with ovarian carcinoma with pegylated liposomal doxorubicin: Analysis of toxicities and predictors of outcome, *Cancer*, 91(1), 90–100, 2001.

15. Gabizon, A., Tzemach, D., Mak, L., Bronstein, M., and Horowitz, A. T., Dose dependency of pharmacokinetics and therapeutic efficacy of pegylated liposomal doxorubicin (DOXIL) in murine models, *J. Drug Target*, 10(7), 539–548, 2002.

16. Gregoriadis, G., Leathwood, P. D., and Ryman, B. E., Enzyme entrapment in liposomes, *FEBS Lett.*, 14(2), 95–99, 1971.

17. Allen, T. M. and Chonn, A., Large unilamellar liposomes with low uptake into the reticuloendothelial system, *FEBS Lett.*, 223(1), 42–46, 1987.

18. Gabizon, A. and Papahadjopoulos, D., Liposome formulations with prolonged circulation time in blood and enhanced uptake by tumors, *Proc. Natl Acad. Sci. USA*, 85(18), 6949–6953, 1988.

19. Allen, T. M., Austin, G. A., Chonn, A., Lin, L., and Lee, K. C., Uptake of liposomes by cultured mouse bone marrow macrophages: Influence of liposome composition and size, *Biochim. Biophys. Acta*, 1061(1), 56–64, 1991.

20. Woodle, M. C. and Lasic, D. D., Sterically stabilized liposomes, *Biochim. Biophys. Acta*, 1113(2), 171–199, 1992.

21. Klibanov, A. L., Maruyama, K., Torchilin, V. P., and Huang, L., Amphipathic polyethyleneglycols effectively prolong the circulation time of liposomes, *FEBS Lett.*, 268(1), 235–237, 1990.

22. Allen, T. M., Hansen, C., Martin, F., Redemann, C., and Yau-Young, A., Liposomes containing synthetic lipid derivatives of poly(ethylene glycol) show prolonged circulation half-lives in vivo, *Biochim. Biophys. Acta*, 1066(1), 29–36, 1991.

23. Lasic, D. D. and Martin, F. J., *Stealth Liposomes*, CRC Press, Boca Raton, FL, 1995.

24. Huang, S. K., Lee, K. D., Hong, K., Friend, D. S., and Papahadjopoulos, D., Microscopic localization of sterically stabilized liposomes in colon carcinoma-bearing mice, *Cancer Res.*, 52(19), 5135–5143, 1992.

25. Yuan, F., Leunig, M., Huang, S. K., Berk, D. A., Papahadjopoulos, D., and Jain, R. K., Microvascular permeability and interstitial penetration of sterically stabilized (stealth) liposomes in a human tumor xenograft, *Cancer Res.*, 54(13), 3352–3356, 1994.

26. Gabizon, A. A., Selective tumor localization and improved therapeutic index of anthracyclines encapsulated in long-circulating liposomes, *Cancer Res.*, 52(4), 891–896, 1992.

27. Bally, M. B., Masin, D., Nayar, R., Cullis, P. R., and Mayer, L. D., Transfer of liposomal drug carriers from the blood to the peritoneal cavity of normal and ascitic tumor-bearing mice, *Cancer Chemother. Pharmacol.*, 34(2), 137–146, 1994.

28. Maeda, H., The enhanced permeability and retention (EPR) effect in tumor vasculature: The key role of tumor-selective macromolecular drug targeting, *Adv. Enzyme Regul.*, 41, 189–207, 2001.

29. Goren, D., Horowitz, A. T., Zalipsky, S., Woodle, M. C., Yarden, Y., and Gabizon, A., Targeting of stealth liposomes to erbB-2 (Her/2) receptor: In vitro and in vivo studies, *Br. J. Cancer*, 74(11), 1749–1756, 1996.

30. Gabizon, A., Shmeeda, H., Horowitz, A. T., and Zalipsky, S., Tumor cell targeting of liposome-entrapped drugs with phospholipid-anchored folic acid-PEG conjugates, *Adv. Drug Deliv. Rev.*, 56(8), 1177–1192, 2004.

31. Stohrer, M., Boucher, Y., Stangassinger, M., and Jain, R. K., Oncotic pressure in solid tumors is elevated, *Cancer Res.*, 60(15), 4251–4255, 2000.

32. Harrington, K. J., Mohammadtaghi, S., Uster, P. S., Glass, D., Peters, A. M., Vile, R. G. et al., Effective targeting of solid tumors in patients with locally advanced cancers by radiolabeled pegylated liposomes, *Clin. Cancer Res.*, 7(2), 243–254, 2001.

33. Symon, Z., Peyser, A., Tzemach, D., Lyass, O., Sucher, E., Shezen, E. et al., Selective delivery of doxorubicin to patients with breast carcinoma metastases by stealth liposomes, *Cancer*, 86(1), 72–78, 1999.

34. Northfelt, D. W., Martin, F. J., Working, P., Volberding, P. A., Russell, J., Newman, M. et al., Doxorubicin encapsulated in liposomes containing surface-bound polyethylene glycol: Pharmacokinetics, tumor localization, and safety in patients with AIDS-related Kaposi's sarcoma, *J. Clin. Pharmacol.*, 36(1), 55–63, 1996.

35. Young, R. C., Ozols, R. F., and Myers, C. E., The anthracycline antineoplastic drugs, *N. Engl. J. Med.*, 305(3), 139–153, 1981.

36. Forssen, E. A. and Tokes, Z. A., In vitro and in vivo studies with adriamycin liposomes, *Biochem. Biophys. Res. Commun.*, 91(4), 1295–1301, 1979.

37. Gabizon, A. A., Liposomal anthracyclines, *Hematol. Oncol. Clin. North Am.*, 8(2), 431–450, 1994.

38. Lasic, D. D., Frederik, P. M., Stuart, M. C., Barenholz, Y., and McIntosh, T. J., Gelation of liposome interior. A novel method for drug encapsulation, *FEBS Lett.*, 312(2–3), 255–258, 1992.

39. Haran, G., Cohen, R., Bar, L. K., and Barenholz, Y., Transmembrane ammonium sulfate gradients in liposomes produce efficient and stable entrapment of amphipathic weak bases, *Biochim. Biophys. Acta*, 1151(2), 201–215, 1993.

40. Gabizon, A., Shmeeda, H., and Barenholz, Y., Pharmacokinetics of pegylated liposomal Doxorubicin: Review of animal and human studies, *Clin. Pharmacokinet.*, 42(5), 419–436, 2003.

41. Colbern, G. T., Hiller, A. J., Musterer, R. S. et al., Significant increase in antitumor potency of doxorubicin HCl by its encapsulation in pegylated liposomes, *J. Liposome Res.*, 9, 523–538, 1999.

42. Ewer, M. S., Martin, F. J., Henderson, C., Shapiro, C. L., Benjamin, R. S., and Gabizon, A. A., Cardiac safety of liposomal anthracyclines, *Semin. Oncol.*, 31(6 Suppl. 13), 161–181, 2004.

43. Allen, T. M. and Martin, F. J., Advantages of liposomal delivery systems for anthracyclines, *Semin. Oncol.*, 31(6 Suppl. 13), 5–15, 2004.

44. Vail, D. M., Amantea, M. A., Colbern, G. T., Martin, F. J., Hilger, R. A., and Working, P. K., Pegylated liposomal doxorubicin: Proof of principle using preclinical animal models and pharmacokinetic studies, *Semin. Oncol.*, 31(6 Suppl. 13), 16–35, 2004.

45. Henderson, C. I., Established and emerging roles of liposomal anthracyclines in oncology. Introduction, *Semin. Oncol.*, 31(6 Suppl. 13), 1–4, 2004.

46. Rifkin, R. M., Gregory, S. A., Mohrbacher, A., and Hussein, M. A., PEGylated liposomal doxorubicin, vincristine, and dexamethasone provide significant reduction in toxicity compared with doxorubicin, vincristine, and dexamethasone in patients with newly diagnosed multiple myeloma, *Cancer*, 106(4), 848–858, 2006.

47. Gabizon, A., Catane, R., Uziely, B., Kaufman, B., Safra, T., Cohen, R. et al., Prolonged circulation time and enhanced accumulation in malignant exudates of doxorubicin encapsulated in polyethylene-glycol coated liposomes, *Cancer Res.*, 54(4), 987–992, 1994.

48. Uziely, B., Jeffers, S., Isacson, R., Kutsch, K., Wei-Tsao, D., Yehoshua, Z. et al., Liposomal doxorubicin: Antitumor activity and unique toxicities during two complementary phase I studies, *J. Clin. Oncol.*, 13(7), 1777–1785, 1995.

49. Alberts, D. S., Muggia, F. M., Carmichael, J., Winer, E. P., Jahanzeb, M., Venook, A. P. et al., Efficacy and safety of liposomal anthracyclines in phase I/II clinical trials, *Semin. Oncol.*, 31(6 Suppl. 13), 53–90, 2004.

50. Huang, S. K., Martin, F. J., Jay, G., Vogel, J., Papahadjopoulos, D., and Friend, D. S., Extravasation and transcytosis of liposomes in Kaposi's sarcoma-like dermal lesions of transgenic mice bearing the HIV tat gene, *Am. J. Pathol.*, 143(1), 10–14, 1993.

51. Krown, S. E., Northfelt, D. W., Osoba, D., and Stewart, J. S., Use of liposomal anthracyclines in Kaposi's sarcoma, *Semin. Oncol.*, 31(6 Suppl. 13), 36–52, 2004.

52. Safra, T., Muggia, F., Jeffers, S., Tsao-Wei, D. D., Groshen, S., Lyass, O. et al., Pegylated liposomal doxorubicin (doxil): Reduced clinical cardiotoxicity in patients reaching or exceeding cumulative doses of 500 mg/m^2, *Ann. Oncol.*, 11(8), 1029–1033, 2000.

53. Berry, G., Billingham, M., Alderman, E., Richardson, P., Torti, F., Lum, B. et al., The use of cardiac biopsy to demonstrate reduced cardiotoxicity in AIDS Kaposi's sarcoma patients treated with pegylated liposomal doxorubicin, *Ann. Oncol.*, 9(7), 711–716, 1998.

54. Gabizon, A. A., Lyass, O., Berry, G. J., and Wildgust, M., Cardiac safety of pegylated liposomal doxorubicin (Doxil/Caelyx) demonstrated by endomyocardial biopsy in patients with advanced malignancies, *Cancer Invest.*, 22(5), 663–669, 2004.

55. O'Brien, M. E., Wigler, N., Inbar, M., Rosso, R., Grischke, E., Santoro, A. et al., Reduced cardio-toxicity and comparable efficacy in a phase III trial of pegylated liposomal doxorubicin HCl (CAELYX/Doxil) versus conventional doxorubicin for first-line treatment of metastatic breast cancer, *Ann. Oncol.*, 15(3), 440–449, 2004.

56. Kehrer, D. F., Bos, A. M., Verweij, J., Groen, H. J., Loos, W. J., Sparreboom, A. et al., Phase I and pharmacologic study of liposomal lurtotecan, NX 211: Urinary excretion predicts hematologic toxicity, *J. Clin. Oncol.*, 20(5), 1222–1231, 2002.

57. Gelmon, K. A., Tolcher, A., Diab, A. R., Bally, M. B., Embree, L., Hudon, N. et al., Phase I study of liposomal vincristine, *J. Clin. Oncol.*, 17(2), 697–705, 1999.

58. Soepenberg, O., Sparreboom, A., de Jonge, M. J., Planting, A. S., de Heus, G., Loos, W. J. et al., Real-time pharmacokinetics guiding clinical decisions; phase I study of a weekly schedule of liposome encapsulated paclitaxel in patients with solid tumours, *Eur. J. Cancer*, 40(5), 681–688, 2004.

59. Boulikas, T., Stathopoulos, G. P., Volakakis, N., and Vougiouka, M., Systemic lipoplatin infusion results in preferential tumor uptake in human studies, *Anticancer Res.*, 25(4), 3031–3039, 2005.

60. Gabizon, A., Tzemach, D., Horowitz, A. T., Shmeeda, H., Yeh, Y., and Zalipsky, S., Reduced toxicity and superior therapeutic activity of a Mitomycin C lipid-based prodrug incorporated in Pegylated liposomes. *Clin. Cancer Res.* 12(6), 1913–1920, 2006.

61. Bandak, S., Goren, D., Horowitz, A., Tzemach, D., and Gabizon, A., Pharmacological studies of cisplatin encapsulated in long-circulating liposomes in mouse tumor models, *Anticancer Drugs*, 10(10), 911–920, 1999.

62. Harrington, K. J., Lewanski, C. R., Northcote, A. D., Whittaker, J., Wellbank, H., Vile, R. G. et al., Phase I-II study of pegylated liposomal cisplatin (SPI-077) in patients with inoperable head and neck cancer, *Ann. Oncol.*, 12(4), 493–496, 2001.

63. Meerum Terwogt, J. M., Groenewegen, G., Pluim, D., Maliepaard, M., Tibben, M. M., Huisman, A. et al., Phase I and pharmacokinetic study of SPI-77, a liposomal encapsulated dosage form of cisplatin, *Cancer Chemother. Pharmacol.*, 49(3), 201–210, 2002.

64. Zamboni, W. C., Gervais, A. C., Egorin, M. J., Schellens, J. H., Zuhowski, E. G., Pluim, D. et al., Systemic and tumor disposition of platinum after administration of cisplatin or STEALTH liposomal-cisplatin formulations (SPI-077 and SPI-077 B103) in a preclinical tumor model of melanoma, *Cancer Chemother. Pharmacol.*, 53(4), 329–336, 2004.

65. Mori, A., Kennel, S. J., van Borssum Waalkes, M., Scherphof, G. L., and Huang, L., Characterization of organ-specific immunoliposomes for delivery of $3',5'$-O-dipalmitoyl-5-fluoro-$2'$-deoxyuridine in a mouse lung metastasis model, *Cancer Chemother. Pharmacol.*, 35, 447–456, 1995.

66. Asai, T., Kurohane, K., Shuto, S., Awano, H., Matsuda, A., Tsukada, H. et al., Antitumor activity of $5'$-O-dipalmitoylphosphatidyl $2'$-C-cyano-$2'$-deoxy-1-β-ᴅ-arabino-pentofuranosylcytosine is enhanced by long-circulating liposomalization, *Biol. Pharm. Bull.*, 21(7), 766–771, 1998.

67. Park, J. W., Hong, K., Kirpotin, D. B., Colbern, G., Shalaby, R., Baselga, J. et al., Anti-HER2 immunoliposomes: Enhanced efficacy attributable to targeted delivery, *Clin. Cancer Res.*, 8(4), 1172–1181, 2002.

68. Pastorino, F., Brignole, C., Marimpietri, D., Cilli, M., Gambini, C., Ribatti, D. et al., Vascular damage and anti-angiogenic effects of tumor vessel-targeted liposomal chemotherapy, *Cancer Res.*, 63(21), 7400–7409, 2003.

69. Allen, T. M., Sapra, P., and Moase, E., Use of the post-insertion method for the formation of ligand-coupled liposomes, *Cell Mol. Biol. Lett.*, 7(3), 889–894, 2002.

70. Gabizon, A., Horowitz, A. T., Goren, D., Tzemach, D., Shmeeda, H., and Zalipsky, S., In vivo fate of folate-targeted polyethylene-glycol liposomes in tumor-bearing mice, *Clin. Cancer Res.*, 9(17), 6551–6559, 2003.

71. Harris, L., Batist, G., Belt, R., Rovira, D., Navari, R., Azarnia, N. et al., Liposome-encapsulated doxorubicin compared with conventional doxorubicin in a randomized multicenter trial as first-line therapy of metastatic breast carcinoma, *Cancer*, 94(1), 25–36, 2002.

72. Bomgaars, L., Geyer, J. R., Franklin, J., Dahl, G., Park, J., Winick, N. J. et al., Phase I trial of intrathecal liposomal cytarabine in children with neoplastic meningitis, *J. Clin. Oncol.*, 22(19), 3916–3921, 2004.

73. Meyers, P. A., Schwartz, C. L., Krailo, M., Kleinerman, E. S., Betcher, D., Bernstein, M. L. et al., Osteosarcoma: A randomized, prospective trial of the addition of ifosfamide and/or muramyl tripeptide to cisplatin, doxorubicin, and high-dose methotrexate, *J. Clin. Oncol.*, 23(9), 2004–2011, 2005.

74. Hong, R. L. and Tseng, Y. L., Phase I and pharmacokinetic study of a stable, polyethylene-glycolated liposomal doxorubicin in patients with solid tumors: The relation between pharmacokinetic property and toxicity, *Cancer*, 91(9), 1826–1833, 2001.

75. O'Byrne, K. J., Thomas, A. L., Sharma, R. A., DeCatris, M., Shields, F., Beare, S. et al., A phase I dose-escalating study of DaunoXome, liposomal daunorubicin, in metastatic breast cancer, *Br. J. Cancer*, 87(1), 15–20, 2002.

76. Lu, C., Perez-Soler, R., Piperdi, B., Walsh, G. L., Swisher, S. G., Smythe, W. R. et al., Phase II study of a liposome-entrapped cisplatin analog (L-NDDP) administered intrapleurally and pathologic response rates in patients with malignant pleural mesothelioma, *J. Clin. Oncol.*, 23(15), 3495–3501, 2005.

77. Ozpolat, B., Lopez-Berestein, G., Adamson, P., Fu, C. J., and Williams, A. H., Pharmacokinetics of intravenously administered liposomal all-trans-retinoic acid (ATRA) and orally administered ATRA in healthy volunteers, *J. Pharm. Pharm. Sci.*, 6(2), 292–301, 2003.

78. Matsumura, Y., Gotoh, M., Muro, K., Yamada, Y., Shirao, K., Shimada, Y. et al., Phase I and pharmacokinetic study of MCC-465, a doxorubicin (DXR) encapsulated in PEG immunoliposome, in patients with metastatic stomach cancer, *Ann. Oncol.*, 15(3), 517–525, 2004.

79. Booser, D. J., Esteva, F. J., Rivera, E., Valero, V., Esparza-Guerra, L., Priebe, W. et al., Phase II study of liposomal annamycin in the treatment of doxorubicin-resistant breast cancer, *Cancer Chemother. Pharmacol.*, 50(1), 6–8, 2002.

80. Verschraegen, C. F., Gilbert, B. E., Loyer, E., Huaringa, A., Walsh, G., Newman, R. A. et al., Clinical evaluation of the delivery and safety of aerosolized liposomal 9-nitro-20(s)-camptothecin in patients with advanced pulmonary malignancies, *Clin. Cancer Res.*, 10(7), 2319–2326, 2004.

81. Butts, C., Murray, N., Maksymiuk, A., Goss, G., Marshall, E., Soulieres, D. et al., Randomized phase IIB trial of BLP25 liposome vaccine in stage IIIB and IV non-small-cell lung cancer, *J. Clin. Oncol.*, 23(27), 6674–6681, 2005.

82. Skubitz, K. M. and Anderson, P. M., Inhalational interleukin-2 liposomes for pulmonary metastases: A phase I clinical trial, *Anticancer Drugs*, 11(7), 555–563, 2000.

83. Rudin, C. M., Marshall, J. L., Huang, C. H., Kindler, H. L., Zhang, C., Kumar, D. et al., Delivery of a liposomal c-raf-1 antisense oligonucleotide by weekly bolus dosing in patients with advanced solid tumors: A phase I study, *Clin. Cancer Res.*, 10(21), 7244–7251, 2004.

84. Boulikas, T., Low toxicity and anticancer activity of a novel liposomal cisplatin (lipoplatin) in mouse xenografts, *Oncol. Rep.*, 12(1), 3–12, 2004.

85. Dark, G. G., Calvert, A. H., Grimshaw, R., Poole, C., Swenerton, K., Kaye, S. et al., Randomized trial of two intravenous schedules of the topoisomerase I inhibitor liposomal lurtotecan in women with relapsed epithelial ovarian cancer: A trial of the national cancer institute of Canada clinical trials group, *J. Clin. Oncol.*, 23(9), 1859–1866, 2005.

86. Yoo, G. H., Hung, M. C., Lopez-Berestein, G., LaFollette, S., Ensley, J. F., Carey, M. et al., Phase I trial of intratumoral liposome E1A gene therapy in patients with recurrent breast and head and neck cancer, *Clin. Cancer Res.*, 7(5), 1237–1245, 2001.

87. Colbern, G. T., Dykes, D. J., Engbers, C., Musterer, R., Hiller, A., Pegg, E. et al., Encapsulation of the topoisomerase I inhibitor GL147211C in PEGylated (STEALTH) liposomes: Pharmacokinetics and antitumor activity in HT29 colon tumor xenografts, *Clin. Cancer Res.*, 4(12), 3077–3082, 1998.

88. Beutel, G., Glen, H., Schoffski, P., Chick, J., Gill, S., Cassidy, J. et al., Phase I study of OSI-7904L, a novel liposomal thymidylate synthase inhibitor in patients with refractory solid tumors, *Clin. Cancer Res.*, 11(15), 5487–5495, 2005.

30 Positively-Charged Liposomes for Targeting Tumor Vasculature

Robert B. Campbell

CONTENTS

30.1 Introduction 613
30.2 The Microvascular Endothelium: Highly Accessible Target of Therapeutics 614
30.3 Arguments for Vascular Targeting Approach over Interstitial Drug Targeting 616
30.4 Influence of Cationic Charge on Vascular Targeting in Health and in Disease 617
30.5 Formulation Development: Alternatives to Routine Drug
     Design and Synthesis 617
30.6 Interaction of Cationic Liposomes with Human Endothelial Cells 618
30.7 Stealth (PEGylated) Liposome Technology: A Brief Communication 619
30.8 PEGylated Cationic Liposomes: Are They Suited for
     Targeting Tumor Vessels? 620
30.9 Cationic Liposomes Improve Anti-Cancer Therapy 621
30.10 Validation and Long-Term Tracking of Tumor Response to Therapy 623
30.11 Concluding Remarks 623
References 624

30.1 INTRODUCTION

In order to abolish the function of established vessels and/or suppress the formation of neovascular networks, various drug delivery vehicles have been used to target experimental therapeutics to the tumor vascular surface. Improving site-specific delivery, maximizing tumor response, minimizing unwanted side effects, and overcoming tumor heterogeneity are important considerations.

Clinical strategies are developed in response to unmet clinical needs, but often at the expense of understanding how to effectively exploit notable differences between the tumor and host tissues. For instance, tumor vessels are generally more tortuous, over-express a variety of active genes, and are more permeable to circulating macromolecules (compared to the vasculature of normal tissues). On the other hand, relatively high interstitial fluid pressure and long interstitial transport distances still hinder effective delivery and deep penetration of drugs in solid tumors.[1] For this reason, although the development of interstitial delivery methods represents the focal point of the majority of research efforts, alternative treatment options are being developed and evaluated in parallel.

The justification for parallel research efforts is due to several advantages of endothelium-specific drug targeting over interstitial drug delivery. First, vascular networks in developing tumors make tumors vulnerable to treatment regardless of where the tumor is located in vivo. Second, the endothelium is vulnerable to the effects of circulating agents; this approach is especially useful for treating tumors given the fact that a steady flow of oxygen and nutrients (from the blood to the interstitial compartment) is essential to sustain active angiogenesis and rapid tumor growth. Third, the vast majority of physiological barriers pose little, if any, threat to vascular targeting strategies, and finally, endothelial cells are non-mutagenic. This last advantage deals specifically with the inability of endothelial cells to acquire drug resistance compared to cancer cells that are inherently more prone to mutation and drug resistance. Nonetheless, despite all of these advantages, preservation of normal vascular function must be achieved so that treatment is both efficacious and safe.

The procedures for delivery of drugs to tumor vessels (not the interstitium) have yet to be optimized completely for all human organ systems. Whether the aim is to deliver bulky cytotoxic drugs or dynamic therapeutic genes, peptides, and proteins, minimizing unintentional injury to healthy organ tissues is critical. Until better quality treatments are made available, existing methods will have to be applied with a better understanding of their clinical limitations. Successful strategies need to satisfy a required set of criteria that include:

- The drug should not be toxic to patients;
- Repetitive administration of drug should be well tolerated;
- The drug should be reasonably accessible to the intended cell target;
- The drug should be undetectable by the host-immune system;
- Drug exposure to primary target(s) should yield useful biological product(s); and
- Biological products should be sufficient to elicit a desired clinical response.

Although these guidelines represent the gold standard for a number of routine clinical procedures and given that some human tumors meagerly respond to standard treatment options (yielding inadequate to mediocre responses at best), these routine procedures are simply not sufficient for treating all human cancers.

In general, the endothelium is considered a non-traditional target for FDA approved agents such as doxorubicin, vincristine, and paclitaxel; therefore, much of our understanding in connection with this approach is incomplete.

Cytotoxic agents are typically systemically administered at the maximum tolerated dose (MTD). The goal has long been to deliver *huge* therapeutic doses to eradicate millions of viable cancer cells hidden deep inside tissues. The drugs next pass through a complicated tumor interstitial matrix to prevent the metastatic process from reaching the point of no return. Today, clinical success has been achieved with the use of some interstitial drug targeting practices. The question currently asked is: Can the use of popular MTD strategies be replaced with an approach that focuses on selective delivery of drugs to the tumor vascular compartment rather than to the interstitial tumor compartment? Moreover, will this clinical practice result in a more effective clinical response that is both efficacious and safe? To address these question and related concerns, the structural and functional role of the endothelium should be examined; identification of accessible targets along the endothelium is a good start.

30.2 THE MICROVASCULAR ENDOTHELIUM: HIGHLY ACCESSIBLE TARGET OF THERAPEUTICS

The endothelium is a structural barrier separating the intravascular compartment from the tissue space. Because of wide variability in anatomical structure, the vascular compartment is further grouped into three different subcategories. The subcategories are continuous, fenestrated, or discontinuous endothelia.

Continuous endothelia are the most common type and are found most frequently in blood vessels lining chambers of the heart, walls of capillaries and arterioles in skeletal, skin, cardiac muscle, and connective tissue. They are well known for their relatively tight cellular junctions.[2,3] This particular category of vessels is also critical in the regulation and rapid exchange of ions and solutes.[2,3] Plasmalemmal vesicles involved in endothelial transport are abundant in myocardial endothelia but are far less frequently observed in capillaries of the brain.[4] The actual number of plasmalemmal vesicles existing along the continuous endothelium is heterogeneous, varying as a function of organ and tissue environment. These vesicular structures are also highly sensitive to charge characteristics, favoring associations with anionic over cationic molecules.[5]

Fenestrated vessels are normally found in vessels of organs that secrete (or excrete) biological fluids as in the gastrointestinal mucosa and in the glomerular capillaries of the kidneys. Fenestrae are usually between 50 and 80 nm in size, and they appear either as individual gap openings in the wall of functional vessels or as clusters. Similar to plasmalemmal vesicles, their frequency of occurrence along vessels depends on organ type and microenvironment.

Often, two or more capillaries may join to form post-capillary venules. These newly formed networks are composed of a single lining of endothelial cells with a basement membrane with no smooth muscle cell attachment. These vessels are heavily involved in the exchange of molecules, and they are preferential sites of plasma extravasation as a result of the actions of vasoactive and humoral factors. Fenestrated vessels possess a negatively charged surface density as a result of high heparin sulfate proteoglycan content, and unlike plasmalemmal vesicles of continuous endothelia, they favor interaction with cationic over anionic molecules.[6]

Discontinuous endothelia are found primarily in the liver, spleen, and bone marrow organs.[2] In the liver sinusoids, the endothelia are not continuous, and they possess an average fenestrae size between 100 and 150 nm in diameter with the size of fenestrate often changing in response to local mediators. These changes include, but are not limited to, response to luminal pressures and potent vasodilators such as histamine and bradykinin.[7-10] An investigation into the size of vascular pore openings in animal tumor models revealed gap openings that were significantly larger than those observed along vessels in normal tissues, around 4 μm (4000 nm) in at least one tumor type but normally falling within the range of 0.4–0.6 μm (400–600 nm) for many tumor types.[11,12] Nonetheless, the evidence is overwhelmingly in favor of the development of tumor-targeted delivery of therapeutic carrier molecules that are small enough to enter through tumor vascular pores without passing through openings in normal healthy organ tissues.

The endothelium is responsible for synthesizing a variety of molecules, regulating endothelial cell migration, proliferation, blood vessel maturation, and function. It has been shown to synthesize vascular growth factors, nitric oxide, collagen IV, laminin, glycosaminoglycans, and proteoglycans to highlight several proven functions.[10,13-19] The physical barrier organizes very rapidly to form monolayers, reassembles to form vascular tubes,[20] and can change specific inter- and intra-cellular signaling patterns to meet highly specialized needs of the host. Additionally, endogenous and exogenous mediators of immune and inflammatory response regulate specialized functions at the surface of the endothelium. The endothelium can be considered an effective mediator of organ homeostasis.[10,21]

In many ways, the vascular networks found in solid tumors poorly resemble the more regular, well-defined vascular structure observed in disease-free tissues. Tumors, for example, have a highly chaotic arrangement of vessels compared to vessels in normal tissues. Tumor vessels also have an over abundance of anionic phospholipids in addition to a number of other negatively charged functional groups.[22-25] In view of the negatively charged molecules, glycosaminoglycans carry out important functions in the metastatic disease process and, much like phospholipids, can serve as useful targets of cationic drug carrier molecules. Evaluation of altered proteoglycan expression in human breast tissue revealed a total proteoglycan content that was significantly increased in comparison to that in healthy tissues.[26] Proteoglycans isolated from malignant breast tissue have been shown to stimulate endothelial cell proliferation, and the total glycoprotein content in tumors is produced by many cell types, including cancer and tumor endothelial cells alike.

The vascular networks of tumors have an increased permeability for macromolecules and a higher proliferation rate of endothelial cells compared to vessels in quiescent tissues.[27,28] An estimated 30- to 40-fold increase in the growth rate of endothelial cells lining vessels in tumors compared to that in normal tissues has been demonstrated.[27]

Direct access to intravenously administered agents, rapid proliferation rate of endothelial cells, and over-expression of negatively charged functional groups along vessels are potentially exploitable features of tumors. Endothelium-specific cationic liposomes made more specific through conjugation of antibodies can offer tremendous therapeutic gains.

30.3 ARGUMENTS FOR VASCULAR TARGETING APPROACH OVER INTERSTITIAL DRUG TARGETING

Physiological barriers that prevent effective interstitial delivery and drug transport[29] will likely pose little threat to vascular targeting strategies given that blood vessels have better access to intravenously administered agents. Damage to only a few tumor vessels can result in the death of literally thousands of malignant cells.[30] Limited drug diffusion to hard-to-reach interstitial areas in tumors will ultimately spare clusters of well-differentiated cancer cells from treatment, but direct targeting of cancer cells is not as important for vascular targeting strategies. As a therapeutic approach, vascular targeting is unaffected by convection and diffusion, and it does not require the presence of highly perfused vessels in all tumor regions to be effective. It can be argued that irregular, slow blood flow velocities play a critical role in facilitating interaction of drugs with endothelial cells.[31] Additionally, poorly perfused tumor vessels permit limited access of drugs to cancer cells. Additionally, endothelial cells are non-malignant and are far less likely than cancer cells to acquire the mutations that safeguard them from treatment. Finally, should a drug intended to target functional tumor vessels gain direct access to a population of cancer cells in the process, additional gains for therapy could result. The potential for vascular targeting substances to come in direct contact with cancer cells can be observed in mosaic vessels where tumor cells directly interact with the lumen of vessels and flowing blood and assist in the spread of metastatic disease.[32] This is apparent with large tumor vascular pore openings where drug agents have considerably greater access to malignant cell populations.[11]

Proof-of-principal: A significant amount of experimental evidence can now be used to support vascular targeting as a targeted approach against cancer. To illustrate, tumor-bearing mice treated with doxorubicin–RGD-4C conjugate (tumor-homing peptide), outlived control mice that died from widespread disease.[33] Mice treated every 3 weeks for a total of 12 weeks with doxorubicin-RGD-4C lived 6 months longer compared to doxorubicin-treated mice.[33] An in vitro study showed success with immunoliposome-targeted delivery to ICAM-1-expressing cells in the bronchial epithelium and endothelium.[34] In this study, the extent of ICAM-1 expression was suggested to influence the degree of immunoliposome binding to endothelial cells. The immunotargeting of liposomes to activated vascular endothelial cells of the cardio-vascular system has also been investigated.[35] In this study, murine antibody mAB H18/7 was conjugated to doxorubicin-loaded liposomes, and the extracellular binding domain of E-selectin was targeted to a significant extent over nonactivated human endothelial cells.[35] Thorpe and colleagues characterized TEC-11 antibody and have demonstrated that the anti-class-ricin A-chain immunotoxin produced widespread infarction and tumor regression in treated mice compared to untreated controls.[36] While these groups and others work to demonstrate proof-of-principal, other laboratories have had success with identifying accessible tumor vessel targets. One example of a potentially useful clinical target was studied by Retting and colleagues. They confirmed that the tumor endothelial antigen endosialin (cell glycoprotein) is expressed in tumors and is not expressed in normal healthy tissues.[37] This early finding suggests that, under optimized conditions, the glycoprotein could be used to launch a site-specific, immunological attack on tumor vessels. Multiple lines of evidence from the literature support vascular targeting as an attractive therapeutic approach against cancer disease.

30.4 INFLUENCE OF CATIONIC CHARGE ON VASCULAR TARGETING IN HEALTH AND IN DISEASE

To optimally target drugs to the endothelium, a basic working knowledge of why select carriers accumulate at this site is essential. Extraordinary insights into normal vessel function and pathogenesis now provide some interesting clues. For example, the vessel lumen of capillaries is predominately lined with anionic sites.[22–25] These regions are important in the development and maintenance of normal and abnormal tissues where the latter is observed in inflammation, thrombosis, and in tumors.[24,25,38] Anionic sites (owing to layers of glycoproteins) mediate microvascular permeability and trans-endothelial cell function.[3,6,23–25,30,38] Endothelial cell membrane-associated proteoglycans regulate delivery and uptake of macromolecules. This has been demonstrated in sulfated proteoglycan deficient HeLa cells where cation-mediated delivery of plasmid DNA resulted in a 69% reduction in luciferase expression.[39] The distribution of negatively charged sites is non-uniform,[25,31] varying among vessel types (i.e., arteries, capillaries, or capillary venules) and between young and relatively old mice.[4] Although negatively charged components of mammalian cells are sialoglycoproteins and proteoglycans, some vascular domains (potentially involved in endocytosis and/or trancytosis) are devoid of these functional groups. These specialized areas interact to a lesser extent with cationized molecules, and it has been suggested that they represent regions of anionic molecular transport across capillary networks.[25] The absence of accessible anionic sites can be observed on the vascular surface, specifically on the surface of plasmalemmal vesicles where they readily assist in the transport of circulating macromolecules.[4,6]

Experiments in solid tumors support the notion that success with nanoparticle-assisted delivery of drugs to tumor vessels does not require that drug carrier molecules possess a cationic charge for effective delivery and transport. In fact, when anionic sites were neutralized with the polycation protamine, the blood–brain barrier (BBB) became significantly less guarded against the transport of albumin and inulin; under normal conditions, neither can pass the BBB.[40] Additionally, whereas electron microscopic studies revealed that positively charged electron-dense label cationic ferritin (CF) did not label plasmalemmal vesicles of the mouse endothelium, native anionic and neutralized ferritin molecules had considerably greater access to these sites.[4] Another study showed that following a cerebral insult of a rat brain tumor model, the vasculature was more permeable to horse radish peroxidase, but more importantly, showed a fewer number of negatively charged sites. The mechanism(s) by which anionic surface charge is altered in disease is not completely understood,[40–42] but studies continue to support structural and morphological differentiation of the microvascular endothelium in disease tissues that undoubtedly influences overall vascular targeting efficiency.

The endothelium is structurally and functionally unique, and several studies confirm region-specific, tumor vascular sensitivity to effects of molecular charge.[3,6,18,25,40,43,44] The fact that the endothelium can bind anionic, neutral, and net positively charged molecules supports these observations. The strategies that best exploit the expression of negatively charged functional groups in combination with additional tumor vascular features will achieve the greatest therapeutic success.

30.5 FORMULATION DEVELOPMENT: ALTERNATIVES TO ROUTINE DRUG DESIGN AND SYNTHESIS

The development of novel therapeutics is an expensive and labor-intensive process. There is also no guarantee that a more useful therapeutic agent will result from the time and money invested. The use of combinatorial chemistry, high throughput screening, and drug synthesis has not met all clinical expectations. As a result, serious issues still remain with existing treatment approaches and with a search for a magic bullet.[45] For this reason, strategies that improve tumor endothelial site-specific delivery are being developed in parallel with efforts involving routine drug design and synthesis.

In many instances, the use of nanotechnology represents a wise alternative to routine drug synthesis. This technology creates new opportunities to improve drug access to primary targets and to exploit additional tumor targets as well. Long circulating liposomes, for example, increase deposition of drugs in the perivascular space of tumors.[46,47] When the optimal lipid ratios have been considered, the vasculature may be targeted more selectively compared to free drug.

On the basis of the experimental evidence, a non-toxic phospholipid-based delivery vehicle is a rational choice for targeting many therapeutic agents to tumors. Liposomes are well suited for this purpose. These microparticulates spontaneously form and are generally easier to prepare compared to viral mediated systems.

Liposomes are relatively non-immunogenic; their large-scale manufacturing does not depend on the culturing of living cells, and they are preferred for a variety of practical considerations as well. Liposomes have been shown to increase systemic circulation, reduce drug side effects, enhance cellular uptake, and increase the duration of drug exposure and tumor response to therapy.[16–18] Moreover, the physicochemical properties of liposomes can be manipulated to address physiological considerations such as the effects of size, molecular charge, membrane rigidity, and stability.[48–53]

Developing a target specific, long circulating, stable drug carrier is only half the battle. Ideally, liposomes should also be triggered to release drugs under tumor sensitive conditions. The effect of molecular charge, over-expression of vascular growth factors, sensitivity to pH, and relatively large vascular pore openings are a few physiological features currently used to distinguish tumors from healthy organ tissues.[12,25,31,38,46,53–57]

30.6 INTERACTION OF CATIONIC LIPOSOMES WITH HUMAN ENDOTHELIAL CELLS

One of the first synthetic cationic lipids, developed by Felgner and colleagues in 1987,[58,59] was initially used to improve existing DNA delivery methods. A 1:1 w/w mixture of the new cationic lipid DOTMA (N-[1-(2,3-dioleyloxy) propyl]-N,N,N-trimethylammonium chloride) was combined with the helper lipid DOPE (dioleoylphosphatidylethanolamine). This formulation (aka ∼ * Lipofectin) was found to be 5–100 times more effective than DNA transfection performed with other in vitro methods.[59] A typical eukaryotic cell membrane has a highly negative surface charge density because of the multitude of membrane-bound carbohydrate groups oriented on its surface. Endothelial cells can be targeted on the basis of this surface charge density, and several mechanisms of action have already been proposed for successful delivery of DNA into cells by positively charged liposomes.[60–63]

Studies involving the transfection of human endothelial cells suggest that methods be tailored to specifications of endothelial cells, not more traditional cell types. High transfection efficiency with the use of cationic liposomes is possible when expression is under the control of a strong promoter.[64] Endothelial cells, like most mammalian cells, depend on an optimal DNA/cationic liposome ratio for high efficiency of endothelial cell trafficking and expression of foreign genes. Moreover, compared to other mammalian cell lines, primary endothelial cells are generally more sensitive to constituents of growth medium and passage number.[65] No specific cationic lipid, to date, is accepted for use with all mammalian cell types. Cationic liposomes may represent a suitable alternative to non-viral delivery systems; however, transfection of human endothelial cells by this approach should be optimized to produce the desired therapeutic effect.[64]

Various experimental and clinical applications for cationic lipids have been reported, ranging from optimizing methods for enhanced gene transfer in vitro to treating clinical abnormalities contributing to neoplastic transformations in man.[66] Using cationic lipids to regulate endothelial cell function in vitro or in treatment of disease in highly developed organ systems has been explored. To illustrate, the application of several cell line types have been investigated

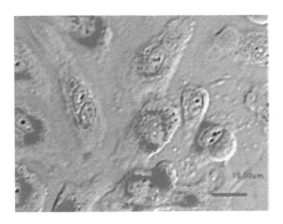

FIGURE 30.1 (See color insert following page 522.) Cationic liposomes are taken up by human endothelial cells. The figure shows a DIC and fluorescence merged image of PCLs associated with human dermal endothelial cells. Perinuclear localization is observed. Captured image is representative of PCLs taken up by other primary endothelial cells derived from various organ tissues. Cells were seeded at 5×10^5 cells/well on cover slips in six well plates. Images were captured 24 h after exposure of cells to PCLs.

(Figure 30.1). Cationic liposome-mediated transfection has been carried out under a broad set of experimental conditions, involving the use of numerous in vitro models of the endothelium in human organ tissues. These in vitro models of the intravascular compartment include the eye,[67] lung,[68,69] heart,[61,70] kidney,[71] and vascular smooth muscle.[72,73]

Site-specific delivery to endothelial cells can be accomplished by coating cationic lipsoemes with peptides and monoclonal antibodies.[65,74] Furthermore, these positively charged vehicles enhanced adeno-associated virus (AAV) uptake by human endothelial and smooth muscle cells. The authors report a highly efficient way to introduce foreign DNA into cells given that modified cationic liposomes (adenosomes) enhanced vector expression by a factor of 10 over adenoviral proteins delivered without liposomes.[75]

Although antibodies have been used to assist gene transfer strategies, cationic liposomes can be optimized to deliver genes to endothelial cells without the use of antibodies. For instance, when cultured bovine corneal endothelial cells (BCECs) were exposed to pUT651 (a plasmid) alone, significant expression was observed and without associated toxicities.[67] In HUVEC cells, reporter gene expression was evident in the absence of select antibodies. In this study, transfection efficiency varied as a function of cell passage number, and a linear reduction in luciferase expression was observed with successive passage numbers.[65] Vascular smooth muscle cells were shown to vary in degree of sensitivity to transfection reagents mediated by cationic lipids, reporting a preference for restenotic lesions compared to normal internal thoracic arteries and atherosclerotic plaques.[73]

30.7 STEALTH (PEGYLATED) LIPOSOME TECHNOLOGY: A BRIEF COMMUNICATION

Rapid elimination of cationic liposomes from circulation by opsonins, lipoproteins, and cell-surface receptors pose an additional threat to effective delivery of drugs to solid tumors.[45,76–80] This particular barrier is different from physiological barriers as it results not from tumors or the influences from the surrounding tissue environment, but it occurs when conventional (relatively short circulating) liposomal therapeutics come in direct contact with circulating proteins in blood.

Stealth liposome technology is one approach used to limit the interaction of conventional liposomes with proteins in blood. The technology involves the coupling of high molecular weight polymers to the liposome surface.[47,50] Surface-bound polymers [i.e., poly(ethylene glycol) (PEG)] provide adequate protection of liposomes from protein molecules that can unfavorably alter their tissue distribution profile. The specific mechanisms underlying delivery mediated by these polymers are unclear, but when PEG is included as a component of liposome preparations, a physical zone is formed around the liposome perimeter. This zone of exclusion reduces liposome–protein interactions as a result of long (PEG) polymer chain constrictions.[47,50,80]

From a therapeutic point-of-view, in addition to favorable kinetics, PEGylated liposomes are usually more active against solid tumors.[47,79] Several lines of evidence support the findings that PEGylated liposomes are generally more stable in blood with only minor leakage of their contents over time compared to conventional liposomal therapeutics.[78]

PEG can also be used to prepare liposomes with significantly higher drug retention under normal physiological pH, or it can be designed to trigger drug release in a low pH environment. In addition to tumor uptake, other organs of relatively high accumulation are the liver, kidney, skin, the female reproductive, and gastrointestinal tract.[77,78]

The extent of modification and molecular mechanics of Stealth liposome technology and exactly how PEG delays reticuloendothelial system (RES) entrapment is not completely understood. However, its benefit(s) in systemic circulation and that the fraction of PEG included in Stealth liposome preparations should reflect preclinical and clinical goals are understood.

Stealth technology might well represent one of the most significant advances in the field of liposome nanotechnology. PEG has provided researchers with an enormous opportunity to exploit the large openings of tumor vessels by enhancing delivery of therapeutic substances through them as a result of the advantage of prolong circulation. In a study performed by Yuan et al., Stealth liposomes accumulated in the perivascular space in tumors 48 h post-administration.[46] This is the region close to tumor vessels but not directly associated with them. The technology is, however, not sufficient to promote vascular-specific delivery; targeting specific physiological feature of vessels (i.e., proteoglycans, endothelial receptors, etc.) in addition to prolonged circulation is also required.

30.8 PEGYLATED CATIONIC LIPOSOMES: ARE THEY SUITED FOR TARGETING TUMOR VESSELS?

The notion of targeting accessible anionic sites along tumor vessels combined with technologies that limit interactions of nanodelivery systems with biological molecules in vivo is promising. PEGylated cationic liposomes (PCLs) have been shown to improve not only oligonucleotide (ODN) loading and delivery, but also the apparent drug solubility profile of various therapeutic molecules as well. In vitro studies conducted with human plasma show that PCLs do not form large molecular aggregates in blood[81,82] or in buffered solutions.[31] These carriers retain most of their physicochemical properties,[31,75,81] and they prolong circulation half-life even in the presence of a net positively charged surface.

Protecting cationic liposomes from serum nucleases enhanced immunostimulatory activity of CpG oligonucleotides motifs compared to free ODN in in vitro and in mouse models.[82] In another study, the anti-HER-2 F (ab′) fragment was coupled to the distal end of PEG chains of cationic liposomes; cellular uptake and nuclear localization increased considerably in cultured SK-BR-3 cells (a human breast cancer cell line) shown to over-expresses HER-2 oncoprotein.[81]

Intravital microscopy has been used to investigate the interactions of PCLs with tumor vessels in vivo.[31] When PCLs were intravenously injected into tumor bearing (LS174T-human colon adenocarcinoma) mice, analysis of spatial distribution of liposomes 24 h post-injection revealed significant vascular targeting in tumors ($\sim$27.5% coverage) compared to vessels in normal tissues ($\sim$3.6% coverage).[31] Moreover, the extent of vascular areas targeted did not vary as a function of tumor type or anatomical location.[31] Several lines of evidence suggest that PEG coating does not inhibit cationic liposomes from binding to negatively charged surfaces. In addition to experimental data supporting this hypothesis, the values of zeta potential (electric potential across a double membrane surface) confirm that a sufficient electrostatic membrane potential still exists in the presence of PEG.[31,52,83]

Vascular targeting can be influenced on the basis of charge content of drug carrier molecules; associations with tumor vessels favor cationic liposomes possessing relatively high cationic charge content.[31,38,63] The abilities to circulate for extended periods in blood and associate with tumor vessels are highly desirable features of an efficient therapeutic approach. For this reason, both the inclusion of PEG and cationic charge content in cationic liposomes are currently being investigated in combination. There are clinical circumstances when relatively long circulating drug carrier molecules (>24 h) accumulating along tumor vessels would be more effective than carriers that are rapidly eliminated from circulation (<0.5 h). For instance, many of the popular chemotherapeutic agents are dependent on phase of the cell cycle; therefore, rapidly dividing tumor endothelial cells could be targeted more effectively with carriers that maintained adequate levels of drug in blood. Such a delivery system would serve less value for agents that can exert drug effects outside the influence of the cell cycle (i.e., membrane targeting agents).

30.9 CATIONIC LIPOSOMES IMPROVE ANTI-CANCER THERAPY

The use of cationic liposomes to deliver experimental therapeutics to solid tumors is a promising alternative to interstitial targeting approaches. However, formulations should be optimized and pre-screened in preclinical models to establish clinical relevance. Given the fact that the architecture of tumor vessels varies significantly in terms of microvascular type, density, vascular surface area, and angiogenic potential, it is understandable why the development of a single therapeutic approach against cancer has yet to be achieved.

When developing cationic liposomal therapeutics, it will be helpful to note that each solid tumor possesses distinctly different organization and specialized structural features. The highly unpredictable nature of tumors further complicates the task of pairing formulations to specific disease states. A common feature among all cationic liposome preparations used in drug delivery is that they must be optimized for site-specific delivery to tumor endothelia and that their intracellular fate should correlate with the drug's mechanism of action (i.e., cationic liposomes should enhance delivery of DNA targeting agents to DNA, and not to irrelevant intracellular targets). This process must be confirmed.

Cationic liposomes can retain drug agents at the tumor vascular site (Figure 30.2), and facilitate interaction of liposomes with subcellular targets prior to releasing their payload. They can also be used to target non-intracellular targets as well as cell-membrane bound molecules other than proteoglycans. This is promising given that many anti-angiogenic agents have been confirmed to exert their action by each mechanism. For instance, SU5416, an inhibitor of tyrosine kinase activity of vascular endothelial growth factor (VEGF), requires direct access to a specific endothelial cell membrane-associated receptor (Flk-1/KDR) in order to suppress neovascular growth of tumors.[84,85] Although SU5416 is suggested to exert long-lasting effects on VEGF phosphorylation and function, cationic liposome-assisted drug delivery could enhance interactions with specific endothelial cell targets.

Effective anti-angiogenic therapy requires the continuous presence of drugs in circulation.[85,86] The inclusion of PEG in cationic liposome preparations can extend circulation half-life of SU5416 compared to SU5416 alone; the goal here is to enhance the duration of drug (SU5416) exposure with tumor target.

Most anti-angiogenic agents suppress neovascularization by blocking activators of angiogenesis and endothelial cell-specific signals. There exist clinical situations where targeting non-tumoral tissues might be equally beneficial. Although the majority of PCLs are recovered by the

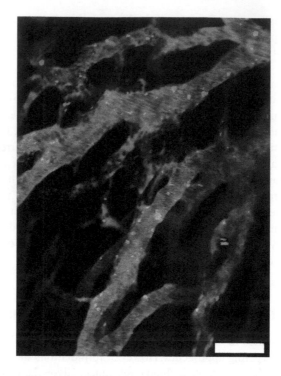

FIGURE 30.2 (See color insert following page 522.) Targeting tumor vessels with cationic liposomes. The image shows vascular targeting of LS174T (human colon adenocarcinoma) with cationic liposomes and was captured by two-photon microscopy 250 μm deep within the tumor. The intrinsic fluorescent property of doxorubicin was used to probe the location of the drug within the tumor. When doxorubicin was loaded in PCLs and administered via tail vein in tumor bearing mice, the drug was shown to be delivered to tumor vessels. Green represents tumor vessels, and orange dots represent doxorubicin. Mag. Bar = 50 μm.

liver post-intraveneous[31] and intrasplenic[87] administration, once these carrier molecules have been loaded with chemotherapeutic agents, they can diminish primary tumor growth in this anatomical location without significantly altering its primary function. To illustrate, poly(I):poly(C)–cationic liposome complex was used to treat metastatic (pancreatic and colorectal) carcinoma cell growth in the liver, and the result was significant delay in tumor growth compared to the administration of poly(I):poly(C), and adriamycin alone.[87]

In a study using PCLs to deliver the chemotherapeutic agent cisplatin (CDDP), CDDP-loaded PCLs suppressed primary tumor growth and distant liver metastasis as a result of tumor and liver-selective uptake.[88] The authors note that the high chondroitin sulfate content of cancer cells was the primary target of PCLs. In addition to this mechanism, another possibility is that drug loaded-PCLs deliver drugs like CDDP to tumor vessels and that cancer cells are not the primary, but the secondary, target of drug-loaded PCLs. Depending on the overall clinical objectives, liver accumulation could result in some unwanted complications; for wide-spread general applicability, high tumor selectivity at the expense of the liver is desired. To minimize uptake of cationic carriers at this anatomical site, additional specialized features of tumor vessels may have to be taken into account.

In addition to targeting tumor vascular targets on the basis of negatively charged functional groups, cationic drug carriers can also be used to enhance the interaction of anti-angiogenic agents with integrins. Integrins are involved in cell invasion and metastasis and in signaling processes that regulate the actions of each.[89,90] Some integrins are preferentially expressed in angiogenic endothelium like integrin $\alpha v\beta 3$.[89] Researchers have since introduced a novel cationic delivery vehicle in order to exploit this endothelial cell receptor.[91] In as study carried out by Hood and colleagues, a cationic lipid based nanoparticle (NP) was coupled to αvβ3-targeting ligand and intravenously injected in tumor bearing mice; the complex (αvβ3–NP) selectively targeted the vasculature and induced tumor regression.[91] Furthermore, sustained regression was achieved without significant accumulation in, and damage to, the liver. This finding is encouraging. It supports the use of positively charged drug carriers with highly specialized molecules to improve vascular targeting at all stages of tumor development.

The knowledge that cationic liposomes deliver drugs to solid tumors as well as the degree to which each tumor compartment is affected by treatment is essential to future advances in nanotechnology. Unfortunately, the overwhelming majority of published works document overall tumor response, animal survival, and drug effects involving the interstitial matrix, yet hidden mechanism(s) underlying tumor vascular damage in connection with treatment is often not a part of routine investigations.

No intravenously administered substances (drug alone or nanoparticle-assisted) can gain access to the interstitial matrix without first passing through the vascular compartment, so drug effects to this tumor compartment must be explored. Cationic liposomes possessing a high cationic lipid content (~ 50 mol%) target the endothelium several-fold over the interstitial environment.[31,38] Growth inhibition and animal survival studies have proven to be extremely helpful in establishing safety conditions for drug administration and predicting drug efficacy in humans. Nonetheless, the specific mechanisms by which cationic liposomes target vessels and alter tumor growth kinetics must be investigated. Studies should probe interstitial and vascular compartments without experimental bias.

30.10 VALIDATION AND LONG-TERM TRACKING OF TUMOR RESPONSE TO THERAPY

It has been established that there are unique molecular and physiological features of tumors that can be used to develop relatively disease-specific treatments. The expression of tumor-specific genes has been used to monitor progression of disease in response to these treatments and others. For instance, VEGF is a potent angiogenic cytokine that is critically linked to angiogenesis; it is so vital to this process that inhibitors are rapidly being developed to suppress VEGF's tumor angiogenesis stimulatory activity.[84,92–94] Normally, when select inhibitors of VEGF bind Flt-1 (VEGF target site) on endothelial cells, proliferation is inhibited among other critical steps involved in tumor angiogenesis.[20,30,84] VEGF expression is a highly regulated process in healthy tissue; elevated expression is observed only under special circumstances, but constitutive expression is observed in tumors that aggressively seek to recruit a new blood supply. Given the fact that the process of developing new blood vessels is fundamentally important to maintaining an aggressive tumor phenotype, VEGF and a variety of other growth factors are often used as molecular and pharmacological endpoints.

Site-specific delivery to tumor vessels and angiogenesis has demonstrated promise in preclinical models, warranting follow-up investigations in clinical settings. Evaluating tumor volume, animal survival and vascular density are useful indicators of tumor response. However, monitoring gene expression in response to therapy can provide deeper insights into the specific mechanisms underlying the disease process.

30.11 CONCLUDING REMARKS

Systemic administration of drugs is a frequently applied strategy used to control local tumor growth and metastasis. The issue of targeting these agents more specifically to tumors while reducing uptake by healthy organ tissues is a formidable challenge. Cationic liposomes have been used to mediate efficient transfection of a variety of different mammalian cell types. The loading of cationic liposomes with chemotherapeutic agents has even greater potential. The relatively new class of liposomes possesses high cationic charge content (and membrane surface charge potential) similar to those used for transfection studies, but they deliver their drug payload to tumor vessels. The development of new synthetic cationic lipids like N-[1-(2,3-dioleyloxy)propyl]-N, N, N-trimethyl-ammonium chloride(DOTMA) and N-[1-(2,3-dioleyloxy)propyl]-N, N, N-trimethylammonium chloride(DOTAP) has helped make this field a reality.

A deeper understanding of how tumors develop relative to the expression of vascular receptors and other targets will create new opportunities for nanotechnologists. These opportunities extend to include more effective treatments at various stages of tumor development and planned attacks against cancer.

REFERENCES

1. Jain, R. K., Vascular and interstitial barriers to delivery of therapeutic agents in tumors, *Cancer Metastasis Rev.*, 9, 253–266, 1990.
2. Simionescu, M., Simionescu, N., and Palade, G. E., Morphometric data on the endothelium of blood capillaries, *J. Cell Biol.*, 60, 128–152, 1974.
3. Simionescu, M., Simionescu, N., and Palade, G. E., Structural basis of permeability in sequential segments of the microvasculature. I. Bipolar microvascular fields in the diaphragm, *Microvasc. Res.*, 15, 1–16, 1978.
4. Simionescu, M., Simionescu, N., Santoro, F., and Palade, G. E., Differentiated microdomains of the luminal plasmalemma of murine muscle capillaries: Segmental variations in young and old animals, *J. Cell Biol.*, 100, 1396–1407, 1985.
5. Ghinea, N. and Simionescu, N., Anionized and cationized hemeundecapeptides as probes for cell surface charge and permeability studies: Differentiated labeling of endothelial plasmalemmal vesicles, *J. Cell Biol.*, 100, 606–612, 1985.
6. Simionescu, M., Simionescu, N., and Palade, G. E., Differentiated microdomains on the luminal surface of the capillary endothelium. I. partial characterization of their anionic sites, *J. Cell Biol.*, 90, 614–621, 1981.
7. Movat, H. Z., The role of histamine and other mediators in microvascular changes in acute inflammation, *Can. J. Physiol. Pharmacol.*, 65, 451–457, 1987.
8. Mylecharane, E. J., Mechanisms involved in serotonin-induced vasodilation, *Blood Vessels*, 27, 116–126, 1990.
9. Vercellotti, G. M. and Tolins, J. P., Endothelial activation and the kidney: Vasomediator modulation and antioxidant strategies, *Am. J. Kidney Dis.*, 21, 331–343, 1993.
10. van Hinsbergh, V. W., The endothelium: Vascular control of haemostasis, *Atherosclerosis*, 95, 198–201, 2001.
11. Hobbs, S. K. et al., Regulation of transport pathways in tumor vessels: Role of tumor type and microenvironment, *Proc. Natl Acad. Sci. USA*, 95, 4607–4612, 1998.
12. Yuan, F. et al., Vascular permeability in a human tumor xenograft: Molecular size dependence and cutoff size, *Cancer Res.*, 55, 3752–3756, 1995.
13. Hudson, B. G., Reeders, S. T., and Trygvason, K., Type IV collagen: Structure, gene organization and role in human diseases, *J. Biol. Chem.*, 268, 26033–26036, 1993.
14. Gorog, P. and Pearson, J. D., Sialic acid moieties on surface glycoproteins protect endothelial cells from proteolytic damage, *J. Pathol.*, 146, 205–212, 1985.
15. Hallmann, R. et al., Expression of function of laminins in the embryonic and mature vasculature, *Physiol. Rev.*, 85, 979–1000, 2005.
16. Sund, M., Xie, L., and Kalluri, R., The contribution of vascular basement membranes and extracellular matrix to the mechanics of tumor angiogenesis, *APMIS*, 112, 450–462, 2004.
17. Ruoslahti, E., Structure and biology of proteoglycans, *Annu. Rev. Cell Biol.*, 4, 229–255, 1988.
18. Ruoslahti, E., Specialization of tumor vasculature, *Nat. Rev. Cancer*, 2, 83–90, 2002.

19. Satoh, A., Toida, T., Yoshida, K., Kojima, K., and Matsumoto, I., New role of glycosaminoglycans on the plasma membrane proposed by their interaction with phosphatidylcholine, *FEBS Lett.*, 477, 249–252, 2000.

20. Folkman, J. and Haudenschild, C., Angiogenesis by capillary endothelial cells in culture, *Trans. Ophthalmol. Soc. UK*, 100, 346–353, 1980.

21. Michiels, C., Arnould, T., and Remacle, J., Endothelial cell responses to hypoxia: Initiation of a cascade of cellular interactions, *Biochim. Biophys. Acta*, 1497, 1–10, 2000.

22. Charonis, A. S. and Wissig, S. L., Anionic sites in basement membranes. Differences in their electro-static properties in continuous and fenestrated capillaries, *Microvasc. Res.*, 25, 265–285, 1983.

23. Roberts, W. G. and Palade, G. E., Neovasculature induced by vascular endothelial growth factor is fenestrated, *Cancer Res.*, 57, 765–772, 1997.

24. Ran, S., Downes, A., and Thorpe, P. E., Increased exposure of anionic phospholipids on the surface of tumor blood vessels, *Cancer Res.*, 62, 6132–6140, 2002.

25. Vincent, S., DePace, D., and Finkelstein, S., Distribution of anionic sites on the capillary endothelium in an experimental brain tumor model, *Microcirc. Endothleium Lymphatics*, 45–67, 1988.

26. Vijayagopal, P., Figueoroa, J. E., and Levine, E. A., Altered composition and increased endothelial cell proliferative activity of proteoglycans isolated from breast carcinoma, *J. Surg. Oncol.*, 68, 250–254, 1998.

27. Denekamp, J. and Hobson, B., Endothelial-cell proliferation in experimental tumours, *Br. J. Cancer*, 46, 711–720, 1982.

28. Denekamp, J., Vasculature as a target for tumor therapy, *Prog. Appl. Microcirc.*, 4, 28–38, 1984.

29. Jain, R. K., The next frontier of molecular medicine: Delivery of therapeutics, *Nat. Med.*, 4, 655–657, 1998.

30. Folkman, J., Tumor angiogenesis: Therapeutic implications, *N. Engl. J. Med.*, 285, 1182–1186, 1971.

31. Campbell, R. B. et al., Cationic charge determines the distribution of liposomes between the vascular and extravascular compartments of tumors, *Cancer Res.*, 62, 6831–6836, 2002.

32. Chang, Y. S. et al., Mosaic blood vessels in tumors: Frequency of cancer cells in contact with flowing blood, *Proc. Natl Acad. Sci. USA*, 97, 14608–14613, 2000.

33. Arap, W., Pasqualini, R., and Ruoslahti, E., Cancer treatment by targeted drug delivery to tumor vasculature in a mouse model, *Science*, 279, 377–380, 1998.

34. Bloemen, P. G. et al., Adhesion molecules: A new target for immunoliposome-mediated drug delivery, *FEBS Lett.*, 357, 140–144, 1995.

35. Spragg, D. D. et al., Immunotargeting of liposomes to activated vascular endothelial cells: A strategy for site-specific delivery in the cardiovascular system, *Proc. Natl Acad. Sci. USA*, 94, 8795–8800, 1997.

36. Thorpe, P. E. and Burrows, F. J., Antibody-directed targeting of the vasculature of solid tumors, *Breast Cancer Res. Treat.*, 36, 237–251, 1995.

37. Rettig, W. J. et al., Identification of endosialin, a cancer surface glycoprotein of vascular endothelial cells in human cancer, *Proc. Natl Acad. Sci. USA*, 89, 10832–10836, 1992.

38. Thurston, G. et al., Cationic liposomes target angiogenic endothelial cells in tumors and chronic inflammation in mice, *J. Clin. Investig.*, 101, 1401–1413, 1998.

39. Mislick, K. A. and Baldeschwieler, J. D., Evidence for the role of proteoglycans in cation-mediated gene transfer, *Proc. Natl Acad. Sci. USA*, 93, 12349–12354, 1996.

40. Hardebo, J. E. and Kahrstrom, J., Endothelial negative surface charge areas and blood–brain barrier function, *Acta Physiol. Scand.*, 125, 495–499, 1985.

41. Nag, S., Role of the endothelial cytoskeleton in blood–brain-barrier permeability to proteins, *Acta Neuropathol. (Berl.)*, 90, 454–460, 1995.

42. Nag, S., Blood–brain barrier permeability using tracers and immunohistochemistry, *Methods Mol. Med.*, 89, 133–144, 2003.

43. Renkin, E. M., Transport of macromolecules: Pores and other endothelial pathways, *Appl. Physiol.*, 58, 315–325, 1985.

44. Renkin, E. M., Multiple pathways of capillary permeability, *Circ. Res.*, 41, 735–743, 1977.

45. Gabizon, A. A., Stealth liposomes and tumor targeting: One step further in the quest for the magic bullet, *Clin. Cancer Res.*, 7, 223–225, 2001.

46. Yuan, F. et al., Microvascular permeability and interstitial penetration of sterically stabilized (Stealth) liposomes in a human tumor xenograft, *Cancer Res.*, 54, 3352–3356, 1994.

47. Papahadjopoulos, D. et al., Sterically stabilized liposomes: Improvements in pharmacokinetics and anti-tumor efficacy, *Proc. Natl Acad. Sci. USA*, 88, 11460–11464, 1991.

48. Campbell, R. B., Balasubramanian, S. V., and Straubinger, R. M., Physical properties of phospho-lipid–cationic lipid interactions: Influences on domain structure, liposome size and cellular uptake, *Biochim. Biophys. Acta*, 1512, 27–39, 2001.

49. Campbell, R. B., Balasubramanian, S. V., and Straubinger, R. M., Influence of cationic lipids on the stability and membrane properties of paclitaxel-containing liposomes, *J. Pharm. Sci.*, 90, 1091–1105, 2001.

50. Torchilin, V. P., Polymer-coated long circulating microparticulate pharmaceuticals, *J. Microencapsul.*, 15, 1–19, 1998.

51. Szoka, F. and Papahadjopoulos, D., Comparative properties and methods of preparation of lipid vesicles (liposomes), *Annu. Rev. Biophys. Bioeng.*, 9, 467–508, 1980.

52. Levchenko, T. S., Rammohan, R., Lukyanov, A. N., Whiteman, K. R., and Torchilin, V. P., Liposome clearance in mice: The effect of a separate and combined presence of surface charge and polymer coating, *Int. J. Pharm.*, 240, 95–102, 2002.

53. Straubinger, R. M., pH-Sensitive liposomes for delivery of macromolecules into cytoplasm of cultured cells, *Methods Enzymol.*, 221, 361–376, 1993.

54. Simberg, G., Weisman, S., Talmon, Y., and Barenholz, Y., DOTAP (and other cationic lipids): Chemistry, biophysics, and transfection, *Crit. Rev. Ther. Drug Carrier Syst.*, 21, 257–317, 2004.

55. Jain, R. K., Munn, L. L., and Fukumura, D., Dissecting tumour pathophysiology using intravital microscopy, *Nat. Rev. Cancer*, 2, 266–276, 2002.

56. Hashizume, H. et al., Openings between defective endothelial cells explain tumor vessel leakiness, *Am. J. Pathol.*, 156, 1363–1380, 2000.

57. Fukumura, D., Yuan, F., Monsky, W. L., Chen, Y., and Jain, R. K., Effect of host microenvironment on the microcirculation of human adenocarcinoma, *Am. J. Pathol.*, 151, 679–688, 1997.

58. Felgne, P. et al., Lipofection: A highly efficient, lipid-mediated DNA-transfection procedure, *Proc. Natl Acad. Sci. USA*, 84, 7413–7417, 1987.

59. Felgner, P. L. and Ringold, G. M., Cationic liposome-mediated transfection, *Nature*, 337, 387–388, 1989.

60. Sakurai, F. et al., Effect of DNA/liposome mixing ratio on the physicochemical characteristics, cellular uptake and intracellular trafficking of plasmid DNA/cationic liposome complexes and sub-sequent gene expression, *J. Control. Release*, 255–269, 2000.

61. Dabbas, S., Kaushik, R., Dandamudi, S., and Campbell, R.B., Physiochemical evaluation of PEG-modified cationic liposomes (PCLs) and properties in vitro using human primary and immortalized microvascular endothelial cells. *J. Pharm. Sci. Submitted.*

62. Dellian, M., Yuan, F., Trubetskoy, V. S., Torchilin, V. P., and Jain, R. K., Vascular permeability in a human tumor xenograft: Molecular charge dependence, *Br. J. Cancer*, 82, 1513–1518, 2000.

63. McLean, J. W. et al., Organ-specific endothelial cell uptake of cationic liposome–DNA complexes in mice, *Am. J. Physiol.*, 273, 387–404, 1997.

64. Tanner, F., Carr, D. P., Nabel, G. J., and Nabel, E. G., Transfection of human endothelial cells, *Cardiovasc. Res.*, 35, 522–528, 1997.

65. Matsumura, J. S., Kim, R., Shively, V. P., MacDonald, R. C., and Pearce, W. H., Characterization of vascular gene transfer using a novel cationic lipid, *J. Surg. Res.*, 85, 339–345, 1999.

66. Nabel, G. J. et al., Direct gene transfer with DNA–liposome complexes in melanoma: Expression, biologic activity, and lack of toxicity in humans, *Proc. Natl Acad. Sci. USA*, 90, 11307–11311, 1993.

67. Pleyer, U. et al., Efficiency and toxicity of liposome-mediated gene transfer to corneal endothelial cells, *Exp. Eye Res.*, 73, 1–7, 2001.

68. Uyechi, L. S., Gagne, L., Thurston, G., and Szoka, F. C. J., Mechanism of lipoplex gene delivery in mouse lung: Binding and internalization of fluorescent lipid and DNA components, *Gene Ther.*, 8, 828–836, 2001.

69. Trubetskoy, V. S., Torchilin, V. P., Kennel, S., and Huang, L., Cationic liposomes enhance targeted delivery and expression of exogenous DNA mediated by N-terminal modified poly(L-lysine)–antibody conjugate in mouse lung endothelial cells, *Biochim. Biophys. Acta*, 1131, 311–313, 1992.

70. Abraham, N. G. et al., Transfection of the human heme oxygenase gene into rabbit coronary micro-vessel endothelial cells: Protective effect against heme hemoglobin toxicity, *Proc. Natl Acad. Sci. USA*, 92, 6798–6802, 1995.

71. Nakatani, K. et al., Endothelial adhesion molecules in glomerular lesions: Association with their severity and diversity in lupus models, *Kidney Int.*, 65, 1290–1300, 2004.

72. Jeschke, M. G. et al., IGF-I gene transfer in thermally injured rats, *Gene Ther.*, 6, 1015–1020, 1999.

73. Pickering, J. G. et al., Liposome-mediated gene transfer into human vascular smooth muscle cells, *Circulation*, 89, 13–21, 1994.

74. Tiukinhoy, S. D. et al., Development of echogenic, plasmid-incorporated, tissue-targeted cationic liposomes that can be used for directed gene delivery, *Invest. Radiol.*, 35, 732–738, 2000.

75. Hong, Z. et al., Enhanced adeno-associated virus vector expression by adenovirus protein–cationic liposome complex, *Chin. Med. J.*, 108, 332–337, 1995.

76. Torchilin, V. P., Liposomes as targetable drug carriers, *CRC Crit. Rev. Ther. Drug Carrier Syst.*, 2, 65–115, 1985.

77. Ishida, T., Harashima, H., and Kiwada, H., Liposome clearance, *Biosci. Reports*, 22, 197–224, 2002.

78. Harrington, K. J., Lewanski, C. R., and Stewart, J. S., Liposomes as vehicles for targeted therapy of cancer. Part 2. Clinical development, *Clin. Oncol.*, 12, 16–24, 2000.

79. Gabizon, A. and Paphadjopoulos, D., Liposome formulations with prolonged circulation time in blood and enhanced uptake in tumours, *Proc. Natl Acad. Sci.*, 85, 6949–6953, 1985.

80. Woodle, M. C. and Lasic, D. L., Sterically stabilized liposomes, *Biochim. Biophys. Acta*, 1113, 171–199, 1992.

81. Meyer, O. et al., Cationic liposomes coated with polyethylene glycol as carriers for oligonucleotides, *J. Biol. Chem.*, 273, 15621–15627, 1998.

82. Gursel, I., Gursel, M., Ishii, K. J., and Klinman, D. M., Sterically stabilized cationic liposomes improved the uptake and immunostimulatory activity of CpG oligonucleotides, *J. Immunol.*, 167, 3324–3328, 2001.

83. Harigai, T. et al., Preferential binding of polyethylene glycol-coated liposomes containing a novel cationic lipid, TRX-20, to human subendothelial cells via chondroitin sulfate, *Pharm. Res.*, 18, 1284–1290, 2001.

84. Strawn, L. M. and Shawver, L. K., Tyrosine kinase in disease: Overview of kinase inhibitors as therapeutic agents and current drugs in clinical trials, *Expert Opin. Investig. Drugs*, 7, 553–573, 1998.

85. Mendel, D. B. et al., The angiogenesis inhibitor SU5416 has long-lasting effects on vascular endothelial growth factor receptor phosphorylation and function, *Clin. Cancer Res.*, 6, 4848–4858, 2000.

86. Brower, V., Tumor angiogenesis-new drugs on the block, *Nat. Biotechnol.*, 17, 963–968, 1999.

87. Hirabayashi, K., Yano, J., Takesue, H., Fujiwara, N., and Irimura, T., Inhibition of metastatic carcinoma cell growth in livers by poly(I):poly(C)/cationic liposome complex (LIC), *Oncol. Res.*, 11, 497–504, 1999.

88. Lee, C. M. et al., Novel chondroitin sulfate-binding cationic liposomes loaded with cisplatin efficiently suppress the local growth and liver metastasis of tumor cells in vivo, *Cancer Res.*, 62, 4282–4288, 2002.

89. Hood, J. D. and Cheresh, D. A., Role of integrins in cell invasion and migration, *Nat. Rev. Cancer*, 2, 91–100, 2002.

90. Akalu, A., Cretu, A., and Brooks, P. C., Targeting integrins for control of tumour angiogenesis, *Expert Opin. Investig. Drugs*, 14, 1475–1486, 2005.

91. Hood, J. D. et al., Tumor regression by targeted gene delivery to the neovasculature, *Science*, 296, 2404–2407, 2002.

92. Shinkarauk, S., Bayle, M., Lain, G., and Deleris, G., Vascular endothelial cell growth factor (VEGF), an emerging target for cancer therapy, *Curr. Med. Chem. Anti-Cancer. Agents*, 3, 95–117, 2003.

93. Senger, D. R. et al., Vascular permeability factor (VPF, VEGF) in tumor biology, *Metastasis Rev.*, 12, 303–324, 1993.

94. Dvorak, H. F., Brown, L. F., Detmar, M., and Dvorak, A. M., Vascular permeability factor/vascular endothelial growth factor, microvascular hyperpermeability and angiogenesis, *Am. J. Pathol.*, 146, 1029–1039, 1995.

31 Cell Penetrating Peptide (CPP)–Modified Liposomal Nanocarriers for Intracellular Drug and Gene Delivery

Vladimir P. Torchilin

CONTENTS

31.1 Introduction 629
31.2 Liposomes for Intracellular Drug Delivery 630
31.3 Transduction 631
31.4 Intracellular Delivery of Pharmaceutical Nanocariers with CPPs 633
31.5 Intracellular Delivery of Liposomes with CPPs 635
References 637

31.1 INTRODUCTION

Although many important pharmacological targets for cancer therapy are located inside cells, membranes pose a serious obstacle to the uptake of charged hydrophilic molecules inside the cells. Many pharmacological proteins or peptides require to be intracellularly delivered to target and modulate cellular functions at the subcellular levels. In this connection, an important task is a proper identification of intracellular targets to be reached and affected. Cancer therapy provides a whole set of good examples to illustrate this importance. In a search of new anti-cancer drugs, a shift is gradually occurring from the semi-empirical approach based on evaluation of efficacy of candidate compounds against cultured cancer cells and animal tumor models to molecular mechanism-based drug discovery.[1] A huge body of information about cellular metabolic and signaling pathways essential for tumorogenesis and tumor cell development allows for the identification of proper targets for interference with the tumor growth. Quite a few molecular targets have already been identified.[1] The creation of a working draft of the human genome sequence[2,3] in combination with high-throughput methods of molecular biology promises continued rapid growth in identifying such targets.[4,5] However, not all drug targets that can be identified and validated by molecular biology tools are considered suitable for drug development, often because of their intracellular location.[1] Myc/Max dimerization, Src homology-2 domain interaction, and Ras/Raf association have been identified as promising interference targets for the inhibition of tumor development.[6,7] Although peptide inhibitors, discovered through the use of phage-display[8] or combinatorial peptide[9] libraries had excellent in vitro activity, they were not considered for drug development because of poor pharmacokinetics and their inability to reach the molecular targets inside the cells. Sometimes, tumors result from malfunctions of tumor suppressors genes and the lack of activity of the proteins they encode.[5,10] In this case, the delivery into tumor cells of working copies of proteins obtained by recombinant methods would provide indispensable tools for validation of gene functions and potential development of protein or gene therapy-based methods of treatment. The use of these proteins for molecular target validation and eventual development of anti-cancer drugs is again hampered by low permeability of cell membranes.

Intracellular transport of biologically active molecules with therapeutic properties is one of the key problems in drug delivery in general. However, the very nature of cell membranes prevents proteins, peptides, and nanoparticulate drug carriers from entering cells unless there is an active transport mechanism that is usually the case for very short peptides.[11] Various vector molecules promote the delivery of associated drugs and drug carriers inside the cells via receptor-mediated endocytosis.[12] This process involves the attachment of the vector molecule and an associated drug carrier to specific ligands on the target cell membranes, followed by the energy-dependent formation of endosomes. The problem, however, is that any molecule/particle entering cells via the endocytic pathway and becoming entrapped into endosome eventually ends in lysosome where active degradation processes take place under the action of numerous lysosomal enzymes. As a result, only a small fraction of unaffected substance appears in the cell cytoplasm. Even if an efficient cellular uptake via endocytosis is observed, the delivery of intact peptides and proteins is compromised by an insufficient endosomal escape and subsequent lysosomal degradation. Enhanced endosomal escape can be achieved through the use of, for example, lytic peptides,[13–15] pH-sensitive polymers,[16] or swellable dendritic polymers.[17] These agents have provided encouraging results in overcoming limitations of endocytosis-based cytoplasmic delivery, but there is still a need for further improvement or consideration of alternative delivery strategies.

31.2 LIPOSOMES FOR INTRACELLULAR DRUG DELIVERY

For almost two decades, liposomes have been considered promising carriers for biologically active substances.[18–20] They are biologically inert and completely biocompatible, and they cause practically no toxic or antigenic reactions. Drugs included into liposomes are protected from the destructive action of the external media, and liposomes are able to deliver their content inside cells and even inside different cell compartments. Water-soluble drugs can be captured by the inner water space of liposomes, whereas lipophilic compounds can be incorporated into the liposomal membrane. The definite drawback of liposomal preparations is their fast elimination from the blood and capture by the cells of the reticulo-endothelial system (RES), primarily, in the liver and spleen that is usually believed to be the result of rapid opsonization of the liposomes.[21] To increase liposome accumulation in the required areas, the use of targeted liposomes has been suggested.[22] Liposomes with a specific affinity for an affected organ or tissue were believed to increase the efficacy of liposomal pharmaceutical agents. Immunoglobulins, primarily of the IgG class, and their fragments are the most promising and widely used targeting moieties for various drugs and drug carriers, including liposomes. Despite evident success in the development of antibody-to-liposome coupling techniques and improvements in the targeting efficacy, the majority of immunoliposomes still ended in the liver that was usually a consequence of insufficient time for the interaction between the target and targeted liposome.

Different methods have been suggested to achieve long circulation of liposomes in vivo, including coating the liposome surface with inert, biocompatible polymers such as polyethylene glycol (PEG) that form a protective layer over the liposome surface and slows down the liposome recognition by opsonins and subsequent clearance.[23,24] Long-circulating liposomes are now widely used in biomedical in vitro and in vivo studies and even found their way into clinical practice.[25,26] Explanations of the phenomenon involve the role of surface charge and hydrophilicity of PEG-coated liposomes,[27] the participation of PEG in the repulsive interactions between PEG-grafted membranes and other particles,[28] and more generally, the decreased rate of plasma protein adsorption on the hydrophilic surface of PEGylated liposomes.[29] Long-circulating immunoliposomes have also been prepared containing on their surface both antibody and PEG, thereby possessing both abilities— i.e., to recognize and bind the target and to circulate long enough to provide high target accumulation.[30]

As previously discussed, most liposomes are internalized by cells via endocytosis and destined to lysosomes for degradation.[31] When one needs to achieve liposome-mediated drug delivery inside cells into a cytoplasmic compartment, pH-sensitive liposomes are frequently used (for one of many reviews, see V. P. Torchilin, F. Zhou, and L. Huang 1993).[32] As noticed in Puyal et al.,[33] cellular drug delivery mediated by pH-sensitive liposomes is not a simple intracellular leakage from the lipid vesicle because the drug has to also cross the endosomal membrane. It is usually supposed that inside an endosome, the low pH and some other factors destabilize liposomal membrane that, in turn, interacts with the endosomal membrane, provoking its secondary destabilization and drug release into the cytoplasm. To prepare pH-sensitive immunoliposomes, the latter were additionally supplied with surface-immobilized antibodies. The advantages of antibody and pH-sensitive liposome combination are mutually additive: cytoplasmic delivery, targetability, and facilitated uptake (i.e., improved intracellular availability) via receptor-mediated endocytosis. Successful application of pH-sensitive immunoliposomes has been demonstrated in the delivery of a variety of molecules including fluorescent dyes, anti-tumor drugs, proteins, and DNA.[34]

31.3 TRANSDUCTION

A promising approach that seems to be the solution of overcoming the cellular barrier has emerged over the last decade. In this approach, certain proteins or peptides can be tethered to the hydrophilic drug of interest, and together, the construct possesses the ability to translocate across the plasma membrane and intracellularly deliver the payload. This process of translocation is called protein transduction. Such proteins or peptides contain domains of less than 20 amino acids, termed as Protein Transduction Domains (PTDs) or CPPs, that are highly rich in basic residues. These peptides have been used for intracellular delivery of various cargoes with molecular weights several times greater than their own.[35] CPPs are a group of peptides, usually containing a cluster of basic residues that have been recognized as promising drug delivery vectors over the last decade. The CPPs are gaining increased attention as they possess the remarkable property of translocating across the hydrophobic cell membrane that forms a formidable barrier to the entry of hydrophilic and high molecular-weight drugs. As a result, different therapeutic moieties that are mostly hydrophilic in nature and/or are of high molecular weight can be tagged to CPPs and transported across the cell membrane to exert their pharmacological actions at the subcellular level. This process of traversing across the biological barrier is called protein transduction, and it is confined to a domain of less than 20 amino acids, termed as PTD or CPP. CPPs are rich in basic residues and are either derived from corresponding transducing proteins or synthesized. The original concept of protein transduction came out from the observation that the trans-activating transcriptional activator, TAT, protein encoded by HIV-1, was efficiently internalized by cells in vitro, resulting in trans-activation of the viral promoter.[36,37] Subsequently, several other proteins and peptides were found to display the translocation activity that encompasses penetratin,[38] VP22,[39] transportan,[40] model amphipathic peptide MAP,[41] signal sequence-based peptides,[42] and synthetic arginine-enriched sequences.[43]

Penetratin is a CPPs derived from the homeodomain of Antennapedia (Drosophila homeoprotein). Homeoproteins are transcription factors that comprise a stretch of 60 amino acids called the homeodomain that is involved in the DNA binding. The homeodomain was shown to traverse through the mammalian nerve cells and accumulate in their nuclei.[44] More specifically, the translocation ability was narrowed down to a 16-mer peptide, termed as penetratin (Antp PTD, 43–58 residues, Arg-Gln-Ile-Lys-Ile-Trp-Phe-Gln-Asn-Arg-Arg-Met-Lys-Trp-Lys-Lys) present in the third helix of the homeodomain.[38] VP22 is a herpes virus type 1 protein associated with the transport between cells where it also ends up in the nuclei.[39] Transportan is a chimeric CPP built of galanin and mastoparan. Inside the cells, transportan is taken up into the nuclei and concentrates in subnuclear structures, probably the nucleoli.[40] MAP is a synthetic 18-mer peptide capable of transporting various cargoes across the cellular membrane. MAP has presented the highest uptake and cargo delivery efficiency among other CPPs.[45] Signal-sequence-based peptides comprise membrane translocating sequences (MTSs) that direct the pre-protein toward the accurate intracellular organelles. Such sequences, when coupled to nuclear localization signals, can be directed to accumulate in the nuclei of the cells.[46] To identify the specific regions of TAT peptide responsible for translocation, synthetic peptides rich in arginines were prepared such as arginine-substituted TAT (R_9-Tat) and D-amino acid substituted TAT (D-TAT).[43] Such peptides were internalized with similar efficiencies as TAT peptide. In addition, various arginine-rich peptides such as RNA-binding peptides derived from proteins HIV-1 Rev, flock house virus coat, and DNA-binding peptides from c-Fos, c-Jun, and yeast GCN4 also displayed translocation properties.

TAT peptide (TATp) remains the most frequently used CPP for drug delivery purposes. TAT is transcriptional activator protein encoded by human immunodeficiency virus type 1 (HIV-1).[47] TAT-mediated transduction was first utilized in 1994 for the intracellular delivery of variety of cargos such as β-galactosidase, horseradish peroxidase, RNase A, and domain III of *Pseudomonas* exotoxin A in vitro.[48] In vivo transduction using TAT peptide (37–72)-conjugated to β-galactosidase resulted in protein delivery to different tissues such as the heart, liver, spleen, lung, and skeletal muscle. Next, attempts were made to narrow down to the specific domain responsible for transduction. For this, synthetic peptides with deletions in the α-helix domain and the basic cluster domain were prepared for investigating their translocation ability.[49] The whole basic cluster from 48 to 60 residues was found accountable for membrane translocation because any deletions or substitutions of basic residues in TAT (48–60) reduced the cellular uptake property. Rothbard et al.[50] studied TAT (48–57) by deletion analysis. They found that deletion of Gly-48 did not affect the transduction efficiency, whereas deletions of Lys-50,51, Arg-55–57 and Gln-54 markedly reduced transduction efficiencies. Therefore, the minimal transduction domain was assigned to TAT (49–57) residues. The commonly studied transduction domain of TAT (TAT PTD) extends from residues 47–57: Tyr-Gly-Arg-Lys-Lys-Arg-Arg-Gln-Arg-Arg-Arg that contain six arginines (Arg) and two lysine residues.[51]

Two types of the endocytic uptake of the CPPs have been proposed: the classical clathrin-mediated endocytosis and the lipid-raft-mediated caveolae endocytosis. Clathrin-mediated endocytosis involves the formation of clathrin-coated membrane pits that pinch off the membrane to form vesicles for subsequent processing.[52] This type of endocytosis was suggested in Console et al. where the TAT peptide showed the co-localization with the classical endocytic marker, transferrin. This was substantiated further in Lundberg, Wilkstrom, and Johansson where TAT PTD and Antp PTD showed uptake only at 37°C. In addition, the internalization required the expression of negatively charged glycosaminoglycans on the cell surface for interaction with CPPs that was followed by endocytosis. Studies also suggested that the PTDs do not provoke a real translocation, but they are only responsible for cell surface adherence that subsequently results in their endocytosis and accumulation in endosomes.[53–57] In fact, direct electrostatic interaction between the positive residues of CPPs and the negative residues of the cell-surface proteoglycans or glycosaminoglycans (such as heparan sulfate, heparin) is required in internalization regardless of the mechanism of cellular uptake[54,58–60] although some studies suggested the lack of correlation between proteoglycans and transduction process.[61]

Another proposed mechanism for internalization is caveolae-mediated clathrin-independent uptake. Caveolae uptake involves the formation of flask-like uncoated invaginations (50–70 nm), principally composed of a subclass of detergent-resistant membrane domains enriched in cholesterol and sphingolipids called as lipid rafts.[62] This type of uptake was suggested in.[63,64] Unlike the rapid uptake of transferrin, a marker for clathrin-mediated endocytosis, the internalization of TAT-cargo was very slow, reaching the plateau after several hours with the co-localization of TATp with the markers of caveolar uptake. The cellular uptake was affected in the presence of drugs that either disrupt lipid rafts or alter caveolar trafficking.

A different mechanism has been proposed for the transport of guanidinium-rich CPPs conjugated to small molecules (MW < 3,000 Da).[65,66] The guanidinium groups of the CPPs form bidentate hydrogen bonds with the negative residues on the cell-surface; the resulting ion pairs then translocate across the cell membrane under the influence of the membrane potential. The ion pair dissociates on the inner side of the membrane, releasing the CPPs into the cytosol. The number of guanidinium groups is critical for translocation with around eight groups being the optimum number for the efficient translocation.

A recent mechanism proposed for CPP-conjugated to large cargos (MW > 30,000 Da) is nonclathrin, noncaveolar endocytosis, called macropinocytosis. Macropinocytosis is a nonspecific form of cellular uptake, brought about by large vesicles known as macropinosomes that are generated from the actin filaments.[67] Because a majority of fusion proteins remained entrapped in macropinosomes, the TAT transduction domain was conjugated to a fusogenic peptide, the N-terminus domain of the influenza virus hemagglutinin protein HA2, to trigger the release of the TAT-fusion protein from endosome, enhancing their nuclear transport. A very recent study also demonstrates the macropinocytosis mechanism for small PTD peptides (1,000–5,000 Da).[68]

Therefore, it looks like that more than one mechanism works for CPP-mediated intracellular delivery of small and large molecules. Individual CPPs or CPP conjugated to small molecules are internalized into cells via electrostatic interactions and hydrogen bonding, whereas CPP conjugated to large molecules occur via the energy-dependent macropinocytosis. However, in both cases, the direct contact between the CPPs and negative residues on cell-surface is a requisite for successful transduction.

Because the majority of the studies believe in endocytosis as a major mechanism for transduction, an important moment is the escape of CPPs from the endosomes and their translocation to the nuclei. Studies suggest that endosomal acidification prior to the disruption is required for CPP escape.[69] Another study suggested that the TAT-fusion proteins enter cells via the endosomal pathway, circumvent lysosomal degradation, and then sequester in the periphery of the nuclei.[70] Overall, the efficiency of nuclear translocation process is limited.

31.4 INTRACELLULAR DELIVERY OF PHARMACEUTICAL NANOCARIERS WITH CPPs

The applications of TAT-mediated delivery extend to nanoparticles delivery inside cells. The concept of TAT-mediated nanoparticle delivery was first realized in 1999 when dextran-coated superparamagnetic iron oxide nanoparticles (size ca. 40 nm) conjugated to TATp (48–57), showing significantly higher uptake in lymphocytes in vitro than control TAT-free particle.[71] The technique showed potential for magnetically labeling cells in order to allow for their magnetic separation or magnetic resonance (MR) imaging. In vivo, TATp effectively transduced iron nanoparticles into hematopoietic and neural progenitor cells for stem cell analysis[72] and into T-cells for MR imaging.[73] Contrary to nuclear translocation of TATp, particulate TATp-iron conjugates were observed in the cytoplasm and not in the nuclei.[74] Because superparamagnetic nanoparticles were shown to be useful MR contrast agents for imaging and for cell labeling and cell tracking, the conjugation of such nanoparticles with TATp provides better signal from the treated cells for MR imaging.[75] For in vivo MRI, cells need to be labeled with magnetic particles through internalizing receptors. The limitation, however, is that most of the cells lack efficient internalization receptors or pathways. This problem was overcome by attaching the CPP such as TAT peptide (48–57) to the dextran-coated superparamagnetic iron oxide particles (CLIO). The average size of the particles was 41 nm, and the conjugate carried average 6.7 TAT moieties per particle. The cellular uptake studies of TAT-CLIO were performed on mouse lymphocytes, human natural killer cells, and HeLa cells. In all the three cell lines tested, the uptake of TAT–CLIO nanoparticles was about 100-fold higher than the nonmodified iron oxide particle. The labeling with TAT–CLIO did not induce toxicity and did not alter the differentiation or proliferation pattern of CD34 + cells. The TAT–CLIO-labeled CD34 + cells and control cells were subsequently intravenously injected into the immunodeficient mice.[75] Around 4% of the injected dose of the cells migrated to the bone marrow (per gram of the tissue), and the labeled and control cells showed similar biodistribution profile. Nonetheless, it was possible to visualize the labeled cells by MRI within mouse bone marrow at the single-cell level. Also, such magnetically labeled cells could be recovered from the bone marrow after in vivo homing using magnetic separation columns.

Similarly, gold nanoparticles were also modified with TATp.[76] As with iron-conjugates, these particles did not reach the cell nuclei in experiments with NIH3T3 and HepG2 cells. The uptake of the particles was found to proceed by endocytosis. TATp (48–57) was also conjugated to FITC-doped silica nanoparticles (FSNPs) for bioimaging purposes.[77] TATp-modified nanoparticles have also been investigated for their capability to deliver the diagnostic and therapeutic agents across the blood–brain barrier.[78] TATp–FSNPs were prepared by microemulsion system and studied for labeling of human lung adenocarcinoma cells (A-549) in vitro. The cells were efficiently labeled with TAT–FSNPs, unlike FSNPs that showed no effective labeling. For in vivo bioimaging potential, TAT–FSNPs were intraarterially administered to the rats' brains. TAT-conjugated FSNPs labeled the brain blood vessels, showing the potential for delivering agents to the brain without compromising the blood–brain barrier.

A new application of CPP-mediated delivery is the labeling of cells with quantum dots using CPP-modified quantum dot-loaded polymeric micelles.[79] Quantum dots are gaining popularity over standard fluorophores for studying tumor pathophysiology because they are photostable and are very bright fluorophores. They can be tuned to a narrow emission spectrum, and they are relatively insensitive to the wavelength of the excitation light. Quantum dots have the ability to distinguish tumor vessels from both the perivascular cells and the matrix with concurrent imaging. Quantum dots were trapped within micelles prepared of PEG–phosphatidyl ethanolamine (PEG–PE) conjugates bearing TAT–PEG–PE linker. TAT-quantum dot conjugates could label mouse endothelial cells in vitro. For in vivo racking, bone marrow-derived progenitor cells were labeled with TAT-bearing quantum dot-containing micelle ex vivo and then injected in the mouse bearing tumor in a cranial window model. It was then possible to track the movement of labeled progenitor cell to tumor endothelium, introducing an attempt toward understanding fine details of tumor neovascularization.

CPPs have also been used to enhance the delivery of genes via solid lipid nanoparticle (SLN).[80] SLN gene vector was modified with dimeric TATp (TAT2) and compared with polyethylenimine (PEI) for gene expression in vitro and in vivo. The presence of TAT2 in SLN gene vector enhanced the gene transfection compared to PEI both in vitro and in vivo. In another study, TATp (47–57) conjugated to nanocage structures showed binding and transduction of the cells in vitro.[81] The shell cross-linked (SCK) nanoparticles were prepared by the micellization of amphiphilic block copolymers of poly(epsilon-caprolactone-b-acrylic acid) and conjugated to TATp that was independently built on a solid support, resulting in TAT-modified nanocage conjugate. Such conjugate was analyzed by confocal microscopy with CHO and HeLa cells. The conjugated nanoparticles showed binding and transduction inside the cells. The authors then characterized the SCK nanoparticles for the optimum number of TATp per particle required to enhance the transduction efficiency.[82] The authors prepared the conjugates with 52, 104, and 210 CPP peptides per particle that were then evaluated for the biocompatibility in vitro and in vivo.[83] In vitro studies showed the inflammatory responses to the conjugates, but in vivo evaluation of the conjugates in mice did not result in major incompatible responses. Therefore, TATp–SCK conjugate can be used as scaffolds for preparing antigen for immunization.

TAT-mediated nanoparticulate delivery is also finding its way in vaccination fields to elicit better immune response. When TAT (1–72)-coated anionic nanoparticles were used to immunize mice, it generated antibodies and T helper type-1 immune response to TAT.[84] In another study, TAT microspheres were used for vaccinations.[85] Anionic microspheres of different compositions, size, and surface charge density were prepared, and all of them adsorbed biologically active TAT protein in a reversible mode. The microspheres were intracellularly delivered by TAT and were not toxic, both in vitro and in vivo.

31.5 INTRACELLULAR DELIVERY OF LIPOSOMES WITH CPPs

CPPs have also augmented the delivery of liposomal drug carriers. TAT peptide (47–57)-modified liposomes could be intracellularly delivered in different cells such as murine Lewis lung carcinoma (LLC) cells, human breast tumor BT20 cells, and rat cardiac myocyte H9C2 cells.[86] The liposomes were tagged with TAT peptide via the spacer p-nitrophenylcarbonyl–PEG–PE at the density of 500 TAT peptide per single liposome vesicle. It was shown that the cells treated with liposomes where TAT peptide–cell interaction was hindered either by direct attachment of TAT peptide to the liposome surface or by the long PEG grafts on the liposome surface shielding the TAT moiety did not show TAT-liposome internalization; however, the preparations of TAT-liposomes that allowed for the direct contact of TAT peptide residues with cells displayed an enhanced uptake by the cells. This suggested that the translocation of TAT peptide (TATp)-liposomes into cells requires direct free interaction of TAT peptide with the cell surface. Further studies on the intracellular trafficking of rhodamine-labeled TATp-liposomes loaded with FITC-dextran revealed that TATp-liposomes remained intact inside the cell cytoplasm within 1 h of translocation; after 2 h, they migrated into the perinuclear zone, and eventually, the liposomes disintegrated there.[87]

The TATp-liposomes were also investigated for their gene delivery ability. For this, TATp-liposomes prepared with the addition of a small quantity of a cationic lipid (DOTAP) were incubated with DNA. The liposomes formed firm noncovalent complexes with DNA. Such TATp-liposome-DNA complexes, when incubated with mouse fibroblast NIH 3T3 and cardiac myocytes H9C2, showed substantially higher transfection in vitro with lower cytotoxicity than the commonly used Lipofectin®. NIH/3T3, BT20, or H9C2 cells were transfected by incubation with TAT peptide–liposome/plasmid complexes in serum-free media for 4 h at 37°C under 6% CO_2. The flow cytometry data demonstrated that the treatment of NIH/3T3 cells with TAT peptide–liposome/pEGFP-N1 complexes results in high fluorescence, i.e., high transfection outcome. Similar results were obtained with all cell lines tested. Confocal microscopy confirmed the transfection of both HCC150 and H9C2 cells with TAT peptide–liposome/DNA complexes (Figure 31.1). From 30 to 50% of both cell types in the field of view show a bright green fluorescence, whereas lower fluorescence was observed in virtually all cells. Under in vivo conditions, the intratumoral injection of TATp-liposome–DNA complexes into the LLC tumor in mice resulted in an efficient transfection of the tumor cells. Histologically, hematoxylin/eosin-stained tumor slices in both control and experimental animals showed a typical pattern of poorly differentiated carcinoma; however, under the fluorescence microscope, samples from control mice (nontreated mice or mice injected with TAT peptide–free liposome/plasmid complexes) showed only a background fluorescence, and slices from tumors injected with TAT peptide–liposome/plasmid complexes contained bright green fluorescence in tumor cells, indicating an efficient TAT peptide-mediated transfection. The study implicated the usefulness of TATp-liposomes for in vitro and localized in vivo gene therapy.

Another study examined the kinetics of uptake of the TAT- and penetratin-modified liposomes.[88] It was found that the translocation of liposomes by TAT peptide or penetratin was

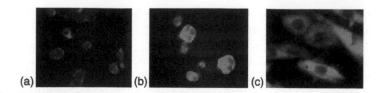

FIGURE 31.1 The enhancement of the liposome-mediated transfection by TAT peptide. HCC1500 cells (human breast carcinoma; a,b) or H9C2 cells (murine myoblasts; c) were incubated with the presence of liposome/pEGFP-N1 plasmid (encoding for the Green Fluorescence Protein, GFP) complexes (10 μg DNA per 100,000 cells) for 4 h at 37°C. The transfection efficiency (the appearance of the green fluorescence of the GFP inside cells) was detected after 72 h. (a), Cells were incubated with control plain (TAT-free) plasmid-bearing liposomes; (b,c), Cells were incubated with plasmid-bearing TAT-liposomes of the same composition. (a), (b) and (c)—fluorescent microscopy with FITC filter. Background transfection can only be seen with controls, whereas the introduction of TAT peptide into the preparation provides a dramatic enhancement of the GFP expression, i.e., increases the transfection efficiency.

proportional to the number of peptide molecules attached to the liposomal surface. A peptide number of as few as five was already sufficient to enhance the intracellular delivery of liposomes. The kinetics of the uptake were peptide- and cell-type dependent. With TATp-liposomes, the intracellular accumulation was time-dependent, and with penetratin-liposomes, the accumulation within the cells was quick to reach the peak within 1 h, after that, it gradually declined.

A study on the similar lines showed that Antp (43–58) and TAT (47–57) peptides coupled to small unilamellar liposomes were accumulated in higher proportions within tumor cells and dendritic cells than unmodified control liposomes.[89] The uptake was time- and concentration-dependent, and at least, 100 PTD molecules per small unilamellar liposomes were required for efficient translocation inside cells. The uptake of the modified liposomes was inhibited by the preincubation of liposomes with heparin, confirming the role of heparan sulfate proteoglycans in CPP-mediated uptake.

In a different approach for improving the transfection and protecting DNA from degradation, thiocholesterol-based cationic lipids (TCL) were used in the formation of nanolipoparticles (NLPs). The NLPs were sequentially modified with TAT peptide that resulted in TAT–NLPs with a zwitterionic surface and higher transfection efficiency than for the cationic NLPs.[90]

TABLE 31.1
Intracellular Delivery of Nanoparticles by CPPs

Particle and size	CPP	Cell
CLIO (MION) particles, 40 nm	TATp	Mouse lymphocytes, human natural killer, HeLa, human hematopoietic CD34+, mouse neural progenitor C17.2, human lymphocytes CD4+, T-cells, B-cells, macrophages
Gold particles, 20 nm	TATp	NIH3T3, HepG2, HeLa, human fibroblast HTERT-BJ1
Quantum dot-loaded polymeric micelles, 20 nm	TATp	Mouse endothelial cells, bone marrow-derived progenitor cells
Sterically stabilized liposomes, 200 nm	TATp	Mouse LLC, human BT20, rat H9C2, LLC tumor in mice
Sterically stabilized liposomes, 65–75 nm	TATp or penetratin	Human bladder carcinoma HTB-9, murine colon carcinoma C26, human epidermoid carcinoma A431, human breast cancer SK-BR-3, MCF7/WT, MCF7/ADR, murine bladder cancer MBT2, dendritic cells

Although initial studies suggested an energy-independent character for the internalization of TATp-liposomes,[86] recent studies have revealed the endocytosis as the main mechanism for the intracellular uptake of TATp-liposomes. As was shown in M. M. Fretz, et al.,[93] the conjugation of TAT peptide to lipoplexes enhanced the gene transfection in primary cell cultures by the endocytic uptake. Similarly, coupling of TAT peptide to the outer surface of liposomes resulted in an enhanced binding and endocytosis of the liposomes in ovarian carcinoma cells.[92] The binding was inhibited in the presence of heparin or dextran sulfate, suggesting that the proteoglycans expressed on the cell surface are also involved in cell binding in this case. In contrast, a new class of transducing peptides, haptides, after binding to the liposomal surface, augmented liposomes penetration through the cell membrane into the cell cytoplasm by a nonreceptor mediated process,[93] confirming that a variety of mechanisms could be involved in CPP-mediated intracellular delivery of nanoparticulates.

CPPs clearly can serve as versatile delivery vectors for intracellular drug delivery. They can bring inside cells a wide range of cargos of different sizes—from small molecules to relatively large nanoparticles (see Table 31.1 for nanoparticles). Of the different CPPs used, TATP remains the most frequently used for this purpose in experimental cancer therapy. Intracellular cytoplasmic delivery of drugs and DNA by CPP-modified pharmaceutical nanocarriers may find applications for in vitro and ex vivo cell treatment as well as in different protocols of local drug application.

REFERENCES

1. Gibbs, J. B., Mechanism-based target identification and drug discovery in cancer research, *Science*, 287, 1969–1973, 2000.
2. Venter, J. C., Adams, M. D., Myers, E. W., Li, P. W., Mural, R. J., Sutton, G. G., Smith, H. O., Yandell, M., Evans, C. A., and Holt, R. A., The sequence of the human genome, *Science*, 291, 1304–1351, 2001.
3. Lander, E. S., Linton, L. M., Birren, B., Nusbaum, C., Zoby, M. C., Baldwin, M.C, Devon, K., Dewar, K., Doyle, M., and Fitzhugh, W., Initial sequencing and analysis of the human genome, *Nature*, 409, 860–921, 2001.
4. Workman, P., New drug targets for genomic cancer therapy: Successes, limitations, opportunities and future challenges, *Curr. Cancer Drug Targets*, 1, 33–47, 2001.
5. Balmain, A., Cancer genetics: From Boveri and Mendel to microarrays, *Nat. Rev. Cancer*, 1, 77–82, 2001.
6. Gibbs, J. B. and Oliff, A., Pharmaceutical research in molecular oncology, *Cell*, 79, 193–198, 1994.
7. Sawyer, T. K., Src homology-2 domains: Structure, mechanisms, and drug discovery, *Biopolymers*, 47, 243–261, 1998.
8. Katz, B. A., Structural and mechanistic determinants of affinity and specificity of ligands discovered or engineered by phage display, *Annu. Rev. Biophys. Biomol. Struct.*, 26, 27–45, 1997.
9. Cortese, R., Ed, *Combinatorial Libraries: Synthesis, Screening and Application Potential*, Walter de Gruyter, Inc., New York, 1995.
10. Hussain, S. P., Hofseth, L. J., and Harris, C. C., Tumor suppressor genes: At the crossroads of molecular carcinogenesis, molecular epidemiology and human risk assessment, *Lung Cancer*, 34 (Suppl. 2), S7–S15, 2001.
11. Egleton, R. D. and Davis, T. P., Bioavailability and transport of peptides and peptide drugs into the brain, *Peptides*, 18, 1431–1439, 1997.
12. Park, J. W., Kirpotin, D. B., Hong, K., Shalaby, R., Shao, Y., Nielsen, U. B., and Marks, J. D., Tumor targeting using anti-her2 immunoliposomes, *J. Controlled Release*, 74, 95–113, 2001.
13. Kamata, H., Yagisawa, H., Takahashi, S., and Hirata, H., Amphiphilic peptides enhance the efficiency of liposome-mediated DNA transfection, *Nucleic Acids Res.*, 22, 536–537, 1994.
14. Midoux, P., Kichler, A., Boutin, V., Maurizot, J. C., and Monsigny, M., Membrane permeabilization and efficient gene transfer by a peptide containing several histidines, *Bioconjugate Chem.*, 9, 260–267, 1998.

15. Mastrobattista, E., Koning, G. A., Van Bloois, L., Filipe, A. C. S., Jiskoot, W., and Storm, G., Functional characterization of an endosome-disruptive peptide and its application in cytosolic delivery of immunoliposome-entrapped proteins, *J. Biol. Chem.*, 277, 27135–27143, 2002.

16. Lackey, C. A., Press, O., Hoffman, A., and Styton, P., A biomimetic pH-responsive polymer directs endosomal release and intracellular delivery of an endocytosed antibody complex, *Bioconjugate Chem.*, 13, 996–1001, 2002.

17. Padilla De Jesus, O. L., Ihre, H. R., Gagne, L., Frechet, J. M., and Szoka, F. C. Jr., Polyester dendritic systems for drug delivery applications: In vitro and in vivo evaluation, *Bioconjugate Chem.*, 13, 453–461, 2002.

18. Gregoriadis, G., Ed, *Liposomes as Drug Carriers*, Wiley, Chinchester, UK, 1988.

19. Lasic, D. D., *Liposomes. From Physics to Applications*, Elsevier, Amsterdam, 1993.

20. Lasic, D. D. and Papahadjopoulos, D., Eds., *Medical Application of Liposomes*, Elsevier, Amsterdam, 1998.

21. Senior, J. H., Fate and behavior of liposomes in vivo: A review of controlling factors, *Crit. Rev. Ther. Drug Carrier Syst.*, 3, 123–193, 1987.

22. Torchilin, V. P., Liposomes as targetable drug carriers, *CRC Crit. Rev. Ther. Drug Carrier Syst.*, 1, 65–115, 1985.

23. Klibanov, A. L., Maruyama, K., Torchilin, V. P., and Huang, L., Amphipatic polyethyleneglycols effectively prolong the circulation time of liposomes, *FEBS Lett.*, 268, 235–238, 1990.

24. Torchilin, V. P. and Trubetskoy, V. S., Which polymers can make nanoparticulate drug carriers long-circulating?, *Adv. Drug Deliv. Rev.*, 16, 141–155, 1995.

25. Martin, F. and Lasic, D., Eds., *Stealth® Liposomes*, CRC Press, Boca Raton, 1995.

26. Gabizon, A. A., Pegylated liposomal doxorubicin: Metamorphosis of an old drug into a new form of chemotherapy, *Cancer Invest.*, 19, 424–436, 2001.

27. Blume, G. and Cevc, G., Molecular mechanism of the lipid vesicle longevity in vivo, *Biochim. Biophys. Acta*, 1146, 157–168, 1993.

28. Kenworthy, A. K., Hristova, K., Needham, D., and McIntosh, T. J., Range and magnitude of the steric pressure between bilayers containing phospholipids with covalently attached poly(ethylene glycol), *Biophys. J.*, 68, 1921–1936, 1995.

29. Woodle, M. C., Newman, M. S., and Cohen, J. A., Sterically stabilized liposomes: Physical and biological properties, *J. Drug Target.*, 5, 397–403, 1994.

30. Torchilin, V. P., Klibanov, A. L., Huang, L., O'Donnell, S., Nossiff, N. D., and Khaw, B. A., Targeted accumulation of polyethelene glycol-coated immunoliposomes in infarcted rabbit myocardium, *FASEB J.*, 6, 2716–2719, 1992.

31. Straubinger, R. M. N., pH-sensitive liposomes mediate cytoplasmic delivery of encapsulated macro-molecules, *FEBS Lett.*, 179, 148–154, 1985.

32. Torchilin, V. P., Zhou, F., and Huang, L., pH-Sensitive liposomes (review), *J. Liposome Res.*, 3, 201–255, 1993.

33. Puyal, C., Milhaud, P., Bienvenue, A., and Philippot, J. R., A new cationic liposome encapsulating genetic material. A potential delivery system for polynucleotides, *Eur. J. Biochem.*, 228, 697–703, 1995.

34. Collins, D. and Huang, L., Cytotoxicity of diphtheria toxin A fragment to toxin-resistant murine cells delivered by pH-sensitive immunoliposomes, *Cancer Res.*, 47, 735–739, 1987.

35. Zorko, M. and Langel, U., Cell-penetrating peptides: Mechanism and kinetics of cargo delivery, *Adv. Drug Deliv. Rev.*, 57, 529–545, 2005.

36. Green, M. and Loewenstein, P. M., Autonomous functional domains of chemically synthesized human immunodeficiency virus tat trans-activator protein, *Cell*, 55, 1179–1188, 1988.

37. Frankel, A. D. and Pabo, C. O., Cellular uptake of the tat protein from human immunodeficiency virus, *Cell*, 55, 1189–1193, 1988.

38. Derossi, D., Joliot, A. H., Chassaing, G., Derossi, D., and Prochiantz, A., The third helix of the Antennapedia homeodomain translocates through biological membranes, *J. Biol. Chem.*, 269, 10444–10450, 1994.

39. Elliot, G. and O'Harre, P., Intercellular trafficking and protein delivery by a herpesvirus structural protein, *Cell*, 88, 223–233, 1997.

40. Pooga, M., Hällbrink, M., Zorko, M., and Langel, Ü, Cell penetration by transportan, *FASEB J.*, 12, 67–77, 1998.
41. Oehlke, J., Scheller, A., Wiesner, B., Krause, E., Beyermann, M., Klauschenz, E., Melzig, M., and Bienert, M., Cellular uptake of an alpha-helical amphipathic model peptide with the potential to deliver polar compounds into the cell interior non-endocytically, *Biochim. Biophys. Acta*, 1414, 127–139, 1998.
42. Lindgren, M., Hällbrink, M., Prochiantz, A., and Langel, U., Cell-penetrating peptides, *Trends Pharmacol. Sci.*, 21, 99–103, 2000.
43. Futaki, S., Suzuki, T., Ohashi, W., Yagami, T., Tanaka, S., Ueda, K., and Sugiura, Y., Arginine-rich peptides. An abundant source of membrane-permeable peptides having potential as carriers for intracellular protein delivery, *J. Biol. Chem.*, 276, 5836–5840, 2001.
44. Joliot, A., Pernelle, C., Deagostini-Bazin, H., and Prochiantz, A., Antennapedia homeobox peptide regulates neural morphogenesis, *Proc. Natl. Acad. Sci. USA*, 88, 1864–1868, 1991.
45. Hällbrink, M., Florén, A., Elmquist, A., Pooga, M., Bartfai, T., and Langel, U., Cargo delivery kinetics of cell-penetrating peptides, *Biochim. Biophys. Acta*, 1515, 101–109, 2001.
46. Lin, Y. Z., Yao, S. Y., Veach, R. A., Torgerson, T. R., and Hawiger, J., Inhibition of nuclear translocation of transcription factor NF-kappa B by a synthetic peptide containing a cell membrane-permeable motif and nuclear localization sequence, *J. Biol. Chem*, 270, 14255–14258, 1995.
47. Calnan, B. J., Tidor, B., Biancalana, S., Hudson, D., and Frankel, A. D., Arginine-mediated RNA recognition: The arginine fork, *Science*, 252, 1167–1171, 1991.
48. Schwarze, S. R., Hruska, K. A., and Dowdy, S. F., Protein transduction: Unrestricted delivery into all cells?, *Trends Cell Biol.*, 10, 290–295, 2000.
49. Mitchell, D. J., Kim, D. T., Steinman, L., Fathman, C. G., and Rothbard, J. B., Polyarginine enters cells more efficiently than other polycationic homopolymers, *J. Pept. Res.*, 56, 318–325, 2000.
50. Rothbard, J. B., Kreider, E., Vandeusen, C. L., Wright, L., Wylie, B. L., and Wender, P. A., Arginine-rich molecular transporters for drug delivery: Role of backbone spacing in cellular uptake, *J. Med. Chem.*, 45, 3612–3618, 2002.
51. Sieczkarski, S. B. and Whittaker, G. R., Dissecting virus entry via endocytosis, *J. Gen. Virol.*, 83, 1535–1545, 2002.
52. Richard, J. P., Melikov, K., Vives, E., Ramos, C., Verbeure, B., Gait, M. J., Chernomordik, L. V., and Lebleu, B., Cell-penetrating peptides. A reevaluation of the mechanism of cellular uptake, *J. Biol. Chem.*, 278, 585–590, 2003.
53. Console, S., Marty, C., Garcia-Echeverria, C., Schwendener, R., and Ballmer-Hofer, K., Antennapedia and HIV transactivator of transcription (TAT) protein transduction domains promote endocytosis of high molecular weight cargo upon binding to cell surface glycosaminoglycans, *J. Biol. Chem.*, 278, 35109–35114, 2003.
54. Lundberg, M., Wilkstrom, S., and Johansson, M., Cell surface adherence and endocytosis of protein transduction domains, *Mol. Ther.*, 8, 143–150, 2003.
55. Leifert, J. A., Harkins, S., and Whitton, J. L., Full-length proteins attached to the HIV tat protein transduction domain are neither transduced between cells, nor exhibit enhanced immunogenicity, *Gene Ther.*, 9, 1422–14428, 2002.
56. Vives, E., Richard, J. P., Rispal, C., and Lebleu, B., TAT peptide internalization: Seeking the mechanism of entry, *Curr. Protein Pept. Sci.*, 4, 125–132, 2003.
57. Tyagi, M., Rusnati, M., Presta, M., and Giacca, M., Internalization of HIV-1 tat requires cell surface heparan sulfate proteoglycans, *J. Biol. Chem.*, 276, 3254–3261, 2001.
58. Sandgren, S., Cheng, F., and Belting, M., Nuclear targeting of macromolecular polyanions by an HIV-Tat derived peptide. Role for cell-surface proteoglycans, *J. Biol. Chem*, 277, 38877–38883, 2002.
59. Mai, J. C., Shen, H., Watkins, S. C., Cheng, T., and Robbins, P. D., Efficiency of protein transduction is cell type-dependent and is enhanced by dextran sulfate, *J. Biol. Chem.*, 277, 30208–30218, 2002.
60. Violini, S., Sharma, V., Prior, J. L., Dyszlewski, M., and Piwnica-Worms, D., Evidence for a plasma membrane-mediated permeability barrier to Tat basic domain in well-differentiated epithelial cells: Lack of correlation with heparan sulfate, *Biochemistry*, 41, 12652–12661, 2002.
61. Anderson, R. G., The caveolae membrane system, *Annu. Rev. Biochem.*, 67, 199–225, 1998.

62. Fittipaldi, A., Ferrari, A., Zoppe, M., Arcangeli, C., Pellegrini, V., Beltram, F., and Giacca, M., Cell membrane lipid rafts mediate caveolar endocytosis of HIV-1 Tat fusion proteins, *J. Biol. Chem.*, 278, 34141–34149, 2003.

63. Ferrari, A., Pellegrini, V., Arcangeli, C., Fittipaldi, A., Giacca, M., and Beltram, F., Caveolae-mediated internalization of extracellular HIV-1 tat fusion proteins visualized in real time, *Mol. Ther.*, 8, 284–294, 2003.

64. Rothbard, J. B., Jessop, T. C., and Wender, P. A., Role of membrane potential and hydrogen bonding in the mechanism of translocation of guanidinium-rich peptides into cells, *J. Am. Chem. Soc.*, 126, 9506–9507, 2004.

65. Rothbard, J. B., Jessop, T. C., and Wender, P. A., Adaptive translocation: The role of hydrogen bonding and membrane potential in the uptake of guanidinium-rich transporters into cells, *Adv. Drug Deliv. Rev.*, 57, 495–504, 2005.

66. Swanson, J. A. and Watts, C., Macropinocytosis, *Trends Cell Biol.*, 5, 424–428, 1995.

67. Kaplan, I. M., Wadia, J. S., and Dowdy, S. F., Cationic TAT peptide transduction domain enters cells by macropinocytosis, *J. Controlled Release*, 102, 247–253, 2005.

68. Potocky, T. B., Menon, A. K., and Gellman, S. H., Cytoplasmic and nuclear delivery of a TAT-derived peptide and a beta-peptide after endocytic uptake into HeLa cells, *J. Biol. Chem.*, 278, 50188–50194, 2003.

69. Caron, N. J., Quenneville, S. P., and Tremblay, J. P., Endosome disruption enhances the functional nuclear delivery of Tat-fusion proteins, *Biochem. Biophys. Res. Commun.*, 319, 12–20, 2004.

70. Zhao, M., Kircher, M. F., Josephson, L., and Weissleder, R., Differential conjugation of tat peptide to superparamagnetic nanoparticles and its effect on cellular uptake, *Bioconjugate Chem.*, 13, 840–844, 2002.

71. Lewin, M., Carlesso, N., and Tung, C. H., Tat peptide-derivatized magnetic nanoparticles allow in vivo tracking and recovery of progenitor cells, *Nat. Biotechnol.*, 18, 410–414, 2000.

72. Dodd, C. H., Hsu, H. C., Chu, W. J., Yang, P., Zhang, H. G., Zhang, J. D. Jr., Zinn, K., Forder, J., Josephson, L., Weissleder, R., Mountz, J. M., and Mountz, J. D., Normal T-cell response and in vivo magnetic resonance imaging of T cells loaded with HIV transactivator-peptide-derived superparamagnetic nanoparticles, *J. Immunol. Methods*, 256, 89–105, 2001.

73. Kaufman, C. L., Williams, M., Ryle, L. M., Smith, T. L., Tanner, M., and Ho, C., Superparamagnetic iron oxide particles transactivator protein-fluorescein isothiocyanate particle labeling for in vivo magnetic resonance imaging detection of cell migration: Uptake and durability, *Transplantation*, 76, 1043–1046, 2003.

74. Zhao, M., Kircher, M. F., Josephson, L., and Weissleder, R., Differential conjugation of tat peptide to superparamagnetic nanoparticles and its effect on cellular uptake, *Bioconjugate Chem.*, 13, 840–844, 2002.

75. Tkachenko, A. G., Xie, H., Liu, Y. L., Coleman, D., Ryan, J., Glomm, W. R., Shipton, M. K., and Feldheim, D. L., Cellular trajectories of peptide-modified gold particle complexes: Comparison of nuclear localization signals and peptide transduction domains, *Bioconjugate Chem.*, 15, 482–490, 2004.

76. Santra, S., Yang, H., Dutta, D., Stanley, J., Holloway, P., Tan, W., Moudgil, B., and Mericle, R., TAT conjugated, FITC doped silica nanoparticles for bioimaging applications, *Chem. Commun. (Camb.)*, 24, 2810–2811, 2004.

77. Santra, S., Yang, H., Stanley, J. T., Holloway, P. H., Moudgil, B. M., Walter, G., and Mericle, R. A., Rapid and effective labeling of brain tissue using TAT-conjugated CdS:Mn/ZnS quantum dots, *Chem. Commun. (Camb.)*, 25, 3144–3146, 2005.

78. Stroh, M., Zimmer, J. P., Zimmer, D. G., Levchenko, T. S., and Cohen, K. S., Quantum dots spectrally distinguish multiple species within the tumor milieu in vivo, *Nat. Med.*, 11, 678–682, 2005.

79. Rudolph, C., Schillinger, U., Ortiz, A., Tabatt, A., Plank, C., Müller, R. H., and Rosenecker, J., Application of novel solid lipid nanoparticle (SLN)-gene vector formulations based on a dimeric HIV-1 TAT-peptide in vitro and in vivo, *Pharm. Res.*, 21, 1662–1669, 2004.

80. Liu, J., Zhang, Q., Remsen, E. E., and Wooley, K. L., Nanostructured materials designed for cell binding and transduction, *Biomacromolecules*, 2, 362–368, 2001.

81. Becker, M. L., Remsen, E. E., Pan, D., and Wooley, K. L., Peptide-derivatized shell-cross-linked nanoparticles. 1. Synthesis and characterization, *Bioconjugate Chem.*, 15, 699–709, 2004.

82. Becker, M. L., Bailey, L. O., and Wooley, K. L., Peptide-derivatized shell-cross-linked nanoparticles. 2. Biocompatibility evaluation, *Bioconjugate Chem.*, 15, 710–717, 2004.

83. Cui, Z., Patel, J., Tuzova, M., Ray, P., Phillips, R., Woodward, J. G., Nath, A., and Mumper, R. J., Strong T cell type-1 immune responses to HIV-1 Tat (1-72) protein-coated nanoparticles, *Vaccine*, 22, 2631–2640, 2004.

84. Caputo, A., Brocca-Cofano, E., Castaldello, A., De Michele, R., Altavilla, G., Marchisio, M., Gavioli, R., Rolen, V., Chiarantini, L., Cerasi, A., Magnani, M., Cafaro, A., Sparnacci, K., Laus, M., Tondalli, L., and Ensoli, B., Novel biocompatible anionic polymeric microspheres for the delivery of the HIV-1 Tat protein for vaccine application, *Vaccine*, 22, 2910–2924, 2004.

85. Torchilin, V. P., Rammohan, R., Weissig, V., and Levchenko, T. S., TAT peptide on the surface of liposomes affords their efficient intracellular delivery even at low temperature and in the presence of metabolic inhibitors, *Proc. Natl. Acad. Sci. USA*, 98, 8786–8796, 2001.

86. Torchilin, V. P., Levchenko, T. S., Rammohan, R., and Volodina, B., Cell transfection in vitro and in vivo with nontoxic TAT peptide-liposome-DNA complexes, *Proc. Natl. Acad. Sci. USA*, 100, 1972–1977, 2003.

87. Tseng, Y. L., Liu, J. J., and Hong, R. L., Translocation of liposomes into cancer cells by cell-penetrating peptides penetratin and tat: A kinetic and efficacy study, *Mol. Pharmacol.*, 62, 864–872, 2002.

88. Marty, C., Meylan, C., Schott, H., Ballmer-Hofer, K., and Schwendener, R. A., Enhanced heparan sulfate proteoglycan-mediated uptake of cell-penetrating peptide-modified liposomes, *Cell Mol. Life Sci.*, 61, 1785–1794, 2004.

89. Huang, Z., Li, W., MacKay, J. A., and Szoka, F. C. Jr., Thiocholesterol-based lipids for ordered assembly of bioresponsive gene carriers, *Mol. Ther.*, 11, 409–417, 2005.

90. Hyndman, L., Lemoine, J. L., Huang, L., Porteous, D. J., and Boyd, A. C., HIV-1 Tat protein transduction domain peptide facilitates gene transfer in combination with cationic liposomes, *J. Controlled Release*, 99, 435–444, 2004.

91. Fretz, M. M., Koning, G. A., Mastrobattista, E., Jiskoot, W., and Storm, G., OVCAR-3 cells internalize TAT-peptide modified liposomes by endocytosis, *Biochim. Biophys. Acta*, 1665, 48–56, 2004.

92. Gorodetsky, R., Levdansky, L., Vexler, A., Shimeliovich, I., Kassis, I., Ben-Moshe, M., Magdassi, S., and Marx, G., Liposome transduction into cells enhanced by haptotactic peptides (Haptides) homologous to fibrinogen C-termini, *J. Controlled Release*, 95, 477–488, 2004.

32 RGD-Modified Liposomes for Tumor Targeting

P. K. Dubey, S. Mahor, and S. P. Vyas

CONTENTS

32.1 Introduction 643
32.2 Tumor Vasculature Targeting 644
    32.2.1 Rationale for Tumor Vasculature Targeting... 644
    32.2.2 Molecular Targets in Angiogenic Tumor Vasculature........................... 646
32.3 Role of RGD in Integrin- and Fibronectin-Mediated Bioevents........................ 647
    32.3.1 Integrins... 647
    32.3.2 Fibronectin... 650
    32.3.3 RGD Mediated Metastasis Inhibition 651
32.4 RGD-Modified Liposomes for Cancer Therapeutics.. 653
32.5 RGD-Modified Liposomes in Cancer Gene Therapy 656
References 657

32.1 INTRODUCTION

Various drug delivery systems have been developed or are under development in order to minimize drug degradation or loss, to prevent harmful side effects, and to increase drug bioavailability and the fraction of the drug accumulated in the required zone. Among drug carriers, one can name microparticles, nanoparticles, synthetic polymers, cell ghosts, lipoproteins, micelles, niosomes, and liposomes. Each of these carriers offers its own advantages and has its own shortcomings; therefore, the choice of a certain carrier for each given case can be made only taking into account the whole bunch of relevent considerations. These carriers can be made slowly biodegradable, stimuli-sensitive, e.g., pH- or temperature sensitive, and even targeted, e.g., by conjugating them with various ligands.

Liposomes-encapsulated anti-cancer drugs reveal their potential for increased therapeutic efficacy and decreased nonspecific toxicities as a result of their ability to enhance the delivery of chemotherapeutic agents selectively or preferentially to tumors (Papahadjopoulos and Gabizon 1995; Martin 1998). However, these liposomes localize in the tumor extracellular compartment and are rarely taken up by the cancer cells located deep within the tumor (Papahadjopoulos et al. 1991a, 1991b; Gabizon and Papahadjopoulos 1992) so that these cancer cells are readily reached with high concentration of drug and are given an opportunity to opt for drug-resistance. To further enhance cytotoxic efficacy, selective delivery of drugs to target cells can be achieved through liposomes appended with antibodies (Maruyama et al. 1990; Lopes de Menezes, Pilarski, and

Allen 1998), serum proteins (Brown and Silvius 1990; Lundberg, Hong, and Papahadjoppoulos 1993), or antibody fragments (Park et al. 1995; Kirpotin et al. 1997) that categorically recognize specific determinants (receptors) on target cells. However, attachment of antibodies and larger proteins to the exterior of liposomes has not been without problems. Chemistries producing phospholipid headgroup-protein coupling are often complex, jeopardizing the subsequent use of such engineered liposomes, especially in living systems. These chemistries can also change the protein's confirmation and do not guarantee the final orientation with respect to either the potential ligand or the phospholipid bilayer, thereby affecting the effectiveness of the targeting system.

An alternative to the coupling of larger proteins is an anchoring of a defined small peptide domain on to the liposome surface (Gyongyossy-Issa, Muller, and Devine 1998). Ligand targeting using small peptides has advantages over the use of large protein molecules such as antibodies. These include ease of preparation, potentially lower antigenicity, and increased stability (Forssen and Willis 1998). RGD peptides have reported as effective in delivering cytotoxic molecules to the tumor vasculature (Arap, Pasqualini, and Ruoslahti 1998; Ellerby et al. 1999; Gerlag et al. 2001; Schiffelers et al. 2003).

32.2 TUMOR VASCULATURE TARGETING

There are many potential barriers to the effective delivery of chemotherapeutic agents, having narrowest therapeutic indices in all medicines to the disseminated malignant tumors. The nonselective toxic effects on normal tissues restrict the dose of the anti-cancer agents. Through selective delivery of chemotherapeutic agents into the malignant tumors, the concentration of drugs in the tumors could be increased. This results in a decrease in the amount and types of nonspecific toxicites and an increased amount of drug that can be effectively delivered to the tumor. Monoclonal antibodies have been used to provide existing targeting opportunities. However, this approach has met with limited success; only a few antigens are known, and their expression on the cells within an individual tumor is not necessarily uniform. Moreover, antibodies against tumor antigens poorly penetrate into solid tumors (Dvorak et al. 1991). Furthermore, antibody targeted drug delivery may be impaired by clonal selection for cells that have lost the tumor antigen because tumor cells are genetically unstable and adopt for mutations advantageous for growth. The targeting of drug delivery systems to the tumor vasculature overcomes some of the problems associated with conventional tumor targeting.

Angiogenesis is an important process in tumor growth as well as in metastasis. Growth and survival of tumor cells depend on oxygen and nutrients supplied by the blood. As a consequence, the size of a tumor is restricted to a few cubic millimeters without the recruitment of new blood vessels. The architecture of a solid tumor is such that numerous layers of tumor cells are fed by one blood vessel. This poses a significant barrier for selective delivery of cytotoxic drugs into malignant tumors that need to extravasate from blood into the tumor tissue to reach the target cell; the larger the chosen carrier, the less accesible the tumor tissue will be. In contrast, endothelial cells lining the tumor vasculature are easily accessible for these macromolecular preparations. The tumor vasculature targeting is a valuable tool for selective delivery of drugs is which either do not reach the target cells insufficiently or are too toxic to non-target cells when administered systemically. Figure 32.1 shows the principle of targeted chemotherapy to tumor blood vessels.

32.2.1 RATIONALE FOR TUMOR VASCULATURE TARGETING

Vascular targeting offers several benefits over conventional targeting to tumor cells (Table 32.1). The endothelium of the tumor vasculature is readily accessible to a circulating probe, whereas a conventional tumor targeting probe has to overcome the high interstitial pressure within the solid

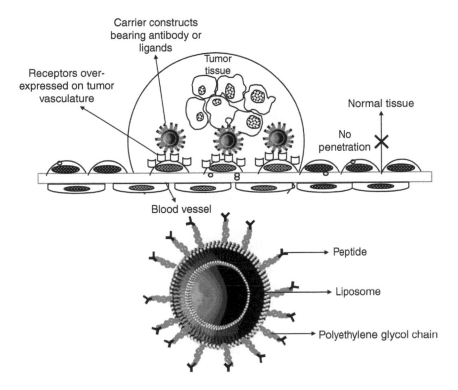

FIGURE 32.1 Principle of targeted chemotherapy to tumor blood vessels.

tumor by penetrating among closely packed tumor cells and dense tumor stroma (Dvorak et al. 1991).

The survival and growth of the tumor cells largely depend on blood supply to them. An anticancer therapy directed against the tumor vasculature does not have to eliminate every endothelial

TABLE 32.1
Comparison Between Conventional Tumor Targeting and Vascular Tumor Targeting

Feature	Conventional Tumor Targeting	Vascular Tumor Targeting
Pharmacokinetic system	Multiple compartment	Single compartment
Primary target cell type	Malignant tumor cells	Non-malignant activated tumor vascular cells
Target cell accessibility	Poor	Good
Receptor	Tumor antigen	Angiogenic markers
Receptor expression	Uneven and heterogeneous	Possibly uniform and homogeneous
Homing moiety	Monoclonal antibodies	Monoclonal antibodies, peptides
Therapeutic moiety	Cytotoxic drug	Cytotoxic drug, angiogenesis inhibitors
Target cell genome	Genetically unstable cells	Genetically stable, diploid cells
Acquired cytotoxic drug resistance	Yes	Not yet reported
Posttargeting amplification loop	Absent	Present
Complete elimination of target cell	Required	Not required
Potential clinical applicability	Limited by antigen heterogeneity	Broad, as angiogenic markers may be shared

cell; partial denuding of the endothelium is likely to lead to the formation of an occlusive thrombus that stops flow to the part of the tumor served by the vessel. In addition, tumor vasculature targeting has an intrinsic amplification mechanism. It has been estimated that elimination of a single endothelial cell can inhibit the growth of approximately 100 tumor cells (Denekamp 1993). Moreover, tumor endothelial cells are diploid and non-malignant, and they are unlikely to lose cell surface target receptors or acquire drug resistance through mutation and clonal evolution.

Oncologists have long recognized that tumors commonly develop resistance to chemotherapy, whereas normal tissues do not. Therefore, toxicity to normal tissues such as chemotherapy-induced myelosuppression continues to occur even after tumor cells have become drug resistant and progress in the face of continued treatment. Endothelial cells, being non-malignant cells, are expected to behave in a manner analogous to bone marrow cells. Long-term antiangiogenic therapy has not been shown to produce drug resistance in experimental animals (Boehm et al. 1997) or in clinical trials (Folkman 1997).

32.2.2 MOLECULAR TARGETS IN ANGIOGENIC TUMOR VASCULATURE

For successful targeting to angiogenesis-associated blood vessels, angiogenic endothelial cells need to be discriminated from the normal quiescent endothelium. In the past decade, using cellular and molecular biological approaches, many receptors, e.g., $\alpha_v\beta_3$ (vitronectin receptor), VEGFR, MMP-2/-9, endoglin (CD105), and aminopeptidase N (CD13), have been identified to be differentially upregulated on tumor endothelial cells. These receptors can be explored as target determinants for drug targeting purposes (Table 32.2).

Several approaches exploiting different target epitopes/determinants were investigated for their potential to interfere with the endothelial neovascularization. VEGFR-I and VEGFR-2 are over-expressed on tumor vasculature, while being present at a low density in the surrounding normal tissues. VEGF-dephtheria toxin conjugate treatment of tumor bearing mice resulted in selective vascular damage in the tumor tissue and inhibited tumor growth (Olson et al. 1997). Histological studies revealed that conjugate treatment spared the blood vessels of normal tissues such as liver, lung, and kidney from being damaged.

Endoglin (CDI05) is a transmembrane glycoprotein, over-expressed in the vasculature of tumors and other tissues undergoing vascular remodelling (Miller et al. 1999). Differential

TABLE 32.2
Potential Target Epitopes on Tumor Vascular Endothelium

Target Epitope	Reference
30.5-kDa antigen	Hagemeier et al. (1986)
CD34	Schlingemann et al. (1990)
VEGF/VEGF receptor complex	Ramakrishnan et al. (1996)
VEGF receptor	Dvorak et al. (1991)
Endosialin	Rettig et al. (1992)
E-selectin	Nguyen et al. (1993)
α_V integrins	Brooks et al. (1994)
Endoglin	Burrows et al. (1995)
Tie-2	Sato et al. (1995)
TNF α receptor	Eggermont et al. (1996)
CD44	Griffioen et al. (1997)
Angiostatin receptor	Moser et al. (1999)
Endostatin receptor	Chang et al. (1999)
MMP-2/MMP-9	Koivunen et al. (1999)

up-regulation of endoglin presents an interesting opportunity to the selective delivery of cytotoxic molecules to the target endothelial cells. Several monoclonal antibodies specific for human endoglin have been produced (Seon et al. 1997). Radioiodinated monoclonal antibodies (10 μi) given to tumor-bearing animals significantly inhibited the tumor growth. This indicates the clinical potential of cytotoxic agent targeting using endoglin (Tabata et al. 1999).

Other receptors over-expressed in tumor vessels include FGF receptors and Tie-2 receptors. Davol and colleagues prepared an endothelial cell-specific cytotoxic conjugate (Davol et al. 1995) by chemically linking the plant-derived ribosomal inhibitory protein saporin to FGF. The FGF-saporin conjugate inhibited proliferation of endothelial cells effectively in vitro. Radioimmunotherapy using ^{213}Bi or ^{131}I targeted to lung vasculature with a monoclonal antibody against thrombomodulin has been reported (Kennel and Mirzadeh 1998). Because thrombomodulin is selectively expressed in the pulmonary blood vessels, the antibody homed to lungs resulted in destruction of small tumor colonies. Cationic liposomes were used to target angiogenic vasculature in mouse pancreatic islet cell tumors. Confocal microscopy demonstrated a 15- to 33-fold preferentially higher uptake of the liposomes by angiogenic than normal endothelial cells (Thurston et al. 1998).

Erkki Ruoslahti and colleagues (1996) developed a novel targeting strategy by using polypeptidal system capable of delivering cytotoxic drugs selectively to integrins. An in vivo selection of phage display libraries identified peptides that specifically home to tumor blood vessels. Ruoslahti's research group identified two major classes of peptides, one containing the RGD motif and the other containing an NGR motif. These polypeptides were then chemically linked to the anti-cancer drug doxorubicin. Treatment of breast carcinoma-bearing mice with the conjugated doxorubicin caused selective vascular damage in the tumors with consequential strong antitumor effect at a 10–40 times lower concentration as compared of free doxorubicin, whereas liver and heart toxicity was significantly reduced compared to that observed with free doxorubicin (Arap, Pasqualini, and Ruoslahti 1998). Their results illustrate the potential of targeting therapeutic agents to integrins expressed on the vasculature of tumors as an effective means of cancer treatment. Other approaches have also been used, e.g., toxic drugs (Arap, Pasqualini, and Ruoslahti 1998; Arora et al. 1999), apoptosis-inducing agents (Ellerby et al. 1999), cytokines (Curnis et al. 2000), and therapeutic genes (Hood et al. 2002) were delivered to the tumor endothelial cells, leading to reduced tumor growth. Furthermore, angiogenesis and tumor growth were inhibited upon occlusion of blood vessels by targeting coagulation factors to tumor vasculature (Huang et al. 1997; Ran et al. 1998). Recently, immune effector cells have successfully been targeted to angiogenic blood vessels, with resultant lysis of tumor endothelial cells and suppression of tumor growth (Niederman et al. 2002).

32.3 ROLE OF RGD IN INTEGRIN- AND FIBRONECTIN-MEDIATED BIOEVENTS

Among the many molecules that are selectively expressed in angiogenic vasculature, the integrins show particular promise in tumor targeting. Saiki and co-workers reported that tumor angiogenesis can be inhibited by blocking the interaction between integrins and the RGD motif-containing extra cellular matrix proteins (Saiki et al. 1990). One of the most studied target epitopes on angiogenic neovasculature is the $\alpha_v\beta_3$ integrin. Studies show that $\alpha_v\beta_3$ and $\alpha_v\beta_5$ integrins are upregulated in angiogenic endothelial cells (Brooks, Clark, and Cheresh, 1994a) and the inhibition of these integrins by antibodies, cyclic RGD peptides (Brooks et al. 1994b) and RGD peptidomimetics (Carron et al. 1998) can block neovascularization.

32.3.1 INTEGRINS

Members of the integrin family of adhesion molecules are non-covalently-associated α/β heterodimers that mediate cell–cell, cell–extracellular matrix, and cell–pathogen interactions by binding to distinct, but often overlapping, combinations of ligands. Eighteen different integrin α-subunits and eight different β-subunits have been identified in vertebrates that form at least 24 α/β hetero-dimers, perhaps making integrins the most structurally and functionally diverse family of cell-adhesion molecules (Hynes 1992; Springer 1994) (Figure 32.2). Half of integrin α-subunits contains inserted (I) domains that are the principal ligand-binding domains (Shimaoka 2002). The complexity and structural and functional diversity of integrins allow this family of adhesion molecules to play a pivotal role in broad contexts of biology, including inflammation, innate and antigen specific immunity, haemostasis, wound healing, tissue morphogenesis, and regulation of cellular growth and differentiation. Conversely, disregulation of integrins is involved in the patho-genesis of many diseased states from autoimmunity to thrombotic vascular diseases to cancer metastasis (Curley, Blum, and Humphries 1999). Therefore, extensive efforts have been directed toward the discovery and development of integrins antagonists for clinical applications.

Significant advances have been made in targeting αIIbβ3 integrins on platelets for inhibiting thrombosis (Scarborough and Gretler 2000), αvβ3 and αvβ5 for blocking tumor metastasis, angio-genesis and bone resorption (Varner and Cheresh 1996) and β2 integrins and α4 integrins on leukocytes for treating autoimmune diseases and other inflammatory disorders (Giblin and Kelly 2001; Yusuf-Makagiansar et al. 2002). Small-molecules, i.e., integrin inhibitors not only interfere with ligand binding but also stabilize particular integrin conformations that have provided insights into integrin structural rearrangements.

The general structural features of all integrins appear to be similar. Both α- and β-subunits are transmembrane glycoproteins with large globular amino-terminal extracellular domains that, together, make up an ellipsoidal head (Figure 32.3). Each subunit provides a relatively thin leg that traverses across the plasma membrane and ends into a relatively small cytoplasmic tail of less than 60 amino acids. The only known integrin that does not fit this general description is β4 integrin that has a cytoplasmic domain of close to 1000 amino acids (Suzuki and Naitoh 1990; Tamura et al. 1990).

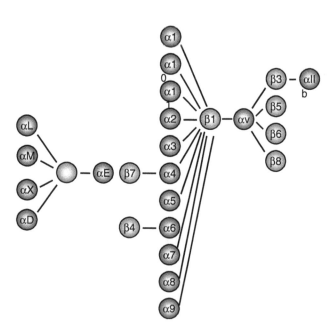

FIGURE 32.2 Integrin heterodimer composition. Integrin α- and β-subunits form 24 heterodimers that recognize distinct, but overlapping, ligands. (Adapted from Hynes, R. O., *Cell*, 69, 11, 1992.)

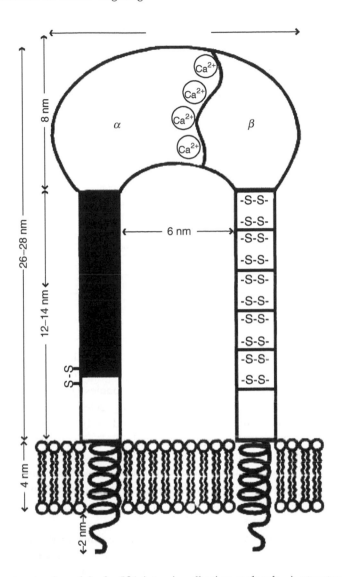

FIGURE 32.3 The structural model of $\alpha 5\beta 1$ integrin adhesion molecule. A structural model based on predicted primary structure and electron microscopy visualization of purified integrin is shown. The $\alpha 5$-subunit consists of part of the globular extracellular domain and one leg that contains predicted 12 β strands (gray box). The $\beta 1$- subunit makes up the remainder of the globular extracellular domain. The $\beta 1$- subunit also contains five cysteine-rich repeats. Both the $\alpha 5$- and $\beta 1$- subunits have transmembrane domains and small cytoplasmic domains. (Adapted from Varner, J. A. and Cheresh, D. A., *Important Adv. Oncol.*, 69, 1996.)

The β_1 integrins that can bind fibronectin include $\alpha_3\beta_1$, $\alpha_4\beta_1$, $\alpha_5\beta_1$, and $\alpha_v\beta_1$. Many integrins, including the $\alpha_3\beta_1$, $\alpha_5\beta_1$, $\alpha_v\beta_1$, $\alpha_v\beta_3$, $\alpha v\beta_5$, $\alpha_v\beta_6$, and the $\alpha_{IIb}\beta_3$, recognize the RGD site located in adhesive proteins (Table 32.3). The fibronectin-specific integrin that consists of an α_5 subunit and a β_1 subunit is the major fibronectin receptor expressed on most of the cells. This integrin mediates such cellular responses to fibronectin as adhesion, migration, assembly of a cytoskeleton, and assembly of the fibronectin extracellular matrix (The $\alpha_5\beta_1$ integrin interacts with the central cell adhesive region of fibronectin and requires both the RGD and synergy sites for maximal binding (Obara and Yoshizato 1995). The major platelet integrin $\alpha_{IIb}\beta_3$ also recognizes a similar synergy site in fibronectin for mediating platelet interactions with fibronectin (Bowditch et al. 1994).

TABLE 32.3
Integrin Receptors and Their Binding Sites

Receptor	Ligand	Receptor-Mediated Action	Amino Acid Sequence Recognized
$\alpha_2\beta_1$	Collagen	Adhesion	DGEA
$\alpha_5\beta_1$	Fibronectin	Adhesion	RGD
$\alpha_6\beta_1$	Laminin	Adhesion	YIGSR
$\alpha_{IIb}\beta_3$	Fibrinogen	Aggregation	KQAGDV or RGD
	Fibronectin	Aggregation	RGD
	von Willebrand factor	Aggregation	RGD
	Vitronectin	Aggregation	RGD
$\alpha_{\varpi}\beta_3$	Vitronectin	Adhesion	RGD
	Fibrinogen	Adhesion	RGD
	Fibronectinvon	Adhesion	RGD
	von Willebrand factor	Adhesion	RGD

32.3.2 FIBRONECTIN

Fibronectin is a large adhesive glycoprotein found in extracellular matrices and body fluids (Mosher 1989; Carsons 1990; Hynes 1990). The primary structure of fibronectin is comprised of three different types of homologous repeating units or modules (Figure 32.4). There are several alternatively spliced forms of fibronectin that result from the deletion or insertion of complete type III modules. In addition, one particular region designated as IIICS can be partially inserted or deleted in some isoforms by alternative splicing. The homologous modules comprising of fibronectin are arranged in protease-resistant domains that are separated by more flexible, protease-susceptible regions. When cleaved from intact fibronectin by partial proteolysis or expressed in bacterial or mammalian cells and individually purified, these domains often retain the specific binding functions of intact fibronectin such as those for he-parin, fibrin, denatured collagen (gelatin), and cell surface receptors.

Fibronectin contains at least two distinct regions that can independently interact with distinct cell surface receptors. The first fibronectin cell-adhesive site to be identified was isolated in the form of protease-resistant fragments of 110–120 kDa, 75 kDa, and 37 kDa (Ruoslahti 1981; Hayashi and Yamada 1983; Zardi et al. 1985; Nagai et al. 1991) derived from the central portion of the protein. Such fragments of fibronectin retained similar cell adhesive activities as those of intact fibronectin. The cell adhesive activity attributed to these fragments was initially localized to the tenth type III module in the form of an 11.5 kDa pepsin fragment (Pierschbacher,

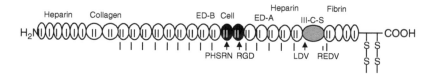

FIGURE 32.4 Structure of fibronectin. Fibronectin is composed of three types of internal repeating modules designated as Type I, Type II, and Type II. The ED-A, ED-B, and III-C-S modeules can be present or absent in some forms of fibronectin as a result of alternative splicing. The binding domains are indicated at the top. The central cell-binding domain consists of ninth and tenth Type III nodules, containing the minimal PHSRN and RGD cell recognition sequence (Adapted from Mosher, D. F., *Fibronectin*, Academic Press, NY, p.474, 1989.)

Hayman, and Ruoslahti 1981) and to a smaller peptide with the sequence Gly-Arg-Gly-Asp-Ser (GRGDS). Although the 11.5 kDa fragment and synthetic peptides containing the RGD sequence can inhibit fibronectin cell-adhesive functions in vitro and in vivo when added as soluble inhibitors (Pierschbacher and Ruoslahti 1984; Lash, Linask, and Yamada 1987), they only poorly promote cell adhesion mediated by the major fibronectin receptor $\alpha_5\beta_1$ integrin, and their affinities are too low to be estimated in direct binding studies (Akiyama and Yamada 1985), suggesting that sequences outside of the tenth type III module are also important for maximal cell binding and adhesion.

This additional fibronectin sequence that is important for maximal cell adhesive activity was identified and characterized using a series of mutants of the central fibronectin cell adhesive region expressed in E. coli (Aota, Nagai, and Yamada 1991) and anti-fibronectin monoclonal antibodies (Nagai et al. 1991). Fragments containing the RGD sequence but truncated approximately 10–14 kDa to the amino-terminal side of the RGD sequence were <4% as active as compared to intact fibronectin in mediating cell adhesion, whereas fragments containing these amino terminal sequences retained >97% of the activity of intact fibronectin. The novel, amino-terminal (non-RGD) site appeared to act synergistically with the RGD site to promote better cell adhesion, leading to its designation as a synergistic adhesive site or synergy site.

The biological function of the synergistic cell adhesive site was characterized by using a panel of anti-fibronectin monoclonal antibodies (mAbs) developed to bind a 37 kDa cell adhesive fibronectin fragment. One of these antibodies, designated 8E3, bound to the ninth type III module at a site approximately 14–16 kDa to the amino-terminal side of the RGD sequence and close to the synergy site identified by mutational studies. Other mAbs that inhibited cell adhesion such as 333 (Akiyama et al. 1985) and 16G3 bound to the tenth type III module and inhibited the RGD site. Interestingly, there was also an antibody, designated 13G12, that bound to fibronectin between the RGD and synergy sites but did not inhibit cell adhesion. Antibodies that bound near the RGD and synergy sites could each individually inhibit cell spreading at high concentrations. Furthermore, mAb 8E3 and mAb 16G3, at concentrations too low to inhibit individually spreading, can be highly inhibitory in combination, underscoring the synergistic nature of the two cell adhesive sites. Antibody inhibition experiments have also shown that both the RGD and synergy sites function in cell migration and cytoskeleton assembly on fibronectin substrates as well as in the assembly of a fibronectin extracellular matrix (Nagai et al. 1991), indicating that both sites are required for a range of fibronectin activities.

32.3.3 RGD-Mediated Metastasis Inhibition

A major cause of morbidity and death as a result of cancer is the metastasis of cells from the primary tumor to distant sites where secondary tumors become established. As schematically shown in Figure 32.5, metastasis is a multistep process. These steps include detachment of cells from the tumor mass, degradation of basement membrane, migration to and invasion into the vascular or lymphatic systems, arrest at a distant site, adhesion to the vascular endothelium, degradation of basement membrane, extravasation, and migration to and proliferation at the secondary site. Many of these steps require cell adhesive interactions or loss of adhesion. Tumor cell adhesion to components of the extracellular matrix and basement membranes is mediated by specific cell surface receptors that bind to extracellular adhesive proteins. The fibronectin-integrin system has provided a valuable model system for the study of molecular mechanisms of ligand-receptor interactions involved in cell adhesive steps in metastasis.

Several different assay systems have been used to directly analyze the role of fibronectin and integrins in in vivo metastasis, but they all fall into two broad classes. The experimental metastasis models mainly involve the intravenous injection of tumor cell suspensions into mice and subsequent quantitation of either the number of metastatic colonies or the sizes of colonies usually in the lungs or liver. Experimental metastasis models, however, exclude the earlier steps of the

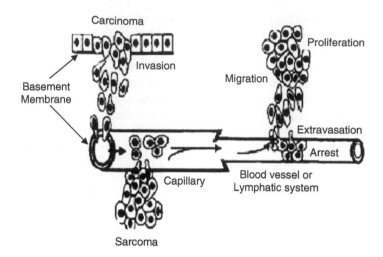

FIGURE 32.5 Major steps in metastasis. Tumor cells are shown invading the circulatory system from a carcinoma by degrading and migrating through a basement membrane or from a sarcoma. Cells separate from the primary tumor, enter the vasculature, and eventually arrest in capillaries where they penetrate and migrate through basement membrane and underlying connective tissue to the metastatic site where colonization and proliferation occur.

metastatic process involving detachment from the tumor mass and invasion into the vasculature and, subsequently, replicate only following the final steps of hematogenous metastasis. Spontaneous metastasis models where tumors or cells are implanted into mice and subsequent metastases to distant organs are quantitated either by counting colonies or measuring colony size are more complicated, but they are required to analyze of the earlier steps of the metastatic cascade.

Synthetic peptides containing the Gly-Arg-Gly-Asp-Ser (GRGDS) sequence derived from fibronectin were assayed by examining their effects on lung colonization of B16-F10 murine melanoma cells in syngeneic C57BL/6 mice using an experimental metastasis model (Humphries, Olden, and Yamada 1986). GRGDS peptide that had a circulatory half-life of approximately eight minutes specifically inhibited lung colonization in a concentration-dependent manner. The major effect of the GRGDS peptide appeared to be to inhibit arrest of melanoma cells in the lungs without affecting the size of either melanoma cell clusters in suspension or lung colonies. The efficacy of peptide treatment was unaltered in animals with impaired platelet function and in animals lacking natural killer cells. Taken together, these results suggest that the GRGDS peptide could inhibit metastasis by disrupting an early adhesive process. As shown in Figure 32.6, a single administration of the GRGDS peptide also had the dramatic effect of increasing the survival of mice when co-injected with melanoma cells (Humphries, Yamada, and Olden 1988).

RGD peptides and RGD mimetics are very promising molecules. RGD peptides not only inhibit metastatic colony formation but a single administration of the peptide can dramatically increase animal survival also however, the use of such agents have some drawbacks. One potential problem is the relatively high concentrations of peptide required for activity. Possible solutions to the problems may include the use of cyclic peptides and designer peptides that reportedly increased the affinity for cell surface integrins or the use of repeating polymers containing RGD units (Saiki et al. 1989; Murata et al. 1991; Yamamoto et al. 1994). Polymers containing RGD sequences have the dual advantages of augmenting the affinity of small peptide sequences through multivalent interactions and the possibility of solving the problem of the relatively short circulatory half-life of GRGDS peptides. An alternative approach would be to develop novel, small, synthetic integrin antagonists. Such compounds that inhibit RGD-dependent and LDV-dependent cell adhesion have already been synthesized and tested in mice (Greenspoon et al. 1993, 1994).

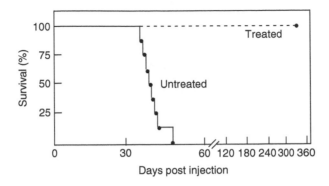

FIGURE 32.6 Effect of GRGDS peptide on survival of C57BL/6 mice injected with B16F10 murine melanoma cells. Mice were co-injected with 3×10^4 B16F10 melanoma cells mixed with 3 mg GRGDS paptide. (Data are from Humphries et al., *J. Clin. Invest.*, 81, 782, 1988.)

32.4 RGD-MODIFIED LIPOSOMES FOR CANCER THERAPEUTICS

Cytotoxic drug incorporation into liposomes has been reported since the mid-1970s. These early reports laid the groundwork for selecting therapeutic agents amenable to incorporation by liposomes as well as determining optimal liposome size and net charge required for effective drug delivery (Gregoriadis 1976). Early reports also noted the problems associated with liposome-mediated drug delivery; liposomes were removed from circulation by fixed macrophages in the RES, particularly in the liver and spleen. Although first-generation liposome-incorporated drugs may be effective in macrophage-related diseases, particularly in the liver and spleen, many other tumors would require other drug-targeting mechanisms. Later studies refined many of the considerations required for effective liposome-mediated drug delivery (Drummond et al. 1999).

In 2005, Pattillo et al. developed RGD-modified immunoliposmes for targeting the antivascular drug combretastatin to irradiated mouse melanomas. Combretastatin was incorporated into liposomes with surfaces modified by the addition of cyclo(Arg-Gly-Asp-D-Phe-Cys) (RGD) to create an immunoliposome. In this transplanted tumor model (B16-F10 melanoma), there was no significant increase in the tumor volume when treated with a single dose of 5-Gy radiation and RGD modified immunoliposomes (14.5 mg/kg of combretastatin) during the initial six days post treatment; all other treatment groups exhibited exponential growth curves after three days. The treatment resulted in a 5.1-day tumor growth delay compared to untreated controls. These findings indicate that preferential targeting of antivascular drugs to irradiated tumors results in significant tumor growth delay.

Various strategies are adopted to target the sites of tumor-associated angiogenesis or other inflammatory environments with liposome-immobilized ligands where adhesion receptors (or ligands) serve as molecular targets (Figure 32.7). Targeting can be achieved using RGD or YIGSR immobilized on liposome surface and interacting with integrin receptors on normal or malignant cells that over-express them. Targeting can be achieved using synthetic antibodies anchored on the liposomes and bind integrins and other cell adhesion molecules molecules by mimicking their natural ligands. Targeting can be achieved by immobilizing specific carbohydrate ligands (sLe^x and sLe^a) on liposomes that show specific interaction with selectin. These strategies are directed either to block the normal recruitment of leucocytes during vasculogenesis or for selective targeting of chemotherapeutic agents to tumor-associated vasculature.

Cyclic RGD peptide, cyclo(Arg-Gly-Asp-Phy-Lys) anchored sterically stabilized liposomes (RGD-SL) were investigated for selective and preferential presentation of carrier contents at angiogenic endothelial cells over-expressing alphavbeta3 integrins on and around tumor tissue and for assessing their targetabilty (Dubey et al. 2004; Dubey 2005). Liposomes were prepared using

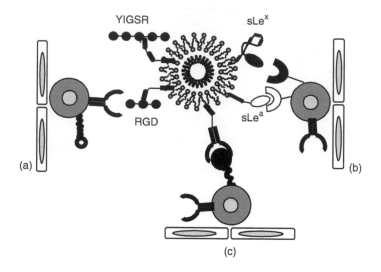

FIGURE 32.7 Liposome-immobilized ligands to target the cell adhesion molecules on tumor vasculature and extracellular matrix. These liposomes serve as artificial leucocyte to mediate interactions with integrins (a), integrins and other CAM molecules (b), and selectins (c) with the help of immobilized ligands or anti-receptor antibody. (Adapted from Vyas, S. P., and Khar, R. K., *Novel Carrier Systems*, CBS publishers, New Delhi, p.535, 2001.)

distearoylphosphatidylcholine (DSPC), cholesterol, and distearoylphosphatidylethanolamine-poly-ethyleneglycol-RGD peptide conjugate (DSPE-PEG-RGD) in a molar ratio 56:39:5. The control RAD peptide anchored sterically stabilized liposomes (RAD-SL), and liposome with 5 mol% PEG (SL) without peptide conjugate that had similar lipid composition were used for comparison. The average size of all liposome preparations prepared was approximately 105 nm and maximum drug entrapment was $10.5 +/- 1.1\%$. In vitro endothelial cell binding of liposomes exhibited 7-fold higher binding of RGD-SL to HUVEC in comparison to the SL and RAD-SL (Figure 32.8a).

Table 32.4 and Figure 32.9 indicate RGD-modified liposomes mediated spontaneous lung metastasis and angiogenesis assays, respectively. RGD peptide-anchored liposomes are significantly ($p < 0.01$) effective in the prevention of lung metastasis and angiogenesis compared to

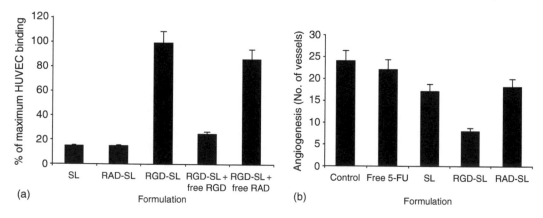

FIGURE 32.8 (a) Binding of RGD-modified liposomes with HUVEC cells. (b) Effect of various formulations on tumor angiogenesis by an intradermal injection of B16F10 melanoma. Results are given as means (S.D.) $N = 5$ ($P < 0.01$). 5-FU, SL, RGD-SL, and RAD-SL indicate free 5-fluorouracil, stealth liposomes (SL), RGD-anchored stealth liposomes (RGD-SL), and RAD-anchored stealth liposomes (RAD-SL), respectively (Adapted from Dubey, P. K., et al., *J. Drug. Target.*, 12, 257, 2004).

TABLE 32.4

Effect of Free 5-FU, SL, RGD-SL, and RAD-SL on Spontaneous Lung Metastasis in BALB/c Mice by an Intra-Footpad Injection of B16F10 Melanoma

Formulations	Average Number of Metastasis/ Mouse	Average Number of Metastasis/ Tumor Bearing Mouse	Number of Metastasis Free Mice (N = 10)
Control	52 ± 15.5	52 ± 15.5	0
Free 5-FU	46 ± 13.6	46 ± 13.6	0
SL	26 ± 8.6	32 ± 7.65	2
RGD-SL	$8 \pm 4.2^*$	16 ± 6.58	5
RAD-SL	28 ± 6.5	31 ± 8.7	1

Source: Dubey et al., *J. Drug. Target.,* 12, 2004. $^*P < 0.01$

free 5-FU, SL, and RAD-SL. In therapeutic experiments, 5-FU, SL, RGD-SL, and RAD-SL were intravenously administered on day 4 at the dose of 10 mg 5-FU/kg body weight to B16F10 tumor bearing BALB/c mice, resulting in effective regression of tumors compared with free 5-FU, SL, and RAD-SL (Dubey et al. 2004). Results indicated that cyclic RGD peptide anchored sterically stabilized liposomes bearing 5-FU were more significantly ($p < 0.01$) active against primary tumor and metastasis than the non-targeted sterically stabilized liposomes and free drug (Figure 32.9). Vascular damage and reduced vascularization were recorded in tumors treated with RGD-SL, but not in the tumor treated with SL (Figure 32.10). These results also confirmed that RGD peptide-anchored liposomes cause a marked improvement in therapeutic efficacy and growth inhibitory effect on tumors.

RGD-modified sterically stabilized liposomes have also been evaluated to improve the anti-tumor efficacy of doxorubicin (Schiffelers et al. 2003; Xiong et al. 2005). RGD peptide was coupled to the distal end of the poly (ethylene glycol)-coated liposomes. RGD peptide anchored sterically

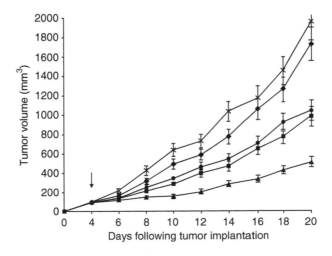

FIGURE 32.9 Anti-tumor efficacy of control ($\times$), free 5-FU ($\blacklozenge$), SL ($\blacksquare$), RAD-SL ($\bullet$), and RGD-SL ($\blacktriangle$) in mice bearing B16F10 melanoma subcutaneously inoculated in the back. The arrow represents the time of injection. Results are given as means (S.D.) N = 10 (P < 0.01). (Adapted from Dubey, P.K. et al., *J. Drug. Target.*, 12, 257, 2004.)

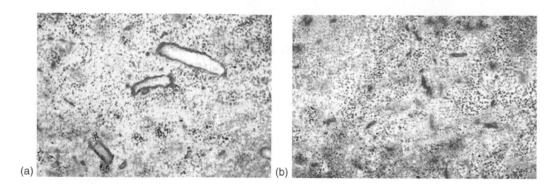

FIGURE 32.10 (See color insert following page 522.) Histopathological analysis (hematoxylin and eosin stain) of B16F10 tumors treated with (a) Stealth liposomes and (b) RGD-modified stealth liposomes (250X). (Adapted from Dubey, P. K., Development and characterization of biochemical based strategies for tumor targeting. Ph. D. thesis, Department of Pharmaceutical Sciences, Dr. H. S. Gour University, Sagar (MP) India, 2005.)

stabilized liposomes (dose 5 mg doxorubicin/kg) have demonstrated prolonged circulation time and increased tumor accumulation and effective retardation in tumor growth compared with sterically stabilized liposomes without RGD modification.

Holig and coworkers have isolated from phage display RGD motif libraries with novel high-affinity cyclic RGD peptides on the basis of their selectivity towards endothelial or melanoma cells (Holig et al. 2004). Although the starting sequences contained only two cysteine residues flanking the RGD motif, several of the isolated peptides possessed four cysteine residues. A high-affinity peptide (RGD10) (DGARYCRGDCFDG) constrained by only one disulfide bond was used to generate novel lipopeptides composed of a lipid anchor, a short flexible spacer, and the peptide ligand conjugated to the spacer end. Incorporation of RGD10 lipopeptides into liposomes resulted in specific and efficient binding of the liposomes to integrin-expressing cells. In vivo experiments applying doxorubicin-loaded RGD10 liposomes in a C26 colon carcinoma mouse model demonstrated improved efficacy compared with free doxorubicin and untargeted liposomes.

Administration of large amounts of synthetic peptides based on the Arg-Gly-Asp (RGD) sequence has been shown to suppress tumor metastasis. To overcome the rapid degradation of peptides in the circulation, an RGD mimetic, L-arginyl-6-aminohexanoic acid (NOK), was synthesized and conjugated with phosphatidylethanolamine (PE) (NOK-PE) for liposomalization (Kurohane, Namba, and Oku 2000). Cell adhesion assays revealed that B16BL6 murine melanoma cells adhered to immobilized NOK-PE. This adhesion was inhibited by the addition of either soluble RGDS or NOK at similar concentrations in a dose-dependent manner. Administration of NOK-PE liposomes (equivalent to approximately 500 microg RGD peptides) via the tail vein completely inhibited lung colonization of B 16BL6 cells. The same dose of soluble NOK was not effective in inhibition of the tumor metastasis. In addition, injection of NOK-PE liposomes via the tail vein inhibited spontaneous lung metastasis of B16BL6 cells from the primary tumor site in the hind footpad. These results suggest that NOK, a structural mimetic of RGD, is capable of suppressing metastasis by blocking the binding of the integrins present on tumor cells to the RGD-containing extracellular matrix.

32.5 RGD-MODIFIED LIPOSOMES IN CANCER GENE THERAPY

Gene therapy, by definition, aims at modifying the genetic program of a cell toward a therapeutic or prophylactic goal. Several gene therapy strategies for tumors are currently under evaluation and some of them include: the modification of the function of oncogenes and tumor suppressor genes;

the modification of the host immune response toward the tumor; the disruption of the tumor neovascularization; the lysis of tumor cells with replication-competent viruses, and suicide gene therapy where an inactive prodrug is converted into a cytotoxic drug by gene-expressed enzymes. Both viral and non-viral gene vectors have been investigated and well documented (Ledley 1995). However, the inefficiency of current gene vectors in infecting targeted cells and their inability to selectively access the diseased cells systemically distributed are two major impediments that have to be overcome for further successful clinical applications.

RGD peptides have been used to taget the lipid-protamine-DNA (LPD) lipopolyplexes to tumor cells (MDA-MB-231), expressing appropriate integrin receptors (Harvie et al. 2003). The incorporation of pegylated lipid into Lipid-Protamine-DNA (LPD-PEG) lipopolyplexes causes a decrease of their in vitro transfection activity. This can be partially attributed to a reduction in particle binding to cells. To restore particle binding and specifically target LPD formulations to tumor cells, the lipid-peptide conjugate DSPE-PEG5K-succinyl-ACDCRGDCFCG-COOH (DSPE-PEG5K-RGD-4C) was generated and incorporated into LPD formulations (LPD-PEG-RGD). LPD-PEG-RGD was characterized with respect to its biophysical and biological properties. The incorporation of DSPE-PEG5K-RGD-4C ligands into LPD formulations results in a 5- and 15-fold increase in the LPD-PEG-RGD binding and uptake, respectively, over an LPD-PEG formulation. Enhancement of binding and uptake resulted in a 100-fold enhancement of transfection activity. Moreover, this transfection enhancement was specific to cells expressing appropriate integrin receptors (MDA-MB-231). Huh7 cells, known for their low level of $\alpha_v\beta_3$ and $\alpha_v\beta_5$ integrin expression, failed to show RGD-mediated transfection enhancement. This transfection enhancement can be abolished in a competitive manner using free RGD peptide, but not an RGE control peptide. Results demonstrated RGD-mediated enhanced LPD-PEG cell binding and transfection in cells expressing the integrin receptor. These formulations provide the basis for effective, targeted, systemic gene delivery.

In 2002, Fahr et al. developed a novel liposomal vector (Artificial Virus Particles; AVPs) for cancer gene therapy. Artificial virus-like particles (AVPs) represent a novel type of liposomal vector, resembling retroviral envelopes. AVPs are serum-resistant and non-toxic and can be endowed with a peptide ligand as a targeting device. These workers investigated if AVPs carrying cyclic peptides with an RGD integrin-binding motif (RGD-AVPs) were suitable for the specific and efficient transduction of human melanoma cells. In the experimentation, they prepared Plasmid DNA-low molecular weight non-linear polyethyleneimine complexes and packaged into anionic liposomes. Transduction efficiencies were determined after transient transfection of different cell lines in serum-free medium using green fluorescent protein or luciferase reporter genes. It was found that RGD-AVPs transduced human melanoma cells with high efficiencies of >60%. Efficient transduction was clearly dependent on the presence of the cyclic RGD ligand and was selective for melanoma cells. Using the melanocyte-specific tyrosinase promoter to drive transgene expression, the specificity of the vector system could be further enhanced.

REFERENCES

Akiyama, S. K. and Yamada, K. M., Synthetic peptides competitively inhibit both direct binding to fibroblasts and functional biological assays for the purified cell-binding domain of fibronectin, *J. Biol. Chem.*, 260, 10402, 1985.

Akiyama, S. K. et al., The interaction of fibronectin fragments with fibroblastic cells, *J. Biol. Chem.*, 260, 13256, 1985.

Aota, S., Nagai, T., and Yamada, K. M., Characterization of regions of fibronectin besides the arginine-glycine-aspartic acid sequence required for adhesive function of the cell-binding domain using site-directed mutagenesis, *J. Biol. Chem.*, 266, 15938, 1991.

Arap, W., Pasqualini, R., and Ruoslahti, E., Cancer treatment by targeted drug delivery to tumor vasculature in a mouse model, *Science*, 279, 377, 1998.

Arora, N. et al., Vascular endothelial growth factor chimeric toxin is highly active against endothelial cells, *Cancer Res.*, 59, 183, 1999.

Boehm, T. et al., Antiangiogenic therapy of experimental cancer does not induce acquired drug resistance, *Nature*, 390, 404, 1997.

Bowditch, R. D. et al., Identification of a novel integrin-binding site in fibronectin: differential utilization by β, integrins, *J. Biol. Chem.*, 269, 10856, 1994.

Brooks, P. C., Clark, R. A., and Cheresh, D. A., Requirement of vascular integrin αvβ3 for angiogenesis, *Science*, 264, 569, 1994a.

Brooks, P. C. et al., Integrin Alpha-v Beta-3 antagonists promote tumor regression by inducing apoptosis of angiogenic blood vessels, *Cell*, 79, 1157, 1994b.

Brown, P. M. and Silvius, J. R., Mechanisms of delivery of liposome-encapsulated cytosine arabinoside to CV-1 cells in vitro: Fluorescence-microscopic and cytotoxicity studies, *Biochim. Biophys. Acta*, 1023, 341, 1990.

Burrows, F. J. et al., Endoglin is an endothelial cell proliferation marker that is selectively expressed in tumor vasculature, *Clin. Cancer Res.*, 1, 1623, 1995.

Carron, C. P. et al., A. peptidomimetic antagonist of the intergin alpha (V) beta 3 inhibits Leydig cell tunour growth and the development of hypercalcemia of malignancy, *Cancer Res.*, 58, 1930, 1998.

Carsons, S. E., *Fibronectin in Health and Disease*, CRC Press, Boca Raton, 1990.

Chang, Z., Choon, A., and Friedl, A., Endostatin binds to blood vessels in situ independent of heparan sulfate and does not compete for fibroblast growth factor-2 binding, *Am. J. Pathol.*, 155, 71, 1999.

Curley, G. P., Blum, H., and Humphries, M. J., Integrin antagonists, *Cell. Mol. Life Sci.*, 56, 427, 1999.

Curnis, F. et al., Enhancement of tumor neaosis factor alpha antitumour immunothecapetic properties by targeted delivery to aminopeptidase N (CD13), *Nat. Biotechnol.*, 18, 1185, 2000.

Davol, P. et al., Saporin toxins directed to basic fibroblast growth factor receptors effectively target human ovarian teratocarcinoma in an animal model, *Cancer*, 76, 79, 1995.

Denekamp, J., Angiogenesis, neovascular proliferation and vascular pathophysiology as targets for cancer therapy, *Br. J. Radiol.*, 66, 181, 1993.

Drummond, D. C. et al., Optimizing liposomes for delivery of chemotherapeutic agents to solid tumors, *Pharmacol. Rev.*, 51, 691, 1999.

Dubey, P. K., Development and characterization of biochemical based strategies for tumor targeting. Ph. D. thesis, Department of Pharmaceutical Sciences, Dr. H. S. Gour University, Sagar (MP) India, 2005.

Dubey, P. K. et al., Liposomes modified with cyclic RGD peptide for tumor targeting, *J. Drug Target*, 12, 257, 2004.

Dvorak, H. F. et al., Distribution of vascular permeability factor (vascular endothelial growth factor) in tumors: Concentration in tumor blood vessels, *J. Exp. Med.*, 174, 1275, 1991.

Eggermont, A. M. et al., Isolated limb perfusion with high-dose tumor necrosis factor-alpha in combination with interferon-gamma and melphalan for nonresectable extremity soft tissue sarcomas: A multicenter trial, *J. Clin. Oncol.*, 14, 2653, 1996.

Ellerby, H. M. et al., Anticancer activity of targeted pro-apoptotic peptides, *Natl. Med.*, 5, 1032, 1999.

Fahr, A. et al., A new colloidal lipidic system for gene therapy, *J. Liposome Res.*, 12, 37, 2002.

Folkman, J., Angiogenesis and angiogenesis inhibition: An overview, *EXS*, 79, 1, 1997.

Forssen, E. and Willis, W., Ligand targeted liposomes, *Adv. Drug Del. Rev.*, 29, 249, 1998.

Gabizon, A. and Papahadjopoulos, D., The role of surface charge and hydrophilic groups on liposome clearance in vivo, *Biochim. Biophys. Acta*, 1103, 94, 1992.

Gerlag, D. M. et al., Suppression of murine collagen-induced arthritis by targeted apoptosis of synovial neovasculature, *Arthritis Res.*, 3, 357, 2001.

Giblin, P. A. and Kelly, T. A., Antagonists of β2 integrin-mediated cell adhesion, *Ann. Rep. Med. Chem.*, 36, 181, 2001.

Greenspoon, N. et al., Structural analysis of integrin recognition and the inhibition of integrin-mediated cell function by novel nonpeptidic surrogates of the Arg-Gly-Asp sequence, *Biochemistry*, 32, 1001, 1993.

Greenspoon, N. et al., Novel ^{f}F-S-CH, peptide-bond replacement and its utilization in the synthesis of nonpeptidic surrogates of the Leu-Asp-Val sequence that exhibit specific inhibitory activities on CD4$^+$T cell binding to fibronectin, *Int. J. Pept. Res.*, 43, 417, 1994.

Gregoriadis, G., The carrier potential of liposomes in biology and medicine (first of two parts), *N. Engl. J. Med.*, 295, 704, 1976.

Griffioen, A. W. et al., CD44 is an activation antigen on human endothelial cells, involvement in tumor angiogenesis, *Blood*, 90, 1150, 1997.

Gyongyossy-Issa, M. I. C., Muller, W., and Devine, D., The covalent coupling of Arg-Gly-Asp-containing peptides to liposomes: Purification and biochemical function of lipopeptide, *Arch. Biochem. Biophys.*, 353, 101, 1998.

Hagemeier, H. H. et al., A monoclonal antibody reacting with endothelial cells of budding vessels in tumors and inflammatory tissues, and non-reactive with normal adult tissues, *Int. J. Cancer*, 38, 481, 1986.

Harvie, P. et al., Targeting of lipid-protamine-DNA (LPD) lipopolyplexes using RGD motifs, *J. Liposome Res.*, 13, 231, 2003.

Hayashi, M. and Yamada, K. M., Domain structure of the carboxy-terminal half of human plasma fibronectin, *J. Biol. Chem.*, 258, 3332, 1983.

Holig, P. et al., Novel RGD lipopeptides for the targeting of liposomes to integrin-expressing endothelial and melanoma cells, *Protein Eng. Des. Sel.*, 17, 433, 2004.

Hood, J. D. et al., Tumor regression by targeted gene delivery to the neovasculature, *Science*, 269, 2404, 2002.

Huang, X. et al., Tumor infarction in mice by antibody-directed targeting of tissue factor to tumor vasculature, *Science*, 275, 547, 1997.

Humphries, M. J., Olden, K., and Yamada, K. M., A synthetic peptide from fibronectin inhibits experimental metastasis of murine melanoma cells, *Science*, 233, 466, 1986.

Humphries, M. J., Yamada, K. M., and Olden, K., Investigation of the biological effects of anti-cell adhesive synthetic peptides that inhibit experimental metastasis of B16-F10 murine melanoma cells, *J. Clin. Investig.*, 81, 782, 1988.

Hynes, R. O., *Fibronectins*, Springer-Verlag, NY, p. 546, 1990.

Hynes, R. O., Integrins: Versatility, modulation, and signaling in cell adhesion, *Cell*, 69, 11, 1992.

Kennel, S. J. and Mirzadeh, S., Vascular targeted radioimmunotherapy with ^{213}Bi-an-particle emitter, *Nucl. Med. Biol.*, 25, 241, 1998.

Kirpotin, D. et al., Sterically stabilized anti-HER2 liposomes: Design and targeting to human breast cancer cells in vitro, *Biochemistry*, 36, 66, 1997.

Koivunen, E. et al., Tumor targeting with a selective gelatinase inhibitor, *Natl. Biotechnol.*, 17, 768, 1999.

Kurohane, K., Namba, Y., and Oku, N., Liposomes modified with a synthetic Arg-Gly-Asp mimetic inhibit lung metastasis of B16BL6 melanoma cells, *Life Sci.*, 68, 273, 2000.

Lash, J. W., Linask, K. K., and Yamada, K. M., Synthetic peptides that mimic the adhesive recognition signal of fibronectin: Differential effects on cell–cell and cell substratum adhesion in embryonic chick cells, *Dev. Biol.*, 123, 411, 1987.

Ledley, F. D., Nonviral gene therapy: the promise of genes as pharmaceutical products, Hum. *Gene. Ther.*, 6, 1129,1995.

Lopes de Menezes, D. E., Pilarski, L. M., and Allen, T. M., In vitro and in vivo targeting of immunoliposomal doxorubicin to human B-cell lymphoma, *Cancer Res.*, 58, 3320, 1998.

Lundberg, B., Hong, K., and Papahadjopoulos, D., Conjugation of apolipoprotein B with liposome and targeting of cells in culture, *Biochim. Biophys. Acta*, 1149, 305, 1993.

Martin, F. J., Clinical pharmacology and antitumor efficacy of DOXIL (pegylated liposomal doxorubicin), In *Medical Applications of Liposomes*, Lasic, D. D. and Papahadjopoulos, D., Eds., Elsevier Science BV, New York, p. 635, 1998.

Maruyama, K. et al., Characterization of in vitro immunoliposome targeting to pulmonary endothelium, *J. Pharm. Sci.*, 79, 978, 1990.

Miller, D.W. et al., Elevated expression of endoglin, a component of the TGF-β-receptor complex, correlates with proliferation of tumor endothelial cells, *Int. J. Cancer*, 81,568,1999.

Moser, T. L. et al., Angiostatin binds ATP synthase on the surface of human endothelial cells, *Proc. Natl Acad. Sci. USA*, 96, 2811, 1999.

Mosher, D. F., *Fibronectin*, Academic Press, New York, 1989.

Murata, J. et al., Molecular properties of poly(RGD) and its binding capacities to metastatic melanoma cells, *Int. J. Pept. Prot. Res.*, 38, 212, 1991.

Nishiya, T. and Sloan, S., Interaction of RGD liposomes with platelets, *Biochem. Biophys. Res. Commun.*, 224, 242, 1996.

Nguyen, M. et al., A role for sialyl Lewis-X/A glycoconjugates in capillary morphogenesis, Nature, 365, 267,1993. Erratum in: Nature, 366, 368, 1993.

Niederman, T.M. et al., Antitumor activity of cytoxic T lymphocytes engineered to target vascular endothelial growth factor receptors, *Proc. Natl. Acad. Sci. USA*, 99, 709, 2002.

Obara, M. and Yoshizato, K., Possible involvement of the interaction of the a, subunit of the α, β, integrin with the synergistic region of the central cell-binding domain of fibronectin in cells to fibronectin binding, *Exp. Cell Res.*, 216, 273, 1995.

Olson, T. A. et al., Targeting the tumor vasculature: inhibition of tumor growth by a vascular endothelial growth factor toxin conjugate, *Int. J. Cancer*, 73, 865, 1997.

Papahadjopoulos, D. and Gabizon, A. A., Sterically stabilized (stealth) liposomes: Pharmacological properties and drug carrying potentialin cancer, In *Liposomes as Tools in Basic Research and Industry*, Philippot, J. R. and Schuber, F., Eds., CRC Press, Boca Raton, p. 177, 1995.

Papahadjopoulos, D. et al., Sterically stabilized liposomes: Improvements in pharmacokinetics and antitumor therapeutic efficacy, *Proc. Natl Acad. Sci. USA*, 88, 11460, 1991a.

Papahadjopoulos, D. et al., Sterically stabilized liposomes: Improvements in pharmacokinetics and antitumor therapeutic efficacy, *Proc. Natl Acad. Sci. USA*, 88, 11460, 1991b.

Park, J. W. et al., Development of anti-p185 HER2 immunoliposomes for cancer therapy, *Proc. Natl Acad. Sci. USA*, 92, 1327, 1995.

Pattillo, C. B. et al., Targeting of the antivascular drug combretastatin to irradiated tumors results in tumor growth delay, *Pharm. Res.*, 22, 1117, 2005.

Pierschbacher, M. D. and Ruoslahti, E., Cell attachment activity of fibronectin can be duplicated by small synthetic fragments of the molecule, *Nature*, 309, 30, 1984.

Pierschbacher, M. D. and Ruoslahti, E., Influence of stereochemistry of the sequence Arg-Gly- Asp-Xaa on binding specificity in cell adhesion, *J. Biol. Chem.*, 262, 17294, 1987.

Pierschbacher, M. D., Hayman, E. G., and Ruoslahti, E., Location of the cell attachment site in fibronectin with monoclonal antibodies and proteolytic fragments of the molecule, *Cell*, 26, 259, 1981.

Ramakrishnan, S. et al., Vascular endothelial growth factor-toxin conjugate specifically inhibits KDR/Flk-1-positive endothelial cell proliferation in vitro and angiogenesis in vivo, *Cancer Res.*, 56, 1324, 1996.

Ran, S. et al., Infarction of solid Hodgkin's tumors in mice by antibody-directed targeting of tissue factor to tumor vasculature, *Cancer Res.*, 58, 4646, 1998.

Rettig, W. J. et al., Identification of endosialin, a cell surface glycoprotein of vascular endothelial cells in human cancer, *Proc. Natl Acad. Sci. USA*, 89, 10832, 1992.

Ruoslahti, E., RGD and other recognition sequences for integrins, Annu. Rev., *Cell Dev. Biol.*, 12, 697, 1996.

Ruoslahti, E. et al., Alignment of biologically active domains in the fibronectin molecule, *J. Biol. Chem.*, 256, 7277, 1981.

Saiki, I. et al., Inhibition of the metastasis of murine malignant melanoma by synthetic polymeric peptides containing core sequences of cell-adhesive molecules, *Cancer Res.*, 49, 3815, 1989.

Saiki, I. et al., Anti-metastatic and anti-invasive effects of polymeric Arg-Gly-Asp (RGD) peptide, poly (RGD), and its analogues, *Jpn. J. Cancer Res.*, 81, 660, 1990.

Sato, T.N. et al., Distinct roles of the receptor tyrosine kinases Tie-1 and Tie-2 in blood vessel formation, *Nature*, 376, 70, 1995.

Scarborough, R. M. and Gretler, D. D., Platelet glycoprotein IIb-IIIa antagonists as prototypical integrin blockers: Novel parenteral and potential oral antithrombotic agents, *J. Med. Chem.*, 43, 3453, 2000.

Schiffelers, R. M. et al., Anti-tumor efficacy of tumor vasculature-targeted liposomal doxorubicin, *J. Control. Release*, 91, 115, 2003.

Schlingemann, R. O. et al., Leukocyte antigen CD34 is expressed by a subset of cultured endothelial cells and on endothelial abluminal microprocesses in the tumor stroma, *Lab. Investig.*, 62, 690, 1990.

Seon, B. K. et al., Long-lasting complete inhibition of human solid tumors in SCID mice by targeting endothelial cells of tumor vasculature with antihuman endoglin immunotoxin, *Clin. Cancer Res.*, 3,1031,1997.

Shimaoka, M., Takagi, J., and Springer, T. A., Conformational regulation of integrin structure and function, *Annu. Rev. Biophys. Biomol. Struct.*, 31, 485, 2002.

Springer, T. A., Traffic signals for lymphocyte recirculation and leukocyte emigration: The multi-step paradigm, *Cell*, 76, 301, 1994.

Suzuki, S. and Naitoh, Y., Amino acid sequence of a novel integrin 34 subunit and primary expression of the mRNA in epithelial cells, *EMBO J.*, 9, 757, 1990.

Tabata, M. et al., Antiangiogenic radioimmunotherapy of human solid tumors in SCID mice using (125)I-labeled anti-endoglin monoclonal antibodies, *Int. J. Cancer*, 82, 737, 1999.

Tamura, R. N. et al., Epithelial integrin a4p6: Complete primary structure of a_6 and variants forms of P_4, *J. Cell Biol.*, 111(1593), 604, 1990.

Thurston, G. et al., Cationic liposomes target angiogenic endothelial cells in tumors and chronic inflammation in mice, *J. Clin. Invest.*, 101,1401, 1998.

Varner, J. A. and Cheresh, D. A., Tumor angiogenesis and the role of vascular cell integrin avβ3, *Important Adv. Oncol.*, 69, 1996.

Vyas, S. P. and Khar, R. K., *Drug Delivery to Tumor, in Targeted and Controlled Drug Delivery, Novel Carrier Systems*, CBS Publishers, New Delhi, 2001.

Xiong, X. B. et al., Enhanced intracellular uptake of sterically stabilized liposomal Doxorubicin in vitro resulting in improved antitumor activity in vivo, *Pharm. Res.*, 22, 933, 2005.

Yamamoto, S. et al., Antimetastatic effects of synthetic peptides containing the core sequence of the type III connecting segment domain (IIICS) of fibronectin, *Anticancer Drugs*, 5, 4424, 1994.

Yusuf-Makagiansar, H. et al., Inhibition of LFA-1/ICAM-1 and VLA-4/VCAM-1 as a therapeutic approach to inflammation and autoimmune diseases, *Med. Res. Rev.*, 22, 146, 2002.

Zardi, L. et al., Elution of fibronectin proteolytic fragments from a hydroxyapatite chromatography column, *Eur. J. BioChem.*, 146, 5711, 1985.

33 Folate Receptor-Targeted Liposomes for Cancer Therapy

Xiaobin B. Zhao, Natarajan Muthusamy,
John C. Byrd, and Robert J. Lee

CONTENTS

33.1 Introduction 663
33.2 The Folate Receptor as a Cancer-Specific Cellular Marker................................ 663
33.3 Targeting FR for Cancer Therapy.. 664
33.4 Liposomes as Drug Carriers.. 665
33.5 Formulation Strategies for FR-Targeted Liposomes 667
33.6 In Vitro Drug Delivery Using FR-Targeted Liposomes 668
33.7 Pharmacokinetic and Biodistribution Properties of FR-Targeted Liposomes 670
33.8 In Vivo Therapeutic Efficacy of FR-Targeted Liposomes 671
33.9 Conclusions and Future Directions.. 672
Acknowledgments 673
References.. 673

33.1 INTRODUCTION

Targeted drug delivery is a promising strategy to improve both the efficacy and safety of treatment. This is especially attractive for cancer therapy as most anti-cancer drugs have only marginal therapeutic index. Various cellular surface receptors and antigens have been targeted through different strategies, among which folate receptor (FR) has been one of the targets extensively investigated. FR is selectively amplified on human malignant cells and can take up folate and its analogs through a receptor-mediated endocytosis process.[1,2] FR-targeted liposomes can be utilized to deliver therapeutic agents to FR positive cells.[3] These liposomes have been evaluated for the targeted delivery of a wide variety of agents. This review will highlight recent developments on FR-targeted liposomes with emphasis on the application of this targeted nanoscale delivery system in cancer therapy. Potential applications of the FR-targeted liposomes in the clinics will also be discussed.

33.2 THE FOLATE RECEPTOR AS A CANCER-SPECIFIC CELLULAR MARKER

FR, also known as folate-binding proteins (FBP), is an *N*-glycosylated protein with high binding affinity to folate. FR has three isoforms, α, β, and γ/γ'. FR-α and FR-β are glycosyl-phosphatidyl-inositol (GPI)-anchored membrane bound proteins,[4,5] whereas FR-γ/γ' are constitutively secreted.[1,6,7] The γ' isoform is a truncated form of FR-γ.[8] These isoforms display different patterns of tissue specificity, and they also present differential ligand stereospecificities.[1] The two membrane receptor subtypes, α and β, share high amino acid sequence identity (70%) but are distinguishable by differential affinities for folic acid and stereoisoforms of reduced folates (affinities for folic acid: FR-α $K_d \sim 0.1$ nM,[9] FR-β $K_d \sim 1$ nM,[10] and FR-γ $K_d \sim 0.4$ nM[6]).

Expression of FR is both tissue-specific and differentiation dependent. The expression of FR has been identified by immuno-histochemical staining, reverse transcription-polymerase chain reaction (RT–PCR), Western blotting, and ^{3}H-folic acid binding both in normal and malignant tissues.[11,12] Functional FR expression is absent in most normal tissues with the exception of the luminal surface of certain epithelial cells[2] where it has limited accessibility to the blood stream. On the other hand, FR-α is consistently expressed in several carcinomas,[12] especially in non-mucinous ovarian carcinomas, uterine carcinomas, testicular choriocarcinomas, ependymomas, and pleural mesotheliomas, and it is less frequently expressed in breast, colon, and renal cell carcinomas.[2,13] Correlation of FR-α expression level and histologic grade has also been shown in ovarian and breast cancers, suggesting FR-α as a possible malignant cellular marker for cancers.[14] Methods for upregulating FR-α expression using anti-estrogens and gluococorticoid agonists have recently been published.[15,16] The effect of these agents is further enhanced by histone deacetylase (HDAC) inhibitors.[16] FR-β is a differentiation marker in the myelomonocytic lineage during neutrophil maturation,[17] and it is amplified in activated monocytes and macrophages.[18] However, FR-β in neutrophils is unable to bind folate as a result of aberrant post-translational modifications.[19] FR-β is also expressed in a functional form in chronic myelogenous leukemia (CML), in 70% of acute myelogenous leukemias (AML),[7,17,20,21] and in activated macrophages associated with rheumatoid arthritis.[22] FR-β expression is regulated by retinoid receptors and can be upregulated by all-trans retinoic acid (ATRA).[23] The effect of ATRA is further enhanced by HDAC inhibitors. The selective amplification of FR-α and FR-β expression in human cancers suggests possible roles for FRs as cancer specific cellular markers and their potential utility as targets for drug and gene delivery to these diseases. FR-γ/γ' are reported to be specific to hematopoietic tissues but are expressed at very low levels.[7] FR-γ/γ' are in secreted form, having limited application for targeted drug delivery but may serve as a serum marker to monitor certain hematological malignancies.[7] The additional possibility of selective FR upregulation in the target cells might further enhance the targeting strategy.

33.3 TARGETING FR FOR CANCER THERAPY

The selective amplification of FR expression in both human solid tumors and leukemia suggests its utility as a potentially valuable target for drug and gene delivery. Both monoclonal antibodies (MAbs) against FR and folate itself have been evaluated as moieties for targeting to the FR.

Two MAbs, MOv18, and MOv19, have been produced against FR-α on ovarian cancers. Both are murine MAbs of the IgG$_1$ class raised against a poorly differentiated ovarian carcinoma, recognizing two distinct epitopes.[24] Clinical studies on radioimmunoscintigraphy using ^{131}I-MOv18 have been carried out in ovarian cancer patients and demonstrated targeting of ovarian cancer.[25] In addition, α particle-emitter ^{211}At conjugated to MOv18 has been found to prolong survival in a murine peritoneal tumor model developed by intraperitoneal inoculation of OVCAR-3 cells.

Folate itself has, in fact, been the focus of recent development of FR targeting ligand. This is possibly because folate is a low molecular weight (MW = 441) ligand and has high affinity to FR, the conjugation chemistry of folate is defined and relatively easy to carry out, folate is a compound of unlimited availability, and folate lacks immunogenicity in contrast to murine MAb. The conjugation of folate via its γ-carboxyl has been shown to preserve its binding affinity to the FRs. Recent studies also found that the glutamate residue of folic acid is not critical for FR recognition, and only pteroic acid is essential for FR binding.[61] Similar to the uptake process of the vitamin folate, FRs mediate the cellular internalization of folate conjugates via receptor-mediated endocytosis;[19,20] therefore, this highly specific event can be utilized to promote uptake of folate-therapeutic agent conjugates in FR positive cancer cells.

Early proof-of-concept in vivo studies were conducted using either radio-labeled or fluoro-chrome-labeled folic acid. The selectivity has also been documented in tumor-bearing mice using radiolabeled folate conjugates such as ^{67}Ga-deferoxamine-folate, ^{99m}Tc-hydrazinonicotinic acid (HYNIC)-folate, and ^{111}In-diethylenetriamine pentaacetic acid (DTPA)-folate as potential folate-based radiopharmaceuticals.[26,27] Among these, ^{111}In-DTPA-folate has been further evaluated in human clinical trials. Concentration of radioactivity in abdominal masses was found in women suspected of ovarian cancer, and only the kidneys and livers in some patients displayed significant retention of ^{111}In-DTPA-folate. This reagent was developed by Endocyte Inc., an Indiana-based biopharmaceutical company, to be a folate-targeted radiopharmaceutical imaging agent (FolateScan™) and is currently in phase II clinical trial designed to evaluate if FolateScan™ can be used to detect and make a disease or pathological assessment of ovarian cancer.

A wide range of therapeutic agents have also been evaluated for enhanced cancer cell selective delivery. These include chemotherapeutic agents,[28] radiopharmaceuticals,[29] prodrug-converting enzymes,[30] protein toxins,[31] T-cell specific antibodies,[32] haptens,[30] gene transfer vectors,[33,34] antisense oligodeoxyribonucleotides (ODNs),[35] polymeric drug carriers,[36] and liposomes[19,37,38,44,49,51,56–60] carrying a variety of therapeutic agents. These conjugates potentially have broad applications in numerous imaging and therapeutic modalities.

Based on the action site of therapeutic agents, application of FR-targeted delivery can be categorized into two types. For hydrophilic compounds with limited permeability through cell membrane, FR offers an efficient drug uptake pathway that can both specify and facilitate the accessibility of therapeutic active agents to cancer cells. Folate-conjugates enter FR positive cells through receptor-mediated endocytosis that is different from passive permeation process. Delivery to cancer cells through a process different from traditional drug transportation has the potential to make the drug more accessible to cancer cells. On the other hand, for drugs that are ready to take effect on the surface of cells such as immunomodulator and prodrug-activating enzymes, FR may also act as a concentrator for these types of drugs. FRs are generally over expressed on the cellular surface of various cancer types, and upon binding of FR drug conjugates, they may enrich these therapeutic agents on the cell surface. Therefore, application of FR-targeted delivery can be exploited for variety of compounds (Figure 33.1).

Taken together, FR targeting has been investigated as a valuable approach to enhance the delivery of variety of agents. Therefore, the interest in exploiting this receptor for targeting application is still increasing as demonstrated by the active researches conducted in both academia and industry. Among these, FR-targeted liposome, first designed and investigated by Lee and Low,[37,38] has been attractive because of the advantages it possesses as a useful delivery system both for hydrophilic and hydrophobic agents.

33.4 LIPOSOMES AS DRUG CARRIERS

Liposomes are spherical, self-closed structures composed of lipophilic bilayers with entrapped hydrophilic core. The development of small size, stable, long-circulating, targeted liposomes has led to an exciting era in the pharmaceutical application of nanotechnology. Giving the unique structure of liposomes compared to other drug delivery systems, both lipophilic and lipophobic therapeutic agents can be delivered by liposomes. The biocompatibility of liposomes, if used with proper lipid composition, also makes liposomes attractive as safe and effective drug carriers. A variety of therapeutic agents have been incorporated into liposomes. Several have reached clinical use. These include liposomal doxorubicin (Doxil™), daunorubicin (Daunoxome™), amphotericin B (Amphotec™, Ambisome™, Abelcet™), cytarabine (Depocyte™), and verteporfin (Visudyne™).

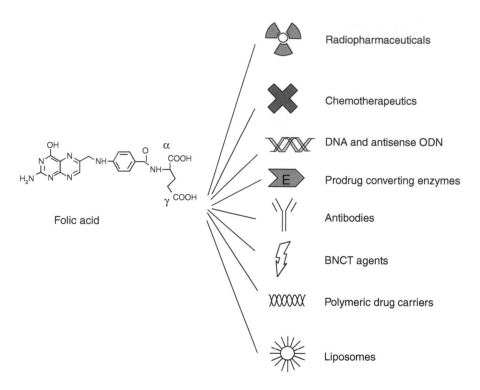

FIGURE 33.1 Folate-conjugates for FR-targeted delivery.

Numerous liposomal formulations are in clinical trial, including vincristine, all-trans retinoid acid, topotecan, and cationic liposome-based therapeutic gene transfer vectors. Many more are in preclinical evaluation. Besides potential use in systemic gene delivery, cationic liposomes are routinely used as transfection reagents for plasmid DNA and oligonucleotides in the laboratory. The use of liposomes can be deliberately engineered to possess unique properties such as long systemic circulation time, pH sensitivity, temperature sensitivity, and target cell specificity. These are achieved by selecting the appropriate lipid composition and surface modification of the liposomes, expanding the application of liposomes for drug delivery.

Liposomal delivery of anti-cancer drugs has been shown to greatly extend their systemic circulation time, reduce toxicity by lowering plasma free drug concentration, and facilitate preferential localization of drugs in solid tumors based on increased endothelial permeability and reduced lymphatic drainage or enhanced permeability and retention (EPR) effect.[39] Clearance of drugs in a liposomal formulation is mediated by phagocytic cells of the reticuloendothelial system (RES), primarily located in the liver and spleen. Drugs can be loaded during liposome preparation through methods such as passive entrapment. Alternatively, they can also be loaded after liposome preparation through a process called remote loading, utilizing the ion gradient across the membrane as in the case of doxorubicin.[31] Compared to free drugs, liposomal drugs are not subjected to rapid renal clearance. Instead, they are cleared by RES, exhibiting prolonged systemic circulation time.[40] RES clearance of liposomes can be reduced by surface coating with polyethyleneglycol (PEG). Passive targeting effect has also been shown in liposomal drug delivery to solid tumors where liposomes preferentially localize in these tumor tissues as a result of the EPR effect.[41]

Liposomes can be targeted to specific cell populations via incorporation of a desired targeting moiety such as a chemical conjugated ligand, antibody, or antibody fragment directed against the surface molecules. Targeted delivery of these liposomes can greatly improve their selectivity of cancer cells and facilitate their cellular uptake and intracellular drug release. Examples of such targeted liposomes are folate-conjugated liposomes targeting to acute myelogenous leukemia, CD19-targeted immunoliposomes for non-Hodgkin's lymphoma (NHL) therapy,[42] and anti-HER2 immunoliposomal doxorubicin targeting to HER2-overexpressing breast cancer cells.[43] A variety of techniques have been described to incorporate targeting moieties to liposomes. First, ligands such as folate can be conjugated to a lipid component such as poly(ethylene glycol)-conjugated distearoyl phosphatidylethanolamine (PEG-DSPE) and PEG-Chol. This amphophilic lipid component can be constructed into the lipid bilayer during the formation of membrane leaving the ligand on the surface of the particles.[38,59] In the second method, a functionalized lipid anchor is included first to the outside of the liposomes after the formation of the bilayer membrane. Therefore, the exposed crosslinker can react with activated targeting moieties such as thiolated proteins (antibody, scFv, transferrin, etc.).[42,43] A useful thiolation reagent is 2-iminothiolane that is also known as Traut's reagent. There are two types of immunoliposomes based on the antibody coupling strategy. Type one involves direct attachment of the antibodies to the lipids, and type two involves linking the antibodies to the terminal ends of reactive PEG derivates. Most of these coupling techniques lead to a random antibody orientation at the liposome surface. A significant contribution to the development of the type two immunoliposome, recently adopted for formulation of HER2-targeted immunoliposomes, is an antibody coupling method called post-insertion method.[43] This method involves formation of micelles of lipid-derivatized antibodies first, followed by transferring the antibody-coupled PEG-lipid micelles to the preformed liposomes, converting the conventional liposomes to immunoliposome through a simple one-step incubation method. Maleimide-terminated PEG–DSPE is a coupling lipid that has been validated to successfully convert commercially available liposomal doxorubicin (Doxil/Caelyx) to targeted sterically stabilized liposomal doxorubicin. The last post-insertion method seems highly promising for future clinical development.

33.5 FORMULATION STRATEGIES FOR FR-TARGETED LIPOSOMES

FR-targeted liposomes can be prepared by including a small fraction (e.g., 0.1 mole%) of a lipophilic folate derivative into the lipid composition.[38] Both folate-PEG (M.W. 3350)-distearoyl phosphatidylethanolamine (folate-PEG–DSPE)[38] and folate-PEG-cholesterol (folate-PEG-chol) have been synthesized and shown to be effective in targeting liposomes to the FR. The synthesis of folate-PEG–DSPE and folate-PEG-Chol can be realized through two stages. First, an intermediate folate-PEG-amine is synthesized by reacting NHS-activated folate and reacted with PEG-bis amine and further purifized using a Sepharose column. This folate-PEG-NH$_2$ is then ready to react with N-succinyl-DSPE to form folate-PEG–DSPE[38] or with cholestreryl chloroformate to form folate-PEG-chol.[59] Another synthetic method has also been described by Gabizon that is to couple FA to readily attainable H$_2$N-PEG–DSPE with the mediation of carbodiimide.[44] These two synthesized lipophilic folate-derivatives have both been shown to be good components to be incorporated to liosomes. Compared to the synthesis of folate-PEG–DSPE, the synthesis of folate-PEG-chol is relatively simple and is less costly. With regard to stability, cholesterol also provides better chemical stability, and it is also known to be able to increase the structural integrity of lipid membranes through tight packing. The molecule of cholesterol is neutrally charged, different from the negatively charged DSPE molecule, and can possibly provide better targeting effect when used as a lipophilic anchor to incorporate the folate molecule. For these reasons, this lab has been routinely using folate-PEG-chol for the formulation of FR-targeted liposomes.

Preparation of folate receptor (FR) targeted liposomes is relatively straightforward since the targeting lipid component can simply be included into the lipid composition during the formation of liposomes. A small percentage (0.1–0.5%, molar ratio) of folate-PEG–DSPE or folate-PEG-chol in the lipid can yield targeting effect to FR. Relatively long linker between folate and the lipid anchor (e.g., PEG$_{3350}$) was found to be necessary for FR binding of the liposomes.[37] FR targeting was compatible with incorporation of 3 mole% of PEG–DSPE that is required for prolonging the in vivo circulation time of the liposomes. Lab scale production of these targeted liposomes can be performed by thin-layer hydration followed by polycarbonate membrane extrusion. For scaling up, alternative methods such as liquid phase emulsification, high pressure homogenization, and tangential flow filtration can be used. Recently, this lab has successfully validated a processing method combining homogenization, ultrafiltration, and remote loading for FR-targeted liposomal doxorubicin at clinical relevant scale quantity. Using this method, three consecutive batches of liposomes with mean particle size around 100 nm curve produced. Another alternative method for folate-PEG-Chol incorporation has not yet been investigated, using post-insertion method similar to HER2-targeted immunoliposome production by incubating micelles of folate-PEG-chol/PEG-chol to pre-formed liposomal drugs. The latter method avoided the ligand stability concern during the liposome production and has the potential to be used for future Current Good Manufacturing Practices (cGMP) grade production of FR-targeted liposomes. Both hydrophilic drugs and hydrophobic drugs can be potentially loaded to the targeted particles through remote loading (doxorubicin, daunorubicin, vincristine, and topetecan) or passive entrapment (paclitaxel and ATRA); therefore, this FR-targeted liposomal drug delivery represents a valuable approach for FR-positive cancer therapy (Figure 33.2).

FR-targeted cationic liposome formulations have also been studied for gene delivery. Several cationic liposome formulations that incorporate a lipopholic folate derivative as a targeting entity have been reported. FR-targeted lipoplexes constructed using a cationic lipid RPR209120 was shown to efficiently transfer gene.[45] Another study using 2% DPPE–PEG$_{3340}$-folate and a cationic dithiol-detergent (dimerized tetradecyl-ornithinly-cysteine) demonstrated efficient FR-dependent cellular uptake and transfection. Therapeutic gene p53 was also delivered by Xu et al., using cationic liposomes conjugated to folate, and this was shown to enhance the anti-tumor efficacy of conventional chemotherapy.[46] FR-targeted cationic liposomes can further form complex with DNA-condensing polymer to form FR-targeted lipopolyplexes, and this was demonstrated by Reddy et al.[54] to have superior transfection activity in FR+M109 murine lung carcinoma cells as well as in L1210A murine lymphocytic leukemia cells. Lee et al. developed FR-targeted type II lipopolyplexes, and these FR-targeted vectors exhibited improved gene transfer activity even in the presence of serum.[47,48] Leamon et al. also investigated the use of FR-targeted liposome constructed with lipid-PEG-pteroate for effective delivery of antisense oligonucleotides into FR-bearing cells in vitro.[61] FR-targeted liposomal drug and gene delivery has also been the subject of another review article.[35]

33.6 IN VITRO DRUG DELIVERY USING FR-TARGETED LIPOSOMES

Cellular uptake of FR-targeted liposomes has been characterized using KB cells, a FR-α (+) human oral carcinoma cell line.[37] Binding of the liposomes has a relatively slow kinetics, occurring over several hours, presumably a result of their size. The specific binding of these liposomes to KB cells was saturable and could be blocked by excess free folate in the binding media.[37] Interestingly, the apparent affinity of the folate-conjugated liposomes to KB cells was much higher than that of free folate.[37] This is due to multivalent binding of the folate ligand on the liposomes with the FRs on the cellular surface.[37] The folate-liposomes are internalized into a low pH compartment via receptor-mediated endocytosis, and the encapsulated drug molecules are released into the cytoplasm to induce cytotoxicity[47] as illustrated in Figure 33.3.

Drug delivery properties of FR-targeted liposomes have been studied in vitro using liposomes loaded with chemotherapeutic agents such as doxorubicin,[38] daunorubicin,[49] and cisplatin.[44] Lee et al. first reported the in vitro effect of doxorubin and showed these targeted liposomes showed ~86-fold greater cytotoxicity in KB cells compared to non-targeted control liposomes.[38] The enhancement in cytotoxicity was correlated with the increase in doxorubicin uptake and could be blocked by excess free folate.[38] Furthermore, a study by Goren et al. showed that the delivery of

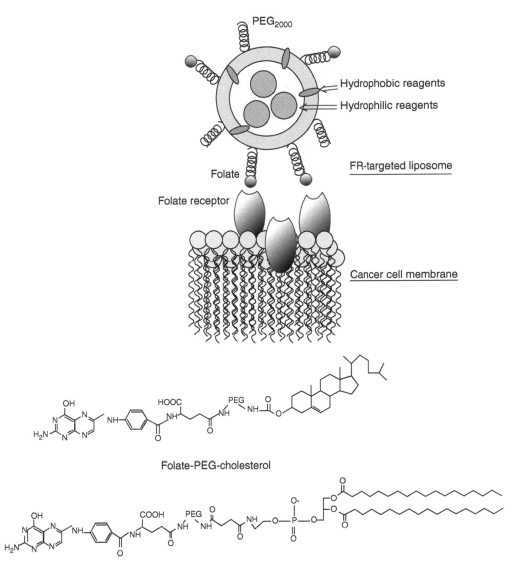

FIGURE 33.2 FR-targeted liposome. The ligand folate is conjugated to a lipophilic anchor (PEG$_{3350}$-DSPE or PEG$_{3350}$-chol) and incorporated to the lipid bilayer. PEG$_{2000}$–DSPE is also included in the construction of the liposome membrane, prolonging the systemic circulation of these particles by avoiding the RES uptake. Hydrophilic anti-cancer reagents can be loaded to the aqueous core through active loading or passive entrapment, and hydrophobic reagents can be loaded to the lipid bilayer through formation of electrostatic complex or passive entrapment. The structures of two synthetic targeting lipids (F-PEG–DSPE and F-PEG-Chol) were also shown.

doxorubicin via FR-targeted liposomes bypassed Pgp-dependent drug efflux in drug resistant FR-α(+) M109 murine lung carcinoma cells in vitro and in an adoptive assay in mice.[50] These data suggest that this delivery modality might overcome drug resistance in these tumor cells.[50] More recently, folate-liposomal doxorubicin was studied in FR-β(+) KG1 human AML cells and similarly showed enhanced cytotoxicity relative to non-targeted control liposomes,[19] and the effect

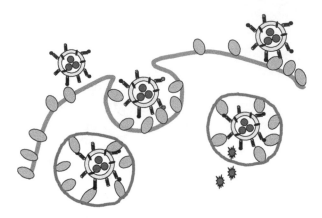

FIGURE 33.3 FR-mediated endocytosis of folate-liposomes.

was enhanced by FR-β upregulation using ATRA.[19] Therefore, both FR-α and FR-β are potential targets for folate-conjugated liposomes.

In addition to chemotherapeutics, a variety of agents have been loaded into FR-targeted liposomes and evaluated in vitro. These include fluorescent[37] probes, BNCT agents,[51] photosensitizers,[52] antisene oligodeoxyribonucleotides against the EGF receptor,[53] and plasmid DNA carrying luciferase reporter gene.[47,54] Similar formulations of FR-targeted lipid nanoparticles containing paclitaxel have also been recently reported.[55] In each of these studies, the FR-targeted formulation was shown to be much more efficient in cellular uptake and biological activity compared to the non-targeted control formulation.

33.7 PHARMACOKINETIC AND BIODISTRIBUTION PROPERTIES OF FR-TARGETED LIPOSOMES

The in vivo clearance rate for folate-conjugated liposomal doxorubicin was faster than non-targeted control liposomes in mice engrafted with FR-overexpressing M109 or human KB carcinoma or mouse J6453 lymphoma.[56] These results also suggested a possible role of FR in the differential rates of clearance.[56] Plasma clearance kinetics of FR-targeted liposomes radiolabled with ^{67}Ga were studied in rats. These liposomes showed enhanced plasma clearance rates and higher uptakes in the liver and spleen compared to non-targeted control liposomes.[56] This effect was even more pronounced in rats that were placed on a folate-free diet, suggesting that FR expression in the RES organs might be responsible for the increase in the rate of liposome clearance.[56] One possible explanation is that FR-targeted liposomes, by virtue of their multivalency, can interact with the FR-β that is present at a relatively low level in the phagocytic cells.

In the biodistribution studies in murine tumor models, FR-targeted liposomes did not show enhanced accumulation in tumors relative to non-targeted liposomes.[59] This might be due to the passive accumulation of the liposomes via the EPR effect was the predominant factor in tumor localization of both targeted and non-targeted liposomes because endothelial extravasation was the rate-limiting step in tumor localization. In addition, FR-mediated enhancement in uptake of the targeted liposomes might be offset by the more rapid plasma clearance of these liposomes as discussed above.[60] Moreover, in solid tumors, FRs on tumor cell surface are not directly exposed to the circulation and not fully accessible by circulating FR-targeted liposomes. Therefore, extravasation, instead of specific binding, is the rate-limiting factor of liposome distribution in solid tumors.

An important observation in vivo is that macrophages also have high uptake of FR-targeted liposomes. In a study done in an ovarian carcinoma ascites mouse model, results showed that macrophages took up FR-targeted liposomes efficiently in the ascites fluid even in the presence of FR positive ascetic tumor cells,[57] although other reports did show increased FR-targeted liposome uptake in FR(+) ascitic tumor cells.[56] The relative uptake of folate-liposomes by these two cell populations was likely determined by the relative FR expression levels.[56]

33.8 IN VIVO THERAPEUTIC EFFICACY OF FR-TARGETED LIPOSOMES

The therapeutic efficacy of folate-liposomal doxorubicin has been evaluated in several murine tumor models. In a KB cell xenograft model of athymic mice on a folate-free diet, folate-liposomal doxorubicin given i.p. at 10 mg/kg×6 injections showed significantly greater tumor inhibition compared to non-targeted liposomes.[58] However, in another study using a M109R-HiFR syngraft model in BALB/c mice on a regular folate-containing diet, folate-liposomal doxorubicin given intravenously at 8 mg/kg×4 injections showed a similar therapeutic efficacy to non-targeted control liposomes.[3] The differences in these studies might be partly due to the different routes of administration, animal diet, and dosages applied.[3] A challenge in demonstrating a therapeutic benefit of folate-liposomal doxorubicin is the excellent anti-tumor activity of PEGylated liposomal doxorubicin used as a control that has optimal pharmacokinetic properties and EPR effect and are much more effective than the free drug.[3] Further studies in additional models are likely needed to confirm possible therapeutic benefits of FR-targeted liposomal doxorubicin.

A possible mechanism for increased therapeutic efficacy for the folate-liposomal doxorubicin in solid tumor is altered intratumoral drug distribution. These liposomes might be more efficiently internalized by FR(+) tumor cells and/or tumor infiltrating macrophages that are FR(+).[57] This, in turn, results in an increase in drug release within the tumor. In contrast, non-targeted liposomal doxorubicin may remain extracellular and be distributed in the interstitial space and releases drug more slowly. This may result in a bioavailable (free) drug concentration that falls below the minimum effective concentration (MEC) for inducing tumor cell apoptosis. Therefore, cellular internalization and associated drug release may be a critical mechanism of liposomal drug action in solid tumors and a rate-limiting factor in overall drug delivery pathway. The relative importance of FR expression on tumor cells and tumor infiltrating macrophages and their role in uptake of folate-coated liposomes warrant further investigation.

Leukemia is a better disease target as compared to solid tumors for this targeted delivery strategy. This potentially is because malignant cells are circulating in leukemia patients and are fully accessible by targeted liposomes. In contrast, these liposomes need to overcome several barriers in solid tumors before reaching cancer cells. Several studies have already reported the anti-tumor efficacy of FR-targeted liposomal anthracyclines in leukemia. In both L1210JF murine lymphocytic leukemia ascites models and a KG-1 human acute myelogenous leukemia xenograft model, FR-targeted liposomal doxorubicin or daunorubicin, when administered i.p,[19] showed improvement in survival of treated mice compared to non-targeted control liposomes or the free drug in the L1210JF model.[19] In both mice models, folate-liposomal doxorubicin given i.p. was also more effective in prolonging animal survival.[19] The relatively greater accessibility of leukemia cells in ascites tumors might have contributed to the superior efficacy of the targeted liposomes.

Interestingly, pretreatment of mice with ATRA, an agent known to upregulate FR-β expression in KG-1 cells in vitro,[19] before the injection of liposomes was also shown to enhance the efficacy. An increase in long-term survival in these mice as a result of FR-targeted liposomal doxorubicin was shown from 12.5% to 60%. Therefore, administering ATRA to upregulate

FR-β expression seems to be a potentially promising therapeutic strategy to be combined with the FR-targeted therapy.

33.9 CONCLUSIONS AND FUTURE DIRECTIONS

Targeted drug delivery is a promising approach to selectively deliver neoplastic agents to cancer cells and to improve the therapeutic index of chemotherapy. FR outstands among various tumor markers because of the specificity and relatively high level of endocytosis, representing a very attractive target for developing targeted delivery for future cancer therapy. FR-targeted liposomal delivery is a promising strategy for treatment of FR(+) tumors and leukemias. Although a variety of agents have been delivered to FR(+) tumor cell in vitro using these liposomes, relatively few reports address the in vivo properties of these liposomes. However, at least some studies indicated a therapeutic advantage of folate-liposomal doxorubicin over the non-targeted control.[19,49,58] Leukemia, that has greater accessibility from the systemic circulation, may present an even better therapeutic target for these liposomes. The concept of sensitizing AML cells with FR-β-upregulating agents such as ATRA may bring further rationale to take FR-targeted liposomal drug delivery to clinical trials.

Several critical factors to be considered for designing in vivo application of FR-targeted liposomes for cancer therapy include FR expression level on tumor cells relative to normal cells, the accessibility of liposomes to these receptors, the intra-tumor diffusion rate of vectors, and the presence of endogenous folate in systemic circulation. The first obstacle of insufficient levels of FR expression on tumor cells in patients could be possibly overcome by inducing FR expression upregulation by co-administration of anti-estrogen or retinoid receptor ligands. To address the second concern of accessibility, optimization of formulation parameter is necessary to reduce the particle size, reduce the plasma protein binding, and avoid the RES and other normal tissue uptake. For solid tumor delivery, priming the vasculature in the tumor to increase the distribution of targeted particles is a possible approach. The concern that the presence of endogenous folate in systemic circulation can block FR binding is not a critical one. It is believed that this should not constitute a barrier as the endogenous form of folate has lower affinity for the FR compared to folic acid and should not significantly impede the binding of multivalent folate-coated liposomes.

Clinical success in targeted drug delivery area has so far been limited by the lack of a suitable cellular target and/or a high degree of difficulty in the production of clinical quantities of targeted drug carriers. FR-targeted liposomes represent an attractive candidate for clinical development with potential application in the treatment of AML and other FR-over expressing solid tumors. Areas of investigation for future studies include the possible role of receptor upregulation in promoting targeted therapy, both for FR-α and FR-β positive tumors; and the cellular population responsible for enhancement in therapeutic response in solid tumors, i.e., relative roles of tumor infiltrating macrophages and the tumor cells themselves.[57] Because activated macrophages have increased FR-β expression,[22] it is possible that folate-conjugates can effectively target tumor infiltrating macrophages in non-FR expressing tumors, extending the FR-targeting strategy to receptor negative tumors.

The area of FR targeting remains an area of great excitement, suggested by large number of recent publications and reports from industrial development. So far, no clinical study targeting FR for therapeutic purposes has been carried out. Given the advantages of the FR-targeting strategy, further preclinical studies are urgently needed to define mechanisms of in vivo targeting and the potential for clinical efficacy. FR-targeted liposomal doxorubicin represents an compelling candidate for clinical development with potential application in the treatment of AML. Furthermore, future application to other tumors with high frequency of FR overexpression such as ovarian and lung carcinomas warrants investigation.

ACKNOWLEDGMENTS

This work was supported, in part, by NCI grant ROI CA095673 to RJL, P01CA95426 to JCB, NSF grants EEC-0425626 to RJL, and Leukemia and Lymphoma Society grant 6113-02 to RJL.

REFERENCES

1. Elnakat, H. and Ratnam, M., Distribution, functionality and gene regulation of folate receptor isoforms: Implications in targeted therapy, *Advanced Drug Delivery Reviews*, 56(8), 1067–1084, 2004.
2. Weitman, S. D. et al., Distribution of the folate receptor GP38 in normal and malignant cell lines and tissues, *Cancer Research*, 52(12), 3396–3401, 1992.
3. Gabizon, A. et al., Tumor cell targeting of liposome-entrapped drugs with phospholipid-anchored folic acid-PEG conjugates, *Advanced Drug Delivery Reviews*, 56(8), 1177–1192, 2004.
4. Lacey, S. W. et al., Complementary DNA for the folate-binding protein correctly predicts anchoring to the membrane by glycosyl-phosphatidylinositol, *The Journal of Clinical Investigation*, 84(2), 715–720, 1989.
5. Ratnam, M. et al., Homologous membrane folate-binding proteins in human placenta: Cloning and sequence of a cDNA, *Biochemistry*, 28(20), 8249–8254, 1989.
6. Shen, F. et al., Folate receptor type gamma is primarily a secretory protein due to lack of an efficient signal for glycosyl-phosphatidylinositol modification: Protein characterization and cell type specificity, *Biochemistry*, 34(16), 5660–5665, 1995.
7. Shen, F. et al., Identification of a novel folate receptor, a truncated receptor, and receptor type beta in hematopoietic cells: cDNA cloning, expression, immunoreactivity, and tissue specificity, *Biochemistry*, 33(5), 1209–1215, 1994.
8. Wang, H., Ross, J. F., and Ratnam, M., Structure and regulation of a polymorphic gene encoding folate receptor type gamma/gamma', *Nucleic Acids Research*, 26(9), 2132–2142, 1998.
9. Kamen, B. A. and Caston, J. D., Properties of a folate-binding protein (FBP) isolated from porcine kidney, *Biochemical Pharmacology*, 35(14), 2323–2329, 1986.
10. da Costa, M. and Rothenberg, S. P., Purification and characterization of folate-binding proteins from rat placenta, *Biochimica Et Biophysica Acta*, 1292(1), 23–30, 1996.
11. Ross, J. F., Chaudhuri, P. K., and Ratnam, M., Differential regulation of folate receptor isoforms in normal and malignant tissues in vivo and in established cell lines, Physiologic and clinical implications. *Cancer*, 73(9), 2432–2443, 1994.
12. Weitman, S. D. et al., Cellular localization of the folate receptor: Potential role in drug toxicity and folate homeostasis, *Cancer Research*, 52(23), 6708–6711, 1992.
13. Garin-Chesa, P. et al., Trophoblast and ovarian cancer antigen LK26. Sensitivity and specificity in immunopathology and molecular identification as a folate-binding protein, *American Journal of Pathology*, 142(2), 557–567, 1993.
14. Toffoli, G. et al., Overexpression of folate-binding protein in ovarian cancers, *International Journal of Cancer*, 74(2), 193–198, 1997.
15. Kelley, K. M. M., Rowan, B. G., and Ratnam, M., Modulation of the folate receptor alpha gene by the estrogen receptor: Mechanism and implications in tumor targeting, *Cancer Research*, 63(11), 2820–2828, 2003.
16. Tran, T. et al., Enhancement of folate receptor alpha expression in tumor cells through the glucocorticoid receptor: A promising means to improved tumor detection and targeting, *Cancer Research*, 65(10), 4431–4441, 2005.
17. Ross, J. F. et al., Folate receptor type beta is a neutrophilic lineage marker and is differentially expressed in myeloid leukemia, *Cancer*, 85(2), 348–357, 1999.
18. Nakamura, M., Nitration and chlorination of folic acid by peroxynitrite and hypochlorous acid, and the selective binding of 10-nitro-folate to folate receptor beta, *Biochemical and Biophysical Research Communications*, 297(5), 1238–1244, 2002.
19. Pan, X. Q. et al., Strategy for the treatment of acute myelogenous leukemia based on folate receptor beta-targeted liposomal doxorubicin combined with receptor induction using all-trans retinoic acid, *Blood*, 100(2), 594–602, 2002.

20. Sadasivan, E. et al., Purification, properties, and immunological characterization of folate-binding proteins from human leukemia cells, *Biochimica Et Biophysica Acta*, 925(1), 36–47, 1987.

21. Sadasivan, E. et al., Characterization of multiple forms of folate-binding protein from human leukemia cells, *Biochimica Et Biophysica Acta*, 882(3), 311–321, 1986.

22. Paulos, C. M. et al., Folate receptor-mediated targeting of therapeutic and imaging agents to activated macrophages in rheumatoid arthritis, *Advanced Drug Delivery Reviews*, 56(8), 1205–1217, 2004.

23. Wang, H. et al., Differentiation-independent retinoid induction of folate receptor type beta, a potential tumor target in myeloid leukemia, *Blood*, 96(10), 3529–3536, 2000.

24. Miotti, S. et al., Characterization of human ovarian carcinoma-associated antigens defined by novel monoclonal antibodies with tumor-restricted specificity, *International Journal of Cancer*, 39(3), 297–303, 1987.

25. Kalofonos, H. P., Karamouzis, M. V., and Epenetos, A. A., Radioimmunoscintigraphy in patients with ovarian cancer, *Acta Oncology*, 40(5), 549–557, 2001.

26. Mathias, C. J., et al., Tumor-selective radiopharmaceutical targeting via receptor-mediated endocytosis of gallium-67-deferoxamine-folate, *Journal of Nuclear Medicine*, 37(6), 1003–1008, 1996.

27. Mathias, C. J., et al., Indium-111-DTPA-folate as a potential folate-receptor-targeted radiopharmaceutical, *Journal of Nuclear Medicine*, 39(9), 1579–1585, 1998.

28. Leamon, C. P. and Reddy, J. A., Folate-targeted chemotherapy, *Advanced Drug Delivery Reviews*, 56(8), 1127–1141, 2004.

29. Ke, C. Y., Mathias, C. J., and Green, M. A., Folate-receptor-targeted radionuclide imaging agents, *Advanced Drug Delivery Reviews*, 56(8), 1143–1160, 2004.

30. Lu, J. Y. et al., Folate-targeted enzyme prodrug cancer therapy utilizing penicillin-V amidase and a doxorubicin prodrug, *Journal of Drug Targeting*, 7(1), 43–53, 1999.

31. Leamon, C. P. and Low, P. S., Selective targeting of malignant cells with cytotoxin-folate conjugates, *Journal of Drug Targeting*, 2(2), 101–112, 1994.

32. Roy, E. J. et al., Folate-mediated targeting of T cells to tumors, *Advanced Drug Delivery Reviews*, 56(8), 1219–1231, 2004.

33. Mislick, K. A. et al., Transfection of folate-polylysine DNA complexes: Evidence for lysosomal delivery, *Bioconjugate Chemistry*, 6(5), 512–515, 1995.

34. Gottschalk, S. et al., Folate receptor mediated DNA delivery into tumor cells: Potosomal disruption results in enhanced gene expression, *Gene Therapy*, 1(3), 185–191, 1994.

35. Zhao, X. B. and Lee, R. J., Tumor-selective targeted delivery of genes and antisense oligodeoxyribonucleotides via the folate receptor, *Advanced Drug Delivery Reviews*, 56(8), 1193–1204, 2004.

36. Shukla, S. et al., Synthesis and biological evaluation of folate receptor-targeted boronated PAMAM dendrimers as potential agents for neutron capture therapy, *Bioconjugate Chemistry*, 14(1), 158–167, 2003.

37. Lee, R. J. and Low, P. S., Delivery of liposomes into cultured KB cells via folate receptor-mediated endocytosis, *Journal of Biological Chemistry*, 269(5), 3198–3204, 1994.

38. Lee, R. J. and Low, P. S., Folate-mediated tumor cell targeting of liposome-entrapped doxorubicin in vitro, *Biochimica Et Biophysica Acta*, 1233(2), 134–144, 1995.

39. Gabizon, A. and Papahadjopoulos, D., Liposome formulations with prolonged circulation time in blood and enhanced uptake by tumors, *Proceedings of the National Academy of Sciences of the United States of America*, 85(18), 6949–6953, 1988.

40. Abraham, S. A. et al., The liposomal formulation of doxorubicin, *Methods in Enzymology*, 391, 71–97, 2005.

41. Uster, P. S. et al., Insertion of poly(ethylene glycol) derivatized phospholipid into pre-formed liposomes results in prolonged in vivo circulation time, *FEBS Letters*, 386(2,3), 243–246, 1996.

42. Allen, T. M., Mumbengegwi, D. R., and Charrois, G. J. R., Anti-CD19-targeted liposomal doxorubicin improves the therapeutic efficacy in murine B-cell lymphoma and ameliorates the toxicity of liposomes with varying drug release rates, *Clinical Cancer Research*, 11(9), 3567–3573, 2005.

43. Park, J. W. et al., Anti-HER2 immunoliposomes: Enhanced efficacy attributable to targeted delivery, *Clinical Cancer Research*, 8(4), 1172–1181, 2002.

44. Gabizon, A., Targeting folate receptor with folate linked to extremities of poly(ethylene glycol)-grafted liposomes: In vitro studies, *Bioconjugate Chemistry*, 10(2), 289–298, 1999.

45. Hofland, H. E. J. et al., Folate-targeted gene transfer in vivo, *Molecular Therapy*, 5(6), 739–744, 2002.

46. Xu, L., Pirollo, K. F., and Chang, E. H., Tumor-targeted p53-gene therapy enhances the efficacy of conventional chemo/radiotherapy, *Journal of Controlled Release: Official Journal of the Controlled Release Society*, 74(1–3), 115–128, 2001.

47. Lee, R. J. and Huang, L., Folate-targeted, anionic liposome-entrapped polylysine-condensed DNA for tumor cell-specific gene transfer, *The Journal of Biological Chemistry*, 271(14), 8481–8487, 1996.

48. Gosselin, M. A., Guo, W., and Lee, R. J., Incorporation of reversibly cross-linked polyplexes into LPDII vectors for gene delivery, *Bioconjugate Chemistry*, 13(5), 1044–1053, 2002.

49. Pan, X. Q. and Lee, R. J., In vivo antitumor activity of folate receptor-targeted liposomal daunorubicin in a murine leukemia model, *Anticancer Research*, 25(1A), 343–346, 2005.

50. Goren, D. et al., Nuclear delivery of doxorubicin via folate-targeted liposomes with bypass of multi-drug-resistance efflux pump, *Clinical Cancer Research*, 6(5), 1949–1957, 2000.

51. Pan, X. Q., Wang, H., and Lee, R. J., Boron delivery to a murine lung carcinoma using folate receptor-targeted liposomes, *Anticancer Research*, 22(3), 1629–1633, 2002.

52. Qualls, M. M. and Thompson, D. H., Chloroaluminum phthalocyanine tetrasulfonate delivered via acid-labile diplasmenylcholine-folate liposomes: Intracellular localization and synergistic phototoxicity, *International Journal of Cancer*, 93(3), 384–392, 2001.

53. Wang, S. et al., Delivery of antisense oligodeoxyribonucleotides against the human epidermal growth factor receptor into cultured KB cells with liposomes conjugated to folate via polyethylene glycol, *Proceedings of the National Academy of Sciences of the United States of America*, 92(8), 3318–3322, 1995.

54. Reddy, J. A. et al., Folate-targeted, cationic liposome-mediated gene transfer into disseminated peritoneal tumors, *Gene Therapy*, 9(22), 1542–1550, 2002.

55. Stevens, P. J., Sekido, M., and Lee, R. J., A folate receptor-targeted lipid nanoparticle formulation for a lipophilic paclitaxel prodrug, *Pharmaceutical Research*, 21(12), 2153–2157, 2004.

56. Gabizon, A. et al., In vivo fate of folate-targeted polyethylene-glycol liposomes in tumor-bearing mice, *Clinical Cancer Research*, 9(17), 6551–6559, 2003.

57. Turk, M. J., Waters, D. J., and Low, P. S., Folate-conjugated liposomes preferentially target macrophages associated with ovarian carcinoma, *Cancer Letters*, 213(2), 165–172, 2004.

58. Pan, X. Q., Wang, H., and Lee, R. J., Antitumor activity of folate receptor-targeted liposomal doxorubicin in a KB oral carcinoma murine xenograft model, *Pharmaceutical Research*, 20(3), 417–422, 2003.

59. Guo, W., Lee, T., Sudimack, J., and Lee, R. J., Receptor targeted delivery of liposomes via folate-PEG-chol, *J. Liposome Res.*, 10, 179–195, 2000.

60. Gabizon, A., Huberty, J., Straubinger, R. M., Price, D. C., and Papahadjopoulos, D., An improved method for in vivo tracing and imaging liposome using a ^{67}Gallium-deferoxamine complex, *Journal of Liposome Research*, 1, 123–135, 1998.

61. Leamon, C. P., Cooper, S. R., and Hardee, G. E., Folate-liposome-mediated antisense oligodeoxy-nucleotide targeting to cancer cells: Evaluation in vitro and in vivo, *Bioconjugate Chemistry*, 14(4), 738–747, 2003.

34 Nanoscale Drug Delivery Vehicles for Solid Tumors: A New Paradigm for Localized Drug Delivery Using Temperature Sensitive Liposomes

David Needham and Ana Ponce

CONTENTS

34.1   Introduction: The Problem and Current Solutions 678
34.2   The Challenges Facing Nanotherapeutic Intervention: A Brief Review
        of Lipid-Based Nanocarriers 682
        34.2.1   Getting the Drug In 682
        34.2.2   Getting the Nanocapsules to the Tumor 684
        34.2.3   Mechanisms of Liposomal Drug Delivery 687
34.3   New Capsular Design for Delivery of Drugs to and Release of Drugs
        in Localized Solid Tumors 691
        34.3.1   Getting the Drug Out 691
        34.3.2   A New Liposome Design 693
        34.3.3   Nanostructures and Membrane Permeability 695
        34.3.4   In Vivo Growth Delay Studies 701
34.4   A New Paradigm for Drug Delivery to Tumor Tissues 705
        34.4.1   Vascular Shutdown 705
        34.4.2   Exploring Other Tumor Models 707
        34.4.3   Regional Deposition of Drugs Observed Using MRI 709
        34.4.4   Clinical Trials and Hyperthermia Devices 710
34.5   Conclusions 711
References 712

34.1 INTRODUCTION: THE PROBLEM AND CURRENT SOLUTIONS

A January 2006 search for the word *cancer* at www.clinicaltrials.gov yields a list of 4,370 federally and privately supported clinical trials in human volunteers. Within these results, a focus on *chemotherapy* reduces the number to 1,745 (~40% of all trials).[1] Therefore, the discovery and testing of chemotherapeutic agents for cancer represents a significant effort in current basic and clinical research. In addition, as cancer is the second leading cause of death in the United States, this research represents a more than significant investment in health and disease.

Chemotherapeutics are designed to kill cancer cells or prevent them from dividing. The major types of cancer drugs include antimetabolites, alkylating agents, antibiotics, antimitotics, hormones, and inorganics (e.g., cisplatin). The National Cancer Institute (NCI) Drug Dictionary contains technical definitions and synonyms for more than 500 agents that are being used in the treatment of patients with cancer or cancer-related conditions.[1] However, the magic bullet,* interpreted here as "a drug that is solely specific for cancer cells," is still not available.[2,3] The drugs that have been developed can kill cancer cells, but most of them can also kill normal cells, resulting in a low therapeutic index (ratio of lethal dose to effective dose). This limits their dosage and, ultimately, their efficacy.[4] Apart from the success of cisplatin in testicular cancer and methotrexate in choriocarcinoma, systemic chemotherapy has not yet exposed solid tumors to sufficiently high drug concentrations for an adequate period of time to cause meaningful tumor regressions.[4] Therefore, surgery† and radiation are the mainstays for local control of tumors.‡ In modern clinical practice, chemotherapy is mostly used as a post-surgical adjuvant to remove microscopic metastases that have spread from the original site. Other applications include neoadjuvant therapy to downsize local tumors before surgery or radiation and palliative care for the incurable patient.[4]

One solution to the low therapeutic index of these non-specific chemotherapeutic drugs is to target the drugs to solid tumors by associating them with microscopic or submicroscopic carriers that possess some affinity for the tumors. Microscopic particles such as microspheres accumulate in tumor capillaries by chemoembolism, whereas submicroscopic carriers such as liposomes accumulate perivascularly through passive extravasation and/or active targeting to cancer cells.[5–10] It is here, in the context of drug delivery, that nanotechnology has already been explored for over 40 years as an approach for reducing toxicity and increasing the accumulation of drugs in tumors when compared with free drug. As depicted in Figure 34.1, attention has principally focused on either self-assembling systems such as polymeric micelles formed from amphiphilic block co-polymers,[11–14] phospholipid-based liposomes,[15,16] and polymer surfactant polymersomes;[17] or covalent-linked structures such as polymer–drug conjugates,[18–21] dendrimers,[22,23] and radioactive antibodies.[24,25]

The primary goal of nanotechnology for cancer therapy is to change the biodistribution of drugs, reducing free drug toxicity and favoring tumor accumulation rather than normal tissue accumulation or excretion. In order to start to achieve this goal, a nanoscopic formulation must satisfy the well-known requirements of a drug delivery system: load and retain the drug in or on a submicroscopic or nanoscopic carrier to prevent extravasation in healthy organs and tissues (e.g., heart and kidneys for doxorubicin)[26,27] and thereby reduce toxicity compared to free drug administration;[28,29] evade the body's defenses (the stealth effect) and avoid premature uptake into the reticuloendothelial system (RES), making the carrier available in the blood stream;[30] and achieve carrier uptake in solid tumors by the enhanced permeability and retention effect (EPR effect or

---

* If an organism is pictured as infected by a certain species of bacterium, researchers can hope to discover a drug with a specific affinity for these bacteria and no affinity for the normal constituents of the body. Such a substance would then be a magic bullet.

† About 90% of cancer patients undergo some kind of surgery, but only about 13% of cancers can be cured by surgery alone.

‡ Although safety has been improved, people are still dying of the side effects of radiation given for cancer such as Mo Mowlan, former Northern Ireland Secretary, Aug 19th, 2005.

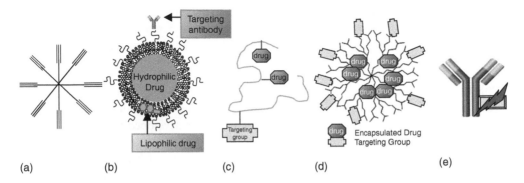

FIGURE 34.1 Schematic illustrations of the most developed nanoparticles for drug delivery: (a) polymer micelles; (b) PEGylated liposome showing encapsulated hydrophilic drug, membrane incorporated hydrophobic drug, and a targeting antibody; (c) a polymer drug conjugate that can also be targeted; (d) a dendrimer with an encapsulated drug as well as targeting ligands; and (e) a radioactive antibody.

passive targeting) that is based on the inherent leakiness of tumor microvessels.[31,32] This accumulation may possibly be aided by active targeting using antibodies.[32,33]

A recent review by Allen and Cullis outlines many of the problems that have been addressed and solved for nanoscale carrier systems, and it counts the successes for 13 systems that have gained FDA approval. If the focus is only on the anti-cancer applications, the systems, manufacturers, indications, and years of approval are:[8]

1. The anti-CD20 radioactive antibodies for non-Hodgkin's lymphoma (NHL), ^{90}Y-ibritumomab tiuxetan (Zevalin by Biogen IDEC in 2002)[34] and ^{131}I-tositumomab (Bexxar by Corixa and GlaxoSmithKline in 2003).
2. An anti-CD33-linked calicheamicin (Mylotarg by Wyeth-Ayerst) for CD33[+] relapsed acute myeloid leukemia (2002).
3. The three liposomal products
   a. Stealth (PEG[*]-stabilized) liposomal doxorubicin (Doxil/Caelyx by ALZA and Schering Plough) approved for Kaposi's Sarcoma (1995), refractory ovarian cancer (1999), and refractory breast cancer (Europe 2003 for patients with cardiac risk)
   b. Liposomal daunorubicin (DaunoXome by NeXstar Pharmaceuticals and Gilead) for advanced, HIV-associated Kaposi's Sarcoma (1996)
   c. Liposomal doxorubicin (Myocet by Elan) for metastatic breast cancer in combination with cyclophosphamide (Europe 2000)
4. Adjuncts including PEG–granulocyte colony stimulating factor (pegfilgrastim/Neulasta by Amgen) for reduction of febrile neutropenia associated with chemotherapy (2002)

In further depth, how are some of these nanotech-formulated drugs faring in terms of clinical use and outcome? For the anti-CD20 antibody Zevalin, a randomized, controlled phase III trial supported its accelerated FDA approval for NHL.[35] This study included 143 subjects with relapsed or refractory, low grade or follicular NHL, or transformed B-cell NHL. An overall response rate of 80% was obtained in subjects receiving the Zevalin therapeutic regimen (73 subjects) compared to 56% for subjects receiving Rituxan alone (70 subjects). Complete response rates (with no detectable cancer) were 30% for the Zevalin group and 16% for Rituxan alone. The side effects of Zevalin

---

[*] PEG, Poly(ethylene glycol).

include thrombocytopenia and neutropenia. This bone marrow suppression, sometimes persistent in nature, has led to severe hemorrhaging and infection in a small number of patients.[35,36] Since its approval in 2002, further studies have supported the efficacy of Zevalin for NHL with response rates greater than 80%; however, the mean time to progression for these patients has still been limited to only 9–12 months.[25,37–39]

With over 40 years of research dedicated to liposomes alone and over 20,000 papers in the literature, the question can be asked: has liposomal drug delivery been successful? The first liposomal carrier, AmBisome (liposomal amphotericin B), was developed for severe fungal infections[40] with a potential market of approximately 100,000 cases per year in the United States. AmBisome was originally approved in Europe in 1990 (Gilead and NeXstar Pharmaceuticals), and it was cleared for use in the United States in 1997 (Gilead and Astellas). It is now available in more than 44 countries worldwide.[41] This formulation then is one of the first successes in therapeutic nanotechnology. Clinical trials are now underway to examine the expanded use of AmBisome for prophylaxis against systemic fungal infection following bone marrow transplantation and other anti-cancer treatments. However, as discussed by Becker et al., although the efficacy of high-dose AmBisome is equal to or better than free amphotericin B, failure rates still give cause for concern.[42] One explanation for this might be the limited immediate bioavailability of amphotericin B when it is liposome-encapsulated. This is a good example of a common problem facing all nanotech drug delivery systems; that is, the challenge of reducing toxicity of free drug is solved, but the flip side is compromised bioavailability.

As for liposomes against cancer, the FDA has approved two anthracycline formulations, Daunoxome (liposomal daunorubicin) and Doxil (PEGylated liposomal doxorubicin). These drugs were originally tested and approved for Kaposi's Sarcoma, an AIDS-related malignancy with an incidence of only 0.02%–0.05% in the U.S.[43] Doxil has also recently been indicated for refractory ovarian cancer that accounts for up to 70% of the 25,500 annual cases of ovarian cancer in the U.S.[44] Response rates of 12%–20% have been reported for Doxil in this patient group, proving it to be one of the most effective salvage therapies for ovarian cancer patients, especially for those that have developed platinum resistance.[45,46] However, complete responses are rare, and chronic recurrence leads to an average survival time of only 12–24 months after therapy.[44,47] One of Doxil's successes is that it is associated with lower incidences of cardiotoxicity, bone marrow suppression, and nausea than free doxorubicin, but it can cause other side effects, including hand-foot syndrome and stomatitis.[28,48,49] Aside from the formulations currently approved, are new liposomes in development or are there new indications for previously approved liposomes? A search for cancer, chemotherapy, and liposome at www.clinicaltrials.gov produced 25 currently recruiting trials, mainly testing liposomal doxorubicin in combination with other free drugs for a variety of cancers as listed in Table 34.1.[50]

Liposomes were discovered in 1965[51] and explored almost immediately in applications for drug delivery, principally for cancer.[52] They now represent the most advanced nanoscale drug delivery system in clinical use. Apart from radioactive antibodies,* other nanoscale molecules and particles such as polymer–drug conjugates and polymer–drug micelles are being developed but have not yet progressed to clinical trials. However, even the liposomal nanotechnology effort accounts for only 25/4370 or 0.6% of all cancer-related clinical trials. This review of the literature and current clinical trials is illuminating and shows that there is a huge opportunity for nanotechnology in the form of nanomedicines to impact cancer therapy. If successful (and failed) designs, tests, and trials are collectively built upon, existing nanotech systems may be integrated with new nanotech ideas may be integrated to achieve more optimal outcomes for the delivery of non-specific drugs—that is, a reduced toxicity compared to free drug and an enhanced accumulation and bioavailability within the tumor. It is here that the thermal sensitive liposome (LTSL)

---

* A search for monoclonal, antibody, and cancer found 297 trials, but only one phase I, using radio-labeled antibodies.

TABLE 34.1
Cancers Under Investigation for Therapy with Liposomal Doxorubicin

Breast cancer
Cervical cancer
Endometrial cancer
Male breast cancer
   Ovarian epithelial cancer
Ovarian germ cell tumor
Fallopian tube cancer
   Peritoneal cavity cancer
Prostate cancer
Non-small cell lung cancer
Small cell lung cancer
Adult hepatocellular carcinoma
Cutaneous T-cell lymphoma
Adult NHL
Mycosis fungoides/sezary syndrome
Adult Hodgkin's lymphoma
Lymphocyte depleted
Mixed cellularity
Nodular sclerosing
Adult large cell lymphoma
Anaplastic
Recurrent diffuse
Recurrent immunoblastic
Multiple myeloma

offers a new paradigm for local drug delivery and could provide the motivation and experience for other nanotechnologies that not only encapsulate drugs but also release them, making them bioavailable at the tumor site.

This chapter will compare and contrast a number of contemporary liposome formulations (e.g., Doxil, Myocet, DaunoXome, and the currently trialed Onco-TCS). In particular, it will discuss the rationale and events that have led to the phase I/II testing of the doxorubicin-containing, temperature-sensitive liposome (ThermoDox) in prostate, breast, and liver cancer. This a particular example of how old and new concepts, especially ones that are bio-inspired, may be combined to face the challenges of cancer nanotherapeutics. The new capsular design addresses an unmet need in liposomal drug delivery, that of getting the drug out of the nanocarrier and thereby improving bioavailability.[53] This temperature sensitive liposome formulation containing lysolipid in the membrane (designated in the literature as LTSL) has been specifically designed to rapidly release drugs in combination with mild hyperthermic temperatures. In preclinical trials, the LTSL successfully delivers doxorubicin to solid tumors in bioavailable quantities that cannot be achieved by free drug administration or passive liposome accumulation, even with stealth liposomes. As depicted in Figure 34.2, this formulation is establishing a new paradigm for localized drug delivery to tumor tissue—one that moves nanotechnology for cancer therapy from the requirements of drug retention and carrier extravasation to the accomplishment of drug release in the blood stream of the tumor, thereby making the drug bioavailable. This gives the drug access to endothelial cells as well as neoplastic cells, offering the potential for anti-vascular as well as anti-cancer drug therapies.[54]

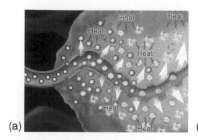
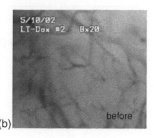
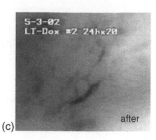

(a) (b) before (c) after

FIGURE 34.2 (a) A new paradigm for drug delivery, i.e., drug is released from circulating temperature-triggered liposomes in the blood stream of the tumor; drug gains access to tumor endothelia as well as to the neoplasm. (b and c) Video micrograph images of tumor vasculature before (b) and 24 h after (c) treatment with temperature-sensitive liposomes containing doxorubicin (Dox–LTSL) with an almost complete elimination of functioning blood vessels. (From Kong, G., et al., *Cancer Res.*, 60 (24), 6950–6957, 2000; Chen, Q., et al., *Mol. Cancer Ther.*, 3 (10), 1311–1317, 2004. With permission.)

34.2 THE CHALLENGES FACING NANOTHERAPEUTIC INTERVENTION: A BRIEF REVIEW OF LIPID-BASED NANOCARRIERS

34.2.1 GETTING THE DRUG IN

Small molecules are on the order of a few nanometers. A liposomal carrier by definition must be larger than the drug it is carrying. Furthermore, a larger liposome can encapsulate a much greater volume, resulting in a higher drug to lipid ratio (for water-soluble drugs).[*] This is important because an overload of carrier material can impair liver metabolism and phagocytosis (RES uptake) and can cause granulomatous inflammation of the liver. In fact, large doses of empty liposomes were used in the past to block the RES and promote longer liposome-circulation half-lives.[55,56] Figure 34.3 shows that although only two molecules thick (~ 5 nm), the lipid bilayer[†] becomes a significant fraction of the total liposome volume at diameters of 200 nm and smaller. For reasons to be later discussed, liposomes are traditionally ~ 100 nm in diameter. At this size, the membrane already takes up 30% of the total liposome volume. Because of mechanical restrictions imposed by membrane thickness and bending stiffness of the lipid bilayer itself (limited curvature), the smallest possible liposome has a diameter of ~ 30 nm, exhibiting a hydrophobic volume fraction of $\sim 80\%$ and a limited capacity to carry water soluble drugs

Therefore, for these nanoscale carriers, how much drug could be loaded in a 100 nm capsule? The number of molecules of hydrophilic drug per liposome (N_d) depends on the solubility (S_d) of the drug as well as the internal radius (R_i) of the liposome

$$N_d = \frac{4}{3} \pi R_i^3 S_d A_0$$

Note that Avogadro's number is included in the calculation. For cisplatin with a solubility of 5.8 mM at 25°C, a 100 nm liposome is capable of encapsulating only 1,340 molecules of this drug. By comparison, the 100 nm liposome is composed of about $\sim 140,000$ molecules of lipid, giving a drug to lipid ratio of only 0.005 (w/w).

The drug doxorubicin, on the other hand, is amphiphilic with a protonatable amine (pK_A 8.4);[57] therefore, depending on the pH, a fraction of the drug is soluble in the bilayer. This fraction can transit the bilayer if an electrochemical gradient is imposed, facilitating the active loading method

---

[*] This chapter does not consider hydrophobic drugs.

[†] The hydrophobic bilayer thickness is ~ 4 nm.

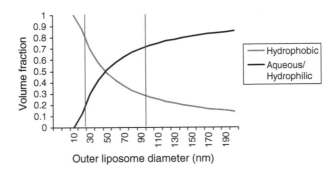

FIGURE 34.3 Plot of hydrophobic volume fraction as a function of liposome diameter for single bilayer vesicles. For the traditional 100 nm diameter liposome, 70% of the liposome volume is available for aqueous content. However, for the smallest liposome possible (30 nm because of limited curvature), there is considerably more hydrophobic volume per liposome ($\sim 80\%$).

that has revolutionized liposomal drug loading technology for weak acids (Figure 34.4).[58,59] This method results in extensive loading of up to 61,000 molecules of doxorubicin per liposome (drug to lipid ratio of 0.3) with the formation of a complex of drug crystals inside the liposome. In considering new nanoscale drug delivery systems, these kinds of simple calculations are necessary to determine carrying capacity and drug dosage. As previously mentioned and will be later discussed, nanoscopic carriers are readily identified by the immune system as foreign and can become toxic to the RES in high doses. There is, therefore, a delicate balance between efficacy and toxicity, not just for the drug but also for the nanoscopic carrier.

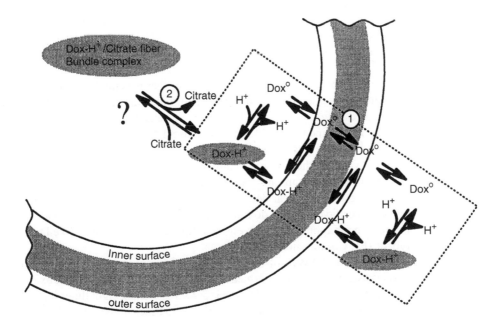

FIGURE 34.4 Remote loading of doxorubicin into liposomes using a pH gradient. Doxorubicin, an amphiphilic drug, crosses the lipid bilayer in neutral form. When it encounters the acidic interior of the liposome, an amine is protonated to form Dox-H+. This compound is retained on the interior, forming fiber bundles by electrostatic interaction with citrate ions.

34.2.2 Getting the Nanocapsules to the Tumor

The next issue to address for any nanocarrier is the probability that it will reach the tumor microvessels after injection into the blood stream. Considerations here include circulation through the tumor and pharmacokinetics of the carrier. A large fraction of the human cardiac output circulates through the kidneys and liver, an this explains the rapid clearance of particles by these organs. It is estimated that only a small fraction of the cardiac output actually goes to the peripheral circulation (and to many tumors).[60] Within this blood, what concentration of liposomal drug is actually present? Liposome extrusion methods are limited to a lipid concentration of 100 mM, above which the high viscosity of the lipid suspension limits processing. Assuming a clinical injection volume of 20 mLs, the lipids would be diluted 250 times (in 5 L blood) to a plasma concentration of 0.4 mM lipid. If drugs such as doxorubicin can be loaded into liposomes at internal drug concentrations of 300 mM, then the actual plasma drug concentration (still encapsulated) is now only 0.216 mM. This is about 200 times the amount required for a log kill of cells in culture (~ 1 μM Dox). However, this assumes complete mixing and no clearance (see below). As the complexities of drug transport become apparent, it is clear that delivery (and not just reduction of toxicity) is the major goal of nanotherapeutics.

Given these limits on actual blood levels available for delivery, how does the nanoparticle accumulate in the tumor? Where does it actually collect within the tumor, and does it stay there? Either the carrier must lodge in the microvascular bed of the tumor as in chemoembolism strategies, or it must leak out of the tumor vasculature (by the EPR effect) as in most nanoparticle strategies. The vascular permeability of a drug (across the endothelial lining) is limited by the size and charge of the drug as well as the structures of the vessel wall, particularly the basement membrane.[61] For small therapeutic drugs such as doxorubicin, permeabilities through normal vasculature are on the order of 10^{-4} cm/s. Similarly, sulforhodamine has been measured to have a permeability of 3.4×10^{-5} cm/s.[62] As the size of a nanoparticle increases beyond 6 nm, its ability to extravasate through the endothelial lining drops precipitously.[63] For example, the microvascular permeability of albumin (radius of gyration 7 nm) in normal tissues is 25×10^{-8} cm/s, and that for 150,000 MW dextran (radius of gyration 11 nm) is 7.3×10^{-8} cm/s.[62,64–66] Therefore, these polymers and proteins are well retained in the blood. For 100 nm liposomes, the normal vascular permeability is also limited at 9×10^{-8} cm/s.[65] Therefore, liposomes are over 1,000 times less permeable across normal vessels than free doxorubicin. It is this physical barrier that is actually responsible for the reduction in toxicity, dramatically altered biodistribution, and somewhat extended circulation half-life of liposome-encapsulated drug compared to free drug. For example, conventional liposomal daunorubicin (Daunoxome) is retained in the blood with a circulation half-life of 4 h compared to 0.8 h for free daunorubicin.[67] As the size of the nanoparticle is increased, motivated by the need to create a large encapsulated volume, the liposome gains the unique ability to keep the drug in the bloodstream and away from healthy organs and tissues such as the heart and kidneys. This was one of the major successes of liposomes, a clear demonstration that nanotechnology could impact the therapeutic toxicity of non-specific drugs like doxorubicin.[28,48,49] In fact, improved toxicity profiles have been the primary reason for approval of conventional liposome formulations like Daunoxome (FDA 1996) and Myocet (EU Commission 2000).

But how does this aid in the delivery of drugs to tumors? Fortunately, as already intimated, tumors have leaky blood vessels that are more permeable than normal blood vessels. For example, the tumor microvascular permeability for 150 kDa dextran is 5.7×10^{-7} cm/s, eight times higher than in normal tissue,[64] and for albumin, tumor vessel permeability is 25 times higher (1×10^{-7} cm/s).[66] It has been shown that this permeability extends to nanoparticles such as colloidal carbon[68] and liposomes up to 400 nm in diameter.[69] Therefore, one favored mechanism of targeting tumors has been to take advantage of the natural leakiness of tumor blood vessels to certain nanoparticles such as drug–polymer complexes and liposomes.[9,70,71] These drug delivery systems achieve localization simply by passive accumulation. Dosimetry results for the conventional liposome Daunoxome show that approximately 2.84 ng/mg (5 μM) of daunorubicin accumulates in the tumor in the first 24 h after an injected dose of 2 mg/kg body weight.[68] This concentration of drug is actually fairly successful at killing tumor cells in culture (IC50 2.8 μg/ml).[72] So why does this formulation show limited success in the clinic? The problem is probably not just inadequate liposome accumulation but also the inability to make the drug bioavailable in a fast enough time period, as will be discussed later.

Returning to a discussion of dosimetry, an important challenge was to determine what limits liposome accumulation in the tumor and how this uptake may be increased. It was soon discovered that the reason for limited tumor uptake is simply the limited circulation half-life of conventional liposomes (2–4 h).[73] These liposomes are labeled by complement and other opsonins as foreign and are subsequently removed by the phagocytic RES of the liver and spleen.[74,75] Incidentally, particles larger than 0.6 μm are cleared by the spleen regardless of opsonization.[76] This is another major lesson for any nanotechnology-based drug carrier system—that the surface chemistry of particles (especially nanoparticles) is crucial in the foreign-body reaction. Essentially, the early researchers and companies (The Liposome Company/Elan and Nexstar), intending to target cancer, unwittingly created an artificial parasite. Some formulations, however, have cleverly taken advantage of this natural localization to the Kupffer cells of the liver for treatment of the parasite Leishmania that resides in these same cells.[77–79] By creating a highly immunogenic (opsonizable) liposome containing negatively charged lipids, these formulations were rapidly removed from circulation by the RES. A similar liver-targeting technique was put into clinical practice when Ambisome was approved for use against leishmaniasis.[80,81]

The solution to the clearance problem was to favorably modify the liposome surface chemistry. Inspired by the very long (120 day) half-life of the red blood cell, Allen and Papahadjopoulos[75] explored the idea of incorporating GM1, a red blood cell membrane glycolipid, into the liposomal membranes (10 mol%). This conferred longevity to the liposomes, increasing plasma half life to ∼24 h. However, GM1 is an expensive molecule to purify from whole blood. Therefore, they[82] and others[83–86] turned to the chemical synthesis and development of polyethylene glycol derivatized lipid (e.g., DSPE–PEG*) that similarly increased circulation time. This conjugated lipid had a single negative charge because the derivatization left the normal zwitterionic PE charged at the phosphate. Although the incorporation of the negatively charged PS was highly immunogenic, their original hypothesis was that when incorporated into the bilayer at a few mol%, this shielded negative charge (similar to that of GM1) was the source of resistance to opsonization (Lasic, D., Personal Communication. Liposome Technology, Inc., Menlo Park, CA, 1992). Subsequently, evidence has been accumulated that indicates that the mechanism is primarily one of steric stability because the electrostatic potential lies somewhat underneath the steric barrier profile provided by PEG as demonstrated by McIntosh et al.[87,88] The mechanism for the long circulation lifetime conferred by incorporating GM1 is still unknown, although Huang postulated a dysopsonization for this ninja liposome.[89,90]

Preclinical animal studies demonstrated the dramatic changes in circulation half-life, tumor accumulation, and anti-tumor efficacy that resulted from these somewhat bioinspired surface chemical strategies.[30,82,91–96] The longer circulation time of PEG-grafted lipids allowed for increased extravasation from leaky tumor microvessels, successfully exhibiting the EPR paradigm of passive targeting. The new so-called "Stealth" technology was developed by Sequus and incorporated into a PEGylated liposomal doxorubicin formulation, Doxil, by Johnson and Johnson. Doxil has achieved circulation half lives of 20 h in mice[82] and ∼50 h in humans.[97] Furthermore, at 48 h after injection, the median tumor concentration of Doxil (in Kaposi's sarcoma lesions in humans) was 19 times higher than in normal skin.[97] In mice, about 7 ng/mg of Dox accumulated in FaDu xenografts (human squamous cell carcinoma) after injection of 5 mg/kg Doxil (Figure 34.24).[98]

---

* DSPE–PEG, Distearoylphosphatidylcholine grafted with polyethylene glycol.

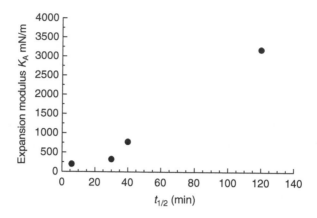

FIGURE 34.5 Plasma half life of conventional liposomes versus elastic modulus of the membrane (for normal RES). Membranes with a higher elastic modulus circulate for longer times. That is, a low resistance to mechanical expansion of the bilayer interface is correlated with liposome destruction in the blood stream. This points to a mechanical mechanism for the presumed opsonization of liposomes. A similar trend holds when the RES is pre-saturated with liposomes, but half-lives are on the order of hours.

An alternative mechanism to resist opsonization that also relies on reducing protein adsorption and binding at the bilayer head group region was to modify the membrane mechanically rather than to chemically modify the surface. This was achieved by using a lipid composition that creates a very tight interface among the headgroups as measured by a high area dilation modulus for the bilayer.[99] This composition utilizes lipids with saturated acyl chains and incorporates high concentrations (~50 mol%) of cholesterol in the membranes. Taking data from the literature on circulation half lives and comparing them to the elastic area dilation modulus measured using the micropipette technique for similar lipid compositions, a direct correlation was found between circulation half life and bilayer elasticity as shown in Figure 34.5.[87]

This shows that resistance to bilayer area expansion is a key property in achieving a long circulation performance for these nanocarriers. That is, a high resistance to mechanical expansion of the bilayer interface is correlated with reduced opsonization in the blood stream. This points to a nanomechanical mechanism for the presumed opsonization of liposomes where the molecular insertion that underlies opsonization must create area expansion in the bilayer. A similar trend holds when the RES is pre-saturated with liposomes. Here, half-lives are on the order of hours, and clearance may be due to destruction in the blood stream. Again, low membrane compliance resists this and is the property that correlates with long circulation performance.

A particular example is the sphingomyelin/cholesterol bilayer with a high area dilation modulus of ~2,200 mN/m[99] that has been utilized in Onco-TCS, the liposomal vincristine* formulation of Inex.[100] Onco-TCS achieves a prolonged circulation half life (6.6 h) when compared to free vincristine (1.36 h) in preclinical studies.[101] Therefore, a low bilayer membrane compliance can be used to increase circulation time. Onco-TCS also exhibits a higher concentration in tumors and demonstrates less neurotoxicity and greater efficacy in mice bearing L-1210 and P-388 leukemias. This formulation has been shown to have a similar MTD when compared with free vincristine, and it is currently in phase II/III clinical trials for aggressive recurrent NHL.[8,102–104]

---

* First extracted from the Madagascar periwinkle.

34.2.3 MECHANISMS OF LIPOSOMAL DRUG DELIVERY

As the study of liposomes continued, it was clear that in order to address mechanistic questions about drug delivery such as the EPR effect and the distribution of liposomes and other nanoparticles within the tumor, a direct in vivo measurement in real time was required. The skin flap tumor window chamber, first used in the 1940s for observation of tumor microvasculature, offered this capacity.[105] This research team[54,71,106] and others[9,66] have extensively utilized the window chamber to measure liposomal circulation half-life, accumulation in tumors, and tumor distribution both with and without hyperthermia. For the conventional, stealth, and temperature-sensitive liposomes, this technique has been invaluable in the direct observation of accumulation and/or triggered release of encapsulated drug in the tumor tissue. The window chamber is shown in Figure 34.6.

The procedure involves implanting tumor cells onto the exposed fascia and covering the surgical site with a glass window. After the tumor grows, the vasculature is visible by intravital microscopy with transmitted white light. In addition, using epifluorescence, it is possible to quantitatively evaluate fluorescent or fluorescently labeled molecules within the vasculature and within the tumor tissue over time. With this model, complex drug delivery data such as pharmacokinetics, vessel wall permeability, and intratumoral distribution have been quantified.[105] Using small fluorescent molecules fluorescently labeled albumin,[62] and fluorescently labeled liposomes,[65] the following have been shown:

1. Small molecules such as sulforhodamine (606 Da) and proteins such as albumin (66,430 Da) are readily extravasated into tumor tissue, supporting the established permeability of tumor vasculature.[62]
2. Conventional liposomes also extravasate but the extent of accumulation in the perivascular space is severely limited by fast RES uptake and disappearance from circulation (Figure 34.7, middle panels and right graph).[65]
3. Stealth liposomes have an extended circulation half-life and accumulate to a greater extent as shown in the epifluorescent images of the tumor vasculature in Figure 34.7 (left panel and left graph).[65] This supports the early efficacy studies by Vaage and

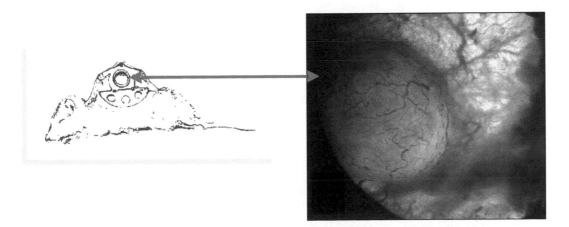

FIGURE 34.6 (See color insert following page 522.) Schematic of rat with a window chamber surgically attached to the dorsal skin flap (left). The skin flap is a 200 μm thick section of dermis with implanted tumor cells. As these cells grow, the tumor is visible via trans-illumination (right) or epi-illumination. This model was used to quantify complex data such as tumor vascular permeability and liposome accumulation. (From Huang, Q., et al., *Nat. Biotechnol.*, 17, 1033–1035, 1999. With permission.)

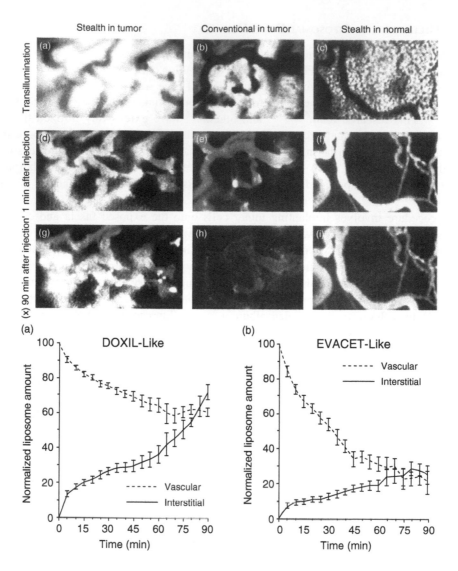

FIGURE 34.7 Video micrographs showing the window chamber vasculature via trans-illumination (top row) and epi-illumination (second and third rows) after injection of stealth liposomes in tumor (left column), conventional liposomes in tumor (middle), or stealth liposomes in normal tissue (right). The second row shows 1 min after injection, and the third row shows 90 min after injection. The graphs show the relative amounts of liposomes in the blood stream and in the tumor versus time (stealth on left and conventional on right) as calculated from the fluorescent images. (From Wu, N. Z., et al., *Cancer Res.*, 53 (16), 3765–3770, 1993. With permission.)

Metcalfe that showed that stealth, but not conventional, liposomes produced tumor efficacy in mouse tumor models.[93–96]

What this data[99,107,108] clearly shows is that the stealth liposomes out-perform the conventional liposomes with respect to circulation half-life and accumulation in the heterogeneously leaky tumor (left and center panels). Furthermore, as shown in the right panel, normal blood vessels (in the same window chamber) maintain a seal against liposome extravasation. The small 100 nm diameter of the liposome and the incorporation of PEG-lipid (4–5 mol%) work to preferentially increase the amount of encapsulated

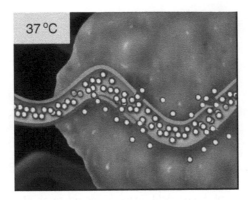

FIGURE 34.8 Schematic showing the traditional paradigm for liposomal (and polymer) drug delivery. In this paradigm, extended circulation time takes advantage of leaky tumor vasculature to preferentially deliver carriers to the tumor tissue.

drug being delivered to the tumor while normal tissue is protected. This behavior establishes the current paradigm of the EPR effect for liposomes (and other macro-molecules and nanoparticles) as depicted in Figure 34.8; that is, the basis for preferential drug delivery is this disparity in vascular permeability to even relatively large 100 nm particles, resulting in a fraction of the injected dose gaining access to the perivascular space of the tumor tissue.

4. The vascular pore cut-off size for liposome extravasation in the human ovarian carcinoma SKOV3 was found to be between 7 and 100 nm diameter at 37°C;[71] other studies have shown a cut-off size of up to 400 nm in the human colon adenocarcinoma LS174T.[70]

5. Mild local hyperthermia (HT) of 42°C could dramatically increase the permeability of tumor vessels to stealth (non-thermal sensitive) liposomes up to 400 nm in diameter with the most dramatic effect seen at 100 nm.[71] In one experiment, the amount of liposome extravasation was increased by 50-fold with HT compared to normothermic controls for the relatively impermeable SKOV3 tumor.[109]

6. In spontaneous tumors in pet cats, HT dramatically increased the accumulation of non-thermal sensitive technetium-99-labeled liposomes in the tumor tissue by 2–16-fold, further supporting the utility of HT in spontaneous tumors.[106]

These results demonstrate the important benefits of mild hyperthermia for liposome distribution in the EPR paradigm (even with non-thermal sensitive liposomes). However, they also point to a number of flaws in relying on the EPR effect alone to deliver nanoparticles and their drug. It is apparent that not all tumors are permeable to nanoparticles on the scale of 100 nm. In addition, throughout the window chamber studies, researchers have observed that liposomes and other nanoparticles only accumulate in the perivascular space after extravasation.[9,71,109] Even antibody-targeted liposomes show only limited interstitial diffusion.

As for the visualization of liposome distribution, Figure 34.9 shows magnified videomicro-graph images of the microvasculature of a mammary adenocarcinoma model in a skin flap window chamber.[65,110] On the left, a trans-illuminated image shows blood vessels of approximately 30 μm in diameter. The smallest (arrow) is about 10 μm, allowing only slow passage of red cells in single file. On the right is an epi-illuminated image of fluorescent liposome distribution 90 min after injection of NBD-labeled liposomes. This characteristic hot spot pattern illustrates the gross hetero-geneity of tumor vascular permeability and the limited transport of fluorescently labeled liposomes into the packed cellular bed of the tumor tissue. This image very dramatically demonstrates that

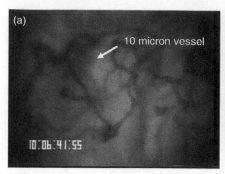
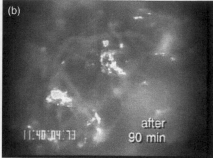

FIGURE 34.9 (a) Videomicrograph bright-field image of the microvasculature of a mammary adenocarcinoma in a skin flap window chamber. (b) Videomicrograph epifluorescent image of the same window chamber, showing a hot spot pattern of stealth liposomes 90 min after tail vein injection. The focal deposits of liposomes are limited to the perivascular space. (From Chen, Q., et al., *Mol. Cancer Ther.*, 3 (10), 1311–1317, 2004. With permission.)

liposomes only accumulate in the perivascular space of the tumor and do not diffuse to cells deep within the tumor. In fact, the hot spots are close enough to the circulation that a portion of the liposomes can actually wash back into the vessels.[111] The important conclusion here is that the liposome-encapsulated drug does not reach the majority of the tumor cells based on the EPR effect alone. This is also expected for other nanoparticles that aim to deliver drugs to tumors; they will not reach beyond the perivascular space unless some additional chemical or external influence is exerted.

Whereas distribution of nanoscopic particles only to the perivascular space is clearly a limitation, a bigger question is if the drug is actually bioavailable when delivered to even these regions. How does the drug get to its target once the carrier has accumulated in the tumor? How do the properties of the carrier and the local tumor environment contribute to this process? This chapter previously described how the currently approved liposomal nanotechnologies achieve a significant decrease in drug toxicity and a modest accumulation of drug in the tumor. However, the design requirement of retaining drug in the carrier during circulation results in most of the drug remaining inside the carrier if and when it localizes in the tumor tissue. It is expected that the drug will passively leak across the membrane, but it seems that it is often not bioavailable in high concentrations on the time scale necessary for therapeutic action. In particular, non-cell cycle specific drugs like doxorubicin should ideally be available to the intracellular targets (including DNA and RNA) as a sharp peak of concentration, maximizing drug effect in a short time period to prevent cell recovery and resistance. This inherent slow release may actually be an advantage for drugs that have cell cycle specificity such as the vincristine formulation Onco-TCS.[104] In this scenario, the slow release and long circulation half life of the sphingomyelin/cholesterol liposomes make free vincristine available to as many cancer cells as possible as they progress through the cell cycle to the G2/M stage.

Overall, the controlled release of the drug from carriers at the tumor site is a problem that all polymer, liposomal, micro-particle, and capsular nanotechnologies still face.[8] For polymer–drug conjugates, it has been fairly straightforward to convert the covalent linkers into biodegradable linkers (cleaved by endogenous enzyme action or by simple bond hydrolysis). Preclinical studies using this method have demonstrated drug release specifically at the tumor with a resulting increase in anti-tumor efficacy.[112,113] However, for liposomes, as mentioned above, the main focus of modern carrier design has been to lessen leakage from the liposomes in the circulation phase rather than to trigger release at the tumor site.

34.3 NEW CAPSULAR DESIGN FOR DELIVERY OF DRUGS TO AND RELEASE OF DRUGS IN LOCALIZED SOLID TUMORS

This section will briefly review the events leading up to the invention of the temperature sensitive liposome (LTSL), a potential solution to this paradox of drug retention in circulation and rapid release at the tumor site. It will also discuss studies on the composition, structure, and property relationships of membranes. It was this wealth of fundamental lipid bilayer knowledge that allowed a liposome to be forward engineered to perform a number of functions: survive the physical and chemical conditions of the blood stream, accumulate in the tumor, and respond to stimuli (in this case, mild hyperthermia) to trigger the release of encapsulated drug. The key to the design and its functional performance, as in all nanotechnologies for drug delivery, is to understand the materials relationships of the carrier.

34.3.1 GETTING THE DRUG OUT

After 35 years of research, the long-circulating liposomes (e.g., Doxil) have fulfilled the requirements of loading and retention of drug, evasion of the body's defenses, and accumulation in tumors by EPR. However, it is widely recognized that liposome performance is often hampered by the very lipid compositions that create low bilayer permeability and allow retention of the encapsulated drug during the blood-borne transport phase. The remaining goal of releasing a drug such as doxorubicin fast enough to kill tumor cells was not part of the original conventional (Myocet, Daunoxome) or stealth (Doxil) designs. One solution that now comprises the main topic for the rest of this chapter has been to take a materials engineering approach to liposome design and to explore nanoscale material defects in these 100 nm carriers. The task undertaken for this particular system was to characterize the relationships between lipid composition (lipids and lysolipids), bilayer nanostructure (grain boundary defects), and membrane properties (mechanical, thermal, and permeability) in order to design a liposome that releases drug rapidly upon an externally applied trigger, in this case, mild heat of around 42°C.[114,115]

The functions (load, retain, target, and release) have been defined, and there is an established wealth of materials data, prototype tests, production methods, and performance studies for liposomes in the literature. Using these guidelines, the goal is to trigger release of the liposomes after extravasation by EPR. Based on size alone, liposomes greater than 600 nm are taken up by the spleen, and those less than 70 nm are taken up by hepatocytes.[76,116] In addition, 400 nm is the cut-off for tumor extravasation.[70] Therefore, the optimal liposome size for EPR-based drug delivery is about 100 nm. In addition, 100 nm liposomes have demonstrated the greatest increase in tumor uptake when combined with hyperthermia.[109] As for surface design, conventional formulations only provide a relatively short circulation half life (2–4 h)[73], but PEGylated lipids with their surface repulsive potentials[87] tend to avoid RES uptake. Therefore, the starting point was the 100 nm stealth formulation, shown schematically in Figure 34.10.

Given this liposome, researchers must remember that many physical limitations still hamper effective drug delivery. As discussed above, only a small fraction of the circulation passes through the tumor. Some tumors have low permeability, preventing uptake of even 100 nm nanoparticles.[71] Upon extravasation, liposomes accumulate only in the perivascular space, and the liposomes that do accumulate achieve only slow release of bioavailable drug. In 1995, even though liposomal formulations were being tested and approved, the current technology was not achieving the kind of bioavailable drug levels necessary for increased efficacy. Through collaboration with the hyperthermia program at Duke Medical Center, this team found that local hyperthermia may be used to address many of these problems. Before discussing the materials aspects of the liposome that promote triggered release, the effects of hyperthermia on liposomal drug delivery must be considered.

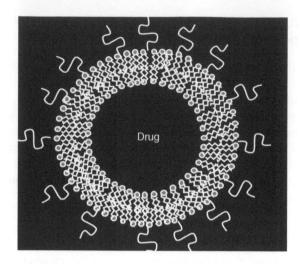

FIGURE 34.10 A tentative liposome design was a 100 nm diameter stealth liposome composed of a DPPC bilayer with 4–5 mol% PEG–DSPE (only shown on the outside) and encapsulating the drug doxorubicin via remote loading.

The mechanisms of synergy between hyperthermia (HT) and liposomal chemotherapy are manifold (Figure 34.11). HT improves vascular perfusion, thereby increasing blood flow through tumors that usually have significant areas of poor perfusion. Furthermore, HT dramatically enhances transvascular permeability, increasing pore size between endothelial cells to enhance liposome extravasation. Tumor microvessels are particularly sensitive to the effects of HT, providing added selectivity that protects the normal tissues.[117,118] These changes last for several hours following HT treatment and have been exploited to augment liposomal drug delivery to tumors up to 50-fold in otherwise impermeable tumor vessels as described above.[109] In addition, hyperthermia provides direct cytotoxicity to tumor cells and synergism with many drugs, including doxorubicin and cisplatin.[117,119] A number of liposomal drugs have been tested in a variety of

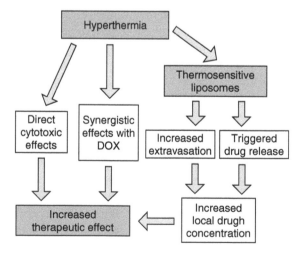

FIGURE 34.11 Schematic showing the mechanisms of synergism between hyperthermia and liposomal drug delivery. HT alone has some direct cytotoxic effects as well as synergism with many chemotherapeutic agents. In addition, HT enhances liposome extravasation and local tumor perfusion. The new concept was rapid triggered drug release from liposomes at mild hyperthermic temperatures.

tumor models, and these reports consistently prove that HT increases liposome accumulation in tumors and enhances anti-tumor effect.[117] Finally, with new advances, the current HT technology has the ability to administer controlled local heat to superficial or deep locations with accurate thermal dosimetry.[118] The remaining question, then, is if the performance of stealth liposomes can be improved by releasing the drug using externally applied heat. The answer lies in tailoring the materials composition, nanostructure, and properties of the lipid bilayer itself.

It is important to note here that the traditional target for chemotherapeutic drug delivery has always been distant metastases; very few primary tumors have been treated, much less cured, with chemotherapy alone. Therefore, these studies deviated from the traditional target for liposomes, took a bold and possibly unconventional track, and focused on local tumors. Whereas metastases are clearly the most important problem, an inability to control the local growth of solid tumors leads to significant morbidity and mortality in cancer patients. In the U.S., over 60% of therapies eventually fail with local recurrence.[120] Furthermore, improved locoregional control has been shown to enhance survival in many common cancers, including breast, prostate, lung, and cervical cancer.[121,122] What is needed here is a parenteral drug delivery system that can dramatically reduce the tumor size, to debulk for subsequent surgery or radiation, or maybe even achieve complete regression.

What this team set out to do was to design, create, and test a liposome that would respond to clinically attainable mild hyperthermia temperatures in the range of 41°C–42°C. The design required an intravenously injected liposome that released its drug only upon entering the heated tumor mass. The great surprise, in the very first series of preclinical tumor growth delay studies, was a response of 11 out of 11 complete regressions.[123] In addition, measurements of drug accumulation in the tumor tissue showed concentrations three times higher than with the combination of Doxil-like liposomes and HT[98] in spite of similar plasma half lives.[*]

34.3.2 A New Liposome Design

The lysolipid-containing temperature-sensitive liposome (LTSL) was invented in 1996, offering an effective way to achieve targeted tumor drug release for cancer chemotherapy.[98,114] This formulation is now in phase 1 human clinical trials.[53] Preclinical studies in mice, rats, and dogs have shown promising efficacy, reducing tumor growth in many of the tumors tested.[50,98,123] However, some variations in tumor response and subtleties in mechanism are still to be investigated. As will be discussed at the end of this section, attempts to correlate tumor growth delay with measurable parameters such as hypoxic fraction, intratumoral drug concentration, and tumor drug sensitivity reveal a complex relationship between these parameters. The current discourse will focus on how this liposome achieves the performance requirement of rapid triggered release of doxorubicin at clinically attainable, mild hyperthermic temperatures.

As demonstrated ~35 years ago by de Kruijff[124] and Papahadjopoulos[125] and subsequently modeled by Mouritsen,[126] the permeability of a lipid bilayer membrane to small ions, water, and even small drug molecules is dramatically increased at its gel-to-liquid melting phase transition when compared to its gel state. At the melting phase transition temperature (T_m), gel and liquid domains coexist, and solid phase lipids form grain structures to be described below. At the grain boundaries, leaky interfacial regions are present.[126–129] The increased permeability may be attributed to these interfacial regions as well as the much higher lateral compressibility of the membrane in the phase transition region.[130,131] For HT-triggered drug release in the body, a lipid was needed that remained solid and impermeable at body temperature (37°C) but became leaky to its encapsulated drug when heated just above this temperature. The lipid of choice was relatively obvious, DPPC[†] has a phase transition temperature of 41.9°C[132] and so has been used in many studies that aimed to explore temperature-sensitive drug delivery.[117,133]

---

[*] The LTSL, if anything, showed a shorter half-life.

[†] DPPC, Dipalmitoylphosphatidylcholine.

Thermal sensitive liposomes (TSLs) were not new in 1996. Previous formulations composed of DPPC mixed with other lipids like DSPC*,[133] DPPG†,[134] or HSPC‡:Cholesterol,[135] all required a temperature of 43°C–45°C to reach the peak in membrane permeability. The two- (or more) component phase diagrams show the origin of this transition temperature elevation, an ideal thermodynamics of mixing (entropic) that broadens the transition and raises it compared to the pure DPPC bilayer.[136–138] As a result, these formulations were not well matched to the clinical setting as hyperthermia temperatures greater than 43°C are not clinically attainable and may cause discomfort and vascular damage to the patient.[118] Also, even at the transition temperature, the rates of drug release for these lipid mixtures were still relatively slow (15–60 min). This is much longer than the transit time of a liposome through the tumor circulation, so these early TSLs relied on the EPR effect to position the liposomes in the tumor before releasing the drug. Unfortunately, the stealth technology had not yet been invented when these TSLs were being tested, and they did not accumulate in the tumor enough to be efficacious. Paradoxically, the stealth liposomes achieved accumulation but did not release their drug. Because many development agendas for conventional and stealth liposomes were already in place, TSLs received less attention for many years. With a renewed interest in thermally triggered systems and with the benefit of hindsight, this team began the LTSL design in 1996 by combining old and new technologies (thermal sensitivity and stealth PEG). In addition, experience from many years of studying, measuring, and modeling the material properties of lipid bilayers using micropipette manipulation techniques was applied.

The micropipette manipulation system has been developed and used by researchers to measure the mechanical properties of lipid bilayer vesicles,[99,107,108] permeability of lipid membranes,[139–141] molecular exchange with lysolipids,[142] and other mechanical and thermomechanical properties[143,144] in order to characterize bimolecular membranes as previously reviewed.[145–147] The micromechanical experiments on lipid vesicles established several key findings that influenced the new design.

1. Experiments that measured the resistance of lipid vesicle bilayers to area expansion and so derived the bilayer elastic expansion modulus showed that liquid state (L_α) lipids such as egg PCs plus cholesterol provided sufficient strength to retain drug and remain intact in the blood stream.
2. The more saturated PCs§ and sphingomyelin plus cholesterol exhibited enormous elastic moduli approaching traditional plastics[99] and were even more impermeable to water[139] and especially drugs.
3. The inclusion of cholesterol (to achieve strength and impermeability) abolished the melting phase transition and so attenuated any possible thermal sensitivity for DPPC (e.g., DPPC:Cholesterol in a molar ratio of 1:1).[135]
4. Even non-cholesterol-containing PC bilayers had moderate strength as gel phase membranes that were essentially thin (5 nm) layers of wax.[107]

Therefore, a liposome using gel phase lipids was designed. The design was built around DPPC with its T_m of 41.9°C plus a PEG¶-conjugated lipid (DSPE–PEG). The inclusion of DSPE–PEG did not change the transition permeability compared to pure DPPC.[148] Ideally, a liposome with a slightly lower T_m and higher membrane permeability than that of pure DPPC is desired. The additional component that achieved these goals will be discussed later in this section. First, the

---

* DSPC, Distearoylphosphatidylcholine.

† DPPG, Dipalmitoylphosphatidylglycerol.

‡ HSPC, Hydrogenated Soy Phosphatidylcholine.

§ Phosphatidylcholines.

¶ Poly(ethylene glycol)

structural features that apparently made the liposomes leak their drug, i.e., the grains and grain boundaries, must be discussed.

34.3.3 Nanostructures and Membrane Permeability

In 1987, working with giant lipid vesicles (20–30 μm diameter), this team showed that a single lipid vesicle could be heated or cooled through its main phase transition at the tip of a micropipette.[107,143] This allowed the transition to be characterized in terms of the bilayer area change reflecting a $\sim 25\%$ change in cumulative area per molecule (from 65Å^2 in the liquid phase to $\sim 41 \text{Å}^2$ in the $P_\beta{}'$ solid phase). Under no applied membrane tension, the solid vesicle was crumpled and faceted, whereas under controlled low tension (suction), it formed flat plate-like domains. Upon melting, these lipid 2D icebergs flowed past the pipet tip as they moved in a sea of melting lipid. In the gel (solid) state, it was shown that the rippled nanosuperstructure that characterized the $P_\beta{}'$ phase could be pulled flat.[143] Then, subjecting the vesicle to shear deformation, a yield shear and shear viscosity for the membrane were measured. The values for these material properties were on the order of simple waxes and plastics just below their melting temperatures. The membrane behaved like a Bingham plastic, yielding after an elastic shear deformation and then flowing with a velocity that was proportional to the excess stress (provided by the micropipette suction pressure) above the yield point. Both the yield shear and shear viscosity increased with decreasing temperature below the lipid melting temperature, showing that the more solid a membrane, the more difficult it was to deform in shear. It has been established that, for crystalline materials in general, the yield shear and shear viscosity reflect density and mobility of crystal defects in the solid structure. In 1987, this team postulated that grain boundaries and/or intra-grain dislocations allowed the shear deformation for the 2-molecule thick wax membrane.

It was expected that because crystalline solids nucleate in liquids upon cooling, the solid lipid membrane should form as grains (analogous to panels in a soccer ball; see Figure 34.12) that meet at grain boundaries (stitches in the soccer ball). This grain-faceted nanostructure was observed in subsequent electron micrograph images of the 100 nm diameter liposome formulation containing

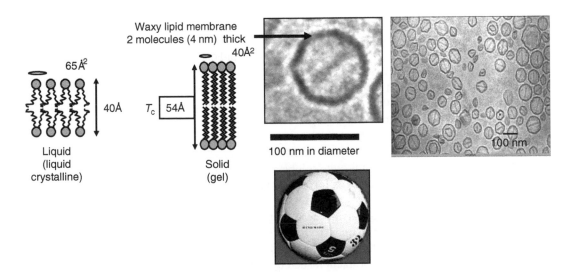

FIGURE 34.12 Schematic showing the soccer ball-like structure of the waxy liposome (center). The electron micrograph images show solid-phase liposomes, 100 nm in diameter, that are loaded with doxorubicin. The nanoscale grains and grain boundaries are evident from the faceted appearance. The phoshpholipid bilayer is 2 molecules thick (~ 4 nm), and there is a difference in molecular area and thickness between solid and liquid phase lipids (left). (From Ipsen, J. H. and O. G. Mouritsen, *Biochim. Biophys. Acta*, 944 (2), 121–134, 1988.)

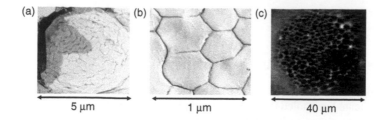

FIGURE 34.13 Solid lipid monolayers on gas microbubbles also show grain structure as revealed by freeze fracture electron microscopy (a and b, 5 μm and 1 μm scale). (c) Fluorescence micrograph images show that Bodipy lipids included in the lipid formulation at ~1 mol% are excluded from the solid grains and forced to accumulate at the grain boundaries. (From von Dreele, P. H., *Biochemistry*, 17 (19), 3939–3943, 1978. With permission.)

doxorubicin (Figure 34.12).[149] It was confirmed that this grain structure was evident at the nanoscale (100 nm) and for the 2-molecule thick membrane. Also shown in Figure 34.12 is a schematic of the solid/liquid interface, illustrating the mismatch in lipid molecular area and thickness between solid and liquid states. It is this nanoscale defect that provides the anomalous membrane permeability that has been observed and modeled in the transition region[125,126,150] and that is the underlying key to the LTSL drug release mechanism.

Additional evidence for the grain structure came from micromechanical and ultrastructural studies of solid lipid monolayers at microgas bubble interfaces (using similar saturated PC lipids). This work aimed to create and characterize lipid-stabilized gas-filled ultrasound contrast agents.[151] It was observed that the yield shear and shear viscosity were consistent with a polycrystalline monolayer, and both freeze fracture electron microscopy and fluorescence microscopy showed the grain structure in these micro-lipid shells (Figure 34.13).

With the grain structure now characterized, the goal of enhancing the permeability at the phase transition (compared to pure DPPC and other lipid mixtures) was now the priority. A series of micropipette experiments measured and modeled the exchange of molecules that entered the bilayer interface (tannic acid)[152] or incorporated into or even crossed the bilayer (peptides[153] and lysolipids)[142,154] and could then be washed out with bathing media. By measuring the area change of a single vesicle and by knowing the areas per molecule of the SOPC* bilayer lipid ($65A^2$) and the MOPC† lysolipid ($50A^2$), the mol fraction of lysolipid entering the bilayer from the aqueous solution was measured, down to a fraction of a mol%, was measured. Importantly, the desorption of lysolipid upon wash-out with lysolipid-free media was also measured, finding that it leaves the outer monolayer rapidly (within seconds)[142]. This team then discussed the possibility that the membrane permeability to small drugs might be increased if a certain amount of water-soluble lysolipids were incorporated in the solid bilayer and then desorbed from the membrane upon heating through the transition. The lysolipid MPPC‡ is structurally and chemically compatible with the bilayer lipid DPPC§, but it is quite soluble in water, especially as micelles. It was expected that MPPC would leave the bilayer (or form defects) at the T_m, making the membrane leaky to encapsulated drug.

Therefore, this team arrived at the new tentative liposome design: a 100 nm liposome comprised of a DPPC bilayer (underlying phase transition at 41.9°C) with 4 mol% of DSPE–PEG (for extended circulation half life) and 10 mol% of the lysolipid MPPC (to enhance the permeability), containing doxorubicin loaded by remote pH loading. The key to the design was

---

* SOPC, Stearolyoleoylphosphatidylcholine.

† MOPC, Monooleoylphosphatidylcholine.

‡ MPPC, Monopalmitoylphosphatidylcholine.

§ DPPC, Dipalmitoylphosphatidylcholine.

to work with the recognized anomalous permeability at the T_m (an apparent grain boundary effect in the melting solid membranes) and to enhance it with a second component (lysolipid). Ultimately, if this change in composition could change the membrane property of permeability (when heated), the drug would be freely available, and the performance of the temperature-sensitive liposome in vivo could be enhanced.

This lysolipid-based design was tested in 100 nm liposomes, and it characterized their ability to become permeable to small molecules, ions, and the drug doxorubicin. By incorporating 10 mol% of MPPC or MSPC* into the gel phase bilayers of DPPC liposomes, the permeability rate to carboxyfluorescein (CF)[148] and dithionite[115] as well as the rate of release of encapsulated drugs like doxorubicin[115,123,155] was greatly enhanced over that of the pure DPPC bilayer at its phase transition. For a liposome formulation composed of DPPC:MPPC:DSPE–PEG-2000† (molar ratio, 90:10:4), 60% of the entrapped content of CF was released in 1 min at 40°C, whereas pure DPPC liposomes released less than 20% CF even in 5 min at 40°C. Using the dithionite permeability assay,[156,157] several lipid systems were compared, providing some very interesting results. In this assay, the addition of dithionite quenches the fluorescence and produces a decay in the absorbance spectrum of NBD lipids‡ that are incorporated in the lipid bilayers at low (1 mol%) concentration (Figure 34.14).[156] In these experiments, the interaction between dithionite and the NBD-lipid head groups was recorded as a decay in absorbance at 465 nm in order to avoid attenuation of fluorescence because of repeated exposure to exciting wavelengths. If the bilayer is permeable, the absorbance shift will take place on both sides of the bilayer. Otherwise, if the bilayer is impermeable (e.g., in its gel state), only the outer monolayer is affected. The change in relative absorbance of the outside monolayer upon addition of dithionite matched its relative lipid distribution (i.e., 54% outside versus 46% remaining inside), reflecting the high 50 nm radius of curvature of the liposome.[115] As expected, the permeability for pure DPPC was somewhat enhanced at 40, 42, and 43°C in the region of its phase transition compared to lower temperatures of 30 and 37°C in the gel phase region (Figure 34.14). With the inclusion of 10 mol% lysolipid in the bilayers, the absorbance shift occurred much faster than for pure DPPC alone, showing that the permeability at the phase transition was dramatically enhanced.[115]

Figure 34.15 compares dithionite permeability for several lipid compositions, represented as permeability rates (per second). The permeability for DPPC at its T_m (41.9°C) was enhanced compared to its gel state, but it was surprisingly only slightly higher than that of a liquid phase lipid (POPC§) at 41.9°C.[115] At temperatures above 41.9°C, both DPPC and POPC exhibited permeability rates greater that that of DPPC at its phase transition. In contrast, dramatic increases in the phase transition permeability rate were observed when 10 mol% MPPC or MSPC were incorporated in the DPPC liposome membranes. The incorporation of these acyl chain-compatible lysolipids only shifted the bilayer phase transition temperature slightly from 41.9°C for DPPC alone to 41.0°C with MPPC and 41.4°C with MSPC.[115] Therefore, the incorporation of 10 mol% lysolipid had achieved the goals of enhancing the transition permeability and slightly lowering the transition temperature. In line with the expected influence of the acyl chain, the slightly longer MSPC lipid reduced the T_m less than the MPPC. Differential scanning calorimetry (DSC) studies showed that the transition region was still very sharp (i.e., not broadened as with the previous additions of DSPC and Cholesterol).[148]

A comparison of the DSC data with the permeability rate data confirms that the grain boundaries are responsible for the drug release. As shown in Figure 34.16, the permeability rate rises on the low temperature shoulder of the excess heat flow curve, i.e., before any significant mass

---

* MSPC, Monostearoylphosphatidylcholine.

† DSPE–PEG, Distearoylphosphoethanolamine–Polyethyleneglycol 2000.

‡ Lipids conjugated with the fluorescent molecule NBD, N-(7-nitro-2,1,3-benzoxadiazol).

§ Palmitoyloleoylphosphatidylcholine.

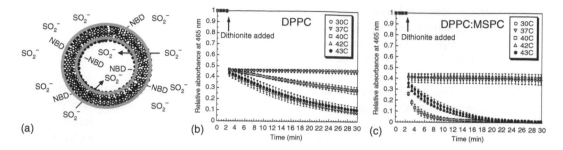

FIGURE 34.14 The dithionite assay for membrane permeability. (a) Dithionite quenches the fluorescence or, in this case, shifts the absorbance for NBD lipids incorporated in the bilayer. The outer monolayer is immediately changed, but the inner monolayer is only affected when the membrane becomes permeable to this ion, i.e., in the phase transition region. Both compositions show only outer monolayer shifts at temperatures in their gel state. Pure DPPC (b) shows the anomalous permeability associated with the transition, but for DPPC:MSPC, (c) the change in absorbance occurs much faster. (From Evans, E. and Needham, D., *J. Phys. Chem.*, 91 (16), 4219–4228, 1987. With permission.)

of solid phase lipid material has melted. Also, the maximum permeability rate coincides with the midpoint of the transition where it is expected that the grain boundary area is also at its maximum.[126]

As for the specific mechanism for this dramatic enhancement of temperature-triggered permeability by including 10 mol% lysolipid, it was initially hypothesized that desorption of the water-soluble lysolipid would leave defects in the membrane. However, Mills et al. recently showed that membrane permeability to dithionite in the transition region was still enhanced following dialysis against lysolipid-free buffer (with one exchange of buffer) at temperatures above T_m for 48 h (Figure 34.17).[115] Under these conditions, with a massive concentration gradient between liposomes and dialysate, there was a strong driving force for lysolipid desorption. This study demonstrated that, contrary to the original hypothesis, lysolipids did not completely desorb from

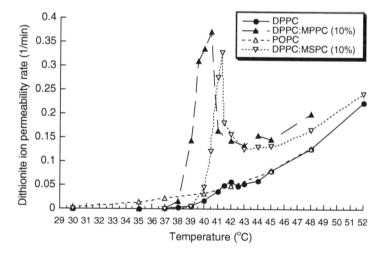

FIGURE 34.15 Dithionite ion permeability rates for pure and lysolipid-containing TSL membranes in vitro. At the phase transition temperature, permeability rates are ∼10-fold higher for the lysolipid-containing TSLs (LTSLs) when compared to the pure DPPC bilayer. (From Mills, J. K. and Needham, D., *Biochim. Biophys. Acta*, 1716 (2), 77–96, 2005. With permission.)

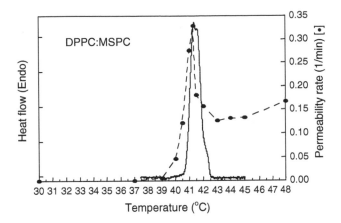

FIGURE 34.16 Comparison between the DSC thermal profile and the dithionite permeability rate for LTSL (DPPC:MSPC). This comparison shows that the onset of increased permeability (dashed line) occurs at the low temperature shoulder of the excess heat curve (solid line). That is, the content release occurs before a large fraction of the lipid has melted. (From Mills, J. K. and Needham, D., *Biochim. Biophys. Acta*, 1716 (2), 77–96, 2005. With permission.)

the membrane upon heating to and through the transition; rather, they largely stayed in the membrane, apparently forming lysolipid-stabilized defects in the bilayer.[115] Because lysolipids (MPPC and MSPC) have a conical molecular shape as a result of a single acyl chain, they prefer the micellized state (Figure 34.17b). Therefore, it seems reasonable that they would accumulate at

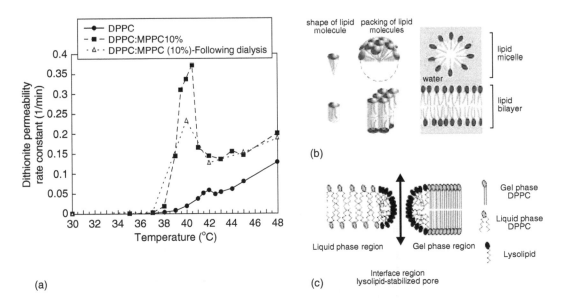

FIGURE 34.17 (a) Dithionite ion permeability as a function of temperature for freshly prepared LTSLs and for LTSLs that have been dialysed for 48 h at 45°C–48°C (liquid state). (b) The maintenance of enhanced (although slightly reduced) permeability after dialysis shows that some lysolipid must remain in the membrane. Pure DPPC is shown for comparison. (c) Lysolipids normally exist in aqueous solution as micelles (left), but they appear to remain present in liposomes at the grain boundaries, possibly stabilizing pores (right). (From Mills, J. K. and Needham, D., *Biochim. Biophys. Acta*, 1716 (2), 77–96, 2005. With permission.)

melting grain boundaries, forming semispherical porous edges (Figure 34.17c). The resulting lysolipid-stabilized pores would form only upon heating to the phase transition region, facilitating rapid temperature-sensitive release of contents. Previous micropipet studies by Zhelev and Needham on giant lipid vesicles support this hypothesis, suggesting that lysolipid can stabilize membrane nanopores large enough to pass glucose, but not sucrose, and that this pore formation does not compromise mechanical stability.[158] Therefore, the incorporation of lysolipids (MPPC or MSPC) is essential for drug release enhancement for temperature sensitive liposomes.

For the LTSL formulation, doxorubicin (Dox) was readily loaded into preformed gel phase liposomes by remote pH gradient methods (as described in Mayer, Bally, and Cullis and Madden, et al. but used now at 35°C).[115,159,160] The liposomes were then heated, and Dox release was measured as a function of time and temperature. The same triggered permeability and enhanced release effects were seen for Dox as with carboxyfluorescein and dithionite. For liposome compositions of DPPC:MPPC:DSPE–PEG-2000 (90:10:4) and DPPC:MSPC:DSPE–PEG2000 (90:10:4), ~100% and ~90%, respectively, of the encapsulated Dox was released within 2 min (at 40 and 41.3°C, respectively).[115] This team's most recent work has improved the experimental conditions in order to more closely observe drug release in the first few minutes.[160,161] With the incorporation of 8 mol% MPPC in the lipid bilayers, 100% of the drug is released in 5 min whereas with 12 mol% MPPC, it takes less than 20 s (Figure 34.18).

The all-important temperature dependence for Dox release was also characterized using 10 mol% lysolipid as shown in Figure 34.19. At 30, 37, and 39°C, very little drug was released. At 40°C, there was a moderate level of release and at 41.3°C, 100% of the drug was released in 4 min. If this 4 min time point is used to plot the amount released versus temperature (Figure 34.20),[148] then the optimal temperature range for triggered release is 39°C–42°C—a temperature window that is attainable using regional hyperthermia, even for deep-seated tumors such as ovarian and prostate tumors.[118] This figure also verifies that, as expected, the stealth liposomes do not release drug in a temperature-triggered fashion. Finally, when the Dox release curve is compared to the DSC curve for the same liposomes (Figure 34.21), it can be seen that drug release occurs on the low temperature shoulder of the phase transition, before a significant mass of lipid is melted. As with the dithionite data shown earlier, this suggests a grain boundary mechanism for the drug permeability.

Importantly, this in vitro data shows that the release time for the lysolipid-containing temperature sensitive liposome (LTSL) is likely to be faster than the transit time of the liposome through the tumor circulation. Moreover, release can be triggered at temperatures in the mild clinical

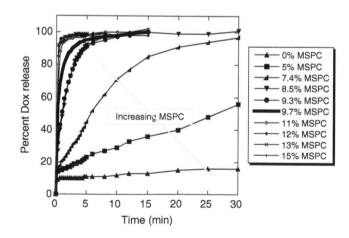

FIGURE 34.18 Drug release as a function of the molar fraction of lysolipid in the bilayer (from 0 mol% to 15 mol%) at 41.5°C. Complete content release occurs within 20 s with 12 mol% MSPC. (From Wright, A., In *Mechanical Engineering and Materials Science*. Duke University, Durham, NC, 2006. With permission.)

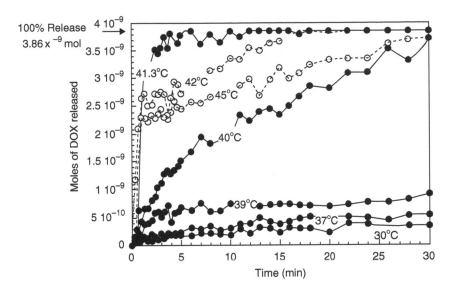

FIGURE 34.19 Drug release as a function of temperature. For liposomes with 10 mol% MSPC, the rate of Dox release increases as a function of temperature with maximum release occurring at the Tm of the lipid mixture. (From Needham, D. and Dewhirst, M. W., *Adv. Drug Deliv.Rev.*, 53 (3), 285–305, 2001; Needham, D., et al., *Cancer Res.*, 60 (5), 1197–1201, 2000. With permission.)

hyperthermia range. These advantages make the LTSL a much more feasible system for tumor drug release than any previous liposome. With this in vitro success, this team was very interested to see what effects the LTSL would have on tumor drug accumulation and tumor growth delay in vivo.

34.3.4 In Vivo Growth Delay Studies

The efficacy of the Dox–LTSL[*] was first evaluated in a human squamous cell carcinoma (FaDu) model, showing unprecedented local control for this tumor (in two studies, respectively, showing 11 out of 11 and 6 out of 9 complete regressions, Figure 34.23).[98,123] The FaDu xenografts were grown in the flanks of nude athymic mice. All animals received an equivalent single dose of 5 mg/kg i.v. doxorubicin (maximally tolerated dose) and one hour of local hyperthermia at 42°C. In addition to Dox–LTSL + HT, a number of control treatments were tested for comparison. As shown in Figure 34.22, with saline control, the FaDu tumors took an average of 10 days to reach the end point of 5 times initial tumor volume, representing a fairly fast growing tumor.[98] Administration of free Dox showed an average growth time of 12 days, delaying the growth by ~2 days. Mild hyperthermia (HT) alone, in the absence of any drug, exhibited a growth time of 18 days. When free Dox was combined with HT, the growth time was extended by another 4 days to an average of 22 days.[98] Therefore, these experiments demonstrate the anti-tumor activity of mild hyperthermia alone as well as its ability to enhance drug efficacy (as depicted schematically in Figure 34.11).

In the same FaDu studies, doxorubicin was encapsulated in Doxil-like, non-temperature sensitive liposomes (NTSL) as well as in the lysolipid-containing temperature sensitive liposomes (LTSL). Figure 34.23a shows the extended growth time of the NTSL in combination with mild HT (35 days) with all tumors decreasing rapidly in size but eventually growing back.[123,†] Compared

---

[*] Dox–LTSL, doxorubicin-encapsulating lysolipid-containing temperature sensitive liposome.

[†] The liposome formulations alone, without heat, did not perform much better than the free drug controls, showing that reduced accumulation and well-entrapped drug is hardly that efficacious.

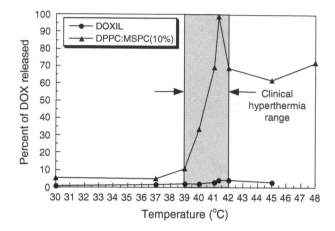

FIGURE 34.20 Percent of Dox released within 4 min as a function of temperature for DOXIL-like liposomes and LTSL (DPPC:MSPC 10%). For LTSL, triggered release occurred in the clinically attainable temperature range of 39°C–42°C. Over the same temperature range, the Doxil-like liposome (non-temperature-sensitive) did not release any drug. (From Needham, D. and Dewhirst, M. W., *Adv. Drug Deliv. Rev.*, 53 (3), 285–305, 2001; Needham, D., et al., *Cancer Res.*, 60 (5), 1197–1201, 2000. With permission.)

to NTSL without heat (21 days, not shown), this again proves the synergism of HT and liposomal drug delivery through increased perfusion and extravasation as demonstrated in previous studies.[106,109,117,162] The tumor regrowth pattern with NTSL indicates that drug was certainly delivered, but some cells were able to survive the treatment, perhaps as a result of the relatively slow leakage of bioavailable drug. The most dramatic result was observed for the LTSL in combination with HT with complete regressions in all 11 animals for up to 60 days (Figure 34.23b).[123],* The importance of rapid drug release was quite apparent. The focus on composition, structure, and properties for the liposome design and the testing of in vitro performance of the LTSL translated into a significant improvement in tumor response in vivo.

Following the success of these anti-tumor efficacy studies, additional FaDu experiments quantified drug delivery to the tumor and to the actual intracellular molecular target. Total drug deposition, tissue distribution, and fraction bound to DNA/RNA were determined for free Dox and three different liposome formulations (Doxil-like NTSL[†], LTSL, and a more traditional TSL with a slightly higher melting point).[98] To do this, two extraction procedures were carried out on tumors harvested just one hour after injection. Extraction with chloroform and silver nitrate identified all drug in the tumor tissue. Extraction of the tumor tissue with chloroform only identified drug that was free, protein-bound, or still trapped inside the liposomes. The difference represented the drug that was bound to the polynucleic acids. Dox concentration was measured using high performance liquid chromatography (HPLC). The result was that only the LTSL formulation showed any doxorubicin bound to DNA and RNA ($\sim$50% of the total amount in the tumor, as shown in Figure 34.24). The total tumor drug delivery using LTSL plus HT was 25 ng doxorubicin per mg tissue ($\sim$50 μM). The other liposomes showed 7–8 ng/mg at 42°C and $\sim$5 ng/mg at 34°C.[98] Growth time directly correlated with the amount of drug deposited. Tumor tissue sections were also visualized for doxorubicin fluorescence to determine the local distribution of the drug in the tumor and confirm the relative drug concentrations.[98] The fluorescence intensity was relatively low at 34°C

---

* The second study again showed superior results with six out of nine complete regressions (one tumor disease free at 40 days) and only two tumors did not respond.

† NTSL, non-temperature sensitive liposomes.

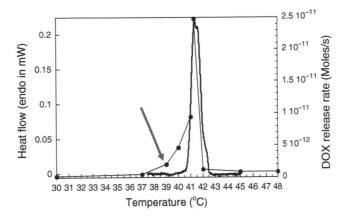

FIGURE 34.21 Comparison between the Dox release rate and the DSC curve for DPPC:MSPC (compare Figure 34.16). Drug permeability again occurs on the low temperature shoulder of the phase transition before significant melting of the lipid mass occurs, indicating a grain boundary mechanism for permeability.

after free drug or any liposomal treatment group. At 42°C, the LTSL showed extensive fluorescence (3.09 arbitrary fluorescence units) with the TTSL 3 times less (1.4) and the Doxil-like NTSL 10 times less (0.31). These fluorescence images show both unencapsulated and encapsulated drug. For the LTSL, the Dox distribution was pervasive throughout these tumor sections (Figure 34.24, right). Overall, the LTSL formulation demonstrated increased drug accumulation and triggered drug bioavailability, leading to complete tumor regressions after a single dose.

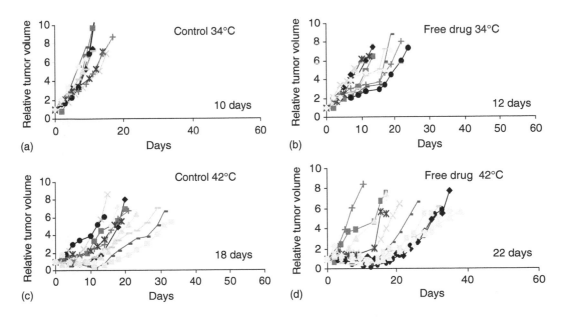

FIGURE 34.22 Tumor growth curves (relative tumor volumes) for flank-implanted FaDu xenografts in nude mice. The treatment groups were (a) control at normothermia (34°C); (b) free doxorubicin at 34°C (maximally tolerated dose of 5 mg/kg); (c) 1 h of hyperthermia at 42°C; and d) injection of free drug with 1 h HT at 42°C. The endpoint was time to 5 times initial tumor volume or 60 days. Average growth times are shown for each group. Overall, hyperthermia shows independent anti-tumor activity and enhancement of drug activity. (From Kong, G., et al., *Cancer Res.*, 60 (24), 6950–6957, 2000. With permission.)

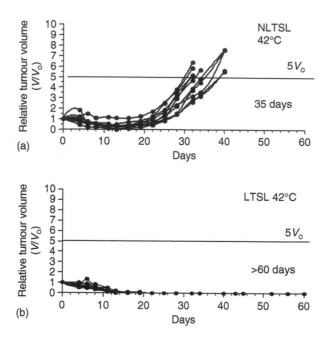

FIGURE 34.23 Tumor growth curves (relative tumor volumes) for flank-implanted FaDu xenografts in nude mice (as in Figure 34.22). All animals received 5 mg/kg doxorubicin and 1 h of local hyperthermia at 42°C. The treatment groups were (a) non-thermal sensitive Doxil-like liposomes plus HT and (b) Dox–LTSL + HT. The Dox–LTSL showed superior anti-tumor efficacy with 11 of 11 complete regressions (up to 60 days). (From Needham, D., et al., *Cancer Res.*, 60(5), 1197–1201, 2000. With permission.)

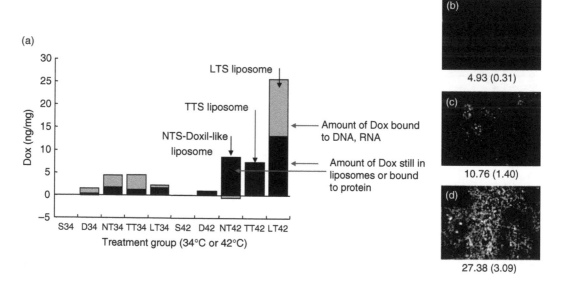

FIGURE 34.24 (a) Dox levels in tumor tissue after i.v. infusion as measured by HPLC for each of the treatment groups. Hatched bars show the amount still encapsulated or bound to protein, and solid bars show the amount bound to DNA/RNA. Only the Dox–LTSL formulation plus HT (42°C) shows significant amounts of Dox bound to DNA/RNA (~50%). The total tumor [Dox] after LTSL delivery was ~50 uM. (right) Doxorubicin fluorescence in a representative tumor section immediately prepared after treatment with 1 h HT in combination with (b) NTSL, (c) TTSL, or (d) LTSL. Fluorescent intensities are denoted in arbitrary units. (From Kong, G., et al., *Cancer Res.*, 60 (24), 6950–6957, 2000. With permission.)

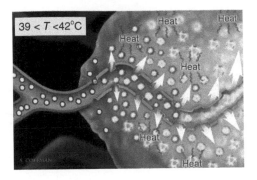

FIGURE 34.25 A new paradigm for local therapy (compare Figure 34.8). Thermal sensitive liposomes release drug RAPIDLY ($<$20 s) when heated to 42°C, resulting in drug release in the blood stream. Therefore, LTSL drug delivery does not require liposome extravasation to deliver free drug to the tumor cell. (From Kong, G., et al., *Cancer Res.*, 60 (24), 6950–6957, 2000. With permission.)

Because all three liposomes were 100 nm in size and coated with PEG, they would be expected to extravasate to the same extent. However, both the total tumor HPLC results and the histologic sections showed additional accumulation of drug (encapsulated plus free) in the tumor with the LTSL. This suggested that the greater delivery achieved by LTSL may be due to release of the drug in the blood stream of the heated tumor as schematically depicted in Figure 34.25. The Dox concentration for even the stealth NTSL ($\sim$15 μM) greatly exceeded the 1.8 μM IC_{50} (media concentration required to kill 50% of cells in culture) of free Dox measured in the multi-drug resistant human cervical carcinoma line KB-85.[163] However, the limited NTSL efficacy in these studies together with the scarcity of complete responses to Doxil in the clinic, suggest that the EPR effect alone has some major limitations in drug delivery. As previously mentioned, the in vitro drug release time for LTSL is shorter than the transit time of a liposome through tumor microvasculature; therefore, additional local delivery through intravascular release would result in free drug diffusion into the tumor tissue. Furthermore, endothelial cells lining the tumor microvasculature would also be exposed to free doxorubicin. This pointed to a new mechanism for drug delivery, one that does not completely rely on the EPR paradigm of liposome extravasation. The possibility of anti-vascular as well as anti-neoplastic drug action can now be considered as another aspect of this remarkable paradigm shift.

34.4 A NEW PARADIGM FOR DRUG DELIVERY TO TUMOR TISSUES

34.4.1 VASCULAR SHUTDOWN

A series of preclinical studies in window chamber and flank tumors have established that the mechanism of Dox–LTSL action is actually one of vascular shut down. This is due to the unique ability of LTSL to intravascularly release Dox as it passes through the blood vessels of the mildly heated tumor (Figure 34.25). Utilizing the window chamber for in situ red blood cell velocity measurements, Yuan et al. demonstrated that treatment with LTSL plus HT (1 h at 42°C) results in significantly reduced blood flow immediately after treatment and complete vascular shutdown within 24 h in all FaDu tumors (Figure 34.26).[54] Before treatment, blood flow velocity was $\sim$0.4 mm/s, and at only 6 h after treatment, blood velocity had slowed to $\sim$0.002 mm/s. Minimal changes in tumor blood flow were seen in all controls, including free drug and empty LTSL both with and without HT.

As shown in Figure 34.27, direct observation of the blood vessels in the tumors before and after treatment showed a startling result—not only had blood flow dramatically slowed, but many of the tumor microvessels had also disappeared. Within 24 h of LTSL plus the 1 h HT treatment, microvessel density in the tumor was reduced from 3.9 mm/mm^2 to 0.9 mm/mm^2, whereas the control

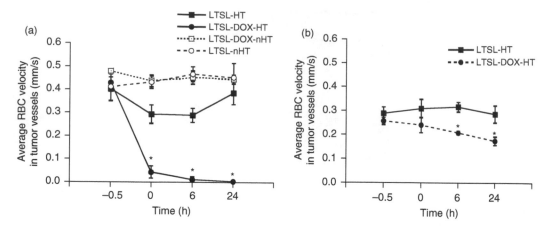

FIGURE 34.26 Red blood cell (RBC) velocity in FaDu tumor microvessels after injection of DiI-labeled fluorescent RBCs. (a) For Dox–LTSL plus HT, blood flow was significantly reduced after therapy and was essentially zero 24 h later. All controls (empty LTSL ± HT, free Dox ± HT) showed little effect on blood flow velocity. b) For normal vasculature, Dox–LTSL + HT had a much smaller effect. (From Chen, Q, et al., *Mol. Cancer Ther.*, With permission.)

treatments again showed little change (Figure 34.27c).[54] Moreover, these effects were only observed in tumor microvessels; heated normal microvessels were spared. Therefore, this represents a unique method of functional targeting of therapeutic agents. Measurements of VEGF and bFGF expression also showed a 50%–60% reduction in growth factor concentrations in treated tumors.[54] Overall, this study provided very strong evidence for the paradigm shift from EPR to intravascular release, leading to anti-vascular effects. The EPR effect alone (i.e., extravasation out of the blood stream) could not achieve this kind of vascular shut down. The unique ability of the lysolipid-containing bilayer to rapidly transition from a solid wax-like membrane into an extremely permeable melting state has made possible this rapid intravascular release of drug triggered by the application of only mild hyperthermia. This opens up the possibility for drug delivery patterns that target both tumor cells and tumor vasculature.

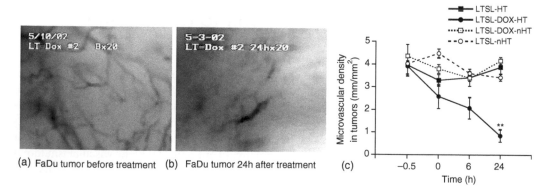

(a) FaDu tumor before treatment (b) FaDu tumor 24h after treatment (c)

FIGURE 34.27 Videomicrographs of tumor vasculature showing the tumor blood vessels before (a) and 24 h after (b) the LTSL plus 1 h HT treatment. After the treatment, almost all blood vessels disappeared with indications of thrombus formation. (c) A plot of tumor microvessel density versus time after the treatment comparing LTSL + HT to the same controls as in Figure 34.26. These results point to an anti-vascular mechanism found only in tumor vessels. Therefore, the Dox–LTSL + HT therapy appears to be anti-vascular as well as anti-neoplastic. (From Chen, Q., et al., *Mol. Cancer Ther.*, 3 (10), 1311–1317, 2004. With permission.)

34.4.2 Exploring Other Tumor Models

Dox–LTSL plus mild HT was tested in a number of other tumor models. As with most anti-cancer therapies, it was found that different tumors responded to LTSL in different ways. Five tumor lines were treated with saline, HT alone, LTSL alone, or LTSL plus HT.[164] Each group received equivalent dosing of 6 mg/kg i.v. Dox and/or 1 h HT at 42°C. The mouse mammary carcinoma 4T07 was implanted in normal BalbC mice, and four human tumor xenograft lines were implanted in nude athymic mice: SKOV-3 (ovarian carcinoma), HCT116 (colon carcinoma), FaDu (head and neck carcinoma), and PC-3 (prostate carcinoma).[164]

All control tumors grew rapidly, reaching 5x tumor volume (5xV) in 10–15 days. HT alone had minimal efficacy in delaying tumor growth, as did LTSL alone without heat. The Dox–LTSL plus HT groups showed superior tumor growth delay times in all tumor types when compared to all other treatment groups ($p < 0.006$), but the degree of prolongation was highly variable from one tumor line to the next (Figure 34.28). For Dox–LTSL + HT, the mean tumor growth times (to 5xV) for each model were, in order of increasing response, 4T07 (19.8 days) < HCT116 (28.1 days) < FaDu (41.8 days) < SKOV3 (51.9 days) < PC3(60 days).[164] This trend correlated to the intrinsic growth rates of each untreated control tumor line, i.e., 4T07 and HCT116 were the fastest growing tumors and PC was the slowest growing tumor. The fraction of animals remaining disease-free at 60 days post treatment was also highly variable among tumor lines. The PC3 tumor line showed the greatest response with 15 out of 20 complete regressions. Of note, the FaDu tumor type showed fewer complete regressions (1 out of 10) than in previous studies.[98,123]

Potential reasons for these variations in LTSL efficacy were explored by measuring parameters describing the tumor microenvironment, tumor cell sensitivity, and overall drug delivery.[164] For each tumor type, these parameters were compared to the relative growth delay time and the fraction of animals showing complete regression up to 60 days.

Measurements of pH and pO_2, reflecting the interstitial microenvironment, were performed in five mice for each tumor line. The hypothesis that pH might be important was based on the fact that tumors produce excess acid.[165] Because Dox is a weak base and is increasingly less hydrophobic with decreasing pH, transport of this drug across cell membranes could, in principle, be inhibited by low extracellular pH. However, all tumors had the same slightly acidic interstitial pH (6.8–7.0), so this could not explain the variations among different tumor lines. The pO_2 histograms (100–200 measurements per tumor) were obtained using the Oxylite probe as described by Braun and colleagues.[166] It has been reported that hypoxia can reduce the cytotoxicity of Dox and may contribute to different tumor responses in vivo.[167] In this team's studies, there were significant

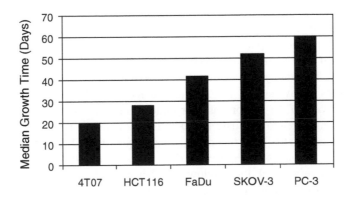

FIGURE 34.28 Median growth time (days to five time tumor volume) for each of five tumor types in response to Dox–LTSL + HT. Ranking of median growth time (MGT) reflects the intrinsic tumor growth rates of these tumor types. (From Zhao, Y., Yarmolenko, P.S., et al. [In preparation]. With permission.)

differences in the levels of hypoxia among these tumor lines, ranging from 34 to 64% hypoxic fraction. It is expected that the least hypoxic (i.e., most oxygenated) tumors would respond better to therapy. This trend was shown in HCT116, SKOV3, and PC3, but FaDu and 4TO7 deviated from this pattern (well oxygenated but non-responding). Therefore, the Dox–LTSL efficacy in a given tumor line was not completely related to its degree of hypoxia.[164]

The amount of drug delivered to tumors is known to be an important factor in anti-tumor efficacy. Therefore, Dox concentration was measured immediately after LTSL+HT treatment in eight tumors from each tumor line, using previously published HPLC methods.[98] Drug concentration correlated very well for four out of the five tumors, but HCT116 showed the highest drug concentration (0.045 µg/mg tumor tissue) along with the poorest response. In addition, cellular sensitivity to Dox was determined for each tumor line using the clonogenic assay after 24 h of exposure to LTSL+HT.[164] It was found that the in vivo responses of three tumors correlated well with in vitro sensitivity to Dox, but two of the most sensitive tumors in vitro (HCT116 and FaDu) showed poor response. Altogether, there was not a significant relationship between efficacy and either drug concentration or in vitro drug sensitivity.[164]

Overall, the four parameters (extracellular pH, hypoxic fraction, intratumoral drug concentrations, and in vitro drug sensitivity) that were examined did not yield a clear pattern to explain the differences in Dox–LTSL efficacy among the tumor types. However, searching for positive correlations, there were some interesting aspects. PC3 achieved the second highest drug concentration, aligning with its high number of complete regressions. In addition, PC3 had the slowest intrinsic (control) growth rate and was the most sensitive tumor line in vitro. On the other hand, HCT116 had the highest drug concentration, but it yielded only one complete regression (out of 20). The 4T07 tumor yielded the most consistent explanation for its relative lack of efficacy (no regressions); it had the highest IC10 (least sensitive), lowest drug concentration, and fastest intrinsic growth rate among all the tumor lines studied. New results from Yuan et al. also show that the relative amount of LTSL induced vascular damage is much greater in the FaDu line than in the 4T07 line (although 4T07 did show some vascular damage, indicating intravascular release of drug from liposomes is still an important mechanism). Accordingly, the relative growth delay time (compared with control) for FaDu was much higher than for 4T07 (FaDu: 41.8 days/12.1 days = 3.5 vs. 4T07: 19.8/7.5 = 2.6). Therefore, multiple potential explanations for the relative resistance of 4T07 to this combined therapy are now available.

These somewhat inconclusive results demonstrate the complexity of liposomal therapeutics for cancer therapy, and this is expected to extend to any other nanotherapeutic agent. For the Dox–LTSL, in particular, a new paradigm of drug delivery has now been established, leading to important questions regarding mechanisms and methods for a given tumor line and even an individual subject. In addition to the parameters already studied, which describe tumor microenvironment, drug concentration, and sensitivity of neoplastic cells, it is believed that individual tumor response may more critically depend on parameters that specifically affect intravascular drug release and vascular shutdown. It is now imperative to evaluate such parameter, including tumor perfusion, vascular density, the heatability of vessels, the temperature profile, the location of drug release (extravascular versus intravascular), the spatial and temporal distribution of the drug, and the relative drug sensitivity of the tumor vascular endothelium (the new target). A comprehensive approach is now being taken to evaluate the Dox–LTSL as this is necessary to optimize clinical therapy. One challenge that has been faced (that, again, extends to all nanotherapeutics) is the lack of tools for real-time evaluation of liposomal drug delivery and drug release at the scale of the whole tumor. Though very useful, the window chamber model has been limited to 2D evaluations and is not applicable to the clinical setting. This team sought to develop a non-invasive method (such as MRI*) to image liposomal drug concentration distributions in vivo.

---

* MRI, magnetic resonance imaging.

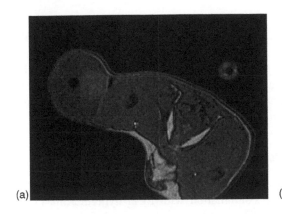
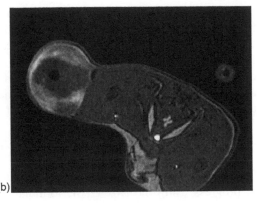

(a)   (b)

FIGURE 34.29 Median growth time (days to five times tumor volume) for each of five tumor types in response to various treatments. The tumor response to Dox–LTSL was greater than all other treatments for all tumor types. Ranking of median growth time (MGT) is the same regardless of treatment type for these tumors, reflecting intrinsic tumor growth rates. MGTs bound by rectangles are not significantly different from each other (Wilcoxon). (From Madden, T. D., et al., *Chem. Phys. Lipids*, 53 (1), 37–46, 1990. With permission.)

34.4.3 Regional Deposition of Drugs Observed Using MRI

Through the use of liposomes containing both drug (Dox) and paramagnetic contrast agents (manganese), Viglianti et al. showed that LTSL content release and distribution within the tumor can be monitored in vivo using T1-weighted MRI.[168] Furthermore, manganese (Mn) and Dox concentrations can be calculated using T1 relaxation times and Mn:Dox encapsulation ratios.[168] This method has been validated in a rat fibrosarcoma model, indicating a concordant linear relationship between T1-based [Dox] and invasive [Dox] as measured by HPLC or histologic fluorescence.[169]

The potential of this type of imaging for real-time evaluation of therapeutic protocols is currently being evaluated, with correlation of drug delivery to outcome on an individual subject basis.[170] Figure 34.29 shows representative axial MR images from the rat fibrosarcoma model before (a) and after (b) administration of i.v. Mn/Dox–LTSL plus local HT. The flank tumor is located in the upper left corner.[169] The catheter at the center of the tumor was used to deliver local hyperthermia (via heated water circulation).* Liposomes were injected after the tumor had reached steady state temperatures of 40°C–42°C. The tumor enhancement pattern shows that this protocol achieved release of contrast agent (and therefore drug) in the peripheral vasculature, i.e., in the feeding vessels as soon as liposomes entered the tumor.. This therapeutic schedule also resulted in rapid accumulation of large amounts of drug, likely due to extensive intravascular release (according to the new paradigm discussed above). Indeed, the kinetics of drug deposition observed here were much faster (<10 min to maximum [Dox]) than the kinetics of liposome extravasation with HT (hours to maximum [Dox] using non-thermally sensitive liposomes).[109] This is not surprising, since LTSL releases contents within 20 s at 41.3°C.[160] Thus, this protocol (inject liposomes during HT) targets the rapid drug release specifically to the feeding blood vessels of the tumor, which could maximize the anti-vascular effects. Therefore, we hypothesize that the most effective tumor response may be achieved with this schedule of therapy.

Current experiments are evaluating different schedules of LTSL plus HT in terms of drug delivery (spatial and temporal tumor distribution) and tumor response (growth delay time) in the rat fibrosarcoma model.[170] These results from this kind of study will be important for future clinical

---

*This is unlikely to be a clinical protocol, but it is a convenient and MR-compatible method to deliver and control local heating in animal experiments.

trials, and preliminary indications are that it may be more beneficial to deliver Dox–LTSL in the hyperthermia facility during steady-state heating. Being able to target not only the tumor but also the vascular source represents a direct clinical application of this new drug delivery paradigm. Furthermore, an imageable temperature-sensitive liposome could be applied in the clinic to monitor content release and so facilitate the control of drug distribution and prediction of tumor response for a given patient. This new multimodal liposome now provides an anti-neoplastic and anti-vascular therapy that can be monitored by MRI in real time.

34.4.4 Clinical Trials and Hyperthermia Devices

Based on the in vitro testing and in vivo pre-clinical success described above, clinical trials are now in progress using the Dox–LTSL formulation in combination with local hyperthermia (ThermoDox™, Celsion Corp.).[53] Clinical HT may be delivered to superficial or deep tumors using microwave, radiofrequency, or ultrasound devices (Figure 34.30).[118] A phase I trial is currently evaluating Dox–LTSL in combination with radiofrequency thermal ablation (RFA) for liver cancer (Figure 34.30a). In RFA, the center of the tumor is heated to cytotoxic temperatures greater than 80°C. Whereas this treatment clearly kills the center of the tumor, it is found that tumors often recur at the margins after RFA. This occurs because the temperature profile at the margins is only in the mild hyperthermia range. Because the Dox–LTSL releases at 39°C–42°C, it is used in this scenario primarily to increase cytotoxicity against these boundary cells. In addition, a phase I/II trial will soon test Dox–LTSL in the setting of chest wall recurrence of breast cancer. This trial will use a

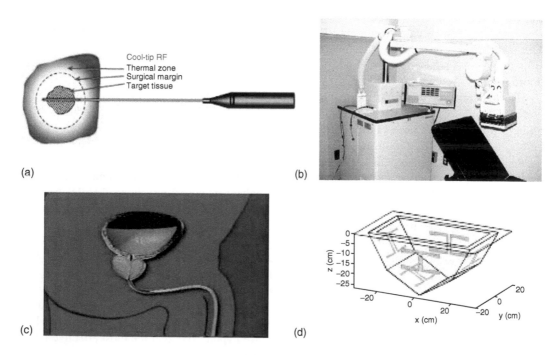

FIGURE 34.30 (**See color insert following page 522.**) Example of clinical hyper hyperthermia devices. (a) The radiofrequency ablation device is percutaneously inserted into the liver tumor under image guidance. (b) The superficial microwave applicator may be used for cutaneous or subcutaneous tumors (from http://www. ucsfhyperthermia.org/patientinfo/hyperucsf.htm). (c) The Prolieve transurethral microwave applicator was developed for benign prostatic hyperplasia but may also be applied for prostate cancer. (d) The RF phased array device for breast cancer may be integrated into a table so that the patient may be positioned prone above the device.

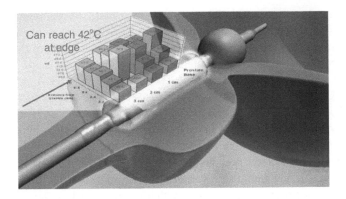

FIGURE 34.31 (See color insert following page 522.) The temperature profile for Celsion's Prolieve transurethral microwave applicator shows that this device can deliver 42°C at the periphery of the prostate. This enables the device to be used in combination with Dox–LTSL for prostate cancer that typically occurs in the periphery.

superficial microwave applicator to apply mild local HT (BSD500™; Figure 34.30b). Other potential clinical applications include prostate cancer (using the Prolieve™ transurethral microwave applicator, Figure 34.30c) and locally advanced breast cancer (using a radiofrequency phased array, Figure 34.30d). The Prolieve™ applicator that was originally designed for benign prostatic hyperplasia has demonstrated an ideal temperature profile for the release of LTSL in the periphery of the prostate where cancer typically occurs (Figure 34.31). That is, temperatures required for the release of drug from circulating liposomes can be attained even at the edges of the prostate by heating from the urethra.

34.5 CONCLUSIONS

This chapter narrated the development of the lysolipid-containing temperature sensitive liposome (LTSL) from bench to bedside in an attempt to highlight many of the formulation, clinical, cellular, and in vivo barriers that must be overcome by any nanotechnologically designed cancer therapeutic. It has shown that this formulation met an unmet need for bioavailable drug by combining previous liposome technologies with a detailed and fundamental knowledge of the materials science and engineering (i.e., composition, structure, properties, and processing) of lipid vesicle membranes. Historically, the successes of FDA-approved liposomes included reduced toxicity compared to free drug administration and a passive tumor accumulation through the EPR effect (especially for long-circulating Doxil). However, there were still many barriers to effective drug action, and clinical studies have shown limited improvements in tumor response. Reasoning that high peaks in bioavailable drug concentration were required for many drugs, this team set out to design a liposome with rapid triggered release of contents. In the absence of reliable endogenous triggerable chemistry (perhaps excluding phospholipases),[171] of an external trigger (e.g., hyperthermia) was utilized with a focus was on local disease rather than global metastases.

A number of studies contributed to the rationale and design for using hyperthermia to trigger liposome release were reviewed. In the window chamber tumor model, the synergistic interactions between hyperthermia and liposomal drug delivery were observed, initially for non-thermal sensitive liposomes. In addition, many years of fundamental lipid bilayer research revealed important relationships among liposome composition, nanostructures (such as crystalline defects in the solid state lipid bilayer), and properties (such as enhanced permeability to drugs at the acyl chain melting transition). Through the incorporation of lysolipids into the lipid membrane, a dramatic increase in permeability at the melting phase transition (41°C) was achived, resulting in complete release of encapsulated liposomal contents within 20 seconds was reached. The release mechanism appears to consist of lysolipid-stabilized pore structures at the interfaces between solid and liquid phase lipids at the crystalline grain boundaries. This is supported by the preservation of liposome permeability after dialysis and by observations of the material structure and properties of giant liposomes.

The preclinical studies that have led to clinical testing of the Dox–LTSL in phase I/II trials were reviewed. The Dox–LTSL achieved dramatic increases in tumor drug accumulation, up to thirty times over free Dox, in mice bearing human squamous cell carcinoma (FaDu) xenografts. This resulted in extensive anti-tumor efficacy (17 out of 20 regressions in two initial studies). Mechanistic studies in window chamber tumors revealed that Dox–LTSL + HT treatment resulted in dramatic reduction in tumor blood flow and complete vascular shutdown for tumors tested. This antivascular action was likely due to intravascular release of Dox from the LTSL, representing a brand new paradigm for liposomal drug delivery that has previously relied upon extravasation of the carrier via the EPR effect. Recent studies in other tumor lines have revealed the complexity and heterogeneity of the tumor response to Dox–LTSL + HT. To monitor liposome release and distribution in vivo, MRI methods using imageable liposomes have been developed. Ongoing research will continue to provide insights as this temperature-sensitive nanotechnology moves forward into clinical trials.

The lessons learned throughout this 15-year effort in triggerable drug delivery are likely to extend to other nanotechnological therapeutics. Previous studies have shown that it is possible to retain drug in a submicroscopic carrier or attach it to a nanomolecule. These drug delivery systems must evade the body's defenses in order to increase circulation lifetime and tumor uptake. However, this chapter has categorically shown that it is also essential to release the drug in free form, either in the blood stream of the tumor or in the perivascular space, so that it is bioavailable to exert its anti-neoplastic effects as well as important anti-vascular effects.

REFERENCES

1. National Library of Medicine. ClinicalTrials.gov [cited Feb 2006]; Available from: http://www. clinicaltrials.gov/.
2. Gabizon, A. A., Stealth liposomes and tumor targeting: One step further in the quest for the magic bullet, *Clin. Cancer Res.*, 7(2), 223–225, 2001.
3. Riethmiller, S., From atoxyl to salvarsan: Searching for the magic bullet, *Chemotherapy*, 51(5), 234–242, 2005.
4. Holland, J. et al., Principles of medical oncology, In *Cancer Medicine*, Bast, R. et al., Eds., B.C. Decker Inc., Hamilton, Ont., 2003.
5. Willmott, N. and Daly, J., *Microspheres and Regional Cancer Therapy*, CRC Press Inc., Boca Raton, FL, 1994.
6. Almond, B. A. et al., Efficacy of mitoxantrone-loaded albumin microspheres for intratumoral chemotherapy of breast cancer, *J. Control. Release*, 91(1–2), 147–155, 2003.
7. Rhines, L. D. et al., Local immunotherapy with interleukin-2 delivered from biodegradable polymer microspheres combined with interstitial chemotherapy: A novel treatment for experimental malignant glioma, *Neurosurgery*, 52(4), 872–879, 2003. discussion 879–880.
8. Allen, T. M. and Cullis, P. R., Drug delivery systems: Entering the mainstream, *Science*, 303(5665), 1818–1822, 2004.
9. Yuan, F. et al., Microvascular permeability and interstitial penetration of sterically stabilized (stealth) liposomes in a human tumor xenograft, *Cancer Res.*, 54(13), 3352–3356, 1994.
10. Hobbs, S. K. et al., Regulation of transport pathways in tumor vessels: Role of tumor type and microenvironment, *Proc. Natl Acad. Sci. U.S.A.*, 95(8), 4607–4612, 1998.
11. Torchilin, V. P., Targeted polymeric micelles for delivery of poorly soluble drugs, *Cell. Mol. Life Sci.*, 61(19–20), 2549–2559, 2004.

12. Wang, J., Mongayt, D., and Torchilin, V. P., Polymeric micelles for delivery of poorly soluble drugs: Preparation and anticancer activity in vitro of paclitaxel incorporated into mixed micelles based on poly(ethylene glycol)–lipid conjugate and positively charged lipids, *J. Drug Target.*, 13(1), 73–80, 2005.

13. Kabanov, A. V., Batrakova, E. V., and Alakhov, V. Y., Pluronic block copolymers as novel polymer therapeutics for drug and gene delivery, *J. Control. Release*, 82(2–3), 189–212, 2002.

14. Rangel-Yagui, C. O., Pessoa, A. Jr., and Tavares, L. C., Micellar solubilization of drugs, *J. Pharm. Pharm. Sci.*, 8(2), 147–165, 2005.

15. Allen, T. M., Liposomal drug formulations. Rationale for development and what we can expect for the future, *Drugs*, 56(5), 747–756, 1998.

16. Drummond, D. C. et al., Optimizing liposomes for delivery of chemotherapeutic agents to solid tumors, *Pharmacol. Rev.*, 51(4), 691–743, 1999.

17. Discher, B. M. et al., Polymersomes: Tough vesicles made from diblock copolymers, *Science*, 284 (5417), 1143–1146, 1999.

18. Dharap, S. S. et al., Molecular targeting of drug delivery systems to ovarian cancer by BH3 and LHRH peptides, *J. Control. Release*, 91(1–2), 61–73, 2003.

19. Kopecek, J. et al., Water soluble polymers in tumor targeted delivery, *J. Control. Release*, 74(1–3), 147–158, 2001.

20. Duncan, R. et al., Polymer–drug conjugates, PDEPT and PELT: Basic principles for design and transfer from the laboratory to clinic, *J. Control. Release*, 74(1–3), 135–146, 2001.

21. Vasey, P. A. et al., Phase I clinical and pharmacokinetic study of PK1 [*N*-(2--hydroxypropyl)methacrylamide copolymer doxorubicin]: First member of a new class of chemotherapeutic agents–drug–polymer conjugates. Cancer Research Campaign Phase I/II Committee, *Clin. Cancer Res.*, 5(1), 83–94, 1999.

22. Patri, A. K., Kukowska-Latallo, J. F., and Baker, J. R. Jr., Targeted drug delivery with dendrimers: Comparison of the release kinetics of covalently conjugated drug and non-covalent drug inclusion complex, *Adv. Drug Deliv. Rev.*, 57(15), 2203–2214, 2005.

23. Choi, Y. and Baker, J. R. Jr., Targeting cancer cells with DNA-assembled dendrimers: A mix and match strategy for cancer, *Cell Cycle*, 4(5), 669–671, 2005.

24. Wong, J. Y. et al., A Phase I trial of 90Y-anti-carcinoembryonic antigen chimeric T84.66 radio-immunotherapy with 5-fluorouracil in patients with metastatic colorectal cancer, *Clin. Cancer Res.*, 9(16 Pt 1), 5842–5852, 2003.

25. Juweid, M. E., Radioimmunotherapy of B-cell non-hodgkin's lymphoma: From clinical trials to clinical practice, *J. Nucl. Med.*, 43(11), 1507–1529, 2002.

26. Gabizon, A. A. et al., Cardiac safety of pegylated liposomal doxorubicin (Doxil/Caelyx) demonstrated by endomyocardial biopsy in patients with advanced malignancies, *Cancer Investig.*, 22(5), 663–669, 2004.

27. Berry, G. et al., The use of cardiac biopsy to demonstrate reduced cardiotoxicity in AIDS Kaposi's sarcoma patients treated with pegylated liposomal doxorubicin, *Ann. Oncol.*, 9(7), 711–716, 1998.

28. Harris, L. et al., Liposome-encapsulated doxorubicin compared with conventional doxorubicin in a randomized multicenter trial as first-line therapy of metastatic breast carcinoma, *Cancer*, 94(1), 25–36, 2002.

29. Working, P. K. et al., Reduction of the cardiotoxicity of doxorubicin in rabbits and dogs by encapsulation in long-circulating, pegylated liposomes, *J. Pharmacol. Exp. Ther.*, 289(2), 1128–1133, 1999.

30. Papahadjopoulos, D. et al., Sterically stabilized liposomes: improvements in pharmacokinetics and antitumor therapeutic efficacy, *Proc. Natl Acad. Sci. U.S.A.*, 88(24), 11460–11464, 1991.

31. Maeda, H. et al., Tumor vascular permeability and the EPR effect in macromolecular therapeutics: A review, *J. Control. Release*, 65(1–2), 271–284, 2000.

32. Allen, T. M. et al., Adventures in targeting, *J. Liposome Res.*, 12(1–2), 5–12, 2002.

33. Bendas, G., Immunoliposomes: A promising approach to targeting cancer therapy, *BioDrugs*, 15(4), 215–224, 2001.

34. Zevalin Injection (Ibritumomab Tiuxetan) Drug Information.[cited; Available from: http://www.drugs.com/, 2005.

35. Witzig, T. E. et al., Randomized controlled trial of yttrium-90-labeled ibritumomab tiuxetan radio-immunotherapy versus rituximab immunotherapy for patients with relapsed or refractory low-grade, follicular, or transformed B-cell non-Hodgkin's lymphoma, *J. Clin. Oncol.*, 20(10), 2453–2463, 2002.

36. Witzig, T. E. et al., Safety of yttrium-90 ibritumomab tiuxetan radioimmunotherapy for relapsed low-grade, follicular, or transformed non-hodgkin's lymphoma, *J. Clin. Oncol.*, 21(7), 1263–1270, 2003.

37. Nademanee, A. et al., A phase 1/2 trial of high-dose yttrium-90-ibritumomab tiuxetan in combination with high-dose etoposide and cyclophosphamide followed by autologous stem cell transplantation in patients with poor-risk or relapsed non-Hodgkin lymphoma, *Blood*, 106 (8), 2896–2902, 2005.

38. Schilder, R. et al., Follow-up results of a phase II study of ibritumomab tiuxetan radioimmu-notherapy in patients with relapsed or refractory low-grade, follicular, or transformed B-cell non-Hodgkin's lymphoma and mild thrombocytopenia, *Cancer Biother. Radiopharm.*, 19(4), 478–481, 2004.

39. Wiseman, G. A. et al., Ibritumomab tiuxetan radioimmunotherapy for patients with relapsed or refractory non-Hodgkin lymphoma and mild thrombocytopenia: A phase II multicenter trial, *Blood*, 99(12), 4336–4342, 2002.

40. Gibbs, W. J., Drew, R. H., and Perfect, J. R., Liposomal amphotericin B: Clinical experience and perspectives, *Expert Rev. Anti. Infect. Ther.*, 3(2), 167–181, 2005.

41. Astellas Pharma, US., Inc. [cited Feb 2006]; Available from: http://www.astellas.us/press_room/docs/corp_brochure033105.pdf.

42. Becker, M. J. et al., Enhanced antifungal efficacy in experimental invasive pulmonary aspergillosis by combination of AmBisome with Fungizone as assessed by several parameters of antifungal response, *J. Antimicrob. Chemother.*, 49, 813–820, 2002.

43. Schwartz, R. A., Lambert, W. C., Kaposi Sarcoma, Voorhees, A. V. et al., Eds., eMedicine.com, Inc, 2005.

44. Armstrong, D. K., Relapsed ovarian cancer: Challenges and management strategies for a chronic disease, *Oncologist*, 7(suppl. 5), 20–28, 2002.

45. Johnston, S. R. and Gore, M. E., Caelyx: Phase II studies in ovarian cancer, *Eur. J. Cancer*, 37(suppl. 9), S8–S14, 2001.

46. Gordon, A. N. et al., Recurrent epithelial ovarian carcinoma: A randomized phase III study of pegylated liposomal doxorubicin versus topotecan, *J. Clin. Oncol.*, 19(14), 3312–3322, 2001.

47. Berek, J. S. and Bast, R. C., Ovarian cancer, In *Cancer Medicine*, Bast, R. et al., Eds., B.C. Decker Inc., Hamilton, Ont., 2003.

48. O'Brien, M. E. et al., Reduced cardiotoxicity and comparable efficacy in a phase III trial of pegylated liposomal doxorubicin HCl (CAELYX/Doxil) versus conventional doxorubicin for first-line treatment of metastatic breast cancer, *Ann. Oncol.*, 15(3), 440–449, 2004.

49. Muggia, F. M., Clinical efficacy and prospects for use of pegylated liposomal doxorubicin in the treatment of ovarian and breast cancers, *Drugs*, 54(suppl. 4), 22–29, 1997.

50. Hauck, M. L., Larue, S. M., et al. Phase I trial of doxorubicin-containing low-temperature sensitive liposomes in spontaneous canine tumors, *Clin. Cancer Res.*, 12(13) 4004–4010, 2006.

51. Bangham, A. D., Standish, M. M., and Watkins, J. C., Diffusion of univalent ions across the lamellae of swollen phospholipids, *J. Mol. Biol.*, 13(1), 238–252, 1965.

52. Gregoriadis, G. et al., Drug-carrier potential of liposomes in cancer chemotherapy, *Lancet*, 1(7870), 1313–1316, 1974.

53. Celsion Corp. [cited Feb 2006]; Available from: www.celsion.com/technology/thermodox.cfm.

54. Chen, Q. et al., Targeting tumor microvessels using doxorubicin encapsulated in a novel thermo-sensitive liposome, *Mol. Cancer Ther.*, 3(10), 1311–1317, 2004.

55. Allen, T. M. and Smuckler, E. A., Liver pathology accompanying chronic liposome administration in mouse, *Res. Commun. Chem. Pathol. Pharmacol.*, 50(2), 281–290, 1985.

56. Allen, T. M. et al., Chronic liposome administration in mice: Effects on reticuloendothelial function and tissue distribution, *J. Pharmacol. Exp. Ther.*, 229(1), 267–275, 1984.

57. Gallois, L., Fiallo, M., and Garnier-Suillerot, A., Comparison of the interaction of doxorubicin, daunorubicin, idarubicin and idarubicinol with large unilamellar vesicles. Circular dichroism study, *Biochim. Biophys. Acta*, 1370(1), 31–40, 1998.

58. Nichols, J. W. and Deamer, D. W., Catecholamine uptake and concentration by liposomes maintaining p/gradients, *Biochim. Biophys. Acta*, 455(1), 269–271, 1976.

59. Mayer, L. D., Bally, M. B., and Cullis, P. R., Uptake of adriamycin into large unilamellar vesicles in response to a pH gradient, *Biochim. Biophys. Acta*, 857(1), 123–126, 1986.

60. Costanzo, L., *Board Review Series, Physiology*, 2nd ed., Lippincott, Williams and Wilkins, New York, 1998.

61. Truskey, G. A., Yuan, F., and Katz, D. F., *Transport Phenomena in Biological Systems*, Pearson Prentice Hall, Upper Saddle River, NJ, 2004.

62. Wu, N. Z. et al., Measurement of material extravasation in microvascular networks using fluorescence video-microscopy, *Microvasc. Res.*, 46(2), 231–253, 1993.

63. Takakura, Y., Mahato, R. I., and Hashida, M., Extravasation of macromolecules, *Adv. Drug Deliv. Rev.*, 34(1), 93–108, 1998.

64. Gerlowski, L. E. and Jain, R. K., Effect of hyperthermia on microvascular permeability to macromolecules in normal and tumor tissues, *Int. J. Microcirc. Clin. Exp.*, 4(4), 363–372, 1985.

65. Wu, N. Z. et al., Increased microvascular permeability contributes to preferential accumulation of stealth liposomes in tumor tissue, *Cancer Res.*, 53(16), 3765–3770, 1993.

66. Monsky, W. L. et al., Augmentation of transvascular transport of macromolecules and nanoparticles in tumors using vascular endothelial growth factor, *Cancer Res.*, 59(16), 4129–4135, 1999.

67. DaunoXome Injection Drug Information. [cited; Available from: http://www.drugs.com/], 2005.

68. Dvorak, H. F. et al., Identification and characterization of the blood vessels of solid tumors that are leaky to circulating macromolecules, *Am. J. Pathol.*, 133(1), 95–109, 1988.

69. Yuan, F., Transvascular drug delivery in solid tumors, *Semin. Radiat. Oncol.*, 8(3), 164–175, 1998.

70. Yuan, F. et al., Vascular permeability in a human tumor xenograft: Molecular size dependence and cutoff size, *Cancer Res.*, 55(17), 3752–3756, 1995.

71. Kong, G., Braun, R. D., and Dewhirst, M. W., Hyperthermia enables tumor-specific nanoparticle delivery: effect of particle size, *Cancer Res.*, 60(16), 4440–4445, 2000.

72. Slater, L. M. et al., Cyclosporin A corrects daunorubicin resistance in ehrlich ascites carcinoma, *Br. J. Cancer*, 54(2), 235–238, 1986.

73. Rivera, E., Liposomal anthracyclines in metastatic breast cancer: Clinical update, *Oncologist*, 8 (Suppl 2), 3–9, 2003.

74. Bradley, A. J. and Devine, D. V., The complement system in liposome clearance: Can complement deposition be inhibited?, *Adv. Drug Deliv. Rev.*, 32(1–2), 19–29, 1998.

75. Allen, T. M., Hansen, C., and Rutledge, J., Liposomes with prolonged circulation times: Factors affecting uptake by reticuloendothelial and other tissues, *Biochim. Biophys. Acta*, 981(1), 27–35, 1989.

76. Moghimi, S. M. et al., Non-phagocytic uptake of intravenously injected microspheres in rat spleen: Influence of particle size and hydrophilic coating, *Biochem. Biophys. Res. Commun.*, 177(2), 861–866, 1991.

77. Alving, C. R. et al., Therapy of leishmaniasis: Superior efficacies of liposome-encapsulated drugs, *Proc. Natl Acad. Sci. U.S.A.*, 75(6), 2959–2963, 1978.

78. New, R. R. et al., Antileishmanial activity of antimonials entrapped in liposomes, *Nature*, 272(5648), 55–56, 1978.

79. Weldon, J. S. et al., Liposomal chemotherapy in visceral leishmaniasis: An ultrastructural study of an intracellular pathway, *Z. Parasitenkd.*, 69(4), 415–424, 1983.

80. Berman, J. D., U.S. Food and Drug Administration approval of AmBisome (liposomal amphotericin B) for treatment of visceral leishmaniasis, *Clin. Infect. Dis.*, 28(1), 49–51, 1999.

81. Catania, S. et al., Visceral leishmaniasis treated with liposomal amphotericin B, *Pediatr. Infect. Dis. J.*, 18(1), 73–74, 1999.

82. Allen, T. M. et al., Liposomes containing synthetic lipid derivatives of poly(ethylene glycol) show prolonged circulation half-lives in vivo, *Biochim. Biophys. Acta*, 1066(1), 29–36, 1991.

83. Gabizon, A. and Martin, F., Polyethylene glycol-coated (pegylated) liposomal doxorubicin. Rationale for use in solid tumours, *Drugs*, 54(suppl. 4), 15–21, 1997.

84. Gabizon, A. et al., In vivo fate of folate-targeted polyethylene–glycol liposomes in tumor-bearing mice, *Clin. Cancer Res.*, 9(17), 6551–6559, 2003.

85. Bedu-Addo, F. K. et al., Effects of polyethyleneglycol chain length and phospholipid acyl chain composition on the interaction of polyethyleneglycol–phospholipid conjugates with phospholipid: Implications in liposomal drug delivery, *Pharm. Res.*, 13(5), 710–717, 1996.

86. Torchilin, V. P. et al., Targeted accumulation of polyethylene glycol-coated immunoliposomes in infarcted rabbit myocardium, *FASEB J.*, 6(9), 2716–2719, 1992.

87. Needham, D., McIntosh, T. J., and Lasic, D. D., Repulsive interactions and mechanical stability of polymer-grafted lipid membranes, *Biochim. Biophys. Acta*, 1108(1), 40–48, 1992.

88. Needham, D., McIntosh, T. J., and Zhelev, D. V., Surface chemistry of the sterically stabilized PEG-liposome: General principles, In *Liposomes: Rational Design*, Janoff, A., Ed., Marcel Dekker, Inc., New York, pp. 13–62, 1999.

89. Mori, A. and Huang, L., "Ninja" liposomes avoiding reticuloendothelial system: Development and application, In *First Shizuoka Drug Delivery System Symposium. Progress in Drug Delivery Systems*, University of Shizuoka, Japan, 1992.

90. Mori, A. et al., Influence of the steric barrier activity of amphipathic poly(ethyleneglycol) and ganglioside GM1 on the circulation time of liposomes and on the target binding of immunoliposomes in vivo, *FEBS Lett.*, 284(2), 263–266, 1991.

91. Allen, C. et al., Controlling the physical behavior and biological performance of liposome formulations through use of surface grafted poly(ethylene glycol), *Biosci. Rep.*, 22(2), 225–250, 2002.

92. Papahadjopoulos, D. and Gabizon, A., Liposomes designed to avoid the reticuloendothelial system, *Prog. Clin. Biol. Res.*, 343, 85–93, 1990.

93. Vaage, J. et al., Chemoprevention and therapy of mouse mammary carcinomas with doxorubicin encapsulated in sterically stabilized liposomes, *Cancer*, 73(9), 2366–2371, 1994.

94. Vaage, J. et al., Therapy of human ovarian carcinoma xenografts using doxorubicin encapsulated in sterically stabilized liposomes, *Cancer*, 72(12), 3671–3675, 1993.

95. Vaage, J. et al., Tissue distribution and therapeutic effect of intravenous free or encapsulated liposomal doxorubicin on human prostate carcinoma xenografts, *Cancer*, 73(5), 1478–1484, 1994.

96. Vaage, J. et al., Tumour uptake of doxorubicin in polyethylene glycol-coated liposomes and therapeutic effect against a xenografted human pancreatic carcinoma, *Br. J. Cancer*, 75 (4), 482–486, 1997.

97. Doxil Injection Drug Information. [cited; Available from: http://www.drugs.com/], 2005.

98. Kong, G. et al., Efficacy of liposomes and hyperthermia in a human tumor xenograft model: Importance of triggered drug release, *Cancer Res.*, 60(24), 6950–6957, 2000.

99. Needham, D. and Nunn, R. S., Elastic deformation and failure of lipid bilayer membranes containing cholesterol, *Biophys. J.*, 58(4), 997–1009, 1990.

100. INEX and Enzon announce FDA acceptance of Onco TCS New Drug Application. [cited Feb 2006]; Available from: http://www.drugs.com/NDA/onco_tcs_040521.html, 2004.

101. Krishna, R. et al., Liposomal and nonliposomal drug pharmacokinetics after administration of liposome-encapsulated vincristine and their contribution to drug tissue distribution properties, *J. Pharmacol. Exp. Ther.*, 298(3), 1206–1212, 2001.

102. Gelmon, K. A. et al., Phase I study of liposomal vincristine, *J. Clin. Oncol.*, 17(2), 697–705, 1999.

103. Sarris, A. H. et al., Liposomal vincristine in relapsed non-Hodgkin's lymphomas: Early results of an ongoing phase II trial, *Ann. Oncol.*, 11(1), 69–72, 2000.

104. Vincristine liposomal—INEX: Lipid-encapsulated vincristine, onco TCS, transmembrane carrier system–vincristine, vincacine, vincristine sulfate liposomes for injection, VSLI. Drugs R D, 5(2): p. 119–23, 2004.

105. Huang, Q. et al., Noninvasive visualization of tumors in rodent dorsal skin window chambers, *Nat. Biotechnol.*, 17, 1033–1035, 1999.

106. Matteucci, M. L. et al., Hyperthermia increases accumulation of technetium-99m-labeled liposomes in feline sarcomas, *Clin. Cancer Res.*, 6(9), 3748–3755, 2000.

107. Evans, E. and Needham, D., Physical-properties of surfactant bilayer-membranes–thermal transitions, elasticity, rigidity, cohesion, and colloidal interactions, *J. Phys. Chem.*, 91(16), 4219–4228, 1987.

108. Kwok, R. and Evans, E., Thermoelasticity of large lecithin bilayer vesicles, *Biophys. J.*, 35(3), 637–652, 1981.
109. Kong, G., Braun, R. D., and Dewhirst, M. W., Characterization of the effect of hyperthermia on nanoparticle extravasation from tumor vasculature, *Cancer Res.*, 61(7), 3027–3032, 2001.
110. Wu, N. Z. et al., Simultaneous measurement of liposome extravasation and content release in tumors, *Microcirculation*, 4(1), 83–101, 1997.
111. Fung, V. W., Chiu, G. N., and Mayer, L. D., Application of purging biotinylated liposomes from plasma to elucidate influx and efflux processes associated with accumulation of liposomes in solid tumors, *Biochim. Biophys. Acta*, 1611(1–2), 63–69, 2003.
112. Satchi-Fainaro, R. et al., PDEPT: Polymer-directed enzyme prodrug therapy. 2. HPMA copolymer-beta-lactamase and HPMA copolymer-C-Dox as a model combination, *Bioconjug. Chem.*, 14(4), 797–804, 2003.
113. Satchi, R., Connors, T. A., and Duncan, R., PDEPT: Polymer-directed enzyme prodrug therapy. I. HPMA copolymer-cathepsin B and PK1 as a model combination, *Br. J. Cancer*, 85(7), 1070–1076, 2001.
114. Needham, D. and Dewhirst, M. W., The development and testing of a new temperature-sensitive drug delivery system for the treatment of solid tumors, *Adv. Drug Deliv. Rev.*, 53(3), 285–305, 2001.
115. Mills, J. K. and Needham, D., Lysolipid incorporation in dipalmitoylphosphatidylcholine bilayer membranes enhances the ion permeability and drug release rates at the membrane phase transition, *Biochim. Biophys. Acta*, 1716(2), 77–96, 2005.
116. Rensen, P. C. et al., Determination of the upper size limit for uptake and processing of ligands by the asialoglycoprotein receptor on hepatocytes in vitro and in vivo, *J. Biol. Chem.*, 276 (40), 37577–37584, 2001.
117. Kong, G. and Dewhirst, M. W., Hyperthermia and liposomes, *Int. J. Hyperthermia*, 15(5), 345–370, 1999.
118. Dewhirst, M. et al., Hyperthermia, In *Cancer Medicine*, Bast, R. et al., Eds., B.C. Decker Inc., Hamilton, Ont., 2003.
119. Herman, T. S. et al., Effect of heating on lethality due to hyperthermia and selected chemotherapeutic drugs, *J. Natl Cancer Inst.*, 68(3), 487–491, 1982.
120. Estimated incidence, mortality, and site of failure of the most common types of cancer in the United States in 2001. Cancer Facts and Figure 2001 in Atlanta: American Cancer Society. 2001.
121. Suit, H., Assessment of the impact of local control on clinical outcome, *Front Radiat. Ther. Oncol.*, 29, 17–23, 1996.
122. Schmidt-Ullrich, R., Local tumor control and survival: Clinical evidence and tumor biologic basis, *Surg. Oncol. Clin. N. Am.*, 9(3), 401–414, 2000. See also page vii.
123. Needham, D. et al., A new temperature-sensitive liposome for use with mild hyperthermia: Characterization and testing in a human tumor xenograft model, *Cancer Res.*, 60(5), 1197–1201, 2000.
124. Haest, C. W. et al., Fragility of the permeability barrier of Escherichia coli, *Biochim. Biophys. Acta*, 288(1), 43–53, 1972.
125. Papahadjopoulos, D. et al., Phase transitions in phospholipid vesicles. Fluorescence polarization and permeability measurements concerning the effect of temperature and cholesterol, *Biochim. Biophys. Acta*, 311(3), 330–348, 1973.
126. Mouritsen, O. G., Jorgensen, K., and Honger, T., Permeability of lipid bilayers near the phase transition, In *Permeability and Stability of Lipid Bilayers*, Disalvo, E. A. and Simon, S. A., Eds., CRC Press, Boca Raton, pp. 137–160, 1995.
127. Mouritsen, O. G. and Zuckermann, M. J., Model of interfacial melting, *Phys. Rev. Lett.*, 58(4), 389–392, 1987.
128. Marsh, D., Watts, A., and Knowles, P. F., Evidence for phase boundary lipid. Permeability of tempo-choline into dimyristoylphosphatidylcholine vesicles at the phase transition, *Biochemistry*, 15, 3570–3578, 1976.
129. Corvera, E. et al., The permeability and the effect of acyl-chain length for phospholipid bilayers containing cholesterol: Theory and experiment, *Biochim. Biophys. Acta*, 1107, 261–270, 1991.
130. Nagle, J. F. and Scott, H. L. Jr., Lateral compressibility of lipid mono- and bilayers. Theory of membrane permeability, *Biochim. Biophys. Acta*, 513(2), 236–243, 1978.

131. Doniach, S., Thermodynamic fluctuations in phospholipid bilayers, *J. Chem. Phys.*, 68, 4912–4916, 1978.

132. Jacobson, K. and Papahadjopoulos, D., Phase transitions and phase separations in phospholipid membranes induced by changes in temperature, pH, and concentration of bivalent cations, *Biochemistry*, 14(1), 152–161, 1975.

133. Yatvin, M. B. et al., Design of liposomes for enhanced local release of drugs by hyperthermia, *Science*, 202(4374), 1290–1293, 1978.

134. Magin, R. L. and Niesman, M. R., Temperature-dependent permeability of large unilamellar liposomes, *Chem. Phys. Lipids*, 34(3), 245–256, 1984.

135. Gaber, M. H., Effect of bovine serum on the phase transition temperature of cholesterol-containing liposomes, *J. Microencapsul.*, 15(2), 207–214, 1998.

136. Ipsen, J. H. and Mouritsen, O. G., Modelling the phase equilibria in two-component membranes of phospholipids with different acyl-chain lengths, *Biochim. Biophys. Acta*, 944(2), 121–134, 1988.

137. Marsh, D., *CRC Handbook of Lipid Bilayers*, CRC Press, Boca Raton, 1990.

138. von Dreele, P. H., Estimation of lateral species separation from phase transitions in nonideal two-dimensional lipid mixtures, *Biochemistry*, 17(19), 3939–3943, 1978.

139. Bloom, M., Evans, E., and Mouritsen, O. G., Physical properties of the fluid lipid-bilayer component of cell membranes: A perspective, *Q. Rev. Biophys.*, 24(3), 293–397, 1991.

140. Zhelev, D. V. and Needham, D., Tension-stabilized pores in giant vesicles: Determination of pore size and pore line tension, *Biochim. Biophys. Acta*, 1147(1), 89–104, 1993.

141. Olbrich, K., Water permeability and mechanical properties of unsaturated lipid membranes and sarcolemma vesicles, In *Mechanical Engineering and Materials Science*. Duke University: Durham, NC, 1997.

142. Needham, D., Stoicheva, N., and Zhelev, D. V., Exchange of monooleoylphosphatidylcholine as monomer and micelle with membranes containing poly(ethylene glycol)-lipid, *Biophys. J.*, 73(5), 2615–2629, 1997.

143. Needham, D. and Evans, E., Structure and mechanical properties of giant lipid (DMPC) vesicle bilayers from 20 degrees C below to 10 degrees C above the liquid crystal-crystalline phase transition at 24 degrees C, *Biochemistry*, 27(21), 8261–8269, 1988.

144. Needham, D., McIntosh, T. J., and Evans, E., Thermomechanical and transition properties of dimyristoylphosphatidylcholine/cholesterol bilayers, *Biochemistry*, 27(13), 4668–4673, 1988.

145. Kim, D. and Needham, D. Lipid bilayers and monolayers: Characterization using micropipet manipulation techniques. In *Encyclopedia of Surface and Colloid Science*, Somasundaran, P., Hubbard, A., Et al., Eds., Marcel Dekker, New York, pp. 3057–3086, 2002.

146. Needham, D. and Zhelev, D., The mechanochemistry of lipid vesicles examined by micropipet manipulation techniques, In *Vesicles*, Rosoff, B., Ed., Macel Dekker, New York, 1996.

147. Needham, D. and Zhelev, D., Use of micropipet manipulation techniques to measure the properties of giant lipid vesicles, In *Giant Vesicles*, Luisi, P. L., Walde, P., Eds., Wiley, New York, 2000.

148. Anyarambhatla, G. R. and Needham, D., Enhancement of the phase transition permeability of DPPC liposomes by incorporation of MPPC: A new temperature-sensitive liposome for use with mild hyperthermia, *J. Liposome Res.*, 9(4), 491–506, 1999.

149. Ickenstein, L. M. et al., Disc formation in cholesterol-free liposomes during phase transition, *Biochim. Biophys. Acta*, 1614(2), 135–138, 2003.

150. Kruijff, B. D., Demel, R. A., and Deenen, L. L. M. V., The effect of cholesterol and epicholesterol incorporation on the permeability and on the phase transition of intact Acholeplasma laidlawii cell membranes and derived liposomes, *Biochim. Biophys. Acta*, 255, 331–347, 1972.

151. Kim, D. H. et al., Mechanical properties and microstructure of polycrystalline phospholipid monolayer shells: Novel solid microparticles, *Langmuir*, 19(20), 8455–8466, 2003.

152. Simon, S. A. et al., Increased adhesion between neutral lipid bilayers: Interbilayer bridges formed by tannic acid, *Biophys. J.*, 66(6), 1943–1958, 1994.

153. Zhelev, D. V. et al., Interaction of synthetic HA2 influenza fusion peptide analog with model membranes, *Biophys. J.*, 81(1), 285–304, 2001.

154. Zhelev, D. V., Exchange of monooleoylphosphatidylcholine with single egg phosphatidylcholine vesicle membranes, *Biophys. J.*, 71(1), 257–273, 1996.

155. Wright, A. M., Ickenstein, L. M., and Needham, D. Optimization of a thermosensitive liposome formulation for doxorubicin, in preparation.

156. Langner, M. and Hui, S. W., Dithionite penetration through phospholipid bilayers as a measure of defects in lipid molecular packing, *Chem. Phys. Lipids*, 65(1), 23–30, 1993.

157. Langner, M. and Hui, S., Effect of free fatty acids on the permeability of 1,2-dimyristoyl-sn-glycero-3-phosphocholine bilayer at the main phase transition, *Biochim. Biophys. Acta*, 1463(2), 439–447, 2000.

158. Zhelev, D. V. and Needham, D., Pore formation and pore dynamics in bilayer membranes, *ASME. Adv. Heat Mass Transfer Biotechnol.*, 34, 47–49, 1996.

159. Madden, T. D. et al., The accumulation of drugs within large unilamellar vesicles exhibiting a proton gradient: A survey, *Chem. Phys. Lipids*, 53(1), 37–46, 1990.

160. Wright, A., Drug loading and release from a thermally sensitive liposome. In *Mechanical Engineering and Materials Science*. Duke University, Durham, NC, 2006.

161. Wright, A. M., Ickenstein, L. M., and Needham, D., Optimization of a thermosensitive liposome formulation for doxorubicin (in preparation Feb 2006).

162. Gaber, M. H. et al., Thermosensitive liposomes: Extravasation and release of contents in tumor microvascular networks, *Int. J. Radiat. Oncol. Biol. Phys.*, 36(5), 1177–1187, 1996.

163. Gaber, M. H., Modulation of doxorubicin resistance in multidrug-resistance cells by targeted liposomes combined with hyperthermia, *J. Biochem. Mol. Biol. Biophys.*, 6(5), 309–314, 2002.

164. Zhao, Y. et al., Comparative effects of a novel thermosensitive doxorubicin-containing liposome in human and mouse tumors. *Mol. Cancer Ther.* (to be submitted March 2006).

165. Owen, C. S., Dependence of proton generation on aerobic or anaerobic metabolism and implications for tumour pH, *Int. J. Hyperthermia*, 12(4), 495–499, 1996.

166. Braun, R. D. et al., Comparison of tumor and normal tissue oxygen tension measurements using OxyLite or microelectrodes in rodents, *Am. J. Physiol. Heart Circ. Physiol.*, 280(6), H2533–H2544, 2001.

167. Teicher, B. A. et al., Classification of antineoplastic treatments by their differential toxicity toward putative oxygenated and hypoxic tumor subpopulations in vivo in the FSaIIC murine fibrosarcoma, *Cancer Res.*, 50(11), 3339–3344, 1990.

168. Viglianti, B. L. et al., In vivo monitoring of tissue pharmacokinetics of liposome/drug using MRI: Illustration of targeted delivery, *Magn. Reson. Med.* 51(6), 1153–62, 2004.

169. Viglianti, B. L. et al., Chemodosimetry of in-vivo tumor liposomal drug concentration using MRI, *J. Magn. Reson. Imaging*, in press.

170. Ponce, A. M. et al., Temperature-sensitive liposome release observed by MRI: Drug dose painting with hyperthermia impacts anti-tumor effect. *Clin. Cancer Res.* in press.

171. Jensen, S. S. et al., Secretory phospholipase A2 as a tumor-specific trigger for targeted delivery of a novel class of liposomal prodrug anticancer etherlipids, *Mol. Cancer Ther.*, 3(11), 1451–1458, 2004.

172. Liposome. 1999 [cited Feb 2006]; Available from: http://www.uic.edu/classes/bios/bios100/lectf03am/liposome.jpg.

173. O'Driscoll, C., and Duncan, R., Chembytes e-zine: Stealth drugs take off 2000 [cited Feb 2006]; Available from: http://www.chemsoc.org/chembytes/ezine/2000/odriscoll_nov00.htm.

174. Heather Bullen Research Group, Northern Kentucky University. [cited Feb 2006]; Available from: http://www.nku.edu/~bullenh1/research.htm.

175. Kukowska-Latallo, J. F. et al., Nanoparticle targeting of anticancer drug improves therapeutic response in animal model of human epithelial cancer, *Cancer Res.*, 65(12), 5317–5324, 2005.

176. Klement, A., TARA News, High-Tech. Behandlung bei malignem Lymphom: Zevalin. 2004 [cited; Available from: http://www.oeaz.at/zeitung/3aktuell/2004/17/serie/serie17_2004tara.html.

177. Hyperthermia at University of California—San Francisco. [cited Feb 2006.]; Available from: http://www.ucsfhyperthermia.org/patientinfo/hyperucsf.htm.

Section 7

Other Lipid Nanostructures

35 Nanoemulsion Formulations for Tumor-Targeted Delivery

Sandip B. Tiwari and Mansoor M. Amiji

CONTENTS

35.1 Introduction 723
35.2 Composition of Nanoemulsions 725
35.3 Preparation of Nanoemulsions 726
35.4 Characterization of Nanoemulsion Systems 727
    35.4.1 Particle Size 727
    35.4.2 Surface Characteristics 728
35.5 Fate of Nanoemulsions In Vivo 729
    35.5.1 Effect of Particle Size on Disposition and Clearance 729
    35.5.2 Effect of Surface Characteristics on Disposition and Clearance 729
35.6 Applications of Nanoemulsions in Cancer Therapy 730
    35.6.1 Improved Delivery of Lipophilic Drugs 730
    35.6.2 Positively Charged Anoemulsions in Cancer Therapy 731
    35.6.3 Photodynamic Therapy of Cancer 732
    35.6.4 Neutron Capture Therapy of Cancer 733
    35.6.5 Perfluorochemical Nanoemulsions in Cancer Therapy 734
35.7 Conclusions and Future Perspectives 735
References 736

35.1 INTRODUCTION

An emulsion is generally described as a heterogonous system composed of two immiscible liquids: one dispersed uniformly as fine droplets throughout the other. The majority of conventional emulsions in pharmaceutical use have dispersed particles ranging in diameter from 0.1 to 100 µm. As with other dispersed systems such as suspensions, emulsions are thermodynamically unstable as a result of the excess free energy associated with the surface of the droplets in the internal phase. The dispersed droplets strive to come together and reduce the surface area. In addition to this flocculation effect, the dispersed particles can coalesce, or fuse, and this can result in the eventual destruction (phase separation) of the emulsion. To minimize this effect, a third component, the emulsifying agent or surface active agent (surfactant), is added to the system to improve the stability. The liquid present as small globules is called the dispersed or internal phase, whereas the liquid where the droplets are dispersed is called the dispersion, continuous, or external phase. In general, the immiscible liquids are described as oil and water phases. The oil phase may be any hydrophobic non-polar liquid, and the water phase may be any hydrophilic polar liquid.

The term macroemulsion is sometimes employed to distinguish the ordinary emulsions defined above from microemulsions, nanoemulsions, and micellar systems. Whereas nanoemulsions are two-phase systems where the dispersed phase droplet size has been made in the nanometer size range, the microemulsions, and micellar systems are single-phase systems.[1–3] All these systems are a result of interfacial phenomenon brought out by surface active agents. Therefore, it is essential to understand the molecular mechanisms responsible for the creation of these systems and distinctions between them. Surface active compounds contain hydrophilic and hydrophobic moieties in the same molecule. When these surfactants are added to the water phase or the oil phase, one portion of the molecule is always incompatible with the other solvent molecules. Surfactants overcome this incompatibility by adsorption at interfaces (air–water, oil–water, or solid–water) and/or by formation of aggregates called micelles. Micelles are colloidal aggregates that form above a specific concentration, the so-called critical micelle concentration (CMC). When the solvent is water, a micellar aggregate exhibit the structure as shown in Figure 35.1a where the surfactant molecules are oriented in such a way that the hydrophilic groups are in contact with the water molecules, and the hydrophobic tails are located in the micellar core away from the aqueous environment. In this case, the core of a micelle resembles a tiny pool of liquid hydrocarbon where compounds that are poorly soluble in water but soluble in non-polar solvents can be solubilized. When the solvent is oil phase, inverse (or reverse) micelles are formed where the surfactant's hydrophilic groups are located inside (Figure 35.1b). In this case, the compounds that are soluble in water phase can be solubilized in the micellar core. The process of solubilization takes place well above the CMC, and because of solubilization, micelles become swollen and may attain the size of a small droplet (~ 100 nm). Consider, for example, that water is solvent, the surfactant concentration is well above CMC, and the oil phase is also present; the micelles would solubilize large amounts of the oil phase and become swollen until they start interacting through a phenomenon called percolation. Such packed swollen micelle structures that could solubilize large amounts of both oil- and water-soluble compounds have been called microemulsions because they were first thought to be extremely small droplet emulsions.[4] This definition of microemulsions is still controversial with some scientists referring microemulsions in similar context of emulsions systems (two phase systems, e.g., nanoemulsions) while others referring to it as a single phase system.[1,4,5] There is, however, general agreement that microemulsions are optically isotropic, thermodynamically stable systems and are not amenable to dilution with the external phase as normal emulsions. Application of microemulsions is usually limited to dermal and peroral application because of their high surfactant concentration that tends to prove toxic through other routes of administration. Also, they exist in narrow regions of the phase diagrams and, as such, they are very sensitive to quantitative changes in formulation. A notable

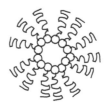

a. Micelle structure
(Dispersion medium is water phase)

b. Inverse micelle
(Dispersion medium is oil phase)

Polar head group

Non-polar tail group

Emulsier or surfactant molecule

FIGURE 35.1 (a) Micelles and (b) inverse micelles.

example of the microemulsion system in pharmaceutical application is Sandimmune Optoral™ and Neoral™ pre-concentrate that contains cyclosporin-A as the active moiety indicated for immune suppression in organ transplantation.[6]

Emulsion formulations can be used orally, parenterally, ophthalmically, or topically, and they can be used by respiratory route.[4,7] Commercial formulations of nanoemulsions are presented in Table 35.1. Major application areas of nanoemulsions are in the field of total parenteral nutrition and as carriers for various drugs, anesthetics, and image contrast-enhancement agents. In this chapter, the discussion on application of nanoemulsion systems is restricted to cancer therapy. Two types of emulsion systems are possible depending on if the dispersed droplets are oil or water (i.e., oil-in-water (O/W) or water-in-oil (W/O) emulsion). O/W nanoemulsion systems are of particular pharmaceutical importance from a parenteral drug delivery point of view. W/O emulsions are also parenterally used; however, their applications are limited for obtaining the prolonged release of the water soluble compounds on intramuscular application. Multiple emulsion or double emulsion systems are also possible where preformed emulsion is re-emulsified with either oil or water. For example, multiple emulsions of oil-in-water-in-oil (O/W/O) are W/O emulsions where the water droplets contain dispersed oil globules. Similarly, water-in-oil-in-water (W/O/W) multiple emulsions are those where the internal and external water phases are separated by an oil phase (Figure 35.2).

35.2 COMPOSITION OF NANOEMULSIONS

Nanoemulsions contain oil phases, surfactants or emulsifiers, active pharmaceutical ingredients (drugs or diagnostic agents), and additives (Table 35.2).[5,8-10] The oil phases are mainly natural or synthetic lipids, fatty acids, oils such as medium or long chain triglycerides, or perflurochemicals. Many oils, in particular, those of vegetable origin, are liable to auto-oxidation, and their use in pharmaceutical formulations requires the addition of an antioxidant. The most widely used oils for parenteral applications are purified soybean, corn, castor, peanut, cottonseed, sesame, and safflower oils. Table 35.1 lists some of the oils that can be used for formulating nanoemulsions. Squalene has been reported to be the choice of oil for formulating stable nanoemulsions with smallest droplet size.[11] Squalene, biocompatible oil, is a linear hydrocarbon precursor of cholesterol found in many tissues, notably the livers of sharks (*Squalus*) and other fishes.[12] Squalane, a derivative of squalene, is prepared by hydrogenation of squalene and is fully saturated that means that it is not subject to auto-oxidation, an important issue from a stability point of view. Purified mineral oil is used in some W/O emulsion preparations.[10] Emulsified perflurochemicals are considered acceptable for

TABLE 35.1
Commercial Nanoemulsion (Sub-Micron Emulsion) Formulations

Drug	Brand	Manufacturer	Therapeutic Indication
Propofol	Diprivan	Astra Zeneca	Anesthetic
Dexamethasone	Limethason	Mitsubishi Pharmaceutical, Japan	Steroid
Palmitate			
Alprostadil	Liple	Mitsubishi Pharmaceutical, Japan	Vasodilator platelet inhibitor
Flurbiprofen axetil	Ropion	Kaken Pharmaceuticals, Japan	Nonsteroidal analgesic
Vitamins A, D, E, K	Vitalipid	Fresenius Kabi, Europe	Parenteral nutrition

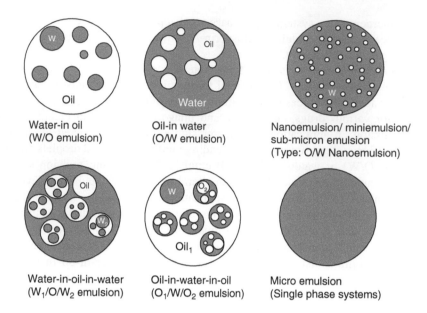

FIGURE 35.2 The different types of colloidal systems. (Adapted from Salager, J. L., *Pharmaceutical Emulsions and Suspensions*, Marcel Dekker, New York, pp. 19–72, 2000.)

intravenous application, and many formulations involving the use of perflurochemicals have reached the marketplace.[13]

The most widely used oils for oral emulsions are various fixed oils of vegetable origin, (corn, cotton seed, or peanut oil), fish liver oil, and mineral oil. The most commonly used emulsifiers for parenteral application include either natural lecithins derived from plant or animal sources [e.g., egg yolk or soy phosphatidylcholine (PC)] or modified lecithins (e.g., diethylenetriaminepentaacetic acid (DTPA) conjugated phosphatidylethanolamine (PE) or poly(ethylene glycol) (PEG)-modified lecithin derivatives). The poly(ethylene oxide) (PEO)-containing block copolymers, in particular Pluronics® or poloxamers, Tetronic® or poloxamines, polysorbates, PEG-conjugated castor oil derivatives (Cremophore EL®), polyglycolized as well as glycerides, stearylamine (SA), 1, 2-dioleoyl-3-trimethylammonium-propane, N-(1, 2-Dioleyl-dihydroxypropyl)-N-(dihydroxy-propyl) (DOTAP) or other positively charged lipids are also used as emulsifiers or co-emulsifiers in the nanoemulsion formulations. Other pharmaceutical additives such as tonicity modifiers, pH adjustment agents, antioxidants, sweeteners, flavors, and preservatives may also be included in the final formulation, depending on the need.

35.3 PREPARATION OF NANOEMULSIONS

As previously discussed, because nanoemulsions are dispersions of fine droplets (in nanometer range) in a dispersion medium, they are thermodynamically unstable and require considerable mechanical energy for their preparation.[14–16] The mechanical energy can be supplied in the form of high pressure homogenizer (e.g., Avestin® homogenizer, Avestin Inc., Ottawa, Canada), Microfluidizer® (MFIC Corporation, Newton, MA, U.S.A.), or an ultrasonic generator (e.g., VC-505, Sonics, and Materials Inc., CT, U.S.A.). Stable preparation of nanoemulsions involves the following steps. First, the active hydrophobic pharmaceutical agent is dissolved in the oil phase and mixed with the water phase, and the mixture is then emulsified with a high shear homogenizer for a short period. The coarse oil-in-water emulsion formed is then further homogenized using a high pressure homogenizer or Microfludizer®. The emulsifier (or surfactant) is either mixed with the water phase or the oil phase. Incorporation of the surfactant in the oil phase has added advantage

TABLE 35.2
Components of Nanoemulsion Formulations

Oils	Emulsifiers
Castor oil	Natural lecithins from plant or animal sources
Coconut oil	(phospholipids)
Corn oil	PEG- phospholipids
Cottonseed oil	Poloxamers (e.g. F68)
Evening primrose oil	Polysorbates
Fish oil	Polyoxyethylene castor oil derivatives
Jojoba oil	Polyglycolized glycerides
Lard oil	Stearlyamine
Linseed oil	Oleylamine
Mineral oil	
Olive oil	*Additives*
Peanut oil	Antioxidants (α-tocopherol, ascorbic acid)
PEG-vegetable oil	Tonicity modifiers (glycerol, sorbitol, xylitol)
Perflurochemicals	pH adjustment agents (NaOH or HCl)
Pine nut oil	Preservatives
Safflower oil	
Sesame oil	
Soybean oil	
Squalene	
Squalane	
Sunflower oil	
Wheatgerm oil	

Source: From Eccleston, G., *Encyclopedia of Pharmaceutical Technology*, Marcel Dekker, New York, pp. 137–188, 1992; Chung, H., Kim, T. W., Kwon, M., Kwon, I. C., and Jeong, S. C., *J. Control Release*, 71, 339–350, 2001; Allison, A. D., *Methods*, 19, 87–93, 1999; Klang, S. and Benita, S., *Submicron Emulsions in Drug Targeting and Delivery*, Academic Publishers, Harwood, pp. 119–152, 1998; Lundberg, B. B., *J. Pharm. Pharmacol.*, 49(1), 16–21, 1997; Campbell, S. E., Kuo, C. J., Hebert, B., Rakusan, K., Marshall, H. W., and Faithfull, N. S., *Can. J. Cardiol.*, 7(5), 234–244, 1991.

as it may aid in enhancing solubility of the lipophillic compound in the oil phase. The size of nanoemulsion droplets can be manipulated by using appropriate pressure settings and number of cycle runs. Nanoemulsions can also be formulated using direct sonication of the oil phase with water phase in the presence of emulsifying agent. In this case, the intensity, mode, and duration of sonication determine the final droplet size. It should, however, be noted that the size of nanoemulsion droplets is also dependant on the nature of oil (e.g., oil composition and viscosity) and the emulsifying phase (e.g., type of emulsifier and its concentration).

35.4 CHARACTERIZATION OF NANOEMULSION SYSTEMS

Nanoemulsion systems are routinely characterized for particle size and surface properties (surface charge and morphology).

35.4.1 Particle Size

The size of nanoemulsion droplets determines their behavior both in vitro and in vivo. Particle size of nanoemulsion droplets can be measured using an ensemble (e.g., spectroscopic methods such as light scattering), counting (e.g., microscopy such as freeze fracture electron microscopy) or separation method (e.g., analytical ultracentrifugation).[17,18] Each method has its own advantages and disadvantages. For example, when the particle size measurement of standard nanoemulsions (made with pine nut oil and egg PC) was performed by this laboratory using an ensemble and counting method, the results indicated some discrepancy. The average particle size of the nanoemulsion droplets was 110 ± 13 nm by dynamic light scattering method using ZetaPALS, 90Plus (Brookhaven Instruments Corporation, Holtsville, NY, U.S.A.). However, the freeze fracture electron microscopy images indicated the size of the droplets in the range of 25–40 nm with some particle aggregates in the size range ~ 100–150 nm (Figure 35.3). It is recommended that, in order to obtain better information on particle size distribution, it is essential that particle size analysis be performed by more than one method. The particle size of the commercial nanoemulsions intended for total parenteral nutrition is reported to be in the range of 100–400 nm.[19]

35.4.2 SURFACE CHARACTERISTICS

Similar to particle size, surface charge of the nanoemulsion droplets has marked effect on the stability of the emulsion system and the droplets' in vivo disposition and clearance. Conventionally, the surface charge on the emulsion droplets has been expressed in terms of zeta potential (ζ). As the emulsion droplets are a result of interfacial phenomenon brought out by surface active agents, their zeta potential is dependent on the extent of ionization of these surface active agents. According to DLVO electrostatic theory, the stability of the colloid is a balance between the attractive van der Waals' forces and the electrical repulsion because of the net surface charge.[20] If the zeta potential

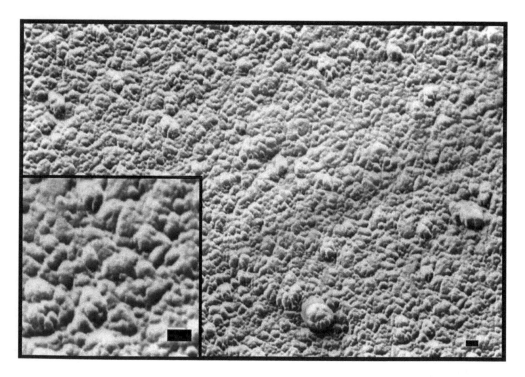

FIGURE 35.3 Freeze fracture electron microscopy image of paclitaxel nanoemulsions formulated using 20% oil phase (pine nut oil) and egg phosphatidylcholine as surfactant and deoxycholic acid as co-surfactant. The inset shows microscopy image at higher magnification. Nanoemulsion droplets were in the size range of 25–40 nm with some particle aggregates in the size range of ~ 100–150 nm. Scale bar in both figures represents 100 nm.

falls below a certain level, the emulsion droplets will aggregate as a result of the attractive forces. Conversely, a high zeta potential (either positive or negative), typically more than 30 mV, maintains a stable system.[8,15,19] The zeta potential of the nanoemulsion droplets is routinely measured using a Zetasizer (Malvern Instruments, U.K.) or the ZetaPlus instrument (Brookhaven Instruments Corporation, Holtsville, NY, U.S.A.). The zeta potential of the commercial nanoemulsions used for total parenteral nutrition is reported to be around -40 to -50 mV at pH 7.[19] The net negative surface charge is thought to be attributed to the anionic components of the emulsions, mainly the phospholipids.[21] One can impart positive charge to the nanoemulsions as well using cationic lipids [e.g., SA,[22–24] oleylamine,[25] 2,3-dioleoyloxypropyl-1-trimethylammonium bromide, DOTMA,[26] dimethylaminoethane carbamoyl cholesterol (DC-cholesterol)],[27] polymers (e.g., chitosan),[28,29] and cationic surfactants (e.g., cetyltrimethylammonium bromide).[30]

35.5 FATE OF NANOEMULSIONS IN VIVO

Similar to liposomal systems, the oil droplets of a conventional O/W nanoemulsion are rapidly cleared after intravenous injection by the reticuloendothelial system (RES) organs such as the liver, spleen, and bone marrow. Alternatively, the emulsion droplets would bind to the apolipoproteins from the blood stream, followed by their metabolism similar to chylomicrons. The clearance rate of emulsion droplets from the systemic circulation is dependant on the physicochemical properties of the emulsion system (size and surface properties in particular). By appropriate control of these physicochemical properties, one can manipulate the in vivo behavior of the nanoemulsion systems.

35.5.1 EFFECT OF PARTICLE SIZE ON DISPOSITION AND CLEARANCE

Small-sized emulsion droplets are cleared slower than larger emulsion droplets. It is reported that small-sized emulsions (100–110 nm) were successful in enhancing the plasma concentration of rhizoxin although the drug was also distributed to the peripheral tissues.[31] On the other hand, emulsions with droplet size >200 nm prevented the drug from entering the bone marrow, small intestine, and other organs not associated with the RES where toxicities of many cytotoxic compounds are expected. Size of nanoemulsion systems can be controlled by processing conditions (e.g., number of homogenization cycles or duration and intensity of sonication) and also by composition of the emulsifier system. Lundberg has reported that it is possible to control the particle size of the nanoemulsion system using appropriate co-surfactant, e.g., use of 0.12 wt% of polysorbate 80 in castor oil: egg PC (1:0.4 wt%) emulsion system was found to reduce the size of the droplets from 50 to 100 nm without using any co-surfactant.[32] The use of Pluronic® F-108 as one of the co-surfactants also decreased the size of the emulsion droplets. Particle size is also dependant on the type of the oil phase used for the preparation of the nanoemulsion systems. For example, it was reported that when squalene was used as the oil phase, smallest nanoemulsion droplets (190 nm) were observed.[11] Castor, safflower, and lard oils yielded emulsion droplets in the size range of 270–285 nm, whereas linseed oil yielded droplets with size of around 355 nm.

35.5.2 EFFECT OF SURFACE CHARACTERISTICS ON DISPOSITION AND CLEARANCE

As previously discussed, emulsion droplets are rapidly taken up by the cells of the RES. The use of PEG-modified phospholipids as emulsifies were shown to reduce the uptake by the cells of the RES and to improve the blood circulation times. For example, it was reported that inclusion of PEG derivatives such as Tween® 80 or PEG–PE into the emulsions composed of castor oil and PC decreases RES uptake and improves systemic circulation time.[33] Hodoshima et al.[34] reported that lipid nanoparticles of lipophillic prodrug 4′-O-tetrahydropyranyl doxorubicin with synthetic PEG (1,900)-lipid derivative and PC enhanced the circulation half of the drug to 4.3 h after intravenous administration in BALB/c mice. A similar approach has been used to improve tumor delivery of rhizoxin.[31,35] When lecithin was replaced with hydrophilic poloxamer 338 in the emulsion formu-lation, it was possible to avoid the normal disposition of droplets to the liver and spleen. In fact, one study reported that one could totally eliminate the liver and spleen uptake by using a nonionic co-emulsifier, the block copolymer poloxamine (P 908).[36] Emulsions prepared with Pluronics® as co-emulsifiers also enhanced the circulation time of the formulation in vivo. The nature of lipid component also alters the biodistribution profile of emulsion systems. For example, it was reported that cholesterol-coated emulsions were rapidly removed from circulation compared to those coated with sphingomyelin.[37] The surface charge of the nanoemulsion droplets also affects the clearance rate in vivo. Negatively charged nanoemulsions are cleared faster and showed high RES organ uptake than neutral or positively charged nanoemulsions.[8] On the other hand, positively charged nanoemulsions showed an initial accumulation in the lungs followed by redistribution to the liver and spleen.[15] This natural fate of positively charged nanoemulsions might be important in devel-oping a suitable system for delivery of drugs targeted specifically to the lungs in case of transplant rejection therapy, lung infections, or lung cancer. In brief, the surface modification of nanoemulsion systems offer an opportunity for prolonging the blood circulation time of the drug, enhancing the possibility of interaction with target cells of interest. This is, in particular, important for delivery of anti-cancer drugs as prolonged exposure of the drug is critical in the treatment of multi-drug-resistant tumors.

35.6 APPLICATIONS OF NANOEMULSIONS IN CANCER THERAPY

35.6.1 Improved Delivery of Lipophilic Drugs

The advantages of formulating various lipophilic anti-cancer drugs in submicron O/W emulsion are obvious. The oil phase of the emulsion systems can act as a solubilizer for the lipophilic compound. Therefore, solubility of lipophilic drugs can be significantly enhanced in an emulsion system, leading to smaller administration volumes compared to an aqueous solution. In addition, because lipophilic drugs are incorporated within the innermost oil phase, they are sequestered from direct contact with body fluids and tissues. Lipid emulsions can minimize the pain associated with intravenously administered drugs by exposing the tissues to lower concentrations of the drug or by avoiding a tissue irritating vehicle. This has been demonstrated with propofol,[38] diazepam,[39] methohexital,[40] clarithromycin,[41] etomidate,[42] and a novel cytotoxic agent.[43] Also, because these systems are similar to chylomicrons, they are well-tolerated and have low incidences of side effects compared to systems based on organic solvents, pH adjustments, and surface active agents (e.g., Cremophor) because there is less chance of drug precipitation upon administration. If the drug is susceptible to hydrolysis or oxidation, it is protected by the non-aqueous environment. Further-more, incorporation of anti-cancer drugs in submicron emulsions (with droplet size of 50–200 nm) with long circulation properties are expected to enhance the tumor accumulation of the drug by passive targeting through the enhanced permeability and retention effect.[44,45] It possible to enhance the tumor accumulation of nanoemulsions with appropriate modification of size or surface func-tionalization as previously discussed. Oil-in-water submicron emulsions appear to be a viable alternative for the intravenous administration of various lipophilic cytotoxic drugs. Several groups of researchers have reported the submicron emulsion formulations of anti-cancer drugs for improved efficacy and/or reduced toxicity.

Paclitaxel, a highly potent anti-cancer agent initially extracted from the bark of the Pacific yew, was entrapped in lipid emulsion droplets with triolein as oil core and dipalmitoylphosphatidylcho-line (DPPC) as the principal emulsifier. The emulsion was stabilized with polysorbate 80 and PEG-dipalmitoyl PE. The results of in vitro anti-proliferative activity of paclitaxel against T47D cells was retained by the drug emulsion with IC_{50} (drug concentration effective in reducing proliferation by 50%) value of 7 nM as compared to 10 and 35 nM for paclitaxel in liposomes and Cremophor® EL-ethanol (50:50) formulations, respectively.[46] The incorporation of PEG-derivatized phospholipid is expected to enhance the in vivo circulation half life of the formulation, thereby enhancing the exposure of the drug to the targeted tumor mass.

Another study reported the formulation of filter sterilizable emulsion formulation of paclitaxel using α-tocopherol as the oil phase and α-tocopherylpolyethyleneglycol-1,000 succinate (TGPS) and poloxamer 407 as emulsifiers.[47] The emulsion droplet size was >200 nm with high levels of paclitaxel loading (8–10 mg/mL). The formulation exhibited better efficacy and was more tolerable when studied in B16 melanoma tumor model in mice. In addition, another group of researchers developed a submicron emulsion (droplet diameter 150 nm) of paclitaxel with oil blend (triglycerols), egg PC, Tween® 80, and glycerol solution. The formulation showed cytotoxicity against HeLa cells with an IC_{50} at 30 nM. An anti-tumor agent, valinomycin, was formulated in the emulsion form using the commercially-available Intralipid 10% soybean oil emulsion used in parenteral nutrition. Evaluation of this formulation in vivo indicated that the emulsion formulation produced similarly shaped dose-response curve to that of an aqueous suspension, but the emulsion formulation required a 30-fold lower dose than the suspension to produce therapeutic similar effects.[7]

In formulation of nanoemulsions, selection of the appropriate oil phase is important because most of the anti-cancer compounds exhibit poor solubility in the oil phase, especially those with highly lipophilic oils. Kan et al. determined the solubility of paclitaxel in various oils such as tributyrin, tricaproin, tricaprylin, corn, soyabean, cotton seed, and mineral oil, and they found that triacylglycerols with short fatty acid chains (tributyrin and tricaproin) were good solvents for paclitaxel with solubility of more than 9.00 mg/g as compared to other vegetable oils (range 0.14–0.23 mg/gm).[48] Another approach to enhance the oil solubility of the anti-cancer compounds is the chemical modification or prodrug formation. Prodrugs with increased oil solubility have been obtained with such anti-cancer drugs as teniposide, etoposide, camptothecin, and paclitaxel, whereas amphiphilic derivatives have been prepared from fluorodeoxyuridine.[49] Esterification with long-chain fatty acids (i.e., oleic acid) has also been reported to increase the oil solubility of many anti-cancer drugs.

Using submicron oil-in-water systems, one can engineer a multifunctional therapeutic system by combining several drugs in the same carrier (i.e., with an oil soluble drug in the core and an amphiphilic one in the surface monolayer) for more effective treatment of tumors or other diseased conditions. Emulsion formulations also show promise in cancer chemotherapy as vehicles for prolonging the drug release after intramuscular and intratumoral injection (W/O systems) and as a means of enhancing the transport of anti-cancer drugs via the lymphatic system.[10]

35.6.2 Positively Charged Anoemulsions in Cancer Therapy

Positively charged nanoemulsions systems are expected to interact with negatively charged cell surfaces more efficiently, and this aspect of the positively charged nanoemulsions has been explored for possibility of oligonucleotide delivery to cancer cells.[50–53] Because oligonucleotides molecules display a polyanioinc character and present a large molecular structure, their ability to cross cell membranes remains very low. Teixeira et al.[50] formulated cationic O/W emulsions of oligonucleotides (pdT_{16}) using medium chain triglycerides, PC, poloxamer, and either monocationic SA lipid PC–SA emulsion or a polycationic lipid, RPR C_{18} (PC–RPRC$_{18}$ emulsion). They studied the biodistribution in leukemic P388/ADR cells grown in the peritoneal cavity after intraperitoneal administration in mice (Table 35.3). The fate of both the pdT and the marker of the oil phase (cholesteryl oleate) were determined in the fluid (devoid of cancer cells) and in the P388/ADR cell pellet. The results indicated that pdT in solution remained only in the fluid and did not associate at all with the tumor cells. However, pdT injected as an emulsion formulation was detectable in the cancer cells pellet even after 24 h and in very high proportions (up to 18% of the injected dose). When the area-under-the-curve values of the concentration versus time profiles for different formulations were compared, it was observed that RPR C_{18} formulations favored an

TABLE 35.3
Area Under the Percent Tissues–Time Curve After Intraperitoneal Administration Into Mice Bearing P388/ADR Ascites of pdT$_{16}$ in Solution or Associated with PC/SA or PC/RPRC$_{18}$ Emulsion

	AUC$_{0-24\,h}$ (% Injected Dose/min)					
Preparations	Blood	Liver	Kidneys	Lungs	Spleen	Cell Pellet
Solution	3,870	16,411	4,725	1,231	2,343	3,095
PC/SA	3,393	16,505	4,214	1,254	2,438	13,634
PC/RPRC	2,984	1,167	2,771	850	2,200	22,592

PC/SA, phosphatidylcholine-stearylamine emulsion; PC/RPRC$_{18}$, Posphatidylcholine-polycationic lipid RPRC$_{18}$ emulsion; pdT$_{16}$, Oligohexadecathymidylate.
Source: Adapted from Teixeira, H., Dubernet, C., Chacun, H., Rabinovich, L., Boutet, V., Deverre, J. R., Benita, S., and Couvreur, P., *J. Control Release*, 89(3), 473–482, 2003.

increased association of pdT to the tumor cells compared to SA (Table 35.3). Because the distribution of the oil marker in the tumor fluid and the P388/ADR cell pellet could not be correlated with that of pdT, the authors postulated that pdT was probably not taken through the endocytosis of the oil droplets but by positive charges of the emulsion that probably increased membrane permeability and allowed the pdT molecules to more efficiently enter the cells. In another study, cationic emulsions composed of 3β [N-(N′,N′-DC-Chol)] and dioleoylphosphatidyl ethanolamine, castor oil, and Tween® 80 efficiently delivered plasmid DNA into various cancer cells with low toxicity. Furthermore, the use of chitosan as a condensing agent (for DNA) and subsequent complexation with cationic emulsion composed of DC-chol enhanced the transfection efficiency in vitro compared to DNA/emulsion complexes with the same formulation with chitosan.[54] In vivo study in mice showed that with chitosan enhanced emulsion complexes, the GFP mRNA expression was prolonged in liver and lung until day 6. Goldstein et al.[55] formulated a monoclonal antibody, AMB8LK-, cationic emulsion for targeted delivery to the tumors that overexpress H-ferritin (that is recognized by AMB8LK). The results of cell culture studies indicated that the coupling of AMB8LK-Fab′ fragment to the cationic emulsion increased the cells uptake by 50% as compared to non-Fab′ conjugated cationic emulsion. Similar conjugation of B-cell lymphoma monoclonal antibody, LL2, to the surface of polyethylene glycol-PE-based lipid emulsion globules has been reported with improved binding to the B-lymphoma cells.[56] The results from these studies indicate significant potential of cationic emulsion systems as effective drug/gene delivery vehicles to tumor cells.

35.6.3 PHOTODYNAMIC THERAPY OF CANCER

Photodynamic therapy (PDT) of cancer is based on the concept that certain photosensitizers can be localized in the neoplastic tissue, and subsequently, these photosensitizers can be activated with the appropriate wavelength (energy) of light to generate active molecular species such as free radicals and singlet oxygen ($^1O_2$) that are toxic to cells and tissues.[57–59] PDT is a binary therapy, and it has a potential advantage in its inherent dual selectivity. First, selectivity is achieved by an increased concentration of the photosensitizer in target tissue; second, the irradiation can be limited to a specified volume. Provided that the photosensitizer is nontoxic, only the irradiated areas will be affected even if the photosensitizer does bind to normal tissues. Selectivity can be further enhanced by binding photosensitizers to molecular delivery systems that have high affinity for target tissue. Many photosensitizers, however, are characterized by a high level of hydrophobicity. The poor water solubility of most photosensitizers often prevents their direct administration into the bloodstream. To overcome this problem, photosensitizers have to be either chemically modified or incorporaed into lipid delivery systems such as liposomes and emulsions.

Morgan et al. utilized emulsions prepared with the ethoxy castor oil Cremophor®-EL (CR) and DPPC liposomes for the administration of several hydrophobic purpurins to rats with an implanted urothelial tumor.[60] PDT studies showed that tin (IV) etiopurpurin (SnET2) at a dose of 1 mg kg-~induced 100% tumor cure with both delivery vehicles. At lower doses, SnET2 in CR was more effective than SnET2 in liposomes.[61] Therapeutic doses of SnET2 in CR caused no phototoxicity to footpad tissue irradiated 24 h after administration although some phototoxicity was observed with liposomes. A higher PDT efficacy was observed by delivering a ketochlorin in CR than in Tween® 80. The better response of the RIF tumor (murine mammary sarcoma) to PDT was ascribed to an increased tumor uptake of the ketochlorin, probably caused by a longer persistence of the drug in the circulation.[62] Similar conclusions were drawn by comparing the pharmacokinetics of several germanium (IV)-octabutoxy–phthalocyanines (GePc) incorporated into DPPC liposomes or in CR emulsion. The Cremophor-delivered GePcs were cleared from the blood circulation at a much slower rate than the liposome-delivered GePcs.[63] At the same time, Cremophor induced a slower and reduced uptake of the GePcs in the liver and spleen, and it greatly enhanced the uptake in the tumor as compared to liposomes. Furthermore, CR induced a four-fold higher accumulation of the Pcs in the MS-2 fibrosarcoma, yielding in all cases two to three times higher tumor-to-muscle (peritumoral tissue) and tumor-to-skin drug concentration ratios.[64] Almost similar results have been reported with silicon (IV)-phthalocyanine (SiPc) CR or liposome formulation in C57B1/6 mice Lewis lung carcinoma.[65]

In brief, various PDT therapies have reported two different vehicles for photosensitizers, a Cremophor oil emulsion and DPPC liposomal vesicles. The reported pharmacokinetic studies clearly indicate that the former vehicle yields a significantly larger selectivity of tumor targeting, mainly as a consequence of an enhanced accumulation in the malignant lesion, and no appreciable differences in the uptake by healthy tissues is observed with Cremophor or DPPC liposomes (except higher uptake by brain tissues with CR formulation). The greater accumulation of Cremophor-delivered photosensitizers has been thought to be attributed to the tendency of this vehicle to release the associated photosensitizing drug to serum low-density lipoproteins (LDLs) in a highly preferential amount, and DPPC liposomes transfer the drug to all the components of the lipoprotein class in aliquots that are proportional to the relative concentration of the individual proteins.[61] LDLs are known to display a preferential interaction with a variety of neoplastic cells through a receptor-mediated endocytotic process. The tumor cells often express an increased number of LDL receptors as compared with normal cells.

35.6.4 Neutron Capture Therapy of Cancer

Neutron Capture Therapy (NCT) is a binary radiation therapy modality that brings together two components that when kept separate, have only minor effects on the cells. The first component is a stable isotope of boron or gadolinium (Gd) that can be concentrated in tumor cells by a suitable delivery vehicle. The second is a beam of low-energy neutrons. Boron or Gd in or adjacent to the tumor cells disintegrates after capturing a neutron, and the high energy heavy charged particles produced through this interaction destroy only the cancer cells in close proximity to it, leaving adjacent normal cells largely unaffected.[66] The success of NCT relies on the targeting of boron and Gd-based compounds to the tumor mass and to achieve desirable intracellular concentrations of these agents. At the present time, there are two targets with NCT, namely glioblastoma (malignant brain tumor) and malignant melanoma.

Lu and co-workers developed and evaluated a very low-density lipoprotein (VLDL), resembling phospholipid-submicron emulsion as a carrier system for new cholesterol-based boronated compound, BCH (anti-cancer boron neutron capture therapy compound), for targeted delivery to cancer cells.[67] Emulsions were formulated by sonication method using triolein, cholesterol,

cholesterol oleate as lipids, egg PC, lysophosphatidylcholine as surfactants, and BCH as active pharmaceutical ingredient. The BCH was solubilized in the inner oily core of the emulsion droplets (155 ± 5 nm). The formulation was designed based on the fact that LDL and high-density lipoproteins (HDL) are natural carriers of cholesteryl esters in the body, and certain human and animal tumor types have been shown to have elevated LDL-receptor activity primarily because the rapidly dividing cancer cells require higher amounts of cholesterol to build new cell membranes. It is expected that VLDL-resembling formulation may mimic the VLDL–LDL biological process for targeted drug delivery to cancer cells. Cell culture data showed sufficient uptake of BCH in rat 9L glioma cells (> 50 mg boron/g cells). The authors proposed that such VLDL-resembling formulation may serve as a targeted drug delivery system to cancerous cells in vivo.

Miyamoto et al. prepared the nanoemulsions of Gd complexed with DTPA-distearylamide (SA) and DTPA-distearylester (SE) using soybean oil and hydrogenated PC and various co-surfactants (Cremophore®-RH60, Myrj® 53, Myrj® 59, or Brij® 700).[68,69] The biodistribution of the Gd was studied in melanoma-bearing hamsters after intraperitoneal administration of the formulations. The co-surfactant-containing emulsions were more rapidly and completely cleared from the abdominal cavity than the plain emulsion without co-surfactants. When the effect of the co-surfactants on the biodistribution of Gd from Gd–DTPA–SA-containing emulsions in the standard-Gd formulation were compared, the Cremophore®-RH60 emulsion exhibited the highest Gd accumulation in the tumor (107 µg Gd/g tumor), possibly resulting from its small particle size (78 nm) and the stable coat on the particle surfaces with PEO. The Brij® 700-containing emulsion allowed the highest blood Gd concentration for a prolonged period. However, it exhibited slower Gd accumulation in the tumor, only reaching an identical level in comparison with the Cremophore®-RH60 emulsion. Gd–DTPA–SE emulsions exhibited poor tumor accumulation of Gd.

The same group of researchers evaluated the effect of intravenous administration of Gd-nanoemulsions on biodistribution and tumor accumulation.[70] The biodistribution data revealed that the intravenous injection had three advantages over the ip injection, namely, a faster and higher accumulation of Gd and a more extended retention time of Gd. Two intravenous injections of the standard-Gd-nanoemulsion (1.5 mg Gd/ml) with a 24 h interval doubled the tumor accumulation of Gd, resulting in 49.7 mg Gd/g wet tumor 12 h after administration. By using a two-fold Gd-enriched formulation (3.0 mg Gd/ml) in the repeated administration schedule, the accumulation was doubled again, reaching 101 mg Gd/g wet tumor. This level was comparable to the maximum level in the single ip injection previously reported. The authors proposed that intravenous injection could be an alternative to ip injection as an administration route for Gd.

35.6.5 Perfluorochemical Nanoemulsions in Cancer Therapy

Perfluorocarbon emulsions are being clinically evaluated as artificial oxygen carriers to reduce allogeneic blood transfusions or to improve tissue oxygenation.[13] Perfluorochemicals are chemically inert synthetic molecules that primarily consist of carbon and fluorine atoms and are clear, colorless liquids. They have the ability to physically dissolve significant quantities of many gases, including oxygen and carbon dioxide. Perfluorochemicals are hydrophobic and are not miscible with water. Perfluorochemicals have to be emulsified for intravenous use. To mimic the natural oxygen carrying cells (RBCs), the droplet size of perfluorocarbon emulsions is maintained in sub-micron range (median diameter < 0.2 µm). Egg phospholipid has been used as an emulsifier of choice in these formulations. The examples of the commercial perfluorocarbon emulsions are Oxygent™ (Alliance Pharmaceutical Corporation, San Diego, CA, U.S.A.), Oxyfluor® (Hemagen Inc., St Louis, MO, U.S.A.) and Fluosol-DA (Alpha Therapeutic Corp., Los Angeles, CA, U.S.A.).

The perfluorochemical nanoemulsions (PFCE) have opened interesting opportunities in cancer therapy. It is suggested that fluorocarbon emulsions might find a role in photodynamic therapy, both as carriers for sensitizing dyes and also to maintain tissue oxygenation in hypoxic regions of solid tumors. The high solubility of oxygen in fluorocarbon emulsions maintains solution oxygen tension, optimizing photo-oxidative damage. The hydrophobic anti-cancer drugs can be delivered to the tumor mass by dissolving them in a hydrophobic core of the emulsion. Furthermore, PFCE can be used as an adjuvant to radiation therapy and/or chemotherapy in the treatment of solid tumors.[71,72] The rationale for the use of PFCE in this therapeutic setting is that solid tumor masses contain areas of hypoxia that are therapeutically resistant. Since x-rays and many chemotherapeutic agents require oxygen to be maximally cytotoxic and most normal tissues are well-oxygenated, the additional oxygen put in circulation by the PFCE should not increase the normal tissue toxicities produced by the various therapies. The preclinical studies have shown very positive effects with single dose and fractionated radiation in several rodent solid tumor models. Many widely used anti-cancer drugs, including anti-tumor alkylating agents and doxorubicin, have shown improved response by PFCE coadministration.[73] Also, local application of toxic doses of PFCEs resulted in the necrosis of cancer cells. This is especially promising in the treatment of cancers of the head and neck regions that are currently difficult to treat.[74]

Lanza and co-workers developed a perfluorocarbon emulsion that can be used for ultrasound and magnetic resonance imaging (MRI) detection of fibrin and thrombus.[75] Liquid perfluorocarbon inside the nanoemulsion core makes it stable and acoustically reflective—a benefit for ultrasound imaging. For MRI, the nanoemulsion can carry Gd molecules. The authors estimates that one nanoemulsion droplet can carry up to 100,000 Gd molecules, enough for a large MR signal and effective imaging of the targeted area. The fluorine component allows quantification of drug amounts to the targeted area using MR spectroscopy. The authors speculated that one could attach radionuclides such as technetium-99m for nuclear imaging or radio-opaque compounds for contrast enhancement in computed tomography in the nanoemulsion core. The same group of researchers have also developed an $\alpha_v\beta_3$ integrin-targeted paramagnetic nanoemulsions that can detect and characterize sparse integrin expression on tumor neovasculature induced by nascent melanoma xenografts.[76] Alpha$_v$-beta$_3$ integrin is an excellent marker for angiogenesis and, therefore, is used as the target to dock the nanoemulsion particles. When the emulsion particles are injected into the body, the homing ligand causes the droplets to seek out $\alpha_v\beta_3$ integrin. Detecting $\alpha_v\beta_3$ integrin provides a direct and early measure of tumor vascularity. Angiogenesis is associated with a wide variety of pathologies, and by targeting angiogenesis, the nanoemulsions can identify solid tumors and lymphoid cancers at an early stage. Because angiogenesis also occurs with atherosclerosis and inflammatory diseases, authors feel that such nanoemulsions can be used for additional diagnostic applications, including imaging early signs of atherosclerosis, restenosis following angioplasty, and inflammatory diseases such as rheumatoid arthritis.

35.7 CONCLUSIONS AND FUTURE PERSPECTIVES

Nanoemulsion formulations offer several advantages for the delivery of drugs, biologicals, or diagnostic agents. Traditionally, nanoemulsions have been used in clinics for more than four decades as total parenteral nutrition fluids. Several other products for drug delivery applications such as Diprivan®, Liple®, and Ropion® have also reached the marketplace. Although nanoemulsions are chiefly seen as vehicles for administering aqueous insoluble drugs, they have more recently received increasing attention as colloidal carriers for targeted delivery of various anti-cancer drugs, photosensitizers, neutron capture therapy agents, or diagnostic agents. Because of their sub-micron size, they can be easily targeted to the tumor area. Moreover, the possibility of surface functionalization with a targeting moiety has opened new avenues for targeted delivery of drugs, genes, photosensitizers, and other molecules to the tumor area. Research with perflurochemical nanoemulsions has shown promising results for the treatment of cancer in conjugation with other treatment modalities and targeted delivery to the neovasculature. It is expected that further research and development work will be carried out in the near future for clinical realization of these targeted delivery vehicles.

REFERENCES

1. Salager, J. L., Formulation concepts for the emulsion maker, In *Pharmaceutical Emulsions and Suspensions*, Nielloud, F. and Marti-Mestres, G., Eds., Marcel Dekker, New York, pp. 19–72, 2000.
2. Tenjarla, S., Microemulsions: An overview and pharmaceutical applications, *Crit. Rev. Ther. Drug Carrier Syst.*, 16(5), 461–521, 1999.
3. Keipert, S., Siebenbrodt, I., Luders, F., and Bornschein, M., Microemulsions and their potential pharmaceutical application, *Pharmazie*, 44(7), 433–444, 1989.
4. Schoot, H., Colloidal dispersions, In *Remington: The Science and Practice of Pharmacy*, Gennaro, A., Lippincott, R., Ed., 20th ed., Williams & Witkins, Maryland, pp. 288–315, 2000.
5. Sarker, D. K., Engineering of nanoemulsions for drug delivery, *Curr. Drug Deliv.*, 2(4), 297–310, 2005.
6. Jores, K., Lipid nanodispersions as drug carrier systems -a physicochemical characterization, http://sundoc.bibliothek.uni-halle.de/diss-online/04/04H310/prom.pdf, accessed date, Aug. 16, 2005.
7. Buszello, K. and Muller, B., Emulsions as drug delivery systems, In *Pharmaceutical Emulsions and Suspensions*, Nielloud, F. and Marti-Mestres, G., Eds., Marcel Dekker, New York, pp. 191–228, 2000.
8. Yang, S. C. and Benita, S., Enhanced absorption and drug targeting by positively charged submicron emulsions, *Drug Dev. Res.*, 50, 476–486, 2000.
9. Inactive ingredient search for approved drug products, http://www.accessdata.fda.gov/scripts/cder/iig/index.cfm, date accessed, Nov. 05, 2005.
10. Eccleston, G., Emulsions, In *Encyclopedia of Pharmaceutical Technology*, Swarbrick, J. and Boylan, J., Eds., Marcel Dekker, New York, pp. 137–188, 1992.
11. Chung, H., Kim, T. W., Kwon, M., Kwon, I. C., and Jeong, S. C., Oil components modulate physical characteristics and function of the natural oil emulsions as drug or gene delivery system, *J. Control Release*, 71, 339–350, 2001.
12. Allison, A. D., Squelene and squalane emulsions as adjuvants, *Methods*, 19, 87–93, 1999.
13. Lane, T. A., Perfluorochemical-based artificial oxygen carrying red cell substitutes, *Transfus. Sci.*, 16(1), 19–31, 1995.
14. Benita, S. and Levy, M. Y., Submicron emulsions as colloidal drug carrieres for intravenous administration: Comprehensive physicochemical characterization, *J. Pharm. Sci.*, 82, 1069–1079, 1993.
15. Klang, S. and Benita, S., Design and evaluation of submicron emulsions as colloidal drug carriers for intravenous administration, In *Submicron Emulsions in Drug Targeting and Delivery*, Benita, S., Ed., Academic Publishers, Harwood, pp. 119–152, 1998.
16. Klang, S., Frucht-Pery, J., Hoffman, A., and Benita, S., Physicochemical characterization and acute toxicity evaluation of a positively charged submicron emulsion vehicle, *J. Pharm. Pharmacol.*, 46, 986–993, 1998.
17. Haskell, R., Nanotechnology for drug delivery, www.banyu-zaidan.or.jp/symp/soyaku/haskell.pdf, date accessed, Oct. 06, 2005.
18. Haskell, R. J., Characterization of submicron systems via optical methods, *J. Pharm. Sci.*, 87(2), 125–129, 1998.
19. Wabel, C., Influence of lecithin on structure and stability of parenteral fat emulsions, http://www2.chemie.uni-erlangen.de/services/dissonline/data/dissertation/Christoph_Wabel/html/index.html, accessed date Oct. 09, 2005.
20. Salager, J. L., Emulsion properties and relate know-how and attain them, In *Pharmaceutical Emulsions and Suspensions*, Nielloud, F. and Marti-Mestres, G., Eds., Marcel Dekker, New York, pp. 73–125, 2000.
21. Gaysorn, C., Lyons, R. T., Patel, M. V., and Hem, S. L., Effect of surface charge on the stability of oil/water emulsions during steam sterilization, *J. Pharm. Sci.*, 88(4), 454–458, 1999.
22. Elbaz, E., Zeevi, A., Klang, S., and Benita, S., Positively charged submicron emulsions- a new type of colloidal drug carrier, *Int. J. Pharm.*, 96, R1–R6, 1993.
23. Klang, S. H., Siganos, C. S., Benita, S., and Frucht-Pery, J., Evaluation of a positively charged submicron emulsion of piroxicam on the rabbit corneum healing process following alkali burn, *J. Control Release*, 57(1), 19–27, 1999.
24. Klang, S. H., Frucht-Pery, J., Hoffman, A., and Benita, S., Physicochemical characterization and acute toxicity evaluation of a positively-charged submicron emulsion vehicle, *J. Pharm. Pharmacol.*, 46(12), 986–993, 1994.

25. Gershanik, T., Benzeno, S., and Benita, S., Interaction of a self-emulsifying lipid drug delivery system with the everted rat intestinal mucosa as a function of droplet size and surface charge, *Pharm. Res.*, 15(6), 863–869, 1998.
26. Oku, N., Tokudome, Y., Namba, Y., Saito, N., Endo, M., Hasegawa, Y., Kawai, M., Tsukada, H., and Okada, S., Effect of serum protein binding on real-time trafficking of liposomes with different charges analyzed by positron emission tomography, *Biochim. Biophys. Acta*, 1280(1), 149–154, 1996.
27. Gao, X. and Huang, L., A novel cationic liposome reagent for efficient transfection of mammalian cells, *Biochem. Biophys. Res. Commun.*, 179(1), 280–285, 1991.
28. Calvo, P., Alonso, M. J., Vila-Jato, J. L., and Robinson, J. R., Development of positively charged colloidal drug carrieres: Chitosan coated polyester nanocapsules and sub-micron emulsions, *Colloid Polm. Sci.*, 275, 46–53, 1997.
29. Jumaa, M. and Muller, B. W., Physicochemical properties of chitosan-lipid emulsions and their stability during the autoclaving process, *Int. J. Pharm.*, 183(2), 175–184, 1999.
30. Samama, J. P., Lee, K. M., and Biellmann, J. F., Enzymes and microemulsions: Activity and kinetic properties of liver alcohol dehydrogenase in ionic water in oil microemulsions, *Eur. J. Biochem.*, 163, 609–617, 1987.
31. Kurihara, A., Shibayama, Y., Mizota, A., Yasuno, A., Ikeda, M., and Hisaoka, M., Pharmacokinetics of highly lipophilic antitumor agent palmitoyl rhizoxin incorporated in lipid emulsions in rats, *Biol. Pharm. Bull.*, 19(2), 252–258, 1996.
32. Lundberg, B. B., Preparation of drug carrier emulsions stabilized with phosphatidylcholine surfactant mixtures, *J. Pharm. Sci.*, 83, 72–75, 1994.
33. Liu, F. and Liu, D., Long-circulating emulsions (oil-in-water) as carriers for lipophilic drugs, *Pharm. Res.*, 12(7), 1060–1064, 1995.
34. Hodoshima, N., Udagawa, C., Ando, T., Fukuyasu, H., Watanabe, T., and Nakabayashi, S., Lipid nanoparticles for delivering antitumor drugs, *Int. J. Pharm.*, 146, 81–92, 1997.
35. Kurihara, A., Shibayama, Y., Mizota, A., Yasuno, A., Ikeda, M., Sasagawa, K., Kobayashi, T., and Hisaoka, M., Enhanced tumor delivery and antitumor activity of palmitoyl rhizoxin using stable lipid emulsions in mice, *Pharm. Res.*, 13(2), 305–310, 1996.
36. Davis, S. S., Washington, C., West, P., Illum, L., Liversidge, G., Sternson, L., and Kirsh, R., Lipid emulsions as drug delivery systems, *Ann. N.Y. Acad. Sci.*, 507, 75–88, 1987.
37. Arimoto, I., Matsumoto, C., Tanaka, M., Okuhira, K., Saito, H., and Handa, T., Surface composition regulates clearance from plasma and triolein lipolysis of lipid emulsions, *Lipids*, 33(8), 773–779, 1998.
38. Krobbuaban, B., Diregpoke, S., Kumkeaw, S., and Tanomsat, M., Comparison on pain on injection of a small particle size-lipid emulsion of propofol and standard propofol with or without lidocaine, *J. Med. Assoc. Thai.*, 88(10), 1401–1405, 2005.
39. Thorn-Alquist, A. M., Parenteral use of diazepam in an emulsion formulation. A clinical study, *Acta Anaesthesiol. Scand.*, 21(5), 400–404, 1977.
40. Westrin, P., Jonmarker, C., and Werner, O., Dissolving methohexital in a lipid emulsion reduces pain associated with intravenous injection, *Anesthesiology*, 76(6), 930–934, 1992.
41. Lovell, M. W., Johnson, H. W., Hui, H. W., Cannon, J. B., Gupta, P. K., and Hsu, C. C., Less-painful emulsion formulations for intravenous administration of clarithomycin, *Int. J. Pharm.*, 109, 45–57, 1994.
42. Stuttmann, H., Doenicke, A., Kugler, J., and Laub, M., A new formulation of etomidate in lipid emulsion: Bioavailability and venous irritation, *Anaesthesist*, 38, 421–423, 1989.
43. Glen, J. B. and Hunter, S. C., Pharmacology of an emulsion formulation of ICI 35 868, *Br. J. Anaesth.*, 56(6), 617–626, 1984.
44. Jang, S. H., Wientjes, M. G., Lu, D., and Au, J. L., Drug delivery and transport to solid tumors, *Pharm. Res.*, 20(9), 1337–1350, 2003.
45. Mayhew, E. G., Lasic, D., Babbar, S., and Martin, F. J., Pharmacokinetics and antitumor activity of epirubicin encapsulated in long-circulating liposomes incorporating a polyethylene glycol-derivatized phospholipid, *Int. J. Cancer*, 51(2), 302–309, 1992.
46. Lundberg, B. B., A submicron lipid emulsion coated with amphipathic polyethylene glycol for parenteral administration of paclitaxel (Taxol), *J. Pharm. Pharmacol.*, 49(1), 16–21, 1997.

47. Constantinides, P. P., Lambert, K. J., Tustian, A. K., Schneider, B., Lalji, S., Ma, W., Wentzel, B., Kessler, D., Worah, D., and Quay, S. C., Formulation development and antitumor activity of a filter-sterilizable emulsion of paclitaxel, *Pharm. Res.*, 17(2), 175–182, 2000.

48. Kan, P., Chen, Z. B., Lee, C. J., and Chu, I. M., Development of nonionic surfactant/phospholipid o/w emulsion as a paclitaxel delivery system, *J. Control Release*, 58(3), 271–278, 1999.

49. Lundberg, B. B., Sub-micron lipid emulsions as carrieres for anticancer drugs, http://www.cmbl.org.pl/lundberg.pdf, accessed date, Sept. 25, 2005.

50. Teixeira, H., Dubernet, C., Chacun, H., Rabinovich, L., Boutet, V., Deverre, J. R., Benita, S., and Couvreur, P., Cationic emulsions improves the delivery of oligonucleotides to leukemic P388/ADR cells in ascite, *J. Control Release*, 89(3), 473–482, 2003.

51. Teixeira, H., Dubernet, C., Puisieux, F., Benita, S., and Couvreur, P., Submicron cationic emulsions as a new delivery system for oligonucleotides, *Pharm. Res.*, 16(1), 30–36, 1999.

52. Teixeira, H., Dubernet, C., Rosilio, V., Laigle, A., Deverre, J. R., Scherman, D., Benita, S., and Couvreur, P., Factors influencing the oligonucleotides release from O-W submicron cationic emulsions, *J. Control Release*, 70(1–2), 243–255, 2001.

53. Teixeira, H., Rosilio, V., Laigle, A., Lepault, J., Erk, I., Scherman, D., Benita, S., Couvreur, P., and Dubernet, C., Characterization of oligonucleotide/lipid interactions in submicron cationic emulsions: Influence of the cationic lipid structure and the presence of PEG-lipids, *Biophys. Chem.*, 92(3), 169–181, 2001.

54. Lee, M. K., Chun, S. K., Choi, W. J., Kim, J. K., Choi, S. H., Kim, A., Oungbho, K., Park, J. S., Ahn, W. S., and Kim, C. K., The use of chitosan as a condensing agent to enhance emulsion-mediated gene transfer, *Biomaterials*, 26(14), 2147–2156, 2005.

55. Goldstein, D., Nassar, T., Lambert, G., Kadouche, J., and Benita, S., The design and evaluation of a novel targeted drug delivery system using cationic emulsion-antibody conjugates, *J. Control Release*, 108(2–3), 418–432, 2005.

56. Lundberg, B. B., Griffiths, G., and Hansen, H., Conjugation of an anti-B-cell lymphoma monoclonal antibody, LL2, to long-circulating drug-carrier lipid emulsions, *J. Pharm. Pharmacol.*, 51(10), 1099–1105, 1999.

57. Waldow, S. M., Henderson, B. W., and Dougherty, T. J., Hyperthermic potentiation of photodynamic therapy employing Photofrin I and II: Comparison of results using three animal tumor models, *Lasers Surg. Med.*, 7(1), 12–22, 1987.

58. Dougherty, T. J., Photosensitizers: therapy and detection of malignant tumors, *Photochem. Photobiol.*, 45(6), 879–889, 1987.

59. Potter, W. R., Mang, T. S., and Dougherty, T. J., The theory of photodynamic therapy dosimetry: Consequences of photo-destruction of sensitizer, *Photochem. Photobiol.*, 46(1), 97–101, 1987.

60. Morgan, A. R., Garbo, G. M., Keck, R. W., and Selman, S. H., New photosensitizers for photodynamic therapy: Combined effect of metallopurpurin derivatives and light on transplantable bladder tumors, *Cancer Res.*, 48(1), 194–198, 1988.

61. Reddi, E., Role of delivery vehicles for photosensitizers in the photodynamic therapy of tumours, *J. Photochem. Photobiol. B*, 37(3), 189–195, 1997.

62. Woodburn, K., Chang, C. K., Lee, S., Henderson, B., and Kessel, D., Biodistribution and PDT efficacy of a ketochlorin photosensitizer as a function of the delivery vehicle, *Photochem. Photobiol.*, 60(2), 154–159, 1994.

63. Soncin, M., Polo, L., Reddi, E., Jori, G., Kenney, M. E., Cheng, G., and Rodgers, M. A., Effect of the delivery system on the biodistribution of Ge(IV) octabutoxy-phthalocyanines in tumour-bearing mice, *Cancer Lett.*, 89(1), 101–106, 1995.

64. Soncin, M., Polo, L., Reddi, E., Jori, G., Kenney, M. E., Cheng, G., and Rodgers, M. A., Effect of axial ligation and delivery system on the tumour-localising and -photosensitising properties of Ge(IV)-octabutoxy-phthalocyanines, *Br. J. Cancer*, 71(4), 727–732, 1995.

65. Wohrle, D., Muller, S., Shopova, M., Mantareva, V., Spassova, G., Vietri, F., Ricchelli, F., and Jori, G., Effect of delivery system on the pharmacokinetic and phototherapeutic properties of bis(methyloxyethyleneoxy) silicon-phthalocyanine in tumor-bearing mice, *J. Photochem. Photobiol. B*, 50(2–3), 124–128, 1999.

66. The basics of boron neutron capture therapy, http://web.mit.edu/nrl/www/bnct/info/description/description.html, date accessed, Nov. 27, 2005.

67. Shawer, M., Greenspan, P., Ole, S., and Lu, D. R., VLDL-resembling phospholipid-submicron emulsion for cholesterol-based drug targeting, *J. Pharm. Sci.*, 91(6), 1405–1413, 2002.

68. Miyamoto, M., Hirano, K., Ichikawa, H., Fukumori, Y., Akine, Y., and Tokuuye, K., Biodistribution of gadolinium incorporated in lipid emulsions intraperitoneally administered for neutron-capture therapy with tumor-bearing hamsters, *Biol. Pharm. Bull.*, 22(12), 1331–1340, 1999.

69. Miyamoto, M., Hirano, K., Ichikawa, H., Fukumori, Y., Akine, Y., and Tokuuye, K., Preparation of Gadolinium-containing emulsions stabilized with phosphatidylcholine surfactant mixtures for neutron capture therapy, *Chem. Pharm. Bull. (Tokyo)*, 47(2), 203–208, 1999.

70. Watanabe, T., Ichikawa, H., and Fukumori, Y., Tumor accumulation of gadolinium in lipid-nanoparticles intravenously injected for neutron-capture therapy of cancer, *Eur. J. Pharm. Biopharm.*, 54(2), 119–124, 2002.

71. Teicher, B. A., Use of perfluorochemical emulsions in cancer therapy, *Biomater. Artif. Cells Immobil. Biotechnol.*, 20(2–4), 875–882, 1992.

72. Teicher, B. A., Herman, T. S., and Frei, E., Perfluorochemical emulsions: Oxygen breathing in radiation sensitization and chemotherapy modulation, *Important Adv. Oncol.*, 39–59, 1992.

73. Teicher, B. A., Holden, S. A., Ara, G., Ha, C. S., Herman, T. S., and Northey, D., A new concentrated perfluorochemical emulsion and carbogen breathing as an adjuvant to treatment with antitumor alkylating agents, *J. Cancer Res. Clin. Oncol.*, 118(7), 509–514, 1992.

74. Rockwell, S., Irvin, C. G., and Nierenburg, M., Effect of a perfluorochemical emulsion on the development of artificial lung metastases in mice, *Clin. Exp. Metastasis*, 4(1), 45–50, 1986.

75. Sisk, J., Imaging cancer one nanometer at a time, http://www.radiologytoday.net/imagingcancer.htm, accessed date, Sept. 06, 2005.

76. Schmieder, A. H., Winter, P. M., Caruthers, S. D., Harris, T. D., Williams, T. A., Allen, J. S., and Lacy, E. K., Molecular MR imaging of melanoma angiogenesis with alphanubeta-3-targeted paramagnetic nanoparticles, *Magn. Reson. Med.*, 53(3), 621–627, 2005.

36 Solid Lipid Nanoparticles for Anti-Tumor Drug Delivery

Ho Lun Wong, Yongqiang Li, Reina Bendayan,
Mike Andrew Rauth, and Xiao Yu Wu

CONTENTS

36.1 Introduction 742
36.2 Composition and Structure of SLN 743
    36.2.1 Lipids Used in SLN 744
    36.2.2 Surfactants Used in SLN.. 747
    36.2.3 Other Agents Used in SLN 748
    36.2.4 Types, Proposed Structure, and Morphology of SLN 748
        36.2.4.1 Classical SLN 748
        36.2.4.2 Nanostructured Lipid Carrier (NLC) 749
        36.2.4.3 Polymer–Lipid Hybrid Nanoparticles (PLN) 749
36.3 Manufacturing Methods 751
    36.3.1 High Pressure Homogenization 751
    36.3.2 Microemulsion.. 751
    36.3.3 Solvent Evaporation 753
    36.3.4 w/o/w Double Emulsion Method 753
36.4 Characteristic Properties of SLN 753
    36.4.1 Drug Loading 753
    36.4.2 Drug Release 754
    36.4.3 Stability of SLN 755
36.5 SLN for Anti-Tumor Drug Delivery 756
    36.5.1 Cancer Chemotherapy 756
    36.5.2 Delivery of Cytotoxic Compounds Using SLN................................ 757
    36.5.3 Poorly Water-Soluble Compounds 759
        36.5.3.1 General Principle.. 759
        36.5.3.2 Topoisomerase I Inhibitors (Camptothecin, Irinothecan) 759
        36.5.3.3 Paclitaxel 760
        36.5.3.4 Other Poorly Water-Soluble Anti-Tumor Compounds
                (Etoposide, All-Trans Retinoic Acid, Butyric Acid)........................ 761
    36.5.4 Water-Soluble Ionic Salts 761
        36.5.4.1 General Principle.. 761
        36.5.4.2 Doxorubicin and Idarubicin 761
        36.5.4.3 Water-Soluble, Non-Ionic Drug Molecule 762
            36.5.4.3.1 General Principle 762
            36.5.4.3.2 Fluorouracil 763
            36.5.4.3.3 Overall Remarks on Cytotoxic Drug Delivery 763

36.5.5   Delivery of Chemosensitizing Agents Using SLN... 764
        36.5.5.1   Multidrug Resistance and Chemosensitizers 764
        36.5.5.2   Rationales for Delivery of Chemosensitizers Using SLN................. 764
        36.5.5.3   Current Use of SLN for Chemosensitizer Delivery 765
                36.5.5.3.1   Verapamil and Quinidine.. 765
                36.5.5.3.2   Elacridar (GG918) 765
                36.5.5.3.3   Overall Remarks on Chemosensitizer Delivery 766
    36.5.6   Effect of SLN Formulation on In Vitro Anti-Tumor Toxicity 766
        36.5.6.1   In Vivo Efficacy of SLN Delivered Antineoplastic Agents.............. 768
        36.5.6.2   Improved Pharmacokinetics and Drug Distribution 768
36.6   Conclusions and Future Perspectives.. 770
References 771

36.1 INTRODUCTION

Solid lipid nanoparticles (SLN) are colloidal particles of a lipid matrix that is solid at body temperature. Since their first introduction by Müller et al. in 1991,[1] SLN have attracted increasing interest as a carrier system for therapeutic and cosmetic applications.[2–7] As a drug delivery system, SLN have been investigated in the last ten years for pharmacological and dermatological formulation development. They can be administered through a number of routes including parenteral, peroral, dermal and rectal.[2,4,6,7] Improved bioavailability and targeting capacity have been observed[8,9] and enhanced cytotoxicity against multidrug resistant cancer cells[10,11] have been evidenced when SLN are used as the delivery vehicles.

SLN have been proposed as an alternative to other controlled drug delivery systems (CDDS) such as lipid emulsion, liposome, and polymeric nanoparticles as a result of their several advantages. For instance, in comparison to lipid emulsion, the solid lipid matrix of SLN makes sustained drug release possible. The solid lipids also immobilize drug molecules, thereby protecting the labile and sensitive drugs from coalescence and degradation,[12] and reduce drug leakage that are commonly seen in many other CDDS such as liposomes. Compared with some polymeric nanoparticles, SLN are generally less toxic because physiological and biocompatible lipids are utilized. Meanwhile, all of the less toxic surfactants that have been applied to other CDDS are equally applicable for SLN preparation. Other appealing features of SLN include the feasibility for mass production,[13–16] flexibility in sterilization,[4] and avoidance of organic solvents in a typical SLN preparation process. It should be noted that SLN are also a versatile formulation. Both lipophilic and hydrophilic compounds can be encapsulated and delivered by SLN with modification in the formulation.

The aforementioned useful qualities of SLN make them particularly attractive for the delivery of cancer chemotherapeutic agents. Anti-tumor drugs, especially the cytotoxic compounds that are used in conventional chemotherapy, are unique when compared to other classes of drugs in a number of areas such as the strong toxicity that is typical of cytotoxic drugs often compromises their therapeutic effects; poor specificity of their drug action in general; and the frequent occurrence of drug resistance during chemotherapeutic treatment. These issues may all at least be partly tackled by delivering anti-tumor compounds with a suitable drug carrier system. SLN is potentially a valuable choice for this purpose. The favorable physicochemical characteristics, controlled release kinetics, and site-specific drug delivery of SLN mean most of the anti-tumor drugs can be efficiently loaded and delivered to the cancerous tissues at reasonable rates while keeping the healthy tissues relatively unexposed to these highly toxic agents[17,18] In addition, many compounds that can be used to reduce drug resistance in cancer cells are also lipophilic (to be discussed later). SLN may also be used for the delivery of these compounds to further improve the effectiveness of chemotherapy of cancer normally resistant to cytotoxic drugs.

In order to fully optimize the performance of SLN for anti-tumor drug delivery, it is critical to gain some understanding of their fundamental aspects. This starts from knowing their main constituents that define the physicochemical properties of the SLN system they form, including the particle size, surface charge, and lipid crystallinity. These, in turn, have a strong impact on the SLN performance. Drug loading, release kinetics, tumor targeting ability, and pharmacokinetics in vivo may all be affected. For example, drug encapsulation by SLN is inversely correlated to the degree of crystallinity of lipids that is related to the choice of lipids.[19–21] Surfactants are applied to stabilize the colloids[22] to delay lipid crystallization[23,24] and modulate the degradation rate of lipid.[25,26] The tumor targeting capacity of SLN derives from their particle size, colloidal morphology, and surface modification. The leaky vasculature and poor lymphatic drainage of a tumor generally facilitate preferential penetration by and retention of particulate matters of small size.[27] The submicron size of SLN allows them to passively target tumor tissues. Further manipulation of the in vivo distribution of SLN can be achieved by modifying their surface charges or hydrophobicity. All in all, by understanding the relationship among the SLN constituents, the physicochemical properties, and particle behaviors, the anti-cancer performance of SLN can be optimized.[28,29]

SLN are not without their limitations. In fact, some drawbacks of SLN such as the low loading capacity for hydrophilic drugs, biphasic release kinetics, and formulation stability issues are still barriers against their popular use.[30,31] Fortunately, this class of drug carriers is also rapidly evolving. A number of strategies have been developed to overcome these limitations. Some of these strategies lead to new generations of SLN. In this chapter, an overall review of the fundamental aspects of SLN, the strategies employed for efficient encapsulation and controlled release of various anti-cancer drugs by SLN, and their current use for anti-tumor drug delivery is presented. It is believed that the information described here will contribute to the design and development of new SLN formulations for improved delivery of anti-tumor compounds.

36.2 COMPOSITION AND STRUCTURE OF SLN

The typical ingredients of SLN consist of solid lipid(s), surfactant(s) (plus co-surfactants if required), and incorporated active ingredients. To date, the lipids applied in the preparation of SLN include the following categories based on their structural diversity: fatty acids with different lengths of hydrocarbon chain; fatty esters; fatty alcohol; glycerides with different structures; and mixtures of glyceryl esters. The frequently used anionic and non-ionic lipids and their structures are listed in Table 36.1a and the cationic lipids in Table 36.1b. Waxes such as Cutina CP (cetylpalmitate), solid paraffin, and beewax are also sometimes used for SLN preparation. The surfactants used include amphoteric, ionic, and non-ionic ones (Table 36.2). Although ionic surfactants such as sodium dodecyl sulphate will allow the SLN to possess narrow size distribution and better stability, some of them are associated with undesirable toxicity. In addition to the conventional lipid-based SLN, the synthesized *para*-acyl-calix[4]arene-based SLN was also recently developed.[32]

SLN have the inherent advantage to poorly encapsulate water-soluble agents because of the lipophilicity of solid lipids. However, to incorporate hydrophilic drugs into SLN, some additional measures have to be taken. Counterions such as polymers and esters (organic salts) are required to neutralize the charge of the drugs (both listed in Table 36.2). This may increase the apparent partition coefficient of the selected drug in the lipid phase and enhance its loading. The encapsulation of specific hydrophilic anti-tumor compounds is further discussed in Section 36.5.2.2 and Section 36.5.2.3.

According to the drug incorporation mechanisms, SLN-based systems can be classified into the following categories, i.e., classical SLN, polymer–lipid hybrid SLN (PLN), nanostructured lipid carriers (NLC), and lipid-drug conjugate (LDC). Their differences are summarized in Table 36.3.

TABLE 36.1a

Structures of Lipids Used in the Preparation of Solid Lipid Nanoparticles (SLN)

Chemical Structure	Name of Lipid
$H_3C{-}{[}CH_2{]}_n{-}COOH$	Dodecanoic acid ($n = 10$)
	Myristic acid ($n = 12$)
	Palmitic acid ($n = 14$)
	Stearic acid ($n = 16$)
triglyceride structure	Caprylate triglycerides ($n = 6$)
	Caprate triglycerides ($n = 8$)
	Trilaurin ($n = 10$)
	Tripalmitin ($n = 14$)
	Tristearin ($n = 16$)
	Tribehenin ($n = 20$)
glyceryl monostearate structure	Glyceryl monostearate ($n = 16$)
glyceryl hydroxystearate structure	Glyceryl hydroxystearate ($n = 10$, $m = 5$)
glyceryl behenate (mono-) structure	Glyceryl behenate (mono-, $n = 20$)
glyceryl behenate (di-) structure	Glyceryl behenate (di-, $n = 20$)
cetylpalmitate structure $H_{31}C_{15}{-}CO{-}O{-}C_{16}H_{33}$	Cetylpalmitate

36.2.1 Lipids Used in SLN

The choice of lipids is critical for SLN to achieve the desired drug loading capacity, stability, and sustained release behavior. Different lipid phases lead to different apparent partition coefficients and different loading capacities for the same drug. Polymorphism of lipids also affects the properties of a SLN system. In addition, hydrophobicity of lipids varies with the balance of the hydrophobic and hydrophilic functional groups on the lipid molecules. For instance, a fatty acid

TABLE 36.1b
Cationic Lipids Used for Preparation of Cationic SLN

Stearylamine (SA)

N,N-di-(β-steaorylethyl)-*N,N*-dimethylammonium chloride (EQ1)

Benzalkonium chloride (BA)

Cetrimide (CTAB)

Dimethyldioctadecylammonium bromide (DDAB)

N-[1-(2,3-dioleoyloxy)propyl-*N,N,N*-trimethylammonium chloride (DOTAP)

Chloroquine phosphate (CQ)

Cetylpyridinium chloride (CPC)

Dimethyldioctadecylammonium bromide (DDAB)

TABLE 36.2
A List of Surfactants, Co-Surfactants, Counterions, and Surface-Modifying Agents Commonly Used in the Formulation of SLN-Based Systems

Surfactants and Co-Surfactants

Amphoteric surfactants

Egg phosphatidylcholine (EggPC)

Egg lecithin (Lipoid E80)

Soy phosphatidylcholine (SP)

(Epikuron 200, 95% SP)

(Lipoid S100)

(Lipoid S75, 68% SP)

(Lipoid S75, 68% SP)

(Phospholipon 90G, 90%)

Ionic surfactants

Sodium cholate

Sodium cocoamphoacetate

Sodium dodecyl sulfate (SDS)

Sodium glycocholate

Sodium oleate

Sodium taurocholate

Sodium taurodeoxycholate

Non-ionic surfactants

Brij78

Poloxamer 188

Poloxamer 407

Poloxamine 908

Polyglycerol methyl glucose distearate

(Tego care 450)

Solutol HS15

Span 85

Trehalose

Tween 20

Tween 80

Tyloxapol

Co-surfactants

Butanol

Butyric acid

Counterions

Organic salts

Monodecylphosphate

Monohexadecylphosphate

Mono-octylphosphate

Sodium hexadecylphosphate

Ionic polymers

Dextran sulfate sodium salt

Hydrolyzed, polymerized epoxidized soy bean oil (HPESO)

Surface-modifying Agents For long-circulating SLN

Dipalmitoylphosphatidyl-ethanolamine conjugated with polyethylene glycol 2000 (DPPE-PEG 2000)

Distearoyl-phosphatidylethanolamine–*N*-poly(ethylene glycol) 2000 (DSPE–PEG)

stearic acid-PEG 2000

TABLE 36.3

Categories of Solid Lipid Nanoparticles (SLN)

SLN	1. Solid lipid matrix as carrier
	2. Lipophilic drug or hydrophilic drug–ester complex is molecularly incorporated into the solid lipid matrix
NLC	1. Binary mixture of solid lipid and spatially different liquid oil as carrier
	2. Lipophilic drug or hydrophilic drug–ester complex is molecularly incorporated into the solid lipid matrix
PLN	1. Solid lipid as carrier
	2. Hydrophilic drug–polymer complex is incorporated into the solid lipid matrix
LDC	1. Hydrophilic drug–lipid conjugate as the homogeneous matrix

becomes more hydrophobic with the increase in the length of its hydrocarbon chain. A fatty ester is more hydrophobic than the fatty acid of the same chain length because the hydrophilic carboxyl group of the fatty acid is replaced by the more hydrophobic ester group. Triglycerides are more hydrophobic than mono- and di-glycerides because all three hydrophilic hydroxyl groups are substituted by fatty ester. The lipid composition essentially determines the overall hydrophobicity of SLN.

Thus far, there has not been a definite rule available for choice of lipids for SLN preparation though some empirical guidelines are proposed. Measurement of solubility of a given drug was suggested for oral drug delivery by SLN.[33] Empirical equations were used to mathematically predict the partition of a drug between a lipid or oil phase and an aqueous phase based on physical interactions of the solute (drug) and solvents (lipid or aqueous phase). For parenteral SLN formulations, certain requirements need to be met in the selection of lipids such as the suitability of producing particulates in a submicron range, sterilization by autoclaving, long-term storage stability either in dispersion or after lyophilization, and acceptable toxicity.[34]

In general, high drug loading in SLN can only be obtained when the drug has a high solubility in the lipid melt or a high partition coefficient between the molten lipid and aqueous phase. Based on the general principle of like-dissolve-like, Li et al. proposed a method to rationally screen the lipid carrier for SLN to obtain high drug loading capacity by a combination of the theoretical calculations and experimental partition coefficient measurements.[35] The theoretical calculations include total solubility parameter, polarity, and mixing enthalpy. This method has already been used for the screening of 16 lipids from different categories, and it has shown good predictive power.

The occurrence of lipid polymorphism is a phenomenon fairly unique to SLN as compared to other lipid formulations. Lipid polymorphism is referred to the multiple crystalline forms of solid lipids. One of these polymorphic forms, usually the one that forms perfect crystalline lattice, is more thermodynamically stable. For example, triglyceride has three forms, i.e., α-form, β-form, and β'-form. Among them, only the β-form is stable.[36] The relatively less stable or meta-stable forms would eventually transform to the stable polymorphic form, leading to a number of challenges in the development of SLN formulations. Drug molecules are primarily loaded into SLN in the defects of the lattices of solid lipids. When the lipid molecules are converted from the meta-stable forms into the stable form, they become more orderly packed, and some of the defects in the lattices disappear. As a result, drug expulsion from the lipid core to the particle surface may occur, contributing to high initial burst release and drug leakage during storage.

Therefore, one additional criterion for choosing lipids for SLN formulation is the tendency of these lipids to form perfect crystals, or more precisely, their polymorphic transition rates. Once again, there is no strict rule to follow. However, in general, lipids with longer fatty acid chains have a slower transition rate than those with shorter chains.[22] SLN made from waxes as lipid matrix exhibited significant drug expulsion phenomenon because of the more crystalline structure although wax-based SLN are usually more physically stable.[19] In addition to the type of lipids, some other factors may also affect the lipid crystallinity, including the storage condition and SLN production methods.[37–39] For example, rapid cooling process will be beneficial to maintain the lipid matrix in a meta-stable form.[37] The choice of surfactant and co-surfactant plays a crucial role as well.[24] The crystallization rate of Dynasan 114-based SLN was shown much faster when poloxamer was used as the surfactant than those stabilized by the sodium salt of cholic acid.[40] An increased amount of the meta-stable form of stearic acid was also found when a high concentration of co-surfactant such as butanol was included in the SLN preparation.[25]

To avoid the aforementioned crystallinity and polymorphism problems, lipid oils, e.g., Miglyol 812 (caprylic/capric triglycerides) or oleic acid, were incorporated into solid lipids or a binary mixture of physically incompatible solid lipids was used, attempting to disrupt the crystallinity of the solid lipid matrix. The resulting modified form of SLN is often referred to as nanostructured lipid carrier (NLC) (see more description in Section 36.2.4.2).[41–43] This method may also provide additional advantages such as increased drug payload.

Rather than avoiding the complexity caused by lipid polymorphism and crystallinity, some researchers attempted to exploit this phenomenon, and they turned it into a useful feature. For instance, the polymorphic conversion of lipid matrix from the meta-stable β' form to the stable β form, as confirmed by form wide-angle x-ray scattering pattern, altered drug release rate,[44] suggesting that by careful design, release rate of SLN can be modulated by controlling the polymorphic interconversion.

Cationic SLN formulations containing positively charged lipids were developed for gene delivery[45] as the positive surface charge may enhance in vivo transfection efficiency of genes. Some of the catioic lipids are listed in Table 36.1b. From the point of view of cytotoxicity, two-tailed cationic lipids are usually preferable to the one-tailed cationic lipids (i.e., CPC and CTAB) and cationic surfactants.

36.2.2 Surfactants Used in SLN

The main functions of surfactants in SLN are to disperse the melt lipid into aqueous phase in the SLN preparation process and then to stabilize the nanoparticles after cooling. A surfactant molecule has a hydrophilic head and lipophilic tail to reduce the surface tension between two phases. The hydrophile–lipophile balance (HLB) value of surfactants represents the relative proportion of the hydrophilic and lipophilic parts of the molecule.

As illustrated in Table 36.2, surfactants applied in SLN include four categories: anionic, e.g., sodium stearate and sodium dodecyl sulfate; cationic, dodecyltrimethyl ammonium bromide (DTAB); non-ionic, e.g., members of the Pluronics® or Tweens® family; and amphoteric, e.g., egg lecithin. In general, ionic surfactants employ the electrostatic approach, and non-ionic surfactants rely on steric repulsion to stabilize particles. Among them, most non-ionic surfactants consist of a hydrophobic moiety (i.e., hydrocarbon chain) and a hydrophilic part (i.e., ethylene oxide group). Amphoteric surfactants possess both positively and negatively charged groups and exhibit cationic surfactant's feature at low pH and anionic at high pH.

Several factors should be considered when choosing surfactants for SLN preparation, namely HLB value, route of administration, toxicity, and the effect on lipid modification and particle size. Surfactants with HLB values in the range of 8–18 are suitable for the preparation of oil-in-water dispersion. Non-ionic surfactants are preferred for oral and parenteral preparations over ionic ones because of their lower toxicity and irritancy. In general, the order of the toxicity of surfactants is: cationic > anionic > non-ionic > amphoteric.

One important factor in the choice of a surfactant for SLN should be considered: different surfactants have variable influences on the in vivo biodegradation of the lipid matrix. For instance, non-ionic surfactants are more effective for inhibiting the degradation of lipid matrix in vivo.

Poly(ethylene oxide) (PEO) chains of non-ionic surfactants provided steric hindrance of the anchoring of lipase/co-lipase complex. Therefore, in vivo degradation rate of SLN may be tailored by adjusting the density of PEO chains on the particle surface. For SLN made of the same lipid but different non-ionic surfactants, the inhibitory activity against lipid degradation is in the order: cholic acid, sodium salt (NaCh) < lipoid E80 < Tween® 80 < poloxamer 407.[25,40] Another study shows that Dynasan 114-based SLN have the fastest degradation rate when stabilized by lecithin and NaCh; intermediate rate using Tween® 80; and the slowest rate using poloxamer 407.[26] Mixing NaCh and poloxamer does not affect the degradation rate.

Finally, the choice of surfactants and co-surfactants also affects the particle size of SLN. SLN made of the same lipids may have different sizes because of the use of different surfactants.[25] Zhang et al. prepared two types of long-circulating SLN using Brij® 78 or Pluronic® F68 as the surfactant, and they found that different surfactants led to the SLNs with totally different particle size.[46] Bocca et al. showed that the average diameters of SLN systems produced by using glycocholate as the co-surfactant were larger than those using taurocholate.[47] The authors attributed this phenomenon to the difference in pKa values of the co-surfactants, i.e., 4.4 for glycocholate and 1.4 for taurocholate.

36.2.3 Other Agents Used in SLN

In addition to lipids and surfactants (and sometimes co-surfactants), in some modified forms of SLN, other agents may also be used. For example, for the encapsulation of cationic, water-soluble compounds, a source of counterions is needed that includes organic anions and anionic polymers (see Table 36.2).

SLN can also be surface-modified. When a drug carrier was coated with hydrophilic polymers such as poloxamers, poloxamines, or poly(ethylene glycol) (PEG), this so-called stealth or long-circulating carrier may reduce the capture by reticuloendothelial system (RES), and it stays longer in systemic circulation.[48,49] For a detailed description of the stealth technique, refer to other chapters in the book that deal with long-circulating nanoplaforms. This technique has been equally valuable to minimize clearance of SLN by phagocytosis.[47] Because of the high surface hydrophobicity of SLN, the hydrophilic coating agents are usually pre-conjugated to lipid moieties to form amphiphilic molecules (see Table 36.2) so their attachment on to the particle surface can be more secure. For the effects of SLN surface modification on the pharmacokinetics and biodistribution of anti-tumor agents, refer to Section 36.5.6.

36.2.4 Types, Proposed Structure, and Morphology of SLN

The structure and morphology of SLN are affected by several factors such as the physicochemical characteristics of the SLN ingredients and the production method adopted. To date, there has been no concrete data revealing the detailed structure of SLN. Most SLN researchers believe that SLN are probably spherical because a spherical subject has the smallest possible surface area-to-volume ratio. This may minimize the surface tension between the lipids and the aqueous phase, thereby leading to thermodynamic stability. Nevertheless, it should be noted that the platelet shape of SLN and NLC was also claimed by some groups, and they have provided evidence as well.[50,51]

36.2.4.1 Classical SLN

Based on the difference of melting points (MP) between drug and lipid matrix together with the consideration of the release kinetics of SLN, Mühlen et al. proposed three hypothetical models for classical SLN: drug-enriched core, drug-enriched shell, and solid solution[52] as illustrated in Figure 36.1. Atomic force microscopy (AFM) and DSC studies of prednisolone-loaded SLN showed that the particles had an inner crystalline core (hard) and an outside amorphous layer (soft).[53] The core-enriched model is based on the fact that the compound with a higher MP will precipitate first during cooling. In contrast, if the drug has a lower MP than the lipid matrix,

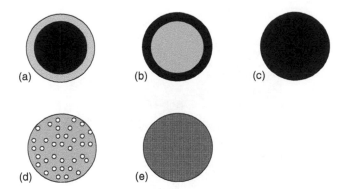

FIGURE 36.1 Proposed structures of solid lipid nanoparticles (SLN): (a) drug-enriched core model; (b) drug-enriched shell model; and (c) drug molecularly dispersed model. (Adopted from zur Mühlen, A., Schwarz, C., and Mehnert, W., *Eur. J. Pharm. Biopharm.*, 45, 149–155, 1998.); and nanostructured lipid carrier (NLC): (d) micro-compartment model; (e) liquid oil molecularly dispersed model.

a drug-enriched shell model will fit the particles. If the drug has a similar MP to the lipid matrix, a solid solution model will best describe the particle. Although there is no formal study on the structure of LDC, it should be similar to classical SLN with the loaded drug evenly distributed within the lipid matrix because they are directly bonded to the lipid molecules.

36.2.4.2 Nanostructured Lipid Carrier (NLC)

To overcome the difficulties of classical SLN, NLC was introduced. In this case, the lipid matrix is composed of the binary mixture of a solid lipid and the medium chain triglycerides or liquid oil.

Based on the composition of the lipid matrix of NLC, Jenning et al. proposed two drug incorporation models[54] as demonstrated in Figure 36.1. In the first model, liquid oil is molecularly dispersed within the solid lipid matrix when the concentration of the liquid oil is below its solubility in the lipid. In the second model, liquid oil is distributed in the solid lipid in droplet form. Physically separate phases (solid, liquid) coexist in this model.

36.2.4.3 Polymer–Lipid Hybrid Nanoparticles (PLN)

In order to deliver the hydrophilic drugs with a high drug loading capacity and simultaneously control the release kinetics of SLN, a new variation of SLN, PLN was recently developed.[55,56] The schematic illustration of proposed the formation mechanism and structure of PLN is presented in Figure 36.2. An ionic drug, e.g., doxorubicin HCl, forms complex with a counter ionic polymer, e.g., dextran sulfate that then partitions into the lipid phase because of higher hydrophobicity, enabling incorporation of water soluble drugs in the lipid nanoparticles. PLN have a spherical

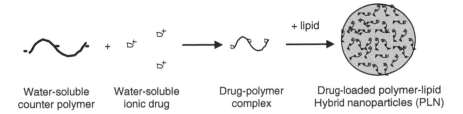

FIGURE 36.2 Proposed formation mechanism and structure of polymer–lipid hybrid nanoparticles (PLN).

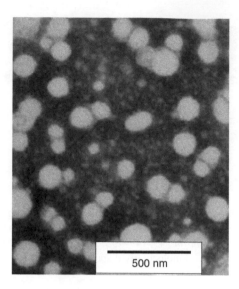

FIGURE 36.3 Transmission electron microscope (TEM) image of polymer–lipid hybrid nanoparticles (PLN) loaded with doxorubicin. (From Wong et al., *Pharm. Res.*, in press.)

morphology as shown in Figure 36.3. Modest initial burst release and subsequent sustained release profiles have been observed using this system. However, if the drug–polymer complex is uniformed distributed in the solid lipid matrix as shown in Figure 36.2 or just attached to the surface of particles needs further verification.

Li et al. applied differential scanning calorimetry (DSC) and powder x-ray diffraction (PXRD) crystallography to examine crystallinity and crystal structures of PLN made of verapamil HCl, dextran sulfate, and dodecanoic acid.[35] The DSC and the PXRD results (Figure 36.4) showed that

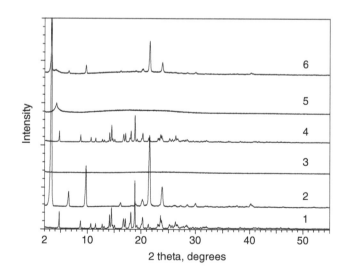

FIGURE 36.4 Overlaid powder x-ray diffraction crystallographs of (1) verapamil HCl (VRP); (2) dodecanoic acid; (3) dextran sulfate sodium salt (DS); (4) physical mixture of verapamil HCl and dextran sulfate sodium salt; (5) VRP–DS complex; and (6) VRP–DS complex incorporated dodecanoic acid. (From Li et al., *Pharm. Res.*, 2005.)

the crystal structure of the lipid remained, but its melting temperature decreased a bit as the drug became amorphous in the complex form. This finding suggests that the presence of drug–polymer complex in PLN may be similar to the liquid oils that reduce the lipid crystallinity in NLC. Nevertheless, the exact significance of the finding will require further investigations.

36.3 MANUFACTURING METHODS

Currently, many techniques are available for SLN preparation. These include high pressure homogenization,[57] microemulsion,[58] and solvent diffusion.[59,60] The more commonly used methods will be briefly described in this chapter. For detailed descriptions of SLN preparation, please refer to the up-to-date reviews[30,38,57–60] where the optimization of the processing variables is also discussed.

36.3.1 HIGH PRESSURE HOMOGENIZATION

High pressure homogenization methods include hot and cold approaches. The main advantages of these methods include narrow size distribution of nanoparticle produced and the possibility of scale-up production. The cold approach minimizes the exposure of drug to heat (although brief period of heating is still required) and is more suitable for labile drugs sensitive to heating. However, this approach requires more tedious steps, and it results in particles of more variable sizes. The general scheme of high pressure homogenization method is shown in Figure 36.5 (hot method) and Figure 36.6 (cold method).

36.3.2 MICROEMULSION

A warm microemulsion is prepared under stirring before it is dispersed in a large volume of cold water at 2–4°C. The typical volume ratio of emulsion to cold water is in the range of 1–20. The procedure of microemulsion method for SLN preparation is illustrated in Figure 36.5.

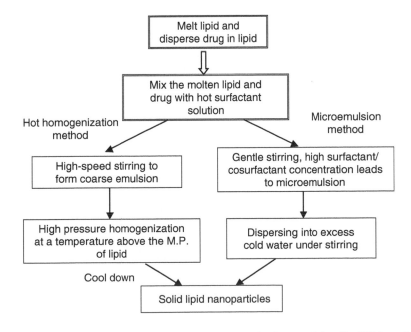

FIGURE 36.5 Schematic procedures of hot homogenization and microemulsion for SLN production.

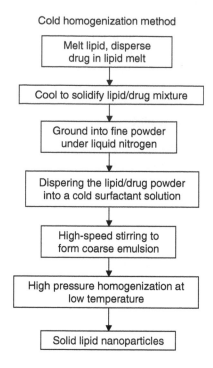

FIGURE 36.6 Schematic procedure of cold homogenization for SLN production.

Using this method, it is noted that the particle size, size distribution, and stability of SLN is correlated with the properties of microemulsion such as the lipids and surfactants applied, the concentration of surfactant, and the drug/lipid ratio.

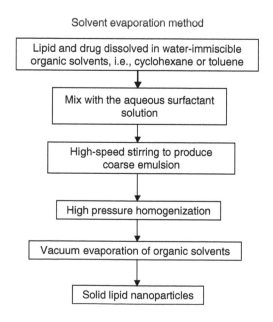

FIGURE 36.7 Schematic procedure of solvent evaporation method for SLN production.

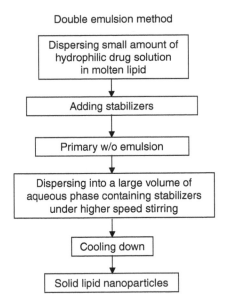

Double emulsion method

FIGURE 36.8 Schematic procedure of w/o/w double emulsion method for SLN production.

36.3.3 Solvent Evaporation

As demonstrated in Figure 36.7, the lipid is dissolved in water-immiscible organic solvent first, and then the hydrophobic drug is added. Dispersing the drug–lipid–organic solvent mixture into an aqueous solution of emulsifiers produces an o/w emulsion. SLN are obtained by removing the organic solvent with evaporation under low pressure.

This method totally avoids heating. Its main disadvantage is the inclusion of organic solvent in preparation. Residues of organic solvent may raise the bio-safety concern.

36.3.4 w/o/w Double Emulsion Method

This method is designated for the delivery of hydrophilic drugs by using SLN.[61] Highly concentrated drug solution is dispersed into a lipid phase under vigorous stirring and stabilized by a surfactant to form the primary w/o emulsion. The primary emulsion is subsequently dispersed into a large volume of aqueous solution containing another kind of surfactant to form w/o/w emulsion. The nanoparticles are obtained by ultrafiltration or solvent evaporation. The detailed procedure is illustration in Figure 36.8.

36.4 CHARACTERISTIC PROPERTIES OF SLN

36.4.1 Drug Loading

For classical SLN, NLC, and PLN, the drug loading process involves partitioning and re-partitioning of the drug between the lipid and aqueous phase during the heating and subsequent cooling process. No covalent bonding or ionic bonding between the molecules of the drug and lipid occurs. On the other hand, drug incorporation by LDC, another modified form of SLN, involves direct covalent or ionic bonding of the drug molecules to the lipids.

Two models are currently proposed to describe how drug molecules are incorporated in SLN, NLC, and PLN. In the first model, drug molecules are accommodated in the flaws of the lipid crystal lattice. This model has been widely used to interpret the drug loading behaviors and the biphase

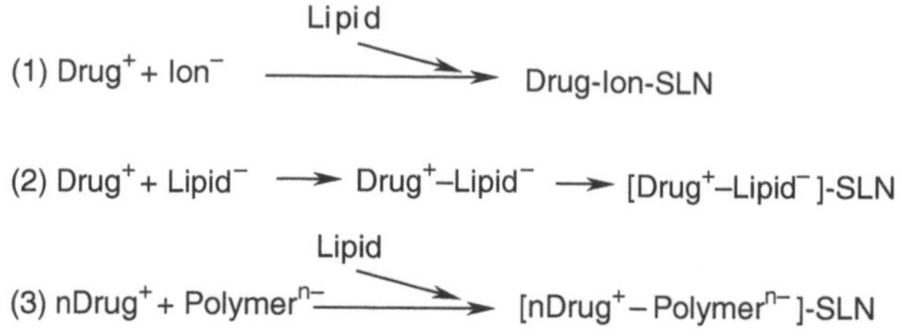

$$Lipid$$
$$(1) \; Drug^+ + Ion^- \longrightarrow Drug\text{-}Ion\text{-}SLN$$

$$(2) \; Drug^+ + Lipid^- \longrightarrow Drug^+\text{--}Lipid^- \longrightarrow [Drug^+\text{--}Lipid^-]\text{-}SLN$$

$$Lipid$$
$$(3) \; nDrug^+ + Polymer^{n-} \longrightarrow [nDrug^+ - Polymer^{n-}]\text{-}SLN$$

FIGURE 36.9 Strategies to load cationic, water-soluble anti-tumor compounds into SLN-based systems. (1) Ion-pair formation method; (2) lipid drug conjugate (LDC); and (3) polymer–lipid hybrid nanoparticles (PLN).

release kinetics of SLN and NLC. In the second model, the physical and chemical compatibilities (affinity, miscibility) between the drug–polymer complex and the solid lipid phase are used to predict the drug loading by PLN.[35]

For water-soluble, ionic anti-tumor compounds, to allow their efficient encapsulation by SLN, three possible strategies have so far been employed. These strategies include mixing the anti-tumor compound with an organic salt carrying opposite charges and packaging the resulting ion-pairs into lipids;[62–65] complexing the ionic anti-tumor compound to lipids carrying opposite charges and directly using the drug–lipid complexes for SLN preparation (i.e., LDC preparation); and complexing the ionic anti-tumor compound to polymer molecules carrying opposite charges, i.e., polyelectrolytes and dispersing these drug–polymer complexes into lipids to form SLN (i.e., PLN preparation). Figure 36.9 summarizes these three feasible strategies. These strategies are similar in terms of their underlying principle—to neutralize the charge on the drug with a counter ion to facilitate drug partitioning into the lipids.

When an organic salt or ionic polymer interacts with the drug molecules of opposite charge, the resulting complex loses charges and gains in hydrophobicity that favors their incorporation into the solid lipid matrix. The ionic molar ratio between the organic salt or polymer and the drug plays a critical role in this process. By employing isothermal titration calorimetry in conjunction with partition experiments, Li et al. found that at an equal ionic molar ratio of verapamil HCl (VRP) to dextran sulfate (DS), the VRP-DS complex reached maximum hydrophobicity and partition in the lipid phase.[35]

36.4.2 Drug Release

The release kinetics of SLN usually exhibit a biphasic feature: a strong initial burst release (usually > 50% of the loaded drug in the first several minutes or hours) followed by a subsequent prolonged release. It is a consensus that the release behavior of SLN is determined by the physicochemical properties of the constituents and the morphology of SLN. By correlating the drug release kinetics to the physicochemical properties of the drug and lipids, the hypothetical structures of SLN can be deduced (as mentioned in Section 36.2.4.1). In general, sustained release profiles are correlated to the drug-enriched core model, and both the drug-enriched shell model and the solid solution model will have fast release kinetics.

High drug loading capacity can be achieved in NLC. However, the existence of the micro-compartments of oil droplet in the solid lipid matrix may compromise the control over the release kinetics. The control of release can be improved in the NLC model with liquid oil molecularly dispersed in the solid lipid matrix.

In comparison to SLN and NLC, PLN offer an additional mechanism to regulate the drug release kinetics. In addition to the regular diffusion control, the need for ion exchange between the drug–polymer complex and the surrounding medium for the drug to be released may also be exploited for better drug release control. The synergistic effect of both approaches will possibly lead to reduced

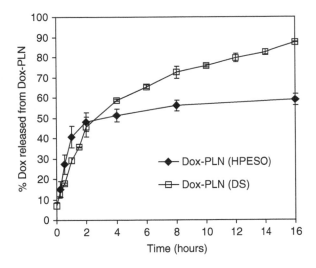

FIGURE 36.10 Cumulative percent of doxorubicin released versus time for PLN in 0.05 M CaCl$_2$ solution at 37°C. (◆) Dextran sulfate (DS) sodium salt as the counter polymer; and (▱) hydrolyzed polymer of expoxidized soybean oil (HPESO). (Adapted from Wong et al., *Pharm. Res.*, in press.)

initial burst release and a subsequent sustained release. Figure 36.10 presents representative release profiles of PLN. Moderate initial releases (50% w/w of drug released in first few hours in both dextran sulfate-based or HPESO-based PLN loaded with doxorubicin) were achieved when 0.05M CaCl$_2$ was used as the release medium, and sustained releases were observed in the late phase of release.

36.4.3 STABILITY OF SLN

The practical usage of SLN depends on their long-term physical and chemical stability. Typically, the shelf life of one viable formulation should be at least one year. The criteria for the assessment of physical stability of SLN include average particle size, zeta potential, and drug leakage during storage. For labile compounds, it is critical to evaluate their chemical stability as well.

The physicochemical properties of the materials used to prepare SLN such as the lipids, drugs, and surfactants play a crucial role in determining the stability of SLN. Waxes usually lead to slower particle growth and aggregation than glycerides.[19] Glycerides that include a fraction of monoglycerides are usually more stable than those without (e.g., Dynasan 116 > Compritol 888 ATO > Imwitor 900). Ionic surfactants often stabilize SLN better than non-ionic surfactants.

In addition to the physicochemical properties of the SLN material, the stability of SLN is also strongly affected by the various processes the SLN subject to after their preparation. For instance, sterilization is strictly needed for parenteral SLN formulation. When compared to most other drug delivery systems, SLN are surprisingly stable when challenged by the harsh conditions (compressed, saturated steam at 121°C) during autoclaving. Cavalli et al. found no significant changes occurred for SLN in terms of the particle size, polydispersity (PI) index, and zeta potential before and after sterilization.[66] However, there are conflicting reports. Venkateswarlu et al.[39] demonstrated an increase of SLN size almost 2–3-fold, a shift of zeta potential from positive to negative, and a slight drop in drug loading after autoclaving of SLN. The differences between the two cases are probably attributed to the different lipids, surfactants, and drugs used in their studies.

Lyophilization is another process that has a strong influence on the formulation stability. Lyophilization, or freeze drying, is necessary to prevent SLN aggregation or hydrolysis in aqueous medium. In general, carbohydrate cryoprotectants such as trehalose may stabilize a SLN formulation during lyophilization.[67] Proper freezing velocity and re-dispersion method (e.g., sonication) are also required. It has been demonstrated that with the optimized parameters for lyophilization, the reconstituted SLN are feasible for i.v. injection;[68] although, there are also studies showing that some deterioration in the particle quality, particularly the particle size, may occur.[69] Even in these studies, the reconstituted SLN remained sufficiently small in size and stable for oral administration.

During storage, the stability of SLN will also be affected by the energy introduced from outside sources such as storage temperature, light, and packing materials.[70] Strong light and high temperature will accelerate the decrease of the particles' zeta potential and, in turn, induce particle aggregation because colloids with zeta potential less negative than -15 mV are prone to aggregation. Some packing materials such as siliconized vials also exhibit statistically significant effects on the particle size growth in comparison to that packed in untreated glass vials.

Most aforementioned stability data derive from non-biological studies. There have been more in vivo SLN studies recently on this issue. This is important because the environment of the systemic circulation is actually harsh on SLN. For example, the electrolytes in the plasma will lead to the shrink of double layer, the rapid fall off of zeta potential, and the subsequent coagulation of particles occurs.[71] Particles may also grow in size and subsequently form semi-solid gels through adding mono-, di-, and trivalent ions.[72] In addition, the biodegradation of the lipid matrix caused by enzymes in vivo will accelerate drug release from SLN and expose the agents to the various degradative systems before they can reach the disease site. In brief, even when performing SLN studies in vitro, it will be advisable to at least use media that simulate the composition of body fluids such as Hanker's buffer salt solution (HBSS) for the experimental works. This may help the researchers to gain insight into the behaviors of SLN in vivo.

Finally, it must be re-emphasized that the various properties of SLN are strongly affected by the occurrence of lipid polymorphism, and these include stability as well. During storage of SLN, drug expulsion and subsequent drug leakage may occur as a result of lipid polymorphic transition, causing re-distribution of drug molecules within particles. For details of polymorphism, refer to Section 36.2.1. All in all, the factors that accelerate particle size growth as a result of the loss of surface charge, including high storage temperature and strong light, will also speed up polymorphic transition rate.

36.5 SLN FOR ANTI-TUMOR DRUG DELIVERY

36.5.1 CANCER CHEMOTHERAPY

Although hormonal therapy and immunotherapy are useful for certain types of cancer, chemotherapy with cytotoxic drugs remains the most established and commonly used form of drug therapy. Sometimes, chemotherapy with cytotoxic drugs is the only treatment a patient receives. More often, cytotoxic drugs are used in addition to other modalities (e.g., surgery, radiotherapy) to improve their effectiveness and prevent cancer recurrence.[73] Until recently, most of the SLN systems for cancer treatment are designed for cytotoxic drug delivery. Therefore, unless otherwise specified, in the following discussions the terms *cytotoxic drug*, *anti-cancer drug* and *anti-tumor* drug will be interchangeably used, even though the latter two terms actually carry broader meaning.

Cytotoxic drugs treat cancers by preferentially causing death or arrest in growth of cancer cells. However, their cytotoxicity is not highly specific for cancer cells. The healthy cells, especially those rapidly dividing, almost inevitably suffer from the chemotherapeutic treatment as well.[74] The major goal in cancer chemotherapy is, therefore, to achieve reasonably high drug concentration for effective cancer cell kill without causing strong toxicity. In addition, clinical resistance to chemotherapy drug therapy often occurs in cancer management.[75] One prominent form of drug resistance is related to the expression of ATP-dependent membrane-associated drug transporters in cancer cells. Cells with over expression of these drug transporters are resistant to a broad range of structurally diverse anti-cancer compounds, a phenotype often referred to as multi-drug resistance

(MDR).[75,76] Cytotoxic drugs can be actively transported out of the cancer cells by these transporters, rendering the chemotherapy ineffective even when an adequate dose of drug is administered.

36.5.2 Delivery of Cytotoxic Compounds Using SLN

SLN have been quickly gaining ground as a carrier of anti-tumor drugs.[10,11,55,62,77–90] Some of these SLN systems have been evaluated for their anti-tumor activities, stabilities, and/or biodistribution in biological systems such as cultured cell lines and animal models. These systems and

TABLE 36.4
SLN Formulations Used for Delivery of Drugs with Anti-Cancer Properties and a Brief Summary of Their Works

Drugs	Groups[a]	F/C[b]	In Vitro	In Vivo	P/B[c]	Stealth	Comments/References
			Focus of Studies				
				Poorly Water-Soluble			
Camptotecan	Yang	Y			Y(mice)		Stability study/[91,92]
Cholesteryl butyrate	Serpe[73]		Y				In colon cancer cell line/[73]
Etoposide	Murthy	Y			Y		Routes of admin compared/[93]
Irinotecan	Unger	Y		Y	Y		SN38, irinotecan prodrug used, stability studied/[94,95]
Paclitaxel	Gasco	Y					
	Lee	Y	Y			Y	Paclitaxel prodrug used/[97]
	Muller	Y					[98]
	Serpe		Y				In HT28 colon cancer cell line/[73]
	Zhang	Y			Y(mice)		2 formulations using different surfactants compared/[99]
Retinoic acid	Kim		Y				In various human cancer cell lines/[90]
				Ionic, water-soluble			
Idarubicin	Gasco	Y			Y (rabbit)		Duodenal route/[101]
Doxorubicin	Gasco	Y			Y (rat), (rabbit)	Y	[8,100,102]
	Serpe		Y				In colon cancer cell line/[73]
	Wu	Y	Y				In MDR breast cancer cells, cytotoxic-chemosensitizer combination/[10,60]
				Non-ionic, water-soluble			
5-Flurouracil	Wang	Y			Y(mice)		FudR, derivative of 5-fluouracil used/[103]

[a] According to corresponding author.
[b] F/C- formulation and characterization.
[c] P/B Pharmacokinetics and/or biodistribution studies presented—pharmacokinetics.

related work are listed in Table 36.4. To this point, studies of SLN systems have demonstrated the following properties of SLN that are desirable for anti-tumor drug delivery:

1. Versatility and modifiability of the carrier that allow encapsulation of cytotoxic agents of vastly diverse physicochemical properties;
2. Improved anti-cancer drug stability;
3. Capability to load more than one drug in one carrier system that opens up the possibility of a drug combination delivery (e.g., a cytotoxic agent and a chemosensitizer);
4. Improved in vitro cytotoxicity against cancer cells, including those normally refractory to chemotherapy;
5. Enhanced drug efficacy in animal models; and
6. Improved pharmacokinetics and in vivo drug distribution.

In the following sections, the use of properties (1) and (2) by SLN for lipophilic anti-tumor drug delivery will be jointly discussed in Section 36.5.2.1, and Section 36.5.2.2 through 36.5.2.5 will focus on properties (3) to (6), respectively.

Anti-tumor drugs can be categorized based on their solubility properties. In general, the majority of anti-tumor agents, particularly the cytotoxic compounds, are fairly lipophilic and poorly water-soluble.[91] There are several exceptions. Some anti-tumor drugs are water-soluble because their molecules are small and contain multiple hydrophilic groups. Examples of these water-soluble drugs include 5-fluorouracil (5-FU) (Figure 36.11a, MW $= 130.1$) and mito-mycin-C (MW $= 334.3$). In addition, there are some lipophilic anti-tumor molecules that can be converted into water-soluble, ionic salts because of their ionizable basic functional groups. These salts are much more commonly used in clinical settings than their corresponding water-insoluble free bases because they can be conveniently dissolved and administered using standard aqueous diluents such as 5% dextrose and 0.9% saline. For example, the hydrochloride salts of anthracy-clines, e.g., doxorubicin and idarubicin (Figure 36.11b), and sulfates of alkaloids, e.g., vincristine and vinblastine, are all commercially available and frequently used in cancer chemotherapy.

As previously discussed, the efficiency and extent of drug incorporation into SLN are strongly influenced by the partition behavior of the drug in a lipid–water two-phase system. The lipophilic, uncharged anti-tumor drugs may be encapsulated with relative ease using classical SLN preparation methods. The other two classes (uncharged, hydrophilic compounds and charged, lipophilic compounds) of water-soluble anti-tumor agents, on the other hand, require new strategies to enhance their incorporation into SLN. Means to control the release rates of these compounds

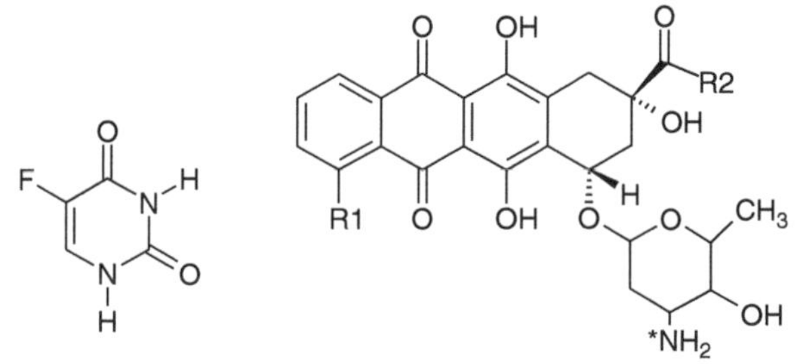

(a) 5-fluorouracil   (b) Doxorubicin /Idarubicin

FIGURE 36.11 Examples of the molecular structures of two different types of water-soluble anti-cancer compounds. (a) 5-fluorouracil (5-FU); and (b) doxorubicin ($R_1 = OCH_3$, $R_2 = CH_2OH$) or idarubicin ($R_1 = H$, $R_2 = CH_3$). * Basic amine group that is protonated when forming the water-soluble hydrochloride salt of doxorubicin or idarubicin.

from SLN are also necessary because they may rapidly diffuse from the drug carriers into the surrounding aqueous medium.

36.5.3 Poorly Water-Soluble Compounds

36.5.3.1 General Principle

There have been several SLN systems successfully formulated for encapsulation of poorly water-soluble anti-tumor compounds. These include camptothecin-based topoisomerase I inhibitors, including camptothecin and irinotecan and their prodrugs; taxanes, mainly paclitaxel; and other lipophilic compounds with anti-tumor properties. In general, these poorly water-soluble compounds partition reasonably well in the lipid phase and can be efficiently encapsulated by the conventional SLN without much nanoparticle modifications required. These agents, however, carry other problems such as poor stability and likelihood of precipitation. In fact, each group of the compounds has its specific problems to be solved before they can be properly delivered using SLN.

36.5.3.2 Topoisomerase I Inhibitors (Camptothecin, Irinothecan)

Camptothecin is a plant alkaloidal compound extracted from *Camptotheca acuminata*.[92] It serves as the prototype compound of a relatively new class of anti-tumor agents known as topoisomerase I inhibitors (e.g., irinotecan, topotecan). The molecules of camptothecin and its related compounds all possess extensive aromatic structures and are quite lipophilic (Figure 36.12a).[93] The anti-tumor activity of camptothecin is strongly correlated to the functionality of the lactone ring of the drug molecule. However, in an aqueous environment, particularly at basic pH, this lactone ring is vulnerable to hydrolysis, leading to the formation of therapeutically inactive carboxylate. The poor water solubility and chemical stability of this drug has seriously diminished its practical value. These two shortcomings can both be overcome by the use of SLN for their capabilities to efficiently load lipophilic compounds and provide a lipid environment to prevent hydrolytic degradation. Yang et al.[79] described a SLN system for camptothecin delivery. A very high encapsulation efficiency (99.6%) of camptothecin was achieved. The same study also showed that the drug mostly remained in its active lactone form until it was released, confirming the protective effect of SLN against hydrolytic degradation.

Irinotecan (or Camptosar or CPT-11) is a newer member in the topoisomerase I inhibitor family. It is relatively hydrophilic and stable compared to camptothecin and has been put to clinical use for colon cancer chemotherapy. However, compared to its parent drug SN-38 (Figure 36.12b), the anti-tumor activity of irinotecan is nearly a thousand-fold weaker.[94] SN-38 is prevented from being put to clinical use mainly because of its extreme hydrophobicity. Moreover, hydrolytic opening of the labile lactone ring of SN-38 and irinotecan is still possible, though less likely. To

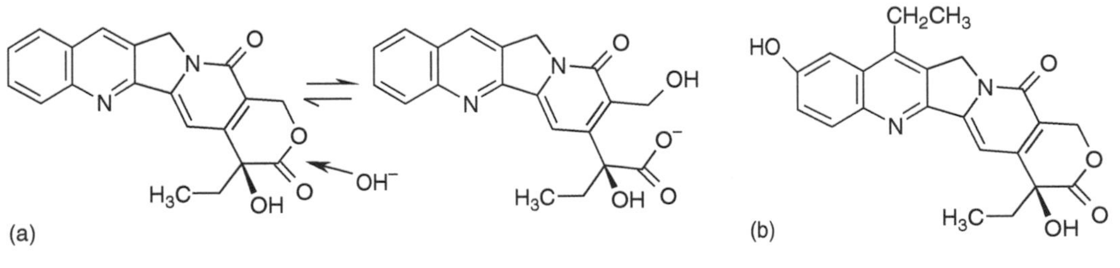

FIGURE 36.12 Molecular structures of two topoisomerase I inhibitors. (a) The hydrolytic degradation of camptothecin from its active lactone form (left) to its inactive carboxylate form (right). (b) SN-38, a more potent lipophilic parent drug of irinotecan.

improve its water solubility and chemical stability, SLN formulation of SN-38 was prepared.[81] The concentrations of SN-38 in the SLN formulation reached as high as 1.0 mg/ml. The lactone ring stability of the compound was also improved. After 3 h incubation in human serum albumin, the lactone concentration in SN-38 loaded-SLN remained 80%. This was superior to the retentions of only 40% lactone for unformulated SN-38 and irinotecan and even more so when compared to free camptothecin that lost more than 95% of its active lactone form in 90 min.

In brief, SLN are a useful drug carrier for camptothecin-based anti-tumor compounds. Both of the studies mentioned above have demonstrated the value of SLN for drug solubilization and preservation of drug stability.

36.5.3.3 Paclitaxel

Paclitaxel (Taxol) is an anti-microtubule agent with broad spectrum anti-tumor activities. It is gaining popularity among oncologists for its effectiveness against several types of malignancies, even when used alone.[95] Paclitaxel is a poorly water-soluble drug. At present, it is commercially available as a non-aqueous micellar solution containing a polyoxyethylated castor oil Cremophor EL and 49.7% dehydrated ethanol. Cremophor EL is known to cause serious hypersensitivity reactions and nephrotoxicity in human subjects.[96] In order to solubilize paclitaxel without the need for Cremophor EL, a few SLN formulations have been developed and studied.[62,82–86] In general, the drug loading in these SLN formulations ranged from 2 to 5%. In these SLN formulations, surfactants or stabilizers such as Pluronic® F68 (i.e., poloxamer 188), Brij-78, and phosphatidylcholine were used. In comparison to Cremophor EL, all of these excipients are commonly included in parenteral formulations with better track records of safety.

Another major issue associated with the conventional paclitaxel formulation is its poor physical stability. Paclitaxel is routinely diluted to a concentration of 0.6–1.2 mg/ml with saline or dextrose for clinical use. The dilution can destabilize the formulation and drug precipitation frequently occurs, necessitating filtration of the drug solution prior to administration. This extra step adds to the labor cost and may affect the drug concentration. This problem is apparently not found in paclitaxel-SLN formulations. In one of these studies,[62] the SLN formulation of paclitaxel was even diluted to nanomolar range using cell culture medium for in vitro cytotoxicity evaluation. There was no report of drug precipitation.

Drug precipitation also happens when paclitaxel is too quickly discharged from a carrier system and the concentration of the released drug exceeds the saturation point. One solution is to limit the rate of drug release. As demonstrated in the paclitaxel-loaded SLN system by Cavelli et al.,[82] no drug precipitation was observed because only 0.1% of paclitaxel was released in pseudo-first order in 2 h.

Besides SLN, researchers have also successfully encapsulated paclitaxel in other drug delivery systems, particularly those that are lipid emulsion based.[97–99] However, all of these formulations are prone to drug precipitation during storage. It is believed that paclitaxel has limited solubility in lipid as well, and the limited solubilities of paclitaxel in both aqueous and lipid phases leads to the strong tendency of drug precipitation when the formulations are stored for a long term. Stevens et al.[84] described the use of a lipophilic paclitaxel prodrug (paclitaxel-2'-carbonyl-cholesterol) to overcome this difficulty. Despite the strong lipophilicity of the prodrug, this group was able to demonstrate over 10% of drug release into bovine serum albumin in a relatively linear fashion within 2 h. In addition, even without resorting to the use of prodrug, Cavalli et al. showed that their formulation containing the unaltered paclitaxel was autoclavable at 121°C without a significant reduction in the amount of the incorporated drug.[82] The same formulation was also stable for over 18 months after lyophilization, and the lyophilized nanoparticles appeared easily dispersable when cryoprotectant (2% trehalose) was added.

Therefore, the main shortcomings of paclitaxel, including a high risk of hypersensitivity and frequent drug precipitation, are correctable using SLN. A number of new, more cancer-type-specific paclitaxel analogs have been synthesized in the past decade.[100,101] It will be interesting to see if SLN are equally useful for these compounds.

36.5.3.4 Other Poorly Water-Soluble Anti-Tumor Compounds (Etoposide, All-Trans Retinoic Acid, Butyric Acid)

Etoposide, all-trans retinoic acid (ATRA), and butyric acid are all lipophilic molecules that have anti-tumor properties. Out of these three compounds, etoposide (Vepesid, VP-16) has the longest history in cancer chemotherapy and is still in clinical use for the treatment of several forms of cancer. Etoposide is a plant alkaloid. It inhibits topoisomerase II and produces reactive derivatives that can lead to DNA strand break.[102] This drug was successfully encapsulated in SLN made of tripalmitin.[80] The authors quoted that the particles were stable after four months of preparation with excellent redispersibility. However, probably because the focus of that study was on the evaluation of drug biodistribution and in vivo anti-tumor efficacy, few details in this area was revealed. The SLN system demonstrated an excellent encapsulation efficiency of 98.96% with 4% payload of etoposide obtained.

Both ATRA and butyric acid remained at the experimental stage in terms of cancer therapy. ATRA is well documented for its chemical instability.[103] It is sensitive to heat, light, and oxidation, and it may isomerize into isotretinoin or be oxidized to form inactive products such as all-trans-4-oxo. This compound should be well incorporated into SLN for its high lipophilicity; although, in the only SLN system documented, the drug payload attempted was merely 2.5 mg ATRA per g of SLN powder.[77] The chemical stability of ATRA in SLN during storage was evaluated. More than 90% of the encapsulated drug remained intact after one month of storage at 4°C versus less than 60% when the drug was stored in the forms of methanol sol ution or Tween 80 solution under the same conditions. Butyric acid is a short chain fatty acid that shows in vitro inhibitory effect against colon cancer cell growth. However, its in vivo activity is very low because of its rapid metabolism.[99] It was believed that by using its prodrug cholesteryl butyrate and by delivering this prodrug using SLN, the anti-tumor activity could be preserved. The prodrug was successfully loaded in SLN, and the prodrug demonstrated good in vitro activity.[62,104,105] Although the in vivo activity has not yet been confirmed, the studies provided another example of the value of SLN to protect a labile chemical.

36.5.4 WATER-SOLUBLE IONIC SALTS

36.5.4.1 General Principle

The anti-tumor drug salts typically have lipophilic molecular structures. The main obstacle against these salts from efficient loading into SLN is their ionic charges. For details of the drug loading mechanism of this class of compounds into SLN-based systems, refer to Section 36.4.1. In essence, it is about neutralizing the charge on the ionic drug salt with a counter ion. It may be argued that these extra steps can simply be avoided by using the free bases of these agents. However, the use of free base compounds will essentially lead to the same scenario described previously in Section 36.5.1.1. These poorly water-soluble free base compounds will likely be very slowly released, and the released drug concentrations will not be as high as compared to when the salt forms are used. The free base compounds may also require dissolution in organic solvents first during the preparation process to allow even mixing with the lipids. In general, it is preferable to use the water-soluble salt of an anti-tumor drug for SLN formulation whenever it is available.

36.5.4.2 Doxorubicin and Idarubicin

Although there are quite a few anti-tumor drugs available in ionic salt forms for improved water solubility, only two of the anthracyclines have been studied for delivery by SLN-based formulations. These include doxorubicin and idarubicin. These two compounds are similar in structures except that doxorubicin is more hydrophilic because of the presence of additional polar substituent groups (Figure 36.11b). Each of these two drug molecules carry an amine group and can be converted into hydrochloride salts. To facilitate the loading of these salts into SLN, they need to be made lipid-soluble based on one of the above-mentioned strategies. Currently, there is still no LDC system designed for anti-tumor drug delivery. The other two strategies described in Section 36.4.1 (ion-pair, PLN) have both been used and tested.

Gasco's group used decyl phosphate or hexadecyl phosphate to form ion pairs with doxorubicin hydrochloride and idarubicin hydrochloride to enhance their loading of into SLN made of stearic acid and egg lecithin.[87] The ion pair formation resulted in increases in lipophilicity as defined by apparent partition coefficients between water and stearic acid at 70°C by more than 1000-fold for doxorubicin and more than 300-fold for idarubicin. SLN prepared using these lipophilic ion pairs carried payloads of doxorubicin and idarubicin up to 7% and 8.4%, respectively. Probably because of the high lipophilicity, the drug also released very slowly. For both drugs, less than 0.1% drug was released after 2 h. With these release rates, there lies a theoretical risk that only the cancer cells in physical contacts with the SLN may be exposed to sufficiently high drug concentrations and properly treated. In addition, as demonstrated in several studies, chronic exposure to sub-lethal concentrations of cytotoxic drugs can induce expression of Pgp.[106] Cancer cells that are farther away from the SLN could possibly develop resistance to the future chemotherapeutic treatment.

Another feasible approach to encapsulate anthracycline salts is to prepare PLN instead. With the inclusion of dextran sulfate to provide counterions, PLN of doxorubicin hydrochloride were successfully prepared.[55] The encapsulation efficiency of doxorubicin in the PLN, depending on the drug payload, was generally over 70% in the presence of dextran sulfate versus approximately 40% in its absence. More recently, another doxorubicin-loaded PLN formulation that used a more lipophilic HPESO polymer (hydrolyzed and polymerized epoxidized soybean oil) was formulated.[10] Payload over 6% was achieved. The drug releases from both PLN formulations were fast compared to ion pair-based SLN (see Figure 36.10). The lower drug encapsulation efficiency and faster drug release compared to the ion pair-based SLN indicate that unlike ion pair formation, the drug–polymer complexation probably does not completely neutralize all of the charges on the polymer molecules. These residual charges might help draw the water molecules into the lipid matrix to accelerate its disintegration, and they may lead to faster and more complete drug release from the nanoparticles. If this will lead to more effective cancer therapy still requires further investigations.

36.5.4.3 Water-Soluble, Non-Ionic Drug Molecule

36.5.4.3.1 General Principle

All of the strategies mentioned in Section 36.4.1 rely on neutralization of the charge on the drug molecules. For small, non-ionic hydrophilic molecules, these strategies cannot be applied. In fact, researchers have already encountered similar technical challenges in the formulation of SLN for the treatment of non-cancer diseases. For instance, 3′azido-3′deoxythymidine (AZT) is a non-ionic antiviral drug with low molecular weight (MW = 267.24). This drug is too water soluble to be formulated into SLN. One way to overcome this difficulty is to prepare a lipophilic drug derivative. In the case of AZT, researchers conjugated a palmitate group to the AZT molecule, and the drug loading was shown as significantly improved subsequent to the increased lipophilicity.[107]

Hydrophilic cytotoxic compounds are relatively few. Both 5-FU and mitomycin-C fall into this category. Cisplatin and its analog carboplatin can also be considered as water-soluble small drugs. Until recently, only 5-FU has been attempted for SLN encapsulation.

36.5.4.3.2 Fluorouracil

The drug-derivatization strategy was adopted for SLN encapsulation of 5-FU. The water solubility of 5-FU is 12.2 mg/ml that makes it very difficult to be loaded in SLN. Wang et al. reduced the water solubility by conjugating two octanoyl groups to the 5-FU molecule,[90] resulting in 3′,5′-dioctanoyl-5-fluoro-2′-deoxyuridine (FuDR). This lipophilic drug derivative could be loaded into SLN with an encapsulation efficiency at over 90%. This strategy, however, requires tedious procedures of chemical synthesis. The chemical purity and identity of the final product have to be confirmed, and the toxicity and efficacy of the derivative also needs to be evaluated. All these steps demand substantial amounts of work. Drugs like 5-FU can be easily diluted and administered without the concern for drug precipitation such as in the cases of the lipophilic compounds, this probably provides very little motivation for the researchers to go through all these tedious procedures. However, one must keep in mind that SLN provide more than just drug solubilization and stabilization. They also allow controlled release, and they possibly improve the anti-tumor activities, side effect profiles, and biodistribution patterns of the loaded drugs. Successful development of SLN formulations for this class of anti-tumor drugs can still be rewarding.

36.5.4.3.3 Overall Remarks on Cytotoxic Drug Delivery

Cytotoxic anti-tumor drugs are a class of compounds well known for their heterogeneity in terms of molecular structures and physicochemical properties. Although a number of strategies have been devised to allow encapsulation of water-soluble salts of cytotoxic compounds, so far, only two anthracyclines have been successfully encapsulated. It is obvious that there is more to explore in this area. For low molecular weight and neutral water-soluble compounds, the major limitation of SLN—inefficient loading of hydrophilic molecules—is still a concern. Until recently, only limited efforts using the drug-derivatization strategy have been implemented. The various strategies that may be suitable for encapsulation of the different classes of cytotoxic compounds by SLN are summarized in Table 36.5. Some of the newer SLN-based systems have not been tested for their use for anti-tumor drug delivery (e.g., NLC and LDC), but based on their general properties, it is expected that they should be able to adequately handle the suggested drug types. It is expected that even more well-designed systems will be developed in the near future to allow more specific and efficient cytotoxic drug delivery.

TABLE 36.5
A Summary of the Possible SLN-Based Systems Currently Available That Can Be Used for the Delivery of Cytotoxic Drugs and Chemosensitizers

Drug	Physicochemical Properties	Possible Systems
Cytotoxic drugs	Lipophilic, water-insoluble molecules	SLN, NLC
	Water-soluble, ionic salts	Ion-pair-SLN, LDC, PLN
	Water-soluble, hydrophilic small molecules	SLN or NLC with lipophilic drug derivative
Chemosensitizers	Lipophilic, amphiphilic molecules in salt forms	SLN, NLC similar to cytotoxic drugs

SLN, classical SLN and microemulsion-based SLN; LD, lipid-drug conjugate lipid nanoparticles; PLN, polymer lipid hybrid nanoparticles.

36.5.5 DELIVERY OF CHEMOSENSITIZING AGENTS USING SLN

36.5.5.1 Multidrug Resistance and Chemosensitizers

As previously mentioned, MDR phenotype in cancer cells presents a significant obstacle to anti-tumor therapy at a cellular level. MDR was demonstrated when tumor cells that have been exposed to one cytotoxic agent develop cross resistance to a broad range of structurally and functionally unrelated compounds.[108] The cytotoxic drugs that are most frequently associated with MDR are typically hydrophobic, amphipathic products, including vinca alkaloids (vincristine, vinblastine), taxanes (paclitaxel, docetaxel), epipodophyllotoxins (etoposide, teniposide), anthracyclines (doxorubicin, daunorubicin, epirubicin), topotecan, and mitomycin C.[109,110]

MDR mediated by membrane-bound drug efflux transporters such as P-glycoprotein (Pgp) is probably the most studied form of drug resistance.[76,111] If MDR as a result of Pgp expression could be reduced, it would possibly increase the success rate of cancer chemotherapy. One approach is to use chemicals with Pgp modulatory activity to restore the drug sensitivity of cancer-demonstrating MDR. These agents, including verapamil, cyclosporin A, quinidine, tamoxifen, and several calmodulin antagonists, were identified in the 1980s,[110] and they were often referred as chemosensitizers or multi-drug resistance reversal agents. Newer and more potent analogs of the older chemosensitizers, including Valspodar (PSC833), Biricodar, and later, the more Pgp-specific synthetic compounds such as XR9576,[112] LY335979,[113] and Elacridar (GG918/GF120918)[114] were successively developed.

36.5.5.2 Rationales for Delivery of Chemosensitizers Using SLN

When tested in clinical settings, the earlier chemosensitizers often produced disappointing results, primarily because of their low affinities for Pgp. High doses of these agents are required to achieve the desired chemosensitizing effect, and this results in unacceptable toxicity[115] Many of these chemosensitizers are also substrates for other enzymes and transporters. This further leads to unpredictable pharmacokinetic interactions when used with other anti-tumor agents. Although the recently developed chemosensitizers are more potent, less toxic, and more Pgp-specific, pharmacokinetic interactions still occur[115–117] even when the newest agents such as GG918 were chosen.[118]

Like most Pgp substrates, chemosensitizers are usually hydrophobic molecules. Many chemosensitizers also carry basic, ionizable groups like doxorubicin. Figure 36.13a and Figure 36.13b show the molecular structures of two examples of chemosensitizers. These properties allow them to be incorporated into SLN at high efficiency. By delivering chemosensitizers using SLN, it is hoped

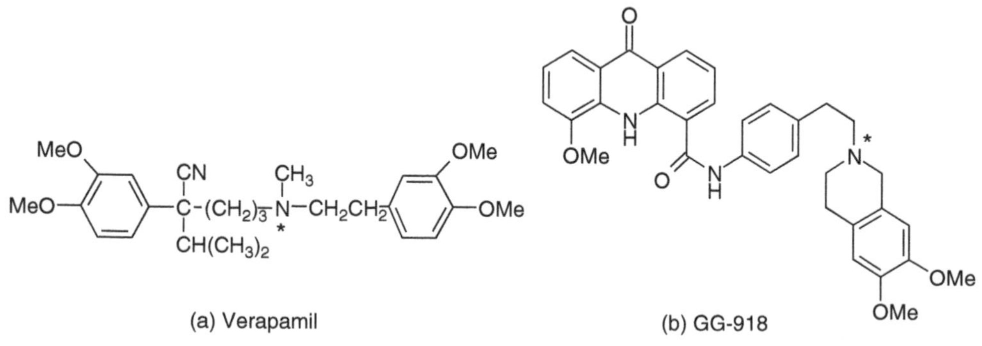

(a) Verapamil   (b) GG-918

FIGURE 36.13 Examples of molecular structures of two chemosensitizers. (a) Verapamil; and (b) GG918. * Basic amine group that is ionizable by protonation.

that the risk of pharmacokinetic drug interactions will be further minimized. Moreover, even without any surface-modifications, SLN are able to improve tumoral drug concentration (to be discussed later), so the chemosensitizing effects could potentially be improved, and the non-cellular drug resistance that is contributed by subtherapeutic tumoral drug concentration may also be alleviated.

36.5.5.3 Current Use of SLN for Chemosensitizer Delivery

So far, only a handful of chemosensitizers have been encapsulated by SLN. These include cyclosporin-A, verapamil, quinidine, and GG918. The cyclosporin-A SLN formulation was actually intended for immunomodulation, the official indication of cyclosporin-A, rather than for drug resistance reversal.[119] The discussion will, therefore, be focused on the other three compounds.

36.5.5.3.1 Verapamil and Quinidine

Verapamil and quinidine are both lipophilic molecules commonly available in ionic salt forms (verapamil hydrochloride and quinidine sulfate) that can both partition into a lipophilic phase when complexed to an anionic polymer. It was determined that the encapsulation efficiencies of the salts of verapamil and quinidine in lipid nanoparticles were improved from 19.6 to 69.3% and 29.5 to 58.3%, respectively, when dextran sulfate was co-loaded to form PLN.[55] When verapamil and quinidine were simultaneously loaded, those encapsulation efficiencies slightly dropped to 50.3% and 55.8%, still significantly higher than without the addition of an anionic polymer. The drug release rates of both drugs from the dual-loaded PLN system were similar to when the drugs were separately loaded. The study shows that it is possible to simultaneously load and deliver two chemosensitizers by PLN with only marginal interferences between each other. This could be therapeutically valuable as it was reported that combinations of two chemosensitizers, even both at suboptimal concentrations, were able to achieve synergistic or additive effect to overcome MDR because of Pgp-over expression.[120]

PLN dual-loaded with verapamil and doxorubicin were also evaluated.[55] The drug loading properties and drug release kinetics of the dual-loaded system was similar to the single-drug systems. Nevertheless, the systemic cytotoxicity of verapamil is very high ($LD_{50} = 163$ µg/g in mice). Considering this drug requires a concentration in the range of µM to mM to be effective for multidrug resistance (MDR) reversal,[121] there is too much risk to put it in clinical use even it is in encapsulated form. The study, however, reveals the possibility of using SLN-based system for multiple drug delivery that is critical in terms of cancer chemotherapy as drug combinations are typically used.

36.5.5.3.2 Elacridar (GG918)

GG918 has strong inhibitory activity on Pgp and breast cancer resistance protein,[122,123] two transporters that can contribute to cellular drug resistance. It is more potent and safer than verapamil.[124,125] The use of SLN for simultaneous delivery of GG918 and doxorubicin is still at experimental stage.[126] It was reported that the two drugs could be efficiently loaded in the same PLN formulation (co-loaded with HPESO polymer), and the resulting dual-drug system is effective against Pgp-over expressing cancer cell lines. It is noteworthy that the dual-drug system was also significantly more effective than when the two drugs were both in solution form or both in separate single-drug PLN systems. In initial results,[126] the anti-tumor effects of various formulations and their combinations were as follows:

doxorubicin/GG918-PLN > free doxorubicin/GG918 in solution > doxorubicin-PLN + GG918 solution ≈ doxorubicin solution + GG918-PLN > doxorubicin-PLN + GG918-PLN (a mixture of two single-drug PLN)

In other words, synergistic effect in drug-resistant cancers may be achieved using this type of dual-drug SLN (or PLN) system. The drug combination appears to require proper spatial distributions to optimally function. When the two drugs are separately encapsulated, the synergistic effect is lost. These findings have laid the foundation for using SLN-based system for combinational drug chemotherapy.

36.5.5.3.3 Overall Remarks on Chemosensitizer Delivery

The study of SLN systems for the delivery of chemosensitizers is still in its infancy. Only limited information in this field has been revealed thus far. Given the good lipophilicities of chemosensitizers in general, it is not surprising to see that the few chemosensitizers studied until now are well incorporated in PLN. It is expected that other free-base chemosensitizers can be loaded into even unmodified, classical SLN (See Table 36.5). Formulation of PLN dual-loaded with two chemosensitizers or a chemosensitizer/cytotoxic agent combination was also shown to be possible. This, once again, demonstrates the versatility of SLN-based delivery systems.

36.5.6 EFFECT OF SLN FORMULATION ON IN VITRO ANTI-TUMOR TOXICITY

The primary goals of cytotoxic drug therapy are to kill as many cancer cells as possible and to prevent their proliferation. Therefore, it is important to ensure that the drug released from a SLN system preserves its cytotoxicity, and the drug delivered by SLN is at least as cytotoxic to cancer cells as the conventional free drug solution. These two issues appear identical but are, in fact, quite different. Drug releases from SLN are rarely 100% complete. In fact, in many SLN formulations to be discussed later, only fractions of the loaded anti-tumor drugs are releasable. However, these formulations all demonstrated cancer cytotoxicity comparable or even superior to the corresponding free drugs. These findings suggest that, in addition to simply unloading the drug outside the cancer cells and killing the cells with this fraction of free, released drug, the portion of drug that remains associated with the SLN is still cytotoxic to various extents. SLN quite possibly have additional mechanisms to deliver this unreleased drug to achieve cancer cell kill or growth suppression. It is important to evaluate the cytotoxicity of a drug-loaded SLN formulation as an integral unit and, whenever possible, to also investigate the underlying cytotoxicity mechanisms that derive from the use of SLN.

Only until recently, a limited number of in vitro cytotoxicity studies on anti-tumor drug-loaded SLN have been conducted. Miglietta et al. performed trypan blue exclusion assays on breast cancer cell line MCF-7 and leukemia cell lines (HL60) treated with SLN incorporating doxorubicin or paclitaxel for 72 h.[83] The results of the assays demonstrated improved cytotoxicity with both formulations compared to the corresponding free drugs. A similar study using the same methodology was later carried out to evaluate the cytotoxicity of SLN formulations carrying doxorubicin, paclitaxel, or cholesteryl butyrate on a colorectal cancer cell line HT-28.[62] SLN of doxorubicin and cholesteryl butyrate were both significantly more cytotoxic than the corresponding free drugs as indicated by the lower drug concentrations needed for the SLN to achieve 50% cell non-viability (IC_{50} of SLN versus solution: doxorubicin: 81.87 nM versus 126.57 nM, butyrate: 0.3 mM versus > 0.6 mM;). Paclitaxel SLN were similarly cytotoxic compared to the drug solution. This was possibly caused by the extremely slow release of paclitaxel from the SLN.[82]

One interesting finding in Serpe et al.'s study is about the combination use of anti-tumor drugs.[62] The authors reported that the combination of low concentrations of cholesteryl butyrate-loaded SLN and doxorubicin or paclitaxel in solution exerted synergistic reduction in cancer cell viability, whereas no such reduction was demonstrated using the same combinations of drugs both as solution. Although the cytotoxicities appear low and only trypan blue exclusion assays were employed, the findings nonetheless support the possibility that combination drug therapy involving SLN formulations could be more effective than the conventional free drug treatment.

Some nanoparticle systems and liposomal formulations have been shown to be effective against cancer cells with MDR phenotype.[127–130] These systems were all made of lipids or lipophilic substances, and many of them were coated with non-ionic surfactants. Also with lipid-based formulations that often use non-ionic surfactants, it is reasonable to expect MDR reversal activities in at least some SLN systems. In the aforementioned studies, even though HL-60 and HT-28 cell lines are not specifically drug-resistant cell lines, they are moderately refractory to doxorubicin, and the SLN formulations were still able to effectively treat them. In another study, PLN dispersed using Pluronic F68, a non-ionic block co-polymer, were able to enhance the cytotoxicity of doxorubicin in Pgp-over expressing breast cancer cell line MDA435/LCC6/MDR1.[10] Doxorubicin-loaded PLN also led to significantly stronger reductions in cell viability in trypan blue exclusion assays than free doxorubicin. In addition, the use of clonogenic assays, a more sensitive method to evaluate the effectiveness of cancer proliferation suppression, showed a larger than 8-fold increase in the anti-tumor activity using PLN formulation compared to free doxorubicin (Figure 36.14). It is interesting to observe that blank particles did not enhance the anti-tumor activity of doxorubicin solution, and doxorubicin–polymer (Dox–HPESO) was not more effective than free doxorubicin. The findings of this study further reveal the potential of SLN for drug-resistant cancer treatment, and they support the earlier claim that encapsulated cytotoxic compound is still effective. In fact, a cytotoxic compound may need to be encapsulated to achieve that enhanced activity in drug-resistant cancer.

Initial work has been undertaken to delineate the underlying MDR reversal mechanisms by SLN.[11] Increased drug uptake and more noticeably, prolonged drug retention, were observed when doxorubicin-loaded PLN were used instead of free drug. This was further supported by fluorescence microscope images.[11] In Figure 36.15, cancer cells were treated for 2 h and re-incubated in fresh medium for an additional 2 h. A high level of doxorubicin that has fluorescent property was intracellularly detected when the drug was delivered in form of PLN. Another fluorescence microscopy study also revealed that lipids were associated with the loaded drug molecules in the cells. All these findings suggest that some of the nanoparticles may carry doxorubicin into the cell,

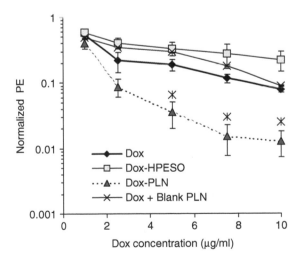

FIGURE 36.14 Clonogenic assay experiments for the toxicity of Dox human P-glycoprotein overexpressing MDA435/LCC6/MDR1 breast cancer cell line. The normalized plating efficiencies (normalized PE) after 4 h exposure to doxorubicin (Dox) solution, Dox-polymer (Dox-HPESO) aggregates, Dox-loaded lipid nanoparticles (Dox-PLN), or Dox solution + blank PLN are shown. Results are expressed as mean ± SD of the measurements obtained in three separate experiments ($n = 6$ in each experiment). *, $p < 0.05$, significantly different from Dox solution group. (Adapted from Wong, H. L. et al., *Pharm. Res.*, in press.)

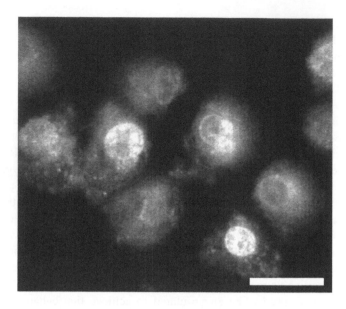

FIGURE 36.15 Fluoroescence microscope images demonstrating the effect of a PLN formulation containing doxorubicin on drug retention and intracellular drug distribution in Pgp-over expressing breast cancer cells MDA435/LCC6/MDR1. Cells were incubated with nanoparticles containing doxorubicin for 2 h, washed, and re-incubated in fresh medium for two additional hours. $\lambda_{ex} = 540$ nm and $\lambda_{em} = 590$ nm. Magnification of objective $= 100\times$. Bar represents 20 μm. (Adapted from Wong, H. L. et al., *Pharmacol. Exp. Ther.*, in press.)

possibly by endocytosis, bypassing the membrane-bound drug efflux mechanisms. These drug-loaded nanoparticles could be staying intracellularly to serve as mini-drug depot systems and chronically inflicting damages to cancer cells.

36.5.6.1 In Vivo Efficacy of SLN Delivered Antineoplastic Agents

In terms of cancer therapy, particularly with regard to solid tumors, in vitro data frequently may only serve as a good starting point. It is because there are several drug resistance mechanisms, sometimes referred as non-cellular mechanisms, that exist only in in vivo situations. A treatment with strong in vitro anti-tumor activity thus may not be working in tumor-bearing animals. However, information on the in vivo efficacy of SLN formulations of anti-tumor drugs remains limited. Most studies that included animal models focused on drug biodistribution. To our knowledge, only one study with in vivo efficacy component has been published. In this study, mice model xenografted with HT-28 tumor was used.[62] Tumors treated with the SLN formulation of SN-38 took longer or comparable time to reach the cut off tumor weight at lower drug dose. The study provided the first evidence that the ability of SLN to enhance anti-tumor drug effectiveness is not limited in cultured cells, but may also be applicable to in vivo systems.

36.5.6.2 Improved Pharmacokinetics and Drug Distribution

As discussed in Section 36.2.3, SLN can be coated with stealth agents such as PEG to minimize the clearance by RES to achieve extended systemic circulation time. Using stearic acid PEG2000, long-circulating SLN formulations of doxorubicin and paclitaxel have been formulated.[9,84] A number of pharmacokinetic and biodistribution studies were based on these SLN systems.

In principle, the unmodified, non-stealth SLN should be rapidly cleared from the systemic circulation by RES. In reality, unmodified SLN were able to remain in the bloodstream for long times in the animal studies performed. Table 36.6 summarizes the area-under-the-curve (AUC)

TABLE 36.6

A Comparison of the Area-Under-the-Curve Values of Anti-Cancer Drugs Administered in Solution to Drugs Encapsulated in Unmodified SLN or Long-Circulating SLN

Study Group[a] Drug[b] AUC[c]	Funaro DOX	Yang CTT	Wang FuDR	Zara 99 DOX	Zara 02 DOX	Zara 02 IDA
Drug solution	83.7	66	44.5	98.0	47.6	31.4, 108.5[d]
Unmodified SLN	814.5	324	138.4	1713.4	123.1	674.2, 458.9
Long-circulating SLN	1121.1	—	—	—	259.7[e]	—
					318.2	
					433.3	

[a] Groups named under the first authors of the studied.

[b] DOX, doxorubicin; CTT, camptothecin; FuDR, 3′,5′-dioctanoyl-5-fluoro-2′-deoxyuridine; IDA, idarubicin.

[c] AUC, area under curve, units in min.(g.ml^{-1} except the studies by Yang (h.ng.g^{-1}) and Wang (h.mg.ml^{-1}).

[d] First figure and second figure correspond to AUC of IDA after intravenous and duodenal administration, respectively.

[e] SLN with three different levels of PEG2000-stearic acid.First, second, and third figures correspond to 0.15%, 0.30%, and 0.45% PEG2000 content in lipid microemulsion for SLN preparation.

values of several anti-tumor compounds obtained in pharmacokinetic studies. Drugs in solution, unmodified SLN, or long-circulating SLN were administered. In all of the studies listed, unmodified SLN were able to increase the AUC values by 3-fold up to over 20-fold, indicating low RES clearance. Nevertheless, long-circulating SLN still have an advantage over the unmodified SLN. In the two studies that compared unmodified and long-circulating formulations, the AUC of drugs delivered by the long-circulating SLN were approximately 25% to over 3-fold higher.

In addition to surface modification, the routes of administration also appear to affect the pharmacokinetics and drug biodistribution. In the etoposide-SLN study,[80] improvements in tumoral drug accumulation were observed when the etoposide formulations were intraperitoneally or subcutaneously injected. Subcutaneous administration even led to multiple-fold increases in the tumor drug concentrations 24 h post-injection. The authors suggested that the slower and progressive penetration of the nanoparticles (and the loaded drug) from the subcutaneous injection site into the tumor may result in more favorable patterns of drug distribution. In the study by Zara et al.,[88] duodenal administration of idarubicin-loaded SLN also led to a higher AUC than when the SLN were intravenously administered. It is evident that the biodistribution of an anti-tumor drug delivered by SLN may be further manipulated to achieve the desired therapeutic goal. The manner by which a SLN drug formulation is administered will be a key aspect to consider when designing animal or clinical studies of SLN for anti-tumor drug delivery.

Regardless of how long an anti-tumor drug can stay in the systemic circulation, the drug has to finally accumulate in the tumor and avoid the non-cancerous tissues to exercise its therapeutic effects without causing toxicity to the patient. To confirm the tumor drug accumulation, an in vivo tumor model is required. Only one SLN study employing tumor-bearing animals has been performed.[80] Nearly 67% and 30% increases of tumor drug concentrations were measured 1 h and 24 h post-injection, respectively. This is promising in terms of cancer chemotherapy. More SLN biodistribution studies involving tumor models will further strengthen the case.

In many of the biodistribution studies of anti-tumor drug-loaded SLN, even without surface modifications, the drug accumulations in the liver, spleen, and kidney, three RES organs, were reduced.[8,9,80,88] The role of RES clearance in the biodistribution of SLN may need to be re-examined. It is also noteworthy to point out the high brain concentrations of SLN delivered drugs in many of these studies. Cytotoxic drugs do not usually accumulate in the brain tissue because of the presence of the blood–brain barrier. This barrier presents a serious challenge to chemotherapeutic treatment of central nervous system tumors (refer to Chapter 17). The most prominent mechanism leading to the blood–brain barrier is the high expression of Pgp. The ability of SLN to carry drugs across the Pgp-rich blood–brain barrier is consistent with the previously described findings of SLN in Pgp-over expressing cancer cell lines (Section 36.5.3). This Pgp-bypassing feature of SLN may be useful for cancer chemotherapy if properly exploited.

In most of the SLN biodistribution studies, the drug concentrations in the heart were significantly lower using SLN formulations.[8,9,80,88] This is important for anthracycline delivery as this class of anti-tumor compounds are notorious in causing irreversible cardiotoxicity. By selectively reducing the anthracycline concentration in the cardiac tissues, higher dose intensity may be given to achieve better therapeutic outcomes.[131]

Overall, SLN are useful for the improvement of the pharmacokinetics and biodistribution profiles of anti-tumor drugs. Drugs administered tend to stay longer, penetrate the tumor sites better, and avoid some vulnerable organs. These may be further improved by the use of long-circulating SLN systems.

36.6 CONCLUSIONS AND FUTURE PERSPECTIVES

SLN are a promising drug delivery device for rational chemotherapy. After just slightly longer than a decade of time, SLN already have gone through several stages of development. Although the ingredients of SLN are relatively simple—basically only lipids and surfactants—there are a diverse number of parameters involved in the preparation processes that can be manipulated to alter the SLN properties. These include the physicochemical characteristics of SLN compositions, production methods, and storage conditions. Significant progresses in the improvement of drug loading capacity, mass-production, modulation of release kinetics, stability in storage, bioavailability, and active targeting have been made. In addition, the early, classical SLN that were made of highly crystallized lipids have quickly evolved into several variations, including microemulsion-based SLN, NLC, LDC, PLN, and surface-modified SLN. Each of these variations corrects a specified set of problems commonly encountered during the early stage of SLN development. Currently not only limited to lipophilic compounds, hydrophilic and charged agents can also be efficiently encapsulated into SLN-based systems, stored with good stability, and released under a certain degree of control. This makes SLN an excellent choice for the delivery of anti-tumor drugs, a highly heterogeneous class of drugs with diverse molecular structures and physical properties and are usually chemical reactive and strongly toxic.

This relatively new class of delivery systems still has a lot of untapped potential as well as limitations to overcome. Because of the complexity of such drug delivery systems and complex interactions existed among various causal factors, the fundamental problems hampering the practical application of SLN such as drug loading and release mechanisms, the correlation between the constituents of SLN and the properties of SLN have not been completely solved yet. The current development of SLN still mainly depends on the trial-and-error approach that leads to occasional controversial results. It is expected that with better understanding of the structure of SLN and their mechanisms of drug loading and release, the overall performance of SLN can be further optimized.

The biological aspects of SLN for cancer treatment, including SLN-cell interactions and in vivo SLN-tumor interactions are areas largely unexplored. In addition, there are a number of new applications of SLN that have just started developing, e.g., the use of SLN for delivery of drug combinations such as cytotoxic drug—chemosensitizer. The potentials of SLN for gene therapy and molecular targeting are also emerging.[84,132] It will not be surprising to witness significant research progress in the field of SLN in the near future, providing more efficacious, less toxic, and more specific chemotherapy treatments for cancer.

REFERENCES

1. Müller, R. H., Colloidal carriers for controlled drug delivery and targeting: Modification, characterization and in vivo distribution, CRC Press, 45–55, 1991.
2. Wissing, S. A., Kayser, O., and Müller, R. H., Solid lipid nanoparticles for parenteral drug delivery, *Adv. Drug Deliv. Rev.*, 56, 1257–1272, 2004.
3. Müller, R. H. et al., Solid lipid nanoparticles (SLN)-An alternative colloidal carrier system for controlled drug delivery, *Eur. Pharm. Biopharm.*, 41, 62–69, 1995.
4. Schwarz, C. et al., Solid lipid nanoparticles (SLN) for controlled drug delivery. I. Production, characterization and sterilization, *J. Control. Release*, 30, 83–96, 1994.
5. Almeida, A. J., Runge, S., and Müller, R. H., Peptide-loaded solid lipid nanoparticles (SLN): Influence of production parameters, *Int. J. Pharm.*, 149, 255–265, 1997.
6. Müller, R. H., Redtke, M., and Wissing, S. A., Solid lipid nanoparticles (SLN) and nanostructured lipid carriers (NLC) in cosmetic and dermatological preparations, *Adv. Drug Deliv. Rev.*, 54, S131–S155, 2002.
7. Wissing, S. A. and Müller, R. H., Solid lipid nanoparticles as carrier for sunscreens: In vitro release and in vivo skin penetration, *J. Control. Release*, 81, 345–355, 2002.
8. Zara, G. P. et al., Intravenous administration to rabbits of non-stealth and stealth doxorubicin-loaded solid lipid nanoparticles at increasing concentration of stealth agent: Pharmacokinetics and distribution in brain and other tissues, *J. Drug Target.*, 10, 327–335, 2002.
9. Fundarò, A. et al., Non-stealth and stealth solid lipid nanoparticles (SLN) carrying doxorubicin: Pharmacokinetics and tissue distribution after I.V. administration to rat, *Pharmacol. Res.*, 42, 337–343, 2000.
10. Wong, H. L. et al., A new polymer–lipid hydrid nanoparticle system increases cytotoxicity of doxorubicin against multidrug resistant human breast cancer cells, *Pharm. Res.*, 23, 7, 1574–1585, 2006.
11. Wong, H. et al., A Mechanistic study of enhanced doxorubicin uptake and retention in multidrug resistant breast cancer cells using a polymer–lipid hybrid nanoparticle (PLN) system, *J. Pharmcol. Exp. Ther.*, (in press).
12. Westesen, K., Siekmann, B., and Koch, M. H. J., Investigations on the physical state of lipid nanoparticles by synchrotron radiation X-ray diffraction, *Int. J. Pharm.*, 93, 189–199, 1993.
13. Dingler, A. and Gohla, S., Production of solid lipoid nanoparticles (SLN): Scaling up feasibilities, *J. Microencapsul.*, 19, 11 16, 2002.
14. Marengo, E. et al., Scale-up of the preparation process of solid lipid nanoparticles. Part I, *Int. J. Pharm.*, 205, 3–13, 2000.
15. Jenning, V., Lippacher, A., and Gohla, S. H., Medium scale production of solid lipid nanoparticles (SLN) by high pressure homogenization, *J. Microencapsul.*, 19, 1–10, 2002.
16. Gohla, S. H. and Dingler, A., Scaling up feasibility of the production of solid lipid nanoparticles (SLN), *Pharmazie*, 56, 61–63, 2001.
17. Allen, T. M. and Cullis, P. R., Drug delivery systems: Entering the mainstream, *Science*, 303, 1818–1822, 2004.
18. Moses, M. A., Brem, H., and Langer, R., Advancing the field of drug delivery: Taking aim at cancer, *Cancer Cell*, 4, 337–341, 2003.
19. Jenning, V. and Gohla, S., Comparison of wax and glyceride solid lipid nanoparticles (SLN), *Int. J. Pharm.*, 196, 219–222, 2000.
20. Jenning, V. and Gohla, S. H., Encapsulation of retinoids in solid lipid nanoparticles (SLN), *J. Microencapsul.*, 18, 149–158, 2001.
21. Bunjes, H. et al., Incorporation of the model drug ubidecarenone into solid lipid nanoparticles, *Pharm. Res.*, 18, 287–293, 2001.
22. Aulton, M. E., *Pharamaceutics: The Science of Dosage form Design.*, Churchill Livingstone, Edinburgh. 81–118, 1988.
23. Bunjes, H., Koch, M. H. J., and Westesen, K., Influence of emulsifiers on the crystallization of solid lipid nanoparticles, *J. Pharm. Sci.*, 92, 1509–1520, 2002.
24. Aquilano, D., Cavalli, R., and Gasco, M. R., Solid lipospheres obtained from hot microemulsions in the presence of different concentration of cosurfactant: The crystallinizatiion of stearic acid polymorphs, *Thermochim. Acta.*, 230, 29–37, 1993.

25. Olbrich, C. and Müller, R. H., Enzymatic degradation of SLN-effect of surfactant and surfactant mixtures, *Int. J. Pharm.*, 180, 31–39, 1999.

26. Olbrich, C., Kayser, O., and Müller, R. H., Enzymatic degradation of Dynasan 114 SLN—effect of surfactants and particle size, *J. Nanoparticle Res.*, 4, 121–129, 2002.

27. Matsumura, Y. and Maeda, H., A new concept for macromolecular therapeutics in cancer chemotherapy: Mechanism of tumoritropic accumulation of proteins and the antitumor agent SMANCS, *Cancer Res.*, 6, 193–210, 1986.

28. Müller, R. H., Mader, K., and Gohla, S., Solid lipid nanoparticles (SLN) for controlled drug delivery—a review of the state of art, *Eur. J. Pharm. Biopharm.*, 50, 161–177, 2000.

29. Kawashima, Y., Nanoparticulate systems for improved drug delivery, *Adv. Drug Deliv. Rev.*, 47, 1–2, 2001.

30. Mehnert, W. and Mader, K., Solid lipid nanoparticles, production, characterization and applications, *Adv. Drug Deliv. Rev.*, 47, 165–196, 2001.

31. Westesen, K., Bunjes, H., and Koch, M. H. J., Physicochemical characterization of lipid nanoparticles and evaluation of their drug loading capacity and sustained release potential, *J. Control. Release*, 48, 223–236, 1997.

32. Shahgaldian, P. et al., Para-acyl-calix-arene based solid lipid nanoparticles (SLNs): A detailed study of preparation and stability parameters, *Int. J. Pharm.*, 253, 23–28, 2003.

33. Bummer, P. M., Physical chemical considerations of lipid-based oral drug delivery-solid lipid nanoparticles, *Crit. Rev. Ther. Drug Carrier Syst.*, 21, 1–20, 2004.

34. Manjunath, K., Reddy, J. S., and Venkateswarlu, V., Solid lipid nanoparticles as drug delivery systems, *Methods Find. Exp. Clin. Pharmacol.*, 27, 127–144, 2005.

35. Li, Y., Taulier, N., Rauth, A. M., and Wu, X. Y., American Association of Pharmaceutical Scientists (AAPS) conference. TN, USA, Nov. 7–11, 2005.

36. Chapman, D., The polymorphism of glycerides, *Chem. Rev.*, 62, 433–456, 1961.

37. Bunjes, H., Koch, M. H. J., and Westesen, K., Influence of emulsifiers on the crystallization of solid lipid nanoparticles, *J. Pharm. Sci.*, 92, 1509–1520, 2003.

38. Hou, D. et al., The production and characteristics of solid lipid nanoparticles (SLNs), *Biomaterials*, 24, 1781–1785, 2003.

39. Venkateswarlu, V. and Manjunath, K., Preparation, characterization and in vitro release kinetics of clozapine solid lipid nanoparticles, *J. Control. Release*, 95, 627–638, 2004.

40. Olbrich, C., Kayser, O., and Müller, R. H., Lipase degradation of Dynasan 114 and 116 solid lipid nanoparticles (SLN)-effect of surfactants, storage time and crystallinity, *Int. J. Pharm.*, 237, 119–128, 2002.

41. Souto, E. B. et al., Development of a controlled release formulation based on SLN and NLC for topical clotrimazole delivery, *Int. J. Pharm.*, 278, 71–77, 2004.

42. Müller, R. H., Radtke, M., and Wissing, S. A., Nanostructured lipid matrices for improved micro-encapsulation of drugs, *Int. J. Pharm.*, 242, 121–128, 2002.

43. Jenning, V., Thunemann, A. F., and Gohla, S. H., Characterization of a novel solid lipid nanoparticle carrier system based on binary mixtures of liquid and solid lipids, *Int. J. Pharm.*, 199, 167–177, 2000.

44. Jenning, V., Korting, M. S., and Gohla, S., Vitamin A-loaded solid lipid nanoparticles for topical use: Drug release properties, *J. Control. Release*, 66, 115–126, 2000.

45. Tabatt, K. et al., Effect of cationic lipid and matrix lipid composition on solid lipid nanoparticle-mediated gene transfer, *Eur. J. Pharm. Biopharm.*, 57, 155–162, 2004.

46. Chen, D. N. et al., In vitro and in vivo study of two types of ion-circulating solid lipid nanoparticles containing paclitaxel, *Chem. Pharm. Bull.*, 49, 1444–1447, 2001.

47. Bocca, C. et al., Phagocytic uptake of fluorescent stealth and non-stealth solid lipid nanoparticles, *Int. J. Pharm.*, 175, 185–193, 1998.

48. Gref, R. et al., The controlled intravenous delivery of drugs using PEG-coated sterically stabilized nanoparticles, *Adv. Drug Deliv. Rev.*, 16, 215–233, 1995.

49. Illum, L. et al., The organ distribution and circulation time of intravenous injected colloidal carriers sterically stabilized with a block copolymer–Poloxamine 908, *Life Sci.*, 40, 367–374, 1987.

50. Jores, K., Mehnert, W., and Mader, K., Physicochemical investigations on solid lipid nanoparticles and on oil-loaded solid lipid nanoparticles: A nuclear magnetic resonance and electron spin resonance study, *Pharm. Res.*, 20, 1274–1283, 2003.

51. Jores, K. et al., Investigation on the structure of solid lipid nanoparticles (SLN) and oil-loaded solid lipid nanoparticles by photon correlation spectroscopy, field-flow fractionation and transmission electron microscopy, *J. Control. Release*, 95, 217–227, 2004.

52. zur Mühlen, A., Schwarz, C., and Mehnert, W., Solid lipid nanoparticles for controlled drug delivery—drug release and release mechanism, *Eur. J. Pharm. Biopharm.*, 45, 149–155, 1998.

53. zur Mühlen, A. et al., Atomic force microscopy studies of solid lipid nanoparticles, *Pharm. Res.*, 13, 1411–1416, 1996.

54. Jenning, V., Mader, K., and Gohla, S. H., Solid lipid nanoparticles (SLNTM) based on binary mixtures of liquid and solid lipids: A 1H-NMR study, *Int. J. Pharm.*, 205, 15–21, 2000.

55. Wong, H. et al., Development of solid lipid nanoparticles containing ionically complexed chemo-metherapeutic drugs and chemosensitizers, *Int. J. Pharm.*, 93, 1993–2008, 2004.

56. Li, Y., Taulier, N., Rauth, A. M., and Wu, X. Y., Screening of lipid carriers and characterization of drug–polymer–lipid interactions for the rational design of polymer–lipid hybrid nanoparticles (PLN), *Pharm. Res.*, (Submitted).

57. Müller, R. H. et al., Solid lipid nanoparticles (SLN)-an alternative colloidal carrier system for controlled drug delivery, *Eur. J. Pharm. Biopharm.*, 41, 62–69, 1995.

58. Gasco, M. R., Method for producing solid lipid microspheres having a narrow size distribution, US Patent No. 5250236, 1993.

59. Trotta, M., Debernardi, F., and Caputo, O., Preparation of solid lipid nanoparticles by a solvent emulsification-diffusion technique, *Int. J. Pharm.*, 257, 153–160, 2003.

60. Hu, F. Q. et al., Preparation of solid lipid nanoparticles with clobetasol propionate by a novel solvent diffusion method in aqueous systems and physicochemical characterization, *Int. J. Pharm.*, 239, 121–128, 2002.

61. Cortesi, R. et al., Production of lipospheres as carriers for bioactive compounds, *Biomaterials*, 23, 2283–2294, 2002.

62. Serpe, L. et al., Cytotoxicity of anticancer drugs incorporated in solid lipid nanoparticles on HT-29 colorectal cancer cell line, *Eur. J. Pharm. Biopharm.*, 58, 673–680, 2004.

63. Cavalli, R. et al., Solid lipid nanoparticles (SLN) as ocular delivery system for tobramycin, *Int. J. Pharm.*, 238, 241–245, 2002.

64. Cavalli, R. et al., Preparation and evaluation in vitro of colloidal lipospheres containing pilocarpine as ion pair, *Int. J. Pharm.*, 117, 243–246, 1995.

65. Gasco, M. R., Cavalli, R., and Carlotti, M. E., Timolol in lipospheres, *Pharmazie*, 47, 119–121, 1992.

66. Cavalli, R. et al., Sterilization and freeze-drying of drug-free and drug loaded solid lipid nanoparticles, *Int. J. Pharm.*, 148, 47–54, 1997.

67. Shahgaldian, P. et al., A study of the freeze-drying conditions of calixarene based solid lipid nanoparticles, *Eur. J. Pharm. Biopharm.*, 55, 181–184, 2003.

68. Zimmermann, E., Müller, R. H., and Mäder, K., Influence of different parameters on reconstitution of lyophilized SLN, *Int. J. Pharm.*, 196, 211–213, 2000.

69. Schwarz, C. and Mehnert, W., Freeze-drying of drug-free and drug-loaded solid lipid nanoparticles (SLN), *Int. J. Pharm.*, 157, 171–179, 1997.

70. Freitas, C. and Müller, R. H., Effect of light and temperature on zeta potential and physical stability in solid lipid nanoparticle (SLNTM) dispersions, *Int. J. Pharm.*, 168, 221–229, 1998.

71. Martin, A., *Physical Pharmacy: Physical Chemical Principles in the Pharmaceutical Sciences*, 4th ed., Lea & Febiger, Philadelphia, PA pp. 362–388, 1993.

72. Freitas, C. and Müller, R. H., Stability determination of solid lipid nanoparticles (SLN) in aqueous dispersion after addition of electrolyte, *J. Microencapsul.*, 16, 59–71, 1999.

73. Skeel, R. T., Selection of treatment for the patient with cancer, In *Handbook of Cancer Chemotherapy*, Skeel, R. T., Ed., Lippincott, Williams, and Wilkins, Philadelphia, PA, pp. 46–52, 2003.

74. Tipton, J. M., Side effects of cancer chemotherapy, In *Handbook of Cancer Chemotherapy*, Skeel, R. T., Ed., Lippincott, Williams, and Wilkins, Philadelphia, PA, pp. 561–580, 2003.

75. Gottesman, M. M., Mechanisms of cancer drug resistance, *Ann. Rev. Med.*, 53, 615–627, 2002.

76. Endicott, J. A. and Ling, V., The biochemistry of P-glycoprotein-mediated multidrug resistance, *Ann. Rev. Biochem.*, 58, 137–171, 1989.

77. Lim, S. J. and Kim, C. K., Formulation parameters determining the physicochemical characteristics of solid lipid nanoparticles loaded with all-trans retinoic acid, *Int. J. Pharm.*, 243, 135–146, 2002.

78. Yang, S. C. et al., Body distribution in mice of intravenously injected camptothecin solid lipid nanoparticles and targeting effect on brain, *J. Control. Release*, 59, 299–307, 1999.

79. Yang, S. C. and Zhu, J. B., Preparation and characterization of camptothecan solid lipid nanoparticles, *Drug. Dev. Ind. Pharm.*, 28, 265–274, 2002.

80. Reddy, L. H. et al., Influence of administration route on tumor uptake and biodistribution of etoposide loaded solid nanoparticles in Dalton's lymphoma tumor bearing mice, *J. Control. Release*, 105, 185–198, 2005.

81. Williams, J. et al., Nanoparticle drug delivery system for intravenous delivery of topoisomerase inhibitors, *J. Control. Release*, 91, 167–172, 2003.

82. Cavalli, R., Caputo, O., and Gasco, M. R., Preparation and characterization of solid lipid nanospheres containing paclitaxel, *Eur. J. Pharm. Sci.*, 10, 305–309, 2000.

83. Miglietta, A. et al., Cellular uptake and cytotoxicity of solid lipid nanospheres (SLN) incorporating doxorubicin or paclitaxel, *Int. J. Pharmaceut.*, 210, 61–67, 2000.

84. Stevens, P. J., Sekido, M., and Lee, R. J., A folate-receptor-targeted lipid nanoparticle formulation for a lipophilic paclitaxel prodrug, *Pharm. Res.*, 21, 2153–2157, 2004.

85. Videira, M. A., Almeida, A. J., and Muller, R. H., Formulation and physical stability assessment of SLN containing paclitaxel. Incorporation of paclitaxel in SLN: Assessment of drug–lipid interaction, *Proc. 3rd World Meeting Pharm. Technol.*, 453–454, 2000.

86. Chen, D. B. et al., In vitro and in vivo study of two types of long-circulating solid lipid nanoparticles containing paclitaxel, *Chem. Pharm. Bull.*, 49, 1444–1447, 2001.

87. Cavalli, R., Caputo, O., and Gasco, M. R., Solid lipospheres of doxorubicin and idarubicin, *Int. J. Pharm.*, 89, R9–R12, 1993.

88. Zara, G. P. et al., Pharmacokinetics and tissue distribution of idarubicin-loaded solid lipid nanoparticles after duodenal administration to rats, *J. Pharm. Sci.*, 91, 1324–1333, 2002.

89. Zara, G. P. et al., Pharmacokinetics of doxorubicin incorporated in solid lipid nanospheres (SLN), *Pharm. Res.*, 40, 281–286, 1989.

90. Wang, J. X., Sun, X., and Zhang, Z. R., Enhanced brain targeting by synthesis of 3-,5-dioctanoyl-5-fluoro-2'-deoxyuridine and incorporation into solid lipid nanoparticles, *Eur. J. Pharm. Biopharm.*, 54, 285–290, 2002.

91. Pratt, W. B. et al., *The Anticancer Drugs*, Oxford University Press, New York.

92. Wall, M. E. et al., Plant antitumor agents. I. The isolation and structure of camptothecin, a novel alkaloidal leukemia and tumour inhibitor from camptotheca acuminate, *J. Am. Chem. Soc.*, 88, 3888–3890, 1966.

93. Gottlieb, J. A. et al., Preliminary pharmacologic and clinical evaluatin of camptothecin sodium (NSC-100880), *Cancer Chemother. Rep.*, 54, 461–470, 1970.

94. Takimoto, C. H. and Arbuck, S. G., Topoisomerase I targeting agents: The camptothecins, In *Cancer Chemotherapy and Biotherapy: Principles and Practice*, Chabner, B. A. and Longo, D. L., Eds., Lippincott, Williams, and Wilkins, Philadelphia, PA, pp. 579–646, 2001.

95. Rowinsky, E. K. and Donehower, R. C., Paclitaxel (taxol), *N. Engl. J. Med.*, 332, 1004–1014, 1995.

96. Singla, A. K., Garg, A., and Aggarwal, D., Paclitaxel and its formulations, *Int. J. Pharm.*, 235, 179–192, 2002.

97. Constantinides, P. P. et al., Formulation development and antitumor activity of a filter-sterilizable emulsion of paclitaxel, *Pharm. Res.*, 17, 175–182, 2000.

98. Kan, P. et al., Development of nonionic surfactant/phospholipid o/w emulsion as a paclitaxel delivery system, *J. Controll. Release*, 58, 271–278, 1999.

99. Bernsdorff, C., Reszka, R., and Winter, R., Interaction of the anticancer agent taxol (paclitaxel) with phopholipid layers, *J. Biomed. Mater. Res.*, 46, 141–149, 1999.

100. Clarke, S. J. and Rivory, L. P., Clinical pharmacokinetics of docetaxel, *Clin. Pharmacokinet.*, 36, 99–114, 1999.

101. Ojima, I. and Geney, R., BMS-184476 Bristol-Myers Squibb, *Curr. Opin. Investig. Drugs*, 4, 732–736, 2003.

102. Baldwin, E. L. and Osheroff, N., Etoposide, topoisomerase II and cancer, Curr, *Med. Chem. Anti-Cancer Agents*, 5, 363–372, 2005.

103. Brisaert, M., Gabriels, M., and Plaizier-Vercammen, J., Investigation of the chemical stability of an erythromycin-tretinoin lotion by the use of an optimization system, *Int. J. Pharm.*, 243, 135–146, 2000.

104. Pouillart, P. R., Role of butyric acid and its derivatives in the treatment of colorectal cancer and hemoglobinopathies, *Life Sci.*, 63, 1739–1760, 1998.

105. Pellizzaro, C. et al., Cholesteryl butyrate in solid lipid nanoparticles as an alternative approach for butyric acid delivery, *Anticancer Res.*, 19 (5B), 3921–3926, 1999.

106. Hahn, S. M. et al., A multidrug-resistant breast cancer line induced by weekly exposure to doxorubicin, *Int. J. Oncol.*, 14, 273–279, 1999.

107. Heiati, H. et al., Solid lipid nanoparticles as drug carriers. I. Incorporation and retention of the lipphilic prodrug 3'-azido-3'-deoxythymidine palmitate, *Int. J. Pharm.*, 146, 123–131, 1997.

108. Biedler, J. L. and Riehm, H., Cellular resistance to actinomycin D in Chinese hamster cells in vitro: Cross-resistance, radioautographic, and cytogenic studies, *Cancer Res.*, 30, 1174–1184, 1970.

109. Ambudkar, S. V., Dey, S., and Hrycyna, C. A., Biochemical, cellular and pharmacological aspects of the multidrug transporter, *Ann. Rev. Pharmacol. Toxicol.*, 39, 361–398, 1999.

110. Krishna, R. and Mayer, L. D., Multidrug resistance (MDR) in cancer. Mechanisms, reversal using modulators of MDR and the role of MDR modulators in influencing the pharmacokinetics of anticancer drugs, *Eur. J. Pharm. Sci.*, 11, 265–283, 2000.

111. Juliano, R. L. and Ling, V., A surface glycoprotein modulating drug permeability in Chinese hamster ovary cell mutants, *Biochim. Acta.*, 455, 152–162, 1976.

112. Roe, M. et al., Reversal of P-glycoprotein mediated multidrug resistance by novel anthranilamide derivatives, *Bioorg. Med. Chem. Lett.*, 9, 595–600, 1999.

113. Dantzig, A. H. et al., Reversal of P-glycoprotein-mediated multidrug resistance by a potent cyclo-propyldibenzosuberane modulator, LY335979, *Cancer Res.*, 56, 4171–4179, 1996.

114. Wallstab, A. et al., Selective inhibition of MDR1 P-glycoprotein-mediated transport by the acridone carboxamide derivative GG918, *Br. J. Cancer*, 79, 1053–1060, 1999.

115. Robert, J. and Jarry, C., Multidrug resistance reversal agents, *J. Med. Chem.*, 46, 4805–4817, 2003.

116. Thomas, H. and Coley, H. M., Overcoming multidrug resistance in cancer: An update on the clinical strategy of inhibiting P-glycoprotein, *Cancer Control*, 10, 159–165, 2003.

117. Wandel, C. et al., P-glycoprotein and cytochrome P450 3A inhibition: dissociation of inhibitory potencies, *Cancer Res.*, 59, 3944–3948, 1999.

118. Planting, A. S. T. et al., A phase I and pharmacologic study of the MDR converter GF120918 in combination with doxorubicin in patients with advanced solid tumors, *Cancer Chemother. Pharmacol.*, 55, 91–999, 2005.

119. Ugazio, E., Cavalli, R., and Gasco, M. R., Incorporation of cyclosporine A in solid lipid nanoparticles (SLN), *Int. J. Pharm.*, 241, 341–344, 2002.

120. Huang, M. et al., Effect of combinations of suboptimal concentrations of P-glycoprotein blockers on the proliferation of MDR gene expressing cells, *Int. J. Cancer*, 65, 389–397, 1996.

121. Belpomme, D., Gauthier, S., and Pujade-Lauraine, E., Verapamil increases the survival of patients with anthracycline-resistant metastatic breast carcinoma, *Ann. Oncol.*, 11, 1471–1476, 2000.

122. Hyafil, F. et al., In vitro and in vivo reversal of multidrug resistance by GF120918, an acridone-carboxamide derivative, *Cancer Res.*, 53, 4595–4602, 1993.

123. Maliepaard, M. et al., Circumvention of breast cancer resistance protein (BCRP)-mediated resistance to camptothecins in vitro using non-substrate drugs or the BCRP inhibitor GF120918, *Clin. Cancer Res.*, 7, 935–941, 2001.

124. den Ouden, D. et al., In vitro effect of GF120918, a novel reversal agent of multidrug resistance, on acute leukemia and multiple myeloma cells, *Leukemia*, 10, 1930–1936, 1996.

125. Sparreboom, A. et al., Clinical pharmacokinetics of doxorubicin in combination with GF120918, a potent inhibitor of MDR1 P-glycoprotein, *Anticancer Drugs*, 10, 719–728, 1999.

126. Wong, H. L. et al., Formulation, characterization and in vitro cytotoxicity of solid lipid nanoparticles carrying polymer-linked doxorubicin and GG918, a novel P-glycoprotein inhibitor, *AAPS J.*, 6, Abstract T3208, 2004.

127. de Verdiere, A. C. et al., Reversion of multidrug resistance with polyalkylcyanoacrylate nanoparticles: Towards a mechanism of action, *Br. J. Cancer*, 76, 198–205, 1997.

128. Soma, C. E. et al., Ability of doxorubicin-loaded nanoparticles to overcome multidrug resistance of tumor cells after their capture by macrophages, *Pharm. Res.*, 16, 1710–1716, 1999.

129. Thierry, A. R. et al., Modulation of doxorubicin resistance in multidrug-resistant cells by liposomes, *FASEB J.*, 7, 572–579, 1993.

130. Romsicki, Y. and Sharom, F. J., The membrane lipid environment modulates drug interactions with the P-glycoprotein multidrug transporter, *Biochemistry*, 38, 6887–6896, 1999.

131. Moore, M. J. and Erlichman, C., Pharmacology of anticancer drugs, In *The Basic Science of Oncology*, Tannock, I. F. and Hill, R. F., Eds., McGraw-Hill, Toronto, pp. 370–391, 1998.

132. Tabatt, K. et al., Transfection with different colloidal systems: Comparison of solid lipid nanoparticles and liposomes, *J. Control. Release*, 97, 321–332, 2004.

37 Lipoprotein Nanoparticles as Delivery Vehicles for Anti-Cancer Agents

Andras G. Lacko, Maya Nair, and Walter J. McConathy

CONTENTS

37.1 Introduction 777
37.2 Lipoproteins as Vehicles for the Targeted Delivery of Anti-Cancer Agents 777
37.3 LDL as a Delivery Vehicle for Anti-Cancer Agents 779
37.4 HDL as a Delivery Vehicle for Anti-Cancer Agents 779
37.5 Conclusions 781
References 782

37.1 INTRODUCTION

Nanotechnology and its biomedical applications have exploded in the last several years. Because of initiatives in Europe[1] and the United States,[2,3] increased research funding and venture capital[4] have been available to facilitate these developments. Nanotechnology, nanoscience, or nanoengineering encompass pharmaceutical formulations, devices, chemical processes, or that may be encapsulated into a compartment of less than 1,000 nm in diameter. This volume is dedicated to the description of a wide variety of drug delivery devices and their application to cancer therapy.

Although lipoproteins have long been considered as potential vehicles to deliver the "magic bullet" for targeted cancer chemotherapy,[5,6] they have yet to be considered as prime candidates for clinical applications as recent reviews on targeted drug delivery[7–9] contain only a limited amount of discussion on lipoproteins[7] or none at all.[8,9] The purpose of this chapter is to evaluate lipoproteins as delivery agents for anti-cancer drugs and to discuss the challenges that have so far prevented clinical applications of lipoprotein-based formulations.

37.2 LIPOPROTEINS AS VEHICLES FOR THE TARGETED DELIVERY OF ANTI-CANCER AGENTS

Plasma lipoproteins are macromolecular complexes composed of specific protein and lipid components. The major physiological role of lipoproteins is to transport water-insoluble lipids from their point of origin to their respective destinations. Lipoproteins contain an outer shell that is made up of phospholipids, apolipoproteins, and unesterified cholesterol (Figure 37.1a) and an interior core compartment, accommodating water-insoluble lipids (triacylglycerols and cholesteryl esters).

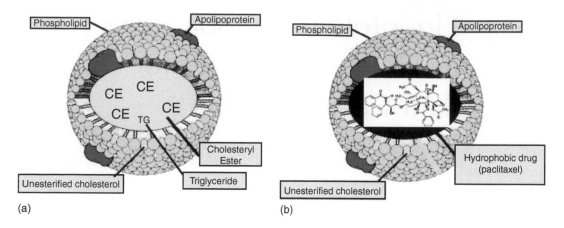

(a)   (b)

FIGURE 37.1 Loading of a hydrophobic drug (paclitaxel) into the core region of a generic lipoprotein (Adapted from Wainberg, P. B., *Hosp. Pract.*, 22, 223–227, 1987.)

This arrangement provides considerable stability to the overall structure of lipoproteins and makes them particularly suitable for the transport of hydrophobic drugs (Figure 37.1b).

Because of the opportunity to convert hydrophilic anti-cancer agents to highly effective hydrophobic pro-drugs via chemical modifications,[10–12] there is essentially no structural limit to the type and kind of pharmaceutical agents that may be transported by lipoproteins. Plasma lipoproteins are spherical particles that range from 5 to over 1,000 nm in diameter (Figure 37.2). Because of their superior stability and smaller size, the classes of low density lipoproteins (LDL) and high density lipoproteins (HDL) qualify as drug carrying nanoparticles and will be the subject of this review.

The outstanding features that render lipoproteins suitable carriers of anti-cancer agents may be summarized as follows:

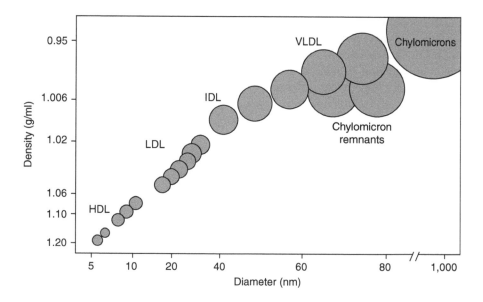

FIGURE 37.2 Buoyant density and size distribution of plasma lipoproteins.

- Lipoproteins represent natural, biologically compatible components providing enhanced safety and efficacy.
- Apolipoproteins provide a signal for the receptor mediated uptake of the whole particle by endocytosis[13] or the selective uptake of the drug/core component.[14]
- Lipoproteins nearly perfectly exemplify the "magic bullet" concept[6] via the receptor-mediated uptake of the encapsulated drug. Because of the over expression of lipoprotein receptors by malignant cells,[15–17] there is an opportunity to attack malignant cells selectively without substantially impacting normal cells. The targeting potential of lipoproteins may be further enhanced by attaching ligands to the respective lipid or protein components.

37.3 LDL AS A DELIVERY VEHICLE FOR ANTI-CANCER AGENTS

The application of LDL as a targeted drug delivery vehicle was initiated by Krieger et al.[18,19] who replaced the core of native LDL with cholesteryl linoleate and suggested that hydrophobic compounds, including drugs, may be incorporated into LDL for diagnostic or therapeutic purposes in a similar manner.[18] Subsequently, Gal et al.[20] proposed an LDL-based drug delivery system for treating gynecological neoplasms. To date, numerous studies have been performed with LDL particles as targeted drug carriers; nevertheless, the enthusiasm for LDL-based drug delivery systems has been modest as indicated by the paucity of approved therapeutic formulations or ongoing clinical trials at present.

The major challenge in the preparation of reconstituted LDL (rLDL) nanoparticles is the availability of apolipoprotein B-100 (apoB-100), the major protein component of LDL.[13] In addition, the high molecular weight of apoB-100 and its tendency to aggregate upon delipidation present additional difficulties in the preparation of rLDL-based drug formulations. Early attempts to prepare rLDL/drug complexes were undertaken by Lundberg,[21–23] combining egg yolk phosphatidylcholine (EYPC), a hydrophobic drug (cytotoxic mustard carbamate), and apoB-100.[20] The resultant nanoparticle had a diameter of about 23 nm and was metabolized by fibroblasts in a manner similar to native LDL.[21] A series of studies by Van Berkel et al. developed unilamellar liposome-like particles, containing apolipoprotein E that were taken up by tissues via the LDL receptor upon injection into rats.[24] Van Berkel et al. also encapsulated a lipophilic derivative of daunorubicin into these nanoparticles that were subsequently taken up by B16 tumors[10] via the LDL receptor.[25]

Maranhao et al. developed an alternate approach to prepare a drug carrying, protein-free microemulsion (LDE) by combining egg yolk phosphatidyl choline, triolein, cholesterol, and cholesteryl oleate.[26] The LDE particles apparently associated with apoE upon injection into rats and had clearance rates similar to LDL. These LDE particles were subsequently loaded with either oleoyl-etoposide[12] or oleoyl-paclitaxel[27] to explore their respective therapeutic potentials. Studies with oleoyl-paclitaxel in mice showed a nine-fold increase in LD_{50} of oleoyl-paclitaxel encapsulated into LDE vs. the free drug.[27] A novel approach for preparing rLDL/drug complexes was reported by Owens et al. who utilized peptides representing the receptor binding region of apoB instead of full length apo B.[28] This microemulsion resembling LDL was able to support the growth of U937 cells similar to native LDL.[29]

37.4 HDL AS A DELIVERY VEHICLE FOR ANTI-CANCER AGENTS

The utilization of lipoprotein nanoparticles for the delivery of anti-cancer agents is based on the hypothesis that rapidly proliferating cells (including cancer cells) have a higher expression of lipoprotein receptors[15–17] to meet their increased need for cholesterol (new cell membrane synthesis). This view is consistent with the lower plasma cholesterol levels found in cancer patients.[30] Clinical studies showed that HDL cholesterol levels were lower in cancer patients versus normal controls than their LDL-cholesterol.[31] Similar findings were reported for patients with hematologic malignancies,[32–34] small cell lung cancer,[35] and colorectal adenoma.[36] Recent findings with the HDL receptors (scavenger receptor B1 [SR-B1], CD36, and LIMPII analogous-1 [CLA-1]) show that breast cancer cells have substantially higher expression of this HDL receptor than the surrounding normal cells.[37] These findings suggest that targeted delivery of anti-cancer drugs using an HDL delivery system is likely to be highly effective.[38]

The proliferation of adenocarcinoma[39] and other cancer cells[40,41] were shown to be enhanced by HDL or HDL components. In addition, Pussinen et al. have shown that breast cancer cells take up cholesterol from HDL via the CLA-1 receptor pathway.[42] These findings show that HDL receptors may play a key role in controlling the proliferative capacity of cancer cells, and they establish the concept of selective targeting of anti-cancer agents delivered via reconstituted HDL (rHDL). These findings are consistent with findings from this laboratory, showing that cancer cells produced substantially stronger immunoblots with an SR-BI antibody than normal fibroblasts.[43] It has also been shown that the respective uptakes of paclitaxel and cholesteryl esters were highly correlated ($r^2 = 0.88$; $p = 0.04$) when delivered as core components of rHDL nanoparticles, indicating that the paclitaxel was taken up by a receptor-mediated pathway.[43]

The delivery of anti-cancer agents via native HDL is exceedingly challenging because of the labor involved in the large scale isolation of human plasma HDL. Concerns about bio-safety would likely render the development of the HDL/drug formulations perhaps even more costly than those employing LDL.[44] So far, there have been only limited attempts to incorporate drugs into native HDL[45] and none with the ultimate purpose of therapeutic applications. Here, rHDL has proven to be a much more effective drug delivery vehicle than native HDL.

The nomenclature currently used in the literature is confusing as to what rHDL actually represents. The terms *reconstituted* and *recombinant* are both being used to describe artificially generated HDL-type nanoparticles although the former designation is becoming more predominant. The nomenclature is further confounded by the naming of rHDL nanoparticles based on the

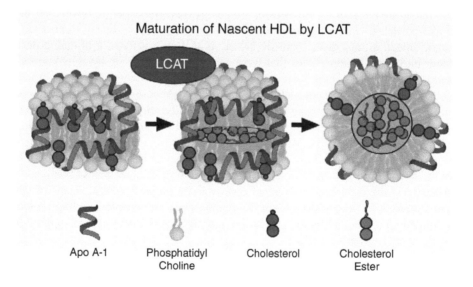

FIGURE 37.3 Transformation of the discoidal precursor (nascent) HDL to its stable, spherical configuration via the accumulation of cholesteryl esters, catalyzed by lecithin: cholesterol acyltransferase (LCAT). (From Alexander, E. T., et al., *Biochemistry*, 44(14), 5409–5419, 2005. With permission.)

different stages in the maturation of the circulating HDL particle (Figure 37.3).[46,47] HDL metabolism commences by the secretion of discoidal, primarily phosphatidyl choline (PC) and apolipoprotein A-I (apoA-I) containing particles by the liver[48] and the intestine. These HDL precursors are subsequently converted to spherical (mature) HDL[49] by the action of lecithin:cholesterol acyltransferase (LCAT) by expanding the particle's cholesteryl ester core (Figure 37.3). Scientific reports and patent disclosures have referred to *both discoidal and spherical nanoparticles* as "rHDL."[50–54]

For the purpose of formulating drug delivery complexes, spherical nanoparticles have an advantage over other configurations as their core region can accommodate the drug to be transported (Figure 37.1). Van Berkel et al. prepared spherical neo-high density lipoproteins as drug delivery vehicles[54] by incorporating the lipophilic derivative of iododeoxyuridine into neo-HDL, resulting in a nanoparticle resembling circulating apoE-free HDL.[11] Earlier work in this laboratory yielded stable rHDL/drug nanoparticles composed of Taxol® and the components of circulating HDL.[43] The paclitaxel from the rHDL/Taxol® nanoparticles was apparently taken up by cancer cells via a receptor mediated mechanism.[43] An enhanced rHDL formulation containing paclitaxel has been recently developed with increased drug carrying capacity and cytotoxicity toward several cancer cell lines.[55]

37.5 CONCLUSIONS

Lipoprotein-based nanoparticles have the potential to provide a superior performance compared to most alternative formulations for drug delivery agents because of their natural, biodegradable ingredients that also help to protect against rapid removal from the circulation by the reticuloendothelial system.[44,56] Additional important issues in favor of lipoproteins nanoparticles are the biological compatibility of lipoprotein based formulations,[44,56,57] many that have already been safely injected into human subjects.[58–62] Lipoprotein nanoparticles have substantially smaller molecular diameter than most liposomal drug preparations, a considerable advantage, as even smaller liposomes have been considered to have superior pharmacokinetic properties.[63] Another important issue favoring lipoprotein nanoparticles over alternate drug delivery systems is their targeting potential via receptor-mediated mechanisms that are over expressed in cancer cells vs. normal cells.[15,17,37] This type of lipoprotein-based (LDL) drug delivery mechanism has been shown to be superior in the delivery of omega-3 fatty acids to cancer cells over serum albumin.[64]

One of the major advantages of lipoprotein nanoparticles in cancer chemotherapy is that their targeting potential may be substantially enhanced by attaching ligands to their surface lipid or protein components. Targeted drug delivery is of major interest in cancer chemotherapy where several monoclonal antibodies have already been employed as targeting agents.[65,66] However, there have been only limited attempts to explore this drug delivery model; it has essentially unlimited potential in cancer chemotherapy. Earlier attempts to enhance the targeting of the lipoprotein nanoparticles was reported by Van Berkel et al. by using lactosylated derivatives of LDL[67,68] and neo-HDL.[69] These modified lipoprotein nanoparticles were shown to be preferentially targeted to parenchymal liver cells[67,68] that provided a model for converting water soluble drugs to pro-drugs for encapsulaton into lipoproteins[69] and a potential drug delivery vehicle for treating hepatitis B.[70] Additional attempts for enhancing the targeting of lipoprotein nanoparticles involved the attachment of monoclonal antibodies (MABs) to LDL to target histocompatibility antigens.[45] Recent studies in this laboratory led to the development of an rHDL particle with folate residues attached to the apo A-I component. These folate-bearing nanoparticles promoted a substantially enhanced uptake of paclitaxel by OVCAR-3 cells from the modified rHDL/paclitaxel complex.

Despite many potential advantages over other drug delivery models, the application of lipoprotein nanoparticles for cancer chemotherapy currently lags behind other drug delivery systems. Although lipoprotein nanoparticles are biologically compatible and offer superb targeting potential via receptor-mediated mechanisms, the challenge of obtaining the apolipoprotein starting material has so far been a major obstacle in the manufacturing of lipoprotein-based pharmaceutical formulations.

In summary, lipoprotein-based nanoparticles have the potential to perform as superior drug delivery agents for anti-cancer drugs because of their natural biodegradable components that protect against their rapid removal from the circulation and their targeting potential via receptors that are over expressed in cancer cells vs. normal cells.

REFERENCES

1. European Science Foundation Policy Briefing, ESF scientific forward look on nanomedicine. www. esf.org, 2005.
2. U.S. Department of Health and Human Services, National Institutes of Health, and National Cancer Institute, Cancer NANOTECHNOLOGY plan, *A strategic initiative to transform clinical oncology and basic research through the directed application of nanotechnology*, 2004.
3. NCI Alliance for Nanotechnology in Cancer Awards. http://nano.cancer.gov/alliance_awards/
4. Paull, R., Wolfe, J., Hebert, P., and Sinkula, M., Investing in nanotechnology, *Nat. Biotechnol.*, 21, 1144–1147, 2003.
5. Counsell, R. E. and Pohland, R. C., Lipoproteins as potential site-specific delivery systems for diagnostic and therapeutic agents, *J. Med. Chem.*, 25, 1115–1120, 1982.
6. Sörgel, F., The return of Ehrlich's 'Therapia magna sterilisans' and other Ehrlich concepts. Series of papers honoring Paul Ehrlich on the occasion of his 150th birthday, *Chemotherapy*, 50, 6–10, 2004.
7. Dubowchik, G. M. and Walker, M. A., Receptor-mediated and enzyme-dependent targeting of cytotoxic anticancer drugs, *Pharmacol. Ther.*, 83, 67–123, 1999.
8. Allen, T. M. and Cullis, P. R., Drug delivery systems: Entering the mainstream, *Science*, 303, 1818–1822, 2004.
9. Guillemard, V. and Saragovi, H. U., Novel approaches for targeted cancer therapy, *Curr. Cancer Drug Targets*, 4, 313–326, 2004.
10. Versluis, A. J., Rump, E. T., Rensen, P. C., van Berkel, T. J., and Bijsterbosch, M. K., Stable incorporation of a lipophilic daunorubicin prodrug into apolipoprotein E-exposing liposomes induces uptake of prodrug via low-density lipoprotein receptor in vivo, *J. Pharmacol. Exp. Ther.*, 289, 1–7, 1999.
11. de Vrueh, R. L., Rump, E. T., Sliedregt, L. A., Biessen, E. A., van Berkel, T. J., and Bijsterbosch, M. K., Synthesis of a lipophilic prodrug of 9-(2-phosphonyl methoxyethyl)adenine (PMEA) and its incorporation into a hepatocyte-specific lipidic carrier, *Pharm. Res.*, 16, 1179–1185, 1999.
12. Azevedo, C. H., Carvalho, J. P., Valduga, C. J., and Maranhao, R. C., Plasma kinetics and uptake by the tumor of a cholesterol-rich microemulsion (LDE) associated to etoposide oleate in patients with ovarian carcinoma, *Gynecol. Oncol.*, 97, 178–182, 2005.
13. Segrest, J. P., Jones, M. K., De Loof, H., and Dashti, N., Structure of apolipoprotein B-100 in low density lipoproteins, *J. Lipid Res.*, 42, 1346–1367, 2001.
14. Williams, D. L., Temel, R. E., and Connelly, M. A., Roles of scavenger receptor BI and APO A-I in selective uptake of HDL cholesterol by adrenal cells, *Endocr. Res.*, 26, 639–651, 2000.
15. Tosi, M. R. and Tugnoli, V., Cholesteryl esters in malignancy, *Clin. Chim. Acta*, 359, 27–45, 2005.
16. Ho, Y. K., Smith, R. G., Brown, M. S., and Goldstein, J. L., Low-density lipoprotein (LDL) receptor activity in human acute myelogenous leukemia cells, *Blood*, 52, 1099–1114, 1978.
17. Imachi, H., Murao, K., Sayo, Y., Hosokawa, H., Sato, M., Niimi, M., Kobayashi, S., Miyauchi, A., Ishida, T., and Takahara, J., Evidence for a potential role for HDL as an important source of cholesterol in human adrenocortical tumors via the CLA-1 pathway, *Endocr. J.*, 46, 27–34, 1999.
18. Krieger, M., Brown, M. S., Faust, J. R., and Goldstein, J. L., Replacement of endogenous cholesteryl esters of low density lipoprotein with exogenous cholesteryl linoleate. Reconstitution of a biologically active lipoprotein particle, *J. Biol. Chem.*, 253, 4093–4101, 1978.
19. Krieger, M., Goldstein, J. L., and Brown, M. S., Receptor-mediated uptake of low density lipoprotein reconstituted with 25-hydroxycholesteryl oleate suppresses 3-hydroxy-3-methylglutaryl-coenzyme. A reductase and inhibits growth of human fibroblasts, *Proc. Natl Acad Sci. U.S.A.* 75, 5052–5056, 1978.

20. Gal, D., Ohashi, M., MacDonald, P. C., Buchsbaum, H. J., and Simpson, E. R., Low-density lipoprotein as a potential vehicle for chemotherapeutic agents and radionucleotides in the management of
gynecologic neoplasms, *Am. J. Obstet. Gynecol.*, 139, 877–885, 1981.

21. Lundberg, B., Preparation of drug-low density lipoprotein complexes for delivery of antitumoral
drugs via the low density lipoprotein pathway, *Cancer Res.*, 47, 4105–4108, 1987.

22. Lundberg, B. and Suominen, L., Preparation of biologically active analogs of serum low density
lipoprotein, *J. Lipid Res.*, 25, 550–558, 1984.

23. Lundberg, B., Cytotoxic activity of two new lipophilic steroid nitrogen carbamates incorporated into
low-density lipoprotein, *Anticancer Drug Des.*, 9, 471–476, 1994.

24. Rensen, P. C., Schiffelers, R. M., Versluis, A. J., Bijsterbosch, M. K., Van Kuijk-Meuwissen, M. E.,
and Van Berkel, T. J., Human recombinant apolipoprotein E-enriched liposomes can mimic low-
density lipoproteins as carriers for the site-specific delivery of antitumor agents, *Mol. Pharmacol.*, 52,
445–455, 1997.

25. Versluis, A. J., Rensen, P. C., Rump, E. T., Van Berkel, T. J., and Bijsterbosch, M. K., Low-density
lipoprotein receptor-mediated delivery of a lipophilic daunorubicin derivative to B16 tumours in mice
using apolipoprotein E-enriched liposomes, *Br. J. Cancer*, 78, 1607–1614, 1998.

26. Maranhão, R. C., Cesar, T. B., Pedroso-Mariani, S. R., Hirata, M. H., and Mesquita, C. H., Metabolic
behavior in rats of a nonprotein microemulsion resembling low-density lipoprotein, *Lipids*, 28,
691–696, 1993.

27. Rodrigues, D. G., Maria, D. A., Fernandes, D. C., Valduga, C. J., Couto, R. D., Ibañez, O. C., and
Maranhão, R. C., Improvement of paclitaxel therapeutic index by derivatization and association to a
cholesterol-rich microemulsion: In vitro and in vivo studies, *Cancer Chemother. Pharmacol.*, 55,
565–576, 2005.

28. Owens, M. D., Baillie, G., and Halbert, G. W., Physicochemical properties of microemulsion analogues of low density lipoprotein containing amphiphatic apoprotein B receptor sequences, *Int.
J. Pharm.*, 228, 109–117, 2001.

29. Baillie, G., Owens, M. D., and Halbert, G. W., A synthetic low density lipoprotein particle capable of
supporting U937 proliferation in vitro, *J. Lipid Res.*, 43, 69–73, 2002.

30. Markel, A. and Brook, G. J., Cancer and hypocholesterolemia, *Isr. J. Med. Sci.*, 30, 787–793, 1994.

31. Fiorenza, A. M., Branchi, A., and Sommariva, D., Serum lipoprotein profile in patients with cancer.
A comparison with non-cancer subjects, *Int. J. Clin. Lab Res.*, 30, 141–145, 2000.

32. Dessì, S., Batetta, B., Pulisci, D., Accogli, P., Pani, P., and Broccia, G., Total and HDL cholesterol in
human hematologic neoplasms, *Int. J. Hematol.*, 54, 483–486, 1991.

33. Moschovi, M., Trimis, G., Apostolakou, F., Papassotiriou, I., and Tzortzatou-Stathopoulou, F. J.,
Serum lipid alterations in acute lymphoblastic leukemia of childhood, *Pediatr. Hematol. Oncol.*,
26, 289–293, 2004.

34. Scribano, D., Baroni, S., Pagano, L., Zuppi, C., Leone, G., and Giardina, B., Prognostic relevance of
lipoprotein cholesterol levels in acute lymphocytic and nonlymphocytic leukemia, *Haematologica*,
81, 343–345, 1996.

35. Siemianowicz, K., Gminski, J., Stajszczyk, M., Wojakowski, W., Goss, M., Machalski, M., Telega,
A., Brulinski, K., and Magiera-Molendowska, H., Serum HDL cholesterol concentration in patients
with squamous cell and small cell lung cancer, *Int. J. Mol. Med.*, 6, 307–311, 2000.

36. Bayerdörffer, E., Mannes, G. A., Richter, W. O., Ochsenkühn, T., Seeholzer, G., Köpcke, W.,
Wiebecke, B., and Paumgartner, G., Decreased high-density lipoprotein cholesterol and increased
low-density cholesterol levels in patients with colorectal adenomas, *Ann. Int. Med.*, 118, 481–487,
1993.

37. Cao, W. M., Murao, K., Imachi, H., Yu, H., Abe, H., Yamauchi, A., Niimi, M., Miyauchi, A., Wong,
N. C., and Ishida, T., A mutant high-density lipoprotein receptor inhibits proliferation of human breast
cancer cells, *Cancer Res.*, 64, 1515–1521, 2004.

38. Lacko, A. G., Nair, M. P., Paranjape, S., Johnson, S., and McConathy, W. J., A novel delivery system
for targeted cancer therapy, In *Stem Cell and Targeted Therapy*, Dickey, K. A. and Keating, A., Eds.,
Garden Jennings, Charlottesville, VA, pp. 179–183, 2003.

39. Favre, G., Tazi, K. A., Le Gaillard, F., Bennis, F., Hachem, H., and Soula, G., High density lipopro-
tein3 binding sites are related to DNA biosynthesis in the adenocarcinoma cell line A549, *J. Lipid
Res.*, 34, 1093–1106, 1993.

40. Gospodarowicz, D., Lui, G. M., and Gonzalez, R., High-density lipoproteins and the proliferation of human tumor cells maintained on extracellular matrix-coated dishes and exposed to defined medium, *Cancer Res.*, 42, 3704–3713, 1982.

41. Jozan, S., Faye, J. C., Tournier, J. F., Tauber, J. P., David, J. F., and Bayard, F., Interaction of estradiol and high density lipoproteins on proliferation of the human breast cancer cell line MCF-7 adapted to grow in serum free conditions, *Biochem. Biophys. Res. Commun.*, 133, 105–112, 1985.

42. Pussinen, P. J., Karten, B., Wintersperger, A., Reicher, H., McLean, M., Malle, E., and Sattler, W., The human breast carcinoma cell line HBL-100 acquires exogenous cholesterol from high-density lipoprotein via CLA-1 (CD-36 and LIMPII analogous 1)-mediated selective cholesteryl ester uptake, *Biochem. J.*, 349, 559–566, 2000.

43. Lacko, A. G., Nair, M., Paranjape, S., Johnson, S., and McConathy, W. J., High density lipoprotein complexes as delivery vehicles for anticancer drugs, *Anticancer Res.*, 22, 2045–2049, 2002.

44. De Smidt, P. C. and van Berkel, T. J., LDL-mediated drug targeting, *Crit. Rev. Ther. Drug Carrier. Syst.*, 7, 99–120, 1990.

45. Shaw, J. M. and Shaw, K. V., Key issues in the delivery of pharmacological agents using lipoproteins: Design of a synthetic apoprotein-lipid carrier, *Targeted Diagn. Ther.*, 5, 351–383, 1991.

46. Barter, P. J., Hugh sinclair lecture: The regulation and remodelling of HDL by plasma factors, *Atheroscler. Suppl.*, 3, 39–47, 2002.

47. Linsel-Nitschke, P. and Tall, A. R., HDL as a target in the treatment of atherosclerotic cardiovascular disease, *Nat. Rev. Drug Discov.*, 4, 193–205, 2005.

48. Hamilton, R. L., Williams, M. C., Fielding, C. J., and Havel, R. J., Discoidal bilayer structure of nascent high density lipoproteins from perfused rat liver, *J. Clin. Invest.*, 58, 667–680, 1976.

49. Forte, T., Norum, K. R., Glomset, J. A., and Nichols, A. V., Plasma lipoproteins in familial lecithin: Cholesterol acyltransferase deficiency: Structure of low and high density lipoproteins as revealed by electron microscopy, *J. Clin. Invest.*, 50, 1141–1148, 1971.

50. Lerch, P. G., Fortsch, V., Hodler, G., and Bolli, R., Production and characterization of a reconstituted high density lipoprotein for therapeutic applications, *Vox Sang.*, 71, 155–164, 1996.

51. Levine, D. M., Simon, S. R., Gordon, B. R., Parker, T. S., Saal, S. D., and Rubin, A. L., United States Patent no. 5,128,318.

52. Sparks, D. L., Phillips, M. C., and Lund-Katz, S., The conformation of apolipoprotein A-I in discoidal and spherical recombinant high density lipoprotein particles 13C. NMR studies of lysine ionization behavior, *J. Biol. Chem.*, 267, 25830–25838, 1992.

53. Jonas, A., Wald, J. H., Toohill, K. L., Krul, E. S., and Kézdy, K. E., The conformation of apolipoprotein A-I in discoidal and spherical recombinant high density lipoprotein particles. 13C NMR studies of lysine ionization behavior, *J. Biol. Chem.*, 265, 22123–22129, 1990.

54. Schouten, D., van der Kooij, M., Muller, J., Pieters, M. N., Bijsterbosch, M. K., and van Berkel, T. J., Development of lipoprotein-like lipid particles for drug targeting: Neo-high density lipoproteins, *Mol. Pharmacol.*, 44, 486–492, 1993.

55. Lacko, A. G., Nair, M., Paranjape, S., Mooberry, L., and McConathy, W. J., *Advanced Drug Delivery System for Breast Cancer Chemotherapy Department of Defense, Congressionally Directed Medical Research Program "ERA of Hope" Breast Cancer Research Review*, Philadelphia, Pennsylvania, 2005. http://mrmcweb4.detrick.army.mil/bcrp/era/abstracts2005/0110582_abs.pdf

56. Shaw, J. M., Shaw, K. V., Yanovich, S., Iwanik, M., Futch, W. S., Rosowsky, A., and Schook, L. B., Delivery of lipophilic drugs using lipoproteins, *Ann. NY Acad. Sci.*, 507, 252–271, 1987.

57. Adams, T., Alanazi, F., and Lu, D. R., Safety and utilization of blood components as therapeutic delivery systems, *Curr. Pharm. Biotechnol.*, 4, 275–282, 2003.

58. Nanjee, M. N., Crouse, J. R., King, J. M., Hovorka, R., Rees, S. E., Carson, E. R., Morgenthaler, J. J., Lerch, P., and Miller, N. E., Effects of intravenous infusion of lipid-free apo A-I in humans, *Arterioscler. Thromb. Vasc. Biol.*, 16, 1203–1214, 1996.

59. Nanjee, M. N., Doran, J. E., Lerch, P. G., and Miller, N. E., Acute effects of intravenous infusion of ApoA1/phosphatidylcholine discs on plasma lipoproteins in humans, *Arterioscler. Thromb. Vasc. Biol.*, 19, 979–989, 1999.

60. Bisoendial, R. J., Hovingh, G. K., Levels, J. H., Lerch, P. G., Andresen, I., Hayden, M. R., Kastelein, J. J., and Stroes, E. S., Restoration of endothelial function by increasing high-density lipoprotein in subjects with isolated low high-density lipoprotein, *Circulation*, 107, 2944–2948, 2003.

61. Pajkrt, D., Doran, J. E., Koster, F., Lerch, P. G., Arnet, B., van der Poll, T., ten Cate, J. W., and van Deventer, S. J., Antiinflammatory effects of reconstituted high-density lipoprotein during human endotoxemia, *J. Exp. Med.*, 184, 1601–1608, 1996.

62. Nanjee, M. N., Cooke, C. J., Garvin, R., Semeria, F., Lewis, G., Olszewski, W. L., and Miller, N. E., *J. Lipid Res.*, 42, 1586–1593, 2001.

63. Palatini, P., Disposition kinetics of phospholipid liposomes, *Adv. Exp. Med. Biol.*, 318, 375–391, 1992.

64. Edwards, I. J., Berquin, I. M., Sun, H., O'Flaherty, J. T., Daniel, L. W., Thomas, M. J., Rudel, L. L., Wykle, R. L., and Chen, Y. Q., Differential effects of delivery of omega-3 fatty acids to human cancer cells by low-density lipoproteins versus albumin, *Clin. Cancer Res.*, 10, 8275–8283, 2004.

65. Toi, M., Takada, M., Bando, H., Toyama, K., Yamashiro, H., Horiguchi, S., and Saji, S., Current status of antibody therapy for breast cancer, *Breast Cancer*, 11, 10–14, 2004.

66. Boskovitz, A., Wikstrand, C. J., Kuan, C. T., Zalutsky, M. R., Reardon, D. A., and Bigner, D. D., Monoclonal antibodies for brain tumour treatment, *Expert. Opin. Biol. Ther.*, 4, 1453–1471, 2004.

67. Bijsterbosch, M. K. and Van Berkel, T. J., Uptake of lactosylated low-density lipoprotein by galactose-specific receptors in rat liver, *Biochem. J.*, 270, 233–239, 1990.

68. Bijsterbosch, M. K. and Van Berkel, T. J., Lactosylated high density lipoprotein: A potential carrier for the site-specific delivery of drugs to parenchymal liver cells, *Mol. Pharmacol.*, 41, 404–411, 1992.

69. de Vrueh, R. L., Rump, E. T., Sliedregt, L. A., Biessen, E. A., van Berkel, T. T., and Bijsterbosch, M. K., Synthesis of a lipophilic prodrug of 9-(2-phosphonyl methoxyethyl)adenine (PMEA) and its incorporation into a hepatocyte-specific lipidic carrier, *Pharm. Res.*, 16, 1179–1185, 1999.

70. de Vrueh, R. L., Rump, E. T., van De Bilt, E., van Veghel, R., Balzarini, J., Biessen, E. A., van Berkel, T. J., and Bijsterbosch, M. K., Carrier-mediated delivery of 9-(2-phosphonylmethoxyethyl)adenine to parenchymal liver cells: A novel therapeutic approach for hepatitis B, *Antimicrob. Agents Chemother.*, 44, 477–483, 2000.

38 DQAsomes as Mitochondria-Targeted Nanocarriers for Anti-Cancer Drugs

Shing-Ming Cheng, Sarathi V. Boddapati, Gerard G. M. D'Souza, and Volkmar Weissig

CONTENTS

38.1   Introduction 787
38.2   Apoptosis and Mitochondria 788
38.3   Proapoptotic Drugs Acting on Mitochondria 788
38.4   Mitochondriotropic Vesicles (DQAsomes) 790
38.5   DQAsome-Mediated Delivery of pDNA to Mitochondria in Living
        Mammalian Cells 791
38.6   Encapsulation of Paclitaxel into DQAsomes 793
38.7   DQAsomal-Encapsulated Paclitaxel Triggers Apoptosis In Vitro 794
38.8   Tumor Growth Inhibition Study with Paclitaxel-Loaded DQAsomes In Vivo 796
38.9   Concluding Remarks 797
References 797

38.1 INTRODUCTION

A major challenge in treating cancer is the lack of selectivity of the currently available cytotoxic agents. Many such agents fail to significantly distinguish between cancer cells and healthy cells. Consequently, systemic application of these drugs is prone to cause severe side effects in other tissues, greatly limiting the maximal allowable dose of the drug. Another major hurdle to successfully treating cancer is multi-drug resistance (MDR). There are multiple putative origins of MDR, but one of the most studied today is the efflux pump that serves to remove drug molecules from the cell before they can act at their particular subcellular target inside the cell. Both problems, i.e., non-specificity and MDR, could potentially be solved by the development of highly selective tumor-specific drug delivery systems. Progress in this direction has already been made with the development and subsequent Food and Drug Administration (FDA) approval of Doxil®, the first liposomal anti-cancer drug, and similar systems are under development. However, the DQAsome-based approach described in this chapter takes drug delivery one significant step farther. To efficiently and selectively eradicate carcinoma cells, the cytotoxic drug not only needs to be delivered to the tumor cell, but it must also be delivered to the particular target inside the cell. Tumor-specific subcellular drug delivery constitutes a new approach for the chemotherapy of cancer. This approach becomes increasingly feasible as the particular mechanism of action, i.e., the molecular targets of anti-cancer drugs, is being recognized. Transporting the cytotoxic drug to its intracellular target could potentially overcome MDR by bypassing the p-glycoprotein, i.e., the drug would literally be hidden from the p-glycoprotein inside the delivery system until it becomes selectively released at the particular intracellular site of action. At the same time, a sub-cellular delivery system would significantly increase the subcellular bioavailability of any drug acting inside a cell.

38.2 APOPTOSIS AND MITOCHONDRIA

Apoptosis (programmed cell death) plays a central role in tissue homeostasis, and it is generally recognized that inhibition of apoptosis may contribute to cell transformation.[1] Significant knowledge has been accumulated during the past several years about how the apoptotic machinery is controlled.[2–20] As each new regulatory mechanism had been identified, dysfunction of that mechanism has been linked to one or another type of cancer.[17] Dysregulation of the apoptotic machinery is now generally accepted as an almost universal component of the transformation process of normal cells into cancer cells.

A large body of experimental data demonstrates that mitochondria play a key role in the complex apoptotic mechanism.[18–49] Mitochondria have been shown to trigger cell death via several mechanisms: by disrupting electron transport and energy metabolism, by releasing or activating proteins that mediate apoptosis, and by altering cellular redox potential. A critical event leading to programmed cell death is the mitochondrial membrane permeabilization that is under the control of the permeability transition pore complex (mPTPC), a multiprotein complex formed at the contact site between the mitochondrial inner and outer membranes. The mPTPC is widely accepted as being central to the process of cell death and has accordingly been recommended as a privileged pharmacological target for cytoprotective and for cytotoxic therapies in general.[50] In particular, it has been suggested that targeting specific mPTPC components may overcome bcl-2 mediated apoptosis inhibition in cancer cells.[1] Several studies have already demonstrated the feasibility of eliminating neoplastic cells by selectively inducing apoptosis (reviewed in reference 17). The design of mitochondria-targeted cytotoxic drugs has been formulated as a novel strategy for overcoming apoptosis resistance in tumor cells,[1] which, intriguingly, opens up a whole new avenue for the therapy of cancer by "tricking cancer cells into committing suicide."[17]

38.3 PROAPOPTOTIC DRUGS ACTING ON MITOCHONDRIA

Several conventional anti-cancer drugs, such as doxorubicin, and cisplatin, have no direct effect on mitochondria.[51] These conventional chemotherapeutic agents elicit mitochondrial permeabilization in an indirect fashion by induction of endogenous effectors that are involved in the physiologic control of apoptosis.[1] However, a variety of clinically approved drugs such as paclitaxel,[52–60] VP-16 (etoposide)[61–64] and vinorelbine[58] as well as an increasing number of experimental anti-cancer drugs such as betulinic acid, lonidamine, CD-437 (a synthetic retinoids) and ceramide (reviewed in[1]) have been found to act directly on mitochondria resulting in triggering apoptosis. These agents may induce apoptosis in circumstances in which conventional drugs fail to act because endogenous apoptosis inducing pathways, e.g., such as those involving p53, death receptors or apical caspase activation, are disrupted, leading to the apoptosis-resistance of tumor cells. For example, several in vitro and in vivo studies have shown that the synthetic retinoid CD437 is able to induce apoptosis in human lung, breast, cervical and ovarian carcinoma cells (reviewed in Kaufmann and Gores 2000). It could be demonstrated that in intact cells, CD437-dependent caspase activation is preceded by the release of cytochrome C from mitochondria.[65] Moreover, it was shown that when added to isolated mitochondria, CD437 causes membrane permeabilization and that this effect is prevented by inhibitors of the mPTPC such as cyclosporine A. CD437 constitutes an experimental drug that exerts its cytotoxic effect via the mPTPC, i.e., by acting directly at the surface or inside of mitochondria.

The development of anti-cancer drugs whose cytotoxic effects depend on their direct interaction with mitochondria inside of living cells raises the issue of the intracellular distribution of these drugs after having been taken up by the cell, i.e., the question of their intracellular bioavailability. Independent of their mode of cell entry that should mostly take place via passive diffusion through the cell membrane, drug molecules become randomly distributed among all cell organelles and, depending on the chemical nature of the drug, eventually metabolized. Therefore, anti-cancer drugs that exert their cytotoxic activity by directly acting at or inside of mitochondria in living cells would dramatically benefit from a delivery system that can selectively transport these drugs to and into mitochondria.

One of the major roles of mitochondria in the metabolism of eukaryotic cells is the synthesis of Adenosine triphosphate (ATP) by oxidative phosphorylation via the respiratory chain. According to Mitchell's chemiosmotic hypothesis, electrons from the hydrogens on Nicotine amide adenine dinucleotide (NADH) and Flavin adenine dinucleotide ($FADH_2$) are carried along the respiratory chain at the mitochondrial inner membrane, thereby releasing energy that is used to pump protons across the inner membrane from the mitochondrial matrix into the intermembrane space. This process creates a transmembrane electrochemical gradient that includes contributions from both a membrane potential (negative inside) and a pH difference (acidic outside). The membrane potential of mitochondria in vitro is between 180 and 200 mV, the maximum a lipid bilayer can sustain while maintaining its integrity.[66] Although this potential is reduced in living cells and organism to about 130–150 mV as a result of metabolic processes such as ATP synthesis and ion transport,[67] it is by far the largest within cells.

Most interestingly, carcinoma cells posses a different mitochondrial membrane potential relative to normal cells. It has been found that in many carcinoma cell lines, the mitochondrial membrane potential is higher than in normal epithelial cells (reviewed in reference 68). For example, the difference of the mitochondrial membrane potential between the colon carcinoma cell line CX-1 and the control green monkey kidney epithelial cell line CV-1 has been reported to be approximately 60 mV.[68] Moreover, some carcinoma cells, in particular, human breast adeno-carcinoma-derived cells, have in addition to the higher mitochondrial membrane potential also an elevated plasma membrane potential relative to their normal parent cell lines.[68–75]

The striking difference between normal cells and human adenocarcinoma cells regarding the electrical charge of both plasma and mitochondrial membranes has lead numerous investigators during the 1990s to explore fundamentally new strategies for the selective targeting of cancer cells. Their attempts have been based on compounds with a delocalized charge that have long been known to accumulate in mitochondria of living cells in response to the mitochondrial membrane potential. Many of these DLCs are toxic to mitochondria at high concentration. For example, the rhodacyanine official name of this compound, unknown what the letters stand for (MKT-077[71,76–80]) was the first DLC to be approved by the FDA for clinical trials for the treatment of carcinoma.[81] The trials were discontinued, however, because efficacy in tumor cell killing was not demonstrated at the particular approved dosage and drug regimen.[68] From the phase I clinical trial, it was concluded that it is feasible to target carcinoma cell mitochondria with rhodacyanine analogues if drugs with higher therapeutic indices could be developed.[81]

The DQAsomal-based strategy involving the use of dequalinium chloride, a typical representative of DLCs, for tumor and mitochondria-specific targeting factually starts where these failed attempts of the 1990s have left off. This approach combines the well-proven ability of DLCs to specifically target carcinoma cell mitochondria with the selective delivery of apoptotically active compounds known to trigger programmed cell death via directly acting at the mitochondrial surface. The DQAsomal approach utilizes the mitochondria-specific affinity of DLCs for the delivery of pro-apoptotic drugs to mitochondria. Therefore, the actual cytotoxic effect is caused by apoptosis-triggering drugs. Any inherent cytotoxicity of the carrier system, i.e., the DLCs used, might only add to the overall efficiency of killing carcinoma cells.

38.4 MITOCHONDRIOTROPIC VESICLES (DQASOMES)

Dequalinium chloride (DQA, Figure 38.1a) represents a typical mitochondriotropic delocalized cation that already almost 20 years ago was shown to selectively accumulate in carcinoma cell mitochondria.[73] Most interestingly in the context of this chapter, it was demonstrated recently that dequalinium B, a new boron carrier for neutron capture therapy, also accumulates preferentially in carcinoma cells over non-transformed cells.[82]

Dequalinium is a dicationic compound resembling "bola"-form electrolytes, i.e., it is a symmetrical molecule with two charge centers separated at a relatively large distance. Such symmetric bola-like structures are well known from archaeal lipids that usually consist of two glycerol backbones connected by two hydrophobic chains.[85,86] The self-assembly behavior of bipolar lipid from Archaea has been extensively studied (reviewed in Gambacorta, Gliozi, and De Rosa 1995). Generally, it has been shown that these symmetric bipolar archaeal lipids can self-associate into mechanically, very stable monolayer membranes. The most striking structural difference between dequalinium and archaeal lipids lies in the number of bridging hydrophobic chains between the polar head groups. Contrary to the common arachaeal lipids, in dequalinium there is only one carbohydrate chain that connects the two cationic hydrophilic head groups. Therefore, this type of bola lipids has been named *single-chain bola-amphiphile*.[87,88] The self-association behavior of this single-chain cationic bola amphiphile was investigated using Monte Carlo computer simulations (Figure 38.1c, left panel),[83] several electron microscopic (EM) techniques (Figure 38.1c, right panel) as well as dynamic laser light scattering.[84]

It was found that, upon sonication, dequalinium forms spherical aggregates with diameters between about 70 and 700 nm that were termed *DQAsomes*.[84] Freeze fracture images (Figure 38.1) show both convex and concave fracture faces. These images strongly indicate the

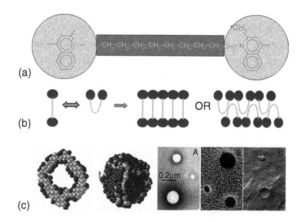

FIGURE 38.1 (a) Chemical structure of dequalinium chloride with overlaid colors indicating, in blue, the hydrophilic part and in yellow the hydrophobic part of the molecule. (b) Theoretical possible conformations of dequalinium chloride, i.e., stretched versus horseshoe conformation, leading to either a monolayer or a bilayer membranous structure following the process of self-assembly. (c) Left panel: Monte Carlo Computer Simulations demonstrate the possible self-assembly of dequalinium chloride into vesicles. (From Weissig, V., Mogel, H. J., Wahab, M., and Lasch, J., *Proceed. Intl. Symp. Control. Rel. Bioact. Mater.*, 25, 312, 1998.) Right panel: Electron microscopic images of vesicles (DQAsomes) made from dequalinium chloride, from left to right: Negatively stained, rotary shadowed, freeze fractured. (From Weissig, V., Lasch, L., Erdos, G., Meyer, H. W., Rowe, T. C., and Hughes, J., *Pharm. Res.*, 15(2), 334–337, 1998. With permission.)

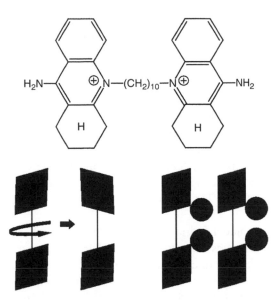

FIGURE 38.2 Top panel: Structure of the cyclohexyl derivative of dequalinium. Bottom panel: Schematic illustration of the stabilizing effect of the cyclohexyl ring system (black circles).

liposome-like aggregation of dequalinium. Negatively stained samples (Figure 38.1) demonstrate that the vesicle is impervious to the stain and appears as a clear area surrounded by stain with no substructure visible. Particle size measurements of DQAsomes stored at room temperature for 24 and 96 h (not shown) do not show any significant changes in their size distribution in comparison to freshly made vesicles measured after one hour.[84] This indicates that DQAsomes do not seem to precipitate, to fuse with each other, or to aggregate in solution over a period of several days. The use of dequalinium derivatives for the preparation of DQAsome-like vesicles leads to vesicles with different size distributions.[89] For example, substituting the methyl group by an aliphatic ring system (Figure 38.2) confers superior vesicle forming properties to this bolaamphiphile. These DQAsome-like vesicles, i.e., vesicles prepared from the cyclohexyl-derivative of DQA shown in Figure 38.2, have a very narrow size distribution of 169 ± 50 nm and can be stored at room temperature for at least five months.

In contrast to vesicles made from dequalinium, bolasomes made from the cyclohexyl derivative are also stable upon dilution of the original vesicle preparation. Whereas dequalinium-based bolasomes upon dilution slowly disintegrate over a period of several hours, bolasomes made from the cyclohexyl compound do not show any change in size distribution following dilution. It appears that bulky aliphatic residues attached to the quinolinium heterocycle favor self-association of the planar ring system. It has, therefore, been speculated that the bulky group sterically prevents the free rotation of the hydrophilic head of the amphiphile around the CH_2-axis (Figure 38.2, bottom panel), contributing to improved intermolecular interactions between the amphiphilic monomers.

38.5 DQASOME-MEDIATED DELIVERY OF pDNA TO MITOCHONDRIA IN LIVING MAMMALIAN CELLS

During efforts in developing a mitochondria-specific DNA delivery vector, it could be demonstrated that DQAsomes are able to selectively deliver plasmid DNA (pDNA) to mitochondria within living mammalian cells. It has been shown, in particular, that DQAsomes stably incorporate pDNA,[84] protect the pDNA from nuclease digestion, and mediate its cellular uptake most likely via non-specific endocytosis.[90] Using membrane-mimicking liposomal membranes and isolated rat liver mitochondria, it was shown that DQAsome/pDNA complexes become destabilized upon contact with mitochondrial membranes, but not at cell plasma membranes.[91,92]

The selective destabilization of DQAsomes at mitochondrial membranes may either be caused by the difference in the lipid composition between cytoplasmic and mitochondrial membranes[91] or by the membrane potential-driven diffusion of individual dequalinium molecules from the DQAsome/pDNA complex into the mitochondrial matrix leading to the disintegration of the complex. However, the fact that DQAsomes do not lose their cargo (pDNA) during their contact with the plasma cell membrane (i.e., during their cellular uptake) but do release the entrapped pDNA upon contact with mitochondria appears as a very attractive feature of DQAsomes as a mitochondria-specific drug delivery system. Any encapsulated drug would potentially be released upon contact with mitochondrial membranes leading to the desired high local drug concentration in the immediate proximity to the mPTPC that is recognized as a major target for apoptotically active experimental drugs.[1,93,94]

Before translocating to mitochondria in response to the mitochondrial membrane potential, however, endocytosed DQAsomes have to be released from endosomes into the cytosol. From studies about the intracellular fate of cationic liposome/DNA complexes (lipoplexes), it is known that cationic lipids exert a destabilizing effect on endosomal membranes, leading to the release of at least a fraction of the lipoplex from early endosomes.[95–98] In agreement with these data, it was found that dequalinium-based DQAsomes also display endosomolytic activity.[99] It was shown that adding DQAsomes to liposomes mimicking the lipid composition of endosomal membranes reproducibly lead to the release of liposomal encapsulated fluorescence marker.[99]

Direct evidence for the ability of DQAsomes to transport pDNA selectively to the site of mitochondria was provided by studying the intracellular distribution of mitochondrial leader sequence peptide -pDNA conjugates in cultured BT20 cells using confocal fluorescence microscopy.[100] Figure 38.3 shows images representative of the confocal fluorescence micrographs obtained in this study. The green and red channels used to generate the overlaid images in the far right column are shown separately in the preceding columns to facilitate a careful comparison of the observed staining patterns.

The characteristic mitochondrial staining pattern seen with the red channel is a strong indicator of mitochondrial viability in the imaged cells. From the composite image obtained by overlaying the green and red channels, it can be seen that a sizeable fraction of the intracellular green fluorescence co-localized with the red mitochondrial fluorescence (depicted as white areas in Figure 38.3c). These observations indicate that, in addition to mediating the cellular uptake of the pDNA conjugate, the use of DQAsomes resulted in a definite association of an appreciable fraction of the internalized conjugate with mitochondria.

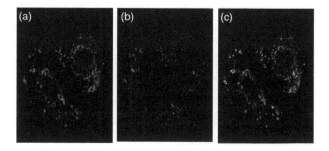

FIGURE 38.3 (**See color insert following page 522.**) Representative confocal fluorescence micrographs of BT20 cells stained with Mitotracker® Red CMXRos (red) after exposure to fluorescein labeled linearized MLS-pDNA conjugate (green) complexed with DQAsomes; (a) red channel, (b) green channel, (c) overlay of red and green channels with white, indicating co-localization of red and green fluorescence.(From D'Souza, G. G., Boddapati, S. V., and Weissig, V., *Mitochondrion*, 5(5), 352–358, 2005. With permission.)

38.6 ENCAPSULATION OF PACLITAXEL INTO DQASOMES

Paclitaxel, generally known as an anti-microtubule agent, has recently been demonstrated to trigger apoptosis by directly acting on mitochondria.[54,55,58] It has been shown that clinically relevant concentrations of paclitaxel directly target mitochondria and trigger apoptosis by inducing cytochrome c (cyt c) release in a permeability transition pore (PTP)-dependent manner.[54] This mechanism of action is known from other pro-apoptotic, directly on mitochondria acting agents.[51] A 24-hour delay between the treatment with paclitaxel or with other PTP inducers and the release of cyt c in cell-free systems compared to intact cells has been explained by the existence of several drug targets inside the cell. Making only a subset of the drug available for mitochondria.[54] Like-wise, other anti-tubulin agents such as vinorelbine or nocodazole could also been shown to trigger the release of cyt c via the direct interaction with mitochondria[101] that subsequently resulted in apoptotic cell death. The detection of tubulin as an inherent component of mitochondrial membranes able to interact with the voltage-dependent anion channel[102] and thereby able to influence the mPTPC seems to make the rather surprising identification of mitochondria as a target for well-established anti-tubulin agents plausible.

For encapsulation of paclitaxel into DQAsomes, dequalinium chloride and paclitaxel were dissolved in methanol followed by removing the organic solvent.[103] After adding buffer, the suspension was sonicated with a probe sonicator until a clear opaque solution was formed. To remove any undissolved material, the sample was centrifuged for 10 min at 3000 rpm. The solubility of paclitaxel in water at 25°C at pH 7.4 is with 0.172 mg/L (0.2 μM) extremely low, making any separation procedure of non-encapsulated paclitaxel unnecessary. However, for control, a paclitaxel suspension was probe sonicated under identical conditions used for the incorporation of paclitaxel into DQAsomes but in the complete absence of dequalinium. As expected, upon centrifugation, no paclitaxel was detectable in the supernatant using UV spectroscopy at 230 nm.[103] Following this procedure, paclitaxel can be incorporated into DQAsomes at a molar ratio paclitaxel to dequalinium of about 0.6. In comparison to the free drug, encapsulation of paclitaxel into DQAsomes increases the drug's solubility by a factor of about 3000.

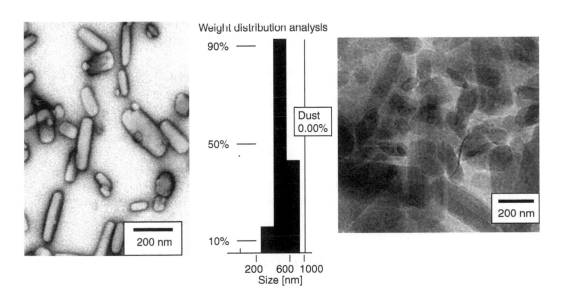

FIGURE 38.4 Paclitaxel encapsulated into DQAsomes. Left panel: transmission electron microscopic image (uranyl acetate staining); middle panel: Size distribution; right panel: Cryo-electron microscopic image. (From Cheng, S. M., Pabba, S., Torchilin, V. P., Fowle, W., Kimpfler, A., Schubert, R. and Weissig, V. *J. Drug Deliv. Sci. Technol.*, 15(1), 81–86, 2005. With permission.)

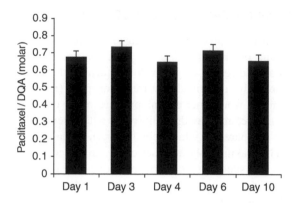

FIGURE 38.5 Paclitaxel remains stably incorporated in DQAsomes: Molar ratio of paclitaxel to DQA in DQAsomes upon storage at 4C. (From Cheng, S. M., Pabba, S., Torchilin, V. P., Fowle, W. Kimpfler, A. Schubert, R., and Weissig, V., *J. Drug Deliv. Sci. Technol.*, 15(1), 81–86, 2005. With permission.)

Considering the known spherical character of DQAsomes, the results of an EM analysis of paclitaxel-loaded DQAsomes seem rather surprising. The transmission EM image (Figure 38.4, left panel) and the cryo-EM image (Figure 38.4, right panel) of an identical sample show with a remarkable conformity worm- or rod-like structures roughly around 400 nm in length, the size of which could also be confirmed by size distribution analysis shown in Figure 38.4 (middle panel). These complexes may represent the formation of worm-like micelles as recently described for self-assembling amphiphilic block co-polymers.[104] Figure 38.5 shows data about the stability of paclitaxel-containing DQAsomes upon storage. Over a period of ten days, the ratio of paclitaxel to dequalinium in the DQAsome preparation remains almost constant, indicating that paclitaxel remains stably incorporated in DQAsomes. However, from a gradual decrease of the total concentration of paclitaxel and of dequalinium over several days and to the same extent (by approximately 20–30% after four days, data not shown) can be concluded that paclitaxel-loaded DQAsomes tend to form larger aggregates that are removable by centrifugation. This tendency to form aggregates is well known from artificial phospholipids vesicles (liposomes) and is generally reversible by shaking or slightly vortexing the preparation.

The stability of paclitaxel-loaded DQAsomes is under physiological conditions in comparison to their shelf-life stability significantly reduced. In general, upon transferring paclitaxel-loaded DQAsomes either into serum-free cell medium (DMEM) at 37°C or into complete serum, they form large aggregates leading to precipitation. However, this process of aggregation seems to be kinetically controlled, giving at least a portion of monomeric paclitaxel-loaded DQAsomes sufficient time to reach their target. The tendency to form aggregates under physiological conditions is common to all cationic carrier systems. Nevertheless, it should be mentioned that cationic lipoplexes and polyplexes are being tested in numerous gene therapeutic clinical trials. Also, cationic colloidal carriers (Catioms) are currently being successfully developed for systemic drug administrations.[105–107]

38.7 DQASOMAL-ENCAPSULATED PACLITAXEL TRIGGERS APOPTOSIS IN VITRO

To test if DQAsomal-encapsulated paclitaxel triggers apoptosis at paclitaxel concentrations where the free drug does not have a significant cytotoxic effect, human colo 205 colon cancer cells were incubated with the free drug, with empty DQAsomes, with a mixture of empty DQAsomes and the free drug, and with the DQAsomal-encapsulated drug. Following the staining of the treated cells

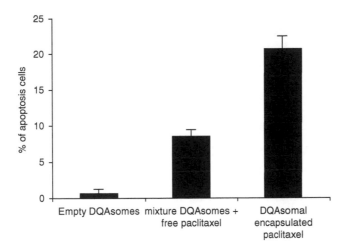

FIGURE 38.6 Human colo 25 colon cancer cells were treated in triplicates with empty DQAsomes (20 nM DQA) with a mixture of empty DQAsomes (20 nM) and free paclitaxel (10 nM) and DQAsomal-encapsulated paclitaxel (20 nM DQA/10 nM paclitaxel). In each case, approximately 400 cells were counted (From Chen, S. M. and Weissig, V., 2006, manuscript in preparation).

with the DNA-binding fluorophore Hoechst 33258, apoptotic nuclei showing the typical apoptotic condensation and fragmentation of chromatin were counted and expressed as percent of the total number of nuclei. Figure 38.6 shows that under identical incubation conditions, 10 nM paclitaxel-encapsulated in DQAsomes more than doubles the number of apoptotic nuclei in comparison to the control whee cells were treated with a mixture of empty DQAsomes and 10 nM free paclitaxel. Likewise, a DNA ladder caused by DNA fragmentation typical for apoptosis could be detected upon incubation of colon cancer cells with 10 nM DQAsomal-encapsulated paclitaxel but not upon incubation with the free drug either alone or in mixture with empty DQAsomes (Figure 38.7). Incubating the cells for the same period of time, the amount of free paclitaxel had to be increased at least 5-fold over the amount of DQAsomal-encapsulated drug in order to generate a DNA ladder (not shown).

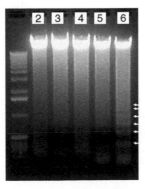

FIGURE 38.7 Human colo 25 colon cancer cells were incubated for 30 h with buffer (lane 2), 20 nM empty DQAsomes (lane 3), 10 nM free paclitaxel (lane 4), a mixture of 20 nM empty DQAsomes and 10 nM free paclitaxel (lane 5), and 20 nM DQAsomes with 10 nM encapsulated paclitaxel (lane 6). White arrow heads indicate the apoptotic DNA ladder (Chen, S. M. and Weissig, V., 2006, manuscript in preparation).

Considering that paclitaxel, generally known as an anti-microtubule agent, has recently been demonstrated to trigger apoptosis by directly acting on mitochondria,[54,55,58] it can be concluded from the data shown in Figure 38.6 and Figure 38.7 that encapsulating paclitaxel into a mitochondria-specific drug delivery system appears to increase the sub-cellular, i.e., mitochondrial, bioavailability of the drug.

38.8 TUMOR GROWTH INHIBITION STUDY WITH PACLITAXEL-LOADED DQASOMES IN VIVO

Paclitaxel-loaded DQAsomes were tested for their ability to inhibit the growth of human colon cancer cells in nude mice.[103] COLO-205 cells were inoculated s.c. into the left flank of nude mice that all formed palpable tumors within seven days after cell injection. For controls with free paclitaxel, the drug was suspended in 100% DMSO at 20 mM, stored at 4°C, and immediately before use, diluted in warm medium. In all controls, the dose of free paclitaxel and empty DQAsomes, respectively, was adjusted according to the dose of paclitaxel and dequalinium given in the paclitaxel-loaded DQAsome sample. Because of the lack of any inhibitory effect on tumor growth, the dose was tripled after 1.5 weeks. Figure 38.8 shows that at concentrations where free paclitaxel and empty DQAsomes do not show any impact on tumor growth, paclitaxel-loaded DQAsomes (with paclitaxel and dequalinium concentrations identical to controls) seem to inhibit the tumor growth by about 50%. Correspondingly, the average tumor weight in the treatment group after sacrificing the animals after 26 days is approximately half of that in all controls (Figure 38.9). Although this result seems to suggest that DQAsomes might be able to increase the therapeutic potential of paclitaxel, the preliminary character of this first in vivo study should be emphasized.

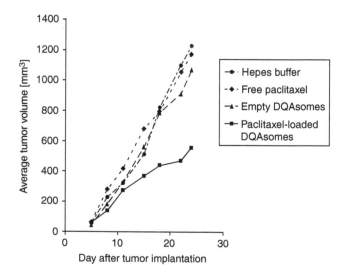

FIGURE 38.8 Tumor growth inhibition study in nude mice implanted with human colon cancer cells. The mean tumor volume from each group was blotted against the number of days. Each group involved eight animals. For clarity, error bars were omitted. Note that after 1.5 weeks, the dose, normalized for paclitaxel, was tripled in all treatment groups. (From Cheng, S. M., Pabba, S., Torchilin, V. P., Fowle, W., Kimpfler, A., Schubert, R., and Weissig, V., *J. Drug Deliv. Sci. Technol.*, 15(1), 81–86, 2005. With permission.)

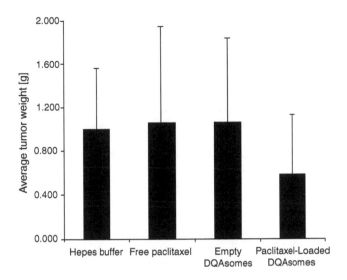

FIGURE 38.9 Average tumor weight ($n = 8$) at time of sacrifice of nude mice implanted with human colon cancer cells. For treatment group (paclitaxel-loaded DQAsomes) versus all three control groups combined $P = 0.054$ (parametric Student t-test). (From Cheng, S. M., Pabba, S., Torchilin, V. P., Fowle, W., Kimpfler, A., Schubert, R., and Weissig, V., *J. Drug Deliv. Sci. Technol.*, 15(1), 81–86, 2005. With permission.)

38.9 CONCLUDING REMARKS

DQAsomes and DQAsome-like vesicles have been established as the first mitochondria-specific cationic drug delivery system potentially able to deliver cytotoxic drugs selectively to mitochondria in cancer cells. However, the encapsulation of antineoplastic drugs into cationic vesicles has already been suggested in 1998.[108] It was found that upon intravenous injection of gold-labeled cationic liposomes, 32% associated with tumor endothelial cells, 53% were internalized into endosomes, and 15% were extravascular 20 min after injection.[108] Following these early experimental data, Munich Biotech AG (Neuried, Germany) is currently developing so-called Catioms for the systemic delivery of anti-cancer drugs to solid tumors.[105–107] However, whereas Munich Biotech is focusing on the targeting of the tumor vasculature, the DQAsomal strategy described in this chapter is aimed at the delivery of pro-apoptotic compounds to and into tumor cell mitochondria.

REFERENCES

1. Costantini, P., Jacotot, E., Decaudin, D., and Kroemer, G., Mitochondrion as a novel target of anticancer chemotherapy, *J. Natl. Cancer Inst.*, 92(13), 1042–1053, 2000.
2. Galati, G., Teng, S., Moridani, M. Y., Chan, T. S., and O'Brien, P. J., Cancer chemoprevention and apoptosis mechanisms induced by dietary polyphenolics, *Drug Metabol. Drug Interact.*, 17(1–4), 311–349, 2000.
3. Ghafourifar, P., Bringold, U., Klein, S. D., and Richter, C., Mitochondrial nitric oxide synthase, oxidative stress and apoptosis, *Biol. Signals Recept.*, 10(1–2), 57–65, 2001.
4. Davis, W., Ronai, Z., and Tew, K. D., Cellular thiols and reactive oxygen species in drug-induced apoptosis, *J. Pharmacol. Exp. Ther.*, 296(1), 1–6, 2001.
5. Creagh, E. M., Sheehan, D., and Cotter, T. G., Heat shock proteins—modulators of apoptosis in tumour cells, *Leukemia*, 14(7), 1161–1173, 2000.
6. Hall, A. G., Review: The role of glutathione in the regulation of apoptosis, *Eur. J. Clin. Invest.*, 29(3), 238–245, 1999.

7. Deigner, H. P. and Kinscherf, R., Modulating apoptosis: Current applications and prospects for future drug development, *Curr. Med. Chem.*, 6(5), 399–414, 1999.

8. Peltenburg, L. T., Radiosensitivity of tumor cells. Oncogenes and apoptosis, *Q. J. Nucl. Med.*, 44(4), 355–364, 2000.

9. Deigner, H. P., Haberkorn, U., and Kinscherf, R., Apoptosis modulators in the therapy of neuro-degenerative diseases, *Expert Opin. Investig. Drugs*, 9(4), 747–764, 2000.

10. Kanduc, D., Mittelman, A., Serpico, R., Sinigaglia, E., Sinha, A. A., Natale, C., Santacroce, R., Di Corcia, M. G. *et al.*, Cell death: Apoptosis versus necrosis (review), *Int. J. Oncol.*, 21(1), 165–170, 2002.

11. Robertson, J. D., Orrenius, S., and Zhivotovsky, B., Review: Nuclear events in apoptosis, *J. Struct. Biol.*, 129(2,3), 346–358, 2000.

12. Tomei, L. D. and Umansky, S. R., Apoptosis and the heart: A brief review, *Ann. N. Y. Acad. Sci.*, 946, 160–168, 2001.

13. Sathasivam, S., Ince, P. G., and Shaw, P. J., Apoptosis in amyotrophic lateral sclerosis: A review of the evidence, *Neuropathol. Appl. Neurobiol.*, 27(4), 257–274, 2001.

14. Tinaztepe, K., Ozen, S., Gucer, S., and Ozdamar, S., Apoptosis in renal disease: A brief review of the literature and report of preliminary findings in childhood lupus nephritis, *Turk. J. Pediatr.*, 43(2), 133–138, 2001.

15. Zusman, I., Gurevich, P., Gurevich, E., and Ben-Hur, H., The immune system, apoptosis and apoptosis-related proteins in human ovarian tumors (a review), *Int. J. Oncol.*, 18(5), 965–972, 2001.

16. Belzacq, A. S., Vieira, H. L., Kroemer, G., and Brenner, C., The adenine nucleotide translocator in apoptosis, *Biochimie*, 84(2,3), 167–176, 2002.

17. Kaufmann, S. H. and Gores, G. J., Apoptosis in cancer: Cause and cure, *Bioessays*, 22(11), 1007–1017, 2000.

18. Olson, M. and Kornbluth, S., Mitochondria in apoptosis and human disease, *Curr. Mol. Med.*, 1(1), 91–122, 2001.

19. Lin, Q. S., Mitochondria and apoptosis, *Sheng Wu Hua Xue Yu Sheng Wu Wu Li Xue Bao (Shanghai)*, 31(2), 116–118, 1999.

20. Pollack, M. and Leeuwenburgh, C., Apoptosis and aging: Role of the mitochondria, *J. Gerontol. A Biol. Sci. Med. Sci.*, 56(11), B475–B482, 2001.

21. Henkart, P. A. and Grinstein, S., Apoptosis: Mitochondria resurrected? *J. Exp. Med.*, 183(4), 1293–1295, 1996.

22. Skulachev, V. P., Why are mitochondria involved in apoptosis? Permeability transition pores and apoptosis as selective mechanisms to eliminate superoxide-producing mitochondria and cell, *FEBS Lett.*, 397(1), 7–10, 1996.

23. Petit, P. X., Zamzami, N., Vayssiere, J. L., Mignotte, B., Kroemer, G., and Castedo, M., Implication of mitochondria in apoptosis, *Mol. Cell. Biochem.*, 174(1,2), 185–188, 1997.

24. Decaudin, D., Marzo, I, Brenner, C., and Kroemer, G., Mitochondria in chemotherapy-induced apoptosis: A prospective novel target of cancer therapy (review), *Int. J. Oncol.*, 12(1), 141–152, 1998.

25. Mignotte, B. and Vayssiere, J. L., Mitochondria and apoptosis, *Eur. J. Biochem.*, 252(1), 1–15, 1998.

26. Susin, S. A., Zamzami, N., and Kroemer, G., Mitochondria as regulators of apoptosis: Doubt no more, *Biochim. Biophys. Acta*, 1366(1,2), 151–165, 1998.

27. Green, D. R. and Reed, J. C., Mitochondria and apoptosis, *Science*, 281(5381), 1309–1312, 1998.

28. Tatton, W. G. and Chalmers-Redman, R. M., Mitochondria in neurodegenerative apoptosis: An opportunity for therapy? *Ann. Neurol.*, 44(3 Suppl. 1), S134–S141, 1998.

29. Gottlieb, R. A., Mitochondria: Ignition chamber for apoptosis, *Mol. Genet. Metab.*, 68(2), 227–231, 1999.

30. Nieminen, A. L., Apoptosis and necrosis in health and disease: Role of mitochondria, *Int. Rev. Cytol.*, 224, 29–55, 2003.

31. Weber, T., Dalen, H., Andera, L., Negre-Salvayre, A., Auge, N., Sticha, M., Lloret, A. *et al.*, Mitochondria play a central role in apoptosis induced by alpha-tocopheryl succinate, an agent with antineoplastic activity: Comparison with receptor-mediated pro-apoptotic signaling, *Biochemistry*, 42(14), 4277–4291, 2003.

32. Richter, C., Oxidative Stress, mitochondria, and apoptosis, *Restor. Neurol. Neurosci.*, 12(2,3), 59–62, 1998.

33. Gulbins, E., Dreschers, S., and Bock, J., Role of mitochondria in apoptosis, *Exp. Physiol.*, 88(Pt 1), 85–90, 2003.

34. Hockenbery, D. M., Giedt, C. D., O'Neill, J. W., Manion, M. K., and Banker, D. E., Mitochondria and apoptosis: New therapeutic targets, *Adv. Cancer Res.*, 85, 203–242, 2002.

35. Heiligtag, S. J., Bredehorst, R., and David, K. A., Key role of mitochondria in cerulenin-mediated apoptosis, *Cell Death Differ.*, 9(9), 1017–1025, 2002.

36. Boichuk, S. V., Minnebaev, M. M., and Mustafin, I. G., Key role of mitochondria in apoptosis of lymphocytes, *Bull. Exp. Biol. Med.*, 132(6), 1166–1168, 2001.

37. Birbes, H., Bawab, S. E., Obeid, L. M., and Hannun, Y. A., Mitochondria and ceramide: Intertwined roles in regulation of apoptosis, *Adv. Enzyme Regul.*, 42, 113–129, 2002.

38. Naoi, M., Maruyama, W., Akao, Y., and Yi, H., Mitochondria determine the survival and death in apoptosis by an endogenous neurotoxin, *N*-methyl(R)salsolinol, and neuroprotection by propargy-lamines, *J. Neural Transm.*, 109(5,6), 607–621, 2002.

39. Roucou, X., Antonsson, B., and Martinou, J. C., Involvement of mitochondria in apoptosis, *Cardiol. Clin.*, 19(1), 45–55, 2001.

40. Wang, X., The expanding role of mitochondria in apoptosis, *Genes Dev.*, 15(22), 2922–2933, 2001.

41. Raha, S. and Robinson, B. H., Mitochondria, oxygen free radicals, and apoptosis, *Am. J. Med. Genet.*, 106(1), 62–70, 2001.

42. Waterhouse, N. J., Goldstein, J. C., Kluck, R. M., Newmeyer, D. D., and Green, D. R., The (Holey) study of mitochondria in apoptosis, *Methods Cell Biol.*, 66, 365–391, 2001.

43. Sordet, O., Rebe, C., Leroy, I., Bruey, J. M., Garrido, C., Miguet, C., Lizard, G., Plenchette, S., Corcos, L., and Solary, E., Mitochondria-targeting drugs arsenic trioxide and lonidamine bypass the resistance of TPA-differentiated leukemic cells to apoptosis, *Blood*, 97(12), 3931–3940, 2001.

44. Gottlieb, R. A., Mitochondria and apoptosis, *Biol. Signals Recept.*, 10(3,4), 147–161, 2001.

45. Shi, Y., A structural view of mitochondria-mediated apoptosis, *Nat. Struct. Biol.*, 8(5), 394–401, 2001.

46. Gottlieb, R. A., Role of mitochondria in apoptosis, *Crit. Rev. Eukaryot. Gene. Expr.*, 10(3,4), 231–239, 2000.

47. Jiang, Z. F., Zhao, Y., Hong, X., and Zhai, Z. H., Nuclear apoptosis induced by isolated mitochondria, *Cell Res.*, 10(3), 221–232, 2000.

48. Brenner, C. and Kroemer, G., Apoptosis. Mitochondria—the death signal integrators, *Science*, 289(5482), 1150–1151, 2000.

49. Desagher, S. and Martinou, J. C., Mitochondria as the central control point of apoptosis, *Trends Cell Biol.*, 10(9), 369–377, 2000.

50. Kroemer, G., Mitochondrial control of apoptosis: An overview, In *Mitochondria and Cell Death*, Brown, G. C., Nicholls, D. G., and Cooper, C. E., Eds., Princton University Press, Princeton, NJ, pp. 1–15, 1999.

51. Fulda, S., Susin, S. A., Kroemer, G., and Debatin, K. M., Molecular ordering of apoptosis induced by anticancer drugs in neuroblastoma cells, *Cancer Res.*, 58(19), 4453–4460, 1998.

52. Ferlini, C., Raspaglio, G., Mozzetti, S., Distefano, M., Filippetti, F., Martinelli, E., Ferrandina, G., Gallo, D., Ranelletti, F. O., and Scambia, G., Bcl-2 down-regulation is a novel mechanism of paclitaxel resistance, *Mol. Pharmacol.*, 64(1), 51–58, 2003.

53. von Haefen, C., Wieder, T., Essmann, F., Schulze-Osthoff, K., Dorken, B., and Daniel, P. T., Paclitaxel-induced apoptosis in BJAB cells proceeds via a death receptor-independent, caspases-3/-8-driven mitochondrial amplification loop, *Oncogene*, 22(15), 2236–2247, 2003.

54. Andre, N., Carre, M., Brasseur, G., Pourroy, B., Kovacic, H., Briand, C., and Braguer, D., Paclitaxel targets mitochondria upstream of caspase activation in intact human neuroblastoma cells, *FEBS Lett.*, 532(1–2), 256–260, 2002.

55. Kidd, J. F., Pilkington, M. F., Schell, M. J., Fogarty, K. E., Skepper, J. N., Taylor, C. W., and Thorn, P., Paclitaxel affects cytosolic calcium signals by opening the mitochondrial permeability transition pore, *J. Biol. Chem.*, 277(8), 6504–6510, 2002.

56. Makin, G. W., Corfe, B. M., Griffiths, G. J., Thistlethwaite, A., Hickman, J. A., and Dive, C., Damage-induced Bax N-terminal change, translocation to mitochondria and formation of Bax dimers/complexes occur regardless of cell fate, *Embo J.*, 20(22), 6306–6315, 2001.

57. Varbiro, G., Veres, B., Gallyas, F., and Sumegi, B., Direct effect of Taxol on free radical formation and mitochondrial permeability transition, *Free Radic. Biol. Med.*, 31(4), 548–558, 2001.

58. Andre, N., Braguer, D., Brasseur, G., Goncalves, A., Lemesle-Meunier, D., Guise, S., Jordan, M. A., and Briand, C., Paclitaxel induces release of cytochrome c from mitochondria isolated from human neuroblastoma cells', *Cancer Res.*, 60(19), 5349–5353, 2000.

59. Carre, M., Carles, G., Andre, N., Douillard, S., Ciccolini, J., Briand, C., and Braguer, D., Involvement of microtubules and mitochondria in the antagonism of arsenic trioxide on paclitaxel-induced apoptosis, *Biochem. Pharmacol.*, 63(10), 1831–1842, 2002.

60. Perkins, C. L., Fang, G., Kim, C. N., and Bhalla, K. N., The role of Apaf-1, caspase-9, and bid proteins in etoposide- or paclitaxel-induced mitochondrial events during apoptosis, *Cancer Res.*, 60(6), 1645–1653, 2000.

61. Itoh, M., Noutomi, T., Toyota, H., and Mizuguchi, J., Etoposide-mediated sensitization of squamous cell carcinoma cells to tumor necrosis factor-related apoptosis-inducing ligand (TRAIL)-induced loss in mitochondrial membrane potential, *Oral Oncol.*, 39(3), 269–276, 2003.

62. Custodio, J. B., Cardoso, C. M., Madeira, V. M., and Almeida, L. M., Mitochondrial permeability transition induced by the anticancer drug etoposide, *Toxicol. In Vitro*, 15(4,5), 265–270, 2001.

63. Fujino, M., Li, X. K., Kitazawa, Y., Guo, L., Kawasaki, M., Funeshima, N., Amano, T., and Suzuki, S., Distinct pathways of apoptosis triggered by FTY720, etoposide, and anti-Fas antibody in human T-lymphoma cell line (Jurkat cells), *J. Pharmacol. Exp. Ther.*, 300(3), 939–945, 2002.

64. Robertson, J. D., Gogvadze, V., Zhivotovsky, B., and Orrenius, S., Distinct pathways for stimulation of cytochrome *c* release by etoposide, *J. Biol. Chem.*, 275(42), 32438–32443, 2000.

65. Marchetti, P., Zamzami, N., Joseph, B., Schraen-Maschke, S., Mereau-Richard, C., Costantini, P., Metivier, D., Susin, S. A., Kroemer, G., and Formstecher, P., The novel retinoid 6-[3-(1-adamantyl)-4-hydroxyphenyl]-2-naphtalene carboxylic acid can trigger apoptosis through a mitochondrial pathway independent of the nucleus, *Cancer Res.*, 59(24), 6257–6266, 1999.

66. Murphy, M. P., Slip and leak in mitochondrial oxidative phosphorylation, *Biochim. Biophys. Acta*, 977, 123–141, 1989.

67. Murphy, M. P. and Smith, R. A., Drug delivery to mitochondria: The key to mitochondrial medicine, *Adv. Drug Deliv. Rev.*, 41(2), 235–250, 2000.

68. Modica-Napolitano, J. S. and Aprille, J. R., Delocalized lipophilic cations selectively target the mitochondria of carcinoma cells, *Adv. Drug Deliv. Rev.*, 49(1,2), 63–70, 2001.

69. Modica-Napolitano, J. S. and Aprille, J. R., Basis for the selective cytotoxicity of rhodamine 123, *Cancer Res.*, 47(16), 4361–4365, 1987.

70. Modica-Napolitano, J. S. and Singh, K. K., Mitochondria as targets for detection and treatment of cancer,*Expert Rev. Mol. Med.* 2002 April 11; 2002:1–19.

71. Modica-Napolitano, J. S., Koya, K., Weisberg, E., Brunelli, B. T., Li, Y., and Chen, L. B., Selective damage to carcinoma mitochondria by the rhodacyanine MKT-077, *Cancer Res.*, 56(3), 544–550, 1996.

72. Manetta, A., Emma, D., Gamboa, G., Liao, S., Berman, M., and DiSaia, P., Failure to enhance the in vivo killing of human ovarian carcinoma by sequential treatment with dequalinium chloride and tumor necrosis factor, *Gynecol. Oncol.*, 50(1), 38–44, 1993.

73. Weiss, M. J., Wong, J. R., Ha, C. S., Bleday, R., Salem, R. R., Steele, G. D., and Chen, L. B., Dequalinium, a topical antimicrobial agent, displays anticarcinoma activity based on selective mitochondrial accumulation, *Proc. Natl. Acad. Sci. USA*, 84(15), 5444–5448, 1987.

74. Christman, J. E., Miller, D. S., Coward, P., Smith, L. H., and Teng, N. N., Study of the selective cytotoxic properties of cationic, lipophilic mitochondrial-specific compounds in gynecologic malignancies, *Gynecol. Oncol.*, 39(1), 72–79, 1990.

75. Davis, S., Weiss, M. J., Wong, J. R., Lampidis, T. J., and Chen, L. B., Mitochondrial and plasma membrane potentials cause unusual accumulation and retention of rhodamine 123 by human breast adenocarcinoma-derived MCF-7 cells, *J. Biol. Chem.*, 260(25), 13844–13850, 1985.

76. Koya, K., Li, Y., Wang, H., Ukai, T., Tatsuta, N., Kawakami, M., Shishido, T., and Chen, L. B., MKT-077, a novel rhodacyanine dye in clinical trials, exhibits anticarcinoma activity in preclinical studies based on selective mitochondrial accumulation, *Cancer Res.*, 56(3), 538–543, 1996.

77. Weisberg, E. L., Koya, K., Modica-Napolitano, J., Li, Y., and Chen, L. B., In vivo administration of MKT-077 causes partial yet reversible impairment of mitochondrial function, *Cancer Res.*, 56(3), 551–555, 1996.

78. Chiba, Y., Kubota, T., Watanabe, M., Matsuzaki, S. W., Otani, Y., Teramoto, T., Matsumoto, Y., Koya, K., and Kitajima, M., MKT-077, localized lipophilic cation: Antitumor activity against human tumor xenografts serially transplanted into nude mice, *Anticancer Res.*, 18(2A), 1047–1052, 1998.

79. Chiba, Y., Kubota, T., Watanabe, M., Otani, Y., Teramoto, T., Matsumoto, Y., Koya, K., and Kitajima, M., Selective antitumor activity of MKT-077, a delocalized lipophilic cation, on normal cells and cancer cells in vitro, *J. Surg. Oncol.*, 69(2), 105–110, 1998.

80. Petit, T., Izbicka, E., Lawrence, R. A., Nalin, C., Weitman, S. D., and Von Hoff, D. D., Activity of MKT 077, a rhodacyanine dye, against human tumor colony-forming units, *Anticancer Drugs*, 10(3), 309–315, 1999.

81. Propper, D. J., Braybrooke, J. P., Taylor, D. J., Lodi, R., Styles, P., Cramer, J. A., Collins, W. C. *et al.*, Phase I trial of the selective mitochondrial toxin MKT077 in chemo-resistant solid tumours, *Ann. Oncol.*, 10(8), 923–927, 1999.

82. Adams, D. M., Ji, W., Barth, R. F., and Tjarks, W., Comparative in vitro evaluation of dequalinium B, a new boron carrier for neutron capture therapy (NCT), *Anticancer Res.*, 20(5B), 3395–3402, 2000.

83. Weissig, V., Mogel, H. J., Wahab, M., and Lasch, J., Computer simulations of DQAsomes, *Proceed. Intl. Symp. Control. Rel. Bioact. Mater.*, 25, 312, 1998.

84. Weissig, V., Lasch, L., Erdos, G., Meyer, H. W., Rowe, T. C., and Hughes, J., DQAsomes: A novel potential drug and gene delivery system made from Dequalinium, *Pharm. Res.*, 15(2), 334–337, 1998.

85. De Rosa, M., Gambacorta, A., and Gliozi, A., Structure, biosynthesis, and physicochemical properties of archaebacterial lipds, *Microbiol. Rev.*, 50, 70–80, 1986.

86. Gambacorta, A., Gliozi, A., and De Rosa, M., Archaeal lipids and their biotechnological applications, *World J. Microbiol. Biotechnol.*, 11, 115–131, 1995.

87. Weissig, V. and Torchilin, V. P., Mitochondriotropic cationic vesicles: A strategy towards mitochondrial gene therapy, *Curr. Pharm. Biotechnol.*, 1(4), 325–346, 2000.

88. Weissig, V. and Torchilin, V. P., Cationic single-chain bolaamphiphiles as new materials for application in medicine and biotechnology. In *Materials research society meeting*, Boston, MA., EE.5.4, p. 173, 1999.

89. Weissig, V., Lizano, C., Ganellin, C. R., and Torchilin, V. P., DNA binding cationic bolasomes with delocalized charge center: A structure-activity relationship study, *STP Pharma Sci.*, 11, 91–96, 2001.

90. Lasch, J., Meye, A., Taubert, H., Koelsch, R., Mansa-ard, J., and Weissig, V., Dequalinium vesicles form stable complexes with plasmid DNA which are protected from DNase attack, *Biol. Chem.*, 380(6), 647–652, 1999.

91. Weissig, V., Lizano, C., and Torchilin, V. P., Selective DNA release from DQAsome/DNA complexes at mitochondria-like membranes, *Drug Deliv.*, 7(1), 1–5, 2000.

92. Weissig, V., D'Souza, G. G., and Torchilin, V. P., DQAsome/DNA complexes release DNA upon contact with isolated mouse liver mitochondria, *J. Control Release*, 75(3), 401–408, 2001.

93. Ravagnan, L., Marzo, I., Costantini, P., Susin, S. A., Zamzami, N., Petit, P. X., and Hirsch, F., Lonidamine triggers apoptosis via a direct, Bcl-2-inhibited effect on the mitochondrial permeability transition pore., *Oncogene*, 18(16), 2537–2546, 1999.

94. Jacotot, E., Costantini, P., Laboureau, E., Zamzami, N., Susin, S. A., and Kroemer, G., Mitochondrial membrane permeabilization during the apoptotic process, *Ann. N. Y. Acad. Sci.*, 887, 18–30, 1999.

95. van der Woude, I., Wagenaar, A., Meekel, A. A., ter Beest, M. B., Ruiters, M. H., Engberts, J. B., and Hoekstra, D., Novel pyridinium surfactants for efficient, nontoxic in vitro gene delivery, *Proc. Natl. Acad. Sci. USA*, 94(4), 1160–1165, 1997.

96. Zuhorn, I. S. and Hoekstra, D., On the mechanism of cationic amphiphile-mediated transfection. To fuse or not to fuse: Is that the question? *J. Membr. Biol.*, 189(3), 167–179, 2002.

97. Wattiaux, R., Jadot, M., Warnier-Pirotte, M. T., and Wattiaux-De Coninck, S., Cationic lipids destabilize lysosomal membrane in vitro, *FEBS Lett.*, 417(2), 199–202, 1997.

98. El Ouahabi, A., Thiry, M., Pector, V., Fuks, R., Ruysschaert, J. M., and Vandenbranden, M., The role of endosome destabilizing activity in the gene transfer process mediated by cationic lipids, *FEBS Lett.*, 414(2), 187–192, 1997.

99. D'Souza, G. G., Rammohan, R., Cheng, S. M., Torchilin, V. P., and Weissig, V., DQAsome-mediated delivery of plasmid DNA toward mitochondria in living cells, *J. Control Release*, 92, 189–197, 2003.

100. D'Souza, G. G., Boddapati, S. V., and Weissig, V., Mitochondrial leader sequence-plasmid DNA conjugates delivered into mammalian cells by DQAsomes co-localize with mitochondria, *Mitochondrion*, 5(5), 352–358, 2005.

101. Braguer, D., Andre, N., Carre, M., Carles, G., Gonzalves, A., and Briand, C., *92nd Annual Meeting of the American Association for Cancer Research*, American Association for Cancer Research, New Orleans, LA, p. 369, 2001.

102. Carre, M., Andre, N., Carles, G., Borghi, H., Brichese, L., Briand, C., and Braguer, D., Tubulin is an inherent component of mitochondrial membranes that interacts with the voltage-dependent anion channel, *J. Biol. Chem.*, 277, 33664–33669, 2002.

103. Cheng, S. M., Pabba, S., Torchilin, V. P., Fowle, W., Kimpfler, A., Schubert, R., and Weissig, V., Towards mitochondria-specific delivery of apoptosis-inducing agents: DQAsomal incorporated paclitaxel, *J. Drug Deliv. Sci. Techno.*, 15(1), 81–86, 2005.

104. Discher, E. D. and Eisenberg, A., Polymer vesicles, *Science*, 297, 967–973, 2002.

105. Schuppenhauer, M., Munich Biotech: Catioms for cancer. *BioCentury, The Bernstein Report on BioBusiness*, May 19, A10, 2003.

106. Schmitt-Sody, M., Strieth, S., Krasnici, S., Sauer, B., Schulze, B., Teifel, M., Michaelis, U., Naujoks, K., and Dellian, M., Neovascular targeting therapy: Paclitaxel encapsulated in cationic liposomes improves antitumoral efficacy, *Clinical Cancer Research*, 9(6), 2335–2341, 2003.

107. Krasnici, S., Werner, A., Eichhorn, M. E., Schmitt-Sody, M., Pahernik, S. A., Sauer, B., and Schulze, B., Effect of the surface charge of liposomes on their uptake by angiogenic tumor vessels, *Int. J. Cancer*, 105(4), 561–567, 2003.

108. Thurston, G., McLean, J. W., Rizen, M., Baluk, P., Haskell, A., Murphy, T. J., Hanahan, D., and McDonald, D. M., Cationic liposomes target angiogenic endothelial cells in tumors and chronic inflammation in mice, *J. Clin. Invest.*, 101(7), 1401–1413, 1998.

Index

A

Abberant vasculature, nanotechnology applications
    targeting strategies and, 30–32
Acidic tumor pH_e, 444
Active targeting, 244
    PEO-*b*-PLAA micelles and, 373
    polymeric micelles and, 342
Agents, other types, solid lipid nanoparticles and,
    748
Alkylcyanoacrylates, chemoembolization and,
    277–278
Alpha-v Beta-3, 294–295
Aminopeptidases, 162, 164
Amphotericin B, 372–373
Angiogenesis
    conjugates for, 163–164
    markers, VEGF receptors, 162
    targeted tumors, imaging and therapy, polymer
        conjugates and, 159–176
    targeting of, 218–220
        integrins, 218–219
        polymeric conjugates, 168–175
        vascular endothelial growth factor receptors,
            218
    tumor-targeted drug delivery and, 218–220
Angiogenic tumor vasculature, molecular targets,
    646–647
Animal tumor model
    Paramagnetic
        HPMA copolymer conjugates and, 204–205
        poly (ι-glutamic acid) conjugates and,
            206–208
        PGA-Mce$_6$ -(Gd-DOTA)conjugates and,
            208–210
Anthracyclines, 187–188
Antiangiogenic gene therapy, 169–170
Antibodies, folate receptors and, 221–222
Anti-cancer
    drugs, polymeric micelles, soluable qualities of,
        409–411
    efficacy, block copolymer micelles and, 338–339
    therapy, cationic liposomes and, 621–623
Antimetabolites, 188
Antisense oligonucleotides, ligand-mediated
    targeting, 483
Antitumor drug delivery, solid lipid nanoparticles
    and, 756–770

Apoptosis, 788
    assessment and, 119–120
Application site, nanoparticles and, 33–34
Aptamers (nucleic acid ligands)
    cancer nanotechnology, 289–306
    cytotoxic T-cell antigen-4, 296
    discovery and usage of, 291–296
    drug conjugates, targeted cancer therapy,
        301–304
    fibrinogen-like domain of tenascin-C, 296
    gold-nanoparticle conjugate, 305
    isolation methods of, 293–294
    nanoparticle conjugates, 296–301
        charge of, 298–299
        polymers, 298
        properties of, 297
        protein detection, 304–306
        size of, 297
        strategies of, 300–301
        surface modification, 299–300
    nucleolin, 296
    physicochemical properties, 292–293
    pigpen, 296
    QD conjugate, protein detection, 304–305
    sialyl lewis X, 295
    targeted delivery of, 294–296
        alpha-v Beta-3, 294–295
        human epidermal growth factor-3, 295
        prostate specific membrane antigen, 295
Avidin targeting, 226

B

Biocompatibility
    nanoparticle characteristics and, 22–23
    pegylated dendrimers and, 535–537
        enhanced biocompatibility of, 535–537
Biodegradable PLGA/PLA nanoparticles, 243–248
Biodegradation, Poly(L-glutamic acid) and, 187
Biodistribution
    alterations, nanoparticles and, surface modifi-
        cations of, 264
    experiments, dendrimer nanocomposites and,
        562–563
    gold nanocomposites and, 577
    nanodevice delivery and, 580
    poly(L-glutamic acid) and, 187
Biological effects, ultrasound and, 425–430

Biosynthetic interface
  modular functionalities
    imaging agents, 61–64
    remote application, 69
    sensing, 64–66
    targeting, 60–61
    therapeutic payloads, 66–68
  multifunctional nanoparticles
    modular functionalities
      imaging agents, 61–64
      remote application, 69
      sensing, 64–66
      targeting, 60–61
      therapeutic payloads, 66–68
Biotin targeting, 226
Block copolymer
  formulations, polymeric micelles and, 323
  micelles, 336–340
    anti-cancer efficacy, 338–339
    doxorubicin formulations, 339
    in vivo biological performance, 336–340
      drug release profiles, 338
      fate of, 337–338
    paclitaxel formulations, 339–340
    pharmacokinetics of, 338–339
    toxicity of, 338–339
Blood contact properties, 125–127
Bone delivery, paramagnetic HPMA copolymer
  conjugates and, 206
Boranes, 87–89
Bornoted dendrimers
  attachment to mAb, 81
  clusters linked to mAb, 80–81
Boron
  delivery agents
    high-molecular-weight types, 79–80
    low-molecular-weight types, 79
    neutron capture cancer therapy and, 78–79
    requirements for, 78–79
  delivery of
    brain tumors, 93–95
      convection-enhanced, 94–95
      direct intracerebral delivery, 94
      drug-transport vectors, 93–94
    dextrans, 92–93
    other compounds, 93
  neutron capture therapy, 494–495
  pegtylated dendrimers and, 544
Boronared dendrimers
  monoclonal antibodies, linked to, 80–81
  other types, 85–86
  receptor ligands, 81–85
    epidermal growth factors, 81, 83
    folate receptor targeting agents, 83
    vascular endothelial growth factor, 83–85

Boronephyenylalanine, 86–87
Brain tumors
  boron delivery to, 93–95
    direct intracerebral delivery, 94
    drug-transport vectors, 93–94
  poly (alkyl cyanocrylate) nanoparticle delivery
    and, 274–275
Breast cancer, poly (alkyl cyanocrylate) nanoparticle
  delivery and, 276–277

C

Camptothecin, 372, 759–760
Cancer
  breast, 276–277
  cells
    nanotechnology applications targeting
      strategies and, 27–30
      cell surface properties, 28–29
      metabolic properties, 29–30
    poly (alkyl cyanocrylate) delivery, resistance
      to, 268–272
    polymeric micelles
      cellular internalization, 332–334
      drug diffusion, 331
      drug resistance, 335–336
      in vitro cytotoxicity, 334
      interactions with, 330–336
    targets, poly (alkyl cyanocrylate) nanoparticle
      delivery and, 266–280
  chemotherapy, 756
    nanoparticles, 525–527
      nanotechnology challenges, 527
    polymer micelles and, 392–393
    problems in, 524–525
  drug delivery systems
    polymeric nanoparticles, 14–15
    solid tumor and, 12–16
      macromolecular and related delivery, 16
      nanotechnology-derived nanoparticles,
        15–16
    targeting and, general principles of, 244
  gene therapy, RGD modified liposomes and,
    656–657
  imaging, targeted nanoparticles and, 225
  multidrug resistance, 444–445
  nanotechnology
    aptamers and, 289–306
    rationale for, 3–8
    targeted drug delivery, 289–291
  polymeric nanoparticles and, 216
  prostate, 224–225
  specific
    cellular marker, 663–664
    types, tumor-targeted drug delivery and,
      223–225

therapeutics, RGD modified liposomes and, 653–656

therapy, 252
    aptamer drug conjugates and, 301–304
    challenges of, 596–508
    multifunctional nanoparticles and, 59–70
    nanoemulsion formulations and, 730–735
    neutron capture, 77–96

Capusular design, lipid-based nanocarriers and, getting drug out, 691–693

Carboranes, 87–89

Carcinoma, haptocellular, 272–273

Carcinoma, lymphatic, 276

Cationic liposomes
    anti-cancer therapy and, 621–623
    human endothelial cells and interaction of, 618–619
    lung targeting, 52–53

Cell
    penetrating peptide, 629–637
        liposomes, 631–632
            delivery of, 635–637
        pharmaceutical nanocarriers, delivery of, 633–635
        transduction, 631–633
    surface properties, cancer cells and, nanotechnology targeting, 28–29

Cellular
    internalization, polymeric micelles and, 332–334
    level, ultrasonic biological effects and, 425–427
    marker, 663–664

Chemistry, Poly(L-glutamic acid) and, 186–187

Chemoembolization
    alkylcyanoacrylates, 277–278
    poly (alkyl cyanoacrylate) delivery and, 277–278

Chemosensitizing agents, delivery by solid lipid nanoparticles, 763–770
    current use of, 765–766
        elacridar, 765
        quinidine, 765
        verapamil, 765

Chemotherapy, 524–525, 756
    polymer micelles and, 392–393

Chromatography, 109

Cisplatin, 370–371

Classical, solid lipid nanoparticles and, 748–749

Clinical trials, polymeric micelles and, 423–425

Clinic-doxil, liposomal anthracyclines and, 602–605

Complications, nanoparticle characteristics and, 25–26

Composite nanodevices, 552–571
    dendrimers, nanocomposites, 559–561
    dendrimers, 555–558
    gold, 558–559
    introduction to, 552–571

radioactive gold, 558–559
    toxicity, 558

Composition, nanoparticle characteristics and, 22–23

Conjugates
    angiogenesis and, 163–164
    copolymer, 204–206
    poly(L-glutamic acid) and, 187–189
        anthracyclines, 187–188
        antimetabolites, 188
        DNA-binding drugs, 188–189

Contrast-enhanced magnetic resonance imaging, 203–204

Convection-enhanced delivery, boron delivery to brain tumors and, 94–95

Copolymer conjugates, 204–206

Co-solvent evaporation method, micellization of PEO-b-PLAA block polymers, 362

Covalent conjugation, 493–494

cRGD encoded
    micelles, 468–469
    SPIO loaded micelles, 472–473

Cyclophosphamide, 369

Cytotoxic
    agents, liposome-entrapped, 605–606
    compound delivery, 757–758
    T-cell antigen-4, 296

Cytotoxicity, 117–118

D

Degradation, nanoparticles and, 260

Delivery agents, Boron, 78–79

Dendrimer, 489–504
    application as drug carrier, 47–48
    boronated, 80–81
    composite nanodevices and, 555–558
    drug
        conjugates, 511–514
        delivery system, PEGylated dendritic nanoparticulate carriers and, 529–534
        delivery, 489–498
        encapsulation and, 514–516
    gene delivery, 498
    history and characteristics, 47–48
    hyperbranched structures and, differences in, 532
    in vivo biodistribution, 573
    nanocomposites, 551–586
        biodistribution experiments, 562–563
        biology of, 571–586
        composite nanodevices, 552–571
        folate targeted devices, 564
        gold composite nanodevice, 564–569
        gold/dendrimer synthesis, 561
        gold, radiation polymerization, 569

nanodevice delivery, 572–580
   targeting, 572–580
   toxicity studies, 580–585
   tritium labeled, 564
nanodevice delivery and, biodistribution,
      575–577
PAMAM, 498–501
pegylated, 534–545
pharmacokinetic
   characteristics of, 47–48
   properties of nanocarriers and, 47–48
polymeric drug carriers, 517–518
polypropylenimine, 501–502
properties of, 80
related delivery agents, 80–86
   properties of, 80
tritium labeled, 564
Dendritic
   gels, pegtylated dendrimers and, 543
   nanoparticulate carriers, PEGylated, 523–545
   nanostructures, 509–518
Derivitization, nanoparticle characteristics and, 23
Detecion, nanoparticle characteristics and, 23–24
Dextrans, boron delivery and, 92–93
Diagnostic agent carrier, Poly(L-glutamic acid) and,
      193–195
Dialysis method, micellization of PEO-b-PLAA
      block polymers, 362
Dispersion polymerization process, 253
DNA-binding drugs, 188–189
Double emulsion method, solid lipid nanoparticles
      manufacturing methods and, 753
Doxorubicin, 327, 369–370, 761–762
   formulations, 339
DQAsomes, 787–797
   mitochondriotropic vesicles, 790–791
   paclitaxel and, 793–794
Drug
   carriers
      dendrimers and, applications of, 47–48
      folate receptor targeted liposomes and, 665–667
      hydrotropic polymer micelles and, 400–404
      linear polymers and, applications of, 45–46
      liposomes and, applications of, 48–50
      polymer micelles and, 391
         applications of, 46–47
   conjugates, dendrimer, 511–514
   delivery system
      dendrimers as, 529–534
         classification of, 532–533
         hyperbranched structures, differences in, 532
         nanodrugs, 533–534
   delivery, 498
      background, 498
      covalent conjugation, 493–494

cytotoxic compound, 757–758
liposomes and, 631–632
nanocarrier-mediated, 43–54
nanoparticles and, 252–280
neutron capture therapy, 494–498
non-covalent encapsulation, 490–492
PEO-b-PLAA micelles and, 363–369
photodynamic therapy, 496–498
PLGA/PLA nanoparticles and, 245–246
polymeric
   micelles and, 410–411, 422–425
   nanoparticles and, 233
      examples of, 235–237
solid lipid nanoparticles and, 756–770
ultrasound and, 425–430
vehicles
   current methods, 678–681
   problems with, 678–681
diffusion, polymeric micelles and, 331
loading
   factors, micelle formulations and, 324–325
   poly (alkyl cyanoacrylate) and, 256–258
   solid lipid nanoparticles characteristic
      properties of, 753–754
partitioning, micelles formulations and, 324
regional deposition, MRI observation and,
      709–710
release
   profiles, block copolymer micelles and, 338
   properties, PEO-b-PLAA micelles drug
      delivery and, 368–369
   solid lipid nanoparticles characteristic
      properties of, 754–755
   ultrasound and, 428
resistance, Poly (alkyl cyanoacrylate)
      nanoparticles and, 268–272
resistance, polymeric micelles and, 335–336
resistant tumors, micelle/ultrasound drug delivery
      and, 433–435
targeting
   active, 244
   cancer, general principles of, 244
   integrin $\alpha_v\beta_3$-mediated, 466–467
   passive, 244
   PEO-b-PLAA micelles and, 373
Drug-sensitive tumors, micelle/ultrasound drug
      delivery and, 432–433
Drug-transport vectors, boron delivery to brain
      tumors and, 93–94

E

EGFR, 91–92
Elacridar, 765
Electric charge, effect of, pharmacokinetics
      macromolecule tissue distribution and, 51

Emulsion polymerization process, 254
Encapsulation, drug delivery, 490–492
Enzyme influence, poly (alkyl cyanocrylate)
    nanoparticles and, 260–261
Epidermal growth factors, 81, 83
EPR effect, passive targeting and, 217–218
Evaporation methods, micellization of PEO-*b*-
    PLAA block polymers, 362
Extravasation, barriers to, solid tumors and, 12

F

FDA (Food and Drug Administration)
    experience in nanotechnology regulation,
        140–143
    initiatives in nanotechnology, 143–144
    nanotechnology
        definition of, 143
        regulatory considerations and, 145–146
        research and, 145
        scientific considerations and, 147–149
            nanoparticle characterization, 147–149
        safety of, 149
    regulatory considerations
        existing, 146–147
        jurisdiction of, 145–146
Fibrinogen-like domain of tenascin-C, 296
Fibronectin, 650–651
Fluorouracil, 762–763
Food and Drug Administration. *See* FDA.
Folate receptor, 220–223
    antibodies, 221–222
    cancer
        specific cellular marker, 663–664
        therapy and, 664–665
    mediated targeting, 54
    targeted liposomes, 90–91, 663–672
        biodistribution properties, 670–671
        drug carriers, 665–667
        formulation strategies, 667–668
        in vitro drug delivery, 668–670
        in vivo therapeutic efficacy, 671–672
        pharmacokinetic properties, 670–671
    targeting agents, 83
Folate targeted
    devices, 564
    nanoparticles, gene delivery, 222–223

G

Gadolinium neutron capture therapy, 495
Galactose liposomes, liver PC targeting, 53
Gastric carcinoma, poly (alkyl cyanocrylate)
    nanoparticle delivery and, 276–277

Gene
    delivery
        background, 498
        dendrimers, 498
            other types, 502–504
        folate-targeted nanoparticles and, 222–223
        nanocarrier-mediated, 43–54
            liposomes, 52
                cationic for lung targeting, 52–53
                folate receptor-mediated targeting, 54
                galactose for liver PC targeting, 53
                mannose for liver NPC targeting, 53
                polymers, 52
            nanoparticles and, 233–234
            PAMAM dendrimer, 498–501
            PEO-*b*-PLAA micelles and, 374–376
            PLGA nanoparticles and, 247–248
            polymeric nanoparticles and, examples of,
                237–238
            polypropylenimine dendrimer, 501–502
        therapy, RGD modified liposomes and, 656–657
Genexol-PM, 329–330
Gliomas, 78
Gold, 558–559
    composite nanodevice, 564–569
        biologic differences, 577–579
        in vivo distribution of, 573
    content determination, instrumental neutron
        activation analysis method and, 575
    dendrimer nanocomposites, 561
        radiation polymerization, 569
    nanocomposites, biodistribution of, 577

H

Hapatocellular carcinoma, poly (alkyl cyanocrylate)
    nanoparticle delivery and, 272–273
High density liproteins, 778
    delivery vehicle, 779–781
High pressure homoegenization, solid lipid
    nanoparticles manufacturing methods and,
        751
High-molecular-weight Boron delivery agents,
        79–80
Human endothelial cells, cationic liposomes and,
        interaction of, 618–619
Human epidermal growth factor-3, 295
Hydrotropic polymer micelles, 385–408
    anti-cancer drug carrier, 400–404
    design strategy of, 395–400
    pharmaceuticals and, 393–395
    solubilization studies, 395
Hyperbranched structures, dendrimers and,   **Q1**
        differences in, 5432
Hyperthermia devices, 710–711

I

Idarubicin, 761–762
Imaging agents
  biosynthetic interface and, 61–64
  integrins and, 220
Immunogencity, nanoparticle evaluation and,
  128
Immunoliposomes, 89–90
Immunotoxicity
  blood contact properties, 125–127
  nanoparticle
    evaluation, 125–127
    preclinical characterization and, 124–128
In vitro
  antitumor toxicicity, solid lipid nanoparticles,
    effectiveness on, 766–770
  characterization, nanoparticles and, 107
  cytotoxicity, polymeric micelles and, 334
  drug delivery, folate receptor targeted liposomes
    and, 668–670
  pharmaceutical assessment, nanoparticle
    preclinical characterization and, 115–121
  target-organ toxicity, 116–117
  toxicological assessment, nanoparticle preclinical
    characterization and, 115–121
In vivo
  applications, micellar antisense oligonucleotide
    delivery systems and, 481–482
  biodistribution, nanodevice delivery and, 573
  biological performance, block copolymer
    micelles and, 336–340
    drug release profiles, 338
    fate of, 337–338
  characterization, nanoparticles and, 107
  delivery, paramagnetic polymer conjugates and,
    201–212
  distribution
    gold composite nanodevices and, 573
    nanoparticles and, 261–264
  evaluation, micellar tumor targeting modality and,
    430–437
  growth delay studies, liposome designs and,
    701–705
  pharmacokinetic assessment, 121–124
  therapeutic efficiacy, folate receptor targeted
    liposomes and, 671–672
  toxicological assessment, 121–124
  nanoemulsion formulations and, 729–730
Indomethacin, 373
Inherent targeting, nanoparticle characteristics and,
  24–25
Instrumental neutron activation analysis method,
  574–575
  gold content determination, 575

Instrumentation, 107–115
  chromatography, 109
  microscopy, 110–111
  nanomaterial
    composition, 113–114
    functionality, 112–113
    purity, 113–114
    stability, 114–115
  size and size distribution, 110–112
  spectroscopy, 108–109
  surface characteristics, 112
Integrins, 166–167, 218–219, 647–649
  for imaging, 220
  $\alpha_v\beta_3$-mediated drug targeting, 466–467
Interstitial drug targeting vs vascular targeting, 616
Intracellular
  drug uptake, ultrasound and, 428–430
  fate, nanoparticles and, 34–35
  uptake, nanoparticles and, 34–35
Intracerebral delivery, boron delivery to convection-
  enhanced, 94–95
Ionic salts, 761–763
Irinothecan, 759–760

J

Jurisdiction, FDA regulatory considerations and,
  145–146

K

KRN5500, 371–372

L

Ligand
  based drug delivery, pegtylated dendrimers and,
    543–544
  coupling, polymeric micelles and, 343–344
  mediated targeting, antisense oligonucleotides
    and, 483
  modification, pharmacokinetics macromolecule
    tissue distribution and, 51
  targeted delivery, tumor angiogenesis and, 162
Linear polymers
  applications as drug carrier, 45–46
  history and characteristics of, 44–45
  pharmacokinetic properties and, 44–46
    nanocarriers and, 44–46
Lipid
  nanoparticles, 741–770
  solid lipid nanoparticles and, 744–747
Lipid-based nanocarriers
  capsular design, 691
    getting drug out, 691–693
  liposomal drug delivery, mechanisms of, 687–690

liposome design, 693–695
review of, 682–690
transporting
drug to destination, 682–683
nanocapsules to tumor, 684–686
Lipolhilic drug delivery, 730
Lipoprotein nanoparticles, 777–782
high density liproteins, 778
low density liproteins, 778
Liposomal
anthracyclines, 600–605
drug delivery systems, 595–607
entrapped cytotoxic agents, 605–606
future of, 606
liposomal anthracyclines, 600–602
low-molecular weight drugs, pharmacokinetics, 597–598
pharmacokinetics, 597–598
rationale for use of, 597
Liposomes, 12–14
application as drug carrier, 48–50
boron delivery agents and, 86–92
cationic for lung targeting, 52–53
cell penetrating peptide and, delivery of, 635–637
designs
in vivo growth delay studies, 701–705
lipid-based nanocarriers and, 693–695
membrane permeability, 695–701
nanostructures, 695–701
drug delivery and, 631–632
pharmacokinetics, 597–598
cytotoxic agents, 605–606
folate receptor
mediated targeting, 54
targeted, 663–672
formulation, liposome pharmacokinetics-stealth liposomes, correlation between, 598–600
galactose for liver PC targeting, 53
history and characteristics, 48–50
mannose for liver NPC targeting, 53
nanocarrier-mediated gene delivery and, 52
nontargeted, 86–89
overview of, 86
pharmacokinetic
characteristics of, 48–50
properties of nanocarriers and, 48–50
stealth liposomes, liposome formulation and, correlation between, 598–600
RGD modified, 643–657
targeted, 89–92
Lipsomal
anthracyclines, clinic-doxil, 602–605
drug delivery, mechanism of, 687–690

Liver
NPC targeting, mannose liposomes and, 53
C targeting, galactose liposomes and, 53
Long-circulating polymeric nanoparticles, 231–239
Low density liproteins, 778
delivery vehicle, 779
Low-molecular weight
delivery agents, 79
drugs delivery systems and, pharmacokinetics, 597–598
Lung targeting, cationic liposomes and, 52–53
Lymphatic carcinoma, poly (alkyl cyanocrylate) nanoparticle delivery and, 276
mAb, boronated dendrimer
attached to, 81
clusters linked to, 80–81

M

Macromolecular and related cancer drug delivery, 16
Macromolecule tissue distribution, pharmacokinetics and, 50–51
electric charge, 51
ligand modification, 51
size, 51
Magnetic resonance imaging
contrast-enhanced, 203–204
paramagnetic
HPMA copolymer conjugates, 204–206
poly (L-glutamic acid) conjugates, 206–210
animal tumor model, 206–208
PGA-Mce$_6$ -(Gd-DOTA)conjugates, animal tumor model, 208–210
Poly(L-glutamic acid) and, 193–195
principles of, 202–203
Mannose liposomes, liver NPC targeting, 53
Manufacturing methods, solid lipid nanoparticles and, 751–753
Markers, tumor angiogenesis and, 162
Matrix metalloproteinases, 164–166
MDR reversal, tumor pH targeting and, 455–460
Membrane permeability, 695–701
Metabolic
environment, nanotechnology applications targeting strategies and, 32
properties, cancer cells and, nanotechnology targeting, 29–30
Methotrexate, 371
Micellar
antisense oligonucleotide delivery systems, 478–482
in vivo applications, 481–482
delivery systems, targeted antisense oligonucleotide, 477–484

dimensions, PEO-*b*-PLAA micelles drug delivery
   and, 363–367
encapsulated drugs
   micellar tumor targeting modality and,
      430–432
   therapy, 421–438
      polymeric micelles, drug delivery,
         422–425
means, peglyated dendrimers and, 537–538
stability, PEO-*b*-PLAA micelles drug delivery
   and, 367–368
tumor targeting modality, in vivo evaluation,
   430–437
tumor targeting modality
   in vivo evaluation
      micellar-encapsulated drugs, 430–432
      polymeric micelles biodistribution, 430
      ultrasound-induced drug release, 430
      ultrasound application timing, 432
      micelle/ultrasound drug delivery, tumor growth
         inhibition, 432–437
Micelle
   cRGD encoded SPIO loaded, 472–473
   formulations
      drug loading factors, 324–325
      polymeric micelles and, 323–326
      properties of, 323–324
         drug partitioning, 324
   hydrotropic polymer, 385–408
   MRI-visible polymeric, 465–484
   PEO-modified poly(L-amino acid micelles),
      357–376
   polyHis/PEG copolymer, 446–447
      PLLA/PEG, mixtures of, 448–451
   polyhis-peg-based pH sensitive, 445–451
   polymeric, 409–417
Micelle/ultrasound drug delivery
   bioeffect mechanisms, 436–437
   drug
      resistant tumors, 433–435
      sensitive tumors, 432–433
   poorly vascularized tumors, 435–436
   tumor growth inhibition, 432–437
Micellization, PEO-*b*-PLAA block polymers,
   361–362
Microcopy, 110–111
Microemulsion, solid lipid nanoparticles
   manufacturing methods and, 751–752
Microvascular endothelium, tumor vasculature and,
   614–616
Mitochondria, 788
   proapotptotic drugs acting on, 788–790
   targeted nanocarriers, 787–797
      apoptosis, 788
Mitochondrial dysfunction, 119–120

Mitochondriotropic vesicles, DQAsomes and,
   790–791
Modular functionalities, biosynthetic interface
   imaging agents, 61–64
   application, 69
   sensing, 64–66
   targeting and, 60–61
   therapeutic payloads, 66–68
Molecular targets, tumor vasculature targeting and,
   646–647
Monoclonal antibodies, boronated dendrimers and,
   links of, 80–81
Morphology
   polymer micelles physico-chemical properties
      and, 321–322
   solid lipid nanoparticles and, 748
MR T2 contrast agents, superparamagnetic iron
   oxide and, 470
MRI
   contrast agents, pegtylated dendrimers and,
      544–545
   observation, drug regional deposition and,
      709–710
   visible polymeric micelles, 465–484
      cRGD encoded SPIO loaded micelles, 472–473
      cRGD-encoded micelles, 468–469
      development of, 470–472
      integrin $\alpha_v\beta_3$-mediated drug targeting, 466–467
      non-cRGD micelles, 468–469
      non-invasive imaging methods, pharmaco-
         kinetic studies, 469–470
      superparamagnetic iron oxide nanoparticles
         (SPIO), 470
Multidrug resistance, cancer and, 444–445
Multifunctional nanoparticles
   biosynthetic interface, modular functionalities,
      60–69
      imaging agents, 61–64
      remote application, 69
      sensing, 64–66
      targeting, 60–61
      therapeutic payloads, 66–68
   cancer therapy and, 59–70
   future of, 70

N

Nanocapsules, 256
   transporting to tumors, 685–686
Nanocarrier
   mediated gene delivery, pharmacokinetics and,
      43–54
      liposomes, 52
      polymers, 52
   pharmacokinetic properties of, 44–50

dendrimers, 47–48
linear polymers, 44–46
liposomes, 48–50
polymeric micelles, 46–47
Nanocomposites
biodistribution, size, importance of, 579–580
dendrimer, 551–586
gold/dendrimer, 561
Nanodevice delivery
biodistribution summary, 580
dendrimer
biodistribution of, 575–577
in vivo biodistribution, 573
gold
composite nanodevices, biologic differences,
577–579
in vivo distribution of, 573
nanocomposites, biodistribution of, 577
instrumental neutron activation analysis method,
574–575
nanocomposite biodistribution, size, importance
of, 579–580
Nanodrugs, dendrimers as, 533–534
Nanoemulsion formulations, 723–735
applications of, 730–735
lipolhilic drugs delivery, 730
neutron capture therapy, 733–734
perfluorochemical, 734–735
photodynamic therapy, 732–733
positively charged anoemulsions, 731–732
background, 723–725
chararacterization of, 727–729
particle size, 727–728
surface, 728–729
composition of, 725–725
in vivo, 729–730
preparation of, 726–727
Nanomaterial
composition, instrumentation and, 113–114
functionality, instrumentation and, 112–113
purity, instrumentation and, 113–114
stability, instrumentation and, 114–115
Nanoparticles
cancer chemotherapy and, 525–527
nanotechnology challenges, 527
characteristics
biocompatibility, 22–23
composition, 22–23
derivitization, 23
detection, 23–24
nanotechnology applications and, 20–24
targeting, 24–27
complicating aspects, 25–26
inherent, 24–25
safety issues, 26–27

characterization of, 258–261
degradation, 260
FDA scientific considerations and, 147–149
physiochemical, 258–260
poly (alkyl cyanocrylate), enzyme influence,
260–261
conjugates, aptamers, 396–401
charge of, 298–299
polymers, 298
properties of, 297
protein detection, 304–306
size of, 297
surface modification, 299–300
strategies of, 300–301
drug delivery systems, 252–280
in vivo distribution of, 261–264
polyalkyl cyanoacrylate, 264–266
evaluation
immunogenicity, 128
immunotoxicity, 125–127
blood contact properties, 125–127
fate of
initial application site, 33–34
intracellular uptake, 34–35
nanotechnology applications targeting
strategies and, 32–35
specific organ distribution, 34
gene delivery, 233–234
in vitro characterization, 107
biodistribution alterations, surface
modifications, 264
lipoprotein, 777–782
multifunctional, 59–70
physicochemical characterization, 107–115
instrumentation, 107–115
strategies, 108
PLGA polymer, 244–245
poly (alkyl cyanocrylate), 251–280
delivery, cancer cell targets, 266–280
drug
loading, 256–258
elease from, 261
polymeric, 14–15, 253–256
preclinical characterization, 105–129
immunotoxicity, 124–128
in vitro pharmaceutical assessment,
115–121
in vitro toxicological assessment, 115–121
in vivo pharmacokinetic and toxicological
assessment, 121–124
solid lipid, 741–770
Nanoscale drug delivery vehicles, 677–712
clinical trials, 710–711
hyperthermia devices, 710–711
lipid-based nanocarriers, review of, 682–690

Q1

regional deposition of drug, 709–710
use on other tumor types, 707–708
vascular shutdown, 705–707
Nanostructures
  dendritic, 509–518
  liposome designs and, 695–701
Nanotechnology
  applications
    nonoparticle characteristics, 20–24
    targeting strategies, 19–36
      cancer cells, 27–30
        cell surface properties, 28–29
        metabolic properties, 29–30
      fate of nanoparticles, 32–35
      strategies, tumors, 30–32
  ancer, 3–8
  challenges, 527
  derived nanoparticles, 15–16
  FDA
    definition of, 143
    initiatives, 143–144
    research, 145
    regulatory considerations and, 145–146
    scientific considerations and, 147–149
  regulatory perspectives, 139–150
    FDA experience, 140–143
Near-infrared fluorescence optical imaging,
    poly(L-glutamic acid) and, 195
Neutron
  activation analysis method, instrumental,
    574–575
  capture cancer therapy, 77–96, 733–734
    boron delivery agents, 494–495, 544
      dextrans, 92–93
      liposomes, 86–92
      requirement for, 78–79
    dendrimer-related delivery agents, 80–86
    drug delivery and, 494–498
    gadolinium neutron capture therapy, 495
    gliomas, 78
NK105, 330
NK911, 327–328
Noble metal dendrimer nanocomposites, toxicity
    studies and, 582
Non-covalent encapsulation, drug delivery and,
    490–492
Non-cRGD micelles, 468–469
Non-invasive
  imaging methods, MRI-visible polymeric
    micelles and, pharmacokinetic studies,
    469–470
  visualization, in vivo delivery of, paramagnetic
    polymer conjugates and, 201–212
Non-ionic drug molecules, 762–763
  fluorouracil, 762–763

Nontargeted liposomes, 86–89
  boranes, 87–89
  boronephenylalanine, 86–87
  carboranes, 87–89
  sodium borocaptate, 86–87
Nuclear imaging, 170–175
Nucleic acid ligands. See aptamers.
Nucleolin, 296

O

Optical imaging, near-infrared fluorescence, 195
Organ distribution, nanoparticles and, 34
Oxidative stress, 118–119

P

Paclitaxel, 329, 369–370, 760
  DQAsomes and, 793–794
  formulations, 339–340
  loaded DQAsomes, tumor growth inhibition and,
    796
PAMAM dendrimer, 498–501
Paramagnetic
  HPMA copolymer conjugates
    animal tumor model, 204–205
    bone delivery, 206
    magnetic resonance imaging of, 204–206
  poly (L-glutamic acid) conjugates, magnetic
    resonance imaging and, 206–210
    animal tumor model, 206–208
  in vivo delivery, noninvasive visualization,
    201–212
  magnetic resonance imaging, 202–203
Particle size, nanoemulsion formulations and,
    727–728
Passive targeting, 244
  EPR effect, 217–218
Pegylated dendrimers
  applications of, 534–545
    boron neutron capture therapy, 544
    dendritic gels, 543
    enhanced biocompatibility of, 535–537
    enhancing solubility micellar means, 537–538
    ligand-based drug delivery, 543–544
    MRI contrast agents, 544–545
    pharmacokinetic properties, 538–539
    photodynamic therapy, 543
    stimuli responsive carriers, 540–541
    toxicity alteration, 541–543
    transfection reagents, 539–540
  dendritic nanoparticulate carriers, 523–545
    dendrimers, drug delivery system, 529–534
    poly (ethylene glycol) conjugation, 527–529
  liposome technology, 619–621
    tumor vessels, effectiveness of, 620–621

PEO shell, PEO-*b*-PLAA micelles drug delivery
and, 363
PEO-*b*-PLAA
block polymers
micellization of, 361–362
co-solvent evaporation method, 362
dialysis method, 362
solvent evaporation method, 362
synthesis of, 358–361
micelles
active drug targeting, 373
anti-cancer types, 369–372
camptothecin, 372
cisplatin, 370–371
cyclophosphamide, 369
doxorubicin, 369–370
KRN5500, 371–372
methotrexate, 371
paclitaxel, 369–370
drug delivery, 363–369
micellar dimensions, 363–367
micellar stability, 367–368
PEO shell, 363
PLAA core, 363
release properties, 368–369
gene delivery, 374–376
other therapeutic agents, 372–373
amphotericin B, 372–373
indomethacin, 373
PEO-modified poly(L-amino acid micelles),
357–376
PEO-*b*-PLAA block polymers, synthesis of,
358–361
Peptide
anti-cancer drugs, poly (alkyl cyanocrylate)
nanoparticle delivery and, 278–279
cell penetrating, 629–637
Perfluorochemical nanoemulsions, 734–735
PGA-Mce$_6$ -(Gd-DOTA)conjugates, animal tumor
model, 208–210
PG-CPT, development and testing of, 192–193
PG-TXL
development and testing of, 189–192
radiotherapy and, 193
Pharmaceutical
assessment, nanoparticle preclinical characterization and, 115–121
hydrotropic polymer micelles and, 393–395
nanocarriers, cell penetrating peptide and,
delivery of, 633–635
Pharmacokinetics
assessment, in vivo, 121–124
block copolymer micelles and, 338–339
liposomes and, 597–598
low-molecular weight drugs and, 597–598

macromolecule tissue distribution and, 50–51
effect of size, 51
effect of electric charge, 51
ligand modification, 51
nanocarrier-mediated drug delivery, 43–54
liposomes, 52
polymers, 52
properties
folate receptor targeted liposomes and,
670–671
linear polymers, 44–46
nanocarriers and, 44–50
dendrimers, 47–48
linear polymers, 44–46
liposomes, 48–50
polymeric micelles, 46–47
pegylated dendrimers and, 538–539
studies, MRI-visible polymeric micelles and,
469–470
Photodynamic therapy, 342, 496–498, 732–733
pegylated dendrimers and, 543
PLGA nanoparticles and, 248
pH-response systems, polymeric micelles and,
344–345
Physico-chemical
characterization, strategies of, nanoparticles and,
108
properties
polymer micelles and, 321–323
morphology, 321–322
size and size distribution, 321
stability, 322–323
Physiochemical characterization, nanoparticles and,
106–115, 258–260
instrumentation, 107–115
Pigpen, 296
PLAA core, PEO-*b*-PLAA micelles drug delivery
and, 363
PLGA
nanoparticles
gene delivery, 247–248
photodynamic therapy, 248
polymer nanoparticles, 244–245
PLGA/PLA nanoparticles, drug delivery, 245–246
PLLA/PEG micelle, polyhis/PEG micelle, mixture
of, 448–451
Poly (alkyl cyanoacrylate), 254–255
brain tumors, 274–275
breast cancer, 275–276
cancer cell targets, 266–268
delivery
chemoembolization, 277–278
gastric carcinoma, 276–277
hapatocellular carcinoma, 272–273
lymphatic carcinoma, 276

peptide anti-cancer drugs, 278–279
    resistant cancer cell targets, 268–272
nanocapsules, 256
nanoparticles, 251–280
    drug loading, 256–258
    drug release from, 261
    enzyme influence, 260–261
    toxicity of drugs associated with, 266
Poly (ethylene glycol) conjugation
    activation of, 528
    derivatization of, 528
    PEGylated dendritic nanoparticulate carriers and,
        527–529
    properties of, 528–529
Poly(L-glutamic acid), 185–196
    anti-cancer drug conjugates, 187–189
    biodegradation, 187
    biodistribution, 187
    chemistry of, 186–187
    conjugates
        anthracyclines, 187–188
        antimetabolites, 188
        DNA-binding drugs, 188–189
    diagnostic agent carrier, 193–195
        magnetic resonance imaging, 193–195
        near-infrared fluorescence optical imaging,
            195
    PG-CPT, development and testing of, 192–193
    PG-TXL, development and testing of, 189–192
    properties of, 186–187
    synthesis of, 186–187
Polyalkyl cyanoacrylate nanoparticles, toxicity of,
        264–266
Polyhis PEG based pH sensitive micelle
    polyHis/PEG copolymer micelle, 446–447
    synthesis of, 445–451
Polyhis PEG copolymer micelle, 446–447
Polyhis PEG micelle, PLLA/PEG micelle, mixture
        of, 448–451
Polymer
    conjugates
        angiogenesis targeted tumors, 168–175
            imaging and therapy, 159–176
        antiangiogenic gene therapy, 169–170
        nuclear imaging, 170–175
        paramagnetic, 201–212
        radiotherapy, 170–175
        tumor angiogenesis, 161
    lipid hybrid nanoparticles, 749–751
    micelles
        drug carriers, review of, 387–391
        application as drug carrier, 46–47
        cancer chemotherapy, 392–393
        drug carrier, 391
        general properties of

building blocks, 319–321
    physico-chemical, 321–323
history and characteristics, 46
pharmacokinetic
    characteristics of, 46–47
    properties of nanocarriers and, 46–47
physico-chemical properties
    morphology, 321–322
        size and size distribution, 321
        stability, 322–323
    solubilization qualities of, 391
    targeting system, 392–393
nanoparticles, aptamers and, 298
Polymeric drug carriers, dendrimer and, 517–518
Polymeric micelles, 317–345, 409–417, 465–484
    active targeting, 342
        ligand coupling, 343–344
        pH-responsive systems, 344–345
    biodistribution, micellar tumor targeting modality
        and, 430
    block copolymer, 336–340
    cancer cells
        cellular internalization, 332–334
        drug
            diffusion, 331
            resistance, 335–336
        in vitro cytotoxicity, 334
        interactions with, 330–336
    drug delivery and, 410–411, 422–425
        advantages and shortcomings of, 422–423
        clinical trials, 423–425
    drug release from, 325–326
    evolution of, 340
    formulation of, 326–330
        Genexol-PM, 329–330
        NK105, 330
        NK911, 327–328
        Paclitaxel, 329
        SP1049C, 328–329
    doxorubicin, 327
    general properties of, 319–326
        block copolymer formulations, 323
        micelle formulations, 323–326
    photodynamic therapy, 342
    polyhis-peg-based pH sensitive micelle, 445–451
    sparingly-soluable anti-cancer drugs, 409–411
    targeting tumor pH, 443–461
    thermosensitive systems, 341–342
    tumor targeting, 411–417
    ultrasonic irradiation, 340–341
Polymeric nanoparticles, 14–15, 232, 253–256
    drug delivery, 233
        examples of, 235–237
    folate-targeted, gene delivery, 222–223
    gene delivery, examples of, 237–238

long-circulating, 231–239
poly (alkyl cyanoacrylate), 254–255
polymerizaton process, 253–254
targeted
  cancer, 216
    imaging, 225
  folate receptors, 220–223
  tumor-targeted drug delivery and, 215–226
Polymerization process, 253–254
dispersion, 253
emulsion, 254
Polymers, nanocarrier-mediated gene delivery and, 52
Polypropylenimine dendrimer, 501–502
Poorly water-soluable compounds, 759–761
others, 761
paclitaxel, 760
topoisomerase I inhibitors, 759–760
Positively charged
anoemulsions, 731–732
liposomes
  cationic, 621–623
    human endothelial cells, 618–619
  formulations of, 617–618
  pegylated liposome technology, 619–620
  therapy effectiveness, 623
  tumor vasculature, 613–624
Preclinical characterization, nanoparticles and,
  in vitro pharmaceutical assessment,115–121
  in vivo pharmacokinetic and toxicological
    assessment, 121–124
Proapoptotic drugs, mitochondria and, 788–790
Prostate specific membrane antigen, 295
Protein detection, aptamer
gold-nanoparticle conjugate, 305
nanoparticle conjugates and, 304–306
QD conjugate, 304–305
Proteomics, 121

Q

Quinidine, 765

R

Radiation polymerization, dendrimer
  nanocomposites and gold, 569
Radioactive gold, 558–559
Radiotherapy, 170–175
  PG-TXL and, 193
Receptor ligands
boronated dendrimers and, 81–85
epidermal growth factors, 81, 83
folate receptor targeting agents, 83
vascular endothelial growth factor, 83–85

Regulatory
considerations, nanotechnology and FDA,
  145–147
perspectives
  nanotechnology and, 139–150
  FDA experience, 140–143
Remote application, biosynthetic interface and, 69
RGD
mediated metastasis inhibition, 651–652
modified liposomes, 643–657
  cancer
    gene therapy, 656–657
    therapeutics, 653–656
  fibronectin, 650–651
  integrin, 647–649
  mediated metastasis inhibition, 651–652
  tumor vasculature targeting, 644
multimers, 167

S

Safety
FDA nanotechnology scientific considerations
  and, 149
issues, nanoparticle characteristics and, 26–27
Salts, ionic, 761–763
Science, FDA nanotechnology considerations and,
  147–149
Sensing, biosynthetic interface and, 64–66
Sialyl Lewis X, 295
Size
effect, pharmacokinetics macromolecule tissue
  distribution and, 51
importance of, nanocomposite biodistribution
  and, 579–580
polymer micelles physico-chemical properties
  and, 321
and size distribution, instrumentation and,
  110–112
Sodium borocaptate, 86–87
Solid lipid nanoparticles, 741–770
antitumor drug delivery, 756–770
cancer chemotherapy, 756
characteristic properties of, 753–756
  drug
    loading, 753–754
    release, 754–755
    stability of, 755–756
chemosensitizing agents, 763–770
composition and structure of, 743–751
  classical SLN, 748–749
    lipids, 744–747
    morphology, 748
    other agents, 748
  polymer lipid hybrid nanoparticles, 749–751

surfactants, 747–748
types, 748
cytotoxic compound delivery, 757–758
in vitro antitumor toxicity, effectiveness on,
766–770
manufacturing methods, 751–753
double emulsion method, 753
high pressure homoegenization, 751
microemulsion, 751–752
solvent evaporation, 752–753
poorly water-soluable compounds, 759–761
structure of, 743–751
water soluable
ionic salts, 761–763
non-ionic drug molecules, 762–763
Solid tumors
barriers to extravasation, 12
cancer drug delivery systems, 12–16
liposomes, 12–14
macromolecular and related delivery, 16
nanotechnology-derived nanoparticles, 15–16
polymeric nanoparticles, 14–15
passive targeting of, 11–17
Solubility micellar means, pegylated dendrimers
and, 537–538
Solubilization qualities, polymer micelles and, 391
Solvent evaporation
method, micellization of PEO-*b*-PLAA block
polymers, 362
solid lipid nanoparticles manufacturing methods
and, 752–753
SP1049C, 328–329
Spectroscopy, 108–109
SPIO. *See* superparamagnetic iron oxide
nanoparticles.
Stability
polymer micelles physico-chemical properties
and, 322–323
solid lipid nanoparticles characteristic properties
of, 755–756
Stimuli responsive carriers, peglyated dendrimers
and, 540–541
Superparamagnetic iron oxide nanoparticles, MR T2
contrast agents, 470
Surface
characteristics, instrumentation and, 112
modification, aptamers and, 399–300
nanoemulsion formulations and, 728–729
Surfactants, solid lipid nanoparticles and, 747–748
Systemic levels, ultrasonic biological effects and,
427–428

T

Targeting
active polymeric micelles and, 342

angiogenesis
integrins, 218–219
polymeric conjugates, 168–175
vascular endothelial growth factor receptors,
218
antisense oligonucleotide micellar delivery
systems, 477–484
ligand-mediated targeting of, 483
biosynthetic interface and, 60–61
cancer therapy, aptamer drug conjugates and,
301–304
delivery
aptamers and, 294–296
ligands and, 162
dendrimer nanocomposites and, 572–580
drug delivery, 289–291
folate receptors, 220–223
liposomes
EGFR, 91–92
folate receptor-targeted, 90–91
immunoliposomes, 89–90
nanoparticle
characteristics, 24–24
complicating aspects, 25–26
inherent, 24–25
safety issues, 26–27
cancer imaging and, 225
nanotechnology applications and, 19–36
cancer cells, 27–30
cell surface properties, 28–29
metabolic properties, 29–30
tumors, 30–32
abberant vasculature, 30–32
metabolic environment, 32
fate of nanoparticles, 32–35
system, polymer micelles and, 392–393
tumor pH, MDR reversal, 455–460
tumor pH$_e$, 451–455
Tenascin-C, 296
Therapeutic payloads, biosynthetic interface and,
66–68
Thermosensitive systems, 341–342
Topoisomerase I inhibitors, 759–760
camptothecin, 759–760
irinothecan, 759–760
Toxicity
alteration, pegtylated dendrimers and,
541–543
block copolymer micelles and, 338–339
composite nanodevices and, 558
polyalkyl cyanoacrylate nanoparticles and,
264–266
studies
dendrimer nanocomposites and, 580–585
noble metal dendrimer nanocomposites, 582

tritiated dendrimer nanoparticles, 580
    tumor treatment, 582–585
Toxicogenomics, 121
Toxicological assessment
    apoptosis, 119–120
    cytotoxicity, 117–118
    in vitro target-organ toxicity, 116–117
    in vivo, 121–124
    mitochondrial dysfunction, 119–120
    nanoparticle preclinical characterization and,
        115–121
    oxidative stress, 118–119
    proteomics, 121
    special considerations, 115–116
    toxicogenomics, 121
Transduction, cell penetrating peptide and, 631–633
Transfection reagents, pegylated dendrimers and,
    539–540
Tritiated dendrimer nanoparticles, toxicity studies
    of, 580
Tritium labeled dendrimers, 564
Tumor
    angiogenesis, 161
        ligands, targeted delivery to, 162
        markers, 162
    brain, 274–275
    drug resistant, micelle/ultrasound drug delivery
        and, 433–435
    drug-sensitive, micelle/ultrasound drug delivery
        and, 432–433
    growth inhibition
        micelle/ultrasound drug delivery and, 432–437
        paclitaxel loaded DQAsomes and, 796
    nanocapsules to, 685–686
    nanotechnology applications targeting strategies
        and, 30–32
        abberant vasculature, 30–32
        metabolic environment, 32
    pH
        acidic tumor pH_e, 444
        multidrug resistance and cancer, 444–445
        targeting of, polymeric micelles and, 443–461
    pH_e targeting, 451–455
    poorly vascularized, micelle/ultrasound drug
        delivery and, 435–437
    solid, passive targeting of, 11–17
    targeting, polymeric micelles and, 411–417
    treatment, toxicity studies and, 582–585
    vasculature
        microvascular endothelium, 614–616
        positively charged liposomes and, 613–624
        targeting, 644

angiogenic tumor, molecular targets,
    646–647
    vs interstitial drug targeting, 616
    rationale for, 644–646
Tumor-targeted drug delivery
    angiogenesis, 218–220
    avidin, 226
    biotin, 226
    cancer, specific types, 223–225
    other targets, 225–226
    passive targeting, EPR effect, 217–218
    polymeric nanoparticles and, 215–226

U

Ultrasonic irradiation, 340–341
Ultrasound
    application timing, micellar tumor targeting
        modality and, 432
    biological effects of, 425–430
        cellular level, 425–427
        drug release qualities, 428
        intracellular drug uptake, 428–430
        systemic levels, 427–428
    combining of, 437
    drug delivery, 425–430
    induced drug release, micellar tumor targeting
        modality and, 430
    therapy, 421–438

V

Vascular
    cationic charge, 617
    endothelial growth factor, 83–85
        receptors, 218
    vs interstitial drug targeting, 616
    nanoscale drug delivery vehicles and, 705–707
Vascularized tumors, micelle/ultrasound drug
    delivery and, 435–437
VEGF receptors, 162
    aminopeptidases, 162, 164
    integrins, 166–167
    matrix metalloproteinases, 164–166
    RGD multimers, 167
Verapamil, 765

W

Water soluable ionic salts, 761–763
    doxorubicin, 761–762
    idarubicin, 761–762
    non-ionic drug molecules, 762–763

Author Queries

Q1   Please check the page range.